Probability Theory and Stochastic Modelling

Volume 83

The **Probability Theory and Stochastic Modelling** series is a merger and continuation of Springer's two well established series Stochastic Modelling and Applied Probability and Probability and Its Applications series. It publishes research monographs that make a significant contribution to probability theory or an applications domain in which advanced probability methods are fundamental. Books in this series are expected to follow rigorous mathematical standards, while also displaying the expository quality necessary to make them useful and accessible to advanced students as well as researchers. The series covers all aspects of modern probability theory including

- Gaussian processes
- Markov processes
- Random fields, point processes and random sets
- Random matrices
- Statistical mechanics and random media
- Stochastic analysis

as well as applications that include (but are not restricted to):

- Branching processes and other models of population growth
- Communications and processing networks
- Computational methods in probability and stochastic processes, including simulation
- Genetics and other stochastic models in biology and the life sciences
- Information theory, signal processing, and image synthesis
- Mathematical economics and finance
- Statistical methods (e.g. empirical processes, MCMC)
- Statistics for stochastic processes
- Stochastic control
- Stochastic models in operations research and stochastic optimization
- Stochastic models in the physical sciences

More information about this series at http://www.springer.com/series/13205

René Carmona • François Delarue

Probabilistic Theory of Mean Field Games with Applications I

Mean Field FBSDEs, Control, and Games

Springer

René Carmona
ORFE Department
Program in Applied and
Computational Mathematics
Princeton University
Princeton, NJ, USA

François Delarue
Institut Universitaire de France
& Laboratoire J.A. Dieudonné
Université Nice Sophia Antipolis
Nice, France

ISSN 2199-3130 ISSN 2199-3149 (electronic)
Probability Theory and Stochastic Modelling
ISBN 978-3-030-13260-6 ISBN 978-3-319-58920-6 (eBook)
DOI 10.1007/978-3-319-58920-6

Mathematics Subject Classification (2010): 91A13, 91A06, 91A55, 49N70, 49N90, 60H10, 60H15, 60H30

Printed on acid-free paper

This Springer imprint is published by Springer Nature
The registered company is Springer International Publishing AG
The registered company address is: Gewerbestrasse 11, 6330 Cham, Switzerland

Foreword

Since its inception about a decade ago, the theory of Mean Field Games has rapidly developed into one of the most significant and exciting sources of progress in the study of the dynamical and equilibrium behavior of large systems. The introduction of ideas from statistical physics to identify approximate equilibria for sizeable dynamic games created a new wave of interest in the study of large populations of competitive individuals with "mean field" interactions. This two-volume book grew out of series of lectures and short courses given by the authors over the last few years on the mathematical theory of Mean Field Games and their applications in social sciences, economics, engineering and finance. While this is indeed the object of the book, by taste, background, and expertise, we chose to focus on the probabilistic approach to these game models.

In a trailblazing contribution, Lasry and Lions proposed in 2006 a methodology to produce approximate Nash equilibria for stochastic differential games with symmetric interactions and a large number of players. In their models, a given player *feels* the presence and the behavior of the other players through the empirical distribution of their private states. This type of interaction was extensively studied in the statistical physics literature under the name of *mean field* interaction, hence the terminology Mean Field Game coined by Lasry and Lions. The theory of these new game models was developed in lectures given by Pierre-Louis Lions at the Collège de France which were video-taped and made available on the internet. Simultaneously, Caines, Huang, and Malhamé proposed a similar approach to large games under the name of Nash Certainty Equivalence principle. This terminology fell from grace and the standard reference to these game models is now Mean Field Games.

While slow to pick up momentum, the subject has seen a renewed wave of interest over the last seven years. The mean field game paradigm has evolved from its seminal principles into a full-fledged field attracting theoretically inclined investigators as well as applied mathematicians, engineers, and social scientists. The number of lectures, workshops, conferences, and publications devoted to the subject has grown exponentially, and we thought it was time to provide the applied mathematics community interested in the subject with a textbook presenting the state of the art, as we see it. Because of our personal taste, we chose to focus on what we like to call the probabilistic approach to mean field games. While a significant portion of the text is based on original research by the authors, great care was taken to include models and results contributed by others, whether or not they were aware of the fact they were working with mean field games. So the book should feel and read like a textbook, not a research monograph.

Most of the material and examples found in the text appear for the first time in book form. In fact, a good part of the presentation is original, and the lion's share of the arguments used in the text have been designed especially for the purpose of the book. Our concern for pedagogy justifies (or at least explains) why we chose to divide the material in two volumes and present mean field games without a common noise first. We ease the introduction of the technicalities needed to treat models with a common noise in a crescendo of sophistication in the complexity of the models. Also, we included at the end of each volume four extensive indexes (author index, notation index, subject index, and assumption index) to make navigation throughout the book seamless.

Acknowledgments

First and foremost, we want to thank our wives Debbie and Mélanie for their understanding and unwavering support. The intensity of the research collaboration which led to this two-volume book increased dramatically over the years, invading our academic lives as well as our social lives, pushing us to the brink of sanity at times. We shall never be able to thank them enough for their patience and tolerance. This book project would not have been possible without them: our gratitude is limitless.

Next we would like to thank Pierre-Louis Lions, Jean-Michel Lasry, Peter Caines, Minyi Huang, and Roland Malhamé for their incredible insight in introducing the concept of mean field games. Working independently on both sides of the pond, their original contributions broke the grounds for an entirely new and fertile field of research. Next in line is Pierre Cardaliaguet, not only for numerous private conversations on game theory but also for the invaluable service provided by the notes he wrote from Pierre-Louis Lions' lectures at the Collège de France. Although they were never published in printed form, these notes had a tremendous impact on the mathematical community trying to learn about the subject, especially at a time when writings on mean field games were few and far between.

We also express our gratitude to the organizers of the 2013 and 2015 conferences on mean field games in Padova and Paris: Yves Achdou, Pierre Cardaliaguet, Italo Capuzzo-Dolcetta, Paolo Dai Pra, and Jean-Michel Lasry.

While we like to cast ourselves as proponents of the probabilistic approach to mean field games, it is fair to say that we were far from being the only ones following this path. In fact, some of our papers were posted essentially at the same time as papers of Bensoussan, Frehse, and Yam, addressing similar questions, with the same type of methods. We benefitted greatly from this stimulating and healthy competition.

We also thank our coauthors, especially Jean-François Chasagneux, Dan Crisan, Jean-Pierre Fouque, Daniel Lacker, Peiqi Wang, and Geoffrey Zhu. We used our joint works as the basis for parts of the text which they will recognize easily.

Also, we would like to express our gratitude to the many colleagues and students who gracefully tolerated our relentless promotion of this emerging field of research through courses, seminar, and lecture series. In particular, we would like to thank Jean-François Chasagneux, Rama Cont, Dan Crisan, Romuald Elie, Josselin Garnier, Marcel Nutz, Huyen Pham, and Nizar Touzi for giving us the opportunity to do just that.

Finally, we would like to thank Paul-Eric Chaudru de Raynal, Mark Cerenzia, Christy Graves, Dan Lacker, Victor Marx, and Peiqi Wang for their help in proofreading an earlier version of the manuscript.

The work of R. Carmona was partially supported by NSF DMS-1211928 and DMS-1716673, and by ARO W911NF-17-1-0578. The work of F. Delarue was partially supported by the Institut Universitaire de France and by the Agence Nationale de la Recherche (project MFG, ANR-16-CE40-0015-01).

Princeton, NJ, USA | René Carmona
Nice, France | François Delarue
July 29, 2016

Preface to Volume I

This first volume of the book is entirely devoted to the theory of mean field games in the absence of a source of random shocks common to all the players. We call these models *games without a common noise*. This volume is divided into two main parts. Part I is a self-contained introduction to mean field games, starting from practical applications and concrete illustrations, and ending with ready-for-use solvability results for mean field games without a common noise. While Chapters 1 and 2 are mostly dedicated to games with a finite number of players, the asymptotic formulation which constitutes the core of the book is introduced in Chapter 3. For the exposition to be as pedagogical as possible, we chose to defer some of the technical aspects of this asymptotic formulation to Chapter 4 which provides a complete toolbox for solving forward-backward stochastic differential equations of the McKean-Vlasov type. Part II has a somewhat different scope and focuses on the main principles of analysis on the Wasserstein space of probability measures with a finite second moment, which plays a key role in the study of mean field games and which will be intensively used in the second volume of the book. We present the mathematical theory in Chapter 5, and we implement its results in Chapter 6 with the analysis of stochastic mean field control problems, which are built upon a notion of equilibrium different from the search for Nash equilibria at the root of the definition of mean field games. Extensions, including infinite time horizon models and games with finite state spaces, are discussed in the epilogue of this first volume.

The remainder of this preface expands, chapter by chapter, the short content summary given above. A diagram summarizing the connections between the different components of the book is provided on page xix.

The first chapter sets the stage for the introduction of mean field games with a litany of examples of increasing complexity. Starting with one-period deterministic games with a large number of players, we introduce the mean field game paradigm. We use examples from domains as diverse as finance, macroeconomics, population biology, and social science to motivate the introduction of mean field games in different mathematical settings. Some of these examples were studied in the literature before the introduction of, and without any reference to, mean field games. We chose them because of their powerful illustrative power and the motivation they offer for the introduction of new mathematical models. The examples of *bank runs* modeled as mean field games of timing are a case in point. For pedagogical reasons, we highlight practical applications where the interaction between the players does not necessarily enter the model through the empirical distributions of the states of the players, but via the empirical distributions of the actions of the players, or even the joint empirical distributions of the states and the controls of the players. Most of these examples will be revisited and solved throughout the book.

Chapter 2 offers a crash course on the mathematical theory of stochastic differential games with a finite number of players. The material of this chapter is not often found in book form, and since we make extensive use of its notations and results throughout the book, we thought it was important to present them early for the sake of completeness and future references. We concentrate on what we call the probabilistic approach to the search for Nash equilibria, and we introduce games with mean field interactions as they are the main object of the book. Explicitly solvable models are few and far between. Among them, linear quadratic (LQ for short) models play a very special role because their solutions, when they exist, can be obtained by solving matrix Riccati equations. The last part of the chapter is devoted to a detailed analysis of a couple of linear quadratic models already introduced in Chapter 1, and for which explicit solutions can be derived. To wit, these models do not require the theory of mean field games since their finite player versions can be solved explicitly. However, they provide a testbed for the analysis of the limit of finite player equilibria when the number of players grows without bound, offering an invaluable opportunity to introduce the concept of mean field game and discover some of its essential features.

The probabilistic approach to mean field games is the main thrust of the book. The underpinnings of this approach are presented in Chapter 3. Stochastic control problems and the search for equilibria for stochastic differential games can be tackled by reformulating the optimization and equilibrium problems in terms of backward stochastic differential equations (BSDEs throughout the book) and forward-backward stochastic differential equations (FBSDEs for short). In this chapter, we review the major forms of FBSDEs that may be used to represent the optimal trajectories of a standard optimization problem: the first one is based on a probabilistic representation of the value function, and the second one on the stochastic Pontryagin maximum principle. Combined with the consistency condition issued from the search for Nash equilibria as fixed points of the best response function, this prompts us to introduce a new class of FBSDEs with a distinctive McKean-Vlasov character. This chapter presents a basic existence result for McKean-Vlasov FBSDEs. This result will be extended in Chapter 4. As a by-product, we obtain early solvability results for mean field games by straightforward implementations of the two forms of the probabilistic approach just mentioned. However, since our primary aim in this chapter is to make the presentation as pedagogical as possible, we postpone the most general versions of the existence results for mean field games to Chapter 4, as some of the proofs are rather technical. Instead, we highlight the role of monotonicity, as captured by the so-called Lasry-Lions monotonicity conditions, in the analysis of uniqueness of equilibria. Finally, we specialize the results of this chapter to the case of linear-quadratic mean field games, which can be handled directly via the analysis of Riccati equations. Most of the results of this chapter will be revisited and extended in the second volume to accommodate a *common noise* which is found in many economic and physical applications.

Chapter 4 starts with a stochastic analysis primer on the theory of FBSDEs. As explained above, in the mean field limit of large games, the fixed point step of the search for Nash equilibria turns the standard FBSDEs derived from optimization problems into equations of the McKean-Vlasov type by introducing the distribution of the solution into the coefficients. These FBSDEs characterize the equilibria. Since this new class of FBSDEs was not studied before the advent of mean field games, one of the main objectives of Chapter 4 is to provide a systematic approach to their solution. We show how to use Schauder's fixed point theorem to prove existence of a solution. The chapter closes with the analysis of the so-called *extended mean field games*, in which the players are interacting not only through the distribution of their states but also through the distribution of their controls. Finally, we demonstrate how the methodology developed in the chapter applies to some of the examples presented in the opening chapter.

Although it contains very few results on mean field games, Chapter 5 plays a pivotal role in the book. It contains all the results on spaces of probability measures which we use throughout the book, including the definitions and properties of the Wasserstein distances, the convergence of the empirical measures of a sequence of independent and identically distributed random variables ... and most importantly, a detailed presentation of the differential calculus on the Wasserstein space introduced by Lions in his unpublished lectures at the Collège de France, and by Cardaliaguet in the notes he wrote from Lions' lectures. Even though the use of this differential calculus in the first volume is limited to the ensuing Chapter 6, the differential calculus on the Wasserstein space plays a fundamental role in the study of the master equation for mean field games, whose presentation and analysis will be provided in detail in the second volume. Still, a foretaste of the master equation is given at the end of this chapter. Its derivation is based on an original form of Itô's formula for functionals of the marginal laws of an Itô process, the proof of which is given in full detail. For the sake of completeness, we also provide a thorough and enlightening discussion of the connections between Lions' differential calculus, which we call L-differential calculus throughout the book, and other forms of differential calculus on the space of probability measures, among which the differential calculus used in optimal transportation theory.

One of the remarkable features of the construction of solutions to mean field game problems is the similarity with a natural problem which did not get much attention from analysts and probabilists: the optimal control of (stochastic) differential equations of the McKean-Vlasov type, which could also be called mean field optimal control. The latter is studied in Chapter 6. Both problems can be interpreted as searches of equilibria for large populations, claim which will be substantiated in Chapter 6 in the second volume of the book. Interestingly, the optimal control of McKean-Vlasov stochastic dynamics is intrinsically a stochastic optimization problem while the search for Nash equilibria in mean field games is more of a fixed point problem than an optimization problem. So despite the strong similarities between the two problems, differences do exist, and we highlight them starting with Chapter 6. There, we show that since the problem at hand is a stochastic control problem, the optimal control of McKean-Vlasov stochastic dynamics can be tackled by means of an appropriate version of the Pontryagin stochastic maximum principle. Following this strategy leads to FBSDEs for which the backward part involves the derivative of the Hamiltonian with respect to a measure argument. This novel feature is handled with the tools provided in Chapter 5. We close the chapter with the discussion of an alternative strategy for solving mean field optimal control problems, based on the notion of relaxed controls. Also, we review several crucial examples, among them potential games. These latter models are mean field games for which the solutions can be reduced to the solutions of mean field optimal control problems, and optimal transportation problems.

Chapter 7 is a capstone which we use to revisit some of the examples introduced in Chapter 1, especially those which are not exactly covered by the probabilistic theory of stochastic differential mean field games developed in the first volume. Indeed, Chapter 1 included a considerable amount of applications hinting at mathematical models with distinctive features which are not accommodated by the models and results of the first part of this first volume. We devote this chapter to presentations, even if only informal, of extensions of the Mean Field Game paradigm to these models. They include extensions to several homogenous populations, infinite horizon optimization, and models with finite state spaces. These mean field game models have a great potential for the quantitative analysis of very important practical applications, and we show how the technology developed in the first volume of the book can be brought to bear on their solutions.

Princeton, NJ, USA René Carmona
Nice, France François Delarue
July 29, 2016

Organization of the Book: Volume I Organigram

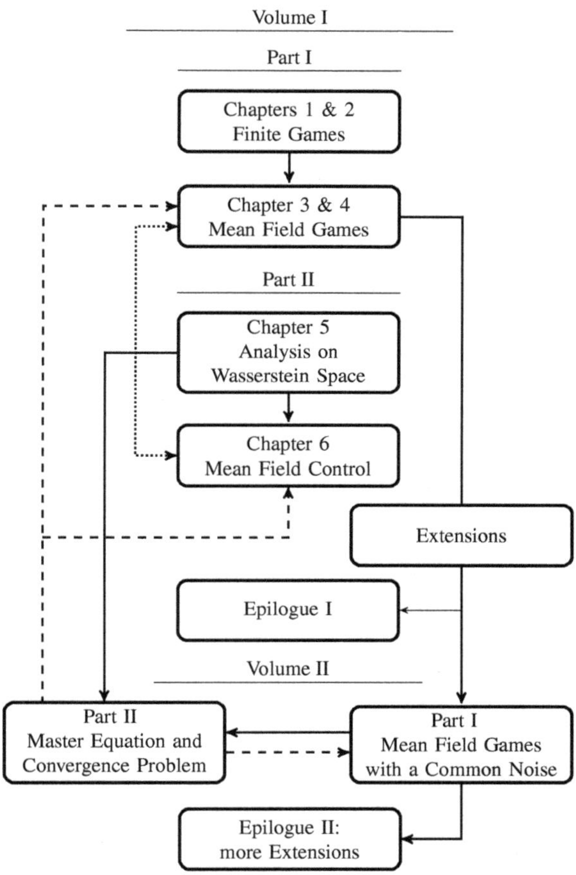

Thick lines indicate the logical order of the chapters. The dotted line between Chapters 3–4 and 6 emphasizes the fact that—in some cases like potential games—mean field games and mean field control problems share the same solutions. Finally, the dashed lines starting from Part II (second volume) point toward the games and the optimization problems for which we can solve approximately the finite-player versions or for which the finite-player equilibria are shown to converge.

References to the second volume appear in the text in the following forms: Chapter (Vol II)-X, Section (Vol II)-$X.x$, Theorem (Vol II)-$X.x$, Proposition (Vol II)-$X.x$, Equation (Vol II)-$(X.x)$, $\ldots$, where X denotes the corresponding chapter in the second volume and x the corresponding label.

Contents

Volume I

Volume II

Part I MFGs with a Common Noise

Part I

The Probabilistic Approach to Mean Field Games

Learning by Examples: What Is a Mean Field Game? 1

Abstract
The goal of this first chapter is to introduce a diverse collection of examples of games whose descriptions help us introduce the notation and terminology as well as some preliminary results at the core of the theory of mean field games. This compendium of models illustrates the great variety of applications which can potentially be studied rigorously by means of this concept. These introductory examples are chosen for their simplicity, and for pedagogical reasons, we take shortcuts and ignore some of their key features. As a consequence, they will need to be revisited before we can subject them to the treatment provided by the theoretical results developed later on in the book. Moreover, some of the examples may not fit well with the typical models analyzed mathematically in this text as the latter are most often in *continuous time*, and of the *stochastic differential* variety. For this reason, we believe that the best way to think about this litany of models is to remember a quote by Box and Draper *"All models are wrong, but some are useful"*.

1.1 Introduction of the Key Ingredients

The analysis of a N-player game involves the *simultaneous* minimization of N functions $J^1, J^2, \cdots, J^N$, the argument of which, say $(\alpha^1, \alpha^2, \cdots, \alpha^N)$, represents the actions taken by the N players in the game. One should think of $J^i(\alpha^1, \alpha^2, \cdots, \alpha^N)$ as the cost/reward to player i when the first player takes the action α^1, ..., and the last player takes the action α^N. Since the mathematical treatments of games where players minimize or maximize are perfectly equivalent, we discuss only the first alternative. An obvious question which will need to be addressed concerns the kind of actions the players are allowed to take, and also, the kind of information they can use to choose their actions. Moreover, since the minimization of $\mathbb{R}^N$-valued

R. Carmona, F. Delarue, *Probabilistic Theory of Mean Field Games with Applications I*, Probability Theory and Stochastic Modelling 83,
DOI 10.1007/978-3-319-58920-6_1

functions can be a touchy business, we also need to be clear on what we mean by simultaneous minimization of the individual functions J^i. This is not too bad of a question and we can settle it from the get-go: we shall consider that the goal of such a simultaneous minimization is achieved when the system of the N players is in a *Nash equilibrium*, concept which we shall define later on. See for example Definition 1.1 below.

By the *simultaneous* nature of the optimization involved in the search for Nash equilibria when N is a finite integer, this problem is typically very challenging, both at the theoretical as well as at the numerical level. So like mathematicians often do, we shall search for simplifications by taking the limit $N \to \infty$, hoping that in such a large game limit, some equations could simplify, providing a form of asymptotic analysis of the model. The rationale of the so-called *Mean Field Games* (MFGs for short) studied in this text is to consider the asymptotic behavior when $N \to \infty$ of games involving a large number of players. The analysis of the asymptotic regime $N \to \infty$ will be our point of contact with Aumann's theory of games with a continuum of players. Motivated by the discussions of this chapter, we show in Chapter (Vol II)-6 how Nash equilibria of N-player games converge in a certain sense as $N \to \infty$ to solutions of MFGs, and conversely, we show how solutions of MFGs can provide approximate Nash equilibria for finite player games.

1.1.1 First Example: A One-Period Deterministic Game

Our first example is mostly intended to motivate the terminology and explain the notation used throughout. We consider a population of N individuals which we denote by $i \in \{1, \cdots, N\}$, individual i having the choice of a point α^i in a space A^i which is assumed to be a compact metric space for the sake of simplicity. This is a simple mathematical model for a game discussed frequently by P.L. Lions in his lectures [265] under the name of "*Where do I put my towel on the beach?*", all the A^i being equal to the same set A representing the beach, and α the location where the player is setting his or her towel.

We are interested in large games, claim which is translated mathematically by the assumption $N \to \infty$. The game nature of the problem comes from the fact that each individual is an optimizer (a minimizer for the sake of definiteness) in the sense that player i tries to minimize a quantity $J^i(\alpha^1, \cdots, \alpha^N)$. In order for simplifications to occur in the limit $N \to \infty$, regularity and symmetry assumptions will be needed. For example, we shall assume that all the spaces A^i are in fact the same compact metric space A. Next, we shall also assume that each function J^i is symmetric in the choices α^j of the players $j \neq i$. For instance, the following cost functions could be used in the description of the example "*Where do I put my towel on the beach?*":

$$J^i(\alpha^1, \cdots, \alpha^N) = \alpha d(\alpha^i, \alpha^0) - \beta \frac{1}{N-1} \sum_{1 \leqslant j \neq i \leqslant N} d(\alpha^i, \alpha^j). \tag{1.1}$$

Here, α and β are two real numbers, d denotes the distance of the metric space A, and $\alpha^0 \in A$ is a special point of interest (say a food stand on the beach). So if the constants α and β are positive, the fact that the players try to minimize this cost function means that the players want to be closer to the point of interest α^0, but at the same time, as far away from each other as possible. Clearly, the above cost functions are of the form:

$$J^i(\alpha^1, \cdots, \alpha^N) = \tilde{J}\left(\alpha^i, \frac{1}{N-1} \sum_{1 \leqslant j \neq i \leqslant N} \delta_{\alpha^j}\right). \tag{1.2}$$

where here and throughout the book, we use the notation δ_x for the unit mass at the point x, for the function $\tilde{J}$ of $\alpha \in A$ and $\mu \in \mathcal{P}(A)$ defined by $\tilde{J}(\alpha, \mu) = \alpha d(\alpha, \alpha_0) - \beta \int_E d(\alpha, \alpha')\mu(d\alpha')$. Also, we use the notation $\mathcal{P}(A)$ for the space of probability measures on A. The special form (1.2) is the epitome of the type of interaction we shall encounter throughout the text. The fact that it involves a function of a measure, an empirical measure in this instance, will be justified by Lemma 1.2 below which gives a rigorous foundation to the rationale for the main assumption of mean field game models. When the cost J^i to player i is a function of α^i and a symmetric function of the other α^j with $j \neq i$, and the dependence of J^i upon each individual α^j is minor, then the function J^i can be viewed as a function of α^i and the empirical distribution of the remaining α^j. In other words, in the limit $N \to \infty$ of large games, the costs J^i can be viewed as functions of α^i and a probability measure. See Lemma 1.2 below for details.

It is customary to use the notation α to denote the actions taken by, or controls used by the players. In this way, the cost to each player appears as a function of the values of the various α^i chosen by the players. This is the notation system used above. On the other hand, it is also customary to use the notation x to denote the state of a system influenced by the actions of the players. Most often in the applications we shall consider, the state of the system comprises individual states x^i attached to each player i, and a few other factors, all of them entering the computation of the costs J^i. Notice that in the above example, $x^i = \alpha^i$ since the controls α^i describe the choice made by the players as well as the states they put themselves in.

Frequently Used Notation

Given an N-tuple $(x^1, \cdots, x^N)$ and the choice of an index $i \in \{1, \cdots, N\}$, we denote by x^{-i} the $(N-1)$-tuple of the x^j with $j \neq i$ and $j \in \{1, \cdots, N\}$. Moreover, with a slight abuse of notation, we will identify (x^i, x^{-i}) with $(x^1, \cdots, x^N)$, and more generally (x, x^{-i}) with the N-tuple $(x^1, \cdots, x^{i-1}, x, x^{i+1}, \cdots, x^N)$ which is nothing but the original $(x^1, \cdots, x^N)$ with the i-th entry x^i replaced by x.

1.1.2 Nash Equilibria and Best Response Functions

The lexicographical order, which is often considered as the most natural notion of order in $\mathbb{R}^N$, is not complete and as a result, the simultaneous minimization of the N scalar functions J^i can be a touchy business. The standard way to resolve the possible ambiguities associated with this simultaneous minimization is to appeal to the notion of Nash equilibrium.

Definition 1.1 *A point* $(\hat{\alpha}^1, \cdots, \hat{\alpha}^N) \in A^N$ *is said to be a Nash equilibrium if for every* $i \in \{1, \cdots, N\}$ *and* $\alpha \in A$,

$$J^i(\hat{\alpha}^1, \cdots, \hat{\alpha}^i, \cdots, \hat{\alpha}^N) \leqslant J^i(\hat{\alpha}^1, \cdots, \hat{\alpha}^{i-1}, \alpha, \hat{\alpha}^{i+1}, \cdots, \hat{\alpha}^N).$$

Note that the above inequality is easier to write and read if one uses the special notation introduced earlier. Indeed, it reads:

$$J^i(\hat{\alpha}^i, \hat{\alpha}^{-i}) = J^i(\hat{\alpha}^1, \cdots, \hat{\alpha}^i, \cdots, \hat{\alpha}^N) \leqslant J^i(\alpha, \hat{\alpha}^{-i}).$$

The notion of Nash equilibrium is best understood in terms of the so-called best response function $B : A^N \to A^N$ defined by:

$$B(\alpha^1, \cdots, \alpha^N) = (\beta^1, \cdots, \beta^N) \text{ if } \forall i \in \{1, \cdots, N\}, \ \beta^i = \arg\min_{\alpha} J^i(\alpha, \alpha^{-i})$$

which is obviously well defined if we assume that there exists a unique minimum of the function in the right of this expression, but which can also be defined in more general situations. With the intuitive concept of best response formalized in this way, a Nash equilibrium appears as a *fixed point* of the best response function B.

1.1.3 Large Games Asymptotics

Solving N player games for Nash equilibria is often difficult, even for one period deterministic games, and the strategy behind the theory of mean field games is to search for simplifications in the limit $N \to \infty$ of large games. Since such simplifications cannot be expected in full generality, we shall need to restrict our attention to games with specific properties. We shall assume:

- a strong symmetry property of the model: typically, we shall assume that each cost function J^i is a symmetric function of the $N-1$ variables α^j for $j \neq i$;
- the influence of each single player on the whole system is diminishing as N gets larger.

We formalize these two properties in precise mathematical terms below in a simple lemma. Its result will be the genesis for many informal arguments used throughout to justify the set-ups of mean field games. But first, we introduce a special notation for empirical distributions. If $n \geqslant 1$ is an integer, given a point $X = (x^1, \cdots, x^n) \in E^n$ we shall denote by $\bar{\mu}^n_X$ the probability measure defined by:

$$\bar{\mu}^n_X = \frac{1}{n} \sum_{i=1}^{n} \delta_{x^i} \tag{1.3}$$

where we use the notation δ_x to denote the unit mass at the point $x \in E$. Notice that $\bar{\mu}^n_X$ belongs to $\mathcal{P}(E)$, the space of probability measures on E. For the purpose of the present discussion, E is assumed to be a compact metric space. Later on in the book, we shall consider more general topological spaces. Accordingly, we assume that the space $\mathcal{P}(E)$ is equipped with the topology of the weak convergence of measures according to which a sequence $(\mu_k)_{k\geqslant 1}$ of probability measures in $\mathcal{P}(E)$ converges to $\mu \in \mathcal{P}(E)$ if and only if $\int_E f d\mu_k \to \int_E f d\mu$ when $k \to \infty$ for every real valued continuous function f on E. The space $\mathcal{P}(E)$ is a compact metric space for this topology and we denote by ρ a distance compatible with this topology. We shall see several examples of such metrics in Chapter 5 where we provide an in-depth analysis of several forms of calculus on spaces of measures.

The following simple result is the basis for our formulation of the limiting problems for games with a large number of players. Its assumptions formalize what we mean when we informally talk about *symmetric function weakly dependent on each of its arguments*.

Lemma 1.2 *For each $n \in \mathbb{N}$, let $u^n : E^n \to \mathbb{R}$ be a symmetric function of its n variables. Let us further assume that we have:*

1. *Uniform boundedness: $\sup_{n\geqslant 1} \sup_{X\in E^n} |u^n(X)| < \infty$;*
2. *Uniform Lipschitz continuity: there exists a finite constant $c > 0$ such that for all $n \geqslant 1$ and all $X, Y \in E^n$, we have:*

$$|u^n(X) - u^n(Y)| \leqslant c\rho(\bar{\mu}^n_X, \bar{\mu}^n_Y).$$

Then, there exists a subsequence $(u^{n_k})_{k\geqslant 1}$ and a Lipschitz continuous map $U : \mathcal{P}(E) \to \mathbb{R}$ such that:

$$\lim_{k\to\infty} \sup_{X\in E^{n_k}} |u^{n_k}(X) - U(\bar{\mu}^{n_k}_X)| = 0.$$

Remark 1.3 *The adjective "symmetric" in the statement of the lemma is redundant. It was included as a matter of emphasis. Indeed, assumption 2 implies that, for each $n \geqslant 1$, u^n is necessarily symmetric and continuous. Choosing Y as a permutation of the entries of X, we get $u^n(X) = u^n(Y)$ by observing that $\bar{\mu}^n_X = \bar{\mu}^n_Y$. Moreover, for any sequence $(X^{n_p})_{p\geqslant 1}$ with values in E^n converging (for the product topology on E^n) toward some X^n, the sequence $(\bar{\mu}^n_{X^{n_p}})_{p\geqslant 1}$ converges toward $\bar{\mu}^n_{X^n}$ implying the continuity of u^n.*

Proof. For each integer $n \geqslant 1$, define the function U^n on $\mathcal{P}(E)$ by:

$$U^n(\mu) = \inf_{X \in E^n} \Big[u^n(X) + c\rho(\bar{\mu}^n_X, \mu) \Big], \qquad \mu \in \mathcal{P}(E). \tag{1.4}$$

These functions are uniformly bounded since the functions $(u^n)_{n\geqslant 1}$ are uniformly bounded, see *1* in the statement, while the function $\mathcal{P}(E)^2 \ni (\mu, \nu) \mapsto \rho(\mu, \nu)$ is bounded since $\mathcal{P}(E)$ is compact. Also, the functions $(U^n)_{n\geqslant 1}$ extend the original functions $(u^n)_{n\geqslant 1}$ to $\mathcal{P}(E)$ in the sense that $u^n(Y) = U^n(\bar{\mu}^n_Y)$ for $Y \in E^n$ and $n \geqslant 1$. Indeed, by choosing $X = Y$ in the infimum appearing in definition (1.4) we get $U^n(\bar{\mu}^n_Y) \leqslant u^n(Y)$. Equality then holds true because, if not, there would exist an $X \in E^n$ such that $u^n(X) + c\rho(\bar{\mu}^n_X, \bar{\mu}^n_Y) < u^n(Y)$ which would contradict assumption 2. Furthermore, these extensions are c-Lipschitz continuous on $\mathcal{P}(E)$ in the sense that:

$$|U^n(\mu) - U^n(\nu)| \leqslant c\rho(\mu, \nu)$$

for all $\mu, \nu \in \mathcal{P}(E)$. To prove this, let $X \in E^n$ and $Y \in E^n$ be such that $U^n(\mu) = u^n(X) + c\rho(\bar{\mu}^n_X, \mu)$ and $U^n(\nu) = u^n(Y) + c\rho(\bar{\mu}^n_Y, \nu)$. The infimum in the definition (1.4) is attained because the space E^n is compact and, for each fixed $\mu \in \mathcal{P}(E)$, the function $E^n \ni X \mapsto u^n(X) + c\rho(\bar{\mu}^n_X, \mu)$ is continuous. Now:

$$\begin{aligned} U^n(\mu) - U^n(\nu) &\leqslant u^n(Y) + c\rho(\bar{\mu}^n_Y, \mu) - u^n(Y) - c\rho(\bar{\mu}^n_Y, \nu) \\ &= c[\rho(\bar{\mu}^n_Y, \mu) - \rho(\bar{\mu}^n_Y, \nu)] \\ &\leqslant c\rho(\mu, \nu). \end{aligned}$$

Similarly, we prove that $U^n(\nu) - U^n(\mu) \leqslant c\rho(\mu, \nu)$ by using X in the infimum defining $U^n(\nu)$. This completes the proof of the c-Lipschitz continuity.

Now since $\mathcal{P}(E)$ is a compact metric space, Arzelà-Ascoli theorem gives the existence of a subsequence $(n_k)_{k\geqslant 1}$ for which U^{n_k} converges uniformly toward a limit U and consequently:

$$\limsup_{k\to\infty} \sup_{X\in E^{n_k}} |u^{n_k}(X) - U(\bar{\mu}^{n_k}_X)| \leqslant \lim_{k\to\infty} \sup_{\mu\in\mathcal{P}(E)} |U^{n_k}(\mu) - U(\mu)| = 0,$$

which follows from the fact that $u^{n_k}(X) = U^{n_k}(\bar{\mu}^{n_k}_X)$ and which concludes the proof. □

The way we wrote formula (1.1) was in anticipation of the result of the above lemma. Indeed given this result, the *take home message* is that symmetric functions of many variables weakly depending on each of its arguments will be conveniently approximated (up to a subsequence) by regular functions of measures evaluated at the empirical distributions of its original arguments.

Returning to the game set-up of the previous subsection, we assume that the cost functions, which we denote by $(J^{N,i})_{i=1,\cdots,N}$ to emphasize the fact that they depend on N variables in A, satisfy:

Assumption (Large Symmetric Cost Functional). For each $N \geqslant 1$, there exists a function $J^N : A^N \to \mathbb{R}$ such that:

(A1) For all $N \geqslant 1$ and $(\alpha^1, \cdots, \alpha^N) \in A^N$,

$$J^{N,i}(\alpha^1, \cdots, \alpha^N) = J^N(\alpha^i, \alpha^{-i}).$$

(A2) $\sup_{N \geqslant 1} \sup_{(\alpha^1,\cdots,\alpha^N)\in A^N} |J^N(\alpha^1, \cdots, \alpha^N)| < \infty$.

(A3) There exists a finite constant $c > 0$ such that for all $N \geqslant 1$, and all $(\alpha^1, \cdots, \alpha^N), (\beta^1, \cdots, \beta^N) \in A^N$,

$$\left|J^N(\alpha^1, \cdots, \alpha^N) - J^N(\beta^1, \cdots, \beta^N)\right| \leqslant c\left[d_A(\alpha^1, \beta^1) + \rho\left(\bar{\mu}^{N-1}_{\alpha^{-1}}, \bar{\mu}^{N-1}_{\beta^{-1}}\right)\right],$$

where d_A is the distance on A and ρ is a distance on $\mathcal{P}(A)$ consistent with the weak convergence of probability measures.

According to the above conventions, the notation $\bar{\mu}^{N-1}_{\alpha^{-i}}$ in **(A3)** stands for:

$$\bar{\mu}^{N-1}_{\alpha^{-i}} = \frac{1}{N-1} \sum_{1 \leqslant j \neq i \leqslant N} \delta_{\alpha^j}.$$

Following Lemma 1.2, for each $N \geqslant 1$, $\alpha \in A$ and $\mu \in \mathcal{P}(A)$, we let:

$$\begin{aligned} &J^{(N)}(\alpha, \mu) \\ &= \inf_{(\alpha^2,\cdots,\alpha^N)\in A^{N-1}} \left[J^{(N)}\left(\alpha, \alpha^2, \cdots, \alpha^N\right) + c\rho\left(\bar{\mu}^{N-1}_{(\alpha^2,\cdots,\alpha^N)}, \mu\right)\right]. \end{aligned} \tag{1.5}$$

Clearly $J^{(N)}$ is c-Lipschitz in α. By Lemma 1.2, it is also c-Lipschitz in μ with respect to ρ. By **(A2)** in assumption **Large Symmetric Cost Functional**, the sequence $(J^{(N)})_{N \geqslant 1}$ is uniformly bounded. Repeating the proof of Lemma 1.2, we deduce that there exist a continuous function J from $A \times \mathcal{P}(A)$ into $\mathbb{R}$ and a subsequence $(N_k)_{k \geqslant 1}$ such that $(J^{(N_k)})_{k \geqslant 1}$ converges to J uniformly. Then,

$$\lim_{k\to\infty} \sup_{\alpha^{N_k} \in A^{N_k}} \left|J^{N_k}\left(\alpha^{N_k,1}, \ldots, \alpha^{N_k,N_k}\right) - J\left(\alpha^{N_k,1}, \bar{\mu}^{N_k-1}_{\alpha^{N_k,-1}}\right)\right| = 0,$$

where we used the notation $\alpha^N = (\alpha^{N,1}, \cdots, \alpha^{N,N})$ to denote N-tuples of elements in A in order to emphasize the dependence over N. As usual, $\alpha^{N,-i}$ is a short notation for $(\alpha^{N,1}, \cdots, \alpha^{N,i-1}, \alpha^{N,i+1}, \cdots, \alpha^{N,N})$.

In order to finalize the formulation of the mean field game problem as an asymptote of finite player games, we now prove the following proposition.

Proposition 1.4 *We assume that for each integer $N \geqslant 1$, $\hat{\alpha}^N = (\hat{\alpha}^{N,1}, \cdots, \hat{\alpha}^{N,N})$ is a Nash equilibrium for the game defined by the cost functions $J^{N,1}, \cdots, J^{N,N}$ which are assumed to satisfy assumption* **Large Symmetric Cost Functional**. *Also, we assume that the metric ρ in* **Large Symmetric Cost Functional** *satisfies, for all $N \geqslant 1$, $\alpha \in A$ and $\mu \in \mathcal{P}(A)$:*

$$\rho\Big(\mu, \frac{N-1}{N}\mu + \frac{1}{N}\delta_\alpha\Big) \leqslant \frac{c}{N}, \tag{1.6}$$

for a constant $c > 0$. Then there exist a subsequence $(N_k)_{k\geqslant 1}$ and a continuous function $J : A \times \mathcal{P}(A) \to \mathbb{R}$ such that the sequence $(\bar{\mu}^{N_k}_{\hat{\alpha}^{N_k}})_{k\geqslant 1}$ converges as $k \to \infty$ toward a probability measure $\hat{\mu} \in \mathcal{P}(A)$, and

$$\lim_{k\to\infty} \sup_{\alpha^{N_k} \in A^{N_k}} \Big| J^{N_k}\big(\alpha^{N_k,1}, \cdots, \alpha^{N_k,N_k}\big) - J\big(\alpha^{N_k,1}, \bar{\mu}^{N_k-1}_{\alpha^{N_k,-1}}\big)\Big| = 0, \tag{1.7}$$

and

$$\int_A J(\alpha, \hat{\mu})\, \hat{\mu}(d\alpha) = \inf_{\mu \in \mathcal{P}(A)} \int_A J(\alpha, \hat{\mu})\mu(d\alpha). \tag{1.8}$$

Remark 1.5 *Notice that property (1.8) is equivalent to the fact that the topological support of the measure $\hat{\mu}$ is contained in the set of minima of the function $\alpha \mapsto J(\alpha, \hat{\mu})$, or in other words that:*

$$\hat{\mu}\big(\{\alpha \in A :\ J(\alpha, \hat{\mu}) \leqslant J(\alpha', \hat{\mu}),\ \text{for all } \alpha' \in A\}\big) = 1. \tag{1.9}$$

To go from (1.8) to (1.9), it suffices to choose $\mu = \delta_{\alpha_0}$ in (1.8) where α_0 is a minimizer of the function $A \ni \alpha \mapsto J(\alpha, \hat{\mu})$. The converse is obvious.

Property (1.9) can be interpreted as a necessary condition for a Nash equilibrium in the limit $N \to \infty$. Obviously, it is tempting to expect it to be a sufficient condition as well, and consequently to be a characterization of a Nash equilibrium in the limit $N \to \infty$.

Remark 1.6 *The bound (1.6) is easily checked in the case of the Kantorovich-Rubinstein distance $\rho = d_{KR}$ introduced in Chapter 5, but it can also be checked directly for other metrics consistent with the weak convergence of measures.*

Remark 1.7 *In (1.8), we used the notation $\mu(d\alpha)$ in the integral, but throughout the text, we use indistinguishably the two notations $\mu(d\alpha)$ and $d\mu(\alpha)$.*

Proof. Since A is compact, we can find a subsequence $(N_k)_{k\geqslant 1}$ such that $(\bar{\mu}^{N_k}_{\hat{\alpha}^{N_k}})_{k\geqslant 1}$ converges toward a probability measure $\hat{\mu}$. Also, by Lemma 1.2, we can assume that (1.7) holds. We just check that $\hat{\mu}$ and J satisfy (1.8). By definition of a Nash equilibrium and by assumption **Large Symmetric Cost Functional**, we have:

$$\delta_{\hat{\alpha}^{N,i}} \in \arg \inf_{\mu \in \mathcal{P}(A)} \int_A J^{(N)}\Big(\alpha, \frac{1}{N-1}\sum_{j \neq i} \delta_{\hat{\alpha}^{N,j}}\Big)\mu(d\alpha),$$

for any $i \in \{1, \cdots, N\}$, where we used the fact that:

$$J^{(N)}\Big(\alpha, \frac{1}{N-1}\sum_{j \neq i} \delta_{\hat{\alpha}^{N,j}}\Big) = J^N(\alpha, \hat{\alpha}^{N,-i}),$$

see (1.5). We now use the fact that:

$$\rho\Big(\mu, \frac{N-1}{N}\mu + \frac{1}{N}\delta_\alpha\Big) \leqslant \frac{c}{N},$$

for all $\mu \in \mathcal{P}(A)$ and $\alpha \in A$. Since the functions $(J^{(N)})_{N \geqslant 1}$ are uniformly Lipschitz continuous, we can find a constant c', independent of N, such that:

$$J^{(N)}(\hat{\alpha}^{N,i}, \bar{\mu}^N_{\hat{\alpha}^N}) \leqslant \inf_{\mu \in \mathcal{P}(A)} \int_A J^{(N)}(\alpha, \bar{\mu}^N_{\hat{\alpha}^N})\mu(d\alpha) + \frac{c'}{N},$$

where $\hat{\alpha}^N = (\hat{\alpha}^{N,1}, \cdots, \hat{\alpha}^{N,N})$.

Summing over $i = 1, \cdots, N$ and dividing by N, we get:

$$\int_A J^{(N)}(\alpha, \bar{\mu}^N_{\hat{\alpha}^N})\bar{\mu}^N_{\hat{\alpha}^N}(d\alpha) \leqslant \inf_{\mu \in \mathcal{P}(A)} \int_A J^{(N)}(\alpha, \bar{\mu}^N_{\hat{\alpha}^N})\mu(d\alpha) + \frac{c'}{N}.$$

Choosing $N = N_k$ and using the fact that the sequence $(J^{(N_k)})_{k \geqslant 1}$ converges to J uniformly, we deduce that there exists a sequence of positive reals $(\epsilon_k)_{k \geqslant 1}$, converging to 0 as k tends to ∞, such that:

$$\int_A J(\alpha, \bar{\mu}^{N_k}_{\hat{\alpha}^{N_k}})\bar{\mu}^{N_k}_{\hat{\alpha}^{N_k}}(d\alpha) \leqslant \inf_{\mu \in \mathcal{P}(A)} \int_A J(\alpha, \bar{\mu}^{N_k}_{\hat{\alpha}^{N_k}})\mu(d\alpha) + \epsilon_k.$$

Letting k tend to ∞ and using the fact that J is Lipschitz continuous on $A \times \mathcal{P}(A)$, we get:

$$\int_A J(\alpha, \hat{\mu})\hat{\mu}(d\alpha) = \inf_{\mu \in \mathcal{P}(A)} \int_A J(\alpha, \hat{\mu})\mu(d\alpha),$$

which completes the proof. □

Remark 1.8 *Below, we shall use a special form of Proposition 1.4. Starting from a limiting cost functional $J : A \times \mathcal{P}(A) \to \mathbb{R}$, we shall reconstruct the N-player cost functionals $J^{N,i}$ by letting:*

$$J^{N,i}(\alpha^1, \cdots, \alpha^N) = J(\alpha^i, \bar{\mu}^{N-1}_{\alpha^{-i}}).$$

In other words, we assume that not only (1.7) *holds in the limit $k \to \infty$, but also for any $k \geqslant 1$.*

Summary of the MFG Strategy

As suggested by the definition of a Nash equilibrium, in order to solve the equilibrium problem in the large game limit, we consider a generic player who tries to minimize its objective function given the choices of the other players. We mimic the construction of the best response function by fixing a probability distribution μ (as a proxy for the choices of the other players assumed to be in large numbers), and solving the minimization problem:

$$\inf_{\alpha \in A} J(\alpha, \mu).$$

In this context, the last step of the search for equilibrium, namely the fixed point argument, amounts to making sure that the support of the measure μ is included in the set of minimizers. So the definition of a Nash point can be rewritten in the form:

$$\mathrm{Supp}(\mu) \subset \arg\inf_{\alpha \in A} J(\alpha, \mu),$$

where $\mathrm{Supp}(\mu)$ denotes the support of μ. Summarizing the above discussion, the search for Nash equilibria via the search for fixed points of the best response function can be recast as the two steps procedure:

- Fix a probability distribution $\mu \in \mathcal{P}(A)$ and solve the minimization problem $\inf_{\alpha \in A} J(\alpha, \mu)$;
- Solve the fixed point step by finding a measure $\hat{\mu}$ concentrated on the arguments of the minimization, or equivalently, satisfying (1.8) of the above proposition.

The above two step problem is what we call the MFG problem. This formulation captures the information of ALL the Nash points as $N \to \infty$. In particular, the aberrations with the weird Nash points which are expected when N is finite should disappear in the limit $N \to \infty$. Clearly, the goal of the analysis of an MFG problem is to find the equilibrium distribution $\hat{\mu}$ for the population, not so much the optimal positions of the individuals.

One of the objectives of the book is to generalize Proposition 1.4 to stochastic differential games. Indeed, the theory of mean field games is grounded in the premise of the convergence of Nash equilibria of large stochastic differential games with a mean field interaction. This is the object of Chapter (Vol II)-6. Interestingly enough, we shall provide in Chapter 7 at the end of this first volume, a counter-example showing that the limiting MFG problem may not capture all the limits of the finite-game equilibria.

Uniqueness

Existence will not be much of a problem when the cost function J is jointly continuous, which we denote by $J \in C(A \times \mathcal{P}(A))$, as long as we still assume that A is compact. Indeed, in this case, a plethora of fixed point theorems for continuous functions on compact spaces can be brought to bear on the model to prove existence.

Uniqueness is typically more difficult. In fact uniqueness is not even true in general. The main sufficient condition used to derive uniqueness (whether we consider the simple case at hand or the more sophisticated stochastic differential games which we study later) will be a monotonicity property identified by Lasry and Lions. We illustrate its workings in the present framework of static deterministic games. It will be studied in a more general context in Section 3.4 of Chapter 3.

Theorem 1.9 *Uniqueness holds if J is strictly monotone in the sense that:*

$$\int_A [J(\alpha, \mu_1) - J(\alpha, \mu_2)] d[\mu_1 - \mu_2](\alpha) > 0, \tag{1.10}$$

whenever $\mu_1 \neq \mu_2$.

Proof. If μ_1 and μ_2 are two solutions of the MFG problem, then:

$$\int J(\alpha, \mu_1) d\mu_1(\alpha) \leqslant \int J(\alpha, \mu_1) d\mu_2(\alpha),$$

because μ_1 is an optimum, and similarly,

$$\int J(\alpha, \mu_2) d\mu_2(\alpha) \leqslant \int J(\alpha, \mu_2) d\mu_1(\alpha),$$

because μ_2 is an optimum. Summing the two last inequalities and computing the left-hand side of (1.10), we find a contradiction unless the two solutions μ_1 and μ_2 are the same. □

Remark on Local Interactions. Earlier in this section, we saw conditions under which the cost functions $J^{N,i} : E^N \to \mathbb{R}$ of symmetric N player games converge as $N \to \infty$ toward a continuous function $J : A \times \mathcal{P}(A) \to \mathbb{R}$ in the sense that:

$$\sup_{\alpha \in A^N} \left| J^{N,i}(\alpha^1, \cdots, \alpha^N) - J\left(\alpha^i, \frac{1}{N-1} \sum_{1 \leqslant j \neq i \leqslant N} \delta_{\alpha^j}\right) \right| \to 0$$

through a fixed subsequence. Typically, the requirement of a weak dependence upon each of the variables could be formalized as the existence of a modulus of continuity uniform in the value of N. The fact that the limiting function J is necessarily continuous with respect to the weak topology restricts greatly the possible choices for such a function. In particular, it cannot be expected to be local. In other words, except for obvious situations with discrete state spaces, $J(\alpha, \mu)$ cannot be a function of the form $J(\alpha, \mu) = g(\alpha, p(\alpha))$ for some real valued function g on $A \times [0, \infty)$ where $p(\alpha)$ denotes the density of μ with respect to a fixed measure, computed at the point $\alpha \in A$.

Coming back to the example of the game "*Where do I put my towel on the beach?*" considered earlier in this section, as we already pointed out, a natural choice for the function $(\alpha, \mu) \mapsto J(\alpha, \mu)$ could be a function of the form:

$$J(\alpha, \mu) = J_0(\alpha) + J_1(\alpha, \mu)$$

where J_0 is a function on A (independent of μ) which any given player would want to minimize if he or she was alone, and J_1 is a scalar function on A and the space of probability measures on A trying to capture the concentration of the measure μ around α. A natural way to choose such a function would be to use a monotone function of the density of μ at α. But since we cannot use functions of the values of Radon-Nykodym densities at α, smoothing by local averaging can offer an alternative. When A is a subset of a Euclidean space, a typical way-out of this dilemma is to use a function of the form:

$$J_1(\alpha, \mu) = f\left(\frac{1}{\text{Leb}(B_\epsilon)}\mu * \mathbf{1}_{B_\epsilon}(\alpha)\right),$$

for some real valued function f on the real line. We used the notation Leb for the Lebesgue measure, and for $\epsilon > 0$, we denoted by B_ϵ the Euclidean ball of radius ϵ around the origin. Consequently, $\mu * \mathbf{1}_{B_\epsilon}(\alpha) = \int_{B_\epsilon} \mu(\alpha - d\alpha')$. So back to the N player game, the argument of f is the proportion of individuals in the ball of radius ϵ around α, so that the cost to player i is given by:

$$J^{N,i}(\alpha^1, \cdots, \alpha^N) = J_0(\alpha^i) + f\left(\frac{\#\{j;\ d(\alpha^i, \alpha^j) < \epsilon\}}{N\text{Leb}(B_\epsilon)}\right),$$

and is a function of the fraction of the population inside the neighborhood in which player i would like to choose α^i. This is a particular case of a smoothing of the measure μ to guarantee the existence of a density: if φ is a positive function with integral 1 which vanishes outside a ball around the origin, one replaces the density $p(\alpha)$, which may not exist, by the quantity $[\mu * \varphi](\alpha)$. The closer to the delta function φ is, the better the approximation of the density. Notice that the function $\mu \mapsto [\mu * \varphi](\alpha)$ is continuous with respect to the topology of the weak convergence of measures whenever φ is continuous. As we already explained, this may play a crucial role in the analysis.

An example which we will use several times in the sequel corresponds to the choice of the function $f(t) = ct^a$ for some positive exponent a and a real constant c. The case $c > 0$ corresponds to aversion to crowds while $c < 0$ indicates the desire to mingle with the crowd. This example offers instances of models for which uniqueness does not hold for some of the values of the parameters.

1.1.4 Potential Games and Informed Central Planner

At the risk of indulging in anticipation of the differentiability concepts introduced later in Chapter 5, we cannot resist the temptation to describe the connection between the theory of MFG problems as outlined above and the general theory of equilibrium in economics (whose applications most often reduce to the solutions of large linear programs). We shall recall the precise definition of potential games in Chapter 2. For the purpose of the present discussion, we identify a specific class of games in the limit $N \to \infty$. We shall say that the large game is a potential game if the cost function $J : A \times \mathcal{P}(A) \to \mathbb{R}$ is of the form:

$$J(\alpha, \mu) = \delta F(\mu)(\alpha), \tag{1.11}$$

for some function $F : \mathcal{P}(A) \to \mathbb{R}$ which is differentiable in the following sense: for every $\mu, \nu \in \mathcal{P}(A)$,

$$\lim_{\epsilon \searrow 0} \frac{F((1-\epsilon)\mu + \epsilon\nu) - F(\mu)}{\epsilon} = \int_A \delta F(\mu)(\alpha)(\nu - \mu)(d\alpha),$$

for some continuous function $\delta F(\mu)$ on A. We shall provide a complete account of the various notions of derivative for functions of measures in Chapter 5. For the time being, it suffices to observe that $(1-\epsilon)\mu + \epsilon\nu$ belongs to $\mathcal{P}(A)$, and thus $F((1-\epsilon)\mu + \epsilon\nu)$ is well defined, when $\epsilon \in [0, 1]$. Also, the right-hand side is consistent with (1.11) when J is continuous in α and A is compact, which is one of the typical assumptions we used so far.

In this case we have the following simple result.

Proposition 1.10 *Every minimum of the potential function F on $\mathcal{P}(A)$ is a solution of the MFG problem. If J is strictly monotone then the solution of the MFG is the unique minimum of F on $\mathcal{P}(A)$.*

Proof. If μ is a minimum of F, the Euler first order condition for the minimization of F reads

$$\int \delta F(\mu)(\alpha) d[\nu - \mu](\alpha) \geqslant 0$$

for all $\nu \in \mathcal{P}(A)$, which is the definition of a solution of the MFG problem. Uniqueness when J is strictly monotone was argued earlier. □

Equilibria obtained in this way by minimization of a global criterion provide a form of decentralization of the problem. This is typical of the economic equilibrium arguments based on the *invisible hand* or the mysterious *informed central planner* also known as the *representative agent*.

1.1.5 A Simple One-Period Stochastic Game

The discussion of the previous section was mostly introductory. Its purpose was to introduce important notation and definitions, and to highlight, already in the case of a simple static deterministic game, the philosophy of the mean field game approach, by finding a more tractable problem than the N-player game when N is large.

The first stochastic game model we present was called *"When does the meeting start?"* when it was first introduced. A meeting is scheduled to start at a time t_0 known from everyone, and participant $i \in \{1, \cdots, N\}$ tries to get to the meeting at time t_i (which is a number completely under the control of player i), except for the fact that despite this desire to get to the meeting at time t_i, due to uncertainties in traffic conditions and public transportations, individual i arrives at the meeting at time X^i which is the realization of a Gaussian random variable with mean t_i and variance $\sigma_i^2 > 0$ which is also random. So in this game model, the control of player i is the time t_i which we shall denote by α^i from now on in order to conform with the notation used throughout the book.

The random variables X^i may not have the same variances because the players may be coming from different locations and facing different traffic conditions. To be specific, we shall assume that for $i \in \{1, \cdots, N\}$, $X^i = \alpha^i + \sigma^i \epsilon^i$ for a sequence $(\epsilon^i)_{1 \leqslant i \leqslant N}$ of independent identically distributed (i.i.d. for short) $N(0, 1)$ random variables, and the $(\sigma^i)_{1 \leqslant i \leqslant N}$ form an i.i.d. sequence of their own which is assumed to be independent of the sequence $(\epsilon^i)_{1 \leqslant i \leqslant N}$. We denote by ν the common distribution of the $(\sigma^i)_{1 \leqslant i \leqslant N}$'s; ν is assumed to be a probability measure on $[0, \infty)$ whose topological support does not contain 0.

If a meeting scheduled to start at time t_0 actually starts at time t, and if agent i arrives at the meeting at time X^i controlled as defined above, the expected overall cost to participant i is defined as:

$$J^i(\alpha^1, \cdots, \alpha^N) = \mathbb{E}[a(X^i - t_0)^+ + b(X^i - t)^+ + c(t - X^i)^+], \tag{1.12}$$

for three positive constants a, b and c, where we use the notation x^+ to denote the positive part $\max(0, x)$ of a real number $x \in \mathbb{R}$. The interpretations of the three terms appearing in the total cost to agent i are as follows:

- the first term $a(X^i - t_0)^+$ represents a *reputation cost* for arriving late (as compared to the announced starting time t_0);
- the term $b(X^i - t)^+$ quantifies the inconvenience for missing the beginning of the meeting;
- $c(t - X^i)^+$ represents the cost of the time wasted for being early and having to wait.

We assume that the actual start time of the meeting is decided algorithmically by computing an agreed upon function of the empirical distribution $\bar{\mu}_X^N$ of the arrival times $X = (X^1, \cdots, X^N)$. In other words, we assume that $t = \tau(\bar{\mu}_X^N)$ for some function $\tau : \mathcal{P}(\mathbb{R}_+) \to \mathbb{R}$. For the sake of illustration, we can think of the case where $\tau(\mu)$ is chosen to be the $100p$-percentile of μ, for some fixed number $p \in [0, 1]$. In other words, the meeting starts when $100p$-percent of the agents already joined the meeting. This is a form of *quorum* rule. Notice that the fact that the choice of the start time t is a function of the empirical distribution $t = \tau(\bar{\mu}_X^N)$ is the source of the interactions between the agents who need to take into account the decisions of the other agents in order to make their own decision on how to minimize (1.12).

As expected, the search for a Nash equilibrium starts with the computation of the *best response function* of each player given the decisions of the other players. So we need, for each agent $i \in \{1, \cdots, N\}$, to assume that the other players have made their choices α^{-i}, and then solve the minimization problem:

$$\inf_{\alpha^i} \mathbb{E}[a(X^i - t_0)^+ + b(X^i - t)^+ + c(t - X^i)^+], \tag{1.13}$$

with $t = t(X^1, \cdots, X^N) = \tau(\bar{\mu}_X^N)$. Observe that, in contrast with the type of cost functional used in Remark 1.8, the empirical measure is here computed over the states instead of the controls, and the computation is performed over all the players as opposed to over the players different from i.

The problem (1.13) is not easy to solve, especially given the fact that the computation of the best response function needs to be followed by the computation of its fixed points in order to identify Nash equilibria. So instead of searching for exact Nash equilibria for the finite player game, we settle for an approximate solution and try to solve the limiting MFG problem. The rationale for the MFG strategy explained earlier was partly based on the convergence of the empirical distributions toward a probability measure μ. Now, because of the randomness of the arrival times and the independence assumption, the intuition behind the MFG strategy is reinforced when N is relatively large, since a form of the *law of large numbers* should kick in, and the empirical measures $\bar{\mu}_X^N$ (which are random) should converge toward a deterministic probability measure μ. Moreover, the sensitivity of $\bar{\mu}_X^N$ with respect to perturbations of α^i for a single individual should not affect the limit, and it seems reasonable to consider the alternate optimization problem obtained by replacing $t = \tau(\bar{\mu}_X^N)$ in (1.13) by $t = \tau(\mu)$. In other words, we assume that the times of arrival of the agents have a statistical distribution μ, and we compute for a representative individual, the best response to this distribution. Since we choose a rule given by a predetermined function τ of the arrival time distribution we, de facto, compute the best response, say $\hat{\alpha}$, to $t = \tau(\mu)$. The fixed point step required to determine Nash equilibria in the classical case, could be now mimicked by searching for a starting value of t which would lead to a best response $\hat{\alpha} = t$. This is exactly the MFG program outlined in the previous section. We now give the details of its implementation in the present set-up.

Proposition 1.11 *If $t \in [t_0, \infty)$ is given and $X = \alpha + \sigma\epsilon$ where $\sigma \sim \nu$ and $\epsilon \sim N(0, 1)$ are independent random variables, then for any set of strictly positive constants a, b, and c, there exists a unique minimizer $\hat{\alpha}$:*

$$\hat{\alpha} = \arg \inf_{\alpha} \mathbb{E}[a(X - t_0)^+ + b(X - t)^+ + c(t - X)^+],$$

which can be identified as the unique solution α of the implicit equation:

$$aF(\alpha - t_0) + (b + c)F(\alpha - t) = c, \tag{1.14}$$

where F denotes the distribution function of the random variable $Z = \sigma\epsilon$.

Proof. Notice that $F(z) = \int \Phi(z/s)\nu(ds)$, and that F is a strictly positive, strictly increasing continuous function satisfying $\lim_{z\searrow -\infty} F(z) = 0$, $\lim_{z\nearrow\infty} F(z) = 1$, and $1 - F(z) = F(-z)$. Here and throughout, Φ denotes the cumulative distribution function of the standard Gaussian distribution $N(0,1)$. Moreover the fact that the topological support of ν does not contain 0 implies that F is differentiable and that its derivative is uniformly bounded over $\mathbb{R}$. The quantity to minimize reads:

$$\begin{aligned}
&\mathbb{E}[a(X-t_0)^+ + b(X-t)^+ + c(t-X)^+] \\
&\quad = \mathbb{E}[a(X-t_0)^+ + (b+c)(X-t)^+ + c(t-X)] \\
&\quad = a\mathbb{E}[(\alpha - t_0 + Z)^+] + (b+c)\mathbb{E}[(\alpha - t + Z)^+] - c(\alpha - t),
\end{aligned}$$

so that, taking the derivative of the above expression with respect to α we get the first order condition of optimality:

$$a\mathbb{P}[\alpha - t_0 + Z > 0] + (b+c)\mathbb{P}[\alpha - t + Z > 0] = c, \tag{1.15}$$

which is exactly (1.14). Given the properties of F identified earlier, this equation has a unique solution $\hat{\alpha}$. □

We take care of the fixed point step in the following proposition.

Theorem 1.12 *Assume that the function $\tau : \mathcal{P}(\mathbb{R}_+) \to \mathbb{R}$ satisfies the following three properties.*

1. $\forall \mu \in \mathcal{P}(\mathbb{R}_+)$, $\tau(\mu) \geqslant t_0$, *in other words, the meeting never starts before* t_0*;*
2. Monotonicity*: if $\mu, \mu' \in \mathcal{P}(\mathbb{R}_+)$ and if $\mu([0,\alpha]) \leqslant \mu'([0,\alpha])$ for all $\alpha \geqslant 0$, then $\tau(\mu) \geqslant \tau(\mu')$;*
3. Sub-additivity*: if $\mu \in \mathcal{P}(\mathbb{R}_+)$, then for all $\alpha \geqslant 0$, $\tau(\mu(\cdot - \alpha)) \leqslant \tau(\mu) + \alpha$;*

If the constants a, b, and c are strictly positive, there exists a unique fixed point for the map $\alpha \mapsto \hat{\alpha}$, as defined in the previous proposition with $t = \tau(F(\cdot - \alpha))$.

In the statement of the theorem as well as in the subsequent proof, when we use the notation $\tau(F(\cdot - \alpha))$, we identify the cumulative distribution function $F(\cdot - \alpha)$ with the distribution of the random variable $\alpha + \sigma\epsilon$.

Proof. We are looking for a fixed point of the map $\alpha \mapsto G(\alpha) = \hat{\alpha}$ defined by:

$$\alpha \mapsto F(\cdot - \alpha) \mapsto t = \tau\Big(F(\cdot - \alpha)\Big) \mapsto \hat{\alpha} = \hat{\alpha}(t),$$

the last step being given by the solution of equation (1.14). Assuming that $x < y$, the monotonicity assumption on τ gives that:

$$\tau\Big(F(\cdot - x)\Big) \leqslant \tau\Big(F(\cdot - y)\Big),$$

and the sub-additivity assumption implies that:

$$\tau\Big(F(\cdot - y)\Big) - \tau\Big(F(\cdot - x)\Big) \leq y - x.$$

Using the special form of equation (1.14), the implicit function theorem implies that when viewing $\hat{\alpha}$ as a function of t, we have:

$$\frac{d\hat{\alpha}}{dt} = \frac{(b+c)F'(\hat{\alpha}(t)-t)}{aF'(\hat{\alpha}(t)-t_0)+(b+c)F'(\hat{\alpha}(t)-t)}$$

which is bounded from above by a constant strictly smaller than 1 because F' is nonnegative. This implies that G is a strict contraction, and that it admits a unique fixed point. □

Remark 1.13 *Notice that the* quorum *rule given by a* $100p$*-percentile of the distribution of the arrival times satisfies the three assumptions stated in the above theorem. As a natural extension of the models considered above, one could envision cases where the cost to each agent depends on more general functionals of the distribution of individual arrival times. In such a case, the optimization problem should be solved for each fixed distribution* μ*, and the fixed point part of the proof should concern* μ *instead of* t*. This is much more involved mathematically as the space of measures* μ *is infinite dimensional.*

1.2 Games of Timing

In the first stochastic game model just presented, the interventions of the players were choices affecting random times. As such, this game could have been called a game of timing. However, we simplified the model by allowing the actions of the players to be limited to the choices of the means of these random times. The standard terminology for game of timing seems to be restricted to situations when as time evolves, the information available to each player increases with time, and the choices of random times are made in a non-anticipative way vis-à-vis the information available at the time of the decision. Mathematically, this means that the random times are in fact *stopping times* for the *filtrations* representing the information available to the players.

In this section, we present two important examples of games of this type. We call them *games of timing*. As expected, we concentrate on models for which the players interact in a mean field fashion. We choose to motivate the mathematical framework with an application to a very important issue in the stability theory of the financial system: liquidity and bank runs. The first model below is intended to present the fundamental economic equilibrium principles which underpin the analysis of liquidity and bank runs in the financial system. It is static in nature, and as such, it may not fully capture the mathematical challenges we want the games of timing to address. Indeed, it reduces the choice of the time of the run on the bank to a binary decision: whether or not to withdraw a deposit, at a predetermined time. Despite the shortcomings of this oversimplification, we describe the model in extensive details to highlight the relevance of the important economic issues which are addressed. The reader is referred to the Notes & Complements at the end of the chapter for references and a simple historical perspective on early works on game models for bank runs.

The second model is fully dynamic. It is set in continuous time. It ports the important stylized facts identified in the static model to a dynamic framework in which the timing decision becomes the major technical difficulty. Since it involves diffusion processes, it is more in line with the theoretical developments presented throughout the book. We argue that if all the investors have the same information, they decide to withdraw their funds at the same time, and will collect their initial deposits *just in time*, before being hurt. This consequence of the full information assumption is highly unrealistic as large losses are always part of bank runs. Anticipating on the terminology used later in the book, this means that for the model to be relevant to the understanding of actual bank runs, it has to include what we call a *common noise*. For this reason, mean field games of timing will only be studied in Chapter (Vol II)-7 at the end of the second volume after the presentation of mean field games with a common noise.

Still, it is important to emphasize that early models of bank runs have emphasized the complementarity property of these game models which have often been called supermodular games. Intuitively speaking, these models formalize mathematically the following feature: if other customers of the bank in which you deposited your savings withdraw their money, it might be good for you to also withdraw your money from the bank.

1.2.1 A Static Model of Bank Run

Our first model is set in a *three dates/two periods* framework often used in the early economic literature on *bank runs*. While some of the assumptions are common to many studies, the gory details of the specific model presented below are borrowed from a study of banking liquidity, attempting to derive policy implications of the models, highlighting among other things, the role of the *lender of last resort*. See the Notes & Complements at the end of the chapter for references.
We first summarize the state of a bank *ex ante* (at time $t = 0$) by its balance sheet:

- D_0 is the amount of uninsured deposits, typically Certificates of Deposits (CDs), which will be repaid as $D = (1 + r)D_0$ upon withdrawal, independently of the withdrawal date, unless the bank fails before;
- E represents the bank's own funds (equity capital).

These funds are used to finance an investment of size I in risky assets like loans, the remaining funds, say M, being held in cash reserves. In particular, $D_0 + E = I + M$. For the sake of convenience, we shall eventually normalize D_0 to 1, that is $D_0 = 1$, in which case $1 + E = I + M$.

The horizon is $t = 2$, at which time the returns on the risky investments are collected, the CDs are repaid, and the stockholders of the bank get the remaining funds whenever there are any left. The returns on the bank's investments are given by the value of a Gaussian random variable $R \sim N(\overline{R}, \sigma_R^2)$ with mean $\bar{R}$ and variance σ_R^2, the value of R being revealed at time $t = 2$. The bank regulator lets the bank operate based on:

- the solvency threshold $R_S = (D - M)/I$; observe that it may be rewritten as:

$$R_S = \frac{D - (D_0 + E - I)}{I} = 1 - \frac{D_0 + E - D}{I},$$

  which becomes $R_S = 1 - (1 + E - D)/I$ when $D_0 = 1$;
- the liquidity ratio $m = M/D$.

The liquidity ratio m should be thought of as the maximum number of CDs which could be redeemed without the bank being forced to seek cash by changing its investment portfolio. We assume that the number N of investors is large and that they handed their investment decisions to fund managers.

At time $t = 1$ early withdrawals are possible. At that time, the N investors/fund managers $i \in \{1, \cdots, N\}$ have access to a private signal $X_i = R + \epsilon_i$ where the ϵ_i's are independent identically distributed Gaussian variables with mean 0 and inverse variance β, i.e., $\epsilon_i \sim N(0, \beta^{-1})$, independent of R. On the basis of the private signal X_i, each investor/fund manager makes a binary decision α^i which will be his or her control: to withdraw and collect D/N in case $\alpha^i = 1$, or do nothing if $\alpha^i = 0$. We denote by n (resp. $\bar{\alpha}^N$) the number (resp. proportion) of investors who decide to withdraw their deposits at time $t = 1$. So $n = \sum_{i=1}^N \alpha^i$ and $\bar{\alpha}^N$ is the mean of the empirical distribution $\bar{\theta}^N$ of the controls α^i. We model a bank run by the refusal of investors to renew their CDs at time $t = 1$. We assume that the investors cannot coordinate their investment strategies: if they could pool their information, they would gain near perfect knowledge of the return R which could then be treated as a deterministic constant instead of a random variable whose outcome is only revealed at time $t = 2$.

- If $\bar{\alpha}^N D \leqslant M$ or equivalently $\bar{\alpha}^N \leqslant m$, the bank remains liquid and can sustain the withdrawals;
- On the other hand, if $\bar{\alpha}^N > m$, the bank is forced to sell some of its assets, for example on the repo market, in order to meet the depositors' withdrawal requests. Let us denote by y the volume of the loans which need to be sold in order to meet the withdrawal demands.
  - If $y > I$, i.e., the bank needs to sell more than it actually owns, the bank fails at time $t = 1$.
  - Otherwise, the bank continues until $t = 2$, and failure occurs at $t = 2$ whenever $R(I - y) < (1 - \bar{\alpha}^N)D$, i.e., if the return collected from the remaining investment is not sufficient to pay back the depositors remaining invested at time $t = 2$.

We now explain how to compute y. When forced to sell, banks cannot get full price for their assets, they can only get a fraction, say $1/(1+\lambda)$ for some $\lambda > 0$, of their value. We assume that the market aggregates efficiently all the private signals, gaining perfect knowledge of the return R. Accordingly, the price at which the loan portfolio can be sold is $P = R/(1+\lambda)$. Since the number of withdrawals is n, the volume of *fire-sales* needed to compensate for the withdrawals is given by:

$$y = \frac{(\bar{\alpha}^N D - M)^+}{P} = (1+\lambda)\frac{(\bar{\alpha}^N D - M)^+}{R},$$

where we use the notation x^+ to denote the positive part $\max(0, x)$ of a real number $x \in \mathbb{R}$.

The bank is close to insolvency when the return R is small or when there is a liquidity shortage and λ is large (the interbank markets are not enough to prevent failure). If the bank needs to close at time $t = 1$, we assume that the liquidation value of its assets is νR for some constant ν much smaller than $1/(1+\lambda)$ modeling the liquidity premium.

Runs and Solvency

Let us summarize, and reorganize the various contingencies, first in a few bullet points.

- if $\bar{\alpha}^N D \leqslant M$: every run is covered without the need to sell assets at $t = 1$.
  - Failure occurs at $t = 2$ if and only if $RI + M < D \iff R < R_S$, where as above $R_S = \frac{D-M}{I} = 1 - \frac{1+E-D}{I}$ if we normalize $D_0 = 1$ and use the fact that $I + M = 1 + E$;
- if $M < \bar{\alpha}^N D \leqslant M + RI/(1+\lambda) = M + IP$: partial sale of assets at $t = 1$ to recover the $\bar{\alpha}^N D - M$ missing;
  - Failure occurs at $t = 2$ if and only if $RI - (1+\lambda)(\bar{\alpha}^N D - M) < (1-\bar{\alpha}^N)D \iff R < R_S + \lambda\frac{\bar{\alpha}^N D - M}{I} = R_S\big[1 + \lambda\frac{\bar{\alpha}^N D - M}{D-M}\big]$, because the quantity $RI - (1+\lambda)(\bar{\alpha}^N D - M)$ appearing on the left is the cash reserves left in the bank after the partial fire sales at time $t = 1$ and failure occurs at time $t = 2$ if this quantity is smaller than what needs to be given to the remaining depositors;
- finally, when $\bar{\alpha}^N D > M + PI$ the bank is closed at $t = 1$ (early closure).

The quantity R_S was defined earlier as the solvency threshold of the bank because if there are no withdrawals at $t = 1$, namely if $\bar{\alpha}^N = 0$, the bank fails at $t = 2$ if and only if $R < R_S$. The threshold R_S is a decreasing function of the solvency ratio E/I.

The second bullet point shows that solvent banks can fail when the number of early withdrawals is too large. Notice however that when the returns are high enough, the bank is *supersolvent* by which we mean $R > (1+\lambda)R_S$, in which case it can never fail, even if everybody withdraws at $t = 1$, i.e., $\bar{\alpha}^N = 1$.

While the ϵ_i appear as idiosyncratic sources of noise, R is a source of randomness common to all the participants. As a result, we expect the behavior of the investors and the failure/non-failure of the bank to depend upon the outcome of this common

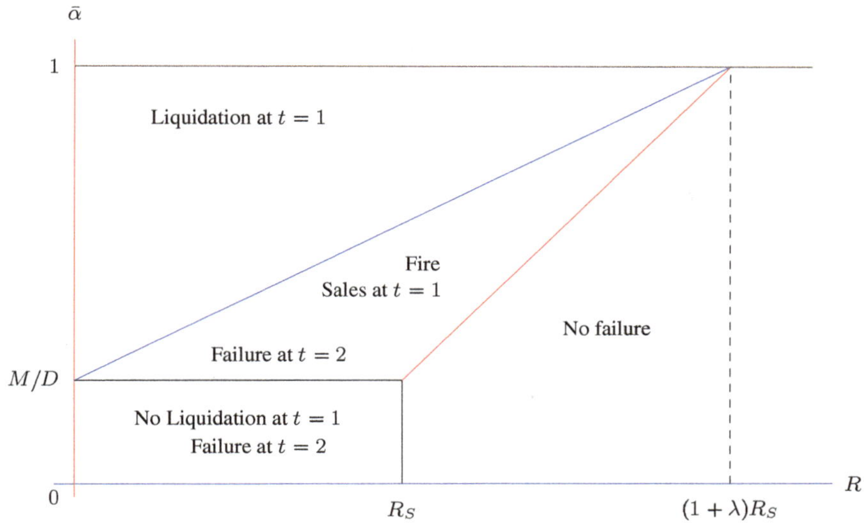

Fig. 1.1 Diagram of the possible outcomes.

noise. For this reason, we express the critical values of R as functions of the natural parameters of the model. Recall that, from the above discussion, we learned that:

- the bank is closed early if $R < R_{EC}(\bar{\alpha}^N)$ where the function R_{EC} is defined on $[0, 1]$ by:

$$R_{EC}(\alpha) = (1 + \lambda)\frac{(\alpha D - M)^+}{I} = R_S(1 + \lambda)\frac{(\alpha - m)^+}{1 - m};$$

- the bank fails if $R < R_F(\bar{\alpha}^N)$ where the function R_F is defined on $[0, 1]$ by:

$$R_F(\alpha) = R_S + \lambda\frac{(\alpha D - M)^+}{I} = R_S\Big(1 + \lambda\frac{(\alpha - m)^+}{1 - m}\Big).$$

Notice that $R_{EC}(\alpha) < R_F(\alpha)$. Our findings are further illustrated in Figure 1.1.

Investor Behavior and Equilibria

Given the parameters of the model, i.e., D, M, I, λ, it is important to realize that we were able to describe all the possible outcomes as functions of the proportion $\bar{\alpha}^N$ of investors attempting to withdraw their funds at time $t = 1$, and the actual value of the return on the risky investment at time $t = 2$. We now switch to the cost/reward analysis driving the behavior of the individual depositors and their fund managers in charge of their investments. Typically, fund managers prefer to renew their investments, and not withdraw early, but they are penalized if they are still invested when the bank fails. So we assume that:

- each fund manager gets a benefit $B > 0$ if they get their money back at time $t = 2$, or if the bank fails and they withdraw their funds at time $t = 1$, nothing otherwise;
- fund managers' compensations are based on the size of their funds more than on their returns, so each withdrawal at time $t = 1$ has a reputation cost $C > 0$.

Recall the information on the basis of which the fund managers make their decisions. At time $t = 1$, fund manager i observes the *private signal* $X_i = R + \epsilon_i$, and decides on the basis of the value of the observation, say $X_i = x$, whether or not to withdraw. In other words, the strategy of player i is a function $x \mapsto \alpha^i(x) \in \{0, 1\}$, with $\alpha^i(x) = 1$ if the decision is to withdraw the funds, and $\alpha^i(x) = 0$ if the decision is to remain invested. Since α^i is a binary function on the range of the signal X_i, it is the indicator function of a set of outcomes of X_i. So given strategies $\alpha^1, \cdots, \alpha^N$, for each player $i \in \{1, \cdots, N\}$, one can define the expected reward $J^i(\alpha^1, \cdots, \alpha^N)$ to player i given the rules and scenarios stated and reviewed in the above bullet points and Figure 1.1. We refrain from writing such an expectation in detail, but notice that it should only depend upon α^i and the average $\bar{\alpha}^N$. Rather than risking to drown in the gory details of the search for equilibria with a finite number of investors, we switch to the mean field game formulation obtained in the limit $N \to \infty$.

Finally, notice that because of the monotonicity properties alluded to in the remark below, we can restrict ourselves to strategies writing as indicator functions of intervals of the form $(-\infty, \ell_i]$, so that $\alpha^i(x) = 1$ if and only if $x \leqslant \ell_i$.

Remark 1.14 *In the case of full information, i.e., when R is common knowledge at time $t = 1$, everybody runs at time $t = 1$ on the event $\{R < R_S\}$ in which case $\bar{\alpha}^N = 1$, while nobody runs on the bank, i.e., $\bar{\alpha}^N = 0$ on the event $\{R > (1 + \lambda)R_S\}$. These two extreme equilibria co-exist in the intermediate regime $\{R_S \leqslant R \leqslant (1 + \lambda)R_S\}$. See the Notes & Complements at the end of the chapter and Section (Vol II)-7.2 of Chapter (Vol II)-7 for further discussion and references.*

Remark 1.15 *If more managers withdraw at time $t = 1$, then the probability of failure conditional on receiving a signal $X_i = x$ increases. This just means that the payoff to a fund manager displays increasing differences with respect to the actions of the other fund managers. This property is known as complementarity, and games with this property are called supermodular games. The equilibrium theory of these games is based on their order structure more than their analytic properties. We shall revisit these issues in Section (Vol II)-7.2 of Chapter (Vol II)-7. For the time being we mention that these games have a largest and a smallest equilibria. Fund managers withdraw in the largest number of scenarios they can in the largest equilibrium, while they withdraw in the smallest number of occasions in the smallest equilibrium. These extremal equilibria act as bounds for the set of equilibria, and the most natural approach to proving uniqueness is to prove that these extremal equilibria coincide.*

Mean Field Game Formulation

To guarantee symmetry, we shall assume that the investors are statistically identical, and in particular, that their initial deposits are the same, say C_0. For the limit $N \to \infty$ of a large number of depositors to make sense, we assume that $C_0 = 1/N$ to be consistent with the previous normalization $D_0 = 1$. So each individual depositor should expect to be repaid $(1+r)C_0$, and in the previous notation, $D = 1 + r$.

The idiosyncratic noise terms ϵ_i being independent and identically distributed, we expect $\bar{\alpha}^N$ to converge as $N \to \infty$ to a limit which we denote $\bar{\alpha}$. However, since the randomness of the return R is common to all the observations X_i, it may not be averaged out in the limit, and the value of $\bar{\alpha}$ is likely to depend upon R. Still, the description of the model given above for a finite number of investors can be used to describe all the possible outcomes to the individual investor and to the bank, as functions of the values of the couple $(R, \bar{\alpha}) \in \mathbb{R} \times [0, 1]$. We restate the conclusions given in the above bullet points in terms of the variables of the mean field limit, and we summarize these outcomes in Figure 1.1.

- if $\bar{\alpha} \leqslant M/D$,
  - at $t = 1$, the runs on the bank can be covered without the need to sell assets;
  - at $t = 2$, failure occurs if and only if $R < R_S$ with R_S;
- if $M/D < \bar{\alpha}$,
  - at $t = 1$ the bank needs to sell $(\bar{\alpha}D - M)(1+\lambda)/R$ worth of its loan investments in order to return $\bar{\alpha}D - M$ to the requests for withdrawal;
    - if $\bar{\alpha}D > M + PI$, i.e., if $\bar{\alpha} > p_1(R)$ with $p_1(R) = R\,I/[D(1+\lambda)] + M/D$, the bank needs to sell more than what it owns, so it is liquidated (early closure);
  - at $t = 2$
    - failure occurs if and only if $RI - (1+\lambda)(\bar{\alpha}D - M) < (1 - \bar{\alpha})D$, i.e., $\bar{\alpha} > p_2(R)$ with $p_2(R) = (R - R_S)(D - M)/(\lambda D R_S) + (M/D)$.

Recall that all these scenarios are visualized in the Figure 1.1.

Remark 1.16 *Notice that, independently of the investors' behaviors:*

- *the bank fails if the returns are too small, specifically, if $R < R_S$;*
- *the bank does not fail if the returns are sufficiently large, specifically, if $R > (1+\lambda)R_S$;*

and in the intermediate regime:

- *if $R_S \leqslant R < (1+\lambda)R_S$, the outcome depends upon how $\bar{\alpha}$ and R relate, specifically, the position of $\bar{\alpha}$ relative to the liquidity ratio $m = M/D$, and the position of the point $(R, \bar{\alpha})$ relative to the lines of equations $\bar{\alpha} = p_1(R)$ and $\bar{\alpha} = p_2(R)$.*

Characterization of the Equilibrium

The limiting formulation of the game model may go as follows. We assume the existence of two independent random variables $R \sim N(\bar{R}, \sigma_R^2)$ and $\epsilon \sim N(0, \sigma_\epsilon^2)$ standing for the return at time $t = 2$ on the risky investment of the bank, and the noise in the private signal $X = R + \epsilon$ of a representative investor. We model the investor's decision whether to run on the bank at time $t = 1$ by a binary function $\alpha : \mathbb{R} \ni x \mapsto \alpha(x) \in \{0, 1\}, \alpha(x) = 1$ if the investor decides to withdraw the deposit after observing the signal x, and $\alpha(x) = 0$ if the decision is to keep the investment until time $t = 2$. As already explained, a simple monotonicity argument can be invoked to prove that we can restrict ourselves to withdrawal strategies α for which the set $\{x \in \mathbb{R} : \alpha(x) = 1\}$ is an interval, and moreover that the decision functions α can be chosen of the form $\alpha = \mathbf{1}_{(-\infty,\ell]}$ for some $\ell \in \mathbb{R}$. The level ℓ can be thought of as the comfort level beyond which the funds may remain invested.

So given the probability distribution θ of the decision function taken by the other investors, the first step of the search for an equilibrium is to solve the optimization problem of a representative investor and find his or her best response to θ. Notice that θ is a probability on the finite set $\{0, 1\}$, so it is characterized by the number $p = \theta(\{1\})$ giving the probability that any one of the *other investors* withdraws the funds at time $t = 1$. However, because of the remark made earlier on the existence of the common noise coming from the randomness of the return R, we expect that θ is in fact a conditional probability, and that p should in fact be a function of R, namely $p = p(R)$.

Let us assume that for each choice of a function p of the form $\mathbb{R} \ni r \mapsto p(r) \in [0, 1]$, one can solve the optimization problem of the individual investor. In other words, we assume that for each possible function p, we can find a strategy $\hat{\alpha}$ in the form of a binary function of the signal value $x \in \mathbb{R}$, say $\hat{\alpha}(x) = \mathbf{1}_{(-\infty,\hat{\ell}^p]}(x)$ for which the expected return is maximal. Given this optimal threshold ℓ^p (which obviously depends upon the function $r \mapsto p(r)$ we started from), we can compute the conditional probability that the investor runs on the bank at time $t = 1$, namely the probability $\hat{p}^p(r) = \mathbb{P}[X \leq \hat{\ell}^p | R = r]$. We use an exponent ^{p} to emphasize the fact that this probability depends upon the choice of the original conditional probability function p. In these conditions, a Nash equilibrium for the mean field game corresponds to a conditional distribution (equivalently a function $r \mapsto p(r)$) such that the resulting probability at the optimum, namely $r \mapsto \hat{p}^p(r)$, is the probability function p we started from.

Notice that when the return R is deterministic or when it is known to all the investors at time $t = 1$, the function p is in fact a constant $p(r) \equiv \bar{p}$ for some $\bar{p} \in [0, 1]$. Indeed, being a conditional probability, it only makes sense on the range of R, namely for the possible values $R = r$ of the random variable R. This is consistent with Remark 1.14.

Remark 1.17 *Once an equilibrium $\hat{p}$ is found, it is possible, at least in principle, to find the probability that the point $(R, \hat{p}(R))$ falls into each single region of the diagram of Figure 1.1 giving the probability that the bank needs to be liquidated* *at time $t = 1$, or fails at time $t = 2$, and if it does, why it fails (e.g., because of a run on the bank from the fund managers, or the poor performance of the risky investments, or the losses due to the fire sales, ...). The values of these probabilities have important policy implications, and depending upon the values of the parameters of the model, regulator intervention can be proposed to alleviate some of the consequences of undesirable scenarios.*

1.2.2 A Diffusion Model of Bank Runs

We now try to capture the most important stylized facts of the above static model in a dynamic setting. We simplify the description of the balance sheet of the bank, as well as the impact of the fire sales, in order to still be able to identify the optimal timing decisions of the investors who decide to run on the bank.

We assume that the market value of the assets of a bank are given at time t by an Itô process:

$$Y_t = Y_0 + \int_0^t b_s ds + \int_0^t \sigma_s dW_s^0,$$

where the value $Y_0 > 0$ is known to everyone, and in particular to the N depositors. We assume that the assets generate a dividend cash flow at rate $\bar{r}$ strictly greater than the risk free rate r. These dividends are not reinvested in the sense that their values are not included in Y_t. The depositors are promised the same interest rate $\bar{r}$ on their deposits. The bank collapses if Y_t reaches 0.

As we did in the treatment of the static model considered above, without any loss of generality, we normalize the aggregate initial deposits to 1. Moreover, since we shall eventually cast the problem as a mean field game, we shall require a strong symmetry in the model, and as a result we shall assume that each initial deposit is in the amount $D_0^i = 1/N$. At any given time t, the liquidation value of the assets of the bank is given by $L(Y_t)$ where $L : y \mapsto L(y)$ is a deterministic continuously differentiable function satisfying:

$$L(0) = 0, \qquad L'(y) \in (0, 1), \qquad \liminf_{y \to \infty} L(y) > 1.$$

Given that the depositors can withdraw their funds at any time, the bank can tap a credit line at interest rate $\bar{r} > r$ to pay off the running depositors. At any given time t, the credit line limit is equal to the liquidation value $L(Y_t)$ of the bank's assets.

The bank is said to be safe if all depositors can be paid in full, even in case of a run. The bank is said to have liquidity problems if the current market value of its assets is sufficient to pay depositors, but the liquidation value is not. Finally, it is said to be insolvent if the current market value of its assets is less than its obligation to depositors. We shall confirm below that in the case of complete information about the solvency of the bank, depositors start to run as soon as the bank starts having liquidity problems, long before the bank is insolvent.

We now introduce an exponential random variable with parameter λ, say τ, independent of the driving Wiener process $\boldsymbol{W}^0 = (W_t^0)_{t\geqslant 0}$ (throughout the book, we often denote processes in boldface type). At time τ, the bank's assets mature and generate a single payoff Y_τ which can be used to pay the credit line and the depositors. Cash flows stop after time τ.

- if $Y_\tau \geqslant 1$, the bank is safe and everybody is paid in full;
- if $Y_\tau < 1$, we talk about an *endogenous default* since the bank cannot pay everybody in full.

The endogenous default is not the only way the bank can default. Indeed there is the possibility of an *exogenous default* at time $t < \tau$ if the mass of running depositors reaches $L(Y_t)$. Let us denote by τ^i the time at which depositor i tries to withdraw his or her deposit and let us denote by $\bar{\nu}^N$ the empirical distribution of these times, i.e.,

$$\bar{\nu}^N = \frac{1}{N}\sum_{i=1}^N \delta_{\tau^i}.$$

Notice that $\bar{\nu}^N([0,t))$ represents the proportion of depositors who tried to withdraw before time t, and that the time of endogenous default is given by:

$$\tau^{\text{endo}} = \inf\{t \in (0,\tau) : \; \bar{\nu}^N([0,t)) > L(Y_t)\}.$$

For the sake of simplicity we assume that once a depositor runs, he or she cannot get back in the game, in other words, his or her decision is irreversible.

Depositor Strategic Behavior

We now explain the strategic behavior of the N depositors. We denote by $\mathbb{F}^i = (\mathcal{F}_t^i)_{t\geqslant 0}$ the information available to player $i \in \{1,\cdots,N\}$. This is a filtration, $\mathcal{F}_t^i$ representing the information available to player i at time t. In the first model we consider, these filtrations are all identical and based on a perfect (though non-anticipative) observation of the signal $\boldsymbol{Y} = (Y_t)_{0\leqslant t\leqslant T}$. We call that *public monitoring*. In a more realistic form of the model, the filtration $\mathbb{F}^i$ will be given by the filtration $\mathbb{F}^{\boldsymbol{X}^{i,N}} = (\mathcal{F}_t^{\boldsymbol{X}^{i,N}})_{t\geqslant 0}$ generated by the process $\boldsymbol{X}^{i,N}$, the life time τ and the process $(\bar{\nu}^N([0,t]))_{t\geqslant 0}$. Here $X_t^{i,N}$ is the *private signal* of depositor i at time t, namely the value of the observation of Y_t he or she can secure at time t. We shall assume that it is of the form:

$$X_t^{i,N} = X_0^{i,N} + Y_t + \sigma_X W_t^i,$$

where $\sigma_X > 0$ and for $i \in \{1,\cdots,N\}$, the processes $\boldsymbol{W}^i = (W_t^i)_{t\geqslant 0}$ are independent Wiener processes (also independent of $\boldsymbol{W}^0$ and τ) representing idiosyncratic noise terms blurring the observations of the exact value Y_t of the assets of the bank. When $\mathbb{F}^i = \mathbb{F}^{\boldsymbol{X}^{i,N}}$ we talk about *private monitoring* of the asset value of the bank. However, for a more realistic form of the model, we shall require that the filtration $\mathbb{F}^{X^{i,N}}$ does not include the information provided by the process $(\bar{\nu}^N([0,t]))_{t\geq 0}$ which involves the private signals of the other depositors. This model will be more challenging mathematically as the individual depositors will have to choose their withdrawal strategies in a distributed manner, using only the information contained in their private signals.

In any case, the filtrations $\mathbb{F}^i$ will be specified in each particular application. Clearly τ^i should be a $\mathbb{F}^i$-stopping time in order to be admissible. In other words, depositors do not have a crystal ball to decide if and when to run.

Given that all the other players $j \neq i$ have chosen their times τ^j to try to withdraw their deposits, the payoff $P^i_t(\tau^{-i})$ at time t to depositor i for trying to run on the bank at time t (i.e., for $\tau^i = t$) can be written as:

$$P^i_t(\tau^{-i}) = D^i_0 \wedge \Big(L(Y_t) - \bar{\nu}^N([0,t))\Big)^+ \mathbf{1}_{[0,\tau)}(t),$$

if $L(Y_s) - \frac{1}{N}\sum_{j=1}^N \mathbf{1}_{[0,s)}(\tau^j) > 0$ for all $s < t$, and $P^i_t(\tau^{-i}) = 0$ otherwise. The problem of depositor i is then to choose for τ^i, the $\mathbb{F}^i$-stopping time θ solving the maximization problem:

$$J^i(\tau^{-i}) = \max_{0\leqslant\theta\leqslant\tau} \mathbb{E}\Big[e^{(\bar{r}-r)\theta} P^i_\theta(\tau^{-i})\Big]$$

which is an optimal stopping problem. Finding a set of stopping times τ^i for $i = 1,\dots,N$ satisfactory to all the players simultaneously will be achieved by finding a Nash equilibrium for this game.

Solution in the Case of Public Monitoring Through Perfect Observation

In this subsection we assume that $\sigma_X = 0$, and that the filtration $\mathbb{F}^i$ giving the information available to depositor i is given by the filtration $\mathbb{F}^Y = (\mathcal{F}^Y_t)_{t\geqslant 0}$ generated by the asset value process and the random time τ. If all the depositors have access to this information, in other words if at time t each depositor knows the past up to time t of the asset value Y_s for $s \leqslant t$, as well as whether or not τ occurred before time t, and if all the depositor decisions (to run or not to run) are based only on this information, then for each $i \in \{1,\cdots,N\}$, $\bar{\nu}^N([0,t]) \in \mathcal{F}^Y_t$ since this information is known by depositor i at time t.

Proposition 1.18 *In the case of public information, if we define the stopping time $\hat{\theta}$ by:*

$$\hat{\theta} = \Big(\tau \wedge \inf\Big\{t \geq 0 : L(Y_t) = \frac{N-1}{N}\Big\}\Big),$$

then a Nash equilibrium is when all the depositors decide to run at time $\hat{\theta}$.

So a bank run occurs as soon as the bank has liquidity problems, even if this is long before it is insolvent. Notice also that according to this proposition, all the depositors experience full recovery of their deposits, which is in flagrant contrast with typical bank runs in which depositors usually experience significant losses.

Proof. We argue that we have indeed identified a Nash equilibrium. If all other depositors but i choose the strategy given by the running time $\hat{\theta}$, we show that player i cannot do better than choosing to also run at time $\hat{\theta}$. If $L(Y_0) < (N-1)/N$, all the others depositors run immediately, and his or her only hope to get something out of his or her deposit is to run at time 0 as well. Similarly, if $L(Y_0) = (N-1)/N$ and all the other depositors run, depositor i needs to run at that time as well. Now if $L(Y_0) > (N-1)/N$, no depositor has a reason to run while $L(Y_t) > (N-1)/N$ since, by not running for a small time interval while $L(Y_t)$ is still strictly greater than $(N-1)/N$, he or she can earn the superior interest $\bar{r} > r$ without facing any risk. This proves that every depositor using $\hat{\theta}$ as time to run is a Nash equilibrium. □

The Mean Field Game Formulation

We now consider an asymptotic regime corresponding to a large number of depositors, and we track the behavior of a representative depositor after normalizing his or her deposit to 1 (while it was $1/N$ before). When N is large, we expect $\bar{\nu}^N$ to approach a probability measure ν. If the Itô process $\boldsymbol{Y}$ giving the asset value of the bank is not deterministic, this probability measure ν is likely to be random through its dependence upon the time evolution of $(Y_t)_{t \geqslant 0}$ and the value of the terminal time τ. If such a probability measure ν is fixed, one defines the individual payoff $P^\nu(t, y)$ of a withdrawal attempt at time t when the value of the assets of the bank is y and the terminal time is τ:

$$P^\nu(t, y) = 1 \wedge \Big(L(y) - \nu([0, t))\Big)^+ \mathbf{1}_{[0,\tau)}(t),$$

as long as $L(Y_s) - \mu[0, s)$ is positive for all $s < t$; otherwise the payoff is null. Then, the optimal time for a representative depositor to claim his or her deposit back is given by the stopping time (for his or her own information filtration) solving the optimal stopping problem:

$$\hat{\theta} = \arg \max_{0 \leqslant \theta \leqslant \tau} \mathbb{E}\Big[e^{(\bar{r}-r)\theta} P^\nu(\theta, Y_\theta)\Big],$$

where the argument θ is required to be a stopping time with respect to the filtration describing the available information to the typical player and where for the sake of definiteness we choose $\hat{\theta}$ to be the smallest of the optimal stopping times when uniqueness of the maximizer does not hold. As explained before, the representative player bases his/her own decision on the observation of a private signal of the form:

$$X_t = X_0 + Y_t + \sigma_X W_t, \quad t \geq 0,$$

where $\boldsymbol{W} = (W_t)_{t \geq 0}$ is independent of $(\boldsymbol{W}^0, \tau)$, σ_X being now non-zero. As $\boldsymbol{Y}$ and τ are common to all the players, the aggregate withdrawal may be captured through the conditional law $\mathcal{L}(\hat{\theta}|\boldsymbol{Y},\tau)$. Hence this creates a map $\nu \mapsto \mathcal{L}(\hat{\theta}|\boldsymbol{Y},\tau)$, and the final step of the mean field game approach is to find a fixed point μ for this map.

We shall continue the discussion of this model in Section (Vol II)-7.2 of Chapter (Vol II)-7.

1.3 Financial Applications

1.3.1 An Explicitly Solvable Toy Model of Systemic Risk

This example will stay with us throughout. We solve the finite player game explicitly in Chapter 2, both for open and Markovian closed loop equilibria, see Chapter 2 for precise definitions. In the following Chapter 3, we identify the limits as $N \to \infty$ of the solutions to the finite player games and solve the corresponding limiting problem as a mean field game. Finally, in Chapter (Vol II)-4 we revisit this example one more time to check the so-called *master equation* by explicit computations.

We describe the model as a network of N banks and we denote by $X_t^{(i)}$ the logarithm of the cash reserves of bank $i \in \{1,\cdots,N\}$ at time t. The following simple model for borrowing and lending between banks through the drifts of their log-cash reserves, while unrealistic, will serve perfectly our pedagogical objectives. For independent Wiener processes $\boldsymbol{W}^i = (W_t^i)_{0\leqslant t\leqslant T}$ for $i = 0,1,\ldots,N$ and a positive constant $\sigma > 0$ we assume that:

$$
\begin{aligned}
dX_t^i &= \frac{a}{N}\sum_{j=1}^{N}(X_t^j - X_t^i)\,dt + \sigma dB_t^i \\
&= a(\bar{X}_t - X_t^i)\,dt + \sigma dB_t^i, \quad i = 1,\ldots,N,
\end{aligned}
\tag{1.16}
$$

where:

$$
dB_t^i = \sqrt{1-\rho^2}dW_t^i + \rho dW_t^0,
$$

for some $\rho \in [-1,1]$. In other words, we assume that the log-cash reserves are Ornstein-Uhlenbeck (OU) processes reverting to their sample mean $\bar{X}_t$ at a rate $a > 0$. This sample mean represents the interaction between the various banks. We also consider a negative constant $D < 0$ which represents a critical liability threshold under which a bank is considered in a state of default.

A remarkable feature of this model is the presence of the Wiener process $\boldsymbol{W}^0$ in the dynamics of all the log-cash reserve processes $\boldsymbol{X}^i$. While these state processes are usually correlated through their empirical distribution, when $\rho \neq 0$, the presence of this *common noise* $\boldsymbol{W}^0$ creates an extra source of dependence which makes the solution of mean field games much more challenging. Models with a common noise will only be studied in the second volume because of their high level of technicality. However, we shall see that the present model can be solved explicitly whether or not $\rho = 0$!

The following are easy consequences of the above assumptions. Summing up equations (1.16) shows that the sample mean $(\bar{X}_t)_{0\leqslant t\leqslant T}$ is a Wiener process with volatility $\sigma/\sqrt{N}$. Simple Monte Carlo simulations or simple computations show that stability of the system is easily achieved by increasing the rate a of borrowing and lending given by the parameter a. Moreover, it is plain to compute analytically the loss distribution, i.e., the distribution of the number of firms whose log-cash reserves cross the level D, and large deviations estimates (which are mere Gaussian tail probability estimates in the present situation) show that increasing a increases systemic risk understood as the simultaneous default of a large number of banks.

While attractive, these conclusions depend strongly on the choice of the model and its specificities. We now consider a modification of the model which will hopefully lead to an equilibrium in which we expect the same conclusions to hold. We consider the new dynamics:

$$dX_t^i = \left[a(\bar{X}_t - X_t^i) + \alpha_t^i\right] dt + \sigma dB_t^i, \quad i = 1, \cdots, N,$$

where $\boldsymbol{\alpha}^i$ is understood as the control of bank i, say the amount of lending and borrowing outside of the N bank network (e.g., issuing debt, borrowing at the Fed window, etc). In this modified model, firm i tries to *minimize*:

$$J^i(\boldsymbol{\alpha}^1, \cdots, \boldsymbol{\alpha}^N)$$
$$= \mathbb{E}\left[\int_0^T \left(\frac{1}{2}|\alpha_t^i|^2 - q\alpha_t^i(\bar{X}_t - X_t^i) + \frac{\epsilon}{2}(\bar{X}_t - X_t^i)^2\right) dt + \frac{c}{2}(\bar{X}_T - X_T^i)^2\right],$$

for some positive constants ϵ and c which balance the individual costs of borrowing and lending with the average behavior of the other banks in the network. The parameter $q > 0$ weighs the contributions of the relative sizes of these components imposing the choice of the sign of α_t^i and the decision whether to borrow or lend. The choice of q is likely to be the regulator's prerogative. Notice that:

- If X_t^i is small relative to the empirical mean $\bar{X}_t$, bank i will want to borrow and choose $\alpha_t^i > 0$;
- If X_t^i is large, then bank i will want to lend and set $\alpha_t^i < 0$.

Throughout the analysis, we shall assume that:

$$q^2 \leqslant \epsilon \tag{1.17}$$

to guarantee the convexity of the running cost functions. In this way, the problem is an instance of a linear-quadratic game (LQ for short). We shall solve explicitly several forms of this model in Chapter 2.

Remark 1.19 *We used the notation* $\bar{x} = (x^1 + \cdots + x^N)/N$ *for the empirical mean of* all *the* x^j*s. So in the above specification of the model, each bank interacts with the* *empirical mean of the states of* all *the banks, including its own state. As we often do when we discuss games with mean field interactions (see for example Section 2.3 of Chapter 2), it may be more natural to formalize the interactions as taking place between the state of a given bank with the empirical mean of the states of the* other *banks. Doing so would require to replace* $(\bar{X}_t - X_t^i)$ *by:*

$$\overline{X_t^{-i}} - X_t^i = \frac{1}{N-1} \sum_{1 \leqslant j \neq i \leqslant N} [X_t^j - X_t^i]$$
$$= \frac{1}{N-1} \sum_{1 \leqslant j \leqslant N} [X_t^j - X_t^i]$$
$$= \frac{N}{N-1} (\bar{X}_t - X_t^i).$$

This shows that the model would be exactly the same as long as we multiply the constants a and q by $(N-1)/N$ *and the constants* ϵ *and c by* $(N-1)^2/N^2$*. So for N fixed, the qualitative properties of the models should be the same, and in any case, the quantitative differences should disappear in the limit* $N \to \infty$*.*

1.3.2 A Price Impact Model

The model presented in this section is of great importance in modern financial engineering because it is used as input to many optimal execution engines in the high frequency electronic markets. Our interest in the model is that it presents an instance of stochastic differential game in which individuals interact through the empirical distribution of their controls instead of the empirical distributions of their private states, which is in stark contrast with what we will find in most of the examples studied in this book. These models are sometime called *extended mean field games*. We devote Section 4.6 of Chapter 4 to their analysis, and we shall revisit the present example of price impact in Subsection 4.7.1.

The Market Model

We analyze the interaction between N traders. Trader i controls its inventory X_t^i, i.e., the number of shares owned at time t by its *rate of trading* α_t^i through a stochastic differential equation of the form:

$$dX_t^i = \alpha_t^i \, dt + \sigma^i dW_t^i, \quad t \in [0, T],$$

where the $\boldsymbol{W}^i = (W_t^i)_{t \geqslant 0}$ are independent standard Wiener processes and the volatilities $\sigma^i \geqslant 0$ are assumed to be constant for the sake of simplicity, and all equal to the same positive number $\sigma > 0$ for symmetry reasons. All the agents trade the same stock whose mid-price at time t is denoted by S_t. The amount of cash held by trader i at time t is denoted by K_t^i. It evolves according to:

$$dK_t^i = -[\alpha_t^i S_t + c(\alpha_t^i)]\, dt,$$

where the function $\alpha \mapsto c(\alpha)$ is a nonnegative convex function satisfying $c(0) = 0$, representing the *cost for trading at rate* α. For example, using for trading cost function $c(\alpha) = c\alpha^2$ for some constant $c > 0$ would correspond to a flat order book.

The actual price impact is encapsulated in the following formula:

$$dS_t = \frac{1}{N}\sum_{i=1}^{N} h(\alpha_t^i)\, dt + \sigma_0 dW_t^0, \quad t \in [0, T],$$

for the changes over time of the mid-price. Here we assume that $\alpha \mapsto h(\alpha)$ is a deterministic function known to everyone, $\sigma_0 > 0$ is a constant, and $\mathbf{W}^0 = (W_t^0)_{t \geqslant 0}$ is a standard Wiener process independent of the family $(\mathbf{W}^i)_{1 \leqslant i \leqslant N}$. This is a particular case of the classical model of Almgren and Chriss for permanent price impact. Since the drift of the mid-price is the integral of the function h with respect to the empirical measure $\bar{\mu}_{\alpha_t}^N$ of the controls α_t^i, we see that in this model, each participant interacts with the empirical distribution of the controls of the other participants. In order to avoid unruly notation, we shall denote by $\langle h, \bar{\mu}_{\alpha_t}^N \rangle$ this integral.

Note that S_t follows an arithmetic Brownian motion with a drift which depends on the accumulated impacts of previous trades. The function h is sometimes called the *instantaneous market impact function*. Since a buy is expected to increase the price of the stock and a sell will tend to decrease the stock price, the function h should satisfy $h(\alpha)\alpha \geqslant 0$. Linear, power-law and logarithmic functions are often used in practice for that reason. Price impact models are most of the time used in optimal execution problems for high frequency trading. Therefore the fact that $(S_t)_{0 \leqslant t \leqslant T}$ can become negative is not a real issue in practice since the model is most often used on a time scale too short for the mid-price to become negative.

The Trader's Optimization Problem

The wealth V_t^i of trader i is defined as the sum of the cash held by the trader and the value of the inventory as marked to the mid-price:

$$V_t^i = K_t^i + X_t^i S_t.$$

If we use the standard self-financing condition of Black-Scholes' theory (see the Notes & Complements at the end of the chapter for references to alternatives), the changes over time of the wealth V_t^i are given by the equation:

$$\begin{aligned} dV_t^i &= dK_t^i + X_t^i\, dS_t + S_t\, dX_t^i \\ &= \Big[- c(\alpha_t^i) + X_t^i \frac{1}{N}\sum_{j=1}^{N} h(\alpha_t^j) \Big] dt + \sigma S_t dW_t^i + \sigma_0 X_t^i dW_t^0. \end{aligned} \tag{1.18}$$

When the controls are square integrable and the instantaneous market impact function h has at most linear growth, the processes X_t^i and S_t are square integrable and the stochastic integrals in the wealth's dynamic are martingales. We assume that the traders are subject to a running liquidation constraint modeled by a function c_X of the shares they hold, and to a terminal liquidation constraint at maturity T represented by a scalar function g. Usually c_X and g are convex nonnegative quadratic functions in order to penalize unwanted inventories. If as usual we denote by J^i the expected costs of trader i as a function of the controls of all the traders we have:

$$J^i(\boldsymbol{\alpha}^1, \cdots, \boldsymbol{\alpha}^N) = \mathbb{E}\biggl[\int_0^T c_X(X_t^i)dt + g(X_T^i) - V_T^i\biggr]. \tag{1.19}$$

By (1.18), we can write the expected cost to trader i as:

$$J^i(\boldsymbol{\alpha}^1, \cdots, \boldsymbol{\alpha}^N) = \mathbb{E}\biggl[\int_0^T f(t, X_t^i, \bar{\mu}^N_{\alpha_t}, \alpha_t^i)dt + g(X_T^i)\biggr], \tag{1.20}$$

where as before, we use the notation $\bar{\mu}^N_{\alpha_t}$ for the empirical distribution of the N components of α_t, which is a probability measure on the space A in which the controls take their values. Here, the running cost function f is defined by:

$$f(t, x, \theta, \alpha) = c(\alpha) + c_X(x) - x\langle h, \theta\rangle, \tag{1.21}$$

for $0 \leqslant t \leqslant T$, $x \in \mathbb{R}^d$, $\theta \in \mathcal{P}(A)$, and $\alpha \in A$. This model is a perfect archetype of a N-player stochastic differential game with interactions through the empirical distribution of the controls. We shall not be able to solve the finite player game problem. We shall approach its solution via the analysis of the mean field game formulation based on the intuition (and the results proved earlier) of the limit $N \to \infty$ of a large number of players.

Remark 1.20 *It is important to emphasize the crucial role played by the innocent looking assumption that the traders are risk neutral as they choose to minimize the expectation of their cost (as opposed to a nonlinear function of the costs). Indeed, the quadratic variation terms disappear in the computation of the expected cost, and so doing, the common noise* $\mathbf{W}^0$ *as well as the mid-price* $\mathbf{S}$ *disappear from the expression of the individual expected costs. The game would be much more difficult to solve it they didn't.*

Since the common noise disappeared from the model, if we restrict the rates of trading to functions of the inventories, or in the case of open loop models, if we assume that they are adapted to the filtrations generated by the $\mathbf{W}^i$ and are independent of the common noise, the independence of the random shocks dW_t^i suggest that in the limit $N \to \infty$ the empirical measures $\bar{\mu}^N_{\alpha_t}$ converge, provided that the rates are sufficiently symmetric, toward a deterministic measure θ_t which, in equilibrium, should be the distribution of the optimal rate of trading of a generic trader. We switched from the notation μ_t to θ_t in order to emphasize the fact that the interaction between the players is now through the empirical distribution of the controls, hence a probability measure in $\mathcal{P}(A)$ instead of a probability measure in $\mathcal{P}(\mathbb{R}^d)$. Consequently, the MFG approach to this game is to fix a deterministic flow $\boldsymbol{\theta} = (\theta_t)_{t\geqslant 0}$ of measures on the space A of controls, solve the optimal control problem:

$$\begin{cases} \inf_{\boldsymbol{\alpha}} \mathbb{E}\Big[\int_0^T f(t, X_t, \theta_t, \alpha_t)dt + g(X_T)\Big] \\ dX_t = \alpha_t dt + \sigma dW_t, \quad t \in [0, T], \end{cases} \tag{1.22}$$

for a given Wiener process $\boldsymbol{W}$ and then, try to find a flow of measures $\boldsymbol{\theta} = (\theta_t)_{t\geqslant 0}$ so that $\theta_t = \mathcal{L}(\hat{\alpha}_t)$ where $\hat{\boldsymbol{\alpha}} = (\hat{\alpha}_t)_{0\leqslant t\leqslant T}$ is an optimal control for the above problem.

As a general rule, this form of mean field game is more difficult to solve than a standard MFG problem where the interaction between the players is through the empirical distribution of their states. As already explained at the beginning of this subsection, we shall call these models *extended mean field games*. We study them in Section 4.6 of Chapter 4, and we solve this particular model of price impact in Subsection 4.7.1.

1.4 Economic Applications

In this section, we introduce several general equilibrium economic growth models. While the first is structured in the spirit of Aghion and Howitt's model, the next two examples can be regarded as variations on the theme of incomplete markets and uninsured idiosyncratic labor income risk as proposed by Aiyagari.

1.4.1 Pareto Distributions and a Macro-Economic Growth Model

Our first instance of economic application was originally introduced as an example of mean field game with a common noise. Here, we review its main features as we plan to use it as an example for which the so-called master equation can be stated and solved. See Chapter (Vol II)-4 for details.

Background

As for many models in the economic literature, the problem was set for an infinite time horizon ($T = \infty$) with a positive discount rate $r > 0$, but to be consistent with the rest of the text, we shall frame it with a finite horizon T. In this model, the private states of the individual agents are not subject to idiosyncratic shocks. They react to a common source of noise given by a one-dimensional Wiener process $\boldsymbol{W}^0 = (W_t^0)_{0\leqslant t\leqslant T}$. We denote by $\mathbb{F}^0 = (\mathcal{F}_t^0)_{0\leqslant t\leqslant T}$ its filtration. We also assume that the volatilities of these states are linear as given by a function $x \mapsto \sigma x$ for some positive constant σ, and that each player controls the drift of his or her own state so that the dynamics of the state of player i read:

$$dX_t^i = \alpha_t^i dt + \sigma X_t^i dW_t^0, \quad t \in [0, T]. \tag{1.23}$$

We shall restrict ourselves to Markovian controls of the form $\alpha_t^i = \alpha(t, X_t^i)$ for a deterministic function $(t, x) \mapsto \alpha(t, x)$, which will be assumed to be nonnegative and Lipschitz in the variable x. See Chapter 2 for a detailed discussion of the use of these special controls. Under these conditions, for any player, say player i, $X_t^i \geqslant 0$ at all times $t > 0$ if $X_0^i \geqslant 0$, and for any two players, say players i and j, the homeomorphism property of Lipschitz stochastic differential equations (SDE for short) implies that $X_t^i \leqslant X_t^j$ at all times $t > 0$ if $X_0^i \leqslant X_0^j$.

For later purposes we notice that when the Markovian control is of the form:

$$\alpha(t, x) = \gamma_t x, \tag{1.24}$$

then,

$$X_t^j = X_t^i + (X_0^j - X_0^i) e^{\int_0^t \gamma_s ds - (\sigma^2/2)t + \sigma W_t^0}. \tag{1.25}$$

We assume that $k > 0$ is a fixed parameter, and we introduce a special notation for the family of scaled Pareto distributions with decay parameter k. For any real number $q > 0$, we denote by $\mu^{(q)}$ the Pareto distribution:

$$\mu^{(q)}(dx) = k \frac{q^k}{x^{k+1}} \mathbf{1}_{[q,\infty)}(x) dx. \tag{1.26}$$

Notice that for any random variable X, $X \sim \mu^{(1)}$ is equivalent to $qX \sim \mu^{(q)}$. We shall use the notation μ_t for the conditional distribution of the state X_t of a generic player at time $t \geqslant 0$ conditioned by the knowledge of the past up to time t as given by $\mathcal{F}_t^0$. Under the prescription (1.24), we claim that, if $\mu_0 = \mu^{(1)}$, then $\mu_t = \mu^{(q_t)}$ where $q_t = e^{\int_0^t \gamma_s ds - (\sigma^2/2)t + \sigma W_t^0}$. In other words, conditioned on the history of the common noise, the distribution of the states of the players remains Pareto with parameter k if it starts that way, and the left most point of the support of the distribution, say q_t, can be understood as a sufficient statistic characterizing the distribution μ_t. This remark is an immediate consequence of formula (1.25) applied to $X_t^i = q_t$, in which case $q_0 = 1$, and $X_t^j = X_t$, implying that $X_t = X_0 q_t$. So if $X_0 \sim \mu^{(1)}$, then $\mu_t \sim \mu^{(q_t)}$. This simple remark provides an explicit formula for the time evolution of the (conditional) marginal distributions of the states. As we shall see, this time evolution is generally difficult to come by, and requires the solution of a forward Partial Differential Equation (PDE for short) known as forward Kolmogorov equation or forward Fokker-Planck equation, which in the particular case at hand should be a Stochastic Partial Differential Equation because of the presence of the common noise. More details will be given in Chapters 2 and 4 (second volume).

Optimization Problems

We now introduce the reward functions of the individual agents and define their optimization problems. The technicalities required to describe the interactions between the agents at a rigorous mathematical level are a hindrance to the intuitive understanding of the nature of these interactions. Indeed, the reward functions would have to be defined in such a way to accommodate empirical distributions for the fact that the latter do not have densities with respect to the Lebesgue measure. Overcoming this technical difficulty would force us to *jump through hoops*, which we consider as an unnecessary distraction at this stage of our introduction to mean field games. For the time being, we define the running reward function f by:

$$f(x,\mu,\alpha) = c\frac{x^a}{[(d\mu/dx)(x)]^b} - \frac{E}{p}\frac{\alpha^p}{[\mu([x,\infty))]^b},$$

for $x, \alpha \geqslant 0$ and $\mu \in \mathcal{P}(\mathbb{R}_+)$ and for some positive constants a, b, c, E and $p > 1$ whose economic meanings are discussed in the references provided in the Notes & Complements at the end of the chapter. We use the convention that the density is the density of the absolutely continuous part of the Lebesgue's decomposition of the measure μ, and that in the above sum, the first term is set to 0 when this density is not defined or is itself 0. Similarly, the second term is set to 0 when μ does not charge the interval $[x, \infty)$.

Solutions of the MFG problem as formulated in this section will be given in Section 4.5.2 of Chapter (Vol II)-4.

1.4.2 Variation on Krusell-Smith's Macro-Economic Growth Model

We first consider a version of a stochastic growth model originally proposed by Aiyagari and later extended by Krusell and Smith. Our presentation follows the lines of the paper of Krusell and Smith whose contribution is twofold. First they added an aggregate source of random shocks (what we shall call common noise throughout the book) to the idiosyncratic shocks originally considered by Aiyagari, and second, they proposed an original algorithm leading to actual computations of equilibrium statics. While we shall not comment on the reliability and/or merits of the numerical algorithm, we notice and we find quite remarkable that the description of its components clearly outlines, step by step, the mean field game strategy articulated in this chapter. The version we present here differs from the original contributions of Aiyagari and Krusell Smith in two respects: we consider the finite horizon case, and we set up the model in continuous time. Also, we formulate the problem for a finite number of agents instead of directly modeling the economy as a continuum using a nonatomic measure space with measure one for the space of agents.

One major difference with the growth model discussed in the previous subsection is the fact that, on the top of the common noise affecting all the states, we also consider idiosyncratic random shocks specific to each individual agent in the economy. While the random shocks are assumed to be independent and identically distributed with a common distribution in Aiyagari's model, for the sake of definiteness, we first discuss the approach of Krusell and Smith in which the shocks are kept discrete and finite in nature for the purpose of numerical implementation. In the next subsection, we change the nature of the random shocks by introducing Wiener process to recast the model in the framework of stochastic differential games.

Modeling the Uncertainty in the Economy

As usual, we denote by N the number of agents in the economy. For the purpose of the present discussion, we can think that they are consumers. The randomness in the model is given by a set of N continuous time Markov chains $(z_t, \eta_t^i)_{t\geqslant 0}$ which, at any time t, can take four possible values $(1-\Delta_z, 0)$, $(1-\Delta_z, 1)$, $(1+\Delta_z, 0)$, and $(1+\Delta_z, 1)$ for some constant $\Delta_z \geqslant 0$. We shall assume that, given the knowledge of the $\boldsymbol{z} = (z_t)_{t\geq 0}$-component, the $\boldsymbol{\eta}^i$ are independent for $i = 1, \cdots, N$. The economic interpretation of these random sources is the following. The shocks $\boldsymbol{z} = (z_t)_{t\geq 0}$ capture the health of the overall economy, like an aggregate productivity measure, so $z_t = 1 + \Delta_z$ in good times, and $z_t = 1 - \Delta_z$ in bad times. Clearly the case $\Delta_z = 0$ corresponds to the absence of common random shocks. The second component $\boldsymbol{\eta}^i$ is specific to the consumer, $\eta_t^i = 1$ when consumer i is employed, and $\eta_t^i = 0$ whenever he or she is unemployed.

Remark 1.21 *In the absence of the* common noise z, *the model was originally proposed by Aiyagari. It consists of a more traditional game with independent idiosyncratic noise terms given by the individual changes in employment.*

The production technology is modeled by a Cobb-Douglas production function in the sense that the per-capita output is given by:

$$Y_t = z_t K_t^{\alpha} \, (\bar{\ell} L_t)^{1-\alpha} \tag{1.27}$$

where K_t and L_t stand for per-capita capital and employment rates respectively. Here the constant $\bar{\ell}$ can be interpreted as the number of units of labor produced by an employed individual. In such a model, two quantities play an important role: the capital rent r_t and the wage rate w_t. In equilibrium, these marginal rates are defined as the partial derivatives of the per-capita output Y_t with respect to capital and employment rate respectively. So,

$$r_t = r(K_t, L_t, z_t) = \alpha z_t \left(\frac{K_t}{\bar{\ell} L_t} \right)^{\alpha - 1}, \tag{1.28}$$

and

$$w_t = w(K_t, L_t, z_t) = \bar{\ell}(1-\alpha)z_t\left(\frac{K_t}{\bar{\ell}L_t}\right)^{\alpha}. \tag{1.29}$$

Individual Agent's Optimization Problem

Each agent i controls his or her consumption c_t^i at time t, and maximizes his or her expected utility of overall consumption:

$$\mathbb{E}\left[\int_0^T \beta^t U(c_t^i)dt + \tilde{U}(c_T^i)\right],$$

for some utility function U and scrap function $\tilde{U}$. The discount factor $\beta \in (0,1]$ does not play any significant role in the finite horizon version of the model so we take it equal to 1 in this case. However, it is of crucial importance in the infinite horizon version for which the optimization is to maximize:

$$\mathbb{E}\left[\int_0^\infty \beta^t U(c_t^i)dt\right],$$

in which case $\beta \in (0,1)$. To conform with most of the economic literature on the subject, we use the power utility function:

$$U(c) = \frac{c^{1-\gamma}-1}{1-\gamma}, \tag{1.30}$$

also known as CRRA (short for Constant Relative Risk Aversion) utility function and its limit as $\gamma \to 1$ given by the logarithmic utility function.

The constraints on the individual consumption choices are given by the values of the individual capitals k_t^i at time t which need to remain nonnegative at all times. The changes in capital over time are given by the equation:

$$dk_t^i = \left[(r_t - \delta)k_t^i + [(1-\tau_t)\bar{\ell}\eta_t^i + \bar{\delta}(1-\eta_t^i)]w_t\right]dt - c_t^i dt. \tag{1.31}$$

Here, the constant $\delta > 0$ represents a depreciation rate. The second term in the above right-hand side represents the wages earned by the consumer. It is equal to $\bar{\delta}w_t$ when the consumer is unemployed, a quantity which should be understood as an unemployment benefit rate. On the other hand, it is equal to $(1-\tau_t)\bar{\ell}w_t$ after adjustment for taxes, when he or she is employed. Here,

$$\tau_t = \frac{\bar{\delta}u_t}{\bar{\ell}L_t},$$

where $u_t = 1 - L_t$ is the unemployment rate.

The above form (1.31) of the dynamics of the state variables k_t^i is rather deceiving as they partially hide the coupling between the equations. The main source of coupling comes from the quantities r_t and w_t which not only depend upon the common noise z_t, but also depend upon the aggregate capital K_t which is in some sense the average of the individuals $(k_t^i)_{1\leqslant i\leqslant N}$!

Mean Field Game Formulations

Let us denote by $\bar{\mu}_k^N$ and for each $i \in \{1, \cdots, N\}$ by $\bar{\mu}_k^{-i,N}$ the empirical distributions of the capital and of the capitals endowments of all the consumers other than i:

$$\bar{\mu}_k^N = \frac{1}{N}\sum_{i=1}^N \delta_{k_t^i}. \quad \text{and} \quad \bar{\mu}_k^{-i,N} = \frac{1}{N-1}\sum_{j=1, j\neq i}^N \delta_{k_t^j}, \quad i = 1, \cdots, N.$$

Because of de Finetti's law of large numbers, we expect that these empirical measures converge when the size N of the economy tends to ∞. It is clear that if these limits exist, they have to be the same. Let us denote by μ_t^z the common limit as $N \to \infty$. This limit will give the *conditional distribution* of capital k_t given the history of the states $(z_s)_{0\leq s\leq t}$ of the economy, hence our notation. Clearly, μ_t^z is independent of z in the case $\Delta_z = 0$ of no common noise. If one searches for Nash equilibria in the limit $N \to \infty$, assuming the knowledge of the empirical distribution of the (other) individual capitals is reduced to assuming the knowledge of the flow $(\mu_t^z)_{0\leqslant t\leqslant T}$ of probability measures. In the case $\Delta_z > 0$, since z can only take two values $1-\Delta_z$ and $1+\Delta_z$, the same assumption amounts to the knowledge of deterministic flows of measures $(\mu_t^\zeta)_{\geq 0}$ parameterized by paths ζ with values in $\{d, u\}$, where d and u are the possible values of z, say down and up, namely $d = 1-\Delta_z$ and $u = 1+\Delta_z$.

Once the flow of conditional measures is assumed to be known, the computation of the best response of a representative agent reduces to the solution of the optimal control problem:

$$\max_c \mathbb{E}\left[\int_0^T \beta^t U(c_t)dt + \tilde{U}(c_T)\right],$$

under the constraints $k_t \geqslant 0$ and:

$$dk_t = \left[(r(K_t, L_t, z_t) - \delta)k_t + [(1-\tau_t)\bar{\ell}\eta_t + \bar{\delta}(1-\eta_t)]w(K_t, L_t, z_t)\right]dt - c_t dt.$$

Here, $(z_t, \eta_t)_{0\leqslant t\leqslant T}$ is a continuous time Markov chain with the same law as any of the $(z_t, \eta_t^i)_{0\leqslant t\leqslant T}$ introduced earlier, the rental rate function r and the wage level function w are as in (1.28) and (1.29), and $K_t = \bar{k}_t^z$ is the mean of the conditional measure μ_t^z, namely:

$$K_t = \int_{[0,\infty)} k\mu_t^z(dk),$$

and where L_t is as above. This is the source of the mean field interaction in the model.

1.4.3 A Diffusion Form of Aiyagari's Growth Model

In this form of the model, the private state at time t of agent i is a two-dimensional vector $X_t^i = (Z_t^i, A_t^i)$. As before, the agents $i \in \{1, \cdots, N\}$ can be viewed as the workers comprising the economy: Z_t^i gives the labor productivity of worker i, and A_t^i his or her wealth at time t. The time evolutions of the states are given by stochastic differential equations:

$$\begin{cases} dZ_t^i &= \mu_Z(Z_t^i)dt + \sigma_Z(Z_t^i)dW_t^i, \\ dA_t^i &= [w_t^i Z_t^i + r_t A_t^i - c_t^i]dt, \end{cases} \tag{1.32}$$

for some functions $\mu_Z, \sigma_Z : \mathbb{R} \to \mathbb{R}$. Here, the random shocks are given by N independent Wiener processes $\boldsymbol{W}^i = (W_t^i)_{t\geqslant 0}$, for $i = 1, \cdots, N$, r_t is the interest rate at time t, w_t^i represents the wages of worker i at time t, and the consumption $\boldsymbol{c}^i = (c_t^i)_{t\geqslant 0}$ is the control of player i. Notice that in this form of the model, the random shocks are idiosyncratic, in other words, the model does not include a common noise.

Remark 1.22 *In many economic applications, a borrowing limit $A_t^i \geqslant \underline{a}$ is imposed with $\underline{a} \leqslant 0$. Moreover, the processes $\boldsymbol{Z}^i = (Z_t^i)_{t\geqslant 0}$ are also restricted by requiring that they are ergodic, or even restricted to an interval $[\underline{z}, \bar{z}]$ for some finite constants $0 \leqslant \underline{z} < \bar{z} < \infty$.*

In this model, given processes $\boldsymbol{r} = (r_t)_{t\geqslant 0}$ and $\boldsymbol{w}^i = (w_t^i)_{t\geqslant 0}$ for $i = 1, \cdots, N$, each worker tries to maximize:

$$J^i(\boldsymbol{c}^1, \cdots, \boldsymbol{c}^N) = \mathbb{E}\int_0^\infty e^{-\rho t} U(c_t^i)dt. \tag{1.33}$$

As usual in economic applications, the model is set up in infinite horizon, and U is an increasing concave utility function common to all the workers. We now explain how the workers interact in the economy, and how the interest rate and the wage processes are determined in equilibrium. As before, the aggregate production in the economy is given by a production function $Y = F(K, L)$, the total capital K_t supplied in the economy at time t being given by the aggregate wealth:

$$K_t = \int a\bar{\mu}_{X_t}^N(dz, da) = \frac{1}{N}\sum_{i=1}^N A_t^i, \tag{1.34}$$

while the total amount of labor L_t supplied in the economy at time t is normalized to 1. In a competitive equilibrium the interest rate and the wages are given by the partial derivatives of the production function:

$$\begin{cases} r_t &= [\partial_K F](K_t, L_t)|_{L_t=1} - \delta, \\ w_t &= [\partial_L F](K_t, L_t)|_{L_t=1}, \end{cases}$$

where $\delta \geqslant 0$ is the rate of capital depreciation, so that $r_t + \delta$ can be viewed as the user cost of capital. So in equilibrium, the interaction between the agents in the economy is through the mean K_t of the empirical distribution $\bar{\mu}^N_{\mathbf{A}_t}$ of the workers' wealth $(A^i_t)_{1 \leqslant i \leqslant N}$. More generally, we capture the state of the economy at time t by the empirical measure:

$$\bar{\mu}^N_{\mathbf{X}_t} = \frac{1}{N} \sum_{i=1}^N \delta_{X^i_t}.$$

Practical Application

We shall revisit this diffusion model several times in the sequel. It will be one of the testbeds we use to apply the tools developed in the book. We first solve the model in Subsection 3.6.3 of Chapter 3 as an example of Mean Field Game (MFG) without common noise. We revisit this solution in Subsection 6.7.4 of Chapter 6 in light of our analysis of the optimal control of McKean-Vlasov dynamics. In order to check the assumptions under which our results are proven, we use a specific model for the mean reverting labor productivity process $\mathbf{Z} = (Z_t)_{t \geqslant 0}$. We choose an Ornstein-Uhlenbeck process for the sake of definiteness. As already mentioned, we use the CRRA isoelastic utility function with constant relative risk aversion:

$$U(c) = \frac{c^{1-\gamma} - 1}{1 - \gamma}, \tag{1.35}$$

for some $\gamma > 0$, with $U(c) = \ln(c)$ if $\gamma = 1$, in which case:

$$U'(c) = c^{-\gamma} \quad \text{and} \quad (U')^{-1}(y) = y^{-1/\gamma}. \tag{1.36}$$

As before, we use the Cobb-Douglas production function:

$$F(K, L) = \bar{a}\, K^{\alpha}\, L^{1-\alpha},$$

for some constants $\bar{a} > 0$ and $\alpha \in (0, 1)$. With this choice, in equilibrium, we have:

$$r_t = \alpha \bar{a} K_t^{\alpha-1} L_t^{1-\alpha} - \delta, \quad \text{and} \quad w_t = (1-\alpha)\bar{a} K_t^{\alpha} L_t^{-\alpha},$$

and since we normalized the aggregate supply of labor to 1,

$$r_t = \frac{\alpha \bar{a}}{K_t^{1-\alpha}} - \delta, \qquad \text{and} \qquad w_t = (1-\alpha)\bar{a}K_t^{\alpha}, \tag{1.37}$$

where K_t is given by (1.34) and provides the mean field interaction.

The unconstrained version (i.e., without any sign constraint on $\boldsymbol{Z}$ or $\boldsymbol{A}$) of this model will be solved as a mean field game in Subsection 3.6.3 in Chapter 3 where we provide numerical illustrations of some of the properties of the solution.

1.4.4 Production of Exhaustible Resources

In the first part of this section, we present a macro-economic model for exhaustible resources. We consider N oil producers in a competitive economy. We denote by $x_0^1, \cdots, x_0^N$ the initial oil reserves of the N producers. Each producer tries to control his or her own rate of production so that, if we denote by X_t^i the oil reserves of producer i at time t, the changes in reserves are given by equations of the forms:

$$dX_t^i = -\alpha_t^i dt + \sigma X_t^i dW_t^i, \quad t \geqslant 0, \tag{1.38}$$

where $\sigma > 0$ is a volatility level common to all the producers, the nonnegative adapted and square integrable processes $(\boldsymbol{\alpha}^i = (\alpha_t^i)_{t\geq 0})_{i=1,\cdots,N}$ are the controls exerted by the producers, and the $(\boldsymbol{W}^i = (W_t^i)_{t\geq 0})_{i=1,\cdots,N}$ are independent standard Wiener processes. The interpretation of X_t^i as an oil reserve requires that it remains nonnegative. However, we shall not say that much on this sign constraint for the purpose of the present discussion. If we denote by P_t the price of one barrel of oil at time t, and if we denote by $C(\alpha) = \frac{b}{2}\alpha^2 + a\alpha$ the cost of producing α barrels of oil, then producer i tries to maximize:

$$J^i(\boldsymbol{\alpha}^1, \cdots, \boldsymbol{\alpha}^N) = \sup_{\boldsymbol{\alpha}:\alpha_t \geq 0, X_t \geq 0} \mathbb{E}\left[\int_0^\infty [\alpha_t^i P_t - C(\alpha_t^i)]e^{-rt}dt\right]. \tag{1.39}$$

The price P_t is the source of coupling between the producer strategies. The model is set up over an infinite horizon with a discount factor $r > 0$. In this competitive economy model, the price is given by a simple equilibrium argument forcing supply to match demand. The demand at time t, when the price is p, is given as $D(t,p)$ by a demand function D.

Mean Field Formalization

We follow the approach of Guéant, Lasry and Lions who suggest to consider the mean field game problem, short-circuiting in this way the difficulties of the finite player games. They work with the demand function $D(t,p) = we^{\rho t}p^{-\gamma}$ where $we^{\rho t}$ gives the total wealth in the economy, and γ represents the elasticity of substitution between oil and other goods. We shall also use the inverse demand function D^{-1} characterized by the fact that $q = D(t,p) \Longleftrightarrow p = D^{-1}(t,q)$.

In the mean field limit, we fix a deterministic flow $\boldsymbol{\mu} = (\mu_t)_{t\geqslant 0}$ of probability measures, interpreting μ_t as the distribution of the oil reserves of a generic producer at time t. Notice that the knowledge of $(\mu_t)_{t\geqslant 0}$ determines the price $(P_t)_{t\geqslant 0}$ since:

$$P_t = p(t, \boldsymbol{\mu}) = D^{-1}\left(t, -\frac{d}{dt}\int x\mu_t(dx)\right), \quad t \geqslant 0, \tag{1.40}$$

which is a mere statement that the quantity produced is given by the negative of the change in reserves. The optimization problem which needs to be solved to determine the best response to this flow of distributions is based on the running reward function of a representative producer. It is given by:

$$f(t, x, \boldsymbol{\mu}, \alpha) = [\alpha p(t, \boldsymbol{\mu}) - C(\alpha)]e^{-rt},$$

where the quantity $p(t, \boldsymbol{\mu})$ is given by (1.40). Accordingly, the value function of the representative producer is defined by:

$$u^{\boldsymbol{\mu}}(t, x) = \sup_{(\alpha_s)_{s\geq t}:\alpha_s\geq 0, X_s\geq 0} \mathbb{E}\left[\int_t^\infty [\alpha_s P_s - C(\alpha_s)]e^{-r(s-t)}\right]ds,$$

the supremum being computed under the dynamical constraint:

$$dX_s = -\alpha_s ds + \sigma X_s dW_s, \qquad X_t = x. \tag{1.41}$$

As usual, once the best response is found by solving this optimization problem for a given initial condition x_0 at time 0, the fixed point argument amounts to finding a measure flow $\boldsymbol{\mu} = (\mu_t)_{t\geqslant 0}$ so that the marginal distributions of the optimal paths $(X_t)_{t\geqslant 0}$ are exactly the distributions $(\mu_t)_{t\geqslant 0}$ we started from. In particular, μ_0 must be equal to δ_{x_0}. This model was originally treated in a Partial Differential Equation (PDE for short) formalism which we shall call *the analytic approach* in this book: the optimization problem is solved by computing the value function $u^{\boldsymbol{\mu}}$ of the problem as the solution of the corresponding Hamilton-Jacobi-Bellman (HJB for short) equation, and the fixed point property is enforced by requiring $(\mu_t)_{t\geqslant 0}$ to satisfy a forward Kolmogorov (or Fokker-Planck) PDE to guarantee that $(\mu_t)_{t\geqslant 0}$ is indeed the distribution of the optimal paths. Kolmogorov's equation is linear and forward in time while the HJB equation is nonlinear and backward in time. These two equations are highly coupled and, more than the nonlinearities, the opposite directions of the time are the major source of difficulty in solving these equations. The authors did not solve them analytically. Numerical approximations were provided as illustration. As we shall see, the occurrence of a forward-backward system is a characteristic feature of the analysis of mean field games. It will be a mainstay of our probabilistic approach as we start emphasizing in Chapter 3.

Remark 1.23 *The reader will notice that the dependence of P_t upon $\boldsymbol{\mu}$ in* (1.40) *is not of the form $P_t = p(t, \mu_t)$ but of the more complicated form $P_t = p(t, \boldsymbol{\mu})$ where $p(t, \cdot)$ is a functional of the path $\boldsymbol{\mu} = (\mu_s)_{s\geqslant 0}$ (at least when regarded in a neighborhood of t).*

In order to fit the formulation we have used so far, we may notice by taking the expectation in (1.41) *that:*

$$\frac{d}{dt}\mathbb{E}[X_t] = -\mathbb{E}[\alpha_t], \quad t \geqslant 0,$$

which shows that:

$$\frac{d}{dt}\int x\mu_t(dx) = -\mathbb{E}[\alpha_t], \quad t \geqslant 0,$$

when $\boldsymbol{\mu}$ is an equilibrium. Therefore, we can recast the search for an equilibrium as the search for a fixed point on the marginal distributions of the controls (and not of the states). Precisely, a deterministic flow of probability measures $\boldsymbol{\theta} = (\theta_t)_{t\geq 0}$ forms an equilibrium if each θ_t, for $t \geqslant 0$, coincides with the marginal distribution at time t of an optimal control process in the optimization problem:

$$\sup_{\boldsymbol{\alpha}:\alpha_t\geq 0, X_t\geq 0} \mathbb{E}\left[\int_0^\infty [\alpha_t P_t^{\boldsymbol{\theta}} - C(\alpha_t)]e^{-rt}\right]dt,$$

the supremum being computed under the dynamical constraint:

$$dX_t = -\alpha_t dt + \sigma X_t dW_t, \qquad t \geqslant 0, \tag{1.42}$$

and $(P_t^{\boldsymbol{\theta}})_{t\geqslant 0}$ being given by:

$$P_t^{\boldsymbol{\theta}} = D^{-1}\left(t, \int x\theta_t(dx)\right), \quad t \geqslant 0.$$

Remark 1.24 *It is not too much of a stretch to imagine that the above mean field formulation can be tweaked to include terms in the maximization which incentivize producers to avoid being the last to produce, the effects of externalities, the impact of new entrants producing from alternative energy sources, ...*

Cournot and Bertrand Variations on the Same Model

In a subsequent study (see the Notes & Complements at the end of the chapter for references), it was suggested to look at dynamics:

$$dX_t^i = -\alpha_t^i dt + \sigma dW_t^i, \qquad t \geq 0, \tag{1.43}$$

with absorption at 0 to guarantee that the reserves of a generic oil producer do not become negative, and to assume that, as in most models for Cournot games, the price P_t^i experienced by each producer, is given by a linear inverse demand function of the rates of productions of all the other players in the form:

$$P_t^i = 1 - \alpha_t^i - \frac{\epsilon}{N-1} \sum_{1 \leqslant j \neq i \leqslant N} \alpha_t^j,$$

so that if we denote by θ the distribution of the rate of extraction α, the running reward function in the mean field game regime becomes:

$$f(t, x, \theta, \alpha) = \big(\alpha[1 - \alpha - \epsilon\bar{\theta}] - C(\alpha)\big)\, e^{-rt}\mathbf{1}_{x>0}.$$

1.5 Large Population Behavior Models

Biologists, social scientists, and engineers have a keen interest in understanding the behavior of schools of fish, flocks of birds, animal herds, and human crowds. Understanding collective behavior resulting from the aggregation of individual decisions and actions is a time honored intellectual challenge which has only received partial answers. In this section, we introduce a small sample of mathematical models which have been proposed for the analysis and simulation of the behavior of large crowds from individual *rules of conduct*.

1.5.1 The Cucker-Smale Model of Flocking

In a groundbreaking contribution, Cucker and Smale gave a complete mathematical analysis of a form of Vicsek model which they propose for the behavior of flocks of birds. Their model is now known as the Cucker Smale model of flocking. According to these authors, the collective motion of the birds can be captured mathematically by a high dimensional deterministic dynamical system, namely a system of Ordinary Differential Equations (ODEs for short) where the state at time t of each bird i in a flock of N birds is described by a 6-dimensional vector $X_t^i = [x_t^i, v_t^i]$ where x_t^i represents its position and v_t^i its velocity at time t. As per their model, the time evolution of the states is given by the system:

$$\begin{cases} dx_t^i = v_t^i dt, \\ dv_t^i = \sum_{j=1}^N w_{i,j}(t)[v_t^j - v_t^i]dt, \quad t \geqslant 0, \end{cases}$$

for a family of weights defined as:

$$w_{i,j}(t) = w(|x_t^i - x_t^j|) = \frac{\tilde{\kappa}}{(1 + |x_t^i - x_t^j|^2)^\beta}, \tag{1.44}$$

for some constants $\tilde{\kappa} > 0$ and $\beta \geqslant 0$. The first equation is a mere consistency condition since it states that the velocity is the time derivative of the position. The second equation says that the changes in the velocity of a bird are given by a weighted average of the differences between its velocity and the velocities of the other birds in the flock, a form of mean reversion toward the mean velocity. Given the form of the weights posited in (1.44), the further apart are the birds, the smaller the weights. Notice that the exact value of the mean reversion constant $\tilde{\kappa}$ does not play much role when N is fixed. However, when the size of the flock increases, namely when N grows, it is natural to expect that κ should be of order $1/N$ for the system to remain stable. More on that later on.

The fundamental result of Cucker and Smale's original mathematical analysis is that if N is fixed and $0 \leqslant \beta < 1/2$, then:

$$\begin{cases} \lim_{t\to\infty} v_t^i = \overline{v}_0^N, \quad \text{for } i = 1, \cdots, N, \\ \sup_{t\geqslant 0} \max_{i,j=1,\cdots,N} |x_t^i - x_t^j| < \infty, \end{cases} \tag{1.45}$$

irrespective of the initial configuration. The first bullet point states that for large times, all the birds in the flock eventually align their velocities, while the second bullet point implies that the birds remain bunched, hence the relevance of the analysis of this system of ODEs to the biological theory of flocking phenomena. When $\beta > 1/2$, flocking can still occur depending upon the initial configuration. Since the publication of the original paper of Cucker and Smale, many extensions and refinements appeared, including a treatment of the case $\beta = 1/2$. See the Notes & Complements at the end of the Chapter for details and references.

The extension which teased our curiosity was proposed by Nourian, Caines, and Malhamé in the form of an equilibrium problem. Instead of positing a phenomenological description of the behavior of the birds in the flock, the idea is to let the birds decide of the macroscopic behavior of the flock by making rational decisions at the microscopic level. By rational decision, we mean resulting from a careful risk-reward optimization. So in this new formulation, the behavior of the flock of N birds will still be captured by their individual states $X_t^i = [x_t^i, v_t^i]$ which have the same meanings as before, but whose dynamics are now given by Stochastic Differential Equations (SDEs):

$$\begin{cases} dx_t^i = v_t^i dt, \\ dv_t^i = \alpha_t^i dt + \sigma dW_t^i, \quad t \geqslant 0. \end{cases}$$

While the first equation has the same obvious interpretation, the second just says that except for random shocks given by the increments of a Wiener process σdW_t^i proper to the specific bird, each bird can control the changes in its velocity through the term $\alpha_t^i dt$. However, this control comes at a cost which each bird will try to minimize. To be specific, for a given strategy profile $\boldsymbol{\alpha} = (\boldsymbol{\alpha}^1, \cdots, \boldsymbol{\alpha}^N)$ giving the control choices of all the birds over time, the cost to bird i is given by:

$$J^i(\boldsymbol{\alpha}) = \lim_{T\to\infty} \frac{1}{T} \int_0^T \left(\frac{1}{2}|\alpha_t^i|^2 + \frac{1}{2}\left| \sum_{j=1}^N w_{i,j}(t)[v_t^j - v_t^i] \right|^2 \right) dt. \tag{1.46}$$

The special form of these cost functionals is very intuitive and can be justified in the following way: by trying to minimize this cost, each bird tries to save energy (minimization of the contribution from the first term) in order to be able to go far, and tries to align its velocity with those close to him in order to remain in the pack and avoid becoming an easy prey to aggressive predators. While introducing the infinite horizon model stated in (1.46), the authors noticed that in the case $\beta = 0$, the nonlinear weights $w_{i,j}(t)$ are independent of i, j, and t, and the model reduces to a Linear Quadratic (LQ) game which can be solved. They also suggest to approach the case $\beta > 0$ by perturbation techniques for $\beta \ll 1$ (i.e., β small), but fall short of the derivation of asymptotic expansions which could be used to analyze the qualitative properties of the model.

For the purpose of illustration, we recast their model in the finite horizon set-up, even though this will presumably prevent us from feeling the conclusions (1.45) of the deterministic model which account for large time properties of the dynamical system. Our reason to work on a finite time interval is to conform with the notation and the analyses which the reader will find throughout the book. So the dynamics of the velocity become:

$$dv_t^i = \alpha_t^i dt + \sigma dW_t^i, \quad t \in [0, T],$$

and we rewrite the individual costs in the form:

$$J^i(\boldsymbol{\alpha}) = \mathbb{E}\left[\int_0^T f(t, X_t^i, \bar{\mu}_t^N, \alpha_t^i) dt \right], \tag{1.47}$$

with:

$$f(t, X, \mu, \alpha) = \frac{1}{2}|\alpha|^2 + \frac{\kappa^2}{2}\left| \int_{\mathbb{R}^6} \frac{v' - v}{(1 + |x - x'|^2)^\beta} \mu(dx', dv') \right|^2 \tag{1.48}$$

where $t \in [0, T]$, $X = [x, v]$, $\mu \in \mathcal{P}(\mathbb{R}^6)$ and $\alpha \in \mathbb{R}^3$.

Remark 1.25 *We explain how the constant κ should relate to the constant $\tilde{\kappa}$ introduced earlier in* (1.44). *If we want the probability measure $\bar{\mu}_t^N$ appearing in formula* (1.47) *to be the empirical measure of the states of the other birds, namely the $X_t^j = [x_t^j, v_t^j]$ for $j \neq i$, then we need to choose $\kappa^2 = (N-1)^2\tilde{\kappa}^2$. On the other hand, if we want this probability measure to be the empirical measure of all the bird states, namely the $X_t^j = [x_t^j, v_t^j]$ for $j = 1, \cdots, N$ including $j = i$, then we need to choose $\kappa^2 = N^2\tilde{\kappa}^2$.In any case, since we already noticed that in the large flock limit, the constant $\tilde{\kappa}$ should be of order $1/N$, κ should be viewed as a* dimensionless *constant independent of the size of the flock.*

One of the main challenges of this model is the fact that the *running cost* function f is not convex for $\beta > 0$. This complicates significantly the solution of the optimization problem. In particular, such a running cost function will not satisfy the typical assumptions under which we provide solutions for this type of models. Moreover, many population biologists have argued that more general interactions (e.g., involving quantiles of the empirical distribution) are needed for the model to have any biological significance. Finally, restricting the random shocks affecting the system to idiosyncratic shocks *attached* to individual birds is highly unrealistic, as an ambient source of noise common to all individuals should be present in the physical environment in which the birds are evolving.

Remark 1.26 *In this example, like in many other examples, each individual interacts, inside the running cost, with the empirical distribution of the states of the other individuals involved in the game.*

The particular case $\beta = 0$ will be solved explicitly in Section 2.4 of Chapter 2 for a finite number of birds, and in Section 3.6.1 of Chapter 3 in the mean field game limit when the number of birds is infinite. The general case will be analyzed with the tools developed for the probabilistic approach to the solution of mean field games in Subsection 4.7.3 of Chapter 4. There, we propose a solution to a slightly modified model, and we provide numerical illustrations.

1.5.2 An Attraction-Repulsion Model

Another popular model of large population behavior is the self-propelling friction and attraction-repulsion model defined by the time evolution:

$$\begin{cases} dx_t^i = v_t^i dt, \\ dv_t^i = \Big[(a - b|v_t^i|^2)v_t^i - \dfrac{1}{N}\displaystyle\sum_{j=1}^N \nabla U(x_t^i - x_t^j)\Big]dt, \quad t \geqslant 0, \end{cases}$$

where a and b are nonnegative parameters and $U : \mathbb{R}^3 \to \mathbb{R}$ is a given potential function modeling the short range repulsion and long range attraction between the individual members of the population. One often uses the Morse potential defined by:

$$U(x) = -C_A e^{-|x|/\ell_A} + C_R e^{-|x|/\ell_R}, \tag{1.49}$$

where C_A, C_R, and ℓ_A, ℓ_R are the strengths and the typical lengths of attraction and repulsion respectively. As in the case of the Cucker-Smale model of flocking, we turn this deterministic descriptive model into a stochastic differential game with mean field interactions by defining the same controlled dynamics as before:

$$\begin{cases} dx_t^i = v_t^i dt, \\ dv_t^i = \alpha_t^i dt + \sigma dW_t^i, \quad t \geqslant 0, \end{cases}$$

each individual trying to minimize the cost:

$$J^i(\boldsymbol{\alpha}) = \mathbb{E}\left[\int_0^T f(t, X_t, \bar{\mu}_t^N, \alpha_t^i)dt\right],$$

with:

$$f(t, X, \mu, \alpha) = \frac{1}{2}|\alpha|^2 + \frac{1}{2}\left|\int_{\mathbb{R}^6} U(x - x')\mu(dx', dv')\right|^2,$$

where $X = [x, v]$ as before. Notice that the value of the running cost function does not depend upon the v-component of the state X, and that it only depends upon the x-marginal of the probability distribution of the state.

1.5.3 Congestion Models for Crowd Behavior

This subsection is an attempt to introduce, in the spirit and with the notation of this chapter, several models of crowd behavior. Part of our motivation is to consider models with different groups of individuals for which the mean field limit and the mean field game strategy apply separately to each group. But first, as a motivation, we start with a single group as we did so far.

Congestion Model for a Single Group of Individuals

For the sake of simplicity, we do not model position and velocity separately as before. We assume that the changes in position of individual $i \in \{1, \cdots, N\}$ are given by an equation of the form:

$$dX_t^i = \alpha_t^i dt + \sigma dW_t^i, \quad t \geqslant 0, \tag{1.50}$$

where the $(\boldsymbol{W}^i)_{1 \leqslant i \leqslant N}$ are N independent standard Wiener processes, $\sigma > 0$, and $\boldsymbol{\alpha}^i$ is a square integrable process giving individual i a form of control on the evolution of its position over time. In other words, each individual chooses its own velocity, at least up to the idiosyncratic noise shocks σdW_t^i. If individual i is at x at time t, its velocity is α, and the empirical distribution of the other individuals is μ, then it faces a running cost given by the value of the function:

$$f(t, x, \mu, \alpha) = \frac{1}{2}|\alpha|^2\left(\int_{\mathbb{R}^3} \rho(x - x')d\mu(x')\right)^a + e^{-rt}k(t, x). \tag{1.51}$$

Here ρ is a smooth density with a support concentrated around 0, and the function k models the effect of panic which frightens individuals depending on where they are, though this effect is dampened with time through the *actualization factor* e^{-rt} for which we assume that $r \geqslant 0$. The power $a \geqslant 0$ is intended to penalize congestion since large positive values of a penalize the kinetic energy, and make it difficult to move where the highest density of population can be found.

We shall generalize the model to the case of two subpopulations (which are often called species in biological applications) in Chapter 7.

Forced Exit from a Room

In this example, we assume that the individuals need to leave a room in a rush for the exit. We model the geometry of the room by a bounded domain D in $\mathbb{R}^d$, and we assume that the exit is only possible through a subset Γ of the boundary ∂D of the domain. Individuals hitting the boundary ∂D away from the exit Γ will bounce back inside the room in a motion of panic. Inside the room, we assume that the individuals control their motion according to the same law as above in (1.50).

From a mathematical point of view, this example poses a certain number of new challenges which were not present in the models discussed so far. While the boundedness of the domain D could make our search for compactness easier, especially when we seek out Nash fixed points, this assumption requires us to confine the stochastic dynamics of the state to D forcing us to specify the boundary behavior of the state processes controlled by the individual players. Typically, we shall impose reflecting boundary conditions on $\partial D \setminus \Gamma$, and Dirichlet boundary condition on Γ to model the fact that the individuals disappear when they hit the door.

These technicalities can be overwhelming mathematically, so we spend a significant amount of time in Subsection 4.7.2 of Chapter 4 to explain how they can be handled with appropriate stochastic analysis tools. There, we provide theoretical solutions and numerical illustrations showing the impact of the congestion parameter on the time a typical individual takes to exit the room.

1.6 Discrete State Game Models

The purpose of this section is to introduce a special class of models which, while not central to the subject matter of the book, still play a crucial role in many practical applications of great importance. For these models, although the time variable varies continuously like with all the subsequent developments in the book, the states controlled (or at least influenced) by the players are restricted to a discrete set which we shall assume to be finite in some cases. These models cannot be cast as stochastic differential games, and they require a special treatment which we provide in Section 7.2 of Chapter 7 and extend to games with major and minor players in Subsection 7.1.9 of Chapter 7 in Volume II.

1.6.1 A Queuing-Like Limit Order Book Model

In this model, we assimilate the Limit Order Book (LOB for short) of an electronic trading exchange, to a set of M different queues, so that at each time t, the state of the limit order book is given by the lengths of these queues. When one of the N agents, typically a trading program, arrives and is ready to trade, the value X_t^i of its private state is kept to 0 if it decides to leave and not enter any queue, or to $j \in \{1, \cdots, M\}$ if it decides to enter the j-th queue. So, in this particular instance, the space in which the states of the players evolve is the finite set $E = \{0, 1, \cdots, M\}$ instead of the Euclidean space $\mathbb{R}^d$ in most of the examples treated in this book. While $X_t = (X_t^1, \cdots, X_t^N)$ gives the states of the N market participants at time t, the empirical distribution:

$$\bar{\mu}_{X_t}^N = \frac{1}{N}\sum_{i=1}^N \delta_{X_t^i} = \frac{1}{N}\sum_{k=0}^M \Big(\sum_{i: X_t^i = k} \delta_k \Big) = \sum_{k=0}^M \frac{\#\{i : X_t^i = k\}}{N} \delta_k,$$

contains the histogram of the relative lengths of the individual queues. For the sake of convenience, we shall denote by $m_t = (m_t^1, \cdots, m_t^M)$ the relative lengths of these queues, namely $m_t^k = \#\{i : X_t^i = k\}/N$. So because the private states of the individual agents can only take finitely many values at any given time, 0, 1, $\cdots$, M in this instance, the marginal distribution of the system can be identified to an element of $\mathbb{R}^{M+1}$, or an element of the M-dimensional probability simplex in this Euclidean space if we wanted to be more specific.

So if we want to capture a property of the game via a real valued function $U(t, x, m)$ of time t, of the state x of an individual player at time t, and of the statistical distribution of this state at time t, say m, such a function can be viewed as a function $u : [0, T] \times \mathbb{R}^{M+1} \to \mathbb{R}^{M+1}$ with the convention $[u(t, m)]_x = u_x(t, m) = U(t, x, m)$, the value of the individual state determining which component of the vector u we use. So if the function U were to appear as the solution of a complex equation in its three variables, or even an infinite dimensional Partial Differential Equation (PDE), such an equation could be rewritten as an equation for the function u which would appear as the solution of a simpler system (say a system of scalar ordinary differential equations, ODEs for short). Such a function U will be introduced in Chapter 7 as a solution of what we shall call the master equation, equation which will be studied in full detail in Chapter 4 of Volume II.

1.6.2 A Simple Cyber Security Model

We now discuss a second instance of departure from the great majority of models treated in the book which happen to have a continuum state space. In this example, not only do we consider a finite state space, but we also break the symmetry among the players as we single out one of them. Note that this special player could also be a small number of players which we would bundle together into what we often call a *major player*. This special player faces a large number of *opponents* which will be assumed to be statistically similar, restoring the framework of mean field game models in this large group. We shall use the terminology *minor player* to refer to each element of this homogeneous group. Stochastic differential mean field games with major and minor players will be studied in Section 7.1 of Chapter 7 in Volume II, the special case of games with finite state spaces being discussed in Subsection 7.1.9 at the end of that same section.

The example we choose to illustrate these two new features is inspired by academic works on cyber security, to which an extensive literature on two-player games has been devoted. Typically, the first player, characterized as the *attacker*, tries to infect, or take control of, the computers of a network administered and protected by a *defender*. The connectivity of the network and the relative importance of the nodes dictate the defensive measures implemented by the administrator of the network. The costs incurred as a result of the attacks and the implementation of defensive measures, cast the model as a *zero-sum game* since the cost to the network parallels the reward to the attacker. Zero-sum games are very popular in the mathematical literature. This stems mostly from the fact that, since the analysis reduces to the study of one single value function (as opposed to one value function per player), the techniques of stochastic control can be extended with a minimal overhead.

In the model which we study in detail in Chapter (Vol II)-7, we consider an interconnected network of N computers labeled by $i \in \{1, \cdots, N\}$, which we identify to their owners or users, and whose levels of security depend upon the levels of security of the other computers in the same network. For the sake of definiteness, we shall assume that each computer can be in one of four possible states: DI for "defended infected"; DS for "defended and susceptible to infection"; UI for "unprotected and infected"; and finally US for "unprotected and susceptible to infection." Each computer user makes an investment trying to keep its machine secure by installing anti-virus filters, setting up firewalls, ... and pays a cost for this investment. On the other hand, the attacker will be rewarded for taking control of computers in the network, and pay a cost for the implementation of attacks on the network, the intensity of its attack and the associated cost depending upon the proportion of computers in the network already infected. This last feature is what guarantees the mean field nature of the model.

1.6.3 Searching for Knowledge

The purpose of this subsection is to present a model inspired by the works of Duffie and collaborators on information percolation as an illustration of a game model for which the state space is countable. These authors model how individuals improve their own knowledge of an unknown random quantity by meeting with other individuals and exchanging information. One of the main features of these models is to involve a continuum of participants so that, when two of them meet, to be sure, neither one previously met any of the persons the other one met in the past. This assumption guarantees enough independence to provide sufficient statistics reducing the complexity of the game. For example, when agents try to guess the value of a random variable (say a Gaussian $N(0, 1)$ random variable) by sharing information when they meet, the number of past encounters happens to be a sufficient statistic whenever the information of each agent is in the form of a private signal which is also a Gaussian random variable.

In Chapter 3 Section 3.7, we provide the mathematical setting needed to make rigorous sense of the continuum of independent random variables without losing measurability requirements, and we prove a weak form of the exact law of large numbers at the core of this modeling assumption.

Our stylized version of the model also assumes that each single one of N agents can improve the level of his or her information only by meeting other agents and sharing information. It would be easy to add a few bells and whistles to allow each individual to increase his or her information on his or her own, but this would add significantly to the complexity of the notation without adding to the crux of the matter, or to its relevance to the theory of mean field games. So we shall assume that, in order to improve his or her knowledge of an unknown random quantity, each agent tries to meet other agents, that the private information of agent i at time t can be represented by an integer X_t^i, and that the sharing of information can be modeled by the fact that if agent i meets agent j at time t, we have $X_t^i = X_t^j = X_{t-}^i + X_{t-}^j$. In other words, the state X_t^i is an integer representing the precision at time t of the best guess player i has of the value of a random variable of common interest.

Each agent controls his or her own search intensity as follows. For each $i \in \{1, \cdots, N\}$, there exists a (measurable) function $[0, T] \times \mathbb{N} \ni (t, n) \mapsto c_t^i(n) \in \mathbb{R}_+$, for a finite time horizon $T > 0$, so that $c_t^i(n)$ represents the intensity with which agent i searches when his or her state is n at time t (i.e., when $X_t^i = n$). Typically, each c^i takes values in a bounded subinterval $[c_L, c_U]$ of $[0, \infty)$. We then model the dynamics of the state $(X_t = (X_t^1, \cdots, X_t^N))_{0 \leqslant t \leqslant T}$ of the information of the set of N agents in the following way:

$$X_t = X_0 + \sum_{1 \leqslant i \neq j \leqslant N} \int_{[0,t]} \int_{[0,\infty)} \varphi_{i,j}(s, X_{s-}, v) M_{i,j}(ds, dv), \tag{1.52}$$

where the $(\boldsymbol{M}_{i,j})_{1 \leqslant i \neq j \leqslant N}$ are independent homogeneous Poisson random measures on $\mathbb{R}_+ \times \mathbb{R}_+$ with mean measures proportional to the 2-dimensional Lebesgue measure Leb_2. More precisely,

$$\mathbb{E}\big[M_{i,j}(A \times B)\big] = \frac{1}{2(N-1)} \mathrm{Leb}_2(A \times B),$$

for $i \neq j$ and A, B two Borel subsets of $\mathbb{R}_+$. The functions $(\varphi_{i,j})_{1 \leqslant i \neq j \leqslant N}$ are given by:

$$\varphi_{i,j}(t,x,v) = \begin{cases} (0,\cdots,0) & \text{if } \kappa_{i,j}(t,x) < v \\ y & \text{otherwise,} \end{cases}$$

where $y = (y^1, \cdots, y^N)$ is defined by $y^k = 0$ if $k \neq i$ and $k \neq j$, and $y^i = x^j$, and $y^j = x^i$, and the function $\kappa_{i,j}$ is defined by:

$$\kappa_{i,j}(t,x) = \kappa_{i,j}(t,x^1,\cdots,x^N) = c_t^i(x^i)c_t^j(x^j).$$

As before, c^i and c^j are the respective search intensities of agents i and j.

The goal of player k is to minimize the quantity:

$$J^k(\alpha^1,\cdots,\alpha^N) = \mathbb{E}\left[\int_0^T K(\alpha_t^k)dt + g(X_T^k)\right],$$

where the controls are given by the feedback functions $((\alpha_t^i = c_t^i(X_t^i))_{0\leq t\leq T})_{i=1,\cdots,N}$ and $K : [c_L, c_U] \ni c \mapsto K(c)$ is a bounded measurable function representing the cost for an individual searching with intensity c. It is natural to assume that this latter function is increasing and convex. The terminal cost function g represents the penalty for ending the game with a given level of information. Typically, $g(n)$ will be chosen to be inversely proportional to $(1+n)$, but any convex function decreasing with n would do as well.

To understand the behavior the system, we describe the dynamics of the state $(X_t^k)_{0\leqslant t\leqslant T}$ of the information of agent k by extracting the k-th component of both sides of (1.52).

In order to proceed, we fix a coordinate $k \in \{1,\cdots,N\}$ and we provide an alternative representation of $\boldsymbol{X}^k$, obtained by choosing $(\boldsymbol{M}_{i,k})_{i\neq k}$ and $(\boldsymbol{M}_{k,j})_{j\neq k}$ in a relevant way. For any time $t \in [0,T]$, our construction of $(M_{i,k}(dt,\cdot))_{i\neq k}$ and $(M_{k,j}(dt,\cdot))_{j\neq k}$, is based on a suitable coupling with the past of the whole system up until time t. For two independent homogeneous Poisson random measures $\tilde{\boldsymbol{M}}^1$ and $\tilde{\boldsymbol{M}}^2$ on $\mathbb{R}_+ \times \mathbb{R}_+ \times (0,1]$ with $\frac{1}{2}\text{Leb}_3$ as intensity, where Leb_3 is the 3-dimensional Lebesgue measure, we choose $M_{i,k}(dt,dv)$ and $M_{k,j}(dt,dv)$ as:

$$M_{i,k}(dt,dv) = \int_{[0,1]} \mathbf{1}_{\frac{\sigma_t(i)-1}{N-1} < w \leqslant \frac{\sigma_t(i)}{N-1}} \tilde{M}^1(dt,dv,dw), \quad i \neq k,$$

$$M_{k,j}(dt,dv) = \int_{[0,1]} \mathbf{1}_{\frac{\sigma_t(j)-1}{N-1} < w \leqslant \frac{\sigma_t(j)}{N-1}} \tilde{M}^2(dt,dv,dw), \quad j \neq k,$$

where $(\sigma_t)_{0\leqslant t\leqslant T}$ is a predictable process with values in the set of one-to-one mappings from $\{1,\cdots,N\} \setminus \{k\}$ onto $\{1,\cdots,N-1\}$. The precise form of $(\sigma_t)_{0\leqslant t\leqslant T}$ will be specified later on.

Recall that:

$$X_t^k = X_0^k + \sum_{1 \leqslant i \neq j \leqslant N} \int_{[0,t]} \int_{[0,\infty)} \varphi_{i,j}^k(X_{s-}, v) M_{i,j}(ds, dv).$$

Since $\varphi_{i,j}^k(x, v) = 0$ if $k \notin \{i, j\}$, we get:

$$\begin{aligned}
X_t^k &= X_0^k \\
&+ \sum_{i=1, i\neq k}^{N} \int_{[0,t]} \int_{[0,\infty)} \int_{[0,1]} \varphi_{i,k}^k(s, X_{s-}, v) \mathbf{1}_{\frac{\sigma_s(i)-1}{N-1} < w \leqslant \frac{\sigma_s(i)}{N-1}} \tilde{M}^1(ds, dv, dw) \\
&+ \sum_{j=1, j\neq k}^{N} \int_{[0,t]} \int_{[0,\infty)} \int_{[0,1]} \varphi_{k,j}^k(s, X_{s-}, v) \mathbf{1}_{\frac{\sigma_s(j)-1}{N-1} < w \leqslant \frac{\sigma_s(j)}{N-1}} \tilde{M}^2(ds, dv, dw).
\end{aligned}$$

Using the fact that $\varphi_{i,k}^k(x, v) = \varphi_{k,i}^k(x, v)$, we obtain:

$$\begin{aligned}
X_t^k &= X_0^k \\
&+ \int_{[0,t]} \int_{[0,\infty)} \int_{[0,1]} \left(\sum_{i=1, i\neq k}^{N} \varphi_{i,k}^k(s, X_{s-}, v) \mathbf{1}_{\frac{\sigma_s(i)-1}{N-1} < w \leqslant \frac{\sigma_s(i)}{N-1}} \right) \tilde{M}(ds, dv, dw),
\end{aligned} \tag{1.53}$$

where $\tilde{\boldsymbol{M}} = \tilde{\boldsymbol{M}}^1 + \tilde{\boldsymbol{M}}^2$ is a homogeneous Poisson measure on $\mathbb{R}_+ \times \mathbb{R}_+ \times (0, 1]$ with Leb_3 as intensity.

We now recall the form of $\varphi_{i,k}^k(s, X_{s-}, v)$:

$$\varphi_{i,k}^k(s, X_{s-}, v) = X_{s-}^i \mathbf{1}_{[v,\infty)} \big(c_s^i(X_{s-}^i) c_s^k(X_{s-}^k) \big),$$

which shows that the integrand appearing in (1.53) is equal to:

$$\begin{aligned}
&\sum_{i=1, i\neq k}^{N} \varphi_{i,k}^k(s, X_{s-}, v) \mathbf{1}_{\frac{\sigma_s(i)-1}{N-1} < w \leqslant \frac{\sigma_s(i)}{N-1}} \\
&= \sum_{i=1, i\neq k}^{N} X_{s-}^i \mathbf{1}_{[v,\infty)} \big(c_s^i(X_{s-}^i) c_s^k(X_{s-}^k) \big) \mathbf{1}_{\frac{\sigma_s(i)-1}{N-1} < w \leqslant \frac{\sigma_s(i)}{N-1}} .
\end{aligned} \tag{1.54}$$

If we now restrict ourselves to symmetric Nash equilibria, in the sense that the control profiles $(\boldsymbol{\alpha}^1, \cdots, \boldsymbol{\alpha}^N)$ are exchangeable at equilibrium, we can assume that the search intensities of all the other agents are given by the same feedback function, say $[0, T] \times \mathbb{N} \ni (t, x^i) \mapsto c_t^i(x^i) = \tilde{c}_t^k(x^i) \in \mathbb{R}_+$ for $i \neq k$, for some function $\tilde{c}^k : [0, T] \times \mathbb{N} \to \mathbb{R}_+$. In such a case, the above sum becomes:

$$\begin{aligned}&\sum_{i=1,i\neq k}^{N} \varphi_{i,k}^k(s, X_{s-}, v)\mathbf{1}_{\frac{\sigma_s(i)-1}{N-1}<w\leqslant\frac{\sigma_s(i)}{N-1}} \\ &\quad = \sum_{i=1,i\neq k}^{N} X_{s-}^i \mathbf{1}_{[v,\infty)}\big(\tilde{c}_s^k(X_{s-}^i)c_s^k(X_{s-}^k)\big)\mathbf{1}_{\frac{\sigma_s(i)-1}{N-1}<w\leqslant\frac{\sigma_s(i)}{N-1}}.\end{aligned} \tag{1.55}$$

Now here is the thrust of our formulation. We choose σ_s as a one-to-one mapping from $\{1,\cdots,N\}\setminus\{k\}$ onto $\{1,\cdots,N-1\}$, such that $(X_{s-}^{\sigma_s^{-1}(i)})_{1\leq i\leq N-1}$ coincides with the order statistics of $(X_{s-}^i)_{1\leq i\neq k\leq N}$. The order statistics are denoted by $(X_{s-}^{(i),-k})_{1\leq i\leq N-1}$. This is especially convenient since for any function $f:\mathbb{N}\to\mathbb{R}$,

$$\sum_{i=1,i\neq k}^{N} f(X_{s-}^i)\mathbf{1}_{\frac{\sigma_s(i)-1}{N-1}<w\leqslant\frac{\sigma_s(i)}{N-1}} = \sum_{i=1}^{N-1} f\big(X_{s-}^{(i),-k}\big)\mathbf{1}_{\frac{i-1}{N-1}<w\leqslant\frac{i}{N-1}}.$$

We now call $\bar{\mu}_{s-}^{-k}$ the empirical measure of X_{s-}^{-k} and $\bar{F}_{s-}^{-k}(\cdot)$ the associated empirical distribution function and we denote by $\bar{Q}_{s-}^{-k}(\cdot)$ its pseudo-inverse. We recall that for any $w\in(0,1]$ and $i\in\{1,\cdots,N-1\}$:

$$\frac{i-1}{N-1} < w \leqslant \frac{i}{N-1} \Longrightarrow \bar{Q}_{s-}^{-k}(w) = X_{s-}^{(i),-k}.$$

We finally get:

$$\sum_{i=1,i\neq k}^{N} f(X_{s-}^i)\mathbf{1}_{\frac{\sigma_s(i)-1}{N-1}<w\leqslant\frac{\sigma_s(i)}{N-1}} = f\big(\bar{Q}_{s-}^{-k}(w)\big), \quad w\in(0,1].$$

Returning to (1.55), we deduce:

$$\begin{aligned}&\sum_{i=1,i\neq k}^{N} \varphi_{i,k}^k(s, X_{s-}, v)\mathbf{1}_{\frac{\sigma_s(i)-1}{N-1}<w\leqslant\frac{\sigma_s(i)}{N-1}} \\ &\qquad = \bar{Q}_{s-}^{-k}(w)\mathbf{1}_{[v,\infty)}\big(\tilde{c}_s^k(\bar{Q}_{s-}^{-k}(w))c_s^k(X_{s-}^k)\big).\end{aligned}$$

Inserting in (1.53), we finally get the following representation for X_t^k:

$$\begin{aligned}X_t^k &= X_0^k \\ &+ \int_{[0,t]}\int_{[0,\infty)}\int_{[0,1]} \bar{Q}_{s-}^{-k}(w)\mathbf{1}_{[v,\infty)}\big(\tilde{c}_s^k(\bar{Q}_{s-}^{-k}(w))c_s^k(X_{s-}^k)\big)\tilde{M}(ds,dv,dw).\end{aligned}$$

This writing will serve as a starting point for the formulation of the game with a large number of agents as a mean field game.

The Mean Field Game Formalization

The mean field game formulation of the problem is to first fix a flow $[0, T] \ni t \mapsto (c_t, \mu_t)$. Here c_t is a function $c_t : \mathbb{N} \ni n \mapsto c_t(n)$ representing the search intensity of a representative participant at time $t \in [0, T]$. As before, we shall assume that this function takes values in $[c_L, c_U]$. On the other hand, μ_t is a probability measure on $\mathbb{N}$ giving the distribution at time t of the state (which it sometimes called the *precision*) of a generic participant. Once such a flow is fixed, we fix an individual, and find the best response to the population behavior whose evolution over time is captured by $(c_t, \mu_t)_{0\leqslant t\leqslant T}$. This best response will be obtained by solving the optimal control problem:

$$\inf_{\alpha\in\mathbb{A}} \mathbb{E}\left[\int_0^T K(\alpha_t)dt + g(X_T)\right], \tag{1.56}$$

under the constraint that:

$$X_t = X_0 + \int_{[0,t]}\int_{[0,\infty)}\int_{[0,1]} \varphi(s, \alpha_s, v, w)\tilde{M}(ds, dv, dw). \tag{1.57}$$

where:

$$\varphi(t, \alpha, v, w) = \mathbf{1}_{[v,\infty)}\big(\alpha c_t(Q_t(w))\big)Q_t(w),$$

$\tilde{M}$ denoting a Poisson random measure on $\mathbb{R}_+ \times \mathbb{R}_+ \times [0, 1]$ with intensity given by the Lebesgue measure Leb_3 and Q_t denoting the quantile function of μ_t. The set $\mathbb{A}$ of admissible controls is chosen to be the set of feedback controls of the form $\alpha_t = \phi_t(X_{t-})$ for some deterministic function $\phi : [0, T] \times \mathbb{N} \ni (t, x) \mapsto \phi_t(x) \in [c_L, c_U]$.

When $(\alpha_t = \phi_t(X_{t-}))_{0\leqslant t\leqslant T}$, the instantaneous jump rate at time $t \in [0, T]$ is, conditional on the event $\{X_{t-} = n\}$, equal to:

$$\int_0^\infty \int_0^1 \mathbf{1}_{[v,\infty)}\big(\phi_t(n)c_t(Q_t(w))\big)\, dvdw = \phi_t(n)\int_{\mathbb{N}} c_t d\mu_t = \phi_t(n)\bar{c}_t,$$

with:

$$\bar{c}_t = \int_{\mathbb{N}} c_t d\mu_t = \sum_{n'\in\mathbb{N}} c_t(n')\mu_t(\{n'\}).$$

In other words, $(X_t)_{t\geqslant 0}$ is a pure-jump process whose jump-arrival intensity at time t is given by the function $\bar{c}_t\phi_t(\,\cdot\,)$. The jump-size distribution at time t is then given by:

$$\mathbb{N} \ni n \mapsto \frac{c_t(n)\mu_t(\{n\})}{\bar{c}_t}.$$

Also, if we denote by $(L_t(n,y))_{n\in\mathbb{N},y\in\mathbb{N}^*}$ the jump rate kernel at time t of the state process $\boldsymbol{X}$, that is $\mathbb{P}[X_{t+dt}=n+y\,|\,X_{t-}=n]=L_t(n,y)dt+o(dt)$, then,

$$L_t(n,y)=\phi_t(n)c_t(y)\mu_t(\{y\}),\quad y\in\mathbb{N}^*. \tag{1.58}$$

Now, the last step of the mean field game approach is to find a flow $\boldsymbol{\mu}=(\mu_t)_{0\le t\le T}$ of distributions on $\mathbb{N}$ and a function $c:[0,T]\times\mathbb{N}\ni(t,n)\mapsto c_t(n)\in[c_L,c_U]$ such that the stochastic control problem (1.56)–(1.57) has an optimizer, given by a family $(\phi_t)_{0\le t\le T}$ of feedback functions from $\mathbb{N}$ to $[c_L,c_U]$ and admitting $\boldsymbol{\nu}=(\nu_t)_{0\le t\le T}$ as flow of marginal distributions of the optimal states, such that the following fixed point condition holds true: $\phi_t=c_t$ and $\nu_t=\mu_t$, for $t\in[0,T]$.

We can characterize the equilibria as the solutions of a nonlinear Fokker-Planck-Kolmogorov equation. To do so, the following notation will be useful. If $\theta:\mathbb{N}\to\mathbb{R}$ is a bounded function and $\boldsymbol{\mu}=(\mu_t)_{0\le t\le T}$ is a flow of probability measures on $\mathbb{N}$, we denote by μ_t^θ the measure with density θ with respect to the measure μ_t, for any $t\in[0,T]$. In other words:

$$\mu_t^\theta(\{n\})=\theta(n)\mu_t(\{n\}),\qquad n=0,1,2,\cdots.$$

Now, for a given flow $\boldsymbol{\mu}=(\mu_t)_{0\le t\le T}$ of probability measures, we write the Fokker-Planck-Kolmogorov equation satisfied by the flow of marginal distributions $\boldsymbol{\nu}=(\nu_t)_{t\geqslant 0}$ of the pure jump process admitting $(L_t)_{0\le t\le T}$ in (1.58) as jump rate kernel. We find this equation by computing:

$$\frac{d}{dt}\mathbb{E}[f(X_t)]=\frac{d}{dt}\langle f,\nu_t\rangle,\quad t\in[0,T],$$

for a bounded function $f:\mathbb{N}\to\mathbb{R}$, with $(X_t)_{0\le t\le T}$ as in (1.57). This gives:

$$\begin{aligned}
\frac{d}{dt}\langle\nu_t,f\rangle &= \sum_{n\in\mathbb{N}}\nu_t(\{n\})\sum_{m\geqslant 1}[f(n+m)-f(n)]L_t(n,m)\\
&=\sum_{n\in\mathbb{N}}\nu_t(\{n\})\sum_{m\in\mathbb{N}}[f(n+m)-f(n)]\phi_t(n)c_t(m)\mu_t(\{m\})\\
&=\sum_{n,m\in\mathbb{N}}\phi_t(n)\nu_t(\{n\})\,c_t(m)\mu_t(\{m\})f(n+m)\\
&\qquad\qquad -\sum_{n,m\in\mathbb{N}}f(n)\phi_t(n)\nu_t(\{n\})\,c_t(m)\mu_t(\{m\})\\
&=\langle\nu_t^{\phi_t}*\mu_t^{c_t},f\rangle-\langle\mu_t^{c_t},1\rangle\,\langle\nu_t^{\phi_t},f\rangle,
\end{aligned}$$

where for two probability measures μ and ν on $\mathbb{N}$, the convolution $\mu*\nu$ is given by:

$$(\nu*\mu)(\{n\})=\sum_{k=0}^{n}\nu(\{k\})\mu(\{n-k\}),\quad n\in\mathbb{N}.$$

This shows that the (linear) Fokker-Planck-Kolmogorov equation reads:

$$\frac{d}{dt}\nu_t = \mu_t^{c_t} * \nu_t^{\phi_t} - \langle \mu_t^{c_t}, 1 \rangle \, \nu_t^{\phi_t}, \quad t \in [0, T].$$

Notice that, if the last step of the mean field game approach can be performed, in other words, if the fixed point problem can be solved, then $\phi_t = c_t$ and $\nu_t = \mu_t$ and the (linear) Fokker-Planck-Kolmogorov equation for the state distribution becomes the nonlinear McKean-Vlasov equation:

$$\frac{d}{dt}\mu_t = \mu_t^{c_t} * \mu_t^{c_t} - \langle \mu_t^{c_t}, 1 \rangle \, \mu_t^{c_t}, \quad t \in [0, T], \tag{1.59}$$

which is exactly the finite horizon analog of the equation obtained in the literature (see the Notes and Complements below) since we do not have entrance and exit of agents at exponential random times.

1.7 Notes & Complements

The notion of Nash equilibrium in game theory goes back to the seminal works by Nash [288] and [289]. Throughout the book, we consider game models with finitely many players, and mathematical objects capturing the limits of features of these models when the number of players grows without bound. In some sense, the mean field game models we formulate and solve pertain to an infinite, though countable, population of players. For a long time, economists have used alternative models for which the set of players is a measurable space equipped with a nonatomic probability measure. Even though we chose not to use this framework, we recognize that it is an attempt to abstract stylized facts from the same finite player game models which we study. The reader interested in Aumann's theory of games with a continuum of players is referred to [27, 28] and to Section 3.7 of Chapter 3.

We borrowed the idea of grounding the formulation of a mean field game problem on the elementary Lemma 1.2 and Proposition 1.4 from P.L. Lions' lectures [265] as explained in Cardaliaguet's presentation [83]. These simple results provide a rigorous foundation for the fundamental assumptions we make on the coefficients of a mean field game. These sources also prompted us to introduce early the notion of potential game and its connections to centralized decision making in lieu of Nash equilibrium. This idea will be revisited several times in Chapter 2 and Chapter 6 for example, with increasingly more sophisticated models and analysis tools.

The discussion of the model *"When does the meeting start?"* given in the text is borrowed from [189]. In fact, the deterministic and stochastic one period examples, as well as the stochastic differential game models used to illustrate the management of exhaustible resources and the economic growth model are all borrowed from the survey [189] by Guéant, Lasry and Lions. We chose not to include a discussion of the Mexican wave (fixture of most crowd behaviors at soccer games all over the world) even though it is one of P.L. Lions' favorite examples of mean field games. It is explained in detail in [189].

Early game theoretic models for the banking system are due to Bryant [73] and Diamond and Dybvig whose fundamental paper [136] initiated a wave of interest leading to a series of papers with increasing realism. In their original analysis, Diamond and Dybvig proposed a banking model in the form of a game played by depositors. A distinctive feature of the model is that there always exist at least a good equilibrium and a bad equilibrium. Many generalizations were proposed, for example to model random returns. The first of the two models discussed in the text is in this line of research, analyzing a static model of the inter-banking system. It is borrowed from the paper [319] of Rochet and Vives. There, the authors use the methodology of global games proposed by Morris and Shin in [286], and the differences in opinions among investors to prove existence and uniqueness of a Nash equilibrium. They go on to analyze the economic and financial underpinnings of bank runs and propose a benchmark for the role of lenders of last resort. We gave a detailed account of their set-up because of the mean field game nature of their approach, despite the fact that their model is de facto static. The theory of games with strategic complementarities goes back to the original works [339] of Vives and [284] of Milgrom and Roberts. An application to games with mean field interactions can be found in the paper [8] by Adlakha and Johari.

The second model of bank run presented in the text ports the most important stylized facts of the first model to a dynamic set-up in continuous time. It is based on a diffusion model for the value of the assets of a bank, and for that reason, it is more in line with the theoretical developments presented later in the text. It was inspired by a lecture given by Olivier Gossner at a PIMS Workshop on Systemic Risk in July 2014. This model builds on an earlier paper [197] by He and Xiong modeling staggered debt maturities in continuous time. Section (Vol II)-7.2 of Chapter (Vol II)-7 is devoted to the discussion and the solution of more general games of timing.

The toy model of systemic risk introduced in Subsection 1.3.1 is borrowed from the paper [102] of Carmona, Fouque, and Sun to which we refer the interested reader for the interpretation of the results in terms of systemic risk.This simple model is remarkable because it can be used as a testbed for all the theoretical tools developed in the text. It is solved explicitly in Chapter 2 to illustrate the differences between open loop and closed loop equilibria for finite player games. It is used in Chapter 3 as an example for which the limit $N \to \infty$ of large games can be performed leading to a solvable mean field game in the limit. It will also be revisited in Chapter (Vol II)-4 to illustrate how the master equation can appear in the limit of finite player games with a common noise. The model was recently extended in [101] to include delay in the control in order to make the model more realistic and more in line with the way interbank borrowing and lending actually occur.

Price impact models as mean field games have been considered by Gomes and Saude in [182] where the problem is approached from a PDE perspective, and by Carmona and Lacker in [103] where it is treated within the framework of the weak formulation. The model presented in Subsection 1.3.2 is borrowed from a technical report by Aghbal and Carmona where it is treated by adapting the tools developed later on in the text. The model of price impact used in the text is due to Almgren and Chriss. See for example [18]. It is relatively easy to calibrate it to high frequency data, hence its popularity among practitioners. As we see in the solution proposed in Subsection 4.7.1 of Chapter 4, this model of price impact can lead to tractable solutions, hence its popularity in the financial engineering literature. Carmona and Webster proved in [107] that the self-financing condition of the classical Black-Scholes theory is not always appropriate to account for frictions in high frequency trading data. They propose an alternative including price impact and adverse selection and it would be interesting to solve the global equilibrium problem in the mean field set-up they propose in [108].

Games for which the interaction between the players occurs through the distribution of the controls of the players have been called extended mean field games by Gomes and Saude in [182]. They will be studied in Section 4.6 of Chapter 4.

The first macro-economic growth model presented in Subsection 1.4.1 is borrowed from [189]. The discussion of the second growth model of Subsection 1.4.2 is patterned after the original paper [241] by Krusell and Smith where the authors propose a formulation which leads to approximations and computational algorithms for approximate solutions. We learned about the version of Aiyagari's model presented in Subsection 1.4.3 from a talk by B. Moll at the Institute of Mathematics and its Applications, Minneapolis MA in November 2012. This continuous time version of the original discrete time model proposed by Aiyagari is also discussed in the review [1] by Achdou, Buera, Lasry, Lions, and Moll of partial differential equation models in macroeconomics. There, it is stated that a mathematical solution of such a model is not known. We shall provide such a mathematical solution in Subsection 3.6.3 of Chapter 3 together with numerical illustrations of the properties of the solution.

The mean field game model for exhaustible resources presented in Subsection 1.4.4 is borrowed from [189]. The last subsection is adapted from a recent technical report [113] by Chan and Sircar.

Our presentation of the deterministic model of flocking is based on the original paper [126] of Cucker and Smale. Soon after the publication of the original treatment [126], Cucker and Mordecki proposed in [125] a generalization in which the dynamics of the velocities are perturbed by a mean zero stationary Gaussian process. In [125], the authors compute a lower bound for the probability that the velocities eventually align in terms of the various parameters of the model. The idea of using a mean field game model to formulate the flocking problem as the search for equilibrium in a stochastic game model with mean field interactions is due to Nourian, Caines, and Malhamé in [285]. In this paper, the authors recognize that the case $\beta = 0$ leads to a linear quadratic mean field game which could be solved, and they suggest that an asymptotic expansion for β small could provide a reasonable approximation to the solution. In this book, we solve the particular case $\beta = 0$ in Section 2.4 of Chapter 2 explicitly for a finite number of birds, and in Section 3.6.1 of Chapter 3 in the mean field game limit when the number of birds increases without bound. Furthermore, we give a theoretical solution and numerical illustrations for the general case $\beta \neq 0$ in Section 4.7.3 of Chapter 4.

The models of crowds behaviors mentioned briefly in Subsection 1.5.3 will be used to show how versatile the mean field game models can be in the analysis of large populations. They provide a rigorous mathematical framework to try to understand complex phenomena like schooling, flocking, hurdling, etc. on the basis of the rational behavior of individuals optimizing their own interest within a large population. Our models of pedestrian dynamics with congestion effects were inspired by the paper [253] of Lachapelle and Wolfram who provided numerical evidence of the explanatory power of these models.

The idea of using a mean field game model for the analysis of the exit from a room in the presence of congestion is borrowed from the paper [5] by Achdou and Laurière who provided numerical illustrations of the expected properties of the model at the end of a paper actually devoted to the optimal control of McKean-Vlasov dynamics which we study in Chapter 6. We propose a theoretical solution as well as numerical illustrations of the properties of the solutions in Subsection 4.7.2 of Chapter 4.

In an effort to model trading on a limit order book, Lachapelle, Lasry, Lehalle, and Lions introduced in [251] a simple model for the trade-off between trading as slow as possible to avoid being detected and becoming a prey, and trading too fast and moving the price in an unfavorable direction, a form of adverse selection. In [251], the authors propose two different models. The most sophisticated of these models involves a mix of heterogenous agents: on one hand, they consider institutional investors trading large quantities, and on the other hand, High Frequency Traders (HFTs) faster than the institutional investors, using smaller orders, showing patience if needed, and relying on Smart Order Routing (SOR) algorithms. For the sake of simplicity, we only discussed the simpler of their models, dealing only with a homogenous class of agents interested in selling only, the other class of market participants, characterized as *impatient buyers*, being modeled exogenously. More recently, Gayduk and Nadtochiy proposed in [172] a mean field game model encapsulating some of the features of the set-up proposed by Lachapelle, Lasry, Lehalle, and Lions, while deriving an endogenous construction of the limit order book in the spirit of Carmona and Webster in [108].

We chose a model of cyber-security introduced recently by Kolokoltsov and Bensoussan in [235] to motivate the analysis of games for which the state of the system can only take finitely many values. While the book is almost exclusively devoted to stochastic differential games, the last couple of examples of this chapter provide the motivation for the analysis of stochastic games with a discrete state space. The mathematical analysis of these models will be presented in Section 7.2 of Chapter 7 where we show how to adapt the results and tools developed in the book, including the master equation, to the special class of continuous-time dynamic-game models exhibiting mean field interactions. The extension to mean field games with major and minor players will be featured in Subsection (Vol II)-7.1.9 of Chapter (Vol II)-7. An early model of computer network security can be found in the work of Lye and Wing [268]. The interested reader may also want to look at a more abstract network model in the conference proceedings [290] by Nguyen, Alpcan, and Basar where the authors frame the problem as a zero-sum game between an attacker and the defender of the network.

Our discussion of the propagation of knowledge in Subsection 1.6.3 was inspired by the work of Duffie, Giroux, Malamud, and Manso, see for example [144–147] which are based on a form of *continuous law of large numbers* proven in [148]. In keeping with the spirit of this first chapter, we used slightly different assumptions to formulate a finite player game with mean field interactions. However, in order to remain faithful to the original works of Duffie et al. we use $\mathbb{N}$ as the state space of the system and as a result, model the idiosyncratic sources of random shocks by Poisson processes, which will prevent us from applying directly the main existence results of the book to these models.

In a couple of interesting papers, [199] and [200], Horst discusses stochastic games with a form of weak interaction between players. There, the author includes peer and neighborhood effects in a dynamic analysis of equilibrium, very much in the spirit of an earlier work of Bisin, Horst, and Özgür of 2002 [59] eventually published in 2006, in which the authors proved the existence of rational expectations equilibria of random economies with locally interacting agents under the assumption that the interaction between different players is weak. The weak interaction approach suggested in this paper provides a unified framework for integrating strategic behavior into dynamic models of social interactions. However, the weak interactions of the models considered in this book are of a different nature. They are defined in an asymptotic sense when the number of players tends to ∞. This is different from Horst's notion of weak interaction for which the weakness of the interaction is localized to the neighbors of the given agent. Other examples of application may be found in the survey by Djehiche, Tcheukam Siwe and Tembine [137].

Probabilistic Approach to Stochastic Differential Games

2

Abstract
This chapter offers a crash course on the theory of nonzero sum stochastic differential games. Its goal is to introduce the jargon and the notation of this class of game models. As the main focus of the text is the probabilistic approach to the solution of stochastic games, we review the strategy based on the stochastic Pontryagin maximum principle and show how BSDEs and FBSDEs can be brought to bear in the search for Nash equilibria. We emphasize the differences between open loop and closed loop or Markov equilibria and we illustrate the results of the chapter with detailed analyses of some of the models introduced in Chapter 1.

2.1 Introduction and First Definitions

Despite the fact that we discussed in Chapter 1 a couple of examples of one and two period games, the book is devoted to continuous time models. As before, we consider N players, and we label them by the integers $1, \cdots, N$. They act at time t on a system whose state X_t they influence through their actions. As the title of the chapter suggests, the dynamics of the state of the system are given by an Itô stochastic differential equation. As can be seen from the examples introduced in Chapter 1, some of the state dynamics appearing naturally in these examples are not easily amenable to stochastic differential equations. For this reason, we devote an entire section to games with finite state spaces in Chapter 7. Moreover, we shall add remarks when appropriate, to indicate how to handle more general Markov dynamics, possibly including jumps, and we shall give precise references in the Notes & Complements sections at the end of all the chapters.

R. Carmona, F. Delarue, *Probabilistic Theory of Mean Field Games with Applications I*, Probability Theory and Stochastic Modelling 83,
DOI 10.1007/978-3-319-58920-6_2

As explained in Remark 2.1 below, in order to avoid alternating between he, his, ... *and* she, her, ..., *we decided to make the players genderless and use the pronouns* it, its, ... *throughout the book. This will sound strange at times but in the end, the discussions of players' behaviors will be consistent.*

We assume that the Itô process used to specify the state dynamics is driven by an M-dimensional Wiener process $\boldsymbol{W} = (W_t)_{0 \leqslant t \leqslant T}$ defined on a probability space $(\Omega, \mathcal{F}, \mathbb{P})$, $\mathbb{F} = (\mathcal{F}_t)_{0 \leqslant t \leqslant T}$ being the completion of the natural filtration of the Wiener process.

We denote by $A_1, \cdots, A_N$ the sets of actions that players $1, \cdots, N$ can take at any point in time. The sets A_i are typically compact metric spaces or subsets of an Euclidean space, say $A_i \subset \mathbb{R}^{k_i}$, and we denote by $\mathcal{A}_i = \mathcal{B}(A_i)$ their Borel σ-fields, where, throughout the book $\mathcal{B}(E)$ denotes the Borel σ-field on any metric space (E, d). We use the notation $\mathbb{A}^{(N)}$ for the set of admissible strategy profiles. The elements $\boldsymbol{\alpha}$ of $\mathbb{A}^{(N)}$ are N-tuples $\boldsymbol{\alpha} = (\boldsymbol{\alpha}^1, \cdots, \boldsymbol{\alpha}^N)$ where each $\boldsymbol{\alpha}^i = (\alpha^i_t)_{0 \leqslant t \leqslant T}$ is a progressively measurable A_i-valued process. Most often, these individual strategies will have to satisfy extra conditions (e.g., measurability and integrability constraints). These conditions change with the application. In most of the cases considered in this book, we assume that these constraints can be defined player by player, independently of each other. To be more specific, we often assume that $\mathbb{A}^{(N)} = \mathbb{A}_1 \times \cdots \times \mathbb{A}_N$ where, for each $i \in \{1, \cdots, N\}$, $\mathbb{A}_i$ is the space of control strategies which are deemed admissible to player i, irrespective of what the other players do. In most applications, $\mathbb{A}_i$ will be a space of A_i-valued progressively measurable processes $\boldsymbol{\alpha}^i = (\alpha^i_t)_{0 \leqslant t \leqslant T}$, either bounded, or satisfying an integrability condition such as $\mathbb{E}\int_0^T |\alpha^i_t|^2 dt < \infty$. We will also add measurability conditions specifying the kind of information each player can use in order to choose its course of action at any given time. Finally, we shall use the notation $A^{(N)} = A_1 \times \cdots \times A_N$ for the set of actions $\alpha = (\alpha^1, \cdots, \alpha^N)$ available to all the players at any given time.

For each choice of strategy profile $\boldsymbol{\alpha} = (\alpha_t)_{0 \leqslant t \leqslant T} \in \mathbb{A}^{(N)}$, it is assumed that the time evolution of the state $\boldsymbol{X} = \boldsymbol{X}^{\boldsymbol{\alpha}}$ of the system satisfies:

$$\begin{cases} dX_t = B(t, X_t, \alpha_t)dt + \Sigma(t, X_t, \alpha_t)dW_t & 0 \leqslant t \leqslant T, \\ X_0 = x_0, \end{cases} \tag{2.1}$$

for some $x_0 \in \mathbb{R}^D$, where $(B, \Sigma) : [0, T] \times \Omega \times \mathbb{R}^D \times A^{(N)} \to \mathbb{R}^D \times \mathbb{R}^{D \times M}$ satisfies:

Assumption (Games).

(A1) For all $S \in [0, T]$, the function $[0, S] \times \Omega \times \mathbb{R}^D \times A^{(N)} \ni (t, \omega, x, \alpha) \mapsto (B, \Sigma)(t, \omega, x, \alpha)$ is $\mathcal{B}([0, S]) \otimes \mathcal{F}_S \otimes \mathcal{B}(\mathbb{R}^D) \otimes \mathcal{B}(A^{(N)})/\mathcal{B}(\mathbb{R}^D) \otimes \mathcal{B}(\mathbb{R}^{D \times M})$-measurable.

(A2) There exists a constant $c > 0$ such that, for all $t \in [0, T]$, $\omega \in \Omega$, $x, x' \in \mathbb{R}^D$ and $\alpha, \alpha' \in A^{(N)}$,

(continued)

$$|B(t,\omega,x,\alpha) - B(t,\omega,x',\alpha')| + |\Sigma(t,\omega,x,\alpha) - \Sigma(t,\omega,x',\alpha')|$$
$$\leqslant c\big(|x-x'| + |\alpha-\alpha'|\big).$$

(A3) For any $\boldsymbol{\alpha} \in \mathbb{A}^{(N)}$, it holds $\mathbb{E}\int_0^T \big(|B(t,0,\alpha_t)|^2 + |\Sigma(t,0,\alpha_t)|^2\big)dt < \infty$.

As usual, we omit ω from the notation whenever possible. Nevertheless, notice that in many cases, the coefficients B and Σ will be independent of ω, and as a result, be deterministic functions on $[0,T]\times\mathbb{R}^D\times A^{(N)}$. Notice also that, in (2.1), the initial condition X_0 may be assumed to be random, in which case the filtration $\mathbb{F}$ has to be enlarged accordingly: Of course, X_0 is $\mathcal{F}_0$-measurable and is thus independent of $\boldsymbol{W}$; most of the time, it is assumed to be square-integrable. Hence (2.1) is uniquely solvable and its solution satisfies $\mathbb{E}[\sup_{0\leqslant t\leqslant T}|X_t|^2] < \infty$.

Remark 2.1 ***Gender of the Players.*** *This book is about mathematical models, their theories and solutions. Practical applications are used as motivations for the introduction of models, and numerical results are given as illustrations of the power and the limitations of the methods developed in the book. Given this emphasis, we do not feel that* political correctness *should be an issue forcing us to choose the way we address the individual* players, *as* he *or* she. *Clearly, their biological genders have no bearing on what we are interested in, and keeping track of grammatical genders can only be a hindrance and a distraction. So as already stated in the warning at the beginning of this introductory section, we decided that, for the sake of definiteness, we shall refer to the individuals involved in the game models we study as neutral from a grammatical standpoint. As a result, we shall treat the players as* genderless.

Remark 2.2 ***Frequently Used Notation.*** *Given a complete filtered probability space $(\Omega,\mathcal{F},\mathbb{F} = (\mathcal{F}_t)_{0\leqslant t\leqslant T},\mathbb{P})$, for each integer $n \geqslant 1$, we denote by $\mathbb{H}^{2,n}$ the space:*

$$\mathbb{H}^{2,n} = \Big\{\mathbf{Z} \in \mathbb{H}^{0,n} : \mathbb{E}\int_0^T |Z_s|^2 ds < \infty\Big\},$$

where $\mathbb{H}^{0,n}$ stands for the collection of all $\mathbb{R}^n$-valued progressively measurable processes on $[0,T]$. The set $\mathbb{H}^{2,n}$ is a Hilbert space for the inner product obtained by polarization of the double integral appearing in the definition. We shall also denote by $\mathbb{S}^{2,n}$ the space of all the continuous processes $\boldsymbol{S} = (S_t)_{0\leqslant t\leqslant T}$ in $\mathbb{H}^{0,n}$ such that $\mathbb{E}[\sup_{0\leqslant t\leqslant T}|S_t|^2] < +\infty$, the square root of this quantity providing $\mathbb{S}^{2,n}$ with a norm which we shall use repeatedly in the sequel.

Information Structures and Admissible Actions and Strategies
As we already saw in Chapter 1, the information structure of a stochastic game can be very complex as each player may have its own filtration $\mathbb{F}^i = (\mathcal{F}^i_t)_{0\leqslant t\leqslant T}$ formalizing the information it has access to at time t in order to choose its action $\alpha^i_t \in A_i$. In general, there is no reason why different players should have access to the same information. The search for solutions and the analytical tools which can be brought to bear in this search depend strongly upon the kind of information each player has access to at any given time, and which subset of this information it is allowed to use to take action.

The subtleties introduced by the interaction and competition between players make game theory much more intricate than the theory of stochastic control. Instead of aiming at the greatest generality, we shall limit ourselves to the definitions needed for the models we treat in the book.

2.1.1 A Typical Set-Up for Stochastic Differential Games

Quite often, the state of the system is the mere aggregation of private states of individual players, so that $X_t = (X^1_t, \cdots, X^N_t)$ where $X^i_t \in \mathbb{R}^{d_i}$ can be interpreted as the private state of player $i \in \{1, \cdots, N\}$. Here $D = d_1 + \cdots + d_N$ and, consequently, $\mathbb{R}^D = \mathbb{R}^{d_1} \times \cdots \times \mathbb{R}^{d_N}$. Moreover, the dynamics of the private states will be assumed to be given by stochastic differential equations driven by separate Wiener processes $\boldsymbol{W}^i = (W^i_t)_{0\leqslant t\leqslant T}$ which are most often assumed to be independent of each other, even though we already saw many examples in Chapter 1 for which this assumption failed to hold. For us, the typical example for which this assumption fails will be given by models with random shocks which are common to all the players. We shall identify these examples as games with a common noise. We focus on these models in Chapters (Vol II)-2 and (Vol II)-3. Typically for these models, we assume that the state dynamics are of the form:

$$dX^i_t = b^i(t, X_t, \alpha_t)dt + \sigma^i(t, X_t, \alpha_t)dW^i_t + \sigma^0(t, X_t, \alpha_t)dW^0_t,$$

for $0 \leqslant t \leqslant T$ and $i = 1, \cdots, N$, where for $i = 0, \cdots, N$, the $(N+1)$ processes $\boldsymbol{W}^i = (W^i_t)_{0\leqslant t\leqslant T}$ are m_i-dimensional independent Wiener processes giving the components of $\boldsymbol{W} = (W_t)_{0\leqslant t\leqslant T}$, and where:

$$(b^i, \sigma^i) : [0, T] \times \Omega \times \mathbb{R}^D \times A^{(N)} \to \mathbb{R}^{d_i} \times \mathbb{R}^{d_i \times m_i},$$

for $i \in \{1, \cdots, N\}$, satisfy the same assumptions as before, and σ^0 satisfies the same assumptions as the $(\sigma^i)_{1\leqslant i\leqslant N}$'s. The Wiener processes $\boldsymbol{W}^i$ with $i = 1, \ldots, N$ represent idiosyncratic random shocks while $\boldsymbol{W}^0$ is used to model what we call the *common noise*. It is important to notice that, even in the absence of a common noise (i.e., when $\sigma^0 \equiv 0$), these N dynamical equations are coupled by the fact that all the private states and all the actions enter into the coefficients of these N equations.

The common noise models described above are not the most general (see the discussion in the Notes & Complements at the end of Chapter 2 in Volume II). They are the ones we introduce in Chapter (Vol II)-2 and solve in Chapter (Vol II)-3 despite the fact that, as we saw in Chapter 1, and especially in the section on macroeconomic models, some of the instances of common noise are not covered by this additive intervention of the common noise term.

The popularity of the formulation described in this subsection is due to the ease with which we can define the information structures and admissible strategy profiles of some specific games of interest. For example, in a game where each player can only use the information of the state of the system at time t when making a strategic decision at that time, the admissible strategy profiles will be of the form $\alpha_t^i = \phi^i(t, X_t)$ for some deterministic function ϕ^i which we often call a feedback function. These strategies are said to be closed loop in feedback form, or Markovian. Moreover, if the information which can be used by player i at time t can only depend upon the state of player i at time t, then the admissible strategy profiles will be of the form $\alpha_t^i = \phi^i(t, X_t^i)$. Such strategies are usually called distributed.

2.1.2 Cost Functionals and Notions of Optimality

Under assumption **Games**, we assume further that each player faces instantaneous and running costs. So for each $i \in \{1, \cdots, N\}$, player i has cost coefficients:

(A4) A running cost function $f^i : [0,T] \times \Omega \times \mathbb{R}^D \times A^{(N)} \to \mathbb{R}$ such that, for all $S \in [0,T]$, the function $[0,S] \times \Omega \times \mathbb{R}^D \times A^{(N)} \ni (t,\omega,x,\alpha) \mapsto f^i(t,\omega,x,\alpha)$ is $\mathcal{B}([0,S]) \otimes \mathcal{F}_S \otimes \mathcal{B}(\mathbb{R}^D) \otimes \mathcal{B}(A^{(N)})/\mathcal{B}(\mathbb{R})$-measurable.

(A5) A terminal cost function $g^i : \Omega \times \mathbb{R}^D \to \mathbb{R}$ which is assumed to be $\mathcal{F}_T \otimes \mathcal{B}(\mathbb{R}^D)/\mathcal{B}(\mathbb{R})$-measurable

Both (**A4**) and (**A5**) will be regarded as part of assumption **Games**.

From this, we define its overall expected cost as given by a cost functional J^i defined for all admissible strategy profiles $\boldsymbol{\alpha} = (\boldsymbol{\alpha}^1, \cdots, \boldsymbol{\alpha}^N) \in \mathbb{A}^{(N)}$ by:

$$J^i(\boldsymbol{\alpha}) = \mathbb{E}\left[\int_0^T f^i(s, X_s, \alpha_s)ds + g^i(X_T)\right], \tag{2.2}$$

where we implicitly assume that the expectation in the above right-hand side is well defined. For instance so is the case if the cost coefficients are at most of quadratic growth in x, uniformly in the other variables, and, for any $\boldsymbol{\alpha} \in \mathbb{A}^{(N)}$,

$$\mathbb{E}\left[\int_0^T |f^i(t, 0, \alpha_t)|dt\right] < \infty.$$

Notice that, in the general situation considered here, the cost to a given player depends upon the strategies used by the other players indirectly through the values of the state X_t over time, but also directly as the specific actions α_t^j taken by the other players may appear explicitly in the expression of the running cost f^i of player i.

While the notion of Pareto optimality is natural in problems of optimal allocation of resources and, as a result, very popular in the economic literature and in operations research applications, as explained in Chapter 1, we shall use the notion of optimality associated with the concept of Nash equilibrium. For the sake of convenience, we repeat a definition already stated in Chapter 1.

Definition 2.3 *A set of admissible strategy profiles* $\hat{\boldsymbol{\alpha}} = (\hat{\boldsymbol{\alpha}}^1, \cdots, \hat{\boldsymbol{\alpha}}^N) \in \mathbb{A}^{(N)}$ *is said to be a Nash equilibrium for the game if:*

$$\forall i \in \{1, \cdots, N\}, \quad \forall \boldsymbol{\alpha}^i \in \mathbb{A}_i, \qquad J^i(\hat{\boldsymbol{\alpha}}) \leqslant J^i(\boldsymbol{\alpha}^i, \hat{\boldsymbol{\alpha}}^{-i}), \tag{2.3}$$

where $(\boldsymbol{\alpha}^i, \hat{\boldsymbol{\alpha}}^{-i})$ *stands for the strategy profile* $(\hat{\boldsymbol{\alpha}}^1, \cdots, \hat{\boldsymbol{\alpha}}^{i-1}, \boldsymbol{\alpha}^i, \hat{\boldsymbol{\alpha}}^{i+1})$*, in which the player* i *chooses the strategy* $\boldsymbol{\alpha}^i$ *while the others, indexed by* $j \in \{1, \cdots, N\} \setminus \{i\}$*, keep the original ones* $\hat{\boldsymbol{\alpha}}^j$*.*

The existence and uniqueness (or lack thereof) of Nash equilibria, as well as the properties of the corresponding optimal strategy profiles, strongly depend upon the information structures available to the players, and the types of actions they are allowed to take. In particular, the above definition of a Nash equilibrium can only make sense once we have properly defined the nature of the *frozen strategies* $\hat{\boldsymbol{\alpha}}^{-i}$ in the Nash condition: it is indeed necessary to specify how the players compute (we could even say "update") their strategies when one of them uses $\boldsymbol{\alpha}^i$ instead of $\hat{\boldsymbol{\alpha}}^i$. So rather than referring to a single game with several information structures and admissible strategy profiles for the players, we shall often talk about models, e.g., the open loop model for the game or the closed loop model, or even the Markovian model for the game. We give precise definitions below.

Open Loop Nash Equilibria

Definition 2.4 *If the strategy profile* $\hat{\boldsymbol{\alpha}} = (\hat{\boldsymbol{\alpha}}^1, \cdots, \hat{\boldsymbol{\alpha}}^N) \in \mathbb{A}^{(N)}$ *satisfies the conditions of Definition 2.3, without further restriction on the strategy* $\boldsymbol{\alpha}^i$ *and under the prescription that* $\hat{\boldsymbol{\alpha}}^{-i}$ *is the process with the same trajectories as the strategy profile* $(\hat{\boldsymbol{\alpha}}^1, \ldots, \hat{\boldsymbol{\alpha}}^{i-1}, \hat{\boldsymbol{\alpha}}^{i+1}, \ldots, \hat{\boldsymbol{\alpha}}^N)$*, even after player* i *changes strategy from* $\hat{\boldsymbol{\alpha}}^i$ *to* $\boldsymbol{\alpha}^i$*, then we say that* $\hat{\boldsymbol{\alpha}}$ *is an open loop Nash equilibrium for the game, or equivalently, a Nash equilibrium for the open loop game model.*

This arcane definition is best understood when the filtration $\mathbb{F}$ is generated by the Wiener process $\boldsymbol{W}$, except possibly for the presence of independent events in $\mathcal{F}_0$. In this case, the strategy profiles used in an open loop game model are given by controls of the form:

$$\alpha_t^i = \phi^i(t, X_0, W_{[0,t]}), \tag{2.4}$$

for some deterministic (measurable) functions $\phi^1, \cdots, \phi^N$, where we use the notation $W_{[0,t]}$ for the path of the Wiener process between time 0 and time t, regarded as a random variable with values in the space $\mathcal{C}([0,t];\mathbb{R}^M)$ of continuous functions from $[0,t]$ to $\mathbb{R}^M$, and it is enlightening to express the above definition in terms of the functions ϕ^ℓ instead of the actual strategies $\boldsymbol{\alpha}^\ell$. Indeed, the content of the definition of an open loop Nash equilibrium given above can be restated in the following way:

> *The strategy profile $\hat{\boldsymbol{\alpha}} = ((\hat{\alpha}_t^1)_{0 \leqslant t \leqslant T}, \cdots, (\hat{\alpha}_t^N)_{0 \leqslant t \leqslant T})$ where each strategy $\hat{\boldsymbol{\alpha}}^\ell$ is of the form $\hat{\alpha}_t^\ell = \hat{\phi}^\ell(t, X_0, W_{[0,t]})$ for some measurable function $\hat{\phi}^\ell$ as above, is an open loop Nash equilibrium if each time a player $i \in \{1, \cdots, N\}$ uses a different strategy $\boldsymbol{\alpha}^i$ given by a function ϕ^i possibly different from $\hat{\phi}^i$ while the other players keep using the same functions $\hat{\phi}^j$ for $j \neq i$, then this player i is not better off in the sense that $J^i(\hat{\boldsymbol{\alpha}}) \leqslant J^i(\boldsymbol{\alpha}^i, \hat{\boldsymbol{\alpha}}^{-i})$.*

A similar definition can be used when the functions $\hat{\phi}^\ell$ and ϕ^i fail to depend upon the Wiener process $\boldsymbol{W}$. This leads to the notion of deterministic open loop Nash equilibrium. In some sense, this merely amounts to redefining the sets $\mathbb{A}_i$ and $\mathbb{A}^{(N)}$ of admissible controls and strategy profiles as sets of deterministic processes, i.e., functions which do not dependent upon $\boldsymbol{W}$.

Definition 2.5 *If the strategy profile $\hat{\boldsymbol{\alpha}} = (\hat{\boldsymbol{\alpha}}^1, \cdots, \hat{\boldsymbol{\alpha}}^N) \in \mathbb{A}^{(N)}$ satisfies the conditions of Definition 2.3 with the restriction that the strategies $\hat{\boldsymbol{\alpha}}^j$, for $j = 1, \cdots, N$, and $\boldsymbol{\alpha}^i$ are deterministic functions of time and the initial state, then we say that $\hat{\boldsymbol{\alpha}}$ is a deterministic open loop Nash equilibrium for the game.*

To be more specific, in the search for a deterministic equilibrium, the strategy profiles are given by controls of the form:

$$\alpha_t^i = \phi^i(t, X_0),$$

for deterministic measurable functions $\phi^1, \cdots, \phi^N$. In other words, in this definition, the tuple $\hat{\boldsymbol{\alpha}} = (\hat{\boldsymbol{\alpha}}^1, \cdots, \hat{\boldsymbol{\alpha}}^N)$ in (2.3) reads as $(\hat{\phi}^1(t, X_0), \cdots, \hat{\phi}^N(t, X_0))$, while $(\boldsymbol{\alpha}^i, \hat{\boldsymbol{\alpha}}^{-i})$ has the form:

$$\begin{aligned} &\big(\phi^i(t, X_0), \hat{\phi}^{-i}(t, X_0)\big) \\ &\quad = \big(\hat{\phi}^1(t, X_0), \cdots, \hat{\phi}^{i-1}(t, X_0), \phi^i(t, X_0), \hat{\phi}^{i+1}(t, X_0), \cdots, \hat{\phi}^N(t, X_0)\big). \end{aligned}$$

Definitions 2.4 and 2.5 are consistent with the definitions of open loop equilibria used in the standard literature on deterministic games. They accommodate models with and without sources of randomness.

Closed Loop Nash Equilibria

In contrast with (2.4), the strategy profiles used in the search for a closed loop equilibrium are given by controls of the form:

$$\alpha_t^i = \phi^i(t, X_{[0,t]}), \tag{2.5}$$

for some deterministic (measurable) functions $\phi^1, \cdots, \phi^N$, $\boldsymbol{X} = (X_t)_{0 \leqslant t \leqslant T}$ denoting the solution of (2.1) and $X_{[0,t]}$ being regarded as a random variable with values in $\mathcal{C}([0,t];\mathbb{R}^D)$.

So, the expression $\alpha_t^i = \phi^i(t, X_{[0,t]})$ is merely an implicit equation for α_t^i, as $\boldsymbol{X}$ then depends upon $\boldsymbol{\alpha}^i$ itself. In that case, $\boldsymbol{X}$ has to be found as the solution of a stochastic differential equation with random and path-dependent coefficients. If one restricts oneself to functions ϕ^i that are Lipschitz in the path argument with respect to the uniform topology, uniformly in t, and that are locally bounded, then the stochastic differential equation for $\boldsymbol{X}$ is well posed under assumption **Games**.

This leads to the following definition of a Nash equilibrium.

Definition 2.6 *A strategy profile $\hat{\boldsymbol{\alpha}} = (\hat{\boldsymbol{\alpha}}^1, \cdots, \hat{\boldsymbol{\alpha}}^N)$ of the form $\hat{\alpha}_t^\ell = \hat{\phi}^\ell(t, \hat{X}_{[0,t]})$ for measurable functions $\hat{\phi}^\ell$ for $\ell = 1, \cdots, N$ is said to be a closed loop Nash equilibrium for the game if, for every $i \in \{1, \cdots, N\}$ and any feedback function ϕ^i, one has $J^i(\hat{\boldsymbol{\alpha}}) \leqslant J^i(\boldsymbol{\alpha})$ whenever:*

- *$\hat{\boldsymbol{\alpha}} = (\hat{\boldsymbol{\alpha}}^1, \cdots, \hat{\boldsymbol{\alpha}}^N)$ with $\hat{\alpha}_t^i = \hat{\phi}^i(t, \hat{X}_{[0,t]})$ where $\hat{\boldsymbol{X}} = (\hat{X}_t)_{0 \leqslant t \leqslant T}$ is the solution of the state equation* (2.1) *in which we use the controls $\alpha_t = (\hat{\phi}^1(t, \hat{X}_{[0,t]}), \cdots, \hat{\phi}^N(t, \hat{X}_{[0,t]}))$;*
- *$\boldsymbol{\alpha} = (\boldsymbol{\alpha}^1, \cdots, \boldsymbol{\alpha}^N)$ with $\alpha_t^i = \phi^i(t, X_{[0,t]})$ and $\alpha_t^j = \hat{\phi}^j(t, X_{[0,t]})$ for $j \neq i$, where $\boldsymbol{X} = (X_t)_{0 \leqslant t \leqslant T}$ is now the solution of the same state equation* (2.1) *in which we use the control α_t given by:*

$$\Big(\hat{\phi}^1(t, X_{[0,t]}), \cdots, \hat{\phi}^{i-1}(t, X_{[0,t]}), \phi^i(t, X_{[0,t]}), \\ \hat{\phi}^{i+1}(t, X_{[0,t]}), \cdots, \hat{\phi}^N(t, X_{[0,t]})\Big).$$

This definition may seem rather cumbersome and pedantic, but we chose to spell out the details needed to understand the subtle differences between open and closed loop equilibria.

Important Differences.

It is crucial to realize the major differences between the notions of open and closed loop Nash equilibria. When checking that a strategy profile $\hat{\boldsymbol{\alpha}} = (\hat{\boldsymbol{\alpha}}^1, \cdots, \hat{\boldsymbol{\alpha}}^N)$ is an open loop equilibrium, the fact that a given player i changes its strategy from $\hat{\boldsymbol{\alpha}}^i$ to $\boldsymbol{\alpha}^i$ does not affect the strategies $\hat{\boldsymbol{\alpha}}^j$ for $j \neq i$ of the other players. This comes from the fact that the open loop controls $\hat{\alpha}_t^j$ are functions of the trajectories of the

Wiener process and, as long as $j \neq i$, they do not change when player i changes its own function ϕ^i of the Wiener process. Even if this change affects the state of the system, it does not change the strategies $\hat{\boldsymbol{\alpha}}^j$ of the other players.

However, things are different in the case of closed loop equilibria. Indeed, if player i changes its strategy from $\hat{\boldsymbol{\alpha}}^i = (\hat{\phi}^i(t, X_{[0,t]}))_{0 \leqslant t \leqslant T}$ to a new control strategy $\boldsymbol{\alpha}^i = (\phi^i(t, X_{[0,t]}))_{0 \leqslant t \leqslant T}$, this change is likely to affect the trajectory of the state of the system, and even if the other players $j \neq i$ still use the same feedback functions $\hat{\phi}^j$, their controls $\hat{\boldsymbol{\alpha}}^j = (\hat{\phi}^j(t, X_{[0,t]}))_{0 \leqslant t \leqslant T}$ will change because of the changes in the value of the state!

To put it another way, the prescription that the players other than i keep using the same feedback functions $\hat{\phi}^{-i}$ to compute their controls allows their controls to take into account the new state of the system if player i changes its strategy.

Also, notice that writing $\alpha_t^i = \phi^i(t, X_{[0,t]})$ supposes that player i has *perfect* observation of the state of the whole system. In many practical applications, it has only *partial* observation, meaning that player i can only observe (possibly indirectly) the states of some of the players in the system.

Finally, we give the definition of closed loop Nash equilibria in feedback form. In the search for a closed loop equilibrium in feedback form, the strategy profiles are given by controls of the form:

$$\alpha_t^i = \phi^i(t, X_0, X_t), \tag{2.6}$$

for some deterministic (measurable) functions $\phi^1, \cdots, \phi^N$. Similar to Definition 2.6, we thus have the following definition.

Definition 2.7 *A strategy profile $\hat{\boldsymbol{\alpha}} = (\hat{\boldsymbol{\alpha}}^1, \cdots, \hat{\boldsymbol{\alpha}}^N)$ of the form $\hat{\alpha}_t^\ell = \hat{\phi}^\ell(t, X_0, \hat{X}_t)$ for measurable feedback functions $\hat{\phi}^\ell$ for $\ell = 1, \cdots, N$ is said to be a closed loop Nash equilibrium in feedback form for the game if, for every $i \in \{1, \cdots, N\}$ and any feedback function ϕ^i, one has $J^i(\hat{\boldsymbol{\alpha}}) \leqslant J^i(\boldsymbol{\alpha})$ if*

- *$\hat{\boldsymbol{\alpha}} = (\hat{\boldsymbol{\alpha}}^1, \cdots, \hat{\boldsymbol{\alpha}}^N)$ with $\hat{\alpha}_t^i = \hat{\phi}^i(t, X_0, \hat{X}_t)$ where $\hat{\boldsymbol{X}} = (\hat{X}_t)_{0 \leqslant t \leqslant T}$ is the solution of the state equation* (2.1) *in which we use $\alpha_t = (\hat{\phi}^1(t, X_0, \hat{X}_t), \cdots, \hat{\phi}^N(t, X_0, \hat{X}_t))$;*
- *$\boldsymbol{\alpha} = (\boldsymbol{\alpha}^1, \cdots, \boldsymbol{\alpha}^N)$ with $\alpha_t^i = \phi^i(t, X_0, X_t)$ and $\alpha_t^j = \hat{\phi}^j(t, X_0, X_t)$ for $j \neq i$, where $\boldsymbol{X} = (X_t)_{0 \leqslant t \leqslant T}$ is the solution of the same state equation* (2.1) *in which we use:*

$$\alpha_t = \big(\hat{\phi}^1(t, X_0, X_t), \cdots, \hat{\phi}^{i-1}(t, X_0, X_t), \\ \phi^i(t, X_0, X_t), \hat{\phi}^{i+1}(t, X_0, X_t), \cdots, \hat{\phi}^N(t, X_0, X_t)\big).$$

The most important example of application of this notion of equilibrium concerns the case of deterministic drift and volatility coefficients B and Σ, and cost functions f^i and g^i. This case will be discussed in Subsection 2.1.4 treating Markovian diffusions and we shall strengthen the notion of closed loop equilibrium in feedback form into the notion of Markovian equilibrium.

Remark 2.8 *It is important to emphasize that in open loop models, when a player makes its decisions, it may not be able to take into account the plays of its opponents since its decision can only be a function of the history of the random shocks. On the other hand, in closed loop models, past plays impact the values of the state, and in that way, become part of the common knowledge of the players. Though less realistic, open loop equilibria are more mathematically tractable than closed loop equilibria. Indeed, players need not consider how their opponents would react to deviations from the equilibrium path. With this in mind, one should expect that when the impact of players on their opponents' costs/rewards is small, open loop and closed loop equilibria should be the same. It is often conjectured that this should be the case for large games. We shall see instances of this claim at the end of the chapter in the large N limit of the Linear Quadratic (LQ) models for flocking with* $\beta = 0$*, and for systemic risk introduced in Subsections 1.5.1 and 1.3.1 of Chapter 1. More generally, we shall address this question in a more systematic way in Chapter (Vol II)-6.*

As a final remark on the definitions of Nash equilibria, we insist that the interpretation stressed in Chapter 1 in terms of 1) a sequence of individual optimal control problems to construct the best response function; 2) the search for fixed points of this best response map, still holds in any of the cases discussed above.

2.1.3 Players' Hamiltonians

For each player $i \in \{1, \cdots, N\}$, we define its Hamiltonian as the function H^i:

$$
\begin{aligned}
[0,T] \times \Omega \times \mathbb{R}^D \times \mathbb{R}^D \times \mathbb{R}^{D\times M} \times A^{(N)} \ni & (t,x,y,z,\alpha) \\
& \mapsto H^i(t,x,y,z,\alpha) \in \mathbb{R}
\end{aligned}
\tag{2.7}
$$

defined by:

$$
H^i(t,x,y,z,\alpha) = \underbrace{B(t,x,\alpha)\cdot y}_{\substack{\text{inner product of}\\ \text{state drift } B\\ \text{and covariable } y}} + \underbrace{\text{trace}\,[\Sigma(t,x,\alpha)^\dagger z]}_{\substack{\text{inner product of}\\ \text{state volatility } \Sigma\\ \text{and covariable } z}} + \underbrace{f^i(t,x,\alpha)}_{\substack{\text{running cost}\\ \text{of player } i}}. \tag{2.8}
$$

Pay attention that, at some point below, the inner product $\text{trace}[\Sigma(t,x,\alpha)^\dagger z]$ in the space of matrices is just denoted by $\Sigma(t,x,\alpha)\cdot z$.

When the volatility is uncontrolled, namely when Σ is independent of α, we usually do not include the second term in the above right-hand side, and talk about reduced Hamiltonian instead.

The following definition is motivated by the generalization to stochastic differential games of the Hamilton-Jacobi-Bellman equation and the stochastic maximum principle which we give below.

Definition 2.9 *We say that the generalized Isaacs (minmax) condition holds if there exists a function:*

$$\hat{\alpha} : [0,T] \times \mathbb{R}^D \times (\mathbb{R}^D)^N \times (\mathbb{R}^{D\times M})^N \ni (t,x,y,z) \mapsto \hat{\alpha}(t,x,y,z) \in A^{(N)}$$

satisfying for every $i \in \{1,\cdots,N\}$, and for all $t \in [0,T]$, $x \in \mathbb{R}^D$, $y = (y^1,\cdots,y^N) \in (\mathbb{R}^D)^N$, and $z = (z^1,\cdots,z^N) \in (\mathbb{R}^{D\times M})^N$:

$$H^i(t,x,y^i,z^i,\hat{\alpha}(t,x,y,z)) \leqslant H^i\big(t,x,y^i,z^i,(\alpha^i,\hat{\alpha}(t,x,y,z)^{-i})\big), \tag{2.9}$$

for all $\alpha^i \in A_i$.

Notice that in this definition, the function $\hat{\alpha}$ could be allowed to depend upon the random scenario $\omega \in \Omega$ if the Hamiltonians H^i did. In words, this definition says that for each set of dual variables $y = (y^1,\cdots,y^N)$ and $z = (z^1,\cdots,z^N)$, for each time t and state x at time t, and possibly random scenario ω, one can find a set of actions $\hat{\alpha} = (\hat{\alpha}^1,\cdots,\hat{\alpha}^N)$ depending on these quantities, such that, if we fix $N-1$ of these actions, say $\hat{\alpha}^{-i}$, then the remaining one $\hat{\alpha}^i$ minimizes the i-th Hamiltonian in the sense that:

$$\hat{\alpha}^i \in \arg\inf_{\alpha^i \in A^i} H^i\big(t,x,y^i,z^i,(\alpha^i,\hat{\alpha}^{-i})\big), \qquad \text{for all } i \in \{1,\cdots,N\}. \tag{2.10}$$

Once again, the notation can be lightened slightly when the volatility is not controlled. Indeed, minimizing the Hamiltonian gives the same $\hat{\alpha}$ as minimizing the reduced Hamiltonian. In this case, the argument $\hat{\alpha}$ of the minimization is independent of z. So when the volatility is not controlled we say that the generalized Isaacs (minmax) condition holds if there exists a function:

$$\hat{\alpha} : [0,T] \times \mathbb{R}^D \times (\mathbb{R}^D)^N \ni (t,x,y) \mapsto \hat{\alpha}(t,x,y) \in A^{(N)}$$

satisfying:

$$\begin{aligned} &\forall i \in \{1,\cdots,N\},\ t \in [0,T],\ x \in \mathbb{R}^D,\ y = (y^1,\cdots,y^N) \in (\mathbb{R}^D)^N, \\ &\quad H^i(t,x,y^i,\hat{\alpha}(t,x,y)) \leqslant H^i(t,x,y^i,(\alpha^i,\hat{\alpha}(t,x,y)^{-i})) \quad \text{for all } \alpha^i \in A_i, \end{aligned} \tag{2.11}$$

where H^i stands for the reduced Hamiltonian of player i. Notice that we use the same letter H for the full-fledged and for the reduced Hamiltonians. We are confident that, at least at this stage, there is no possible confusion between the two because of the context and the variables appearing as arguments. In particular, if there is only one adjoint variable in H, then H should be the reduced Hamiltonian. Alternatively if there are more than one adjoint variable, then H should be understood as the full-fledged Hamiltonian. Still, in the second volume, we shall make the distinction between the two forms of Hamiltonians by writing $H^{(r)}$ for the reduced Hamiltonian.

2.1.4 The Case of Markovian Diffusion Dynamics

In many applications of interest, the coefficients of the state dynamics (2.1) depend only upon the *present value* X_t of the state instead of the entire past $X_{[0,t]}$ of the state of the system, or of the Wiener process driving the evolution of the state. In this case, the dynamics of the state are given by a diffusion-like equation:

$$dX_t = B(t, X_t, \alpha_t)dt + \Sigma(t, X_t, \alpha_t)dW_t, \qquad 0 \leqslant t \leqslant T, \tag{2.12}$$

with initial condition $X_0 = x$, and drift and volatility coefficients given by deterministic functions:

$$(B, \Sigma) : [0, T] \times \mathbb{R}^D \times A^{(N)} \ni (t, x, \alpha) \mapsto (B(t, x, \alpha), \Sigma(t, x, \alpha)) \in \mathbb{R}^D \times \mathbb{R}^{D \times M}.$$

In this setting, it is natural to use strategy profiles which are deterministic functions of time and the current value of the state to force the controlled state process to be a Markov diffusion. Furthermore, we also assume that the running and terminal cost functions f^i and g^i are Markovian in the sense that, like B and Σ, f^i and g^i do not depend upon the random scenario $\omega \in \Omega$, but only upon the current values of the state and the actions taken by the players, so that $f^i : [0, T] \times \mathbb{R}^D \times A^{(N)} \ni (t, x, \alpha) \mapsto f^i(t, x, \alpha) \in \mathbb{R}$, and $g^i : \mathbb{R}^D \ni x \mapsto g^i(x) \in \mathbb{R}$ with (at most) quadratic growth. So in the case of Markovian diffusion dynamics, the cost functional is of the form:

$$J^i(\boldsymbol{\alpha}) = \mathbb{E}\left[\int_0^T f^i(t, X_t, \alpha_t)dt + g^i(X_T)\right], \qquad \boldsymbol{\alpha} \in \mathbb{A}^{(N)}, \tag{2.13}$$

and we tailor the notion of equilibrium to this situation by considering closed loop strategy profiles in feedback forms which provide *simultaneously* Nash equilibria for all the games starting at times $t \in [0, T]$ (i.e., over the time periods $[t, T]$) and all the possible initial conditions $X_t = x$ as long as they share the same state drift and volatility coefficients B and Σ, and cost functions f^i and g^i. More precisely:

Definition 2.10 *A set $\phi^* = (\phi^{*1}, \cdots, \phi^{*N})$ of N deterministic (measurable) functions $\phi^{*i} : [0, T] \times \mathbb{R}^D \to A_i$ for $i = 1, \cdots, N$ is said to be a Markovian Nash equilibrium, or a Nash equilibrium for the Markovian game model, if for each $(t, x) \in [0, T] \times \mathbb{R}^D$, the strategy profile $\boldsymbol{\alpha}^* = (\boldsymbol{\alpha}^{*1}, \cdots, \boldsymbol{\alpha}^{*N}) \in \mathbb{A}^{(N)}$ defined for $s \in [t, T]$ by $\alpha_s^{*i} = \phi^{*i}(s, X_s^{t,x})$ where $\boldsymbol{X}^{t,x} = (X_s^{t,x})_{t \le s \le T}$ is the unique solution of the stochastic differential equation (which is implicitly required to be well-posed):*

$$dX_s = B\big(s, X_s, \phi^*(s, X_s)\big)ds + \Sigma\big(s, X_s, \phi^*(s, X_s)\big)dW_s, \quad t \le s \le T,$$

with initial condition $X_t = x$, satisfies the conditions of Definition 2.3 with:

- *the restriction that the strategy $\boldsymbol{\alpha}^i$ is also given by a deterministic function ϕ^i on $[t,T] \times \mathbb{R}^D$,*
- *the prescription that $(\boldsymbol{\alpha}^i, \hat{\boldsymbol{\alpha}}^{-i})$ in (2.3) uses the functions ϕ^i and $(\phi^{*j})_{1 \leqslant j \neq i \leqslant N}$.*

The strategy profiles used in the above definition are called Markovian strategy profiles and the deterministic functions ϕ^* and ϕ^i feedback functions. Extra regularity assumptions on the functions ϕ^* and ϕ^i may be needed for the stochastic differential equations giving the dynamics of the controlled state to have a unique strong solution. Under assumption **Games**, the coefficients B and Σ are Lipschitz in (x, α) uniformly in $t \in [0,T]$, so that, assuming that the feedback function ϕ^* is locally bounded and Lipschitz in x uniformly in t is enough. However in some cases, requiring that the feedback functions are Lipschitz continuous may be an overkill somehow. Indeed, using the Markovian nature of the dynamics, the stochastic differential equation for $\boldsymbol{X}$ is known to be well posed, regardless of the smoothness of ϕ^*, when coefficients are bounded and the volatility is Lipschitz-continuous, uniformly nondegenerate and uncontrolled. We shall use this fact in Chapter (Vol II)-6 in order to identify an instance of uniqueness for Markovian Nash equilibria in Proposition (Vol II)-6.27.

We shall still use the same notation (2.8) for the players' Hamiltonians. However, their roles and their interpretations will be slightly different than in the search for open loop Nash equilibria. Indeed, using Markovian strategy profiles instead of merely *state-insensitive* adapted processes for controls will dramatically affect the dependence upon the state variable x. In order to illustrate this last point, we abandon momentarily the notation $\boldsymbol{\alpha}$ for the control processes, and we use the notation $\phi = (\phi^1, \cdots, \phi^N)$ for deterministic functions $\phi^i : [0.T] \times \mathbb{R}^D \to A_i$ to emphasize the Markovian nature of the strategy profiles. The controlled dynamics of the state $\boldsymbol{X} = \boldsymbol{X}^{\boldsymbol{\alpha}}$ solve the Markovian stochastic differential equation:

$$dX_t = B(t, X_t, \phi(t, X_t))\, dt + \Sigma(t, X_t, \phi(t, X_t))\, dW_t.$$

Hence, the state of the system is a Markov process with infinitesimal generator $\mathcal{L}^\phi$ defined by:

$$\mathcal{L}_t^\phi = \frac{1}{2} \sum_{p,q=1}^{D} A_{pq}^\phi(t,x) \frac{\partial^2}{\partial x_p \partial x_q} + \sum_{p=1}^{D} B_p^\phi(t,x) \frac{\partial}{\partial x_p},$$

with, at least when the components of the Wiener processes $\boldsymbol{W}$ are independent,

$$B_p^\phi(t,x) = B_p(t, x, \phi(t,x)),$$

and

$$A^{\phi}_{pq}(t,x) = \sum_{\ell=1}^{m} \Sigma_{p\ell}(t,x,\phi(t,x))\Sigma_{q\ell}(t,x,\phi(t,x)).$$

Let us assume momentarily that the Markovian strategy profile $\phi^* = (\phi^{*i})_{i=1,\cdots,N}$ is a Markovian Nash equilibrium. Being a fixed point of the best response function, the Markovian control ϕ^{*i} of player i, for each fixed $i \in \{1,\cdots,N\}$, minimizes $J^i(\phi^i,\phi^{*-i})$ over Markovian controls ϕ^i. So if we assume that the Markovian strategy profiles ϕ^{*-i} are *frozen*, we can think of ϕ^{*i} as the solution of a standard stochastic control problem. We discuss the consequences of this fact in light of the stochastic maximum principle in Subsection 2.2.2 below, but for the time being, we review its implications in terms of the standard analytic approach to the control of Markovian diffusions based on the solution of Partial Differential Equations (PDEs).

The value function V^i of player i is defined as:

$$V^i(t,x) = \inf_{\alpha^i \in \mathbb{A}^i} \mathbb{E}\Big[\int_t^T f^i\big(s,X_s,(\alpha^i_s,\phi^{*-i}(s,X_s))\big)ds + g^i(X_T)\,\big|\,X_t = x\Big]. \tag{2.14}$$

It depends upon the feedback functions ϕ^{*-i} of the other players. It is expected to satisfy the Hamilton-Jacobi-Bellman (HJB) equation:

$$\partial_t V^i + L^{i*}\big(t,x,\partial_x V^i(t,x),\partial^2_{xx}V^i(t,x)\big) = 0, \tag{2.15}$$

where the operator symbol L^{*i} is defined by:

$$L^{i*}(t,x,y,z) = \inf_{\alpha \in A^i} L^i(t,x,y,z,\alpha)$$

and the function L^i by:

$$\begin{aligned} L^i(t,x,y,z,\alpha) =& f^i\big(t,x,(\alpha,\phi^{*-i}(t,x))\big) + y\cdot B\big(t,x,(\alpha,\phi^{*-i}(t,x))\big) \\ &+ \frac{1}{2}\mathrm{trace}\big[z\Sigma\big(t,x,(\alpha,\phi^{*-i}(t,x))\big)\Sigma\big(t,x,(\alpha,\phi^{*-i}(t,x))\big)^{\dagger}\big]. \end{aligned}$$

The HJB equation (2.15) does not stand on its own since the $N-1$ feedback functions ϕ^{*-i} enter the definition of the operator symbol L^{*i}. Indeed, the fact that a Nash equilibrium is a fixed point of the best response map is the source of intricate dependencies between the optimal controls of the individual players, creating strong couplings between these HJB equations. In many applications of interest, the equilibrium feedback functions ϕ^{*i} are given in terms of the gradients $\partial_x V^j(t,x)$ and possibly of the Hessians $\partial^2_{xx}V^j(t,x)$ of the individual players' value functions. In particular, we shall study examples for which $\phi^{*j}(t,x) = -\partial_x V^j(t,x)$, or a simple function of the gradient of V. In these cases, the value functions $(V^i)_{i=1,\cdots,N}$ appear as the solution of a system of N strongly coupled HJB equations, though very difficult to solve in most cases.

Notations. Throughout the book, gradients of scalar valued functions will be regarded, when needed, as column vectors. Derivatives of vector valued functions will be regarded as matrices, the number of lines being given by the dimension of the arrival vector space and the number of columns being given by the number of directions in the differentiation. Hence, if $v : \mathbb{R}^d \to \mathbb{R}$, $\partial_x v = (\partial_{x_i} v(x))_{1 \leqslant i \leqslant d}$ is regarded as d-dimensional column vector, while, if $v = (v^1, \cdots, v^n) : \mathbb{R}^d \to \mathbb{R}^n$, $n \geqslant 2$, $\partial_x v = (\partial_{x_j} v^i(x))_{1 \leqslant i \leqslant n, 1 \leqslant j \leqslant d}$ is regarded as a matrix of dimension $n \times d$.

Uncontrolled Volatilities

We now argue that the fact that the feedback function ϕ^{*i} is a function of $\partial_x V$ happens often when the volatility Σ is not controlled, in which case the operator symbol L^i is, up to the second order term, identical to the reduced Hamiltonian H^i of player i. So when the second order term is independent of the controls, it is equivalent to search for a minimizer of L^i or of H^i. Consequently in this case, the HJB equation (2.15) can be equivalently written using the minimized reduced Hamiltonian H^{*i}:

$$\partial_t V^i(t,x) + \frac{1}{2}\text{trace}\big[\Sigma(t,x)\Sigma(t,x)^\dagger \partial^2_{xx} V^i(t,x)\big] + H^{*i}\big(t,x,\partial_x V^i(t,x)\big) = 0, \qquad (2.16)$$

where the minimized reduced Hamiltonian H^{*i} of player i is defined by:

$$H^{*i}(t,x,y) = \inf_{\alpha \in A^i} H^i\big(t,x,y,(\alpha, \phi^{*-i}(t,x))\big).$$

We shall often write the HJB equation using the minimized reduced Hamiltonian as in (2.16) above.

The form of the system (2.16) can be made more explicit when Isaacs condition is in force, see Definition 2.9. In that case, the minimizer $\hat{\alpha}$ only depends on t, x, and y and is independent of z since the volatility is not controlled. Then, the guess is that ϕ^{*j} should be given by $\phi^{*j}(t,x) = \hat{\alpha}^j(t,x,\partial_x V(t,x))$, in which case (2.16) would take the form:

$$\begin{aligned} &\partial_t V^i(t,x) + \frac{1}{2}\text{trace}\big[\Sigma(t,x)\Sigma(t,x)^\dagger \partial^2_{xx} V^i(t,x)\big] \qquad (2.17) \\ &\quad + H^i\big(t,x,\partial_x V^i(t,x), \hat{\alpha}(t,x,\partial_x V(t,x))\big) = 0, \quad (t,x) \in [0,T] \times \mathbb{R}^D, \end{aligned}$$

for all $i \in \{1,\dots,N\}$, with the terminal condition:

$$V^i(T,x) = g^i(x), \quad x \in \mathbb{R}^D, \quad i \in \{1,\dots,N\}. \qquad (2.18)$$

System (2.17) is called the *Nash system* associated with the game. It is especially useful as a verification tool for the existence of Markovian Nash equilibria.

Proposition 2.11 *Assume that $\mathbb{A}^{(N)} \subset \mathbb{H}^{2,K}$, with $K = k_1 + \cdots + k_N$, and that there exists a classical solution of the Nash system* (2.17) *with the property that, for any given initial condition $x_0 \in \mathbb{R}^D$, the stochastic differential equation:*

$$dX_t^* = B\big(t, X_t^*, \hat{\alpha}(t, X_t^*, \partial_x V(t, X_t^*))\big)dt + \Sigma(t, X_t^*)dW_t, \quad t \in [0, T],$$

is uniquely solvable, with $(\hat{\alpha}(t, X_t^*, \partial_x V(t, X_t^*)))_{0 \leqslant t \leqslant T}$ belonging to $\mathbb{A}^{(N)}$, and with the property that for any other controlled dynamics:

$$dX_t = B(t, X_t, \alpha_t)dt + \Sigma(t, X_t)dW_t, \quad t \in [0, T],$$

with the same initial condition and with $\boldsymbol{\alpha} \in \mathbb{A}^{(N)}$, the expectation:

$$\mathbb{E}\int_0^T \big|\Sigma^\dagger(t, X_t)\partial_x V(t, X_t)\big|^2 dt$$

is finite. Then, provided that the costs to $(\hat{\alpha}(t, X_t^*, \partial_x V(t, X_t^*)))_{0 \leqslant t \leqslant T}$ are well defined, the tuple $(\phi^{*1}, \cdots, \phi^{*N}) = (\hat{\alpha}^1(\cdot, \cdot, \partial_x V(\cdot, \cdot)), \cdots, \hat{\alpha}^N(\cdot, \cdot, \partial_x V(\cdot, \cdot)))$ is a Markovian Nash equilibrium over strategy profiles ϕ^i, for $i \in \{1, \cdots, N\}$, for which the SDE:

$$dX_t = B\big(t, X_t, (\phi^i(t, X_t), \hat{\alpha}(t, X_t, \partial_x V(t, X_t))^{-i})\big)dt + \Sigma(t, X_t)dW_t,$$

for $t \in [0, T]$ and with $X_0 = x_0$, is uniquely solvable and its solution satisfies $(\phi^i(t, X_t), \hat{\alpha}(t, X_t, \partial_x V(t, X_t))^{-i})_{0 \leq t \leq T} \in \mathbb{A}^{(N)}$ and has finite costs.

Remark 2.12 *In fact, the proof shows that $V^i(0, x_0)$ is the cost to player i under the Markovian Nash equilibrium*

$$(\phi^{*1}, \ldots, \phi^{*N}) = (\hat{\alpha}^1(\cdot, \cdot, \partial_x V(\cdot, \cdot)), \cdots, \hat{\alpha}^N(\cdot, \cdot, \partial_x V(\cdot, \cdot))).$$

In particular, the tuple $(V^1, \ldots, V^N)$ reads as the value function of the game under the equilibrium $(\phi^{*1}, \cdots, \phi^{*N})$.

Proof. The proof is a mere application of Itô's formula. For a given $i \in \{1, \ldots, N\}$ and a given ϕ^i such that the SDE:

$$dX_t = B\big(t, X_t, (\phi^i(t, X_t), \hat{\alpha}(t, X_t, \partial_x V(t, X_t))^{-i})\big)dt + \Sigma(t, X_t)dW_t, \quad t \in [0, T],$$

with $X_0 = x_0$, is uniquely solvable, its solution satisfying $(\phi^i(t, X_t), \hat{\alpha}(t, X_t, \partial_x V(t, X_t))^{-i})_{0 \leq t \leq T} \in \mathbb{A}^{(N)}$ and having finite costs, we expand $(V^i(t, X_t))_{0 \leqslant t \leqslant T}$ by Itô's formula. Thanks to with the system (2.17), we get:

$$
\begin{aligned}
d\Big[V^i(t,X_t) &+ \int_0^t f^i\big(s,X_s,(\phi^i(s,X_s),\hat{\alpha}(s,X_s,\partial_x V(s,X_s))^{-i})\big)ds\Big] \\
&= \Big[H^i\big(t,X_t,\partial_x V^i(t,X_t),(\phi^i(t,X_t),\hat{\alpha}(t,X_t,\partial_x V(t,X_t))^{-i})\big) \\
&\quad - H^i\big(t,X_t,\partial_x V^i(t,X_t),\hat{\alpha}(t,X_t,\partial_x V(t,X_t))\big)\Big]dt \\
&\quad + \partial_x V^i(t,X_t)\Sigma(t,X_t)dW_t, \quad 0 \leqslant t \leqslant T,
\end{aligned}
$$

Taking the expectation and implementing Isaacs' condition (2.9), we deduce that:

$$
\mathbb{E}\Big[V^i(T,X_T) + \int_0^T f^i\big(t,X_t,(\phi^i(t,X_t),\hat{\alpha}(t,X_t,\partial_x V(t,X_t))^{-i})\big)dt\Big] \geq V^i(0,x_0),
$$

with equality when $\phi^i(t,x) = \hat{\alpha}^i(t,x,\partial_x V(t,x))$. □

Here is a specific set of assumptions under which the conclusion of the above proposition holds true.

Assumption (*N*-Nash System). The set $A^{(N)}$ is bounded. Moreover,

(A1) The function B is bounded on $[0,T] \times \mathbb{R}^D \times A^{(N)}$ and is Lipschitz continuous in $\alpha \in A^{(N)}$, uniformly in $(t,x) \in [0,T] \times \mathbb{R}^D$.

(A2) The function Σ is bounded and continuous on $[0,T] \times \mathbb{R}^D$ and is Lipschitz continuous in $x \in \mathbb{R}^D$, uniformly in time $t \in [0,T]$. Moreover the matrix-valued function $\Sigma\Sigma^\dagger$ is uniformly nondegenerate, that is the lowest eigenvalue is bounded from below by a positive constant, uniformly on $[0,T] \times \mathbb{R}^D$.

(A3) For each $i \in \{1,\ldots,N\}$, the function f^i is bounded on $[0,T] \times \mathbb{R}^D \times A^{(N)}$ and is Lipschitz continuous in $\alpha \in A^{(N)}$, uniformly in $(t,x) \in [0,T] \times \mathbb{R}^D$. The function g^i is bounded and Lipschitz continuous.

(A4) The minimizer in the Isaacs condition is Lipschitz continuous in $y \in \mathbb{R}^{D\times M}$, uniformly in $(t,x) \in [0,T] \times \mathbb{R}^D$.

The following result is taken from the Partial Differential Equation (PDE) literature:

Proposition 2.13 *Under assumption N-***Nash System***, the Nash system* (2.17)–(2.18) *has a unique solution V in the space of $\mathbb{R}^N$-valued bounded and continuous functions on $[0,T] \times \mathbb{R}^D$ that are differentiable in space on $[0,T) \times \mathbb{R}^D$, with a bounded and continuous gradient on $[0,T) \times \mathbb{R}^D$, and that have generalized time first-order and space second-order derivatives in $L^p_{\text{loc}}([0,T) \times \mathbb{R}^D)$, for any $p \geqslant 1$, where the index* loc *is used to indicate that integrability holds true on compact subsets only.*

Moreover, for any bounded and measurable function ϕ from $[0,T] \times \mathbb{R}^D$ into $A^{(N)}$ and for any initial condition $x_0 \in \mathbb{R}^D$, the stochastic differential equation:

$$dX_t = B(t, X_t, \phi(t, X_t))dt + \Sigma(t, X_t)dW_t, \quad t \in [0,T],$$

with $X_0 = x_0$ is uniquely solvable, and the conclusion of Proposition 2.11 holds true without any further restriction on the profile strategies used in the definition of a Markovian Nash equilibrium.

Pay attention that, although the conclusion of Proposition 2.11 is claimed to hold under assumption N-**Nash System**, the Nash system is not claimed to have a classical solution. The solution V provided by Proposition 2.13 is usually said to be *strong*.

Proof. The first part of the statement is a standard result from the PDE literature; references are given in the Notes & Complements at the end of the chapter. What really matters is the fact that the volatility coefficient is uniformly nondegenerate.

The second part of the statement is also standard in stochastic analysis. Again, references are provided in the Notes & Complements below.

The third part of the statement follows from the same argument as that used in the proof of Proposition 2.11, except that an extension of Itô's formula is needed to overcome the lack of continuity of the first order time derivative and second order space derivatives of V. Such an extension is covered by the so-called Itô-Krylov formula for Itô processes driven by a bounded drift and a bounded and uniformly nondegenerate volatility coefficient. Once again, references are given in the Notes & Complements. □

2.2 Game Versions of the Stochastic Maximum Principle

Generalizations of the Pontryagin maximum principle to stochastic games are not as straightforward as one would like. While many versions are possible, we limit ourselves to open loop and Markovian equilibria for the sake of simplicity. We treat them separately to emphasize the differences.

Throughout this section, we assume that assumption **Games** is in force together with:

Assumption (Games SMP). Each A_i, for $i = 1, \cdots, N$, is a convex subset of $\mathbb{R}^{k_i}$. Moreover, the coefficients satisfy:

(A1) The drift and volatility functions B and Σ, as well as the running and terminal cost functions $(f^i)_{i=1,\cdots,N}$ and $(g^i)_{i=1,\cdots,N}$ are locally bounded deterministic functions which are continuously differentiable with respect to (x, α).

(A2) For $i = 1, \cdots, N$, the partial derivatives $\partial_x f^i$ and $\partial_\alpha f^i$ (respectively $\partial_x g^i$) for $i = 1, \cdots, N$ are at most of linear growth in (x, α) (respectively in x), uniformly in $t \in [0,T]$.

Observe from the Lipschitz property of B and Σ that the partial derivatives $\partial_x B$, $\partial_\alpha B$, $\partial_x \Sigma$ and $\partial_\alpha \Sigma$ are also bounded.

Notice that, since B takes values in $\mathbb{R}^D$ and $x \in \mathbb{R}^D$, $\partial_x B$ is an element of $\mathbb{R}^{D\times D}$, in other words a $D \times D$ matrix whose entries are the partial derivatives of the components B^i of B with respect to the components x^j of x. Analog statements can be made concerning $\partial_x \Sigma$ which has the interpretation of a tensor.

In this section, we allow the volatility Σ to depend upon the control parameter $\alpha \in A^{(N)}$. Also, we assume that $\mathbb{A}^{(N)} = \prod_{i=1}^N \mathbb{A}_i$ with $\mathbb{A}_i \subset \mathbb{H}^{2,k_i}$.

2.2.1 Open Loop Equilibria

The generalization of the stochastic Pontryagin maximum principle to open loop stochastic games can be approached in a very natural way. Forms of the sufficient condition for the existence and identification of an open loop Nash equilibrium can be used in the case of linear quadratic models. Specific examples are given below. We also refer to the Notes & Complements at the end of the chapter for further references.

Definition 2.14 *Given an admissible strategy profile $\boldsymbol{\alpha} \in \mathbb{A}^{(N)}$ and the corresponding controlled state $\boldsymbol{X} = \boldsymbol{X}^{\boldsymbol{\alpha}}$ of the system, a set of N couples of processes $((\boldsymbol{Y}^{i,\boldsymbol{\alpha}}, \boldsymbol{Z}^{i,\boldsymbol{\alpha}}) = (Y_t^{i,\boldsymbol{\alpha}}, Z_t^{i,\boldsymbol{\alpha}})_{0\leqslant t\leqslant T})_{i=1,\cdots,N}$ in $\mathbb{S}^{2,D} \times \mathbb{H}^{2,D\times M}$ for each $i = 1,\cdots,N$, is said to be a set of adjoint processes associated with $\boldsymbol{\alpha} \in \mathbb{A}^{(N)}$ if, for each $i \in \{1,\cdots,N\}$, they satisfy the Backward Stochastic Differential Equation (BSDE):*

$$\begin{cases} dY_t^{i,\boldsymbol{\alpha}} = -\partial_x H^i(t, X_t, Y_t^{i,\boldsymbol{\alpha}}, Z_t^{i,\boldsymbol{\alpha}}, \alpha_t)dt + Z_t^{i,\boldsymbol{\alpha}} dW_t, \quad t \in [0,T], \\ Y_T^{i,\boldsymbol{\alpha}} = \partial_x g^i(X_T^{\boldsymbol{\alpha}}). \end{cases} \tag{2.19}$$

Given $\boldsymbol{\alpha} \in \mathbb{A}^{(N)}$ and the corresponding state $\boldsymbol{X} = \boldsymbol{X}^{\boldsymbol{\alpha}}$ defined in (2.1), equation (2.19) can be viewed as a BSDE with random coefficients, a terminal condition in L^2 and a bounded variation term $-\psi^i$ with the following p-th component, for each $p \in \{1,\cdots,D\}$:

$$\begin{aligned}\psi^{i,p}(t,\omega,y,z) &= \partial_{x_p} B\big(t, X_t(\omega), \alpha_t(\omega)\big) \cdot y \\ &\quad + \mathrm{trace}\big[\partial_{x_p} \Sigma\big(t, X_t(\omega), \alpha_t(\omega)\big)^\dagger z\big] + \partial_{x_p} f^i\big(t, X_t(\omega), \alpha_t(\omega)\big),\end{aligned}$$

which is an affine function of y and z. Also, as $\boldsymbol{X} = \boldsymbol{X}^{\boldsymbol{\alpha}} \in \mathbb{S}^{2,D}$ and $\boldsymbol{\alpha}^i \in \mathbb{H}^{2,k_i}$ for each $i \in \{1,\cdots,N\}$, it holds:

$$\mathbb{E}\int_0^T |\psi^i(t, X_t, Y_t^{i,\boldsymbol{\alpha}}, Z_t^{i,\boldsymbol{\alpha}}, \alpha_t)|^2 dt < \infty.$$

In the stochastic analysis literature, the bounded variation part of a BSDE (up to the minus sign) is often called the driver of the equation and denoted by the letters ψ or f. We shall not use f for the driver because we use the letter f for the running cost. So, for each $i \in \{1, \cdots, N\}$, existence and uniqueness of a solution follow from standard results on BSDEs. See the Notes & Complements at the end of the chapter for references and Chapter 4 for precise statements of these results.

Necessary Part of the Stochastic Maximum Pontryagin Principle

The following result is the open loop *game analog* of the necessary part of the stochastic maximum principle of stochastic control in the case of convex sets of admissible controls.

Theorem 2.15 *Under the above conditions, if $\hat{\boldsymbol{\alpha}} \in \mathbb{A}^{(N)}$ is an open loop Nash equilibrium, if we denote by $\hat{\mathbf{X}} = (\hat{X}_t)_{0 \le t \le T}$ the corresponding controlled state of the system, and the adjoint processes by $\hat{\mathbf{Y}} = (\hat{\mathbf{Y}}^1, \cdots, \hat{\mathbf{Y}}^N)$ and $\hat{\mathbf{Z}} = (\hat{\mathbf{Z}}^1, \cdots, \hat{\mathbf{Z}}^N)$, then the generalized min-max Isaacs conditions hold along the optimal paths in the sense that, for each $i \in \{1, \cdots, N\}$,*

$$
\begin{aligned}
&H^i\big(t, \hat{X}_t, \hat{Y}_t^i, \hat{Z}_t^i, \hat{\alpha}_t\big) \\
&\quad = \inf_{\alpha^i \in A^i} H^i\big(t, \hat{X}_t, \hat{Y}_t^i, \hat{Z}_t^i, (\alpha^i, \hat{\alpha}_t^{-i})\big), \quad \mathrm{Leb}_1 \otimes \mathbb{P}\ a.e.,
\end{aligned}
\tag{2.20}
$$

provided that the mapping $A_i \ni \alpha \mapsto H^i(t, \hat{X}_t, \hat{Y}_t^i, \hat{Z}_t^i, (\alpha, \hat{\alpha}^{-i}))$ is convex, $\mathrm{Leb}_1 \otimes \mathbb{P}$ *almost-everywhere, where* Leb_1 *is the one-dimensional Lebesgue measure.*

Proof. We only provide a sketch of the proof since we exclusively use this result as a rationale behind our search strategy for a function satisfying the min-max Isaacs condition.

The proof is a consequence of the stochastic maximum principle of stochastic control whose statement is recalled in Theorem 3.27 and which is proven in the greater generality of the control of McKean-Vlasov equation in Chapters 6 and (Vol II)-1. See for example Theorem 6.14 in Chapter 6. For a given $i \in \{1, \ldots, N\}$, in order to find the best response to the strategies $\hat{\boldsymbol{\alpha}}^{-i}$ we may consider the optimal control problem consisting in minimizing the cost functional $J^i(\boldsymbol{\alpha}^i, \hat{\boldsymbol{\alpha}}^{-i})$ over control strategies $\boldsymbol{\alpha}^i \in \mathbb{A}_i$ and controlled Itô processes:

$$
dX_t = B\big(t, X_t, (\alpha_t^i, \hat{\alpha}_t^{-i})\big)dt + \Sigma\big(t, X_t, (\alpha_t^i, \hat{\alpha}_t^{-i})\big)dW_t, \quad t \in [0, T].
$$

Since $\boldsymbol{\alpha}^i = \hat{\boldsymbol{\alpha}}^i$ is a minimizer, we can use the necessary part of the standard stochastic maximum principle of stochastic control. The proof is completed by adapting the proofs of Theorems 6.14 and (Vol II)-1.59 to the current setting, under which the coefficients are random and the volatility is controlled. □

Sufficiency Part of the Stochastic Maximum Pontryagin Principle

We now state and prove the sufficient condition which we shall use repeatedly.

Theorem 2.16 *For an admissible strategy profile $\hat{\boldsymbol{\alpha}} \in \mathbb{A}^{(N)}$, with $\hat{\boldsymbol{X}} = (\hat{X}_t)_{0 \leqslant t \leqslant T}$ as corresponding controlled state and $(\hat{\boldsymbol{Y}}, \hat{\boldsymbol{Z}}) = ((\hat{\boldsymbol{Y}}^1, \cdots, \hat{\boldsymbol{Y}}^N), (\hat{\boldsymbol{Z}}^1, \cdots, \hat{\boldsymbol{Z}}^N))$ as corresponding adjoint processes, assume that, for each $i \in \{1, \cdots, N\}$,*

- $\mathbb{R}^D \times A^i \ni (x, \alpha^i) \mapsto H^i(t, x, \hat{Y}_t^i, \hat{Z}_t^i, (\alpha^i, \hat{\alpha}_t^{-i}))$ *is a convex function,* $\mathrm{Leb}_1 \otimes \mathbb{P}$ *almost-everywhere,*
- g^i *is convex,*

and

$$H^i(t, \hat{X}_t, \hat{Y}_t^i, \hat{Z}_t^i, \hat{\alpha}_t) = \inf_{\alpha^i \in A^i} H^i(t, \hat{X}_t, \hat{Y}_t^i, \hat{Z}_t^i, (\alpha^i, \hat{\alpha}_t^{-i})), \ \ \mathrm{Leb}_1 \otimes \mathbb{P} \ a.e., \tag{2.21}$$

then $\hat{\boldsymbol{\alpha}}$ is an open loop Nash equilibrium.

Proof. We fix $i \in \{1, \cdots, N\}$, a generic $\boldsymbol{\alpha}^i \in \mathbb{A}_i$, and for the sake of simplicity, we denote by $\boldsymbol{X}$ the state process $\boldsymbol{X}^{(\alpha^i, \hat{\alpha}^{-i})}$ controlled by the strategies $(\boldsymbol{\alpha}^i, \hat{\boldsymbol{\alpha}}^{-i})$. The function g^i being convex, using the form of the terminal condition of the adjoint equations and integration by parts, we get:

$$\begin{aligned}
g^i(\hat{X}_T) - g^i(X_T) &\leqslant \partial_x g^i(\hat{X}_T) \cdot (\hat{X}_T - X_T) \\
&= \hat{Y}_T^i \cdot (\hat{X}_T - X_T) \\
&= \int_0^T (\hat{X}_t - X_t) \cdot d\hat{Y}_t^i + \int_0^T \hat{Y}_t^i \cdot d(\hat{X}_t - X_t) \\
&\quad + \int_0^T \mathrm{trace}\big\{\big[\Sigma(t, \hat{X}_t, \hat{\alpha}_t) - \Sigma\big(t, X_t, (\alpha_t^i, \hat{\alpha}_t^{-i})\big)\big]^\dagger \hat{Z}_t^i\big\} dt \\
&= -\int_0^T (\hat{X}_t - X_t) \cdot \partial_x H^i(t, \hat{X}_t, \hat{Y}_t^i, \hat{Z}_t^i, \hat{\alpha}_t)\, dt \\
&\quad + \int_0^T \hat{Y}_t^i \cdot \big[B(t, \hat{X}_t, \hat{\alpha}_t) - B\big(t, X_t, (\alpha_t^i, \hat{\alpha}_t^{-i})\big)\big]\, dt \\
&\quad + \int_0^T \mathrm{trace}\big\{\big[\Sigma(t, \hat{X}_t, \hat{\alpha}_t) - \Sigma\big(t, X_t, (\alpha_t^i, \hat{\alpha}_t^{-i})\big)\big]^\dagger \hat{Z}_t^i\big\} dt + M_T,
\end{aligned}$$

where $(M_t)_{0 \leqslant t \leqslant T}$ is a martingale with $M_0 = 0$. Taking expectations of both sides and plugging the result into:

$$J^i(\hat{\boldsymbol{\alpha}}) - J^i((\boldsymbol{\alpha}^i, \hat{\boldsymbol{\alpha}}^{-i})) = \mathbb{E}\bigg[\int_0^T \big[f^i(t, \hat{X}_t, \hat{\alpha}_t) - f^i\big(t, X_t, (\alpha_t^i, \hat{\alpha}_t^{-i})\big)\big] dt + g^i(\hat{X}_T) - g^i(X_T)\bigg],$$

we get:

$$\begin{aligned}
&J^i(\hat{\boldsymbol{\alpha}}) - J^i\big((\boldsymbol{\alpha}^i, \hat{\boldsymbol{\alpha}}^{-i})\big) \\
&= \mathbb{E}\bigg[\int_0^T \big[H^i(t, \hat{X}_t, \hat{Y}_t^i, \hat{Z}_t^i, \hat{\alpha}_t) - H^i\big(t, X_t, \hat{Y}_t^i, \hat{Z}_t^i, (\alpha_t^i, \hat{\alpha}_t^{-i})\big)\big]dt\bigg] \\
&\quad - \mathbb{E}\bigg[\int_0^T \hat{Y}_t^i \cdot \big[B(t, \hat{X}_t, \hat{\alpha}_t) - B\big(t, X_t, (\alpha_t^i, \hat{\alpha}_t^{-i})\big)\big]dt\bigg] \\
&\quad - \mathbb{E}\bigg[\int_0^T \mathrm{trace}\big\{\big[\Sigma(t, \hat{X}_t, \hat{\alpha}_t) - \Sigma\big(t, X_t, (\alpha_t^i, \hat{\alpha}_t^{-i})\big)\big] \cdot \hat{Z}_t^i\big\}dt\bigg] \\
&\quad + \mathbb{E}\big[g^i(\hat{X}_T) - g^i(X_T)\big] \\
&\leqslant \mathbb{E}\bigg[\int_0^T \big[H^i(t, \hat{X}_t, \hat{Y}_t^i, \hat{Z}_t^i, \hat{\alpha}_t) - H^i\big(t, X_t, \hat{Y}_t^i, \hat{Z}_t^i, (\alpha_t^i, \hat{\alpha}_t^{-i})\big) \\
&\qquad\qquad - (\hat{X}_t - X_t) \cdot \partial_x H^i(t, \hat{X}_t, \hat{Y}_t^i, \hat{Z}_t^i, \hat{\alpha}_t)\big]\, dt\bigg] \\
&\leqslant 0,
\end{aligned} \tag{2.22}$$

because the above integrand is non-positive for $\mathrm{Leb}_1 \otimes \mathbb{P}$ almost every $(t, \omega) \in [0, T] \times \Omega$. The last claim is easily seen by using the convexity of H^i together with the fact that $\hat{\alpha}_t$ is a critical point. Indeed, by convexity of A_i and by the generalized Isaacs condition (2.21), we have, $dt \otimes d\mathbb{P}$ a.s., for all $\alpha^i \in A_i$,

$$\big(\alpha^i - \hat{\alpha}_t^i\big) \cdot \partial_{\alpha^i} H^i(t, \hat{X}_t, \hat{Y}_t^i, \hat{Z}_t^i, \hat{\alpha}_t) \geqslant 0,$$

which completes the proof. □

Implementation Strategy

We shall use this sufficient condition in the following manner. Under assumptions **Games** and **Games SMP**, we shall search for a deterministic function:

$$\big(t, x, (y^1, \cdots, y^N), (z^1, \cdots, z^N)\big) \mapsto \hat{\alpha}\big(t, x, (y^1, \cdots, y^N), (z^1, \cdots, z^N)\big)$$

defined on $[0, T] \times \mathbb{R}^D \times (\mathbb{R}^D)^N \times (\mathbb{R}^{D\times M})^N$, with values in $A^{(N)}$ and satisfying the generalized Isaacs conditions. Next, we replace the *adapted* controls $\boldsymbol{\alpha}$ by:

$$\hat{\alpha}\big(t, X_t, (Y_t^1, \cdots, Y_t^N), (Z_t^1, \cdots, Z_t^N)\big)$$

both in the forward and backward equations. This creates a large FBSDE comprising a forward equation in dimension D, and N backward equations in dimension D. The couplings between these equations may be highly nonlinear, and this system may be very difficult to solve. However, if we find processes $\boldsymbol{X}$, $(\boldsymbol{Y}^1, \cdots, \boldsymbol{Y}^N)$, $(\boldsymbol{Z}^1, \cdots, \boldsymbol{Z}^N)$ solving this FBSDE, namely:

$$
\left\{
\begin{aligned}
dX_t &= B\Big(t, X_t, \hat{\alpha}\big(t, X_t, (Y_t^1, \cdots, Y_t^N), (Z_t^1, \cdots, Z_t^N)\big)\Big)dt \\
&\qquad + \Sigma\big(t, X_t, \hat{\alpha}\big(t, X_t, (Y_t^1, \cdots, Y_t^N), (Z_t^1, \cdots, Z_t^N)\big)\big)dW_t, \\
dY_t^1 &= -\partial_x H^1\Big(t, X_t, Y_t^1, Z_t^1, \hat{\alpha}\big(t, X_t, (Y_t^1, \cdots, Y_t^N), (Z_t^1, \cdots, Z_t^N)\big)\Big)dt \\
&\qquad + Z_t^1 dW_t, \\
\vdots \quad &\vdots \qquad\qquad\qquad\qquad \vdots \\
dY_t^N &= -\partial_x H^N\Big(t, X_t, Y_t^N, Z_t^N, \hat{\alpha}\big(t, X_t, (Y_t^1, \cdots, Y_t^N), (Z_t^1, \cdots, Z_t^N)\big)\Big)dt \\
&\qquad + Z_t^N dW_t,
\end{aligned}
\right.
\tag{2.23}
$$

for $t \in [0, T]$, with the initial condition $X_0 = x_0 \in \mathbb{R}^D$ for the forward equation and with the terminal conditions $Y_T^i = \partial_x g^i(X_T)$ for the backward equations $i \in \{1, \cdots, N\}$, then, provided that the convexity assumptions in the statement of Theorem 2.16 are satisfied, the above sufficient condition says that the strategy profile $\hat{\boldsymbol{\alpha}}$ defined by $\hat{\alpha}_t = \hat{\alpha}(t, X_t, (Y_t^1, \cdots, Y_t^N), (Z_t^1, \cdots, Z_t^N))$, for $t \in [0, T]$, forms an open loop Nash equilibrium.

For instance, the convexity assumptions are satisfied whenever, for each $i \in \{1, \ldots, N\}$, g^i is convex and, for any $(t, y, z) \in [0, T] \times (\mathbb{R}^D)^N \times (\mathbb{R}^{D\times M})^N$ and $\alpha^{-i} \in A^{-i} = \prod_{j\neq i} A_j$, the function $\mathbb{R}^D \times A_i \ni (x, \alpha^i) \mapsto H^i(t, x, y, z, (\alpha^i, \alpha^{-i}))$ is convex.

In the last two sections of the chapter, we implement this strategy in the cases of the flocking model with $\beta = 0$, and the systemic risk toy model introduced in Chapter 1. We refer to Chapter 4 for general solvability results for forward-backward systems.

2.2.2 Markovian Nash Equilibria

In the search for Markovian Nash equilibria, despite the strong appeal of the HJB equation based PDE approach reviewed earlier, we may want to use a version of the stochastic maximum principle to tackle the individual control problems entering the construction of the best response function.

If we choose to do so, we may directly invoke, in the spirit of the sketch of proof of Theorem 2.15, the usual stochastic maximum principle for standard optimal control problems. As explained in the Notes & Complements below, a detailed review of this usual version of the stochastic maximum principle is provided in the next chapters, but anticipating on the sequel, we use now some of this material.

In full analogy with the derivation of the Nash system in Subsection 2.1.4, we may indeed regard a Nash equilibrium $(\phi^{*1}, \ldots, \phi^{*N})$, say in closed loop feedback form, as a partial optimizer. Once the Markovian feedback functions ϕ^{*-i} are given to the players $j \neq i$, the feedback function ϕ^{*i} reads as a minimizer of the cost J^i to player i. Throughout the subsection, we use the convenient notation $J^i(\phi^i, \phi^{*-i})$

to denote the cost to player i when using the Markovian feedback function ϕ^i while the others use the Markovian feedback functions $(\phi^{*j})_{j\neq i}$.

Then, the Hamiltonian associated with the minimization of the cost $J^i(\phi^i, \phi^{*-i})$ over Markovian feedback functions ϕ^i reads:

$$
\begin{aligned}
H^{-i}(t,x,y,z,\alpha) &= B\big(t,x,(\alpha,\phi^{*-i}(t,x))\big)\cdot y \\
&\quad + \mathrm{trace}\big\{\Sigma\big(t,x,(\alpha,\phi^{*-i}(t,x))\big)^\dagger z\big\} + f^i\big(t,x,(\alpha,\phi^{*-i}(t,x))\big) \\
&= H^i\big(t,x,y,z,(\alpha,\phi^{*-i}(t,x))\big),
\end{aligned}
\tag{2.24}
$$

for $(t,x,y,z) \in [0,T]\times\mathbb{R}^D\times\mathbb{R}^D\times\mathbb{R}^{D\times M}$. Recalling that the adjoint equation in the stochastic maximum principle is driven by the negative of $\partial_x H^{-i}$, see for instance Subsection 3.3.2, the p-th component of the driver in the adjoint BSDE reads:

$$
\begin{aligned}
\partial_{x_p} H^{-i}(t,x,y,z,\alpha) &= \partial_{x_p} H^i\big(t,x,y,z,(\alpha,\phi^{*-i}(t,x))\big) \\
&\quad + \sum_{j=1,j\neq i}^{N} \partial_{\alpha^j} H^i\big(t,x,y,z,(\alpha,\phi^{*-i}(t,x))\big)\,\partial_{x_p}\phi^{*j}(t,x).
\end{aligned}
\tag{2.25}
$$

Notice that the last line of the above formula is not present when we use the stochastic maximum principle to search for open loop Nash equilibria. Obviously, this new line corresponds to the fact that, in closed loop equilibria, the strategies chosen by the players are sensitive to the states of the others.

Let us assume that $\phi = (\phi^1, \cdots, \phi^N)$ is a jointly measurable function from $[0,T]\times\mathbb{R}^D$ into $A^{(N)} = A_1\times\cdots\times A_N$ which is locally bounded and differentiable in $x\in\mathbb{R}^D$, for $t\in[0,T]$ fixed, with derivatives that are uniformly bounded in $(t,x)\in[0,T]\times\mathbb{R}^D$. Recalling that the drift and volatility functions B and Σ are Lipschitz in (x,α) uniformly in $t\in[0,T]$, we denote by $\boldsymbol{X}^\phi$ the unique strong solution of the state equation:

$$
dX_t = B\big(t,X_t,\phi(t,X_t)\big)dt + \Sigma\big(t,X_t,\phi(t,X_t)\big)dW_t, \quad t\in[0,T],
\tag{2.26}
$$

with initial condition $X_0 = x_0$.

Definition 2.17 *A set of N couples $((\boldsymbol{Y}^{\phi,i}, \boldsymbol{Z}^{\phi,i}) = (Y_t^{\phi,i}, Z_t^{\phi,i})_{0\leq t\leq T})_{i=1,\dots,N}$ of processes in $\mathbb{S}^{2,D}\times\mathbb{H}^{2,D\times M}$ is said to be a set of adjoint processes associated with the Markovian feedback functions $\phi = (\phi^1,\cdots,\phi^N)$, if for each $i\in\{1,\cdots,N\}$, they satisfy the BSDE:*

$$
\begin{cases}
dY_t^{\phi,i} = -\Big[\partial_x H^i(t, X_t^\phi, Y_t^{\phi,i}, Z_t^{\phi,i}, \phi(t, X_t^\phi)) \\
\qquad + \displaystyle\sum_{j=1, j\neq i}^{N} \partial_{\alpha^j} H^i(t, X_t^\phi, Y_t^{\phi,i}, Z_t^{\phi,i}, \phi(t, X_t^\phi))\partial_x \phi^j(t, X_t^\phi)\Big]dt \\
\qquad + Z_t^{\phi,i} dW_t, \quad t \in [0, T], \\
Y_T^{\phi,i} = \partial_x g^i(X_T^\phi).
\end{cases}
\tag{2.27}
$$

We shall often drop the superscript ϕ when no confusion is possible. Given the current assumptions on the coefficients of the model and the functions ϕ^i, the existence and uniqueness of the adjoint processes follow from the same argument as in the open loop case.

We now state and prove the sufficient condition for the existence of a Markovian equilibrium.

Theorem 2.18 *For a Markovian feedback function $\phi^* = (\phi^{*1}, \cdots, \phi^{*N})$, continuously differentiable in $x \in \mathbb{R}^D$ with a bounded gradient, we denote by $\boldsymbol{Y}^* = (\boldsymbol{Y}^{*1}, \cdots, \boldsymbol{Y}^{*N})$ and $\boldsymbol{Z}^* = (\boldsymbol{Z}^{*1}, \cdots, \boldsymbol{Z}^{*N})$ the adjoint processes associated with ϕ^*. If, for each $i \in \{1, \cdots, N\}$,*

- *the (random) function:*

$$
\mathbb{R}^D \times A_i \ni (x, \alpha^i) \mapsto h^i(x, \alpha^i) = H^i\big(t, x, Y_t^{*i}, Z_t^{*i}, (\alpha^i, \phi^{*-i}(t, x))\big)
$$

  is convex $\mathrm{Leb}_1 \otimes \mathbb{P}$ *a.e.,*
- *g^i is convex,*

and

$$
\begin{aligned}
&H^i\big(t, X_t^*, Y_t^{*i}, Z_t^{*i}, \phi^*(t, X_t^*)\big) \\
&\quad = \inf_{\alpha^i \in A^i} H^i\big(t, X_t^*, Y_t^{*i}, Z_t^{*i}, (\alpha^i, \phi^{*-i}(t, X_t^*))\big)
\end{aligned}
\tag{2.28}
$$

$\mathrm{Leb}_1 \otimes \mathbb{P}$ *a.e., where $\boldsymbol{X}^* = \boldsymbol{X}^{\phi^*}$, then ϕ^* is a Markovian Nash equilibrium.*

Proof. The proof is essentially the same as in the case of open loop equilibria. The differences will become clear below. As before, we fix $i \in \{1, \cdots, N\}$ together with a generic feedback function $(t, x) \mapsto \psi(t, x) \in A_i$ and, for the sake of simplicity, we denote by $\boldsymbol{X}$ the solution $\boldsymbol{X}^{(\psi, \phi^{*-i})}$ of (2.26) with the feedback function (ψ, ϕ^{*-i}) in lieu of ϕ. Starting from:

$$
\begin{aligned}
&J^i(\phi^*) - J^i((\psi, \phi^{*-i})) \\
&= \mathbb{E}\Big[\int_0^T \big[f^i(t, X_t^*, \phi^*(t, X_t^*)) - f^i\big(t, X_t, (\psi(t, X_t), \phi^{*-i}(t, X_t))\big)\big]dt \\
&\quad + g^i(X_T^*) - g^i(X_T)\Big],
\end{aligned} \tag{2.29}
$$

we use the definition of the Hamiltonian H^i to replace f^i in the above expression. We get:

$$
\begin{aligned}
&J^i(\phi^*) - J^i((\psi, \phi^{*-i})) \\
&= \mathbb{E}\Big[\int_0^T \big[H^i(t, X_t^*, Y_t^{*i}, Z_t^{*i}, \phi^*(t, X_t^*)) \\
&\qquad\qquad - H^i\big(t, X_t, Y_t^{*i}, Z_t^{*i}, (\psi(t, X_t), \phi^{*-i}(t, X_t))\big)\big]dt\Big] \\
&\quad - \mathbb{E}\Big[\int_0^T \big[B(t, X_t^*, \phi^*(t, X_t^*)) - B\big(t, X_t, (\psi(t, X_t), \phi^{*-i}(t, X_t))\big)\big] \cdot Y_t^{*i}\, dt\Big] \\
&\quad - \mathbb{E}\Big[\int_0^T \mathrm{trace}\Big[\big[\Sigma(t, X_t^*, \phi^*(t, X_t^*)) - \Sigma\big(t, X_t, (\psi(t, X_t), \phi^{*-i}(t, X_t))\big)\big]^\dagger Z_t^{*i}\Big]dt\Big] \\
&\quad + \mathbb{E}\big[g^i(X_T^*) - g^i(X_T)\big].
\end{aligned} \tag{2.30}
$$

We bound this last expectation using the convexity of g^i which implies:

$$
\begin{aligned}
&g^i(X_T^*) - g^i(X_T) \\
&\leqslant \partial_x g^i(X_T^*) \cdot (X_T^* - X_T) \\
&= Y_T^{*i} \cdot (X_T^* - X_T) \\
&= \int_0^T (X_t^* - X_t) \cdot dY_t^{*i} + \int_0^T Y_t^{*i} \cdot d(X_t^* - X_t) \\
&\qquad + \int_0^T \mathrm{trace}\Big[\big[\Sigma(t, X_t^*, \phi^*(t, X_t^*)) - \Sigma\big(t, X_t, (\psi(t, X_t), \phi^{*-i}(t, X_t))\big)\big]^\dagger Z_t^{*i}\Big]dt,
\end{aligned}
$$

where we used the special form (2.27) of the adjoint BSDE in order to compute the bracket in the last equality. Expanding dY_t^{*i} in the third line, we obtain:

$$
\begin{aligned}
&g^i(X_T^*) - g^i(X_T) \\
&\leqslant - \int_0^T (X_t^* - X_t) \cdot \Big[\partial_x H^i\big(t, X_t^*, Y_t^{*i}, Z_t^{*i}, \phi^*(t, X_t^*)\big) \\
&\qquad + \sum_{j=1, j\neq i}^{N} \partial_{\alpha^j} H^i\big(t, X_t^*, Y_t^{*i}, Z_t^{*i}, \phi^*(t, X_t^*)\big) \partial_x \phi^{*j}(t, X_t^*)\Big]\, dt \qquad (2.31) \\
&\quad + \int_0^T \big[B\big(t, X_t^*, \phi^*(t, X_t^*)\big) - B\big(t, X_t, (\psi(t, X_t), \phi^{*-i}(t, X_t))\big)\big] \cdot Y_t^{*i}\, dt \\
&\quad + \int_0^T \mathrm{trace}\Big[\big[\Sigma\big(t, X_t^*, \phi^*(t, X_t^*)\big) - \Sigma\big(t, X_t, (\psi(t, X_t), \phi^{*-i}(t, X_t))\big)\big]^\dagger Z_t^{*i}\Big] dt \\
&\quad + M_T,
\end{aligned}
$$

where $(M_t)_{0 \leqslant t \leqslant T}$ is a martingale starting from 0.

Putting together (2.30) and (2.31), we get:

$$
\begin{aligned}
&J^i(\phi^*) - J^i\big((\psi, \phi^{*-i})\big) \\
&\leqslant \mathbb{E}\Big[\int_0^T \Big(H^i\big(t, X_t^*, Y_t^{*i}, Z_t^{*i}, \phi^*(t, X_t^*)\big) \\
&\qquad - H^i\big(t, X_t, Y_t^{*i}, Z_t^{*i}, (\psi(t, X_t), \phi^{*-i}(t, X_t^*))\big) \qquad (2.32) \\
&\quad - (X_t^* - X_t) \cdot \big[\partial_x H^i\big(t, X_t^*, Y_t^{*i}, Z_t^{*i}, \phi^*(t, X_t^*)\big) \\
&\qquad + \sum_{j=1, j\neq i}^{N} \partial_{\alpha^j} H^i\big(t, X_t^*, Y_t^{*i}, Z_t^{*i}, \phi^*(t, X_t^*)\big) \partial_x \phi^{*j}(t, X_t^*)\big]\Big) dt\Big].
\end{aligned}
$$

We conclude by using the convexity of the function $(x, \alpha^i) \mapsto h^i(x, \alpha^i)$ for (t, ω) fixed, where for the sake of notation we defined this function as:

$$
h^i(x, \alpha^i) = H^i\big(t, x, Y_t^{*i}(\omega), Z_t^{*i}(\omega), (\alpha^i, \phi^{*-i}(t, x))\big).
$$

We use the convexity assumption in the form:

$$
h^i(x, \alpha^i) - h^i(\tilde{x}, \tilde{\alpha}^i) - (x - \tilde{x}) \cdot \partial_x h^i(x, \alpha^i) - (\alpha^i - \tilde{\alpha}^i) \cdot \partial_{\alpha^i} h^i(x, \alpha^i) \leqslant 0,
$$

which we apply to $x = X_t^*$, $\tilde{x} = X_t$, $\alpha^i = \phi^{*i}(t, X_t^*)$ and $\tilde{\alpha}^i = \psi(t, X_t)$. Since the minimum of the Hamiltonian is attained along the (candidate for the) optimal path, notice also that:

$$
\forall \beta \in A^i, \quad \big(\phi^i(t, X_t^*) - \beta\big) \cdot \partial_{\alpha^i} h^i\big(X_t^*, \phi^{*i}(t, X_t^*)\big) \leqslant 0.
$$

Altogether, this gives:

$$
\begin{aligned}
&H^i\big(t, X_t^*, Y_t^{*i}, Z_t^{*i}, \phi^*(t, X_t^*)\big) - H^i\big(t, X_t, Y_t^{*i}, Z_t^{*i}, (\psi(t, X_t), \phi^{*-i}(t, X_t^*))\big) \\
&\qquad - (X_t^* - X_t) \cdot \Big[\partial_x H^i\big(t, X_t^*, Y_t^{*i}, Z_t^{*i}, \phi^*(t, X_t^*)\big) \\
&\qquad\qquad + \sum_{j=1, j\neq i}^{N} \partial_{\alpha^j} H^i\big(t, X_t^*, Y_t^{*i}, Z_t^{*i}, \phi^*(t, X_t^*)\big) \partial_x \phi^{*j}(t, X_t^*)\Big] \\
&\leqslant 0,
\end{aligned}
$$

$\text{Leb}_1 \otimes \mathbb{P}$ *a.e.*, which shows that:

$$
J^i(\phi^*) - J^i\big((\psi, \phi^{*-i})\big) \leqslant 0.
$$

This concludes the proof that ϕ^* is a Markovian Nash equilibrium. □

Implementation Strategy

As one can expect, the systematic use of the above sufficient condition to construct Markovian Nash equilibria is much more delicate than in the open loop case. Under assumptions **Games** and **Games SMP**, we should, as in the open loop case, search for a deterministic function $\hat{\alpha}$:

$$
\big(t, x, (y^1, \cdots, y^N), (z^1, \cdots, z^N)\big) \mapsto \hat{\alpha}\big(t, x, (y^1, \cdots, y^N), (z^1, \cdots, z^N)\big)
$$

defined on $[0, T] \times \mathbb{R}^D \times (\mathbb{R}^D)^N \times (\mathbb{R}^{D\times M})^N$, with values in $A^{(N)}$ and satisfying the generalized Isaacs conditions. Let us assume for example that such a function is found and that it does not depend upon $z = (z^1, \cdots, z^N)$. As in the case of the open loop models, we would like to replace the instances of the controls in the forward dynamics of the state as well as in the adjoint BSDEs by $\hat{\alpha}(t, X_t, (Y_t^1, \cdots, Y_t^N))$, looking for an FBSDE which could be solved. Unfortunately, while this idea was reasonable in the open loop case, it cannot be implemented in a straightforward manner for Markov games because the adjoint equations require the derivatives of the controls.

However, taking advantage of the deterministic structure of the coefficients of such an FBSDE, we may expect that there exists a smooth function $(u^1, \ldots, u^N)$: $[0, T] \times \mathbb{R}^D \to (\mathbb{R}^D)^N$ such that that $Y_t^i = u^i(t, X_t)$ and $Z_t^i = \partial_x u^i(t, X_t)\Sigma(t, X_t)$. Such a function is called *decoupling field*; we refer to the first section in Chapter 4 for a review on this notion. Basically, it is understood as the space derivative of the solution to the Nash system (2.17).

So we may want to use $\phi^*(t, x) = \hat{\alpha}(t, x, u(t, x))$, and in the BSDE giving the adjoint processes, the quantity $\partial_x \hat{\alpha}(t, x, u(t, x)) + \partial_y \hat{\alpha}(t, x, u(t, x))\partial_x u(t, x)$ instead of the term $\partial_x \phi^*(t, x)$. As in the case of the open loop models, this creates a large FBSDE which we need to solve in order to obtain a Markov Nash equilibrium. In the last two sections of the chapter, we do just that in the cases of the flocking model with $\beta = 0$ and the systemic risk toy model introduced in Chapter 1.

2.3 N-Player Games with Mean Field Interactions

In this section, we specialize the results of the first part of the chapter to the class of models at the core of the book. As explained in Chapter 1, these models are characterized by strong symmetry properties in the coefficients and cost functions, and the interactions between players need to be such that the influence of each individual player on the rest of the population disappears as the size of the game grows without bound.

2.3.1 The N-Player Game

Stochastic differential game models with the strong symmetry property alluded to above would require the dynamics of the private states of the N players $i \in \{1, \cdots, N\}$ to be given by Itô stochastic differential equations of the form:

$$dX_t^i = b^i(t, X_t^i, X_t^{-i}, \alpha_t^i, \alpha_t^{-i})dt + \sigma^i(t, X_t^i, X_t^{-i}, \alpha_t^i, \alpha_t^{-i})dW_t^i, \tag{2.33}$$

for $i = 1, \cdots, N$ and $t \in [0, T]$, where the m-dimensional Wiener processes $\mathbf{W}^i = (W_t^i)_{0 \leqslant t \leqslant T}$ are assumed to be independent for the time being. We shall consider the case of dependent Wiener processes when we study models in which states are subject to a common source of random shocks. See Chapters (Vol II)-2 and (Vol II)-3.

The symmetry requirement forces the dimensions d_i of all the private states X_t^i to be the same. We denote by d their common value, so that now $D = Nd$. The common value m of all the second dimensions of the $(\sigma^i)_{1 \leq i \leq N}$'s is now as well the common dimension of the Wiener processes $(\mathbf{W}^i)_{1 \leq i \leq N}$, so that $M = Nm$. For the same reason, all the control sets A_i should be the same. We shall denote by A this common set, so that $A^{(N)}$ is now $A \times \cdots \times A = A^N$. Also, the functions b^i and σ^i should satisfy specific properties to account for the symmetry requirement. First, these functions should not depend upon i and hence be the same for all the players. Moreover, their dependence upon the $N-1$ other states in X_t^{-i}, and/or the $N-1$ controls in α_t^{-i} should be symmetric. The same considerations apply to the running cost functions f^i, but for the sake of simplicity, we only argue the case of the coefficients of the dynamics. So if we want symmetry and if we want each player to have a vanishing influence when the number of players grows, then in light of Lemma 1.2 in Chapter 1, it is reasonable to assume that, at least for large games, these coefficients can be well approximated by functions of one private state, say X_t^i, the corresponding control α_t^i, and the empirical distribution $\bar{\mu}^{N-1}_{(X_t,\alpha_t)^{-i}}$ of the other couples private states/controls.

So for the sake of our mathematical analysis, we shall assume that, instead of (2.33), the dynamics of the private states of the individual players are given by Itô's stochastic differential equations of the form:

$$dX_t^i = b(t, X_t^i, \bar{\mu}^{N-1}_{(X_t,\alpha_t)^{-i}}, \alpha_t^i)dt + \sigma(t, X_t^i, \bar{\mu}^{N-1}_{(X_t,\alpha_t)^{-i}}, \alpha_t^i)dW_t^i, \tag{2.34}$$

for $i = 1, \cdots, N$, where $(b, \sigma) : [0, T] \times \mathbb{R}^d \times \mathcal{P}(\mathbb{R}^d \times A) \times A \to \mathbb{R}^d \times \mathbb{R}^{d\times m}$ is a deterministic measurable function satisfying specific assumptions which we spell out later on. Recall the notation $\bar{\mu}^{N-1}_{(X_t^{-i},\alpha_t^{-i})}$ for the empirical measure:

$$\bar{\mu}^{N-1}_{(X_t,\alpha_t)^{-i}} = \frac{1}{N-1} \sum_{1\leqslant j\neq i\leqslant N} \delta_{(X_t^j,\alpha_t^j)}, \tag{2.35}$$

Remark 2.19 *Dependencies between the Wiener processes will be introduced in Chapters (Vol II)-2 and (Vol II)-3 by adding a term of the form $\sigma^0(\cdots)dW_t^0$ to the right-hand side of the state dynamics given by* (2.34)*. For obvious reasons, the increments dW_t^0 of the Wiener process $\mathbf{W}^0$ will be called common noise as opposed to the increments of the Wiener processes $\mathbf{W}^i$ for $i = 1, \cdots, N$ which are intrinsic to the private states and called idiosyncratic noises.*

Remark 2.20 *In most of the applications considered in the book, the private states of the players interact through the empirical distributions of the states themselves. However, as we saw in Chapter 1, in many applications of interest, the interactions are built into the models through the empirical distribution of the controls, or even through the empirical distribution of the couple of state and control as posited in* (2.34) *above. These last two classes of problems are more difficult to solve than the first one, and we shall not try to offer a systematic presentation of their analyses. We shall only discuss these models in the limit $N \to \infty$ in Section 4.6 of Chapter 4 where we call them extended mean field games. But as a general rule, we shall mostly restrict our discussions of extended models to particular cases for which a solution can be derived with a low overhead from the theoretical results we provide for the models with interactions through the states only.*

As explained in the previous section, for each player, the choice of a strategy is driven by the desire to minimize an expected cost over a period $[0, T]$, each individual cost being a combination of running and terminal costs. Based on the above discussion of the form of the drift and volatility coefficients in the private state dynamics, we shall assume that for each $i \in \{1, \cdots, N\}$, the running cost to player i is given by a measurable function $f : [0, T] \times \mathbb{R}^d \times \mathcal{P}(\mathbb{R}^d \times A) \times A \to \mathbb{R}$ and the terminal cost by a measurable function $g : \mathbb{R}^d \times \mathcal{P}(\mathbb{R}^d \times A) \to \mathbb{R}$ in such a way that if the N players use the strategy profile $\boldsymbol{\alpha} = (\boldsymbol{\alpha}^1, \cdots, \boldsymbol{\alpha}^N) \in \mathbb{A}^N$, the expected total cost to player i is:

$$J^i(\boldsymbol{\alpha}) = \mathbb{E}\left[\int_0^T f(t, X_t^i, \bar{\mu}_t^{N-1}, \alpha_t^i)dt + g(X_T^i, \bar{\mu}_T^{N-1})\right], \tag{2.36}$$

where we use the short notation $\bar{\mu}_t^{N-1}$ for the empirical measures (2.35). Also, we denote by $\mathbb{A}^N$ the product of N copies of $\mathbb{A}$. In some instances, we shall also use the notation $J^{N,i}$ instead of J^i when we want to emphasize the dependence upon the number N of players. This will be the case when we study the limit $N \to \infty$ in Chapter (Vol II)-6. Notice that even though only α_t^i appears in the formula giving the cost to player i, this cost depends upon the strategies used by the other players indirectly, as these strategies affect not only the private state X_t^i, but also the empirical distribution $\bar{\mu}_t^{N-1}$.

2.3.2 Hamiltonians and the Stochastic Maximum Principle

We revisit our notation system in order to take advantage of the current emphasis on the decomposition of the state of the system as the aggregation of the private states of the individual players, and the strong symmetry conditions we impose on the dynamics and the costs. In particular, we pay special attention to the players' Hamiltonians introduced in (2.8). The state variable, denoted by the bold face letter $\boldsymbol{x}$, reads as an N-tuple $\boldsymbol{x} = (x^1, \ldots, x^N) \in (\mathbb{R}^d)^N$, describing the private states of the players, the first dual variable $\boldsymbol{y}$ reads as an N-tuple $(\boldsymbol{y}^1, \ldots, \boldsymbol{y}^N) \in [(\mathbb{R}^d)^N]^N$, each $\boldsymbol{y}^i$ being itself an N-tuple $(y^{i,1}, \ldots, y^{i,N}) \in (\mathbb{R}^d)^N$ and the second dual variable $\boldsymbol{z}$ reads as an N-tuple $(\boldsymbol{z}^1, \ldots, \boldsymbol{z}^N) \in [(\mathbb{R}^{d\times m})^{N\times N}]^N$, each $\boldsymbol{z}^i$ denoting an $N \times N$-tuple $(z^{i,1,1}, \ldots, z^{i,N,N}) \in (\mathbb{R}^{d\times m})^{N\times N}$. Finally, the control variable $\boldsymbol{\alpha}$ reads as an N-tuple $(\alpha^1, \ldots, \alpha^N) \in A^N$ of possible actions by the players.

With these conventions, the Hamiltonian of player i is given by:

$$
\begin{aligned}
H^i\big(t, \boldsymbol{x}, \boldsymbol{y}^i, \boldsymbol{z}^i, \boldsymbol{\alpha}\big) &= \sum_{j=1}^{N} b\big(t, x^j, \bar{\mu}_{(\boldsymbol{x},\boldsymbol{\alpha})^{-j}}^{N-1}, \alpha^j\big) \cdot y^{i,j} \\
&\quad + \sum_{j=1}^{N} \sigma\big(t, x^j, \bar{\mu}_{(\boldsymbol{x},\boldsymbol{\alpha})^{-j}}^{N-1}, \alpha^j\big) \cdot z^{i,j,j} + f\big(t, x^i, \bar{\mu}_{(\boldsymbol{x},\boldsymbol{\alpha})^{-i}}^{N-1}, \alpha^i\big),
\end{aligned}
$$

where in full analogy with (2.35), $\bar{\mu}_{(\boldsymbol{x},\boldsymbol{\alpha})^{-j}}^{N-1}$ is the empirical measure defined by:

$$
\bar{\mu}_{(\boldsymbol{x},\boldsymbol{\alpha})^{-j}}^{N-1} = \frac{1}{N-1} \sum_{1 \leqslant j \neq i \leqslant N} \delta_{(x^i, \alpha^i)}. \tag{2.37}
$$

The necessary part of the stochastic maximum principle suggests the minimization of H^i when $t, \boldsymbol{x}, \boldsymbol{y}^i, \boldsymbol{z}^i$ and $\boldsymbol{\alpha}^{-i}$ are frozen. Interestingly, whenever the interaction in the coefficients is through the state only as explained in Remark 2.20, that is $\bar{\mu}_{(\boldsymbol{x},\boldsymbol{\alpha})^{-j}}^{N-1}$ and $\bar{\mu}_{(\boldsymbol{x},\boldsymbol{\alpha})^{-i}}^{N-1}$ are replaced by $\bar{\mu}_{\boldsymbol{x}^{-j}}^{N-1}$ and $\bar{\mu}_{\boldsymbol{x}^{-i}}^{N-1}$, the Hamiltonian has a distributed additive structure, in the sense that each given control variable appears separately from the others in each of the terms of the above sum. As a by-product, the partial optimization procedure over α^i whenever all the other variables $t, \boldsymbol{x}, y^i, z^i$ and $\boldsymbol{\alpha}^{-i}$ are frozen reduces to the optimization problem:

$$\inf_{\alpha \in A} \left[b\big(t, x^i, \bar{\mu}^{N-1}_{\boldsymbol{x}^{-i}}, \alpha\big) \cdot y^{i,i} + \sigma\big(t, x^i, \bar{\mu}^{N-1}_{\boldsymbol{x}^{-i}}, \alpha\big) \cdot z^{i,i,i} + f\big(t, x^i, \bar{\mu}^{N-1}_{\boldsymbol{x}^{-i}}, \alpha\big) \right],$$

of the type we have been dealing with so far. The symmetry assumed in this section implies that one can focus on the Hamiltonian of one single player. By symmetry, one can drop the superscript i in the notation of the Hamiltonian as it is enough to consider $H(t, x^i, \bar{\mu}^{N-1}_{\boldsymbol{x}^{-i}}, y^{i,i}, z^{i,i,i}, \alpha^i)$, where H is defined as:

$$H\big(t, x, \mu, y, z, \alpha\big) = b\big(t, x, \mu, \alpha\big) \cdot y + \sigma\big(t, x, \mu, \alpha\big) \cdot z + f\big(t, x, \mu, \alpha\big),$$

for $t \in [0, T]$, $x \in \mathbb{R}^d$, $\mu \in \mathcal{P}(\mathbb{R}^d)$, $y \in \mathbb{R}^d$, $z \in \mathbb{R}^{d \times m}$ and $\alpha \in A$. For that very reason, the minimizer $\hat{\alpha}$ of $H(t, x, \mu, y, z, \cdot)$ plays an important role in the subsequent analysis.

Definition 2.21 *Assume that the interaction in the coefficients is through the state only, as explained in Remark 2.20. Then, given $t \in [0, T]$, $x \in \mathbb{R}^d$, $\mu \in \mathcal{P}(\mathbb{R}^d)$, $y \in \mathbb{R}^d$ and $z \in \mathbb{R}^{d \times m}$, we use the generic notation $\hat{\alpha}(t, x, \mu, y, z)$ to denote a minimizer of the function $A \ni \alpha \mapsto H(t, x, \mu, y, z, \alpha)$, in other words:*

$$\hat{\alpha}(t, x, \mu, y, z) \in \text{argmin}_{\alpha \in A} H(t, x, \mu, y, z, \alpha).$$

If the above minimizer is well-defined, then the function $[0, T] \times (\mathbb{R}^d)^N \times ((\mathbb{R}^d)^N)^N \times ((\mathbb{R}^{d \times m})^{N \times N})^N \ni (t, \boldsymbol{x}, (\boldsymbol{y}^1, \cdots, \boldsymbol{y}^N), (\boldsymbol{z}^1, \cdots, \boldsymbol{z}^N)) \mapsto (\hat{\alpha}(t, x^i, \bar{\mu}^{N-1}_{\boldsymbol{x}^{-i}}, y^{i,i}, z^{i,i,i}))_{1 \leqslant i \leqslant N}$ satisfies the Isaacs condition, see Definition 2.9.

Remark 2.22 *In the most desirable situations, the minimizer in Definition 2.21 is uniquely defined. This is for instance the case when A is convex and the Hamiltonian is strictly convex in the variable α. We shall often restrict ourselves to this situation, although it is rather restrictive since it requires the drift b to be linear in α. As already used before, whenever A is convex and H is differentiable in α, we have:*

$$\forall \beta \in A, \quad (\beta - \alpha) \cdot \partial_\alpha H\big(t, x, \mu, y, z, \hat{\alpha}(t, x, \mu, y, z)\big) \geqslant 0,$$

and, when $\hat{\alpha}(t, x, \mu, y, z)$ is in the interior of A,

$$\partial_\alpha H\big(t, x, \mu, y, z, \hat{\alpha}(t, x, \mu, y, z)\big) = 0.$$

When H is strictly convex, the implicit function theorem may be used to transfer the smoothness properties of H in the directions x, μ, y and z into smoothness properties of $\hat{\alpha}$. We shall make this important remark precise in several lemmas in the sequel.

2.3.3 Potential Stochastic Differential Games

The notion of potential game introduced in Chapter 1 in the particular case of one period models can be generalized to the setting of stochastic differential games. Here, we specialize this notion to the case of N-player games with mean field interactions, and we concentrate on N-player open loop games for the sake of definiteness. We leave the discussion of potential mean field games to Chapter 6 where we emphasize the connection with the control of McKean-Vlasov stochastic differential equations.

Recall the definitions of the running and terminal cost functions of the players entering the definition of the stochastic differential game (2.34)–(2.36), and in particular the fact that J^i is the cost functional of player $i \in \{1, \cdots, N\}$.

Definition 2.23 *The game is said to be a potential game if there exists a functional* $(\boldsymbol{\alpha}^1, \cdots, \boldsymbol{\alpha}^N) \mapsto J(\boldsymbol{\alpha}^1, \cdots, \boldsymbol{\alpha}^N)$*, from* $\mathbb{A}^N$ *to* $\mathbb{R}$*, satisfying*

$$J(\boldsymbol{\alpha}^i, \boldsymbol{\alpha}^{-i}) - J(\boldsymbol{\beta}, \boldsymbol{\alpha}^{-i}) = J^i(\boldsymbol{\alpha}^i, \boldsymbol{\alpha}^{-i}) - J^i(\boldsymbol{\beta}, \boldsymbol{\alpha}^{-i}) \tag{2.38}$$

for all $i \in \{1, \cdots, N\}$ *and admissible control strategies* $(\boldsymbol{\alpha}^1, \cdots, \boldsymbol{\alpha}^N) \in \mathbb{A}^N$ *and* $\boldsymbol{\beta} \in \mathbb{A}$.

In other words, a game is a potential game if one can find a single function J of the set of player strategy profiles, such that any change in the value of this function J when one (and only one) strategy is perturbed, equals the corresponding change in the cost functional J^i of the player i whose strategy is perturbed, for the same change in strategies. This special property makes it possible to replace the search for a Nash equilibrium by the search for a minimum of this function, problem which is usually simpler! Assume indeed that $\hat{\boldsymbol{\alpha}} = (\hat{\alpha}_t^1, \cdots, \hat{\alpha}_t^N)_{0 \leqslant t \leqslant T}$ is an argument of the minimization of J. Then, one readily checks that $\hat{\boldsymbol{\alpha}} = (\hat{\alpha}_t^1, \cdots, \hat{\alpha}_t^N)_{0 \leqslant t \leqslant T}$ is a Nash equilibrium for the game because

$$0 \geqslant J(\hat{\boldsymbol{\alpha}}^i, \hat{\boldsymbol{\alpha}}^{-i}) - J(\boldsymbol{\alpha}^i, \hat{\boldsymbol{\alpha}}^{-i}) = J^i(\hat{\boldsymbol{\alpha}}^i, \hat{\boldsymbol{\alpha}}^{-i}) - J^i(\boldsymbol{\alpha}^i, \hat{\boldsymbol{\alpha}}^{-i}),$$

for any $i \in \{1, \ldots, N\}$.

The following example illustrates the power of the concept. We consider a stochastic differential game (2.34)–(2.36) with uncoupled private state dynamics in the sense that the drift and volatility coefficients do not depend upon the measure argument responsible for the interactions. In other words, we assume that they are of the form:

$$b(t, x, \mu, \alpha) = b(t, x, \alpha) \quad \text{and} \quad \sigma(t, x, \mu, \alpha) = \sigma(t, x, \alpha).$$

The fact that we use the same notation for different functions should not create ambiguities. Next, we assume that the running cost function $f : [0, T] \times \mathbb{R}^d \times \mathcal{P}(\mathbb{R}^d \times A) \times A \to \mathbb{R}$ is in fact of the form:

$$f(t, x, \mu, \alpha) = \frac{1}{2}|\alpha|^2 + \tilde{f}(t, x, \mu_x)$$

for some function $\tilde{f} : [0, T] \times \mathbb{R}^d \times \mathcal{P}(\mathbb{R}^d) \to \mathbb{R}$ and where $\mu_x \in \mathcal{P}(\mathbb{R}^d)$ denotes the marginal in $x \in \mathbb{R}^d$ of $\mu \in \mathcal{P}(\mathbb{R}^d \times A)$, in other words, the projection of μ onto $\mathbb{R}^d$. Similarly, we assume that the terminal cost function $g : \mathbb{R}^d \times \mathcal{P}(\mathbb{R}^d \times A) \to \mathbb{R}$ is of the form:

$$g(x, \mu) = \tilde{g}(x, \mu_x).$$

Under these conditions, we have the following result.

Proposition 2.24 *If on top of the above assumptions on the form of the coefficients, there exist functions* $[0, T] \times \mathcal{P}(\mathbb{R}^d) \ni (t, \mu) \mapsto F(t, \mu)$ *and* $\mathcal{P}(\mathbb{R}^d) \ni \mu \mapsto G(\mu)$ *satisfying:*

$$\begin{aligned}\tilde{f}(t, x, \bar{\mu}_X^{N-1}) &- \tilde{f}(t, x', \bar{\mu}_X^{N-1}) \\ &= F\Big(t, \frac{1}{N}\delta_x + \frac{N-1}{N}\bar{\mu}_X^{N-1}\Big) - F\Big(t, \frac{1}{N}\delta_{x'} + \frac{N-1}{N}\bar{\mu}_X^{N-1}\Big)\end{aligned}$$

and

$$\begin{aligned}\tilde{g}(x, \bar{\mu}_X^{N-1}) &- \tilde{g}(x', \bar{\mu}_X^{N-1}) \\ &= G\Big(\frac{1}{N}\delta_x + \frac{N-1}{N}\bar{\mu}_X^{N-1}\Big) - G\Big(\frac{1}{N}\delta_{x'} + \frac{N-1}{N}\bar{\mu}_X^{N-1}\Big)\end{aligned}$$

for every $x, x' \in \mathbb{R}^d$ *and* $X = (x^1, \cdots, x^{N-1}) \in \mathbb{R}^{d(N-1)}$, *then the game is a potential game and the function J defined by:*

$$J(\boldsymbol{\alpha}^1, \cdots, \boldsymbol{\alpha}^N) = \mathbb{E}\Big[\int_0^T \Big[\frac{1}{2}\sum_{i=1}^N |\alpha_t^i|^2 + F(t, \bar{\mu}_{X_t}^N)\Big]dt + G(\bar{\mu}_{X_T}^N)\Big] \tag{2.39}$$

can be used in Definition 2.23.

Here and in the following we use the notation $\bar{\mu}_Z^N$ for the empirical measure of the N-tuple $Z = (z^1, \cdots, z^N)$ as defined by formula (1.3) in Chapter 1. As usual, it is implicitly required that the expectation in (2.39) is well-defined for $(\boldsymbol{\alpha}^1, \cdots, \boldsymbol{\alpha}^N) \in \mathbb{A}^N$.

It is crucial to emphasize the practical importance of this seemingly innocuous result. While it is typically very difficult to identify Nash equilibria for stochastic games, especially for large games, it appears that if the cost functions are of the special form identified in the above statement, the search for Nash equilibria reduces to a single optimization problem. This type of result had many applications in economics where this single optimization is often referred to as the central planner, or representative agent or even the invisible hand optimization.

Proof. The proof is based on a direct verification argument relying on the specific form of the cost functions.

For a given $\hat{\boldsymbol{\alpha}} = (\hat{\boldsymbol{\alpha}}^1, \cdots, \hat{\boldsymbol{\alpha}}^N) \in \mathbb{A}^N$, we denote by $\hat{\boldsymbol{X}} = (\hat{X}_t)_{0 \leqslant t \leqslant T}$ the corresponding state process. For a given $i \in \{1, \cdots, N\}$ and for another $\boldsymbol{\alpha}^i \in \mathbb{A}$, we change $\hat{\boldsymbol{\alpha}}^i$ into $\boldsymbol{\alpha}^i$. Thanks to the special form of the forward dynamics, this does not affect $\hat{\boldsymbol{X}}^{-i}$. We then denote by $\boldsymbol{X}^i$ the state process to player i associated with $\boldsymbol{\alpha}^i$. Also, we let $\boldsymbol{X} = (\boldsymbol{X}^i, \hat{\boldsymbol{X}}^{-i})$. Importantly, we can write, for any $t \in [0, T]$,

$$\bar{\mu}^N_{X_t} = \frac{1}{N}\delta_{X^i_t} + \frac{N-1}{N}\bar{\mu}^{N-1}_{\hat{X}^{-i}_t}.$$

In particular,

$$F(t, \bar{\mu}^N_{X_t}) - F(t, \bar{\mu}^N_{\hat{X}_t}) = \tilde{f}(t, X^i_t, \bar{\mu}^N_{\hat{X}^{-i}_t}) - \tilde{f}(t, \hat{X}^i_t, \bar{\mu}^N_{\hat{X}^{-i}_t}).$$

and similarly for G.

Now, by (2.39),

$$\begin{aligned}
&J(\boldsymbol{\alpha}^i, \hat{\boldsymbol{\alpha}}^{-i}) - J(\hat{\boldsymbol{\alpha}}) \\
&\quad = \mathbb{E}\Big[\int_0^T \Big[\frac{1}{2}(|\alpha^i_t|^2 - |\hat{\alpha}^i_t|^2) + F(t, \bar{\mu}^N_{X_t}) - F(t, \bar{\mu}^N_{\hat{X}_t})\Big]dt + G(\bar{\mu}^N_{X_T}) - G(\bar{\mu}^N_{\hat{X}_T})\Big] \\
&\quad = \mathbb{E}\Big[\int_0^T \Big[\frac{1}{2}(|\alpha^i_t|^2 - |\hat{\alpha}^i_t|^2) + \tilde{f}(t, X^i_t, \bar{\mu}^N_{\hat{X}^{-i}_t}) - \tilde{f}(t, \hat{X}^i_t, \bar{\mu}^N_{\hat{X}^{-i}_t})\Big]dt \\
&\qquad + \tilde{g}(X^i_T, \bar{\mu}^{N-1}_{\hat{X}^{-i}_T}) - \tilde{g}(\hat{X}^i_T, \bar{\mu}^{N-1}_{\hat{X}^{-i}_T})\Big] \\
&\quad = J^i(\boldsymbol{\alpha}^i, \hat{\boldsymbol{\alpha}}^{-i}) - J^i(\hat{\boldsymbol{\alpha}}^i, \hat{\boldsymbol{\alpha}}^{-i}),
\end{aligned}$$

which completes the proof. □

Notice that the last statement of the proof is specific to the open loop nature of the problem as it does not hold any longer if the controls are closed loop.

Remark 2.25 *We shall revisit this result about potential games in the case of the asymptotic regime $N \to \infty$ of large games. In this asymptotic regime, the search for Nash equilibria will amount to solving for a mean field game equilibrium, and the central planner optimization problem will reduce to the solution of an optimal control problem for McKean-Vlasov's stochastic differential equations. For this reason, it is instructive to use the stochastic maximum principle to get an idea of* *what the controls forming the Nash equilibrium look like. We do just that in the analysis of an example below. Notice that this exercise sheds some new and different light on the above argument.*

Examples.

1. The first obvious situation in which the assumptions of the above proposition are satisfied is when there exist functions F and G such that:

$$\tilde{f}(t, x, \bar{\mu}_X^{N-1}) = F\left(t, \frac{1}{N}\delta_x + \frac{N-1}{N}\bar{\mu}_X^{N-1}\right),$$

and

$$\tilde{g}(x, \bar{\mu}_X^{N-1}) = G\left(\frac{1}{N}\delta_x + \frac{N-1}{N}\bar{\mu}_X^{N-1}\right).$$

2. A more interesting example is provided by functions $\tilde{f}$ and $\tilde{g}$ of the form:

$$\tilde{f}(t, x, \mu) = [h(t, \cdot) * \mu](x), \quad \text{and} \quad \tilde{g}(x, \mu) = [k * \mu](x),$$

for some smooth even functions $h(t, \cdot)$ and k. Indeed, if we define the functions F and G by:

$$F(t, \mu) = \frac{N^2}{2(N-1)}\langle h(t, \cdot) * \mu, \mu\rangle, \quad \text{and} \quad G(\mu) = \frac{N^2}{2(N-1)}\langle k * \mu, \mu\rangle$$

a straightforward computation shows that the assumptions of Proposition 2.24 are satisfied. Indeed, with the same notations as in the statement of Proposition 2.24, we have:

$$\begin{aligned}
&F\left(t, \frac{1}{N}\delta_x + \frac{N-1}{N}\bar{\mu}_X^{N-1}\right) - F\left(t, \frac{1}{N}\delta_{x'} + \frac{N-1}{N}\bar{\mu}_X^{N-1}\right) \\
&= \frac{N^2}{2(N-1)}\Bigg[\frac{1}{N^2}\left(Nh(t,0) + 2\sum_{i=1}^{N} h(t, x - x^i) + \sum_{i,j=1}^{N-1} h(t, x^i - x^j)\right) \\
&\qquad\qquad - \frac{1}{N^2}\left(Nh(t,0) + 2\sum_{i=1}^{N} h(t, x' - x^i) + \sum_{i,j=1}^{N-1} h(t, x^i - x^j)\right)\Bigg] \\
&= \frac{1}{N-1}\left[\sum_{i=1}^{N} h(t, x - x^i) - \sum_{i=1}^{N} h(t, x' - x^i)\right] \\
&= \tilde{f}(t, x, \bar{\mu}_X^{N-1}) - \tilde{f}(t, x', \bar{\mu}_X^{N-1}),
\end{aligned}$$

and similarly for G.

The reader may wonder about the scaling used in the cost coefficients F and G, which grow linearly with N. Indeed, such a scaling may seem to contradict our objective to investigate the asymptotic behavior of the equilibria as N tends to ∞. Actually, when it comes to potential games, what really matters is the fact that Nash equilibria coincide with minima of some collective cost functional J. Of course, those minima remain the same if J is replaced by J/N. In our specific case, J/N has the right scaling: it represents the average cost to the society. We shall come back to this question in Chapters 6 and (Vol II)-6 when we face optimal control problems for McKean-Vlasov diffusion processes.

For the purpose of illustration, we show that, at least for this particular example, the classical version of the Pontryagin stochastic maximum principle when applied to the standard stochastic control problem of the minimization problem of the central planner leads to the same FBSDE as the game version of the stochastic maximum principle when applied to the above potential N-player game. Let us assume for example that the dynamics of the private states of the N players are given by:

$$dX_t^i = \alpha_t^i dt + \sigma dW_t^i, \quad t \in [0, T].$$

For the sake of simplicity, we also assume that $d = 1$, $A = \mathbb{R}$ and h is independent of t. The reduced Hamiltonian of the central planner optimization problem with $F(\mu) = (1/2)\langle h * \mu, \mu\rangle$ and $G(\mu) = (1/2)\langle k * \mu, \mu\rangle$ reads:

$$H(t, \boldsymbol{x}, \boldsymbol{y}, \boldsymbol{\alpha}) = \sum_{i=1}^{N} \alpha^i y^i + \frac{1}{2}\sum_{i=1}^{N} |\alpha^i|^2 + \frac{1}{2(N-1)} \sum_{i,j=1}^{N} h(x^i - x^j),$$

for $\boldsymbol{x} = (x^1, \cdots, x^N) \in \mathbb{R}^N$, $\boldsymbol{y} = (y^1, \cdots, y^N) \in \mathbb{R}^N$ and $\boldsymbol{\alpha} = (\alpha^1, \ldots, \alpha^N) \in \mathbb{R}^N$. The partial minimizer in α is $\hat{\alpha}(t, x, y) = -y$ and then, implementing the stochastic maximum principle for standard optimal control exposed in Chapter 3, a straightforward computation of $\partial_{x^i} H$ shows that any optimal control $(\hat{\alpha}_t)_{0\leqslant t\leqslant T}$ satisfies $(\hat{\alpha}_t^i = -Y_t^i)_{0\leqslant t\leqslant T}$ where $(Y_t)_{0\leqslant t\leqslant T}$ is the backward component of the FBSDE system:

$$\begin{cases} dX_t^j &= -Y_t^j dt + \sigma dW_t^j, \quad j = 1, \cdots, N, \\ dY_t^i &= -\dfrac{1}{N-1}\Big[\displaystyle\sum_{j=1, j\neq i}^{N} h'(X_t^i - X_t^j)\Big]dt + \sum_{j=1}^{N} Z_t^{ij} dW_t^j, \end{cases} \tag{2.40}$$

for $i = 1, \cdots, N$ and $t \in [0, T]$, with $Y_T^i = \frac{1}{N-1}[\sum_{j\neq i} k'(X_T^i - X_T^j)]$. Note that we used the fact that h and k are even functions in the computation of the drift term and terminal condition of $\boldsymbol{Y}^i$. Whenever h and k are convex, the Hamiltonian H is convex in (x, α) (regarded as a variable in $(\mathbb{R}^N)^2$) and the terminal cost $(x^1, \ldots, x^N) \mapsto G(\frac{1}{N}\sum_{i=1}^N x^i)$ is also convex, in which case the stochastic maximum principle is not only a necessary but also a sufficient condition of optimality, see Subsection 3.3.2

We can check that the application of the game version of the stochastic maximum principle gives the same result. Indeed, in the case of the N player game with $\tilde{f}$ and $\tilde{g}$ as above, the (reduced) Hamiltonian of player i reads:

$$H^i(t,\boldsymbol{x},\boldsymbol{y}^i,\boldsymbol{\alpha}) = \sum_{j=1}^N \alpha^j y^{i,j} + \frac{1}{2}|\alpha^i|^2 + \frac{1}{N-1}\sum_{j=1,j\neq i}^N h(x^i - x^j),$$

and the necessary part of the Pontryagin stochastic maximum principle identifies the same candidate $\hat{\alpha}_t^i = -Y_t^{i,i}$ for the equilibrium controls, so that the forward dynamics of the state $(X_t)_{0\leq t\leq T}$ are still given by the first equation in the system (2.40):

$$dX_t^i = -Y_t^{i,i}dt + \sigma dW_t^i, \quad t \in [0,T].$$

The backward component of the FBSDE which provides a necessary condition for any Nash equilibrium includes:

$$\begin{aligned}
dY_t^{i,i} & \\
&= -\partial_{x^i} H^i\big(t, X_t, (Y_t^{i,1},\cdots,Y_t^{i,N}), -(Y_t^{i,1},\cdots,Y_t^{i,N})\big)dt + \sum_{j=1}^N Z_t^{i,i,j} dW_t^j \\
&= -\frac{1}{N-1}\Big[\sum_{j=1,j\neq i}^N h'(X_t^i - X_t^j)\Big]dt + \sum_{j=1}^N Z_t^{i,i,j} dW_t^j, \quad t \in [0,T],
\end{aligned}$$

which is exactly the same equation as the second equation in (2.40) with the same exact terminal conditions if we identify Y_t^i and $Y_t^{i,i}$, and $Z_t^{i,j}$ with $Z_t^{i,i,j}$.

Remark 2.26 *As a final remark, and anticipating on the discussion of the differentiability of functions of measures in Chapter 5, we emphasize the fact that a crucial role is played by the identity $\delta F(t,\mu)(x) = \frac{N^2}{N-1}\tilde{f}(t,x,\mu)$ μ-a.e. which we will prove in Chapter 5. Here, $\delta F(t,\mu)$ is as in (1.11). Such a derivative is a function and, in the notation $\delta F(t,\mu)(x)$, this function is evaluated at $x \in \mathbb{R}$.*

2.3.4 Linear Quadratic Games with Mean Field Interactions

Linear quadratic (LQ) game models are popular because their solutions reduce to systems of ordinary differential equations, the only nonlinearity appearing in a matrix Riccati equation. Because these equations are not always solvable in the multivariate case, we refrain from dwelling on a discussion of the general form of LQ games, and instead, we restrict our attention to those linear quadratic games with mean field interactions. For such models, the dynamics of the state of player i are given by a linear equation of the form:

$$dX_t^i = \left(b_1(t)X_t^i + \bar{b}_1(t)\bar{X}_t^{-i} + b_2(t)\alpha_t^i \right) dt + \sigma dW_t^i, \tag{2.41}$$

where as usual the $(\boldsymbol{W}^i = (W_t^i)_{0 \leqslant t \leqslant T})_{1 \leqslant i \leqslant N}$'s are N independent standard Wiener processes of dimension m and where:

$$\bar{X}_t^{-i} = \frac{1}{N-1} \sum_{j=1, j \neq i}^{N} X_t^j = \int_{\mathbb{R}^d} x d\bar{\mu}_{X_t^{-i}}^{N-1}(x)$$

denotes the sample mean of the states of the players $j \neq i$. Here $\boldsymbol{b}_1 = (b_1(t))_{0 \leqslant t \leqslant T}$, $\bar{\boldsymbol{b}}_1 = (\bar{b}_1(t))_{0 \leqslant t \leqslant T}$ and $\boldsymbol{b}_2 = (b_2(t))_{0 \leqslant t \leqslant T}$ are deterministic continuous functions of $t \in [0, T]$ with values in $\mathbb{R}^{d \times d}$, $\mathbb{R}^{d \times d}$ and $\mathbb{R}^{d \times k}$ respectively, while σ is a constant matrix in $\mathbb{R}^{d \times m}$. As explained before, the volatility does not need to be constant. We make this assumption for the sake of simplicity. In terms of the notation used in this chapter, we have:

$$b(t, x, \mu, \alpha) = b_1(t)x + \bar{b}_1(t)\bar{\mu} + b_2(t)\alpha, \qquad \text{and} \qquad \sigma(t, x, \mu, \alpha) \equiv \sigma,$$

where we use the notation $\bar{\mu}$ for the mean $\int x d\mu(x)$ of the probability measure μ (which, in contrast with what we have done so far, is thus required to have a finite first-order moment). Clearly, the continuity assumption is stronger than what we need to assume on the coefficients. Again, it is here for the sake of simplicity. Using the same notation f and g for the running and terminal cost functions, we assume that these costs are of the form:

$$f(t, x, \mu, \alpha) = \frac{1}{2} \left(x^\dagger q(t) x + (x - s(t)\bar{\mu})^\dagger \bar{q}(t)(x - s(t)\bar{\mu}) + \alpha^\dagger r(t) \alpha \right),$$

and

$$g(x, \mu) = \frac{1}{2} \left(x^\dagger q x + (x - s\bar{\mu})^\dagger \bar{q}(x - s\bar{\mu}) \right),$$

where the symbol $\dagger$ is used for the transposition so that $x^\dagger q(t) x$ stands for the inner product $x \cdot (q(t)x)$ (and similarly for the others) and where $q, \bar{q}, s \in \mathbb{R}^{d \times d}$, $\boldsymbol{q} = (q(t))_{0 \leqslant t \leqslant T}$, $\boldsymbol{r} = (r(t))_{0 \leqslant t \leqslant T}$, $\boldsymbol{s} = (s(t))_{0 \leqslant t \leqslant T}$ and $\bar{\boldsymbol{q}} = (\bar{q}(t))_{0 \leqslant t \leqslant T}$ are deterministic continuous functions of $t \in [0, T]$ with values in $\mathbb{R}^{d \times d}$, $\mathbb{R}^{k \times k}$, $\mathbb{R}^{d \times d}$ and $\mathbb{R}^{d \times d}$ respectively. Moreover, we assume that q, $\bar{q}$, $q(t)$, and $\bar{q}(t)$ are symmetric and nonnegative semi-definite, which guarantees that f and g are convex in the direction x (which, in fact, would be true under the weaker assumption that $q+\bar{q}$ and $q(t)+\bar{q}(t)$ are nonnegative semi-definite), while $r(t)$ is assumed to be symmetric and strictly positive definite, hence invertible, which guarantees that f is strictly convex in α. Obviously, f is convex in the pair (x, α). All the matrices we consider in this book are real. Even though the notion of nonnegative definiteness in the sense of complex vector spaces implies that the matrices in question are Hermitian (hence symmetric since their entries are real), we shall make explicit the fact that we are considering symmetric matrices in order to avoid any possible ambiguity. Eigenvalues of the symmetric nonnegative matrices q, $\bar{q}$, $q(t)$, and $\bar{q}(t)$ are nonnegative. Moreover, all of the eigenvalues of $r(t)$ are strictly positive. So $r(t)$ is invertible, and by continuity, its inverse is also symmetric, with strictly positive eigenvalues, and is a continuous function of time. The form chosen above for the running cost f is not the most general. Indeed, we could have included cross terms in α and x. Similar results could be obtained for such an extension, at the cost of slightly more complicated formulas, and extra assumptions on the coefficients of the cross terms. We refrain from going to this level of generality as it does not bring anything new to the understanding of the models.

Unless specified otherwise, when dealing with LQ game models, we shall implicitly assume that the set A of possible control values is the whole Euclidean space $\mathbb{R}^k$ and the space $\mathbb{A}$ of admissible strategies is the Hilbert space $\mathbb{H}^{2,k}$. These assumptions can be relaxed when needed. We make them for the sake of simplicity.

The matrix $r(t)$ being invertible, one sees that the players' reduced Hamiltonians:

$$
\begin{aligned}
H^i(t, \boldsymbol{x}, \boldsymbol{y}^i, \boldsymbol{\alpha}) &= \sum_{j=1}^{N} \big(b_1(t)x^j + \bar{b}_1(t)\bar{\boldsymbol{x}}^{-j} + b_2(t)\alpha^j\big) \cdot y^{i,j} \\
&\quad + \frac{1}{2}\big((x^i)^\dagger q(t) x^i + (x^i - s(t)\bar{\boldsymbol{x}}^{-i})^\dagger \bar{q}(t)(x^i - s(t)\bar{\boldsymbol{x}}^{-i}) + (\alpha^i)^\dagger r(t)\alpha^i\big),
\end{aligned}
$$

with $\bar{\boldsymbol{x}}^{-i} = \frac{1}{N-1}\sum_{j=1, j\neq i}^{N} x^j$, for $(t, \boldsymbol{x}, \boldsymbol{y}, \boldsymbol{\alpha}) \in [0, T] \times (\mathbb{R}^d)^N \times (\mathbb{R}^d)^N \times (\mathbb{R}^k)^N$ and $i \in \{1, \cdots, N\}$, are strictly convex in the control variables and the generalized Isaacs conditions are satisfied with $\hat{\alpha}^i = -r(t)^{-1} b_2(t)^\dagger y^{i,i}$. We could write the large system of forward and backward equations by substituting these values for the controls appearing in the forward dynamics of the states and in the adjoint equations and try to solve the resulting high dimensional FBSDE by reducing it to a set of ordinary differential equations and a matrix Riccati equation. Unfortunately, matrix Riccati equations are not always well posed, and their analysis can be involved. So we refrain from pursuing the search for solutions at the present level of generality to avoid unnecessary technicalities.

However, we show in the last two sections of this chapter that linear quadratic games with mean field interactions can be explicitly solvable. We substantiate this claim by solving completely two of the models introduced in Chapter 1. We first treat the particular case of the flocking model in the case $\beta = 0$, and next, we solve the systemic risk model.

2.4 The Linear Quadratic Version of the Flocking Model

As a first application of the theory developed in this chapter, we consider the special case of the flocking model introduced in Chapter 1 corresponding to the particular choice $\beta = 0$ of the parameter. In that case, the position x_t^i of the bird at time t does not appear in the cost function and as a result, the mathematical analysis can be focused on the velocity component of the state. So for the purpose of this section, $X_t^i \in \mathbb{R}^3$ represents the velocity at time t of bird i, and its dynamics are given by:

$$dX_t^i = \alpha_t^i dt + \sigma dW_t^i, \quad 0 \leqslant t \leqslant T, \qquad i = 1, \cdots, N,$$

where the 3-dimensional standard Wiener processes $\boldsymbol{W}^i = (W_t^i)_{0 \leqslant t \leqslant T}$ are independent for $i = 1, \cdots, N$, and where $\sigma > 0$ is assumed to be a scalar for the sake of simplicity. If we specialize the discussion of Subsection 1.5.1 of Chapter 1 to the case $\beta = 0$, we find that if $\boldsymbol{\alpha} = (\boldsymbol{\alpha}^1, \cdots, \boldsymbol{\alpha}^N)$ is a strategy profile for the flock, bird i will want to minimize the expected cost:

$$J^i(\boldsymbol{\alpha}) = \mathbb{E}\left[\int_0^T f^i\big(t, (X_t^1, \cdots, X_t^N), (\alpha_t^1, \cdots, \alpha_t^N)\big) dt\right],$$

where the running cost function is given by:

$$f^i\big(t, (x^1, \cdots, x^N), (\alpha^1, \cdots, \alpha^N)\big) = \frac{\kappa^2}{2}|x^i - \bar{x}|^2 + \frac{1}{2}|\alpha^i|^2,$$

for $\boldsymbol{x} = (x^1, \cdots, x^N) \in \mathbb{R}^{3N}$ and $\boldsymbol{\alpha} = (\alpha^1, \cdots, \alpha^N) \in \mathbb{R}^{3N}$, and $\kappa \geqslant 0$. Recall that there is no terminal cost in the model as we stated it.

Here, we could choose to have each bird interact with the empirical distribution of the velocities of the other birds. We decided to have each bird interact with the empirical distribution of all the velocities, including its own. So, we use the notation $\bar{x} = (x^1 + \cdots + x^N)/N$ for the sample mean of the states of all the birds. As explained in Remark 1.25 of Chapter 1, having bird i pay or be rewarded by the difference between its state and the mean of the states of the other birds $j \neq i$ would simply amount to multiplying the constant κ by a quantity which converges to 1 as $N \to \infty$. So for the sake of definiteness, we use the empirical mean of all the states.

We first consider the open loop equilibrium problem. From now on, we use the terms *bird* and *player* interchangeably.

2.4.1 Open Loop Nash Equilibria

We use reduced Hamiltonians since the volatility is constant. Since we concentrate on the open loop problem, for each player, the set of admissible strategies $\mathbb{A}$ is the space $\mathbb{H}^{2,3}$ of $\mathbb{R}^3$-valued square integrable adapted processes. For each $i \in \{1, \cdots, N\}$, the reduced Hamiltonian of player i reads:

$$H^i(\boldsymbol{x}, \boldsymbol{y}^i, \boldsymbol{\alpha}) = \sum_{j=1}^{N} \alpha^j \cdot y^{i,j} + \frac{\kappa^2}{2}|\bar{\boldsymbol{x}} - x^i|^2 + \frac{1}{2}|\alpha^i|^2,$$

where $\boldsymbol{x} = (x^1, \ldots, x^N) \in \mathbb{R}^{3N}$, $\bar{\boldsymbol{x}} = \frac{1}{N}\sum_{i=1}^N x^i$ and $\boldsymbol{y}^i = (y^{i,1}, \ldots, y^{i,N}) \in \mathbb{R}^{3N}$. The value of α^i minimizing this reduced Hamiltonian with respect to α^i, when all the other variables, including α^j for $j \neq i$, are fixed, is given by:

$$\begin{aligned} \hat{\alpha}^i = \hat{\alpha}^i(\boldsymbol{x}, \boldsymbol{y}^i) = -y^{i,i}, \quad \boldsymbol{x} &= (x^1, \ldots, x^N) \in \mathbb{R}^{3N}, \\ \boldsymbol{y}^i &= (y^{i,1}, \ldots, y^{i,N}) \in \mathbb{R}^{3N}. \end{aligned} \tag{2.42}$$

Now, given an admissible strategy profile $\boldsymbol{\alpha} = (\alpha_t^1, \cdots, \alpha_t^N)_{0\le t\le T}$ and the corresponding controlled state $\boldsymbol{X} = \boldsymbol{X}^{\boldsymbol{\alpha}}$, the adjoint processes associated with $\boldsymbol{\alpha}$ are the processes $\boldsymbol{Y} = (\boldsymbol{Y}^1, \cdots, \boldsymbol{Y}^N)$ and $\boldsymbol{Z} = (\boldsymbol{Z}^1, \cdots, \boldsymbol{Z}^N)$, each $\boldsymbol{Y}^i$ being $(\mathbb{R}^3)^N$-valued and each $\boldsymbol{Z}^i$ being $(\mathbb{R}^3)^{N\times N}$-valued, solving the system of BSDEs:

$$\begin{aligned} dY_t^{i,j} &= -\partial_{x^j} H^i\big(t, X_t, (Y_t^{i,1}, \ldots, Y_t^{i,N}), \alpha_t\big)dt + \sum_{\ell=1}^{N} Z_t^{i,j,\ell} dW_t^\ell, \\ &= -\kappa^2(\delta_{i,j} - \frac{1}{N})(X_t^i - \bar{X}_t)\, dt + \sum_{\ell=1}^{N} Z_t^{i,j,\ell} dW_t^\ell, \qquad t \in [0, T], \end{aligned}$$

for $i,j = 1, \cdots, N$, with terminal conditions $Y_T^{i,j} = 0$. Notice that the controls do not appear explicitly in the adjoint equations. According to the strategy outlined earlier, we replace all the occurrences of the controls α_t^i in the forward dynamics by $\hat{\alpha}^i(X_t, Y_t^i) = -Y_t^{i,i}$, and we try to solve the resulting system of forward-backward equations. If we manage to do so, the strategy profile $\boldsymbol{\alpha} = (\alpha_t^1, \cdots, \alpha_t^N)_{0\le t\le T}$ defined by:

$$\alpha_t^i = \hat{\alpha}^i\big(X_t, Y_t^i\big) = \hat{\alpha}^i\big(X_t, (Y_t^{i,1}, \ldots, Y_t^{i,N})\big) = -Y_t^{i,i}, \quad t \in [0, T], \tag{2.43}$$

will provide an open loop Nash equilibrium. Notice indeed that the stochastic maximum principle here provides both a necessary and sufficient condition of equilibrium since the Hamiltonian H^i is convex in $(\boldsymbol{x}, \boldsymbol{\alpha})$. In the present situation, the system of FBSDEs reads:

$$\begin{cases} dX_t^i = -Y_t^{i,i} dt + \sigma dW_t^i, \\ dY_t^{i,j} = -\kappa^2(\delta_{i,j} - \frac{1}{N})(X_t^i - \bar{X}_t)\, dt + \sum_{k=1}^{N} Z_t^{i,j,\ell} dW_t^\ell, \qquad t \in [0, T], \\ Y_T^{i,j} = 0, \quad i,j = 1, \ldots, N. \end{cases} \tag{2.44}$$

This is a system of affine FBSDEs, so we expect $Y_t^{i,j}$ to be an affine function of X_t^i, or equivalently, using the terminology introduced in Chapter 4, we expect the decoupling field to be affine. In other words, we expect that the backward components Y_t will be given by an affine function of X_t. In the present situation, since the couplings between all these equations depend only upon quantities of the form $X_t^i - \bar{X}_t$, we search for a solution of the form:

$$Y_t^{i,j} = \eta_t\big(\delta_{i,j} - \frac{1}{N}\big)(X_t^i - \bar{X}_t) \tag{2.45}$$

for some smooth deterministic function $[0,T] \ni t \mapsto \eta_t \in \mathbb{R}$ to be determined. With such an ansatz, for any $i \in \{1, \ldots, N\}$ and $t \in [0,T]$, the forward dynamics become:

$$dX_t^i = -\eta_t\big(1 - \frac{1}{N}\big)(X_t^i - \bar{X}_t)dt + \sigma dW_t^i, \tag{2.46}$$

so that, by summing over i, we get:

$$d(X_t^i - \bar{X}_t) = -\eta_t\big(1 - \frac{1}{N}\big)(X_t^i - \bar{X}_t)dt + \sigma\Big(dW_t^i - \frac{1}{N}\sum_{\ell=1}^N dW_t^\ell\Big).$$

Therefore, computing the differential $dY_t^{i,j}$ from the ansatz (2.45), we get:

$$\begin{aligned} dY_t^{i,j} &= \big(\delta_{i,j} - \frac{1}{N}\big)(X_t^i - \bar{X}_t)\Big[\dot{\eta}_t - \eta_t^2\big(1 - \frac{1}{N}\big)\Big]dt \\ &\quad + \sigma\eta_t\big(\delta_{i,j} - \frac{1}{N}\big)\Big(dW_t^i - \frac{1}{N}\sum_{\ell=1}^N dW_t^\ell\Big). \end{aligned} \tag{2.47}$$

Identifying this differential with the right-hand side of the backward component of (2.44) we get:

$$Z_t^{i,j,\ell} = \sigma\eta_t\big(\delta_{i,j} - \frac{1}{N}\big)\big(\delta_{i,\ell} - \frac{1}{N}\big), \quad \ell = 1, \cdots, N,$$

and

$$\dot{\eta}_t = \big(1 - \frac{1}{N}\big)\eta_t^2 - \kappa^2, \tag{2.48}$$

with terminal condition $\eta_T = 0$. This is a scalar Riccati equation. Since we shall encounter this type of equation frequently in the sequel, we state a standard existence result for the sake of future reference.

Scalar Riccati Equations. Let us assume that A, B, γ and C are real numbers such that $B \neq 0$, $\gamma B \geqslant 0$ and $BC > 0$. Then, the scalar Riccati equation:

$$\dot{\eta}_t = 2A\eta_t + B\eta_t^2 - C, \quad t \in [0, T], \tag{2.49}$$

with terminal condition $\eta_T = \gamma$, has a unique solution:

$$\eta_t = \frac{-C\big(e^{(\delta^+-\delta^-)(T-t)} - 1\big) - \gamma\big(\delta^+ e^{(\delta^+-\delta^-)(T-t)} - \delta^-\big)}{\big(\delta^- e^{(\delta^+-\delta^-)(T-t)} - \delta^+\big) - \gamma B\big(e^{(\delta^+-\delta^-)(T-t)} - 1\big)}, \quad t \in [0, T], \tag{2.50}$$

where $\delta^\pm = -A \pm \sqrt{R}$, with $R = A^2 + BC > 0$.

Formula (2.50) follows from the standard change of variable:

$$A + B\eta_t = -\frac{\dot{\theta}_t}{\theta_t}, \quad t \in [0, T],$$

which transforms the nonlinear equation (2.49) into the second order linear ordinary differential equation:

$$\ddot{\theta}_t = (A^2 + BC)\theta_t, \quad t \in [0, T].$$

The crucial point is to observe that $\delta^+ > 0$ and $\delta^- < 0$, so that the denominator in (2.50) does not vanish, which excludes any possibility of blow-up. Similarly, it is clear that the numerator does not vanish except maybe in $t = T$, so that the function $\eta : [0, T] \ni t \mapsto \eta_t$ takes values in $[0, +\infty)$ if $\gamma \geqslant 0$ and in $(-\infty, 0]$ if $\gamma \leqslant 0$.

In the particular case at hand we find that the unique solution of the Riccati equation (2.48) is:

$$\eta_t = \kappa \sqrt{\frac{N}{N-1}} \frac{e^{2\kappa\sqrt{(N-1)/N}(T-t)} - 1}{e^{2\kappa\sqrt{(N-1)/N}(T-t)} + 1}, \quad t \in [0, T]. \tag{2.51}$$

Notice that $\eta_t > 0$ if $t < T$. For this specific function $[0, T] \ni t \mapsto \eta_t \in \mathbb{R}$, the sufficiency part of the Pontryagin stochastic maximum principle, recall Theorem 2.18, says that the strategy profile $\hat{\boldsymbol{\alpha}} = (\hat{\boldsymbol{\alpha}}^1, \cdots, \hat{\boldsymbol{\alpha}}^N)$ where $\hat{\boldsymbol{\alpha}}^i = (\hat{\alpha}_t^i)_{0 \leqslant t \leqslant T}$ is given by:

$$\hat{\alpha}_t^i = -\big(1 - \frac{1}{N}\big)\eta_t(X_t^i - \bar{X}_t), \tag{2.52}$$

obtained by plugging the ansatz (2.45) into (2.43), is an open loop Nash equilibrium. Notice that the controls (2.52) are in feedback form since they only depend upon the current value of the state X_t at time t. Note also that in equilibrium, the state X_t is Gaussian, and more precisely, the dynamics of the states $((X_t^i)_{0\leq t\leq T})_{i=1\cdots,N}$ of the individual birds are given by the stochastic differential equations (2.46), which show that the velocities of the individual birds are Ornstein-Uhlenbeck processes mean reverting toward the sample average of the velocities in the flock.

Important Remark. The strategy profile given by (2.52) was constructed in order to satisfy the sufficient condition for an open loop Nash equilibrium given by the stochastic maximum principle, and as such, it is indeed an open loop Nash equilibrium. However, even though it is in closed loop form, or even Markovian form, there is a priori no reason, except possibly wishful thinking, to believe that it could also be a closed loop or a Markovian Nash equilibrium, simply because of the definition we chose of Markovian equilibria.

2.4.2 Markovian Nash Equilibrium by the Stochastic Maximum Approach

In this subsection, we search for a set $\phi = (\phi^1, \cdots, \phi^N)$ of $\mathbb{R}^3$-valued feedback functions ϕ^i forming a Nash equilibrium for the Markov model of the game. For each player $i \in \{1, \cdots, N\}$, the reduced Hamiltonian (recall that the volatility depends neither on the state nor the controls) reads:

$$H^{-i}(\boldsymbol{x}, \boldsymbol{y}^i, \alpha) = \sum_{j=1, j\neq i}^{N} \phi^j(t, \boldsymbol{x}) \cdot y^{i,j} + \alpha \cdot y^{i,i} + \frac{\kappa^2}{2}|x^i - \bar{\boldsymbol{x}}|^2 + \frac{1}{2}|\alpha|^2,$$

for $\boldsymbol{x} = (x^1, \ldots, x^N) \in \mathbb{R}^{3N}$, $\boldsymbol{y}^i = (y^{i,1}, \ldots, y^{i,N}) \in \mathbb{R}^{3N}$ and $\alpha \in \mathbb{R}^3$. We refer to (2.24) for the definition of H^{-i}. The value of α minimizing this Hamiltonian (when all the other variables are fixed) is the same as before: $\hat{\alpha} = -y^{i,i}$. Added to the fact that the adjoint equations will lead to an affine FBSDE where the couplings depend only upon quantities of the form $X_t^i - \bar{X}_t$, we propose a similar ansatz for the Markov feedback functions. To be specific, we search for equilibrium feedback functions ϕ^i in the form:

$$\phi^i(t, \boldsymbol{x}) = \big(1 - \frac{1}{N}\big)\eta_t(x^i - \bar{\boldsymbol{x}}), \qquad (t, \boldsymbol{x}) \in [0, T] \times \mathbb{R}^{3N}, \quad i = 1, \cdots, N, \qquad (2.53)$$

for some deterministic function $[0,T] \ni t \mapsto \eta_t \in \mathbb{R}$. Even though we use the same notation $(\eta_t)_{0\leq t\leq T}$, this function may differ from the one identified above in the case of open loop equilibria. Using the special form of the Hamiltonian H^{-i}, which is convex in $(\boldsymbol{x},\alpha)$ since ϕ is linear in $\boldsymbol{x}$, we get as FBSDE derived from the stochastic maximum principle for Markovian equilibria:

$$
\begin{cases}
dX_t^i = -Y_t^{i,i}dt + \sigma dW_t^i, \\
dY_t^{i,j} = -\Big[\sum_{\ell=1,\ell\neq i}^{N} \big(\partial_{x^j}\phi^\ell(t,X_t)\big)^\dagger Y_t^{i,\ell} + \kappa^2(\delta_{i,j} - \frac{1}{N})(X_t^i - \bar{X}_t)\Big]dt \\
\qquad\qquad + \sum_{\ell=1}^{N} Z_t^{i,j,\ell} dW_t^\ell, \qquad t \in [0,T], \\
Y_T^{i,j} = 0,
\end{cases}
\tag{2.54}
$$

for $i,j = 1,\cdots,N$. Each $\boldsymbol{X}^i$, each $\boldsymbol{Y}^{i,j}$ and each $\boldsymbol{Z}^{i,j,\ell}$ is $\mathbb{R}^3$-valued. For the particular choice (2.53) of feedback functions, we have:

$$
\partial_{x^j}\phi^\ell(t,\boldsymbol{x}) = \big(\delta_{j,\ell} - \frac{1}{N}\big)\big(1 - \frac{1}{N}\big)\eta_t I_3,
$$

where I_3 denotes the 3×3 identity matrix. The backward component of (2.54) can be rewritten as:

$$
dY_t^{i,j} = -\Big[\big(1 - \frac{1}{N}\big)\eta_t \sum_{\ell=1,\ell\neq i}^{N} \big(\delta_{\ell,j} - \frac{1}{N}\big)Y_t^{i,\ell} + \kappa^2\big(\delta_{i,j} - \frac{1}{N}\big)(X_t^i - \bar{X}_t)\Big]dt
$$

$$
+ \sum_{\ell=1}^{N} Z_t^{i,j,\ell} dW_t^\ell, \tag{2.55}
$$

for $t \in [0,T]$, and $i,j = 1,\cdots,N$. For the same reasons as in the open loop case (couplings depending only upon $\bar{X}_t - X_t^i$), we make the same ansatz (2.45) on the form of $Y_t^{i,j}$, and we search for a solution of the FBSDE (2.54) in the form (2.45) with the same function $[0,T] \ni t \mapsto \eta_t \in \mathbb{R}$ as in (2.53). Evaluating the right-hand side of the BSDE part of (2.55) using the ansatz (2.45), we get:

$$
\begin{aligned}
dY_t^{i,j} &= -\Big[(1-\frac{1}{N})\eta_t \sum_{\ell=1,\ell\neq i}^{N} (\delta_{\ell,j}-\frac{1}{N})\eta_t(\delta_{i,\ell}-\frac{1}{N})(X_t^i-\bar{X}_t) \\
&\quad + \kappa^2(\delta_{i,j}-\frac{1}{N})(X_t^i-\bar{X}_t)\Big]dt + \sum_{\ell=1}^{N} Z_t^{i,j,\ell}dW_t^\ell \\
&= -\Big[(1-\frac{1}{N})\eta_t^2(X_t^i-\bar{X}_t) \sum_{\ell=1,\ell\neq i}^{N} (\delta_{\ell,j}-\frac{1}{N})(\delta_{i,\ell}-\frac{1}{N}) \\
&\quad + \kappa^2(\delta_{i,j}-\frac{1}{N})(X_t^i-\bar{X}_t)\Big]dt + \sum_{\ell=1}^{N} Z_t^{i,j,\ell}dW_t^\ell \\
&= -\Big[\frac{1}{N}(1-\frac{1}{N})\eta_t^2+\kappa^2\Big](\delta_{i,j}-\frac{1}{N})(X_t^i-\bar{X}_t)dt + \sum_{\ell=1}^{N} Z_t^{i,j,\ell}dW_t^\ell,
\end{aligned}
$$

where, to pass from the second to the third equality, we used the identity:

$$
\sum_{\ell=1,\ell\neq i}^{N} (\delta_{\ell,j}-\frac{1}{N})(\delta_{\ell,i}-\frac{1}{N}) = \frac{1}{N}(\delta_{i,j}-\frac{1}{N}). \tag{2.56}
$$

Equating with the differential $dY_t^{i,j}$ obtained in (2.47) from the ansatz (remember that (2.47) only depends upon the form of the ansatz and not on the nature of the equilibrium), we get the same identification for the $Z_t^{i,j,\ell}$ and the following Riccati equation for η_t:

$$
\dot{\eta}_t = (1-\frac{1}{N})^2\eta_t^2 - \kappa^2, \quad t\in[0,T], \tag{2.57}
$$

with the same terminal condition $\eta_T = 0$. This equation is very similar, but still different from the Riccati equation (2.48) obtained in the search for open loop equilibria. By (2.50), we get an explicit formula for the solution. In the present case, $\delta^+ = \kappa(1-1/N)$ and $\delta^- = -\delta^+$ and consequently:

$$
\eta_t = \kappa\frac{N}{N-1}\frac{e^{2\kappa(1-1/N)(T-t)}-1}{e^{2\kappa(1-1/N)(T-t)}+1}, \qquad 0\leqslant t\leqslant T. \tag{2.58}
$$

As in the case of the open loop problem, the equilibrium dynamics of the state ($\boldsymbol{X} = (\boldsymbol{X}^1,\cdots,\boldsymbol{X}^N)$) are given by an $\mathbb{R}^{3N}$-valued Ornstein-Uhlenbeck process reverting toward the sample mean $(\bar{X}_t = \frac{1}{N}\sum_{j=1}^N X_t^j)_{0\leqslant t\leqslant T}$.

Remark 2.27 *Clearly, the function* $[0, T] \ni t \mapsto \eta_t \in \mathbb{R}$ (2.58) *obtained in our search for Markov Nash equilibria is different from the function giving the open loop Nash equilibrium found in* (2.51)*. Notice that both functions converge toward the same limit as* $N \to \infty$*, this common limit solving the Riccati equation:*

$$\dot{\eta}_t = \eta_t^2 - \kappa^2.$$

and as a consequence, being given explicitly by the formula:

$$\eta_t = \kappa \frac{e^{2\kappa(T-t)} - 1}{e^{2\kappa(T-t)} + 1}, \qquad 0 \leqslant t \leqslant T, \tag{2.59}$$

obtained by passing to the limit $N \to \infty$ *in* (2.58).

2.5 The Coupled OU Model of Systemic Risk

We now present a second example of finite player game which can be solved explicitly, and for which the open loop and Markovian equilibria, though similar and given by feedback functions, differ as long as the number of players remains finite. As in the previous section, the model is of the linear quadratic type and the interactions are of a mean field nature. The computations will be more involved than in the previous section, but the analysis will remain very similar. Our interest in this example lies mostly in the fact that it contains a common noise and cross terms in the running cost function.

The example is based on the model of systemic risk introduced in Subsection 1.3.1 in Chapter 1. It is a particular case of linear quadratic game with mean field interactions as introduced in Subsection 2.3.4 above. Like in the previous section, we solve the model completely, and illustrate how the several versions of the stochastic maximum principle presented in this chapter can lead to different Nash equilibria. Moreover, we also implement the analytic approach based on the solution of a large system of coupled Hamilton-Jacobi-Bellman partial differential equations, if only to show that the Markovian equilibrium found in this way does coincide with the Markov equilibrium found via the Pontryagin stochastic maximum principle.

As in the case of the flocking model analyzed earlier, we choose to work with the form of the interaction where each player interact with the empirical mean of the states of all the banks, as opposed to the empirical mean of the states of all the other banks. As already explained for the flocking model, switching from one form of interaction to the other, simply amounts to multiplying the constants of the model by functions of N which tend to 1 as $N \to \infty$.

2.5.1 Open Loop Nash Equilibria

In this model, we assume that the log-cash reserves $X_t = (X_t^1, \cdots, X_t^N)$ of N banks are Ornstein-Uhlenbeck (OU) processes reverting to their sample mean $\bar{X}_t = (X_t^1 + \cdots + X_t^N)/N$ at a rate $a > 0$. To be specific, we assume that the dynamics of the log-reserves of the banks are given by equations of the form:

$$dX_t^i = \left[a(\bar{X}_t - X_t^i) + \alpha_t^i\right] dt + \sigma dB_t^i, \quad i = 1, \cdots, N, \tag{2.60}$$

where:

$$dB_t^i = \sqrt{1-\rho^2}dW_t^i + \rho dW_t^0,$$

for some $\rho \in [-1, 1]$. The major fundamental difference between this model and the flocking model considered in the previous section is the presence of the Wiener process $\boldsymbol{W}^0$ in the dynamics of all the log-cash reserve processes $\boldsymbol{X}^i$. The state processes are usually correlated through their empirical distribution, but when $\rho \neq 0$, the presence of this *common noise* $\boldsymbol{W}^0$ creates an extra source of dependence. The process $\boldsymbol{\alpha}^i$ is understood as the control of bank i.

In this model, bank i tries to *minimize*:

$$J^i(\boldsymbol{\alpha}^1, \cdots, \boldsymbol{\alpha}^N)$$
$$= \mathbb{E}\left[\int_0^T \left(\frac{\epsilon}{2}(\bar{X}_t - X_t^i)^2 - q\alpha_t^i(\bar{X}_t - X_t^i) + \frac{1}{2}|\alpha_t^i|^2\right) dt + \frac{c}{2}(\bar{X}_T - X_T^i)^2\right],$$

where ϵ, c, and q are positive constants.

As before, we use reduced Hamiltonians since the volatility depends neither upon the state X_t, nor the control α_t. For each player, the set of admissible strategies is the space $\mathbb{H}^2 = \mathbb{H}^{2,1}$ of real valued, square integrable adapted processes. For each $i \in \{1, \cdots, N\}$, the reduced Hamiltonian of player i reads:

$$H^i(\boldsymbol{x}, \boldsymbol{y}^i, \boldsymbol{\alpha}) = \sum_{j=1}^N [a(\bar{\boldsymbol{x}} - x^j) + \alpha^j]y^{i,j} + \frac{\epsilon}{2}(\bar{\boldsymbol{x}} - x^i)^2 - q\alpha^i(\bar{\boldsymbol{x}} - x^i) + \frac{1}{2}(\alpha^i)^2,$$

where $\boldsymbol{x} = (x^1, \cdots, x^N) \in \mathbb{R}^N$, $\bar{\boldsymbol{x}} = \frac{1}{N}\sum_{i=1}^N x_i$, $\boldsymbol{y}^i = (y^{i,1}, \cdots, y^{i,N}) \in \mathbb{R}^N$ and $\boldsymbol{\alpha} = (\alpha^1, \cdots, \alpha^N) \in \mathbb{R}^N$. The value of α^i minimizing this reduced Hamiltonian with respect to α^i, when all the other variables, including α^j for $j \neq i$, are fixed, is given by:

$$\hat{\alpha}^i = \hat{\alpha}^i(\boldsymbol{x}, \boldsymbol{y}^i) = -y^{i,i} + q(\bar{\boldsymbol{x}} - x^i), \tag{2.61}$$

for $\boldsymbol{x} = (x^1, \cdots, x^N) \in \mathbb{R}^N$, and $\boldsymbol{y}^i = (y^{i,1}, \cdots, y^{i,N}) \in \mathbb{R}^N$. Now, given an admissible strategy profile $\boldsymbol{\alpha} = (\alpha_t^1, \cdots, \alpha_t^N)_{0 \le t \le T}$ and the corresponding controlled state $\boldsymbol{X} = \boldsymbol{X}^{\boldsymbol{\alpha}}$, the adjoint processes associated with $\boldsymbol{\alpha}$ are the processes $\boldsymbol{Y} = (\boldsymbol{Y}^1, \cdots, \boldsymbol{Y}^N)$ and $\boldsymbol{Z} = (\boldsymbol{Z}^1, \cdots, \boldsymbol{Z}^N)$, each $\boldsymbol{Y}^i$ being $\mathbb{R}^N$-valued and each $\boldsymbol{Z}^i$ being $\mathbb{R}^{N\times(N+1)}$-valued, solving the system of BSDEs:

$$
\begin{aligned}
dY_t^{i,j} &= -\partial_{x^j} H^i\big(t, X_t, (Y_t^{i,1}, \cdots, Y_t^{i,N}), \alpha_t\big)dt + \sum_{\ell=0}^{N} Z_t^{i,j,\ell} dW_t^\ell, \\
&= -\Big[\sum_{\ell=1}^{N} a(\frac{1}{N} - \delta_{\ell,j})Y_t^{i,\ell} - q\alpha_t^i(\frac{1}{N} - \delta_{i,j}) + \epsilon(\bar{X}_t - X_t^i)(\frac{1}{N} - \delta_{i,j})\Big]dt \\
&\quad + \sum_{\ell=0}^{N} Z_t^{i,j,\ell} dW_t^\ell, \qquad t \in [0,T],
\end{aligned}
\tag{2.62}
$$

for $i,j = 1, \cdots, N$ with terminal conditions $Y_T^{i,j} = c(\bar{X}_T - X_T^i)(\frac{1}{N} - \delta_{i,j})$. According to the strategy outlined earlier, we replace all the occurrences of the controls α_t^i, in the forward equations giving the dynamics of the states, and in the backward adjoint equations, by $\hat{\alpha}^i(X_t, Y_t^i) = -Y_t^{i,i} + q(\bar{X}_t - X_t^i)$. Then, we try to solve the resulting system of forward-backward equations. If we succeed, the strategy profile $\boldsymbol{\alpha} = (\alpha_t^1, \cdots, \alpha_t^N)_{0\le t\le T}$ defined by:

$$
\alpha_t^i = \hat{\alpha}^i\big(X_t, (Y_t^{i,1}, \dots, Y_t^{i,N})\big) = -Y_t^{i,i} + q(\bar{X}_t - X_t^i), \quad t \in [0,T], \tag{2.63}
$$

will provide an open loop Nash equilibrium. Notice that the condition $\epsilon \geqslant q^2$ implies that H^i is convex in $(\boldsymbol{x}, \boldsymbol{\alpha})$. In the present situation, the FBSDEs read:

$$
\begin{cases}
dX_t^i = \Big[(a+q)(\bar{X}_t - X_t^i) - Y_t^{i,i}\Big]dt + \sigma\rho dW_t^0 + \sigma\sqrt{1-\rho^2}dW_t^i, \\
dY_t^{i,j} = -\Big[a\sum_{\ell=1}^{N}(\frac{1}{N} - \delta_{\ell,j})Y_t^{i,\ell} + q[Y_t^{i,i} - q(\bar{X}_t - X_t^i)](\frac{1}{N} - \delta_{i,j}) \\
\qquad\qquad + \epsilon(\bar{X}_t - X_t^i)(\frac{1}{N} - \delta_{i,j})\Big]dt \\
\qquad\qquad + \sum_{\ell=0}^{N} Z_t^{i,j,\ell} dW_t^\ell, \quad t \in [0,T], \\
Y_T^{i,j} = c(\bar{X}_T - X_T^i)(\frac{1}{N} - \delta_{i,j}), \quad i,j = 1, \cdots, N.
\end{cases}
\tag{2.64}
$$

This is a system of affine FBSDEs, so we expect that the backward components Y_t at time t will be given by an affine function of X_t. However, since the couplings between all these equations depend only upon quantities of the form $\bar{X}_t - X_t^i$, we search for a solution of the form:

$$
Y_t^{i,j} = \eta_t(\bar{X}_t - X_t^i)(\frac{1}{N} - \delta_{i,j}), \tag{2.65}
$$

for some smooth deterministic function $[0, T] \ni t \mapsto \eta_t \in \mathbb{R}$ to be determined. With such an ansatz, it holds that, for any $i \in \{1, \ldots, N\}$ and $t \in [0, T]$:

$$dX_t^i = \Big[a + q + (1 - \frac{1}{N})\eta_t\Big](\bar{X}_t - X_t^i)dt + \sigma\rho dW_t^0 + \sigma\sqrt{1-\rho^2}dW_t^i,$$

from which it easily follows that:

$$\begin{aligned} d(\bar{X}_t - X_t^i) \\ &= -\Big[a + q + (1 - \frac{1}{N})\eta_t\Big](\bar{X}_t - X_t^i)dt + \sigma\sqrt{1-\rho^2}\Big(\frac{1}{N}\sum_{\ell=1}^N dW_t^\ell - dW_t^i\Big). \end{aligned}$$

Therefore, computing the differential $dY_t^{i,j}$ from the ansatz (2.65), we get:

$$\begin{aligned} dY_t^{i,j} &= (\frac{1}{N} - \delta_{i,j})(\bar{X}_t - X_t^i)\Big[\dot{\eta}_t - \eta_t\Big(a + q + (1 - \frac{1}{N})\eta_t\Big)\Big]dt \\ &\quad + \sigma\sqrt{1-\rho^2}\eta_t(\frac{1}{N} - \delta_{i,j})\Big(\frac{1}{N}\sum_{\ell=1}^N dW_t^\ell - dW_t^i\Big). \end{aligned} \tag{2.66}$$

Evaluating the right-hand side of the BSDE part of (2.64) using the ansatz (2.65) we get:

$$\begin{aligned} dY_t^{i,j} = -\Big[a\sum_{\ell=1}^N (\frac{1}{N} - \delta_{\ell,j})\eta_t(\bar{X}_t - X_t^i)(\frac{1}{N} - \delta_{\ell,i}) \\ +q\big[\eta_t(\bar{X}_t - X_t^i)(\frac{1}{N} - 1) - q(\bar{X}_t - X_t^i)\big](\frac{1}{N} - \delta_{i,j}) \\ +\epsilon(\bar{X}_t - X_t^i)(\frac{1}{N} - \delta_{i,j})\Big]dt + \sum_{\ell=0}^N Z_t^{i,j,\ell}dW_t^\ell. \end{aligned}$$

Similar to (2.56), we observe that, for any $\boldsymbol{x} = (x^1, \cdots, x^N) \in \mathbb{R}^N$,

$$\sum_{\ell=1}^N (\frac{1}{N} - \delta_{\ell,j})\eta_t(\bar{\boldsymbol{x}} - x^i)(\frac{1}{N} - \delta_{\ell,i}) = -\eta_t(\bar{\boldsymbol{x}} - x^i)(\frac{1}{N} - \delta_{i,j}). \tag{2.67}$$

Therefore, for any $t \in [0, T]$, we have:

$$\begin{aligned} dY_t^{i,j} \\ &= (\frac{1}{N} - \delta_{i,j})(\bar{X}_t - X_t^i)\Big[(a + q(1 - \frac{1}{N}))\eta_t + q^2 - \epsilon\Big]dt + \sum_{\ell=0}^N Z_t^{i,j,\ell}dW_t^\ell. \end{aligned} \tag{2.68}$$

Identifying the two Itô decompositions of $Y_t^{i,j}$ given in (2.66) and (2.68) we get, as a necessary condition for (2.65):

$$Z_t^{i,j,0} = 0, \quad Z_t^{i,j,\ell} = \sigma\sqrt{1-\rho^2}\eta_t(\frac{1}{N} - \delta_{i,j})(\frac{1}{N} - \delta_{i,\ell}), \quad \ell = 1, \cdots, N,$$

and

$$\dot{\eta}_t - \eta_t\left(a + q + (1 - \frac{1}{N})\eta_t\right) = (a+q)\eta_t - \frac{1}{N}q\eta_t + q^2 - \epsilon,$$

which we rewrite as a standard scalar Riccati's equation:

$$\dot{\eta}_t = \left[2(a+q) - \frac{1}{N}q\right]\eta_t + (1 - \frac{1}{N})\eta_t^2 + q^2 - \epsilon, \tag{2.69}$$

with terminal condition $\eta_T = c$. Under the condition $\epsilon \geqslant q^2$, the existence result which we recalled in the previous section says that this Riccati equation has a unique solution, given by:

$$\eta_t = \frac{-(\epsilon - q^2)\left(e^{(\delta^+ - \delta^-)(T-t)} - 1\right) - c\left(\delta^+ e^{(\delta^+ - \delta^-)(T-t)} - \delta^-\right)}{\left(\delta^- e^{(\delta^+ - \delta^-)(T-t)} - \delta^+\right) - c(1 - 1/N)\left(e^{(\delta^+ - \delta^-)(T-t)} - 1\right)}, \tag{2.70}$$

with:

$$\begin{aligned} \delta^{\pm} &= -\left(a + q - \frac{q}{2N}\right) \pm \sqrt{R}, \\ \text{and} \quad R &= \left(a + q - \frac{q}{2N}\right)^2 + \left(1 - \frac{1}{N}\right)(\epsilon - q^2) > 0. \end{aligned} \tag{2.71}$$

Figure 2.1 gives the plots of the solution for a few values of the parameters.

With such a function $[0,T] \ni t \mapsto \eta_t \in \mathbb{R}$ in hand, the sufficiency part of the Pontryagin stochastic maximum principle given in Theorem 2.18 implies that the strategy profile given by:

$$\alpha_t^i = \left[q + (1 - \frac{1}{N})\eta_t\right](\bar{X}_t - X_t^i), \quad t \in [0,T], \tag{2.72}$$

obtained by plugging the value (2.65) of $Y_t^{i,j}$ in (2.63), is an open loop Nash equilibrium. Notice that the controls (2.72) are in feedback form since they only depend upon the current value of the state X_t at time t. Note also that in equilibrium, the dynamics of the state $\boldsymbol{X}$ are given by the stochastic differential equations:

$$dX_t^i = \left[a + q + (1 - \frac{1}{N})\eta_t\right](\bar{X}_t - X_t^i)dt + \sigma\rho dW_t^0 + \sigma\sqrt{1-\rho^2}dW_t^i, \tag{2.73}$$

for $i = 1, \cdots, N$, which are exactly the uncontrolled versions of the equations we started from, except for the fact that the mean reversion coefficient a is replaced by the time dependent mean reversion rate $a + q + (1 - \frac{1}{N})\eta_t$.

Same Remark as Before. Even though the strategy profile given by (2.72) is in closed loop form, we can only claim that it is an open loop Nash equilibrium.

2.5.2 Markovian Nash Equilibrium by the Stochastic Maximum Approach

We now search for a set $\phi = (\phi^1, \cdots, \phi^N)$ of feedback functions ϕ^i forming a Nash equilibrium for the Markov model of the game. For each player $i \in \{1, \cdots, N\}$, the reduced Hamiltonian (recall that the volatility depends neither on the state nor the controls) reads:

$$\begin{aligned} H^{-i}(\boldsymbol{x}, \boldsymbol{y}^i, \alpha) = & \sum_{\ell=1,\ell\neq i}^{N} [a(\bar{x} - x^\ell) + \phi^\ell(t, \boldsymbol{x})]y^{i,\ell} + [a(\bar{x} - x^i) + \alpha]y^{i,i} \\ & + \frac{\epsilon}{2}(\bar{x} - x^i)^2 - q\alpha(\bar{x} - x^i) + \frac{1}{2}\alpha^2, \end{aligned}$$

for $\boldsymbol{x} = (x^1, \ldots, x^N) \in \mathbb{R}^N$, $\boldsymbol{y}^i = (y^{i,1}, \ldots, y^{i,N}) \in \mathbb{R}^N$ and $\alpha \in \mathbb{R}$. The value of α minimizing this Hamiltonian (when all the other variables are fixed) is again the value $\hat{\alpha}$ given by (2.61). Using this formula and the fact that the adjoint equations will lead to an affine FBSDE where the couplings depend only upon quantities of the form $(\bar{X}_t - X_t^i)_{0\leq t\leq T}$, we search, as in the open loop case, for equilibrium feedback functions ϕ^i in the form:

$$\phi^i(t, \boldsymbol{x}) = \left[q + \left(1 - \frac{1}{N}\right)\eta_t\right](\bar{x} - x^i), \qquad (t, \boldsymbol{x}) \in [0, T] \times \mathbb{R}^N, \tag{2.74}$$

for $i = 1, \cdots, N$ and for some deterministic function $[0, T] \ni t \mapsto \eta_t \in \mathbb{R}$ to be determined, and we try to find such a function in order for these feedback functions ϕ^i to form a Markovian Nash equilibrium. Importantly, with these feedback functions in hand, H^{-i} is convex in $(\boldsymbol{x}, \alpha)$.

Using formula (2.25) for the partial derivative of the Hamiltonian, we can solve the Markovian model by means of the Pontryagin principle, which leads to the FBSDE:

$$
\begin{cases}
dX_t^i = \big[(a+q)(\bar{X}_t - X_t^i) - Y_t^{i,i}\big]dt + \sigma\rho dW_t^0 + \sigma\sqrt{1-\rho^2}dW_t^i, \\
dY_t^{i,j} = -\Big[a\sum_{\ell=1}^{N}(\frac{1}{N} - \delta_{\ell,j})Y_t^{i,\ell} + \sum_{\ell=1,\ell\neq i}^{N} \partial_{x^j}\phi^\ell(t,X_t)Y_t^{i,\ell} \\
\qquad +q[Y_t^{i,i} - q(\bar{X}_t - X_t^i)](\frac{1}{N} - \delta_{i,j}) + \epsilon(\bar{X}_t - X_t^i)(\frac{1}{N} - \delta_{i,j})\Big]dt \\
\qquad +\sum_{\ell=0}^{N} Z_t^{i,j,\ell}dW_t^\ell, \qquad t\in[0,T], \\
Y_T^{i,j} = c(\bar{X}_T - X_T^i)(\frac{1}{N} - \delta_{i,j}), \qquad i,j = 1,\cdots,N.
\end{cases}
\tag{2.75}
$$

For the particular choice (2.74) of feedback functions, we have:

$$
\partial_{x^j}\phi^\ell(t,\boldsymbol{x}) = \big(\frac{1}{N} - \delta_{j,\ell}\big)\big[q + (1-\frac{1}{N})\eta_t\big],
$$

and the backward component of the BSDE rewrites:

$$
\begin{aligned}
dY_t^{i,j} = &-\Big[a\sum_{\ell=1}^{N}(\frac{1}{N} - \delta_{\ell,j})Y_t^{i,\ell} + \sum_{\ell=1,\ell\neq i}^{N}(\frac{1}{N} - \delta_{\ell,j})[q + \eta_t(1-\frac{1}{N})]Y_t^{i,\ell} \\
&+ q[Y_t^{i,i} - q(\bar{X}_t - X_t^i)](\frac{1}{N} - \delta_{i,j}) + \epsilon(\bar{X}_t - X_t^i)(\frac{1}{N} - \delta_{i,j})\Big]dt \\
&+\sum_{\ell=0}^{N} Z_t^{i,j,\ell}dW_t^\ell, \qquad t\in[0,T], \quad i,j = 1,\cdots,N.
\end{aligned}
\tag{2.76}
$$

For the same reasons as in the open loop case (couplings depending only upon $\bar{X}_t - X_t^i$), we make the same ansatz on the form of $Y_t^{i,j}$, namely $Y_t^{i,j} = \eta_t(\bar{X}_t - X_t^i)(\frac{1}{N} - \delta_{i,j})$, and search for a solution of the FBSDE (2.75) in the form (2.65). Evaluating the right-hand side of the BSDE part of (2.76) using the ansatz (2.65), we get:

$$
\begin{aligned}
dY_t^{i,j} = &-\Big[a\sum_{\ell=1}^{N}\big(\frac{1}{N}-\delta_{\ell,j}\big)\eta_t(\bar{X}_t-X_t^i)\big(\frac{1}{N}-\delta_{\ell,i}\big) \\
&\quad+\sum_{\ell=1,\ell\neq i}^{N}\big(\frac{1}{N}-\delta_{\ell,j}\big)\big[q+\eta_t(1-\frac{1}{N})\big]\eta_t(\bar{X}_t-X_t^i)\big(\frac{1}{N}-\delta_{\ell,i}\big) \\
&\quad+q\big[\eta_t(\bar{X}_t-X_t^i)\big(\frac{1}{N}-1\big)-q(\bar{X}_t-X_t^i)\big]\big(\frac{1}{N}-\delta_{i,j}\big) \\
&\quad+\epsilon(\bar{X}_t-X_t^i)\big(\frac{1}{N}-\delta_{i,j}\big)\Big]dt \\
&+\sum_{\ell=0}^{N}Z_t^{i,j,\ell}dW_t^\ell, \qquad t\in[0,T], \quad i,j=1,\cdots,N.
\end{aligned}
$$

Following (2.56), we observe that, for any $\boldsymbol{x}=(x^1,\dots,x^N)\in\mathbb{R}^N$,

$$
\begin{aligned}
&\sum_{\ell=1,\ell\neq i}^{N}\big(\frac{1}{N}-\delta_{\ell,j}\big)\big[q+\eta_t(1-\frac{1}{N})\big]\eta_t(\bar{\boldsymbol{x}}-x^i)\big(\frac{1}{N}-\delta_{\ell,i}\big) \\
&\qquad=\frac{1}{N}\big[q+\eta_t(1-\frac{1}{N})\big]\big(\delta_{i,j}-\frac{1}{N}\big)\eta_t(\bar{\boldsymbol{x}}-x^i).
\end{aligned}
$$

Using (2.67) to handle the first line in $dY_t^{i,j}$, we get:

$$
\begin{aligned}
dY_t^{i,j} = &\big(\frac{1}{N}-\delta_{i,j}\big)(\bar{X}_t-X_t^i)\Big[(a+q)\eta_t+\frac{1}{N}\big(1-\frac{1}{N}\big)\eta_t^2+q^2-\epsilon\Big]dt \\
&+\sum_{\ell=0}^{N}Z_t^{i,j,\ell}dW_t^\ell.
\end{aligned}
$$

Identifying this Itô decomposition with the differential $dY_t^{i,j}$ obtained in (2.66) from the ansatz, we get the same identification for the $Z_t^{i,j,k}$ and the following Riccati equation for η_t:

$$
\dot{\eta}_t = 2(a+q)\eta_t+\big(1-\frac{1}{N^2}\big)\eta_t^2+q^2-\epsilon, \quad t\in[0,T], \tag{2.77}
$$

with the same terminal condition $\eta_T = c$ as before. This equation has a unique solution since $\epsilon \geqslant q^2$ and (2.50) gives for any $t\in[0,T]$:

$$
\eta_t = \frac{-(\epsilon-q^2)\big(e^{(\delta^+-\delta^-)(T-t)}-1\big)-c\big(\delta^+e^{(\delta^+-\delta^-)(T-t)}-\delta^-\big)}{\big(\delta^-e^{(\delta^+-\delta^-)(T-t)}-\delta^+\big)-c(1-1/N^2)\big(e^{(\delta^+-\delta^-)(T-t)}-1\big)}, \tag{2.78}
$$

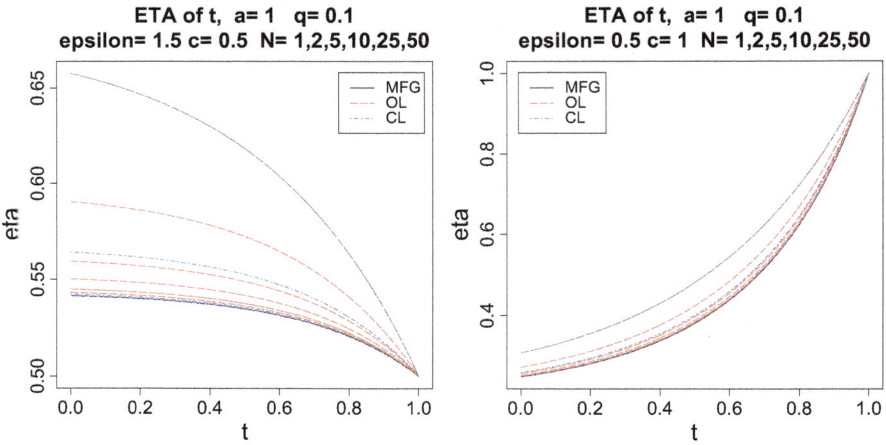

Fig. 2.1 Plot of the solution η_t of the Riccati equations (2.69) and (2.80) for several values of the parameters and numbers of players N increasing from 1 to 50.

with:

$$\delta^{\pm} = -(a+q) \pm \sqrt{R}, \qquad \text{and} \quad R = (a+q)^2 + \Big(1 - \frac{1}{N^2}\Big)(\epsilon - q^2) > 0. \tag{2.79}$$

Clearly, the function $[0,T] \ni t \mapsto \eta_t \in \mathbb{R}$ obtained in our search for Markov Nash equilibria is different from the function giving the open loop Nash equilibrium found in (2.70) and (2.71).

Notice that both functions converge toward the same limit as $N \to \infty$, this common limit solving the Riccati equation:

$$\dot{\eta}_t = 2(a+q)\eta_t + \eta_t^2 + q^2 - \epsilon, \qquad \eta_T = c. \tag{2.80}$$

Figure 2.1 gives the plots of the solutions for the two types of equilibria and for a few values of the parameters. We indeed observe from the plots that, as N increases, the two functions $[0,T] \ni t \mapsto \eta_t \in \mathbb{R}$ decrease to their common limit as $N \to \infty$. In the limit of large games ($N \to \infty$) the open loop and the closed loop (Markovian) Nash equilibria found with the Pontryagin stochastic maximum principle coincide. The fact that the differences between open and closed loop equilibria disappear in the limit of large games is expected. It is part of the *game theory folklore*. We will elaborate further on that limit $N \to \infty$ in Chapter 3 when we discuss Mean Field Games (MFGs), and at the end of the Notes & Complements section of that chapter where we give references to papers and book chapters discussing this claim. These references include Chapter 6 of the second volume, which is dedicated to the passage from games with finitely many players to mean field games.

2.5.3 Markovian Nash Equilibria by PDE Methods

For the sake of completeness, we show that the analytic approach based on the solution of a system of coupled partial differential equations of the Hamilton-Jacobi-Bellman (HJB for short) type can also be implemented, and that it gives exactly the same Markovian Nash equilibrium as the stochastic maximum principle approach implemented in the previous subsection. Notice that the present set-up fits the setting used in Subsection 2.1.4, with $B(t,\boldsymbol{x},\boldsymbol{\alpha}) = (B_i(t,\boldsymbol{x},\boldsymbol{\alpha}))_{1\leqslant i\leqslant N}$ and $\Sigma(t,\boldsymbol{x},\boldsymbol{\alpha}) = (\Sigma_{i,j}(t,\boldsymbol{x},\boldsymbol{\alpha}))_{1\leqslant i\leqslant N,0\leqslant j\leqslant N+1}$, given by:

$$B_i(t,\boldsymbol{x},\boldsymbol{\alpha}) = a(\bar{x} - x^i) + \alpha^i, \quad i \in \{1,\cdots,N\},$$

$$\Sigma_{i,j}(t,\boldsymbol{x}) = \begin{cases} \sigma\rho & \text{if } i = j \neq 0, \\ \sigma\sqrt{1-\rho^2} & \text{if } j = 0, \\ 0 & \text{otherwise,} \end{cases}$$

for $\boldsymbol{x},\boldsymbol{\alpha} \in \mathbb{R}^N$, where, as above, we use the notation $\bar{x}$ for the mean $\bar{x} = (x^1 + \cdots + x^N)/N$. Accordingly, the noise in (2.12) is regarded as an $(N+1)$-dimensional Wiener process $\boldsymbol{W} = (W_t)_{0\leqslant t\leqslant T} = (W_t^0, W_t^1, \cdots, W_t^N)_{0\leqslant t\leqslant T}$.

Recall that, given an N-tuple $(\phi^i)_{1\leqslant i\leqslant N}$ of functions from $[0,T]\times\mathbb{R}$ into $\mathbb{R}$, we define, for each $i \in \{1,\cdots,N\}$, the related value function V^i by:

$$V^i(t,x^1,\ldots,x^N) = \inf_{(\alpha_s^i)_{t\leqslant s\leqslant T}} \mathbb{E}\biggl[\int_t^T f\bigl(s,X_s^i,\bar{\mu}_s^N,\alpha_s^i\bigr)ds + g(X_T^i,\bar{\mu}_T^N) \,\big|\, X_t = \boldsymbol{x}\biggr],$$

with $\boldsymbol{x} = (x^1,\ldots,x^N) \in \mathbb{R}^N$ and with the same cost functions f and g as before. Here the dynamics of $(X_s^1,\ldots,X_s^N)_{t\leqslant s\leqslant T}$ are given by (2.60) with $X_t^j = x^j$ for $j \in \{1,\ldots,N\}$ and $\alpha_s^j = \phi^j(s,X_s)$ for $j \neq i$. By dynamic programming, each scalar function V^i, for $i = 1,\cdots,N$, must satisfy the HJB equation:

$$\begin{aligned} &\partial_t V^i(t,\boldsymbol{x}) \\ &\quad + \inf_{\alpha\in\mathbb{R}} \bigl\{\bigl(a(\bar{x}-x^i)+\alpha\bigr)\partial_{x^i}V^i(t,\boldsymbol{x}) + \frac{1}{2}\alpha^2 - q\alpha\bigl(\bar{x}-x^i\bigr)\bigr\} + \frac{\epsilon}{2}(\bar{x}-x^i)^2 \\ &\quad + \sum_{j=1,j\neq i}^N \bigl(a(\bar{x}-x^j)+\phi^j(t,x^j)\bigr)\partial_{x^j}V^i(t,\boldsymbol{x}) \\ &\quad + \frac{\sigma^2}{2}\sum_{j=1}^N\sum_{k=1}^N \bigl(\rho^2 + \delta_{j,k}(1-\rho^2)\bigr)\,\partial^2_{x^jx^k}V^i(t,\boldsymbol{x}) = 0, \end{aligned}$$

for $(t, \boldsymbol{x}) \in [0, T] \times \mathbb{R}^N$, with the terminal condition $V^i(T, \boldsymbol{x}) = c(\bar{\boldsymbol{x}} - x^i)^2/2$. The infima in these HJB equations can be computed explicitly:

$$\inf_{\alpha \in \mathbb{R}} \left\{ \left(a(\bar{\boldsymbol{x}} - x^i) + \alpha\right) \partial_{x^i} V^i(t, \boldsymbol{x}) + \frac{1}{2}\alpha^2 - q\alpha\left(\bar{\boldsymbol{x}} - x^i\right) \right\}$$
$$= a(\bar{\boldsymbol{x}} - x^i)\partial_{x^i} V^i(t, \boldsymbol{x}) - \frac{1}{2}\left[q\left(\bar{\boldsymbol{x}} - x^i\right) - \partial_{x^i} V^i(t, \boldsymbol{x})\right]^2,$$

the infima being attained for

$$\hat{\alpha} = q\left(\bar{\boldsymbol{x}} - x^i\right) - \partial_{x^i} V^i(t, \boldsymbol{x}).$$

Therefore, the Markovian strategies $(\phi^i)_{1 \leqslant i \leqslant N}$ will form a Nash equilibrium if $\phi^i(t, \boldsymbol{x}) = q(\bar{\boldsymbol{x}} - x^i) - \partial_{x^i} V^i(t, \boldsymbol{x})$, which suggests that we need to solve the system of N coupled HJB equations:

$$\begin{aligned}
&\partial_t V^i(t, \boldsymbol{x}) + \sum_{j=1}^{N} \left[(a + q)\left(\bar{\boldsymbol{x}} - x^j\right) - \partial_{x^j} V^j(t, \boldsymbol{x})\right] \partial_{x^j} V^i(t, \boldsymbol{x}) \\
&\quad + \frac{\sigma^2}{2} \sum_{j=1}^{N} \sum_{k=1}^{N} \left(\rho^2 + \delta_{j,k}(1 - \rho^2)\right) \partial^2_{x^j x^k} V^i(t, \boldsymbol{x}) \\
&\quad + \frac{1}{2}(\epsilon - q^2)\left(\bar{\boldsymbol{x}} - x^i\right)^2 + \frac{1}{2}\left(\partial_{x^i} V^i(t, \boldsymbol{x})\right)^2 = 0,
\end{aligned} \tag{2.81}$$

for $i = 1, \cdots, N$, with the same terminal condition as above. In (2.17), we called the system (2.81) the Nash system of the game. If and when this system is solved, the feedback functions $\phi^i(t, \boldsymbol{x}) = q(\bar{\boldsymbol{x}} - x^i) - \partial_{x^i} V^i(t, \boldsymbol{x})$ should give the equilibrium Markovian strategies. Generally speaking, these systems of HJB equations are difficult to solve. Here, because of the particular forms of the couplings and the terminal conditions, we can solve the system by inspection, checking that a solution can be found in the form:

$$V^i(t, \boldsymbol{x}) = \frac{\eta_t}{2}(\bar{\boldsymbol{x}} - x^i)^2 + \chi_t, \quad (t, \boldsymbol{x}) \in [0, T] \times \mathbb{R}^N, \tag{2.82}$$

for some deterministic scalar functions $[0, T] \ni t \mapsto \eta_t \in \mathbb{R}$ and $[0, T] \ni t \mapsto \chi_t \in \mathbb{R}$ satisfying $\eta_T = c$ and $\chi_T = 0$ in order to match the terminal conditions for the functions $(V^i)_{1 \leqslant i \leqslant N}$. With this ansatz, the partial derivatives $\partial_{x^j} V^i$ and $\partial_{x^j x^k} V^i$ read:

$$\partial_{x^j} V^i(t, \boldsymbol{x}) = \eta_t\left(\frac{1}{N} - \delta_{i,j}\right)\left(\bar{\boldsymbol{x}} - x^i\right),$$
$$\partial^2_{x^j x^k} V^i(t, \boldsymbol{x}) = \eta_t\left(\frac{1}{N} - \delta_{i,j}\right)\left(\frac{1}{N} - \delta_{i,k}\right),$$

and plugging these expressions into (2.81), and identifying term by term, we see that the system of HJB equations is solved if and only if:

$$\begin{cases} \dot{\eta}_t = 2(a+q)\eta_t + \left(1 - \frac{1}{N^2}\right)\eta_t^2 - (\epsilon - q^2), \\ \dot{\chi}_t = -\frac{1}{2}\sigma^2(1-\rho^2)\left(1 - \frac{1}{N}\right)\eta_t, \end{cases} \tag{2.83}$$

for $t \in [0,T]$, with the terminal conditions $\eta_T = c$ and $\chi_T = 0$. As already explained earlier, the Riccati equation is scalar and can be solved explicitly. Here it coincides with (2.77), and following (2.78), we get:

$$\eta_t = \frac{-(\epsilon - q^2)\left(e^{(\delta^+ - \delta^-)(T-t)} - 1\right) - c\left(\delta^+ e^{(\delta^+ - \delta^-)(T-t)} - \delta^-\right)}{\left(\delta^- e^{(\delta^+ - \delta^-)(T-t)} - \delta^+\right) - c(1 - 1/N^2)\left(e^{(\delta^+ - \delta^-)(T-t)} - 1\right)}, \tag{2.84}$$

provided we set:

$$\delta^{\pm} = -(a+q) \pm \sqrt{R}, \qquad \text{with} \qquad R = (a+q)^2 + \left(1 - \frac{1}{N^2}\right)(\epsilon - q^2) > 0. \tag{2.85}$$

Once η_t is identified, one solves for χ_t (remember that $\chi_T = 0$) and finds:

$$\chi_t = \frac{1}{2}\sigma^2(1-\rho^2)\left(1 - \frac{1}{N}\right)\int_t^T \eta_s \, ds. \tag{2.86}$$

For the record, we note that the optimal Markovian strategies read:

$$\hat{\alpha}_t^i = q(\bar{X}_t - X_t^i) - \partial_{x^i} V^i(t, X_t) = \left(q + (1 - \frac{1}{N})\eta_t\right)(\bar{X}_t - X_t^i), \tag{2.87}$$

for $t \in [0,T]$, and the optimally controlled dynamics:

$$dX_t^i = \left(a + q + (1 - \frac{1}{N})\eta_t\right)(\bar{X}_t - X_t^i)dt + \sigma\left(\sqrt{1-\rho^2}dW_t^i + \rho dW_t^0\right), \tag{2.88}$$

for $t \in [0,T]$. As announced, we recover the solution found by the Pontryagin stochastic maximum principle.

2.6 Notes & Complements

The main purpose of this chapter was to present background material and notation for the analysis of finite player stochastic differential games. The published literature on general nonzero sum stochastic differential games is rather limited, especially in textbook form. Moreover, the terminology varies from one source to the next. In particular, there is no clear consensus on the names to give to the many notions of admissibility for strategy profiles and for the corresponding equilibria. The definitions we use in this text reflect our own personal biases. They are borrowed from Carmona's recent text [94]. The reader is referred to Chapter 5 of this book for proofs of the necessary part of the stochastic Pontryagin maximum principle, and detailed discussions of linear quadratic game models and applications to predatory trading.

The formulation of the Isaacs condition as given in Definition 2.9 is credited to Isaacs in the case of two-player ($N = 2$) zero-sum games, and to Friedman in the general case of noncooperative N-player games. Earlier results on the solvability of the Nash system (2.17)–(2.18) in the classical or strong sense and with bounded controls may be found in the monograph by Ladyzenskaja et al. [258] and in the paper by Friedman [163]. We also refer to the series of papers by Bensoussan and Frehse [45, 48, 49] for refined solvability properties and estimates for parabolic or elliptic Nash systems allowing for Hamiltonians of quadratic growth. For other monographs on semilinear PDEs, we refer to Friedman [162] and Lieberman [264]. The solvability property of the Nash system used in the proof of Proposition 2.13 may be explicitly found in Delarue and Guatteri [134]. The unique solvability of the SDE appearing in the same proof is taken from the seminal work by Veretennikov [336]. The Itô-Krylov formula is due to Krylov, see Chapter II in his monograph [242].

The stochastic maximum principle for stochastic differential games was used in the linear quadratic setting by Hamadène [193] and [194]. Generalizations have been considered by several authors, among which generalizations to games with stochastic dynamics including jumps or with partial observation. We refer the interested reader to [22] and [23] and the references therein. For further details on the stochastic maximum principle for stochastic optimal control problems, from which the stochastic maximum principle for games may be derived, we refer the reader to the subsequent Chapters 3, 4, 6, and (Vol II)-1: The standard version with deterministic coefficients is exposed in Chapters 3 and 4, while the case with random coefficients is addressed in Chapter (Vol II)-1; Chapter 6 is dedicated to the optimal control of McKean-Vlasov diffusion processes.

Our reasons to present the case $\beta = 0$ of the flocking model, and the systemic risk toy model (whose discussion is based on the paper by Carmona, Fouque, and Sun [102]), are mainly pedagogical. Indeed, in both cases, the open and closed loop forms of the models can be solved explicitly, and the large game limits appear effortlessly. So in this sense, they offer a perfect introduction to the discussion of mean field games, hence our decision to present them in full detail, despite their possible shortcomings. Indeed, the LQ form of the flocking model is rather unrealistic, and when viewed as a model for systemic risk in an interbank system, our toy model of systemic risk is very naive. Indeed, despite the strong case made in [102] for the relevance of the model to systemic risk of the banking system, it remains that according to this model, banks can borrow from each other without having to repay their debts, and even worse, the case for the model is further weakened by the fact that the liabilities of the banks are not included in the model. As already mentioned in the Notes & Complements of Chapter 1, the realism of the model was recently improved in [101] by Carmona, Fouque, Moussavi, and Sun who included delayed terms in the drift of the state to account for the fact that the decision to borrow or lend at a given time will have an impact down the road on the ability of a bank to borrow or lend.

Stochastic Differential Mean Field Games

3

Abstract

The goal of this chapter is to propose solutions to asymptotic forms of the search for Nash equilibria for large stochastic differential games with mean field interactions. We implement the Mean Field Game strategy, initially developed by Lasry and Lions in an analytic set-up, in a purely probabilistic framework. The roads to solutions go through a class of standard stochastic control problems followed by fixed point problems for flows of probability measures. We tackle the inherent stochastic optimization problems in two different ways. Once by representing the value function as the solution of a backward stochastic differential equation (reminiscent of the so-called weak formulation approach), and a second time using the Pontryagin stochastic maximum principle. In both cases, the optimization problem reduces to the solutions of a Forward-Backward Stochastic Differential Equation (FBSDE for short). The search for a fixed flow of probability measures turns the FBSDE into a system of equations of the McKean-Vlasov type where the distribution of the solution appears in the coefficients. In this way, both the optimization and interaction components of the problem are captured by a single FBSDE, avoiding the twofold reference to Hamilton-Jacobi-Bellman equations on the one hand, and to Kolmogorov equations on the other hand.

3.1 Notation, Assumptions, and Preliminaries

Here, we recall the basic results and ingredients from stochastic analysis and optimal stochastic control theory which we use throughout the chapter. We leverage the resources of Chapter 2 to formalize what we mean by a mean field game problem.

R. Carmona, F. Delarue, *Probabilistic Theory of Mean Field Games with Applications I*, Probability Theory and Stochastic Modelling 83,
DOI 10.1007/978-3-319-58920-6_3

3.1.1 The N Player Game

We consider a stochastic differential game with N players. As usual the players are denoted by the integers $i \in \{1, \cdots, N\}$. Each player is controlling its own private state $X_t^i \in \mathbb{R}^d$ at time $t \in [0, T]$ by taking an action α_t^i in a closed convex set $A \subset \mathbb{R}^k$. We assume that the dynamics of the private states of the individual players are given by Itô's stochastic differential equations of the form:

$$dX_t^i = b(t, X_t^i, \bar{\mu}_{X_t^{-i}}^{N-1}, \alpha_t^i)dt + \sigma(t, X_t^i, \bar{\mu}_{X_t^{-i}}^{N-1}, \alpha_t^i)dW_t^i, \qquad t \in [0, T], \tag{3.1}$$

for $i = 1, \cdots, N$, where the $(\boldsymbol{W}^i = (W_t^i)_{0 \leqslant t \leqslant T})_{i=1,\cdots,N}$ are m-dimensional independent Wiener processes. Often in the text, we choose $m = d$ for simplicity. The function $(b, \sigma) : [0, T] \times \mathbb{R}^d \times \mathcal{P}(\mathbb{R}^d) \times A \to \mathbb{R}^d \times \mathbb{R}^{d \times m}$ is deterministic and satisfies assumptions to be spelled out later on, and $\bar{\mu}_{X_t^{-i}}^{N-1}$ denotes the empirical distribution of the states X_t^{-i}, namely the states X_t^j for $j \neq i$. This is a slight departure from our earlier discussion of finite player games with mean field interactions in which the empirical distribution was typically assumed to be the empirical distribution of the couples "state/control." We saw in Chapter 1 several instances of models for which the interactions appeared through the empirical distributions of the controls, or even through the empirical distributions of the couples "state/control." We shall provide in Section 4.6 of Chapter 4 insight and tools to handle some of these more general classes of mean field games which we call extended mean field games.

Recall that the symmetry and small individual influence conditions articulated in Chapters 1 and 2 have been incorporated in the model through the choice of the form of the coefficients of the states dynamics. Indeed, the dimensions of the states and the random shocks, as well as the drift and volatility coefficients b and σ are the same for all the players. Moreover, since we want the influence of the players $j \neq i$ on the state of player i to be symmetric and diminishing quickly as the number of players grows, we used the intuition behind the result of Lemma 1.2 to assume that the coefficients are given by functions of measures. In this way, the state of player i is influenced by the empirical distribution of the states of the other players.

Remark 3.1 *Later on, we shall add a term of the form* $\sigma^0(t, X_t^i, \bar{\mu}_{X_t^{-i}}^{N-1}, \alpha_t^i)dW_t^0$ *to the right-hand side of the state dynamics given by* (3.1). *For obvious reasons, the Wiener process* $\boldsymbol{W}^0$ *will be called a common noise as opposed to the Wiener processes* $\boldsymbol{W}^i$ *for* $i = 1, \cdots, N$ *which are intrinsic to the private states and called idiosyncratic noises.*

In this chapter, we concern ourselves with both open and closed loop equilibria, without paying much attention to the differences between the two cases since, in our framework, the asymptotic formulations are expected to be the same in the limit $N \to \infty$. So, whatever the type of the equilibrium, each player chooses a strategy in the space $\mathbb{A}$ of progressively measurable A-valued stochastic processes $\boldsymbol{\alpha} = (\alpha_t)_{0 \leqslant t \leqslant T}$ satisfying the admissibility condition:

$$\mathbb{E}\left[\int_0^T |\alpha_t|^2 dt\right] < +\infty. \tag{3.2}$$

As explained in Chapter 2, the choice of a strategy is driven by the desire to minimize an expected cost over a period $[0, T]$, each individual cost being a combination of running and terminal costs. For each $i \in \{1, \cdots, N\}$, the running cost to player i is given by a measurable function $f^i : [0, T] \times \mathbb{R}^d \times \mathcal{P}(\mathbb{R}^d) \times A \to \mathbb{R}$ and the terminal cost by a measurable function $g^i : \mathbb{R}^d \times \mathcal{P}(\mathbb{R}^d) \to \mathbb{R}$ in such a way that if the N players use the strategy profile $\boldsymbol{\alpha} = (\boldsymbol{\alpha}^1, \cdots, \boldsymbol{\alpha}^N) \in \mathbb{A}^N$, the expected total cost to player i is:

$$J^i(\boldsymbol{\alpha}) = \mathbb{E}\left[\int_0^T f^i\bigl(t, X_t^i, \bar{\mu}_{X_t^{-i}}^{N-1}, \alpha_t^i\bigr) dt + g^i\bigl(X_T^i, \bar{\mu}_{X_T^{-i}}^{N-1}\bigr)\right]. \tag{3.3}$$

Quite often, we denote by $\mathbb{A}^N$ the product of N copies of $\mathbb{A}$. We shall also use the notation $J^{N,i}$ when we want to emphasize the dependence upon the number N of players. This will be the case when we study the limit $N \to \infty$ in Chapter (Vol II)-6. Notice that even though only α_t^i appears in the formula giving the cost to player i, this cost depends upon the strategies used by the other players indirectly, as these strategies affect not only the private state X_t^i, but also the empirical distribution $\bar{\mu}_{X_t^{-i}}^{N-1}$ of the private states of the other players. As emphasized in Chapter 1, we restrict ourselves to games with strong symmetry properties and our models require that the behaviors of the players be *statistically identical* when driven by controls which are *statistically invariant* under permutation, imposing that the running and terminal cost functions f^i and g^i, like the drift and volatility coefficients, do not depend upon i. We denote them by f and g respectively.

The final remark of this introductory subsection is related to the actual definition of the mean field interaction between finitely many players. In accordance with earlier discussions, the empirical measure which appears in (3.1) and (3.3) is the empirical measure of the *other states*, namely of the variables X_t^j for $j \neq i$. However, as we already explained in several instances, if we were to use instead the empirical measure $\bar{\mu}_t^N$ of *all the states* X_t^j including $j = i$, the results would be qualitatively the same, though possibly different quantitatively. This was highlighted in Remark 1.19 and Remark 1.25 of Chapter 1 where the net effect of switching from one empirical measure to the other amounts to applying multiplicative factors on the parameters of the models. We also argued that these multiplicative factors were converging to 1 as $N \to \infty$. Since the mean field game problems which we study throughout the book are essentially limits as $N \to \infty$ of N-player stochastic differential games with mean field interactions (see Chapter (Vol II)-6 for rigorous proofs), the convention we use for the empirical measures in the finite player games should not matter in the end. For this reason, we shall often start from the empirical measure $\bar{\mu}_t^N$ of the states of all the players when we motivate the formulation of mean field game models.

3.1.2 The Mean Field Game Problem

We now formalize the definition of the Mean Field Game problem without a common noise. For this purpose, we start with a complete filtered probability space $(\Omega, \mathcal{F}, \mathbb{F} = (\mathcal{F}_t)_{0 \leqslant t \leqslant T}, \mathbb{P})$, the filtration $\mathbb{F}$ supporting a d-dimensional Wiener process $\boldsymbol{W} = (W_t)_{0 \leqslant t \leqslant T}$ with respect to $\mathbb{F}$ and an initial condition $\xi \in L^2(\Omega, \mathcal{F}_0, \mathbb{P}; \mathbb{R}^d)$. As announced in Subsection 3.1.1, we thus choose $\boldsymbol{W}$ of the same dimension as the state variable. This is for convenience only. Most of the time, the filtration $\mathbb{F}$ will be chosen as the filtration generated by $\mathcal{F}_0$ and $\boldsymbol{W}$. As usual, the law of ξ is denoted by $\mathcal{L}(\xi)$. From a practical point of view, $\mu^0 = \mathcal{L}(\xi)$ should be understood as the initial distribution of the population. Following the notations introduced in the previous subsection, we shall denote by $\mathbb{A}$ the set of $\mathbb{F}$-progressively measurable A-valued stochastic processes $\boldsymbol{\alpha} = (\alpha_t)_{0 \leqslant t \leqslant T}$ that satisfy the square-integrability condition (3.2).

In the present context, the mean field game problem derived from the finite player game model introduced in the previous section is articulated in the following way:

(i) For each fixed deterministic flow $\boldsymbol{\mu} = (\mu_t)_{0 \leqslant t \leqslant T}$ of probability measures on $\mathbb{R}^d$, solve the standard stochastic control problem:

$$
\begin{gathered}
\inf_{\boldsymbol{\alpha} \in \mathbb{A}} J^{\boldsymbol{\mu}}(\boldsymbol{\alpha}) \text{ with } J^{\boldsymbol{\mu}}(\boldsymbol{\alpha}) = \mathbb{E}\left[\int_0^T f(t, X_t^{\boldsymbol{\alpha}}, \mu_t, \alpha_t) dt + g(X_T^{\boldsymbol{\alpha}}, \mu_T) \right], \\
\text{subject to} \\
\begin{cases} dX_t^{\boldsymbol{\alpha}} = b(t, X_t^{\boldsymbol{\alpha}}, \mu_t, \alpha_t) dt + \sigma(t, X_t^{\boldsymbol{\alpha}}, \mu_t, \alpha_t) dW_t, & t \in [0, T], \\ X_0^{\boldsymbol{\alpha}} = \xi. \end{cases}
\end{gathered}
\tag{3.4}
$$

(ii) Find a flow $\boldsymbol{\mu} = (\mu_t)_{0 \leqslant t \leqslant T}$ such that $\mathcal{L}(\hat{X}_t^{\boldsymbol{\mu}}) = \mu_t$ for all $t \in [0, T]$, if $\hat{\boldsymbol{X}}^{\boldsymbol{\mu}}$ is a solution of the above optimal control problem.

Notice that here, X_t represents the private state of a representative player, not the whole system as before. Recasting these two steps in the set-up of finite player games and the concept of Nash equilibrium, we see that the first step provides the best response of a given player interacting with the statistical distribution of the states of the other players if this statistical distribution is assumed to be given by μ_t, while the second step solves a specific fixed point problem in the spirit of the search for fixed points of the best response function. The strategy outlined by these two steps parallels exactly what needs to be done to construct Nash equilibria for finite player games. Once these two steps have been taken successfully, if the *fixed-point* optimal control $\hat{\boldsymbol{\alpha}}^{\boldsymbol{\mu}}$ identified in step (ii) is in feedback form, in the sense that it is of the form $\alpha_t^{\boldsymbol{\mu}} = \phi(t, \hat{X}_t^{\boldsymbol{\mu}}, \mu_t)$ for some deterministic function ϕ on $[0, T] \times \mathbb{R}^d \times \mathcal{P}(\mathbb{R}^d)$, where $\boldsymbol{\mu} = (\mu_t = \mathcal{L}(\hat{X}_t^{\boldsymbol{\mu}}))_{0 \leqslant t \leqslant T}$ is the flow of marginal distributions at the fixed point, we expect that the prescription $\hat{\alpha}_t^i = \phi(t, X_t^i, \mu_t)$, if used by the players $i = 1, \cdots, N$ of a large game, should form an approximate Nash equilibrium. This fact will be proven rigorously in Chapter (Vol II)-6, where we also quantify the accuracy of the approximation.

Remark 3.2 *Throughout the book, we shall consider the case when the optimization problem* $\inf_{\boldsymbol{\alpha}\in\mathbb{A}} J^{\boldsymbol{\mu}}(\boldsymbol{\alpha})$ *has a unique minimizer for any input* $\boldsymbol{\mu}$*. In that case, we denote by* $\hat{\boldsymbol{X}}^{\boldsymbol{\mu}}$ *the unique optimal trajectory under the input* $\boldsymbol{\mu}$ *and* $\boldsymbol{\mu}$ *is said to be an equilibrium (or a solution of the mean field game) if* $\mu_t = \mathcal{L}(\hat{X}_t^{\boldsymbol{\mu}})$ *for all* $t \in [0, T]$*.*

When the optimization problem $\inf_{\boldsymbol{\alpha}\in\mathbb{A}} J^{\boldsymbol{\mu}}(\boldsymbol{\alpha})$ *has several solutions, the two steps (i) and (ii) may be reformulated as follows. Denoting the set of minimizing controls by* $\hat{\mathbb{A}}^{\boldsymbol{\mu}} = \operatorname{argmin}_{\boldsymbol{\alpha}\in\mathbb{A}} J^{\boldsymbol{\mu}}(\boldsymbol{\alpha})$*,* $\boldsymbol{\mu}$ *is said to be an equilibrium if there exists* $\hat{\boldsymbol{\alpha}} \in \hat{\mathbb{A}}^{\boldsymbol{\mu}}$ *such that, for all* $t \in [0, T]$*,* $\mu_t = \mathcal{L}(X_t^{\hat{\boldsymbol{\alpha}}})$*. However, we shall not consider this level of generality in the book.*

3.1.3 An Alternative Description of the Mean Field Game Problem

If our goal is to study the limiting MFG problem more than solving the finite player games from which it is issued, a possible alternative introduction of the problem may be useful. We motivate this approach by the limit of finite player games, but it should be understood that the finite player games we are about to introduce are different from the games we started from to derive the MFG problem.

We framed the search for Nash equilibria as a search for fixed points of the best response map. We exploit this point of view systematically throughout the book and our formulation of the mean field game problem was strongly influenced by this approach. It naturally leads to a search for fixed points on flows of probability measures. The present discussion will remain informal, as we do not spend much effort providing explicit definitions of all the objects we manipulate. As before, we assume that the N players use controls $(\boldsymbol{\alpha}^i)_{i=1,\cdots,N}$ given by deterministic functions $(\phi^i)_{i=1,\cdots,N}$. We shall denote by $\bar{\mu}_t^N$ the empirical distribution of $X_t = (X^1, \cdots, X_t^N)$ at time t, and assume that the controls used by the players are of the form:

$$\alpha_t^i = \phi^i(t, X_t^i, \bar{\mu}_t^N), \quad i = 1, \cdots, N,$$

when we search for distributed Markovian equilibria, or $\alpha_t^i = \phi^i(t, X_{[0,t]}^i, \bar{\mu}_t^N)$ when we aim at distributed feedback controls, or $\alpha_t^i = \phi^i(t, W_{[0,t]}^i, \bar{\mu}_t^N)$ in the case of distributed open loop controls. Given the form of the coefficients, we expect that in the limit $N \to \infty$, all the players will use the same functions ϕ^i in equilibrium. So instead of fixing a specific player, say $i \in \{1, \cdots, N\}$, and try to find its best response to the other players $j \neq i$ assuming they chose their strategies $\boldsymbol{\alpha}^{-i}$, because of the symmetries among the players and the fact that we shall eventually consider the case $N \to \infty$, we might as well assume that all the N players have chosen their strategies $\boldsymbol{\alpha} = (\boldsymbol{\alpha}^1, \cdots, \boldsymbol{\alpha}^N)$ given by the same function ϕ, and we try to determine the best response, say $\bar{\boldsymbol{\alpha}}$ of a virtual $(N+1)$-th player (which we will refer to using the index 0), interacting with the empirical distribution of the states of the N players. In other words

- We fix a feedback function $\phi : (t, x, \mu) \mapsto \phi(t, x, \mu)$ to determine the strategies $\alpha_t^i = \phi(t, X_t^i, \bar{\mu}_t^N)$, or $\alpha_t^i = \phi(t, X_{[0,t]}^i, \bar{\mu}_t^N)$, or $\alpha_t^i = \phi(t, W_{[0,t]}^i, \bar{\mu}_t^N)$ depending upon the case we study.
- Next we solve the system of N coupled stochastic differential equations

$$dX_t^i = b(t, X_t^i, \bar{\mu}_t^N, \alpha_t^i)dt + \sigma(t, X_t^i, \bar{\mu}_t^N, \alpha_t^i)dW_t^i, \quad i = 1, \cdots, N,$$

  and we treat $X_t^1, \cdots, X_t^N$ as the states of the N players of the game.
- Given the flow $(\bar{\mu}_t^N)_{0 \leq t \leq T}$ of (random) empirical measures of the $X_t^1, \cdots, X_t^N$ just computed, we now introduce a virtual player which looks for a control $\boldsymbol{\alpha}^0$ given by a feedback function ϕ^0 which bears to $\boldsymbol{\alpha}^0$ the same relationship as ϕ bears to $\boldsymbol{\alpha}^i$ for $i = 1, \cdots, N$, in order to minimize:

$$J(\boldsymbol{\alpha}^0) = \mathbb{E}\left[\int_0^T f(t, X_t, \bar{\mu}_t^N, \alpha_t^0)dt + g(X_T, \bar{\mu}_T^N)\right]$$

  under the dynamical constraint

$$dX_t = b(t, X_t, \bar{\mu}_t^N, \alpha_t^0)dt + \sigma(t, X_t, \bar{\mu}_t^N, \alpha_t^0)dW_t,$$

  for a Wiener process $\boldsymbol{W}$ independent of the Wiener processes $\boldsymbol{W}^i$ for $i = 1, \cdots, N$. Here, we assume that the feedback function ϕ^0 bears to $\boldsymbol{\alpha}^0$ the same relationship as ϕ bears to $\boldsymbol{\alpha}^i$ for $i = 1, \cdots, N$ with $\boldsymbol{X}$ and $\boldsymbol{W}$ in lieu of $\boldsymbol{X}^i$ and $\boldsymbol{W}^i$.
- Denoting by $\boldsymbol{\alpha}^{0,*}(\boldsymbol{\alpha})$ the set of optimal Markov controls $\boldsymbol{\alpha}^0$ (or equivalently by $\phi^{0,*}(\phi)$ the set of optimal feedback functions ϕ^0) for the virtual player, the set of controls $\boldsymbol{\alpha}^{0,*}(\boldsymbol{\alpha})$ (resp. $\phi^{0,*}(\phi)$) plays the role of the best response to the control $\boldsymbol{\alpha}$ (resp. ϕ) of the virtual player interacting with the flow of empirical distributions $(\bar{\mu}_t^N)_{0 \leq t \leq T}$ of the N players $i = 1, \cdots, N$.

We now explain how in the limit $N \to \infty$ of large games, the above steps lead to the mean field game paradigm introduced in the previous subsection.

First we notice that the first two bullet points do not involve any optimization. Anticipating on the several discussions of the propagation of chaos which we give in Chapters 4, 5, (Vol II)-2 and (Vol II)-7, we realize that when $N \to \infty$, the $(\boldsymbol{X}^i)_{i=1,\cdots,N}$ become independent of each other, and their marginal distributions converge toward the law of the solution of the McKean-Vlasov equation:

$$d\tilde{X}_t = b(t, \tilde{X}_t, \mathcal{L}(\tilde{X}_t), \tilde{\alpha}_t)dt + \sigma(t, \tilde{X}_t, \mathcal{L}(\tilde{X}_t), \tilde{\alpha}_t)d\tilde{W}_t,$$

where $\mathcal{L}(\tilde{X}_t)$ denotes the law of the random element $\tilde{X}_t$, $\tilde{\boldsymbol{W}}$ is a Wiener process, and $\tilde{\boldsymbol{\alpha}} = (\tilde{\alpha}_t)_{0 \leq t \leq T}$ is computed from the function ϕ and the processes $\tilde{\boldsymbol{X}}$ and $\tilde{\boldsymbol{W}}$. In some sense, in the limit $N \to \infty$, the role of the first two bullet points is to associate, to each control $\boldsymbol{\alpha}$ or function ϕ, a flow $\boldsymbol{\mu} = (\mu_t)_{0 \leqslant t \leqslant T}$ of probability given by the marginal laws $\mu_t = \mathcal{L}(\tilde{X}_t)$ of the solution of the above McKean-Vlasov stochastic differential equation.

Once this flow of measures is obtained, the third bullet point proposes a standard optimal control problem (still not a game) in which a virtual player minimizes its expected costs in interaction with the flow of distributions. This optimization problem is exactly the same as (3.4) except for the fact that the input flow of probability measures $\boldsymbol{\mu} = (\mu_t)_{0 \leqslant t \leqslant T}$ is not arbitrary. Instead, this flow is given by the marginal laws of the solution of a McKean-Vlasov stochastic differential equation whose coefficients are determined by the choice of a control $\boldsymbol{\alpha}$ or a function ϕ. Clearly, if the fixed point problem can be solved, its solution provides a solution to the mean field game problem as articulated in the previous subsection. Conversely, any solution to the mean field game problem of the previous section provides a solution to the problem stated in the above bullet points.

The present formulation of the mean field game problem exhibits two useful features.

1. It offers an alternative to the fixed point step by formulating it in a space of controls instead of flows of measures.
2. It highlights from the start the fact that, because we are interested in large games and mean field interactions, the state dynamics are necessarily of a McKean-Vlasov nature, fact which is not clear from the previous formulation.

3.1.4 The Hamiltonian

As already emphasized, we shall assume (unless stated otherwise) that A is a closed convex subset of $\mathbb{R}^k$. This makes it easier to minimize the Hamiltonian when the running cost f is convex in α. Here, the full-fledged Hamiltonian has the form:

$$H(t, x, \mu, y, z, \alpha) = b(t, x, \mu, \alpha) \cdot y + \sigma(t, x, \mu, \alpha) \cdot z + f(t, x, \mu, \alpha),$$

for $t \in [0, T]$, $x, y \in \mathbb{R}^d$, $z \in \mathbb{R}^{d \times d}$, $\alpha \in A$ and $\mu \in \mathcal{P}(\mathbb{R}^d)$. Above, the dots '$\cdot$' stand for the inner products in $\mathbb{R}^d$ and $\mathbb{R}^{d \times d}$ respectively.

In order to lighten the notation and avoid unwanted technicalities at this early stage of the discussion, we also assume throughout the chapter (and actually throughout most of the book) that the volatility is uncontrolled (*i.e.* does not depend upon the value of the control). In other words we assume that:

$$\sigma(t, x, \mu, \alpha) = \sigma(t, x, \mu).$$

In fact, for some of the derivations in this chapter, we shall sometimes assume that the volatility is also independent of μ or, even, that it is a constant matrix $\sigma \in \mathbb{R}^{d \times d}$

The fact that the volatility is uncontrolled allows us to use the *reduced Hamiltonian* defined as:

$$H(t, x, \mu, y, \alpha) = b(t, x, \mu, \alpha) \cdot y + f(t, x, \mu, \alpha), \tag{3.5}$$

for $t \in [0, T]$, $x, y \in \mathbb{R}^d$, $\alpha \in A$ and $\mu \in \mathcal{P}(\mathbb{R}^d)$. As in Chapter 2, we use the same letter H for the full-fledged and the reduced Hamiltonians. The context and the variables appearing as arguments specify which Hamiltonian we are using: it is the reduced Hamiltonian when there is only one single adjoint variable, and it is the regular Hamiltonian otherwise.

Our first task will be to minimize the reduced Hamiltonian with respect to the control parameter; in other words, to search for a function $(t, x, \mu, y) \mapsto \hat{\alpha}(t, x, \mu, y)$ satisfying:

$$\hat{\alpha}(t, x, \mu, y) \in \mathrm{argmin}_{\alpha \in A} H(t, x, \mu, y, \alpha), \tag{3.6}$$

and understand how such a minimizer depends upon its variables.

It will be convenient to use the following spaces of probability measures. Some of them already appeared in earlier chapters. They will be studied in great detail in Chapter 5. Here and in the following, whenever E is a separable metric space and $p \geqslant 1$, we denote by $\mathcal{P}_p(E)$ the subspace of $\mathcal{P}(E)$ of probability measures of order p, having a finite moment of order p meaning that the p-th power of the distance to a fixed point of E is integrable. Obviously, the specific choice of this fixed point is irrelevant. When E is a normed space, $\mu \in \mathcal{P}_p(E)$ if $\mu \in \mathcal{P}(E)$ and

$$M_{p,E}(\mu) = \left(\int_E \|x\|_E^p d\mu(x) \right)^{1/p} < \infty. \tag{3.7}$$

We simply write M_p whenever E is a Euclidean space. Bounded subsets of $\mathcal{P}_p(E)$ are defined as sets of probability measures with uniformly bounded moments of order p, in other words we call bounded subsets of $\mathcal{P}_p(E)$, subsets on which $M_{p,E}$ is bounded.

Most of the time, the measure argument in the coefficients b, f and g is taken in $\mathcal{P}_2(\mathbb{R}^d)$ instead of $\mathcal{P}(\mathbb{R}^d)$, which is precisely what we do in the next set of assumptions, under which we will be able to minimize the Hamiltonian and control the properties of the minimizer:

Assumption (Minimization of the Hamiltonian). The coefficients b and f are defined on $[0, T] \times \mathbb{R}^d \times \mathcal{P}_2(\mathbb{R}^d) \times \mathbb{R}^k$ and $[0, T] \times \mathbb{R}^d \times \mathcal{P}_2(\mathbb{R}^d) \times A$ respectively and satisfy:

(continued)

(A1) The drift b is an affine function of α of the form

$$b(t,x,\mu,\alpha) = b_1(t,x,\mu) + b_2(t)\alpha, \tag{3.8}$$

where $[0,T] \ni t \mapsto b_2(t) \in \mathbb{R}^{d\times k}$ is measurable and bounded, and the mapping $[0,T] \times \mathbb{R}^d \times \mathcal{P}_2(\mathbb{R}^d) \ni (t,x,\mu) \mapsto b_1(t,x,\mu) \in \mathbb{R}^d$ is measurable and bounded on bounded subsets of $[0,T] \times \mathbb{R}^d \times \mathcal{P}_2(\mathbb{R}^d)$.

(A2) There exist two constants $\lambda > 0$ and $L \geqslant 0$ such that for any $t \in [0,T]$ and $\mu \in \mathcal{P}_2(\mathbb{R}^d)$, the function $\mathbb{R}^d \times A \ni (x,\alpha) \mapsto f(t,x,\mu,\alpha) \in \mathbb{R}$ is once continuously differentiable with respect to α, the derivative being L- Lipschitz-continuous in x and α. Moreover, it satisfies the following strong form of convexity:

$$f(t,x,\mu,\alpha') - f(t,x,\mu,\alpha) - (\alpha' - \alpha)\cdot \partial_\alpha f(t,x,\mu,\alpha) \geqslant \lambda|\alpha' - \alpha|^2. \tag{3.9}$$

Finally, f and $\partial_\alpha f$ are locally bounded over $[0,T] \times \mathbb{R}^d \times \mathcal{P}_2(\mathbb{R}^d) \times A$.

The minimization of the Hamiltonian is taken care of by the following result.

Lemma 3.3 *If A is closed and convex and assumption* **Minimization of the Hamiltonian** *is in force, then for each $(t,x,\mu,y) \in [0,T] \times \mathbb{R}^d \times \mathcal{P}_2(\mathbb{R}^d) \times \mathbb{R}^d$, there exists a unique minimizer $\hat\alpha(t,x,\mu,y)$ of H in A. Moreover, the function $[0,T] \times \mathbb{R}^d \times \mathcal{P}_2(\mathbb{R}^d) \times \mathbb{R}^d \ni (t,x,\mu,y) \mapsto \hat\alpha(t,x,\mu,y) \in A$ is measurable, locally bounded and Lipschitz-continuous with respect to (x,y), uniformly in $(t,\mu) \in [0,T] \times \mathcal{P}_2(\mathbb{R}^d)$, with a Lipschitz constant depending only upon λ, the supremum norm of b_2 and the Lipschitz constant of $\partial_\alpha f$ in x. In fact, an explicit upper bound for $\hat\alpha$ reads:*

$$\begin{aligned} &\forall (t,x,\mu,y) \in [0,T] \times \mathbb{R}^d \times \mathcal{P}_2(\mathbb{R}^d) \times \mathbb{R}^d, \\ &\qquad |\hat\alpha(t,x,\mu,y)| \leqslant \lambda^{-1}\big(|\partial_\alpha f(t,x,\mu,\beta_0)| + |b_2(t)|\,|y|\big) + |\beta_0|, \end{aligned} \tag{3.10}$$

where β_0 is any arbitrary point in A.

Proof. For any given (t,x,μ,y), the function $A \ni \alpha \mapsto H(t,x,\mu,y,\alpha)$ is once continuously differentiable and strictly convex so that $\hat\alpha(t,x,\mu,y)$ appears as the unique solution of the variational inequality:

$$\forall \beta \in A, \quad \big(\beta - \hat\alpha(t,x,\mu,y)\big)\cdot \partial_\alpha H\big(t,x,\mu,y,\hat\alpha(t,x,\mu,y)\big) \geqslant 0. \tag{3.11}$$

By strict convexity, measurability of the minimizer $\hat\alpha(t,x,\mu,y)$ is a consequence of the *gradient descent* algorithm with convex constraints. See the Notes & Complements at the end of the chapter for references. Local boundedness of $\hat{\alpha}(t,x,\mu,y)$ also follows from strict convexity since by (3.9), for any arbitrary point $\beta_0 \in A$,

$$\begin{aligned}H(t,x,\mu,y,\beta_0) &\geqslant H(t,x,\mu,y,\hat{\alpha}(t,x,\mu,y))\\&\geqslant H(t,x,\mu,y,\beta_0)\\&\quad + \big(\hat{\alpha}(t,x,\mu,y)-\beta_0\big)\cdot\partial_\alpha H(t,x,\mu,y,\beta_0) + \lambda\big|\hat{\alpha}(t,x,\mu,y)-\beta_0\big|^2,\end{aligned}$$

so that:

$$\big|\beta_0 - \hat{\alpha}(t,x,\mu,y)\big| \leqslant \lambda^{-1}\big(|\partial_\alpha f(t,x,\mu,\beta_0)| + |b_2(t)|\,|y|\big),$$

and consequently:

$$\big|\hat{\alpha}(t,x,\mu,y)\big| \leqslant \lambda^{-1}\big(|\partial_\alpha f(t,x,\mu,\beta_0)| + |b_2(t)|\,|y|\big) + |\beta_0|,$$

which proves the local boundedness claim since β_0 is arbitrary, $\partial_\alpha f$ is locally bounded, and b_2 is bounded.

The smoothness of $\hat{\alpha}$ with respect to x and y follows from a suitable adaptation of the implicit function theorem to variational inequalities driven by coercive functionals. Indeed, for $x,x',y,y' \in \mathbb{R}^d$ and $(t,\mu) \in [0,T]\times\mathcal{P}_2(\mathbb{R}^d)$, we have the two inequalities:

$$\begin{aligned}&\big(\hat{\alpha}(t,x',\mu,y')-\hat{\alpha}(t,x,\mu,y)\big)\cdot\partial_\alpha H\big(t,x,\mu,y,\hat{\alpha}(t,x,\mu,y)\big) \geqslant 0,\\&\big(\hat{\alpha}(t,x,\mu,y)-\hat{\alpha}(t,x',\mu,y')\big)\cdot\partial_\alpha H\big(t,x',\mu,y',\hat{\alpha}(t,x',\mu,y')\big) \geqslant 0.\end{aligned}$$

Summing these inequalities, we get:

$$\begin{aligned}&\big(\hat{\alpha}(t,x',\mu,y')-\hat{\alpha}(t,x,\mu,y)\big)\\&\qquad\cdot\Big(\partial_\alpha H\big(t,x',\mu,y',\hat{\alpha}(t,x',\mu,y')\big)-\partial_\alpha H\big(t,x,\mu,y,\hat{\alpha}(t,x,\mu,y)\big)\Big) \leqslant 0,\end{aligned}$$

that is:

$$\begin{aligned}&\big(\hat{\alpha}(t,x',\mu,y')-\hat{\alpha}(t,x,\mu,y)\big)\\&\qquad\cdot\Big(\partial_\alpha H\big(t,x,\mu,y,\hat{\alpha}(t,x',\mu,y')\big)-\partial_\alpha H\big(t,x,\mu,y,\hat{\alpha}(t,x,\mu,y)\big)\Big)\\&\quad\leqslant \big(\hat{\alpha}(t,x',\mu,y')-\hat{\alpha}(t,x,\mu,y)\big)\\&\qquad\cdot\Big(\partial_\alpha H\big(t,x,\mu,y,\hat{\alpha}(t,x',\mu,y')\big)-\partial_\alpha H\big(t,x',\mu,y',\hat{\alpha}(t,x',\mu,y')\big)\Big).\end{aligned}$$

Exchanging the roles of α and α' in (3.9) and summing the resulting bounds, we check that for any $\alpha,\alpha' \in A$,

$$(\alpha'-\alpha)\cdot\big(\partial_\alpha f(t,x,\mu,\alpha')-\partial_\alpha f(t,x,\mu,\alpha)\big) \geqslant 2\lambda|\alpha'-\alpha|^2.$$

Using the two previous inequalities together with the fact that b is linear in α, we deduce that:

$$
\begin{aligned}
2\lambda \big|\hat{\alpha}(t,x',\mu,y') - \hat{\alpha}(t,x,\mu,y)\big|^2
&\leqslant \big(\hat{\alpha}(t,x',\mu,y') - \hat{\alpha}(t,x,\mu,y)\big) \\
&\qquad \cdot \Big(\partial_\alpha H\big(t,x,\mu,y,\hat{\alpha}(t,x',\mu,y')\big) - \partial_\alpha H\big(t,x',\mu,y',\hat{\alpha}(t,x',\mu,y')\big)\Big) \\
&\leqslant C\big|\hat{\alpha}(t,x',\mu,y') - \hat{\alpha}(t,x,\mu,y)\big|\big(|x'-x| + |y'-y|\big),
\end{aligned}
$$

where C only depends upon the bound for b_2 and the Lipschitz-constant of $\partial_\alpha f$ as a function of x. □

Remark 3.4 *Various generalizations of Lemma 3.3 to cases for which b_2 is allowed to depend upon x and μ are possible. For the sake of simplicity, we refrain from giving such generalizations as we shall most often focus on the case where b_2 is a function of the sole variable t. Indeed, in this case, the whole drift b in* (3.8) *is Lipschitz continuous in the variable x whenever b_1 is itself Lipschitz continuous in x. This assumption on b_2 may be rather restrictive for some practical applications. In order to consider more general models, the reader may want to reformulate some of the results proven in this chapter (and the next one) in the more general case where b_2 is a function of (t,x,μ) (or of (t,μ)). In doing so, he/she must pay particular attention to the regularity of the whole drift b.*

3.1.5 The Analytic Approach to MFGs

Going back to the program (i)–(ii) articulated in Subsection 3.1.2, the first step consists in solving a standard stochastic control problem when the deterministic flow $\boldsymbol{\mu} = (\mu_t)_{0\leqslant t\leqslant T}$ of probability measures is given and *frozen*. A natural route is to express the value function of the optimization problem (3.4) as the solution of the corresponding Hamilton-Jacobi-Bellman (HJB for short) equation. This is the keystone of the analytic approach to the MFG theory, the matching problem (ii) being resolved by coupling the HJB equation with a Kolmogorov equation intended to identify the flow $\boldsymbol{\mu} = (\mu_t)_{0\leqslant t\leqslant T}$ with the flow of marginal distributions of the optimal states. With the same notation $\hat{\alpha}$ as above for the minimizer of the reduced Hamiltonian H, the resulting system of PDEs can be written as:

$$
\begin{cases}
\partial_t V(t,x) + \dfrac{1}{2}\mathrm{trace}\Big[\big(\sigma\sigma^\dagger\big)(t,x,\mu_t)\partial^2_{xx}V(t,x)\Big] \\
\qquad\qquad + H\Big(t,x,\mu_t,\partial_x V(t,x),\hat{\alpha}(t,x,\mu_t,\partial_x V(t,x))\Big) = 0, \\
\partial_t \mu_t - \dfrac{1}{2}\mathrm{trace}\Big[\partial^2_{xx}\Big(\big(\sigma\sigma^\dagger\big)(t,x,\mu_t)\mu_t\Big)\Big] \\
\qquad\qquad + \mathrm{div}_x\Big(b\big(t,x,\mu_t,\hat{\alpha}(t,x,\mu_t,\partial_x V(t,x))\big)\mu_t\Big) = 0,
\end{cases}
\tag{3.12}
$$

in $[0,T]\times\mathbb{R}^d$, with $V(T,\cdot) = g(\cdot,\mu_T)$ as terminal condition for the first equation and $\mu_0 = \mu^0$ as initial condition for the second (recall (3.4) for the meaning of μ^0).

The first equation is the HJB equation of the stochastic control problem when the flow $\boldsymbol{\mu} = (\mu_t)_{0\leqslant t\leqslant T}$ is frozen, see for instance Lemma 4.47. Notice that, as pointed out earlier, this equation can be written using the reduced Hamiltonian H instead of the usual minimized operator symbol because the volatility is not controlled and because H is assumed to have a minimizer. The existence of $\hat{\alpha}$ is especially useful as it provides the form of the optimal feedback function, which reads $[0,T] \times \mathbb{R}^d \ni (t,x) \mapsto \hat{\alpha}(t,x,\partial_x V(t,x))$. The second equation is the Kolmogorov (sometimes referred to Fokker-Planck) equation giving the time evolution of the flow $\boldsymbol{\mu} = (\mu_t)_{0\leqslant t\leqslant T}$ of measures dictated by the dynamics (3.4) of the *state of the system* once we have implemented the optimal feedback function. These two equations are coupled by the fact that the Hamiltonian appearing in the HJB equation is a function of the measure μ_t at time t and the drift appearing in the Kolmogorov equation is a function of the gradient of the value function V. Notice that the first equation is a backward equation to be solved from a terminal condition, while the second equation is forward in time, starting from an initial condition.

The resulting system thus reads as a two-point boundary value problem, notorious for being difficult to solve. In other words, the system (3.12) is nothing but a forward-backward deterministic differential system in infinite dimension. From experience with the analysis of forward-backward stochastic differential systems in finite dimension, we expect that Cauchy-Lipschitz like theory, when it can be applied, will only provide solutions in small time. One of the major difficulties of mean field games is to identify sufficient conditions under which existence and/or uniqueness of a solution hold over a time-interval of arbitrary length. Moreover, it is also to be expected that, for systems of the same type as (3.12), ellipticity of the diffusion matrix σ cannot suffice to decouple the two equations as the forward component is entirely deterministic. On this last point, we refer to Subsection 3.2.3 below for a more detailed account.

As we shall see next, the crux of our approach is to recast the infinite dimensional deterministic forward-backward system (3.12) into a finite dimensional stochastic forward-backward system of the McKean-Vlasov type. The fact that the probabilistic point of view yields a finite dimensional system should not be a surprise. The infinite dimensional feature is in fact hidden in the McKean-Vlasov component.

3.2 Why and Which FBSDEs?

We do not intend to solve MFG problems in the analytic approach described above. We presented it for the sake of completeness, and to give an enlightening perspective to the different ways we shall approach these problems. For this reason, we revisit the formulation of the MFG problem with a view toward the probabilistic approaches we intend to follow. As before, this section is rather informal, favoring ideas and strategies over precise quantitative statements and proofs. The latter will come later in this chapter and in the next chapter as well.

In the search for a Nash equilibrium, a typical player has to compute its best response to all the other players, assuming that they have already chosen their own strategies. This search for the best response amounts to the solution of an optimal control problem whereby the typical player seeks a control strategy in order to minimize its expected cost, assuming that all the other players have chosen their own strategies and play the game without deviating from these choices. If we work with models of stochastic differential games with mean field interactions, such an optimal control problem can be written in the same form as in (3.4) (with, in full generality, σ possibly depending on α):

$$\inf_{\alpha\in\mathbb{A}} \mathbb{E}\left[\int_0^T f(t, X_t, \mu_t, \alpha_t)dt + g(X_T, \mu_T)\right]$$

subject to (3.13)

$$\begin{cases} dX_t = b(t, X_t, \mu_t, \alpha_t)dt + \sigma(t, X_t, \mu_t, \alpha_t)dW_t, \quad t \in [0, T], \\ X_0 = \xi. \end{cases}$$

Here, X_t is the private state of the typical player, α_t the action it chooses to take at time t, and μ_t represents the impact of the strategies chosen by the other players. For the purpose of this control problem, μ_t is regarded as an input: it is fixed. We saw in Chapter 2 that, in the case of games with finitely many players, μ_t should be the empirical distribution of the states (or the actions, or both states and actions) of the other players. This can be viewed as a random measure with finite support. In the case of mean field games, μ_t is typically deterministic. It is the marginal distribution at time t of the state of a generic player in the population. However in the case of mean field games with a common noise discussed in Chapter (Vol II)-2, μ_t is a random measure representing the conditional marginal distribution of a generic state given the realization of the common noise. In that case and as already alluded to in Remark 3.1, the state dynamics appearing in the stochastic control problem (3.13) contain an extra diffusion term involving the common noise.

The solutions of the optimal control problems leading to the best responses of individual players form an important component of the search for Nash equilibria. However, they are not the whole story. Since Nash equilibria are the fixed points of the best response function, the second step needs to be the search for fixed points of this function, in accordance with step (ii) in the program articulated in Subsection 3.1.2. In our context, this will involve the search for particular flows of probability measures $\boldsymbol{\mu} = (\mu_t)_{0\leqslant t\leqslant T}$ which, if used as input, need to be recovered as output.

While the analytic methods discussed in the previous section have underpinned the first works on mean field games, our contention is that a probabilistic approach should bring new insights and allow for more general and possibly less regular models to be solved. The purpose of this section is to explain why and how Forward-Backward Stochastic Differential Equations (FBSDEs) appear naturally in the solutions of the mean field game problems, and to develop the tools necessary for their analyses.

3.2.1 FBSDEs and Control Problems

We first consider the optimal control step of the formulation of the mean field game problems described earlier. Probabilists have a two pronged approach to these optimal control problems. We proceed to their descriptions when the input $\boldsymbol{\mu} = (\mu_t)_{0 \leqslant t \leqslant T}$ is deterministic and fixed. We refer the reader to Chapter (Vol II)-1 for a review of the corresponding optimization problem in a random environment given by a stochastic input. As announced, we assume that the volatility function σ appearing in the state dynamics part of the stochastic control problem (3.13) is independent of the control α.

1. The first method is closer in spirit to the analytic approach based on the Hamilton-Jacobi-Bellman (HJB) equation derived from the dynamic programming principle. The crux of this method is to give a probabilistic representation of the value function of the optimization problem as the solution of a Backward Stochastic Differential Equation (BSDE). Assuming that the volatility σ is an invertible matrix, this BSDE reads:

$$dY_t = -f\big(t, X_t, \mu_t, \alpha_t\big)dt + Z_t \cdot dW_t, \quad t \in [0, T], \tag{3.14}$$

with terminal condition $Y_T = g(X_T, \mu_T)$, where $\boldsymbol{X} = (X_t)_{0 \leqslant t \leqslant T}$ is the controlled process obtained by choosing for $\boldsymbol{\alpha} = (\alpha_t)_{0 \leqslant t \leqslant T}$ the specific control:

$$\alpha_t = \hat{\alpha}\big(t, X_t, \mu_t, \sigma(t, X_t, \mu_t)^{-1\dagger} Z_t\big), \quad t \in [0, T].$$

In (3.14), the process $\boldsymbol{Y} = (Y_t)_{0 \leqslant t \leqslant T}$ is scalar valued while $\boldsymbol{Z} = (Z_t)_{0 \leqslant t \leqslant T}$ takes values in $\mathbb{R}^d$. For that reason, the stochastic integration is denoted under the form of an inner product. This is in contrast with the notation used when $\boldsymbol{Y}$ is vector valued, in which case the stochastic integration is written under the form of a matrix multiplication. Moreover, $\sigma(t, X_t, \mu_t)^{-1}$ is the inverse of $\sigma(t, X_t, \mu_t)$ and $\sigma(t, X_t, \mu_t)^{-1\dagger}$ denotes its transpose. Also, the function $\hat{\alpha}$ is the minimizer of the Hamiltonian in the sense that:

$$\hat{\alpha}(t, x, \mu, y) \in \operatorname{argmin}_{\alpha \in A} H(t, x, \mu, y, \alpha),$$

where H is the reduced Hamiltonian function introduced when σ is uncontrolled:

$$H(t, x, \mu, y, \alpha) = y \cdot b(t, x, \mu, \alpha) + f(t, x, \mu, \alpha).$$

Lemma 3.3 provides conditions for existence, uniqueness and regularity of such a function $\hat{\alpha}$. A first remark is that, even before we replace α_t, the state X_t appears in the driver of the BSDE (3.14), by which we mean the coefficient, up to the sign $-$, in front of the dt. So this BSDE has random coefficients. But if, as we require, the player has to use the specific control α_t given by the function $\hat{\alpha}$ for the system to be at the optimum, then the term Z_t appears in the forward dynamics of the state of the control problem (3.13). Equations giving dX_t and dY_t are now strongly coupled, and instead of a BSDE with random coefficients (due to the dependence of the driver upon X_t), we now need to solve a fully coupled FBSDE, whose coefficients are deterministic and depend upon the measure flow $\boldsymbol{\mu} = (\mu_t)_{0 \leqslant t \leqslant T}$.

2. The second prong of the probabilistic approach is based on the extension of the Pontryagin maximum principle to the control of stochastic differential equations. It does not require invertibility of the volatility matrix σ, but it requires differentiability of the coefficients. It is based on a probabilistic representation of the derivative of the value function (the so-called adjoint variable or adjoint process) as a solution of a BSDE called the adjoint equation:

$$dY_t = -\partial_x H^{\text{full}}(t, X_t, \mu_t, Y_t, Z_t, \alpha_t)dt + Z_t dW_t, \quad t \in [0, T], \tag{3.15}$$

with terminal condition $Y_T = \partial_x g(X_T, \mu_T)$, where as before, the control α_t is chosen to be the specific control $\alpha_t = \hat{\alpha}(t, X_t, \mu_t, Y_t)$. Above, $\boldsymbol{Y}$ takes values in $\mathbb{R}^d$ and $\boldsymbol{Z}$ in $\mathbb{R}^{d \times d}$. In agreement with our previous remark, the stochastic integration is written under the form of a matrix acting on a vector and the Hamiltonian $H^{\text{full}}(t, x, \mu, y, z, \alpha)$ is equal to $H(t, x, \mu, y, \alpha) + \sigma(t, x, \mu) \cdot z$, which also admits $\hat{\alpha}(t, x, \mu, y)$ as minimizer in α.

The equation (3.15) is a BSDE with random coefficients because of the presence of X_t in the expression giving the driver. But as before, replacing α_t by $\hat{\alpha}(t, X_t, \mu_t, Y_t)$ in the forward dynamics of the control problem (3.13) creates a strong coupling between dX_t and dY_t and the solution of the control problem reduces to the solution of an FBSDE with deterministic coefficients.

Remark 3.5 *This remark complements Remark 3.1 on mean field games with a common noise. Indeed, the above discussion makes a strong case for the use of FBSDEs in the solution of mean field game problems. However, the FBSDEs touted above cannot be used to handle mean field games with a common noise which we study later in the book. Indeed, as seen in some of the models introduced in Chapter 1 (see for instance paragraphs 1.3.2 and 1.4.1), the forward SDEs giving the dynamics of the state should contain an extra term in dW_t^0 accounting for a common source of random shocks. Accordingly, the input $\boldsymbol{\mu} = (\mu_t)_{0 \leqslant t \leqslant T}$ should be random and stand for the conditional distribution of a generic state given the realization of the common noise. As a result, the BSDE should also have a term $Z_t^0 dW_t^0$ (or $Z_t^0 \cdot dW_t^0$ depending on the dimension of the backward equation). However, as explained in detail in Chapter (Vol II)-2, although $\boldsymbol{\mu} = (\mu_t)_{0 \leqslant t \leqslant T}$ is expected to be adapted to the filtration generated by $\mathbf{W}^0 = (W_t^0)_{0 \leqslant t \leqslant T}$, it may happen that $\boldsymbol{\mu}$ involves additional sources of randomness, as it is the case in the construction of weak solutions of stochastic differential equations which often end up not being adapted to the underlying Brownian filtration. Therefore, randomness in the measure $\boldsymbol{\mu}$ may prevent us from assuming that the filtrations satisfy the martingale representation theorem. As a result, we should be prepared to face cases for which the extra martingale term forced on us by the presence* *of a common noise, may not be a stochastic integral of the form* $Z_t^0 dW_t^0$*. Instead, this martingale term should be of the more general form* $Z_t^0 dW_t^0 + dM_t$ *for some martingale* $\boldsymbol{M} = (M_t)_{0 \leq t \leq T}$ *orthogonal to* $(\boldsymbol{W}, \boldsymbol{W}^0) = (W_t, W_t^0)_{0 \leq t \leq T}$*. This class of FBSDEs is not standard, and as far as we know, was little studied in the existing literature. We call them FBSDEs in a random environment (due to the randomness of* $\boldsymbol{\mu}$*), or simply FBSDEs with random coefficients. We develop their theory (or at least what we need for the purpose of the analysis of mean field games with a common noise) in Section (Vol II)-1.1.*

3.2.2 FBSDEs of McKean-Vlasov Type and MFGs

The second step of the search for Nash equilibria is the construction of the fixed points (if any) of the best response map, see for instance step (ii) in the program articulated in Subsection 3.1.2. In the present context of mean field games, this step amounts to finding particular flows of probability measures $\boldsymbol{\mu} = (\mu_t)_{0 \leq t \leq T}$ which, if used as input to the stochastic control problem, will force the marginal distribution at time t of the optimal state of the controlled problem to coincide with the original μ_t we started from. In other words, these fixed points will force $(\mu_t)_{0 \leq t \leq T}$ to be the flow of marginal distributions of the $\boldsymbol{X} = (X_t)_{0 \leq t \leq T}$-component of the solution of the FBSDE associated with the control problem. Replacing μ_t by $\mathcal{L}(X_t)$ in the FBSDE turns the family of standard FBSDEs parameterized by the flow of measures $\boldsymbol{\mu}$ into an FBSDE of McKean-Vlasov type.

Remark 3.6 *Strictly speaking, the above discussion applies to the mean field game models solved in this chapter. In the presence of a common noise, as we shall see in Chapter (Vol II)-2, the measures* $(\mu_t)_{0 \leq t \leq T}$ *are random since they depend upon the realizations of the common noise. In that case, the fixed point argument says that* μ_t *needs to be identified to the conditional distribution of* X_t *given the common noise. So technically speaking, the solution of the mean field game becomes equivalent to the solution of an FBSDE of conditional McKean-Vlasov type. These new FBSDEs of McKean-Vlasov type will also appear in Chapter (Vol II)-6 when we study mean field games with a major and minor players. The solvability of FBSDEs of conditional McKean-Vlasov type is addressed in Chapter (Vol II)-3.*

Remark 3.7 *As explained above,* μ_t *should hopefully be identified with* $\mathcal{L}(X_t)$*. Since solutions of generic stochastic differential equations have finite moments, we shall work from the get-go with probability measures* μ_t *already in the so-called Wasserstein space* $\mathcal{P}_2(\mathbb{R}^d)$ *of probability measures over* $\mathbb{R}^d$ *with a finite second moment. The space* $\mathcal{P}_2(\mathbb{R}^d)$ *is equipped with the Wasserstein distance* W_2*. For the convenience of the reader and for the present chapter to be self-contained, we give a precise definition of the Wasserstein distances. They are studied in full detail in Chapter 5. Recall that if* E *is a complete separable metric space (quite often* $\mathbb{R}^d$ *in the book), for any* $p \geq 1$ *we denote by* $\mathcal{P}_p(E)$ *the subspace of* $\mathcal{P}(E)$ *of the probability* *measures of order p, namely those probability measures which integrate the p-th power of the distance to a fixed point whose choice is irrelevant in the definition of* $\mathcal{P}_p(E)$*. If* $\mu, \mu' \in \mathcal{P}_p(E)$*, the p-Wasserstein distance* $W_p(\mu, \mu')$ *is defined by:*

$$W_p(\mu, \mu') = \inf_{\pi \in \Pi_p(\mu,\mu')} \left[\int_{E\times E} d(x, y)^p \, \pi(dx, dy) \right]^{1/p}, \tag{3.16}$$

where $\Pi_p(\mu, \mu')$ *denotes the set of probability measures in* $\mathcal{P}_p(E\times E)$ *with marginals* μ *and* μ'*. It is customary to talk about Wasserstein space and Wasserstein distance (without referring to p) when p is assumed to be equal to* 2*.*

Recasting the two prongs of the probabilistic approach into a single formulation, and leaving for later the introduction of an additional common noise $\boldsymbol{W}^0$ as explained in Remark 3.5, we see that the optimal control part leads in both cases to the analysis of an FBSDE of the form:

$$\begin{cases} dX_t = B(t, X_t, \mu_t, Y_t, Z_t)dt + \Sigma(t, X_t, \mu_t)dW_t, \\ dY_t = -F(t, X_t, \mu_t, Y_t, Z_t)dt + Z_t dW_t, \quad t \in [0, T], \\ Y_T = G(X_T, \mu_T), \end{cases} \tag{3.17}$$

where $\boldsymbol{W} = (W_t)_{0\leqslant t\leqslant T}$ is a Wiener processes in $\mathbb{R}^d$, B and F are functions from $[0, T] \times \mathbb{R}^d \times \mathcal{P}_2(\mathbb{R}^d) \times \mathbb{R}^m \times \mathbb{R}^{m\times d}$ into $\mathbb{R}^d$ and $\mathbb{R}^m$ respectively, and Σ is a function from $[0, T] \times \mathbb{R}^d \times \mathcal{P}_2(\mathbb{R}^d)$ into $\mathbb{R}^{d\times d}$. Observe that the integer m used for the dimension of the backward component has nothing to do with the integer m used in Subsection 3.1.1 to denote the dimension of the noise; recall indeed that $\boldsymbol{W}$ is assumed to be d-dimensional. The flow $\boldsymbol{\mu} = (\mu_t)_{0\leqslant t\leqslant T}$ accounts for the input, which is deterministic for the time being. It is assumed to take values (continuously) in the Wasserstein space $\mathcal{P}_2(\mathbb{R}^d)$ of probability measures of order 2. A solution of such an FBSDE comprises progressively measurable processes $\boldsymbol{X} = (X_t)_{0\leqslant t\leqslant T}$, $\boldsymbol{Y} = (Y_t)_{0\leqslant t\leqslant T}$ and $\boldsymbol{Z} = (Z_t)_{0\leqslant t\leqslant T}$ with values in $\mathbb{R}^d$, $\mathbb{R}^m$ and $\mathbb{R}^{m\times d}$ respectively. We refer to Section 4.1 for a more detailed account of FBSDEs.

To be more specific, when implementing approach 1 based on the representation of the value function as the solution of a BSDE, we end up with:

$$\begin{aligned} B(t, x, \mu, y, z) &= b\big(t, x, \mu, \hat\alpha(t, x, \mu, \sigma(t, x, \mu)^{-1\dagger} z)\big), \\ F(t, x, \mu, y, z) &= f\big(t, x, \mu, \hat\alpha(t, x, \mu, \sigma(t, x, \mu)^{-1\dagger} z)\big), \end{aligned} \tag{3.18}$$

where $\sigma(t, x, \mu)^{-1\dagger}$ is the transpose of $\sigma(t, x, \mu)^{-1}$.

On the other hand, when implementing approach 2 based on the stochastic version of the Pontryagin maximum principle, we end up with:

$$\begin{aligned} B(t,x,\mu,y,z) &= b(t,x,\mu,\hat{\alpha}(t,x,\mu,y)), \\ F(t,x,\mu,y,z) &= \big(\partial_x H^{\text{full}}\big)\big(t,x,\mu,y,z,\hat{\alpha}(t,x,\mu,y)\big). \end{aligned} \tag{3.19}$$

In the search for an MFG equilibrium, the flow $\boldsymbol{\mu}$ is required to match the flow of marginal distributions $(\mathcal{L}(X_t))_{0\leqslant t\leqslant T}$ of the forward process $\boldsymbol{X} = (X_t)_{0\leqslant t\leqslant T}$ in (3.17). The resulting equation is the epitome of an FBSDE of McKean-Vlasov type:

$$\begin{cases} dX_t = B\big(t,X_t,\mathcal{L}(X_t),Y_t,Z_t\big)dt + \Sigma\big(t,X_t,\mathcal{L}(X_t)\big)dW_t, \\ dY_t = -F\big(t,X_t,\mathcal{L}(X_t),Y_t,Z_t\big)dt + Z_t dW_t, \quad t \in [0,T], \\ Y_T = G\big(X_T,\mathcal{L}(X_T)\big). \end{cases} \tag{3.20}$$

We shall provide a systematic analysis of this new class of FBSDEs in Section 4.3.

Remark 3.8 *Our discussion highlighted a one-to-one correspondence between optimal stochastic control problems and a specific class of FBSDEs. Yet, the FBSDEs of McKean-Vlasov type introduced for the purpose of solving mean field game problems are not directly associated with optimization problems. However, as we shall see after developing a special differential calculus for measures in Chapter 5, some of the FBSDEs of McKean-Vlasov type are in fact associated with optimal control problems. The latter correspond to the optimal control of dynamics given by stochastic differential equations of McKean-Vlasov type. This theory is developed in Chapter 6.*

Remark 3.9 *When the dimension m of the backward component in* (3.17) *is equal to* 1, *the process $\boldsymbol{Z} = (Z_t)_{0\leqslant t\leqslant T}$ takes values in $\mathbb{R}^{1\times d}$. As already explained, it will be more convenient to regard it as a process with values in $\mathbb{R}^d$ and to write the product $Z_t dW_t$ as an inner product in $\mathbb{R}^d$. For that reason, in the special case $m = 1$, we often write $Z_t \cdot dW_t$ instead of $Z_t dW_t$, Z_t being understood as an element of $\mathbb{R}^d$.*

3.2.3 Road Map to the Solution of FBSDEs of the McKean-Vlasov Type

The above reformulation of the MFG problems is screaming for the investigation of the solvability of forward-backward SDEs of the McKean-Vlasov type. Most of Chapter 4 will be devoted to this specific question, while Chapter (Vol II)-3 will address the same problem in the presence of a common noise. Here, we try to provide new insight on the nature of the technical difficulties we are about to face, and the tools that we shall bring to bear to overcome them.

The most challenging feature of these equations is the twofold structure of the boundary condition. To wit, a forward-backward equation is a *two-point boundary value problem*. One of the simplest examples we may think of for this type of equation is a pair of two coupled ODEs (with values in $\mathbb{R}$), one being set forward and the other one being set backward:

$$\begin{cases} \dot{x}_t = b(t, x_t, y_t), \\ \dot{y}_t = -f(t, x_t, y_t), \quad t \in [0, T], \end{cases} \tag{3.21}$$

with a given initial condition $x_0 \in \mathbb{R}$ for $\boldsymbol{x} = (x_t)_{0 \leqslant t \leqslant T}$ and terminal condition $y_T = g(x_T)$ for $\boldsymbol{y} = (y_t)_{0 \leqslant t \leqslant T}$. In most cases of interest to us, the forward-backward system is stochastic as the forward equation is forced by a random noise, and is driven by coefficients depending on an infinite dimensional variable since the state variable at time t comprises both the private state X_t and the *collective state* $\mathcal{L}(X_t)$. Quite remarkably, the purely analytic formulation of the MFG problems presented in Subsection 3.1.5 also consists in a forward-backward system: the forward equation is the Fokker-Planck equation for the evolution of the population, while the backward equation is the Hamilton-Jacobi-Bellman equation for the value function. In that case, the forward-backward system is clearly infinite dimensional, though deterministic, since both the forward and backward components are of infinite dimension.

As we already mentioned, one of the major drawbacks of forward-backward systems, even those of the simplest form (3.21), is that Cauchy-Lipschitz theory fails except possibly in small time. There are very simple examples of systems of the form (3.21), with coefficients b and f Lipschitz continuous in the variables x and y, for which existence and/or uniqueness fail. We shall present one of them in Subsection 4.3.4 in order to enlighten the fact that the same difficulties plague FBSDEs of the McKean-Vlasov type. This is clearly a bad omen as we cannot expect to solve systems like (3.20) by a standard Picard fixed point argument, except possibly when T is small enough or equivalently, when the coupling between the forward and the backward components is weak. Henceforth, we must seek alternative strategies for proving existence and/or uniqueness of solutions to (3.20); one of the major objectives of the book is to present some of them.

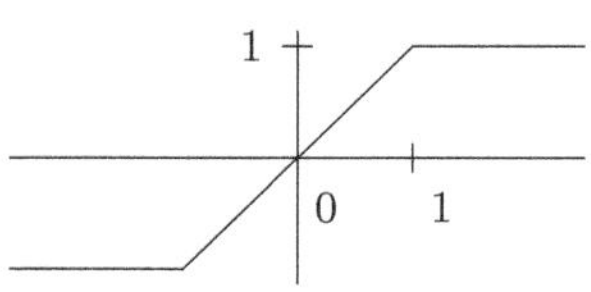

Plot of $x \mapsto G(x) = (-1) \vee x \wedge 1$

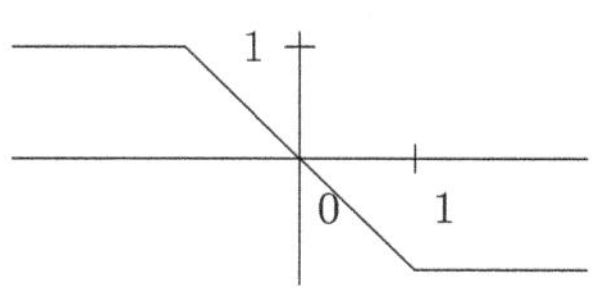

Plot of $x \mapsto G(x) = -(-1) \vee x \wedge 1$

How Can We Solve an FBSDE?

We first consider the case of systems like (3.20) without McKean-Vlasov interaction. But even then, the discussion of the deterministic example (3.21) will show that despite the simplification, solvability of the forward-backward system may still be a *touchy business*. As a case in point, we concentrate on the following specific examples:

$$\begin{cases} \dot{x}_t = -y_t, & x_0 \in \mathbb{R}, \\ \dot{y}_t = 0, & t \in [0, T], \\ y_T = G(x_T). \end{cases} \tag{3.22}$$

with terminal condition functions $G(x) = \pm(-1) \vee x \wedge 1$ whose plots are reproduced below.

When $T = 1$ and the leading sign in G is $+$, it is easy to check that the solution to (3.22) is given by $(x_t = x_0(1 - t/2), y_t = x_0/2)_{0 \leqslant t \leqslant T}$ when $x_0 \in [-2, 2]$. We plot paths of the x-component of the solution in the left-hand side below. When $T = 1$ and the leading sign in G is $-$, the solutions to (3.22) are $(x_t = x_0 + t\mathrm{sign}(x_0), y_t = -\mathrm{sign}(x_0))_{0 \leqslant t \leqslant T}$ if $x_0 \neq 0$, but when $x_0 = 0$, all the curves $(x_t = at, y_t = -a)_{0 \leqslant t \leqslant T}$, for $a \in [-1, 1]$, are solutions. Plots of the x-components are given in the right-hand side below, the solutions with $x_0 = 0$ and sign $= -$ being in red.

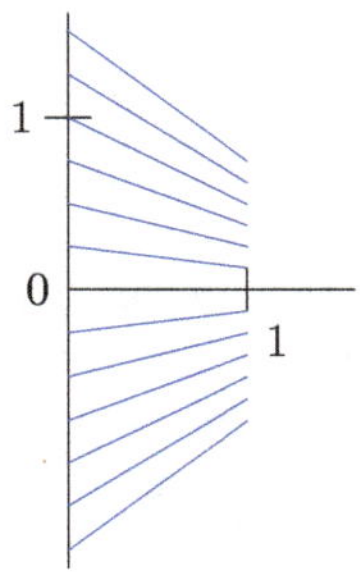

$T = 1$ and $+$ sign for G.

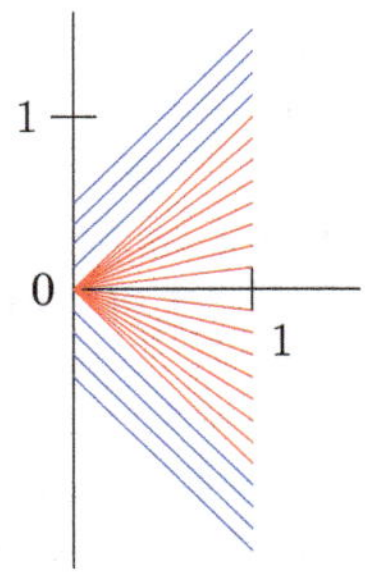

$T = 1$ and $-$ sign for G.

This example is quite enlightening as it shows that the monotonicity properties of the coefficients (here of the function giving the terminal condition) may play a key role in the properties of the forward-backward system.

In order to gain a better understanding of the meaning of this monotonicity property, we may view equation (3.22) as a particular case of (3.15), with $\sigma \equiv 0$ and without input $\boldsymbol{\mu}$. Indeed, a straightforward computation based on formula (3.19) shows that (3.22) is an instance of adjoint equation (3.15) for $(\alpha_t = \hat{\alpha}(t, X_t, \mu_t, Y_t))_{0 \leqslant t \leqslant T}$ when $A = \mathbb{R}$, $b(t, x, \mu, \alpha) = \alpha$, $f(t, x, \mu, \alpha) = \alpha^2/2$ and $g(x) = \int_0^x G(r)dr$. We then understand the monotonicity property of G as a convexity property of the terminal cost function g.

This observation is crucial. As we shall see in Subsection 3.3.2 below, the forward-backward system obtained by coupling (3.15) with the corresponding forward controlled equation is well posed when the underlying optimization problem has a full-fledged convex structure in both α and x. This is a prime strategy to solve an FBSDE. In the sequel, we shall use it in order to solve the FBSDE derived from the stochastic Pontryagin principle when convexity holds.

Although very helpful, this first solvability result will not suffice. As the reader may have already noticed from the examples given in Chapter 1, convexity does not always hold, even in simple models. However, as highlighted by Lemma 3.3, convexity in the direction α can play a crucial role as it guarantees that $\hat{\alpha}$ inherits the smoothness properties of the coefficients. The challenge for us will be to relax the convexity condition in the direction x.

In regard to our preliminary discussion of the breakdown of the Cauchy-Lipschitz theory for forward-backward systems, our goal is to replace the convexity assumption by another sufficient condition guaranteeing the well posedness of the FBSDE. In order to do so, we shall invoke another key ingredient in the theory of forward-backward SDEs. When driven by deterministic coefficients, these forward-backward systems can be viewed as systems of characteristics of nonlinear PDEs. Indeed, going back to system (3.22) and assuming that existence and uniqueness hold true, and that there exists a smooth function $u : [0,T] \times \mathbb{R} \to \mathbb{R}$ such that $y_t = u(t, x_t)$ for all $t \in [0,T]$, condition $\dot{y}_t = 0$ can be rewritten as $\partial_t u(t,x_t) + \partial_x u(t,x_t)\dot{x}_t = 0$. Recalling that $\dot{x}_t = -y_t$, we end up with:

$$\partial_t u(t,x_t) - u(t,x_t)\partial_x u(t,x_t) = 0, \quad t \in [0,T].$$

If existence and uniqueness hold true for any initial condition $x_0 \in \mathbb{R}$ at any initialization time $t_0 \in [0,T]$, then u must solve the nonlinear equation:

$$\partial_t u(t,x) - u(t,x)\partial_x u(t,x) = 0, \quad t \in [0,T], \ x \in \mathbb{R}, \tag{3.23}$$

which is the backward Burgers' equation with terminal condition $u(T,\cdot) = G$. The plots given earlier are then the plots of the characteristics of the Burgers equation, and the meaning of the monotonicity property of G may be reformulated as follows: if G is nondecreasing, the equation is in a *dilation* regime (when described in the backward sense, the characteristics diverge from each other); if G is nonincreasing, the equation is in a *compression* regime (when described in the backward sense, the characteristics get closer to each other). In the dilation regime, the Burgers equation has a solution which is Lipschitz in space, while, in the compression regime, solutions develop a singularity. We conclude that upward monotonicity (i.e., convexity when regarded at the level of the optimal control problem) has a regularizing effect onto the whole system.

Another way to regularize a nonlinear PDE such as the Burgers equation is to force it by a diffusive term. Irrespective of the sign of G, (3.23) with $(1/2)\partial^2_{xx}u(t,x)$ added to the left-hand side has a classical solution. Thanks to the regularizing effect of the heat kernel, the solution cannot develop singularities. We now revisit the discussion of the characteristics, and explain how the inclusion of the Laplacian in the equation manifests in the form of a Brownian motion! In that case, the system (3.22) becomes stochastic:

$$\begin{cases} dX_t = -Y_t + dW_t, \\ dY_t = Z_t dW_t, \quad t \in [0,T], \\ X_0 = x_0 \in \mathbb{R}, \quad Y_T = G(X_T), \end{cases} \tag{3.24}$$

where $\boldsymbol{W} = (W_t)_{0 \leqslant t \leqslant T}$ is a one-dimensional Wiener process. Here the martingale term in the backward equation is needed to force the solution to be adapted to the filtration of the noise. It turns out that the well posedness of the viscous Burgers equation transfers to the forward-backward SDE (3.24), which is also uniquely solvable. Moreover, the solution $(X_t, Y_t, Z_t)_{0 \leqslant t \leqslant T}$ admits the representation:

$$\mathbb{P}\Big[\forall t \in [0,1],\ Y_t = u(t, X_t)\Big] = 1,$$

u being the solution of the viscous Burgers equation, and $Z_t = \partial_x u(t, X_t)$ almost everywhere under $\mathrm{Leb}_1 \otimes \mathbb{P}$, where Leb_1 denotes the one-dimensional Lebesgue measure. In the theory of FBSDEs, u is called the *decoupling field* of the forward-backward system.

This provides still another avenue to solve FBSDEs. Indeed, when the diffusion coefficient (or volatility) driving the noise term is nondegenerate and the coefficients are bounded in the space variable, it may be shown that the Cauchy-Lipschitz theory still holds true, see the Notes and Complements at the end of the chapter. We shall state and use this result in Chapter 4, see Theorem 4.12. The need for the boundedness of coefficients has been documented in the literature with examples of linear forward-backward systems with an additive nondegenerate noise for which existence and/or uniqueness fail.

Implementing Schauder's Fixed Point Theorem

The results discussed above have the potential to be very helpful. Indeed, they provide effective tools for investigating the well posedness of the equation (3.17) driven by an input $\boldsymbol{\mu} = (\mu_t)_{0 \leqslant t \leqslant T}$. However, they fall short of being sufficient for our purposes, since we are interested in the solvability of the McKean-Vlasov version (3.20). As explained earlier and documented later in Chapter 4, there is no Cauchy-Lipschitz theory for McKean-Vlasov FBSDEs, except in small time. Once again, we shall need innovative technologies to bypass this roadblock.

Our strategy for solving the equation will rely on two main ingredients:

1. The first one is a suitable notion of upward monotonicity for functionals depending upon a measure argument, in full analogy with the upward monotonicity

property of G in the example 3.22 discussed above. This notion is due to Lasry and Lions and will be shown in Subsection 3.4, to play a key role in the uniqueness of equilibria for mean field games.

2. The second one is a systematic use of Schauder's fixed point theorem in order to prove the existence (though not the uniqueness) of solutions to FBSDEs of the McKean-Vlasov type and subsequently, of equilibria to the corresponding mean field games. The implementation of Schauder's theorem is discussed in detail in Chapter 4 where it is applied to the function mapping the input $\boldsymbol{\mu} = (\mu_t)_{0 \leqslant t \leqslant T}$ in (3.17) onto the output flow of marginal laws $(\mathcal{L}(X_t))_{0 \leqslant t \leqslant T}$ formed by the forward component $\boldsymbol{X} = (X_t)_{0 \leqslant t \leqslant T}$ of the solution. This approach works well because we can easily imbed the input $\boldsymbol{\mu}$ and the output $(\mathcal{L}(X_t))_{0 \leqslant t \leqslant T}$ in a topological space to which simple compactness criteria can be applied. This is crucial as Schauder's theorem is based on compactness arguments.

For the sake of illustration, we provide below the statement of one of the solvability results proven in Chapter 4 by means of Schauder's theorem. This statement is given here for pedagogical reasons, in anticipation of the discussion of next subsection where we use it to prove our first results of existence of equilibria for mean field games. It will be generalized in Chapter 4.

The statement given below addresses the existence (but not the uniqueness) of a solution to a fully coupled McKean-Vlasov forward-backward system of the type (3.20), namely of the form:

$$\begin{cases} dX_t = B\big(t, X_t, \mathcal{L}(X_t), Y_t, Z_t\big)dt + \Sigma\big(t, X_t, \mathcal{L}(X_t), Y_t\big)dW_t \\ dY_t = -F\big(t, X_t, \mathcal{L}(X_t), Y_t, Z_t\big)dt + Z_t dW_t, \quad t \in [0, T], \\ Y_T = G\big(X_T, \mathcal{L}(X_T)\big), \end{cases} \tag{3.25}$$

with initial condition $X_0 = \xi$ for some $\xi \in L^2(\Omega, \mathcal{F}_0, \mathbb{P}; \mathbb{R}^d)$. The unknown processes $(\boldsymbol{X}, \boldsymbol{Y}, \boldsymbol{Z})$ are of dimensions d, m, and $m \times d$ respectively. The coefficients are assumed to be deterministic. The functions B and F map $[0, T] \times \mathbb{R}^d \times \mathcal{P}_2(\mathbb{R}^d) \times \mathbb{R}^m \times \mathbb{R}^{m \times d}$ into $\mathbb{R}^d$ and $\mathbb{R}^m$ respectively, while the coefficient Σ maps $[0, T] \times \mathbb{R}^d \times \mathcal{P}_2(\mathbb{R}^d) \times \mathbb{R}^m$ into $\mathbb{R}^{d \times d}$. The function G giving the terminal condition maps $\mathbb{R}^d \times \mathcal{P}_2(\mathbb{R}^d)$ into $\mathbb{R}^m$. All these functions are assumed to be Borel-measurable. The space $\mathcal{P}_2(\mathbb{R}^d)$ is equipped with the 2-Wasserstein distance W_2, see Remark 3.7. For $\mu \in \mathcal{P}_2(\mathbb{R}^d)$, we call $M_2(\mu)$ the second moment:

$$M_2(\mu) = \left(\int_{\mathbb{R}^d} |x|^2 d\mu(x) \right)^{1/2}. \tag{3.26}$$

For the sake of definiteness, we state formally the precise assumptions under which the existence result will be proven.

Assumption (Nondegenerate MKV FBSDE). There exists a constant $L \geqslant 1$ such that

(A1) For any $t \in [0,T]$, $x, x' \in \mathbb{R}^d$, $y, y' \in \mathbb{R}^m$, $z, z' \in \mathbb{R}^{m\times d}$ and $\mu \in \mathcal{P}_2(\mathbb{R}^d)$,

$$|(B,F,G,\Sigma)(t,x',\mu,y',z') - (B,F,G,\Sigma)(t,x,\mu,y,z)| \\ \leqslant L|(x,y,z) - (x',y',z')|.$$

Moreover, for any $(t,x,y,z) \in [0,T]\times\mathbb{R}^d\times\mathbb{R}^m\times\mathbb{R}^{m\times d}$, the coefficients $B(t,x,\cdot,y,z)$, $F(t,x,\cdot,y,z)$, $\Sigma(t,x,\cdot,y)$ and $G(x,\cdot)$ are continuous in the measure argument with respect to the 2-Wasserstein distance.

(A2) The functions Σ and G are bounded by L. Moreover, for any $t \in [0,T]$, $x \in \mathbb{R}^d$, $y \in \mathbb{R}^m$, $z \in \mathbb{R}^{m\times d}$ and $\mu \in \mathcal{P}_2(\mathbb{R}^d)$,

$$|(B,F)(t,x,\mu,y,z)| \leqslant L\big[1 + |y| + |z| + M_2(\mu)\big].$$

(A3) The function Σ is uniformly elliptic in the sense that, for any $t \in [0,T]$, $x \in \mathbb{R}^d$ and $\mu \in \mathcal{P}_2(\mathbb{R}^d)$, the following inequality holds true:

$$\big(\Sigma\Sigma^\dagger\big)(t,x,\mu,y) \geqslant L^{-1}I_d,$$

in the sense of symmetric matrices, where I_d is the d-dimensional identity matrix, and where the exponent † denotes the transpose of a matrix. Moreover, the function $[0,T]\times\mathbb{R}^d\times\mathcal{P}_2(\mathbb{R}^d) \ni (t,x,\mu) \mapsto \Sigma(t,x,\mu)$ is continuous.

We can now state the anticipated existence result whose proof is deferred to Subsection 4.3:

Theorem 3.10 *Under assumption* **Nondegenerate MKV FBSDE**, *for any random variable* $\xi \in L^2(\Omega,\mathcal{F}_0,\mathbb{P};\mathbb{R}^d)$, *the FBSDE* (3.25) *has a solution* $(\boldsymbol{X},\boldsymbol{Y},\boldsymbol{Z}) = (X_t,Y_t,Z_t)_{0\leqslant t\leqslant T}$ *satisfying*

$$\mathbb{E}\bigg[\sup_{0\leqslant t\leqslant T}\big(|X_t|^2 + |Y_t|^2\big) + \int_0^T |Z_t|^2 dt\bigg] < \infty,$$

with $X_0 = \xi$ *as initial condition.*

3.3 The Two-Pronged Probabilistic Approach

We stress that most of the results given in this section are provisional. They will be stated and proven in full generality in the next chapter. The rationale for presenting them at this stage is to supply the reader with a fair understanding of the philosophy of the probabilistic approach.

In this section, we offer a first, mostly pedagogical, approach to the mean field game problem using probabilistic tools in two different ways. In both cases we introduce BSDEs to tackle the stochastic optimization problem, and in both cases, the optimality condition creates a coupling between the forward dynamics of the state and the original BSDE, leading to the solution of an FBSDE. Both approaches are well understood by probabilists working on optimal control problems. The first approach is known as the weak formulation or martingale method, while the second one is known under the name of stochastic maximum approach. We introduce them below. As emphasized in the previous section, the fixed point step in the solution of the mean field game problem, when implemented in each of these two approaches, turns standard FBSDEs into FBSDEs of the McKean-Vlasov type. This unexpected twist to the standard theory will require special attention in this chapter and in the next one for the solution of mean field game problems without common noise, and in Chapter 6 for the control of McKean-Vlasov dynamics.

In this section and the next, all the processes are assumed to be defined on a complete filtered probability space $(\Omega, \mathcal{F}, \mathbb{F} = (\mathcal{F}_t)_{0 \leqslant t \leqslant T}, \mathbb{P})$, the filtration $\mathbb{F}$ satisfying the usual conditions, supporting a d-dimensional Wiener process $\boldsymbol{W} = (W_t)_{0 \leqslant t \leqslant T}$ with respect to $\mathbb{F}$. Recall that for each random variable/vector or stochastic process X, we denote by $\mathcal{L}(X)$ the law (alternatively called the distribution) of X and, for any integer $n \geqslant 1$, by $\mathbb{H}^{2,n}$ the Hilbert space:

$$\mathbb{H}^{2,n} = \left\{ \boldsymbol{Z} \in \mathbb{H}^{0,n} : \mathbb{E} \int_0^T |Z_s|^2 ds < \infty \right\},$$

where $\mathbb{H}^{0,n}$ stands for the collection of all $\mathbb{R}^n$-valued progressively measurable processes on $[0, T]$. We shall also denote by $\mathbb{S}^{2,n}$ the collection of all continuous processes $\boldsymbol{U} = (U_t)_{0 \leqslant t \leqslant T}$ in $\mathbb{H}^{0,n}$ such that $\mathbb{E}[\sup_{0 \leqslant t \leqslant T} |U_t|^2] < \infty$.

3.3.1 The Weak Formulation Approach

Of the two probabilistic approaches which we propose in this section, the Weak Formulation is closest to the analytic approach. Indeed, it follows the strategy based on the search for an equation for the value function of the optimal control problem (3.4) of step (i) of the formulation of a mean field game problem, when the flow $\boldsymbol{\mu} = (\mu_t)_{0 \leqslant t \leqslant T}$ of probability measures is fixed. Recall that, as stated at the beginning of Subsection 3.1.2, we restrict ourselves to deterministic flows in this chapter. The main characteristic (and possibly the main shortcoming) of the weak formulation as described in this chapter, is to be limited to models for which the volatilities of the state dynamics are not controlled and do not depend upon the measure arguments. In the next Chapter 4, we shall propose a more robust formulation that accommodates a diffusion coefficient depending upon the measure argument.

So, in this subsection, we restrict ourselves to models for which:

$$\sigma(t,x,\mu,\alpha) = \sigma(t,x), \tag{3.27}$$

for a locally bounded and measurable function $[0,T]\times\mathbb{R}^d \ni (t,x) \mapsto \sigma(t,x) \in \mathbb{R}^{d\times d}$ which we assume to be Lipschitz in x uniformly in $t \in [0,T]$. This guarantees existence and uniqueness of a strong solution $\boldsymbol{X} = (X_t)_{0\leqslant t\leqslant T}$ of the equation:

$$dX_t = \sigma(t,X_t)dW_t, \quad t\in[0,T]; \qquad X_0 = \xi, \tag{3.28}$$

for any given square integrable random variable ξ with values in $\mathbb{R}^d$. We shall also assume uniform ellipticity, namely that the spectrum of the matrix $\sigma(t,x)\sigma(t,x)^\dagger$ is bounded from below by a strictly positive constant independent of t and x. This implies that $\sigma(t,x)$ is invertible with a uniformly bounded inverse. This remark is important because we plan to use Girsanov's theorem. Indeed, for each continuous measure flow $\boldsymbol{\mu} = (\mu_t)_{0\leqslant t\leqslant T}$ and admissible control $\boldsymbol{\alpha} \in \mathbb{A}$, we define the probability measure $\mathbb{P}^{\boldsymbol{\mu},\boldsymbol{\alpha}}$ on $(\Omega,\mathcal{F}_T)$ by:

$$\frac{d\mathbb{P}^{\boldsymbol{\mu},\boldsymbol{\alpha}}}{d\mathbb{P}} = \mathcal{E}\left(\int_0^{\cdot} \sigma(t,X_t)^{-1} b\,(t,X_t,\mu_t,\alpha_t)\cdot dW_t\right)_T,$$

where we use the notation $\mathcal{E}$ for the Doléans-Dade exponential of a martingale. Recall that if $\boldsymbol{M} = (M_t)_{0\leqslant t\leqslant T}$ is a local martingale, its Doléans-Dade exponential $\mathcal{E}(\boldsymbol{M})$ (also called the stochastic exponential of $\boldsymbol{M}$) is defined by the formula:

$$\mathcal{E}(\boldsymbol{M})_t = \exp\big(M_t - M_0 - \frac{1}{2}[M,M]_t\big),$$

where $([M,M]_t)_{0\leqslant t\leqslant T}$ stands for the quadratic variation of $\boldsymbol{M}$. The process $\boldsymbol{W}^{\boldsymbol{\mu},\boldsymbol{\alpha}}$ defined by:

$$W_t^{\boldsymbol{\mu},\boldsymbol{\alpha}} = W_t - \int_0^t \sigma(s,X_s)^{-1} b\,(s,X_s,\mu_s,\alpha_s)\,ds, \quad t\in[0,T],$$

is a Wiener process under $\mathbb{P}^{\boldsymbol{\mu},\boldsymbol{\alpha}}$, provided that Girsanov's theorem applies. The latter is true if b is bounded, since σ^{-1} is already known to be bounded. In such a case, it holds $\mathbb{P}^{\boldsymbol{\mu},\boldsymbol{\alpha}}$ almost-surely:

$$dX_t = b\,(t,X_t,\mu_t,\alpha_t)\,dt + \sigma(t,X_t)dW_t^{\boldsymbol{\mu},\boldsymbol{\alpha}}, \quad t\in[0,T].$$

That is, under $\mathbb{P}^{\mu,\alpha}$, X is a weak solution of the state equation. Note that $\mathbb{P}^{\mu,\alpha}$ and $\mathbb{P}$ agree on $\mathcal{F}_0$; in particular, the law of $X_0 = \xi$ remains the same. Moreover, ξ and W remain independent under $\mathbb{P}^{\mu,\alpha}$.

Reformulation of the Mean Field Game. Given $\boldsymbol{\alpha} \in \mathbb{A}$, we redefine the cost functional $J^{\mu}(\boldsymbol{\alpha})$ associated with $\boldsymbol{\alpha}$ by:

$$J^{\mu,\text{weak}}(\boldsymbol{\alpha}) = \mathbb{E}^{\mu,\alpha}\left[g(X_T,\mu_T) + \int_0^T f(t,X_t,\mu_t,\alpha_t)dt\right], \tag{3.29}$$

where X solves the driftless equation (3.28) and $\mathbb{E}^{\mu,\alpha}$ denotes the expectation with respect to $\mathbb{P}^{\mu,\alpha}$. It is worth mentioning that $J^{\mu,\text{weak}}(\boldsymbol{\alpha})$ may differ from $J^{\mu}(\boldsymbol{\alpha})$ in (3.4) since the distribution of the pair $(X,\boldsymbol{\alpha})$ under $\mathbb{P}^{\mu,\alpha}$ may be different from the distribution of the pair $(X^{\alpha},\boldsymbol{\alpha})$ under $\mathbb{P}$. However, when b is bounded and σ is bounded and continuous, and when the control $\boldsymbol{\alpha}$ is Markovian in the sense that $\alpha_t = \phi(t,X_t)$ for some Borel-measurable function $\phi : [0,T]\times\mathbb{R}^d \to A$, Stroock and Varadhan uniqueness in law theorem guarantees that $(X,\boldsymbol{\alpha})$, with $\alpha_t = \phi(t,X_t)$ for any $t \in [0,T]$, has the same law under $\mathbb{P}^{\mu,\alpha}$ as $(X_t^{\phi},\phi(t,X_t^{\phi}))_{0\le t\le T}$ where X^{ϕ} is the solution of the SDE:

$$dX_t^{\phi} = b\big(t,X_t^{\phi},\mu_t,\phi(t,X_t^{\phi})\big)dt + \sigma(t,X_t^{\phi})dW_t, \quad t \in [0,T]; \quad X_0^{\phi} = \xi,$$

under $\mathbb{P}$. In particular, when the optimization of J^{μ} is performed over Markovian controls only, the minimal costs to the weak and strong formulations coincide, the strong formulation referring to the one used in (3.4).

At this stage of the book, we shall avoid any further technical discussion about the possible differences between the weak and strong formulations. We shall provide in Chapter 4 a suitable set of assumptions under which both formulations have the same minimization paths. In the current section, we shall reformulate the mean field game problem (3.4) in terms of the weak formulation, namely we will seek a flow $\boldsymbol{\mu} = (\mu_t)_{0\leqslant t\leqslant T}$ such that, under the probability $\mathbb{P}^{\mu,\star}$ associated with the optimal control $\boldsymbol{\alpha}^{\star}$ minimizing $J^{\mu,\text{weak}}$, $(X_t)_{0\leqslant t\leqslant T}$ has exactly $\boldsymbol{\mu}$ as flow of marginal distributions.

Weak Formulation and BSDEs. In order to proceed, we now provide a set of assumptions under which the weak formulation has, for any continuous flow $\boldsymbol{\mu} = (\mu_t)_{0\leqslant t\leqslant T}$ with values in $\mathcal{P}_2(\mathbb{R}^d)$, a unique minimizer.

Assumption (Weak Formulation). The set A is a bounded subset of $\mathbb{R}^k$, but may not be closed nor convex. Moreover, the coefficients b, f, σ and g are defined on $[0,T]\times\mathbb{R}^d\times\mathcal{P}_2(\mathbb{R}^d)\times A$, $[0,T]\times\mathbb{R}^d\times\mathcal{P}_2(\mathbb{R}^d)\times A$, $[0,T]\times\mathbb{R}^d$, and $\mathbb{R}^d\times\mathcal{P}_2(\mathbb{R}^d)$ respectively, and they satisfy for a constant $L \geqslant 1$:

(continued)

(A1) For any $t \in [0,T]$, $x, x' \in \mathbb{R}^d$, $\alpha, \alpha' \in A$, and $\mu \in \mathcal{P}_2(\mathbb{R}^d)$,

$$
\begin{aligned}
|(b,f)(t,x',\mu,\alpha') - (b,f)(t,x,\mu,\alpha)| + |\sigma(t,x') - \sigma(t,x)| \\
+ |g(x',\mu) - g(x,\mu)| \leqslant L|(x,\alpha) - (x',\alpha')|.
\end{aligned}
$$

(A2) The functions b, f, σ and g are bounded, the common bound being also denoted by L.

(A3) The function σ is continuous and uniformly elliptic in the sense that, for any $t \in [0,T]$ and $x \in \mathbb{R}^d$, the following inequality holds:

$$
\sigma(t,x)\big(\sigma(t,x)\big)^{\dagger} \geqslant L^{-1} I_d,
$$

in the sense of symmetric matrices. Here, I_d denotes the d-dimensional identity matrix.

(A4) There exists a function

$$
\hat{\alpha} : [0,T] \times \mathbb{R}^d \times \mathcal{P}_2(\mathbb{R}^d) \times \mathbb{R}^d \ni (t,x,\mu,y) \mapsto \hat{\alpha}(t,x,\mu,y) \in A,
$$

which is L-Lipschitz continuous in (x,y) such that, for each $(t,x,\mu,y) \in [0,T] \times \mathbb{R}^d \times \mathcal{P}_2(\mathbb{R}^d) \times \mathbb{R}^d$, $\hat{\alpha}(t,x,\mu,y)$ is the unique minimizer of $H(t,x,\mu,y,\alpha)$.

As explained in Subsection 3.1.4, the existence of a strict minimizer for the Hamiltonian, as required in **(A4)** right above, is guaranteed under assumption **Minimization of the Hamiltonian** under the additional assumption that A is closed and convex. This requires the Hamiltonian to have a convex structure in α. Importantly, this is not in conflict with the assumption that f is bounded and Lipschitz, since A is assumed to be bounded.

Also, in the rest of this subsection, the filtration $\mathbb{F}$ is required to be generated by $\mathcal{F}_0$ and $\boldsymbol{W}$.

The workhorse of the first probabilistic approach is the representation of the optimal cost provided by the following result:

Proposition 3.11 *Let assumption* **Weak Formulation** *be in force. Recall also the definition* (3.28) *of the process* $\boldsymbol{X}$ *for an initial condition* $\xi \in L^2(\Omega, \mathcal{F}_0, \mathbb{P}; \mathbb{R}^d)$. *Then, for any continuous flow* $\boldsymbol{\mu} = (\mu_t)_{0 \leqslant t \leqslant T}$ *of probability measures on* $\mathbb{R}^d$, *the BSDE:*

$$
\begin{aligned}
dY_t = &-H\big(t, X_t, \mu_t, \sigma(t,X_t)^{-1\dagger} Z_t, \hat{\alpha}(t, X_t, \mu_t, \sigma(t,X_t)^{-1\dagger} Z_t)\big)dt \\
&- Z_t \cdot dW_t, \quad 0 \leqslant t \leqslant T,
\end{aligned}
\tag{3.30}
$$

with terminal condition $Y_T = g(X_T, \mu_T)$, is uniquely solvable. Moreover, the control $\hat{\boldsymbol{\alpha}} = (\hat{\alpha}_t)_{0\leqslant t\leqslant T}$ defined by $\hat{\alpha}_t = \hat{\alpha}(t, X_t, \mu_t, \sigma(t,X_t)^{-1\dagger}Z_t)$ is the unique optimal control over the interval $[0,T]$, and the optimal cost of the problem is given by:

$$\inf_{\alpha\in\mathbb{A}} J^{\mu,\text{weak}} = Y_0. \tag{3.31}$$

Proof.

First Step. We first show that the BSDE (3.30) has a solution. Importantly, the process $\boldsymbol{X}$ is adapted with respect to the filtration $\mathbb{F}$, which is assumed to satisfy the representation martingale theorem. However, the difficulty is that the Hamiltonian H is not Lipschitz continuous in the variable y, and consequently, the driver of the BSDE is not Lipschitz as a function of Z_t. Indeed, when expanding H, (3.30) takes the form:

$$\begin{aligned} dY_t = &-\big[\big(\sigma(t,X_t)^{-1\dagger}Z_t\big)\cdot b\big(t,X_t,\mu_t,\hat{\alpha}(t,X_t,\mu_t,\sigma(t,X_t)^{-1\dagger}Z_t)\big) \\ &+ f\big(t,X_t,\mu_t,\hat{\alpha}(t,X_t,\mu_t,\sigma(t,X_t)^{-1\dagger}Z_t)\big)\big]dt \\ &- Z_t\cdot dW_t, \quad 0\leqslant t\leqslant T, \end{aligned} \tag{3.32}$$

with $Y_T = G(X_T,\mu_T)$. In order to bypass this obstacle, we shall invoke results from the theory of quadratic BSDEs, a short account of which is given in Chapter 4. Thanks to the boundedness of f and g, we first notice from standard results for backward SDEs that, for any solution $(\boldsymbol{Y},\boldsymbol{Z})$ to (3.32), the component $\boldsymbol{Y}$ is bounded in the sense that $\sup_{0\leqslant t\leqslant T}|Y_t|$ is in $L^\infty(\Omega,\mathcal{F},\mathbb{P};\mathbb{R})$. We only provide a sketch of the proof. We can find two constants $c>0$ and $C\geqslant 0$ such that, when applying Itô's formula to $(\exp(ct)|Y_t|^2)_{0\leqslant t\leqslant T}$, we get, with probability 1, for all $t\in[0,T]$,

$$\exp(ct)|Y_t|^2 + \frac{1}{2}\int_t^T \exp(ct)|Z_s|^2 ds \leqslant C + 2\int_t^T \exp(cs)Y_sZ_s\cdot dW_s.$$

Taking conditional expectation given $\mathcal{F}_t$ on both sides, we get an almost sure bound for $|Y_t|$. Since $\boldsymbol{Y}$ is continuous, we easily obtain an almost sure bound for $\sup_{0\leqslant t\leqslant T}|Y_t|$. Existence and uniqueness then follow from Theorem 4.15.

Second Step. Given an admissible control $\boldsymbol{\beta}\in\mathbb{A}$, since the variable y does not appear in the driver of the BSDE, and hence the map $\mathbb{R}\times\mathbb{R}^d \ni (y,z)\mapsto H(t,X_t,\mu_t,\sigma(t,X_t)^{-1\dagger}z,\beta_t)$ is independent of y and uniformly Lipschitz in z (recall that A is assumed to be bounded), existence and uniqueness hold for the following BSDE, whose solution is denoted by $(\boldsymbol{Y}^{\beta},\boldsymbol{Z}^{\beta})$:

$$\begin{cases} dY_t^{\beta} = -H\big(t,X_t,\mu_t,\sigma(t,X_t)^{-1\dagger}Z_t^{\beta},\beta_t\big)dt + Z_t^{\beta}\cdot dW_t, & t\in[0,T], \\ Y_T^{\beta} = g(X_T,\mu_T). \end{cases}$$

Recalling that $\boldsymbol{X} = (X_t)_{0\leqslant t\leqslant T}$ is the solution of the driftless dynamic equation (3.28), we have:

$$
\begin{aligned}
Y_t^{\beta} &= g(X_T, \mu_T) + \int_t^T H\big(s, X_s, \mu_s, \sigma(s, X_s)^{-1\dagger} Z_s^{\beta}, \beta_s\big) ds - \int_t^T Z_s^{\beta} \cdot dW_s \\
&= g(X_T, \mu_T) + \int_t^T \big[f(s, X_s, \mu_s, \beta_s) + \big(\sigma(s, X_s)^{-1\dagger} Z_s^{\beta}\big) \cdot b(s, X, \mu_s, \beta_s)\big] ds \\
&\quad - \int_t^T Z_s^{\beta} \cdot dW_s \\
&= g(X_T, \mu_T) + \int_t^T f(s, X, \mu_s, \beta_s) ds \\
&\quad + \int_t^T Z_s^{\beta} \cdot \big[\sigma(s, X)^{-1} b(s, X, \mu_s, \beta_s) ds - dW_s\big] \\
&= g(X_T, \mu_T) + \int_t^T f(s, X, \mu_s, \beta_s) ds - \int_t^T Z_s^{\beta} \cdot dW_s^{\mu, \beta}.
\end{aligned}
$$

Since the density of $\mathbb{P}^{\mu,\beta}$ with respect to $\mathbb{P}$ has moments of any order, and since $\boldsymbol{Z}^{\beta}$ is square integrable under $\mathbb{P}$, the stochastic integral above is a martingale under $\mathbb{P}^{\mu,\beta}$. So by taking $\mathbb{P}^{\mu,\beta}$-conditional expectation with respect to $\mathcal{F}_t$, we get:

$$
Y_t^{\beta} = \mathbb{E}^{\mathbb{P}^{\mu,\beta}}\left[g(X_T, \mu_T) + \int_t^T f(s, X, \mu_s, \beta_s) ds \,\middle|\, \mathcal{F}_t \right],
$$

and:

$$
\mathbb{E}\big[Y_0^{\beta}\big] = \mathbb{E}^{\mathbb{P}^{\mu,\beta}}\big[Y_0^{\beta}\big] = \mathbb{E}^{\mathbb{P}^{\mu,\beta}}\left[g(X_T, \mu_T) + \int_0^T f(s, X_s, \mu_s, \beta_s) ds \right] = J^{\mu,\text{weak}}(\boldsymbol{\beta}).
$$

In order to conclude the proof, we notice that the solution $(\boldsymbol{Y}, \boldsymbol{Z})$ of the FBSDE (3.30) is the solution of the BSDE with terminal condition $g(X_T, \mu_T)$ and driver Ψ^* defined by:

$$
\Psi^*(t, \omega, y, z) = H\big(t, X_t(\omega), \mu_t, \sigma(t, X_t(\omega))^{-1\dagger} z, \hat{\alpha}(t, X_t(\omega), \mu_t, \sigma(t, X_t(\omega))^{-1\dagger} z)\big)
$$

while $(\boldsymbol{Y}^{\beta}, \boldsymbol{Z}^{\beta})$ is the solution of the BSDE with the same terminal condition $g(X_T, \mu_T)$ and driver Ψ defined by:

$$
\Psi(t, \omega, y, z) = H\big(t, X_t(\omega), \mu_t, \sigma(t, X_t(\omega))^{-1\dagger} z, \beta_t(\omega)\big),
$$

and by criticality of the function $\hat{\alpha}$, we have:

$$
\Psi^*(t, \omega, y, z) \leqslant \Psi(t, \omega, y, z) \quad \mathbb{P} - \text{a.s.}
$$

for every t, y and z. From this, we conclude $\mathbb{E}[Y_0] \leqslant \mathbb{E}[Y_0^{\beta}]$ by the comparison theorem for BSDEs (see the Notes & Complements at the end of the chapter for references, see also Theorem 4.16). Since the comparison theorem for BSDEs is strict and the minimizer of H is strict as well, we have that:

$$\mathbb{E}[Y_0] = \mathbb{E}\big[Y_0^{\beta}\big] \Leftrightarrow \beta_t = \hat{\alpha}_t \quad \text{Leb}_1 \otimes \mathbb{P} \text{ almost-everywhere},$$

with $\hat{\boldsymbol{\alpha}} = (\hat{\alpha}_t = \hat{\alpha}(t, X_t, \sigma(t, X_t)^{-1\dagger} Z_t))_{0 \leqslant t \leqslant T}$. □

Connection with FBSDEs of the McKean-Vlasov Type. We are now in a position to provide a rigorous definition of a solution to the mean field game associated with (3.29).

Definition 3.12 *Under assumption* **Weak Formulation**, *to any continuous flow of measures* $\boldsymbol{\mu} = (\mu_t)_{0 \leqslant t \leqslant T}$ *from* $[0, T]$ *to* $\mathcal{P}_2(\mathbb{R}^d)$, *we associate the solution* $(\boldsymbol{Y}^{\mu}, \boldsymbol{Z}^{\mu})$ *to* (3.30)*. Letting* $\hat{\boldsymbol{\alpha}}^{\mu} = (\hat{\alpha}(t, X_t, \mu_t, \sigma(t, X_t)^{-1\dagger} Z_t^{\mu}))_{0 \leqslant t \leqslant T}$, *we say that* $\boldsymbol{\mu}$ *is a solution to the mean field game (under the weak formulation of the stochastic optimal control problem) if, for any* $t \in [0, T]$:

$$\mathbb{P}^{\mu, \hat{\alpha}^{\mu}} \circ X_t^{-1} = \mu_t.$$

It is important to notice that, under $\mathbb{P}^{\mu, \hat{\alpha}^{\mu}}$, the process $(\boldsymbol{X}, \boldsymbol{Y}^{\mu}, \boldsymbol{Z}^{\mu})$ solves the FBSDE:

$$\begin{cases} dX_t = b\big(t, X_t, \mu_t, \hat{\alpha}(t, X_t, \mu_t, \sigma(t, X_t)^{-1\dagger} Z_t^{\mu})\big)dt \\ \qquad\quad + \sigma(t, X_t) dW_t^{\mu, \hat{\alpha}^{\mu}}, \\ dY_t^{\mu} = -f\big(t, X_t, \mu_t, \hat{\alpha}(t, X_t, \mu_t, \sigma(t, X_t)^{-1\dagger} Z_t^{\mu})\big)dt \\ \qquad\quad + Z_t^{\mu} \cdot dW_t^{\mu, \hat{\alpha}^{\mu}}, \quad t \in [0, T], \end{cases} \tag{3.33}$$

with $X_0 = \xi$ as initial condition and $Y_T^{\mu} = G(X_T, \mu_T)$ as terminal condition. Indeed, owing to Theorem 4.18 in Chapter 4, we know that, for any $p \geqslant 1$,

$$\mathbb{E}\bigg[\bigg(\int_0^T |Z_t|^2 dt\bigg)^p\bigg] < \infty.$$

Since $d\mathbb{P}^{\mu, \hat{\alpha}^{\mu}}/d\mathbb{P}$ is in any $L^p(\Omega, \mathcal{F}_T, \mathbb{P}; \mathbb{R})$, $p \geqslant 1$, we deduce that

$$\mathbb{E}^{\mu, \hat{\alpha}^{\mu}}\bigg[\bigg(\int_0^T |Z_t|^2 dt\bigg)^p\bigg] < \infty,$$

proving that the martingale in the backward component of (3.33) is square-integrable.

We shall prove in Theorem 4.12 (strong uniqueness for (3.33)) and in Theorem (Vol I)-1.33 (version of the Yamada-Watanabe theorem for FBSDEs) that uniqueness in law holds for (3.33). In particular, the law of the solution remains the same if $\boldsymbol{W}^{\mu, \hat{\alpha}^{\mu}}$ is replaced by $\boldsymbol{W}$. Meanwhile, observe that (3.33) fits (3.17) and (3.18).

We have the following characterization of the MFG equilibria:

Proposition 3.13 *Under assumption* **Weak Formulation**, *a continuous flow of measures* $\boldsymbol{\mu} = (\mu_t)_{0 \leqslant t \leqslant T}$ *from* $[0, T]$ *to* $\mathcal{P}_2(\mathbb{R}^d)$ *is an MFG equilibrium if and only if* $\mu_t = \mathcal{L}(\hat{X}_t)$ *for any* $t \in [0, T]$, *where* $(\hat{\boldsymbol{X}}, \hat{\boldsymbol{Y}}, \hat{\boldsymbol{Z}})$ *solves the McKean-Vlasov FBSDE*

$$\begin{cases} d\hat{X}_t = b\big(t, \hat{X}_t, \mathcal{L}(\hat{X}_t), \hat{\alpha}(t, \hat{X}_t, \mathcal{L}(\hat{X}_t), \sigma(t, \hat{X}_t)^{-1\dagger}\hat{Z}_t\big)dt \\ \qquad\quad + \sigma(t, \hat{X}_t)dW_t, \\ d\hat{Y}_t = -f\big(t, \hat{X}_t, \mathcal{L}(\hat{X}_t), \hat{\alpha}(t, \hat{X}_t, \mathcal{L}(\hat{X}_t), \sigma(t, \hat{X}_t)^{-1\dagger}\hat{Z}_t\big)dt \\ \qquad\quad + \hat{Z}_t dW_t, \end{cases} \tag{3.34}$$

with $\hat{X}_0 = \xi$ *as initial condition and* $\hat{Y}_T = G(\hat{X}_T, \mathcal{L}(\hat{X}_T))$ *as terminal condition.*

As we already alluded to, we shall prove in Theorem 4.12 that the system (3.34) has a unique solution when the McKean-Vlasov component is replaced by a mere input $\boldsymbol{\mu} = (\mu_t)_{0 \leqslant t \leqslant T}$. This proves that there is no loss in replacing the noise $\boldsymbol{W}^{\boldsymbol{\mu}, \hat{\alpha}^{\boldsymbol{\mu}}}$ in (3.33) by $\boldsymbol{W}$, as done in (3.34).

Combining with Theorem 3.10, we finally deduce:

Theorem 3.14 *On top of assumption* **Weak Formulation**, *assume that the coefficients* b, f *and* g *are continuous in the measure argument* μ *and that the optimizer* $\hat{\alpha}$ *is also continuous in* μ. *Then, there exists an MFG equilibrium whenever the optimization problem in* (3.4) *is solved through the weak formulation.*

Remark 3.15 *The main shortcoming of Theorem 3.14 above is the restrictive assumption that the set A of possible control values is bounded. However it is possible to extend the application of the formulation based upon the representation of the value function to cases where this assumption is not satisfied.* ***For instance, Theorem 4.44 in Chapter 4 gives a more general solvability result for MFGs with unbounded*** A, σ ***depending upon*** μ ***and the optimal control problem*** (3.4) ***being understood in the strong sense!*** *At the current stage of our presentation of mean field games, we chose not to introduce the technical tools required to overcome the underlying obstacles by fear of obstructing the view of the road to the solution of these problems with too many technicalities.*

Remark 3.16 *The reader may want to compare the PDE system* (3.12) *with the mean field FBSDE* (3.34). *They suggest that the value of the adjoint process* Y_t *at time* t *should be identified with* $V(t, X_t)$. *Accordingly, the value of the representation process* Z_t *should be identified with* $\sigma^\dagger(t, X_t, \mathcal{L}(X_t))\partial_x V(t, X_t)$. *Hence the dynamics of* $\boldsymbol{Y} = (Y_t)_{0 \leqslant t \leqslant T}$ *are directly connected with the dynamics of the value function* V *in* (3.12) *along the optimal paths. Similarly, the distribution of* X_t *at time* t *should be identified to* μ_t *in* (4.70). *We shall revisit this question again in Chapter 4.*

3.3.2 The Stochastic Maximum Principle Approach

The strategy advocated in this subsection is based on the probabilistic description of the optimal states of the optimization problem (3.4) provided by the stochastic maximum principle. Recall that under very general conditions, the necessary condition of this principle identifies the optimal states of the problem (3.4). It posits that the optimally controlled state satisfies the forward dynamics in a characteristic FBSDE, referred to as *the adjoint system* of the stochastic optimization problem, see the end of the subsection for more details. Moreover, the stochastic maximum principle also provides a sufficient condition singling out convexity conditions, under which the forward dynamics of any solution to the adjoint system is guaranteed to be optimal. In this chapter, we use the sufficient condition to prove the existence of solutions to problem (i)–(ii) as stated in Subsection 3.1.2. The challenges posed by the fixed point step (ii) require additional assumptions. We shall assume:

Assumption (SMP). The coefficients b, f, σ and g are defined on $[0,T] \times \mathbb{R}^d \times \mathcal{P}_2(\mathbb{R}^d) \times A$, $[0,T] \times \mathbb{R}^d \times \mathcal{P}_2(\mathbb{R}^d) \times A$, $[0,T] \times \mathbb{R}^d$, and $\mathbb{R}^d \times \mathcal{P}_2(\mathbb{R}^d)$ respectively. They satisfy:

(A1) The drift b is an affine function of (x, α) of the form:

$$b(t,x,\mu,\alpha) = b_0(t,\mu) + b_1(t)x + b_2(t)\alpha,$$

where $b_0 : [0,T] \times \mathcal{P}_2(\mathbb{R}^d) \ni (t,\mu) \mapsto b_0(t,\mu)$, $b_1 : [0,T] \ni t \mapsto b_1(t)$ and $b_2 : [0,T] \ni t \mapsto b_2(t)$ are $\mathbb{R}^d$, $\mathbb{R}^{d\times d}$ and $\mathbb{R}^{d\times k}$ valued respectively, and are measurable and bounded on bounded subsets of their respective domains.

(A2) The function σ is constant.

(A3) There exist two constants $\lambda > 0$ and $L \geqslant 1$ such that the function $\mathbb{R}^d \times A \ni (x,\alpha) \mapsto f(t,x,\mu,\alpha) \in \mathbb{R}$ is once continuously differentiable with Lipschitz-continuous derivatives (so that $f(t,\cdot,\mu,\cdot)$ is $C^{1,1}$), the Lipschitz constant in x and α being bounded by L (so that it is uniform in t and μ). Moreover, it satisfies the following strong form of convexity:

$$\begin{aligned} &f(t,x',\mu,\alpha') - f(t,x,\mu,\alpha) - (x'-x, \alpha'-\alpha) \cdot \partial_{(x,\alpha)} f(t,x,\mu,\alpha) \\ &\geqslant \lambda |\alpha' - \alpha|^2. \end{aligned} \tag{3.35}$$

The notation $\partial_{(x,\alpha)} f$ stands for the gradient in the joint variables (x,α). Finally, f, $\partial_x f$ and $\partial_\alpha f$ are locally bounded over $[0,T] \times \mathbb{R}^d \times \mathcal{P}_2(\mathbb{R}^d) \times A$.

(continued)

(A4) The function $\mathbb{R}^d \times \mathcal{P}_2(\mathbb{R}^d) \ni (x,\mu) \mapsto g(x,\mu)$ is locally bounded. Moreover, for any $\mu \in \mathcal{P}_2(\mathbb{R}^d)$, the function $\mathbb{R}^d \ni x \to g(x,\mu)$ is once continuously differentiable and convex, and has a L-Lipschitz-continuous first order derivative.

Assumption (**A2**) is presumably too restrictive, and the results of this section could still be true under the more general assumption $\sigma(t,x) = \sigma_0(t) + \sigma_1(t)x$ of linearity instead of boundedness of the volatility, see for instance the generalizations in Section (Vol II)-3.4.

Convenient Forms of the Stochastic Maximum Principle. We shall take advantage of the following variations on the standard proof of the stochastic maximum principle. Their formulations are tailored to the needs of this chapter and the next one. They provide not only existence, but quantitative estimates when the uniform convexity assumptions hold.

Theorem 3.17 *Let us assume that assumption* **SMP** *holds and that the mapping* $\boldsymbol{\mu} : [0,T] \ni t \mapsto \mu_t \in \mathcal{P}_2(\mathbb{R}^d)$ *is measurable and bounded. Then, the FBSDE:*

$$\begin{cases} dX_t = b\big(t, X_t, \mu_t, \hat{\alpha}(t, X_t, \mu_t, Y_t)\big)dt + \sigma dW_t, \\ dY_t = -\partial_x H\big(t, X_t, \mu_t, Y_t, \hat{\alpha}(t, X_t, \mu_t, Y_t)\big)dt \\ \qquad\qquad +Z_t dW_t, \quad t \in [0,T], \\ X_0 = \xi, \quad Y_T = \partial_x g(X_T, \mu_T), \end{cases} \tag{3.36}$$

where $\hat{\alpha}$ is the minimizer of the Hamiltonian constructed in Lemma 3.3, has a solution $(\boldsymbol{X}, \boldsymbol{Y}, \boldsymbol{Z}) = (X_t, Y_t, Z_t)_{0\leq t\leq T}$ satisfying:

$$\mathbb{E}\bigg[\sup_{0\leq t\leq T}\big(|X_t|^2 + |Y_t|^2\big) + \int_0^T |Z_t|^2 dt\bigg] < +\infty. \tag{3.37}$$

If we define the control process $\hat{\boldsymbol{\alpha}} = (\hat{\alpha}_t)_{0\leq t\leq T}$ by $\hat{\alpha}_t = \hat{\alpha}(t, X_t, \mu_t, Y_t)$, then for any progressively measurable admissible control $\boldsymbol{\alpha} = (\alpha_t)_{0\leq t\leq T}$ satisfying (3.2), it holds that:

$$J^{\boldsymbol{\mu}}(\hat{\boldsymbol{\alpha}}) + \lambda \mathbb{E}\int_0^T |\alpha_t - \hat{\alpha}_t|^2 dt \leq J^{\boldsymbol{\mu}}(\boldsymbol{\alpha}). \tag{3.38}$$

Remark 3.18 *While $\hat{\boldsymbol{\alpha}}$ will be shown to be progressively measurable with respect to the filtration generated by $\mathcal{F}_0$ and $\boldsymbol{W}$, $\boldsymbol{\alpha}$ in (3.38) may be assumed to be progressively measurable with respect to a larger filtration. Put it differently, the filtration $\mathbb{F}$ used in the definition of $\mathbb{A}$ may not be generated by $\mathcal{F}_0$ and $\boldsymbol{W}$.*

Proof. The proof of the existence of a solution to (3.36) is deferred to Chapter 4; see Lemma 4.56 there. Here we just focus on the proof of inequality (3.38). By Lemma 3.3, $\hat{\boldsymbol{\alpha}} = (\hat{\alpha}_t)_{0 \leqslant t \leqslant T}$ satisfies (3.2), and the proof of the stochastic maximum principle (see for example the proof given in Theorem 2.16) gives:

$$\begin{aligned} J^{\mu}(\boldsymbol{\alpha}) \geqslant J^{\mu}(\hat{\boldsymbol{\alpha}}) + \mathbb{E}\int_0^T \big[H(t, X_t^{\boldsymbol{\alpha}}, \mu_t, Y_t, \alpha_t) - H(t, X_t, \mu_t, Y_t, \hat{\alpha}_t) \\ - (X_t^{\boldsymbol{\alpha}} - X_t) \cdot \partial_x H(t, X_t, \mu_t, Y_t, \hat{\alpha}_t) - (\alpha_t - \hat{\alpha}_t) \cdot \partial_\alpha H(t, X_t, \mu_t, Y_t, \hat{\alpha}_t)\big] dt. \end{aligned}$$

By linearity of b and assumption **(A3)** on f, the Hessian of H satisfies (3.35), so that the required convexity assumption is satisfied. The result easily follows. □

Remark 3.19 *As the proof shows, and which is exactly what we claimed in the previous remark, there is no need for $\mathbb{F}$ to be the filtration generated by $\mathbb{F}_0$ and the Wiener process $\boldsymbol{W} = (W_t)_{0 \leqslant t \leqslant T}$.*

Remark 3.20 *Theorem 3.17 has interesting consequences. First, it says that the optimal control exists and is unique. Second, it also implies uniqueness of the solution of the FBSDE* (3.36)*. Indeed, given two solutions $(\boldsymbol{X}, \boldsymbol{Y}, \boldsymbol{Z})$ and $(\boldsymbol{X}', \boldsymbol{Y}', \boldsymbol{Z}')$ of* (3.36)*, $\mathrm{Leb}_1 \otimes \mathbb{P}$ a.e. it holds by* (3.38) *that:*

$$\hat{\alpha}(t, X_t, \mu_t, Y_t) = \hat{\alpha}(t, X_t', \mu_t, Y_t'),$$

so that $\boldsymbol{X}$ and $\boldsymbol{X}'$ coincide by the Lipschitz property of the coefficients of the forward equation. As a consequence, $(\boldsymbol{Y}, \boldsymbol{Z})$ and $(\boldsymbol{Y}', \boldsymbol{Z}')$ coincide as well.

The bound provided by Theorem 3.17 is sharp in the class of convex models as shown for example by the following slight variation on the same theme. We shall use this form repeatedly in this chapter and the next one.

Proposition 3.21 *Under the assumptions and notation of Theorem 3.17 above, if we consider in addition another measurable and bounded flow $[0, T] \ni t \mapsto \mu_t' \in \mathcal{P}_2(\mathbb{R}^d)$ of probability measures of order 2, and the corresponding controlled state process $\boldsymbol{X}' = (X_t')_{0 \leqslant t \leqslant T}$ defined by:*

$$X_t' = \xi' + \int_0^t b(s, X_s', \mu_s', \alpha_s) ds + \sigma W_t, \quad t \in [0, T],$$

for an initial condition $\xi' \in L^2(\Omega, \mathcal{F}_0, \mathbb{P}; \mathbb{R}^d)$ possibly different from ξ and an arbitrary control $\boldsymbol{\alpha} \in \mathbb{A}$, then:

$$\begin{aligned} J^{\mu}(\hat{\boldsymbol{\alpha}}) + \mathbb{E}\big[(\xi' - \xi) \cdot Y_0\big] + \lambda \mathbb{E}\int_0^T |\alpha_t - \hat{\alpha}_t|^2 dt \\ \leqslant J^{\mu}\big([\boldsymbol{\alpha}, \boldsymbol{\mu}']\big) - \mathbb{E}\bigg[\int_0^T \big(b_0(t, \mu_t') - b_0(t, \mu_t)\big) \cdot Y_t dt\bigg], \end{aligned} \tag{3.39}$$

where the quantity $J^{\mu}([\boldsymbol{\alpha}, \boldsymbol{\mu}'])$ *is defined by:*

$$J^{\mu}([\boldsymbol{\alpha}, \boldsymbol{\mu}']) = \mathbb{E}\left[g(X'_T, \mu_T) + \int_0^T f(t, X'_t, \mu_t, \alpha_t) dt \right].$$

The process $\boldsymbol{X}'$ is the controlled diffusion process driven by the control $\boldsymbol{\alpha}$ and evolving in the environment $\boldsymbol{\mu}'$, but the cost functional is computed under the environment $\boldsymbol{\mu}$.

Proof. As before, we use the same old strategy of the original proof of the stochastic maximum principle, by computing the Itô differential of the process:

$$\left((X'_t - X_t) \cdot Y_t + \int_0^t \big[f(s, X'_s, \mu_s, \alpha_s) - f(s, X_s, \mu_s, \hat{\alpha}_s) \big] ds \right)_{0 \leqslant t \leqslant T},$$

and integrating it between 0 and T. Since the initial conditions ξ and ξ' are possibly different, we get the additional term $\mathbb{E}[(\xi' - \xi) \cdot Y_0]$ in the left-hand side of (3.39). Similarly, since the drift of $\boldsymbol{X}'$ is driven by $\boldsymbol{\mu}' = (\mu'_t)_{0 \leqslant t \leqslant T}$, we get the additional difference of the drifts in order to account for the fact that the drifts are driven by the different flows of probability measures. □

Connection with FBSDEs of the McKean-Vlasov Type. In order to solve the standard stochastic control problem (3.4) using the Pontryagin maximum principle, we minimize the Hamiltonian H with respect to the control variable α, and inject the minimizer $\hat{\alpha}$ into the forward equation of the state as well as the backward equation defining the adjoint processes. Since the minimizer $\hat{\alpha}$ depends upon both the forward state X_t and the adjoint process Y_t, this creates a strong coupling between the forward and backward equations, leading to the FBSDE (3.36). The MFG matching condition (ii) of Subsection 3.1.2 then reads: seek a flow of probability distributions $\boldsymbol{\mu} = (\mu_t)_{0 \leqslant t \leqslant T}$ of order 2 such that the process $\boldsymbol{X}$ solving the forward equation of (3.36) admits $\boldsymbol{\mu} = (\mu_t)_{0 \leqslant t \leqslant T}$ as flow of marginal distributions. As already explained, the decision to consider probability measures of order 2 is not restrictive because standard estimates for solutions of stochastic differential equations imply that the state X_t will necessarily have a finite second order moment, provided that $\xi \in L^2(\Omega, \mathcal{F}_0, \mathbb{P}; \mathbb{R}^d)$. We thus deduce:

Definition 3.22 *Under assumption* **SMP**, *for any continuous flow of measures* $\boldsymbol{\mu} = (\mu_t)_{0 \leqslant t \leqslant T}$ *from* $[0, T]$ *to* $\mathcal{P}_2(\mathbb{R}^d)$, *call* $(\hat{\boldsymbol{X}}^{\mu}, \hat{\boldsymbol{Y}}^{\mu}, \hat{\boldsymbol{Z}}^{\mu})$ *the solution to FBSDE* (3.36) *(which is unique by Remark 3.20). Then, we say that* $\boldsymbol{\mu}$ *is a solution to the mean field game* (3.4) *or an MFG equilibrium if, for any* $t \in [0, T]$,

$$\mathbb{P} \circ (\hat{X}^{\mu}_t)^{-1} = \mu_t.$$

Similar to Definition 3.12, Definition 3.22 captures the essence of the approach to mean field games summarized in Subsection 3.1.2. The crux of this approach is to freeze the probability measure when optimizing the cost. This is in sharp contrast with the study of the control of McKean-Vlasov dynamics investigated in Chapter 6. Indeed in that case, optimization is also performed with respect to the measure argument.

The net result is that, once the flow of probability measures giving the fixed point is injected in the FBSDE, it becomes clear that the stochastic maximum principle approach to the solution of the mean field game problem amounts to the solution of an FBSDE of the McKean-Vlasov type since the marginal distribution of the solution appears in the coefficients of the equation.

In analogy with Proposition 3.13, we claim:

Proposition 3.23 *Under the assumption of Definition 3.22, a continuous flow of measures $\boldsymbol{\mu} = (\mu_t)_{0 \leq t \leq T}$ from $[0, T]$ to $\mathcal{P}_2(\mathbb{R}^d)$ is an MFG equilibrium if and only if $\mu_t = \mathcal{L}(\hat{X}_t)$ for any $t \in [0, T]$, where $(\hat{\boldsymbol{X}}, \hat{\boldsymbol{Y}}, \hat{\boldsymbol{Z}})$ solves the McKean-Vlasov FBSDE*

$$\begin{cases} d\hat{X}_t = b\big(t, \hat{X}_t, \mathcal{L}(\hat{X}_t), \hat{\alpha}(t, \hat{X}_t, \mathcal{L}(\hat{X}_t), \hat{Y}_t)\big)dt + \sigma dW_t, \\ d\hat{Y}_t = -\partial_x H\big(t, \hat{X}_t, \mathcal{L}(\hat{X}_t), \hat{Y}_t, \hat{\alpha}(t, \hat{X}_t, \mathcal{L}(\hat{X}_t), \hat{Y}_t)\big)dt + \hat{Z}_t dW_t, \end{cases} \tag{3.40}$$

with $\hat{X}_0 = \xi$ as initial condition and $\hat{Y}_T = \partial_x g(\hat{X}_T, \mathcal{L}(\hat{X}_T))$ as terminal condition.

With the special form of coefficients chosen in assumption **SMP**, the FBSDE reads:

$$\begin{cases} d\hat{X}_t = \big[b_0\big(t, \mathcal{L}(\hat{X}_t)\big) + b_1(t)\hat{X}_t + b_2(t)\hat{\alpha}\big(t, \hat{X}_t, \mathcal{L}(\hat{X}_t), \hat{Y}_t\big)\big]dt \\ \qquad\qquad + \sigma dW_t, \\ d\hat{Y}_t = -\big[b_1^\dagger(t)\hat{Y}_t + \partial_x f\big(t, \hat{X}_t, \mathcal{L}(\hat{X}_t), \hat{\alpha}(t, \hat{X}_t, \mathcal{L}(\hat{X}_t), \hat{Y}_t)\big)\big]dt \\ \qquad\qquad + \hat{Z}_t dW_t, \end{cases} \tag{3.41}$$

where, as usual, $b_1^\dagger$ denotes the transpose of the matrix b_1.

Existence of a Solution to the MFG Problem. Using Theorem 3.10, we obtain a first solvability result for the McKean-Vlasov FBSDE (3.41).

Theorem 3.24 *On top of assumption* **SMP**, *assume that the set A is bounded, that σ is invertible, that the coefficients b_0, $\partial_x f$ and $\partial_x g$ are globally bounded, that b_1 is zero, and that the coefficients b_0, $\partial_x f$, $\partial_x g$ and $\hat{\alpha}$ are also continuous in the measure argument μ. Then, there exists a solution to the MFG problem.*

Proof. It suffices to apply Theorem 3.10 with:

$$\begin{aligned} B\big(t, X_t, \mathcal{L}(X_t), Y_t, Z_t\big) &= b\big(t, X_t, \mathcal{L}(X_t), \hat{\alpha}(t, X_t, \mathcal{L}(X_t), Y_t)\big), \\ F\big(t, X_t, \mathcal{L}(X_t), Y_t, Z_t\big) &= \partial_x f\big(t, X_t, \mathcal{L}(X_t), \hat{\alpha}(t, X_t, \mathcal{L}(X_t), Y_t)\big), \\ G\big(X_T, \mathcal{L}(X_T)\big) &= \partial_x g\big(X_T, \mathcal{L}(X_T)\big). \end{aligned}$$

Regularity properties of $\hat{\alpha}$ follow from Lemma 3.3, continuity with respect to μ being easily tackled by a compactness argument. □

Remark 3.25 *Clearly, demanding that $\partial_x f$ and $\partial_x g$ are bounded while f and g are already assumed to be convex, is very restrictive. For instance, the Linear Quadratic (LQ) models considered in Section 3.5 are not covered by this result since the driver F is not allowed to have linear growth. We shall revisit the problem under much weaker conditions in Chapter 4. In full analogy with Remark 3.15,* ***the reader will find in Chapter 4 a more general version of Theorem 3.24*** *in which the set A and the coefficients b_0, $\partial_x f$ and $\partial_x g$ are allowed to be unbounded and the coefficient b_1 to be nonzero. We refer to Theorem 4.53 for a precise statement.*

Remark 3.26 *If we recall the content of Remark 3.16, it is enlightening to compare the PDE system* (3.12) *with the mean field FBSDE* (3.40). *Formally, the value of the adjoint process Y_t at time t should be identified with $\partial_x V(t, X_t)$, so that the dynamics of $\boldsymbol{Y}$ are directly connected with the dynamics of the gradient of the value function V in* (3.12) *along the optimal paths. Similarly, the distribution of X_t at time t should be identified to μ_t in* (3.12).

The Stochastic Maximum Principle as a Necessary Condition

The statement of Theorem 3.17 provides a sufficient condition for proving the optimality of the forward component in the forward-backward system (3.36). As explained above, it is usually referred to as the *sufficient condition of the stochastic maximum principle*.

In the proof of Theorem 3.17, the convexity conditions required in assumption **SMP** play a crucial role as they permit to turn the condition (3.36) into a global minimality property. In this regard, the full complete version of the stochastic maximum principle sheds more light on the exact meaning of the condition (3.36). It is the contribution of the *necessary condition of the stochastic maximum principle* to show that the condition (3.36) is in fact a first-order order criticality condition for the minimization problem $\inf_{\boldsymbol{\alpha}\in\mathbb{A}} J^{\mu}(\boldsymbol{\alpha})$.

In order to state properly this necessary condition, there is no need to require the full convexity condition (**A3**). Under the same hypothesis as before that σ is constant, we shall just assume further:

Assumption (Necessary SMP).

(A1) The functions b and f are differentiable with respect to (x, α), the mappings $\mathbb{R}^d \times A \ni (x, \alpha) \mapsto \partial_x(b,f)(t, x, \mu, \alpha)$ and $\mathbb{R}^d \times A \ni (x, \alpha) \mapsto \partial_\alpha(b,f)(t, x, \mu, \alpha)$ being continuous for each $(t, \mu) \in [0, T] \times \mathcal{P}_2(\mathbb{R}^d)$. Similarly, the function g is differentiable with respect to x, the mapping $\mathbb{R}^d \ni (x, \mu) \mapsto \partial_x g(x, \mu)$ being continuous for each $\mu \in \mathcal{P}_2(\mathbb{R}^d)$.

(A2) The functions $[0, T] \ni t \mapsto (b,f)(t, 0, \delta_0, 0_A)$ are uniformly bounded, for some point $0_A \in A$. The derivative $\partial_{(x,\alpha)} b$ is uniformly bounded and, for any $R \geqslant 0$ and any $(t, \mu) \in [0, T] \times \mathcal{P}_2(\mathbb{R}^d)$ such that $M_2(\mu) \leqslant R$, the function $\partial_x f(t, \cdot, \mu, \cdot)$, $\partial_x g(\cdot, \mu)$ and $\partial_\alpha f(t, \cdot, \mu, \cdot)$ are at most of linear growth in (x, α).

Theorem 3.27 *Let* $\boldsymbol{\mu} = (\mu_t)_{0 \leqslant t \leqslant T}$ *be a bounded and measurable function from* $[0, T]$ *into* $\mathcal{P}_2(\mathbb{R}^d)$ *and* $\boldsymbol{\alpha} = (\alpha_t)_{0 \leqslant t \leqslant T} \in \mathbb{A}$ *be an admissible control. Under assumption* **Necessary SMP**, *assume further that the Hamiltonian* H *is convex in* $\alpha \in A$. *If* $\boldsymbol{\alpha}$ *is optimal, then, for the associated controlled state* $\boldsymbol{X}^{\boldsymbol{\alpha}} = (X_t^{\boldsymbol{\alpha}})_{0 \leqslant t \leqslant T}$, *and the corresponding solution* $(\boldsymbol{Y}, \boldsymbol{Z}) = (Y_t, Z_t)_{0 \leqslant t \leqslant T}$ *of the adjoint backward SDE:*

$$\begin{cases} dY_t = -\partial_x H\big(t, X_t^{\boldsymbol{\alpha}}, \mu_t, Y_t, \alpha_t\big)dt + Z_t dW_t, & t \in [0, T], \\ Y_T = \partial_x g\big(X_T^{\boldsymbol{\alpha}}, \mu_T\big), \end{cases} \tag{3.42}$$

we have for all $\alpha \in A$:

$$H\big(t, X_t^{\boldsymbol{\alpha}}, \mu_t, Y_t, \alpha_t\big) \leqslant H\big(t, X_t^{\boldsymbol{\alpha}}, \mu_t, Y_t, \alpha\big) \quad \text{Leb}_1 \otimes \mathbb{P} \ a.e. . \tag{3.43}$$

Since we make little use of Theorem 3.27 in this chapter and the next, we postpone its proof to Chapters 6 and (Vol II)-1, where more general versions are given, including cases where σ is not constant, see Theorem 6.14 for mean field stochastic control problems and Theorem (Vol II)-1.59 for stochastic control problems in a random environment. Also, as indicated in Proposition 6.15, a weaker form holds if convexity of H in α fails. Roughly speaking, the corresponding version says that, instead of (3.43), it holds $\partial_\alpha H(t, X_t, \mu_t, Y_t, \alpha_t) = 0$ when α_t is in the interior of A.

3.4 Lasry-Lions Monotonicity Condition

We postpone to Chapter 4 the detailed proof of Theorem 3.10, which served in the previous subsection as the basic ingredient for establishing the existence of a solution to the MFG problem. For the time being, we address the question of uniqueness and provide a general criterion under which it is guaranteed. So far, very little has been said about uniqueness. We claimed in Subsection 3.2.3 that Cauchy-Lipschitz theory was true only in small time, a fact which will be proved rigorously in Subsection 4.2.3. Accordingly, we based the construction of solutions over time intervals of arbitrary length upon Theorem 3.10, whose proof relies on Schauder's theorem for the existence, though not the uniqueness, of fixed points.

Regarding the interpretation of the solution (i) – (ii) of an MFG equilibrium based on FBSDEs of the McKean-Vlasov type, it would be tempting to adapt the arguments used to ensure uniqueness of solutions to classical FBSDEs to the McKean-Vlasov setting. Inspired by the discussion of Subsection 3.2.3, we can imagine three possible avenues to uniqueness:

1. The first one is to assume that the coupling between the forward and backward equations is weak in the sense that one of the two equations depends on the solution of the other one through coefficients with a small Lipschitz constant. Basically, this amounts to assuming that the time horizon T is small enough.

2. In full analogy with the analysis of the inviscid Burgers equation presented in Subsection 3.2.3, the second one is to make use of monotonicity conditions, but in the direction of the measure argument.
3. Finally, given the role played by the Laplace operator in the viscous Burgers equation, another possibility is to make use of non-degeneracy conditions, but on the space of probability measures this time around.

Again, existence and uniqueness in small time under Lipschitz conditions will be investigated in Subsection 4.2.3.

Adapting the third strategy to the McKean-Vlasov case is much more challenging as the state variable has to be understood as the pair made of X_t, which describes the private state of the player at time t, and of $\mathcal{L}(X_t)$, which stands for the statistical distribution of the states in the population at time t. As we shall see in Chapters (Vol II)-4 and (Vol II)-5, the analogue of the viscous Burgers equation, whose solution is the *decoupling field* of the FBSDE (3.24), is a PDE on the space of probability measures, called the *master equation*. To put it differently, the decoupling field of a McKean-Vlasov FBSDE has to be understood as a function over an infinite-dimensional space. It is thus a rather intricate object. Moreover, it is worth mentioning that, for mean field games without common noise, the dynamics of $(\mathcal{L}(X_t))_{0\leqslant t\leqslant T}$ is entirely deterministic, and for this reason, we cannot invoke a non-degeneracy argument. As we shall see in Chapters (Vol II)-2, (Vol II)-3, (Vol II)-4, and (Vol II)-5, it is only in the presence of a common noise that we may expect these arguments to make sense. Actually, even the framework considered in these four chapters is too restrictive to address the smoothing effect of the common noise in full generality. Indeed, except for a few cases, strict ellipticity cannot hold true if the common noise is of finite dimension, a situation we encounter throughout the book. The few examples which could work are cases where the marginal laws $(\mathcal{L}(X_t))_{0\leqslant t\leqslant T}$ belong to a parametric family. We provide such an example in Subsection (Vol II)-3.5.2 where we manage to prove that the common noise restores uniqueness in some specific cases.

Therefore, at this stage of the discussion, it seems that there is only one possible road to uniqueness, and it has to be based on a structural monotonicity condition. We make this clear in what follows.

3.4.1 A First Notion of Monotonicity

The following definition of monotonicity is taken from the earlier works by Lasry and Lions. We call it *Lasry-Lions monotonicity condition*.

Definition 3.28 *A real valued function h on $\mathbb{R}^d \times \mathcal{P}_2(\mathbb{R}^d)$ is said to be monotone (in the sense of Lasry and Lions), if, for all $\mu \in \mathcal{P}_2(\mathbb{R}^d)$, the mapping $\mathbb{R}^d \ni x \mapsto h(x,\mu)$ is at most of quadratic growth, and, for all $\mu, \mu' \in \mathcal{P}_2(\mathbb{R}^d)$, we have:*

$$\int_{\mathbb{R}^d} [h(x,\mu) - h(x,\mu')] \, d(\mu - \mu')(x) \geqslant 0.$$

Clearly, any linear combination of functions which satisfy the Lasry-Lions Monotonicity condition also satisfies it if the coefficients are nonnegative. A first set of examples of monotone functions will be provided in Subsection 3.4.2. More properties of monotone functions, including convexity, will be discussed in Chapter 5, see for instance Remark 5.75.

We now introduce what turns out to be the most popular set of assumptions under which uniqueness has been proven to hold in the existing literature. It goes back to the earlier works of Lasry and Lions on mean field games. With the same notation as in Subsection 3.1.2, it reads as follows.

Assumption (Lasry-Lions Monotonicity).

(A1) The coefficients b and σ do not depend upon the measure argument. They thus read as mappings $b : [0,T] \times \mathbb{R}^d \times A \to \mathbb{R}^d$ and $\sigma : [0,T] \times \mathbb{R}^d \to \mathbb{R}^{d \times d}$.

(A2) The running cost f has a separated structure of the form:

$$f(t,x,\mu,\alpha) = f_0(t,x,\mu) + f_1(t,x,\alpha), \qquad t \in [0,T],\ x \in \mathbb{R}^d,\ \alpha \in A,\ \mu \in \mathcal{P}_2(\mathbb{R}^d),$$

where f_0 is a Borel-measurable mapping from $[0,T] \times \mathbb{R}^d \times \mathcal{P}_2(\mathbb{R}^d)$ into $\mathbb{R}$ and f_1 is a Borel-measurable mapping from $[0,T] \times \mathbb{R}^d \times A$ into $\mathbb{R}$. Moreover,

$$|f(t,x,\mu,\alpha)| \leqslant C\big(1 + |x| + M_2(\mu) + |\alpha|\big)^2,$$
$$|g(x,\mu)| \leqslant C\big(1 + |x| + M_2(\mu)\big)^2,$$

for all $(t,x,\mu,\alpha) \in [0,T] \times \mathbb{R}^d \times \mathcal{P}_2(\mathbb{R}^d) \times A$, for a constant $C \geqslant 0$.

(A3) The functions $f_0(t,\cdot,\cdot)$ for $t \in [0,T]$, and g are monotone in the sense of Definition 3.28.

The main result of this section is the following important uniqueness consequence of the monotonicity assumption.

Theorem 3.29 *Let assumption* **Lasry-Lions Monotonicity** *hold, and let us assume that for any deterministic continuous flow* $\boldsymbol{\mu} = (\mu_t)_{0 \leqslant t \leqslant T}$ *from* $[0,T]$ *to* $\mathcal{P}_2(\mathbb{R}^d)$, *the optimal control problem* (3.4) *has a unique minimizer* $\hat{\boldsymbol{\alpha}}^{\boldsymbol{\mu}} \in \mathbb{A}$. *Call* $\hat{\boldsymbol{X}}^{\boldsymbol{\mu}}$ *the corresponding optimal path.*

Then there exists at most one flow $\boldsymbol{\mu} = (\mu_t)_{0 \leqslant t \leqslant T}$ *so that:*

$$\forall t \in [0,T], \quad \mathcal{L}(\hat{X}_t^{\boldsymbol{\mu}}) = \mu_t. \tag{3.44}$$

In other words, there exists at most one MFG equilibrium.

Implicitly, we here require that, for any $\boldsymbol{\mu}$, $\hat{\boldsymbol{X}}^{\boldsymbol{\mu}}$ satisfies $\mathbb{E}[\sup_{0 \leqslant t \leqslant T} |\hat{X}_t^{\boldsymbol{\mu}}|^2] < \infty$.

Proof. Assume that there are two different MFG equilibria $\boldsymbol{\mu} = (\mu_t)_{0 \leqslant t \leqslant T}$ and $\boldsymbol{\mu}' = (\mu'_t)_{0 \leqslant t \leqslant T}$. Then, the processes $\hat{\boldsymbol{\alpha}}^{\boldsymbol{\mu}}$ and $\hat{\boldsymbol{\alpha}}^{\boldsymbol{\mu}'}$ must differ as otherwise $\hat{\boldsymbol{X}}^{\boldsymbol{\mu}}$ and then $\hat{\boldsymbol{X}}^{\boldsymbol{\mu}'}$ would be the same and then, by (3.44), $\boldsymbol{\mu}$ and $\boldsymbol{\mu}'$ would be the same as well. Therefore, by uniqueness of the minimizer of the cost functionals $J^{\boldsymbol{\mu}}$ and $J^{\boldsymbol{\mu}'}$, we have:

$$J^{\boldsymbol{\mu}}(\hat{\boldsymbol{\alpha}}^{\boldsymbol{\mu}}) - J^{\boldsymbol{\mu}}(\hat{\boldsymbol{\alpha}}^{\boldsymbol{\mu}'}) < 0 \quad \text{and} \quad J^{\boldsymbol{\mu}'}(\hat{\boldsymbol{\alpha}}^{\boldsymbol{\mu}'}) - J^{\boldsymbol{\mu}'}(\hat{\boldsymbol{\alpha}}^{\boldsymbol{\mu}}) < 0.$$

Adding the two inequalities, we get:

$$J^{\boldsymbol{\mu}}(\hat{\boldsymbol{\alpha}}^{\boldsymbol{\mu}}) - J^{\boldsymbol{\mu}'}(\hat{\boldsymbol{\alpha}}^{\boldsymbol{\mu}}) - \Big(J^{\boldsymbol{\mu}}(\hat{\boldsymbol{\alpha}}^{\boldsymbol{\mu}'}) - J^{\boldsymbol{\mu}'}(\hat{\boldsymbol{\alpha}}^{\boldsymbol{\mu}'})\Big) < 0. \tag{3.45}$$

Now, we use the fact that the coefficients b and σ are independent of μ. So in the environment $\boldsymbol{\mu}$, the controlled path driven by $\hat{\boldsymbol{\alpha}}^{\boldsymbol{\mu}'}$ is exactly $\hat{\boldsymbol{X}}^{\boldsymbol{\mu}'}$. Similarly, in the environment $\boldsymbol{\mu}'$, the controlled path driven by $\hat{\boldsymbol{\alpha}}^{\boldsymbol{\mu}}$ is exactly $\hat{\boldsymbol{X}}^{\boldsymbol{\mu}}$. Therefore,

$$\begin{aligned}
J^{\boldsymbol{\mu}}(\hat{\boldsymbol{\alpha}}^{\boldsymbol{\mu}}) - J^{\boldsymbol{\mu}'}(\hat{\boldsymbol{\alpha}}^{\boldsymbol{\mu}}) = \mathbb{E}\bigg[& \int_0^T \Big(f_0\big(t, \hat{X}_t^{\boldsymbol{\mu}}, \mu_t\big) - f_0\big(t, \hat{X}_t^{\boldsymbol{\mu}}, \mu'_t\big)\Big) dt \\
& + \int_0^T \Big(f_1\big(t, \hat{X}_t^{\boldsymbol{\mu}}, \hat{\alpha}_t^{\boldsymbol{\mu}}\big) - f_1\big(t, \hat{X}_t^{\boldsymbol{\mu}}, \hat{\alpha}_t^{\boldsymbol{\mu}}\big)\Big) dt + g(\hat{X}_T^{\boldsymbol{\mu}}, \mu_T) - g(\hat{X}_T^{\boldsymbol{\mu}}, \mu'_T)\bigg],
\end{aligned}$$

and we observe that the first term in the second line is zero. Thanks to (3.44), we deduce that:

$$\begin{aligned}
J^{\boldsymbol{\mu}}(\hat{\boldsymbol{\alpha}}^{\boldsymbol{\mu}}) - J^{\boldsymbol{\mu}'}(\hat{\boldsymbol{\alpha}}^{\boldsymbol{\mu}}) = & \int_0^T \int_{\mathbb{R}^d} (f_0(t,x,\mu_t) - f_0(t,x,\mu'_t)) d\mu_t(x) dt \\
& + \int_{\mathbb{R}^d} \big[g(x,\mu_T) - g(x,\mu'_T)\big] d\mu_T(x).
\end{aligned}$$

Similarly,

$$\begin{aligned}
J^{\boldsymbol{\mu}}(\hat{\boldsymbol{\alpha}}^{\boldsymbol{\mu}'}) - J^{\boldsymbol{\mu}'}(\hat{\boldsymbol{\alpha}}^{\boldsymbol{\mu}'}) = & \int_0^T \int_{\mathbb{R}^d} (f_0(t,x,\mu_t) - f_0(t,x,\mu'_t)) d\mu'_t(x) dt \\
& + \int_{\mathbb{R}^d} \big[g(x,\mu_T) - g(x,\mu'_T)\big] d\mu'_T(x).
\end{aligned}$$

Taking differences as in (3.45), we get:

$$
\begin{aligned}
J^{\mu}(\hat{\alpha}^{\mu}) &- J^{\mu'}(\hat{\alpha}^{\mu}) - \left(J^{\mu}(\hat{\alpha}^{\mu'}) - J^{\mu'}(\hat{\alpha}^{\mu'})\right) \\
&= \int_0^T \int_{\mathbb{R}^d} \left(f_0(t,x,\mu_t) - f_0(t,x,\mu_t')\right) d\left(\mu_t - \mu_t'\right)(x) dt \\
&\quad + \int_{\mathbb{R}^d} \left[g(x,\mu_T) - g(x,\mu_T')\right] d\left(\mu_T - \mu_T'\right)(x).
\end{aligned}
$$

By **(A3)** in assumption **Lasry-Lions Monotonicity**, the right-hand side is nonnegative, which contradicts (3.45) and concludes the proof. □

Remark 3.30 *It should be noticed that the Lasry-Lions monotonicity condition also guarantees uniqueness whenever the optimization problem is understood in the weak sense. Indeed, with the same definition of the cost functional $J^{\mu,\text{weak}}$ as in* (3.29)*, we have, for two optimal controls $\hat{\boldsymbol{\alpha}}^{\mu}$ and $\hat{\boldsymbol{\alpha}}^{\mu'}$ in $\mathbb{A}$,*

$$
\begin{aligned}
J^{\mu,\text{weak}}\left(\hat{\boldsymbol{\alpha}}^{\mu}\right) - J^{\mu',\text{weak}}\left(\hat{\boldsymbol{\alpha}}^{\mu}\right) &= \mathbb{E}^{\mu,\hat{\alpha}^{\mu}}\left[g(X_T,\mu_T) + \int_0^T f(t,X_t,\mu_t,\hat{\alpha}_t^{\mu}) dt\right] \\
&\quad - \mathbb{E}^{\mu',\hat{\alpha}^{\mu}}\left[g(X_T,\mu_T') + \int_0^T f(t,X_t,\mu_t',\hat{\alpha}_t^{\mu}) dt\right].
\end{aligned}
$$

Recalling that b and σ are independent of μ under assumption **Lasry-Lions Monotonicity**, *we observe that:*

$$
\mathbb{P}^{\mu,\hat{\alpha}^{\mu}} = \mathbb{P}^{\mu',\hat{\alpha}^{\mu}},
$$

so that:

$$
\begin{aligned}
J^{\mu,\text{weak}}&\left(\hat{\boldsymbol{\alpha}}^{\mu}\right) - J^{\mu',\text{weak}}\left(\hat{\boldsymbol{\alpha}}^{\mu}\right) \\
&= \mathbb{E}^{\mu,\hat{\alpha}^{\mu}}\left[g(X_T,\mu_T) - g(X_T,\mu_T') + \int_0^T \left(f_0(t,X_t,\mu_t) - f_0(t,X_t,\mu_t')\right) dt\right].
\end{aligned}
$$

Exploiting the fact that $\mathbb{P}^{\mu,\hat{\alpha}^{\mu}} \circ X_t^{-1} = \mu_t$, we then deduce that:

$$
\begin{aligned}
J^{\mu,\text{weak}}\left(\hat{\boldsymbol{\alpha}}^{\mu}\right) - J^{\mu',\text{weak}}\left(\hat{\boldsymbol{\alpha}}^{\mu}\right) &= \int_{\mathbb{R}^d} \left[g(x,\mu_T) - g(x,\mu_T')\right] d\mu_T(x) \\
&\quad + \int_0^T \int_{\mathbb{R}^d} \left[f_0(t,x,\mu_t) - f_0(t,x,\mu_t')\right] d\mu_t(x) dt,
\end{aligned}
$$

which suffices to repeat the proof of Theorem 3.29.

3.4.2 Examples

We now provide several examples of real valued functions h on $\mathbb{R}^d \times \mathcal{P}_2(\mathbb{R}^d)$ which are Lasry-Lions monotone, i.e., monotone in the sense of Definition 3.28.

Example 1. If h does not depend upon μ and is of quadratic growth in x, then it satisfies the requirements of Definition 3.28.

Example 2. If h does not depend upon x, then it also satisfies the requirements of Definition 3.28. Indeed, for any function $h : \mathcal{P}_2(\mathbb{R}^d) \to \mathbb{R}$ and for all $\mu, \mu' \in \mathcal{P}_2(\mathbb{R}^d)$,

$$\int_{\mathbb{R}^d} (h(\mu) - h(\mu'))d\big(\mu - \mu'\big)(x) = \big(h(\mu) - h(\mu')\big)\big(\mu - \mu'\big)(\mathbb{R}^d) = 0.$$

Example 3. Let $h : \mathbb{R}^d \times \mathcal{P}_2(\mathbb{R}^d) \to \mathbb{R}$ be given by:

$$h(x, \mu) = a\, x \cdot \bar{\mu}, \quad \text{with} \quad \bar{\mu} = \int_{\mathbb{R}^d} y d\mu(y),$$

for some $a > 0$. Then h satisfies the requirements of Definition 3.28. Indeed, for all $\mu, \mu' \in \mathcal{P}_2(\mathbb{R}^d)$,

$$\int_{\mathbb{R}^d} \big(h(x, \mu) - h(x, \mu')\big)d\big(\mu - \mu'\big)(x) = a \int_{\mathbb{R}^d} \int_{\mathbb{R}^d} x \cdot y d\big(\mu - \mu'\big)(y) d\big(\mu - \mu'\big)(x).$$

Therefore,

$$\int_{\mathbb{R}^d} \big(h(x, \mu) - h(x, \mu')\big)d\big(\mu - \mu'\big)(x) = a \left| \int_{\mathbb{R}^d} x d\big(\mu - \mu'\big)(x) \right|^2.$$

This example may be useful in linear-quadratic optimization problems, see Subsection 3.5 below.

Example 4. Let $h : \mathbb{R}^d \times \mathcal{P}_2(\mathbb{R}^d) \to \mathbb{R}$ be given by:

$$h(x, \mu) = \int_{\mathbb{R}^d} \ell(x - y) d\mu(y),$$

for some Borel-measurable odd function ℓ satisfying $|\ell(x)| \leqslant C(1 + |x|^2)$ for some $C \geqslant 0$ and all $x \in \mathbb{R}$. Then, h is also covered by Definition 3.28. Indeed, for all $\mu, \mu' \in \mathcal{P}_2(\mathbb{R}^d)$,

$$\begin{aligned}
&\int_{\mathbb{R}^d} (h(x,\mu) - h(x,\mu'))d(\mu - \mu')(x) \\
&\qquad = \int_{\mathbb{R}^d}\int_{\mathbb{R}^d} \ell(x-y)d\big(\mu-\mu'\big)(y)d(\mu-\mu')(x) \\
&\qquad = \frac{1}{2}\int_{\mathbb{R}^d}\int_{\mathbb{R}^d} \ell(x-y)d\big(\mu-\mu'\big)(y)d(\mu-\mu')(x) \\
&\qquad\quad - \frac{1}{2}\int_{\mathbb{R}^d}\int_{\mathbb{R}^d} \ell(y-x)d\big(\mu-\mu'\big)(y)d(\mu-\mu')(x) \\
&\qquad = 0,
\end{aligned}$$

where we used the fact that $\ell(x-y) = -\ell(y-x)$ to pass from the second to the third line. This form of function h is well adapted to our discussion of potential games in Chapter 6.

Example 5. Let $h : \mathbb{R}^d \times \mathcal{P}_2(\mathbb{R}^d) \to \mathbb{R}$ be given by:

$$h(x,\mu) = \int_{\mathbb{R}^d} \ell(x-y)d\mu(y),$$

for some symmetric function $\ell : \mathbb{R}^d \to \mathbb{R}_+$ writing

$$\ell(r) = \int_{\mathbb{R}^d} \exp(ir\cdot s)d\lambda(s), \quad r \in \mathbb{R}^d,$$

where λ is a symmetric positive finite measure on $\mathbb{R}^d$ and $i^2 = -1$.

Then, for all $\mu, \mu' \in \mathcal{P}_2(\mathbb{R}^d)$,

$$\begin{aligned}
&\int_{\mathbb{R}^d} \big(h(x,\mu) - h(x,\mu')\big)d\big(\mu - \mu'\big)(x) \\
&\qquad = \int_{\mathbb{R}^d}\int_{\mathbb{R}^d} \ell(x-y)d\big(\mu-\mu'\big)(y)d\big(\mu-\mu'\big)(x) \\
&\qquad = \int_{\mathbb{R}^d}\int_{\mathbb{R}^d}\int_{\mathbb{R}^d} \exp\big(i(x-y)\cdot s\big)d\lambda(s)d\big(\mu-\mu'\big)(y)d\big(\mu-\mu'\big)(x) \\
&\qquad = \int_{\mathbb{R}^d}\left|\int_{\mathbb{R}^d} \exp(ix\cdot s)d\big(\mu-\mu'\big)(x)\right|^2 d\lambda(s)
\end{aligned}$$

Taking λ as the Gaussian or Cauchy distributions, we deduce that monotonicity holds true with $\ell(x) = \exp(-\frac{1}{2}|x|^2)$ or $\ell(x) = \exp(-|x|)$.

Example 6. Let $d = 1$ and h be given by:

$$h(x,\mu) = \mu\big((-\infty,x)\big) + \frac{1}{2}\mu\big(\{x\}\big), \quad x \in \mathbb{R},\ \mu \in \mathcal{P}_2(\mathbb{R}).$$

Then, h satisfies the requirements of Definition 3.28. Notice that, when μ has no atoms, $h(x, \mu)$ coincides with the cumulative distribution function of μ at point x.

Indeed, using the sign function ($\text{sign}(x) = 1$ if $x > 0$, -1 if $x < 0$ and 0 if $x = 0$), we have

$$\begin{aligned}\int_{\mathbb{R}} \text{sign}(x-y)d\mu(y) &= \mu\big((-\infty, x)\big) - \mu\big((x, \infty)\big)\\ &= 2\mu\big((-\infty, x)\big) + \mu\big(\{x\}\big) - 1 = 2h(x, \mu) - 1.\end{aligned}$$

By the third example, we deduce that $2h - 1$ satisfies the Lasry-Lions monotonicity condition. By linearity, h satisfies it as well.

Example 7. Let $h : \mathbb{R}^d \times \mathcal{P}_2(\mathbb{R}^d) \to \mathbb{R}$ be given by

$$h(x, \mu) = \int_{\mathbb{R}^d} L\big(z, \rho * \mu(z)\big)\rho(x-z)dz,$$

where ρ is a bounded even smooth probability density function over $\mathbb{R}^d$ such that $\int_{\mathbb{R}^d} |x|^2 \rho(x)dx < \infty$ and $L : \mathbb{R}^d \times [0, \infty) \to [0, \infty)$ is nondecreasing in the second variable and satisfies, for any $r > 0$ and all $\varrho \in [-r, r]$, $|L(z, \varrho)| \leqslant C_r(1 + |\varrho|^2)$ for some constant $C_r \geqslant 0$. The notation $\rho * \mu$ denotes the standard convolution product. Notice in particular that the function $\rho * \mu$ is bounded. Then, h satisfies Definition 3.28. While the fact that h is at most of quadratic growth is easily checked, monotonicity may be proved as follows. For all $\mu, \mu' \in \mathcal{P}_2(\mathbb{R}^d)$,

$$\begin{aligned}&\int_{\mathbb{R}^d} \big(h(x, \mu) - h(x, \mu')\big)d\big(\mu - \mu'\big)(x)\\ &= \int_{\mathbb{R}^d} \Big[\Big(L\big(z, \rho * \mu(z)\big) - L\big(z, \rho * \mu'(z)\big)\Big) \int_{\mathbb{R}^d} \rho(x-z)\big(d\mu(x) - d\mu'(x)\big)\Big]dz\\ &= \int_{\mathbb{R}^d} \Big(L\big(z, \rho * \mu(z)\big) - L\big(z, \rho * \mu'(z)\big)\Big)\big(\rho * \mu(z) - \rho * \mu'(z)\big)dz \geqslant 0,\end{aligned}$$

the last line following from the fact that L is nondecreasing in the second variable. When h is understood as a cost functional, it increases at a point x as the mass of μ in the neighborhood of x increases.

3.4.3 Another Form of Monotonicity

Although Definition 3.28 is the most frequently used notion of monotonicity, we introduce a variation on the same idea. Even though the concept captured by Definition 3.31 below is slightly different, it will also lead to a useful sufficient condition for uniqueness.

Definition 3.31 *A real valued function H from $\mathbb{R}^d \times \mathcal{P}_2(\mathbb{R}^d)$ into $\mathbb{R}^d$ is said to be* L-*monotone if, for any $\mu \in \mathcal{P}_2(\mathbb{R}^d)$, the mapping $\mathbb{R}^d \ni x \mapsto H(x,\mu) \in \mathbb{R}^d$ is at most of linear growth and, for any two $\mathbb{R}^d$-valued square-integrable random variables X and X′ defined on some probability space $(\Omega, \mathcal{F}, \mathbb{P})$, we have:*

$$\mathbb{E}\big[\big(H\big(X, \mathcal{L}(X)\big) - H\big(X', \mathcal{L}(X')\big)\big) \cdot (X - X')\big] \geqslant 0.$$

Importantly, observe that the above definition does not depend upon the choice of the probability space $(\Omega, \mathcal{F}, \mathbb{P})$, provided that $(\Omega, \mathcal{F}, \mathbb{P})$ is assumed to be rich enough to carry, for any joint distribution $\pi \in \mathcal{P}_2((\mathbb{R}^d \times \mathbb{R}^d)^2)$, a pair of random variables (X, X') with π as distribution. We shall address this latter point in detail in Chapter 5.

Examples of L-monotone functions will be given in Subsection 3.4.3. The reasons for the terminology "L-monotone" will be made clear in Subsection 5.7.1. Therein, we shall show that, surprisingly, the two notions of monotonicity have different origins.

We now provide another sufficient condition for uniqueness using the notion of L-monotonicity introduced in Definition 3.31:

Assumption (L-Monotonicity).

(A1) The coefficient σ is constant and the coefficient b does not depend upon the measure argument and reads, for all $(t, x, \alpha) \in [0, T] \times \mathbb{R}^d \times A$,

$$b(t, x, \alpha) = b_0(t) + b_1(t)x + b_2(t)\alpha,$$

for some bounded measurable deterministic functions b_0, b_1 and b_2 with values in $\mathbb{R}^d$, $\mathbb{R}^{d\times d}$, $\mathbb{R}^{d\times k}$.

(A2) The running cost f has a separated structure of the form:

$$f(t, x, \mu, \alpha) = f_0(t, x, \mu) + f_1(t, x, \alpha),$$
$$t \in [0, T],\ x \in \mathbb{R}^d,\ \alpha \in A,\ \mu \in \mathcal{P}_2(\mathbb{R}^d),$$

where f_0 is a Borel-measurable mapping from $[0, T] \times \mathbb{R}^d \times \mathcal{P}_2(\mathbb{R}^d)$ into $\mathbb{R}$ and f_1 is a Borel-measurable mapping from $[0, T] \times \mathbb{R}^d \times A$ into $\mathbb{R}$.

(A3) For t and μ fixed, the functions $f_0(t, \cdot, \mu)$ and $g(\cdot, \mu)$ are continuously differentiable in x, the partial derivative $\partial_x f_0$ and $\partial_x g$ being at most of linear growth in x, uniformly in (t, μ). The function f_1 is continuously differentiable in (x, α) for t fixed, the derivative being at most linear growth in (x, α), uniformly in $t \in [0, T]$.

The function $[0, T] \ni t \mapsto f_1(t, 0, 0)$ is bounded and the function $[0, T] \times \mathcal{P}_2(\mathbb{R}^d) \ni (t, \mu) \mapsto f_0(t, 0, \mu)$ is bounded on bounded subsets.

(continued)

(A4) The functions $\partial_x f_0(t,\cdot,\cdot)$ for $t \in [0,T]$, and $\partial_x g$ are L-monotone in the sense of Definition 3.31. Moreover, the function f_1 satisfies the following strong form of convexity:

$$\begin{aligned} &f_1(t,x',\alpha') - f_1(t,x,\alpha) - (x'-x,\alpha'-\alpha)\cdot\partial_{(x,\alpha)}f_1(t,x,\alpha) \\ &\quad \geqslant \lambda|\alpha'-\alpha|^2, \end{aligned}$$

for some $\lambda > 0$. The notation $\partial_{(x,\alpha)}f_1$ stands for the gradient in the joint variables (x,α).

Here is the uniqueness result announced earlier.

Theorem 3.32 *If assumption* **L-Monotonicity** *holds and for any deterministic continuous flow* $\boldsymbol{\mu} = (\mu_t)_{0\leqslant t\leqslant T}$ *in* $\mathcal{P}_2(\mathbb{R}^d)$*, the optimal control problem* (3.4) *has a unique minimizer* $\hat{\boldsymbol{\alpha}}^{\boldsymbol{\mu}} \in \mathbb{A}$*, then there exists at most one MFG equilibrium.*

Proof. The proof depends upon the equilibrium criticality condition based on the necessary from of the Pontryagin stochastic maximum principle.

First Step. Owing to Theorem 3.27, the Pontryagin FBSDE of McKean-Vlasov type satisfied by any equilibrium takes the form:

$$\begin{cases} dX_t = b\big(t,X_t,\hat{\alpha}_t\big)dt + \sigma dW_t, \quad t\in[0,T], \\ dY_t = -\partial_x H\big(t,X_t,\mathcal{L}(X_t),Y_t,\hat{\alpha}_t\big)dt + Z_t dW_t, \quad t\in[0,T], \\ Y_T = \partial_x g\big(X_T,\mathcal{L}(X_T)\big), \end{cases} \tag{3.46}$$

where $\hat{\alpha}_t = \hat{\alpha}(t,X_t,\mathcal{L}(X_t),Y_t)$, with $\hat{\alpha}(t,x,\mu,y)$ minimizing the function $A \ni \alpha \mapsto H_1(t,x,y,\alpha)$, where H_1 is the reduced Hamiltonian:

$$H_1(t,x,y,\alpha) = \big(b_2(t)\alpha\big)\cdot y + f_1(t,x,\alpha), \quad (t,x,y,\alpha)\in[0,T]\times\mathbb{R}^d\times\mathbb{R}^d\times A.$$

Observe that, in the Pontryagin system (3.46),

$$\partial_x H(t,x,\mu,y,\alpha) = b_1(t)^\dagger y + \partial_x f_0(t,x,\mu) + \partial_x f_1(t,x,\alpha).$$

Let us assume that $\boldsymbol{X}' = (X'_t)_{0\leqslant t\leqslant T}$ is the optimal path of another equilibrium. The Pontryagin FBSDE of McKean-Vlasov type for $\boldsymbol{X}'$ takes a similar form as long as we replace $(X_t,Y_t,Z_t,\hat{\alpha}_t)_{0\leqslant t\leqslant T}$ with $(X'_t,Y'_t,Z'_t,\hat{\alpha}'_t)_{0\leqslant t\leqslant T}$.

Second Step. Like in the derivation of the stochastic Pontryagin principle, we compute:

$$\begin{aligned}
&d\,(X_t' - X_t) \cdot (Y_t' - Y_t) \\
&= \Big[\big(b(t, X_t', \hat{\alpha}_t') - b(t, X_t, \hat{\alpha}_t)\big) \cdot (Y_t' - Y_t) \\
&\quad - \big(\partial_x H\big(t, X_t', \mathcal{L}(X_t'), Y_t', \hat{\alpha}_t'\big) - \partial_x H\big(t, X_t, \mathcal{L}(X_t), Y_t, \hat{\alpha}_t\big)\big) \cdot (X_t' - X_t) \Big] dt \\
&\quad + dM_t,
\end{aligned}$$

where $(M_t)_{0 \leqslant t \leqslant T}$ is a martingale. Therefore,

$$\begin{aligned}
&d\,(X_t' - X_t) \cdot (Y_t' - Y_t) \\
&= \Big[\Big(b_2(t)(\hat{\alpha}_t' - \hat{\alpha}_t)\Big) \cdot (Y_t' - Y_t) \\
&\quad - \Big(\partial_x f_0\big(t, X_t', \mathcal{L}(X_t')\big) - \partial_x f_0\big(t, X_t, \mathcal{L}(X_t)\big)\Big) \cdot (X_t' - X_t) \\
&\quad - \Big(\partial_x f_1\big(t, X_t', \hat{\alpha}_t'\big) - \partial_x f_1\big(t, X_t, \hat{\alpha}_t\big)\Big) \cdot (X_t' - X_t) \Big] dt + dM_t,
\end{aligned} \tag{3.47}$$

Third Step. Since, for any realization, $\hat{\alpha}_t$ is a minimizer of the function $A \ni \alpha \mapsto H(t, X_t, Y_t, \alpha)$, we have, for all $\alpha \in A$,

$$(\alpha - \hat{\alpha}_t) \cdot \big(b_2(t)^\dagger Y_t + \partial_\alpha f_1(t, X_t, \hat{\alpha}_t)\big) \geqslant 0.$$

So by joint convexity of f_1 in the variable (x, α), we have:

$$\begin{aligned}
&\big(b_2(t)\hat{\alpha}_t'\big) \cdot Y_t + f_1\big(t, X_t', \hat{\alpha}_t'\big) \\
&\quad \geqslant \big(b_2(t)\hat{\alpha}_t\big) \cdot Y_t + f_1\big(t, X_t, \hat{\alpha}_t\big) + (X_t' - X_t) \cdot \partial_x f_1\big(t, X_t, \hat{\alpha}_t\big) + \lambda |\hat{\alpha}_t' - \hat{\alpha}_t|^2.
\end{aligned}$$

Similarly,

$$\begin{aligned}
&\big(b_2(t)\hat{\alpha}_t\big) \cdot Y_t' + f_1\big(t, X_t, \hat{\alpha}_t\big) \\
&\quad \geqslant \big(b_2(t)\hat{\alpha}_t'\big) \cdot Y_t' + f_1\big(t, X_t', \hat{\alpha}_t'\big) + (X_t - X_t') \cdot \partial_x f_1\big(t, X_t', \hat{\alpha}_t'\big) + \lambda |\hat{\alpha}_t' - \hat{\alpha}_t|^2,
\end{aligned}$$

and summing the two inequalities, we get:

$$\begin{aligned}
&\big(b_2(t)\hat{\alpha}_t'\big) \cdot (Y_t - Y_t') \\
&\quad \geqslant \big(b_2(t)\hat{\alpha}_t\big) \cdot (Y_t - Y_t') - (X_t' - X_t) \cdot \Big(\partial_x f_1\big(t, X_t', \hat{\alpha}_t'\big) - \partial_x f_1\big(t, X_t, \hat{\alpha}_t\big)\Big) \\
&\qquad + 2\lambda |\hat{\alpha}_t' - \hat{\alpha}_t|^2,
\end{aligned}$$

and consequently:

$$\begin{aligned}&\Big(b_2(t)(\hat{\alpha}_t' - \hat{\alpha}_t)\Big) \cdot (Y_t' - Y_t) - (X_t' - X_t) \cdot \Big(\partial_x f_1\big(t, X_t', \hat{\alpha}_t'\big) - \partial_x f_1\big(t, X_t, \hat{\alpha}_t\big)\Big) \\ &\quad \leqslant -2\lambda |\hat{\alpha}_t' - \hat{\alpha}_t|^2.\end{aligned}$$

Plugging into (3.47) and using in addition the L-monotonicity of f_0, we deduce that:

$$\mathbb{E}\big[(Y_T' - Y_T) \cdot (X_T' - X_T)\big] + 2\lambda \mathbb{E} \int_0^T |\hat{\alpha}_t' - \hat{\alpha}_t|^2 dt \leqslant 0.$$

Using now the terminal condition, and again the L-monotonicity condition, we get:

$$\begin{aligned}&\mathbb{E}\big[(Y_T' - Y_T) \cdot (X_T' - X_T)\big] \\ &= \mathbb{E}\big[\big(\partial_x g\big(X_T', \mathcal{L}(X_T')\big) - \partial_x g\big(X_T, \mathcal{L}(X_T)\big)\big) \cdot (X_T' - X_T)\big] \geqslant 0,\end{aligned}$$

proving that:

$$\mathbb{E} \int_0^T |\hat{\alpha}_t - \hat{\alpha}_t'|^2 dt = 0,$$

which completes the proof. □

Remark 3.33 *As made clear by the proof of Theorem (Vol II)-1.59 in Chapter (Vol II)-1, the result easily extends to the case when σ takes the form $\sigma(t,x) = \sigma_0(t) + \sigma_1(t)x$.*

Example. Let h be a continuously differentiable even convex function from $\mathbb{R}^d$ into $\mathbb{R}$ whose gradient ∂h is at most of linear growth. Then, the x-derivative of the function:

$$f(x,\mu) = \int_{\mathbb{R}^d} h(x - x') d\mu(x'), \quad x \in \mathbb{R}^d, \ \mu \in \mathcal{P}_2(\mathbb{R}^d),$$

is L-monotone.

Proof. We have:

$$\partial_x f(x,\mu) = \int_{\mathbb{R}^d} \partial h(x - x') d\mu(x'), \quad x \in \mathbb{R}^d, \ \mu \in \mathcal{P}_2(\mathbb{R}^d).$$

Then, for two square-integrable random variables X and X' with values in $\mathbb{R}^d$, we have:

$$\begin{aligned}&\mathbb{E}\Big[\big(\partial_x f(X,\mu) - \partial_x f(X',\mu')\big) \cdot (X - X')\Big] \\ &= \mathbb{E}\Big[\big(\partial h(X - Y) - \partial h(X' - Y')\big) \cdot (X - X')\Big],\end{aligned}$$

where (Y, Y') has the same distribution as, and is independent of, (X, X'). Since ∂h is odd, we have:

$$\begin{aligned}
&\mathbb{E}\Big[\big(\partial h(X-Y) - \partial h(X'-Y')\big)\cdot(X-X')\Big] \\
&= \mathbb{E}\Big[\big(\partial h(Y-X) - \partial h(Y'-X')\big)\cdot(Y-Y')\Big] \\
&= -\mathbb{E}\Big[\big(\partial h(X-Y) - \partial h(X'-Y')\big)\cdot(Y-Y')\Big].
\end{aligned}$$

Therefore,

$$\begin{aligned}
&\mathbb{E}\Big[\big(\partial_x f(X,\mu) - \partial_x f(X',\mu')\big)\cdot(X-X')\Big] \\
&= \frac{1}{2}\mathbb{E}\Big[\big(\partial h(X-Y) - \partial h(X'-Y')\big)\cdot\big(X-Y-(X'-Y')\big)\Big] \geqslant 0,
\end{aligned}$$

the last inequality following from the fact that h is convex. This proves that $\partial_x f$ is L-monotone. □

Importantly, we shall prove in Lemma 5.73 of Chapter 5 that f is not monotone in the sense of Definition 3.28. Therefore, this provides an example where Theorem 3.32 applies but Theorem 3.29 does not! We shall revisit this fact in Subsection 5.7.1 through *ad hoc* notions of convexity for functionals defined on the space of probability measures.

3.5 Linear Quadratic Mean Field Games

Our first application of the strategy and the results obtained in this chapter concerns the Linear Quadratic (LQ for short) models. The linearity of the coefficients and the convexity of the costs are screaming for the use of the stochastic maximum approach, as the weak formulation approach cannot take advantage of these features as easily.

With A being equal to the entire $\mathbb{R}^k$, we use the notation and assumptions introduced in Subsection 2.3.4 of Chapter 2, but with $\boldsymbol{W}$ of dimension d and σ a constant matrix in $\mathbb{R}^{d\times d}$. Since

$$b(t,x,\mu,\alpha) = b_1(t)x + \bar{b}_1(t)\bar{\mu} + b_2(t)\alpha,$$

for deterministic continuous matrix functions b_1, b_2 and $\bar{b}_1$, **(A1)** in assumption **SMP** is satisfied. Since

$$f(t,x,\mu,\alpha) = \frac{1}{2}\Big(x^\dagger q(t)x + (x - s(t)\bar{\mu})^\dagger \bar{q}(t)(x - s(t)\bar{\mu}) + \alpha^\dagger r(t)\alpha\Big),$$

assumption (**A3**) is also satisfied as the matrices $q(t)$ and $\bar{q}(t)$ are symmetric nonnegative semi-definite and continuous in time $t \in [0,T]$ and the matrix $r(t)$ is symmetric strictly positive definite and continuous in $t \in [0,T]$; in particular, $r(t)$ is strictly positive definite, uniformly in $t \in [0,T]$. Finally, since

$$g(x,\mu) = \frac{1}{2}\Big(x^\dagger q x + (x - s\bar{\mu})^\dagger \bar{q}(x - s\bar{\mu})\Big),$$

with q and $\bar{q}$ symmetric nonnegative semi-definite, assumption (**A4**) is satisfied for the same reasons.

However, although assumption **SMP** is satisfied, Theorem 3.24 does not apply. Obviously, the reason is that the set A is unbounded and the coefficients b_0 (which is $\bar{b}_1$ in the present situation), $\partial_x f$ and $\partial_x g$ are unbounded as well. In Chapter 4 we prove an extension of Theorem 3.24, namely Theorem 4.53, which covers a large class of linear quadratic mean field games. It is proven under a set of assumptions called **MFG Solvability SMP**. In particular, assumptions (**A5**) and (**A6**) in assumption **MFG Solvability SMP** are easily checked in the present situation. However, assumption (**A7**) is not always satisfied. It is satisfied when the matrices $\bar{q}(t)s(t)$ and $\bar{q}s$ are non-positive semi-definite. So clearly, a very large class of linear quadratic mean field games are covered by the results of Chapter 4.

Still, and even if assumption **MFG Solvability SMP** is satisfied, it is instructive to know that one can solve LQ mean field games directly, without appealing to the abstract existence and uniqueness results proven in this chapter and the next. In the present set-up, the two main steps of the mean field game strategy articulated in Subsection 3.1.2 above read as:

(i) For each fixed deterministic function $[0,T] \ni t \mapsto \bar{\mu}_t \in \mathbb{R}^d$, solve the standard stochastic control problem

$$\begin{aligned}
&\inf_{\alpha \in \mathbb{A}} \mathbb{E}\bigg[\frac{1}{2}\Big(X_T^\dagger q X_T + (X_T - s\bar{\mu}_T)^\dagger \bar{q}(X_T - s\bar{\mu}_T)\Big) \\
&\quad + \frac{1}{2}\int_0^T \Big(X_t^\dagger q(t) X_t + (X_t - s(t)\bar{\mu}_t)^\dagger \bar{q}(t)(X_t - s(t)\bar{\mu}_t) + \alpha_t^\dagger r(t)\alpha_t\Big)dt\bigg]
\end{aligned}$$

subject to (3.48)

$$\begin{cases}
dX_t = \Big[b_1(t)X_t + b_2(t)\alpha_t + \bar{b}_1(t)\bar{\mu}_t\Big]dt + \sigma dW_t, \\
X_0 = \xi.
\end{cases}$$

(ii) Determine a function $[0,T] \ni t \mapsto \bar{\mu}_t \in \mathbb{R}^d$ so that, for all $t \in [0,T]$, $\mathbb{E}[X_t] = \bar{\mu}_t$, where $(X_t)_{0 \leqslant t \leqslant T}$ is the optimal path of the optimal control problem in the environment $(\bar{\mu}_t)_{0 \leqslant t \leqslant T}$.

The Hamiltonian of the control problem defined in (i) is given by:

$$H(t,x,\mu,y,\alpha) = [b_1(t)x + \bar{b}_1(t)\bar{\mu} + b_2(t)\alpha] \cdot y + \frac{1}{2}\Big(x^\dagger q(t)x + (x - s(t)\bar{\mu})^\dagger \bar{q}(t)(x - s(t)\bar{\mu}) + \alpha^\dagger r(t)\alpha\Big).$$

This Hamiltonian is minimized for:

$$\hat{\alpha} = \hat{\alpha}(t,x,\mu,y) = -r(t)^{-1}b_2(t)^\dagger y, \tag{3.49}$$

which is independent of the measure argument μ. For each fixed $[0,T] \ni t \mapsto \bar{\mu}_t \in \mathbb{R}^d$, the optimal control problem of step (i) has a unique solution if and only if we can uniquely solve the FBSDE:

$$\begin{cases} dX_t = \Big(b_1(t)X_t - b_2(t)r(t)^{-1}b_2(t)^\dagger Y_t + \bar{b}_1(t)\bar{\mu}_t\Big)dt + \sigma dW_t, \\ dY_t = -\Big(b_1(t)^\dagger Y_t + [q(t) + \bar{q}(t)]X_t - \bar{q}(t)s(t)\bar{\mu}_t\Big)dt + Z_t dW_t, \end{cases} \tag{3.50}$$

with initial condition $X_0 = \xi$ and terminal condition $Y_T = (q + \bar{q})X_T - \bar{q}s\bar{\mu}_T$. By Theorem 3.17, this control problem and this FBSDE are indeed uniquely solvable because of the strict convexity assumption (i.e., $r(t)$ is strictly positive definite).

Assuming that the fixed point step (ii) can be solved, we can substitute $\bar{\mu}_t$ for $\mathbb{E}[X_t]$ in (3.50) and the FBSDE becomes the McKean-Vlasov FBSDE:

$$\begin{cases} dX_t = \Big(b_1(t)X_t - b_2(t)r(t)^{-1}b_2(t)^\dagger Y_t + \bar{b}_1(t)\mathbb{E}[X_t]\Big)dt \\ \qquad +\sigma dW_t, \\ dY_t = -\Big(b_1(t)^\dagger Y_t + [q(t) + \bar{q}(t)]X_t - \bar{q}(t)s(t)\mathbb{E}[X_t]\Big)dt \\ \qquad +Z_t dW_t, \quad t \in [0,T], \end{cases} \tag{3.51}$$

with initial condition $X_0 = \xi$ and terminal condition $Y_T = (q + \bar{q})X_T - \bar{q}s\mathbb{E}[X_T]$, which is a particular case of the FBSDE (3.40) characterizing the solution of an MFG problem. Taking expectations of both sides of (3.51) and using the notation $\bar{x}_t$ and $\bar{y}_t$ for the expectations $\mathbb{E}[X_t]$ and $\mathbb{E}[Y_t]$ respectively, we find that:

$$\begin{cases} d\bar{x}_t = \Big([b_1(t) + \bar{b}_1(t)]\bar{x}_t - b_2(t)r(t)^{-1}b_2(t)^\dagger \bar{y}_t\Big)dt, \\ d\bar{y}_t = \Big(-[q(t) + \bar{q}(t) - \bar{q}(t)s(t)]\bar{x}_t - b_1(t)^\dagger \bar{y}_t\Big)dt, \quad t \in [0,T], \\ \bar{x}_0 = \mathbb{E}[\xi], \quad \bar{y}_T = [q + \bar{q} - \bar{q}s]\bar{x}_T. \end{cases} \tag{3.52}$$

This system of ordinary differential equations is not always easy to solve despite its deceptive simplicity. Its forward/backward nature is the source of difficulty. We shall say more below, especially in the univariate case $d = m = k = 1$. In any case, its properties play a crucial role in the solution of the linear quadratic mean field game. Indeed, we have the following statement.

Theorem 3.34 *Existence and uniqueness of a solution to the LQ MFG problem (i)-(ii) hold if and only there is existence and uniqueness of a solution of* (3.52).

Proof. Clearly, existence of a solution to the MFG problem implies existence of a solution for (3.52). We now prove the analogue, but for uniqueness. To do so, let us assume that there is at most one equilibrium for the MFG problem. For each solution $(\bar{x}_t, \bar{y}_t)_{0 \leqslant t \leqslant T}$ of (3.52), we can solve the system (3.50) with $\bar{x}_t$ in lieu of $\bar{\mu}_t$, and conclude that $\mathbb{E}[X_t] = \bar{x}_t$ and $\mathbb{E}[Y_t] = \bar{y}_t$ for all $t \in [0, T]$. Indeed, forming the difference between (3.50) and (3.52), we see that $(\mathbb{E}[X_t] - \bar{x}_t, \mathbb{E}[Y_t] - \bar{y}_t)_{0 \leqslant t \leqslant T}$ is the solution of a homogeneous linear system of order 1 with zero initial condition; invoking Theorem 3.17, or duplicating its proof, we observe that this homogeneous linear system has zero as unique solution because of the strict convexity assumption implied by the fact that $r(t)$ is strictly positive definite. Therefore, the solution $(X_t, Y_t)_{0 \leqslant t \leqslant T}$ of (3.50) with $\bar{x}_t$ in lieu of $\bar{\mu}_t$ solves (3.51); since (3.51) is uniquely solvable, this shows that $(\bar{x}_t = \mathbb{E}[X_t], \bar{y}_t = \mathbb{E}[Y_t])_{0 \leqslant t \leqslant T}$, is uniquely determined. This concludes the proof of the uniqueness of the solution of (3.52).

Conversely, let us assume existence of a unique solution for the deterministic system (3.52). Recall from Theorem 3.17 that for each fixed deterministic continuous function $[0, T] \ni t \mapsto \bar{\mu}_t \in \mathbb{R}^d$, the FBSDE (3.50) is uniquely solvable. Using the unique solution $(\bar{x}_t)_{0 \leqslant t \leqslant T}$ of (3.52) in lieu of $(\bar{\mu}_t)_{0 \leqslant t \leqslant T}$ as input in (3.50) and forming as above the difference between (3.50) and (3.52), we get by the same argument that $\mathbb{E}[X_t] = \bar{x}_t$ for all $t \in [0, T]$, which proves that the fixed point step of the MFG strategy is also satisfied. Furthermore, the uniqueness of the solution of (3.52) and the uniqueness of the solution of (3.50) for $[0, T] \ni t \mapsto \bar{\mu}_t \in \mathbb{R}^d$ fixed imply the uniqueness of the solution of the MFG equilibrium. □

Consistent with the time honored method to solve affine FBSDEs, we may want to look for a solution of (3.52) in the form $\bar{y}_t = \bar{\eta}_t \bar{x}_t + \bar{\chi}_t$ where $t \mapsto \bar{\eta}_t$ and $t \mapsto \bar{\chi}_t$ are smooth functions with values in $\mathbb{R}^{d \times d}$ and $\mathbb{R}^d$ respectively. For the sake of notation, we rewrite the forward-backward system (3.52) in the form:

$$\begin{cases} \dot{\bar{x}}_t = a_t \bar{x}_t + b_t \bar{y}_t, \\ \dot{\bar{y}}_t = c_t \bar{x}_t + d_t \bar{y}_t, \quad t \in [0, T], \\ \bar{x}_0 = \mathbb{E}[\xi], \quad \bar{y}_T = e \bar{x}_T, \end{cases} \tag{3.53}$$

where:

$$a_t = b_1(t) + \bar{b}_1(t), \quad b_t = -b_2(t) r(t)^{-1} b_2(t)^\dagger,$$

and

$$c_t = -[q(t) + \bar{q}(t) - \bar{q}(t)s(t)], \quad d_t = -b_1(t)^\dagger, \quad e = q + \bar{q} - \bar{q}s.$$

Notice that we use the standard ODE notation of a dot for the time derivative of deterministic functions of time. If we compute the derivative of $\bar{y}_t$ from the ansatz $\bar{y}_t = \bar{\eta}_t\bar{x}_t + \bar{\chi}_t$ and use the forward equation to express $\dot{\bar{x}}_t$, we obtain:

$$\dot{\bar{y}}_t = [\dot{\bar{\eta}}_t + \bar{\eta}_t a_t + \bar{\eta}_t b_t \bar{\eta}_t]\bar{x}_t + \bar{\eta}_t b_t \bar{\chi}_t + \dot{\bar{\chi}}_t, \qquad t \in [0, T].$$

If we now replace $\bar{y}_t$ by the ansatz in the backward equation in (3.53), we obtain:

$$\dot{\bar{y}}_t = [c_t + d_t\bar{\eta}_t]\bar{x}_t + d_t\bar{\chi}_t, \qquad t \in [0, T],$$

and identifying the two forms of the derivative $\dot{\bar{y}}_t$, we find that given the ansatz, the system (3.53) is equivalent to the system:

$$\begin{cases} \dot{\bar{\eta}}_t + \bar{\eta}_t a_t - d_t\bar{\eta}_t + \bar{\eta}_t b_t \bar{\eta}_t - c_t = 0, & \bar{\eta}_T = e, \\ \dot{\bar{\chi}}_t + [\bar{\eta}_t b_t - d_t]\bar{\chi}_t = 0, & \bar{\chi}_T = 0. \end{cases} \tag{3.54}$$

The first equation is a matrix Riccati equation which is not always solvable on a time interval of pre-assigned length. When it is, its solution can be injected into the second equation, which then becomes a first order homogenous linear equation with terminal condition zero, so its solution is identically zero.

Let us assume momentarily that the Riccati equation appearing as the first equation in the system (3.54) has a unique solution which we denote by $(\bar{\eta}_t)_{0\leqslant t\leqslant T}$. Injecting the ansatz $\bar{y}_t = \bar{\eta}_t\bar{x}_t$ into the first equation of the system (3.53), we find that $(\bar{x}_t)_{0\leqslant t\leqslant T}$ has to solve the ODE:

$$\dot{\bar{x}}_t = [a_t + b_t\bar{\eta}_t]\bar{x}_t, \qquad \bar{x}_0 = \mathbb{E}[\xi], \tag{3.55}$$

which is a linear ODE for which existence and uniqueness are guaranteed. Finding the optimal mean function $[0, T] \ni t \mapsto \bar{x}_t$ guarantees the existence of a solution to the MFG problem, but it does not tell much about the optimal state trajectories or the optimal control. The latter can be obtained by plugging the so-obtained $\bar{x}_t$ into the FBSDE (3.51) in lieu of $\mathbb{E}[X_t]$ and solving for $\boldsymbol{X} = (X_t)_{0\leqslant t\leqslant T}$ and $\boldsymbol{Y} = (Y_t)_{0\leqslant t\leqslant T}$. This search reduces to the solution of the affine FBSDE:

$$\begin{cases} dX_t = [\mathfrak{a}_t X_t + \mathfrak{b}_t Y_t + \mathfrak{c}_t]dt + \sigma dW_t, \\ dY_t = [\mathfrak{m}_t X_t - \mathfrak{a}_t^\dagger Y_t + \mathfrak{d}_t]dt + Z_t dW_t, \\ X_0 = \xi, \quad Y_T = \mathfrak{q}X_T + \mathfrak{r}, \end{cases} \tag{3.56}$$

where we set:

$$\mathfrak{a}_t = b_1(t), \qquad \mathfrak{b}_t = -b_2(t)r(t)^{-1}b_2(t)^\dagger, \qquad \mathfrak{c}_t = \bar{b}_1(t)\bar{x}_t,$$

and

$$\mathfrak{m}_t = -[q(t) + \bar{q}(t)], \quad \mathfrak{d}_t = \bar{q}(t)s(t)\bar{x}_t, \quad \mathfrak{q} = q + \bar{q}, \quad \mathfrak{r} = -\bar{q}s\bar{x}_T.$$

The standard theory of FBSDEs suggests that Y_t should be given by a deterministic function of t and X_t, the so-called decoupling field. The affine structure of the FBSDE (3.56) suggests that this decoupling field should be affine. So, again, we search for deterministic differentiable functions η_t and χ_t such that:

$$Y_t = \eta_t X_t + \chi_t, \qquad t \in [0, T]. \tag{3.57}$$

Notice that taking expectations on both sides of this ansatz we get $\mathbb{E}[Y_t] = \eta_t\mathbb{E}[X_t] + \chi_t$, but since both functions η and χ depend upon the function $[0, T] \ni t \mapsto \bar{x}_t$, there is no contradiction with the formula $\bar{y}_t = \bar{\eta}_t\bar{x}_t$ even if, as we are about to find out, the function χ is not identically zero, since the function η may solve a Riccati equation different from the Riccati equation solved by $\bar{\eta}$.

Computing dY_t from ansatz (3.57) by using the expression of dX_t given by the first equation of (3.56), we get:

$$dY_t = [(\dot{\eta}_t + \eta_t\mathfrak{a}_t + \eta_t\mathfrak{b}_t\eta_t)X_t + \dot{\chi}_t + \eta_t\mathfrak{b}_t\chi_t + \eta_t\mathfrak{c}_t]dt + \eta_t\sigma dW_t, \quad t \in [0, T],$$

and identifying term by term with the expression of dY_t given in (3.56), we get:

$$\begin{cases} \dot{\eta}_t + \eta_t\mathfrak{b}_t\eta_t + \mathfrak{a}_t^\dagger\eta_t + \eta_t\mathfrak{a}_t - \mathfrak{m}_t = 0, & \eta_T = \mathfrak{q}, \\ \dot{\chi}_t + (\mathfrak{a}_t^\dagger + \eta_t\mathfrak{b}_t)\chi_t - \mathfrak{d}_t + \eta_t\mathfrak{c}_t = 0, & \chi_T = \mathfrak{r}, \\ Z_t = \eta_t\sigma. \end{cases} \tag{3.58}$$

As before, the first equation is a matrix Riccati equation. If and when it can be solved, the third equation becomes solved automatically, and the second equation becomes a first order linear ODE, though not homogenous this time, which can be solved by standard methods. Notice that the quadratic terms of the two Riccati equations (3.54) and (3.58) are the same since $b_t = \mathfrak{b}_t = -b_2(t)r(t)^{-1}b_2(t)^\dagger$. However, the terminal conditions are different since the terminal condition in (3.58) is given by $\mathfrak{q} = q + \bar{q}$ while it was given by $e = q + \bar{q}(I_d - s)$ in (3.54). Notice also that the first order terms are different as well.

3.5.1 Connection with Deterministic Control Theory

In several instances, we alluded to a strong connection between the solvability of Riccati equations and deterministic LQ optimal control problems. We now make this correspondence precise for the purpose of the discussion of the solvability of the above matrix Riccati equation.

Stochastic Maximum Principle and Riccati Equation. Under the same assumptions on the dynamics, we consider the minimization of the functional:

$$\tilde{J}(\boldsymbol{\alpha}) = \frac{1}{2} x_T^\dagger \tilde{q} x_T + \int_0^T \frac{1}{2} \Big[x_t^\dagger \tilde{q}_t x_t + \alpha_t^\dagger r_t \alpha_t \Big] dt$$

under the dynamic constraint:

$$dx_t = (a_t x_t + b_t \alpha_t)\, dt, \qquad x_0 = \mathbb{E}[\xi],$$

where $\boldsymbol{\alpha} = (\alpha_t)_{0 \le t \le T}$ is a deterministic control in $L^2([0,T];A)$. Pay attention that $(a_t)_{0 \le t \le T}$ and $(b_t)_{0 \le t \le T}$ may differ from the coefficients $(a_t)_{0 \le t \le T}$ and $(b_t)_{0 \le t \le T}$ defined earlier and, similarly, $(r_t)_{0 \le t \le T}$ may differ from $(r(t))_{0 \le t \le T}$. Notice that this problem has a unique solution if we assume as before that the matrix coefficients are continuous functions of the time variable t, and if we also assume that $\tilde{q}$ and $\tilde{q}_t$ are symmetric and nonnegative semi-definite, and that r_t is symmetric and strictly positive definite. The Hamiltonian:

$$H(t,x,y,\alpha) = y^\dagger a_t x + y^\dagger b_t \alpha + \frac{1}{2} x^\dagger \tilde{q}_t x + \frac{1}{2} \alpha^\dagger r_t \alpha$$

is minimized for $\alpha = \hat{\alpha}(t,x,y) = -r_t^{-1} b_t^\dagger y$ so that the forward-backward system given by the maximum principle reads:

$$\begin{cases} \dot{x}_t = a_t x_t - b_t r_t^{-1} b_t^\dagger y_t, \\ \dot{y}_t = -\tilde{q}_t x_t - a_t^\dagger y_t, \quad t \in [0,T], \\ x_0 = \mathbb{E}[\xi], \quad y_T = \tilde{q} x_T. \end{cases} \tag{3.59}$$

Existence and uniqueness of a solution are guaranteed by the maximum principle and the convexity properties of the coefficients of the functional $\tilde{J}$, see Theorem 3.17.

Based on the same ansatz as before, we seek a pair of functions $(\eta_t, \chi_t)_{0 \le t \le T}$, independent of the initial condition x_0, such $y_t = \eta_t x_t + \chi_t$ for all $t \in [0,T]$. Proceeding as above, this prompts us to address the solvability of the matrix Riccati equation:

$$\dot{\eta}_t + a_t^\dagger \eta_t + \eta_t a_t - \eta_t [b_t r_t^{-1} b_t^\dagger] \eta_t + \tilde{q}_t = 0, \qquad \eta_T = \tilde{q}, \tag{3.60}$$

together with the linear equation:

$$\dot{\chi}_t + a_t^\dagger \chi_t - \eta_t [b_t r_t^{-1} b_t^\dagger] \chi_t = 0, \qquad \chi_T = 0. \tag{3.61}$$

Of course, (3.60) is locally uniquely solvable on an interval $[T - \delta, T]$, for some $\delta > 0$. This permits to solve the equation:

$$\dot{x}'_t = a_t x'_t - b_t r_t^{-1} b_t^\dagger \eta_t x'_t - b_t r_t^{-1} b_t^\dagger \chi_t$$

on $[T - \delta, T]$, with some initial condition at $T - \delta$. The above equation may be obtained by inserting the ansatz in the forward equation of the system (3.59). Letting $(y'_t = \eta_t x'_t + \chi_t)_{T-\delta \leqslant t \leqslant T}$, we construct in this way a solution to the forward-backward system (3.59) but with an initial condition at time $T - \delta$ in lieu of 0. By small time uniqueness of the solutions to (3.59), or equivalently by uniqueness of the solutions to (3.59) when initialized at time $T - \delta$ in lieu of 0 for δ small enough, this shows that any solution $(x_t, y_t)_{0 \leqslant t \leqslant T}$ of (3.59) must take the form $y_t = \eta_t x_t + \eta_t$ for $t \in [T - \delta, T]$. Anticipating on the notion of *decoupling field* discussed in detail in Chapter 4, this shows that the function $[T - \delta, T] \times \mathbb{R}^d \ni (t, x) \mapsto \eta_t x + \chi_t$ is the decoupling field of the system (3.59) on $[T - \delta, T]$. Now, the stability property provided by Proposition 3.21 may be used to prove that the decoupling field of (3.59) is Lipschitz in x, uniformly in $t \in [0, T]$, see Lemma 4.56 for details. This provides an a priori bound for $(\eta_t)_{T-\delta \leqslant t \leqslant T}$ and permits to extend by induction the solution of the Riccati equation to the entire $[0, T]$.

As a conclusion, we deduce that, under the standing assumption on the coefficients of $\tilde{J}$, the matrix Riccati equation (3.60) is uniquely solvable on $[0, T]$.

Application to LQ Mean Field Games. If we set:

$$a_t = b_1(t), \quad b_t = b_2(t), \quad r_t = r(t), \quad \tilde{q}_t = q(t) + \bar{q}(t) \quad \text{and} \quad \tilde{q} = q + \bar{q},$$

the Riccati equation (3.60) coincides with the Riccati equation in (3.58) since $\mathfrak{a}_t = c_t$, $\mathfrak{b}_t = b_t$ and $\mathfrak{m}_t = -\tilde{q}_t$ and we can use the equivalence provided by the maximum principle to conclude existence and uniqueness of a solution for the Riccati equation in (3.58).

However, in our study of LQ mean field games, it is also necessary to provide conditions implying existence and uniqueness of a solution to the system (3.52) which, according to Theorem 3.34, is equivalent to the existence and uniqueness of a solution to the LQ mean field game problem. Indeed, as we already remarked, the only issue is the solution of the fixed point problem (ii), since for any continuous function $[0, T] \ni t \mapsto \bar{\mu}_t \in \mathbb{R}^d$ the standard optimal control problem (i) has a unique solution under the above assumptions on the coefficients.

Notice that if we assume that the $\mathbb{R}^d$-valued function $t \mapsto \zeta_t$ satisfies an equation of the form:

$$\dot{\zeta}_t = c_t x_t + d_t \zeta_t, \qquad t \in [0, T],$$

like in the second equation of the system (3.53), and if we define $t \mapsto e_t$ as the unique $\mathbb{R}^{d\times d}$-valued solution of the matrix ODE $\dot{e}_t = \bar{b}_1(t)^\dagger e_t$ with initial condition $e_0 = I_d$, then, the $\mathbb{R}^d$-valued function $t \mapsto \tilde{\zeta}_t$ defined by $\tilde{\zeta}_t = e_t \zeta_t$ satisfies:

$$\dot{\tilde{\zeta}}_t = [e_t c_t] x_t + [\bar{b}_1(t)^\dagger + d_t]\tilde{\zeta}_t, \qquad t \in [0, T],$$

if the matrices e_t and d_t commute in the sense that $e_t d_t = d_t e_t$ for all $t \in [0, T]$. Notice that, while this commutativity property is often satisfied in applications (in particular in the unidimensional case $d = k = 1$), it is still rather restrictive. We make it here for the purpose of the discussion of the assumptions under which the fixed point step (ii) can be solved for LQ models.

The relevance of this remark comes about in the following way. The above maximum principle argument shows that if we set:

$$a_t = b_1(t) + \bar{b}_1(t), \qquad b_t = b_2(t), \qquad r_t = \tilde{r}_t,$$

for some continuous and strictly positive-definite-symmetric-matrix-valued function $[0, T] \ni t \mapsto \tilde{r}_t$ to be determined, then the system:

$$\begin{cases} \dot{x}_t = a_t x_t - b_t \tilde{r}_t^{-1} b_t^\dagger \zeta_t, \\ \dot{\zeta}_t = -\tilde{q}_t x_t - a_t^\dagger \zeta_t, \quad t \in [0, T], \\ x_0 = \mathbb{E}[\xi], \quad \zeta_T = \tilde{q}\xi_T, \end{cases} \tag{3.62}$$

has a unique solution, if, as prescribed above, $[0, T] \ni t \mapsto \tilde{q}_t$ is continuous and takes values in the set of symmetric nonnegative semi-definite matrices and $\tilde{q}$ is also a symmetric nonnegative semi-definite matrix. Setting $\tilde{\zeta}_t = e_t \zeta_t$ with e_t as above and assuming that e_t commutes with $a_t^\dagger$, we conclude that the couple $(x_t, \tilde{\zeta}_t)_{0 \leqslant t \leqslant T}$ satisfies:

$$\begin{cases} \dot{x}_t = [b_1(t) + \bar{b}_1(t)]x_t - b_2(t)\tilde{r}_t^{-1} b_2(t)^\dagger e_t^{-1} \tilde{\zeta}_t, \\ \dot{\tilde{\zeta}}_t = -e_t \tilde{q}_t x_t - b_1(t)^\dagger \tilde{\zeta}_t, \quad t \in [0, T], \\ x_0 = \mathbb{E}[\xi], \quad \tilde{\zeta}_T = e_T \tilde{q} x_T, \end{cases} \tag{3.63}$$

which is nothing but the system (3.52) if we can choose $(\tilde{q}_t)_{0 \leqslant t \leqslant T}$ and $\tilde{q}$ so that:

$$\begin{cases} e_t \tilde{q}_t = q(t) + \bar{q}(t) - \bar{q}(t)s(t), \quad t \in [0, T], \\ e_T \tilde{q} = q + \bar{q} - \bar{q}s, \end{cases}$$

and $\tilde{r}_t$ such that:

$$\tilde{r}_t^{-1} b_2(t)^\dagger e_t = r(t)^{-1} b_2(t)^\dagger,$$

for all $t \in [0,T]$. Checking that $\tilde{q}_t$ is symmetric and nonnegative semi-definite and that $\tilde{r}_t$ is symmetric and strictly positive definite may be difficult. Still, if $\bar{b}_1(t)$ happens to be a multiple of the $d \times d$ identity matrix I_d, e_t will also be a multiple of I_d, with a positive multiplicative constant. As such, e_t will commute with all the $d \times d$ matrices. In this case, we may choose $\tilde{r}_t$ as equal to r_t up to a multiplicative constant. Moreover, $\tilde{q}_t$ will be symmetric if $q(t)s(t)$ is symmetric, and nonnegative semi-definite if $q(t) + \bar{q}(t) - q(t)s(t) \geqslant 0$, which is the case if $q(t)s(t) \leqslant 0$, and similarly for $\tilde{q}$. This matches condition (**A7**) in the set of assumption **MFG Solvability SMP** that we shall introduce in Chapter 4 in order to prove existence of an MFG equilibrium within a larger framework that includes both Theorem 3.24 and the linear-quadratic case.

We deduce:

Proposition 3.35 *Assume, as above, that the matrix coefficients are continuous, q, $\bar{q}$, $q(t)$, and $\bar{q}(t)$ are nonnegative semi-definite, and $r(t)$ is strictly positive definite. Assume also that $\bar{b}_1(t)$ is a multiple of the identity matrix I_d, that for all $t \in [0,T]$, $q(t)s(t)$ is symmetric and $q(t)+\bar{q}(t)-q(t)s(t)$ is nonnegative semi-definite and that qs is symmetric and $q+\bar{q}-qs$ is nonnegative semi-definite. Then, the LQ mean field game problem defined through* (3.48) *has a unique solution.*

3.5.2 The One-Dimensional Case $d = k = 1$

In the one-dimensional case, $d = m = k = 1$, the system (3.58) can be rewritten in the form:

$$\begin{cases} \dot{\eta}_t = -\mathfrak{b}_t\eta_t^2 - 2\mathfrak{a}_t\eta_t + \mathfrak{m}_t, & \eta_T = \mathfrak{q}, \\ \dot{\chi}_t + (\mathfrak{a}_t + \mathfrak{b}_t\eta_t)\chi_t = \mathfrak{d}_t - \mathfrak{c}_t\eta_t, & \chi_T = \mathfrak{r}, \\ Z_t = \eta_t\sigma. \end{cases} \tag{3.64}$$

The first equation is now a scalar Riccati equation, and according to the classical theory of scalar Ordinary Differential Equations (ODEs for short), a straightforward approach is to solve the second order linear equation:

$$-\mathfrak{b}_t\ddot{\theta}_t + [\dot{\mathfrak{b}}_t - 2\mathfrak{a}_t\mathfrak{b}_t]\dot{\theta}_t + \mathfrak{m}_t\mathfrak{b}_t^2\theta_t = 0, \qquad t \in [0,T],$$

with terminal conditions $\theta_T = 1$ and $\dot{\theta}_T = \mathfrak{b}_T\mathfrak{q}$, and set $\eta_t = (\mathfrak{b}_t\theta_t)^{-1}\dot{\theta}_t$, provided one can find a solution $[0,T] \ni t \mapsto \theta_t$ which does not vanish. Once $(\eta_t)_{0\leqslant t\leqslant T}$ in (3.64) is computed, the next step is to plug its value in the third equation to determine $\mathbf{Z} = (Z_t)_{0\leqslant t\leqslant T}$, and in the second equation, which can then be solved by:

$$\chi_t = \mathfrak{r}e^{\int_t^T[\mathfrak{a}_u+\mathfrak{b}_u\eta_u]du} - \int_t^T [\mathfrak{d}_s - \mathfrak{c}_s\eta_s]e^{\int_t^s[\mathfrak{a}_u+\mathfrak{b}_u\eta_u]du}ds, \qquad t \in [0,T]. \tag{3.65}$$

When the Riccati equation is well posed, its solution does not blow up and all the terms above are integrable. Now that the deterministic functions $(\eta_t)_{0\leqslant t\leqslant T}$ and $(\chi_t)_{0\leqslant t\leqslant T}$ are computed, we rewrite the forward stochastic differential equation for the dynamics of the state, see (3.56), using the ansatz (3.57):

$$dX_t = [(\mathfrak{a}_t + \mathfrak{b}_t\eta_t)X_t + \mathfrak{b}_t\chi_t + \mathfrak{c}_t]dt + \sigma dW_t, \qquad X_0 = \xi.$$

Such a stochastic differential equation admits an explicit solution:

$$\begin{aligned} X_t &= X_0 e^{\int_0^t(\mathfrak{a}_u+\mathfrak{b}_u\eta_u)du} + \int_0^t (\mathfrak{b}_u\chi_u + \mathfrak{c}_u)e^{\int_u^t(\mathfrak{a}_v+\mathfrak{b}_v\eta_v)dv}du \\ &\quad + \sigma\int_0^t e^{\int_u^t(\mathfrak{a}_v+\mathfrak{b}_v\eta_v)dv}dW_u, \quad t\in[0,T], \end{aligned} \tag{3.66}$$

which provides the solution to (3.56) in the univariate case once the fixed point condition in the LQ mean field game problem has been solved.

In order to solve the fixed point condition, we may pursue the argument started earlier in the multidimensional case. To do so, we notice that in the one-dimensional case, the function $(e_t)_{0\leq t\leq T}$ used in (3.63) is given explicitly by:

$$e_t = \exp\left(\int_0^t \bar{b}_1(u)du\right), \qquad t\in[0,T],$$

and since the commutativity conditions are automatically satisfied we only need to check the positivity conditions. Since $\tilde{r}_t = e_t^{-1}r(t)$ is strictly positive if and only if $r(t)$ is, which is part of our assumptions, the only requirement we need to guarantee existence of a solution to the MFG problem is the nonnegativity of $\tilde{q}_t$, for $t\in[0,T]$, and of $\tilde{q}$, which amounts to assuming that $q(t)+\bar{q}(t)-\bar{q}(t)s(t)\geqslant 0$, for $t\in[0,T]$, and $q+\bar{q}-\bar{q}s\geqslant 0$.

Remark 3.36 *Whenever we solve the above LQ mean field game problems with a deterministic initial private state $\xi = x_0\in\mathbb{R}$, the equilibrium state process $\boldsymbol{X} = (X_t)_{0\leqslant t\leqslant T}$, its adjoint process $\boldsymbol{Y} = (Y_t)_{0\leqslant t\leqslant T}$ as well as the optimal control process $\boldsymbol{\alpha} = (\alpha_t)_{0\leqslant t\leqslant T}$ are Gaussian processes whose mean and auto-covariance functions can be computed explicitly in terms of the functions $(\eta_t)_{0\leqslant t\leqslant T}$ and $(\chi_t)_{0\leqslant t\leqslant T}$.*

Remark 3.37 *We refer to* (2.49)–(2.50) *for the analysis of the Riccati equation in* (3.64) *when the coefficients $\mathfrak{b}$, $\mathfrak{a}$ and $\mathfrak{m}$ are constant.*

A Very Simple Example

For the sake of illustration, we consider an example frequently used in the early literature on mean field games. In this example, the drift b reduces to the control, namely $b(t,x,\mu,\alpha) = \alpha$, so that $b_1(t) = \bar{b}_1(t) = 0$, and $b_2(t) = 1$, and the state equation reads

$$dX_t = \alpha_t dt + \sigma dW_t, \qquad t \in [0, T]; \qquad X_0 = \xi.$$

We also assume that the running cost is simply the square of the control, i.e., $f(t, x, \mu, \alpha) = \alpha^2/2$ so that $r(t) = 1$ and $q(t) = \bar{q}(t) = 0$. In particular, $\mathfrak{a}_t = \mathfrak{d}_t = \mathfrak{c}_t = 0$ and $\mathfrak{b}_t = -1$, see (3.56). In this example, the interaction between the players occurs only through the terminal cost, which we assume to be of the form $g(x, \mu) = \bar{q}(x - s\bar{\mu})^2/2$ for some $\bar{q} \geqslant 0$ and $s \in \mathbb{R}$ to conform with the setting of this section. Using the notation and the results above, we see that after fixing the mean $\bar{\mu}_t = \mathbb{E}[X_t]$, the FBSDE from the Pontryagin stochastic maximum principle has the simple form:

$$\begin{cases} dX_t = -Y_t dt + \sigma dW_t, \\ dY_t = Z_t dW_t, \quad t \in [0, T], \\ X_0 = \xi, \quad Y_T = \mathfrak{q} X_T + \mathfrak{r}, \end{cases} \tag{3.67}$$

with $\mathfrak{q} = \bar{q}$ and $\mathfrak{r} = -\bar{q}s\bar{\mu}_T$. As explained above, we solve this FBSDE by postulating $Y_t = \eta_t X_t + \chi_t$, and solving for the two deterministic functions $t \mapsto \eta_t$ and $t \mapsto \chi_t$. The ODEs of (3.64) read $\dot{\eta}_t - \eta_t^2 = 0$ and $\dot{\chi}_t - \eta_t \chi_t = 0$ with terminal conditions $\eta_T = \mathfrak{q}$ and $\chi_T = \mathfrak{r}$ respectively. Their solutions are:

$$\eta_t = \frac{\mathfrak{q}}{1 + \mathfrak{q}(T - t)}, \qquad \chi_t = \frac{\mathfrak{r}}{1 + \mathfrak{q}(T - t)},$$

(keep in mind that $\mathfrak{q} \geqslant 0$ so that the functions above are well defined) and plugging these expressions into (3.66) we get:

$$X_t = \xi \frac{1 + \mathfrak{q}(T - t)}{1 + \mathfrak{q}T} - \frac{\mathfrak{r}t}{1 + \mathfrak{q}T} + \sigma[1 + \mathfrak{q}(T - t)] \int_0^t \frac{dW_u}{1 + \mathfrak{q}(T - u)}. \tag{3.68}$$

Notice further that the optimal control α_t and the adjoint process Y_t satisfy:

$$-\alpha_t = Y_t = \frac{\mathfrak{q}}{1 + \mathfrak{q}(T - t)} X_t + \frac{\mathfrak{r}}{1 + \mathfrak{q}(T - t)},$$

and that the only quantity depending upon the fixed mean function $t \mapsto \bar{\mu}_t$ is the constant $\mathfrak{r} = -\bar{q}\, s\, \bar{\mu}_T$, which depends only upon the mean state at the end of the time interval. Recalling that $\mathfrak{q} = \bar{q}$, this makes the search for a fixed point very simple and one easily checks that if

$$\bar{\mu}_T = \frac{\mathbb{E}[\xi]}{1 + \bar{q}(1 - s)T}, \tag{3.69}$$

then the mean at time T of the random variable X_T given by (3.68) is $\bar{\mu}_T$.

From this, we deduce that an equilibrium exists if and only if $1 + \bar{q}(1 - s)T \neq 0$.

Remark 3.38 *Whenever* $1 + \bar{q}(1-s)T < 0$, *the reader may find contradictory the fact that* $\bar{\mu}_T$ *and* $\bar{\mu}_0$ *have opposite signs in equilibrium, while* (3.55) *seemingly implies that* $\mathbb{E}[X_t]$ *has the same sign as* $\mathbb{E}[X_0] = \mathbb{E}[\xi]$ *in equilibrium. To resolve this ostensible contradiction, we must recall that* (3.55) *holds whenever the Riccati equation in* (3.54) *is solvable. Whenever* $1 + \bar{q}(1-s)T < 0$, *the Riccati equation in* (3.54) *is certainly not solvable on* $[0,T]$ *as otherwise there would be a solution on the shorter interval* $[0,T_c]$ *with the same terminal condition at* T_c *in lieu of* T, *where* T_c *is the critical time when* $1 + \bar{q}(1-s)T_c = 0$. *Of course, the Riccati equation cannot be solvable on* $[0,T_c]$ *since* (3.69) *asserts that there is no equilibrium when* $T = T_c$.

3.6 Revisiting Some of the Examples of Chapter 1

We test the results of this chapter on some of the examples introduced in Chapter 1.

3.6.1 A Particular Case of the Flocking Model

We first consider the mean field game problem arising from the flocking model introduced in Subsection 1.5.1 of Chapter 1. As in the discussion of the finite player game given in Section 2.4 of Chapter 2, we first consider the particular case $\beta = 0$. The general case $\beta \neq 0$ will be treated in Chapter 4. As explained in Section 2.4, when $\beta = 0$, the weights (1.44) are identically equal to a constant, and the costs to the individuals depend only upon their velocities. Since the position does not appear in the dynamics of the velocities, it is possible to reframe the model in terms of the velocities only. Doing so, we are facing a linear quadratic mean field game model fitting perfectly the framework discussed in this chapter.

As explained in Section 2.4 of Chapter 2, we use the notation X_t^i to denote the velocity at time t of bird i. This choice of the upper case letter X for the velocity, is made to conform with the notation used throughout Chapter 2. Later on, when we consider the general case in Chapter 4, we shall switch back to the original notation introduced in Chapter 1. Here we focus on the dynamics of the velocity:

$$dX_t = \alpha_t dt + \sigma dW_t,$$

and recalling the form (1.48) of the running cost, the minimization concerns the cost functional:

$$J(\boldsymbol{\alpha}) = \mathbb{E}\left[\int_0^T [\frac{1}{2}|\alpha_t|^2 + \frac{\kappa^2}{2}|X_t - \bar{\mu}_t|^2]dt\right].$$

This model appears as a simple particular case covered by the discussion of Section 3.5 with:

$$b_1(t) = 0, \quad b_2(t) = I_3, \quad \bar{b}_1(t) = 0,$$

and

$$r(t) = I_3, \quad s(t) = I_3, \quad \bar{q}(t) = \kappa^2 I_3, \quad q(t) = 0, \quad q = \bar{q} = s = 0.$$

If we denote by Y the adjoint variable of X, and if we set $\bar{x}_t = \mathbb{E}[X_t]$ and $\bar{y}_t = \mathbb{E}[Y_t]$, system (3.52) becomes:

$$\begin{cases} d\bar{x}_t = -\bar{y}_t dt, & \bar{x}_0 = \bar{\mu}_0, \\ d\bar{y}_t = 0, & \bar{y}_T = 0, \end{cases}$$

which shows that $\bar{x}_t = \bar{\mu}_0$ and $\bar{y}_t = 0$. Theorem 3.34 implies existence and uniqueness of a solution to the mean field game. We identify this solution by solving the usual forward-backward system of ODEs (3.58) which reduces to the following system in the present situation:

$$\begin{cases} \dot{\eta}_t - \eta_t^2 + \kappa^2 I_3 = 0, & \eta_T = 0, \\ \dot{\chi}_t - \eta_t \chi_t - \kappa^2 \bar{\mu}_0 = 0, & \chi_T = 0, \\ Z_t = \eta_t \sigma. \end{cases}$$

The first equation is a $d \times d$ matrix Riccati equation. However, in its present very special form, its solution can clearly be searched for as a scalar multiple of the identity. So if the above matrix valued function $t \mapsto \eta_t$ is of the form $t \mapsto \eta_t I_3$ for some real valued function $t \mapsto \eta_t$, the latter should solve the scalar Riccati equation:

$$\dot{\eta}_t - \eta_t^2 + \kappa^2 = 0, \qquad \eta_T = 0. \tag{3.70}$$

We chose to use the same letter η for both the matrix valued and the scalar functions, not so much because of a shortage of Greek characters, but because we already considered a scalar Riccati equation of the type (3.70) in Section 2.4. In fact, this very equation appeared as the limit $N \to \infty$ of the Riccati equations providing open loop and Markovian Nash equilibria for the N player games, and we explained there that this equation has a unique solution given by (2.59):

$$\eta_t = \kappa \frac{e^{2\kappa(T-t)} - 1}{e^{2\kappa(T-t)} + 1}, \qquad t \in [0, T].$$

Similarly, the components χ_t^i of the vector valued function χ_t are given by:

$$\chi_t^i = -\kappa^2 \bar{\mu}_0^i \int_t^T e^{\int_s^t \eta_u du} ds, \qquad t \in [0, T], \quad i = 1, \cdots, d.$$

So the equilibrium trajectories of the velocity are given by the d-dimensional Gaussian dynamics:

$$dX_t = -(\eta_t X_t + \chi_t) dt + \sigma dW_t, \quad t \in [0, T].$$

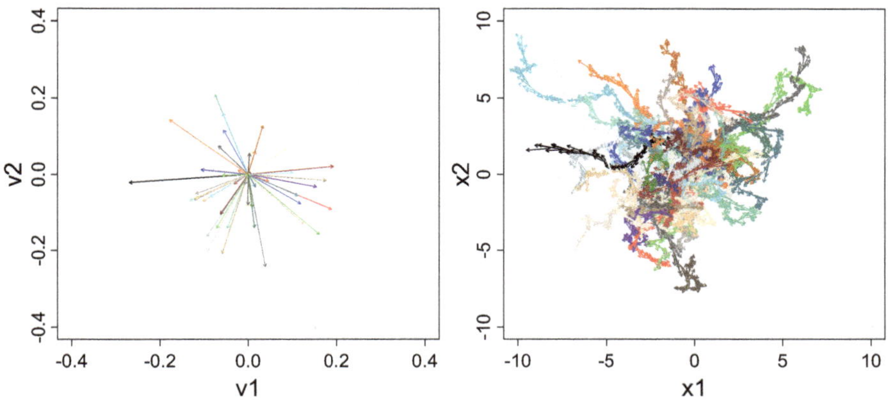

Fig. 3.1 Monte Carlo simulations of $N = 50$ samples with $\kappa = 1$, $\sigma = 1$ and $T = 10$ in dimension $d = 2$. The left pane shows the velocity vectors at time $t = T$, the positions and the (properly rescaled) velocity vectors being shown on the right for $t \in [0, T]$.

These dynamics are mean reverting because the scalar function $t \mapsto \eta_t$ is positive. In fact, the velocity is given by the explicit formula:

$$X_t = e^{-\int_0^t \eta_u du} v_0 - \int_0^t e^{-\int_s^t \eta_u du} \chi_s \, ds + \sigma \int_0^t e^{-\int_s^t \eta_u du} dW_s, \qquad t \in [0, T],$$

where v_0 is the initial velocity, from which we obtain $x_t = x_0 + \int_0^t X_s ds$ for the position at time t of a typical bird in the flock, x_0 denoting the initial position. The left pane of Figure 3.1 shows the results of $N = 50$ Monte Carlo simulations of the model with $\kappa = 1$, $\sigma = 1$ and $T = 10$ in dimension $d = 2$.

3.6.2 Systemic Risk Example Without Common Noise

Even though we were able to solve the finite player game for open and closed loop Nash equilibria, it is instructive to consider the mean field game version of the toy model of systemic risk in the absence of common noise, i.e., when $\rho = 0$. The general case will be discussed later in Chapter (Vol II)-4. When $\rho = 0$, this model is a particular case of the LQ mean field game models considered earlier. Indeed, the MFG strategy is based on the following two steps:

(i) For each fixed deterministic function $[0, T] \ni t \mapsto m_t \in \mathbb{R}$, solve the standard control problem:

$$\inf_{\alpha \in \mathbb{A}} \mathbb{E}\left[\int_0^T \left[\frac{\alpha_t^2}{2} - q\alpha_t(m_t - X_t) + \frac{\epsilon}{2}(m_t - X_t)^2 \right] dt + \frac{c}{2}(m_T - X_T)^2 \right],$$

subject to the dynamics:

$$dX_t = [a(m_t - X_t) + \alpha_t]\,dt + \sigma dW_t,$$

where $\boldsymbol{W} = (W_t)_{0 \leqslant t \leqslant T}$ is a Wiener process independent of the initial value X_0 which may be a square integrable random variable ξ.

(ii) Solve the fixed point problem: find $[0, T] \ni t \mapsto m_t$ so that $m_t = \mathbb{E}[X_t]$ for all $t \in [0, T]$, where $(X_t)_{0 \leqslant t \leqslant T}$ is an optimal trajectory of the optimal control problem in environment $(m_t)_{0 \leqslant t \leqslant T}$.

As stated, this problem is a particular case of the LQ mean field game models discussed above only when $q = 0$. However, in the general case $q \leqslant \epsilon^2$, the arguments used above can be applied mutatis mutandis. The reduced Hamiltonian of the system is given by:

$$H(t, x, y, \alpha) = [a(m_t - x) + \alpha]\,y + \frac{1}{2}\alpha^2 - q\alpha(m_t - x) + \frac{\epsilon}{2}(m_t - x)^2,$$

which is strictly convex in (x, α) under the condition $q^2 \leqslant \epsilon$, and attains its minimum for:

$$\alpha = \hat{\alpha}(t, x, m_t, y) = q(m_t - x) - y.$$

The corresponding *adjoint forward-backward equations* are given by:

$$\begin{cases} dX_t = [(a + q)(m_t - X_t) - Y_t]dt + \sigma dW_t \\ dY_t = \big[(a + q)Y_t + (\epsilon - q^2)(m_t - X_t)\big]\,dt + Z_t dW_t, \quad t \in [0, T], \\ Y_T = c(X_T - m_T). \end{cases} \tag{3.71}$$

This affine FBSDE is of the type considered above and can be solved in the same way. Given our experience with the corresponding finite player game solved in the previous chapter, we make the (educated) ansatz:

$$Y_t = -\eta_t(m_t - X_t), \tag{3.72}$$

and the deterministic function $t \mapsto \eta_t$ must solve the Riccati equation:

$$\dot{\eta}_t = 2(a + q)\eta_t + \eta_t^2 - (\epsilon - q^2), \tag{3.73}$$

with terminal condition $\eta_T = c$. As expected, this equation appears as the limit as $N \to \infty$ of the Riccati equations obtained in the solutions of the N-player games for open and closed loop equilibria, see (2.80). This is a first concrete confirmation of the *folk theorem* according to which the differences between open and closed loop equilibria disappear in the limit $N \to \infty$ of large games. See the Notes & Complements at the end of the chapter for references discussing this claim.

3.6.3 The Diffusion Form of Aiyagari's Growth Model

The diffusion version of Aiyagari's growth model presented in Subsection 1.4.3 of Chapter 1 belongs in the family of MFG models considered in this chapter. In order to satisfy the ergodic property mentioned in Remark 1.22, we model the Z-component of the state (Z_t, A_t) by an Ornstein-Uhlenbeck process, choosing $\mu_Z(z) = 1-z$ and a positive constant for σ_Z. We shall specify the numerical value for σ_Z when we want to compare our numerical results to some computations reported in the literature. So the state dynamics are given by:

$$\begin{cases} dZ_t = -(Z_t - 1)\,dt + \sigma_Z dW_t, \\ dA_t = \big[(1-\alpha)\bar{\mu}_t^{\alpha} Z_t + \big(\alpha\bar{\mu}_t^{\alpha-1} - \delta\big)A_t - c_t\big]dt, \quad t \in [0,T], \end{cases} \tag{3.74}$$

where $(\bar{\mu}_t)_{0\leqslant t\leqslant T}$ denotes the flow of average wealths in the population in equilibrium. We switched to a system of notation used in this chapter, but the reader is warned that this average wealth was denoted K_t when we introduced the model in Chapter 1 using standard notation in the macro-economic literature. In any case, this average wealth is assumed to take (strictly) positive values, both for economic reasons and because of the powers $\alpha \in (0,1)$ and $1-\alpha \in (0,1)$ appearing in the above equation. Observe also that we used the Greek letter α for the exponent in (3.74), although we already used the same letter for the elements of the admissible values for the control processes; clearly, there is no risk of confusion between both in the sequel. In order to make sure that $(Z_t)_{t\geqslant 0}$ is stationary for all times (and not simply "asymptotically stationary"), we can assume that the distribution of Z_0 is the invariant measure $N(1, \sigma_Z^2/2)$ of the process. For our purpose, we just assume that $\mathbb{E}[Z_0] = 1$, Z_0 being independent of $\boldsymbol{W}$. Among other things, this implies that $\mathbb{E}[Z_t] = 1$ for all $t \geqslant 0$, fact which we shall use later on. Last, observe that, in comparison with (1.37) in Chapter 1, we took $\bar{a} = 1$.

The set $\mathbb{A}$ of admissible controls is the set $\mathbb{H}_+^{2,1}$ of real valued square-integrable $\mathbb{F}$-adapted processes $\boldsymbol{c} = (c_t)_{0\leqslant t\leqslant T}$ with nonnegative values, and the cost functional is defined by:

$$J(\boldsymbol{c}) = \mathbb{E}\bigg[\int_0^T (-U)(c_t)dt - \tilde{U}(A_T)\bigg],$$

for the CRRA utility function U given by (1.35), namely $U(c) = (c^{1-\gamma} - 1)/(1-\gamma)$ for $\gamma > 0$ with $U(c) = \ln(c)$ if $\gamma = 1$, and $\tilde{U}(a) = a$. Notice the additional minus signs due to the fact that we want to treat the optimization problem as a minimization problem. Here we chose to take 0 for the discount rate since we are working on a finite horizon. Throughout the analysis, we shall assume that $A_0 > 0$.

Application of the Pontryagin Maximum Principle

The Hamiltonian reads:

$$\begin{aligned} &H(t, z, a, \mu, y_z, y_a, c) \\ &\quad = (1-z)y_z + \big[-c + (1-\alpha)\bar{\mu}^{\alpha} z + (\alpha\bar{\mu}^{\alpha-1} - \delta)a\big]y_a - U(c), \end{aligned}$$

where $\bar{\mu} = \int_{\mathbb{R}^2} a\mu(dz, da)$ denotes the mean of the second marginal of the measure μ. Notice that we use the reduced Hamiltonian because the volatility of Z is constant and the volatility of A is zero. The first adjoint equation reads:

$$\begin{aligned} dY_{z,t} &= -\partial_z H\big(t, Z_t, A_t, \mu_t, Y_{z,t}, Y_{a,t}, c_t\big)dt + \tilde{Z}_{z,t}dW_t \\ &= \big(Y_{z,t} - (1-\alpha)\bar{\mu}_t^{\alpha}\big)dt + \tilde{Z}_{z,t}dW_t, \quad t \in [0, T]. \end{aligned}$$

Since the variables z and y_z do not play any role in the minimization of the Hamiltonian with respect to the control variable c, the process $(Y_{z,t})_{0\leqslant t\leqslant T}$ does not enter the definition of the optimal trajectory. Consequently, we can ignore these variables and not include them in the Hamiltonian. Accordingly, we shall use the (further) reduced Hamiltonian:

$$H(t, a, \mu, y, c) = \big[-c + (\alpha\bar{\mu}^{\alpha-1} - \delta)a\big]y - U(c),$$

y being understood as y_a. Clearly, this Hamiltonian is convex in (a, c) and strictly convex in c. The form (1.36) of the derivative of the utility function implies that the value of the control minimizing the Hamiltonian is $\hat{c} = (-U')^{-1}(y) = (-y)^{-1/\gamma}$. Therefore, the FBSDE derived from the Pontryagin stochastic maximum principle reads:

$$\begin{cases} dA_t = \big[(1-\alpha)\bar{\mu}_t^{\alpha}Z_t + [\alpha\bar{\mu}_t^{\alpha-1} - \delta]A_t - (-Y_t)^{-1/\gamma}\big]dt \\ dY_t = -Y_t[\alpha\bar{\mu}_t^{\alpha-1} - \delta]dt + \tilde{Z}_t dW_t, \quad t \in [0, T]\,; \quad Y_T = -1, \end{cases} \tag{3.75}$$

where we used the notation $(\tilde{Z}_t)_{0\leqslant t\leqslant T}$ to denote the integrand of the backward equation in order to distinguish it from the process $(Z_t)_{0\leqslant t\leqslant T}$ used in the model for the first component of the state. We emphasize that, the utility function U having a singularity at 0, the assumptions of the Pontryagin principle used in this chapter are not satisfied here. However, it is easy to see that the proof of the sufficient part of the Pontryagin principle goes through provided that the adjoint process $(Y_t)_{0\leqslant t\leqslant T}$ lives, with probability 1, in a compact subset of $(-\infty, 0)$.

The crux of our analysis is to notice that the backward equation may be decoupled from the forward equation. Its solution is deterministic and is obtained by solving the backward ordinary differential equation:

$$dY_t = -Y_t[\alpha\bar{\mu}_t^{\alpha-1} - \delta]dt, \quad t \in [0, T]\,; \quad Y_T = -1.$$

Among other things, this shows that the process $(Y_t)_{0\leqslant t\leqslant T}$ is negative valued, whatever the input $(\bar{\mu}_t)_{0\leqslant t\leqslant T}$. Also the optimal trajectory is unique and the optimal consumption $\hat{c}_t = (-Y_t)^{-1/\gamma}$ is also deterministic! Once we know that $(Y_t)_{0\leqslant t\leqslant T}$ is deterministic, taking the expectation in the dynamics of $(A_t)_{0\leqslant t\leqslant T}$, we deduce that the flow $(\bar{\mu}_t)_{0\leqslant t\leqslant T}$ describing the average wealth of the population in equilibrium, if it exists, must solve the deterministic forward-backward system:

$$\begin{cases} d\bar{\mu}_t = \left[\bar{\mu}_t^\alpha - \delta\bar{\mu}_t - (-Y_t)^{-1/\gamma}\right]dt, \\ dY_t = -Y_t[\alpha\bar{\mu}_t^{\alpha-1} - \delta]dt, \quad t \in [0,T]\,; \quad Y_T = -1. \end{cases} \tag{3.76}$$

Above we used the fact that $\mathbb{E}[Z_t] = 1$ for all $t \in [0,T]$. In order to tackle the existence and uniqueness of an MFG equilibrium, it thus suffices to prove that (3.76) admits a unique solution $(\bar{\mu}_t, Y_t)_{0\leqslant t\leqslant T}$ satisfying $\bar{\mu}_t > 0$ (and $Y_t < 0$) for any $t \in [0,T]$. Once (3.76) has been solved, it is indeed straightforward to plug the solution into (3.75) and to solve the forward equation therein.

Looking for an Equilibrium

Obviously, a major difficulty in (3.76) is to guarantee that the solutions have the required signs. In order to proceed, we perturb the original system and *smooth out* the power terms coming from the derivation of the Cobb-Douglas production function. We consider a continuously differentiable concave function φ such that φ is affine on $(-\infty, 0]$ and φ coincides with $x \mapsto x^\alpha$ on $[\epsilon, +\infty)$ for some $\epsilon > 0$. In particular, φ' is bounded and Lipschitz continuous on the whole line. Then, we replace (3.76) by the system:

$$\begin{cases} d\bar{\mu}_t = \left[\varphi(\bar{\mu}_t) - \delta\bar{\mu}_t - (-Y_t)^{-1/\gamma}\right]dt, \\ dY_t = -Y_t[\varphi'(\bar{\mu}_t) - \delta]dt, \quad t \in [0,T]\,; \quad Y_T = -1. \end{cases} \tag{3.77}$$

Figure 3.2 gives the plots of φ and its derivative for the value $\epsilon = 0.01$ which we shall use in the numerical computations below.

Notice that if we assume that $(a_t = \bar{\mu}_t)_{0\leqslant t\leqslant T}$ is given, then the second equation of (3.77) can be solved explicitly. We get:

$$Y_t = -e^{\int_t^T [\varphi'(a_s)-\delta]ds}, \quad t \in [0,T]. \tag{3.78}$$

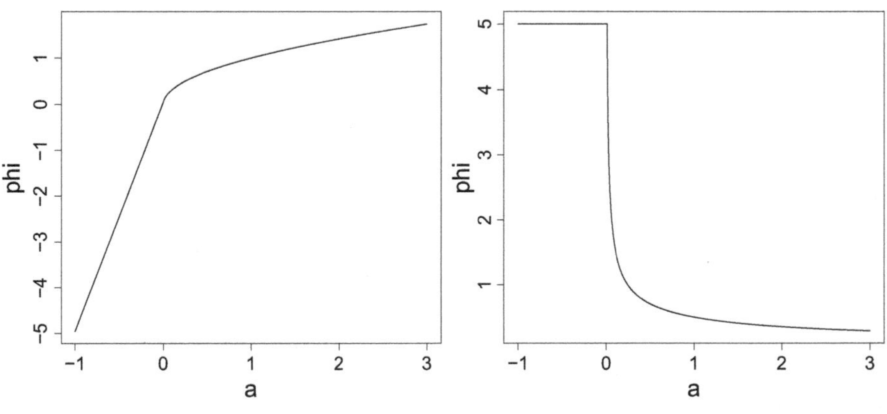

Fig. 3.2 Plots of the regularizing function $a \mapsto \varphi(a)$ (left) and its derivative $a \mapsto \varphi'(a)$ (right) for the values $\epsilon = 0.01$ and $\alpha = 0.5$ of the parameters.

Strangely enough, the system (3.77) is quite easy to solve numerically. Indeed, a simple Picard's iteration converges very quickly to a numerical solution. Typically, we start with $(Y_t = -1)_{0 \leqslant t \leqslant T}$, inject it in the first equation, run a standard Ordinary Differential Equation (ODE) solver to find $(a_t = \bar{\mu}_t)_{0 \leqslant t \leqslant T}$ satisfying this first equation, inject this solution into formula (3.78), retrieve $(Y_t)_{0 \leqslant t \leqslant T}$ which we then inject in the first equation, etc. The process converges after a small number (no more than 5 or 6 depending upon the values of the parameters) of iterations. Figure 3.3 gives the plots of the solutions $(\bar{\mu}_t)_{0 \leqslant t \leqslant T}$ and $(Y_t)_{0 \leqslant t \leqslant T}$ obtained for a few values of the parameters given in the caption.

Figure 3.4 shows how the average wealth $(\bar{\mu}_t)_{0 \leqslant t \leqslant T}$ and the adjoint variable Y_t depend upon the risk aversion level γ of the agents.

As for the mathematical analysis of the system (3.77), we first notice that, quite remarkably, it reads like the forward-backward system obtained from the

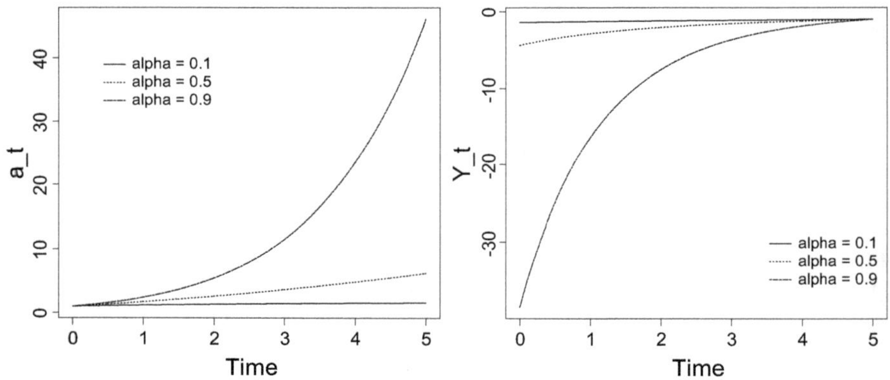

Fig. 3.3 Plots of the solutions $(\bar{\mu}_t)_{0 \leqslant t \leqslant T}$ (left) and $(Y_t)_{0 \leqslant t \leqslant T}$ (right) of the forward/backward system (3.77) for different values of α, and for the values $\epsilon = 0.01$ and $\gamma = 1.5$ for the cut-off and risk aversion parameters.

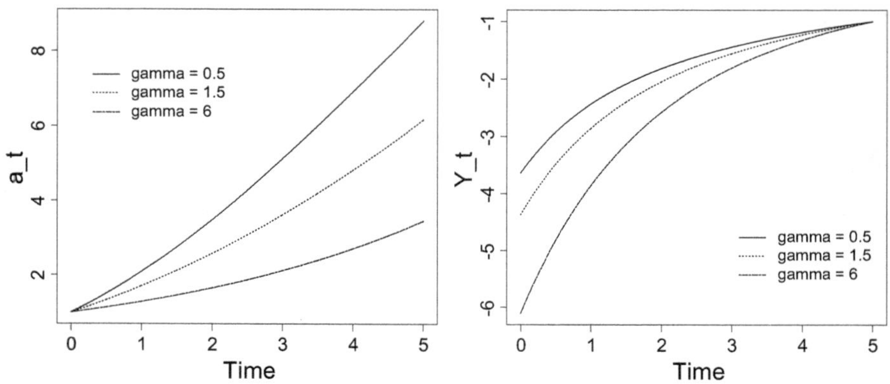

Fig. 3.4 Plots of the solutions $(\bar{\mu}_t)_{0 \leqslant t \leqslant T}$ (left) and $(Y_t)_{0 \leqslant t \leqslant T}$ (right) of the forward/backward system (3.77) for different values of the risk aversion parameter γ, and for the values $\epsilon = 0.01$ and $\alpha = 0.5$ for the cut-off and Cobb-Douglas parameters.

deterministic Pontryagin maximum principle, when applied to the minimization problem:

$$\inf_{c} \bar{J}(\boldsymbol{c}), \qquad \text{with} \qquad \bar{J}(\boldsymbol{c}) = \int_0^T (-U)(c_t)dt - \tilde{U}(a_T),$$

where $\mathbf{c} = (c_t)_{0 \leqslant t \leqslant T}$ is a deterministic integrable path from $[0, T]$ to $\mathbb{R}_+$ and $\boldsymbol{a} = (a_t)_{0 \leqslant t \leqslant T}$ solves:

$$\dot{a}_t = \varphi(a_t) - \delta a_t - c_t. \tag{3.79}$$

Here, the Hamiltonian reads:

$$\bar{H}(a, y, c) = [\varphi(a) - \delta a - c]y - U(c),$$

which is easily seen to be convex in (a, c) when y is restricted to $(-\infty, 0)$. Since φ' is bounded, it is indeed pretty clear that the solution of the backward equation:

$$\dot{y}_t = -y_t\big(\varphi'(a_t) - \delta\big), \quad t \in [0, T]; \quad y_T = -1, \tag{3.80}$$

where $\boldsymbol{a} = (a_t)_{0 \leqslant t \leqslant T}$ solves the controlled equation (3.79), lives in a compact subset in $(-\infty, 0)$. As a byproduct, this shows that, in the minimization of $\bar{J}$, the dual variable y must live in a compact subset of $(-\infty, 0)$. In particular, in (3.77), the singular term $(-Y_t)^{-1/\gamma}$ may be replaced by a bounded and Lipschitz function of Y. With such a prescription, (3.77) may be seen as an FBSDE with Lipschitz coefficients. It is thus uniquely solvable in small time, see Chapter 4. The fact that the backward equation lives in a compact subset of $(-\infty, 0)$, determined by $\|\varphi'\|_\infty$ and δ only, shows that the optimal control, given by $((-Y_t)^{-1/\gamma})_{0 \leqslant t \leqslant T}$, lives a compact subset of $(0, +\infty)$, independently of the initial condition of $(\bar{\mu}_t)_{0 \leqslant t \leqslant T}$. In order to extend inductively the property of unique solvability from small to long time intervals, it suffices to notice that, for the useful values of c and y, the mapping $(a, y, c) \mapsto \bar{H}(a, y, c)$ is convex in (a, c) and is uniformly convex in c. Then, by the Pontryagin maximum principle and using the same argument as in the proof of Lemma 4.56, based on Proposition 3.21, we can control the Lipschitz constant of the decoupling field along the induction used in the extension from small time to long time. This gives the unique solvability of (3.77) for an initial condition $\bar{\mu}_0 > 0$.

We now show that we can replace the function φ and its derivative by the original power function $x \mapsto x^\alpha$ and its derivative, and still solve the system. First we notice that formula (3.78) giving Y_t in terms of a_t implies that:

$$-Y_t \geqslant \exp(\delta(t - T)), \quad t \in [0, T],$$

and therefore:

$$(-Y_t)^{-1/\gamma} \leqslant \exp\Big(\frac{\delta(T - t)}{\gamma}\Big), \quad t \in [0, T].$$

Consider now the solution of the forward ODE:

$$\dot{a}_t = \left[\varphi(a_t) - \delta a_t - (-Y_t)^{-1/\gamma}\right], \quad t \in [0, T].$$

Then,

$$\begin{aligned} a_t &= \exp(-\delta t)\left[a_0 + \int_0^t \exp(\delta s)\varphi(a_s)ds - \int_0^t \exp(\delta s)(-Y_s)^{-1/\gamma}ds\right] \\ &\geqslant \exp(-\delta t)\left[a_0 - \int_0^t \exp(\delta s)\exp\left(\frac{\delta(T-s)}{\gamma}\right)ds\right]. \end{aligned}$$

Therefore, when

$$a_0 = \bar{\mu}_0 > \int_0^T \exp(\delta s)\exp\left(\frac{\delta(T-s)}{\gamma}\right)ds, \tag{3.81}$$

we can choose ϵ small enough so that the solution of (3.77) is also a solution of (3.76). By the same argument, the solution must be unique, since any other solution of (3.76), with the prescribed signed condition (that is $\bar{\mu}_t > 0$ and $Y_t < 0$ for all $t \in [0, T]$), is a solution of (3.77) for a well-chosen ϵ. It is worth mentioning that the solution to (3.75), obtained in the end under the condition (3.81), satisfies $\mathbb{E}[A_t] > 0$ for any $t \in [0, T]$.

Monte Carlo Simulations

Once the equilibrium values of the functions $(a_t = \bar{\mu}_t)_{0\leqslant t\leqslant T}$ and $(Y_t)_{0\leqslant t\leqslant T}$ are obtained, it is easy to run Monte Carlo simulations using the original system (3.74). Indeed, it is straightforward to simulate samples from the Ornstein-Uhlenbeck dynamics given by the first equation, and for each such sample, one can use the second equation to compute the corresponding trajectory of the wealth $(A_t)_{0\leqslant t\leqslant T}$. Doing so, one can keep track of the full distribution of wealth over the entire population. For example, Figure 3.5 shows the histogram of the wealth distribution at the terminal time. It was computed from $N = 20,000$ Monte Carlo scenarios (based on simulations of the Ornstein-Uhlenbeck process $(Z_t)_{0\leqslant t\leqslant T}$) with $T = 5$ year, $\alpha = 0.5$ and $\gamma = 6$, starting from an initial distribution of the wealth, which was chosen to be uniform over the interval $[0, 2]$.

3.7 Games with a Continuum of Players

The rationale for the mean field game models studied in this book is based on the limit as $N \to \infty$ of N-player games with mean field interactions. One of the justifications given in Chapter 1 for the formulation of the mean field game paradigm is that the influence on the game of each individual player vanishes in this limit.

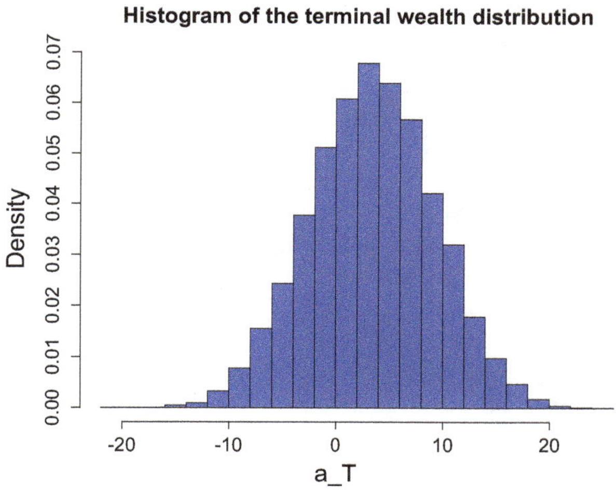

Fig. 3.5 Histogram of the terminal wealth from a sample of $N = 20{,}000$ Monte Carlo scenarios.

Mathematical physicists and economists have been using game models in which the impact of each single player is *insignificant*. They do just that by considering games for which the players are labeled by elements i of an uncountable set I, accounting for a continuum of agents. This set I is equipped with a σ-field $\mathcal{I}$ and a probability measure λ which is assumed to be continuous (i.e., nonatomic). In this way, if $i \in I$ represents a player, the fact that $\lambda(\{i\}) = 0$ accounts for the insignificance of the players in the model. This section is thus intended to be a quick introduction to the framework of games with a continuum of players.

The classical Glivenko-Cantelli form of the Law of Large Numbers (LLN) states that if F denotes the cumulative distribution function of a probability measure on $\mathbb{R}$, if $(X^n)_{n\geqslant 1}$ is an infinite sequence of independent identically distributed random variables on a probability space $(\Omega, \mathcal{F}, \mathbb{P})$ with common distribution μ, and if we use the notation:

$$F_\omega(x) = \limsup_{N\to\infty} \frac{1}{N} \#\big\{n \in \{1, \cdots, N\} : X^n(\omega) \leqslant x\big\}, \quad x \in \mathbb{R},\ \omega \in \Omega, \tag{3.82}$$

for the proportion of $X^n(\omega)$'s not greater than x, then this lim sup is in fact a limit for all $x \in \mathbb{R}$ and $\mathbb{P}$-almost all $\omega \in \Omega$, and $\mathbb{P}[\{\omega \in \Omega : F_\omega(\cdot) = F\}] = 1$.

Switching gears momentarily, recall that, over fifty years ago, economists suggested that the appropriate model for perfectly competitive markets is a model with a continuum of traders represented as elements of a measurable space. In such a set-up, the insignificance of individual traders is captured by the idea of a set with zero measure, and summation or aggregation is generalized by the notion of integral. In games with a continuum of players, the latter are labeled by the elements $i \in I$ of an arbitrary set I (often assumed to be uncountable, and most often chosen to be the unit interval $[0, 1]$) equipped with a σ-field $\mathcal{I}$ and a probability measure λ. In this set-up, if the state of each player $i \in I$ is given by a random variable X^i on a probability space $(\Omega, \mathcal{F}, \mathbb{P})$, in analogy with the countable case leading to formula (3.82), the quantity:

$$F_\omega(x) = \lambda\big(\{i \in I : X^i(\omega) \leqslant x\}\big) \tag{3.83}$$

appears as a natural generalization of the *proportion of* $(X^i(\omega))_{i\in I}$*'s not greater than* x, in other words of the cumulative distribution function of the empirical distribution, and if the $(X^i)_{i\in I}$'s were to be independent with the same distribution, we could have a reasonable generalization of the Law of Large Numbers to this setting. However, as we explain in the next subsection, measurability issues get in the way and such a generalization is, when it does exist, far from trivial.

3.7.1 The Exact Law of Large Numbers

If E is a Polish space, for each probability measure $\mu \in \mathcal{P}(E)$, Kolmogorov's theorem can be used to construct on the product space $\Omega = E^I$ equipped with the σ-field $\mathcal{F}$ obtained as the product of copies of the Borel σ-field of E, the product probability measure $\mathbb{P}$ for which the coordinate projections $(X^i : \Omega \ni \omega \mapsto X^i(\omega) = \omega(i) \in E)_{i\in I}$ become independent and identically distributed random variables with common distribution μ on the probability space $(\Omega, \mathcal{F}, \mathbb{P})$. It is also well known that the sample paths $I \ni i \mapsto X^i(\omega) \in E$ are pretty rough functions since they are (for $\mathbb{P}$-almost $\omega \in \Omega$) nowhere continuous and not even measurable.

Hence, this construction of a continuum of independent identically distributed random variables leads to irregular structures lacking measurability properties. The following definition offers an alternative which keeps most of what is needed from the independence.

Definition 3.39 *If E is a Polish space, the E-valued random variables $(X^i)_{i\in I}$ are said to be essentially pairwise independent if, for λ-almost every $i \in I$, the random variable X^i is independent of X^j for λ-almost every $j \in I$. Accordingly, if the real valued random variables $(X^i)_{i\in I}$ are square integrable, we say that the family $(X^i)_{i\in I}$ is essentially pairwise uncorrelated if, for λ-almost every $i \in I$, the correlation coefficient of X^i with X^j is 0 for λ-almost every $j \in I$.*

One may wonder if essentially pairwise independent families $(X^i)_{i\in I}$ can be constructed on probability spaces $(\Omega, \mathcal{F}, \mathbb{P})$ so that the process $\boldsymbol{X} : I\times\Omega \ni (i, \omega) \mapsto X^i(\omega)$ satisfies relevant measurability properties. To do so, we shall construct such processes on extensions of the product space $(I\times\Omega, \mathcal{I}\otimes\mathcal{F}, \lambda\otimes\mathbb{P})$, which are called Fubini's extensions.

Definition 3.40 *If $\mathcal{I} \boxtimes \mathcal{F}$ is a σ-field containing $\mathcal{I} \otimes \mathcal{F}$ and $\lambda \boxtimes \mathbb{P}$ is a probability measure on $(I \times \Omega, \mathcal{I} \boxtimes \mathcal{F})$, then $(I \times \Omega, \mathcal{I} \boxtimes \mathcal{F}, \lambda \boxtimes \mathbb{P})$ is said to be a Fubini extension of $(I \times \Omega, \mathcal{I} \otimes \mathcal{F}, \lambda \otimes \mathbb{P})$ if, for every measurable and $\lambda \boxtimes \mathbb{P}$-integrable $X : I \times \Omega \ni (i, \omega) \mapsto X^i(\omega) \in \mathbb{R}$, we have:*

1. *for λ-a.e. $i \in I$, $\Omega \ni \omega \mapsto X^i(\omega)$ is a $\mathbb{P}$-integrable random variable, and for $\mathbb{P}$-a.e. $\omega \in \Omega$, $I \ni i \mapsto X^i(\omega)$ is measurable and λ-integrable;*
2. *$I \ni i \mapsto \int_\Omega X^i(\omega) d\mathbb{P}(\omega)$ is measurable and λ-integrable, and $\Omega \ni \omega \mapsto \int_I X^i(\omega) d\lambda(i)$ is a $\mathbb{P}$-integrable random variable, and:*

$$\begin{aligned} \int_I \left(\int_\Omega X^i(\omega) d\mathbb{P}(\omega) \right) d\lambda(i) &= \int_\Omega \left(\int_I X^i(\omega) d\lambda(i) \right) d\mathbb{P}(\omega) \\ &= \int_{I \times \Omega} X^i(\omega) d\big(\lambda \boxtimes \mathbb{P}\big)(i, \omega). \end{aligned} \tag{3.84}$$

In the sequel, we shall use the standard symbol $\mathbb{E}$ for denoting the expectation under the sole probability $\mathbb{P}$.

Measurable essentially pairwise independent processes $\boldsymbol{X}$ are first constructed in such a way that, for each $i \in I$, the law of X^i is the uniform distribution on the unit interval $[0, 1]$. Then, using the tools we develop in Chapter 5, see for example Lemma 5.29, we easily construct measurable essentially pairwise independent Euclidean-valued processes with any given prescribed marginals. So the actual problem is to construct rich product probability spaces in the sense of the following definition.

Definition 3.41 *A Fubini extension $(I \times \Omega, \mathcal{I} \boxtimes \mathcal{F}, \lambda \boxtimes \mathbb{P})$ is said to be rich if there exists a real valued $\mathcal{I} \boxtimes \mathcal{F}$-measurable essentially pairwise independent process $\boldsymbol{X}$ such that the law of X^i is the uniform distribution on $[0, 1]$ for every $i \in I$.*

We refer to the Notes & Complements at the end of the chapter for references to papers giving the construction of essentially pairwise independent measurable processes on Fubini extensions.

The following gives a simple property of rich Fubini extensions.

Lemma 3.42 *If the Fubini extension $(I \times \Omega, \mathcal{I} \boxtimes \mathcal{F}, \lambda \boxtimes \mathbb{P})$ is rich, then λ is necessarily atomless.*

Proof. We shall argue by contradiction. If $A \in \mathcal{I}$, with $\lambda(A) > 0$, is an atom of $(I, \mathcal{I}, \lambda)$, then, for $\mathbb{P}$-a.e. $\omega \in \Omega$, the function $I \ni i \mapsto X^i(\omega)$ is λ-a.e. constant on A. So for $\mathbb{P}$-a.e. $\omega \in \Omega$ and λ-a.e. $i \in A$,

$$X^i(\omega) = \int_A X^j(\omega) \frac{d\lambda(j)}{\lambda(A)},$$

and using the Fubini property (3.84), we deduce that for λ-a.e. $i \in A$, the random variable $\Omega \ni \omega \mapsto X^i(\omega)$ is $\mathbb{P}$-a.e. equal to the random variable $\theta : \Omega \ni \omega \mapsto \int_A X^j(\omega) d\lambda(j)/\lambda(A)$. Also, for any event $B \in \mathcal{F}$,

$$\begin{aligned}\mathbb{P}[\theta \in B] &= \frac{1}{\lambda(A)} \lambda \boxtimes \mathbb{P}\big[(i,\omega) \in A \times \Omega : X^i(\omega) \in B\big] \\ &= \frac{1}{\lambda(A)} \int_A \mathbb{P}[X^i \in B] d\lambda(i) = \mathrm{Leb}_1(B),\end{aligned}$$

proving that θ, as a random variable on $(\Omega, \mathcal{F}, \mathbb{P})$, has the uniform distribution. In particular, $\mathbb{E}[\theta^2] = 1/3$.

On the other hand, we know that, for almost every $i \in I$, the function $I \times \Omega \ni (j,\omega) \mapsto X^i(\omega) X^j(\omega)$ is $\mathcal{I} \boxtimes \mathcal{F}$-measurable. Also, by the Fubini property, the function $I \ni j \mapsto \mathbb{E}[X^i X^j]$ is integrable with respect to λ and

$$\frac{1}{\lambda(A)} \int_A \mathbb{E}[X^i X^j] d\lambda(j) = \mathbb{E}[X^i \theta]. \tag{3.85}$$

Now, we observe that the function $I \times \Omega \ni (i,\omega) \mapsto X^i(\omega)\theta(\omega)$ is also $\mathcal{I} \boxtimes \mathcal{F}$-measurable. Hence, $I \ni i \mapsto \mathbb{E}[X^i \theta]$ is integrable with respect to λ and

$$\frac{1}{\lambda(A)} \int_A \mathbb{E}[X^i \theta] d\lambda(i) = \mathbb{E}[\theta^2] = \frac{1}{3}.$$

The contradiction comes from the fact that, for almost every $i \in I$, X^i is orthogonal to X^j for almost every $j \in I$. In other words, the left-hand side in (3.85) is equal to:

$$\frac{1}{\lambda(A)} \int_A \mathbb{E}[X^i X^j] d\lambda(j) = \frac{1}{4},$$

which gives the desired contradiction. □

Using Lemma 5.29 from Chapter 5 when E is a Euclidean space and an extension of it when E is a more general Polish space, see for instance the Notes & Complements at the end of Chapter 5, we get the following result which we already announced.

Proposition 3.43 *If the Fubini extension $(I \times \Omega, \mathcal{I} \boxtimes \mathcal{F}, \lambda \boxtimes \mathbb{P})$ is rich, if E is a Polish space, and if $\mu : I \to \mathcal{P}(E)$ is $\mathcal{I}$-measurable, then there exists a $\mathcal{I} \boxtimes \mathcal{F}$-measurable E-valued essentially pairwise independent process $\mathbf{Y} : I \times \Omega \to E$ such that for λ-a.e. $i \in I$, $\mathbb{P} \circ Y_i^{-1} = \mu_i$.*

An exact law of large numbers can be proven on Fubini's extensions. In a weak form, this law can be given in the following way.

Theorem 3.44 *Let $\boldsymbol{X} = (X^i)_{i\in I}$ be a measurable square integrable process on a Fubini extension $(I \times \Omega, \mathcal{I} \boxtimes \mathcal{F}, \lambda \boxtimes \mathbb{P})$. The following are equivalent:*

(i) The random variables $(X^i)_{i\in I}$ are essentially pairwise uncorrelated;
(ii) For every $A \in \mathcal{I}$ with $\lambda(A) > 0$, one has for $\mathbb{P}$-almost every $\omega \in \Omega$:

$$\int_A X^i(\omega) d\lambda(i) = \int_A \mathbb{E}[X^i] d\lambda(i).$$

Proof.

First Step. We first check that if $\boldsymbol{Y} = (Y^i)_{i\in I}$ and $\boldsymbol{Z} = (Z^i)_{i\in I}$ are measurable and square integrable processes on the Fubini extension $(I \times \Omega, \mathcal{I} \boxtimes \mathcal{F}, \lambda \boxtimes \mathbb{P})$, and if we set $\tilde{X}^{i,j}(\omega) = Y^i(\omega)Z^j(\omega)$ for $i,j \in I$ and $\omega \in \Omega$, then $\Omega \ni \omega \mapsto \tilde{X}^{i,j}$ is $\mathbb{P}$-integrable for λ-a.e. $i \in I$ and $j \in I$. Now, proceeding as in the proof of Lemma 3.42 and using the Fubini property of the space, we easily check that, for λ-a.e. $i \in I$, the function $I \ni j \mapsto \mathbb{E}[\tilde{X}^{i,j}]$ is λ-integrable, that the function $I \ni i \mapsto \int_I \mathbb{E}[\tilde{X}^{i,j}] d\lambda(j) = \mathbb{E}[Y^i \int_I Z^j d\lambda(j)]$ is λ-integrable, that the function $\Omega \ni \omega \mapsto (\int_I Y^i(\omega) d\lambda(i))(\int_I Z^j(\omega) d\lambda(j))$ is $\mathbb{P}$-integrable and that:

$$\mathbb{E}\Big[\Big(\int_I Y^i(\omega) d\lambda(i)\Big)\Big(\int_I Z^j(\omega) d\lambda(j)\Big)\Big] = \int_I \Big(\int_I \mathbb{E}[\tilde{X}^{i,j}] d\lambda(i)\Big) d\lambda(j). \tag{3.86}$$

Second Step. Let $A, B \in \mathcal{I}$, and let us define the processes $\boldsymbol{Y} = (Y^i)_{i\in I}$ and $\boldsymbol{Z} = (Z^i)_{i\in I}$ by $(Y^i = \mathbf{1}_A(i)(X^i - \mathbb{E}[X^i]))_{i\in I}$ and $(Z^i = \mathbf{1}_B(i)(X^i - \mathbb{E}[X^i]))_{i\in I}$ respectively. Applying (3.86) from the first step we get:

$$\begin{aligned}\int_A \int_B \mathbb{E}\Big[\Big(X^i - \mathbb{E}[X^i]\Big)\Big(X^j - \mathbb{E}[X^j]\Big)\Big] d\lambda(i) d\lambda(j) \\ = \mathbb{E}\Big[\int_A \Big(X^i - \mathbb{E}[X^i]\Big) d\lambda(i) \int_B \Big(X^j - \mathbb{E}[X^j]\Big) d\lambda(j)\Big],\end{aligned} \tag{3.87}$$

and the implication *(i)* ⇒ *(ii)* follows by taking $B = A$. On the other hand, if we assume that *(ii)* holds, equation (3.87) implies that:

$$\int_A \int_B \mathbb{E}\Big[\Big(X^i - \mathbb{E}[X^i]\Big)\Big(X^j - \mathbb{E}[X^j]\Big)\Big] d\lambda(i) d\lambda(j) = 0$$

for all $A, B \in \mathcal{I}$. The set $A \in \mathcal{I}$ being arbitrary, we conclude that:

$$\int_B \mathbb{E}\Big[\Big(X^i - \mathbb{E}[X^i]\Big)\Big(X^j - \mathbb{E}[X^j]\Big)\Big] d\lambda(j) = 0,$$

for λ-a.e. $i \in I$. So for λ-a.e. $i \in I$, $B \in \mathcal{I}$ being arbitrary, we conclude that:

$$\mathbb{E}\Big[\Big(X^i - \mathbb{E}[X^i]\Big)\Big(X^j - \mathbb{E}[X^j]\Big)\Big] = 0$$

for λ-a.e. $j \in I$ which completes the proof. □

Theorem 3.44 provides a form of the weak law of large numbers for essentially pairwise uncorrelated uncountable families of random variables. Here is a stronger form for essentially pairwise independent families of random variables.

Theorem 3.45 *Let E be a Polish space and $\boldsymbol{X} = (X^i)_{i\in I}$ be a measurable E-valued process on a Fubini extension $(I\times\Omega, \mathcal{I}\boxtimes\mathcal{F}, \lambda\boxtimes\mathbb{P})$ such that the random variables $(X^i)_{i\in I}$ are essentially pairwise independent. Then, for $\mathbb{P}$ almost every $\omega\in\Omega$ and for any B in the Borel σ-field $\mathcal{B}(E)$,*

$$\lambda\big[\{i\in I : X^i(\omega)\in B\}\big] = \int_I \mathbb{P}\big[X^i\in B\big]d\lambda(i).$$

Of course, we may choose E as a Euclidean space, in which case we get a strong form of the exact law of large numbers for essentially pairwise independent families of random variables with values in $\mathbb{R}^d$, for some $d\geqslant 1$. By choosing E as a functional space, the same holds true for a continuum of essentially pairwise independent random processes.

Finally, we can also derive conditional versions of these exact laws. We do not give the details here because we want to keep the presentation to a rather nontechnical level since our motivation is merely to connect our approach to mean field games to the existing literature on games with a continuum of players. The interested reader is referred to the Notes & Complements at the end of the chapter for references.

3.7.2 Mean Field Games with a Continuum of Players

We now revisit the introductory discussion of Section 3.1, and especially Subsection 3.1.1 to introduce what would be the analogue with a continuum of players. In other words, we would like to replace the finite set $I=\{1,\cdots,N\}$ of players, by a general probability space $(I,\mathcal{I},\lambda)$ possibly with a continuous measure λ. Under the same assumptions on the drift and volatility functions b and σ, as well as on the running and terminal cost functions f and g, we thus posit that the dynamics of the state process $\boldsymbol{X}^i$ of each player $i\in I$ are given by a stochastic differential equation of the form:

$$dX_t^i = b(t, X_t^i, \rho_t^i, \alpha_t^i)dt + \sigma(t, X_t^i, \rho_t^i, \alpha_t^i)dW_t^i, \qquad t\in[0,T], \tag{3.88}$$

in full analogy with (3.1). Here, each $\boldsymbol{W}^i = (W_t^i)_{0\leqslant t\leqslant T}$, for $i\in I$, is intended to be a Brownian motion with values in $\mathbb{R}^d$, where d is the dimension of the state space, constructed on some common probability space $(\Omega,\mathcal{F},\mathbb{P})$. Accordingly, each $(\alpha_t^i)_{0\leqslant t\leqslant T}$ is intended to be a control process which we require to be progressively measurable with respect to the larger filtration generated by all the noises $(\boldsymbol{W}^i)_{i\in I}$. Similarly, each $(\rho_t^i)_{0\leqslant t\leqslant T}$ is a measure-valued process which we also require to be progressively measurable with respect to the same filtration. We refer to Chapter 5 for the topologies and σ-algebras with which we may equip spaces of measures.

Of course, solving (3.88) for any given $i \in I$ is not a problem as long as we do not ask the family of processes $(\boldsymbol{X}^i)_{i\in I}$ to satisfy any further measurability properties or statistical constraints with respect to the variable i. However, simply stating (3.88) with the additional constraint that the processes $(\boldsymbol{X}^i)_{i\in I}$ must both form a jointly measurable mapping in (i, ω) and be independent with respect to i, which is intuitively what we should expect to be relevant for defining an equilibrium over a continuum of players, immediately raises challenging questions. The purpose of this subsection is to address them in an informal manner, paving the way to a possible rigorous treatment of the issues.

1. Certainly, the starting point is to define a collection $(\boldsymbol{W}^i)_{i\in I}$ of Wiener processes such that the map $\boldsymbol{W} : I\times\Omega \ni (i, \omega) \mapsto \boldsymbol{W}^i(\omega) \in \mathcal{C}([0, T]; \mathbb{R}^d)$ is measurable and the variables $(\boldsymbol{W}^i)_{i\in I}$ are essentially pairwise independent. This major difficulty may be overcome if we use a rich Fubini extension, as defined in the previous subsection.
2. In fact, using such a Fubini extension will kill two birds with one stone. Indeed, the mean field nature of the model is usually guaranteed by the interactions between the private states provided by the presence of the empirical distribution of the states in the coefficients. In other words, the place holder ρ_t^i appearing in (3.88) should be the empirical distribution of all the players, or if not the empirical distribution of the players $j \in I$ with $j \neq i$. According to our earlier discussion, a proxy for the former could be provided by a formula similar to (3.83) as long as $\boldsymbol{X}$ has the right measurability. In such a case, $\rho_t^i(\omega)$, for $t \in [0, T]$ and $(i, \omega) \in I \times \Omega$, becomes independent of i and could be just defined as $\rho_t(\omega) = \lambda \circ (I \ni j \mapsto X_t^j(\omega))^{-1}$, i.e., the push forward of the probability measure λ under the map that associates any player j with its state at time t under the realization ω of the randomness.

The notion of Nash equilibrium could then be defined in a natural way.

Definition 3.46 *On the same rich Fubini extension $(I\times\Omega, \mathcal{I}\boxtimes\mathcal{F}, \lambda\boxtimes\mathbb{P})$ as above, a collection of admissible controls $\boldsymbol{\alpha}^* = (\boldsymbol{\alpha}^{*i})_{i\in I}$ is said to form a Nash equilibrium for the game with a continuum of players whenever the following two conditions are satisfied.*

(i) We can solve the state equation (3.88) in such a way that the state process $\boldsymbol{X}^* : I\times\Omega \ni (i, \omega) \mapsto \boldsymbol{X}^{*i} \in \mathcal{C}([0, T]; \mathbb{R}^d)$ is measurable with respect to $\mathcal{I}\boxtimes\mathcal{F}$ and each $\boldsymbol{X}^{*i}$ solves (3.88) with $\boldsymbol{\alpha}^i = \boldsymbol{\alpha}^{*i}$ and

$$\rho_t^i(\omega) = \rho_t(\omega), \quad \textit{where} \quad \rho_t(\omega) = \lambda \circ \big(I \ni j \mapsto X_t^{*j}(\omega)\big)^{-1},$$

for all $i \in I$.

(ii) If for each player $i \in I$ and any strategy profile $\boldsymbol{\alpha}^i$ for player i, we define the expected cost to player i by the formula:

$$J^i(\boldsymbol{\alpha}) = \mathbb{E}\left[\int_0^T f(t, X_t^i, \rho_t, \alpha_t^i)dt + g(X_T^i, \rho_T)\right],$$

where $\boldsymbol{X}^i$ solves (3.88)*, then, for λ-a.e. $i \in I$ and any admissible control $\boldsymbol{\alpha}^i = (\alpha_t^i)_{0 \leqslant t \leqslant T}$ for player i, we have:*

$$J^i(\boldsymbol{\alpha}^{*i}) \leqslant J^i(\boldsymbol{\alpha}^i).$$

Observe that, in the notation used right above, J^i implicitly depends upon $\boldsymbol{\alpha}^* = (\boldsymbol{\alpha}^{*j})_{j \in I}$ through the empirical distributions $(\rho_t)_{0 \leqslant t \leqslant T}$. The fact that the same $(\rho_t)_{0 \leqslant t \leqslant T}$ is used whenever player i uses $\boldsymbol{\alpha}^i$ in place of $\boldsymbol{\alpha}^{*i}$ is fully legitimated by Lemma 3.42, which asserts that the measure λ is necessarily continuous. Indeed, the continuity of the measure λ guarantees that each player is *insignificant*, and in particular that the empirical measures constructed from the strategy profiles $\boldsymbol{\alpha}^*$ and $(\boldsymbol{\alpha}^i, \boldsymbol{\alpha}^{*-i})$ are the same, where as usual, the strategy profile $(\boldsymbol{\alpha}^i, \boldsymbol{\alpha}^{*-i})$ is given by all the players $j \neq i$ using the controls $\boldsymbol{\alpha}^{*j}$ while player i is using control $\boldsymbol{\alpha}^i$.

It is worth noting that the rich Fubini extension is in fact just needed to construct the state process $\boldsymbol{X}^*$ forming the equilibrium. In stark contrast, any unilateral deviation from the equilibrium calls for the redefinition of the trajectories of one single player $i \in I$ only, which can be done on the sole space $(\Omega, \mathcal{F}, \mathbb{P})$.

We now explain why, at least at an intuitive level,

a solution of the mean field game problem stated in Section 3.1 *provides a Nash equilibrium for the mean field game with a continuum of players.*

Given the drift and volatility functions b and σ, and given the running and terminal cost functions f and g, let us assume that the mean field game problem formulated in Subsection 3.1.2 has a solution. We denote by $\hat{\boldsymbol{\alpha}} = (\hat{\alpha}_t)_{0 \leqslant t \leqslant T}$ the equilibrium strategy, as defined in Subsection 3.1.2, which we assume to be a progressively measurable function of the path of the Wiener process $\boldsymbol{W} = (W_t)_{0 \leqslant t \leqslant T}$ on which the game model is based. We also denote by $\boldsymbol{\mu} = (\mu_t)_{0 \leqslant t \leqslant T}$ the corresponding equilibrium flow of probability measures. Recall that the latter is entirely deterministic. Next, we proceed to define the strategy profile $\boldsymbol{\alpha}^* = (\boldsymbol{\alpha}^{*i})_{i \in I}$ for the mean field game with a continuum of players by demanding that, for each $i \in I$, $\boldsymbol{\alpha}^{*i}$ bears to $\boldsymbol{W}^i$ the same relationship as $\hat{\boldsymbol{\alpha}}$ does to $\boldsymbol{W}$. Next, still for each $i \in I$, we consider the process $\boldsymbol{X}^{*i} = (X_t^{*i})_{0 \leqslant t \leqslant T}$ solving (3.88) when we use the deterministic measures $(\mu_t)_{0 \leqslant t \leqslant T}$ in lieu of the possibly random empirical measures $(\rho_t)_{0 \leqslant t \leqslant T}$. Under standard assumption, the state equation (3.88) is strongly solvable when, for each $i \in I$, $(\alpha_t^i)_{0 \leqslant t \leqslant T}$ and $(\rho_t^i)_{0 \leqslant t \leqslant T}$ are given as we just explained. In particular, we claim that there exists a progressively measurable function F from $\mathcal{C}([0,T]; \mathbb{R}^d)$ into itself such that $\boldsymbol{X}^{*i} = F(\boldsymbol{W}^i)$ for all $i \in I$. We deduce that $\boldsymbol{X}^* : I \times \Omega \ni (i, \omega) \mapsto \boldsymbol{X}^{*i}(\omega) \in \mathcal{C}([0,T]; \mathbb{R}^d)$ is measurable with respect to $\mathcal{I} \boxtimes \mathcal{F}$ and that the family $(\boldsymbol{X}^{*i})_{i \in I}$ is essentially pairwise independent. Hence, the exact law of large numbers implies that, for each $t \in [0, T]$, the corresponding empirical measure, defined as $\lambda \circ (I \ni i \mapsto X_t^{*i})^{-1}$, is nonrandom and coincides with μ_t. It then remains to identify J^i, as defined in (3.46), with the cost functional defined in (3.4). The only subtlety when we do so is to note that $\boldsymbol{\alpha}^i$ is adapted to a much larger filtration than the filtration generated by $\boldsymbol{W}^i$. This says that, to conclude, we must assume that $\hat{\boldsymbol{\alpha}}$ in (3.4) is optimal among a class $\mathbb{A}$ of control processes that are progressively measurable with respect to a filtration $\mathbb{F}$ which is strictly larger than the filtration generated by $\boldsymbol{W}$. In fact, so is the case under standard assumptions on the coefficients of the game, like those we use throughout the book.

3.8 Notes & Complements

The formulation of the mean field game problem given in Subsection 3.1.2 as a set of bullet points leads to a family of standard continuous time stochastic control problems followed by a fixed point problem in the space of deterministic measure flows. This is inspired by the presentation of the Nash Certainty Equivalence (NCE) principle by Huang, Caines and Malhamé in [211]. However, our search for a solution is of a probabilistic nature as opposed to the analytic approach identifying the value functions of the control problems as solutions of HJB partial differential equations as described in Subsection 3.1.5. It is tempting to tackle the solutions of both the HJB equations and the fixed point problems by contraction fixed point arguments. This scheme is followed in [211]. Unfortunately, this strategy faces subtle difficulties created by the fact that the two problems have time evolutions in opposite directions, and as a result, it requires strong hypotheses which are difficult to check in practice, and equilibria are only obtained over sufficiently short time intervals. Existence over arbitrary time intervals can be proved for various types of models at the cost of sophisticated PDE arguments. This was first done by Lasry and Lions in [260–262] when σ is equal to the identity. In all these references, the coefficients are allowed to depend upon the distribution of the population through its density, in which case the coupling is said to be local. The note [260] is dedicated to the stationary case presented in Chapter 7, while the finite horizon case is discussed in both [261] and [262]. The arguments are detailed in the video lectures [265]; the reader may also consult the notes by Cardaliaguet [83]. As in our approach, Schauder's theorem is explicitly invoked in [83] to complete the existence proof of a solution to the MFG system (3.12).

Since Lasry and Lions' work, several contributions have addressed the solvability of the MFG system (3.12). Some efforts have been concentrated on the so-called first order case, when the volatility σ in (3.1) is null: We refer to Cardaliaguet [85] and Cardaliaguet and Graber [87] for results with local coupling and to Cardaliaguet, Mészáros, and Santambrogio [92] for cases when the density of the population is constrained. Second order but degenerate cases were addressed by Cardaliaguet, Graber, Porretta, and Tonon [88], while a great attention has been paid by Gomes and his coauthors to the nondegenerate case but with various forms of Hamiltonians and couplings: Gomes, Pimentel, and Sanchez [179] studied mean field games on the torus with sub-quadratic Hamiltonians and power like dependence on the density of the population, such a form of interaction accounting for some aversion to congestion; Gomes and Pimentel [177] addressed the same problem but with a logarithmic dependence on the density; and Gomes, Pimentel, and Sanchez [180] investigated mean field games with super-quadratic Hamiltonians and power like dependence on the density. Similar models, but on the whole space, are discussed in Gomes and Pimentel [178]. In [85,87,88,92], the construction of a solution relies on the connection with mean field optimal control problems as exposed in Chapter 6; in [177,179,180], it is based on a smoothing procedure of the coefficients permitting to apply Lasry and Lions' original results. We refer to Guéant [186] for a subtle change of variable transforming the MFG system, when driven by quadratic Hamiltonians, into a tractable system of two heat equations. For a more complete account, the reader may also consult the monographs by Bensoussan, Frehse and Yam [50], Gomes, Nurbekyan, and Pimentel [176] and Gomes, Pimentel, and Voskanyan [181].

For a complete overview of the theory of stochastic optimal control, we refer the reader to the textbooks by Fleming and Soner [157], Pham [310], Touzi [334], and Yong and Zhou [343]. For a quicker introduction to the subject, the reader may have at a look at the surveys by Borkar [65] and Pham [309].

The theory of backward SDEs goes back to the pioneering works by Pardoux and Peng in the early 90s, see for instance [297, 298]. We refer to the monograph by Pardoux and Răşcanu [299] for a complete overview of the subject and of the bibliography. For a pedagogical introduction on the connection between backward SDEs and stochastic control, the reader may also have a look at Pham's textbook [310]. Actually, existence of a connection between backward SDEs and stochastic control was known before Pardoux and Peng's works as an earlier version of the stochastic maximum principle appeared in Bismut's contributions [60–62]. The standard version of the stochastic maximum principle, as exposed in this chapter, is due to Peng [302], and it is now a standard tool of stochastic optimization. It is featured in many textbooks on the subject, for example Chapter IV of Yong and Zhou's textbook [343] or Chapter 4 of Carmona's lectures [94]. We also refer to the survey by Hu [202]. The sufficient condition can be found in Chapter 6 of Pham's book [310] or in Chapter 10 of Touzi's monograph [334]. We give a complete proof in the more general set-up of stochastic dynamics of the McKean-Vlasov type in Chapter 6. The representation of Hamilton-Jacobi-Bellman equations by means of backward SDEs, as explained in Remark 3.16, is also due to Peng, see [304]; we shall revisit it in the next chapter. Our presentation of the weak formulation in Subsection 3.3.1 is inspired by the articles by Hamadène and Lepeltier [195] and El Karoui, Peng, and Quenez [226]. Earlier formulation of the comparison principle for BSDEs, as used in the proof of Proposition 3.11 on the weak formulation, may be found in [304]; we refer to Chapter 5 in the monograph by Pardoux and Răşcanu [299] for a more systematic presentation or to any textbook on the subject. See for example [94, 310] or [343].

The analysis of fully coupled forward-backward SDEs is a challenging question, which we shall review in the next Chapter 4. A common reference on the subject is the textbook by Ma and Yong [274].

Inspired by the original works of Lasry, Lions and collaborators, BSDEs of mean field type have been studied in several papers by Buckdahn and his coauthors, see [74, 75] for example. Unfortunately, these results are of little use in the analysis of the mean field game problems discussed in this chapter and in the optimal control of McKean-Vlasov stochastic differential equations studied in Chapter 6. Indeed, more than BSDEs, systems of coupled forward-backward stochastic differential equations (FBSDEs for short) of McKean-Vlasov type need to be solved, and, as already explained for the classical case, existence and uniqueness result for BSDEs are much easier to come by than for FBSDEs. To the best of our knowledge, the FBSDEs of McKean-Vlasov type discussed in this chapter had not been considered before their investigation by Carmona and Delarue in [95]. The main result of [95] will be presented and extended in Chapter 4.

The weak formulation (also known as the martingale method) for mean field game problems, as exposed in Subsection 3.3.1, was used first by Carmona and Lacker in [103] in a more general setting than in the present chapter. There, the authors consider interactions through the empirical distributions of the controls as well as the states of the players. Moreover, weaker assumptions on the dependence of the coefficients upon their measure arguments are required, allowing for the handling of more singular interactions. However, these authors required the control space A to be compact, ruling out, among other things, the LQ models.

Uniqueness criterion provided by Theorem 3.29 is the most popular one in the theory of mean field games. It goes back to the original papers by Lasry and Lions [260–262] The monotonicity condition in Definition 3.28 underpinning the statement of Theorem 3.29 is inspired by the theory of evolution equations with a monotone operator. Example 5 of a monotonous function in Subsection 3.4.2 is taken from the paper of Gatheral, Schied, and Slynko [171]. Example 7 is standard in the theory. From a modeling point of view, Example 7 says that, whenever h is regarded as a cost functional, the cost $h(x, \mu)$ when the representative player is in state x and the population is in state μ increases with the local mass of μ in the neighborhood of x. This reads as an aversion to congestion. The other form of monotonicity in Definition 3.31 is less standard; it appeared in Ahuja [11]. These forms of monotonicity will be revisited in Chapter 5, see Subsection 5.7.1, when equipping the space of probability measures with different forms of differential calculus.

The discussion of the Linear Quadratic (LQ) models provided in Section 3.5 is inspired by the contents of the papers [53] of Bensoussan, Sung, Yam, and Yung, and [99] of Carmona, Delarue, and Lachapelle. The results of these two groups of authors were announced essentially at the same time, and are very similar. They demonstrate how to solve linear quadratic mean field games and linear quadratic optimal control problems for McKean-Vlasov dynamics (also called mean field stochastic control problems), and they both argue that, even in the linear quadratic setting, the solutions to these two problems should be expected to be different. McKean-Vlasov control problems are discussed in Chapter 6 where we emphatically make this point. Several sufficient conditions for existence and uniqueness of a solution to the forward-backward system of deterministic ODEs (3.52) are given in [53]. These restrictive assumptions are reminiscent of existence and uniqueness of a solution for small time. The search for a solution given by an ansatz expressing the backward component as an affine function of the forward component is mentioned in [53] and used systematically in [99] in the one-dimensional case. The paper [33] by Bardi provides explicit solutions of LQ Mean Field Games when the objective function minimized by the players is computed as an ergodic average over an infinite horizon.

Multidimensional linear quadratic mean field games have been considered and solved under various sorts of assumptions. For an alternative to the results presented in this chapter, the interested reader may consult the papers [210] and [212] by Huang, Caines, and Malhamé in the book chapter [213] by the same authors.

Affine FBSDEs of the form (3.56) have been studied in [341, 342]. For the sake of completeness, and at the risk of appearing too elementary, we chose to construct a solution from scratch. As a result of the affine structure, LQ control problems as well as LQ Mean Field Games lead to Gaussian solutions. However they are not the only MFG models with this property. Guéant and his coauthors show in [189] that Gaussian equilibria can be obtained even with logarithmic running costs typically used in crowd models.

The solution of the MFG form of the toy model for systemic risk was shown to be given by the solution of a Riccati equation appearing as the common limit of the Riccati equations giving Nash equilibria for finite player game models associated with the open and closed loop information structures. It has been argued that for large games in which the influence to each individual player disappears as the size of the game grows, the differences between open loop and closed loop Nash equilibria disappear as well. See for example the introduction of the paper [164] by Fudenberg and Levine, or Section 4.7.3 of the book by Fudenberg and Tirole [165] for a discussion of this claim.

The analysis of the diffusion form of Aiyagari's growth model given in Subsection 3.6.3 is due to the authors. It appears in print for the first time here.

A weaker formulation of mean field games, based on the notion of relaxed controls, was introduced in Lacker [254]. Relaxed controls are probability measures on the set A of admissible values of the control. In general, using relaxed controls allows for weaker assumptions on the coefficients and turn the space of controls into a compact set when A is bounded. As a result, the existence of a best response to a given state of the population comes for free if the cost functional is lower semi-continuous. However, this best response may not be unique. In the context of the search of an equilibrium in a mean field game, this says that, in the second step of the procedure detailed in Subsection 3.1.2, the quantity $\hat{X}^{\mu}$ may be multivalued. The fixed point condition then reads $\mu \in (\mathcal{L}(\hat{X}_t^{\mu}))_{0 \leqslant t \leqslant T}$, where $(\mathcal{L}(\hat{X}_t^{\mu}))_{0 \leqslant t \leqslant T}$ is understood as the collection of flows of marginal measures generated by the set of optimal paths $\hat{X}^{\mu}$. Instead of Schauder's fixed point theorem, one may invoke Kakutani's fixed point theorem for multivalued functions in order to solve the equilibrium condition $\mu \in (\mathcal{L}(\hat{X}_t^{\mu}))_{0 \leqslant t \leqslant T}$.

We refer the reader to the survey by Borkar [65] and to the monograph by Yong and Zhou for an overview of controlled diffusion processes with relaxed controls. Earlier results in that direction are due to Young [344] and Fleming [158]. In Chapter 6, we shall study a simple example of mean field control problem by means of the notion of relaxed controls.

Examples of mean field games with other different cost functionals, like risk-sensitive cost functionals for instance, may be found in Tembine, Zhu, and Basar [332]. Models leading to mean field games with several populations, mean field games with an infinite time horizon, and mean field games with a finite state space are discussed in Chapter 7. Mean field games with major minor players will be presented in Chapter (Vol II)-7.

The results on convex optimization used in the text can be found in most standard monographs on the subject, see for instance Bertsekas' [55] or Ciarlet's [115] textbooks.

Games models with a continuum of players were introduced by Aumann in 1964 in a breakthrough paper [27]. Our presentation of the exact law of large numbers is modeled after Sun's paper [324]. This law was also used by Duffie and Sun in [148] to model matching from searches. This last work was used to justify the assumptions of the percolation of information model presented in Chapter 1.

FBSDEs and the Solution of MFGs Without Common Noise

4

Abstract

The goal of this chapter is to develop a general methodology for the purpose of solving mean field games using the forward-backward SDE formulations introduced in Chapter 3. We first proceed with a careful analysis of forward-backward mean field SDEs, that is of McKean-Vlasov type, which shows how Schauder's fixed point theorem can be used to prove existence of a solution. As a by-product, we derive two general solvability results for mean field games: first from the FBSDE representation of the value function, and then from the stochastic Pontryagin maximum principle. In the last section, we revisit some of the examples introduced in the first chapter, and illustrate how our general existence results can be applied.

4.1 A Primer on FBSDE Theory

The goal of this section is to provide basic solvability results for standard forward-backward SDEs. These results will serve us well when we try to prove existence of solutions to forward-backward SDEs of the McKean-Vlasov type. Precise references are cited in the Notes & Complements at the end of the chapter for all the results given without proof.

As a general rule, we consider forward-backward SDEs of a slightly more general form than what is really needed in order to implement the program outlined in Chapter 3. Typically, we allow the diffusion coefficient (or *volatility*) to depend upon the backward component of the solution. In doing so, we obtain almost for free, an existence result for FBSDEs of the McKean-Vlasov type for a larger class of models covering mean field games as a particular case. Being able to handle such a larger class will turn out to be handy in Chapter 6 for the study of the optimal control of McKean-Vlasov diffusion processes.

R. Carmona, F. Delarue, *Probabilistic Theory of Mean Field Games with Applications I*, Probability Theory and Stochastic Modelling 83,
DOI 10.1007/978-3-319-58920-6_4

The framework of this section is as follows. Given a complete probability space $(\Omega, \mathcal{F}, \mathbb{P})$, equipped with a complete and right-continuous filtration $\mathbb{F}$, an initial condition $\xi \in L^2(\Omega, \mathcal{F}_0, \mathbb{P}; \mathbb{R}^d)$, and a d-dimensional $\mathbb{F}$-Brownian $\boldsymbol{W} = (W_t)_{0 \leqslant t \leqslant T}$, we consider the forward-backward SDE:

$$
\begin{cases}
dX_t = B\big(t, X_t, Y_t, Z_t\big)dt + \Sigma(t, X_t, Y_t)dW_t, \\
dY_t = -F\big(t, X_t, Y_t, Z_t\big)dt + Z_t dW_t, \qquad t \in [0, T], \\
X_0 = \xi, \quad Y_T = G(X_T),
\end{cases}
\tag{4.1}
$$

where B and F are functions defined on $[0, T] \times \mathbb{R}^d \times \mathbb{R}^m \times \mathbb{R}^{m \times d}$ into $\mathbb{R}^d$ and $\mathbb{R}^m$ respectively, Σ is a function from $[0, T] \times \mathbb{R}^d \times \mathbb{R}^m$ into $\mathbb{R}^{d \times d}$ and G is a function from $\mathbb{R}^d$ into $\mathbb{R}^m$. The coefficients B, F, G and Σ are assumed to be Borel measurable. As in the previous chapter, the dimensions of $\boldsymbol{X}$ and of $\boldsymbol{W}$ are assumed to be the same for convenience.

We call solution any triple $(\boldsymbol{X}, \boldsymbol{Y}, \boldsymbol{Z}) = (X_t, Y_t, Z_t)_{0 \leqslant t \leqslant T}$ of $\mathbb{F}$-progressively measurable processes, with values in $\mathbb{R}^d \times \mathbb{R}^m \times \mathbb{R}^{m \times d}$, such that $\boldsymbol{X}$ and $\boldsymbol{Y}$ have continuous paths,

$$
\mathbb{E}\bigg[\sup_{0 \leqslant t \leqslant T} \big(|X_t|^2 + |Y_t|^2\big) + \int_0^T |Z_t|^2 dt\bigg] < \infty,
$$

and (4.1) holds true $\mathbb{P}$ almost surely. In the next subsections, we address the existence and uniqueness of such solutions.

Remark 4.1 *For the reader who is not familiar with the theory of forward-backward equations, it may sound rather strange to ask for the well posedness of a system with three unknowns but two equations only. Actually, the reader must remember the fact that the triple $(\boldsymbol{X}, \boldsymbol{Y}, \boldsymbol{Z})$ is required to be progressively measurable with respect to $\mathbb{F}$. In particular, it should not anticipate the future of $\boldsymbol{W}$. The role of the process $\boldsymbol{Z}$ is precisely to guarantee the adaptedness of the solution with respect to the filtration $\mathbb{F}$. In the end, the forward-backward system actually consists of two equations and a progressive measurability constraint.*

All the results stated in this section are given in a rigorous form, and all the assumptions they require are given in full detail. However, proofs are frequently skipped. Indeed, while we want the reader to have a good sense of the underpinnings of the theory of FBSDEs, its main achievements as well as its limitations, we fear that too many technical proofs will distract from the thrust of our analysis. We shall only give proofs of results which further the theory of FBSDEs to help us solve the new challenges posed by mean field game models and the control of McKean-Vlasov dynamics.

4.1.1 Lipschitz Setting and Decoupling Field

Throughout this subsection, we assume that the coefficients are Lipschitz continuous in the variables x, y, and z:

Assumption (Lipschitz FBSDE). There exist two nonnegative constants Λ and L such that

(A1) The function $[0,T] \ni t \mapsto (B(t,0,0,0), F(t,0,0,0), \Sigma(t,0,0), G(0))$ is bounded by Λ.

(A2) For each $t \in [0,T]$, the functions $B(t,\cdot,\cdot,\cdot)$, $F(t,\cdot,\cdot,\cdot)$, $\Sigma(t,\cdot,\cdot)$ and G are L-Lipschitz continuous on their own domain.

Small Time Solvability

As announced in Subsection 3.2.3, Cauchy-Lipschitz theory for forward-backward systems provides existence and uniqueness in small time.

Theorem 4.2 *Under assumption* **Lipschitz FBSDE**, *there exists a constant $c > 0$, only depending on L (and not on Λ), such that, for any initial condition $\xi \in L^2(\Omega, \mathcal{F}_0, \mathbb{P}; \mathbb{R}^d)$, equation* (4.1) *has a unique solution as long as $T \leqslant c$.*

Remark 4.3 *Uniqueness is understood in the following pathwise sense. We say that* (4.1) *has a unique solution if, for any two solutions $(\boldsymbol{X}, \boldsymbol{Y}, \boldsymbol{Z}) = (X_t, Y_t, Z_t)_{0\leqslant t\leqslant T}$ and $(\boldsymbol{X}', \boldsymbol{Y}', \boldsymbol{Z}') = (X_t', Y_t', Z_t')_{0\leqslant t\leqslant T}$ defined on the same probabilistic set-up $(\Omega, \mathcal{F}, \mathbb{F}, \mathbb{P})$ as described above, we have:*

$$\mathbb{E}\left[\sup_{0\leqslant t\leqslant T}\left(|X_t - X_t'|^2 + |Y_t - Y_t'|^2\right) + \int_0^T |Z_t - Z_t'|^2 dt\right] = 0. \tag{4.2}$$

In Chapter (Vol II)-1, we shall address another form of uniqueness, namely uniqueness in law. In particular, we shall prove a suitable version of the Yamada-Watanabe theorem for FBSDEs saying that pathwise uniqueness in the sense of (4.2) *implies weak uniqueness.*

We do not give the proof of Theorem 4.2 here. Indeed, the reader will find the proof of a more general statement, including the McKean-Vlasov case, in Subsection 4.2.3. In fact, a careful inspection of the proof provided in Subsection 4.2.3 shows that Theorem 4.2 may be easily turned into a stability property which we also state without proof.

Theorem 4.4 *There exist two constants $c, C \geqslant 0$, only depending on L, such that, for any other set of coefficients (B', F', G', Σ') satisfying assumption* **Lipschitz**

FBSDE *with the same constant L (and a possibly different constant* Λ*) as* (B, F, G, Σ) *and for any other initial condition* $\xi' \in L^2(\Omega, \mathcal{F}_0, \mathbb{P}; \mathbb{R}^d)$*, the solution* $(\boldsymbol{X}', \boldsymbol{Y}', \boldsymbol{Z}') = (X'_t, Y'_t, Z'_t)_{0 \leqslant t \leqslant T}$ *to the FBSDE driven by* (B', F', G', Σ') *and* ξ' *satisfies, with probability 1:*

$$\mathbb{E}\Big[\sup_{0\leqslant t\leqslant T}\Big(|X_t - X'_t|^2 + |Y_t - Y'_t|^2\Big) + \int_0^T |Z_t - Z'_t|^2 dt \,|\, \mathcal{F}_0\Big]$$

$$\leqslant C\mathbb{E}\Big[|\xi - \xi'|^2 + |(G - G')(X_T)|^2 + \int_0^T \big|\big(B - B', F - F', \Sigma - \Sigma'\big)\big(t, X_t, Y_t, Z_t\big)\big|^2 dt \,|\, \mathcal{F}_0\Big].$$

as long as $T \leqslant c$.

Role of the Decoupling Field

We notice that the forward-backward system (4.1) may be solved with respect to the complete and right-continuous filtration generated by the initial condition ξ and the Brownian motion $\boldsymbol{W}$. When T is small enough, Theorem 4.2 says that a solution exists and is unique. We call this solution *the solution constructed on the canonical set-up*. Quite remarkably, this solution is also a solution of the original problem defined on the set-up $(\Omega, \mathcal{F}, \mathbb{F}, \mathbb{P})$. Since Theorem 4.2 ensures that the equation (4.1), when solved on the original set-up $(\Omega, \mathcal{F}, \mathbb{F}, \mathbb{P})$, is uniquely solvable, we deduce that the solution constructed on the canonical set-up coincides with the solution defined on the set-up $(\Omega, \mathcal{F}, \mathbb{F}, \mathbb{P})$. In short, the solution is left invariant under augmentation of the filtration, provided that $\boldsymbol{W}$ remains a Brownian motion under the new filtration. As we shall see in Chapter (Vol II)-1, this result is false when the coefficients B, F, G, and Σ are random and are not adapted to the filtration generated by $\boldsymbol{W}$ and ξ.

Therefore, we can assume without any loss of generality that $\mathbb{F}$ is the complete and right-continuous augmentation of the filtration generated by ξ and $\boldsymbol{W}$. Whenever ξ is deterministic, we know from Blumenthal's zero-one law that $\mathcal{F}_0$ is almost surely trivial, proving (at least when $T \leqslant c$) that the unique solution of (4.1) is almost surely deterministic at time 0. Denoting, for any $x \in \mathbb{R}^d$, by $(\boldsymbol{X}^{0,x}, \boldsymbol{Y}^{0,x}, \boldsymbol{Z}^{0,x}) = (X_t^{0,x}, Y_t^{0,x}, Z_t^{0,x})_{0 \leqslant t \leqslant T}$ the unique solution to (4.1) with $X_0 = x$, this prompts us to define a mapping $x \mapsto u(0, x)$ by:

$$u(0, x) = \mathbb{E}[Y_0^{0,x}], \quad x \in \mathbb{R}^d.$$

We then have:

$$\mathbb{P}\big[Y_0^{0,x} = u(0, x)\big] = 1. \tag{4.3}$$

Now, applying Theorem 4.4 with $(B', F', G', \Sigma') = (B, F, G, \Sigma)$, $\xi = x$ and $\xi' = x'$, we deduce that, for $T \leqslant c$ (without any loss of generality we can assume that the constants c in the statements of Theorems 4.2 and 4.4 are the same), this function is Lipschitz in the sense that:

$$\forall x, x' \in \mathbb{R}^d, \quad |u(0,x) - u(0,x')| \leqslant C|x - x'|. \tag{4.4}$$

The thrust of the notion of decoupling field is that the relationship (4.3) remains true when the initial condition ξ is random; namely, denoting by $(\boldsymbol{X}^{0,\xi}, \boldsymbol{Y}^{0,\xi}, \boldsymbol{Z}^{0,\xi})$ the unique solution to (4.1) when $T \leqslant c$, we claim that:

$$\mathbb{P}\big[Y^{0,\xi} = u(0,\xi)\big] = 1. \tag{4.5}$$

The proof of (4.5) is quite simple. When ξ is a simple random variable of the form $\xi = \sum_{i=1}^N \mathbf{1}_{A_i} x_i$, for some integer $N \geqslant 1$, events $(A_i)_{i=1,\cdots,N} \in \mathcal{F}_0^N$ and points $(x_i)_{i=1,\cdots,N} \in (\mathbb{R}^d)^N$, (4.5) is quite obvious since $Y^{0,\xi} = \sum_{i=1}^N \mathbf{1}_{A_i} Y_0^{0,x_i}$. In the general case, it suffices to approximate ξ by a sequence of simple random variables and to take advantage of the stability result of Theorem 4.4 and the Lipschitz property of $u(0,\cdot)$ in order to pass to the limit in (4.5) along the approximating sequence.

Another key fact is that the decoupling field may be defined, not only at time 0, but at any time $t \in [0,T]$, provided that $T \leqslant c$. It suffices to initialize the forward equation at time t and to let X_t match some $x \in \mathbb{R}^d$. Solving the equation (4.1) on the interval $[t,T]$ instead of $[0,T]$ with respect to the complete filtration generated by $(W_s - W_t)_{t \leqslant s \leqslant T}$, we then get, by the same argument as above, that Y_t is almost surely deterministic. Below, we shall denote by $(\boldsymbol{X}^{t,x}, \boldsymbol{Y}^{t,x}, \boldsymbol{Z}^{t,x}) = (X_s^{t,x}, Y_s^{t,x}, Z_s^{t,x})_{t \leqslant s \leqslant T}$ the unique solution with $X_t = x$ as initial condition. It satisfies:

$$\begin{cases} dX_s^{t,x} = B\big(s, X_s^{t,x}, Y_s^{t,x}, Z_s^{t,x}\big)ds + \Sigma\big(s, X_s^{t,x}, Y_s^{t,x}\big)dW_s, \\ dY_s^{t,x} = -F\big(s, X_s^{t,x}, Y_s^{t,x}, Z_s^{t,x}\big)ds + Z_s^{t,x} dW_s, \qquad s \in [t,T], \\ X_t^{t,x} = x, \quad Y_T^{t,x} = G\big(X_T^{t,x}\big). \end{cases} \tag{4.6}$$

The decoupling field at time t is then defined by letting:

$$u(t,x) = \mathbb{E}[Y_t^{t,x}],$$

so that $\mathbb{P}[Y_t^{t,x} = u(t,x)] = 1$. Following (4.4), u is Lipschitz in space, uniformly in time. As above, the relationship between the backward component of the solution and the decoupling field may be generalized to any random variable ξ in the space $L^2(\Omega, \mathcal{F}_t, \mathbb{P}; \mathbb{R}^d)$. Denoting by $(\boldsymbol{X}^{t,\xi}, \boldsymbol{Y}^{t,\xi}, \boldsymbol{Z}^{t,\xi}) = (X_s^{t,\xi}, Y_s^{t,\xi}, Z_s^{t,\xi})_{t \leqslant s \leqslant T}$ the unique solution to (4.1) with $X_t = \xi$ as initial condition, we have:

$$\mathbb{P}\big[Y_t^{t,\xi} = u(t,\xi)\big] = 1. \tag{4.7}$$

Relationship (4.7) is fundamental. Indeed, if we consider any solution $(\boldsymbol{X}, \boldsymbol{Y}, \boldsymbol{Z})$ to (4.1) with some initial condition at time 0, then $(X_s, Y_s, Z_s)_{t \leqslant s \leqslant T}$ is also a solution to (4.1) but on the interval $[t, T]$ and with X_t as initial condition. Therefore,

$$\mathbb{P}\big[Y_t = u(t, X_t)\big] = 1,$$

provided u is continuous in (t, x), which is the object of the next lemma:

$$\mathbb{P}\big[\forall t \in [0, T], \quad Y_t = u(t, X_t)\big] = 1. \tag{4.8}$$

The next lemma shows that u is not only continuous in space, but also jointly continuous in time and space.

Lemma 4.5 *Under assumption* **Lipschitz FBSDE** *and for $T \leqslant c$ with c as in the statements of Theorem 4.2 and Theorem 4.4, the decoupling field is Lipschitz in space uniformly in time, and $1/2$-Hölder continuous in time locally in space, the Hölder constant growing at most linearly with the space variable.*

Proof. Given $t \in [0, T]$ and $h > 0$ such that $t, t + h \in [0, T]$, and $x \in \mathbb{R}^d$, we have:

$$\begin{aligned} &u(t, x) - u(t + h, x) \\ &= \mathbb{E}\big[u(t, x) - u(t + h, X^{t,x}_{t+h})\big] + \mathbb{E}\big[u(t + h, X^{t,x}_{t+h}) - u(t + h, x)\big] \\ &= \mathbb{E}\big[Y^{t,x}_t - Y^{t,x}_{t+h}\big] + \mathbb{E}\big[u(t + h, X^{t,x}_{t+h}) - u(t + h, x)\big]. \end{aligned} \tag{4.9}$$

By Theorem 4.4 (with $\xi' = 0$, $B' \equiv B$, $F' \equiv F$, $G' \equiv G$, $\Sigma' \equiv \Sigma$), it is readily seen that:

$$\mathbb{E}\bigg[\sup_{t \leqslant s \leqslant T} \Big(|X^{t,x}_s|^2 + |Y^{t,x}_s|^2\Big) + \int_t^T |Z^{t,x}_s|^2 ds\bigg] \leqslant C\big(1 + |x|^2\big),$$

where C is independent of t and x. Plugging this estimate into (4.6), and using the Lipschitz property of u with respect to x, we deduce that the two terms in (4.9) are less than $C(1 + |x|)h^{1/2}$, for a possibly new value of the constant C. □

We end this subsection with two important remarks.

Remark 4.6 *As we already alluded to, we shall prove in Chapter (Vol II)-1 that strong (or pathwise) uniqueness for FBSDEs, as we consider here, implies uniqueness in law. As a by-product, this will show that the decoupling field u is independent of the probabilistic set-up on which the solution is constructed, see also Lemma 4.25.*

Remark 4.7 *The construction of the decoupling field u provided in this subsection relies on the assumption $T \leqslant c$. The role of this condition is to guarantee the unique solvability of* (4.6)*. Clearly, the above construction of the decoupling field is possible on any time interval on which existence and uniqueness are known to hold for any initial condition.*

Furthermore, the analysis of the regularity of u may be carried out for arbitrary times provided that the conclusion of Theorem 4.4 remains true.

4.1.2 Induction Method and Nondegenerate Case

Our goal is now to provide a systematic method to extend the small time existence and uniqueness result. The counter-example of Subsection 3.2.3 shows that such a method cannot work under the sole assumption **Lipschitz FBSDE**. Our strategy is to prove that, as long as we can exhibit a decoupling field which is Lipschitz continuous in the space variable, existence and uniqueness must hold.

Iteration

Our approach relies on the following observation. For an arbitrary time horizon $T > c$, we can first restrict the analysis of the forward-backward system (4.1) to the interval $[T-c, T]$, for c as in the statements of Theorem 4.2 and Theorem 4.4. Then, following the argument of the previous subsection, we know that for any $t \in [T-c, T]$, for any random variable $\xi \in L^2(\Omega, \mathcal{F}_t, \mathbb{P}; \mathbb{R}^d)$, the restriction of the system (4.1) to $[t, T]$ with $X_t = \xi$ as initial condition, is uniquely solvable. We set $(\boldsymbol{X}^{t,\xi}, \boldsymbol{Y}^{t,\xi}, \boldsymbol{Z}^{t,\xi}) = (X_s^{t,\xi}, Y_s^{t,\xi}, Z_s^{t,\xi})_{t \leqslant s \leqslant T}$ for the unique solution and define the decoupling field

$$u : [T-c, T] \times \mathbb{R}^d \ni (t, x) \mapsto u(t, x) = \mathbb{E}[Y_t^{t,x}].$$

For any $t \in [T-c, T]$ and $\xi \in L^2(\Omega, \mathcal{F}_t, \mathbb{R}^d; \mathbb{P})$, we have $\mathbb{P}[\forall s \in [t, T],\ Y_s^{t,\xi} = u(s, X_s^{t,\xi})] = 1$.

Assume now that the forward-backward system (4.1) has a solution $(\boldsymbol{X}, \boldsymbol{Y}, \boldsymbol{Z})$ for some initial condition $\xi \in L^2(\Omega, \mathcal{F}_0, \mathbb{P}; \mathbb{R}^d)$. Then, as in Subsection 4.1.1, we may regard $(X_t, Y_t, Z_t)_{T-c \leqslant t \leqslant T}$ as a solution to (4.1) on the interval $[T-c, T]$ with X_{T-c} as initial condition at time $T-c$. Since the problem set on $[T-c, T]$ with X_{T-c} as initial condition is uniquely solvable, we get that $(X_t, Y_t, Z_t)_{T-c \leqslant t \leqslant T} = (\boldsymbol{X}^{T-c, X_{T-c}}, \boldsymbol{Y}^{T-c, X_{T-c}}, \boldsymbol{Z}^{T-c, X_{T-c}})$. In particular, we may use the decoupling field to represent the solution:

$$\mathbb{P}\big[\forall t \in [T-c, T], \quad Y_t = u(t, X_t)\big] = 1,$$

from which we get that $(X_t, Y_t, Z_t)_{0 \leqslant t \leqslant T-c}$ is a solution of the forward-backward system (4.1) over the interval $[0, T-c]$ with terminal condition function $u(T-c, \cdot)$ instead of G; in other words:

$$\begin{cases} dX_t = B\big(t, X_t, Y_t, Z_t\big)dt + \Sigma(t, X_t, Y_t)dW_t, \\ dY_t = -F\big(t, X_t, Y_t, Z_t\big)dt + Z_t dW_t, \qquad t \in [0, T-c], \\ X_0 = \xi, \quad Y_{T-c} = u(T-c, X_{T-c}). \end{cases} \tag{4.10}$$

Regarding (4.10) as the new forward-backward system, we may apply the same argument as above. Denoting the previous c by c_0, we deduce that there exists a new c_1 such that (4.10) is uniquely solvable on $[T-(c_0+c_1), T-c_0]$. The reason is that the new terminal condition $u(T-c_0,\cdot)$ is Lipschitz continuous, but with a Lipschitz constant possibly differing from L. Of course, when this Lipschitz constant is greater than L, c_1 is smaller than c_0. We shall come back to this crucial point momentarily.

In order to distinguish solutions constructed on the interval $[T-c_0, T]$ from that constructed on $[T-(c_0+c_1), T-c_0]$, we use the following convention. We label the solutions constructed on the interval $[T-c_0, T]$ with a superscript '0' and those constructed on the interval $[T-(c_0+c_1), T-c_0]$ with a superscript '1'. In particular, for any $t \in [T-c_0, T]$ and $\xi \in L^2(\Omega, \mathcal{F}_t, \mathbb{P}; \mathbb{R}^d)$, the previous $(\boldsymbol{X}^{t,\xi}, \boldsymbol{Y}^{t,\xi}, \boldsymbol{Z}^{t,\xi})$ is now denoted by $(\boldsymbol{X}^{0;t,\xi}, \boldsymbol{Y}^{0;t,\xi}, \boldsymbol{Z}^{0;t,\xi})$. Similarly, for any $t \in [T-(c_0+c_1), T-c_0]$ and $\xi \in L^2(\Omega, \mathcal{F}_t, \mathbb{P}; \mathbb{R}^d)$, we call $(\boldsymbol{X}^{1;t,\xi}, \boldsymbol{Y}^{1;t,\xi}, \boldsymbol{Z}^{1;t,\xi})$ the unique solution to (4.10) with $X_t^{1;t,\xi} = \xi$ as initial condition.

Of course, a very natural idea is to patch together the two solutions. For given $t \in [T-(c_0+c_1), T-c_0]$ and $\xi \in L^2(\Omega, \mathcal{F}_t, \mathbb{P}; \mathbb{R}^d)$, we let for any $s \in [t, T]$:

$$
\begin{aligned}
&(X_s, Y_s, Z_s) \\
&= \begin{cases} \big(X_s^{1;t,\xi}, Y_s^{1;t,\xi}, Z_s^{1;t,\xi}\big), & \text{if } s \in [T-(c_0+c_1), T-c_0], \\ \big(X_s^{0;T-c_0,X_{T-c_0}^{1;t,\xi}}, Y_s^{0;T-c_0,X_{T-c_0}^{1;t,\xi}}, Z_s^{0;T-c_0,X_{T-c_0}^{1;t,\xi}}\big), & \text{if } s \in (T-c_0, T], \end{cases}
\end{aligned}
$$

Observe that, $\mathbb{P}$ almost surely,

$$
\begin{aligned}
\lim_{s \searrow T-c_0} X_s &= \lim_{s \searrow T-c_0} X_s^{0;T-c_0,X_{T-c_0}^{1;t,\xi}} = X_{T-c_0}^{1;t,\xi} = X_{T-c_0}, \\
\lim_{s \searrow T-c_0} Y_s &= \lim_{s \searrow T-c_0} Y_s^{0;T-c_0,X_{T-c_0}^{1;t,\xi}} = \lim_{s \searrow T-c_0} u\big(s, X_s^{0;T-c_0,X_{T-c_0}^{1;t,\xi}}\big) \\
&= u\big(T-c_0, X_{T-c_0}^{1;t,\xi}\big) = Y_{T-c_0},
\end{aligned}
$$

which proves that $(X_s, Y_s)_{T-(c_0+c_1) \leqslant s \leqslant T}$ is continuous at $s = T-c_0$. It is then plain to check that the process $(X_s, Y_s, Z_s)_{T-(c_0+c_1) \leqslant s \leqslant T}$ is a solution of the FBSDE (4.1) on $[T-(c_0+c_1), T]$. Quite remarkably, this solution satisfies:

$$
\mathbb{P}\big[\forall t \in [T-(c_0+c_1), T], \quad Y_t = u(t, X_t)\big] = 1.
$$

To summarize, we managed to extend the definition of the decoupling field to the domain $[T-(c_0+c_1), T] \times \mathbb{R}^d$ and to construct a solution to (4.1) for any initial condition (t,ξ) with $t \in [T-(c_0+c_1), T]$ and $\xi \in L^2(\Omega, \mathcal{F}_t, \mathbb{P}; \mathbb{R}^d)$. Notice that u is Lipschitz continuous in x, uniformly in time, and, by the same argument as in Lemma 4.5, that it is time-space continuous.

Obviously, the method can be iterated. We can find a sequence $(c_n)_{n\in\mathbb{N}}$ of positive real numbers such that, as long as $T-(c_0+\cdots+c_n) \geqslant 0$, we can extend the decoupling field u to the domain $[T-(c_0+\cdots+c_n),T]\times\mathbb{R}^d$ and construct for any initial condition (t,ξ) with $t\in[T-(c_0+\cdots+c_n),T]$ and $\xi\in L^2(\Omega,\mathcal{F}_t,\mathbb{P};\mathbb{R}^d)$, a solution to (4.1) satisfying:

$$\mathbb{P}\Big[\forall t\in\big[T-(c_0+\cdots+c_n),T\big],\quad Y_t=u(t,X_t)\Big]=1. \tag{4.11}$$

Moreover, u is jointly continuous in time and space, and Lipschitz continuous in space, uniformly in time.

Although very attractive, this method suffers from a major drawback. Nothing guarantees that the sequence $(\sum_{k=0}^{n}c_k)_{n\in\mathbb{N}}$ goes beyond T. Actually, the counter-example given in Subsection 3.2.3 shows that even in the Lipschitz regime, the sequence $(\sum_{k=0}^{n}c_k)_{n\in\mathbb{N}}$ may not reach T. Indeed, it may happen that the Lipschitz constant of the decoupling field becomes so large along the induction that the sum of the time lengths $\sum_{n\geqslant 0}c_n$ remains less than T. Recall that each time length c_n is determined by the Lipschitz constant (in space) of the decoupling field u at time $T-(c_0+\cdots+c_{n-1})$.

Before we derive sufficient conditions to control the Lipschitz constant of the decoupling field along the induction, we notice that such a construction necessarily implies uniqueness in addition to the existence of a solution.

Proposition 4.8 *On top of assumption* **Lipschitz FBSDE**, *assume that there exists a continuous function* $u:[0,T]\times\mathbb{R}^d\to\mathbb{R}^m$ *which is Lipschitz continuous in space uniformly in time, and such that, for any* $(t,x)\in[0,T]\times\mathbb{R}^d$, *we can find a solution to* (4.1), *with* $X_t=x$ *as initial condition at time t, satisfying:*

$$\mathbb{P}\Big[\forall s\in[t,T],\quad Y_s=u(s,X_s)\Big]=1. \tag{4.12}$$

Then, for any $t\in[0,T]$ *and* $\xi\in L^2(\Omega,\mathcal{F}_t,\mathbb{P};\mathbb{R}^d)$, *there exists a unique solution to* (4.1) *with* $X_t=\xi$ *as initial condition, and this solution satisfies the representation formula* (4.12).

Proof. With the same notation as before, we start with the following observation: there must exist $\delta>0$ such that, for all $n\in\mathbb{N}$, $c_n\geqslant\delta$. Indeed, since on the interval $[T-c_0,T]$ existence and uniqueness hold true for any initial condition, by (4.12), u must coincide with the decoupling field on $[T-c_0,T]$. Since the Lipschitz constant of u in space is bounded from above by a known (fixed) constant by assumption, this implies that c_1 is bounded from below by a known (fixed) constant. Iterating the argument, we realize that the extension of the decoupling field to the domain $[T-(c_0+c_1),T]\times\mathbb{R}^d$ still coincides with u. Therefore, the Lipschitz constant of the decoupling field at time $T-(c_0+c_1)$ is bounded from above by the same known (fixed) constant and hence, c_2 is bounded from below by a known (fixed) constant, etc.

As a result, there exists a finite integer $n\in\mathbb{N}$ such that $T\leqslant c_0+\cdots+c_n$. In other words, the iteration argument presented above needs only a finite number of steps to provide a solution for any initial condition (t,ξ) satisfying the prescribed conditions.

In order to prove uniqueness, it suffices to prove that any solution satisfies the representation formula (4.12). Indeed, once we have the representation formula (4.12), we may compare any two solutions (say starting from some ξ at time 0) on the small interval $[0, T-(c_0+\cdots+c_{n-1})]$, where n is the smallest integer such that $T \leqslant c_0 + \cdots + c_n$. On this small interval, the two solutions are known to satisfy (4.1) with the same terminal condition because of the representation formula. By existence and uniqueness in small time, they coincide on this first interval. Then, we can repeat the argument on $[T-(c_0+\cdots+c_{n-1}), T-(c_0+\cdots+c_{n-2})]$, since the two solutions are now known to restart from the same (new) initial condition at time $T-(c_0+\cdots+c_{n-1})$. Uniqueness follows by induction.

The fact that any solution $(\boldsymbol{X}, \boldsymbol{Y}, \boldsymbol{Z})$ satisfies (4.12) may be proved by a backward induction starting from the last interval $[T-c_0, T]$. The representation property on $[T-c_0, T]$ is indeed a consequence of Theorem 4.2. It permits to identify Y_{T-c_0} with $u(T-c_0, X_{T-c_0})$ and then to repeat the same argument on $[T-(c_0+c_1), T-c_0]$. And so on... □

Stability

Proposition 4.8 may be complemented with the following stability property, which is the long time analogue of Theorem 4.4.

Lemma 4.9 *Let us assume that there is another set of coefficients (B', F', G', Σ') satisfying the same assumption as (B, F, G, Σ) in the statement of Proposition 4.8, with respect to another decoupling field u'.*

Then, there exists a constant C, only depending on T, L and the Lipschitz constants of u and u' in x such that, for any initial conditions $\xi, \xi' \in L^2(\Omega, \mathcal{F}_0, \mathbb{P}; \mathbb{R}^d)$, the two processes $(\boldsymbol{X}, \boldsymbol{Y}, \boldsymbol{Z})$ and $(\boldsymbol{X}', \boldsymbol{Y}', \boldsymbol{Z}')$ obtained by solving (4.1) with ξ and ξ' as respective initial conditions and with (B, F, G, Σ) and (B', F', G', Σ') as respective coefficients, satisfy:

$$\begin{aligned}
&\mathbb{E}\Big[\sup_{0\leqslant t\leqslant T}\Big(|X_t - X'_t|^2 + |Y_t - Y'_t|^2\Big) + \int_0^T |Z_t - Z'_t|^2 dt\Big] \\
&\leqslant C\mathbb{E}\Big[|\xi - \xi'|^2 + |(G - G')(X_T)|^2 \\
&\qquad + \int_0^T \big|(B - B', F - F', \Sigma - \Sigma')(t, X_t, Y_t, Z_t)\big|^2 dt\Big].
\end{aligned}$$

Proof. For small time $T > 0$, this estimate follows immediately from Theorem 4.4. We only need to show that one can extend it to arbitrarily large values of T. We then choose a regular subdivision $0 = T_0 < T_1 < \cdots < T_{N-1} < T_N = T$ so that the common length of the intervals $[T_i, T_{i+1}]$ is small enough in order to apply Theorem 4.4 on each interval $[T_i, T_{i+1}]$ with $u(T_{i+1}, \cdot)$ or $u'(T_{i+1}, \cdot)$ as terminal condition function. For any $i \in \{0, \cdots, N-1\}$, we have:

$$\begin{aligned}
&\mathbb{E}\Big[\sup_{T_i\leqslant t\leqslant T_{i+1}}\Big(|X_t - X'_t|^2 + |Y_t - Y'_t|^2\Big) + \int_{T_i}^{T_{i+1}} |Z_t - Z'_t|^2 dt\Big] \\
&\leqslant C\Big[\mathbb{E}\big[|X_{T_i} - X'_{T_i}|^2 + |(u - u')(T_{i+1}, X_{T_{i+1}})|^2\big] \qquad (4.13) \\
&\qquad + \mathbb{E}\int_{T_i}^{T_{i+1}} \big|(B - B', F - F', \Sigma - \Sigma')(t, X_t, Y_t, Z_t)\big|^2 dt\Big].
\end{aligned}$$

For simplicity, we denote the left-hand side by $\Theta(T_i, T_{i+1})$ and we let:

$$\Delta_t = \big|(B-B', F-F', \Sigma-\Sigma')(t, X_t, Y_t, Z_t)\big|^2.$$

In this proof, we shall also use the notation $\delta_T = \big|(G-G')(X_T)\big|^2$.

We first consider the last interval $[T_{N-1}, T_N]$ corresponding to the case $i = N-1$. Since $T_N = T$ we have $u(T, \cdot) = G$ and $u'(T, \cdot) = G'$, so that:

$$\Theta(T_{N-1}, T) \leqslant C\left(\mathbb{E}\big[|X_{T_{N-1}} - X'_{T_{N-1}}|^2\big] + \delta_T + \int_{T_{N-1}}^T \Delta_t dt\right),$$

this estimate being true for all possible initial conditions for the process $\boldsymbol{X}'$ at time T_{N-1}. In this regard, notice that, while some freedom is allowed in the choice of $X'_{T_{N-1}}$, the initial condition of $\boldsymbol{X}$ is somehow fixed through $(\Delta_t)_{T_{N-1} \leqslant t \leqslant T}$. Note also that C is implicitly assumed to be larger than 1 and that we can allow its value to change from line to line as long as this new value depends only upon T, L in assumption **Lipschitz FBSDE** and the Lipschitz constants in x of u and u'.

Next we freeze the process $\boldsymbol{X}$ but we let $\boldsymbol{X}'$ vary. We use the fact that the decoupling field u' does not depend on the initial condition of $\boldsymbol{X}'$. In particular, we can choose to keep the coefficients (B', F', G', Σ') but set $X'_{T_{N-1}} = X_{T_{N-1}}$. Then the above inequality implies

$$\mathbb{E}\big[|u(T_{N-1}, X_{T_{N-1}}) - u'(T_{N-1}, X_{T_{N-1}})|^2\big] \leqslant C\left(\delta_T + \int_{T_{N-1}}^T \Delta_t dt\right).$$

We can now plug this estimate into inequality (4.13) with $i = N-2$ to get:

$$\Theta(T_{N-2}, T_{N-1}) \leqslant C\left(\mathbb{E}\big[|X_{T_{N-2}} - X'_{T_{N-2}}|^2\big] + \delta_T + \int_{T_{N-2}}^T \Delta_t dt\right).$$

As before, we can write what this estimate gives if we keep (B', F', G', Σ'), and set $X_{T_{N-2}} = X'_{T_{N-2}}$:

$$\mathbb{E}\big[|u(T_{N-2}, X_{T_{N-2}}) - u'(T_{N-2}, X_{T_{N-2}})|^2\big] \leqslant C\left(\delta_T + \int_{T_{N-2}}^T \Delta_t dt\right).$$

Plugging this estimate into inequality (4.13) with $i = N-3$ we get:

$$\Theta(T_{N-3}, T_{N-2}) \leqslant C\left(\mathbb{E}\big[|X_{T_{N-3}} - X'_{T_{N-3}}|^2\big] + \delta_T + \int_{T_{N-3}}^T \Delta_t dt\right).$$

Iterating, we get:

$$\Theta(T_i, T_{i+1}) \leqslant C\left(\mathbb{E}\big[|X_{T_i} - X'_{T_i}|^2\big] + \delta_T + \int_{T_i}^T \Delta_t dt\right). \tag{4.14}$$

As before the value of the constants can change from line to line. From this, we get the desired estimate once we notice that, for each $i \in \{1, \cdots, N\}$, we have:

$$\begin{aligned}\mathbb{E}\big[|X_{T_i} - X'_{T_i}|^2\big] &\leqslant \mathbb{E}\Big[\sup_{T_{i-1}\leqslant t\leqslant T_i} |X_t - X'_t|^2\Big] \\ &\leqslant C\bigg(\mathbb{E}\big[|X_{T_{i-1}} - X'_{T_{i-1}}|^2\big] + \delta_T + \int_{T_{i-1}}^{T} \Delta_t dt\bigg),\end{aligned}$$

from which we easily derive the required bound for $\mathbb{E}[\sup_{0\leq t\leq T} |X_t - X'_t|^2]$ by means of a forward induction. Summing over i in (4.14), we conclude the proof. □

Connection with PDEs

Representation formula (4.8) is reminiscent of the Markov property as it offers a representation of the backward component at time t as a function of the sole position of the state variable at time t. Furthermore, the proof of the representation formula (4.8) is itself reminiscent of the Markov property as it suggests that the solution $(X_s, Y_s, Z_s)_{t\leqslant s\leqslant T}$ after time t only depends upon the present position X_t at time t and the Brownian increments after time t, at least provided that T is small enough. Although quite intuitive, this last assertion will be made entirely rigorous in Chapter (Vol II)-1 with a suitable version for FBSDEs, of the Yamada-Watanabe theorem in Theorem (Vol II)-1.33.

The fact that the forward component of (4.1) is a Markov process whenever existence and uniqueness hold is a strong indication that u solves a partial differential equation.

Actually, this is another feature of forward-backward equations which we already alluded to in Chapter 3. See for instance Remarks 3.16 and 3.26. Forward-backward SDEs (at least when driven by deterministic coefficients) provide a nonlinear version of the Feynman-Kac formula and it should now be clear that the decoupling field should be the core of the connection with PDEs. This is usually checked with a verification argument.

Lemma 4.10 *For a given $T > 0$, assume that on top of assumption* **Lipschitz FBSDE**, *the system of PDEs:*

$$\begin{aligned}&\partial_t u^i + B\big(t, x, u(t,x), \partial_x u(t,x)\Sigma(t,x,u(t,x))\big) \cdot \partial_x u^i(t,x) \\ &\quad + \frac{1}{2}\text{trace}\big[\big(\Sigma\Sigma^\dagger\big)(t,x,u(t,x))\partial^2_{xx}u^i(t,x)\big] \\ &\quad + F^i\big(t,x,u(t,x), \partial_x u(t,x)\Sigma(t,x,u(t,x))\big) = 0,\end{aligned}$$

for $(t,x) \in [0,T] \times \mathbb{R}^d$, $i \in \{1,\cdots,m\}$, with the terminal condition $u(T,x) = G(x)$ for $x \in \mathbb{R}^d$, has a bounded classical solution $u : [0,T] \times \mathbb{R}^d \ni (t,x) \mapsto u(t,x) = (u^1(t,x),\cdots,u^m(t,x))$, continuous on $[0,T] \times \mathbb{R}^d$, once differentiable in time, and twice differentiable in space with jointly continuous derivatives on $[0,T) \times \mathbb{R}^d$ and with bounded first and second order derivatives in space.

Then, for any $t \in [0,T]$ and $\xi \in L^2(\Omega, \mathcal{F}_t, \mathbb{P}; \mathbb{R}^d)$, the FBSDE (4.1) *with initial condition $X_t = \xi$, has a unique solution $(\boldsymbol{X}^{t,\xi}, \boldsymbol{Y}^{t,\xi}, \boldsymbol{Z}^{t,\xi})$, $\boldsymbol{X}^{t,\xi}$ solving the SDE:*

$$dX_s^{t,\xi} = B\Big(s, X_s^{t,\xi}, u(s, X_s^{t,\xi}), \partial_x u(s, X_s^{t,\xi})\Sigma\big(s, X_s^{t,\xi}, u(s, X_s^{t,\xi})\big)\Big)ds \qquad (4.15)$$
$$+ \Sigma\big(s, X_s^{t,\xi}, u(s, X_s^{t,\xi})\big)dW_s, \quad s \in [t, T],$$

and $\boldsymbol{Y}^{t,\xi}$ *and* $\boldsymbol{Z}^{t,\xi}$ *being given by:*

$$Y_s^{t,\xi} = u(s, X_s^{t,\xi}), \quad Z_s^{t,\xi} = \partial_x u(s, X_s^{t,\xi})\Sigma\big(s, X_s^{t,\xi}, u(s, X_s^{t,\xi})\big), \qquad (4.16)$$

for $s \in [t, T]$.

Proof. The proof is quite straightforward. There is no difficulty for solving (4.15). Once this is done, one can define $\boldsymbol{Y}^{t,\xi}$ and $\boldsymbol{Z}^{t,\xi}$ as in (4.16), and prove the desired result, namely that $(\boldsymbol{X}^{t,\xi}, \boldsymbol{Y}^{t,\xi}, \boldsymbol{Z}^{t,\xi})$ satisfies (4.1), by applying Itô's formula to $\boldsymbol{Y}^{t,\xi} = (u(s, X_s^{t,\xi}))_{t \leqslant s \leqslant T}$ and by taking advantage of the fact that the function u solves the above system of PDEs. Uniqueness follows from Proposition 4.8. □

At this stage, the reader may wonder whether the representation of the gradient given by the formula (4.16) can be directly proved, without any use of a PDE argument. A positive answer is given by the following result.

Lemma 4.11 *On top of assumption* **Lipschitz FBSDE**, *assume that for a random variable* $\xi \in L^2(\Omega, \mathcal{F}_0, \mathbb{P}; \mathbb{R}^d)$, *we can find a solution* $(\boldsymbol{X}, \boldsymbol{Y}, \boldsymbol{Z})$ *to* (4.1) *with* $X_0 = \xi$ *as initial condition and a jointly continuous function* $u : [0, T] \times \mathbb{R}^d \to \mathbb{R}^m$, *once differentiable in space with a bounded and jointly continuous derivative, such that*

$$\mathbb{P}\Big[\forall t \in [0, T], \quad Y_t = u(t, X_t)\Big] = 1. \qquad (4.17)$$

Then, for $\mathrm{Leb}_1 \otimes \mathbb{P}$*-almost every* $(t, \omega) \in [0, T] \times \Omega$ *it holds that:*

$$Z_t = \partial_x u(t, X_t)\Sigma\big(t, X_t, u(t, X_t)\big).$$

When u *is merely Lipschitz-continuous in* x *uniformly in time, and* Σ *is bounded, we can find a constant* C *only depending upon the Lipschitz constant of* u *and the bound for* Σ, *such that* $(Z_t)_{0 \leqslant t \leqslant T}$ *is bounded by* C, *almost everywhere for* $\mathrm{Leb}_1 \otimes \mathbb{P}$. *If* $m = 1$ *and if* Σ *is invertible and its inverse is bounded, then* $(\Sigma^{-1\dagger}(t, X_t, Y_t)Z_t)_{0 \leq t \leq T}$ *is essentially bounded by the Lipschitz constant of* u *in* x.

Proof. We first prove the representation formula under the strong assumptions on u. Consider a uniform subdivision $0 = T_0 < T_1 < \cdots < T_N = T$ of step size h together with a simple process $\boldsymbol{\theta}$ of the form:

$$\theta_t = \sum_{i=0}^{N-1} \theta^i \mathbf{1}_{(t_i, t_{i+1}]}(t), \quad t \in [0, T],$$

with $\theta^i \in L^\infty(\Omega, \mathcal{F}_{t_i}, \mathbb{P}; \mathbb{R}^d)$, the variables $(\theta^i)_{i=0,\cdots,N-1}$ being uniformly bounded by some constant K. Next, for any $i = 0, \cdots, N-1$, notice that:

$$\mathbb{E}\Big[\Big(\int_{T_i}^{T_{i+1}} \theta_t \cdot dW_t\Big) Y_{T_{i+1}}\Big] = \mathbb{E}\Big[\Big(\theta_{T_i} \cdot (W_{T_{i+1}} - W_{T_i})\Big) Y_{T_{i+1}}\Big].$$

Using the fact that:

$$Y_{T_{i+1}} = Y_{T_i} - \int_{T_i}^{T_{i+1}} F\big(s, X_s, Y_s, Z_s\big) ds + \int_{T_i}^{T_{i+1}} Z_s dW_s,$$

with:

$$\mathbb{E}\Big[\sup_{0 \leqslant t \leqslant T} \big(|X_s|^2 + |Y_s|^2\big) + \int_0^T |Z_s|^2 ds\Big] < \infty,$$

we easily see that:

$$\mathbb{E}\Big[\Big(\int_{T_i}^{T_{i+1}} \theta_s \cdot dW_s\Big) Y_{T_{i+1}}\Big] = \mathbb{E}\Big[\int_{T_i}^{T_{i+1}} Z_s \theta_s ds\Big] + h\varepsilon(T_i, h), \tag{4.18}$$

where, here and below in the proof, $(\varepsilon(t,h))_{t \in [0,T], h \geqslant 0}$ is a generic notation for a function satisfying:

$$\lim_{h \searrow 0} \sup_{t \in [0,T]} |\varepsilon(t,h)| = 0.$$

Now, we observe that the left-hand side is also equal to:

$$\begin{aligned}
&\mathbb{E}\Big[\Big(\int_{T_i}^{T_{i+1}} \theta_s \cdot dW_s\Big) Y_{T_{i+1}}\Big] \\
&= \mathbb{E}\Big[\Big(\int_{T_i}^{T_{i+1}} \theta_s \cdot dW_s\Big)\Big(u\big(T_{i+1}, X_{T_{i+1}}\big) - u\big(T_{i+1}, X_{T_i}\big)\Big)\Big] \\
&= \mathbb{E}\Big[\Big(\int_{T_i}^{T_{i+1}} \theta_s \cdot dW_s\Big) \\
&\quad \times \Big(\int_0^1 \partial_x u\big(T_{i+1}, r X_{T_{i+1}} + (1-r) X_{T_i}\big)\big(X_{T_{i+1}} - X_{T_i}\big) dr\Big)\Big].
\end{aligned} \tag{4.19}$$

Thanks to the boundedness and the joint continuity of $\partial_x u$, observe that:

$$\begin{aligned}
&\mathbb{E}\Big[\Big| \int_{T_i}^{T_{i+1}} \partial_x u(s, X_s) \Sigma(s, X_s, Y_s) dW_s \\
&\qquad - \int_0^1 \partial_x u\big(T_{i+1}, r X_{T_{i+1}} + (1-r) X_{T_i}\big)\big(X_{T_{i+1}} - X_{T_i}\big) dr\Big|^2\Big]^{1/2} \\
&= h\varepsilon(T_i, h).
\end{aligned}$$

We finally get:

$$\begin{aligned}
&\mathbb{E}\Big[\Big(\int_{T_i}^{T_{i+1}} \theta_s \cdot dW_s\Big) Y_{T_{i+1}}\Big] \\
&= \mathbb{E}\Big[\Big(\int_{T_i}^{T_{i+1}} \theta_s \cdot dW_s\Big)\Big(\int_{T_i}^{T_{i+1}} \partial_x u(s, X_s)\Sigma(s, X_s, Y_s) dW_s\Big)\Big] + h\varepsilon(T_i, h) \qquad (4.20) \\
&= \mathbb{E}\Big[\int_{T_i}^{T_{i+1}} \partial_x u(s, X_s)\Sigma(s, X_s, Y_s)\theta_s ds\Big] + h\varepsilon(T_i, h).
\end{aligned}$$

Identifying (4.18) and (4.20), summing over $i = 0, \cdots, N-1$, and then letting h tend to 0, we finally get:

$$\mathbb{E}\Big[\int_0^T \partial_x u(s, X_s)\Sigma(s, X_s, Y_s)\theta_s ds\Big] = \mathbb{E}\Big[\int_0^T Z_s\theta_s ds\Big].$$

The proof of the first claim is easily completed, using the fact that the class of simple processes $\boldsymbol{\theta}$ is dense within the family of square-integrable $\mathbb{F}$-progressively measurable processes.

When u is Lipschitz continuous in space and not necessarily differentiable in x, we assume that Σ is bounded. We then observe that (4.19) still makes sense. Indeed, since the function $\mathbb{R} \ni r \mapsto u(T_{i+1}, X_{T_i} + r(X_{T_{i+1}} - X_{T_i}))$ is Lipschitz continuous, we can still give a sense to the integral $\int_0^1 \partial_x u(T_{i+1}, rX_{T_{i+1}} + (1-r)X_{T_i})dr$. Also, we can handle the last term in (4.19) by Itô's formula, namely we can prove:

$$\mathbb{E}\Big[\Big|\Big(\int_{T_i}^{T_{i+1}} \theta_s \cdot dW_s\Big)(X_{T_{i+1}} - X_{T_i}) - \int_{T_i}^{T_{i+1}} \Sigma(s, X_s, Y_s)\theta_s ds\Big|\Big] = h\varepsilon(T_i, h).$$

Therefore, we can find a constant C, only depending on the Lipschitz bound for u and on the bound for Σ, such that:

$$\begin{aligned}
\mathbb{E}\Big[\Big(\int_{T_i}^{T_{i+1}} \theta_s \cdot dW_s\Big) Y_{T_{i+1}}\Big] &\leqslant C\mathbb{E}\Big[\Big|\int_{T_i}^{T_{i+1}} \Sigma(s, X_s, Y_s)\theta_s ds\Big|\Big] + h\varepsilon(T_i, h) \\
&\leqslant C\mathbb{E}\Big[\int_{T_i}^{T_{i+1}} |\theta_s| ds\Big] + h\varepsilon(T_i, h),
\end{aligned}$$

the value of C being allowed to increase from line to line. Identifying again with (4.18), summing over $i \in \{0, \cdots, N-1\}$ and letting h tend to 0, we deduce that:

$$\Big|\mathbb{E}\Big[\int_0^T Z_s\theta_s ds\Big]\Big| \leqslant C\mathbb{E}\Big[\int_0^T |\theta_s| ds\Big].$$

Once again, the proof of the second claim is easily completed. The last claim may be proved by changing $(\theta_s)_{0 \le s \le T}$ into $(\Sigma^{-1\dagger}(s, X_s, Y_s)\theta_s)_{0 \le s \le T}$. □

The Nondegenerate Case

Returning to the statement of Lemma 4.10, we understand that independently of any result of existence of a classical solution to the system (4.10), the theory of PDEs might help in another (though related) way. Actually, we may want to take advantage of gradient estimates from the theory of nonlinear PDEs in order to control the Lipschitz constant of the decoupling field in the induction procedure described earlier. Indeed, some of these gradient estimates can be used to derive existence and uniqueness of a solution to (4.1). A typical instance of a successful implementation of this strategy is provided by the nondegenerate models which we already alluded to in Subsection 3.2.3. Relevant to our current discussion is the following result of Delarue which holds under the assumption given below, see the Notes & Complements at the end of the chapter of a precise reference.

Assumption (Nondegenerate FBSDE). There exists a constant $L \geqslant 1$ such that:

(A1) For any $t \in [0,T]$, $x, x' \in \mathbb{R}^d$, $y, y' \in \mathbb{R}^m$, $z, z' \in \mathbb{R}^{m\times d}$,

$$
\begin{aligned}
&|(B,F,G,\Sigma)(t,x',y',z') - (B,F,G,\Sigma)(t,x,y,z)| \\
&\quad \leqslant L|(x,y,z) - (x',y',z')|.
\end{aligned}
$$

(A2) The functions Σ and G are bounded by L. Moreover, for any $t \in [0,T]$, $x \in \mathbb{R}^d$, $y \in \mathbb{R}^m$ and $z \in \mathbb{R}^{m\times d}$,

$$
|(B,F)(t,x,y,z)| \leqslant L\big[1 + |y| + |z|\big].
$$

(A3) The function Σ is uniformly elliptic in the sense that, for any $t \in [0,T]$, $x \in \mathbb{R}^d$ and $y \in \mathbb{R}^m$, the following inequality holds:

$$
\big(\Sigma\Sigma^\dagger\big)(t,x,y) \geqslant L^{-1} I_d,
$$

in the sense of symmetric matrices, where I_d is the d-dimensional identity matrix. Recall that we use the exponent † to denote the transpose of a matrix. Moreover, the function $[0,T]\times\mathbb{R}^d\times\mathbb{R}^m \ni (t,x,y) \mapsto \Sigma(t,x,y)$ is continuous.

Theorem 4.12 *Under assumption* **Nondegenerate FBSDE**, *for any $t \in [0,T]$ and ξ in the space $L^2(\Omega,\mathcal{F}_t,\mathbb{P};\mathbb{R}^d)$, the forward-backward system* (4.1) *with $X_t = \xi$ as initial condition has a unique solution, denoted by $(X_s^{t,\xi}, Y_s^{t,\xi}, Z_s^{t,\xi})_{t\leqslant s\leqslant T}$. Moreover, the decoupling field $u : [0,T]\times\mathbb{R}^d \ni (t,x) \mapsto u(t,x) = Y_t^{t,x} \in \mathbb{R}^m$, obtained by choosing $\xi = x$, is bounded by a constant γ depending only upon T and L, and is $1/2$-Hölder continuous in time and Lipschitz continuous in space in the sense that:*

$$|u(t,x) - u(t',x')| \leqslant \Gamma\big(|t-t'|^{1/2} + |x-x'|\big),$$

for some constant Γ only depending upon T and L. Finally, $Y_s^{t,\xi} = u(s, X_s^{t,\xi})$ for any $t \leqslant s \leqslant T$ and $|Z_s^{t,\xi}| \leqslant \Gamma L$, $\mathrm{Leb}_1 \otimes \mathbb{P}$ almost everywhere.

Remark 4.13 *The above Lipschitz estimate (in the variable x) will be established in Subsection 4.4.2, when σ is independent of y. The proof relies on the theory of quadratic BSDEs, which we present in the next subsection.*

4.1.3 Quadratic Backward SDEs

So far, we have provided results for general FBSDEs of the form (4.1), allowing the diffusion coefficient Σ to depend upon the variable y. Actually, in most of the applications considered in this book, we do not need such a level of generality. In fact, it will suffice to manipulate FBSDEs driven by a diffusion coefficient only depending on the variables t and x.

A crucial insight into this case is the following: When Σ is independent of the backward component, the forward-backward system can be decoupled by means of a Girsanov transformation, at least when Σ is invertible. For instance, if $(\boldsymbol{X}, \boldsymbol{Y}, \boldsymbol{Z})$ is a solution to (4.1), we may let:

$$\frac{d\mathbb{Q}}{d\mathbb{P}} = \mathcal{E}\bigg(- \int_0^{\cdot} \big(\Sigma^{-1}(t, X_t) B(t, X_t, Y_t, Z_t)\big) \cdot dW_t \bigg)_T,$$

where as earlier, $\mathcal{E}$ stands for the Doléans-Dade exponential of a martingale. If we are allowed to apply Girsanov's theorem, this turns the forward-backward system (4.1) into the following system of decoupled equations:

$$\begin{cases} dX_t = \Sigma(t, X_t) dW_t^{\mathbb{Q}}, \\ dY_t = -H\big(t, X_t, Y_t, Z_t\big)dt + Z_t dW_t^{\mathbb{Q}}, \qquad t \in [0,T], \\ X_0 = \xi, \quad Y_T = G(X_T), \end{cases} \tag{4.21}$$

where $\boldsymbol{W}^{\mathbb{Q}} = (W_t^{\mathbb{Q}})_{0 \leqslant t \leqslant T}$ is a Wiener process under $\mathbb{Q}$ and H is given by:

$$H(t,x,y,z) = F(t,x,y,z) + z\big(\Sigma^{-1}(t,x) B(t,x,y,z)\big),$$

for $(t,x,y,z) \in [0,T] \times \mathbb{R}^d \times \mathbb{R}^m \times \mathbb{R}^{m \times d}$, where $z(\Sigma^{-1}(t,x)B(t,x,y,z))$ is the product of an $m \times d$-matrix and a d-vector.

We already used this strategy in the presentation of mean field games under the weak formulation in Subsection 3.3.1. In particular, (4.21) is very close to (3.30), although the functions H in the two definitions do not exactly coincide.

A key observation with (4.21) is that the driver H is most often of quadratic growth in the variable z. This departure from the standard Lipschitz setting creates additional difficulties and requires a special treatment. In order to overcome these new challenges we shall assume that $m = 1$, implying that the backward component $\boldsymbol{Y}$ is one-dimensional. This restrictive assumption will not be too much of an hindrance in what follows. Indeed, in the majority of the cases of interest to us, quadratic BSDEs will be used to represent a cost; in this regard, quadratic BSDEs under consideration will be only in dimension 1.

Below, we provide some of the basic results in the analysis of quadratic BSDEs. We do not give proofs because of their technicalities. We refer to the Notes & Complements at the end of the chapter for references on the subject.

Existence and Uniqueness of Solutions

In order to address existence and uniqueness, we recast the problem in the more general setting of *non-Markovian* backward equations of the form:

$$dY_t = -\Psi(t, Y_t, Z_t)dt + Z_t \cdot dW_t, \quad t \in [0, T]\,; \quad Y_T = \zeta, \tag{4.22}$$

where ζ is a random variable in $L^\infty(\Omega, \mathcal{F}_T, \mathbb{P}; \mathbb{R})$, and Ψ is a random driver satisfying the following assumption:

Assumption (Quadratic BSDE). The terminal condition ζ is a bounded $\mathcal{F}_T$-measurable random variable with values in $\mathbb{R}$, and the driver Ψ is a function from $[0, T] \times \Omega \times \mathbb{R} \times \mathbb{R}^d$ into $\mathbb{R}$ satisfying:

(A1) For any $(y, z) \in \mathbb{R} \times \mathbb{R}^d$, the process $[0, T] \times \Omega \ni (t, \omega) \mapsto \Psi(t, \omega, y, z)$ is $\mathbb{F}$-progressively measurable, where $\mathbb{F}$ is the usual augmentation of the filtration generated by $\boldsymbol{W}$ and by an initial σ-field $\mathcal{F}_0$, independent of $\boldsymbol{W}$.

(A2) For any $t \in [0, T]$, $\omega \in \Omega$, $y, y' \in \mathbb{R}$, and $z, z' \in \mathbb{R}^d$, we have:

$$|\Psi(t, \omega, y, z)| \leqslant L\big(1 + |y| + |z|^2\big),$$
$$|\Psi(t, \omega, y', z') - \Psi(t, \omega, y, z)| \leqslant L\big(|y - y'| + (1 + |z| + |z'|)|z - z'|\big).$$

Remark 4.14 *In accordance with the convention introduced in Chapter 3, the value Z_t at time t of the martingale integrand process $\boldsymbol{Z}$ in a one-dimensional BSDE will be often regarded as a d-dimensional vector (and not as a $1 \times d$-matrix). This justifies the use of the notation $Z_t \cdot dW_t$ instead of $Z_t dW_t$.*

In order to understand the rationale for the boundedness assumption on the terminal value ζ, the reader is referred to Subsection 4.7.3 below where we use the so-called *Cole-Hopf* transformation in order to linearize a quadratic BSDE. Therein, we transform an equation of the type (4.22) into a linear equation by considering the exponential of the solution (or of a multiple of the solution). This analysis requires the exponential of ζ (up to a multiplicative constant) to be sufficiently integrable. A convenient way to guarantee this restrictive integrability property is to assume that ζ is bounded.

Notice that as usual, we dropped the variable ω in (4.22), and we shall do so whenever possible. We give the main existence result without proof.

Theorem 4.15 *Under assumption* **Quadratic BSDE**, *there exists a unique pair of* $\mathbb{F}$*-progressively measurable processes* $(\boldsymbol{Y}, \boldsymbol{Z}) = (Y_t, Z_t)_{0 \leqslant t \leqslant T}$ *with values in* $\mathbb{R}$ *and* $\mathbb{R}^d$ *satisfying* (4.22) *and*

$$\sup_{0 \leqslant t \leqslant T} |Y_t| \in L^\infty(\Omega, \mathcal{F}_T, \mathbb{P}; \mathbb{R}), \quad \mathbb{E} \int_0^T |Z_t|^2 dt < \infty,$$

both quantities being controlled by T, L and the bound for ζ.

In Chapter 3, we already appealed to the following comparison principle.

Theorem 4.16 *Under assumption* **Quadratic BSDE**, *let* $(\zeta', \boldsymbol{F}')$ *be another pair of coefficients satisfying assumption* **Quadratic BSDE**. *Assume that* $\mathbb{P}[\zeta > \zeta'] = 0$ *and*

$$\forall y \in \mathbb{R},\ z \in \mathbb{R}^d,$$
$$\text{Leb}_1 \otimes \mathbb{P}\big[(t, \omega) \in [0, T] \times \Omega : \Psi(t, \omega, y, z) > \Psi'(t, \omega, y, z)\big] = 0.$$

Then,

$$\forall t \in [0, T], \quad \mathbb{P}\big[Y_t > Y'_t\big] = 0,$$

where $(\boldsymbol{Y}', \boldsymbol{Z}')$ *is the unique solution (in the sense of Theorem 4.15) to* (4.22), *when driven by* $(\zeta', \boldsymbol{F}')$.

Bounded-Mean-Oscillation Martingales

One crucial ingredient with quadratic BSDEs is that the martingale process $(\int_0^t Z_s \cdot dW_s)_{0 \leqslant t \leqslant T}$ is of bounded-mean-oscillation in the sense of the following definition.

Definition 4.17 *Given an* $\mathbb{F}$*-progressively measurable square-integrable process* $\boldsymbol{Z} = (Z_t)_{0 \leqslant t \leqslant T}$ *with values in* $\mathbb{R}^d$ *(that is* $\boldsymbol{Z} \in \mathbb{H}^{2,d}$*), the martingale* $(\int_0^t Z_s \cdot dW_s)_{0 \leqslant t \leqslant T}$ *is said to be of bounded-mean-oscillation (BMO for short) if there exists a constant* $K \geqslant 0$ *such that for any* $\mathbb{F}$*-stopping time with values in* $[0, T]$,

$$\mathbb{P}\bigg(\mathbb{E}\bigg[\int_\tau^T |Z_t|^2 dt \,|\, \mathcal{F}_\tau\bigg] \leqslant K^2\bigg) = 1.$$

We call the smallest constant K with this property the BMO norm of the martingale.

An important property of BMO martingales is given in the following standard result of stochastic analysis.

Proposition 4.18 *Let* $\mathbf{Z}$ *be a process in* $\mathbb{H}^{2,d}$ *such that the martingale* $(\int_0^t Z_s \cdot dW_s)_{0\leqslant t\leqslant T}$ *has a finite BMO norm. Then, there exists a constant* $r > 1$*, depending only on the BMO norm of* $(\int_0^t Z_s \cdot dW_s)_{0\leqslant t\leqslant T}$*, such that the Doléans-Dade exponential* $\mathcal{E}(\int_0^\cdot Z_s \cdot dW_s)$ *belongs to* $L^r(\Omega, \mathcal{F}_T, \mathbb{P}; \mathbb{R})$ *and is thus uniformly integrable. Moreover, the* $L^r(\Omega, \mathcal{F}_T, \mathbb{P}; \mathbb{R})$*-norm of* $\mathcal{E}(\int_0^\cdot Z_s \cdot dW_s)$ *only depends on* T *and the BMO norm of* $(\int_0^t Z_s \cdot dW_s)_{0\leqslant t\leqslant T}$*.*

Finally, for any $p \geqslant 1$, $\int_0^T |Z_s|^2 ds$ *belongs to the space* $L^p(\Omega, \mathcal{F}_T, \mathbb{P}; \mathbb{R})$ *and its* $L^p(\Omega, \mathcal{F}_T, \mathbb{P}; \mathbb{R})$*-norm only depends on the BMO norm of* $(\int_0^t Z_s \cdot dW_s)_{0\leqslant t\leqslant T}$*.*

The relevance of BMO martingales to quadratic BSDEs is captured by the following statement.

Theorem 4.19 *Under assumption* **Quadratic BSDE**, *consider the unique solution* $(\mathbf{Y}, \mathbf{Z}) = (Y_t, Z_t)_{0\leqslant t\leqslant T}$ *to* (4.22) *as given in Theorem 4.15. Then, the process* $(\int_0^t Z_s \cdot dW_s)_{0\leqslant t\leqslant T}$ *is a BMO martingale and its BMO norm only depends upon* T*,* L *and the bound for the* L^∞*-norm of* ζ*.*

Remark 4.20 *Part of the statements given here will be revisited in Chapter (Vol II)-1 when handling optimal control problems in random environments.*

4.2 McKean-Vlasov Stochastic Differential Equations

Motivated by the formulation of mean field games developed in Chapter 3, we proceed with the analysis of (forward, backward, and forward-backward) stochastic differential equations whose coefficients depend upon the law of their own solutions. Solvability of these equations may be addressed in two steps. In the first one, flow of probability measures underpinning the equation is treated as an input of the problem, while the second step involves a fixed point argument. We cast the latter as a matching problem which, once solved, replaces the input by the marginal distributions of the solutions. As a result, the coefficients of the equations end up containing the marginal distributions of the solutions. This characteristic is at the origin of the terminology *McKean-Vlasov equation*. The existence and uniqueness theory for forward SDEs of this type is rather standard by now. We introduce it first. The corresponding theory for BSDEs of McKean-Vlasov type is recent. We present it next, still in this section. Existence and uniqueness results for FBSDEs of McKean-Vlasov type are quite new, and much more involved. In this section, we provide an existence and uniqueness result in short time only. Its proof is modeled after the argument behind the original proof of Theorem 4.2 which was not given when we stated this short time existence and uniqueness result in Subsection 4.1.1.

This argument will be used again when we state and prove Theorem 1.45 in Chapter (Vol II)-1 as a generalization of Theorem 4.2 to the case of random coefficients. Solving FBSDEs of the McKean-Vlasov type on arbitrary time intervals is much more delicate and involved. We present a first class of models for which we can actually do that in the following section. Other models will be investigated in the next sections when we return to the existence of solutions for mean field games.

In this section and the next, all the processes are assumed to be defined on a complete filtered probability space $(\Omega, \mathcal{F}, \mathbb{F} = (\mathcal{F}_t)_{0\leqslant t\leqslant T}, \mathbb{P})$ supporting a d-dimensional Wiener process $\boldsymbol{W} = (W_t)_{0\leqslant t\leqslant T}$ with respect to $\mathbb{F}$, the filtration $\mathbb{F}$ satisfying the usual conditions. We recall that, for each random variable/vector or stochastic process X, we denote by $\mathcal{L}(X)$ the law (alternatively called the distribution) of X and for any integer $n \geqslant 1$, by $\mathbb{H}^{2,n}$ the Hilbert space:

$$\mathbb{H}^{2,n} = \Big\{\boldsymbol{Z} \in \mathbb{H}^{0,n} : \mathbb{E}\int_0^T |Z_s|^2 ds < \infty\Big\}, \tag{4.23}$$

where $\mathbb{H}^{0,n}$ stands for the collection of all $\mathbb{R}^n$-valued progressively measurable processes on $[0,T]$. We shall also denote by $\mathbb{S}^{2,n}$ the collection of all continuous processes $\boldsymbol{U} = (U_t)_{0\leqslant t\leqslant T}$ in $\mathbb{H}^{0,n}$ such that $\mathbb{E}[\sup_{0\leqslant t\leqslant T} |U_t|^2] < +\infty$. As for the dependence of the coefficients (and the solutions) upon the measure parameters, we refer the reader to the definition (3.16) of the Wasserstein distance given earlier, and to Section 5.1 of Chapter 5 for a thorough discussion of its properties. We merely highlight a simple property of the 2-Wasserstein distance W_2:

$$W_2(\mathcal{L}(X), \mathcal{L}(X'))^2 \leqslant \mathbb{E}[|X - X'|^2], \tag{4.24}$$

for any $\mathbb{R}^n$-valued square-integrable random variables X and X'.

4.2.1 Forward SDEs of McKean-Vlasov Type

Let us consider a forward stochastic differential equation in a given environment $\boldsymbol{\mu} = (\mu_t)_{0\leqslant t\leqslant T}$. One could think of the forward part of the general equation (3.17) if we assume that it does not depend upon the backward component. Requiring that the environment $\boldsymbol{\mu} = (\mu_t)_{0\leqslant t\leqslant T}$ therein matches the flow $(\mathcal{L}(X_t))_{0\leqslant t\leqslant T}$ of marginal distributions of the solution turns the ordinary SDE into a nonlinear SDE of the form

$$dX_t = B\big(t, X_t, \mathcal{L}(X_t)\big)dt + \Sigma\big(t, X_t, \mathcal{L}(X_t)\big)dW_t, \qquad t \in [0,T]. \tag{4.25}$$

For technical reasons, we allow the coefficients to be random. This means that the drift and diffusion coefficients of the state X_t of the system at time t are given by a pair of (measurable) functions $(B, \Sigma) : [0,T]\times\Omega\times\mathbb{R}^d\times\mathcal{P}_2(\mathbb{R}^d) \to \mathbb{R}^d\times\mathbb{R}^{d\times d}$. The term *nonlinear* used to qualify (4.25) does not refer to the fact that the coefficients B and Σ could be nonlinear functions of x, but instead to the fact that they depend not only on the value of the unknown process X_t at time t, but also on its marginal distribution $\mathcal{L}(X_t)$. We shall use the following assumptions.

Assumption (MKV SDE). There exists a constant $L \geqslant 0$ such that:

(A1) For each $(x,\mu) \in \mathbb{R}^d \times \mathcal{P}_2(\mathbb{R}^d)$, the processes $B(\cdot,\cdot,x,\mu) : [0,T] \times \Omega \ni (t,\omega) \mapsto B(t,\omega,x,\mu)$ and $\Sigma(\cdot,\cdot,x,\mu) : [0,T] \times \Omega \ni (t,\omega) \mapsto \Sigma(t,\omega,x,\mu)$ are $\mathbb{F}$-progressively measurable and belong to $\mathbb{H}^{2,d}$ and $\mathbb{H}^{2,d\times d}$ respectively.

(A2) For any $t \in [0,T]$, $\omega \in \Omega$, $x,x' \in \mathbb{R}^d$ and $\mu,\mu' \in \mathcal{P}_2(\mathbb{R}^d)$,

$$\begin{aligned}&|B(t,x,\mu)-B(t,x',\mu')|+|\Sigma(t,x,\mu)-\Sigma(t,x',\mu')|\\&\leqslant L\big[|x-x'|+W_2(\mu,\mu')\big].\end{aligned}$$

Under the above conditions we have existence and uniqueness of a solution to (4.25).

Theorem 4.21 *Under assumption* **MKV SDE**, *if* $X_0 \in L^2(\Omega,\mathcal{F}_0,\mathbb{P};\mathbb{R}^d)$, *there exists a unique solution* $\boldsymbol{X} = (X_t)_{0\leqslant t\leqslant T}$ *to* (4.25) *in* $\mathbb{S}^{2,d}$. *In particular, this solution satisfies:*

$$\mathbb{E}\Big[\sup_{0\leqslant t\leqslant T}|X_t|^2\Big] < +\infty.$$

Proof. Let $\boldsymbol{\mu} = (\mu_t)_{0\leqslant t\leqslant T} \in \mathcal{C}([0,T];\mathcal{P}_2(\mathbb{R}^d))$ be temporarily fixed. Substituting momentarily μ_t for $\mathcal{L}(X_t)$ for all $t \in [0,T]$ in (4.25) and recalling that X_0 is given, the classical existence result for Lipschitz SDEs guarantees existence and uniqueness of a strong solution of the classical stochastic differential equation with random coefficients:

$$dX_t = B(t,X_t,\mu_t)dt + \Sigma(t,X_t,\mu_t)dW_t, \qquad t \in [0,T]. \tag{4.26}$$

We denote its solution by $\boldsymbol{X}^{\boldsymbol{\mu}} = (X_t^{\boldsymbol{\mu}})_{0\leqslant t\leqslant T}$. This classical existence result also implies that the law of $\boldsymbol{X}^{\boldsymbol{\mu}}$ is of order 2, so that we can define the mapping

$$\begin{aligned}\Phi : \mathcal{C}([0,T];\mathcal{P}_2(\mathbb{R}^d)) \ni \boldsymbol{\mu} \mapsto \Phi(\boldsymbol{\mu}) &= \big(\mathcal{L}(X_t^{\boldsymbol{\mu}})\big)_{0\leqslant t\leqslant T}\\&= \big(\mathbb{P}\circ(X_t^{\boldsymbol{\mu}})^{-1}\big)_{0\leqslant t\leqslant T} \in \mathcal{C}([0,T];\mathcal{P}_2(\mathbb{R}^d)).\end{aligned}$$

Observe that the last term is in $\mathcal{C}([0,T];\mathcal{P}_2(\mathbb{R}^d))$ because $\boldsymbol{X}^{\boldsymbol{\mu}}$ has continuous paths and satisfies $\mathbb{E}[\sup_{0\leqslant t\leqslant T}|X_t^{\boldsymbol{\mu}}|^2] < \infty$.

Since a process $\boldsymbol{X} = (X_t)_{0\leqslant t\leqslant T}$ satisfying $\mathbb{E}[\sup_{0\leqslant t\leqslant T}|X_t|^2] < \infty$ is a solution of (4.25) if and only if its law is a fixed point of Φ, we prove the existence and uniqueness result of the theorem by proving that the mapping Φ has a unique fixed point. Let us choose $\boldsymbol{\mu}$ and $\boldsymbol{\mu}'$ in $\mathcal{C}([0,T];\mathcal{P}_2(\mathbb{R}^d))$. Since $\boldsymbol{X}^{\boldsymbol{\mu}}$ and $\boldsymbol{X}^{\boldsymbol{\mu}'}$ have the same initial conditions, Doob's maximal inequality and the Lipschitz assumption yield, for all $t \in [0,T]$:

$$
\begin{aligned}
&\mathbb{E}[\sup_{0\leqslant s\leqslant t} |X_s^\mu - X_s^{\mu'}|^2] \\
&\leqslant 2\mathbb{E}\left[\sup_{0\leqslant s\leqslant t}\left|\int_0^s [B(r,X_r^\mu,\mu_r) - B(r,X_r^{\mu'},\mu'_r)]dr\right|^2\right] \\
&\qquad + \mathbb{E}\left[\sup_{0\leqslant s\leqslant t}\left|\int_0^s [\Sigma(r,X_r^\mu,\mu_r) - \Sigma(r,X_r^{\mu'},\mu'_r)]dW_r\right|^2\right] \\
&\leqslant c(T)\left(\int_0^t \mathbb{E}\left[\sup_{0\leqslant r\leqslant s}|X_r^\mu - X_r^{\mu'}|^2\right]ds + \int_0^t \left(W_2(\mu_s,\mu'_s)\right)^2 ds\right. \\
&\qquad \left. + \mathbb{E}\left[\int_0^t |\Sigma(r,X_r^\mu,\mu_r) - \Sigma(r,X_r^{\mu'},\mu'_r)|^2 dr\right]\right) \\
&\leqslant c(T)\left(\int_0^t \mathbb{E}\left[\sup_{0\leqslant r\leqslant s}|X_r^\mu - X_r^{\mu'}|^2\right]ds + \int_0^t \left(W_2(\mu_s,\mu'_s)\right)^2 ds\right),
\end{aligned}
$$

for a constant $c(T)$ depending on T and L, $c(T)$ being nondecreasing in T. As usual, and except for the dependence upon T which we keep track of, we use the same notation $c(T)$ even though the value of this constant can change from line to line. Using Gronwall's inequality one concludes that:

$$
\mathbb{E}\left[\sup_{0\leqslant s\leqslant t}|X_s^\mu - X_s^{\mu'}|^2\right] \leqslant c(T)\int_0^t W_2(\mu_s,\mu'_s)^2\, ds, \tag{4.27}
$$

for $t \in [0,T]$. Therefore,

$$
\sup_{0\leqslant s\leqslant t} W_2\big(\Phi(\mu)_s, \Phi(\mu')_s\big)^2 \leqslant c(T)\int_0^t W_2(\mu_s,\mu'_s)^2\, ds.
$$

Iterating this inequality and denoting by Φ^k the k-th composition of the mapping Φ with itself we get that for any integer $k > 1$:

$$
\begin{aligned}
\sup_{0\leqslant s\leqslant T} W_2\big(\Phi^k(\mu)_s, \Phi^k(\mu')_s\big)^2 &\leqslant c(T)^k \int_0^T \frac{(T-s)^{k-1}}{(k-1)!} W_2(\mu_s,\mu'_s)^2\, ds \\
&\leqslant \frac{c(T)^k T^k}{k!} \sup_{0\leqslant s\leqslant T} W_2(\mu_s,\mu'_s)^2,
\end{aligned}
$$

which shows that for k large enough, Φ^k is a strict contraction and hence, Φ admits a unique fixed point as the space $\mathcal{C}([0,T];\mathcal{P}_2(\mathbb{R}^d))$ is complete. □

The McKean-Vlasov dynamics posited in (4.25) are sometimes called of *mean field type*. This is justified by the fact that stochastic differential equations of McKean-Vlasov type first appeared as the infinite particle limits of large systems of particles with mean field interactions. See Section (Vol II)-2.1 for a detailed account of this theory.

Remark 4.22 *The reader may object to the fact that the SDE* (4.25) *does not include all the models suggested by the general form touted in* (3.17)*. Indeed, in order to do so, we should investigate an SDE of the form*

$$dX_t = B\big(t, X_t, \mathcal{L}(X_t, \zeta_t)\big)dt + \Sigma\big(t, X_t, \mathcal{L}(X_t, \zeta_t)\big)dW_t, \qquad t \in [0, T], \tag{4.28}$$

where $(\zeta_t)_{0\leqslant t\leqslant T}$ *denotes a given* $\mathbb{F}$*-adapted process with paths in* $\mathcal{C}([0,T];\mathbb{R}^m)$ *and with square integrable marginals. However, it is easy to check that Theorem 4.21 extends to this slightly more general setting.*

4.2.2 Mean Field BSDEs

This section is devoted to an existence and uniqueness result for BSDEs of McKean-Vlasov type. We consider a backward stochastic differential equation of the form:

$$dY_t = -\Psi\big(t, Y_t, Z_t, \mathcal{L}(\xi_t, Y_t)\big)dt + Z_t dW_t, \qquad t \in [0, T], \tag{4.29}$$

with terminal condition $Y_T = G$. The driver Ψ and the terminal condition function G are measurable and random, with $\Psi : [0, T] \times \Omega \times \mathbb{R}^m \times \mathbb{R}^{m\times d} \times \mathcal{P}_2(\mathbb{R}^d \times \mathbb{R}^m) \to \mathbb{R}^m$ and $G : \Omega \to \mathbb{R}^m$. Moreover, $\boldsymbol{W} = (W_t)_{0\leqslant t\leqslant T}$ is an $\mathbb{R}^d$-valued Brownian motion and $\boldsymbol{\xi} = (\xi_t)_{0\leqslant t\leqslant T}$ is an $\mathbb{R}^d$-valued square-integrable $\mathbb{F}$-adapted process with continuous paths on $[0, T]$, where $\mathbb{F}$ is the usual augmentation of the filtration generated by $\boldsymbol{W}$ and by an initial σ-field $\mathcal{F}_0$, independent of $\boldsymbol{W}$.

Assumption (MKV BSDE).

(A1) For each $(y, z, \nu) \in \mathbb{R}^m \times \mathbb{R}^{m\times d} \times \mathcal{P}_2(\mathbb{R}^d \times \mathbb{R}^m)$, the process $\Psi(\cdot, \cdot, y, z, \nu)$: $[0, T] \times \Omega \ni (t, \omega) \mapsto \Psi(t, \omega, y, z, \nu)$ is $\mathbb{F}$-progressively measurable and belongs to $\mathbb{H}^{2,m}$. Also, $G \in L^2(\Omega, \mathcal{F}_T, \mathbb{P}; \mathbb{R}^m)$.

(A2) There exists a constant $L \geqslant 0$ such that for any $t \in [0, T]$, $\omega \in \Omega$, $y, y' \in \mathbb{R}^m$, $z, z' \in \mathbb{R}^{m\times d}$, $\nu, \nu' \in \mathcal{P}_2(\mathbb{R}^d \times \mathbb{R}^m)$, ν and ν' having the same first marginal on $\mathbb{R}^d$,

$$|\Psi(t, y, z, \nu) - \Psi(t, y', z', \nu')| \leqslant L\big[|y - y'| + |z - z'| + W_2(\nu, \nu')\big],$$

where we use the same notation W_2 for the 2-Wasserstein distance on $\mathcal{P}_2(\mathbb{R}^d)$ and $\mathcal{P}_2(\mathbb{R}^d \times \mathbb{R}^m)$.

Theorem 4.23 *Under assumption* **MKV BSDE**, *there exists a unique solution* $(\boldsymbol{Y}, \boldsymbol{Z}) \in \mathbb{S}^{2,m} \times \mathbb{H}^{2,m\times d}$ *of* (4.29).

Proof. On $\mathbb{H}^{2,m} \times \mathbb{H}^{2,m\times d}$, we introduce the norm $\|\cdot\|_{\mathbb{H},\alpha}$ defined by:

$$\|(\boldsymbol{Y},\boldsymbol{Z})\|_{\mathbb{H},\alpha}^2 = \mathbb{E}\int_0^T e^{\alpha t}(|Y_t|^2 + |Z_t|^2)dt,$$

with a positive constant α to be chosen later in the proof. For any $(\boldsymbol{Y},\boldsymbol{Z}) \in \mathbb{H}^{2,d} \times \mathbb{H}^{2,m\times d}$, we denote by $(\boldsymbol{Y}',\boldsymbol{Z}')$ the unique solution of the BSDE (which is known to exist by standard results from BSDE theory):

$$dY'_t = -\Psi\big(t, Y'_t, Z'_t, \mathcal{L}(\xi_t, Y_t)\big)dt + Z'_t dW_t, \quad t \in [0,T], \qquad Y'_T = G.$$

This defines a map $\Phi : (\boldsymbol{Y},\boldsymbol{Z}) \mapsto (\boldsymbol{Y}',\boldsymbol{Z}') = \Phi(\boldsymbol{Y},\boldsymbol{Z})$ from $\mathbb{H}^{2,m} \times \mathbb{H}^{2,m\times d}$ into itself. Notice that $\boldsymbol{Y}' \in \mathbb{S}^{2,m}$. The proof consists in showing that one can choose α so that the mapping Φ is a strict contraction, its unique fixed point giving the desired solution to the mean field BSDE (4.29). Let us choose $(\boldsymbol{Y}^1,\boldsymbol{Z}^1)$ and $(\boldsymbol{Y}^2,\boldsymbol{Z}^2)$ in $\mathbb{H}^{2,m} \times \mathbb{H}^{2,m\times d}$ and let us set $(\boldsymbol{Y}'^1,\boldsymbol{Z}'^1) = \Phi(\boldsymbol{Y}^1,\boldsymbol{Z}^1)$, $(\boldsymbol{Y}'^2,\boldsymbol{Z}'^2) = \Phi(\boldsymbol{Y}^2,\boldsymbol{Z}^2)$, $(\hat{\boldsymbol{Y}},\hat{\boldsymbol{Z}}) = (\boldsymbol{Y}^2 - \boldsymbol{Y}^1, \boldsymbol{Z}^2 - \boldsymbol{Z}^1)$ and $(\hat{\boldsymbol{Y}}',\hat{\boldsymbol{Z}}') = (\boldsymbol{Y}'^2 - \boldsymbol{Y}'^1, \boldsymbol{Z}'^2 - \boldsymbol{Z}'^1)$. Applying Itô's formula to $(e^{\alpha t}|\hat{Y}'_t|^2)_{0\leq t\leq T}$, we get, for any $t \in [0,T]$,

$$\begin{aligned}
&|\hat{Y}'_t|^2 + \mathbb{E}\bigg[\int_t^T \alpha e^{\alpha(r-t)}|\hat{Y}'_r|^2 dr \,\Big|\, \mathcal{F}_t\bigg] + \mathbb{E}\bigg[\int_t^T e^{\alpha(r-t)}|\hat{Z}'_r|^2 dr \,\Big|\, \mathcal{F}_t\bigg] \\
&\quad = 2\mathbb{E}\bigg[\int_t^T e^{\alpha(r-t)}\hat{Y}'_r \cdot \big[\Psi\big(r, Y'^2_r, Z'^2_r, \mathcal{L}(\xi_r, Y^2_r)\big) \\
&\qquad\qquad - \Psi\big(r, Y'^1_r, Z'^1_r, \mathcal{L}(\xi_r, Y^1_r)\big)\big]dr \,\Big|\, \mathcal{F}_t\bigg].
\end{aligned}$$

From the integrability assumption **(A1)** and the uniform Lipschitz assumption **(A2)** in **(MKV BSDE)**, we deduce that there exists a constant c, depending on L but not on α, such that:

$$\alpha\mathbb{E}\int_0^T e^{\alpha r}|\hat{Y}'_r|^2 dr + \frac{1}{2}\mathbb{E}\int_0^T e^{\alpha r}|\hat{Z}'_r|^2 dr \leq c\mathbb{E}\int_0^T e^{\alpha r}\Big(|\hat{Y}'_r|^2 + |\hat{Y}_r|^2\Big)dr,$$

which gives, for α large enough:

$$\mathbb{E}\int_0^T e^{\alpha t}(|\hat{Y}'_t|^2 + |\hat{Z}'_t|^2)dt \leq \frac{1}{2}\mathbb{E}\int_0^T e^{\alpha t}(|\hat{Y}_t|^2 + |\hat{Z}_t|^2)dt$$

or equivalently $\|(\hat{\boldsymbol{Y}}',\hat{\boldsymbol{Z}}')\|_{\mathbb{H},\alpha} \leq 2^{-1/2}\|(\hat{\boldsymbol{Y}},\hat{\boldsymbol{Z}})\|_{\mathbb{H},\alpha}$. This completes the proof. □

4.2.3 McKean-Vlasov FBSDEs in Small Time

As stated earlier, our goal is to solve fully coupled McKean-Vlasov forward-backward systems of the form:

$$\begin{cases} dX_t = B\big(t, X_t, Y_t, Z_t, \mathcal{L}(X_t, Y_t)\big)dt \\ \qquad\qquad\qquad + \Sigma\big(t, X_t, Y_t, \mathcal{L}(X_t, Y_t)\big)dW_t, \\ dY_t = -F\big(t, X_t, Y_t, Z_t, \mathcal{L}(X_t, Y_t)\big)dt + Z_t dW_t, \qquad t \in [0, T], \end{cases} \tag{4.30}$$

with initial condition $X_0 = \xi \in L^2(\Omega, \mathcal{F}_0, \mathbb{P}; \mathbb{R}^d)$, and terminal condition $Y_T = G(X_T, \mathcal{L}(X_T))$. Here, the unknown processes $(\boldsymbol{X}, \boldsymbol{Y}, \boldsymbol{Z})$ are of dimensions d, m and $m \times d$ respectively, the random coefficients B and F map $[0, T] \times \Omega \times \mathbb{R}^d \times \mathbb{R}^m \times \mathbb{R}^{m\times d} \times \mathcal{P}_2(\mathbb{R}^d \times \mathbb{R}^m)$ into $\mathbb{R}^d$ and $\mathbb{R}^m$ respectively, while the coefficient Σ maps $[0, T] \times \Omega \times \mathbb{R}^d \times \mathbb{R}^m \times \mathcal{P}_2(\mathbb{R}^d \times \mathbb{R}^m)$ into $\mathbb{R}^{d\times d}$, and the random function G giving the terminal condition maps $\Omega \times \mathbb{R}^d \times \mathcal{P}_2(\mathbb{R}^d)$ into $\mathbb{R}^m$. All these functions are assumed to be measurable. Also, the spaces $\mathcal{P}_2(\mathbb{R}^d \times \mathbb{R}^m)$ and $\mathcal{P}_2(\mathbb{R}^d)$ are assumed to be endowed with the topology of the 2-Wasserstein distance W_2.

Our experience with the classical theory of FBSDEs suggests that existence and uniqueness should hold for short time when the coefficients driving both equations are Lipschitz-continuous in the variables x, y, z and ν (or μ according to the dimension). In this subsection, we prove such a result for FBSDEs of McKean-Vlasov type. However we warn the reader that global existence over a time interval of arbitrarily prescribed length requires more restrictive assumptions and more sophisticated arguments, and we refer to Subsection 4.3.4 for a counter-example showing that Cauchy-Lipschitz theory typically fails over an interval of prescribed length.

Assumption (MKV FBSDE in Small Time).

(A1) For each $(x, y, z, \nu) \in \mathbb{R}^d \times \mathbb{R}^m \times \mathbb{R}^{m\times d} \times \mathcal{P}_2(\mathbb{R}^d \times \mathbb{R}^m)$, the processes $B(\cdot, \cdot, x, y, z, \nu) : [0, T] \times \Omega \ni (t, \omega) \mapsto B(t, \omega, x, y, z, \nu)$, $F(\cdot, \cdot, x, y, z, \nu) : [0, T] \times \Omega \ni (t, \omega) \mapsto F(t, \omega, x, y, z, \nu)$, $\Sigma(\cdot, \cdot, x, y, \nu) : [0, T] \times \Omega \ni (t, \omega) \mapsto \Sigma(t, \omega, x, y, \nu)$ are $\mathbb{F}$-progressively measurable and belong to $\mathbb{H}^{2,d}$, $\mathbb{H}^{2,m}$ and $\mathbb{H}^{2,m\times d}$ respectively. Moreover, for any $x \in \mathbb{R}^d$ and $\mu \in \mathcal{P}_2(\mathbb{R}^d)$, $G(x, \mu) \in L^2(\Omega, \mathcal{F}_T, \mathbb{P})$.

(A2) There exists a constant $L \geqslant 0$ such that for any $t \in [0, T]$, $\omega \in \Omega$, $x, x' \in \mathbb{R}^d$, $y, y' \in \mathbb{R}^m$, $z, z' \in \mathbb{R}^{m\times d}$, $\nu, \nu' \in \mathcal{P}_2(\mathbb{R}^d \times \mathbb{R}^m)$, $\mu, \mu' \in \mathcal{P}_2(\mathbb{R}^d)$, with $\mathbb{P}$-probability 1,

$$\begin{aligned} &|(B, F)(t, x, y, z, \nu) - (B, F)(t, x', y', z', \nu')| \\ &\qquad \leqslant L\big[|x - x'| + |y - y'| + |z - z'| + W_2(\nu, \nu')\big], \\ &|\Sigma(t, x, y, \nu) - \Sigma(t, x', y', \nu')| \leqslant L\big[|x - x'| + |y - y'| + W_2(\nu, \nu')\big], \\ &|G(x, \mu) - G(x', \mu')| \leqslant L\big[|x - x'| + W_2(\mu, \mu')\big], \end{aligned}$$

where we use the same notation W_2 for the 2-Wasserstein distance on $\mathcal{P}_2(\mathbb{R}^d)$ and $\mathcal{P}_2(\mathbb{R}^d \times \mathbb{R}^m)$.

Theorem 4.24 *Under assumption* **MKV FBSDE in Small Time**, *there exists a constant $c > 0$, only depending on the parameter L in the assumption, such that for $T \leqslant c$ and for any initial condition $X_0 = \xi \in L^2(\Omega, \mathcal{F}_0, \mathbb{P}; \mathbb{R}^d)$, the FBSDE* (4.30) *has a unique solution $(\boldsymbol{X}, \boldsymbol{Y}, \boldsymbol{Z}) \in \mathbb{S}^{2,d} \times \mathbb{S}^{2,m} \times \mathbb{H}^{2,m\times d}$.*

Proof. Throughout the proof, the initial condition $\xi \in L^2(\Omega, \mathcal{F}_0, \mathbb{P}; \mathbb{R}^d)$ is fixed. For an element $\boldsymbol{X} = (X_t)_{0\leqslant t\leqslant T} \in \mathbb{S}^{2,d}$, $\boldsymbol{X}$ being progressively measurable with respect to the completion of the filtration generated by ξ and $\boldsymbol{W}$, we call $(\boldsymbol{Y}, \boldsymbol{Z}) = (Y_t, Z_t)_{0\leqslant t\leqslant T}$ the solution of the BSDE:

$$dY_t = -F\big(t, X_t, Y_t, Z_t, \mathcal{L}(X_t, Y_t)\big)dt + Z_t dW_t, \quad t \in [0, T], \tag{4.31}$$

with the terminal boundary condition $Y_T = G(X_T, \mathcal{L}(X_T))$. The pair $(\boldsymbol{Y}, \boldsymbol{Z})$ is progressively measurable with respect to the completion of the filtration generated by ξ and $\boldsymbol{W}$. Its existence is guaranteed by Theorem 4.23 if we use the driver:

$$\Psi(t, \omega, y, z, \nu) = F(t, X_t(\omega), y, z, \nu)$$

and $\xi_t = X_t$. With this $(\boldsymbol{Y}, \boldsymbol{Z}) \in \mathbb{S}^{2,m} \times \mathbb{H}^{2,m\times d}$, we associate $\boldsymbol{X}' = (X'_t)_{0\leqslant t\leqslant T}$ the solution of the SDE:

$$dX'_t = B\big(t, X'_t, Y_t, Z_t, \mathcal{L}(X'_t, Y_t)\big)dt + \Sigma\big(t, X'_t, Y_t, \mathcal{L}(X'_t, Y_t)\big)dW_t, \quad t \in [0, T],$$

with $X'_0 = \xi$ as initial condition, see Theorem 4.21 and Remark 4.22. Obviously, $\boldsymbol{X}'$ is progressively measurable with respect to the completion of the filtration generated by ξ and $\boldsymbol{W}$. In this way, we created a map:

$$\Phi : \mathbb{S}^{2,d,(\xi,\boldsymbol{W})} \ni \boldsymbol{X} \mapsto \boldsymbol{X}' \in \mathbb{S}^{2,d,(\xi,\boldsymbol{W})},$$

where $\mathbb{S}^{2,d,(\xi,\boldsymbol{W})}$ denotes the collection of the processes $\boldsymbol{X} \in \mathbb{S}^{2,d}$ which are progressively measurable with respect to the completion of the filtration generated by ξ and $\boldsymbol{W}$, our goal being now to prove that Φ is a contraction when T is small enough.

Given two inputs $\boldsymbol{X}^1$ and $\boldsymbol{X}^2$ in $\mathbb{S}^{2,d,(\xi,\boldsymbol{W})}$, we denote by $(\boldsymbol{Y}^1, \boldsymbol{Z}^1)$ and $(\boldsymbol{Y}^2, \boldsymbol{Z}^2)$ the solutions of the BSDE (4.31) when $\boldsymbol{X}$ is replaced by $\boldsymbol{X}^1$ and $\boldsymbol{X}^2$ respectively. Moreover, we let $\boldsymbol{X}^{1\prime} = \Phi(\boldsymbol{X}^1)$ and $\boldsymbol{X}^{2\prime} = \Phi(\boldsymbol{X}^2)$. Then, we can find a constant $C \geqslant 1$, depending on L in **MKV FBSDE in Small Time** such that, for $T \leqslant 1$:

$$\mathbb{E}\Big[\sup_{0\leqslant t\leqslant T}|Y^1_t - Y^2_t|^2 + \int_0^T |Z^1_t - Z^2_t|^2 dt\Big] \leqslant C\mathbb{E}\Big[\sup_{0\leqslant t\leqslant T}|X^1_t - X^2_t|^2\Big],$$

and

$$\mathbb{E}\Big[\sup_{0\leqslant t\leqslant T}|X^{1\prime}_t - X^{2\prime}_t|^2\Big] \leqslant CT\mathbb{E}\Big[\sup_{0\leqslant t\leqslant T}|Y^1_t - Y^2_t|^2 + \int_0^T |Z^1_t - Z^2_t|^2 dt\Big],$$

so that, increasing the constant C if needed, we get:

$$\mathbb{E}\Big[\sup_{0\leqslant t\leqslant T}|X_t^{1\prime}-X_t^{2\prime}|^2\Big]\leqslant CT\mathbb{E}\Big[\sup_{0\leqslant t\leqslant T}|X_t^1-X_t^2|^2\Big],$$

which proves that Φ is a contraction when T is small enough. □

4.2.4 A Primer on the Notion of Master Field

As we explained in the previous section, the notion of decoupling field plays a major role in the machinery of forward-backward equations. Importantly, this notion remains meaningful in the McKean-Vlasov framework, at least when the coefficients B, Σ, F, and G are deterministic. However, the domain of the decoupling field has to be enlarged to the whole space $\mathbb{R}^d \times \mathcal{P}_2(\mathbb{R}^d)$, due to the fact that the forward component in the McKean-Vlasov system has to be regarded as a process with values in the *enlarged state space* $\mathbb{R}^d\times\mathcal{P}_2(\mathbb{R}^d)$. Throughout the book, we shall call the resulting decoupling field *the master field* of the underlying McKean-Vlasov FBSDE.

Here is a first step toward the analysis of this master field.

Lemma 4.25 *On top of assumption* **MKV FBSDE in Small Time**, *let us assume that the coefficients B, Σ, F and G are deterministic, and let us also assume that, on any probabilistic set-up $(\Omega,\mathcal{F},\mathbb{F}=(\mathcal{F}_t)_{0\leqslant t\leqslant T},\mathbb{P})$, for any $t\in[0,T]$ and $\xi\in L^2(\Omega,\mathcal{F}_t,\mathbb{P};\mathbb{R}^d)$, there exists a unique solution, denoted by $(X_s^{t,\xi},Y_s^{t,\xi},Z_s^{t,\xi})_{t\leqslant s\leqslant T}$, of (4.30) on $[t,T]$ with $X_t^{t,\xi}=\xi$ as initial condition.*

Then, for any $\mu\in\mathcal{P}_2(\mathbb{R}^d)$, there exists a measurable mapping $U(t,\cdot,\mu):\mathbb{R}^d\ni x\mapsto U(t,x,\mu)\in\mathbb{R}^m$ such that:

$$\mathbb{P}\big[Y_t^{t,\xi}=U(t,\xi,\mathcal{L}(\xi))\big]=1.$$

Furthermore:

$$\forall s\in[t,T],\quad \mathbb{P}\big[Y_s^{t,\xi}=U\big(s,X_s^{t,\xi},\mathcal{L}(X_s^{t,\xi})\big)\big]=1.$$

Remarkably, the mapping U is independent of the particular choice of the probabilistic set-up $(\Omega,\mathcal{F},\mathbb{F},\mathbb{P})$ used in its construction.

Proof. Given a probabilistic set-up $(\Omega,\mathcal{F},\mathbb{F},\mathbb{P})$, an initial time $t\in[0,T)$ and an initial condition $\xi\in L^2(\Omega,\mathcal{F}_t,\mathbb{P};\mathbb{R}^d)$, we can solve (4.30) with respect to the augmented filtration $\mathbb{F}^{*}$ generated by ξ and $(W_s-W_t)_{t\leqslant s\leqslant T}$. The resulting solution is also a solution with respect to the larger filtration $\mathbb{F}$, and by uniqueness, it coincides with the solution obtained by solving the FBSDE (4.30) with respect to $\mathbb{F}$. We deduce that $Y_t^{t,\xi}$ coincides a.s. with a $\sigma\{\xi\}$-measurable $\mathbb{R}^d$-valued random variable. In particular, there exists a measurable function $u_\xi(t,\cdot):\mathbb{R}^d\to\mathbb{R}^m$ such that $\mathbb{P}[Y_t^{t,\xi}=u_\xi(t,\xi)]=1$.

We now claim that the law of $(\xi, Y_t^{t,\xi})$ only depends upon the law of ξ (i.e., it depends on ξ through its law only). This directly follows from the version of the Yamada-Watanabe theorem for FBSDEs that we shall prove in Chapter (Vol II)-1, see Theorem (Vol II)-1.33. Since uniqueness holds pathwise, it also holds in law, so that given two initial conditions with the same law, the solutions also have the same laws. Therefore, given another $\mathbb{R}^d$-valued random vector ξ' with the same law as ξ, it holds $(\xi, u_\xi(t,\xi)) \sim (\xi', u_{\xi'}(t,\xi'))$. In particular, for any measurable function $v : \mathbb{R}^d \to \mathbb{R}^m$, the random variables $u_\xi(t,\xi) - v(\xi)$ and $u_{\xi'}(t,\xi') - v(\xi')$ have the same law. Choosing $v = u_\xi(t,\cdot)$, we deduce that $u_{\xi'}(t,\cdot)$ and $u_\xi(t,\cdot)$ are a.e. equal under the probability measure $\mathcal{L}(\xi)$. To put it differently, denoting by μ the law of ξ, there exists an element $U(t,\cdot,\mu) \in L^2(\mathbb{R}^d,\mu)$ such that $u_\xi(t,\cdot)$ and $u_{\xi'}(t,\cdot)$ coincide μ a.e. with $U(t,\cdot,\mu)$. Identifying $U(t,\cdot,\mu)$ with one of its version, this proves that:

$$\mathbb{P}\big[Y_t^{t,\xi} = U(t,\xi,\mu)\big] = 1.$$

When $t > 0$, we notice that, for any $\mu \in \mathcal{P}_2(\mathbb{R}^d)$, there exists an $\mathcal{F}_t$-measurable random variable ξ such that $\mu = \mathcal{L}(\xi)$. As a result, the procedure we just described permits to define $U(t,\cdot,\mu)$ for any $\mu \in \mathcal{P}_2(\mathbb{R}^d)$. The situation may be different when $t = 0$ as $\mathcal{F}_0$ may reduce to events of measure zero or one. In such a case, $\mathcal{F}_0$ can be enlarged without any loss of generality in order to support $\mathbb{R}^d$-valued random variables with arbitrary distributions.

The fact that U is independent of the choice of the probabilistic set-up $(\Omega,\mathcal{F},\mathbb{F},\mathbb{P})$ directly follows from the uniqueness in law property. □

Remark 4.26 *Notice that the additional variable $\mathcal{L}(\xi)$ is "for free" in the above writing since we could set $v(t,\cdot) = U(t,\cdot,\mathcal{L}(\xi))$ and then have $Y_t^{t,\xi} = v(t,\xi)$. In fact, this additional variable $\mathcal{L}(\xi)$ is specified to emphasize the non-Markovian nature of the equation over the state space $\mathbb{R}^d$: the decoupling fields are not the same if the laws of the initial conditions are different. Indeed, it is important to keep in mind that, in the Markovian framework, the decoupling field is the same for all possible initial conditions, thus yielding the connection with partial differential equations. Here the Markov property holds, but over the enlarged space $\mathbb{R}^d \times \mathcal{P}_2(\mathbb{R}^d)$, justifying the use of the extra variable $\mathcal{L}(\xi)$.*

Remark 4.27 *The notion of master field will be revisited in Subsection 5.7.2, and used in a more systematic way in Chapters (Vol II)-4 and (Vol II)-5. The main challenge will be to prove that the master field solves a partial differential equation on the enlarged state space $[0,T] \times \mathbb{R}^d \times \mathcal{P}_2(\mathbb{R}^d)$, this partial differential equation being referred to as the* master equation.

4.3 Solvability of McKean-Vlasov FBSDEs by Schauder's Theorem

The goal of this section is to provide a general existence result for McKean-Vlasov FBSDEs over an arbitrarily prescribed time interval. The result of this section was announced and appealed to in Chapter 3 in order to provide first existence results of MFG equilibria. It will be revisited in Subsection 4.5 below and in Chapter 6 in order to cover models which elude some of the assumptions made in this section. There, the FBSDEs arise from optimization problems, and by taking advantage of the assumptions specific to the applications to mean field games and control of McKean-Vlasov dynamics respectively, we shall be able to extend the coverage of the existence result of this section. These assumptions include, for example, strong convexity of the cost functions, linearity of the drift, $\cdots$ Instead we here require the diffusion matrix to be nondegenerate and the coefficients to be bounded in the space variable.

The motivation of this section comes from the short time restriction in Theorem 4.24. This restriction is not satisfactory for practical applications, hence the need for conditions under which solutions exist on an arbitrary time intervals. The non-degeneracy condition used in this section is borrowed from the theory of standard FBSDEs, and part of the proof is based upon a result of unique solvability for these equations, see Theorem 4.12.

We emphasize once more that all the regularity properties with respect to the probability measure argument μ are understood in the sense of the 2–Wasserstein's distance W_2 whose definition was given in (3.16), and whose properties will be discussed in detail in Section 5.1 of Chapter 5. We use the same notation as in Section 4.2, see for instance (4.23) and (4.24).

4.3.1 Notation, Assumptions, and Statement of the Existence Result

Our goal is to prove existence (but not necessarily uniqueness) of a solution to a fully coupled McKean-Vlasov forward-backward system of the same form as in (4.30):

$$\left\{\begin{aligned} dX_t &= B\big(t, X_t, Y_t, Z_t, \mathcal{L}(X_t, Y_t)\big)dt \\ &\qquad\qquad + \Sigma\big(t, X_t, Y_t, \mathcal{L}(X_t, Y_t)\big)dW_t, \\ dY_t &= -F\big(t, X_t, Y_t, Z_t, \mathcal{L}(X_t, Y_t)\big)dt + Z_t dW_t, \qquad t \in [0, T], \end{aligned}\right. \tag{4.32}$$

with initial condition $X_0 = \xi$ for some $\xi \in L^2(\Omega, \mathcal{F}_0, \mathbb{P}; \mathbb{R}^d)$, and terminal condition $Y_T = G(X_T, \mathcal{L}(X_T))$. As in Subsection 4.2.3, the unknown processes $\boldsymbol{X}$, $\boldsymbol{Y}$ and $\boldsymbol{Z}$ are of dimensions d, m and $m \times d$ respectively. However, the coefficients are now assumed to be deterministic. The functions B and F map $[0, T] \times \mathbb{R}^d \times \mathbb{R}^m \times \mathbb{R}^{m\times d} \times \mathcal{P}_2(\mathbb{R}^d \times \mathbb{R}^m)$ into $\mathbb{R}^d$ and $\mathbb{R}^m$ respectively, while the coefficient Σ maps $[0, T] \times \mathbb{R}^d \times \mathbb{R}^m \times \mathcal{P}_2(\mathbb{R}^d \times \mathbb{R}^m)$ into $\mathbb{R}^{d\times d}$. The function G giving the terminal condition maps $\mathbb{R}^d \times \mathcal{P}_2(\mathbb{R}^d)$ into $\mathbb{R}^m$. All these functions are assumed to be Borel-measurable. Once again, the spaces $\mathcal{P}_2(\mathbb{R}^d \times \mathbb{R}^m)$ and $\mathcal{P}_2(\mathbb{R}^d)$ are assumed to be endowed with the topology of the 2-Wasserstein distance W_2.

The reader should notice that the system (4.32) is a generalization of the system (3.25) introduced in Chapter 3. Here the McKean-Vlasov constraint involves the full-fledged distribution of the process $(X_t, Y_t)_{0\leqslant t\leqslant T}$.

The following standing assumptions extend those stated in Chapter 3.

Assumption (Nondegenerate MKV FBSDE). There exists a constant $L \geqslant 1$ such that:

(A1) For any $t \in [0,T]$, $x, x' \in \mathbb{R}^d$, $y, y' \in \mathbb{R}^m$, $z, z' \in \mathbb{R}^{m\times d}$, $\nu \in \mathcal{P}_2(\mathbb{R}^d \times \mathbb{R}^m)$ and $\mu \in \mathcal{P}_2(\mathbb{R}^d)$,

$$
\begin{aligned}
&|(B,F)(t,x',y',z',\nu) - (B,F)(t,x,y,z,\nu)| \leqslant L|(x,y,z) - (x',y',z')|,\\
&|\Sigma(t,x',y',\nu) - \Sigma(t,x,y,\nu)| \leqslant L|(x,y) - (x',y')|,\\
&|G(x',\mu) - G(x,\mu)| \leqslant L|x - x'|.
\end{aligned}
$$

Moreover, for any $(t,x,y,z) \in [0,T] \times \mathbb{R}^d \times \mathbb{R}^m \times \mathbb{R}^{m\times d}$, the coefficients $B(t,x,y,z,\cdot)$, $F(t,x,y,z,\cdot)$, $\Sigma(t,x,y,\cdot)$ and $G(x,\cdot)$ are continuous in the measure argument with respect to the 2-Wasserstein distance.

(A2) The functions Σ and G are bounded by L. Moreover, for any $t \in [0,T]$, $x \in \mathbb{R}^d$, $y \in \mathbb{R}^m$, $z \in \mathbb{R}^{m\times d}$ and $\nu \in \mathcal{P}_2(\mathbb{R}^d \times \mathbb{R}^m)$,

$$
\begin{aligned}
&|B(t,x,y,z,\nu)| \leqslant L\big[1 + |y| + |z| + M_2(\nu)\big],\\
&|F(t,x,y,z,\nu)| \leqslant L\big[1 + |y| + |z| + M_2(\nu \circ \pi^{-1})\big], \quad \text{with } \pi(x,y) = y.
\end{aligned}
$$

(A3) The function Σ is uniformly elliptic in the sense that, for any $t \in [0,T]$, $x \in \mathbb{R}^d$, $y \in \mathbb{R}^m$ and $\nu \in \mathcal{P}_2(\mathbb{R}^d \times \mathbb{R}^m)$, the following inequality holds:

$$
(\Sigma\Sigma^{\dagger})(t,x,y,\nu) \geqslant L^{-1} I_d,
$$

in the sense of symmetric matrices, where I_d is the d-dimensional identity matrix. Moreover, the function $[0,T]\times\mathbb{R}^d\times\mathbb{R}^m\times\mathcal{P}_2(\mathbb{R}^d\times\mathbb{R}^m) \ni (t,x,y,\nu) \mapsto \Sigma(t,x,y,\nu)$ is continuous.

Remark 4.28 *Recall that $M_2(\nu)$ denotes the square root of the second moment of μ, see* (3.7). *We also notice that* **(A2)** *may be rewritten as:*

$$
\begin{aligned}
&\big|B\big(t,x,y,z,\mathcal{L}(X,Y)\big)\big| \leqslant L\big[1 + |y| + |z| + \mathbb{E}[|X|^2 + |Y|^2]^{1/2}\big],\\
&\big|F\big(t,x,y,z,\mathcal{L}(X,Y)\big)\big| \leqslant L\big[1 + |y| + |z| + \mathbb{E}[|Y|^2]^{1/2}\big],
\end{aligned}
$$

for any square-integrable random variables X and Y. The fact that F is uniformly bounded with respect to $\mathbb{E}[|X|^2]^{1/2}$ will be explicitly used in the analysis below.

Recall that throughout the book, we use the superscript † to denote the transpose of a matrix. We can now state the main result of this section. Notice that it extends Theorem 3.10.

Theorem 4.29 *Under assumption* **Nondegenerate MKV FBSDE**, *for any random variable* $\xi \in L^2(\Omega, \mathcal{F}_0, \mathbb{P}; \mathbb{R}^d)$, *the FBSDE* (4.32) *has a solution* $(\boldsymbol{X}, \boldsymbol{Y}, \boldsymbol{Z}) \in \mathbb{S}^{2,d} \times \mathbb{S}^{2,m} \times \mathbb{H}^{2,d\times m}$ *with* $X_0 = \xi$ *as initial condition.*

Reminiscing about the discussion of the notion of decoupling field presented in Subsections 4.1.1 and 4.2.4, we expect Y_t and X_t to be connected by a deterministic relationship of the form $Y_t = \varphi(t, X_t)$, φ being a function from $[0,T] \times \mathbb{R}^d$ into $\mathbb{R}^m$. If that is indeed the case, the law of the pair (X_t, Y_t) is entirely determined by the law of X_t since the distribution $\mathcal{L}(X_t, Y_t)$ of (X_t, Y_t) is equal to $(I_d, \varphi(t,\cdot))(\mathcal{L}(X_t)) = (\mathcal{L}(X_t)) \circ (I_d, \varphi(t,\cdot))^{-1}$, depending upon which measure theory notation the reader is familiar with. For a probability measure μ in $\mathbb{R}^d$ and for a measurable mapping ψ from $\mathbb{R}^d$ into $\mathbb{R}^m$, we shall denote by $\psi \diamond \mu$ the image of μ under the map $(I_d, \psi) : \mathbb{R}^d \ni x \mapsto (x, \psi(x)) \in \mathbb{R}^d \times \mathbb{R}^m$, that is $\psi \diamond \mu = \mu \circ (I_d, \psi)^{-1}$. With this notation in hand, it is natural to look for a function $\varphi : [0,T] \times \mathbb{R}^d \to \mathbb{R}^m$ and a flow $[0,T] \ni t \mapsto \mu_t \in \mathcal{P}_2(\mathbb{R}^d)$ such that:

$$\begin{cases} dX_t = B\big(t, X_t, Y_t, Z_t, \varphi(t,\cdot) \diamond \mu_t\big)dt \\ \qquad\qquad + \Sigma\big(t, X_t, Y_t, \varphi(t,\cdot) \diamond \mu_t\big)dW_t, \\ dY_t = -F\big(t, X_t, Y_t, Z_t, \varphi(t,\cdot) \diamond \mu_t\big)dt + Z_t dW_t, \qquad t \in [0,T], \end{cases} \tag{4.33}$$

under the constraints that $Y_t = \varphi(t, X_t)$ and $\mu_t = \mathcal{L}(X_t)$ for $t \in [0,T]$, and with the boundary conditions $X_0 = \xi$ and $Y_T = G(X_T, \mathcal{L}(X_T))$. The strategy we use below consists in recasting the stochastic system (4.33) into a fixed point problem over the arguments $(\varphi, (\mu_t)_{0 \leqslant t \leqslant T})$. The first step is to use $\varphi(t,\cdot) \diamond \mu_t$ as an input, and solve (4.33) as a standard FBSDE. In order to do so, we should be able to use some of the known existence results for standard FBSDEs which we reviewed earlier.

Remark 4.30 *Theorem 4.29 could be extended to the more general case when, in the McKean-Vlasov argument, the joint law* $\mathcal{L}(X_t, Y_t)$ *of* X_t *and* Y_t *is replaced by the joint law* $\mathcal{L}(X_t, Y_t, Z_t)$ *of* X_t, Y_t, *and* Z_t *in* B *and* F. *Indeed, in the nondegenerate setting,* Z_t *is also given by a continuous function of* X_t *in the same way as* Y_t *is, namely* $Z_t = v(t, X_t)$ *with* $v(t,x) = \partial_x u(t,x)\Sigma(t, x, u(t,x), u(t,\cdot) \diamond \mathcal{L}(X_t))$ *whenever* $Y_t = u(t, X_t)$ *(i.e.,* $u \equiv \varphi$ *with the notations used above), see Lemmas 4.10 and 4.11. However, since the proof would require a careful analysis of the smoothing properties of the operator driving the forward component of the equation, we refrain from tackling this question here.*

Remark 4.31 *The assumption that* Σ *is independent of* $(Z_t)_{0 \leqslant t \leqslant T}$ *should not be underestimated. Indeed, if* Σ *depends upon* Z_t, *even in the classical (i.e., non-McKean-Vlasov) case, the arguments needed to prove existence are much more involved if this assumption is not satisfied. In essence, they try to recreate via specific monotonicity assumptions the role played by convexity in the analysis of the so-called* adjoint *FBSDEs arising in optimal stochastic control.*

Below, the fixed point problem is solved by means of Schauder's fixed point theorem. This provides existence of a fixed point from compactness arguments. However, it is important to keep in mind that it does not say anything about uniqueness. The way we implement Schauder's theorem is quite typical of the strategy we shall use later on to solve mean field games. In this regard, Theorem 4.29 serves as a good testbed for our technology, although so far, nothing has been said about the connection with mean field games.

We here recall the statement of Schauder's theorem for the sake of completeness.

Theorem 4.32 *Let $(V, \|\cdot\|)$ be a normed linear vector space and E be a nonempty closed convex subset of V. Then, any continuous mapping from E into itself which has a relatively compact range has a fixed point.*

Preliminary Step: Structure of the Solution for a Given Input

Our fixed point argument relies on a reformulation of the results described in Section 4.1. Lemma 4.33 below is a reformulation of Theorem 4.12, while Lemma 4.34 bears the same relationship to Lemma 4.9.

Lemma 4.33 *Fix $T \geqslant 0$ and on top of assumption* **Nondegenerate MKV FBSDE**, *assume that, instead of* **(A2)**, *B and F have the following growth property:*

$$|(B,F)(t,x,y,z,\nu)| \leqslant L\big[1+|y|+|z|\big],$$

for all $(t,x,y,z,\nu) \in [0,T] \times \mathbb{R}^d \times \mathbb{R}^m \times \mathbb{R}^{m\times d} \times \mathcal{P}_2(\mathbb{R}^d \times \mathbb{R}^m)$.

Then, given a deterministic continuous function $\boldsymbol{\nu} : [0,T] \ni t \mapsto \nu_t \in \mathcal{P}_2(\mathbb{R}^d \times \mathbb{R}^m)$, a probability $\mu \in \mathcal{P}_2(\mathbb{R}^d)$, and an initial condition $(t,\xi) \in [0,T] \times L^2(\Omega, \mathcal{F}_t, \mathbb{P}; \mathbb{R}^d)$, the forward-backward system

$$\begin{cases} dX_s = B\big(s, X_s, Y_s, Z_s, \nu_s\big)ds + \Sigma(s, X_s, Y_s, \nu_s)dW_s, \\ dY_s = -F\big(s, X_s, Y_s, Z_s, \nu_s\big)ds + Z_s dW_s, \qquad s \in [t,T], \end{cases} \tag{4.34}$$

with $X_t = \xi$ as initial condition and $Y_T = G(X_T, \mu)$ as terminal condition, has a unique solution, denoted by $(X_s^{t,\xi}, Y_s^{t,\xi}, Z_s^{t,\xi})_{t\leqslant s\leqslant T}$. Moreover, the decoupling field $u : [0,T] \times \mathbb{R}^d \ni (t,x) \mapsto u(t,x) = Y_t^{t,x} \in \mathbb{R}^m$ obtained by choosing $\xi = x$ is bounded by a constant γ depending only upon T and L, and is $1/2$-Hölder continuous in time and Lipschitz continuous in space in the sense that:

$$|u(t,x) - u(t',x')| \leqslant \Gamma\big(|t-t'|^{1/2} + |x-x'|\big),$$

for some constant Γ only depending upon T and L. In particular, both γ and Γ are independent of $\boldsymbol{\nu}$ and μ. Finally, it holds that $Y_s^{t,\xi} = u(s, X_s^{t,\xi})$ for any $t \leqslant s \leqslant T$ and $|Z_s^{t,\xi}| \leqslant \Gamma L$, $ds \otimes \mathbb{P}$ almost everywhere.

For the time being, we use this existence result in the following way. We start with a bounded continuous function φ from $[0,T] \times \mathbb{R}^d$ into $\mathbb{R}^m$, and a flow of probability measures $\boldsymbol{\mu} = (\mu_t)_{0\leqslant t\leqslant T}$ in $C([0,T];\mathcal{P}_2(\mathbb{R}^d))$, which we want to think of as the flow of marginal laws $(\mathcal{L}(X_t))_{0\leqslant t\leqslant T}$ of the solution. We apply the above existence result for (4.34) to $\mu = \mu_T$ and $\nu_t = \varphi(t,\cdot) \diamond \mu_t$ for $t \in [0,T]$ and solve:

$$\begin{cases} dX_t = B\big(t, X_t, Y_t, Z_t, \varphi(t,\cdot) \diamond \mu_t\big)dt \\ \qquad\qquad + \Sigma(t, X_t, Y_t, \varphi(t,\cdot) \diamond \mu_t)dW_t, \\ dY_t = -F\big(t, X_t, Y_t, Z_t, \varphi(t,\cdot) \diamond \mu_t\big)dt + Z_t dW_t, \qquad t \in [0,T], \end{cases} \tag{4.35}$$

with terminal condition $Y_T = G(X_T, \mu_T)$ and initial condition $X_0 = \xi$. The following estimate will be instrumental in the proof of the main result.

Lemma 4.34 *Under the same assumptions as in Lemma 4.33, there exists a positive constant C, depending on T and L only, such that for any initial conditions $\xi, \xi' \in L^\infty(\Omega, \mathcal{F}_0, \mathbb{P}; \mathbb{R}^d)$ and any inputs $(\varphi, \boldsymbol{\mu})$ and $(\varphi', \boldsymbol{\mu}')$ as above, the processes $(\boldsymbol{X}, \boldsymbol{Y}, \boldsymbol{Z})$ and $(\boldsymbol{X}', \boldsymbol{Y}', \boldsymbol{Z}')$ obtained by solving (4.35) with ξ and ξ' as respective initial conditions and $(\varphi, \boldsymbol{\mu})$ and $(\varphi', \boldsymbol{\mu}')$ as respective inputs, satisfy:*

$$\begin{aligned} &\mathbb{E}\big[\sup_{0\leqslant t\leqslant T} |X_t - X_t'|^2\big] + \mathbb{E}\big[\sup_{0\leqslant t\leqslant T} |Y_t - Y_t'|^2\big] + \mathbb{E}\int_0^T |Z_t - Z_t'|^2 dt \\ &\leqslant C\Big[\mathbb{E}[|\xi - \xi'|^2] + \mathbb{E}\int_0^T \big|(B, F, \Sigma)\big(t, X_t, Y_t, Z_t, \varphi(t,\cdot) \diamond \mu_t\big) \\ &\qquad\qquad - (B, F, \Sigma)\big(t, X_t, Y_t, Z_t, \varphi'(t,\cdot) \diamond \mu_t'\big)\big|^2 dt\Big]. \end{aligned} \tag{4.36}$$

Notice that Lemma 4.34 is the analogue of Lemma 4.9.

Remark 4.35 *We shall use the following bound:*

$$\begin{aligned} &W_2(\varphi(t,\cdot) \diamond \mu_t, \varphi'(t,\cdot) \diamond \mu_t') \\ &\leqslant C\Big[W_2\Big(\mu_t \circ \big(I_d, \varphi(t,\cdot)\big)^{-1}, \mu_t' \circ \big(I_d, \varphi(t,\cdot)\big)^{-1}\Big) \\ &\quad + \Big(\int_{\mathbb{R}^d} |(\varphi - \varphi')(t,x)|^2 d\mu_t'(x)\Big)^{1/2}\Big]. \end{aligned} \tag{4.37}$$

to estimate the integral in the right-hand side of (4.36)

We are now in a position to implement the fixed point part of the strategy touted for the construction of solutions to McKean-Vlasov FBSDEs in arbitrary time.

4.3.2 Fixed Point Argument in the Bounded Case

Some of the arguments in this subsection rely on properties of spaces of measures that are discussed in detail in Chapter 5. The reader may want to consult Section 5.1 of that chapter if he/she is not familiar with the results we take for granted.

In this subsection **we assume that the coefficients B and F are bounded by the constant L. In addition, we assume that the initial condition ξ of the forward component is also bounded in the sense that it belongs to $L^\infty(\Omega, \mathcal{F}_0, \mathbb{P}; \mathbb{R}^d)$.** For any bounded continuous function $\varphi : [0,T] \times \mathbb{R}^d \to \mathbb{R}^m$ and for any flow of probability measures $\boldsymbol{\mu} = (\mu_t)_{0 \leqslant t \leqslant T} \in \mathcal{C}([0,T]; \mathcal{P}_2(\mathbb{R}^d))$, the map $[0,T] \ni t \mapsto \varphi(t,\cdot) \diamond \mu_t \in \mathcal{P}_2(\mathbb{R}^d \times \mathbb{R}^m)$ is continuous. So by Lemma 4.33, there exists a unique triplet $(X_t, Y_t, Z_t)_{0 \leqslant t \leqslant T}$ satisfying (4.35) with $X_0 = \xi$ as initial condition. Moreover, there exists a bounded and continuous mapping u from $[0,T] \times \mathbb{R}^d$ into $\mathbb{R}^m$ such that $Y_t = u(t, X_t)$, the bound for u being denoted by γ. This maps the input $(\varphi, \boldsymbol{\mu})$ into the output $(u, (\mathcal{L}(X_t))_{0 \leqslant t \leqslant T})$ and our goal is to find a fixed point for this map. We shall take advantage of the a priori L^∞ - bound on u to restrict the choice of the functions φ to the set:

$$\mathcal{E}_1 = \{\varphi \in \mathcal{C}([0,T] \times \mathbb{R}^d; \mathbb{R}^m) : \quad \forall (t,x) \in [0,T] \times \mathbb{R}^d, \quad |\varphi(t,x)| \leqslant \gamma\}. \tag{4.38}$$

Similarly, since ξ is in $L^\infty(\Omega, \mathcal{F}_0, \mathbb{P}; \mathbb{R}^d)$ and the drift B and the volatility Σ are uniformly bounded, the fourth moment of the supremum $\sup_{0 \leqslant t \leqslant T} |X_t|$ is bounded by a constant depending only upon the bounds of ξ, B and Σ. Consequently, we shall choose the input measure $\boldsymbol{\mu}$ in the set:

$$\mathcal{E}_2 = \left\{ \boldsymbol{\mu} \in \mathcal{C}\big([0,T]; \mathcal{P}_4(\mathbb{R}^d)\big) : \sup_{0 \leqslant t \leqslant T} \int_{\mathbb{R}^d} |x|^4 \, d\mu_t(x) \leqslant \gamma' \right\}, \tag{4.39}$$

for γ' appropriately chosen in such a way that $\mathbb{E}[\sup_{0 \leqslant t \leqslant T} |X_t|^4] \leqslant \gamma'$ for any input $(\varphi, \boldsymbol{\mu})$. We then denote by $\mathcal{E}$ the Cartesian product $\mathcal{E} = \mathcal{E}_1 \times \mathcal{E}_2$. We view $\mathcal{E}$ as a subset of the product vector space $V = V_1 \times V_2$, where $V_1 = C_b([0,T] \times \mathbb{R}^d; \mathbb{R}^m)$ stands for the space of bounded continuous functions from $[0,T] \times \mathbb{R}^d$ into $\mathbb{R}^m$, and $V_2 = \mathcal{C}([0,T]; \mathcal{M}_f^1(\mathbb{R}^d))$ for the space of continuous functions from $[0,T]$ into the space $\mathcal{M}_f^1(\mathbb{R}^d)$ of finite signed measures μ on $\mathbb{R}^d$ such that $\mathbb{R}^d \ni x \mapsto |x|$ is integrable under $|\mu|$. On V_1, we use the exponentially weighted supremum-norm:

$$\|h\|_{V_1} = \sup_{(t,x) \in [0,T] \times \mathbb{R}^d} e^{-|x|} |h(t,x)|,$$

and on V_2 the supremum over a variant of the Kantorovich-Rubinstein norm:

$$\|\boldsymbol{\mu}\|_{V_2} = \sup_{t \in [0,T]} \|\mu_t\|_{\mathrm{KR}\star},$$

$$\text{with} \quad \|\mu\|_{\mathrm{KR}\star} = |\mu(\mathbb{R}^d)| + \sup \left\{ \int_{\mathbb{R}^d} \ell(x) d\mu(x); \quad \ell \in \mathrm{Lip}_1(\mathbb{R}^d), \ \ell(0) = 0 \right\}.$$

Here, $\mathrm{Lip}_1(\mathbb{R}^d)$ stands for the space of Lip-1 functions on $\mathbb{R}^d$. As we shall show in Corollary 5.4, when restricted to $\mathcal{P}_1(\mathbb{R}^d)$, the distance induced by the Kantorovich-Rubinstein norm $\|\cdot\|_{\mathrm{KR}\star}$ coincides with the 1-Wasserstein metric W_1 as already defined in (3.16) earlier in Chapter 3 and studied in detail in Chapter 5.

The fact that $\|\cdot\|_{V_2}$ is a norm on V_2 may be easily checked. It suffices to check that $\|\cdot\|_{\mathrm{KR}\star}$ is a norm on $\mathcal{M}^1_f(\mathbb{R}^d)$. While the triangular inequality and the homogeneity are easily verified, the property $\|\mu\|_{\mathrm{KR}\star} = 0 \Rightarrow \mu = 0$ may be proved as follows. If $\|\mu\|_{\mathrm{KR}\star} = 0$ then $\mu^+(\mathbb{R}^d) = \mu^-(\mathbb{R}^d)$, where μ^+ and μ^- are the positive and negative parts of μ. If $\mu^+(\mathbb{R}^d) > 0$, we may assume without any loss of generality that μ^+ and μ^- are two probability measures and $\|\mu^+ - \mu^-\|_{\mathrm{KR}\star}$ is equal to:

$$\begin{aligned}\|\mu^+ - \mu^-\|_{\mathrm{KR}\star} &= \sup\left\{ \int_{\mathbb{R}^d} \ell(x) d\big(\mu^+ - \mu^-\big)(x); \quad \ell \in \mathrm{Lip}_1(\mathbb{R}^d),\ \ell(0) = 0\right\} \\ &= \sup\left\{ \int_{\mathbb{R}^d} \ell(x) d\big(\mu^+ - \mu^-\big)(x); \quad \ell \in \mathrm{Lip}_1(\mathbb{R}^d)\right\},\end{aligned}$$

the passage from the first to the second line following from the fact that any ℓ as in the second line can be replaced by $\ell(\cdot) - \ell(0)$. By Corollary 5.4, we get $\|\mu^+ - \mu^-\|_{\mathrm{KR}\star} = W_1(\mu^+, \mu^-)$. Since W_1 is a distance, $\|\mu^+ - \mu^-\|_{\mathrm{KR}\star}$ is equal to 0, which shows that $\mu^+ = \mu^-$.

We emphasize that $\mathcal{E}_1$ is a convex closed bounded subset of V_1. Moreover, we notice that the convergence for the norm $\|\cdot\|_{V_1}$ of a sequence of functions in $\mathcal{E}_1$ is equivalent to the uniform convergence on compact subsets of $[0,T] \times \mathbb{R}^d$. Similarly, $\mathcal{E}_2$ is a convex closed bounded subset of V_2 since the space $\mathcal{P}_1(\mathbb{R}^d)$ is closed under $\|\cdot\|_{\mathrm{KR}\star}$ and since, as shown in Theorem 5.5, the convergence of probability measures for $\|\cdot\|_{\mathrm{KR}\star}$ implies weak convergence of measures, which guarantees that the mapping $\mathcal{P}_1(\mathbb{R}^d) \ni \mu \mapsto \int_{\mathbb{R}^d} |x|^4 d\mu(x) \in [0, +\infty]$ is lower semicontinuous for the distance induced by the Kantorovich-Rubinstein norm. We now claim:

Lemma 4.36 *In addition to assumption* **Nondegenerate MKV FBSDE**, *assume that B and F are bounded and that the initial condition ξ lies in $L^\infty(\Omega, \mathcal{F}_0, \mathbb{P}; \mathbb{R}^d)$. Then, the mapping $\Phi : \mathcal{E} \ni (\varphi, \boldsymbol{\mu}) \mapsto (u, (\mathcal{L}(X_t))_{0 \leq t \leq T}) \in \mathcal{E}$ defined above is continuous and has a relatively compact range.*

Proof. We first check the continuity of Φ. Given a sequence $(\varphi^n, \boldsymbol{\mu}^n)$ in $\mathcal{E}$ converging towards $(\varphi, \boldsymbol{\mu}) \in \mathcal{E}$ with respect to the product norm on $V_1 \times V_2$, and given the corresponding solutions $(\boldsymbol{X}^n, \boldsymbol{Y}^n, \boldsymbol{Z}^n)$ and $(\boldsymbol{X}, \boldsymbol{Y}, \boldsymbol{Z})$ obtained by solving (4.35) with (φ^n, μ^n) and (φ, μ) respectively, we have (compare with (4.37)): (i) for any $t \in [0,T]$, $W_2(\mu_t, \mu^n_t) \to 0$ as $n \to +\infty$ since $(\mu^n_t)_{n \geq 1}$ converges weakly towards μ_t and the moments of order 4 of the measures $(\mu^n_t)_{n \geq 1}$ are uniformly bounded by γ', see Theorem 5.5; (ii) by continuity and boundedness of φ, and by a similar argument, $W_2(\mu_t \circ (I_d, \varphi(t,\cdot))^{-1}, \mu^n_t \circ (I_d, \varphi(t,\cdot))^{-1})$ converges toward 0 as $n \to +\infty$; (iii) since the sup-norms of all the $(\varphi^n)_{n \geq 1}$ are not greater than γ, the tightness of the measures $(\mu^n_t)_{n \geq 1}$, for any $t \in [0,T]$, together with the uniform convergence of $(\varphi^n)_{n \geq 1}$ towards φ on compact sets can be used to prove that:

$$\forall t \in [0,T], \quad \lim_{n\to\infty} \int_{\mathbb{R}^d} |(\varphi - \varphi^n)(t,y)|^2 d\mu_t^n(y) = 0.$$

Therefore, by (4.37),

$$\forall t \in [0,T], \quad \lim_{n\to\infty} W_2(\varphi(t,\cdot) \diamond \mu_t, \varphi^n(t,\cdot) \diamond \mu_t^n) = 0,$$

so that, by Lebesgue's dominated convergence theorem,

$$\lim_{n\to\infty} \mathbb{E}\bigg[\int_0^T \Big|(B,F,\Sigma)\big(t,X_t,Y_t,Z_t,\varphi(t,\cdot)\diamond\mu_t\big) \\ - (B,F,\Sigma)\big(t,X_t,Y_t,Z_t,\varphi^n(t,\cdot)\diamond\mu_t^n\big)\Big|^2 dt\bigg] = 0.$$

Similarly, $W_2(\mu_T, \mu_T^n) \to 0$ as n tends to $+\infty$ and:

$$\lim_{n\to\infty} \mathbb{E}\big[\big|G(X_T,\mu_T) - G(X_T,\mu_T^n)\big|^2\big] = 0.$$

From (4.36), we obtain

$$\lim_{n\to\infty} \bigg[\mathbb{E}\Big(\sup_{0\leqslant t\leqslant T} |X_t - X_t^n|^2\Big) + \mathbb{E}\Big(\sup_{0\leqslant t\leqslant T} |Y_t - Y_t^n|^2\Big) + \mathbb{E}\int_0^T |Z_t - Z_t^n|^2 dt\bigg] = 0,$$

from which we deduce that $(\mathcal{L}(X_t^n))_{0\leqslant t\leqslant T}$ converges towards $(\mathcal{L}(X_t))_{0\leqslant t\leqslant T}$ as n tends to $+\infty$, in the Wasserstein metric W_1 uniformly in $t \in [0,T]$, and thus in the topology associated with the norm $\|\cdot\|_{V_2}$. Denoting by u^n the FBSDE decoupling field, which is a function from $[0,T]\times\mathbb{R}^d$ into $\mathbb{R}^m$ such that $Y_t^n = u^n(t,X_t^n)$, and by u the FBSDE decoupling field for which $Y_t = u(t,X_t)$, we deduce that:

$$\lim_{n\to\infty} \sup_{0\leqslant t\leqslant T} \mathbb{E}\big[|u^n(t,X_t^n) - u(t,X_t)|^2\big] = 0.$$

By Lemma 4.33, we know that all the mappings $(u^n)_{n\geqslant 1}$ are bounded and Lipschitz continuous with respect to x, uniformly with respect to n. Therefore,

$$\lim_{n\to\infty} \sup_{0\leqslant t\leqslant T} \mathbb{E}\big[|u^n(t,X_t) - u(t,X_t)|^2\big] = 0.$$

Moreover, by Arzèla-Ascoli's theorem and by Lemma 4.33 again, the sequence $(u^n)_{n\geqslant 1}$ is relatively compact for the uniform convergence on compact sets, so denoting by $\hat{u}$ the limit of a subsequence converging for the norm $\|\cdot\|_{V_1}$, we deduce that, for any $t \in [0,T]$, $\hat{u}(t,\cdot) = u(t,\cdot)$ $\mathcal{L}(X_t)$-almost surely. By Stroock and Varadhan's support theorem for diffusion processes, $\mathcal{L}(X_t)$ has a full support for any $t \in (0,T]$, so that, by continuity, $\hat{u}(t,\cdot) = u(t,\cdot)$ for any $t \in (0,T]$. By continuity of u and $\hat{u}$ on the whole $[0,T]\times\mathbb{R}^d$, equality holds at $t = 0$ as well. This shows that $(u^n)_{n\geqslant 1}$ converges towards u for $\|\cdot\|_{V_1}$ and completes the proof of the continuity of Φ.

We now prove that $\Phi(\mathcal{E})$ is relatively compact for the product norm of $V_1 \times V_2$. Given $(u,\boldsymbol{\mu}') = \Phi(\varphi,\boldsymbol{\mu})$ for some $(\varphi,\boldsymbol{\mu}) \in \mathcal{E}$, we know from Lemma 4.33 that u is bounded by

γ and $(1/2, 1)$-Hölder continuous with respect to (t, x), the Hölder constant being bounded by Γ. In particular, u remains in a compact subset of $\mathcal{C}([0, T] \times \mathbb{R}^d; \mathbb{R}^m)$ for the topology of uniform convergence on compact sets as $(\varphi, \boldsymbol{\mu})$ varies over $\mathcal{E}$. Similarly, $\boldsymbol{\mu}'$ remains in a compact set when $(\varphi, \boldsymbol{\mu})$ varies over $\mathcal{E}$. Indeed, if $(\mathcal{L}(X_t))_{0 \leqslant t \leqslant T} = \boldsymbol{\mu}'$ is associated with $(\varphi, \boldsymbol{\mu})$, the moments of the measures $(\boldsymbol{\mu}'_t)_{0 \leqslant t \leqslant T}$ can be easily controlled from the fact B and Σ are bounded by constants independent of φ and μ. Using Corollary 5.6 which will be proven in Chapter 5, this implies that all the $(\mu'_t)_{0 \leqslant t \leqslant T}$ live in a compact subset of $\mathcal{P}_1(\mathbb{R}^d)$ equipped with the distance W_1, independently of the input $(\varphi, \boldsymbol{\mu}) \in \mathcal{E}$. Moreover, it is clear that there exists a constant C, independent of the input $(\varphi, \boldsymbol{\mu}) \in \mathcal{E}$, such that:

$$W_1(\mu'_t, \mu'_s) \leqslant C|t - s|^{1/2}, \quad s, t \in [0, T],$$

which proves, by Arzelà-Ascoli theorem, that the path $([0, T] \ni t \mapsto \mu'_t)_{0 \leqslant t \leqslant T}$ lives in a compact subset of $\mathcal{C}([0, T], \mathcal{P}_1(\mathbb{R}^d))$, independently of the input $(\varphi, \boldsymbol{\mu}) \in \mathcal{E}$. □

We completed all the steps needed in the proof of the main result of this subsection.

Proposition 4.37 *In addition to assumption* **Nondegenerate MKV FBSDE,** *assume that B and F are bounded and that $\xi \in L^\infty(\Omega, \mathcal{F}_0, \mathbb{P}; \mathbb{R}^d)$. Then, equation* (4.32) *has a solution $(\boldsymbol{X}, \boldsymbol{Y}, \boldsymbol{Z}) \in \mathbb{S}^{2,d} \times \mathbb{S}^{2,m} \times \mathbb{H}^{2,m\times d}$ with $X_0 = \xi$ as initial condition.*

Proof. By Schauder's fixed point Theorem 4.32, Φ has a fixed point $(\varphi, \boldsymbol{\mu})$. As explained in our description of the strategy of the proof, solving (4.35) with this $(\varphi, \boldsymbol{\mu})$ as input, and denoting by $(X_t, Y_t, Z_t)_{0 \leqslant t \leqslant T}$ the resulting solution, by definition of a fixed point, we have $Y_t = \varphi(t, X_t)$ for any $t \in [0, T]$, a.s., and $(\mathcal{L}(X_t))_{0 \leqslant t \leqslant T} = (\mu_t)_{0 \leqslant t \leqslant T}$. In particular, $\varphi(t, \cdot) \diamond \mu_t$ coincides with $\mathcal{L}((X_t, Y_t))$. We conclude that $(X_t, Y_t, Z_t)_{0 \leqslant t \leqslant T}$ satisfies (4.32). □

4.3.3 Relaxing the Boundedness Condition

We now complete the proof of Theorem 4.29 when the coefficients only satisfy assumption **Nondegenerate MKV FBSDE**. The proof consists in approximating the initial condition ξ and the coefficients B and F by a sequence of initial conditions $(\xi^n)_{n \geqslant 1}$ and sequences of coefficients $(B^n)_{n \geqslant 1}$ and $(F^n)_{n \geqslant 1}$, such that each $(\xi^n, B^n, F^n, \Sigma, G)$, for $n \geqslant 1$, satisfies the assumptions of Proposition 4.37.

Proof. We first construct the approximating sequences. For any $n \geqslant 1$, $t \in [0, T]$, $x \in \mathbb{R}^d$, $y \in \mathbb{R}^m$, $z \in \mathbb{R}^{m\times d}$, $\nu \in \mathcal{P}_2(\mathbb{R}^d \times \mathbb{R}^m)$ and $\mu \in \mathcal{P}_2(\mathbb{R}^d)$, we set:

$$(B^n, F^n)(t, x, y, z, \nu) = (B, F)\big(t, \Pi_n^{(d)}(x), \Pi_n^{(m)}(y), \Pi_n^{(m\times d)}(z), \nu \circ (\Pi_n^{(d+m)})^{-1}\big),$$

where for any integer $k \geqslant 1$, $\Pi_n^{(k)}$ is the orthogonal projection from $\mathbb{R}^k$ onto the k-dimensional ball of radius n centered at the origin, and for any probability measure ν on $\mathbb{R}^k$, $\nu \circ (\Pi_n^{(k)})^{-1}$ denotes the push-forward of ν by $\Pi_n^{(k)}$. Finally, for each $n \geqslant 1$, we define $\xi^n = \Pi_n^{(d)}(\xi)$.

For each $n \geqslant 1$, the assumptions of Proposition 4.37 are satisfied with ξ^n instead of ξ as initial condition and $(B^n, F^n, \Sigma^n, G^n)$ instead of (B, F, Σ, G) as coefficients. We denote by $(\boldsymbol{X}^n, \boldsymbol{Y}^n, \boldsymbol{Z}^n)$ the solution of (4.32) given by Proposition 4.37 when the system (4.32) has ξ^n as initial condition and is driven by the coefficients B^n, F^n, Σ^n, and G^n. As explained in the previous subsection, the process $\boldsymbol{Y}^n$ satisfies $Y_t^n = u^n(t, X_t^n)$, for any $t \in [0, T]$, for some deterministic function u^n.

The next step of the proof is to provide a uniform bound on the decoupling fields $(u^n)_{n\geqslant 1}$. Applying Itô's formula and using the specific growth condition **(A2)** in assumption **Nondegenerate MKV FBSDE**, we get:

$$\forall t \in [0, T], \quad \mathbb{E}\big[|Y_t^n|^2\big] \leqslant C + C \int_t^T \mathbb{E}\big[|Y_s^n|^2\big] ds,$$

for some constant C depending on T and L only, and whose value may vary from line to line at our convenience. By Gronwall's inequality, we deduce that the quantity $\sup_{0\leqslant t\leqslant T} \mathbb{E}[|Y_t^n|^2]$ can be bounded in terms of T and L only. Injecting this estimate into **(A2)** shows that $(-F^n(t, X_t^n, Y_t^n, Z_t^n, \mathcal{L}(X_t^n, Y_t^n)))_{0\leqslant t\leqslant T}$ is bounded by $(C(1 + |Y_t^n| + |Z_t^n|))_{0\leqslant t\leqslant T}$, which fits the growth condition in Lemma 4.33. Moreover, repeating the Itô expansion of $(|Y_t^n|^2)_{0\leqslant t\leqslant T}$, we also have:

$$\mathbb{E}\bigg[\int_0^T |Z_s^n|^2 ds\bigg] \leqslant C. \tag{4.40}$$

Plugging the bounds for $(\sup_{0\leqslant t\leqslant T} \mathbb{E}[|Y_t^n|^2])_{n\geqslant 1}$ and for $(\mathbb{E}\int_0^T |Z_t^n|^2 dt)_{n\geqslant 1}$ into the forward equation, we obtain in a similar way:

$$\forall t \in [0, T], \quad \mathbb{E}\big[|X_t^n|^2\big] \leqslant C + C \int_0^t \mathbb{E}\big[|X_s^n|^2\big] ds,$$

which proves that:

$$\sup_{0\leqslant t\leqslant T} \mathbb{E}\big[|X_t^n|^2\big] \leqslant C.$$

The crucial fact is that C is independent of n. Injecting this new estimate into **(A2)** shows that the drift $(B^n(t, X_t^n, Y_t^n, Z_t^n, \mathcal{L}(X_t^n, Y_t^n)))_{0\leqslant t\leqslant T}$ is bounded by $(C(1 + |Y_t^n| + |Z_t^n|))_{0\leqslant t\leqslant T}$, which also fits the growth condition in Lemma 4.33.

By Lemma 4.33, we deduce that the processes $(\boldsymbol{Y}^n)_{n\geqslant 1}$ and $(\boldsymbol{Z}^n)_{n\geqslant 1}$ are uniformly bounded by a constant C that depends upon T and L.

The next step of the proof is to establish the relative compactness of the family of functions $([0, T] \ni t \mapsto \mathcal{L}(X_t^n, Y_t^n) \in \mathcal{P}_2(\mathbb{R}^d \times \mathbb{R}^m))_{n\geqslant 1}$, $\mathcal{P}_2(\mathbb{R}^d \times \mathbb{R}^m)$ being equipped with the 2-Wasserstein metric W_2. Thanks to the uniform bounds we have for $(\boldsymbol{Y}^n)_{n\geqslant 1}$ and $(\boldsymbol{Z}^n)_{n\geqslant 1}$, we see that the driver:

$$\Big(F^n\big(t, X_t^n, Y_t^n, Z_t^n, \mathcal{L}(X_t^n, Y_t^n)\big)\Big)_{0\leqslant t\leqslant T}$$

of the backward equation is bounded by C, for a possibly new value of C. In particular, using the fact that $(\boldsymbol{Z}^n)_{n\geqslant 1}$ is uniformly bounded, we get:

$$\forall s, t \in [0, T], \quad \mathbb{E}\big[|Y_t^n - Y_s^n|^2\big] \leqslant C|t - s|. \tag{4.41}$$

Similarly, using the bound **(A2)** for B^n, we easily deduce that:

$$\mathbb{E}\big[\sup_{0 \leqslant t \leqslant T} |X_t^n|^4 | \mathcal{F}_0\big]^{1/2} \leqslant C\big(1 + |\xi|^2\big), \tag{4.42}$$

and that:

$$\mathbb{E}\big[|X_t^n - X_s^n|^2\big] \leqslant C|t - s|, \tag{4.43}$$

for all $s, t \in [0, T]$. From (4.41) and (4.43), we deduce that, for all $n \geqslant 1$,

$$\forall s, t \in [0, T], \quad W_2\big(\mathcal{L}(X_t^n, Y_t^n), \mathcal{L}(X_s^n, Y_s^n)\big) \leqslant C|t - s|^{1/2}. \tag{4.44}$$

Moreover, (4.42) implies:

$$\mathbb{E}\big[\sup_{0 \leqslant t \leqslant T} |X_t^n|^2\big] \leqslant C\big(1 + \mathbb{E}[|\xi|^2]\big), \tag{4.45}$$

and using (4.42) once again, we obtain, for any event $D \in \mathcal{F}$:

$$\begin{aligned}
\forall \varepsilon > 0, \quad \mathbb{E}\big[\sup_{0 \leqslant t \leqslant T} |X_t^n|^2 \mathbf{1}_D\big] &\leqslant \mathbb{E}\Big[\mathbb{E}\big[\sup_{0 \leqslant t \leqslant T} |X_t^n|^4 | \mathcal{F}_0\big]^{1/2} \big[\mathbb{P}(D|\mathcal{F}_0)\big]^{1/2}\Big] \\
&\leqslant C\mathbb{E}\Big[\big(1 + |\xi|^2\big)\big[\mathbb{P}(D|\mathcal{F}_0)\big]^{1/2}\Big] \\
&\leqslant C\Big(\varepsilon \mathbb{E}\big[1 + |\xi|^2\big] + \frac{1}{\varepsilon}\mathbb{E}\Big[\big(1 + |\xi|^2\big)\mathbb{P}(D|\mathcal{F}_0)\Big]\Big) \\
&= C\Big(\varepsilon \mathbb{E}\big[1 + |\xi|^2\big] + \frac{1}{\varepsilon}\mathbb{E}\big[\big(1 + |\xi|^2\big)\mathbf{1}_D\big]\Big).
\end{aligned} \tag{4.46}$$

We deduce that:

$$\limsup_{\delta \searrow 0} \sup_{n \geqslant 1} \sup_{D \in \mathcal{F}: \mathbb{P}(D) \leqslant \delta} \mathbb{E}\big[\sup_{0 \leqslant t \leqslant T} |X_t^n|^2 \mathbf{1}_D\big] = 0.$$

Using (4.45), this shows that:

$$\lim_{a \to \infty} \sup_{n \geqslant 1} \mathbb{E}\big[\sup_{0 \leqslant t \leqslant T} |X_t^n|^2 \mathbf{1}_{\{\sup_{0 \leqslant t \leqslant T} |X_t^n|^2 \geqslant a\}}\big] = 0.$$

In particular, by uniform boundedness of the $(\boldsymbol{Y}^n)_{n \geqslant 1}$, this implies that:

$$\lim_{a \to \infty} \sup_{n \geqslant 1} \mathbb{E}\big[\sup_{0 \leqslant t \leqslant T} |(X_t^n, Y_t^n)|^2 \mathbf{1}_{\{\sup_{0 \leqslant t \leqslant T} |(X_t^n, Y_t^n)|^2 \geqslant a\}}\big] = 0. \tag{4.47}$$

By Corollary 5.6 in Chapter 5, we deduce that the family $((\mathcal{L}(X_t^n, Y_t^n))_{0 \leqslant t \leqslant T})_{n \geqslant 1}$ lives in a compact subset of $\mathcal{P}_2(\mathbb{R}^d \times \mathbb{R}^m)$. From (4.44) and Arzéla-Ascoli theorem, we finally obtain that the mappings $([0, T] \ni t \mapsto \mathcal{L}(X_t^n, Y_t^n) \in \mathcal{P}_2(\mathbb{R}^d \times \mathbb{R}^m))_{n \geqslant 1}$ are in a compact subset of $\mathcal{C}([0, T]; \mathcal{P}_2(\mathbb{R}^d) \times \mathbb{R}^m)$.

For the last step of the proof, we denote by $\boldsymbol{\nu} = (\nu_t)_{0\leqslant t\leqslant T}$ a limit point of $(\boldsymbol{\nu}^n = (\nu_t^n)_{0\leqslant t\leqslant T})_{n\geqslant 1}$, with $\nu_t^n = \mathcal{L}(X_t^n, Y_t^n)$, and we call $(\boldsymbol{X}, \boldsymbol{Y}, \boldsymbol{Z})$ the solution to the FBSDE (4.34) with ξ as initial condition, with $\boldsymbol{\nu}$ as input flow of measures, and with $\mu = \nu_T \circ (\mathbb{R}^d \times \mathbb{R}^d \ni (x, y) \mapsto x)^{-1}$ as input terminal measure. From Lemma 4.34 (injecting in **(A2)** the bounds we have on the moments of the solutions in order to fit the framework of Lemma 4.33), we deduce that, possibly modulo the extraction of a subsequence,

$$\lim_{n\to\infty} \mathbb{E}\Big[\sup_{0\leqslant t\leqslant T} |X_t^n - X_t|^2 + \sup_{0\leqslant t\leqslant T} |Y_t^n - Y_t|^2 + \int_0^T |Z_t^n - Z_t|^2 dt \Big] = 0.$$

Therefore, for all t $\in [0, T]$, $(\mathcal{L}(X_t^n, Y_t^n))_{n\geqslant 1}$ converges in W_2 to $\mathcal{L}(X_t, Y_t) = \nu_t$. From there, we easily conclude that $(\boldsymbol{X}, \boldsymbol{Y}, \boldsymbol{Z})$ satisfies (4.32). □

4.3.4 Uniqueness does not Always Hold: A Counter-Example

We close this discussion with a counter-example showing that uniqueness cannot hold in general under assumption **Nondegenerate MKV FBSDE** even with an additional Cauchy-Lipschitz property in the measure argument (with respect to the 2-Wasserstein distance), and even in the case $d = m = p = 1$. Indeed, let us consider the forward-backward system:

$$\begin{cases} dX_t = B(\mathbb{E}[Y_t])dt + dW_t, \\ dY_t = -F(\mathbb{E}[X_t])dt + Z_t dW_t, \qquad t \in [0, T], \\ X_0 = x_0, \quad Y_T = G(\mathbb{E}[X_T]), \end{cases} \tag{4.48}$$

where B, F and G are real valued bounded and Lipschitz-continuous functions on the real line satisfying $B(x) = F(x) = G(x) = x$ for $|x| \leqslant R$. For $T = \pi/4$ and for any $a \in \mathbb{R}$, the pair:

$$x_t = a\sin(t), \quad y_t = a\cos(t), \quad 0 \leqslant t \leqslant T = \frac{\pi}{4},$$

satisfies $\dot{x}_t = y_t$, $\dot{y}_t = -x_t$, for $t \in [0, T]$, with $y_T = x_T$ as terminal condition and $x_0 = 0$ as initial condition. Therefore, for $|a| \leqslant R$, $(a\sin(t), a\cos(t))_{0\leqslant t\leqslant T}$ is a solution of the deterministic forward-backward system:

$$\begin{cases} \dot{x}_t = B(y_t), \\ \dot{y}_t = -F(x_t), \end{cases}$$

with initial condition $x_0 = 0$ and terminal condition $y_T = G(x_T)$ over the interval $[0, T]$. For such a value of a, we now set:

$$X_t = x_t + W_t, \quad Y_t = y_t, \qquad t \in [0, T].$$

Then, $(\boldsymbol{X}, \boldsymbol{Y}, 0)$ solves:

$$\begin{cases} dX_t = B(\mathbb{E}[Y_t])dt + dW_t, \\ dY_t = -F(\mathbb{E}[X_t])dt + 0\, dW_t, \end{cases}$$

with $X_0 = 0$ and $Y_T = G(\mathbb{E}[X_T])$, proving that uniqueness fails.

Remark 4.38 *The reason for the failure of uniqueness can be explained as follows. In the classical FBSDE framework, uniqueness holds because of the smoothing effect of the diffusion operator in the spatial direction. However, in the McKean-Vlasov setting, the smoothing effect of the diffusion operator is ineffective in the direction of the measure variable.*

4.3.5 Focusing on FBSDEs arising from MFG Problems

Theorem 4.29 can be applied directly to some of the FBSDEs of the McKean-Vlasov type describing equilibria of mean field games. However, its setting is somewhat too general for what is actually needed for the solution of MFG problems, and one should be able to do better under weaker assumptions to solve for MFG equilibria.

As we already explained, FBSDEs of the McKean-Vlasov type underpinning mean field games are of the simpler form (at least for games for which the volatility is independent of the control parameter):

$$\begin{cases} dX_t = B\big(t, X_t, \mathcal{L}(X_t), Y_t, Z_t\big)dt + \Sigma\big(t, X_t, \mathcal{L}(X_t)\big)dW_t, \\ dY_t = -F\big(t, X_t, \mathcal{L}(X_t), Y_t, Z_t\big)dt + Z_t dW_t, \qquad t \in [0, T], \end{cases} \tag{4.49}$$

with a terminal condition of the form $Y_T = G(X_T, \mathcal{L}(X_T))$.

In contrast with the notation used in the previous section, we changed the order in which the arguments appear in the coefficients. Most noticeably, the measure argument now appears right after the state x argument, to conform with the notation used in Chapter 3 and earlier in this chapter when we discussed mean field games.

In comparison with (4.32), the McKean-Vlasov constraint integrates the marginal law of the sole forward process instead of the joint law of both the forward and backward components. Moreover, the volatility coefficient is independent of the process $(Y_t)_{0 \leqslant t \leqslant T}$. Clearly, these two facts should make things easier, and in fact, they do!

1. The form of the assumption **Nondegenerate MKV FBSDE** is predicated by the full-fledged coupled structure of (4.32). Parts may be relaxed when handling FBSDEs of the type (4.49). For instance, we shall drop the non-

degeneracy condition when dealing with the stochastic Pontryagin principle in Subsection 4.5.1 or when revisiting the model of flocking in Subsection 4.7.3. In both cases, we shall also relax the growth conditions on the coefficients and allow the coefficients to be unbounded in the space variable.

2. The implementation of Schauder's fixed point theorem in Subsection 4.3.2 relies on a compactness proof wherein we argue that both the mapping φ and the flow of measures $\boldsymbol{\mu}$ may be chosen in compact sets. But clearly, when dealing with FBSDEs of the simpler form (4.49), it suffices to focus on the flow of measures $\boldsymbol{\mu}$.

Based on these observations, we can revisit the proof of Theorem 4.29 and establish, under the new set of assumptions spelled out below, a version of the existence result for a system of the type (4.49). See Theorem 4.39 for the statement.

Assumption (MKV FBSDE for MFG). The coefficient Σ is independent of the variable y and the measure argument of all the coefficients is a probability measure in $\mathcal{P}_2(\mathbb{R}^d)$. Moreover, there exists a constant $L \geqslant 1$ such that:

(A1) Condition **(A1)** in assumption **Nondegenerate MKV FBSDE** holds;

(A2) The function Σ is bounded by L. Moreover, the coefficients B, F and G are of at most linear growth in x, y, z and μ, uniformly in t (linear growth in μ being understood as a bound by $1 + M_2(\mu)$, with $M_2(\mu)$ as in (3.26)).

Introducing the non-McKean-Vlasov parameterized version of (4.49):

$$\begin{cases} dX_t = B\big(t, X_t, \mu_t, Y_t, Z_t\big)dt + \Sigma\big(t, X_t, \mu_t\big)dW_t, \\ dY_t = -F\big(t, X_t, \mu_t, Y_t, Z_t\big)dt + Z_t dW_t, \qquad t \in [0, T], \end{cases} \tag{4.50}$$

with a terminal condition of the form $Y_T = G(X_T, \mu_T)$, where the parameter $\boldsymbol{\mu} = (\mu_t)_{0 \leqslant t \leqslant T}$ is in $\mathcal{C}([0, T]; \mathcal{P}_2(\mathbb{R}^d))$, we also assume:

(A3) For any $(t_0, x) \in [0, T] \times \mathbb{R}^d$, the FBSDE (4.50) over the time interval $[t_0, T]$ with $X_{t_0} = x$ as initial condition at time t_0 has a unique solution $(X_t^{t_0,x}, Y_t^{t_0,x}, Z_t^{t_0,x})_{t_0 \leqslant t \leqslant T}$.

(A4) There exists a continuous mapping $u : [0, T] \times \mathbb{R}^d \to \mathbb{R}^m$, Lipschitz continuous in x uniformly in $t \in [0, T]$, such that for any initial condition $(t_0, x) \in [0, T] \times \mathbb{R}^d$, it holds with probability 1 under $\mathbb{P}$:

$$\mathbb{P}\Big[\forall t \in [t_0, T], \quad Y_t^{t_0,x} = u(t, X_t^{t_0,x})\Big] = 1.$$

(continued)

(A5) The Lipschitz constant of u in x and the supremum norm of the function $[0,T] \ni t \mapsto u(t,0)$ may be bounded independently of $\boldsymbol{\mu} = (\mu_t)_{0 \leqslant t \leqslant T}$.

Here is the result announced earlier.

Theorem 4.39 *Under assumption* **MKV FBSDE for MFG**, *for any random variable ξ belonging to $L^2(\Omega, \mathcal{F}_0, \mathbb{P}; \mathbb{R}^d)$, the FBSDE* (4.49) *has a solution $(\boldsymbol{X}, \boldsymbol{Y}, \boldsymbol{Z}) \in \mathbb{S}^{2,d} \times \mathbb{S}^{2,m} \times \mathbb{H}^{2,d\times m}$ with $X_0 = \xi$ as initial condition.*

Proof. For $\xi \in L^2(\Omega, \mathcal{F}_0, \mathbb{P}; \mathbb{R}^d)$ and $\boldsymbol{\mu} \in \mathcal{C}([0,T]; \mathcal{P}_2(\mathbb{R}^d))$, we know from **(A3)**, **(A4)** and Proposition 4.8 that (4.50) has a unique solution $(\boldsymbol{X}^{\mu}, \boldsymbol{Y}^{\mu}, \boldsymbol{Z}^{\mu}) = (X_t^{\mu}, Y_t^{\mu}, Z_t^{\mu})_{0 \leqslant t \leqslant T}$. We thus define the mapping $\Phi : \mathcal{C}([0,T], \mathcal{P}_2(\mathbb{R}^d)) \ni \boldsymbol{\mu} \mapsto \Phi(\boldsymbol{\mu}) = (\mathcal{L}(X_t^{\mu}))_{0 \leqslant t \leqslant T}$. The goal is to apply Schauder's Theorem 4.32 in order to prove that Φ has a fixed point.

Following the argument used in Subsection 4.3.2, we apply Schauder's fixed point theorem in the space $\mathcal{C}([0,T]; \mathcal{M}_f^1(\mathbb{R}^d))$ of continuous functions $\boldsymbol{\mu} = (\mu_t)_{0 \leqslant t \leqslant T}$ from $[0,T]$ into the space of finite signed measures μ over $\mathbb{R}^d$ such that $\mathbb{R}^d \ni x \mapsto |x|$ is integrable under $|\mu|$, equipped with the supremum of the Kantorovich-Rubinstein norm:

$$\|\boldsymbol{\mu}\| = \sup_{t \in [0,T]} \|\mu_t\|_{\mathrm{KR}\star},$$

with:

$$\|\mu\|_{\mathrm{KR}\star} = |\mu(\mathbb{R}^d)| + \sup\left\{ \int_{\mathbb{R}^d} \ell(x) d\mu(x); \quad \ell \in \mathrm{Lip}_1(\mathbb{R}^d),\ \ell(0) = 0 \right\}.$$

As already mentioned, the norm $\|\cdot\|_{\mathrm{KR}\star}$ is known to coincide with the Wasserstein distance W_1 on $\mathcal{P}_1(\mathbb{R}^d)$. This fact will be proven rigorously in Chapter 5.

We prove existence by proving that there exists a closed convex subset $\mathcal{E}$ included in $\mathcal{C}([0,T]; \mathcal{P}_2(\mathbb{R}^d)) \subset \mathcal{C}([0,T]; \mathcal{M}_f^1(\mathbb{R}^d))$ which is stable under Φ, such that $\Phi(\mathcal{E})$ is relatively compact, and Φ is continuous on $\mathcal{E}$. But first, we establish a priori estimates for the solution of (4.49).

The key point is to notice that the coefficients B, Σ, F and G being Lipschitz in the variable (x,y,z) and the decoupling field satisfying **(A5)**, Lemma 4.11 implies that $|Y_t^{\mu}| \leqslant C(1 + |X_t^{\mu}|)$ and $|Z_t^{\mu}| \leqslant C$, $\mathrm{Leb}_1 \otimes \mathbb{P}$ almost everywhere. Plugging these bounds into the forward part of (4.49) and using **(A2)**, standard L^p estimates for SDEs imply that there exists a constant C, independent of $\boldsymbol{\mu}$ (but possibly depending on $\mathbb{E}[|\xi|^2]$), such that:

$$\mathbb{E}\Big[\sup_{0 \leqslant t \leqslant T} |X_t^{\mu}|^4 |\mathcal{F}_0\Big]^{1/2} \leqslant C\Big(1 + |\xi|^2\Big). \tag{4.51}$$

Following (4.46) and allowing the constant C to change from line to line, we deduce that, for any $\varepsilon > 0$ and $a \geqslant 1$,

$$\mathbb{E}\Big[\sup_{0\leqslant t\leqslant T}|X_t^{\mu}|^2\mathbf{1}_{\{\sup_{0\leqslant t\leqslant T}|X_t^{\mu}|\geqslant a\}}\Big]\leqslant C\Big(\varepsilon+\varepsilon^{-1}\mathbb{E}\big[(1+|\xi|^2)\mathbf{1}_{\{\sup_{0\leqslant t\leqslant T}|X_t^{\mu}|\geqslant a\}}\big]\Big)$$
$$\leqslant C\Big(\varepsilon+\varepsilon^{-1}\sup_{D\in\mathcal{F}:\mathbb{P}(D)\leqslant Ca^{-2}}\mathbb{E}\big[(1+|\xi|^2)\mathbf{1}_D\big]\Big),$$

where we used the fact that $\mathbb{E}[\sup_{0\leqslant t\leqslant T}|X_t^{\mu}|^2]\leqslant C$, which is implied by (4.51). Minimizing over $\varepsilon>0$, we get:

$$\mathbb{E}\Big[\sup_{0\leqslant t\leqslant T}|X_t^{\mu}|^2\mathbf{1}_{\{\sup_{0\leqslant t\leqslant T}|X_t^{\mu}|\geqslant a\}}\Big]\leqslant C\sup_{D\in\mathcal{F}:\mathbb{P}(D)\leqslant Ca^{-2}}\mathbb{E}\big[(1+|\xi|^2)\mathbf{1}_D\big]^{1/2}.$$

Now, for any $D\in\mathcal{F}$ such that $\mathbb{P}(D)\leqslant Ca^{-2}$, with $a\geqslant 1$, we have that:

$$\mathbb{E}\big[(1+|\xi|^2)\mathbf{1}_D\big]\leqslant 2Ca^{-1}+\mathbb{E}\big[|\xi|^2\mathbf{1}_{\{|\xi|\geqslant a^{1/2}\}}\big],$$

so that:

$$\mathbb{E}\Big[\sup_{0\leqslant t\leqslant T}|X_t^{\mu}|^2\mathbf{1}_{\{\sup_{0\leqslant t\leqslant T}|X_t^{\mu}|\geqslant a\}}\Big]\leqslant C\Big(a^{-1}+\mathbb{E}\big[|\xi|^2\mathbf{1}_{\{|\xi|\geqslant a^{1/2}\}}\big]\Big).$$

This prompts us to consider the restriction of Φ to the subset $\mathcal{E}$ defined as:

$$\mathcal{E}=\Big\{\mu\in\mathcal{C}([0,T];\mathcal{P}_2(\mathbb{R}^d)):$$
$$\forall a\geqslant 1,\ \sup_{0\leqslant t\leqslant T}\int_{\{|x|>a\}}|x|^2d\mu_t(x)\leqslant C\Big(a^{-1}+\mathbb{E}\big[|\xi|^2\mathbf{1}_{\{|\xi|\geqslant a^{1/2}\}}\big]\Big)\Big\}.$$

Clearly, $\mathcal{E}$ is convex and closed in $\mathcal{C}([0,T];\mathcal{M}_f^1(\mathbb{R}^d))$ equipped with $\|\cdot\|$. Also Φ maps $\mathcal{E}$ into itself.

Returning to the dynamics of X^{μ}, observe that (4.51) implies that, for any $\mu\in\mathcal{E}$ and $0\leqslant s\leqslant t\leqslant T$:

$$\mathbb{E}\big[|X_t^{\mu}-X_s^{\mu}|^2\big]\leqslant C(t-s),$$

so that:

$$W_2\big([\Phi(\mu)]_t,[\Phi(\mu)]_s\big)=W_2\big(\mathcal{L}(X_t^{\mu}),\mathcal{L}(X_s^{\mu})\big)\leqslant C(t-s)^{1/2},$$

which essentially says that the family $\{[0,T]\ni t\mapsto[\Phi(\mu)]_t;\ \mu\in\mathcal{E}\}$ is equicontinuous. We already know that there exists a compact subset $\mathcal{K}\subset\mathcal{P}_2(\mathbb{R}^d)$ such that for any $\mu\in\mathcal{E}$, $[\Phi(\mu)]_t\in\mathcal{K}$ for any $t\in[0,T]$. By the above bound and by Arzelà-Ascoli theorem, we deduce that $\Phi(\mathcal{E})$ is a relatively compact subset of $\mathcal{C}([0,T];\mathcal{P}_2(\mathbb{R}^d))$, and thus of $\mathcal{C}([0,T];\mathcal{P}_1(\mathbb{R}^d))$.

Finally, the continuity of Φ on $\mathcal{E}$ follows from the stability properties of FBSDEs with a Lipschitz decoupling field, see Lemma 4.9, and from the fact that any $\mathcal{E}$-valued sequence $(\mu^n)_{n\geqslant 1}$ converging in $\mathcal{C}([0,T];\mathcal{M}_f^1(\mathbb{R}^d))$ with respect to $\|\cdot\|$ converges in $\mathcal{C}([0,T];\mathcal{P}_2(\mathbb{R}^d))$. This latter observation is a consequence of Theorem 5.5 and of the uniform integrability property upon which the definition of $\mathcal{E}$ is based. □

Remark 4.40 *Notice that Corollary 5.6 in Chapter 5 implies that the set:*

$$\left\{\mu \in \mathcal{P}_2(\mathbb{R}^d) : \forall a \geqslant 1, \int_{\{|x|\,>\,a\}} |x|^2 d\mu(x) \leqslant C\Big(a^{-1} + \mathbb{E}\big[|\xi|^2 \mathbf{1}_{\{|\xi|>a^{1/2}\}}\big]\Big)\right\},$$

is a compact subset of $\mathcal{P}_2(\mathbb{R}^d)$. This remark will play a crucial role in the sequel. We already appealed to this type of argument in the third step of the proof given in Subsection 4.3.3.

Remark 4.41 *At this stage, it may be worth mentioning the relevant version of the Arzelà-Ascoli theorem which we use: if $\mathcal{X}$ is a compact Hausdorff space, and $\mathcal{Y}$ is a metric space, then $F \subset \mathcal{C}(\mathcal{X};\mathcal{Y})$ is compact in the compact-open topology if and only if it is equicontinuous, pointwise relatively compact and closed. Here pointwise relatively compact means that for each $x \in \mathcal{X}$, the set $F_x = \{f(x); f \in F\}$ is relatively compact in $\mathcal{Y}$.*

Remark 4.42 *The reader presumably noticed the following difference between the proofs of Theorems 4.29 and 4.39. In the proof of Theorem 4.29, we first assume that ξ is in $L^\infty(\Omega, \mathcal{F}_0, \mathbb{P}; \mathbb{R}^d)$, and then handle the general case where $\xi \in L^2(\Omega, \mathcal{F}_0, \mathbb{P}; \mathbb{R}^d)$ by an approximation argument. In contrast, we work directly with $\xi \in L^2(\Omega, \mathcal{F}_0, \mathbb{P}; \mathbb{R}^d)$ in the proof of Theorem 4.39. As a result, different prescriptions are required for the definition of the set $\mathcal{E}$ in the first step of the proof of Theorem 4.39.*

4.4 Solving MFGs from the Probabilistic Representation of the Value Function

In this section and the next, we provide two general solvability results for the MFG problem described in Subsection 3.1.2 of Chapter 3. We remind the reader of the objective: find a deterministic flow $\boldsymbol{\mu} = (\mu_t)_{0 \leqslant t \leqslant T}$ of probability measures on $\mathbb{R}^d$ such that the stochastic control problem:

$$\begin{aligned}
&\inf_{\boldsymbol{\alpha} \in \mathbb{A}} J^{\boldsymbol{\mu}}(\boldsymbol{\alpha}), \quad \text{with } J^{\boldsymbol{\mu}}(\boldsymbol{\alpha}) = \mathbb{E}\left[\int_0^T f(t, X_t^{\boldsymbol{\alpha}}, \mu_t, \alpha_t)dt + g(X_T^{\boldsymbol{\alpha}}, \mu_T)\right], \\
&\text{subject to} \\
&\begin{cases} dX_t^{\boldsymbol{\alpha}} = b(t, X_t^{\boldsymbol{\alpha}}, \mu_t, \alpha_t)dt + \sigma(t, X_t^{\boldsymbol{\alpha}}, \mu_t, \alpha_t)dW_t, \qquad t \in [0, T], \\ X_0^{\boldsymbol{\alpha}} = \xi, \end{cases}
\end{aligned} \tag{4.52}$$

has an optimally controlled process with $(\mu_t)_{0 \leqslant t \leqslant T}$ as flow of marginal distributions.

We recall that the above problem is set on a complete filtered probability space $(\Omega, \mathcal{F}, \mathbb{F} = (\mathcal{F}_t)_{0 \leqslant t \leqslant T}, \mathbb{P})$ supporting a d-dimensional $\mathbb{F}$-Wiener process $\boldsymbol{W} = (W_t)_{0 \leqslant t \leqslant T}$, and for an initial condition $\xi \in L^2(\Omega, \mathcal{F}_0, \mathbb{P}; \mathbb{R}^d)$. The set $\mathbb{A}$ denotes the collection of square-integrable and $\mathbb{F}$-progressively measurable control processes $\boldsymbol{\alpha} = (\alpha_t)_{0 \leqslant t \leqslant T}$ taking values in a convex closed subset $A \subset \mathbb{R}^k$. Moreover, the state process $\boldsymbol{X} = (X_t)_{0 \leqslant t \leqslant T}$ takes values in $\mathbb{R}^d$. The solvability results which we provide in this section and the next are derived within the two forms of FBSDE-based approaches introduced in Subsection 3.2 to characterize the solutions of an optimal stochastic control problem.

This section is specifically dedicated to the method based upon the FBSDE representation of the value function, in the spirit of the weak formulation approach discussed in Subsection 3.3.1, except for the fact that **the control problem underpinning the mean field game is formulated in the strong sense**.

Throughout the section, we assume that the volatility coefficient σ does not depend upon the control. The coefficients b and f will be regarded as measurable mappings from $[0, T] \times \mathbb{R}^d \times \mathcal{P}_2(\mathbb{R}^d) \times A$ into $\mathbb{R}^d$ and $\mathbb{R}$ respectively, the volatility coefficient σ as a measurable mapping from $[0, T] \times \mathbb{R}^d \times \mathcal{P}_2(\mathbb{R}^d)$ into $\mathbb{R}^{d \times d}$, and g as a measurable function from $\mathbb{R}^d \times \mathcal{P}_2(\mathbb{R}^d)$ into $\mathbb{R}$.

4.4.1 Assumptions and Existence Result

As suggested in Subsection 3.3.1, a first way to tackle the MFG problem is to represent the value function of the optimal stochastic control problem (4.52) computed along the optimal path as the backward component of an FBSDE. We shall do so under the following assumptions.

Assumption (MFG Solvability HJB).

(A1) The volatility σ is independent of the control parameter α.

(A2) There exists a constant $L \geqslant 0$ such that:

$$|b(t, x, \mu, \alpha)| \leqslant L\big(1 + |\alpha|\big), \quad \big|f(t, x, \mu, \alpha)\big| \leqslant L\big(1 + |\alpha|^2\big),$$

$$|(\sigma, \sigma^{-1})(t, x, \mu)| \leqslant L, \quad |g(x, \mu)| \leqslant L,$$

for all $(t, x, \mu, \alpha) \in [0, T] \times \mathbb{R}^d \times \mathcal{P}_2(\mathbb{R}^d) \times A$.

(A3) For any $t \in [0, T]$, $x \in \mathbb{R}^d$, and $\mu \in \mathcal{P}_2(\mathbb{R}^d)$, the functions $b(t, x, \mu, \cdot)$ and $f(t, x, \mu, \cdot)$ are continuously differentiable in α.

(continued)

(A4) For any $t \in [0, T]$, $\mu \in \mathcal{P}_2(\mathbb{R}^d)$ and $\alpha \in A$, the functions $b(t, \cdot, \mu, \alpha)$, $f(t, \cdot, \mu, \alpha)$, $\sigma(t, \cdot, \mu)$ and $g(\cdot, \mu)$ are L-Lipschitz continuous in x; for any $t \in [0, T]$, $x \in \mathbb{R}^d$ and $\alpha \in A$, the functions $b(t, x, \cdot, \alpha), f(t, x, \cdot, \alpha)$, $\sigma(t, x, \cdot)$ and $g(x, \cdot)$ are continuous in the measure argument with respect to the 2-Wasserstein distance.

(A5) For the same constant L and for all $(t, x, \mu, \alpha) \in [0, T] \times \mathbb{R}^d \times \mathcal{P}_2(\mathbb{R}^d) \times A$,

$$|\partial_\alpha b(t, x, \mu, \alpha)| \leqslant L, \quad |\partial_\alpha f(t, x, \mu, \alpha)| \leqslant L(1 + |\alpha|).$$

We shall also require:

(A6) Letting

$$H(t, x, \mu, y, \alpha) = b(t, x, \mu, \alpha) \cdot y + f(t, x, \mu, \alpha),$$

for all $(t, x, \mu, y, \alpha) \in [0, T] \times \mathbb{R}^d \times \mathcal{P}_2(\mathbb{R}^d) \times \mathbb{R}^d \times A$, there exists a unique minimizer $\hat{\alpha}(t, x, \mu, y) \in \text{argmin}_\alpha H(t, x, \mu, y, \alpha)$, continuous in μ and L-Lipschitz continuous in (x, y), satisfying:

$$|\hat{\alpha}(t, x, \mu, y)| \leqslant L(1 + |y|), \tag{4.53}$$

for all $(t, x, \mu, y) \in [0, T] \times \mathbb{R}^d \times \mathcal{P}_2(\mathbb{R}^d) \times \mathbb{R}^d$.

In exactly the same way as for Lemma 3.3, one proves the following sufficient condition ensuring that assumption **(A6)** holds.

Lemma 4.43 *On top of assumptions* **(A1–5)** *above, assume that* $b(t, x, \mu, \alpha)$ *has the form:*

$$b(t, x, \mu, \alpha) = b_0(t, x, \mu) + b_1(t)\alpha,$$

for a bounded function b_1*, that, for any* $t \in [0, T]$*, the function* $\partial_\alpha f(t, \cdot, \cdot, \cdot)$ *is continuous in* μ *and* L*-Lipschitz continuous in* (x, α)*, and finally that* f *satisfies the* λ*-convexity assumption:*

$$f(t, x, \mu, \alpha') - f(t, x, \mu, \alpha) - (\alpha' - \alpha) \cdot \partial_\alpha f(t, x, \mu, \alpha) \geqslant \lambda |\alpha' - \alpha|^2,$$

for all $t \in [0, T]$*,* $(x, \mu) \in \mathbb{R}^d \times \mathcal{P}_2(\mathbb{R}^d)$ *and* $(\alpha, \alpha') \in A \times A$*, for some* $\lambda > 0$*. Then* **(A6)** *in assumption* **MFG Solvability HJB** *holds true for a possibly new value of* L*.*

We now state the main result of this section. It provides a solution to the mean field game problem by solving the appropriate FBSDE associated with the stochastic control problem (4.52). Recall that in the first prong of the probabilistic approach, the variable $z\sigma^{-1}(t,x,\mu)$ is substituted for the dual variable y appearing in the coefficients b and f (and hence the Hamiltonian H) and the minimizer $\hat{\alpha}$.

Theorem 4.44 *Let assumption* **MFG Solvability HJB** *be in force. Then, for any initial condition* $\xi \in L^2(\Omega, \mathcal{F}_0, \mathbb{P}; \mathbb{R}^d)$*, the McKean-Vlasov FBSDE:*

$$\begin{cases} dX_t = b\big(t, X_t, \mathcal{L}(X_t), \hat{\alpha}\big(t, X_t, \mathcal{L}(X_t), \sigma(t, X_t, \mathcal{L}(X_t))^{-1\dagger} Z_t\big)\big)dt \\ \qquad\qquad\qquad\qquad\qquad\qquad +\sigma(t, X_t, \mathcal{L}(X_t))dW_t, \\ dY_t = -f\big(t, X_t, \mathcal{L}(X_t), \hat{\alpha}\big(t, X_t, \mathcal{L}(X_t), \sigma(t, X_t, \mathcal{L}(X_t))^{-1\dagger} Z_t\big)\big)dt \\ \qquad\qquad\qquad\qquad\qquad\qquad +Z_t \cdot dW_t, \end{cases} \tag{4.54}$$

for $t \in [0,T]$*, with* $Y_T = g(X_T, \mathcal{L}(X_T))$ *as terminal condition, is solvable.*

Moreover, the flow $(\mathcal{L}(X_t))_{0\leqslant t\leqslant T}$ *given by the marginal distributions of the forward component of any solution is an equilibrium of the MFG problem associated with the stochastic control problem* (4.52)*.*

Following Remark 4.14, the martingale part in (4.54) is denoted by $Z_t \cdot dW_t$ in order to account for the fact that the backward equation is one-dimensional. Put it differently, Z_t is regarded as a d-dimensional vector while it is regarded as a $1 \times d$-random matrix if we use the notation $Z_t dW_t$. Observe that this remark fully justifies the fact that the coefficients depend upon $\sigma(t, X_t, \mu_t)^{-1\dagger} Z_t$ instead of $Z_t \sigma(t, X_t, \mu_t)^{-1}$.

The remainder of this section is devoted to the proof of Theorem 4.44.

4.4.2 FBSDE Representation in the Strong Formulation of the Control Problem

As a preliminary step, we revisit the FBSDE representation of the value function of a stochastic control problem introduced in Chapter 3. In contrast with Subsection 3.3.1, our objective here is to provide a strong representation as opposed to the representation in weak form given there.

Given an input $\boldsymbol{\mu} = (\mu_t)_{0\leqslant t\leqslant T} \in \mathcal{C}([0,T]; \mathcal{P}_2(\mathbb{R}^d))$, the FBSDE used to represent the value function has the form:

$$\begin{cases} dX_t = b\big(t, X_t, \mu_t, \hat{\alpha}\big(t, X_t, \mu_t, \sigma(t, X_t, \mu_t)^{-1\dagger} Z_t\big)\big)dt + \sigma(t, X_t, \mu_t)dW_t, \\ dY_t = -f\big(t, X_t, \mu_t, \hat{\alpha}\big(t, X_t, \mu_t, \sigma(t, X_t, \mu_t)^{-1\dagger} Z_t\big)\big)dt + Z_t \cdot dW_t, \end{cases} \tag{4.55}$$

for $t \in [0,T]$, with the initial condition $X_0 = \xi \in L^2(\Omega, \mathcal{F}_0, \mathbb{P}; \mathbb{R}^d)$ and the terminal condition $Y_T = g(X_T, \mu_T)$.

Remark that (4.55) differs from equation (3.30) appearing in the statement of Proposition 3.11 articulating the so-called *weak formulation* of the optimal stochastic control problem. As demonstrated by the proof of Proposition 3.11, we may go (at least formally) from (3.30) to (4.55) by means of Girsanov's theorem. Part of the argument in the proof of Proposition 3.11 is precisely to check that it is indeed legal to invoke Girsanov's theorem in that context.

Here we work with (4.55) instead of (3.30) in order to avoid any Girsanov transformation, and in so doing, get a direct representation of the solution of the stochastic control problem instead of a weaker one. We give a precise statement now, and we postpone the proof to Subsection 4.4.3.

Theorem 4.45 *For the same input $\boldsymbol{\mu}$ as above and under assumption* **MFG Solvability HJB**, *the FBSDE* (4.55) *with $X_0 = \xi$ as initial condition at time* 0 *has a unique solution $(X_t^{0,\xi}, Y_t^{0,\xi}, Z_t^{0,\xi})_{0\leq t\leq T}$ with $(Z_t^{0,\xi})_{0\leq t\leq T}$ being bounded by a deterministic constant, almost everywhere for* $\mathrm{Leb}_1 \otimes \mathbb{P}$ *on* $[0,T]\times\Omega$.

Moreover, there exists a continuous mapping $u : [0,T]\times\mathbb{R}^d \to \mathbb{R}$, Lipschitz continuous in x uniformly with respect to $t \in [0,T]$ and to the input $\boldsymbol{\mu}$, such that, for any initial condition $\xi \in L^2(\Omega,\mathcal{F}_0,\mathbb{P};\mathbb{R}^d)$, the unique solution $(X_t^{0,\xi}, Y_t^{0,\xi}, Z_t^{0,\xi})_{0\leq t\leq T}$ to the FBSDE (4.55) *with $X_0 = \xi$ as initial condition at time* 0, *satisfies:*

$$\mathbb{P}\Big[\forall t \in [0,T], \quad Y_t^{0,\xi} = u\big(t, X_t^{0,\xi}\big)\Big] = 1.$$

Also, the process $(\sigma(t, X_t^{0,\xi}, \mu_t)^{-1\dagger} Z_t^{0,\xi})_{0\leq t\leq T}$ is bounded by the Lipschitz constant of u in x. Finally, the process $(X_t^{0,\xi})_{0\leq t\leq T}$ is the unique solution of the optimal control problem (4.52). *In particular,* $\mathbb{E}[u(0,\xi)] = J^{\boldsymbol{\mu}}(\hat{\boldsymbol{\alpha}})$ *for* $\hat{\boldsymbol{\alpha}} = (\hat{\alpha}(t, X_t^{0,\xi}, \mu_t, \sigma(t, X_t^{0,\xi}, \mu_t)^{-1\dagger} Z_t^{0,\xi}))_{0\leq t\leq T}$.

Remark 4.46 *Observe that the driver in the backward component of* (4.55) *is not assumed to be Lipschitz continuous, see* **(A5)** *above. This explains why the analysis of* (4.55) *requires a special treatment.*

Connection with the HJB Equation

Recall that, according to our terminology, u in the statement of Theorem 4.45 is the decoupling field of the FBSDE (4.55). As one can expect, a straightforward application of Itô's formula gives the following verification result.

Lemma 4.47 *For the same input $\boldsymbol{\mu}$ as above and under assumption* **MFG Solvability HJB**, *assume that the HJB equation:*

$$\begin{aligned}&\partial_t V(t,x) + \frac{1}{2}\mathrm{trace}\Big[\big(\sigma\sigma^\dagger\big)(t,x)\partial_{xx}^2 V(t,x)\Big]\\&\quad + H\Big(t,x,\mu_t,\partial_x V(t,x),\hat{\alpha}(t,x,\mu_t,\partial_x V(t,x))\Big) = 0, \quad (t,x)\in[0,T]\times\mathbb{R}^d,\end{aligned}$$

with $V(T,x) = g(x,\mu_T)$ *as terminal condition, has a classical solution* V*, once differentiable in time and twice in space with jointly continuous derivatives, and such that* $\partial_x V$ *and* $\partial^2_{xx} V$ *are bounded. Then, the process* $(\boldsymbol{X},\boldsymbol{Y},\boldsymbol{Z}) = (X_t,Y_t,Z_t)_{0\leqslant t\leqslant T}$ *where* $\boldsymbol{X}$ *is the solution of the SDE:*

$$dX_t = b\big(t,X_t,\mu_t,\hat{\alpha}(t,X_t,\mu_t,\partial_x V(t,X_t))\big)dt + \sigma(t,X_t,\mu_t)dW_t, \quad X_0 = \xi,$$

and:

$$Y_t = V(t,X_t), \qquad \text{and} \qquad Z_t = \sigma(t,X_t,\mu_t)^\dagger \partial_x V(t,X_t), \qquad 0 \leqslant t \leqslant T,$$

solves (4.55)*. It is the unique solution for which the process* $(Z_t)_{0\leqslant t\leqslant T}$ *is bounded. Moreover, the assumption of Proposition 4.51 below are satisfied by taking* $u = V$*.*

Observe that the representation of Z_t is fully justified by the fact that Z_t is here understood as a random vector with values in $\mathbb{R}^d$ and that, V being $\mathbb{R}$-valued, $\partial_x V$ is also regarded as an $\mathbb{R}^d$-valued function. When Z_t and $\partial_x V$ are regarded as taking values in $\mathbb{R}^{1\times d}$, Z_t takes the form $\partial_x V(t,X_t)\sigma(t,X_t,\mu_t)$.

We give an example of an application of Lemma 4.47 in Subsection 4.7.3.

Proof. We only provide a sketch of the proof. The fact that (4.55) is satisfied is a straightforward application of Itô's formula to compute $dY_t = dV(t,X_t)$ given the fact that V solves the HJB equation, and the SDE satisfied by $\boldsymbol{X}$, which is solvable under the standing assumption.

Uniqueness of the solution can be proved in two ways. First, one can observe that for solutions with a bounded $(Z_t)_{0\leqslant t\leqslant T}$, the equation may be rewritten as an equation with Lipschitz coefficients. Since we have identified V with a Lipschitz decoupling field, there must be at most one solution by Proposition 4.8. Another strategy is to expand $(V(t,X'_t))_{0\leqslant t\leqslant T}$, for any other solution $(X'_t,Y'_t,Z'_t)_{0\leqslant t\leqslant T}$, and to check that the pair process $(V(t,X'_t),\sigma(t,X'_t,\mu_t)^\dagger\partial_x V(t,X'_t))_{0\leqslant t\leqslant T}$ satisfies a BSDE with random Lipschitz coefficients that is also satisfied by $(Y'_t,Z'_t)_{0\leqslant t\leqslant T}$. By Cauchy-Lipschitz theory for BSDEs (and not FBSDEs), we get $Y'_t = V(t,X'_t)$ and $Z'_t = \sigma(t,X'_t,\mu_t))^\dagger\partial_x V(t,X'_t)$, for $t \in [0,T]$, which shows that $(X'_t)_{0\leqslant t\leqslant T}$ solves the same SDE as $(X_t)_{0\leqslant t\leqslant T}$. Uniqueness then follows. □

Remark 4.48 *We shall not discuss existence results for classical solutions of the HJB equation. We just emphasize that, whenever the coefficients (obtained after composition with* $\boldsymbol{\mu}$*) are Hölder continuous in time and satisfy the conditions required in assumption* **MFG Solvability HJB** *and the terminal condition* g *is smooth in* x*, classical solutions are known to exist. We refer to the Notes & Complements at the end of the chapter for references.*

The following lemma shows that, in the framework of Lemma 4.47, the conclusion of Theorem 4.45 can be checked by a standard verification argument.

Lemma 4.49 *Under the assumptions of Lemma 4.47, the process* $\boldsymbol{X} = (X_t)_{0\leqslant t\leqslant T}$ *identified in the statement is the unique optimal solution of the stochastic control problem* (4.52) *with* $X_0 = \xi$ *as initial condition.*

Proof. The proof consists in another application of Itô's formula. Indeed, consider the process:

$$dX_t^{\beta} = b(t, X_t^{\beta}, \mu_t, \beta_t)dt + \sigma(t, X_t^{\beta}, \mu_t)dW_t, \qquad t \in [0, T],$$

controlled by an $\mathbb{F}$-progressively measurable and square integrable control process $\boldsymbol{\beta} = (\beta_t)_{0\leqslant t\leqslant T}$ with values in A. Using Itô's formula to expand the process $(V(t, X_t^{\beta}) + \int_0^t f(s, X_s^{\beta}, \mu_s, \beta_s))_{0\leqslant t\leqslant T}$, and using the form of the HJB equation to modify the expansion, we get:

$$\begin{aligned} &d\Big(V(t, X_t^{\beta}) + \int_0^t f\big(s, X_s^{\beta}, \mu_s, \beta_s\big)ds\Big) \\ &= \partial_x V\big(t, X_t^{\beta}\big) \cdot \Big(\sigma(t, X_t^{\beta})dW_t\Big) \\ &\quad + \Big[H\Big(t, X_t^{\beta}, \mu_t, \partial_x V(t, X_t^{\beta}), \beta_t\Big) \\ &\qquad - H\Big(t, X_t^{\beta}, \mu_t, \partial_x V(t, X_t^{\beta}), \hat{\alpha}\big(t, X_t^{\beta}, \mu_t, \partial_x V(t, X_t^{\beta})\big)\Big)\Big]dt, \end{aligned}$$

for $0 \leqslant t \leqslant T$. Since $\hat{\alpha}(t, x, \mu, y)$ is the unique minimizer of $A \ni \alpha \mapsto H(t, x, \mu, y, \alpha)$, we deduce, by taking the expectation and by recalling the terminal condition $V(T, \cdot) = g(\cdot, \mu_T)$, that

$$J^{\mu}(\boldsymbol{\beta}) \geqslant \mathbb{E}\big[V(0, \xi)\big] = J^{\mu}(\hat{\boldsymbol{\alpha}}),$$

with equality if and only if $\beta_t = \hat{\alpha}(t, X_t^{\beta}, \mu_t, \partial_x V(t, X_t^{\beta}, \mu_t))$ for all $t \in [0, T]$, where, in the right-hand side above, $\hat{\boldsymbol{\alpha}} = (\hat{\alpha}(t, X_t, \mu_t, Z_t))_{0\leqslant t\leqslant T}$ denotes the control process given by the solution of the FBSDE (4.55). The result easily follows. □

Remark 4.50 *It is important to emphasize that the non-degeneracy condition required on* σ *can be relaxed. Indeed, if for instance* A *is a closed convex subset of* $\mathbb{R}^d$ *and*

$$b(t, x, \mu, \alpha) = b_1(t, x, \mu) + \sigma(t, x, \mu)\alpha,$$

with $b_1(t, x, \mu)$ *as in Lemma 4.43, then the Hamiltonian takes the form:*

$$H_{\sigma}(t, x, \mu, y, \alpha) = b_1(t, x, \mu) \cdot y + \big(\sigma^{\dagger}(t, x, \mu)y\big) \cdot \alpha + f(t, x, \mu, \alpha),$$

and its minimizer $\hat{\alpha}_{\sigma}$ *in the variable* α *is given by the formula:*

$$\hat{\alpha}_{\sigma}(t, x, \mu, y) = \hat{\alpha}_{I_d}(t, x, \mu, \sigma(t, x, \mu)^{\dagger} y), \tag{4.56}$$

where $\hat{\alpha}_{I_d}$ *is the minimizer of the Hamiltonian:*

$$H_{I_d}(t,x,\mu,y,\alpha) = b_1(t,x,\mu)\cdot y + \alpha\cdot y + f(t,x,\mu,\alpha).$$

It is important to notice that because of the assumptions on the function b_1, the minimizer $\hat{\alpha}_{I_d}$ has all the smoothness properties of the minimizers corresponding to nondegenerate volatility matrices σ since it corresponds to the case $\sigma = I_d$. Now, as explained already several times, the FBSDE relevant to the first prong of the probabilistic approach is obtained by replacing the dual variable y by $\sigma(t,x,\mu_t)^{-1\dagger}z$ in the minimizer $\hat{\alpha}$. In the present situation, this substitution gives $\hat{\alpha}_\sigma(t,X_t,\mu_t,\sigma(t,X_t,\mu_t)^{-1\dagger}Z_t)$, but the latter is in fact equal to $\hat{\alpha}_{I_d}(t,X_t,\mu_t,Z_t)$ because of the equality (4.56)*. We claim that the conclusion of Theorem 4.45 remains true for this new FBSDE.*

For a proof, the reader may try to adapt the arguments given in Chapter (Vol II)-1 to the present situation, or have a look at Subsection 4.7.3 for an illustration.

4.4.3 Proof of the Representation Theorem for the Strong Formulation

We now turn to the proof of Theorem 4.45 as we aim at proving that the FBSDE:

$$\begin{cases} dX_t = b\big(t,X_t,\mu_t,\hat{\alpha}(t,X_t,\mu_t,\sigma(t,X_t,\mu_t)^{-1\dagger}Z_t)\big)dt + \sigma(t,X_t,\mu_t)dW_t, \\ dY_t = -f\big(t,X_t,\mu_t,\hat{\alpha}(t,X_t,\mu_t,\sigma(t,X_t,\mu_t)^{-1\dagger}Z_t)\big)dt + Z_t\cdot dW_t, \end{cases} \tag{4.57}$$

for $t \in [0,T]$, with initial condition $X_0 = \xi \in L^2(\Omega,\mathcal{F}_0,\mathbb{P};\mathbb{R}^d)$ and terminal condition $Y_T = g(X_T,\mu_T)$, is uniquely solvable within the class of processes $(\boldsymbol{Y},\boldsymbol{Z})$ with $\boldsymbol{Z}$ bounded. In the process, we identify the forward component as the unique solution of the stochastic control problem (4.52).

We start with the following proposition, which provides a sufficient condition for this identification.

Proposition 4.51 *On top of assumption* **MFG Solvability HJB**, *assume that there exists $R \geqslant 0$ such that, for any initial condition $(t_0,x) \in [0,T]\times\mathbb{R}^d$, the FBSDE* (4.57)*, with $X_{t_0} = x$ as initial condition at time t_0, has a unique solution $(X_t^{t_0,x},Y_t^{t_0,x},Z_t^{t_0,x})_{t_0\leqslant t\leqslant T}$ satisfying $|\sigma(t,X_t^{t_0,x},\mu_t)^{-1\dagger}Z_t^{t_0,x}| \leqslant R$ for almost every $(t,\omega) \in [t_0,T]\times\Omega$ under* $\mathrm{Leb}_1 \otimes \mathbb{P}$*. Assume also that there exists a continuous mapping $u:[0,T]\times\mathbb{R}^d \to \mathbb{R}$, Lipschitz continuous in x uniformly in $t\in[0,T]$, such that for any $(t_0,x)\in[0,T]\times\mathbb{R}^d$:*

$$\mathbb{P}\Big[\forall t\in[t_0,T],\quad Y_t^{t_0,x} = u(t,X_t^{t_0,x})\Big] = 1.$$

Then, the FBSDE (4.57) *with $X_0 = \xi$ as initial condition at time* 0 *has a solution $(X_t,Y_t,Z_t)_{0\leqslant t\leqslant T}$. Also, $(\sigma(t,X_t,\mu_t)^{-1\dagger}Z_t)_{0\leqslant t\leqslant T}$ is bounded by the Lipschitz constant* *of u in x and* $(X_t, Y_t, Z_t)_{0 \leqslant t \leqslant T}$ *is in fact the unique solution of* (4.57) *with a bounded martingale integrand* $(Z_t)_{0 \leqslant t \leqslant T}$. *Finally,* $(X_t)_{0 \leqslant t \leqslant T}$ *is the unique optimal path of the stochastic control problem* (4.52). *In particular,* $\mathbb{E}[u(0,\xi)] = J^{\mu}(\hat{\boldsymbol{\alpha}})$, *for* $\hat{\boldsymbol{\alpha}} = (\hat{\alpha}(t, X_t, \mu_t, \sigma(t, X_t, \mu_t)^{-1\dagger} Z_t))_{0 \leqslant t \leqslant T}$.

Remark 4.52 *While the result is true in full generality, the proof provided below uses the fact that the filtration* $\mathbb{F}$ *is generated by* $\mathcal{F}_0$ *and the Wiener process* $\mathbf{W}$. *This is due to the use of BSDEs with* $\mathbb{F}$*-progressively measurable coefficients in the proof below. A proof of the result in the general case will be provided in Theorem (Vol II)-1.57 in Chapter (Vol II)-1 when we discuss FBSDEs with random coefficients. A key ingredient in that proof will be the so-called Kunita-Watanabe decomposition which we do not want to introduce at this stage.*

Proof. We divide the proof in four separate steps.

First Step. We first prove the unique solvability of (4.57) when $X_0 = \xi$. Notice that since:

$$|\sigma(t, X_t^{t_0,x}, \mu_t)^{-1\dagger} Z_t^{t_0,x}| \leqslant R,$$

for any $t \in [t_0, T]$ and any $(t_0, x) \in [0, T] \times \mathbb{R}^d$, we may replace the driver

$$\Psi(t,x,z) = f\big(t, x, \mu_t, \hat{\alpha}(t, x, \mu_t, \sigma(t, x, \mu_t)^{-1\dagger} z)\big)$$

of the backward component of the FBSDE (4.57) by $\psi(z)\Psi(t,x,z)$ for a smooth cut-off function $\psi : \mathbb{R}^d \to [0,1]$ such that $\psi(z) = 1$ when $|z| \leqslant LR$ and $\psi(z){=}0$ when $|z| \geqslant 2LR$. In this way, we may regard (4.57) as an FBSDE driven by Lipschitz continuous coefficients. By Proposition 4.8, we deduce that the FBSDE driven by $\psi(z)\Psi(t,x,z)$ has a unique solution $(\boldsymbol{X}^{0,\xi}, \boldsymbol{Y}^{0,\xi}, \boldsymbol{Z}^{0,\xi})$ with $X_0 = \xi$. By Lemma 4.11, $\boldsymbol{Z}^{0,\xi}$ is bounded by a deterministic constant. Without any loss generality, we can assume that this constant is LR itself. Therefore, $(\boldsymbol{X}^{0,\xi}, \boldsymbol{Y}^{0,\xi}, \boldsymbol{Z}^{0,\xi})$ is also a solution of (4.57). Uniqueness in the class of processes $(\boldsymbol{Y}, \boldsymbol{Z})$ with $\boldsymbol{Z}$ bounded is proved in the same way, using the same truncation argument.

Second Step. We now return to the control problem (4.52). Given another controlled path $(\boldsymbol{X}^{\boldsymbol{\beta}}, \boldsymbol{\beta})$, the control $\boldsymbol{\beta}$ being bounded by some deterministic constant, we consider, on the original probabilistic set-up and with the same cut-off function ψ as above, the BSDE:

$$\begin{aligned} dY_t^{\beta} &= -\psi(Z_t^{\beta}) f(t, X_t^{\beta}, \mu_t, \hat{\alpha}_t^{\beta}) dt \\ &\quad + Z_t^{\beta} \cdot \Big[(\sigma^{-1} b)(t, X_t^{\beta}, \mu_t, \beta_t) - (\sigma^{-1} b)(t, X_t^{\beta}, \mu_t, \hat{\alpha}_t^{\beta}) \Big] dt \\ &\quad + Z_t^{\beta} \cdot dW_t, \end{aligned} \tag{4.58}$$

with $\hat{\alpha}_t^{\beta} = \hat{\alpha}(t, X_t^{\beta}, \mu_t, \sigma(t, X_t^{\beta}, \mu_t)^{-1\dagger} Z_t^{\beta})$ and terminal condition $Y_T^{\beta} = g(X_T^{\beta}, \mu_T)$. Here, we use the notation $(\sigma^{-1}b)(t,x,\mu,\alpha)$ for $\sigma(t,x,\mu)^{-1} b(t,x,\mu,\alpha)$ despite the fact that σ^{-1} and b do not have the same arguments. Equation (4.58) is a quadratic BSDE and we can solve it using Theorem 4.15. Let $(\mathcal{E}_t^{\beta})_{0 \leqslant t \leqslant T}$ be the Doléans-Dade exponential of the stochastic integral:

$$\left(- \int_0^t \left[(\sigma^{-1}b)\big(s, X_s^{\beta}, \mu_s, \beta_s\big) - (\sigma^{-1}b)\big(s, X_s^{\beta}, \mu_s, \hat{\alpha}_s^{\beta}\big) \right] \cdot dW_s \right)_{0 \leqslant t \leqslant T}.$$

We observe that the integrand, at time t, is bounded by $C(1 + |Z_t^{\beta}|)$. Since the martingale $(\int_0^t Z_s^{\beta} \cdot dW_s)_{0 \leqslant t \leqslant T}$ is of bounded mean oscillation, so is the stochastic integral above. By Proposition 4.18, $(\mathcal{E}_t^{\beta})_{0 \leqslant t \leqslant T}$ is a true martingale and we can define the probability measure $\mathbb{P}^{\beta} = \mathcal{E}_T^{\beta} \cdot \mathbb{P}$. Under $\mathbb{P}^{\beta}$, the process:

$$\left(W_t^{\beta} = W_t + \int_0^t \left[(\sigma^{-1}b)\big(s, X_s^{\beta}, \mu_s, \beta_s\big) - (\sigma^{-1}b)\big(s, X_s^{\beta}, \mu_s, \hat{\alpha}_s^{\beta}\big) \right] ds \right)_{0 \leqslant t \leqslant T},$$

is a d-dimensional Brownian motion. We show, at the end of the proof, that $\mathbb{E}^{\mathbb{P}^{\beta}} \int_0^T |Z_t^{\beta}|^2 dt < \infty$ and that, under $\mathbb{P}^{\beta}$, $(\boldsymbol{X}^{\beta}, \boldsymbol{Y}^{\beta}, \boldsymbol{Z}^{\beta})_{0 \leqslant t \leqslant T}$ is a solution of the FBSDE (4.57) when driven by $\psi(z)\Psi(t,x,z)$ instead of $\Psi(t,x,z)$, and $\boldsymbol{W}^{\beta}$ instead of $\boldsymbol{W}$. Therefore, taking these facts for granted momentarily, we infer (by strong and thus weak uniqueness) that the law of $(\boldsymbol{X}^{\beta}, \hat{\boldsymbol{\alpha}}^{\beta})$ under $\mathbb{P}^{\beta}$ is the same as the law of the pair $(\boldsymbol{X}, \hat{\boldsymbol{\alpha}} = (\hat{\alpha}(t, X_t, \mu_t, \sigma(t, X_t, \mu_t)^{-1\dagger} Z_t))_{0 \leqslant t \leqslant T})$ under $\mathbb{P}$, which proves in particular that:

$$J^{\mu}(\hat{\boldsymbol{\alpha}}) = \mathbb{E}^{\mathbb{P}^{\beta}} \left[g(X_T^{\beta}, \mu_T) + \int_0^T f(t, X_t^{\beta}, \mu_t, \hat{\alpha}_t^{\beta}) dt \right],$$

and that $(\sigma(t, X_t^{\beta}, \mu_t)^{-1\dagger} Z_t^{\beta})_{0 \leqslant t \leqslant T}$ is bounded by R, $\mathrm{Leb}_1 \otimes \mathbb{P}^{\beta}$ almost everywhere, and thus $\mathrm{Leb}_1 \otimes \mathbb{P}$ almost everywhere. As a byproduct, by (4.58), $\psi(Z_t^{\beta})$ is equal to 1. Moreover, $\mathbb{E}^{\mathbb{P}^{\beta}}[Y_0^{\beta}]$ is equal to the above right-hand side, and thus to $J^{\mu}(\hat{\boldsymbol{\alpha}})$. Since

$$\mathbb{E}^{\mathbb{P}^{\beta}}[Y_0^{\beta}] = \mathbb{E}^{\mathbb{P}}[\mathcal{E}_T^{\beta} Y_0^{\beta}] = \mathbb{E}^{\mathbb{P}}[\mathcal{E}_0^{\beta} Y_0^{\beta}] = \mathbb{E}^{\mathbb{P}}[Y_0^{\beta}],$$

we have:

$$\begin{aligned} J^{\mu}(\hat{\boldsymbol{\alpha}}) - J^{\mu}(\boldsymbol{\beta}) &= \mathbb{E}^{\mathbb{P}}[Y_0^{\beta}] - J^{\mu}(\boldsymbol{\beta}) \\ &= \mathbb{E}^{\mathbb{P}} \bigg[\int_0^T \big[H\big(t, X_t^{\beta}, \mu_t, \sigma(t, X_t^{\beta}, \mu_t)^{-1\dagger} Z_t^{\beta}, \hat{\alpha}_t^{\beta}\big) \\ &\qquad - H\big(t, X_t^{\beta}, \mu_t, \sigma(t, X_t^{\beta}, \mu_t)^{-1\dagger} Z_t^{\beta}, \beta_t\big) \big] dt \bigg], \end{aligned} \tag{4.59}$$

so that $J^{\mu}(\hat{\boldsymbol{\alpha}}) \leqslant J^{\mu}(\boldsymbol{\beta})$.

For a generic $\boldsymbol{\beta}$ satisfying $\mathbb{E} \int_0^T |\beta_t|^2 dt < \infty$, we can apply the previous inequality with $\boldsymbol{\beta}$ replaced by $\boldsymbol{\beta}^n = (\beta_t \mathbf{1}_{|\beta_t| \leqslant n})_{0 \leqslant t \leqslant T}$. Using the continuity and growth assumptions on the coefficients, it is plain to prove that $J^{\mu}(\boldsymbol{\beta}^n)$ converges to $J^{\mu}(\boldsymbol{\beta})$ as n tends to ∞, and deduce that $\hat{\boldsymbol{\alpha}}$ is a control minimizing the cost.

Third Step. Since $\hat{\alpha}(t, x, \mu, y)$ is a strict minimizer of $H(t, x, \mu, y, \cdot)$, we have that, for any bounded control $\boldsymbol{\beta}$, $J^{\mu}(\boldsymbol{\beta}) = J^{\mu}(\hat{\boldsymbol{\alpha}})$ if and only if $\boldsymbol{\beta} = \hat{\boldsymbol{\alpha}}^{\beta}$ $\mathrm{Leb}_1 \otimes \mathbb{P}$ almost everywhere. In such a case, the second line in (4.58) vanishes and $(\boldsymbol{X}^{\beta}, \boldsymbol{Y}^{\beta}, \boldsymbol{Z}^{\beta})$ satisfies the FBSDE (4.57) under $\mathbb{P}$ and by uniqueness of solutions with a bounded martingale integrand, we conclude that $\boldsymbol{X}^{\beta} = \boldsymbol{X}$ and $\boldsymbol{\beta} = \hat{\boldsymbol{\alpha}}^{\beta} = \hat{\boldsymbol{\alpha}}$. If $\boldsymbol{\beta}$ is not bounded, we can use the same approximating sequence $(\boldsymbol{\beta}^n)_{n\geqslant 0}$ as above, and since $\boldsymbol{X}^{\beta^n}$ converges to $\boldsymbol{X}^{\beta}$ for the norm $\mathbb{E}^{\mathbb{P}}[\sup_{0\leqslant t\leqslant T}|\cdot|^2]^{1/2}$, we have from (4.59):

$$\begin{aligned}J^{\mu}(\boldsymbol{\beta}) - J^{\mu}(\hat{\boldsymbol{\alpha}}) &= \lim_{n\to\infty} J^{\mu}(\boldsymbol{\beta}^n) - J^{\mu}(\hat{\boldsymbol{\alpha}})\\ &\geqslant \mathbb{E}^{\mathbb{P}}\int_0^T \liminf_{n\to\infty}\Big[\Big(H\big(t,X_t^{\beta^n},\mu_t,\sigma(t,X_t^{\beta^n},\mu_t)^{-1\dagger}Z_t^{\beta^n},\beta_t^n\big)\\ &\qquad - H\big(t,X_t^{\beta^n},\mu_t,\sigma(t,X_t^{\beta^n},\mu_t)^{-1\dagger}Z_t^{\beta^n},\hat{\alpha}_t^{\beta^n}\big)\Big)\Big]dt.\end{aligned}$$

Again, if $\boldsymbol{\beta}$ is not bounded, we can find $R' > L(R+1)+1$ for L such that $|\hat{\alpha}(t,x,\mu,y)| \leqslant L(1+|y|)$, and such that $\mathbb{E}\int_0^T \mathbf{1}_{L(R+1)+1<|\beta_t|<R'}dt \neq 0$. Given $(t,x,\mu) \in [0,T]\times\mathbb{R}^d\times\mathcal{P}_2(\mathbb{R}^d)$, we then let:

$$\Theta(t,x,\mu) = \big\{(y,\beta)\in\mathbb{R}^d\times A:\ |y|\leqslant R,\ |\beta|\leqslant R',\ |\beta-\hat{\alpha}(t,x,\mu,y)|\geqslant 1\big\}.$$

Then,

$$\begin{aligned}&J^{\mu}(\boldsymbol{\beta}) - J^{\mu}(\hat{\boldsymbol{\alpha}})\\ &\geqslant \mathbb{E}^{\mathbb{P}}\int_0^T \liminf_{n\to\infty}\Big[\inf_{(y,\beta)\in\Theta(t,X_t^{\beta^n},\mu_t)}\Big(H\big(t,X_t^{\beta^n},\mu_t,y,\beta\big)\\ &\qquad - H\big(t,X_t^{\beta^n},\mu_t,y,\hat{\alpha}(t,X_t^{\beta^n},\mu_t,y)\big)\Big)\mathbf{1}_{L(R+1)+1\leqslant|\beta_t^n|\leqslant R'}\Big]dt.\end{aligned}$$

By continuity of $H(t,\cdot,\mu_t,\cdot,\cdot)$ and $\hat{\alpha}(t,\cdot,\mu_t,\cdot)$ and by compactness of $\Theta(t,x,\mu)$ for each (t,x,μ), it is plain to deduce that:

$$\begin{aligned}J^{\mu}(\boldsymbol{\beta}) - J^{\mu}(\hat{\boldsymbol{\alpha}}) &\geqslant \mathbb{E}^{\mathbb{P}}\int_0^T \Big[\inf_{(y,\beta)\in\Theta(t,X_t^{\beta},\mu_t)}\Big(H\big(t,X_t^{\beta},\mu_t,y,\beta\big)\\ &\qquad - H\big(t,X_t^{\beta},\mu_t,y,\hat{\alpha}(t,X_t^{\beta},\mu_t,y)\big)\Big)\mathbf{1}_{L(R+1)+1<|\beta_t|<R'}\Big]dt,\end{aligned}$$

which cannot be zero by definition of $\Theta(t,X_t^{\beta},\mu_t)$. This proves that $\boldsymbol{X}$ is the unique minimizing path.

Fourth Step. In order to complete the proof, it remains to check that, for $\boldsymbol{\beta}$ bounded and under $\mathbb{P}^{\beta}$, $(\boldsymbol{X}^{\beta},\boldsymbol{Y}^{\beta},\boldsymbol{Z}^{\beta})$ is a solution of the FBSDE (4.57) when driven by $\psi(z)\Psi(t,x,z)$ instead of $\Psi(t,x,z)$ and by $\boldsymbol{W}^{\beta}$ instead of $\boldsymbol{W}$. We know that, for $t\in[0,T]$,

$$\begin{aligned}dX_t^{\beta} &= b\big(t,X_t^{\beta},\mu_t,\beta_t\big)dt + \sigma\big(t,X_t^{\beta},\mu_t\big)dW_t\\ &= b\big(t,X_t^{\beta},\mu_t,\hat{\alpha}_t^{\beta}\big)dt + \sigma\big(t,X_t^{\beta},\mu_t\big)dW_t^{\beta},\end{aligned}$$

and

$$\begin{aligned} dY_t^\beta &= -\psi(Z_t^\beta) f(t, X_t^\beta, \mu_t, \hat{\alpha}_t^\beta) dt + Z_t^\beta \cdot dW_t^\beta \\ &= -\psi(Z_t^\beta) \Psi(t, X_t^\beta, Z_t^\beta) dt + Z_t^\beta \cdot dW_t^\beta, \end{aligned}$$

with the terminal condition $Y_T^\beta = g(X_T^\beta, \mu_T)$. This shows that the equations in the system (4.57) hold true under $\mathbb{P}^\beta$ with $\boldsymbol{W}$ replaced by $\boldsymbol{W}^\beta$ and with $\psi(z)\Psi(t,x,z)$ as driver in the backward equation. To prove that $(\boldsymbol{X}^\beta, \boldsymbol{Y}^\beta, \boldsymbol{Z}^\beta)$ is indeed a solution of (4.57), it remains to check the appropriate integrability conditions.

By Proposition 4.18, we know that, for any $p \geqslant 1$,

$$\mathbb{E}^{\mathbb{P}}\bigg[\bigg(\int_0^T |Z_t^\beta|^2 ds\bigg)^p\bigg] < \infty.$$

Since $\mathcal{E}_T^\beta$ is in $L^r(\Omega, \mathcal{F}_T, \mathbb{P}; \mathbb{R})$ for some $r > 1$, the above is also true under $\mathbb{P}^\beta$.

By (4.57), we also have $\mathbb{E}^{\mathbb{P}}[\sup_{0 \leqslant t \leqslant T} |Y_t^\beta|^p] < \infty$, and then $\mathbb{E}^{\mathbb{P}^\beta}[\sup_{0 \leqslant t \leqslant T} |Y_t^\beta|^p] < \infty$. Since $\boldsymbol{\beta}$ is bounded, the same holds with $\boldsymbol{X}^\beta$ instead of $\boldsymbol{Y}^\beta$. The proof is easily completed. □

End of the Proof of Theorem 4.45

We now complete the proof of Theorem 4.45.

Proof. The objective is to prove that, for a given deterministic initial condition, the FBSDE (4.57) has a solution with a bounded martingale integrand, and that this solution is unique within the class of solutions with bounded martingale integrands. Meanwhile, we must also construct a decoupling field. As before, we split the proof into successive steps.

First Step. Following the proof of Proposition 4.51, we first focus on a truncated version of (4.57), namely:

$$\begin{cases} dX_t = b\big(t, X_t, \mu_t, \hat{\alpha}\big(t, X_t, \mu_t, \sigma(t, X_t, \mu_t)^{-1\dagger} Z_t\big)\big) dt + \sigma(t, X_t, \mu_t) dW_t, \\ dY_t = -\psi(Z_t) f\big(t, X_t, \mu_t, \hat{\alpha}\big(t, X_t, \mu_t, \sigma(t, X_t, \mu_t)^{-1\dagger} Z_t\big)\big) dt + Z_t \cdot dW_t, \end{cases} \tag{4.60}$$

for $t \in [0, T]$, with the terminal condition $Y_T = g(X_T, \mu_T)$, for a cut-off function $\psi : \mathbb{R}^d \to [0, 1]$, equal to 1 on the ball of center 0 and radius R, and equal to 0 outside the ball of center 0 and radius $2R$, such that $\sup |\psi'| \leqslant 2/R$. For the time being, $R > 0$ is an arbitrary real number. Its value will be fixed later on.

By Theorem 4.12, we know that, for any initial condition $(t_0, x) \in [0, T] \times \mathbb{R}^d$, (4.60) is uniquely solvable. We denote the unique solution by $(\boldsymbol{X}^{R;t_0,x}, \boldsymbol{Y}^{R;t_0,x}, \boldsymbol{Z}^{R;t_0,x}) = (X_t^{R;t_0,x}, Y_t^{R;t_0,x}, Z_t^{R;t_0,x})_{t_0 \leqslant t \leqslant T}$. Thanks to the cut-off function ψ, the driver of (4.60) is indeed Lipschitz-continuous in the variable z. Moreover, the solution can be represented through a continuous decoupling field u^R, Lipschitz continuous in the variable x, uniformly in time. Also, the martingale integrand $\boldsymbol{Z}^{R;t_0,x}$ is bounded by L times the Lipschitz constant of u^R, with L as in assumption **MFG Solvability HJB**. See also Lemma 4.11. Therefore, the proof boils down to showing that we can bound the Lipschitz constant of the decoupling field independently of the cut-off function ψ in (4.60).

Second Step. In this step, we fix the values of $(t_0, x) \in [0, T] \times \mathbb{R}^d$ and $R > 0$, and we use the notation $(\boldsymbol{X}, \boldsymbol{Y}, \boldsymbol{Z})$ for $(\boldsymbol{X}^{R;t_0,x}, \boldsymbol{Y}^{R;t_0,x}, \boldsymbol{Z}^{R;t_0,x})$. We then let $(\mathcal{E}_t)_{0 \leqslant t \leqslant T}$ be the Doléans-Dade exponential of the stochastic integral:

$$\left(- \int_0^t \left[(\sigma^{-1} b)(s, X_s, \mu_s, \hat{\alpha}_s) \right] \cdot dW_s \right)_{0 \leqslant t \leqslant T},$$

where $\hat{\alpha}_s = \hat{\alpha}(s, X_s, \mu_s, \sigma(s, X_s, \mu_s)^{-1\dagger} Z_s)$. As earlier, we write $(\sigma^{-1} b)(t, x, \mu, \alpha)$ for $\sigma(t, x, \mu)^{-1} b(t, x, \mu, \alpha)$ despite the fact that σ^{-1} and b do not have the same arguments. Since the integrand is bounded, $(\mathcal{E}_t)_{0 \leqslant t \leqslant T}$ is a true martingale, and we can define the probability measure $\mathbb{Q} = \mathcal{E}_T \cdot \mathbb{P}$. Under $\mathbb{Q}$, the process:

$$\left(W_t^{\mathbb{Q}} = W_t + \int_0^t (\sigma^{-1} b)(s, X_s, \mu_s, \hat{\alpha}_s) ds \right)_{0 \leqslant t \leqslant T}$$

is a d-dimensional Brownian motion. Following the proof of Proposition 4.51, we learn that under $\mathbb{Q}$, $(X_t, Y_t, Z_t)_{0 \leqslant t \leqslant T}$ is a solution of the forward-backward SDE:

$$\begin{cases} dX_t = \sigma(t, X_t, \mu_t) dW_t^{\mathbb{Q}}, \\ dY_t = -\psi(Z_t) f\big(t, X_t, \mu_t, \hat{\alpha}\big(t, X_t, \mu_t, \sigma(t, X_t, \mu_t)^{-1\dagger} Z_t\big)\big) dt \\ \qquad -Z_t \cdot (\sigma^{-1} b)\big(t, X_t, \mu_t, \hat{\alpha}\big(t, X_t, \mu_t, \sigma(t, X_t, \mu_t)^{-1\dagger} Z_t\big)\big) dt \\ \qquad +Z_t \cdot dW_t^{\mathbb{Q}}, \end{cases} \tag{4.61}$$

over the interval $[t_0, T]$, with the same terminal condition as before. Since $\boldsymbol{Z}$ is bounded, the forward-backward SDE (4.61) may be regarded as an FBSDE with Lipschitz-continuous coefficients. By the FBSDE version of Yamada-Watanabe theorem proven in Theorem (Vol II)-1.33 of Chapter (Vol II)-1, any other solution with a bounded martingale integrand, with the same initial condition but constructed with respect to another Brownian motion, has the same distribution. Therefore, we can focus on the version of (4.61) obtained by replacing $\boldsymbol{W}^{\mathbb{Q}}$ by $\boldsymbol{W}$. If, for this version, the backward component $\boldsymbol{Y}$ can be represented in the form $Y_t = V(t, X_t)$, for all $t \in [t_0, T]$, with V being Lipschitz continuous in space, uniformly in time, and with $\boldsymbol{Z}$ bounded, then $V(t_0, x)$ must coincide with $u^R(t_0, x)$. Repeating the argument for any $(t_0, x) \in [0, T] \times \mathbb{R}^d$, we then have $V \equiv u^R$.

Third Step. The strategy is now as follows. We consider the same FBSDE as in (4.61), but with $\boldsymbol{W}^{\mathbb{Q}}$ replaced by the original $\boldsymbol{W}$:

$$\begin{cases} dX_t = \sigma(t, X_t, \mu_t) dW_t, \\ dY_t = -\psi(Z_t) f\big(t, X_t, \mu_t, \hat{\alpha}\big(t, X_t, \mu_t, \sigma(t, X_t, \mu_t)^{-1\dagger} Z_t\big)\big) dt \\ \qquad -Z_t \cdot (\sigma^{-1} b)\big(t, X_t, \mu_t, \hat{\alpha}\big(t, X_t, \mu_t, \sigma(t, X_t, \mu_t)^{-1\dagger} Z_t\big)\big) dt \\ \qquad +Z_t \cdot dW_t, \qquad t \in [0, T], \end{cases}$$

with $Y_T = g(X_T, \mu_T)$. This BSDE may be regarded as a quadratic BSDE. In particular, Theorem 4.15 applies and guarantees that it is uniquely solvable. However, since the driver in the backward equation is not Lipschitz continuous, we cannot make use of the tools developed in Subsection 4.1.2 for FBSDEs with Lipschitz continuous coefficients. To bypass this obstacle, we shall modify the form of the equation and focus on the following version:

$$\begin{cases} dX_t = \sigma(t, X_t, \mu_t)dW_t, \\ dY_t = -\psi(Z_t)f\big(t, X_t, \mu_t, \hat{\alpha}\big(t, X_t, \mu_t, \sigma(t, X_t, \mu_t)^{-1\dagger}Z_t\big)\big)dt \\ \qquad -\psi(Z_t)Z_t \cdot (\sigma^{-1}b)\big(t, X_t, \mu_t, \hat{\alpha}\big(t, X_t, \mu_t, \sigma(t, X_t, \mu_t)^{-1\dagger}Z_t\big)\big)dt \\ \qquad +Z_t \cdot dW_t, \qquad t \in [0, T]. \end{cases} \tag{4.62}$$

Notice that the cut-off function ψ now appears on the third line. Our objective being to prove that (4.62) admits a solution for which $\boldsymbol{Z}$ is bounded independently of R, when R is large, the presence of the cut-off does not make any difference.

Now, (4.62) may be regarded as both a quadratic and a Lipschitz FBSDE. For any initial condition (t_0, x), we may again denote the solution by $(\boldsymbol{X}^{R;t_0,x}, \boldsymbol{Y}^{R;t_0,x}, \boldsymbol{Z}^{R;t_0,x})$. This is the same notation as in the first step although the equation is different. Since the steps are completely separated, there is no risk of confusion. We denote the corresponding decoupling field by V^R. By Theorem 4.12, it is bounded (the bound possibly depending on R at this stage of the proof) and $\boldsymbol{Z}^{R;t_0,x}$ is bounded.

For the sake of simplicity, we assume that $t_0 = 0$ and we drop the indices R and t_0 in the notation $(\boldsymbol{X}^{R;t_0,x}, \boldsymbol{Y}^{R;t_0,x}, \boldsymbol{Z}^{R;t_0,x})$. We just denote it by $(\boldsymbol{X}^x, \boldsymbol{Y}^x, \boldsymbol{Z}^x)$. Similarly, we just denote V^R by V.

The goal is then to prove that there exists a constant C, independent of R and of the cut-off functions, such that, for all $x, x' \in \mathbb{R}^d$,

$$\big|\mathbb{E}\big[Y_0^{x'} - Y_0^x\big]\big| \leqslant C|x' - x|, \tag{4.63}$$

from which we will deduce that, for all $x, x' \in \mathbb{R}^d$,

$$|V(0, x') - V(0, x)| \leqslant C|x' - x|,$$

which is exactly the Lipschitz control we need on the decoupling field.

Fourth Step. We now proceed with the proof of (4.63). Fixing the values of x and x' and letting

$$\big(\delta X_t, \delta Y_t, \delta Z_t\big) = \big(X_t^{x'} - X_t^x, Y_t^{x'} - Y_t^x, Z_t^{x'} - Z_t^x\big), \quad t \in [0, T],$$

we can write:

$$d\delta X_t = \big[\delta\sigma_t \delta X_t\big]dW_t, \quad t \in [0, T], \tag{4.64}$$

where $\delta\sigma_t\delta X_t$ is the $d \times d$ matrix with entries:

$$\big(\delta\sigma_t\delta X_t\big)_{i,j} = \sum_{\ell=1}^d \big(\delta\sigma_t\big)_{i,j,\ell}\big(\delta X_t\big)_\ell, \quad i,j \in \{1, \cdots, d\}^2,$$

where $(\delta X_t)_\ell$ is the ℓth coordinate of δX_t and

$$(\delta\sigma_t)_{i,j,\ell} = \frac{\sigma_{i,j}(t, X_t^{\ell-1;x\leftrightsquigarrow x'}, \mu_t) - \sigma_{i,j}(t, X_t^{\ell;x\leftrightsquigarrow x'}, \mu_t)}{(\delta X_t)_\ell} \mathbf{1}_{(\delta X_t)_\ell \neq 0},$$

with:

$$X_t^{\ell;x\leftrightsquigarrow x'} = \big((X_t^x)_1, \cdots, (X_t^x)_\ell, (X_t^{x'})_{\ell+1}, \cdots, (X_t^{x'})_d\big).$$

From the Lipschitz property of σ in x, the process $(\delta\sigma_t)_{0\leqslant t\leqslant T}$ is bounded by a constant C only depending upon L in the assumption. Notice that in the notation $\delta\sigma_t\delta X_t$, $(\delta\sigma_t\delta X_t)_{i,j}$ appears as the inner product of $((\delta\sigma_t)_{i,j,\ell})_{1\leqslant\ell\leqslant d}$ and $((\delta X_t)_\ell)_{1\leqslant\ell\leqslant d}$. Because of the presence of the additional indices (i,j), we chose not to use the inner product notation in this definition. This warning being out of the way, we may use the inner product notation when convenient.

Indeed, in a similar fashion, the pair $(\delta Y_t, \delta Z_t)_{0\leqslant t\leqslant T}$ satisfies a backward equation of the form:

$$\begin{aligned}\delta Y_t &= \delta g_T \cdot \delta X_T \\ &\quad + \int_t^T \big(\delta F_s^{(1)} \cdot \delta X_s + \delta F_s^{(2)} \cdot \delta Z_s\big) ds - \int_t^T \delta Z_s \cdot dW_s, \qquad t \in [0,T],\end{aligned} \tag{4.65}$$

where δg_T is an $\mathbb{R}^d$-valued random variable bounded by C and $\delta \boldsymbol{F}^{(1)} = (\delta F_t^{(1)})_{0\leqslant t\leqslant T}$ and $\delta \boldsymbol{F}^{(2)} = (\delta F_t^{(2)})_{0\leqslant t\leqslant T}$ are progressively measurable $\mathbb{R}^d$-valued processes, which are bounded, the bounds possibly depending upon the function ψ. Here, "$\cdot$" denotes the inner product of $\mathbb{R}^d$. Notice that, as a uniform bound on the growth of $\delta \boldsymbol{F}^{(1)}$ and $\delta \boldsymbol{F}^{(2)}$, we have:

$$\begin{aligned}|\delta F_t^{(1)}| &\leqslant C\big(1 + |Z_t^x|^2 + |Z_t^{x'}|^2\big) \\ |\delta F_t^{(2)}| &\leqslant C\big(1 + |Z_t^x| + |Z_t^{x'}|\big)\end{aligned}, \qquad t \in [0,T], \tag{4.66}$$

the constant C only depending on the constant L appearing in the assumption and where we used the assumption $\sup|\psi'| \leqslant 2/R$.

Since $\delta \boldsymbol{F}^{(2)}$ is bounded, we may introduce a probability $\mathbb{Q}$ (again this is not the same $\mathbb{Q}$ as that appearing in the second step, but, since the two steps are completely independent, there is no risk of confusion), equivalent to $\mathbb{P}$, under which $(W_t^{\mathbb{Q}} = W_t - \int_0^t \delta F_s^{(2)} ds)_{0\leqslant t\leqslant T}$ is a Brownian motion. Then,

$$\big|\mathbb{E}(\delta Y_0)\big| = \big|\mathbb{E}^{\mathbb{Q}}(\delta Y_0)\big| = \left|\mathbb{E}^{\mathbb{Q}}\left[\delta g_T \cdot \delta X_T + \int_0^T \delta F_s^{(1)} \cdot \delta X_s \, ds\right]\right|. \tag{4.67}$$

In order to handle the above right-hand side, we need to investigate $d\mathbb{Q}/d\mathbb{P}$. This requires to go back to (4.66) and to (4.62).

Fifth Step. The backward equation in (4.62) may be regarded as a quadratic BSDE satisfying assumption **Quadratic BSDE**, uniformly in R. By Theorem 4.19, the stochastic integral $(\int_0^t Z_s^x \cdot dW_s)_{0\leqslant t\leqslant T}$ is of Bounded Mean Oscillation and its BMO norm is independent of x and R. Without any loss of generality, we may assume that it is less than C.

Coincidentally, the same holds true if we replace Z_s^x by $\delta F_s^{(2)}$ from (4.65), as $|\delta F_s^{(2)}| \leqslant C(1 + |Z_s^x| + |Z_s^{x'}|)$. By Proposition 4.18, we deduce that there exists an exponent $r > 1$, only depending on L and T, such that (allowing the constant C to increase from line to line):

$$\mathbb{E}\Big[\Big(\frac{d\mathbb{Q}}{d\mathbb{P}}\Big)^r\Big] \leqslant C.$$

Now (4.64) implies that, for any $p \geqslant 1$, there exists a constant C'_p, independent of the cut-off functions ψ, such that $\mathbb{E}[\sup_{0\leqslant t\leqslant T} |\delta X_s|^p]^{1/p} \leqslant C'_p|x - x'|$. Therefore, applying Hölder's inequality, (4.67) and the bound for the r-moment of $d\mathbb{Q}/d\mathbb{P}$, we obtain:

$$\big|\mathbb{E}(\delta Y_0)\big| \leqslant C|x - x'|\bigg\{1 + \mathbb{E}\bigg[\bigg(\int_0^T (|Z_s^x|^2 + |Z_s^{x'}|^2)ds\bigg)^{r'}\bigg]^{1/r'}\bigg\}, \tag{4.68}$$

for some $r' > 1$. In order to estimate the right-hand side, we invoke Theorem 4.18 again. We deduce that:

$$\big|\mathbb{E}(\delta Y_0)\big| \leqslant C'|x - x'|,$$

for a constant C' that only depends upon L and T. This proves the required estimate for the Lipschitz constant of the decoupling field associated with the system (4.62). □

4.4.4 Conclusion

We now return to the mean field game associated with the stochastic control problem (4.52).

By Theorem 4.45, any solution to the McKean-Vlasov FBSDE:

$$\begin{cases} dX_t = b\big(t, X_t, \mathcal{L}(X_t), \hat{\alpha}\big(t, X_t, \mathcal{L}(X_t), \sigma(t, X_t, \mathcal{L}(X_t))^{-1\dagger} Z_t\big)\big)dt \\ \qquad\qquad\qquad\qquad\qquad\qquad + \sigma(t, X_t, \mathcal{L}(X_t))dW_t, \\ dY_t = -f\big(t, X_t, \mathcal{L}(X_t), \hat{\alpha}\big(t, X_t, \mathcal{L}(X_t), \sigma(t, X_t, \mathcal{L}(X_t))^{-1\dagger} Z_t\big)\big)dt \\ \qquad\qquad\qquad\qquad\qquad\qquad + Z_t \cdot dW_t, \end{cases} \tag{4.69}$$

for $t \in [0, T]$, with $X_0 = \xi \in L^2(\Omega, \mathcal{F}_0, \mathbb{P}; \mathbb{R}^d)$ as initial condition and $Y_T = g(X_T, \mathcal{L}(X_T))$ as terminal condition, provides a solution to the MFG problem derived from (4.52) since, under assumption **MFG Solvability HJB**, Equation (4.69) but with a frozen interaction falls within the scope of Theorem 4.45.

The goal now is to invoke Theorem 4.39 to solve (4.69). Using the fact that the bound for the process $\mathbf{Z}^{0;\xi}$ in the statement of Theorem 4.45 is independent of μ, we can recover the setting of Theorem 4.39: By a truncation argument, we can indeed assume without any loss of generality that assumption **MKV FBSDE for MFG** holds true. Combining the results of Theorems 4.39 and 4.45, we complete the proof of Theorem 4.44, proving the existence of an equilibrium to the MFG problem.

4.5 Solving MFGs by the Stochastic Pontryagin Principle

4.5.1 Statement of the Solvability Results

We now present another strategy for proving the existence of an MFG equilibrium, based upon the stochastic Pontryagin maximum principle. We already presented this alternative method in Subsection 3.3.2, but we now address the solvability of the underpinning McKean-Vlasov FBSDE under weaker conditions than in Theorem 3.24. Recall that the FBSDE takes the form:

$$\begin{cases} dX_t = b\big(t, X_t, \mathcal{L}(X_t), \hat{\alpha}(t, X_t, \mathcal{L}(X_t), Y_t)\big)dt + \sigma dW_t, \\ dY_t = -\partial_x H\big(t, X_t, \mathcal{L}(X_t), Y_t, \hat{\alpha}(t, X_t, \mathcal{L}(X_t), Y_t)\big)dt + Z_t dW_t, \end{cases} \tag{4.70}$$

for $t \in [0, T]$, with the initial condition $X_0 = \xi \in L^2(\Omega, \mathcal{F}_0, \mathbb{P}; \mathbb{R}^d)$, together with the terminal condition $Y_T = \partial_x g(X_T, \mathcal{L}(X_T))$. For convenience purposes, we recall the assumption introduced in Chapter 3.

Assumption (SMP). The coefficients b, f, σ and g are defined on $[0, T] \times \mathbb{R}^d \times A \times \mathcal{P}_2(\mathbb{R}^d)$, $[0, T] \times \mathbb{R}^d \times A \times \mathcal{P}_2(\mathbb{R}^d)$, $[0, T] \times \mathbb{R}^d$ and $\mathbb{R}^d \times \mathcal{P}_2(\mathbb{R}^d)$ respectively and they satisfy, for two constants $\lambda > 0$ and $L \geqslant 1$,

(A1) The drift b is an affine function of (x, α) in the sense that it is of the form:

$$b(t, x, \mu, \alpha) = b_0(t, \mu) + b_1(t)x + b_2(t)\alpha, \tag{4.71}$$

where $[0, T] \times \mathcal{P}_2(\mathbb{R}^d) \ni (t, \mu) \mapsto b_0(t, \mu)$, $b_1 : [0, T] \ni t \mapsto b_1(t)$ and $b_2 : [0, T] \ni t \mapsto b_2(t)$ are $\mathbb{R}^d$, $\mathbb{R}^{d \times d}$ and $\mathbb{R}^{d \times k}$ valued respectively, and are measurable and bounded on bounded subsets of their respective domains.

(A2) The function σ is constant.

(A3) The function $\mathbb{R}^d \times A \ni (x, \alpha) \mapsto f(t, x, \mu, \alpha) \in \mathbb{R}$ is once continuously differentiable with Lipschitz-continuous derivatives (so that $f(t, \cdot, \mu, \cdot)$ is $C^{1,1}$), the Lipschitz constant in x and α being bounded by L (so that it is uniform in t and μ). Moreover, it satisfies the following form of λ-convexity:

$$\begin{aligned} &f(t, x', \mu, \alpha') - f(t, x, \mu, \alpha) \\ &\quad - (x' - x, \alpha' - \alpha) \cdot \partial_{(x,\alpha)} f(t, x, \mu, \alpha) \geqslant \lambda |\alpha' - \alpha|^2. \end{aligned} \tag{4.72}$$

The notation $\partial_{(x,\alpha)} f$ stands for the gradient in the joint variables (x, α). Finally, f, $\partial_x f$ and $\partial_\alpha f$ are locally bounded over $[0, T] \times \mathbb{R}^d \times \mathcal{P}_2(\mathbb{R}^d) \times A$.

(continued)

(A4) The function $\mathbb{R}^d \times \mathcal{P}_2(\mathbb{R}^d) \ni (x,\mu) \mapsto g(x,\mu)$ is locally bounded. Moreover, for any $\mu \in \mathcal{P}_2(\mathbb{R}^d)$, the function $\mathbb{R}^d \ni x \mapsto g(x,\mu)$ is once continuously differentiable and convex, and has an L-Lipschitz-continuous first order derivative.

On top of assumption **SMP**, we shall use the following assumptions to solve the matching problem (ii) in (3.4). Recall the notation $M_2(\mu)^2$ for the second moment of the measure μ introduced in (3.26). The following assumptions are stated using a fixed point $\alpha_0 \in A$. Clearly, the assumptions do not depend upon the particular choice of this control value in A.

Assumption (MFG Solvability SMP). On top of assumption **SMP**, the coefficients b, f and g satisfy for some $\alpha_0 \in A$ and with the same constant L:

(A5) The functions $[0,T] \ni t \mapsto f(t,0,\delta_0,\alpha_0)$, $[0,T] \ni t \mapsto \partial_x f(t,0,\delta_0,\alpha_0)$ and $[0,T] \ni t \mapsto \partial_\alpha f(t,0,\delta_0,\alpha_0)$ are bounded by L, and for all $t \in [0,T]$, $x,x' \in \mathbb{R}^d$, $\alpha,\alpha' \in A$ and $\mu,\mu' \in \mathcal{P}_2(\mathbb{R}^d)$, it holds:

$$\begin{aligned}
&\big|f(t,x',\mu',\alpha') - f(t,x,\mu,\alpha)\big| + \big|g(x',\mu') - g(x,\mu)\big| \\
&\quad \leqslant L\big[1 + |(x',\alpha')| + |(x,\alpha)| + M_2(\mu) + M_2(\mu')\big] \\
&\qquad\qquad \times \big[|(x',\alpha') - (x,\alpha)| + W_2(\mu',\mu)\big].
\end{aligned}$$

Moreover, b_0, b_1 and b_2 in (4.71) are bounded by L and b_0 satisfies, for any $\mu,\mu' \in \mathcal{P}_2(\mathbb{R}^d)$,

$$|b_0(t,\mu') - b_0(t,\mu)| \leqslant L W_2(\mu,\mu').$$

(A6) For all $t \in [0,T]$, $x \in \mathbb{R}^d$ and $\mu \in \mathcal{P}_2(\mathbb{R}^d)$, $\quad |\partial_\alpha f(t,x,\mu,\alpha_0)| \leqslant L$.

(A7) For all $(t,x) \in [0,T] \times \mathbb{R}^d$,

$$x \cdot \partial_x f(t,0,\delta_x,\alpha_0) \geqslant -L(1+|x|), \qquad \text{and} \qquad x \cdot \partial_x g(0,\delta_x) \geqslant -L(1+|x|).$$

Assumption **(A5)** provides Lipschitz continuity while condition **(A6)** controls the smoothness of the running cost f with respect to α uniformly in the other variables. The most unusual assumption is certainly condition **(A7)**. We refer to it as a *weak mean-reverting* condition as it looks like a standard mean-reverting condition for recurrent diffusion processes, even though the notion of recurrence does not make much sense in our case since we are working on a finite time interval. Still, as shown by the proof of Theorem 4.53 below, its role is to control the expectation of the solution of the forward equation in (4.70), providing an *a priori* bound for it. The latter will play a crucial role in the proof of compactness.

Here is the main result of this section:

Theorem 4.53 *Under assumption* **MFG Solvability SMP**, *when* $A = \mathbb{R}^k$ *and* $\alpha_0 = 0$, *the forward-backward system* (4.70) *has a solution. Hence, so does the MFG problem associated with the stochastic control problem* (4.52).

Moreover, for any solution $(\boldsymbol{X}, \boldsymbol{Y}, \boldsymbol{Z}) = (X_t, Y_t, Z_t)_{0 \leq t \leq T}$ *of* (4.70), *there exists a function* $u : [0,T] \times \mathbb{R}^d \to \mathbb{R}^d$ *satisfying the growth and Lipschitz properties:*

$$\forall t \in [0,T], \quad \forall x, x' \in \mathbb{R}^d, \quad \begin{cases} |u(t,x)| \leqslant C(1+|x|), \\ |u(t,x) - u(t,x')| \leqslant C|x - x'|, \end{cases} \tag{4.73}$$

for some constant $C \geqslant 0$, *and such that* $\mathbb{P}$*-a.s.* $Y_t = u(t, X_t)$ *for all* $t \in [0,T]$. *In particular, for any* $p \geqslant 1$, $\mathbb{E}[\sup_{0 \leqslant t \leqslant T} |X_t|^p] < +\infty$.

In line with the terminology used so far, the function u will be referred to as the decoupling field of the FBSDE when the environment $\boldsymbol{\mu} = (\mu_t = \mathcal{L}(X_t))_{0 \leqslant t \leqslant T}$ is fixed.

Remark 4.54 *An interesting example which should be kept in mind is a particular case of the class of linear-quadratic models which we already studied in detail in Section 3.5. Indeed, assume that* b_0, f, *and* g *have the form:*

$$b_0(t,\mu) = b_0(t)\bar{\mu}, \quad g(x,\mu) = \frac{1}{2}\big|qx + \bar{q}\,\bar{\mu}\big|^2,$$

$$f(t,x,\mu,\alpha) = \frac{1}{2}\big|m(t)x + \bar{m}(t)\bar{\mu}\big|^2 + \frac{1}{2}|n(t)\alpha|^2,$$

where q, $\bar{q}$, $m(t)$ *and* $\bar{m}(t)$ *are elements of* $\mathbb{R}^{d' \times d}$, *for some* $d' \geqslant 1$, $n(t)$ *is an element of* $\mathbb{R}^{k' \times k}$, *for some* $k' \geqslant k$, *and* $\bar{\mu}$ *stands for the mean of* μ. *Assumption* **MFG Solvability SMP** *is satisfied when* $b_0(t) \equiv 0$ *(so that* b_0 *is bounded as required in* **(A5)**), $\bar{q}^\dagger q \geqslant 0$, $\bar{m}(t)^\dagger m(t) \geqslant 0$, *and* $n(t)^\dagger n(t) \geqslant \lambda I_k$ *in the sense of quadratic forms so that* **(A7)** *and* **(A3)** *hold. In the one-dimensional case* $d = m = 1$, **(A7)** *says that* $q\bar{q}$ *and* $m(t)\bar{m}(t)$ *must be nonnegative and* **(A3)** *says that* $n(t)^2$ *must be greater than* λ. *As we saw in Section 3.5, these conditions are not optimal for existence when* $d = m = 1$, *as we showed that* (4.70) *is indeed solvable when* $[0,T] \ni t \mapsto b_0(t)$ *is a (possibly nonzero) continuous function,* $q(q+\bar{q}) \geqslant 0$ *and* $m(t)(m(t) + \bar{m}(t)) \geqslant 0$. *Obviously, the gap between these conditions is the price to pay for treating general systems within a single framework.*

Another example investigated in Section 3.5 is $b_0 \equiv 0$, $b_1 \equiv 0$, $b_2 \equiv 1$, $f \equiv \alpha^2/2$, *with* $d = m = 1$. *When* $g(x,\mu) = \bar{q}(x - s\bar{\mu})^2/2$, *with* $\bar{q} \geqslant 0$ *and* $s \in \mathbb{R}$, *assumption* **MFG Solvability SMP** *is satisfied when* $\bar{q}s \leqslant 0$ *(so that* **(A7)** *holds). The optimal condition given in Section 3.5 is* $1 + \bar{q}(1-s)T \neq 0$.

Notice that assumption **MFG Solvability SMP** *is satisfied when* $g(x,\mu) = \bar{q}(x - s\gamma(\bar{\mu}))^2/2$ *for a bounded Lipschitz-continuous function* γ *from* $\mathbb{R}$ *into itself.*

Remark 4.55 *The reader may want to apply Theorem 4.39 in order to prove Theorem 4.53, but, as made clear in the proof below, the difficulty is that* (**A5**) *in assumption* **MKV FBSDE for MFG** *may not be satisfied.*

The remainder of the section is dedicated to the proof of Theorem 4.53. It split into four main steps.

Preliminary Analysis of the SMP

The first step is to prove, for an initial condition ξ in $L^2(\Omega, \mathcal{F}_0, \mathbb{P}; \mathbb{R}^d)$, the solvability of the FBSDE:

$$\begin{cases} dX_t = b\big(t, X_t, \mu_t, \hat{\alpha}(t, X_t, \mu_t, Y_t)\big)dt + \sigma dW_t, \\ dY_t = -\partial_x H\big(t, X_t, \mu_t, Y_t, \hat{\alpha}(t, X_t, \mu_t, Y_t)\big)dt + Z_t dW_t, \qquad t \in [0, T], \\ X_0 = \xi, \quad Y_T = \partial_x g(X_T, \mu_T), \end{cases} \tag{4.74}$$

whose analysis was left open in the proof of Theorem 3.17.

To this end, we need the following lemma.

Lemma 4.56 *Given a continuous flow* $\boldsymbol{\mu} = (\mu_t)_{0 \leqslant t \leqslant T}$ *from* $[0, T]$ *to* $\mathcal{P}_2(\mathbb{R}^d)$, *the FBSDE* (4.74) *is uniquely solvable under assumption* **SMP**, *A being a general closed convex subset of* $\mathbb{R}^k$. *If we denote its solution by* $(X_t^{0,\xi}, Y_t^{0,\xi}, Z_t^{0,\xi})_{0 \leqslant t \leqslant T}$, *then there exist a constant* $C > 0$, *only depending upon the parameters in* **SMP** *and thus independent of* $\boldsymbol{\mu}$, *and a locally bounded measurable function* $u : [0, T] \times \mathbb{R}^d \to \mathbb{R}^d$, *depending on* $\boldsymbol{\mu}$, *such that*

$$\forall t \in [0, T], \ \forall x, x' \in \mathbb{R}^d, \quad |u(t, x') - u(t, x)| \leqslant C|x' - x|,$$

and $\mathbb{P}$*-a.s., for all* $t \in [0, T]$, $Y_t^{0,\xi} = u(t, X_t^{0,\xi})$.

The proof of Lemma 4.56 is based on Lemma 3.3 and Proposition 3.21 from Chapter 3.

Proof. From the definition of the reduced Hamiltonian, $\partial_x H$ reads $\partial_x H(t, x, \mu, y, \alpha) = b_1(t)^\dagger y + \partial_x f(t, x, \mu, \alpha)$, so that, by Lemma 3.3, the driver $[0, T] \times \mathbb{R}^d \times \mathbb{R}^d \ni (t, x, y) \mapsto \partial_x H(t, x, \mu_t, \hat{\alpha}(t, x, \mu_t, y))$ of the backward equation in (4.70) is Lipschitz continuous in the variables (x, y), uniformly in t. Therefore, by Theorem 4.2, existence and uniqueness hold when T is small enough. Equivalently, when T is arbitrary, there exists $\delta > 0$, depending on the Lipschitz constant of the coefficients in the variables x and y such that unique solvability holds on $[T - \delta, T]$, that is when the initial condition $\xi \in L^2(\Omega, \mathcal{F}_{t_0}, \mathbb{P}; \mathbb{R}^d)$ of the forward process is prescribed at some time $t_0 \in [T - \delta, T]$. The solution is then denoted by $(X_t^{t_0,\xi}, Y_t^{t_0,\xi}, Z_t^{t_0,\xi})_{t_0 \leqslant t \leqslant T}$. Following Subsection 4.1.2, existence and uniqueness can be established on the whole $[0, T]$ by iterating the unique solvability property in short time provided we have:

$$\forall x_0, x_0' \in \mathbb{R}^d, \quad |Y_{t_0}^{t_0,x_0} - Y_{t_0}^{t_0,x_0'}|^2 \leqslant C|x_0 - x_0'|^2, \tag{4.75}$$

for some constant C independent of t_0 and δ. By (3.39), we have:

$$\hat{J}^{t_0,x_0} + (x_0' - x_0) \cdot Y_{t_0}^{t_0,x_0} + \lambda \mathbb{E} \int_{t_0}^{T} |\hat{\alpha}_t^{t_0,x_0} - \hat{\alpha}_t^{t_0,x_0'}|^2 dt \leqslant \hat{J}^{t_0,x_0'}, \tag{4.76}$$

where $\hat{J}^{t_0,x_0} = J^{\mu}((\hat{\alpha}_t^{t_0,x_0})_{t_0 \leqslant t \leqslant T})$ and $\hat{\alpha}_t^{t_0,x_0} = \hat{\alpha}(t, X_t^{t_0,x_0}, \mu_t, Y_t^{t_0,x_0})$ (with similar definitions for $\hat{J}^{t_0,x_0'}$ and $\hat{\alpha}_t^{t_0,x_0'}$ by replacing x_0 by x_0'). Exchanging the roles of x_0 and x_0' in (4.76) and adding the resulting inequality to (4.76), we deduce that:

$$2\lambda \mathbb{E} \int_{t_0}^{T} |\hat{\alpha}_t^{t_0,x_0} - \hat{\alpha}_t^{t_0,x_0'}|^2 dt \leqslant (x_0' - x_0) \cdot (Y_{t_0}^{t_0,x_0'} - Y_{t_0}^{t_0,x_0}). \tag{4.77}$$

Moreover, by standard SDE estimates first and then by standard BSDE estimates, there exists a constant C (the value of which may vary from line to line), independent of t_0 and δ, such that:

$$\begin{aligned} &\mathbb{E}\big[\sup_{t_0 \leqslant t \leqslant T} |X_t^{t_0,x_0} - X_t^{t_0,x_0'}|^2\big] + \mathbb{E}\big[\sup_{t_0 \leqslant t \leqslant T} |Y_t^{t_0,x_0} - Y_t^{t_0,x_0'}|^2\big] \\ &\quad \leqslant C\mathbb{E} \int_{t_0}^{T} |\hat{\alpha}_t^{t_0,x_0} - \hat{\alpha}_t^{t_0,x_0'}|^2 dt. \end{aligned}$$

Plugging (4.77) into the above inequality completes the proof of (4.75). As explained in Section 4.1, the function u is then defined as $u : [0,T] \times \mathbb{R}^d \ni (t,x) \mapsto Y_t^{t,x}$ and the representation property of $\boldsymbol{Y}$ in terms of $\boldsymbol{X}$ follows from (4.8). Local boundedness of u follows from the Lipschitz continuity in the variable x together with the obvious inequality:

$$\sup_{0 \leqslant t \leqslant T} |u(t,0)| \leqslant \sup_{0 \leqslant t \leqslant T} \Big[\mathbb{E}\big[|u(t, X_t^{0,0}) - u(t,0)|\big] + \mathbb{E}\big[|Y_t^{0,0}|\big] \Big] < \infty.$$

□

4.5.2 Existence under Additional Boundedness Conditions

We first prove existence under an extra boundedness assumption. When, in addition, A is bounded, σ is invertible, b does not depend on x and the coefficients of (4.74) are continuous in the measure argument, existence is already known from Theorem 3.24. In the proof below, we derive the a priori estimates needed for compactness in the statement of Schauder's theorem from the strong convexity of the running cost and the convenient form of the stochastic maximum principle recalled earlier.

Proposition 4.57 *The system* (4.70) *is solvable if, in addition to assumption* **MFG Solvability SMP**, *we also assume that* $\mathbb{E}[|\xi|^4] < \infty$ *and that* $\partial_x f$ *and* $\partial_x g$ *are uniformly bounded, i.e., for some constant* $c > 0$

$$\begin{aligned} &\forall t \in [0,T],\ x \in \mathbb{R}^d,\ \mu \in \mathcal{P}_2(\mathbb{R}^d),\ \alpha \in A, \\ &\qquad |\partial_x g(x,\mu)| + |\partial_x f(t,x,\mu,\alpha)| \leqslant c. \end{aligned} \tag{4.78}$$

Notice that (4.78) implies **(A7)** in assumption **MFG Solvability SMP**.

Proof. Throughout the proof, we denote by $(\boldsymbol{X}^{\xi;\boldsymbol{\mu}}, \boldsymbol{Y}^{\xi;\boldsymbol{\mu}}, \boldsymbol{Z}^{\xi;\boldsymbol{\mu}})$ the solution of (4.74) with $\boldsymbol{\mu}$ as input. Also, we denote by $u^{\boldsymbol{\mu}}$ the corresponding decoupling field.

For a given $\xi \in L^4(\Omega, \mathcal{F}_0, \mathbb{P}; \mathbb{R}^d)$, we consider the map $\boldsymbol{\mu} \mapsto \Phi(\boldsymbol{\mu}) = (\mathcal{L}(X_t^{\xi;\boldsymbol{\mu}}))_{0 \leqslant t \leqslant T}$ and try to apply Schauder's Theorem 4.32 in order to prove that Φ has a fixed point, very much in the spirit of the proofs of Theorems 4.29 and 4.39. See also Remark 4.42 for a comment on the difference between the two proofs. Following Subsections 4.3.2 and 4.3.5, we apply Schauder's fixed point theorem in the space $\mathcal{C}([0,T]; \mathcal{M}_f^1(\mathbb{R}^d))$ of continuous functions $\boldsymbol{\mu} = (\mu_t)_{0 \leqslant t \leqslant T}$ from $[0,T]$ into the space of finite signed measures over $\mathbb{R}^d$, equipped with the supremum of the Kantorovich-Rubinstein norm:

$$\|\boldsymbol{\mu}\| = \sup_{t \in [0,T]} \|\mu_t\|_{\mathrm{KR}\star},$$

with

$$\|\mu\|_{\mathrm{KR}\star} = |\mu(\mathbb{R}^d)| + \sup\left\{ \int_{\mathbb{R}^d} \ell(x) d\mu(x); \quad \ell \in \mathrm{Lip}_1(\mathbb{R}^d),\ \ell(0) = 0 \right\}.$$

As already explained several times, $\| \cdot \|_{\mathrm{KR}\star}$ coincides with the Wasserstein distance W_1 on $\mathcal{P}_1(\mathbb{R}^d)$. See the Notes & Complements at the end of Chapter 5 for details and references.

We prove existence by proving that there exists a closed convex subset $\mathcal{E}$ included in $\mathcal{C}([0,T]; \mathcal{P}_2(\mathbb{R}^d))$ which, when viewed as a subset of $\mathcal{C}([0,T]; \mathcal{M}_f^1(\mathbb{R}^d))$, is stable for Φ, with a relatively compact range, Φ being continuous on $\mathcal{E}$.

First Step. We first establish several a priori estimates for the solution of (4.74). The coefficients $\partial_x f$ and $\partial_x g$ being bounded, the terminal condition in (4.74) is bounded, and the growth of the driver is controlled by:

$$|\partial_x H(t, x, \mu_t, y, \hat{\alpha}(t, x, \mu_t, y))| \leqslant c + L|y|.$$

By expanding $(|Y_t^{\xi;\boldsymbol{\mu}}|^2)_{0 \leqslant t \leqslant T}$ as the solution of a one-dimensional BSDE, we can compare it with the solution of a deterministic BSDE with a constant terminal condition. This implies that there exists a constant C, only depending upon c, L and T, such that, for any $\boldsymbol{\mu} \in \mathcal{C}([0,T]; \mathcal{P}_2(\mathbb{R}^d))$, $\mathbb{P}$-almost surely,

$$\forall t \in [0,T], \quad |Y_t^{\xi;\boldsymbol{\mu}}| \leqslant C. \tag{4.79}$$

Notice that the value of the constant C will vary from line to line. By (3.10) in the statement of Lemma 3.3, and by **(A6)** in assumption **MFG Solvability SMP**, we deduce that:

$$\forall t \in [0,T], \quad \left|\hat{\alpha}\left(t, X_t^{\xi;\boldsymbol{\mu}}, \mu_t, Y_t^{\xi;\boldsymbol{\mu}}\right)\right| \leqslant C. \tag{4.80}$$

Plugging this bound into the forward part of (4.74), and again, allowing the constant C to increase when necessary, standard L^p estimates for SDEs imply:

$$\mathbb{E}\big[\sup_{0 \leqslant t \leqslant T} |X_t^{\xi;\boldsymbol{\mu}}|^4\big] \leqslant C\Big(1 + \mathbb{E}[|\xi|^4]\Big). \tag{4.81}$$

We consider the restriction of Φ to the subset $\mathcal{E}$ of continuous flows of probability measures whose fourth moments are not greater than $C(1+\mathbb{E}[|\xi|^4])$, i.e.,

$$\mathcal{E} = \left\{ \mu \in \mathcal{C}([0,T];\mathcal{P}_4(\mathbb{R}^d)) : \sup_{0\leqslant t\leqslant T} \int_{\mathbb{R}^d} |x|^4 d\mu_t(x) \leqslant C\big(1+\mathbb{E}[|\xi|^4]\big) \right\}.$$

Clearly, $\mathcal{E}$ is convex and closed for the 1-Wasserstein distance, and Φ maps $\mathcal{E}$ into itself.

Second Step. By (4.80), we get for any $\mu \in \mathcal{E}$ and $0 \leqslant s \leqslant t \leqslant T$:

$$|X_t^{\xi;\mu} - X_s^{\xi;\mu}| \leqslant C\big[(t-s)\big(1+\sup_{0\leqslant r\leqslant T}|X_r^{\xi;\mu}|\big) + |W_t - W_s|\big],$$

so that, by (4.81),

$$W_2\big([\Phi(\mu)]_t,[\Phi(\mu)]_s\big) = W_2\big(\mathcal{L}(X_t^{\xi;\mu}),\mathcal{L}(X_s^{\xi;\mu})\big) \leqslant C(t-s)^{1/2},$$

where C is now allowed to depend on ξ.

Using (4.81) and Corollary 5.6 in Chapter 5, we deduce that there exists a compact subset $\mathcal{K} \subset \mathcal{P}_2(\mathbb{R}^d)$ such that, for any $\mu \in \mathcal{E}$, $[\Phi(\mu)]_t \in \mathcal{K}$ for any $t \in [0,T]$. By the above bound and by Arzelà-Ascoli theorem, we deduce that $\Phi(\mathcal{E})$ is a relatively compact subset of $\mathcal{C}([0,T];\mathcal{P}_2(\mathbb{R}^d))$ and thus of $\mathcal{C}([0,T];\mathcal{P}_1(\mathbb{R}^d))$; see Chapter 5.

Third Step. We finally check that Φ is continuous on $\mathcal{E}$. Given another flow of measures $\mu' \in \mathcal{E}$, (3.39) in Proposition 3.21 implies that:

$$\begin{aligned} &J^{\mu}(\hat{\boldsymbol{\alpha}}) + \lambda\mathbb{E}\int_0^T |\hat{\alpha}'_t - \hat{\alpha}_t|^2 dt \\ &\qquad \leqslant J^{\mu}\big([\hat{\boldsymbol{\alpha}}',\mu']\big) - \mathbb{E}\int_0^T \big(b_0(t,\mu'_t) - b_0(t,\mu_t)\big)\cdot Y_t^{\xi;\mu} dt, \end{aligned} \tag{4.82}$$

where $\hat{\alpha}_t = \hat{\alpha}(t, X_t^{\xi;\mu}, \mu_t, Y_t^{\xi;\mu})$ for $t \in [0,T]$, $\hat{\alpha}'_t$ being defined in a similar way by replacing μ by μ'. By optimality of $\hat{\alpha}'$ for the cost functional $J^{\mu'}(\cdot)$, we claim:

$$J^{\mu}\big([\hat{\boldsymbol{\alpha}}',\mu']\big) \leqslant J^{\mu'}(\hat{\boldsymbol{\alpha}}) + J^{\mu}\big([\hat{\boldsymbol{\alpha}}',\mu']\big) - J^{\mu'}(\hat{\boldsymbol{\alpha}}'),$$

so that (4.82) yields:

$$\begin{aligned} \lambda\mathbb{E}\int_0^T |\hat{\alpha}'_t - \hat{\alpha}_t|^2 dt &\leqslant J^{\mu'}(\hat{\boldsymbol{\alpha}}) - J^{\mu}(\hat{\boldsymbol{\alpha}}) + J^{\mu}\big([\hat{\boldsymbol{\alpha}}',\mu']\big) - J^{\mu'}(\hat{\boldsymbol{\alpha}}') \\ &\qquad - \mathbb{E}\int_0^T \big(b_0(t,\mu'_t) - b_0(t,\mu_t)\big)\cdot Y_t^{\xi;\mu} dt. \end{aligned} \tag{4.83}$$

We now compare $J^{\mu'}(\hat{\boldsymbol{\alpha}})$ with $J^{\mu}(\hat{\boldsymbol{\alpha}})$, and similarly $J^{\mu'}(\hat{\boldsymbol{\alpha}}')$ with $J^{\mu}([\hat{\boldsymbol{\alpha}}',\mu'])$. We notice that $J^{\mu}(\hat{\boldsymbol{\alpha}})$ is the cost associated with the flow of measures $\mu = (\mu_t)_{0\leqslant t\leqslant T}$ and the diffusion process $X^{\xi;\mu}$, whereas $J^{\mu'}(\hat{\boldsymbol{\alpha}})$ is the cost associated with the flow of measures $\mu' = (\mu'_t)_{0\leqslant t\leqslant T}$ and the controlled diffusion process $\boldsymbol{U} = (U_t)_{0\leqslant t\leqslant T}$ satisfying:

$$dU_t = \big[b_0(t,\mu'_t) + b_1(t)U_t + b_2(t)\hat{\alpha}_t\big]dt + \sigma dW_t, \quad t \in [0,T]; \quad U_0 = \xi.$$

By Gronwall's lemma, we can modify the value of C (which is now allowed to depend on ξ) in such a way that:

$$\mathbb{E}\big[\sup_{0\leqslant t\leqslant T}|X_t^{\xi;\mu} - U_t|^2\big] \leqslant C\int_0^T W_2^2(\mu_t,\mu_t')dt.$$

Since $\boldsymbol{\mu}$ and $\boldsymbol{\mu}'$ are in $\mathcal{E}$, we deduce from **(A5)** in assumption **MFG Solvability SMP**, (4.80) and (4.81) that:

$$J^{\boldsymbol{\mu}'}(\hat{\boldsymbol{\alpha}}) - J^{\boldsymbol{\mu}}(\hat{\boldsymbol{\alpha}}) \leqslant C\bigg[\bigg(\int_0^T W_2^2(\mu_t,\mu_t')dt\bigg)^{1/2} + W_2(\mu_T,\mu_T')\bigg].$$

A similar bound holds for $J^{\boldsymbol{\mu}}([\hat{\boldsymbol{\alpha}}',\boldsymbol{\mu}'])-J^{\boldsymbol{\mu}'}(\hat{\boldsymbol{\alpha}}')$, the argument being even simpler as the costs are driven by the same processes. So from (4.83) and (4.79) again, together with Gronwall's lemma to go back to the controlled SDEs,

$$\begin{aligned}
&\mathbb{E}\int_0^T|\hat{\alpha}_t' - \hat{\alpha}_t|^2dt + \mathbb{E}\big[\sup_{0\leqslant t\leqslant T}|X_t^{\xi;\mu} - X_t^{\xi;\mu'}|^2\big] \\
&\leqslant C\bigg[\bigg(\int_0^T W_2^2(\mu_t,\mu_t')dt\bigg)^{1/2} + W_2(\mu_T,\mu_T')\bigg],
\end{aligned}$$

where C is also allowed to depend on λ.

As probability measures in $\mathcal{E}$ have bounded moments of order 4, Cauchy-Schwartz inequality yields:

$$\begin{aligned}
\sup_{0\leqslant t\leqslant T} W_1\big([\Phi(\mu)]_t,[\Phi(\mu')]_t\big) &\leqslant C\bigg[\bigg(\int_0^T W_2(\mu_t,\mu_t')^2dt\bigg)^{1/4} + W_2(\mu_T,\mu_T')^{1/2}\bigg] \\
&\leqslant C\bigg[\bigg(\int_0^T W_1(\mu_t,\mu_t')^{1/2}dt\bigg)^{1/4} + W_1(\mu_T,\mu_T')^{1/8}\bigg],
\end{aligned}$$

showing that Φ is continuous on $\mathcal{E}$. The last inequality follows from Hölder's inequality $\mathbb{E}[|X|^2] = \mathbb{E}[|X|^{1/2}|X|^{3/2}] \leqslant \mathbb{E}[|X|]^{1/2}\mathbb{E}[|X|^3]^{1/2} \leqslant \mathbb{E}[|X|]^{1/2}\mathbb{E}[|X|^4]^{3/8}$, for any random variable X. □

4.5.3 Approximation Procedure

Obviously, examples of functions f and g which are convex in x and such that $\partial_x f$ and $\partial_x g$ are bounded are rather limited in number and scope. Also, boundedness of $\partial_x f$ and $\partial_x g$ fails in the typical case when f and g are quadratic with respect to x. In order to overcome this limitation, we propose to approximate the cost functions f and g by two sequences $(f^n)_{n\geqslant 1}$ and $(g^n)_{n\geqslant 1}$, referred to as approximated cost functions, satisfying assumption **MFG Solvability SMP** uniformly with respect to $n \geqslant 1$, and such that, for any $n \geqslant 1$, equation (4.70), with $(\partial_x f, \partial_x g)$ replaced by $(\partial_x f^n, \partial_x g^n)$ and with $\xi \in L^4(\Omega,\mathcal{F}_0,\mathbb{P};\mathbb{R}^d)$, has a solution $(\boldsymbol{X}^n,\boldsymbol{Y}^n,\boldsymbol{Z}^n)$. In this framework, Proposition 4.57 says that such approximate FBSDEs are indeed solvable when $\partial_x f^n$ and $\partial_x g^n$ are bounded for any $n \geqslant 1$. Our approximation procedure relies on the following:

Lemma 4.58 *Let us assume that there exist two sequences $(f^n)_{n\geqslant 1}$ and $(g^n)_{n\geqslant 1}$ such that:*

(i) *there exist two parameters L' and $\lambda' > 0$ such that, for any $n \geqslant 1$, f^n and g^n satisfy assumption* **MFG Solvability SMP** *with λ' and L';*
(ii) *f^n (resp. g^n) converges towards f (resp. g) uniformly on bounded subsets of $[0,T] \times \mathbb{R}^d \times \mathcal{P}_2(\mathbb{R}^d) \times A$ (resp. $\mathbb{R}^d \times \mathcal{P}_2(\mathbb{R}^d)$);*
(iii) *for any $n \geqslant 1$, equation* (4.70)*, with $(\partial_x f, \partial_x g)$ replaced by $(\partial_x f^n, \partial_x g^n)$ and with $\xi \in L^4(\Omega, \mathcal{F}_0, \mathbb{P}; \mathbb{R}^d)$ instead of $\xi \in L^2(\Omega, \mathcal{F}_0, \mathbb{P}; \mathbb{R}^d)$, has a solution.*

Then, equation (4.70)*, with the original coefficients and with $\xi \in L^2(\Omega, \mathcal{F}_0, \mathbb{P}; \mathbb{R}^d)$, is solvable.*

Proof. For a sequence of $\mathcal{F}_0$-measurable random variables $(\xi^n)_{n\geqslant 1}$ with values in $\mathbb{R}^d$, such that $|\xi^n| \leqslant |\xi| \wedge n$ for all $n \geqslant 1$ and $\mathbb{E}[|\xi - \xi^n|^2] \to 0$ as n tends to ∞, we denote by $(\boldsymbol{X}^n, \boldsymbol{Y}^n, \boldsymbol{Z}^n)_{n\geqslant 1}$ the sequence of processes obtained by solving (4.70), with $(\partial_x f, \partial_x g)$ replaced by $(\partial_x f^n, \partial_x g^n)$ and ξ by ξ^n.

We establish tightness of the processes $(\boldsymbol{X}^n)_{n\geqslant 1}$ in order to extract a convergent subsequence of $(\boldsymbol{\mu}^n = (\mu^n_t = \mathcal{L}(X^n_t))_{0\leqslant t\leqslant T})_{n\geqslant 1}$. For any $n \geqslant 1$, we consider the approximate Hamiltonian:

$$H^n(t,x,\mu,y,\alpha) = b(t,x,\mu,\alpha)\cdot y + f^n(t,x,\mu,\alpha),$$

together with its minimizer $\hat{\alpha}^n(t,x,\mu,y) = \text{argmin}_\alpha H^n(t,x,\mu,y,\alpha)$. Setting $\hat{\alpha}^n_t = \hat{\alpha}^n(t, X^n_t, \mathcal{L}(X^n_t), Y^n_t)$ for any $t \in [0,T]$ and $n \geqslant 1$, our first step will be to prove that:

$$\sup_{n\geqslant 1} \mathbb{E}\left[\int_0^T |\hat{\alpha}^n_s|^2 ds\right] < +\infty. \tag{4.84}$$

Since $\boldsymbol{X}^n$ is the diffusion process controlled by $\hat{\boldsymbol{\alpha}}^n = (\hat{\alpha}^n_t)_{0\leqslant t\leqslant T}$, we use Theorem 3.17 to compare its behavior to the behavior of a *reference controlled process* $\boldsymbol{U}^n$ whose dynamics are driven by a specific control $\boldsymbol{\beta}^n$. We shall consider two different versions for $\boldsymbol{U}^n$ corresponding to the following choices for $\boldsymbol{\beta}^n$:

$$(i)\ \beta^n_s = \mathbb{E}[\hat{\alpha}^n_s] \quad \text{for } 0 \leqslant s \leqslant T; \qquad (ii)\ \boldsymbol{\beta}^n \equiv 0. \tag{4.85}$$

For each of these controls, we compare its cost to the optimal cost by using the version of the stochastic maximum principle which we proved earlier, and subsequently, derive useful information on the optimal control $\hat{\boldsymbol{\alpha}}^n$.

First Step. We first consider (i) in (4.85). In this case,

$$U^n_t = \xi^n + \int_0^t \big[b_0(s, \mathcal{L}(X^n_s)) + b_1(s)U^n_s + b_2(s)\mathbb{E}(\hat{\alpha}^n_s)\big]ds + \sigma W_t, \quad t \in [0,T]. \tag{4.86}$$

Notice that taking expectations on both sides of (4.86) shows that $\mathbb{E}(U^n_s) = \mathbb{E}(X^n_s)$, for $0 \leqslant s \leqslant T$, and that:

$$\big[U^n_t - \mathbb{E}(U^n_t)\big] = \big[\xi^n - \mathbb{E}(\xi^n)\big] + \int_0^t b_1(s)\big[U^n_s - \mathbb{E}(U^n_s)\big]ds + \sigma W_t, \quad t \in [0,T],$$

from which it easily follows that $\sup_{n\geqslant 1}\sup_{0\leqslant s\leqslant T}\mathrm{Var}(U_s^n) < +\infty$. By Theorem 3.17, with $g^n(\cdot, \mathcal{L}(X_T^n))$ as terminal cost and $(f^n(t,\cdot,\mathcal{L}(X_t^n),\cdot))_{0\leqslant t\leqslant T}$ as running cost, we get:

$$\begin{aligned}
&\mathbb{E}\big[g^n(X_T^n, \mathcal{L}(X_T^n))\big] + \mathbb{E}\int_0^T \big[\lambda'|\hat{\alpha}_s^n - \beta_s^n|^2 + f^n\big(s, X_s^n, \mathcal{L}(X_s^n), \hat{\alpha}_s^n\big)\big]ds \\
&\quad \leqslant \mathbb{E}\Big[g^n\big(U_T^n, \mathcal{L}(X_T^n)\big) + \int_0^T f^n\big(s, U_s^n, \mathcal{L}(X_s^n), \beta_s^n\big)ds\Big].
\end{aligned} \tag{4.87}$$

Using the fact that $\beta_s^n = \mathbb{E}[\hat{\alpha}_s^n]$, the convexity condition in **(A2)** and **(A4)** and Jensen's inequality, we obtain:

$$\begin{aligned}
&g^n\big(\mathbb{E}(X_T^n), \mathcal{L}(X_T^n)\big) + \int_0^T \big[\lambda'\mathrm{Var}(\hat{\alpha}_s^n) + f^n\big(s, \mathbb{E}(X_s^n), \mathcal{L}(X_s^n), \mathbb{E}(\hat{\alpha}_s^n)\big)\big]ds \\
&\quad \leqslant \mathbb{E}\Big[g^n\big(U_T^n, \mathcal{L}(X_T^n)\big) + \int_0^T f^n\big(s, U_s^n, \mathcal{L}(X_s^n), \mathbb{E}(\hat{\alpha}_s^n)\big)ds\Big].
\end{aligned} \tag{4.88}$$

By **(A5)** in assumption **MFG Solvability SMP**, we deduce that there exists a constant C, depending only on λ, L, $\mathbb{E}[|\xi|^2]$ and T, such that (the actual value of C possibly varying from line to line):

$$\begin{aligned}
&\int_0^T \mathrm{Var}(\hat{\alpha}_s^n)ds \leqslant C\big(1 + \mathbb{E}\big[|U_T^n|^2\big]^{1/2} + \mathbb{E}\big[|X_T^n|^2\big]^{1/2}\big)\mathbb{E}\big[|U_T^n - \mathbb{E}(X_T^n)|^2\big]^{1/2} \\
&\quad + C\int_0^T \big(1 + \mathbb{E}\big[|U_s^n|^2\big]^{1/2} + \mathbb{E}\big[|X_s^n|^2\big]^{1/2} + \mathbb{E}\big[|\hat{\alpha}_s^n|^2\big]^{1/2}\big)\mathbb{E}\big[|U_s^n - \mathbb{E}(X_s^n)|^2\big]^{1/2}ds.
\end{aligned}$$

Since $\mathbb{E}(X_t^n) = \mathbb{E}(U_t^n)$ for any $t \in [0,T]$, we deduce from the uniform boundedness of the variance of $(U_s^n)_{0\leqslant s\leqslant T}$ that:

$$\int_0^T \mathrm{Var}(\hat{\alpha}_s^n)ds \leqslant C\Big[1 + \sup_{0\leqslant s\leqslant T}\mathbb{E}[|X_s^n|^2]^{1/2} + \Big(\mathbb{E}\int_0^T |\hat{\alpha}_s^n|^2 ds\Big)^{1/2}\Big]. \tag{4.89}$$

From this, the linearity of the dynamics of $\boldsymbol{X}^n$ and Gronwall's inequality, we deduce:

$$\sup_{0\leqslant s\leqslant T}\mathrm{Var}(X_s^n) \leqslant C\Big[1 + \Big(\mathbb{E}\int_0^T |\hat{\alpha}_s^n|^2 ds\Big)^{1/2}\Big], \tag{4.90}$$

since

$$\sup_{0\leqslant s\leqslant T}\mathbb{E}\big[|X_s^n|^2\big] \leqslant C\Big[1 + \mathbb{E}\int_0^T |\hat{\alpha}_s^n|^2 ds\Big]. \tag{4.91}$$

Bounds like (4.90) allow us to control, for any $0 \leqslant s \leqslant T$, the Wasserstein distance between the distribution of X_s^n and the Dirac mass at the point $\mathbb{E}(X_s^n)$.

Second Step. We now compare $\boldsymbol{X}^n$ to the process controlled by the null control. So we consider case (*ii*) in (4.85), and now:

$$U_t^n = \xi^n + \int_0^t \big[b_0\big(s, \mathcal{L}(X_s^n)\big) + b_1(s)U_s^n\big]ds + \sigma W_t, \quad t \in [0, T].$$

Note that we still denote the solution by $\boldsymbol{U}^n$ although it is different from the one in the first step. By the boundedness of b_0 in **(A5)** in assumption **MFG Solvability SMP**, it holds that $\sup_{n \geqslant 1} \mathbb{E}[\sup_{0 \leqslant s \leqslant T} |U_s^n|^2] < \infty$. Using Theorem 3.17 as before in the derivation of (4.87) and (4.88), we get:

$$\begin{aligned}
&g^n\big(\mathbb{E}(X_T^n), \mathcal{L}(X_T^n)\big) + \int_0^T \big[\lambda' \mathbb{E}\big(|\hat{\alpha}_s^n|^2\big) + f^n\big(s, \mathbb{E}(X_s^n), \mathcal{L}(X_s^n), \mathbb{E}(\hat{\alpha}_s^n)\big)\big]ds \\
&\quad \leqslant \mathbb{E}\bigg[g^n\big(U_T^n, \mathcal{L}(X_T^n)\big) + \int_0^T f^n\big(s, U_s^n, \mathcal{L}(X_s^n), 0\big)ds\bigg].
\end{aligned}$$

By convexity of f^n with respect to α (recall **(A2)** in assumption **MFG Solvability SMP**, together with **(A6)**), we get:

$$\begin{aligned}
&g^n\big(\mathbb{E}(X_T^n), \mathcal{L}(X_T^n)\big) + \int_0^T \big[\lambda' \mathbb{E}\big(|\hat{\alpha}_s^n|^2\big) + f^n\big(s, \mathbb{E}(X_s^n), \mathcal{L}(X_s^n), 0\big)\big]ds \\
&\quad \leqslant \mathbb{E}\bigg[g^n\big(U_T^n, \mathcal{L}(X_T^n)\big) + \int_0^T f^n\big(s, U_s^n, \mathcal{L}(X_s^n), 0\big)ds\bigg] + C\mathbb{E}\int_0^T |\hat{\alpha}_s^n| ds,
\end{aligned}$$

for some constant C, independent of n. Using **(A5)**, we obtain:

$$\begin{aligned}
&g^n\big(\mathbb{E}(X_T^n), \delta_{\mathbb{E}(X_T^n)}\big) + \int_0^T \big[\lambda' \mathbb{E}\big(|\hat{\alpha}_s^n|^2\big) + f^n\big(s, \mathbb{E}(X_s^n), \delta_{\mathbb{E}(X_s^n)}, 0\big)\big]ds \\
&\quad \leqslant g^n\big(0, \delta_{\mathbb{E}(X_T^n)}\big) + \int_0^T f^n\big(s, 0, \delta_{\mathbb{E}(X_s^n)}, 0\big)ds + C\mathbb{E}\int_0^T |\hat{\alpha}_s^n| ds \\
&\qquad + C\big(1 + \sup_{0 \leqslant s \leqslant T} \big[\mathbb{E}[|X_s^n|^2]^{1/2}\big]\big)\big(1 + \sup_{0 \leqslant s \leqslant T} \big[\mathrm{Var}(X_s^n)\big]^{1/2}\big),
\end{aligned}$$

the value of C possibly varying from line to line. From (4.91), Young's inequality yields:

$$\begin{aligned}
&g^n\big(\mathbb{E}(X_T^n), \delta_{\mathbb{E}(X_T^n)}\big) + \int_0^T \big[\frac{\lambda'}{2} \mathbb{E}\big(|\hat{\alpha}_s^n|^2\big) + f^n\big(s, \mathbb{E}(X_s^n), \delta_{\mathbb{E}(X_s^n)}, 0\big)\big]ds \\
&\quad \leqslant g^n\big(0, \delta_{\mathbb{E}(X_T^n)}\big) + \int_0^T f^n\big(s, 0, \delta_{\mathbb{E}(X_s^n)}, 0\big)ds + C\big(1 + \sup_{0 \leqslant s \leqslant T} \big[\mathrm{Var}(X_s^n)\big]\big).
\end{aligned}$$

By (4.90), we obtain:

$$\begin{aligned}
&g^n\big(\mathbb{E}(X_T^n), \delta_{\mathbb{E}(X_T^n)}\big) + \int_0^T \big[\frac{\lambda'}{2} \mathbb{E}\big(|\hat{\alpha}_s^n|^2\big) + f^n\big(s, \mathbb{E}(X_s^n), \delta_{\mathbb{E}(X_s^n)}, 0\big)\big]ds \\
&\quad \leqslant g^n\big(0, \delta_{\mathbb{E}(X_T^n)}\big) + \int_0^T f^n\big(s, 0, \delta_{\mathbb{E}(X_s^n)}, 0\big)ds + C\bigg(1 + \bigg[\int_0^T \mathbb{E}\big(|\hat{\alpha}_s^n|^2\big)ds\bigg]^{1/2}\bigg).
\end{aligned}$$

Young's inequality and the convexity in x of g^n and f^n give:

$$\mathbb{E}(X_T^n) \cdot \partial_x g^n(0, \delta_{\mathbb{E}(X_T^n)}) + \int_0^T \Big[\frac{\lambda'}{4}\mathbb{E}(|\hat{\alpha}_s^n|^2) + \mathbb{E}(X_s^n) \cdot \partial_x f^n(s, 0, \delta_{\mathbb{E}(X_s^n)}, 0)\Big] ds \leqslant C.$$

By **(A7)**, we get $\mathbb{E}\int_0^T |\hat{\alpha}_s^n|^2 ds \leqslant C(1 + \sup_{0\leqslant s\leqslant T} \mathbb{E}[|X_s^n|^2]^{1/2})$, and the bound (4.84) now follows from (4.91), and as a consequence,

$$\mathbb{E}\big[\sup_{0\leqslant s\leqslant T} |X_s^n|^2\big] \leqslant C. \tag{4.92}$$

Using (4.84) and (4.92), we can prove that the processes $(\boldsymbol{X}^n)_{n\geqslant 1}$ are tight in $\mathcal{C}([0,T];\mathbb{R}^d)$. Indeed, there exists a constant C', independent of n, such that, for any $0 \leqslant s \leqslant t \leqslant T$,

$$|X_t^n - X_s^n| \leqslant C'(t-s)^{1/2}\bigg[1 + \Big(\int_0^T \big[|X_r^n|^2 + |\hat{\alpha}_r^n|^2\big] dr\Big)^{1/2}\bigg] + C'|W_t - W_s|,$$

so that tightness follows from (4.84) and (4.92).

Third Step. Let $(n_p)_{p\geqslant 1}$ be an increasing sequence of integers such that the distributions of the processes $(\boldsymbol{X}^{n_p})_{p\geqslant 1}$ on $\mathcal{C}([0,T];\mathbb{R}^d)$ are weakly convergent. We then denote by $(\boldsymbol{\mu}^{n_p} : [0,T] \ni t \mapsto \mu_t^{n_p} = \mathcal{L}(X_t^{n_p}))_{p\geqslant 1}$ the corresponding sequence of flows of marginal distributions. For any $t \in [0,T]$, $\mu_t^{n_p}$ converges in the weak sense to μ_t, where $\boldsymbol{\mu} = (\mu_t)_{0\leqslant t\leqslant T}$ is the flow of marginal distributions of the limit law, which belongs to $\mathcal{P}_2(\mathcal{C}([0,T];\mathbb{R}^d))$ by (4.92). It satisfies $\boldsymbol{\mu} \in \mathcal{C}([0,T];\mathcal{P}_2(\mathbb{R}^d))$. Therefore, by Lemma 4.56, FBSDE (4.74), with $\boldsymbol{\mu}$ as input and on the same space as before, has a unique solution $(X_t, Y_t, Z_t)_{0\leqslant t\leqslant T}$. Moreover, there exists $u : [0,T]\times\mathbb{R}^d \to \mathbb{R}^d$, which is C-Lipschitz in the variable x for the same constant C as in the statement of the lemma, such that $Y_t = u(t, X_t)$ for any $t \in [0,T]$. In particular,

$$\sup_{0\leqslant t\leqslant T} |u(t,0)| \leqslant \sup_{0\leqslant t\leqslant T} \Big[\mathbb{E}\big[|u(t,X_t) - u(t,0)|\big] + \mathbb{E}\big[|Y_t|\big]\Big] < \infty. \tag{4.93}$$

So there exists a constant, still denoted by C', such that $|u(t,x)| \leqslant C'(1 + |x|)$, for $t \in [0,T]$ and $x \in \mathbb{R}^d$. By (3.10) and **(A6)**, we deduce that (for a possibly new value of C') $|\hat{\alpha}(t,x,\mu_t,u(t,x))| \leqslant C'(1+|x|)$. Plugging this bound into the forward SDE satisfied by X in (4.70), we conclude that, for a possibly new value of C',

$$\forall \ell \geqslant 1, \quad \mathbb{E}\big[\sup_{0\leqslant t\leqslant T} |X_t|^{2\ell} | \mathcal{F}_0\big]^{1/\ell} \leqslant C'\big(1 + |\xi|^2\big), \tag{4.94}$$

and, thus,

$$\mathbb{E}\int_0^T |\hat{\alpha}_t|^2 dt < \infty, \tag{4.95}$$

with $\hat{\alpha}_t = \hat{\alpha}(t, X_t, \mu_t, Y_t)$, for $t \in [0,T]$. We can now apply the same argument to any $(X_t^n)_{0\leqslant t\leqslant T}$, for any $n \geqslant 1$. We claim:

$$\forall \ell \geqslant 1, \quad \sup_{n \geqslant 1} \mathbb{E}\Big[\sup_{0 \leqslant t \leqslant T} |X_t^n|^{2\ell} | \mathcal{F}_0 \Big]^{1/\ell} \leqslant C'\big(1 + |\xi|^2\big), \tag{4.96}$$

which is a consequence of the following three observations. First, the constant C in the statement of Lemma 4.56 does not depend on n. Second, the second-order moments of $\sup_{0 \leqslant t \leqslant T} |X_t^n|$ are bounded, uniformly in $n \geqslant 1$ by (4.92). Third, by **(A5)**, the driver of the backward component in (4.74) is at most of linear growth in (x, y, α), so that by (4.84) and standard L^2 estimates for BSDEs, the second-order moments of $\sup_{0 \leqslant t \leqslant T} |Y_t^n|$ are uniformly bounded as well. This shows (4.96) by repeating the proof of (4.94). By (4.94) and (4.96) and by the same uniform integrability argument as in the third step of the proof of Theorem 4.29 in Subsection 4.3.3, we get that $\sup_{0 \leqslant t \leqslant T} W_2(\mu_t^{n_p}, \mu_t) \to 0$ as n tends to $+\infty$. Repeating the proof of (4.83) (see (3.39) for the notations), we have:

$$\begin{aligned} \lambda' \mathbb{E} \int_0^T |\hat{\alpha}_t^n - \hat{\alpha}_t|^2 dt \leqslant\; & J^{n,\mu^n}(\hat{\boldsymbol{\alpha}}) - J^{\mu}(\hat{\boldsymbol{\alpha}}) + J^{\mu}\big([\hat{\boldsymbol{\alpha}}^n, \boldsymbol{\mu}^n]\big) - J^{n,\mu^n}(\hat{\boldsymbol{\alpha}}^n) \\ & - \mathbb{E}\big[(\xi^n - \xi) \cdot Y_0\big] \quad - \mathbb{E} \int_0^T \big(b_0(t, \mu_t^n) - b_0(t, \mu_t)\big) \cdot Y_t dt, \end{aligned} \tag{4.97}$$

where $J^{\mu}(\cdot)$ is given by (4.52) and $J^{n,\mu^n}(\cdot)$ is defined in a similar way, but with (f, g) and $(\mu_t)_{0 \leqslant t \leqslant T}$ replaced by (f^n, g^n) and $(\mu_t^n)_{0 \leqslant t \leqslant T}$. With these definitions at hand, we notice that:

$$\begin{aligned} & J^{n,\mu^n}(\hat{\boldsymbol{\alpha}}) - J^{\mu}(\hat{\boldsymbol{\alpha}}) \\ & \quad = \mathbb{E}\big[g^n(U_T^n, \mu_T^n) - g(X_T, \mu_T)\big] + \mathbb{E} \int_0^T \big[f^n\big(t, U_t^n, \mu_t^n, \hat{\alpha}_t\big) - f\big(t, X_t, \mu_t, \hat{\alpha}_t\big)\big] dt, \end{aligned}$$

where $\boldsymbol{U}^n$ is the controlled diffusion process:

$$dU_t^n = \big[b_0(t, \mu_t^n) + b_1(t) U_t^n + b_2(t) \hat{\alpha}_t\big] dt + \sigma dW_t, \quad t \in [0, T]; \quad U_0^n = \xi^n.$$

By Gronwall's lemma and by convergence of $\boldsymbol{\mu}^{n_p}$ towards $\boldsymbol{\mu}$ for the 2–Wasserstein distance, we claim that $\boldsymbol{U}^{n_p} \to \boldsymbol{X}$ as $p \to +\infty$ for the norm $\mathbb{E}[\sup_{0 \leqslant s \leqslant T} | \cdot_s |^2]^{1/2}$, namely in $\mathbb{S}^{2,d}$. Using on one hand the uniform convergence of f^n and g^n towards f and g on bounded subsets of their respective domains together with the regularity properties of f^n, g^n, f and g, and on the other hand the convergence of μ^{n_p} towards μ together with the bounds (4.94), (4.95) and (4.96)), we deduce that $J^{n_p,\mu^{n_p}}(\hat{\boldsymbol{\alpha}}) \to J^{\mu}(\hat{\boldsymbol{\alpha}})$ as $p \to +\infty$. Similarly, using the bounds (4.84), (4.94) and (4.96), the other differences in the right-hand side in (4.97) tend to 0 along the subsequence $(n_p)_{p \geqslant 1}$ so that $\hat{\boldsymbol{\alpha}}^{n_p} \to \hat{\boldsymbol{\alpha}}$ as $p \to +\infty$ in $L^2([0, T] \times \Omega, \mathrm{Leb}_1 \otimes \mathbb{P})$. We conclude that $\boldsymbol{X}$ is the limit of the sequence $(\boldsymbol{X}^{n_p})_{p \geqslant 1}$ in $\mathbb{S}^{2,d}$. Therefore, $\boldsymbol{\mu}$ matches the flow of marginal laws of $\boldsymbol{X}$, proving that equation (4.70) is solvable. □

4.5.4 Choice of the Approximating Sequences

In order to complete the proof of Theorem 4.53, we must specify the choice of the approximating sequences in Lemma 4.58. Actually, the choice is performed in two steps. We first consider the case when the cost functions f and g are strongly convex in the variables x.

Lemma 4.59 *Assume that, in addition to assumption* **MFG Solvability SMP**, *there exists a constant* $\gamma > 0$ *such that the functions* f *and* g *satisfy (compare with (4.72)):*

$$\begin{aligned} &f(t, x', \mu, \alpha') - f(t, x, \mu, \alpha) \\ &\qquad - (x' - x, \alpha' - \alpha) \cdot \partial_{(x,\alpha)} f(t, x, \mu, \alpha) \geqslant \gamma |x' - x|^2 + \lambda |\alpha' - \alpha|^2, \\ &g(x', \mu) - g(x, \mu) - (x' - x) \cdot \partial_x g(x, \mu) \geqslant \gamma |x' - x|^2. \end{aligned} \tag{4.98}$$

Then, there exist two positive constants λ' *and* L', *depending only upon* λ, L *and* γ, *and two sequences of functions* $(f^n)_{n \geqslant 1}$ *and* $(g^n)_{n \geqslant 1}$ *such that:*

(i) *for any* $n \geqslant 1$, f^n *and* g^n *satisfy* **MFG Solvability SMP** *with the parameters* λ' *and* L' *and* $\partial_x f^n$ *and* $\partial_x g^n$ *are bounded;*
(ii) *for any bounded subsets of* $[0, T] \times \mathbb{R}^d \times \mathcal{P}_2(\mathbb{R}^d) \times \mathbb{R}^k$ *and* $\mathbb{R}^d \times \mathcal{P}_2(\mathbb{R}^d)$, *there exists an integer* n_0, *such that, for any* $n \geqslant n_0$, f^n *and* g^n *coincide with* f *and* g *on these bounded sets.*

Proof. The proof of Lemma 4.59 is a pure technical exercise in convex analysis. We focus on the approximation of the running cost f (the case of the terminal cost g is similar) and we ignore the dependence of f upon t to simplify the notation. For any $n \geqslant 1$, we define f_n as the truncated Legendre transform:

$$f_n(x, \mu, \alpha) = \sup_{|y| \leqslant n} \inf_{z \in \mathbb{R}^d} \bigl[y \cdot (x - z) + f(z, \mu, \alpha) \bigr], \tag{4.99}$$

for $(x, \alpha) \in \mathbb{R}^d \times \mathbb{R}^k$ and $\mu \in \mathcal{P}_2(\mathbb{R}^d)$. By standard properties of the Legendre transform of convex functions,

$$f_n(x, \mu, \alpha) \leqslant \sup_{y \in \mathbb{R}^d} \inf_{z \in \mathbb{R}^d} \bigl[y \cdot (x - z) + f(z, \mu, \alpha) \bigr] = f(x, \mu, \alpha). \tag{4.100}$$

Moreover, by strict convexity of f in x,

$$\begin{aligned} f_n(x, \mu, \alpha) &\geqslant \inf_{z \in \mathbb{R}^d} \bigl[f(z, \mu, \alpha) \bigr] \geqslant \inf_{z \in \mathbb{R}^d} \bigl[\gamma |z|^2 + \partial_x f(0, \mu, \alpha) \cdot z \bigr] + f(0, \mu, \alpha) \\ &\geqslant -\frac{1}{4\gamma} |\partial_x f(0, \mu, \alpha)|^2 + f(0, \mu, \alpha), \end{aligned} \tag{4.101}$$

so that f_n has finite real values. Clearly, it is also n-Lipschitz continuous in x.

First Step. We first check that the sequence $(f_n)_{n \geqslant 1}$ converges towards f, uniformly on bounded subsets of $\mathbb{R}^d \times \mathcal{P}_2(\mathbb{R}^d) \times \mathbb{R}^k$. So for any given $R > 0$, we restrict ourselves to $|x| \leqslant R$, $|\alpha| \leqslant R$, and $\mu \in \mathcal{P}_2(\mathbb{R}^d)$, such that $M_2(\mu) \leqslant R$. By **(A5)** in assumption **MFG Solvability SMP**, there exists a constant $c > 0$, independent of R, such that

$$\sup_{z \in \mathbb{R}^d} \bigl[y \cdot z - f(z, \mu, \alpha) \bigr] \geqslant \sup_{z \in \mathbb{R}^d} \bigl[y \cdot z - c|z|^2 \bigr] - c(1 + R^2) = \frac{|y|^2}{4c} - c(1 + R^2). \tag{4.102}$$

Therefore,

$$\inf_{z\in\mathbb{R}^d}\big[y\cdot(x-z)+f(z,\mu,\alpha)\big]\leqslant R|y|-\frac{|y|^2}{4c}+c(1+R^2). \tag{4.103}$$

By (4.101) and **(A5)** in assumption **MFG Solvability SMP**, $f_n(t,x,\mu,\alpha)\geqslant -c(1+R^2)$, c depending possibly on γ, so that optimization in the variable y in the definition of f_n can be done over points $y^\star$ satisfying:

$$-c(1+R^2)\leqslant R|y^\star|-\frac{|y^\star|^2}{4c}+c(1+R^2),\quad \text{that is}\quad |y^\star|\leqslant c(1+R), \tag{4.104}$$

the constant c being allowed to vary from inequality to another. In particular, for n large enough (depending on R),

$$f_n(x,\mu,\alpha)=\sup_{y\in\mathbb{R}^d}\inf_{z\in\mathbb{R}^d}\big[y\cdot(x-z)+f(z,\mu,\alpha)\big]=f(x,\mu,\alpha). \tag{4.105}$$

So on bounded subsets of $\mathbb{R}^d\times\mathcal{P}_2(\mathbb{R}^d)\times\mathbb{R}^k$, f_n and f coincide for n large enough. In particular, for n large enough, $f_n(0,\delta_0,0)$, $\partial_x f_n(0,\delta_0,0)$ and $\partial_\alpha f_n(0,\delta_0,0)$ exist, coincide with $f(0,\delta_0,0)$, $\partial_x f(0,\delta_0,0)$ and $\partial_\alpha f(0,\delta_0,0)$ respectively, and are bounded by L as in **(A5)**. Moreover, still for $|x|\leqslant R$, $|\alpha|\leqslant R$ and $M_2(\mu)\leqslant R$, we see from (4.100) and (4.104) that optimization in z can be reduced to $z^\star$ satisfying:

$$y^\star\cdot(x-z^\star)+f(z^\star,\mu,\alpha)\leqslant f(x,\mu,\alpha)\leqslant c(1+R^2),$$

the second inequality following from **(A5)**. By strict convexity of f in x, we obtain:

$$-c(1+R)|z^\star|+\gamma|z^\star|^2+\partial_x f(0,\mu,\alpha)\cdot z^\star+f(0,\mu,\alpha)\leqslant c(1+R^2),$$

so that, by **(A5)**, $\gamma|z^\star|^2-c(1+R)|z^\star|\leqslant c(1+R^2)$, in other words:

$$|z^\star|\leqslant c(1+R). \tag{4.106}$$

Second Step. We now investigate the convexity property of $f_n(\cdot,\mu,\cdot)$, for given $\mu\in\mathcal{P}_2(\mathbb{R}^d)$. For any $h\in\mathbb{R}$, $x,e,y,z_1,z_2\in\mathbb{R}^d$ and $\alpha,\beta\in\mathbb{R}^k$, with $|y|\leqslant n$ and $|e|,|\beta|\leq 1$, we deduce from the convexity of $f(\cdot,\mu,\cdot)$:

$$\begin{aligned}
&2\inf_{z\in\mathbb{R}^d}\big[y\cdot(x-z)+f(z,\mu,\alpha)\big]\\
&\leqslant y\cdot\Big((x+he-z_1)+(x-he-z_2)\Big)+2f\bigg(\frac{z_1+z_2}{2},\mu,\frac{(\alpha+h\beta)+(\alpha-h\beta)}{2}\bigg)\\
&\leqslant y\cdot(x+he-z_1)+f(z_1,\mu,\alpha+h\beta)+y\cdot(x-he-z_2)+f(z_2,\mu,\alpha-h\beta)-2\lambda h^2|\beta|^2.
\end{aligned}$$

Taking infimum with respect to z_1,z_2, and supremum with respect to y, we obtain:

$$f_n(x,\mu,\alpha)\leqslant\frac{1}{2}f_n(x+he,\mu,\alpha+h\beta)+\frac{1}{2}f_n(x-he,\mu,\alpha-h\beta)-\lambda h^2|\beta|^2. \tag{4.107}$$

In particular, the function $\mathbb{R}^d \times \mathbb{R}^k \ni (x, \alpha) \to f_n(x, \mu, \alpha) - \lambda|\alpha|^2$ is convex. We prove later on that it is also continuously differentiable so that (4.72) holds.

In a similar way, we can investigate the semi-concavity property of $f_n(\cdot, \mu, \cdot)$. For any $h \in \mathbb{R}$, $x, e, y_1, y_2 \in \mathbb{R}^d$, $\alpha, \beta \in \mathbb{R}^k$, with $|y_1|, |y_2| \leqslant n$ and $|e|, |\beta| \leq 1$,

$$
\begin{aligned}
&\inf_{z \in \mathbb{R}^d} \big[y_1 \cdot (x + he - z) + f(z, \mu, \alpha + h\beta) \big] \\
&\qquad + \inf_{z \in \mathbb{R}^d} \big[y_2 \cdot (x - he - z) + f(z, \mu, \alpha - h\beta) \big] \\
&\quad = \inf_{z \in \mathbb{R}^d} \big[y_1 \cdot (x - z) + f(z + he, \mu, \alpha + h\beta) \big] \\
&\qquad + \inf_{z \in \mathbb{R}^d} \big[y_2 \cdot (x - z) + f(z - he, \mu, \alpha - h\beta) \big].
\end{aligned}
$$

By expanding $f(\cdot, \mu, \cdot)$ up to the first order and by using the Lipschitz regularity of the first order derivatives, we see that:

$$
\begin{aligned}
&\inf_{z \in \mathbb{R}^d} \big[y_1 \cdot (x + he - z) + f(z, \mu, \alpha + h\beta) \big] \\
&\qquad + \inf_{z \in \mathbb{R}^d} \big[y_2 \cdot (x - he - z) + f(z, \mu, \alpha - h\beta) \big] \\
&\quad \leqslant \inf_{z \in \mathbb{R}^d} \big[(y_1 + y_2) \cdot (x - z) + 2f(z, \mu, \alpha) \big] + c|h|^2 \big(|e|^2 + |\beta|^2 \big) \\
&\quad = 2 \inf_{z \in \mathbb{R}^d} \big[\frac{y_1 + y_2}{2} \cdot (x - z) + f(z, \mu, \alpha) \big] + c|h|^2 \big(|e|^2 + |\beta|^2 \big),
\end{aligned}
$$

for some constant c. Taking the supremum over y_1, y_2, we deduce that:

$$
f_n(x + he, \mu, \alpha + h\beta) + f_n(x - he, \mu, \alpha - h\beta) - 2f_n(x, \mu, \alpha) \leqslant c|h|^2 \big(|e|^2 + |\beta|^2 \big).
$$

So for any $\mu \in \mathcal{P}_2(\mathbb{R}^d)$, the function $\mathbb{R}^d \times \mathbb{R}^k \ni (x, \alpha) \mapsto f_n(x, \mu, \alpha) - c[|x|^2 + |\alpha|^2]$ is concave. Therefore, $f_n(\cdot, \mu, \alpha)$ is both convex and semi-concave from which we deduce that it is $\mathcal{C}^{1,1}$, i.e., continuously differentiable with Lipschitz derivatives, the Lipschitz constant of the derivatives being uniform with respect to $n \geqslant 1$ and to $\mu \in \mathcal{P}_2(\mathbb{R}^d)$. Moreover, by definition, the function $f_n(\cdot, \mu, \cdot)$ is n-Lipschitz continuous in the variable x, that is $\partial_x f_n$ is bounded, as required.

Third Step. We now investigate the consequences of **(A5)**. Given $\delta > 0$, $R > 0$ and $n \geqslant 1$, we consider $x \in \mathbb{R}^d$, $\alpha \in \mathbb{R}^k$, $\mu, \mu' \in \mathcal{P}_2(\mathbb{R}^d)$ such that:

$$
\max \big(|x|, |\alpha|, M_2(\mu), M_2(\mu') \big) \leqslant R, \quad W_2(\mu, \mu') \leqslant \delta. \tag{4.108}
$$

By **(A5)** in assumption **MFG Solvability SMP** and (4.106), we can find a constant c' (possibly depending on γ) such that:

$$
\begin{aligned}
f_n(x,\mu',\alpha) &= \sup_{|y|\leqslant n} \inf_{|z|\leqslant c(1+R)} \big[y\cdot(x-z) + f(z,\mu',\alpha) \big] \\
&\leqslant \sup_{|y|\leqslant n} \inf_{z\leqslant c(1+R)} \big[y\cdot(x-z) + f(z,\mu,\alpha) + L(1+R+|z|)\delta \big] \\
&= \sup_{|y|\leqslant n} \inf_{z\in\mathbb{R}^d} \big[y\cdot(x-z) + f(z,\mu,\alpha) \big] + c'(1+R)\delta.
\end{aligned}
\tag{4.109}
$$

This proves local Lipschitz-continuity in the measure argument as in **(A5)**.

In order to prove local Lipschitz-continuity in the variables x and α, we use the $\mathcal{C}^{1,1}$-property. Indeed, for x, μ and α as in (4.108), we know that:

$$
\big|\partial_x f_n(x,\mu,\alpha)\big| + \big|\partial_\alpha f_n(x,\mu,\alpha)\big| \leqslant \big|\partial_x f_n(0,\mu,0)\big| + \big|\partial_\alpha f_n(0,\mu,0)\big| + cR. \tag{4.110}
$$

By (4.105), for any integer $p \geqslant 1$, there exists an integer n_p, such that, for any $n \geqslant n_p$, $\partial_x f_n(0,\mu,0)$ and $\partial_\alpha f_n(0,\mu,0)$ are respectively equal to $\partial_x f(0,\mu,0)$ and $\partial_\alpha f(0,\mu,0)$ for $M_2(\mu) \leqslant p$. In particular, for $n \geqslant n_p$,

$$
\big|\partial_x f_n(0,\mu,0)\big| + \big|\partial_\alpha f_n(0,\mu,0)\big| \leqslant c\big(1+M_2(\mu)\big) \quad \text{whenever} \quad M_2(\mu) \leqslant p, \tag{4.111}
$$

so that (4.110) implies **(A5)** whenever $n \geqslant n_p$ and $M_2(\mu) \leqslant p$. We get rid of these restrictions by modifying the definition of f_n. Given a probability measure $\mu \in \mathcal{P}_2(\mathbb{R}^d)$ and an integer $p \geqslant 1$, we define $\Phi_p(\mu)$ as the push-forward of μ by the mapping $\mathbb{R}^d \ni x \to [\max(M_2(\mu),p)]^{-1}px$ so that $\Phi_p(\mu) \in \mathcal{P}_2(\mathbb{R}^d)$ and $M_2(\Phi_p(\mu)) \leqslant \min(p, M_2(\mu))$. Indeed, if the random variable X has μ as distribution, i.e., $\mathcal{L}(X) = \mu$, then the random variable $X_p = pX/\max(M_2(\mu),p)$ has $\Phi_p(\mu)$ as distribution. It is easy to check that Φ_p is Lipschitz continuous for the 2-Wasserstein distance, uniformly in $n \geqslant 1$. We then consider the approximating sequence:

$$
\hat{f}_p : \mathbb{R}^d \times \mathcal{P}_2(\mathbb{R}^d) \times \mathbb{R}^k \ni (x,\mu,\alpha) \to f_{n_p}\big(x, \Phi_p(\mu), \alpha\big), \quad p \geqslant 1,
$$

instead of $(f_n)_{n\geqslant 1}$ itself. Clearly, on any bounded subset, $\hat{f}_p$ still coincides with f for p large enough. Moreover, the conclusion of the second step is preserved. In particular, the conclusion of the second step together with (4.109), (4.110), and (4.111) say that **(A5)** holds (for a possible new choice of L). From now on, we get rid of the symbol "hat" in $(\hat{f}_p)_{p\geqslant 1}$ and keep the notation $(f_n)_{n\geqslant 1}$ for $(\hat{f}_p)_{p\geqslant 1}$.

Fourth Step. It only remains to check that f_n satisfies the bound **(A6)** and the sign condition **(A7)** in assumption **MFG Solvability SMP**. Since $|\partial_\alpha f(x,\mu,0)| \leqslant L$, the Lipschitz property of $\partial_\alpha f$ implies that there exists a constant $c \geqslant 0$ such that $|\partial_\alpha f(x,\mu,\alpha)| \leqslant c$ for all $(x,\mu,\alpha) \in \mathbb{R}^d \times \mathcal{P}_2(\mathbb{R}^d) \times \mathbb{R}^k$ with $|\alpha| \leqslant 1$. In particular, for any $n \geqslant 1$, it is plain to see that $f_n(x,\mu,\alpha) \leqslant f_n(x,\mu,0) + c|\alpha|$, for any $(x,\mu,\alpha) \in \mathbb{R}^d \times \mathcal{P}_2(\mathbb{R}^d) \times \mathbb{R}^k$ with $|\alpha| \leqslant 1$, so that $|\partial_\alpha f_n(x,\mu,0)| \leqslant c$. This proves **(A6)**.

Finally, we can modify the definition of f_n once more to satisfy **(A7)**. Indeed, for any $R > 0$, there exists an integer n_R, such that, for any $n \geqslant n_R$, $f_n(x,\mu,\alpha)$ and $f(x,\mu,\alpha)$ coincide for $(x,\mu,\alpha) \in \mathbb{R}^d \times \mathcal{P}_2(\mathbb{R}^d) \times \mathbb{R}^k$ with $|x|, |\alpha|, M_2(\mu) \leqslant R$ so that $x \cdot \partial_x f_n(0,\delta_x,0) \geqslant -L(1+|x|)$, for $|x| \leqslant R$ and $n \geqslant n_R$. Next we choose a smooth function $\psi : \mathbb{R}^d \to \mathbb{R}^d$, satisfying $|\psi(x)| \leqslant 1$ for any $x \in \mathbb{R}^d$, $\psi(x) = x$ for $|x| \leqslant 1/2$ and $\psi(x) = x/|x|$ for $|x| \geqslant 1$, and we set $\hat{f}_p(x, \mu, \alpha) = f_{n_p}(x, \Psi_p(\mu), \alpha)$ for any integer $p \geqslant 1$ and $(x, \mu, \alpha) \in \mathbb{R}^d \times \mathcal{P}_2(\mathbb{R}^d) \times \mathbb{R}^k$ where $\Psi_p(\mu)$ is the push-forward of μ by the mapping $\mathbb{R}^d \ni x \to x - \bar{\mu} + p\psi(p^{-1}\bar{\mu})$. Recall that $\bar{\mu}$ stands for the mean of μ. In other words, if X has distribution μ, then $\hat{X}_p = X - \mathbb{E}(X) + p\psi(p^{-1}\mathbb{E}(X))$ has distribution $\Psi_p(\mu)$.

The function Ψ_p is Lipschitz continuous with respect to W_2, uniformly in $p \geqslant 1$. Moreover, for any $R > 0$ and $p \geqslant 2R$, $M_2(\mu) \leqslant R$ implies $|\int_{\mathbb{R}^d} x' d\mu(x')| \leqslant R$ so that $p^{-1}|\int_{\mathbb{R}^d} x' d\mu(x')| \leqslant 1/2$, that is $\Psi_p(\mu) = \mu$ and, for $|x|, |\alpha| \leqslant R$, $\hat{f}_p(x, \mu, \alpha) = f_{n_p}(x, \mu, \alpha) = f(x, \mu, \alpha)$. Therefore, the sequence $(\hat{f}_p)_{p \geq 1}$ is an approximating sequence for f which satisfies the same regularity properties as $(f_n)_{n \geq 1}$. In addition:

$$x \cdot \partial_x \hat{f}_p(0, \delta_x, 0) = x \cdot \partial_x f_{n_p}(0, \delta_{p\psi(p^{-1}x)}, 0) = x \cdot \partial_x f(0, \delta_{p\psi(p^{-1}x)}, 0)$$

for $x \in \mathbb{R}^d$. Finally we choose $\psi(x) = [\rho(|x|)/|x|]x$ (with $\psi(0) = 0$), where ρ is a smooth nondecreasing function from $[0, +\infty)$ into $[0, 1]$ such that $\rho(x) = x$ on $[0, 1/2]$ and $\rho(x) = 1$ on $[1, +\infty)$. If $x \neq 0$, then the above right-hand side is equal to:

$$\begin{aligned} x \cdot \partial_x f(0, \delta_{p\psi(p^{-1}x)}, 0) &= \frac{|p^{-1}x|}{\rho(|p^{-1}x|)} (p\psi(p^{-1}x)) \cdot \partial_x f(0, \delta_{p\psi(p^{-1}x)}, 0) \\ &\geqslant -L \frac{|p^{-1}x|}{\rho(|p^{-1}x|)} (1 + |p\psi(p^{-1}x)|). \end{aligned}$$

For $|x| \leqslant p/2$, we have $\rho(p^{-1}|x|) = |p^{-1}x|$, so that the right-hand side coincides with $-L(1 + |x|)$. For $|x| \geqslant p/2$, we have $\rho(p^{-1}|x|) \geqslant 1/2$ so that:

$$-\frac{|p^{-1}x|}{\rho(|p^{-1}x|)} (1 + |p\psi(p^{-1}x)|) \geqslant -2p^{-1}|x|(1 + |p\psi(p^{-1}x)|) \geqslant -2p^{-1}|x|(1 + p) \geqslant -4|x|.$$

This proves that **(A7)** in assumption **MFG Solvability SMP** holds with a new constant. □

4.5.5 Conclusion

Equation (4.70) is solvable when, in addition to assumption **MFG Solvability SMP**, f and g satisfy the convexity condition (4.98). Indeed, by Lemma 4.59, there exists an approximating sequence $(f^n, g^n)_{n \geq 1}$ satisfying *(i)* and *(ii)* in the statement of Lemma 4.58, and also *(iii)* by Proposition 4.57. When f and g satisfy assumption **MFG Solvability SMP** only, the assumptions of Lemma 4.58 are satisfied with the following approximating sequence:

$$f_n(t, x, \mu, \alpha) = f(t, x, \mu, \alpha) + \frac{1}{n}|x|^2; \quad g_n(x, \mu) = g(x, \mu) + \frac{1}{n}|x|^2,$$

for $(t, x, \mu, \alpha) \in [0, T] \times \mathbb{R}^d \times \mathcal{P}(\mathbb{R}^d) \times \mathbb{R}^k$ and $n \geqslant 1$. Therefore, (4.70) is solvable under assumption **MFG Solvability SMP**. Moreover, given an arbitrary solution to (4.70), the existence of a function u, as in the statement of Theorem 4.53, follows from Lemma 4.56 and (4.93). This completes the proof of Theorem 4.53. □

4.6 Extended Mean Field Games: Interaction Through the Controls

The purpose of this section is to revisit the theory of mean field games when the individual players can also interact via the controls. These models have been referred to as extended mean field games in the PDE literature on mean field games. We saw several examples of this kind in Chapter 1, notably when we discussed models for exhaustible resources as in Subsection 1.4.4. Therein, the inventories of N oil producers are modeled by means of a stochastic differential game in which the cost functional of each player depends upon the empirical mean of the instantaneous rates of production of all the producers. A similar situation appeared when we introduced the price impact model which we shall solve in detail in the next section.

While the N-player game models are usually detailed in the discussions of the practical applications, here, we jump directly to the asymptotic formulation of the mean field games. We work with the usual set-up $(\Omega, \mathcal{F}, \mathbb{F} = (\mathcal{F}_t)_{0 \leqslant t \leqslant T}, \mathbb{P})$, equipped with a d-dimensional $\mathbb{F}$ - Wiener process $\boldsymbol{W} = (W_t)_{0 \leqslant t \leqslant T}$ and an $\mathcal{F}_0$-measurable initial condition $\xi \in L^2(\Omega, \mathcal{F}_0, \mathbb{P}; \mathbb{R}^d)$. We rewrite the matching problem (i)–(ii) of Subsection 3.1.2 as follows:

(i) For each fixed deterministic continuous flow $\boldsymbol{\nu} = (\nu_t)_{0 \leqslant t \leqslant T}$ of probability measures on $\mathbb{R}^d \times A$ (where the closed convex subset $A \subset \mathbb{R}^k$ denotes the set of admissible values for the controls), solve the standard stochastic control problem

$$\inf_{\alpha \in \mathbb{A}} J^{\boldsymbol{\nu}}(\boldsymbol{\alpha}), \quad \text{with } J^{\boldsymbol{\nu}}(\boldsymbol{\alpha}) = \mathbb{E}\left[\int_0^T f(t, X_t^{\boldsymbol{\alpha}}, \nu_t, \alpha_t)dt + g(X_T^{\boldsymbol{\alpha}}, \mu_T)\right],$$

subject to (4.112)

$$\begin{cases} dX_t^{\boldsymbol{\alpha}} = b(t, X_t^{\boldsymbol{\alpha}}, \nu_t, \alpha_t)dt + \sigma(t, X_t^{\boldsymbol{\alpha}}, \nu_t, \alpha_t)dW_t, & t \in [0, T], \\ X_0^{\boldsymbol{\alpha}} = \xi, \end{cases}$$

where μ_T (respectively μ_t for $t \in [0, T]$) denotes the first marginal of ν_T on $\mathbb{R}^d$ (respectively the first marginal of ν_t on $\mathbb{R}^d$).

(ii) Find a flow $\boldsymbol{\nu} = (\nu_t)_{0 \leq t \leq T}$ so that, for all $t \in [0, T]$, $\mathcal{L}(\hat{X}_t^{\nu}, \hat{\alpha}_t^{\nu}) = \nu_t$, if $\hat{\boldsymbol{\alpha}}^{\nu} \in \mathbb{A}$ is a minimizer of $J^{\boldsymbol{\nu}}$ with $\hat{\boldsymbol{X}}^{\nu}$ as optimal path.

Implicitly, the coefficients b, σ, and f are now given as (measurable) mappings from $[0, T] \times \mathbb{R}^d \times \mathcal{P}(\mathbb{R}^d \times A) \times A$ to $\mathbb{R}^d$, $\mathbb{R}^{d \times d}$ and $\mathbb{R}$ respectively. In line with what we have done so far, we shall restrict the discussion to the case when σ is independent of the control: in that case, σ reads as a mapping from $[0, T] \times \mathbb{R}^d \times \mathcal{P}(\mathbb{R}^d \times A)$ into $\mathbb{R}^{d \times d}$.

Notice though that g remains the same, as in the previous chapters: The terminal boundary condition only feels the terminal state and the terminal distribution of the state of a typical agent.

Revisiting the Analytic Approach

Since σ is assumed to be independent of the control, we can work with the *reduced Hamiltonian* instead of the full Hamiltonian associated with the optimization control problem (4.112). The form of the reduced Hamiltonian is the same as in (3.5), except that the variable μ has to be replaced by ν, namely:

$$H(t,x,\nu,y,\alpha) = b(t,x,\nu,\alpha)\cdot y + f(t,x,\nu,\alpha), \tag{4.113}$$

for $t \in [0,T]$, $x, y \in \mathbb{R}^d$, $\alpha \in A$ and $\nu \in \mathcal{P}(\mathbb{R}^d \times A)$. In particular, Lemma 3.3 may be generalized at no cost and under a straightforward adaptation of assumption **Minimization of the Hamiltonian** introduced in Chapter 3, we can find, for any $t \in [0,T]$, $x \in \mathbb{R}^d$ and $\nu \in \mathcal{P}(\mathbb{R}^d \times A)$, a unique minimizer $\hat{\alpha}(t,x,\nu,y) \in \text{argmin}_{\alpha \in A} H(t,x,\nu,y,\alpha)$.

Following Subsection 3.1.5, we denote by V the value function of the optimization problem (i). When the flow of probability measures $\boldsymbol{\nu} = (\nu_t)_{0\leqslant t\leqslant T}$ is fixed, V is the solution of the HJB equation (the reader may want to compare with the first equation (3.12)):

$$\begin{aligned}\partial_t V(t,x) &+ \frac{1}{2}\text{trace}\Big[\big(\sigma\sigma^\dagger\big)(t,x,\nu_t)\partial^2_{xx}V(t,x)\Big] \\ &+ H\Big(t,x,\nu_t,\partial_x V(t,x),\hat{\alpha}\big(t,x,\nu_t,\partial_x V(t,x)\big)\Big) = 0,\end{aligned} \tag{4.114}$$

in $[0,T]\times\mathbb{R}^d$, with $V(T,\cdot) = g(\cdot,\mu_T)$ as terminal condition. The following simple observation will turn out to be instrumental in the subsequent analysis. The optimal feedback associated with the optimization problem has the form:

$$[0,T]\times\mathbb{R}^d \ni (t,x) \mapsto \hat{\alpha}\big(t,x,\nu_t,\partial_x V(t,x)\big),$$

which implies in particular that the optimal control in (i) takes the Markovian form:

$$\hat{\alpha}^{\nu}_t = \tilde{\alpha}\big(t,\hat{X}^{\nu}_t,\nu_t\big), \quad t \in [0,T],$$

for the function $\tilde{\alpha}$ defined as $\tilde{\alpha}(t,x,\nu) = \hat{\alpha}(t,x,\nu,\partial_x V(t,x))$. Therefore, for any $t \in [0,T]$, the law of $(\hat{X}^{\nu}_t,\hat{\alpha}^{\nu}_t)$ appears as the pushed forward image of the law of $\hat{X}^{\nu}_t$ since:

$$\mathcal{L}\big(\hat{X}^{\nu}_t,\hat{\alpha}^{\nu}_t\big) = \mathcal{L}\big(\hat{X}^{\nu}_t\big)\circ\big(I_d,\hat{\alpha}(t,\cdot,\nu_t,\partial_x V(t,\cdot))\big)^{-1}.$$

Now, the equilibrium condition reads:

$$\nu_t = \mathcal{L}\big(\hat{X}^{\nu}_t,\hat{\alpha}^{\nu}_t\big) = \mathcal{L}\big(\hat{X}^{\nu}_t\big)\circ\big(I_d,\hat{\alpha}(t,\cdot,\nu_t,\partial_x V(t,\cdot))\big)^{-1}, \quad t \in [0,T].$$

Consequently, the fixed point condition (ii) for the flow $\boldsymbol{\nu} = (\nu_t)_{0\leqslant t\leqslant T}$ of joint distributions of the state and the control can be rewritten as:

$$\begin{cases} \mu_t = \mathcal{L}(\hat{X}_t^\nu), \\ \nu_t = \mu_t \circ \left(I_d, \hat{\alpha}(t, \cdot, \nu_t, \partial_x V(t, \cdot))\right)^{-1}, \end{cases} \quad t \in [0, T], \tag{4.115}$$

where μ_t is the first marginal of ν_t on $\mathbb{R}^d$. Moreover, the analogue of the forward-backward PDE system (3.12) which characterizes the equilibrium has the form:

$$\begin{cases} \partial_t V(t,x) + \frac{1}{2}\text{trace}\Big[\big(\sigma\sigma^\dagger\big)(t,x,\nu_t)\partial^2_{xx} V(t,x)\Big] \\ \qquad\qquad + H\Big(t,x,\nu_t,\partial_x V(t,x), \hat{\alpha}(t,x,\nu_t,\partial_x V(t,x))\Big) = 0, \\ \partial_t \mu_t - \frac{1}{2}\text{trace}\Big[\partial^2_{xx}\Big(\big(\sigma\sigma^\dagger\big)(t,x,\nu_t)\mu_t\Big)\Big] \\ \qquad\qquad + \text{div}_x\Big(b\big(t,x,\nu_t,\hat{\alpha}(t,x,\nu_t,\partial_x V(t,x))\big)\mu_t\Big) = 0, \end{cases} \tag{4.116}$$

in $[0, T] \times \mathbb{R}^d$, with $V(T, \cdot) = g(\cdot, \mu_T)$ as terminal condition for the first equation, and $\mu_0 = \mathcal{L}(\xi)$ as initial condition for the second equation.

In comparison with the standard case when the mean field interaction is only through the states, the new feature is the second relationship in (4.115), which provides in equilibrium, an implicit expression for the flow $\boldsymbol{\nu} = (\nu_t)_{0 \leqslant t \leqslant T}$ of joint distributions of both the state and the control in terms of the flow $\boldsymbol{\mu} = (\mu_t)_{0 \leqslant t \leqslant T}$ of marginal distributions of the state. Of course, a natural question is to identify cases in which this implicit expression is uniquely solvable. In order to do so, we shall restrict ourselves to flows of probability measures with values in $\mathcal{P}_2(\mathbb{R}^d \times A)$:

Lemma 4.60 *Let assumption* **Minimization of the Hamiltonian** *be in force, the measure argument being in $\mathcal{P}_2(\mathbb{R}^d \times A)$ in lieu of $\mathcal{P}_2(\mathbb{R}^d)$. Assume also that, for any $t \in [0, T]$ and $\nu \in \mathcal{P}_2(\mathbb{R}^d \times A)$, $f(t, \cdot, \nu, \cdot)$ is at most of quadratic growth in (x, α) and that, for any $t \in [0, T]$, $\partial_\alpha f(t, \cdot, \cdot, \cdot)$ is L-Lipschitz continuous in (x, ν, α), for some constant $L \geqslant 0$.*

If, for any $t \in [0, T]$, any probability distribution $\mu \in \mathcal{P}_2(\mathbb{R}^d)$, and any Borel-measurable mappings ψ and ψ' from $\mathbb{R}^d$ into A, it holds:

$$\begin{aligned} &\int_{\mathbb{R}^d} \Big(\big[f\big(t,x,\mu \circ (I_d, \psi(\cdot))^{-1}, \psi(x)\big) - f\big(t,x,\mu \circ (I_d, \psi'(\cdot))^{-1}, \psi(x)\big)\big] \\ &\qquad - \big[f\big(t,x,\mu \circ (I_d, \psi(\cdot))^{-1}, \psi'(x)\big) - f\big(t,x,\mu \circ (I_d, \psi'(\cdot))^{-1}, \psi'(x)\big)\big]\Big) d\mu(x) \\ &\geqslant 0, \end{aligned} \tag{4.117}$$

then, t and μ being fixed, for any Borel-measurable function $\phi : \mathbb{R}^d \to \mathbb{R}^d$ in $L^2(\mathbb{R}^d, \mu; \mathbb{R}^d)$, there exists a unique (up to a μ-negligible Borel subset of $\mathbb{R}^d$) square integrable function $\psi : \mathbb{R}^d \to A$, such that $\nu = \mu \circ (I_d, \psi)^{-1}$ satisfies:

$$\nu = \mu \circ \Big(I_d, \hat{\alpha}\big(t, \cdot, \nu, \phi(\cdot)\big)\Big)^{-1}.$$

Proof. We argue by a continuation argument like in some proofs of existence of solutions to FBSDEs. We prove by induction that, for any $\delta \in [0,1]$, any λ-convex function f as in the statement, any function $f_0 : [0,T]\times\mathbb{R}^d\times A \to \mathbb{R}$ satisfying **(A2)** in assumption **Minimization of the Hamiltonian**, with $f_0(t,\cdot,\cdot)$ being at most of quadratic growth and $\partial_\alpha f_0(t,\cdot,\cdot)$ being L-Lipschitz continuous for any $t \in [0,T]$, and any function $\phi \in L^2(\mathbb{R}^d,\mu;\mathbb{R}^d)$ with values in $\mathbb{R}^d$, the equation:

$$\nu = \mu \circ \Big(I_d, \hat{\alpha}_\delta\big(t,\cdot,\nu,\phi(\cdot)\big)\Big)^{-1}, \tag{4.118}$$

has a unique solution $\nu \in \mathcal{P}_2(\mathbb{R}^d \times A)$ for any $t \in [0,T]$ and $\mu \in \mathcal{P}_2(\mathbb{R}^d)$, where:

$$\forall x,y \in \mathbb{R}^d, \quad \hat{\alpha}_\delta\big(t,x,\nu,y\big) = \text{argmin}_{\alpha\in A}\big[\big(b_2(t)\alpha\big)\cdot y + f_\delta(t,x,\nu,\alpha)\big],$$

and

$$f_\delta(t,x,\nu,\alpha) = \delta f\big(t,x,\nu,\alpha\big) + (1-\delta) f_0\big(t,x,\alpha\big).$$

First Step. We start with the case $\delta = 0$. The result is obviously true since f_0 and thus $\hat{\alpha}_0$ are independent of the argument ν. Therefore, we can just denote $\hat{\alpha}_0(t,x,\nu,y)$ by $\hat{\alpha}_0(t,x,y)$ and the only solution to (4.118) must be given by the function $\psi(\cdot) = \hat{\alpha}_0\big(t,\cdot,\phi(\cdot)\big)$. By Lemma 3.3, such a function ψ is square-integrable because:

$$\forall x \in \mathbb{R}^d, \quad \big|\hat{\alpha}_0\big(t,x,\phi(x)\big)\big| \leqslant c\big(1 + |x| + |\phi(x)|\big), \tag{4.119}$$

and $\phi \in L^2(\mathbb{R}^d,\mu;\mathbb{R}^d)$, the constant c depending upon f_0.

Second Step. Let us assume that for some $\delta \in [0,1)$, for any f_0 and ϕ as above, the equation (4.118) has a unique solution in $\mathcal{P}_2(\mathbb{R}^d \times A)$. Then, for $\epsilon \in [0,1)$ with $\delta + \epsilon \leqslant 1$, we define the mapping Ψ by:

$$\Psi : \mathcal{P}_2(\mathbb{R}^d \times A) \ni \nu \mapsto \nu' \in \mathcal{P}_2(\mathbb{R}^d \times A),$$

where ν' solves the equation:

$$\nu' = \mu \circ \Big(I_d, \hat{\alpha}_\delta^\nu\big(t,\cdot,\nu',\phi(\cdot)\big)\Big)^{-1}, \tag{4.120}$$

where, for a given ν, $\hat{\alpha}_\delta^\nu$ is defined by:

$$\begin{aligned}
&\forall x,y \in \mathbb{R}^d, \ \forall \nu' \in \mathcal{P}_2(\mathbb{R}^d \times A),\\
&\quad \hat{\alpha}_\delta^\nu\big(t,x,\nu',y\big) = \text{argmin}_{\alpha\in A}\big[\big(b_2(t)\alpha\big)\cdot y + f_\delta^\nu\big(t,x,\nu',\alpha\big)\big],
\end{aligned}$$

the function f_δ^ν depending upon the measure ν through the choice of a new function f_0, namely:

$$\begin{aligned}
f_\delta^\nu\big(t,x,\nu',\alpha\big) &= \delta f\big(t,x,\nu',\alpha\big) + \epsilon f\big(t,x,\nu,\alpha\big) + \big(1-(\delta+\epsilon)\big) f_0(t,x,\alpha)\\
&= \delta f\big(t,x,\nu',\alpha\big) + (1-\delta)\tilde{f}_0(t,x,\alpha),
\end{aligned}$$

with

$$\tilde{f}_0(t,x,\alpha) = \frac{\epsilon}{1-\delta} f(t,x,\nu,\alpha) + \frac{1-(\delta+\epsilon)}{1-\delta} f_0(t,x,\alpha).$$

Therefore, f_δ^ν is covered by the induction assumption if we use for f_0 the new function $\tilde{f}_0$.

The induction assumption implies that equation (4.120) has a unique solution in $\mathcal{P}_2(\mathbb{R}^d \times A)$ guaranteeing that the mapping Ψ is well defined. Observe then that any fixed point ν of Ψ satisfies:

$$\nu = \mu \circ \Big(I_d, \hat{\alpha}_\delta^\nu\big(t,\cdot,\nu,\phi(\cdot)\big)\Big)^{-1},$$

providing a solution of the equation (4.118) with δ replaced by $\delta+\epsilon$. Conversely, any solution of the equation (4.118), with δ replaced by $\delta+\epsilon$, is a fixed point of the mapping Ψ. Therefore, in order to prove that (4.118), with δ replaced by $\delta+\epsilon$, is uniquely solvable, it suffices to prove that Ψ is a contraction on $\mathcal{P}_2(\mathbb{R}^d \times A)$ for the Wasserstein distance W_2.

Third Step. We now prove that, for ϵ small enough, the function Ψ is a contraction. Given ν_1 and $\nu_2 \in \mathcal{P}_2(\mathbb{R}^d \times A)$, we call ν_1' and ν_2' their respective images by Ψ, and we denote by X a random variable with distribution μ. Then, by optimality of $\hat{\alpha}_\delta^\nu(t,\cdot,\nu',\phi(\cdot))$, we have:

$$\begin{aligned}
&\big[b_2(t)\hat{\alpha}_\delta^{\nu_1}(t,X,\nu_1',\phi(X))\big]\cdot\phi(X) + f_\delta^{\nu_1}\Big(t,X,\nu_1',\hat{\alpha}_\delta^{\nu_1}(t,X,\nu_1',\phi(X))\Big)\\
&\qquad + \lambda\big|\hat{\alpha}_\delta^{\nu_1}(t,X,\nu_1',\phi(X)) - \hat{\alpha}_\delta^{\nu_2}(t,X,\nu_2',\phi(X))\big|^2\\
&\qquad \leqslant \big[b_2(t)\hat{\alpha}_\delta^{\nu_2}(t,X,\nu_2',\phi(X))\big]\cdot\phi(X) + f_\delta^{\nu_1}\Big(t,X,\nu_1',\hat{\alpha}_\delta^{\nu_2}(t,X,\nu_2',\phi(X))\Big)\\
&\big[b_2(t)\hat{\alpha}_\delta^{\nu_2}(t,X,\nu_2',\phi(X))\big]\cdot\phi(X) + f_\delta^{\nu_2}\Big(t,X,\nu_2',\hat{\alpha}_\delta^{\nu_2}(t,X,\nu_2',\phi(X))\Big)\\
&\qquad + \lambda\big|\hat{\alpha}_\delta^{\nu_1}(t,X,\nu_1',\phi(X)) - \hat{\alpha}_\delta^{\nu_2}(t,X,\nu_2',\phi(X))\big|^2\\
&\qquad \leqslant \big[b_2(t)\hat{\alpha}_\delta^{\nu_1}(t,X,\nu_1',\phi(X))\big]\cdot\phi(X) + f_\delta^{\nu_2}\Big(t,X,\nu_2',\hat{\alpha}_\delta^{\nu_1}(t,X,\nu_1',\phi(X))\Big).
\end{aligned}$$

Summing these two inequalities we get:

$$\begin{aligned}
&f_\delta^{\nu_1}\Big(t,X,\nu_1',\hat{\alpha}_\delta^{\nu_1}(t,X,\nu_1',\phi(X))\Big) + f_\delta^{\nu_2}\Big(t,X,\nu_2',\hat{\alpha}_\delta^{\nu_2}(t,X,\nu_2',\phi(X))\Big)\\
&\qquad + 2\lambda\big|\hat{\alpha}_\delta^{\nu_1}(t,X,\nu_1',\phi(X)) - \hat{\alpha}_\delta^{\nu_2}(t,X,\nu_2',\phi(X))\big|^2\\
&\qquad \leqslant f_\delta^{\nu_1}\Big(t,X,\nu_1',\hat{\alpha}_\delta^{\nu_2}(t,X,\nu_2',\phi(X))\Big) + f_\delta^{\nu_2}\Big(t,X,\nu_2',\hat{\alpha}_\delta^{\nu_1}(t,X,\nu_1',\phi(X))\Big).
\end{aligned}$$

Rearranging the terms we get:

$$\begin{aligned}
& f_\delta^{\nu_1}\Big(t, X, \nu_1', \hat{\alpha}_\delta^{\nu_1}(t, X, \nu_1', \phi(X))\Big) - f_\delta^{\nu_2}\Big(t, X, \nu_2', \hat{\alpha}_\delta^{\nu_1}(t, X, \nu_1', \phi(X))\Big) \\
& \quad + f_\delta^{\nu_2}\Big(t, X, \nu_2', \hat{\alpha}_\delta^{\nu_2}(t, X, \nu_2', \phi(X))\Big) - f_\delta^{\nu_1}\Big(t, X, \nu_1', \hat{\alpha}_\delta^{\nu_2}(t, X, \nu_2', \phi(X))\Big) \\
& \quad + 2\lambda \big|\hat{\alpha}_\delta^{\nu_1}(t, X, \nu_1', \phi(X)) - \hat{\alpha}_\delta^{\nu_2}(t, X, \nu_2', \phi(X))\big|^2 \\
& \leqslant 0.
\end{aligned} \tag{4.121}$$

Now, the expectation of the four terms on the two first lines is equal to:

$$\begin{aligned}
& \mathbb{E}\Big[f_\delta^{\nu_1}\Big(t, X, \nu_1', \hat{\alpha}_\delta^{\nu_1}(t, X, \nu_1', \phi(X))\Big) - f_\delta^{\nu_2}\Big(t, X, \nu_2', \hat{\alpha}_\delta^{\nu_1}(t, X, \nu_1', \phi(X))\Big)\Big] \\
& \qquad + \mathbb{E}\Big[f_\delta^{\nu_2}\Big(t, X, \nu_2', \hat{\alpha}_\delta^{\nu_2}(t, X, \nu_2', \phi(X))\Big) - f_\delta^{\nu_1}\Big(t, X, \nu_1', \hat{\alpha}_\delta^{\nu_2}(t, X, \nu_2', \phi(X))\Big)\Big] \\
& \quad = \delta\Big(\mathbb{E}\Big[f\Big(t, X, \nu_1', \hat{\alpha}_\delta^{\nu_1}(t, X, \nu_1', \phi(X))\Big) - f\Big(t, X, \nu_2', \hat{\alpha}_\delta^{\nu_1}(t, X, \nu_1', \phi(X))\Big)\Big] \\
& \qquad\qquad - \mathbb{E}\Big[f\Big(t, X, \nu_1', \hat{\alpha}_\delta^{\nu_2}(t, X, \nu_2', \phi(X))\Big) - f\Big(t, X, \nu_2', \hat{\alpha}_\delta^{\nu_2}(t, X, \nu_2', \phi(X))\Big)\Big]\Big) \\
& \qquad + \epsilon\Big(\mathbb{E}\Big[f\Big(t, X, \nu_1, \hat{\alpha}_\delta^{\nu_1}(t, X, \nu_1', \phi(X))\Big) - f\Big(t, X, \nu_2, \hat{\alpha}_\delta^{\nu_1}(t, X, \nu_1', \phi(X))\Big)\Big] \\
& \qquad\qquad - \mathbb{E}\Big[f\Big(t, X, \nu_1, \hat{\alpha}_\delta^{\nu_2}(t, X, \nu_2', \phi(X))\Big) - f\Big(t, X, \nu_2, \hat{\alpha}_\delta^{\nu_2}(t, X, \nu_2', \phi(X))\Big)\Big]\Big).
\end{aligned}$$

Since the random vector $(X, \hat{\alpha}^{\nu_1}(t, X, \nu_1', \phi(X)))$ (respectively $(X, \hat{\alpha}^{\nu_2}(t, X, \nu_2', \phi(X)))$) has exactly ν_1' (respectively ν_2') as distribution, we deduce from (4.121) and from the monotonicity property of f that:

$$\begin{aligned}
& 2\lambda \mathbb{E}\Big[\big|\hat{\alpha}_\delta^{\nu_1}(t, X, \nu_1', \phi(X)) - \hat{\alpha}_\delta^{\nu_2}(t, X, \nu_2', \phi(X))\big|^2\Big] \\
& \quad \leqslant \epsilon \Big|\mathbb{E}\Big[f\Big(t, X, \nu_1, \hat{\alpha}_\delta^{\nu_1}(t, X, \nu_1', \phi(X))\Big) - f\Big(t, X, \nu_2, \hat{\alpha}_\delta^{\nu_1}(t, X, \nu_1', \phi(X))\Big)\Big] \\
& \qquad - \mathbb{E}\Big[f\Big(t, X, \nu_1, \hat{\alpha}_\delta^{\nu_2}(t, X, \nu_2', \phi(X))\Big) - f\Big(t, X, \nu_2, \hat{\alpha}_\delta^{\nu_2}(t, X, \nu_2', \phi(X))\Big)\Big]\Big|.
\end{aligned}$$

Thanks to the regularity properties of f, the term between absolute values in the right-hand side is less than:

$$C W_2(\nu_1, \nu_2) \mathbb{E}\Big[\big|\hat{\alpha}_\delta^{\nu_1}(t, X, \nu_1', \phi(X)) - \hat{\alpha}_\delta^{\nu_2}(t, X, \nu_2', \phi(X))\big|^2\Big]^{1/2},$$

for a constant C which only depends on the parameter L in the assumption. In particular, C is independent of δ and of f_0. Therefore,

$$2\lambda\mathbb{E}\Big[\big|\hat{\alpha}_{\delta}^{\nu_1}(t,X,\nu_1',\phi(X)) - \hat{\alpha}_{\delta}^{\nu_2}(t,X,\nu_2',\phi(X))\big|^2\Big]$$
$$\leqslant C\epsilon W_2(\nu_1,\nu_2)\mathbb{E}\Big[\big|\hat{\alpha}_{\delta}^{\nu_1}(t,X,\nu_1',\phi(X)) - \hat{\alpha}_{\delta}^{\nu_2}(t,X,\nu_2',\phi(X))\big|^2\Big]^{1/2}.$$

Allowing the constant C to increase from line to line if necessary, we deduce that:

$$\mathbb{E}\Big[\big|\hat{\alpha}_{\delta}^{\nu_1}(t,X,\nu_1',\phi(X)) - \hat{\alpha}_{\delta}^{\nu_2}(t,X,\nu_2',\phi(X))\big|^2\Big]^{1/2} \leqslant C\epsilon W_2(\nu_1,\nu_2).$$

Using again the fact that $(X,\hat{\alpha}^{\nu_1}(t,X,\nu_1',\phi(X)))$ (respectively $(X,\hat{\alpha}^{\nu_2}(t,X,\nu_2',\phi(X)))$) has exactly ν_1' (respectively ν_2') as distribution, we notice that the left-hand side is greater than $W_2(\nu_1',\nu_2')$, which finally yields:

$$W_2(\nu_1',\nu_2') \leqslant C\epsilon W_2(\nu_1,\nu_2).$$

This shows that the mapping Ψ is a contraction for $C\epsilon < 1$. Therefore, for $C\epsilon \leqslant 1/2$, Equation (4.118), with δ replaced by $\delta+\epsilon$, has a unique solution.

Final Step. Since the constant C, in the condition $C\epsilon \leqslant 1/2$, is independent of δ, we can apply a straightforward induction argument to prove that (4.118) is uniquely solvable for any $\delta \in [0,1]$, which completes the proof. □

Examples

We now provide three important examples of function f satisfying (4.117).

Example 1. If the function f is of the form:

$$f(t,x,\nu,\alpha) = f_0(t,x,\nu) + f_1(t,x,\mu,\alpha),$$

where μ denotes the first marginal of ν on $\mathbb{R}^d$, and $f_0 : [0,T]\times\mathbb{R}^d\times\mathcal{P}_2(\mathbb{R}^d\times A) \to \mathbb{R}$ and $f_1 : [0,T]\times\mathbb{R}^d\times\mathcal{P}_2(\mathbb{R}^d)\times A \to \mathbb{R}$ are sufficiently regular, then assumption (4.117) is satisfied since the left-hand side is identically 0.

Example 2. Consider now a Borel-measurable function $h : [0,T]\times\mathbb{R}^d\times\mathcal{P}_2(\mathbb{R}^d)\times A\times\mathcal{P}_2(A) \to \mathbb{R}$ such that, for any $t\in[0,T]$, $x\in\mathbb{R}^d$ and $\mu\in\mathcal{P}_2(\mathbb{R}^d)$, the function:

$$A\times\mathcal{P}_2(A) \ni (\alpha,\theta) \mapsto h(t,x,\mu,\alpha,\theta)$$

satisfies:

$$|h(t,x,\mu,\alpha,\theta)| \leqslant C\big(1+|x|+|\alpha|+M_2(\mu)+M_2(\theta)\big)^2, \tag{4.122}$$

together with the Lasry-Lions monotonicity condition:

$$\forall \theta,\theta' \in \mathcal{P}_2(A), \qquad \int_A \big[h(t,x,\mu,\alpha,\theta) - h(t,x,\mu,\alpha,\theta')\big] d\big(\theta-\theta'\big)(\alpha) \geqslant 0. \tag{4.123}$$

Then, the function f given by:

$$f(t,x,\nu,\alpha) = h\big(t,x,\mu,\alpha,Q(x,\cdot)\big),$$

where μ denotes the first marginal of ν on $\mathbb{R}^d$, and for μ-almost every $x \in \mathbb{R}^d$, $Q(x,d\alpha)$ is the regular conditional distribution of ν given that the first component is x, also satisfies (4.117). Indeed, it suffices to take $\theta = \delta_{\psi(x)}$ and $\theta' = \delta_{\psi'(x)}$ in (4.123) in order to prove (4.117).

Example 3. Still another example is provided by any function $f : [0,T] \times \mathbb{R}^d \times \mathcal{P}_2(\mathbb{R}^d \times A) \times A \to \mathbb{R}$ satisfying:

$$|f(t,x,\nu,\alpha)| \leqslant C\big(1 + |x| + |\alpha| + M_2(\nu)\big)^2, \tag{4.124}$$

together with the Lasry-Lions monotonicity condition on the whole $\mathbb{R}^d \times A$, namely:

$$\begin{aligned} &\forall \nu, \nu' \in \mathcal{P}_2(\mathbb{R}^d \times A), \\ &\int_{\mathbb{R}^d \times A} [f(t,x,\nu,\alpha) - f(t,x,\nu',\alpha)] d\big(\nu - \nu'\big)(x,\alpha) \geqslant 0. \end{aligned} \tag{4.125}$$

Indeed, condition (4.117) is checked by choosing $\nu = \mu \circ (I_d, \psi)^{-1}$ and $\nu' = \mu \circ (I_d, \psi')^{-1}$.

Recall that examples of functions satisfying the Lasry-Lions monotonicity condition were given in Section 3.4. These examples can easily be adapted to the current framework. Nevertheless, it is more challenging to combine the monotonicity and convexity properties. Examples 1, 2 and 3 of Section 3.4.2 clearly do. However, Examples 4, 5 and 6 do not! As for Example 7 in Section 3.4.2, notice that the function:

$$h(\alpha,\theta) = \int_{\mathbb{R}^k} L\big(\beta, \rho * \theta(\beta)\big)\rho(\alpha - \beta)d\beta, \quad \alpha \in A, \ \theta \in \mathcal{P}_2(A),$$

where $L : \mathbb{R}^k \times [0,+\infty) \ni (x,y) \mapsto L(x,y)$ is twice differentiable and convex in both variables on $A \times [0,+\infty)$, and nondecreasing in the second variable on $[0,+\infty)$, and ρ is nonnegative, even, smooth, with compact support, satisfies:

$$\begin{aligned} \partial^2_\alpha h(\alpha,\theta) = \int_{\mathbb{R}^k} \Big[&\partial^2_{xx} L\big(\beta, \rho * \theta(\beta)\big) + 2\partial^2_{xy} L\big(\beta, \rho * \theta(\beta)\big) \otimes \big(\partial\rho * \theta(\beta)\big) \\ &+ \partial^2_{yy} L\big(\beta, \rho * \theta(\beta)\big)\big(\partial\rho * \theta(\beta)\big)^{\otimes 2}\Big] \rho(\alpha - \beta) d\beta \\ &+ \int_{\mathbb{R}^k} \partial_y L\big(\beta, \rho * \theta(\beta)\big)\big(\partial^2\rho * \theta(\beta)\big)\rho(\alpha - \beta)d\beta. \end{aligned}$$

When the set A is bounded, ρ may be assumed to be convex on the set $A - A = \{\alpha - \beta, \alpha, \beta \in A\}$. Since $\partial_y L$ is nonnegative and θ is supported by A, the matrix $\partial^2_\alpha h$ inherits the convexity properties of L.

FBSDE Formulation

In the spirit of Chapter 3, we may characterize solutions of extended mean field games by means of an FBSDE of the McKean-Vlasov type.

Using the Representation of the HJB Equation. As in the standard case, two strategies are possible. The first one is to represent the value function of the game as in Proposition 3.11 and Theorem 4.44, and the second one is to use the stochastic maximum principle as we did in Proposition 3.21. In both cases, the main issue is the identification of the analogue of relationship (4.115) which provides an implicit expression of ν_t in terms of its first marginal μ_t. When representing the value function of the game in Proposition 3.13, the gradient of the solution of the Hamilton-Jacobi-Bellman equation (4.114) which appears in (4.115), is connected with the martingale integrand $(\hat{Z}_t)_{0\leqslant t\leqslant T}$ which appears in the FBSDE formulation of the game. A quick glance at Proposition 3.13 shows that the analogue of (4.115) should be:

$$\begin{aligned}\nu_t &= \mathcal{L}\Big(\hat{X}_t, \hat{\alpha}\big(t, \hat{X}_t, \nu_t, \sigma(t, X_t, \nu_t)^{-1\dagger}\hat{Z}_t\big)\Big) \\ &= \mathcal{L}\big(\hat{X}_t, \sigma(t, \hat{X}_t, \nu_t)^{-1\dagger}\hat{Z}_t\big) \circ \Big(\mathbb{R}^d \times \mathbb{R}^d \ni (x, y) \mapsto \big(x, \hat{\alpha}(t, x, \nu_t, y)\big)\Big)^{-1},\end{aligned} \tag{4.126}$$

for $t \in [0, T]$, where $(\hat{X}_t, \hat{Y}_t, \hat{Z}_t)_{0\leqslant t\leqslant T}$ denotes the solution of the associated FBSDE:

$$\begin{cases} d\hat{X}_t = b\big(t, \hat{X}_t, \nu_t, \hat{\alpha}\big(t, \hat{X}_t, \nu_t, \hat{\sigma}(t, \hat{X}_t, \nu_t)^{-1\dagger} Z_t\big)\big)dt + \sigma(t, \hat{X}_t, \nu_t)dW_t, \\ d\hat{Y}_t = -f\big(t, \hat{X}_t, \nu_t, \hat{\alpha}\big(t, \hat{X}_t, \nu_t, \sigma(t, \hat{X}_t, \nu_t)^{-1\dagger}\hat{Z}_t\big)\big)dt + \hat{Z}_t \cdot dW_t, \end{cases} \tag{4.127}$$

with $\hat{X}_0 = \xi$ as initial condition and $\hat{Y}_T = g(\hat{X}_T, \mu_T)$ as terminal condition, where μ_T denotes the first marginal of ν_T on $\mathbb{R}^d$.

The next result is given without proof because it can be proven following the steps of the proof of Lemma 4.60.

Lemma 4.61 *Let assumption* **Minimization of the Hamiltonian** *be in force, the measure argument in the coefficients being in $\mathcal{P}_2(\mathbb{R}^d \times A)$ in lieu of $\mathcal{P}_2(\mathbb{R}^d)$. Assume also that, for any $t \in [0, T]$ and $\nu \in \mathcal{P}_2(\mathbb{R}^d \times A)$, $f(t, \cdot, \nu, \cdot)$ is at most of quadratic growth in (x, α) and that, for any $t \in [0, T]$, $\partial_\alpha f(t, \cdot, \cdot, \cdot)$ is L-Lipschitz continuous in (x, ν, α), for some constant $L \geqslant 0$.*

Assume further that, for any $t \in [0, T]$, $\mu \in \mathcal{P}_2(\mathbb{R}^d)$, and transition probability kernels $(Q(x, \cdot))_{x\in\mathbb{R}^d}$ and $(Q'(x, \cdot))_{x\in\mathbb{R}^d}$ from $\mathbb{R}^d$ to A, it holds that:

$$\int_{\mathbb{R}^d}\Big(\int_A \big[f(t,x,\nu,\alpha)-f(t,x,\nu',\alpha)\big]\big[Q(x,d\alpha)-Q'(x,d\alpha)\big]\Big)d\mu(x)\geq 0,$$

where the finite measures ν and ν' are defined by:

$$\nu(D)=\int_{\mathbb{R}^d}\Big[\int_A \mathbf{1}_D(x,\alpha)Q(x,d\alpha)\Big]d\mu(x),$$
$$\nu'(D)=\int_{\mathbb{R}^d}\Big[\int_A \mathbf{1}_D(x,\alpha)Q'(x,d\alpha)\Big]d\mu(x),$$

for $D\in\mathcal{B}(\mathbb{R}^d\times A)$. Then, for any joint distribution π on $\mathbb{R}^d\times\mathbb{R}^d$, there exists a unique distribution $\nu\in\mathcal{P}_2(\mathbb{R}^d\times A)$, which we shall denote $\Pi(t,\pi)$, such that:

$$\nu=\pi\circ\Big(\mathbb{R}^d\times\mathbb{R}^d\ni(x,y)\mapsto\big(x,\hat\alpha(t,x,\nu,y)\big)\in\mathbb{R}^d\times A\Big)^{-1}.$$

The proof of Lemma 4.61 is similar to that of Lemma 4.60. A crucial fact in the proof is that, for any $\pi\in\mathcal{P}_2(\mathbb{R}^d\times\mathbb{R}^d)$ and $\nu\in\mathcal{P}_2(\mathbb{R}^d\times A)$, the first marginal of $\pi\circ(\mathbb{R}^d\times\mathbb{R}^d\ni(x,y)\mapsto(x,\hat\alpha(t,x,\nu,y))\in\mathbb{R}^d\times A)^{-1}$ is equal to the first marginal of π on $\mathbb{R}^d$ and is thus independent of ν. This permits to recover the same framework as in the proof of Lemma 4.60: the first marginal of π, denoted by μ in the proof of Lemma 4.60, is entirely fixed.

When σ only depends on $\nu\in\mathcal{P}_2(\mathbb{R}^d\times A)$ through its first marginal $\mu\in\mathcal{P}_2(\mathbb{R}^d)$ on $\mathbb{R}^d$ and under the assumption of Lemma 4.61, Equation (4.126) has a unique solution:

$$\nu_t=\Pi\Big(t,\mathcal{L}\big(\hat X_t,\sigma(t,\hat X_t,\mathcal{L}(\hat X_t))^{-1\dagger}\hat Z_t\big)\Big),$$

which we may rewrite, without any ambiguity, $\Pi'(t,\mathcal{L}(\hat X_t,\hat Z_t))$. Then, the FBSDE (4.127) rewrites:

$$\begin{cases} d\hat X_t = b\Big(t,\hat X_t,\Pi'(t,\mathcal{L}(\hat X_t,\hat Z_t)), \\ \qquad\qquad \hat\alpha\big(t,\hat X_t,\Pi'(t,\mathcal{L}(\hat X_t,\hat Z_t)),\sigma(t,\hat X_t,\mathcal{L}(\hat X_t))^{-1\dagger}\hat Z_t\big)\Big)dt \\ \qquad\qquad\qquad +\sigma\big(t,\hat X_t,\mathcal{L}(\hat X_t)\big)dW_t, \\ d\hat Y_t = -f\Big(t,\hat X_t,\Pi'(t,\mathcal{L}(\hat X_t,\hat Z_t)), \\ \qquad\qquad \hat\alpha\big(t,\hat X_t,\Pi'(t,\mathcal{L}(\hat X_t,\hat Z_t)),\sigma(t,\hat X_t,\mathcal{L}(\hat X_t))^{-1\dagger}\hat Z_t\big)\Big)dt \\ \qquad\qquad\qquad +\hat Z_t\cdot dW_t, \end{cases} \tag{4.128}$$

with $\hat X_0=\xi$ as initial condition and $\hat Y_T=g(\hat X_T,\mathcal{L}(\hat X_T))$ as terminal condition.

Based on Theorem 4.45, the analog of Proposition 3.11 (the control problem being understood in the strong instead of weak sense) reads:

Proposition 4.62 *Let assumption* **MFG Solvability HJB** *be in force, the measure argument in the coefficients being in $\mathcal{P}_2(\mathbb{R}^d \times A)$ in lieu of $\mathcal{P}_2(\mathbb{R}^d)$ except in σ, which is still assumed to be a function from $[0,T] \times \mathbb{R}^d \times \mathcal{P}_2(\mathbb{R}^d)$ into $\mathbb{R}^{d\times d}$. Then, a continuous flow of measures $\boldsymbol{\nu} = (\nu_t)_{0\leqslant t\leqslant T}$ from $[0,T]$ to $\mathcal{P}_2(\mathbb{R}^d \times A)$ is an MFG equilibrium if and only if*

$$\nu_t = \Pi'(t, \mathcal{L}(\hat{X}_t, \hat{Z}_t)),$$

where $(\hat{\boldsymbol{X}}, \hat{\boldsymbol{Y}}, \hat{\boldsymbol{Z}})$ solves the McKean-Vlasov FBSDE (4.128).

Remark 4.63 *The assumption that σ only depends on ν through μ is well understood. When σ depends on the full measure ν, Equation* (4.126) *becomes:*

$$\nu_t = \Pi\Big(t, \mathcal{L}\big(\hat{X}_t, \sigma(t, \hat{X}_t, \nu_t)^{-1\dagger}\hat{Z}_t\big)\Big),$$

which is still implicit.

A possible way to tackle this problem is to let $\hat{Z}'_t = \sigma(t, \hat{X}_t, \nu_t)^{-1\dagger}\hat{Z}_t$, in which case the fixed point condition becomes:

$$\nu_t = \Pi\big(t, \mathcal{L}(\hat{X}_t, \hat{Z}'_t)\big),$$

and then to represent the backward equation accordingly with:

$$\left(\int_0^t \hat{Z}'_s \cdot \Big(\sigma(s, \hat{X}_s, \nu_s)dW_s\Big)\right)_{0\leqslant t\leqslant T}$$

as martingale part. Obviously, it would require a new analysis.

Using the Stochastic Maximum Principle. When using the Pontryagin stochastic maximum principle, to derive an analog of Proposition 3.23, the gradient of the solution of the Hamilton-Jacobi-Bellman equation is no longer related to the process $\boldsymbol{Z} = (Z_t)_{0\leqslant t\leqslant T}$ appearing in the FBSDE formulation of the game, but with the process $\boldsymbol{Y} = (Y_t)_{0\leqslant t\leqslant T}$ instead. In particular, repeating the above discussion shows that, under the assumption of Lemma 4.61, the analogue of (4.115) reads:

$$\nu_t = \Pi(t, \mathcal{L}(X_t, Y_t)), \tag{4.129}$$

where $(\boldsymbol{X}, \boldsymbol{Y}, \boldsymbol{Z}) = (X_t, Y_t, Z_t)_{0\leqslant t\leqslant T}$ now denotes the solution of the FBSDE:

$$
\begin{cases}
dX_t &= b\Big(t, X_t, \Pi\big(t, \mathcal{L}(X_t, Y_t)\big), \hat{\alpha}\big(t, X_t, \Pi(t, \mathcal{L}(X_t, Y_t)), Y_t\big)\Big)dt \\
&\qquad\qquad\qquad\qquad +\sigma dW_t, \\
dY_t &= -\partial_x H\Big(t, X_t, \Pi\big(t, \mathcal{L}(X_t, Y_t)\big), Y_t, \\
&\qquad\qquad \hat{\alpha}\big(t, X_t, \Pi(t, \mathcal{L}(X_t, Y_t)), Y_t\big)\Big)dt \\
&\qquad\qquad\qquad\qquad +Z_t dW_t,
\end{cases}
\tag{4.130}
$$

with $X_0 = \xi$ as initial condition and $Y_T = \partial_x g(X_T, \mathcal{L}(X_T))$ as terminal condition. Above, we assumed σ to be constant as we did in the statement of Proposition 3.23.

In this case, the analog of Proposition 3.23 reads:

Proposition 4.64 *Under the assumption of Definition 3.22, the measure argument in the coefficients being in $\mathcal{P}_2(\mathbb{R}^d \times A)$, a continuous flow of measures $\boldsymbol{\nu} = (\nu_t)_{0 \leqslant t \leqslant T}$ from $[0, T]$ to $\mathcal{P}_2(\mathbb{R}^d \times A)$ is an MFG equilibrium if and only if $\nu_t = \Pi(t, \mathcal{L}(X_t, Y_t))$ for any $t \in [0, T]$, where $(\boldsymbol{X}, \boldsymbol{Y}, \boldsymbol{Z})$ solves the McKean-Vlasov FBSDE* (4.130).

Example. Following Example 1 right above, we know that, when the running cost f is of the form:

$$
f(t, x, \nu, \alpha) = f_0(t, x, \nu) + f_1(t, x, \mu, \alpha),
$$

where μ denotes the first marginal of ν on $\mathbb{R}^d$, $f_0 : [0, T] \times \mathbb{R}^d \times \mathcal{P}_2(\mathbb{R}^d \times A) \to \mathbb{R}$ and $f_1 : [0, T] \times \mathbb{R}^d \times \mathcal{P}_2(\mathbb{R}^d) \times A \to \mathbb{R}$, the minimizer $\hat{\alpha}$ of the Hamiltonian H depends upon ν through μ only and writing $\hat{\alpha}(t, \cdot, \mu, \cdot)$ instead of $\hat{\alpha}(t, \cdot, \nu, \cdot)$, the fixed point mapping $\Pi(t, \pi)$ has the explicit expression:

$$
\Pi(t, \pi) = \pi \circ \big(\mathbb{R}^d \times \mathbb{R}^d \ni (x, z) \mapsto (x, \hat{\alpha}(t, x, \mu, z)) \in \mathbb{R}^d \times A\big)^{-1},
$$

where μ is here given as the first marginal of π on $\mathbb{R}^d$.

In the application to the model of price impact (recall Chapter 1) which we solve in the next section the situation is even simpler. Indeed in this model, the contributions of ν and α to the drift and the running cost functions are separated as in:

$$
\begin{cases}
b(t, x, \nu, \alpha) = b_0(t, x, \nu) + b_1(t, x, \alpha), \\
f(t, x, \nu, \alpha) = f_0(t, x, \nu) + f_1(t, x, \alpha),
\end{cases}
$$

implying that the minimizer $\hat{\alpha}$ is independent of ν. In this case, the condition $\nu_t = \Pi(t, \mathcal{L}(X_t, Y_t))$ reduces to $\nu_t = \mathcal{L}(X_t, \hat{\alpha}(t, X_t, Y_t))$.

Existence and Uniqueness

We shall not address existence of a solution in full generality. We just notice that Equation (4.130) fits the general form (4.32) investigated in Subsection 4.3.1. This may suffice whenever the (quite demanding) assumptions required in Subsection 4.3.1 are satisfied. We treat below an example which does not satisfy the growth conditions used in Subsection 4.3.1.

We now briefly discuss the uniqueness part. Following Section 3.4, we can prove that, provided that g satisfies the standard Lasry-Lions monotonicity condition and f satisfies the version (4.124)–(4.125) on $\mathbb{R}^d \times A$ (instead of $\mathbb{R}^d$ only), then uniqueness holds. The proof is a straightforward adaptation of that of Theorem 3.29.

Typical Solution

Motivated by the model of price impact which was introduced in Subsection 1.3.2 of Chapter 1, and which we shall solve in the next section, we choose the following specific set of assumptions to illustrate the applicability of the approach based on the stochastic maximum principle discussed above.

Assumption (EMFG). The set A is closed and convex and the coefficients b, σ, f, and g are defined on $[0,T]\times\mathbb{R}^d\times A$, $[0,T]\times\mathbb{R}^d$, $[0,T]\times\mathbb{R}^d\times\mathcal{P}_2(A)\times A$, and $\mathbb{R}^d$ respectively and they satisfy, for two constants $\lambda, L > 0$:

(A1) The volatility σ is constant and the drift is an affine function of (x,α):

$$b(t,x,\alpha) = b_0(t) + b_1(t)x + b_2(t)\alpha,$$

where b_0, b_1 and b_2 are measurable bounded functions from $[0,T]$ into $\mathbb{R}^d$, $\mathbb{R}^{d\times d}$ and $\mathbb{R}^{d\times k}$ respectively.

(A2) The function $f : [0,T]\times\mathbb{R}^d\times\mathcal{P}_2(A)\times A \ni (t,x,\theta,\alpha) \mapsto f(t,x,\theta,\alpha) \in \mathbb{R}$ is of the form:

$$f(t,x,\theta,\alpha) = x\cdot f_0(t,\theta) + f_1(t,x,\alpha),$$

$f_0 : [0,T]\times\mathcal{P}_2(A) \to \mathbb{R}^d$ being measurable, bounded by L and continuous in θ, and $f_1 : [0,T]\times\mathbb{R}^d\times A \to \mathbb{R}$ being measurable and once continuously differentiable with respect to (x,α) and having Lipschitz-continuous derivatives (so that $f(t,\cdot,\theta,\cdot)$ is $C^{1,1}$), the Lipschitz constant in x and α being bounded by L (so that it is uniform in t and θ). Moreover, f_1 satisfies the following strong form of convexity:

$$f_1(t,x',\alpha') - f_1(t,x,\alpha) - (x'-x,\alpha'-\alpha)\cdot\partial_{(x,\alpha)}f_1(t,x,\alpha) \geqslant \lambda|\alpha'-\alpha|^2.$$

(continued)

Recall that the notation $\partial_{(x,\alpha)} f_1$ stands for the gradient in the joint variables (x, α). Finally, f_1, $\partial_x f_1$ and $\partial_\alpha f_1$ are locally bounded over $[0, T] \times \mathbb{R}^d \times A$.

(A3) The function g is differentiable and its derivative is Lipschitz continuous. Moreover, g is convex in the sense that:

$$g(x') - g(x) - \partial_x g(x) \cdot (x' - x) \geqslant 0, \qquad x, x' \in \mathbb{R}^d.$$

Notice that in the present context, the measure $\theta \in \mathcal{P}_2(A)$ should be understood as the distribution of the control, namely the second marginal of the measure ν used throughout this section.

Theorem 4.65 *Under assumption* **EMFG** *and for some* $\xi \in L^2(\Omega, \mathcal{F}_0, \mathbb{P}; \mathbb{R}^d)$ *as initial condition, the extended mean field game has a solution.*

Proof. We make use of Proposition 4.64. Under the standing assumption, the McKean-Vlasov FBSDE (4.130) takes the form:

$$\begin{cases} dX_t &= b\big(t, X_t, \hat{\alpha}(t, X_t, Y_t)\big)dt + \sigma dW_t, \\ dY_t &= -\big[\partial_x H_1\big(t, X_t, Y_t, \hat{\alpha}(t, X_t, Y_t)\big) + f_0\big(t, \mathcal{L}\big(\hat{\alpha}(t, X_t, Y_t)\big)\big)\big]dt + Z_t dW_t, \end{cases}$$

for $t \in [0, T]$, with $X_0 = \xi$ as initial condition and $Y_T = \partial_x g(X_T)$ as terminal boundary condition. Above, H_1 denotes the Hamiltonian:

$$H_1(t, x, y, \alpha) = b(t, x, \alpha) \cdot y + f_1(t, x, \alpha), \quad t \in [0, T], \ x, y \in \mathbb{R}^d, \ \alpha \in A,$$

and $\hat{\alpha}(t, x, y)$ is the unique minimizer of the function $A \ni \alpha \mapsto H_1(t, x, y, \alpha)$.

First Step. We proceed as in the analysis of the system (4.50). For a given continuous flow of measures $\boldsymbol{\theta} = (\theta_t)_{0 \leqslant t \leqslant T}$ with values in $\mathcal{P}_2(A)$, we consider the system:

$$\begin{cases} dX_t &= b\big(t, X_t, \hat{\alpha}(t, X_t, Y_t)\big)dt + \sigma dW_t, \\ dY_t &= -\big[\partial_x H_1\big(t, X_t, Y_t, \hat{\alpha}(t, X_t, Y_t)\big) + f_0(t, \theta_t)\big]dt + Z_t dW_t, \end{cases} \tag{4.131}$$

for $t \in [0, T]$, with $X_0 = \xi$ as initial condition and $Y_T = \partial_x g(X_T)$ as terminal boundary condition.

Following the proof of Lemma 4.56, the system (4.131) is uniquely solvable and we may call $u^{\boldsymbol{\theta}}$ its decoupling field. Also, we can find a constant c, independent of $\boldsymbol{\theta}$, such that, for all $t \in [0, T]$ and $x, x' \in \mathbb{R}^d$,

$$\big|u^{\boldsymbol{\theta}}(t, x') - u^{\boldsymbol{\theta}}(t, x)\big| \leqslant c|x' - x|. \tag{4.132}$$

Second Step. We now consider the system (4.131) but without $(f_0(t,\theta_t))_{0\leqslant t\leqslant T}$ in the backward equation:

$$\begin{cases} dX'_t &= b\big(t, X'_t, \hat{\alpha}(t, X'_t, Y'_t)\big)dt + \sigma dW_t, \\ dY'_t &= -\partial_x H_1\big(t, X'_t, Y'_t, \hat{\alpha}(t, X'_t, Y'_t)\big)dt + Z'_t dW_t, \end{cases} \tag{4.133}$$

for $t \in [0,T]$, with $X'_0 = \xi$ as initial condition and $Y'_T = \partial_x g(X'_T)$ as terminal boundary condition.

Below, we let:

$$\hat{\alpha}_t = \hat{\alpha}(t, X_t, Y_t), \quad \hat{\alpha}'_t = \hat{\alpha}(t, X'_t, Y'_t), \quad t \in [0,T].$$

Computing the Itô differential of the process:

$$\bigg((X'_t - X_t)\cdot Y_t + \int_0^t \big[X'_t \cdot f_0(t,\theta_t) + f_1(s, X'_s, \hat{\alpha}'_s) - X_t \cdot f_0(t,\theta_t) - f_1(s, X_s, \hat{\alpha}_s)\big] ds \bigg)_{0\leqslant t\leqslant T},$$

and following the proof of Proposition 3.21, we can prove that:

$$\begin{aligned} &\mathbb{E}\bigg[\int_0^T \big[X_t \cdot f_0(t,\theta_t) + f_1(t, X_t, \hat{\alpha}_t)\big]dt + g(X_T)\bigg] + \lambda \mathbb{E}\int_0^T |\hat{\alpha}_t - \hat{\alpha}'_t|^2 dt \\ &\quad \leqslant \mathbb{E}\bigg[\int_0^T \big[X'_t \cdot f_0(t,\theta_t) + f_1(t, X'_t, \hat{\alpha}'_t)\big]dt + g(X'_T)\bigg]. \end{aligned} \tag{4.134}$$

Proceeding the other way round, we get:

$$\begin{aligned} &\mathbb{E}\bigg[\int_0^T f_1(t, X'_t, \hat{\alpha}'_t)dt + g(X'_T)\bigg] + \lambda \mathbb{E}\int_0^T |\hat{\alpha}_t - \hat{\alpha}'_t|^2 dt \\ &\quad \leqslant \mathbb{E}\bigg[\int_0^T f_1(t, X_t, \hat{\alpha}_t)dt + g(X_T)\bigg]. \end{aligned} \tag{4.135}$$

Summing the two last inequalities (4.134) and (4.135), we get:

$$\mathbb{E}\int_0^T X_t \cdot f_0(t,\theta_t)dt + 2\lambda \mathbb{E}\int_0^T |\hat{\alpha}_t - \hat{\alpha}'_t|^2 dt \leqslant \mathbb{E}\int_0^T X'_t \cdot f_0(t,\theta_t)dt.$$

Since f_0 is bounded, we easily deduce that:

$$\mathbb{E}\int_0^T |\hat{\alpha}_t - \hat{\alpha}'_t|^2 dt \leqslant c,$$

the constant c being allowed to increase from line to line, as long as it remains independent of $\boldsymbol{\theta}$.

Proceeding as in the proof of Lemma 4.56, we obtain that:

$$\mathbb{E}\big[|Y_0 - Y'_0|^2\big] \leqslant c.$$

It is well checked that the above bound is independent of ξ. We then deduce that $|u^{\theta}(0,0)| \leqslant c$. More generally, we have:

$$\sup_{0 \leqslant t \leqslant T} |u^{\theta}(t,0)| \leqslant c. \tag{4.136}$$

Third Step. We now have all the ingredients to follow the proof of Theorem 4.39. To do so, we call $\boldsymbol{X}^{\theta}$ and $\boldsymbol{Y}^{\theta}$ the forward and backward components of the solution to (4.131).

Proceeding as in the proof of Theorem 4.39, we deduce that there exists a compact subset $\mathfrak{K} \subset \mathcal{C}([0,T], \mathcal{P}_2(\mathbb{R}^d))$ such that, for any input $\boldsymbol{\theta}$ as above, the path $(\mathcal{L}(X_t^{\theta}))_{0 \leqslant t \leqslant T}$ is in $\mathfrak{K}$.

Since u^{θ} satisfies (4.132) and (4.136), we also deduce that, for any $\boldsymbol{\theta}$ as above, there exists a compact subset $\mathcal{K} \subset \mathcal{P}_2(\mathbb{R}^d)$ such that, for any $t \in [0,T]$, $\mathcal{L}(Y_t^{\theta}) \in \mathcal{K}$.

Thanks to (4.132) and (4.136) once again, we may proceed as in (4.9) and deduce that $|u^{\theta}(t,x) - u^{\theta}(s,x)| \leqslant c(1+|x|)|t-s|^{1/2}$, from which we get that:

$$\mathbb{E}\big[|Y_t^{\theta} - Y_s^{\theta}|^2\big] \leqslant c|t-s|,$$

where the constant c is, as we already explained, independent of $\boldsymbol{\theta}$. Following the proof of Theorem 4.39 and modifying if necessary the compact subset $\mathfrak{K} \subset \mathcal{C}([0,T]; \mathcal{P}_2(\mathbb{R}^d))$, we can prove that, for all $\boldsymbol{\theta}$, $(\mathcal{L}(Y_t^{\theta}))_{0 \leqslant t \leqslant T} \in \mathfrak{K}$.

Since $\hat{\alpha}$ is known to be Lipschitz continuous in (x,y) and locally bounded, see Lemma 3.3, we deduce that there exists a compact subset $\mathfrak{K}' \subset \mathcal{C}([0,T], \mathcal{P}_2(\mathbb{R}^k))$ such that, for any $\boldsymbol{\theta}$, $(\mathcal{L}(\hat{\alpha}(t, X_t^{\theta}, Y_t^{\theta})))_{0 \leqslant t \leqslant T}$ belongs to $\mathfrak{K}'$. We then conclude as in the proof of Theorem 4.39 the existence of $\boldsymbol{\theta} \in \mathcal{C}([0,T], \mathcal{P}_2(\mathbb{R}^k))$ such that

$$\forall t \in [0,T], \quad \theta_t = \mathcal{L}\big(\hat{\alpha}(t, X_t^{\theta}, Y_t^{\theta})\big).$$

By construction of $\hat{\alpha}$, it holds that $\theta_t \in \mathcal{P}_2(A)$ for all $t \in [0,T]$. □

4.7 Examples

4.7.1 The Price Impact MFG Model

The price impact model presented in Subsection 1.3.2 of Chapter 1 led to the mean field game model in which for each fixed flow $\boldsymbol{\theta} = (\theta_t)_{0 \leqslant t \leqslant T}$ of probability measures on A, a typical player minimizes the quantity:

$$J(\boldsymbol{\alpha}) = \mathbb{E}\bigg[\int_0^T f(t, X_t, \theta_t, \alpha_t)dt + g(X_T)\bigg] \tag{4.137}$$

under the dynamical constraint:

$$dX_t = \alpha_t dt + \sigma dW_t, \qquad t \in [0,T]; \quad X_0 = \xi,$$

with a running cost function f of the form:

$$f(t,x,\theta,\alpha) = c_\alpha(\alpha) + c_X(x) - x\int_A h d\theta, \qquad t \in [0,T],\ x \in \mathbb{R}^d,\ \theta \in \mathcal{P}(A),\ \alpha \in A, \tag{4.138}$$

where c_α and h are deterministic measurable functions on the control space A. They model the individual price impact and the cost of trading at a given rate. The functions c_X and g are defined on $\mathbb{R}^d$. They model running and terminal liquidation constraints penalizing unwanted inventories. The fixed point step imposes the condition $\theta_t = \mathcal{L}(\hat{\alpha}_t)$ where $\hat{\boldsymbol{\alpha}} = (\hat{\alpha}_t)_{0\leqslant t\leqslant T}$ is the optimal control minimizing (4.137).

We assume that the functions c_X and g are quadratic and that the function c_α is strongly convex in the sense that its second derivative is bounded away from 0. The function h is assumed to be uniformly Lipschitz continuous. In most practical applications, it is even assumed to be linear. Such a particular case is known as the Almgren-Chriss linear price impact model. For the numerical illustrations given below we shall choose (recall that $d = m = k = 1$):

$$f(t,x,\theta,\alpha) = \frac{c_\alpha}{2}\alpha^2 + \frac{c_X}{2}x^2 - x\int_{\mathbb{R}} h d\theta, \quad \text{and} \quad g(x) = \frac{c_g}{2}x^2,$$

and $b(t,x,\alpha) = \alpha$ and $\sigma(t,x) = \sigma > 0$. Here, $t \in [0,T]$, $x \in \mathbb{R}$, $\theta \in \mathcal{P}_2(\mathbb{R})$ and $\alpha \in A = \mathbb{R}$. Above, c_X and c_α are strictly positive constants and m is nonnegative. Also, we assume that $\xi = x^0 \in \mathbb{R}$ is deterministic.

When h is continuous and bounded, assumption **EMFG** is satisfied and one can use Theorem 4.65 to conclude existence of a solution to the mean field game model. We pursue the analysis with the goal of a constructive identification of the solutions and numerical illustrations.

Numerical Results

The Hamiltonian

$$H(t,x,y,\theta,\alpha) = \alpha y + \frac{c_\alpha}{2}\alpha^2 + \frac{c_X}{2}x^2 - x\int_{\mathbb{R}} h d\theta$$

is minimized (in α) for $\hat{\alpha} = -y/c_\alpha$ and the McKean-Vlasov FBSDE we need to solve is:

$$\begin{cases} dX_t = -\frac{1}{c_\alpha} Y_t\, dt + \sigma dW_t, \\ dY_t = \big[- c_X X_t + \mathbb{E}[h(-Y_t/c_\alpha)]\big]dt + Z_t dW_t, \qquad t \in [0,T], \\ X_0 = x^0, \quad Y_T = c_g X_T. \end{cases} \tag{4.139}$$

This form is consistent with Proposition 4.64 since $\nu_t = \Pi(t, \mathcal{L}(X_t, Y_t)) = \mathcal{L}(X_t, -Y_t/c_\alpha)$, recall $\hat{\alpha}(t,x,\nu,y) = -y/c_\alpha$.

We first consider the case of a linear price impact function h, say $h(\alpha) = \bar{h}\alpha$. Under this extra assumption, assumption **EMFG** is not satisfied any longer, but the problem becomes particularly simple because the McKean-Vlasov FBSDE (4.139) is now affine:

$$\begin{cases} dX_t = -\frac{1}{c_\alpha} Y_t\,dt + \sigma dW_t, \\ dY_t = \big[- c_X X_t - \frac{\bar{h}}{c_\alpha}\mathbb{E}[Y_t]\big]dt + Z_t dW_t, \qquad t \in [0,T], \\ X_0 = x^0, \quad Y_T = c_g X_T. \end{cases} \tag{4.140}$$

We could approach existence and uniqueness of a solution to (4.140) by means of our earlier analysis of McKean-Vlasov FBSDEs. However, we shall opt for a direct approach in hope to get solutions via explicit formulas which could then be used for numerical computations. Also, notice that we cannot directly apply the results derived in Section 3.5 because the dependence upon the distribution in (4.140) is through the expectation of the adjoint variable Y_t instead of being through the expectation of the state variable X_t. However, we can still implement the same solution strategy by looking for an affine decoupling field $Y_t = \eta_t X_t + \chi_t$ given by two deterministic functions $(\eta_t)_{0\leqslant t\leqslant T}$ and $(\chi_t)_{0\leqslant t\leqslant T}$. Like in Section 3.5, we first need to compute the mean functions $\bar{x}_t = \mathbb{E}[X_t]$ and $\bar{y}_t = \mathbb{E}[Y_t]$.

Taking expectations on both sides of (4.140), we get:

$$\begin{cases} \dot{\bar{x}}_t = -\frac{1}{c_\alpha}\bar{y}_t, \\ \dot{\bar{y}}_t = -c_X\bar{x}_t - \frac{\bar{h}}{c_\alpha}\bar{y}_t, \qquad t \in [0,T], \\ \bar{x}_0 = x^0 \quad \bar{y}_T = c_g\bar{x}_T. \end{cases} \tag{4.141}$$

This is a particular case of the system (3.53) which we solved by relying on the ansatz $\bar{y}_t = \bar{\eta}_t\bar{x}_t + \bar{\chi}_t$. In the present situation, the functions $\bar{\eta}$ and $\bar{\chi}$ can be identified as the solutions of the system of ODEs:

$$\begin{cases} \dot{\bar{\eta}}_t + \frac{\bar{h}}{c_\alpha}\bar{\eta}_t - \frac{1}{c_\alpha}\bar{\eta}_t^2 + c_X = 0, \\ \dot{\bar{\chi}}_t - [\frac{1}{c_\alpha}\bar{\eta}_t - \frac{\bar{h}}{c_\alpha}]\bar{\chi}_t = 0, \qquad t \in [0,T], \\ \bar{\eta}_T = c_g, \quad \bar{\chi}_T = 0. \end{cases} \tag{4.142}$$

The first equation is a one-dimensional Riccati equation. Recalling that both c_α and c_X are strictly positive and c_g is nonnegative and following (2.49)–(2.50), we get:

$$\bar{\eta}_t = \frac{-C\big(e^{(\delta^+-\delta^-)(T-t)} - 1\big) - c_g\big(\delta^+ e^{(\delta^+-\delta^-)(T-t)} - \delta^-\big)}{\big(\delta^- e^{(\delta^+-\delta^-)(T-t)} - \delta^+\big) - c_g B\left(e^{(\delta^+-\delta^-)(T-t)} - 1\right)}, \tag{4.143}$$

for $t \in [0,T]$, where $A = -\bar{h}/(2c_\alpha)$, $B = 1/c_\alpha$, $C = c_X$, $\delta^\pm = -A \pm \sqrt{R}$, with $R = A^2 + BC > 0$.

The second equation in (4.142) is a first order homogenous linear equation with terminal condition zero, so its solution is identically zero. Consequently the means $\bar{x}_t$ and $\bar{y}_t$ are given by:

$$\bar{x}_t = x^0 e^{-\frac{1}{c_\alpha}\int_0^t \bar{\eta}_u du}, \quad \text{and} \quad \bar{y}_t = x^0 \bar{\eta}_t e^{-\frac{1}{c_\alpha}\int_0^t \bar{\eta}_u du}, \qquad t \in [0,T]. \tag{4.144}$$

We can now go back to the McKean-Vlasov FBSDE (4.140) and solve for the equilibrium processes $\boldsymbol{X}$ and $\boldsymbol{Y}$. As explained above, we use the ansatz $Y_t = \eta_t X_t + \chi_t$ and substitute the quantity $\bar{y}_t$ just computed for $\mathbb{E}[Y_t]$. Computing the stochastic differential of $\boldsymbol{Y}$ using such an ansatz and the equations in (4.140), we find that these functions solve the system of ODEs:

$$\begin{cases} \dot{\eta}_t = \frac{1}{c_\alpha}\eta_t^2 - c_X, \\ \dot{\chi}_t = \frac{1}{c_\alpha}\eta_t \chi_t - \frac{\bar{h}}{c_\alpha}\bar{y}_t, \\ \eta_T = c_g, \qquad \chi_T = 0, \end{cases} \tag{4.145}$$

where we use the notation $\bar{y}_t$ for the expectation $\mathbb{E}[Y_t]$. The first equation is a Riccati equation which can be solved directly. Proceeding as above, we find:

$$\eta_t = -c_\alpha \sqrt{c_X/c_\alpha}\,\frac{c_\alpha\sqrt{c_X/c_\alpha} - c_g - (c_\alpha\sqrt{c_X/c_\alpha} + c_g)e^{2\sqrt{c_X/c_\alpha}(T-t)}}{c_\alpha\sqrt{c_X/c_\alpha} - c_g + (c_\alpha\sqrt{c_X/c_\alpha} + c_g)e^{2\sqrt{c_X/c_\alpha}(T-t)}}, \tag{4.146}$$

for $t \in [0,T]$. Once η is determined, one can inject its value (4.146) into the explicit solution of the second equation in (4.145) which reads:

$$\chi_t = \frac{\bar{h}}{c_\alpha}\int_t^T \bar{y}_s e^{-\frac{1}{c_\alpha}\int_t^s \eta_u du} ds, \qquad t \in [0,T]. \tag{4.147}$$

Observe that $Y_t = \eta_t X_t + \chi_t$ yields $\bar{y}_t = \eta_t \bar{x}_t + \chi_t$, so that $\chi_t = (\bar{\eta}_t - \eta_t)\bar{x}_t$.

As usual in linear quadratic models, the equilibrium state is Gaussian. Here, its dynamics are given by the Ornstein-Uhlenbeck like equation:

$$dX_t = -\frac{1}{c_\alpha}\big(\eta_t X_t + (\bar{\eta}_t - \eta_t)\bar{x}_t\big)dt + \sigma dW_t, \qquad t \in [0,T]\,; \quad X_0 = x^0.$$

Since $Y_t = \eta_t X_t + \chi_t$, the adjoint process and the optimal control process $\hat{\boldsymbol{\alpha}} = (\hat{\alpha}_t = -Y_t/c_\alpha)_{0 \leqslant t \leqslant T}$ are also Gaussian. Notice that $\hat{\alpha}_t \sim N(\mu_t^{(\alpha)}, [\sigma_t^{(\alpha)}]^2)$ with:

$$\mu_t^{(\alpha)} = -\frac{x^0}{c_\alpha}\bar{\eta}_t e^{-\frac{1}{c_\alpha}\int_0^t \bar{\eta}_u\, du} \quad \text{and} \quad [\sigma_t^{(\alpha)}]^2 = \frac{\sigma^2}{c_\alpha^2}\eta_t^2 \int_0^t e^{-\frac{2}{c_\alpha}\int_s^t \eta_u\, du} ds.$$

Figure 4.1 shows the time evolution of the density of the control process.

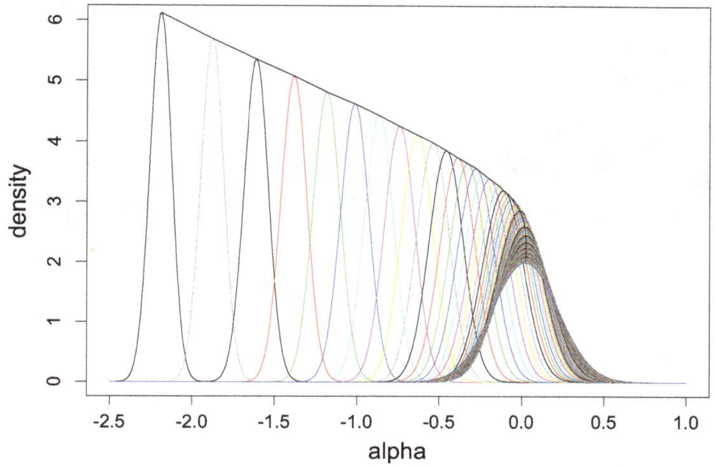

Fig. 4.1 Time evolution (for t ranging from 0.06 to $T = 1$) of the marginal density of the optimal rate of trading $\hat{\alpha}_t$ for a representative trader for the values $x^0 = 1$, $c_X = 0.1$, $c_g = 0.3$, $c_\alpha = 2$, $k = 10$ and $\sigma = 0.7$ of the parameters.

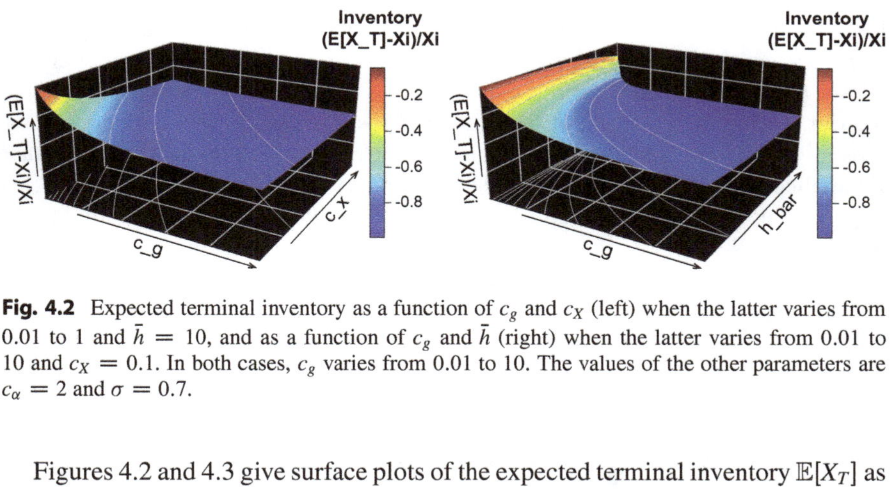

Fig. 4.2 Expected terminal inventory as a function of c_g and c_X (left) when the latter varies from 0.01 to 1 and $\bar{h} = 10$, and as a function of c_g and $\bar{h}$ (right) when the latter varies from 0.01 to 10 and $c_X = 0.1$. In both cases, c_g varies from 0.01 to 10. The values of the other parameters are $c_\alpha = 2$ and $\sigma = 0.7$.

Figures 4.2 and 4.3 give surface plots of the expected terminal inventory $\mathbb{E}[X_T]$ as a function of the various parameters of the model. Clearly, both plots of Figure 4.2 confirm the intuition that large values of the parameter c_g would force this terminal inventory to be small.

Figure 4.3 seems to indicate that the price impact parameter $\bar{h}$ does not have a large influence on the expected terminal inventory for large values of the parameters c_x and c_α. However, for small values of the parameters c_x and c_α, the expected terminal inventory seems to be a decreasing function of the price impact parameter $\bar{h}$.

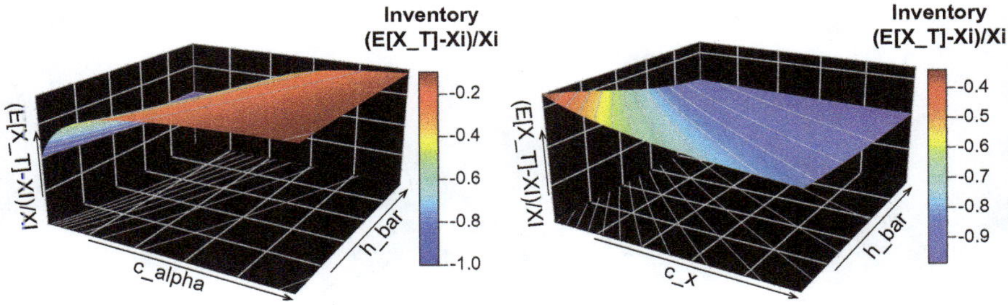

Fig. 4.3 Expected terminal inventory as a function of c_α and $\bar{h}$ (left) when the former varies from 0.01 to 10 and $c_X = 0.1$, and as a function of c_X and $\bar{h}$ (right) when the former varies from 0.01 to 10 and $c_\alpha = 2$. In both cases, $\bar{h}$ varies from 0.01 to 10. The values of the other parameters are $c_g = 1$ and $\sigma = 0.7$.

4.7.2 A Model of Crowd Motion with Congestion

We now study in detail the crowd congestion model introduced in Subsection 1.5.3 of Chapter 1.

We model the behavior of N individuals exiting an enclosed area such as a ballroom, or a theater. The room is modeled as a bounded closed convex polyhedron $D \subset \mathbb{R}^d$, and the exits comprise the connected components of a relatively closed subset E of the boundary ∂D (so that E itself is closed) with a nonempty relative interior. The convexity assumption is mostly for convenience as it is not needed for most of the theoretical arguments we use below. Non-convex models are important for applications. Indeed, domains with holes can be used to model physical obstacles (e.g., barriers, pillars, rows of seats, ...) impeding the motion of the individuals. Also, dumbell-like domains comprising thin corridors connecting convex bodies can provide realistic models for suites of rooms connected by narrow hallways or by staircases.

In the mean field game limit, the dynamics of the position of an individual are assumed to be given by a controlled reflected stochastic differential equation of the form:

$$dX_t = \alpha_t dt + dW_t + dK_t, \quad t \in [0, T], \tag{4.148}$$

where $\boldsymbol{W} = (W_t)_{0 \leqslant t \leqslant T}$ is a d-dimensional Wiener process on a complete filtered probability space $(\Omega, \mathcal{F}, \mathbb{F}, \mathbb{P})$ and $\boldsymbol{K} = (K_t)_{0 \leqslant t \leqslant T}$ is a continuous process with finite variation implementing the normal reflection de facto preventing $\boldsymbol{X}$ from exiting D. We shall give more details on $\boldsymbol{K}$ later, but we already notice that its sample paths only increase at the times of the set $\{t \in [0, T] : X_t \in \partial D\}$. The admissible control processes $\boldsymbol{\alpha} = (\alpha_t)_{0 \leqslant t \leqslant T}$ will be the $\mathbb{F}$-progressively measurable processes with values in a bounded closed convex subset $A \subset \mathbb{R}^d$. Except for the noise term dW_t and the forced reflection produced by dK_t, α_t is the unperturbed velocity and represents how the individual controls its motion through the room. An important quantity we will want to track is the exit time of the room given by the first hitting time of E by the process $\boldsymbol{X}$ defined as:

$$\tau = \inf\{t \in [0, T] : X_t \in E\}. \tag{4.149}$$

When an individual reaches a door, we consider that it is not part of the *game* any longer, and instead of letting the reflection take place, we use a standard procedure in the classical theory of Markov processes to send the individual to a point Δ added to the state space D and called the cemetery. We use the notation Δ for the cemetery to follow the tradition of the classical texts on Markov processes, hoping that it will not be confused with the notation Δ_x used for the Laplacian operator at the end of this subsection. Indeed the cemetery notation will only enter the definition of the extended state space $D^\Delta = D \cup \{\Delta\}$. For the sake of definiteness, we shall assume that $\Delta \in \mathbb{R}^d \setminus D$. We want to apologize to the reader for our willingness to abide by the standard notation and terminology that create this amusing oxymoron: we end up calling cemetery the place the individuals want to reach (since they want to leave the room) in the shortest amount of time !

For the *control step* of the mean field game problem, we fix a continuous flow $\boldsymbol{\mu} = (\mu_t)_{0 \leq t \leq T}$ of probability measures in $\mathcal{C}([0, T]; \mathcal{P}(D^\Delta))$, and to each admissible control $\boldsymbol{\alpha}$, we associate the cost:

$$J^{\mu}(\boldsymbol{\alpha}) = \mathbb{E}\left[\int_0^{T \wedge \tau} \left(\ell(X_t, \mu_t)\frac{|\alpha_t|^2}{2} + f(t)\right)dt\right], \tag{4.150}$$

where ℓ is a continuous function from $\mathbb{R}^d \times \mathcal{P}(D^\Delta)$ equipped with the product of the Euclidean and weak topologies, into a compact subset of $(0, +\infty)$, and f is an integrable function from $[0, T]$ to $\mathbb{R}_+$. Notice that the cost $J^{\mu}(\boldsymbol{\alpha})$ depends only upon the part of the trajectory of $\boldsymbol{X}$ before it reaches the exit, so modifying X_t for $t > \tau$ will have no effect on the cost. The function f penalizes long stays inside the domain D before exiting. It may be chosen to be identical equal to 1, in which case its contribution to the running cost represents the total time spent in the room. Finally, the function ℓ is intended to penalize the amount of energy spent (as given by the term $|\alpha_t|^2$ representing the kinetic energy) where there is congestion. A typical example of interest to us is:

$$\ell(x, \mu) = \varphi(1 + (\mu_{|D} * \rho)(x))$$

for an increasing continuous function φ from $\mathbb{R}_+$ into itself, and a smooth even compactly supported density $\rho : \mathbb{R}^d \to \mathbb{R}_+$. Here, μ is any probability measure on D^Δ, $\mu_{|D}$ denotes its restriction to D, and '*' stands for the standard convolution on Euclidean space.

The *fixed point step* of the mean field game problem can be formulated as follows. If $\boldsymbol{X}$ is the solution of the optimal control problem described above, we define the process X' by $X'_t = X_t$ for $t < \tau$ and $X'_t = \Delta$ whenever $t \geqslant \tau$, and we say that X' (or the flow μ) is a solution of the MFG problem if, for any $t \in [0, T]$, μ_t coincides with the distribution of X'_t (and not of X_t). Notice that in the law of X'_t, the interesting part is the measure $\mu_{t|D} = \mathbf{1}_D \mu_t$ which describes the statistical distribution of the individuals who have not exited yet by time t. This is not necessarily a probability measure since its total mass is the proportion of individuals still in D at time t. We could have avoided the introduction of the cemetery and used flows of sub-probability measures to define the costs and the mean field game equilibrium, but in order to use the tools developed in this book for probability measures we introduced the state process X' and the above definition for the MFG equilibrium. In case when ℓ is given as the convolution of $\mu_{|D}$ with ρ, we may choose for cemetery Δ a point satisfying $\Delta \notin \mathrm{supp}(\rho) + D$. This condition ensures that the convolution of μ and ρ does not feel the mass allocated to the cemetery.

Because the problem is set on a bounded domain instead of the whole Euclidean space $\mathbb{R}^d$, and because of the special type of mixed boundary conditions needed in this model, we cannot use directly the results derived in Subsections 4.4 and 4.5. However, we shall prove that similar arguments may be used to solve the MFG problem as derived from (4.148)–(4.150).

For the purpose of illustration we treat a numerical example at the end of this subsection.

Reflected Brownian Motion

In order to simplify the presentation, we perform the stochastic control step in its weak form, in full analogy with Subsection 3.3.1 in Chapter 3. The main reason is to avoid the technical discussion of (fully coupled) FBSDEs with random terminal times. Also, recall that we assume that the set A is bounded, an assumption which is often in force when using the weak formulation.

Imitating (3.28), we thus consider the uncontrolled dynamics:

$$dX_t = dW_t + dK_t, \quad t \in [0, T]; \qquad X_0 = \xi, \tag{4.151}$$

defined on some filtered complete probability space $(\Omega, \mathcal{F}, \mathbb{F}, \mathbb{P})$ satisfying the usual conditions, for an $\mathcal{F}_0$-measurable initial condition $X_0 = \xi$ whose law μ_0 is assumed to have a bounded density on the η-neighborhood $E^\eta = \{x \in D : d(x, E) < \eta\}$ of E for some $\eta > 0$. Equation (4.151) is the equation for the reflected Brownian motion in the domain D. The process K is a bounded variation process acting in a minimal way to prevent X from exiting D. It is required to satisfy:

$$|K|_t = \int_0^t \mathbf{1}_{\{X_s \in \partial D\}} d|K|_s, \quad K_t = \int_0^t n_s d|K|_s, \qquad t \in [0, T],$$

where for any $t \in [0, T]$, n_t is an inward pointing unit normal vector to ∂D at X_t. The first identity expresses the fact that K only acts when X is on the boundary.

The second one says that $\boldsymbol{K}$ is directed along an inward pointing unit normal to the boundary. Importantly, observe that for any $\mathbb{F}$-progressively measurable process $\boldsymbol{Z} = (Z_t)_{0 \leqslant t \leqslant T}$ with values in D, it holds, for all $t \in [0, T]$,

$$\int_0^t (Z_s - X_s) \cdot dK_s = \int_0^t \mathbf{1}_{\{X_s \in \partial D\}} (Z_s - X_s) \cdot n_s d|K|_s \geqslant 0, \tag{4.152}$$

the last inequality following from the fact that D is convex.

Precise references for the theory and the construction of reflected processes are given in the Notes & Complements at the end of the chapter.

Notice that the process $\boldsymbol{X}$ considered in this section is different from the process $\boldsymbol{X}$ considered earlier in the subsection since it does not contain the controlled drift. However the stopping time τ is defined in the same way as in (4.149), except that the infimum is taken over all $t \geq 0$. Recall also that we use implicitly the convention $\inf \emptyset = \infty$.

We shall need the following technical result whose proof we defer to the final paragraph of this subsection to avoid distracting from the logical steps toward the construction of a solution of the MFG problem.

Lemma 4.66 *For any $\epsilon \in (0, 1/2)$, the cumulative distribution function of $\tau = \inf\{t \geq 0 : X_t \in E\}$, namely the function $\mathbb{R}_+ \ni t \mapsto \mathbb{P}(\tau \leqslant t)$ is $(1/2 - \epsilon)$ Hölder continuous, uniformly in time. Moreover, $\mathbb{P}(\tau < \infty) = 1$ and for any $t > 0$, $\mathbb{P}(\tau \leqslant t) > 0$.*

Weak Formulation

Now, for any measure flow $\boldsymbol{\mu} = (\mu_t)_{0 \leqslant t \leqslant T}$ in $\mathcal{C}([0, T]; \mathcal{P}(D^{\Delta}))$ and any admissible control process $\boldsymbol{\alpha} = (\alpha_t)_{0 \leqslant t \leqslant T}$ (i.e., any $\mathbb{F}$-progressively measurable A-valued process), we define the probability $\mathbb{P}^{\boldsymbol{\mu}, \boldsymbol{\alpha}}$ on $(\Omega, \mathcal{F}_T)$ by:

$$\frac{d\mathbb{P}^{\boldsymbol{\mu}, \boldsymbol{\alpha}}}{d\mathbb{P}} = \exp\left(\int_0^T \alpha_t \cdot dW_t - \frac{1}{2} \int_0^T |\alpha_t|^2 dt \right).$$

Observe that $\mathbb{P}^{\boldsymbol{\mu}, \boldsymbol{\alpha}}$ is in fact independent of $\boldsymbol{\mu}$, the rationale for the exponent $\boldsymbol{\mu}$ being mostly for pedagogical reasons. Notice also that, under $\mathbb{P}^{\boldsymbol{\mu}, \boldsymbol{\alpha}}$, $\boldsymbol{X}$ is not a reflected Brownian motion any longer. It is the result of the reflection of a process which is a Brownian motion plus a drift given by $\int_0^t \alpha_s ds$, which is what we were looking for. Under the weak formulation, the cost associated with $\boldsymbol{\alpha}$ is:

$$J^{\boldsymbol{\mu}, \text{weak}}(\boldsymbol{\alpha}) = \mathbb{E}^{\boldsymbol{\mu}, \boldsymbol{\alpha}} \left[\int_0^{T \wedge \tau} \left[\frac{1}{2} \ell(X_t, \mu_t) |\alpha_t|^2 + f(t) \right] dt \right],$$

where we use the notation $\mathbb{E}^{\boldsymbol{\mu}, \boldsymbol{\alpha}}$ for the expectation with respect to the probability $\mathbb{P}^{\boldsymbol{\mu}, \boldsymbol{\alpha}}$. The reduced Hamiltonian H is independent of the boundary condition. It is given by the same formula:

$$H(t,x,\mu,y,\alpha) = \alpha \cdot y + \frac{1}{2}\ell(x,\mu)|\alpha|^2 + f(t).$$

A straightforward computation shows that the minimizer of the function $A \ni \alpha \mapsto H(t,x,\mu,y,\alpha)$ is equal to the orthogonal projection of $-y/\ell(x,\mu)$ onto the convex set A. With the same notation as above, we thus have $\hat{\alpha}(x,\mu,y) = \Pi_A(-y/\ell(x,\mu))$ where Π_A is the orthogonal projection onto A. Hence, the minimized Hamiltonian H^* is given by:

$$\begin{aligned} H^*(t,x,\mu,y) &= \inf_{\alpha \in A} H(t,x,\mu,y,\alpha) \\ &= \ell(x,\mu)\Big[\Pi_A\Big(-\frac{y}{\ell(x,\mu)}\Big)\cdot\frac{y}{\ell(x,\mu)} + \frac{1}{2}\Big|\Pi_A\Big(-\frac{y}{\ell(x,\mu)}\Big)\Big|^2\Big] + f(t). \end{aligned}$$

In the present context, Proposition 3.11 gives:

Proposition 4.67 *For any continuous flow $\boldsymbol{\mu} = (\mu_t)_{0\leqslant t\leqslant T}$ of probability measures on D^Δ, the BSDE:*

$$dY_t = -\mathbf{1}_{\{t\leqslant\tau\}} H\big(t, X_t, \mu_t, Z_t, \hat{\alpha}(X_t,\mu_t,Z_t)\big)dt + Z_t \cdot dW_t, \tag{4.153}$$

for $t \in [0,T]$, with terminal condition $Y_T = 0$, is uniquely solvable. Moreover, the control $\hat{\boldsymbol{\alpha}} = (\hat{\alpha}_t)_{0\leqslant t\leqslant T}$ defined by $\hat{\alpha}_t = \hat{\alpha}(X_t,\mu_t,Z_t)$ is the unique optimal control over the interval $[0,T]$ and the optimal cost of the problem is given by:

$$\inf_{\boldsymbol{\alpha}\in\mathbb{A}} J^{\boldsymbol{\mu},\text{weak}}(\boldsymbol{\alpha}) = Y_0. \tag{4.154}$$

In order to emphasize the dependence of the optimal control $\hat{\boldsymbol{\alpha}}$ upon $\boldsymbol{\mu}$, we shall denote it by $\hat{\boldsymbol{\alpha}}^{\boldsymbol{\mu}}$. It is worth mentioning that, with $\boldsymbol{Y} = (Y_t)_{0\leqslant t\leqslant T}$ as above, we have with $\mathbb{P}$-probability 1 that $Y_t = 0$ for any $t \in [\tau, T]$. We do not give the proof of Proposition 4.67 as it goes along the very same lines as for Proposition 3.11.

MFG Equilibrium in the Weak Formulation

According to Definition 3.12, solving the MFG problem under the weak formulation consists in finding a flow $\boldsymbol{\mu} = (\mu_t)_{0\leqslant t\leqslant T}$ of probability measures on D^Δ such that

$$\forall t \in [0,T], \quad \mu_t = \mathbb{P}^{\boldsymbol{\mu},\hat{\boldsymbol{\alpha}}^{\boldsymbol{\mu}}} \circ (X'_t)^{-1}, \quad \text{where } X'_t = \begin{cases} X_t & \text{if } t < \tau \\ \Delta & \text{otherwise} \end{cases}. \tag{4.155}$$

Defining the map Φ by:

$$\Phi : \mathcal{C}\big([0,T];\mathcal{P}(D^\Delta)\big) \ni \boldsymbol{\mu} \mapsto \big(\mathbb{P}^{\boldsymbol{\mu},\hat{\boldsymbol{\alpha}}^{\boldsymbol{\mu}}} \circ (X'_t)^{-1}\big)_{0\leqslant t\leqslant T}, \tag{4.156}$$

our strategy is to prove that Φ admits a fixed point by checking that Schauder's theorem can be used in the same way as in Subsection 4.3.2.

We start with the following simple remark. Since D^{Δ} is bounded and closed, $\mathcal{P}(D^{\Delta})$ coincides with $\mathcal{P}_2(D^{\Delta})$ and the topology of weak convergence usually considered on $\mathcal{P}(D^{\Delta})$ is the same as the topology given by the Wasserstein distance W_2 on $\mathcal{P}_2(D^{\Delta})$. Also, $\mathcal{P}(D^{\Delta})$ is a closed compact subset of $\mathcal{P}_2(\mathbb{R}^d)$. These facts were already mentioned in Chapter 1, and they will be discussed in detail in Chapter 5. For this reason, $\mathcal{C}([0,T];\mathcal{P}(D^{\Delta}))$ may be regarded as a closed convex subset of $\mathcal{C}([0,T];\mathcal{P}_2(\mathbb{R}^d))$, and we can work along the same lines as in Section 4.3. Below, we shall write $\mathcal{C}([0,T];\mathcal{P}_2(D^{\Delta}))$ to emphasize the fact that $\mathcal{P}(D^{\Delta})$ is equipped with the 2-Wasserstein distance and that $\mathcal{C}([0,T];\mathcal{P}(D^{\Delta}))$ is equipped with the supremum distance induced by W_2. Henceforth, we aim at applying Schauder's theorem as in Subsection 4.3.2, and for that, it suffices to prove that Φ is continuous and has a relatively compact range. This is proven in Lemmas 4.68 and 4.69 below, proving that the MFG problem (4.155) has a solution.

Lemma 4.68 *There exists a constant C such that, for any $\boldsymbol{\mu} \in \mathcal{C}([0,T];\mathcal{P}_2(D^{\Delta}))$,*

$$\forall s,t \in [0,T], \quad W_2\Big(\mathbb{P}^{\mu,\hat{\alpha}^{\mu}} \circ (X'_t)^{-1}, \mathbb{P}^{\mu,\hat{\alpha}^{\mu}} \circ (X'_s)^{-1}\Big) \leqslant C|t-s|^{1/8}.$$

In particular, Φ maps $\mathcal{C}([0,T];\mathcal{P}_2(D^{\Delta}))$ onto a relatively compact subset of the space $\mathcal{C}([0,T];\mathcal{P}_2(D^{\Delta}))$.

Proof. As we already mentioned, by Girsanov transformation, we know that under $\mathbb{P}^{\mu,\hat{\alpha}^{\mu}}$, $X = (X_t)_{0\leqslant t\leqslant T}$ satisfies the SDE:

$$dX_t = \hat{\alpha}\big(X_t, \mu_t, Z_t\big)dt + dW_t + dK_t.$$

Observe that, for any $0 \leqslant s \leqslant t \leqslant T$,

$$d_t|X_t - X_s|^2 = 2(X_t - X_s)\cdot\Big(\hat{\alpha}\big(X_t,\mu_t,Z_t\big)dt + dW_t + dK_t\Big) + dt.$$

By (4.152), we have:

$$2(X_t - X_s)\cdot dK_t \leqslant 0,$$

from which we obtain

$$d_t|X_t - X_s|^2 \leqslant 2(X_t - X_s)\cdot\Big(\hat{\alpha}\big(X_t,\mu_t,Z_t\big)dt + dW_t\Big) + dt.$$

Taking expectations and using the fact that $\hat{\alpha}$ is bounded, we deduce that there exists a constant C, independent of μ, such that:

$$\forall s,t \in [0,T], \quad \mathbb{E}^{\mu,\hat{\alpha}^{\mu}}\big[|X_t - X_s|^2\big] \leqslant C|t-s|. \tag{4.157}$$

Now, for all $s, t \in [0, T]$, with $s < t$,

$$\begin{aligned}\mathbb{E}^{\mu,\hat{\alpha}^{\mu}}\big[|X'_t - X'_s|^2\big] &\leqslant \mathbb{E}^{\mu,\hat{\alpha}^{\mu}}\big[|X_t - X_s|^2 \mathbf{1}_{\{t<\tau\}}\big] + C\mathbb{P}^{\mu,\hat{\alpha}^{\mu}}\big[s < \tau < t\big] \\ &\leqslant \mathbb{E}^{\mu,\hat{\alpha}^{\mu}}\big[|X_t - X_s|^2\big] + C\mathbb{P}^{\mu,\hat{\alpha}^{\mu}}\big[s < \tau < t\big],\end{aligned}$$

where we allowed the value of the constant C to change from line to line. By (4.157), we get:

$$\mathbb{E}^{\mu,\hat{\alpha}^{\mu}}\big[|X'_t - X'_s|^2\big] \leqslant C(t-s) + C\mathbb{P}^{\mu,\hat{\alpha}^{\mu}}\big[s < \tau < t\big].$$

The density of $\mathbb{P}^{\mu,\hat{\alpha}^{\mu}}$ with respect to $\mathbb{P}$ is defined in terms of $\hat{\boldsymbol{\alpha}}$, which is bounded, independently of $\boldsymbol{\mu}$. Therefore,

$$\mathbb{E}\Big[\Big(\frac{d\mathbb{P}^{\mu,\hat{\alpha}^{\mu}}}{d\mathbb{P}}\Big)^2\Big] \leqslant C,$$

from which we get:

$$\mathbb{E}^{\mu,\hat{\alpha}^{\mu}}\big[|X'_t - X'_s|^2\big] \leqslant C(t-s) + C\mathbb{P}\big[s < \tau < t\big]^{1/2}.$$

The result follows from the fact that according to Lemma 4.66, the cumulative distribution function of τ is Hölder continuous. □

We now investigate the continuity of Φ.

Lemma 4.69 *The function Φ in* (4.156) *is continuous from $\mathcal{C}([0,T];\mathcal{P}_2(D^{\Delta}))$ into itself.*

Proof. Given two continuous flows $\boldsymbol{\mu} = (\mu_t)_{0\leqslant t\leqslant T}$ and $\boldsymbol{\mu}' = (\mu'_t)_{0\leqslant t\leqslant T}$ with values in $\mathcal{P}(D^{\Delta})$, we compare the control processes $\hat{\boldsymbol{\alpha}}^{\mu}$ and $\hat{\boldsymbol{\alpha}}^{\mu'}$. By Proposition 4.15, we may call $(Y_t, Z_t)_{0\leqslant t\leqslant T}$ the solution of the quadratic BSDE (4.153) driven by $\boldsymbol{\mu}$, and $(Y'_t, Z'_t)_{0\leqslant t\leqslant T}$ the solution of the quadratic BSDE (4.153) driven by $\boldsymbol{\mu}'$. Computing the difference between Y_t and Y'_t for $t \in [0, T]$, we get:

$$\begin{aligned}d(Y_t - Y'_t) = -\mathbf{1}_{\{t<\tau\}}\Big[&H\big(X_t, \mu_t, Z_t, \hat{\alpha}(X_t, \mu_t, Z_t)\big) - H\big(X_t, \mu_t, Z'_t, \hat{\alpha}(X_t, \mu_t, Z'_t)\big) \\ &+ H\big(X_t, \mu_t, Z'_t, \hat{\alpha}(X_t, \mu_t, Z'_t)\big) - H\big(X_t, \mu'_t, Z'_t, \hat{\alpha}(X_t, \mu'_t, Z'_t)\big)\Big]dt \\ + \big(Z_t - Z'_t\big)\cdot dW_t.&\end{aligned}$$

Notice now, from the local Lipschitz property of the Hamiltonian H in the variables z and α and from the Lipschitz property of the optimizer $\hat{\alpha}$ in the variables z and α, that we can find a process $\boldsymbol{\zeta} = (\zeta_t)_{0\leqslant t\leqslant T}$ with values in $\mathbb{R}^d$ such that, for some constant $C \geqslant 0$, independent of $\boldsymbol{\mu}$ and $\boldsymbol{\mu}'$,

$$|\zeta_t| \leqslant C\big(1 + |Z_t| + |Z'_t|\big), \quad t \in [0, T],$$

and, for all $t \in [0, T]$,

$$\mathbf{1}_{\{t<\tau\}}\Big[H\big(X_t, \mu_t, Z_t, \hat{\alpha}(X_t, \mu_t, Z_t)\big) - H\big(X_t, \mu_t, Z'_t, \hat{\alpha}(X_t, \mu_t, Z'_t)\big)\Big] = (Z_t - Z'_t) \cdot \zeta_t.$$

Define now the drifted Brownian motion $(W^{\zeta}_t = W_t - \int_0^t \zeta_s ds)_{0 \leqslant t \leqslant T}$ together with the Girsanov transform:

$$\frac{d\mathbb{P}^{\zeta}}{d\mathbb{P}} = \exp\left(\int_0^T \zeta_s \cdot dW_s - \frac{1}{2}\int_0^T |\zeta_s|^2 ds\right).$$

By the BMO property of $\mathbf{Z}$ and $\mathbf{Z}'$, we know from Proposition 4.18 that there exists $r > 1$, independent of $\boldsymbol{\mu}$ and $\boldsymbol{\mu}'$, such that (allowing the constant C to vary from line to line):

$$\mathbb{E}\Big[\Big(\frac{d\mathbb{P}^{\zeta}}{d\mathbb{P}}\Big)^r\Big] \leqslant C. \tag{4.158}$$

Under $\mathbb{P}^{\zeta}$, we have that:

$$\begin{aligned}
d(Y_t - Y'_t) = &-\mathbf{1}_{\{t<\tau\}}\Big[H\big(X_t, \mu_t, Z'_t, \hat{\alpha}(X_t, \mu_t, Z'_t)\big) - H\big(X_t, \mu'_t, Z'_t, \hat{\alpha}(X_t, \mu'_t, Z'_t)\big)\Big]dt \\
&+ (Z_t - Z'_t) \cdot dW^{\zeta}_t,
\end{aligned}$$

for $t \in [0, T]$. Taking the power $2p$ on both sides for some $p \geqslant 1$, we deduce by standard BSDE inequalities that:

$$\begin{aligned}
&\mathbb{E}^{\zeta}\big[\sup_{0\leqslant t \leqslant T} |Y_t - Y'_t|^{2p}\big] + \mathbb{E}^{\zeta}\Big[\Big(\int_0^T |Z_t - Z'_t|^2 dt\Big)^p\Big] \\
&\quad \leqslant C\mathbb{E}^{\zeta}\Big[\Big(\int_0^T \big|H\big(X_t, \mu_t, Z'_t, \hat{\alpha}(X_t, \mu_t, Z'_t)\big) \\
&\qquad\qquad - H\big(X_t, \mu'_t, Z'_t, \hat{\alpha}(X_t, \mu'_t, Z'_t)\big)\big|^2 dt\Big)^p\Big].
\end{aligned}$$

Thanks to (4.158), the above right-hand side is less than:

$$\begin{aligned}
&C\mathbb{E}\Big[\Big(\int_0^T \big|H\big(X_t, \mu_t, Z'_t, \hat{\alpha}(X_t, \mu_t, Z'_t)\big) \\
&\qquad - H\big(X_t, \mu'_t, Z'_t, \hat{\alpha}(X_t, \mu'_t, Z'_t)\big)\big|^2 dt\Big)^{rp/(r-1)}\Big]^{(r-1)/r}.
\end{aligned}$$

Recalling that for any $q \geqslant 1$, $\mathbb{E}[(\int_0^T |Z'_t|^2 dt)^q]$ can be bounded independently of $\boldsymbol{\mu}'$, see again Proposition 4.18, we easily deduce that the above right-hand side tends to 0 as $\boldsymbol{\mu}'$ tends to $\boldsymbol{\mu}$. Therefore, for any $p \geqslant 1$,

$$\mathbb{E}^{\zeta}\Big[\Big(\int_0^T |Z_t - Z'_t|^2 dt\Big)^p\Big], \qquad \text{and thus} \qquad \mathbb{E}^{\zeta}\Big[\Big(\int_0^T |\hat{\alpha}^{\mu}_t - \hat{\alpha}^{\mu'}_t|^2 dt\Big)^p\Big],$$

tend to 0 as $\boldsymbol{\mu}'$ tends to $\boldsymbol{\mu}$ for the topology of uniform convergence on the space $\mathcal{C}([0,T];\mathcal{P}_2(D^{\Delta}))$. Notice however that this result is not entirely satisfactory since $\mathbb{P}^{\zeta}$ depends on $\boldsymbol{\mu}'$. We now prove that the same holds true but under $\mathbb{P}$. To do, observe that, for any $\epsilon > 0$,

$$\lim_{\mu'\to\mu} \mathbb{P}^{\zeta}\bigg[\int_0^T |\hat{\alpha}_t^{\mu} - \hat{\alpha}_t^{\mu'}|^2 dt \geqslant \epsilon\bigg] = 0. \tag{4.159}$$

Therefore, for any $M > 1$,

$$\lim_{\mu'\to\mu} \mathbb{P}\bigg[\frac{d\mathbb{P}^{\zeta}}{d\mathbb{P}} \geqslant \frac{1}{M}, \quad \int_0^T |\hat{\alpha}_t^{\mu} - \hat{\alpha}_t^{\mu'}|^2 dt \geqslant \epsilon\bigg] = 0.$$

It then remains to see that:

$$\begin{aligned}
\mathbb{P}\bigg[\frac{d\mathbb{P}^{\zeta}}{d\mathbb{P}} < \frac{1}{M}\bigg] &= \mathbb{P}\bigg[\int_0^T \zeta_s \cdot dW_s - \frac{1}{2}\int_0^T |\zeta_s|^2 ds < -\ln(M)\bigg] \\
&\leqslant \frac{1}{\ln(M)} \mathbb{E}\bigg[\bigg|\int_0^T \zeta_s \cdot dW_s\bigg| + \frac{1}{2}\int_0^T |\zeta_s|^2 ds\bigg],
\end{aligned}$$

which tends to 0 as M tends to ∞. By combining the two above inequalities we deduce that (4.159) holds with $\mathbb{P}^{\zeta}$ replaced by $\mathbb{P}$. Recalling that A is bounded and thus that $\hat{\boldsymbol{\alpha}}^{\mu}$ and $\hat{\boldsymbol{\alpha}}^{\mu'}$ are bounded independently of $\boldsymbol{\mu}$ and $\boldsymbol{\mu}'$, we obtain:

$$\lim_{\mu'\to\mu} \mathbb{E}\int_0^T |\hat{\alpha}_t^{\mu} - \hat{\alpha}_t^{\mu'}|^2 dt = 0, \tag{4.160}$$

which is the desired result.

We now go back to the expression of $\mathbb{P}^{\mu,\hat{\alpha}^{\mu}}$ and $\mathbb{P}^{\mu',\hat{\alpha}^{\mu'}}$. They are equivalent probability measures and the density of $\mathbb{P}^{\mu',\hat{\alpha}^{\mu'}}$ with respect to $\mathbb{P}^{\mu,\hat{\alpha}^{\mu}}$ is given by:

$$\frac{d\mathbb{P}^{\mu',\hat{\alpha}^{\mu'}}}{d\mathbb{P}^{\mu,\hat{\alpha}^{\mu}}} = \exp\bigg(\int_0^T (\hat{\alpha}_t^{\mu'} - \hat{\alpha}_t^{\mu})\cdot dW_t - \frac{1}{2}\int_0^T (|\hat{\alpha}_t^{\mu'}|^2 - |\hat{\alpha}_t^{\mu}|^2)dt\bigg).$$

Observe that, for any $p \geqslant 1$,

$$\begin{aligned}
&\mathbb{E}\bigg[\exp\bigg(p\int_0^T (\hat{\alpha}_t^{\mu'} - \hat{\alpha}_t^{\mu})\cdot dW_t\bigg)\bigg] \\
&\quad \leqslant \mathbb{E}\bigg[\exp\bigg(2p\int_0^T (\hat{\alpha}_t^{\mu'} - \hat{\alpha}_t^{\mu})\cdot dW_t - 2p^2\int_0^T |\hat{\alpha}_t^{\mu'} - \hat{\alpha}_t^{\mu}|^2 dt\bigg)\bigg]^{1/2} \\
&\qquad\qquad \times \mathbb{E}\bigg[\exp\bigg(2p^2\int_0^T |\hat{\alpha}_t^{\mu'} - \hat{\alpha}_t^{\mu}|^2 dt\bigg)\bigg]^{1/2} \\
&\quad = \mathbb{E}\bigg[\exp\bigg(2p^2\int_0^T |\hat{\alpha}_t^{\mu'} - \hat{\alpha}_t^{\mu}|^2 dt\bigg)\bigg]^{1/2}.
\end{aligned}$$

Recalling that the set A is bounded, we easily deduce that the above right-hand side tends to 1 as $\boldsymbol{\mu}'$ tends to $\boldsymbol{\mu}$. Therefore, for any $p \geqslant 1$,

$$\limsup_{\mu' \to \mu} \mathbb{E}\Big[\Big(\frac{d\mathbb{P}^{\mu',\hat{\alpha}^{\mu'}}}{d\mathbb{P}^{\mu,\hat{\alpha}^{\mu}}}\Big)^p\Big] \leqslant 1.$$

Now,

$$\begin{aligned}\mathbb{E}^{\mu,\hat{\alpha}^{\mu}}\Big[\Big(\frac{d\mathbb{P}^{\mu',\hat{\alpha}^{\mu'}}}{d\mathbb{P}^{\mu,\hat{\alpha}^{\mu}}} - 1\Big)^2\Big] &= \mathbb{E}^{\mu,\hat{\alpha}^{\mu}}\Big[\Big(\frac{d\mathbb{P}^{\mu',\hat{\alpha}^{\mu'}}}{d\mathbb{P}^{\mu,\hat{\alpha}^{\mu}}}\Big)^2\Big] + 1 - 2\mathbb{E}^{\mu,\hat{\alpha}^{\mu}}\Big[\frac{d\mathbb{P}^{\mu',\hat{\alpha}^{\mu'}}}{d\mathbb{P}^{\mu,\hat{\alpha}^{\mu}}}\Big] \\ &= \mathbb{E}^{\mu,\hat{\alpha}^{\mu}}\Big[\Big(\frac{d\mathbb{P}^{\mu',\hat{\alpha}^{\mu'}}}{d\mathbb{P}^{\mu,\hat{\alpha}^{\mu}}}\Big)^2\Big] - 1.\end{aligned} \tag{4.161}$$

Moreover, for any $\eta > 1$

$$\begin{aligned}&\limsup_{\mu' \to \mu} \mathbb{E}^{\mu,\hat{\alpha}^{\mu}}\Big[\Big(\frac{d\mathbb{P}^{\mu',\hat{\alpha}^{\mu'}}}{d\mathbb{P}^{\mu,\hat{\alpha}^{\mu}}}\Big)^2\Big] \\ &= \limsup_{\mu' \to \mu} \mathbb{E}\Big[\frac{d\mathbb{P}^{\mu,\hat{\alpha}^{\mu}}}{d\mathbb{P}}\Big(\frac{d\mathbb{P}^{\mu',\hat{\alpha}^{\mu'}}}{d\mathbb{P}^{\mu,\hat{\alpha}^{\mu}}}\Big)^2\Big] \\ &\leqslant \mathbb{E}\Big[\Big(\frac{d\mathbb{P}^{\mu,\hat{\alpha}^{\mu}}}{d\mathbb{P}}\Big)^{2\eta}\Big]^{1/\eta} \limsup_{\mu' \to \mu} \mathbb{E}\Big[\Big(\frac{d\mathbb{P}^{\mu',\hat{\alpha}^{\mu'}}}{d\mathbb{P}^{\mu,\hat{\alpha}^{\mu}}}\Big)^{2\eta/(\eta-1)}\Big]^{(\eta-1)/\eta} \\ &= \mathbb{E}\Big[\Big(\frac{d\mathbb{P}^{\mu,\hat{\alpha}^{\mu}}}{d\mathbb{P}}\Big)^{2\eta}\Big]^{1/\eta}.\end{aligned}$$

Letting η tend to 1, we get:

$$\limsup_{\mu' \to \mu} \mathbb{E}^{\mu,\hat{\alpha}^{\mu}}\Big[\Big(\frac{d\mathbb{P}^{\mu',\hat{\alpha}^{\mu'}}}{d\mathbb{P}^{\mu,\hat{\alpha}^{\mu}}}\Big)^2\Big] \leqslant 1.$$

And then, by (4.161),

$$\limsup_{\mu' \to \mu} \mathbb{E}^{\mu,\hat{\alpha}^{\mu}}\Big[\Big(\frac{d\mathbb{P}^{\mu',\hat{\alpha}^{\mu'}}}{d\mathbb{P}^{\mu,\hat{\alpha}^{\mu}}} - 1\Big)^2\Big] = 0. \tag{4.162}$$

Finally, for any bounded measurable function F on $\mathcal{C}([0,T];\mathbb{R}^{2d})$, we have:

$$\mathbb{E}^{\mu',\hat{\alpha}^{\mu'}}\big[F(\boldsymbol{X})\big] - \mathbb{E}^{\mu,\hat{\alpha}^{\mu}}\big[F(\boldsymbol{X})\big] = \mathbb{E}^{\mu,\hat{\alpha}^{\mu}}\Big[\Big(\frac{d\mathbb{P}^{\mu',\hat{\alpha}^{\mu'}}}{d\mathbb{P}^{\mu,\hat{\alpha}^{\mu}}} - 1\Big)F(\boldsymbol{X})\Big],$$

and from (4.162), we deduce that:

$$\lim_{\mu' \to \mu} \Big|\mathbb{E}^{\mu',\hat{\alpha}^{\mu'}}\big[F(\boldsymbol{X})\big] - \mathbb{E}^{\mu,\hat{\alpha}^{\mu}}\big[F(\boldsymbol{X})\big]\Big| = 0,$$

the convergence being uniform over measurable mappings F with a supremum norm less than 1.

Observing that, for a given $t \in [0, T]$, the mapping $\mathcal{C}([0, T]; \mathbb{R}^d) \ni \boldsymbol{x} \mapsto x'_t$, with $x'_t = x_t$ if $t < \tau$ and $x'_t = \Delta$ if $t > \tau$ and $\tau = \inf\{t \in [0, T] : x_t \in E\}$, is measurable, we deduce that, for any bounded measurable function $h : D^\Delta \to \mathbb{R}$,

$$\lim_{\mu' \to \mu} \sup_{t \in [0,T]} \left| \mathbb{E}^{\mu', \hat{\alpha}^{\mu'}} \left[h(X'_t) \right] - \mathbb{E}^{\mu, \hat{\alpha}^{\mu}} \left[h(X'_t) \right] \right| = 0,$$

the convergence being uniform over measurable mappings h with a sup-norm less than 1. Expressed in terms of the function Φ used in the statement, this says that:

$$\lim_{\mu' \to \mu} \sup_{\|h\|_\infty \leqslant 1} \sup_{t \in [0,T]} \left| \int_{D^\Delta} h d\big(\Phi(\mu')\big)_t - \int_{D^\Delta} h d\big(\Phi(\mu)\big)_t \right| = 0,$$

that is, the probability measure $(\Phi(\mu'))_t$ weakly converges to $(\Phi(\mu))_t$, uniformly in $t \in [0, T]$. Recall that, since D^Δ is bounded, the metric associated with weak convergence on $\mathcal{P}(D^\Delta)$ is equivalent to the 2-Wasserstein distance. Therefore,

$$\lim_{\mu' \to \mu} \sup_{t \in [0,T]} W_2\Big(\big(\Phi(\mu')\big)_t, \big(\Phi(\mu)\big)_t \Big) = 0,$$

which shows that Φ is a continuous mapping from $\mathcal{C}([0, T]; \mathcal{P}_2(D^\Delta))$ into itself. □

Now, the existence of an MFG equilibrium comes as a consequence of Lemmas 4.68 and 4.69. Indeed, it suffices to regard $\mathcal{E} = \mathcal{C}([0, T]; \mathcal{P}(D^\Delta))$ as a closed convex subset of $\mathcal{C}([0, T]; \mathcal{M}^1_f(\mathbb{R}^d))$ equipped with the same norm $\|\cdot\|$ as in the proof of Theorem 4.39 and to observe any $\mathcal{E}$-valued sequence converging for $\|\cdot\|$ is uniformly convergent with respect to the 2-Wasserstein distance.

Proof of Lemma 4.66

Proof. First, we recall some basic facts about exit times and exit distributions of standard Brownian motion, as well as some properties of reflected Brownian motions. Recall that the boundary ∂D of the domain D is assumed to be piecewise smooth. See the Notes & Complements at the end of the chapter for references to papers providing proofs of these results.

Under the standing assumptions on the domain D, the reflected Brownian motion in D has a fundamental solution $(p(t, x, y))_{t>0, x \in D, y \in D}$, namely:

$$\mathbb{P}\big[X_t \in B | X_0 = x\big] = \int_B p(t, x, y) dy, \quad B \in \mathcal{B}(D).$$

For any $t > 0$, the mapping $D^2 \ni (x, y) \mapsto p(t, x, y)$ is continuous and (strictly) positive. Moreover, there exists a constant C such that, for all $t \in (0, 1)$,

$$\forall x, y \in D, \quad p(t, x, y) \leqslant C t^{-d/2} \exp\Big(-\frac{|x - y|^2}{Ct}\Big), \tag{4.163}$$

and, for all $t \geqslant 1$,

$$\forall x, y \in D, \quad p(t,x,y) \leqslant C. \tag{4.164}$$

We also have that, for all $t \in (0,1)$,

$$\forall x \in D,\ \forall a > 0, \quad \mathbb{P}\Big[\sup_{0 \leqslant s \leqslant t} |X_s - x| \geqslant a \,|\, X_0 = x\Big] \leqslant C \exp\big(-\frac{a^2}{Ct}\big). \tag{4.165}$$

At times, it will be convenient to compare the stopping time τ (defined as the first hitting time of the part E of the boundary ∂D) to the first exit time $\tilde{\tau} = \inf\{t > 0;\ X_t \in \partial D\}$ of the domain D. Although we shall not use this fact, we mention that the joint distribution of the first time of exit and the location of exit, namely $\mathcal{L}(\tilde{\tau}, X_{\tilde{\tau}})$ is absolutely continuous with respect to the measure $dt\,\sigma(dy)$ where $\sigma(dy)$ denotes the surface measure on ∂D. More precisely for any starting point x in the interior of D, we have:

$$\mathbb{P}\big[\tilde{\tau} \in dt,\ X_{\tilde{\tau}} \in dy \,|\, X_0 = x\big] = \frac{1}{2}\frac{\partial}{\partial \boldsymbol{n}_y} p^0(t,x,y)\ dt\,\sigma(dy) \tag{4.166}$$

where $\boldsymbol{n}_y$ denotes the inward pointing unit normal vector to ∂D at $y \in \partial D$, and $p^0(t,x,y)$ is the fundamental solution of the Dirichlet problem in D, namely the density of the Brownian motion killed the first time it hits the boundary ∂D; in other words:

$$\mathbb{P}[X_t \in dy,\ t < \tilde{\tau} \,|\, X_0 = x] = p^0(t,x,y)dy.$$

We used the process $\boldsymbol{X}$ while talking about standard Brownian motion because, at least in distribution, $\boldsymbol{X}$ behaves like a standard Brownian motion up until time $\tilde{\tau}$. See the Notes & Complements for references.

Now, we tackle the proof of the lemma by considering $X_0 = \xi$, as in (4.151). For any $t \geqslant 0$, we denote by ν_t the distribution of X_t. For each $t > 0$, ν_t has a density, which we denote by ρ_t:

$$\rho_t(x) = \int_D p(t,y,x)d\nu_0(y), \qquad x \in D,$$

When $t \in (0,1)$ and $x \in E^{\eta/2}$, we have:

$$\begin{aligned}
\rho_t(x) &\leqslant \int_{E^\eta} p(t,y,x)\rho_0(y)dy + \frac{C}{t^{d/2}}\int_{D\setminus E^\eta} \exp\big(-\frac{|x-y|^2}{Ct}\big)d\nu_0(y) \\
&\leqslant \int_{E^\eta} p(t,y,x)\rho_0(y)dy + \frac{C}{t^{d/2}} \exp\big(-\frac{\eta^2}{4Ct}\big),
\end{aligned}$$

where as before, $E^\eta = \{x \in D : \mathrm{dist}(x,E) < \eta\}$, and we used the notation ρ_0 for the density of the absolutely continuous part of ν_0. Since we assume that ρ_0 is bounded on E^η, we deduce that there exists a constant C_0 such that, for all $t \in (0,1)$,

$$\forall x \in E^{\eta/2}, \quad \rho_t(x) \leqslant C_0. \tag{4.167}$$

The bound remains true when $t \geqslant 1$ because of (4.164).

We now denote by m_t the restriction of the distribution of X'_t to D, namely the sub-probability measure defined by:

$$m_t(B) = \mathbb{P}\big[X_t \in B,\ t < \tau\big], \quad B \in \mathcal{B}(D),$$

the initial condition ξ of X_0 being prescribed. Obviously, we always have $m_t(B) \leqslant \nu_t(B)$. For any $t \geqslant 0$ such that $\mathbb{P}[\tau \geqslant t] > 0$, we have:

$$\mathbb{P}\big[\tau \leqslant t + h \,|\, \tau \geqslant t\big] = \int_D \mathbb{P}\big[\tau \leqslant h \,|\, X_0 = x\big]\, d\nu_t(x).$$

For a given $x \in D$, we deduce from (4.165) that, for all $h \in (0, 1)$,

$$\begin{aligned}\mathbb{P}\big[\tau \leqslant h \,|\, X_0 = x\big] &\leqslant \mathbb{P}\Big[\sup_{0 \leqslant t \leqslant h} |X_t - x| \geqslant \mathrm{dist}(x, E) \,\big|\, X_0 = x\Big] \\ &\leqslant C \exp\Big(-\frac{(\mathrm{dist}(x, E))^2}{Ch}\Big).\end{aligned}$$

Therefore, for $h \in (0, 1)$,

$$\mathbb{P}\big[\tau \leqslant t + h \,|\, \tau \geqslant t\big] \leqslant C \int_D \exp\Big(-\frac{(\mathrm{dist}(x, E))^2}{Ch}\Big)\, d\nu_t(x).$$

We split the integral in the right-hand side into two parts according to the partition of D into $E^{\eta/2}$ and $D \setminus E^{\eta/2}$. Using (4.167) and allowing the constant C to increase from line to line, we get, for any $\epsilon \in (0, 1/2)$,

$$\begin{aligned}&\mathbb{P}\big[\tau \leqslant t + h \,|\, \tau \geqslant t\big] \\ &\leqslant C \int_{E^{\eta/2}} \exp\Big(-\frac{(\mathrm{dist}(x, E))^2}{4Ch}\Big) dx + C \int_{D \setminus E^{\eta/2}} \exp\Big(-\frac{\eta^2}{Ch}\Big) d\nu_t(x) \\ &\leqslant Ch^{(1-\epsilon)/2} \int_{E^{\eta/2}} \frac{dx}{(\mathrm{dist}(x, E))^{1-\epsilon}} dx + C\eta^{-1} h^{1/2}.\end{aligned}$$

Using the fact that:

$$\int_D \frac{1}{(\mathrm{dist}(x, E))^{1-\epsilon}} dx < \infty, \tag{4.168}$$

which we prove below, we can conclude that:

$$\mathbb{P}\big[\tau \leqslant t + h \,|\, \tau \geqslant t\big] \leqslant Ch^{(1-\epsilon)/2},$$

for a constant C independent of t in $[0, +\infty)$ and $h \in (0, 1)$. Therefore,

$$\mathbb{P}\big[\tau \geqslant t + h \,|\, \tau \geqslant t\big] \geqslant 1 - Ch^{(1-\epsilon)/2},$$

and then,

$$\mathbb{P}\big[\tau \geqslant t+h\big] \geqslant \mathbb{P}\big[\tau \geqslant t\big](1 - Ch^{(1-\epsilon)/2}).$$

Since the function $t \mapsto \mathbb{P}[\tau \geqslant t]$ is nonincreasing, this completes the proof of the desired Hölder continuity, as long as we can check that (4.168) holds. This is indeed the case because, writing $E \subset \partial D = \cup_{i=1}^N F_i$, where $(F_i)_{i=1,\cdots,N}$ denote the faces of D, it suffices to prove that, for all $i = 1, \cdots, N$,

$$\int_D \frac{1}{(\mathrm{dist}(x, F_i))^{1-\epsilon}} dx < \infty.$$

Now, the distance from x to F_i is greater than the distance from x to the hyperplane H_i supporting F_i. The result easily follows by changing the coordinates in such a way that, under the new coordinates, $\mathrm{dist}(x, H_i) = x_1$, x_1 denoting the first coordinate of x in the new reference frame.

Finally, we prove that, for any $t > 0$, $\mathbb{P}[\tau \leqslant t] > 0$.

We start with the case when the support of $\nu_0 = \mathcal{L}(\xi)$ is not included in ∂D. Then, we can find $x_0 \in D$ and $\epsilon > 0$ small enough such that the d-dimensional ball $B(x_0, \epsilon)$ is included in the interior of D. Also, we know that E contains a $(d-1)$-dimensional relatively open ball F included in one of the face of ∂D. Thanks to the polyhedral structure of D, we can find, for any $t > 0$, a piecewise linear function $\varrho : [0, t] \to \mathbb{R}^d$ such that $\varrho(0) = x_0$, $\varrho([0, t/2]) \subset D$, $\varrho([0, t/2]) \cap \partial D \subset F$, $\varrho((t/2, t]) \cap D = \emptyset$ and $\mathrm{dist}(\varrho(t), D) \geqslant 1$ (that is $\varrho([0, t])$ crosses ∂D at some point in F). Also, for ϵ small enough, we can a draw a tube $\mathcal{T} = \{x \in \mathbb{R}^d : \inf_{s\in[0,t]} |x - \varrho(s)| \leqslant \epsilon\}$ such that $\mathcal{T} \cap \partial D \subset F$. By support theorem for the Brownian motion, we know that $\mathbb{P}[\forall s \in [0, t],\ X_0 + W_s \in \mathcal{T}] > 0$. Since $X_s = X_0 + W_s$ for $s \leqslant \tilde{\tau}$, we deduce that $\mathbb{P}[\tilde{\tau} \leqslant t, X_{\tilde{\tau}} \in F] > 0$. This concludes the proof since $\mathbb{P}[\tau \leqslant t] > \mathbb{P}[\tilde{\tau} \leqslant t, X_{\tilde{\tau}} \in F]$.

When $X_0 = \xi$ is concentrated on the boundary, we may use the fact that $D^2 \ni (x, y) \mapsto p(t, x, y)$ is strictly positive. In particular, for any $t > 0$, $p(t, x, \cdot)$ must charge a ball in the interior of D. Therefore, when starting from the boundary, there is a positive probability to reach a ball in the interior of D, in any positive time. By the Markov property, we deduce that there is a positive probability to reach E, in any positive time.

In order to prove that τ is almost surely finite, we use the fact that $p(1/2, \cdot, \cdot)$ is bounded from below. Starting from any point, there is a positive probability to belong, at time $1/2$, to the same ball $B(x_0, \epsilon)$ as that constructed right above. Starting from this ball, there is a positive probability to reach E between $t = 1/2$ and $t = 1$. By a standard iteration argument based on the Markov property, we deduce that $\mathbb{P}[\tau < \infty] = 1$. □

The Analytic Approach

Motivated in part by numerical considerations (see numerical example below) we provide the necessary details on the PDE description of the equilibrium. In the present situation, the HJB equation takes the form:

$$\partial_t V(t, x) + \frac{1}{2}\Delta_x V(t, x) + H^*\big(x, \mu_t, \partial_x V(t, x)\big) = 0, \tag{4.169}$$

with 1) a Neumann condition on $\partial D \setminus E$ accounting for the normal reflection present in the dynamics of the position of the typical individual, that is $\partial_x V(t, x) \cdot n(x) = 0$

for $x \in \partial D \setminus E$, where $n(x)$ is the inward normal vector to ∂D at point $x \in \partial D \setminus E$, and 2) a Dirichlet boundary condition on E accounting for the killing at the door, that is $V(t,x) = 0$ for $x \in E$.

Similarly, the equation for the density $x \mapsto m_t(x)$ of the restriction of μ_t to D, or in other words, the mass *remaining* in D at time t, is given by a form of the Kolmogorov/Fokker-Planck equation:

$$\partial_t m_t - \frac{1}{2}\Delta_x m_t + \mathrm{div}_x\big(\tilde{\alpha}(t,x)m_t\big) = 0, \tag{4.170}$$

with mixed boundary conditions:

$$\begin{cases} \frac{1}{2}\partial_x m_t(x) \cdot n(x) - m_t(x)\tilde{\alpha}(t,x) \cdot n(x) = 0, & x \in \partial D \setminus E, \\ m_t(x) = 0, \quad x \in E, \end{cases}$$

where we use the notation $\tilde{\alpha}(t,x) = \hat{\alpha}(x, \mu_t, \partial_x V(t,x))$ for convenience. The intuitive interpretation of $m_t(x)$ is the proportion of individuals who have not yet exited by time t. Since deriving this equation directly involves delicate computations with the singular process $\boldsymbol{K}$ (involving the local time of the process at the relevant part of the boundary), we start from a solution of the above equation and identify it, at least formally, with the distribution of the part of the population still in D by time t. So we pick a time $S \in [0,T]$ and an arbitrary continuous bounded function g on D and we prove that:

$$\int_D g(x)m(S,x)dx = \mathbb{E}\big[g(X_S)\mathbf{1}_{\{S<\tau\}}\big]. \tag{4.171}$$

In order to do so, for S and g given, we consider the solution u of the parabolic Dirichlet-Neumann problem:

$$\partial_t u(t,x) + \frac{1}{2}\Delta_x u(t,x) + \tilde{\alpha}(t,x) \cdot \partial_x u(t,x) = 0, \tag{4.172}$$

for $(t,x) \in [0,S] \times D$, with Neumann boundary condition $\partial_x u(t,x) \cdot n(x) = 0$ for $(t,x) \in [0,S) \times (\partial D \setminus E)$, Dirichlet boundary condition $u(t,x) = 0$ for $t \in [0,S] \times E$, and terminal condition $u(S,x) = g(x)$ for $x \in E$. Consider also the solution $(X_t)_{0 \leqslant t \leqslant T}$ of the reflected SDE:

$$dX_t = \tilde{\alpha}(t,X_t)dt + dW_t + dK_t, \quad t \in [0,T].$$

Introducing as above the first hitting time $\tau = \inf\{t \geqslant 0 : X_t \in E\}$, and assuming that u is smooth enough to apply Itô's formula, we get:

$$\frac{d}{dt}\mathbb{E}\big[u\big(t \wedge \tau, X_{t\wedge\tau}\big)\big] = 0, \quad t \in [0,S]. \tag{4.173}$$

We used the fact that the expectation of the stochastic integral with respect to dW_t is 0 and that the Neumann boundary condition on u kills the integral with respect to dK_t. Now, using the notation $m(t,x)$ for $m_t(x)$ and the equations (4.170) and (4.172) satisfied by m and u, we get:

$$\begin{aligned}
&\frac{d}{dt}\int_D u(t,x)m(t,x)dx \\
&= -\int_D \Big[\frac{1}{2}\Delta_x u(t,x) + \tilde{\alpha}(t,x)\cdot\partial_x u(t,x)\Big]m(t,x)dx \\
&\quad + \int_D \Big[\frac{1}{2}u(t,x)\Delta_x m(t,x) - u(t,x)\mathrm{div}_x\big(\tilde{\alpha}(t,x)m(t,x)\big)\Big]dx.
\end{aligned} \tag{4.174}$$

Using Green's formula we get:

$$\begin{aligned}
&-\int_D \Delta_x u(t,x)m(t,x)dx \\
&= \int_D \partial_x u(t,x)\cdot\partial_x m(t,x)dx + \int_{\partial D} m(t,x)\partial_x u(t,x)\cdot n(x)d\sigma(x), \\
&\int_D u(t,x)\Delta_x m(t,x)dx \\
&= -\int_D \partial_x u(t,x)\cdot\partial_x m(t,x)dx - \int_{\partial D} u(t,x)\partial_x m(t,x)\cdot n(x)d\sigma(x),
\end{aligned}$$

where σ denotes the surface measure. Since $u(t,x) = m(t,x) = 0$ for $x \in E$ and $\partial_x u(t,x)\cdot n(x) = 0$ for $x \in \partial D\setminus E$, we obtain:

$$\begin{aligned}
&-\int_D \Delta_x u(t,x)m(t,x)dx + \int_D u(t,x)\Delta_x m(t,x)dx \\
&= -\int_{\partial D\setminus E} u(t,x)\partial_x m(t,x)\cdot n(x)d\sigma(x).
\end{aligned} \tag{4.175}$$

Similarly, using the divergence theorem, we get:

$$\begin{aligned}
&\int_D \Big[-u(t,x)\mathrm{div}_x\big(\tilde{\alpha}(t,x)m(t,x)\big)\Big]dx \\
&= \int_D \partial_x u(t,x)\cdot\tilde{\alpha}(t,x)m(t,x)dx + \int_{\partial D} u(t,x)m(t,x)\tilde{\alpha}(t,x)\cdot n(x)d\sigma(x) \\
&= \int_D \partial_x u(t,x)\cdot\tilde{\alpha}(t,x)m(t,x)dx + \int_{\partial D\setminus E} u(t,x)m(t,x)\tilde{\alpha}(t,x)\cdot n(x)d\sigma(x).
\end{aligned}$$

By (4.174), (4.175), and the boundary conditions satisfied by m we obtain:

$$\frac{d}{dt}\int_D u(t,x)m(t,x)dx = 0. \tag{4.176}$$

Putting together (4.173) and (4.176) and the fact that $\int_D u(0,x)m(0,x)dx = \mathbb{E}[u(0,X_0)]$, we deduce that:

$$\int_D u(t,x)m(t,x)dx = \mathbb{E}\big[u\big(t\wedge\tau, X_{t\wedge\tau}\big)\big],$$

which gives the desired result (4.171) if we use the terminal condition of u.

Numerical Illustration

For the purpose of illustration, we consider the domain $D = [0,1]\times[0,1]$ in the plane, and two exit doors $E = ([0.95,1]\times\{0\})\cup([0.98,1]\times\{1\})$. They are shown as unions of gray circles on the various panels of Figure 4.5. We implemented the search for an equilibrium density as a simple Picard iteration. At each step of the iteration, we use the values $m(t,x,y)$ obtained from the previous iteration with a simple monotone finite difference Euler scheme to compute the solution of the HJB equation (4.169) backward in time. In the case at hand, this HJB equation becomes:

$$\partial_t V(t,x) + \frac{\sigma^2}{2}\Delta V(t,x) - \frac{\beta}{2(1+m(t,x))^\alpha}|\nabla V|^2 + \delta = 0, \qquad V(T,\cdot)\equiv 0.$$

if we choose $\mathbb{R}^2$ for A and a constant $\delta \geqslant 0$ for the function $f(t)$ appearing in the expression of the loss $J^\mu(\boldsymbol{\alpha})$ in (4.150) penalizing the time spent in the room before exiting, and if we replace the congestion penalty $\ell(x,\mu)$ by a multiple of the quantity $(1+m(x))^\alpha$ where m stands for the density of μ. Once a solution of the HJB equation is found, we solve (again with a simple monotone finite difference Euler scheme) the forward Kolmogorov equation (4.170) which reads:

$$\partial_t m_t - \frac{\sigma^2}{2}\Delta m_t - \beta\,\mathrm{div}\Big(\frac{m_t\nabla V}{(1+m_t)^\alpha}\Big) = 0.$$

For the sake of definiteness (and comparison with numerical studies reported in the still unpublished literature), we chose the values $\sigma = 0.1$, $\beta = 16$, $\delta = 1/320$, and we let time evolve from $t = 0$ to $T = 8$ by increments of size $\Delta_t = 0.02$. Lack of congestion corresponds to the choice $\alpha = 0$. For the purpose of the numerical experiments whose results are reported below, we include minor contagion in the model by choosing $\alpha = 0.1$.

Figure 4.4 shows the time evolution of the total mass of the measure m_t, essentially the number of individuals still in the room at time t, starting from a uniform distribution of individuals in a square at the center of the room. This plot shows that the effect of the congestion is to slow down the exit indeed.

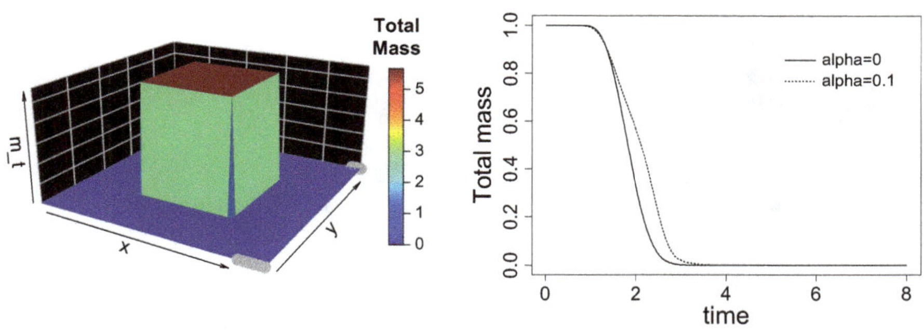

Fig. 4.4 Left: Initial distribution m_0 used for the numerical illustrations. The small exits are marked in gray. Right: Time evolution of the total mass of the distribution m_t of the individuals still in the room at time t for $\alpha = 0$, i.e., absence of congestion (continuous line) and $\alpha = 0.1$, i.e., moderate congestion (dotted line).

Figure 4.5 shows the time evolution of the density $m(t, \cdot)$ starting from a uniform distribution of individuals in a square at the center of the room. Snapshots of the density are given for times $t = 0.42, 1.22$ and $t = 2.42$. For the sake of comparison with numerical studies reported in the unpublished literature, we chose the values $\sigma = 0.2$, $\beta = 16$, and $\alpha = 0.1$.

Clearly the congestion term slows down the exit of the individuals as we see that it takes longer for the same mass of individuals to reach the exit doors, as more individuals are stranded looking for the exit and bouncing off the walls before finding the exit.

4.7.3 A Variant of the Model of Flocking

We revisit the model of flocking presented in Subsection 1.5.1, and offer a theoretical solution as well as numerical illustrations for the general case $\beta \neq 0$. Since the model can be used for schools of fish, flocks of birds, or human crowds, we shall often use the generic terminology *"particles"* for the individual members of the population.

An interesting feature of the model is the form of the state equation. As we already explained in Chapter 1, the state variable contains both the position and the velocity of the particle. While the original flocking example was introduced in the physical dimension $d = 3$, the following discussion will not depend upon the specific value of the dimension d. The state $X_t = (x_t, v_t)$ at time t is an element of the phase space, whose dimension is $2d$, and the dynamics read as a controlled kinetic equation:

$$dx_t = v_t dt, \quad dv_t = \alpha_t dt + \sigma dW_t, \quad t \in [0, T]; \quad X_0 = (x_0, v_0) = \xi, \tag{4.177}$$

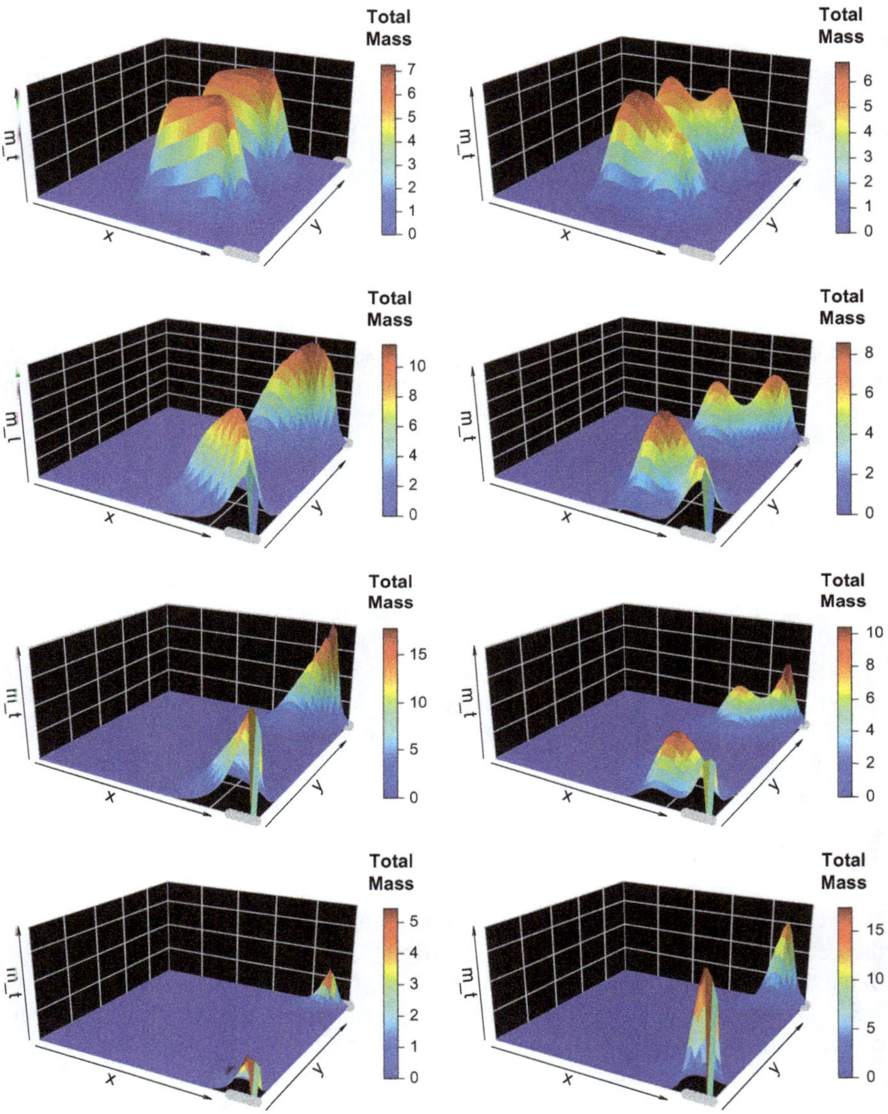

Fig. 4.5 Surface plots of the density m_t for times $t = 0.42, 1.22$ and $t = 2.42$ (from the top) for $\alpha = 0$, i.e., absence of congestion (left column) and $\alpha = 0.1$, i.e., moderate congestion (right column). The exit doors are shown as unions of gray circles.

where $\boldsymbol{W} = (W_t)_{0 \leqslant t \leqslant T}$ is a d-dimensional Brownian motion defined on a filtered complete probability space $(\Omega, \mathcal{F}, \mathbb{F}, \mathbb{P})$ and $\xi \in L^2(\Omega, \mathcal{F}_0, \mathbb{P}; \mathbb{R}^{2d})$. Both $\boldsymbol{x} = (x_t)_{0 \leqslant t \leqslant T}$ and $\boldsymbol{v} = (v_t)_{0 \leqslant t \leqslant T}$ are d-dimensional $\mathbb{F}$-adapted processes. The control $\boldsymbol{\alpha} = (\alpha_t)_{0 \leqslant t \leqslant T}$ is a d-dimensional square-integrable progressively measurable process. Given a fixed flow $\boldsymbol{\mu} = (\mu_t)_{0 \leqslant t \leqslant T} \in \mathcal{C}([0,T];\mathbb{R}^{2d})$ of probability measures on $\mathbb{R}^{2d}$, the cost of a control $\boldsymbol{\alpha}$ is given by:

$$J^{\boldsymbol{\mu}}(\boldsymbol{\alpha}) = \mathbb{E}\left[\int_0^T \left(\frac{1}{2}|\alpha_t|^2 + f\big((x_t, v_t), \mu_t\big)\right)dt\right]. \tag{4.178}$$

Here f is a measurable function from $\mathbb{R}^{2d} \times \mathcal{P}_2(\mathbb{R}^{2d})$ into $\mathbb{R}$. In the original form (1.46) of the cost (4.178) introduced in Chapter 1, and the discussions of the particular case $\beta = 0$ offered in Chapter 2 and Chapter 3, we used the running cost function:

$$f((x,v),\mu) = \frac{\kappa^2}{2}\left|\int_{\mathbb{R}^{2d}} \frac{1}{(1+|x-x'|^2)^\beta}(v-v')\,\mu(dx',dv')\right|^2. \tag{4.179}$$

However, like in the subsequent analysis of the corresponding control problem of McKean-Vlasov dynamics presented in Chapter 6, we shall use the variant:

$$f((x,v),\mu) = \frac{\kappa^2}{2}\int_{\mathbb{R}^{2d}} \frac{|v-v'|^2}{(1+|x-x'|^2)^\beta}\,\mu(dx',dv'). \tag{4.180}$$

Clearly, the parameter $\beta \geqslant 0$ plays a crucial role in both cases. Its role is to quantify how much particles whose positions x are far from the bulk of the positions x' of particles distributed according to the input distribution μ, contribute to the running cost. Interestingly, in the case $\beta = 0$, and for the running cost function (4.179), we saw in Subsection 3.6.1 that the denominator is identically one, and the model reduces to a LQ mean field game which we solved by the methods presented in Section 3.5 of Chapter 3. Existence of equilibria in the case $\beta > 0$ does not follow directly from the results presented so far, mostly because of the degeneracy and because of the lack of convexity of the function f.

In order to simplify somehow the discussion of the mathematical analysis presented below, we shall assume that f is bounded. This comes as a slight restriction in comparison with the original model described in Subsection 1.5.1 as recalled above. In fact, we shall assume much more as we will require that the function f is continuous on $\mathbb{R}^{2d} \times \mathcal{P}_2(\mathbb{R}^{2d})$, $\mathcal{P}_2(\mathbb{R}^{2d})$ being equipped with the 2-Wasserstein distance, and that for any fixed $\mu \in \mathcal{P}_2(\mathbb{R}^{2d})$, f is twice differentiable in $(x,v) \in \mathbb{R}^{2d}$, with derivatives uniformly bounded in μ.

FBSDE Characterizing the MFG

Using $Y = (y^x, y^v) \in \mathbb{R}^{2d}$ for the adjoint variable of the state variable $X = (x,v)$, the reduced Hamiltonian of the model takes the form:

$$H\big(t,(x,v),\mu,(y^x,y^v),\alpha\big) = y^x \cdot v + y^v \cdot \alpha + \frac{1}{2}|\alpha|^2 + f\big((x,v),\mu\big).$$

The minimizer of H with respect to α is:

$$\hat{\alpha}\big(t, (x, v), \mu, (y^x, y^v)\big) = -y^v.$$

Our goal is to use the first prong of the probabilistic approach based on the FBSDE representation of the value function, in the strong formulation as given in (4.55). In this approach, the relevant FBSDE is obtained by replacing the control by its minimizer in both the forward dynamics and the BSDE representation of the value function, and the adjoint variable by the martingale integrand multiplied by the inverse of the volatility. Unfortunately, the $(2d) \times (2d)$ volatility matrix is not invertible in the present situation. Indeed, it is of the form:

$$\sigma(t, x, \mu) = \begin{bmatrix} 0_d & 0_d \\ 0_d & \sigma I_d \end{bmatrix}.$$

However, if we identify the control space to the closed linear subspace $A = \{0_d\} \times \mathbb{R}^d$ of $\mathbb{R}^{2d}$, the current form of the flocking model fits the framework of Remark 4.50 since:

$$b\big(t, (x, v), \mu, (0, \alpha)\big) = (v, 0)^\dagger + \sigma(t, x, \mu)(0, \alpha)^\dagger,$$

and we can use the content of this remark to derive the appropriate version of the FBSDE to be solved. Choosing $\sigma = 1$ in the subsequent analysis, it reads:

$$\begin{cases} dx_t = v_t dt, \quad dv_t = -Z_t dt + dW_t, \\ dY_t = -\Big(\frac{1}{2}|Z_t|^2 + f\big((x_t, v_t), \mu_t\big)\Big)dt + Z_t \cdot dW_t, \end{cases} \tag{4.181}$$

for $t \in [0, T]$, with initial condition $X_0 = (x_0, v_0) = \xi$, and $Y_T = 0$ as terminal condition since the terminal cost is not present in the model. According to the first prong of the probabilistic approach to MFGs, we can solve the MFG problem by solving this FBSDE. Unfortunately, assumption **MFG Solvability HJB** of Subsection 4.4.1 is not satisfied in the present situation. Indeed, the forward dynamics in equation (4.181) are degenerate: only the velocity is randomly forced, and the position component $\boldsymbol{x}$ of $\boldsymbol{X} = (\boldsymbol{x}, \boldsymbol{v})$ can only feel the noise $\boldsymbol{W}$ through the drift. As a consequence, we cannot use Theorem 4.45, which requires σ to be invertible. However, it turns out that this degeneracy is not *fatal* in the sense that it is *hypoelliptic*. Notice that we cannot use the conclusion of Remark 4.50 either, since it requires boundedness of the coefficients in the space variable, which is not the case in the present situation.

As a result, we prove solvability of (4.181) by a direct approach, taking advantage of the quadratic structure of the cost functional J^μ which allows us to implement a form of the so-called *Cole-Hopf transformation*.

Direct Analysis of the FBSDE

Following the same strategy as in Subsection 4.4.1, we first construct the decoupling field U of the FBSDE (4.181). Here we use the upper case U to denote the decoupling field in order to distinguish it from the object which will be referred to by the lower case u later on. Recall that the flow of probability measures $\boldsymbol{\mu}$ is fixed throughout the analysis. Prompted by the method of proof at the crux of the weak formulation, we focus first on the decoupled system:

$$\begin{cases} dx_t = v_t dt, \quad dv_t = dW_t, \\ dY_t = \left(\frac{1}{2}|Z_t|^2 - f\big((x_t, v_t), \mu_t\big)\right)dt + Z_t \cdot dW_t, \qquad t \in [0, T], \\ X_0 = (x_0, v_0) = \xi, \quad Y_T = 0, \end{cases} \tag{4.182}$$

where the control does not appear. We claim:

Proposition 4.70 *For a given input $\boldsymbol{\mu} = (\mu_t)_{0 \leq t \leq T}$ as above, the equation* (4.182) *has a unique solution $(x_t, v_t, Y_t, Z_t)_{0 \leq t \leq T}$, for which $\boldsymbol{Y} = (Y_t)_{0 \leq t \leq T}$ and the martingale integrand $\boldsymbol{Z} = (Z_t)_{0 \leq t \leq T}$ are bounded.*

Moreover, there exists a bounded and continuous function $U : [0, T] \times \mathbb{R}^d \times \mathbb{R}^d \to \mathbb{R}$, differentiable and Lipschitz continuous in the space argument uniformly in time, such that with probability 1, for all $t \in [0, T]$, $Y_t = U(t, x_t, v_t)$ and $Z_t = \partial_v U(t, x_t, v_t)$. For each $x, v \in \mathbb{R}^d$ and $t \in [0, T]$, the quantity $U(t, x, v)$ is given by the formula:

$$U(t, x, v) = \\ -\ln \mathbb{E}\left[\exp\left(-\int_0^{T-t} f\left(\left(x + vs + \int_0^s W_r dr, v + W_s\right), \mu_{s+t}\right)ds\right)\right]. \tag{4.183}$$

Proof. Since the forward and backward components of equation (4.182) are decoupled, existence and uniqueness of a solution $(\boldsymbol{x}, \boldsymbol{v}, \boldsymbol{Y}, \boldsymbol{Z})$ depend only on the BSDE part, and such a result (with $\boldsymbol{Y}$ bounded), follows from Theorem 4.15. The identification of this solution and of the decoupling field relies on the so-called Cole-Hopf transformation. Indeed, by Itô formula, it must hold:

$$d(e^{-Y_t}) = e^{-Y_t} f\big((x_t, v_t), \mu_t\big)dt - e^{-Y_t} Z_t \cdot dW_t, \quad t \in [0, T]\,; \quad e^{-Y_T} = 1.$$

Since $\boldsymbol{Y} = (Y_t)_{0 \leq t \leq T}$ is a bounded process, we deduce that the pair process $(\tilde{\boldsymbol{Y}}, \tilde{\boldsymbol{Z}}) = (e^{-Y_t}, -e^{-Y_t} Z_t)_{0 \leq t \leq T}$ solves the linear BSDE:

$$d\tilde{Y}_t = \tilde{Y}_t f\big((x_t, v_t), \mu_t\big)dt + \tilde{Z}_t \cdot dW_t, \quad t \in [0, T]\,; \quad \tilde{Y}_T = 1,$$

which is uniquely solvable since f is bounded. Call $(\tilde{Y}_t, \tilde{Z}_t)_{0 \leq t \leq T}$ the solution. Then,

$$d\left[\tilde{Y}_t e^{-\int_0^t f((x_s, v_s), \mu_s)ds}\right] = e^{-\int_0^t f((x_s, v_s), \mu_s)ds} \tilde{Z}_t \cdot dW_t.$$

Again, by boundedness of f, the term in the right-hand side must be a martingale. Thus,

$$\mathbb{E}\Big[\tilde{Y}_T e^{-\int_0^T f((x_s,v_s),\mu_s)ds} \mid \mathcal{F}_t\Big] = \tilde{Y}_t e^{-\int_0^t f((x_s,v_s),\mu_s)ds},$$

that is,

$$\tilde{Y}_t = \mathbb{E}\Big[e^{-\int_t^T f((x_s,v_s),\mu_s)ds} \mid \mathcal{F}_t\Big].$$

Because of the Markov property, $\tilde{Y}_t$ can be written as:

$$\tilde{Y}_t = e^{-U(t,x_t,v_t)},$$

with U as in (4.183). Since f is assumed to be differentiable in (x, v), with bounded derivatives, it is easily checked that U is also differentiable in (x, v) with bounded and continuous (in time and space) derivatives. Moreover, by Lemma 4.11, the process $\tilde{\mathbf{Z}}$ may be identified in $L^2([0,T] \times \Omega; \mathbb{R}^d)$ with:

$$\tilde{Z}_t = -\partial_v U(t, x_t, v_t) \exp\big(-U(t, x_t, v_t)\big), \quad t \in [0, T].$$

Recalling that by definition $Y_t = -\ln(\tilde{Y}_t)$ and $Z_t = -\tilde{Z}_t/\tilde{Y}_t$, we conclude that the process $(Y_t, Z_t)_{0\leqslant t\leqslant T}$ satisfies the conditions in the statement of the lemma. □

We further investigate the properties of the function U constructed above.

Lemma 4.71 *The function U constructed in Proposition 4.70 is once differentiable in time and twice continuously differentiable in space, with bounded derivatives. Moreover, the uniform bounds on the derivatives are independent of the flow of probability measures $\boldsymbol{\mu} = (\mu_t)_{0\leqslant t\leqslant T}$.*

Also, U satisfies the HJB equation:

$$\begin{aligned}\partial_t U(t,x,v) + v \cdot \partial_x U(t,x,v) + \frac{1}{2}\partial^2_{vv} U(t,x,v) - \frac{1}{2}|\partial_v U(t,x,v)|^2 \\ + f\big((x,v),\mu_t\big) = 0,\end{aligned}$$

for $(t,(x,v)) \in [0,T] \times \mathbb{R}^{2d}$, with the terminal condition $U(T,\cdot,\cdot) = 0$.

Proof. Existence and continuity of the first and second order derivatives in (x, v) is a straightforward consequence of the regularity of f and of the formula given for U in the statement of Proposition 4.70.

Existence of the first order derivative in time may be proved as follows. For some initial condition $(t,(x,v)) \in [0,T] \times \mathbb{R}^{2d}$, consider the unique solution $(X_s, Y_s, Z_s)_{t\leqslant s\leqslant T}$ of (4.182) with $X_t = (x, v)$ as initial condition at time t. Recall that we use the notation $X_t = (x_t, v_t)$ for $0 \leqslant t \leqslant T$. Then, by applying Itô's formula in the space variable only, we get:

$$\mathbb{E}\big[U(t+h, x_{t+h}, v_{t+h})\big] = U(t+h, x, v)$$
$$+\,\mathbb{E}\int_t^{t+h}\Big(v_s \partial_x U\big(t+h, x_s, v_s\big) + \frac{1}{2}\partial^2_{vv} U\big(t+h, x_s, v_s\big)\Big)ds. \tag{4.184}$$

Recall that we can identify the left-hand side with $\mathbb{E}[Y_{t+h}]$. Going back to the backward equation (4.182) and using the fact that $Z_s = \partial_v U(s, x_s, v_s)$ for $s \in [t, T]$, we see that:

$$\mathbb{E}[Y_{t+h}] = U(t, x, v) + \mathbb{E}\int_t^{t+h}\Big(\frac{1}{2}|\partial_v U(s, x_s, v_s)|^2 - f\big((x_s, v_s), \mu_s\big)\Big)ds. \tag{4.185}$$

Therefore, by identifying the left-hand sides in (4.184) and (4.185) and by taking advantage of the fact that $\partial_x U$ and $\partial^2_{vv} U$ are (jointly) continuous and bounded, we easily deduce that:

$$\lim_{h\searrow 0}\frac{U(t+h, x, v) - U(t, x, v)}{h}$$
$$= -v\cdot\partial_x U(t, x, v) - \frac{1}{2}\partial^2_{vv} U(t, x, v) + \frac{1}{2}|\partial_v U(t, x, v)|^2 - f\big((x, v), \mu_t\big).$$

Since the right-hand side is continuous in t, we conclude that U is differentiable in time and that the time derivative is equal to the right-hand side. □

Using the above two results in the same way we used Lemmas 4.47 and 4.49 earlier, we prove the following existence and uniqueness result for the FBSDE (4.181).

Proposition 4.72 *Equation* (4.181) *has a unique solution* $(X_t, Y_t, Z_t)_{0\leqslant t\leqslant T}$ *for which the process* $\boldsymbol{Y} = (Y_t)_{0\leqslant t\leqslant T}$ *is bounded and* $\boldsymbol{Z} = (Z_t)_{0\leqslant t\leqslant T}$ *is essentially bounded for* $\mathrm{Leb}_1 \otimes \mathbb{P}$. *This solution admits the function U defined in* (4.183) *as decoupling field.*

Moreover, for a given initial condition $X_0 = (x_0, v_0) = \xi \in L^2(\Omega, \mathcal{F}_0, \mathbb{P}; \mathbb{R}^{2d})$, *the process* $\boldsymbol{X} = (\boldsymbol{x}, \boldsymbol{v}) = (x_t, v_t)_{0\leqslant t\leqslant T}$ *is the unique optimal path for the control problem* (4.177)–(4.178).

Matching Reformulation and Solvability of the MFG Problem

Existence of an equilibrium for the MFG problem associated with the stochastic control problem (4.177)–(4.178) is almost a direct consequence of Proposition 4.72 and Theorem 4.39. The only restriction is that, in the statement of Theorem 4.39, coefficients of the McKean-Vlasov BSDE are required to be at most of linear growth in the variable z, which is not the case here. However, this is easily circumvented by using the fact that the process $(Z_t)_{0\leqslant t\leqslant T}$ in the statement of Proposition 4.72 is bounded independently of $\boldsymbol{\mu}$.

Quite remarkably, it turns out that we can say more about the form of the equilibria. Owing to the quadratic structure of the control problem, we are indeed able to provide a direct representation of the law of the optimal paths.

Lemma 4.73 *For a given initial distribution $\mu_0 \in \mathcal{P}_2(\mathbb{R}^{2d})$, we equip the space $\mathcal{C}([0,T];\mathbb{R}^{2d})$ with the image $\mathbb{P}_0$ of the measure $\mu_0 \otimes \mathcal{W}_d$ on $\mathbb{R}^{2d} \times \mathcal{C}([0,T];\mathbb{R}^d)$ by the mapping:*

$$\mathbb{R}^{2d} \times \mathcal{C}([0,T];\mathbb{R}^d) \ni \big((x,v), \boldsymbol{w} = (w_t)_{0 \leqslant t \leqslant T}\big)$$

$$\mapsto \left(x + \int_0^t (v + w_s)ds, v + w_t\right)_{0 \leqslant t \leqslant T},$$

where $\mathcal{W}_d$ denotes the d-dimensional Wiener measure.

Then, for a given flow of probability measures $\boldsymbol{\mu} = (\mu_t)_{0 \leqslant t \leqslant T}$ with μ_0 as initial condition and with paths in $\mathcal{C}([0,T];\mathcal{P}_2(\mathbb{R}^{2d}))$, the law $\mathbb{P}^{\boldsymbol{\mu}}$ on the space $\mathcal{C}([0,T];\mathbb{R}^{2d})$ of the optimal path solving the stochastic control problem (4.177)–(4.178), with an initial condition ξ being distributed according to μ_0, is absolutely continuous with respect to $\mathbb{P}_0$, with the density:

$$\frac{d\mathbb{P}^{\boldsymbol{\mu}}}{d\mathbb{P}_0} = \exp\big(U(0,x_0,v_0)\big) \exp\left(-\int_0^T f\big((x_t,v_t),\mu_t\big)dt\right), \tag{4.186}$$

where $(\boldsymbol{x} = (x_t)_{0 \leqslant t \leqslant T}, \boldsymbol{v} = (v_t)_{0 \leqslant t \leqslant T})$ denotes the canonical process on the space $\mathcal{C}([0,T];\mathbb{R}^{2d})$ and U is as in the statement of Proposition 4.70.

We emphasize that Lemma 4.73 is nothing but a probabilistic reformulation of the Cole-Hopf transformation used in the proof of Proposition 4.70.

Proof. On the canonical space $\mathcal{C}([0,T];\mathbb{R}^{2d})$, equipped with the probability $\mathbb{P}_0$ and with the complete and right-continuous augmentation of the canonical filtration $\mathbb{F}$ generated by the canonical process $(\boldsymbol{x},\boldsymbol{v})$, we let $(w_t = v_t - v_0)_{0 \leqslant t \leqslant T}$. By construction of $\mathbb{P}_0$, $\boldsymbol{w} = (w_t)_{0 \leqslant t \leqslant T}$ is an $\mathbb{F}$-Wiener process, and we can write the dynamics of $\boldsymbol{v}$ in the form:

$$dv_t = -\partial_v U(t,x_t,v_t)dt + d\left(w_t + \int_0^t \partial_v U(s,x_s,v_s)ds\right), \quad t \in [0,T].$$

This prompts us to define the equivalent probability measure (recall that $\partial_v U$ is bounded):

$$\frac{d\mathbb{P}^{\boldsymbol{\mu}}}{d\mathbb{P}_0} = \exp\left(-\int_0^T \partial_v U(s,x_s,v_s) \cdot dw_s - \frac{1}{2}\int_0^T |\partial_v U(s,x_s,v_s)|^2 ds\right).$$

Under $\mathbb{P}^{\boldsymbol{\mu}}$, the process $(w_t^{\boldsymbol{\mu}} = w_t + \int_0^t \partial_v U(s,x_s,v_s)ds)_{0 \leqslant t \leqslant T}$ is an $\mathbb{F}$-Brownian motion. Moreover,

$$dv_t = -\partial_v U(t,x_t,v_t)dt + dw_t^{\boldsymbol{\mu}},$$

so that $(x_t,v_t)_{0 \leqslant t \leqslant T}$ solves, under $\mathbb{P}^{\boldsymbol{\mu}}$, the same SDE as the optimal path of the optimal control problem (4.177)–(4.178), see Propositions 4.70 and 4.72. Therefore, $\mathbb{P}^{\boldsymbol{\mu}}$ is the law of the optimal path.

It only remains to derive the form (4.186) of $d\mathbb{P}^{\mu}/d\mathbb{P}^0$. By Itô's formula, we have:

$$
\begin{aligned}
&U(T, x_T, v_T) \\
&= U(0, x_0, v_0) + \int_0^T \big(\partial_t U(t, x_t, v_t) + v_t \cdot \partial_x U(t, x_t, v_t) + \frac{1}{2}\partial^2_{vv} U(t, x_t, v_t)\big)dt \\
&\qquad + \int_0^T \partial_v U(t, x_t, v_t) \cdot dw_t, \\
&= U(0, x_0, v_0) - \int_0^T f\big((x_t, v_t), \mu_t\big)dt + \int_0^T \frac{1}{2}|\partial_v U(t, x_t, v_t)|^2 dt \\
&\qquad + \int_0^T \partial_v U(t, x_t, v_t) \cdot dw_t,
\end{aligned}
$$

where we have used the HJB equation to pass from the first to the second line. Finally, we complete the proof using the fact that $U(T, \cdot, \cdot) = 0$. □

We can now state the desired characterization of the equilibrium:

Proposition 4.74 *With the same notation as in the statement of Lemma 4.73, any measure flow $\boldsymbol{\mu} = (\mu_t)_{0 \leqslant t \leqslant T} \in \mathcal{C}([0, T]; \mathcal{P}_2(\mathbb{R}^{2d}))$ solving:*

$$
\mu_t = \mathbb{P}^{\mu} \circ e_t^{-1}, \qquad 0 \leqslant t \leqslant T,
$$

where e_t denotes the evaluation map at time t on $\mathcal{C}([0, T]; \mathbb{R}^{2d})$ (i.e., $e_t(\boldsymbol{x}, \boldsymbol{v}) = (x_t, v_t)$) is an equilibrium for the model of flocking.

Computational Implications

Formula (4.186) may be very useful for numerical purposes. Indeed it provides a direct way to simulate the optimal path in the environment $\boldsymbol{\mu} = (\mu_t)_{0 \leqslant t \leqslant T}$ by using an acceptance-rejection method when the supremum norm of f is not too large, or more refined particle method otherwise. In this regard, it is worth noticing that there is no need to solve the HJB equation to simulate the distribution of the optimal path by Monte Carlo methods, even though $U(0, \cdot, \cdot)$ explicitly appears in the density. The reason is that $\exp(U(0, x_0, v_0))$ is nothing but a normalizing constant and simulation methods based on systems of particles do not need to evaluate it. This is the more convenient that the HJB equation is in $2d$ dimensions.

The above discussion could serve as the basis for a strategy to solve MFG problems numerically by means of Picard iterations based on the construction of successive approximations of the measure flow $\boldsymbol{\mu} = (\mu_t)_{0 \leqslant t \leqslant T}$ by the empirical measures of Monte Carlo samples generated according to formula (4.186).

We implemented the Monte Carlo strategy described above to simulate the optimal paths of the solution of the optimal control problem whenever the input measure flow $\boldsymbol{\mu} = (\mu_t)_{0 \leqslant t \leqslant T}$ is fixed. We restricted ourselves to the two-dimensional case, i.e., $d = 2$, and we used the running cost functions f given

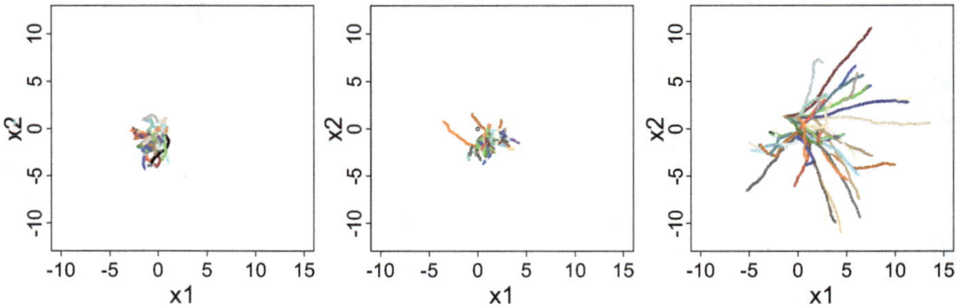

Fig. 4.6 Monte Carlo samples of a system of particles near equilibrium in the case $\beta = 0$ (left), $\beta = 0.1$ (center), and $\beta = 5$ (right).

in (4.180) and (4.179). The qualitative features of the results being the same, we only report on simulations based on the running cost function (4.179):

$$f((x,v),\mu) = \int_{\mathbb{R}^4} \frac{|v - v'|^2}{(1 + |x - x'|^2)^\beta} \mu(dx', dv'), \qquad x, v \in \mathbb{R}^2.$$

We chose $\kappa = \sqrt{2}$ for the sake of definiteness. Figure 4.6 shows Monte Carlo samples of a system near equilibrium (i.e., for an input flow $\boldsymbol{\mu}$ still not a fixed point of the Picard iteration) in the case $\beta = 0$, $\beta = 0.1$, and $\beta = 5$, the other parameters of the model being the same. Even though it may not appear very clearly, each plot contains 2000 Monte Carlo sample trajectories, each of them comprising 100 points, the velocity vector being attached to each of these points. The impact of the size of the parameter β is clear. Indeed, when β is large (right pane of Figure 4.6), positions x far from the bulk of the positions likely to occur according to the flow $\boldsymbol{\mu}$ will create large denominators in the expression of f, and as a consequence, a small overall cost. So one way for the particles to lower the cost is to drift apart, property which we can clearly see from Figure 4.6. As recalled in Subsection 1.5.1 of Chapter 1, the terminology *flocking* was introduced to describe situations in which the birds (particles in our present context) remain in a bounded set as time evolves. From the plots above, it seems clear that flocking is likely for small values of β and highly questionable for large values of β, the threshold separating these regimes being $\beta = 0.5$ in the deterministic (nonequilibrium) dynamical system proposed by Cucker and Smale. One way to prove flocking in the classical deterministic dynamical systems was to prove asymptotic stability of the velocity. We reinforce the points made earlier on the basis of the positions of the particles by looking at the time evolutions of the velocities. Even though we let the time go from $t = 0$ to $T = 4$ in 100 time steps, we see from Figure 4.7 that the velocity vectors seem to remain in a compact set when $\beta = 0$ while they seem to lack stability and diverge when $\beta = 5$, confirming the characterization of flocking touted for deterministic dynamical systems.

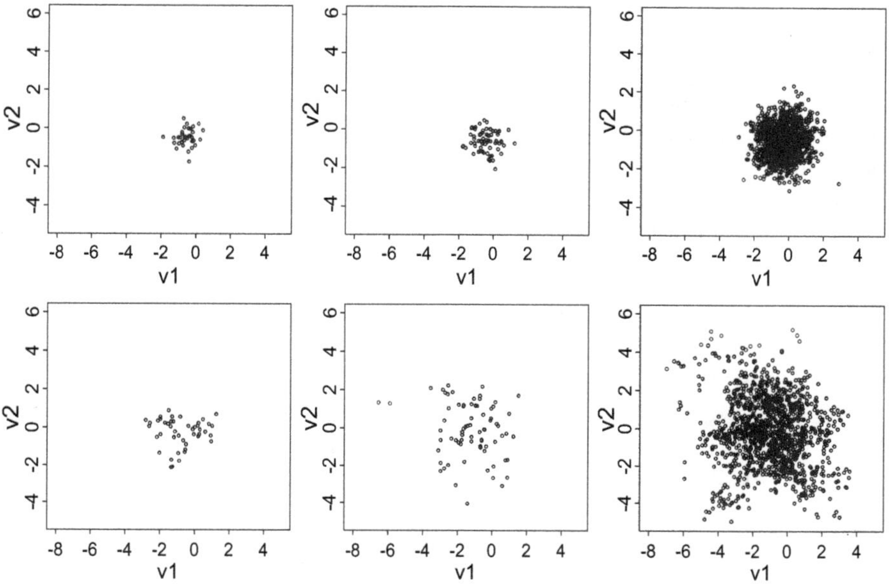

Fig. 4.7 Monte Carlo samples of the two components of the velocities of a system of 2000 particles near equilibrium in the case $\beta = 0$ (top row), $\beta = 5$ (bottom row). The three plots give from left to right, the locations of the atoms of the empirical distributions of the two components of the velocity vectors after 10, 50 and 100 time steps.

4.8 Notes & Complements

The analysis of fully coupled forward-backward SDEs in small time was initiated first by Antonelli in [25]. Since then, several methods have been discussed in order to prove existence and uniqueness over an interval of arbitrary length: Hu and Peng [203] and Peng and Wu [307] implemented a continuation argument under a suitable monotonicity assumption inspired by convexity conditions in the theory of stochastic optimal control (see also Chapter 6 below where we implement this method to solve McKean-Vlasov FBSDEs deriving from the stochastic Pontryagin principle for mean field stochastic control problems), Pardoux and Tang [300] exhibited another type of monotonicity condition that permits to apply the Picard fixed point theorem, and Ma, Protter, and Yong [271] and Delarue [132] made use of the connection with nondegenerate quasilinear PDEs for systems driven by deterministic coefficients. The notion of decoupling field was introduced by Ma and his coauthors in [272]. The reader is referred to the book of Ma and Yong [274] for background material on adjoint equations, FBSDEs and the stochastic maximum principle approach to stochastic optimization problems. The reader is also referred to the papers by Hu and Peng [203] and Peng and Wu [307] for general solvability properties of standard FBSDEs within the same framework of stochastic optimization.

The rehabilitation of the Cauchy-Lipschitz theory when the diffusion coefficient (or volatility) driving the noise term is nondegenerate and the coefficients are bounded in the space variable, see Theorem 4.12, is due to Delarue. The original result can be found in [132], see Theorem 2.6 and Corollary 2.8 therein.

The representation formula for the process $(Z_t)_{0 \leqslant t \leqslant T}$ in the statement of Lemma 4.11 is a standard result in the theory of backward SDEs. It may be seen as a particular case of a more general result permitting to represent $(Z_t)_{0 \leqslant t \leqslant T}$ in terms of the Malliavin derivative of $(Y_t)_{0 \leqslant t \leqslant T}$, see for instance El Karoui, Peng and Quenez [226]. For differentiability properties of BSDEs, as used in the proof of Proposition 4.51, we refer to the seminal paper by Pardoux and Peng [298].

The theory of quadratic BSDEs goes back to the original work of Kobylanski [232], see also the seminal paper by Briand and Hu [71]. We refer the interested reader to Dos Reis' monograph [141] for a quite comprehensive overview of the subject, including a discussion of the differentiability of the flow formed by the solutions.

The most standard reference on the BMO condition is the monograph by Kazamaki [227].

The analysis of SDEs of McKean-Vlasov type has a long history. These equations were first introduced by Henry McKean Jr in [277] and [278] to provide a rigorous treatment of special nonlinear Partial Differential Equations (PDEs). Later on, they were studied for their own sake, and in a more general mathematical setting. The standard reference for existence and uniqueness of solutions to these special SDEs is Sznitman's original set of lectures [325]. See also the paper of Jourdain, Méléard, and Woyczynski [221] for a generalization including jumps. Properties of the solutions have been studied in the framework of the propagation of chaos, as McKean-Vlasov equations appear as effective equations describing the dynamics of large populations of individuals subject to mean field interactions, see again [325] together with Méléard [279]. Propagation of chaos will be revisited in Chapter (Vol II)-2.

As explained in the Notes & Complements of Chapter 3, Backward Stochastic Differential Equations (BSDEs) of mean field type were introduced by Buckdahn and his coauthors, see [74, 75] for example. In these papers, the McKean-Vlasov interaction is of a more restricted form than in Subsection 4.2.2. Also, these results cannot be used in the applications considered in Chapter 3 and in the optimal control of McKean-Vlasov stochastic differential equations studied in Chapter 6. Indeed, we are mostly interested in the analysis of systems of coupled FBSDEs of McKean-Vlasov type. As far as BSDEs (and not FBSDEs) are concerned, the discussion of Subsection 4.2.2 is inspired by Carmona's lecture notes [94]. The proof is adapted from the original existence and uniqueness result of Pardoux and Peng [298] for standard BSDEs.

To the best of our knowledge, fully coupled FBSDEs of McKean-Vlasov type were first investigated by Carmona and Delarue in [95]. In particular, Theorem 4.29 is taken from Carmona and Delarue [95], with a slight simplification in the assumption: We here assume that B is bounded in x whereas it is allowed to be of linear growth in [95]. The argument used to relax the boundedness condition in Subsection 4.3.3 is also different: It is based on PDE arguments in [95] whereas we here use a stability property for FBSDEs. As noticed in Remark 4.55, Theorem 4.39 may not suffice to prove Theorem 4.53; indeed, the analysis of linear-quadratic games performed in Chapter 3 shows that (**A5**) in **MKV FBSDE for MFG** may fail as, in that case, the feedback function may grow up linearly with the mean state of the population. So, Theorem 4.39 does not cover some of the linear-quadratic games already studied in the literature [33,53,99,212]. Most of the technical results of this chapter are thus devoted to the extension of this existence result to coefficients with linear growth. The reader could skip some of these derivations in a first reading. Our approximation and convergence arguments in Subsection 4.5 are taken from the paper [96] by Carmona and Delarue. As we rely on the stochastic maximum principle, we find it natural to derive the *a priori* estimates needed in the proof from convexity properties of the coefficients of the game. The solvability results obtained in this chapter will be extended to more general cases in Chapter (Vol II)-3, when dealing with mean field games with a common noise; by specializing some of them to the case without common noise, we shall prove existence of a solution under weaker assumptions than those used in this chapter, see for instance Section (Vol II)-3.4.

Throughout the analysis of the McKean-Vlasov FBSDEs occurring in mean field games, a significant amount of effort is devoted to the construction of the decoupling field of the FBSDE. The latter expresses the solution of the backward equation as a function of the solution of the forward dynamics. The existence of this function is crucial for the formulation and the proofs of the approximation results given in Section (Vol II)-6.1 of Chapter (Vol II)-6. In the present set-up, the decoupling field is constructed as a function of the sole time and space variables; as a matter of fact, the flow of marginal distributions of the forward component does not appear explicitly although the decoupling field actually depends on it. In Chapter (Vol II)-4, we shall show that, when uniquely solvable, McKean-Vlasov FBSDEs admit an infinite dimensional decoupling field, called *master field*, independent of the flow of marginal measures of the forward equation, but permitting to express the current value of the backward solution in terms of the current realization and marginal distribution of the forward component, see Subsection 4.2.4 for a primer.

The version of Schauder's fixed point theorem used in this chapter, see Theorem 4.32, may be found in many textbooks on fixed point theorems, see for instance Granas and Dugundji [184], Reed and Simon [318] (although the statement therein is limited to the case when E is compact), and Zeidler [345]. The original version by Schauder [323] does not apply here since it requires the space V to be complete. The extension to non-complete normed spaces (or more generally to locally convex spaces) is sometimes called Schauder-Tychonoff's fixed point theorem. For the definition and the properties of the Kantorovich-Rubinstein norm set on the space of finite measures, as used in Subsection 4.3.2 to implement Schauder's theorem, we refer to Chapter 8 in Bogachev [64].

As for classical solutions of HJB equations as cited in Remark 4.48, we refer to monographs on semilinear parabolic PDEs, see for instance Friedman [162], Ladyzenskaja, Solonnikov, and Ural'ceva [258] and Lieberman [264]. Regarding the Cole-Hopf transformation cited in Subsection 4.1.3, the reader may consult Guéant's paper [186] for a quite systematic use within the MFG framework.

The price impact model proposed in Subsection 4.7.1 is based on the Almgren-Chriss linear model of the influence of the trading intensity on prices. See [18]. Not unlike most papers on optimal execution, the control of each trader is the rate of trading. However, the presence of the Wiener processes $(W_t^i)_{0 \leqslant t \leqslant T}$, $i = 1, \cdots, N$, prevents the traders' inventories from being differentiable. This is in accordance with a recent trend in the literature on option trading in the presence of price impact. See for example the works [112] by Cetin, Soner, and Touzi, and [107] by Carmona and Webster, this latter paper containing empirical evidence of the infinite variation nature of the inventories. Existence of a solution for the mean field game in the weak formulation was proven by Carmona and Lacker in [103], though in a nonconstructive way and under more restrictive assumptions including compactness of the space A of controls. The results presented in Subsection 4.7.1 are from the unpublished technical report by Aghbal and Carmona [9]. The solution given in the text is based on the original approach to extended mean field games proposed by the authors in Subsection 4.6. The terminology *extended mean field games* is borrowed from Gomes and Saude [182] and Gomes and Voskanyan [183], although the formulation therein is slightly different and is limited to dynamics without volatility. The uniqueness criterion proven in [183] is similar to that obtained in this chapter. Therein, the construction of a solution relies on the assumption that an equation similar to (4.115) is solvable.

As explained in the Notes & Complements of Chapter 1, our discussion of models of crowd aversion and congestion is inspired by the paper [253] of Lachapelle and Wolfram. The room exit problem is borrowed from the paper [5] by Achdou and Laurière who provided numerical illustrations without proofs. The paper [43] of Benamou, Carlier, and Bonne proposes numerical schemes which are far more sophisticated than the naive ones used to produce the numerical results reproduced in the text. They are tailor-made to handle more realistic domains to exit, including dumbell-shaped domains.

We believe that the theoretical solution provided in Subsection 4.7.2 is original. The properties of the joint distribution of the exit time and location of a Brownian motion from a regular domain can be found in the paper [13] by Aizenman and Simon. By regular domain, we mean a bounded set with a smooth boundary, or at least a piecewise smooth boundary like a convex polygon.

Concerning the reflected Brownian motion used in the application to crowd exits discussed in Subsection 4.7.2, we refer the interested reader to the seminal papers by Tanaka [329], and by Lions and Sznitman [267] for its construction in a bounded domain of an Euclidean space. This construction is based on the use of the so-called Skorokhod map whose theory was enhanced in a more recent paper [81] by Burdzy, Kang, and Ramanan. The properties of Brownian motion reflected in a convex domain which we used in the proof of Lemma 4.66 can be found for instance in the paper [39] of Bass and Hsu, [109] by Carmona and Zheng, and [129] by Davies.

The theoretical solution of the flocking model provided in Subsection 4.7.3 is original.

Part II

Analysis on Wasserstein Space and Mean Field Control

Spaces of Measures and Related Differential Calculus

5

Abstract

The goal of the present chapter is to present in a self-contained manner, elements of differential calculus and stochastic analysis over spaces of probability measures. Such a calculus will play a crucial role in the sequel when we discuss stochastic control of dynamics of the McKean-Vlasov type, and various forms of the master equation for mean field games. After reviewing the standard metric theory of spaces of probability measures, we introduce a notion of differentiability of functions of measures tailor-made to our needs. We provide a thorough analysis of its properties, and relate it to different notions of differentiability which have been used in the existing literature, in particular the geometric notion of Wasserstein gradient. Finally, we derive a first form of chain rule (Itô's formula) for functions of flows of measures, and we illustrate its versatility on a couple of applications.

5.1 Metric Spaces of Probability Measures

In the previous chapters, we investigated special forms of large symmetric games introduced as early as in Chapter 1. A simple limiting argument was used to show that the analysis of symmetric functions of a large number of variables could be advantageously replaced by the analysis of functions of probability measures. The continuity properties of these functions, and the characterization of compact sets of measures played an important role, and they were studied by classical measure theoretical tools. In this first section, we collect some of the notation and concepts already used. This will set the stage for the introduction of the more sophisticated tools needed in the investigation of the differentiability properties of functions of probability measures.

R. Carmona, F. Delarue, *Probabilistic Theory of Mean Field Games with Applications I*, Probability Theory and Stochastic Modelling 83,
DOI 10.1007/978-3-319-58920-6_5

Recall from Chapter 1 that if $(E, \mathcal{E})$ is a measurable space, we use the notation $\mathcal{P}(E)$ for the space of probability measures on $(E, \mathcal{E})$, assuming that the σ-field $\mathcal{E}$ on which the measures are defined is understood.

5.1.1 Distances Between Measures

Most of the results mentioned in this subsection are classical. They exist in book form so we state them without proof. We refer the reader to the Notes & Complements at the end of the chapter for references and a detailed bibliography.

Throughout the section, we denote by (E, d) a complete separable metric space. It is important to keep in mind that what we are about to do depends upon the choice of the distance. Indeed, the metric structure of E is more important than the topology determined on E by d. Indeed, the same topology could be obtained from other metrics, for example a bounded metric, but the results of our analysis would have to be modified. Quite often, we use a somewhat abusive terminology by saying that (E, d) is a Polish space, in which case, we implicitly assume that d is one of the distances on E making E a complete and separable space. The σ-field $\mathcal{E}$ equipping E is always assumed to be the Borel σ-field $\mathcal{B}(E)$ which does not depend upon the choice of the particular compatible distance.

Lévy-Prokhorov Distance

The weak convergence of probability measures on E appears as the convergence in the sense of the so-called Lévy-Prokhorov distance d_{LP} on $\mathcal{P}(E)$ defined in the following way:

$$\begin{aligned} &d_{\mathrm{LP}}(\mu, \nu) \\ &\quad = \inf\bigl\{\epsilon > 0 : \; \forall A \in \mathcal{B}(E),\; \mu(A) \leqslant \nu(A^\epsilon) + \epsilon, \;\; \text{and} \;\; \nu(A) \leqslant \mu(A^\epsilon) + \epsilon\bigr\}, \end{aligned}$$

where we use the notation A^ϵ for the ϵ-neighborhood $A^\epsilon = \{x \in E : \exists y \in A, d(x, y) < \epsilon\}$ of a set A.

A famous result of Strassen (later extended by Dudley to more general metric spaces) asserts that the Lévy-Prokhorov distance between μ and ν can be represented in terms of *couplings* between μ and ν:

$$d_{\mathrm{LP}}(\mu, \nu) = \inf\left\{\epsilon > 0 : \inf_{\pi \in \Pi(\mu,\nu)} \int_{E\times E} \mathbf{1}_{\{d(x,y)>\epsilon\}} d\pi(x, y) < \epsilon\right\}, \tag{5.1}$$

where $\Pi(\mu, \nu)$ denotes the set of probability measures on $E \times E$ with μ and ν as respective first and second marginals. These probability measures are often called couplings (or transport plans) between μ and ν. It is easy to check that if $x \in E$ and $y \in E$, then

$$d_{\mathrm{LP}}(\delta_x, \delta_y) = 1 \wedge d(x, y). \tag{5.2}$$

More generally, for $X = (x^1, \cdots, x^N)$ and $Y = (y^1, \cdots, y^N)$ in E^N, we may want to compute the Lévy-Prokhorov distance between the empirical measures $\bar{\mu}_X^N$ and $\bar{\mu}_Y^N$, where $\bar{\mu}_X^N$ was defined in Chapter 1 in formula (1.3) as:

$$\bar{\mu}_X^N = \frac{1}{N}\sum_{j=1}^{N} \delta_{x^j},$$

and similarly for $\bar{\mu}_Y^N$. Using (5.1), we get the obvious rough bound:

$$d_{\mathrm{LP}}(\bar{\mu}_X^N, \bar{\mu}_Y^N) \leqslant \Big(\max_{i=1,\cdots,N} d(x^i, y^i)\Big) \wedge 1.$$

Actually, the proof of the result by Strassen and Dudley (see the Notes & Complements at the end of the chapter for a precise reference) shows that:

$$\begin{aligned} d_{\mathrm{LP}}&(\bar{\mu}_X^N, \bar{\mu}_Y^N) \\ &= \inf\left\{\epsilon > 0 : \inf_{\sigma \in \mathcal{S}_N} \left(\frac{\sharp\{i \in \{1,\cdots,N\};\ d(x^i, y^{\sigma(i)}) > \epsilon\}}{N}\right) < \epsilon\right\}, \end{aligned} \tag{5.3}$$

where $\mathcal{S}_N$ denotes the set of all the permutations of $\{1, \cdots, N\}$. The argument relies on the so-called pairing theorem (whose statement may be found in the same reference), applied with $\{1, \cdots, N\} \times \{1, \cdots, N+k\}$, where k is the floor part of $N\epsilon$ for $\epsilon > d_{\mathrm{LP}}(\bar{\mu}_X^N, \bar{\mu}_Y^N)$. Two elements $i \in \{1, \cdots, N\}$ and $j \in \{1, \cdots, N+k\}$ are said to be connected if $j \leq N$ and $d(x^i, y^j) < \epsilon$ or if $j \geq N+1$. Since $\epsilon > d_{\mathrm{LP}}(\bar{\mu}_X^N, \bar{\mu}_Y^N)$, the pairing theorem implies the existence of a 1-1 mapping ς from $\{1, \cdots, N\}$ into $\{1, \cdots, N+k\}$ such that i and $\varsigma(i)$ are connected for all $i \in \{1, \cdots, N\}$. Using this ς, we may construct an element $\sigma \in \mathcal{S}_N$ as in (5.3).

Observe that, in comparison with (5.1), the representation formula (5.3) is solely based on the couplings of $\bar{\mu}_X^N$ and $\bar{\mu}_Y^N$ that are of the form $\bar{\mu}_X^N \circ (I_E, \varphi)^{-1}$ where I_E is the identity on E and φ is a measurable mapping from E into itself. Such couplings are said to be *deterministic*. Whenever both $x^1, \cdots, x^N$ and $y^1, \cdots, y^N$ are pairwise distinct, φ must induce a one-to-one mapping from $\{x^1, \cdots, x^N\}$ onto $\{y^1, \cdots, y^N\}$ by restriction.

Connection with the Total Variation Distance. Observe that by choosing:

$$\epsilon = \sup_{A \in \mathcal{B}(E)} |\mu(A) - \nu(A)|$$

for $\mu, \nu \in \mathcal{P}(E)$ in the definition of $d_{\mathrm{LP}}(\mu, \nu)$, we get the bound:

$$d_{\mathrm{LP}}(\mu, \nu) \leqslant \frac{1}{2} d_{\mathrm{TV}}(\mu, \nu),$$

where $d_{\mathrm{TV}}(\mu, \nu) = 2\sup_{A \in \mathcal{B}(E)} |\mu(A) - \nu(A)|$ denotes the total variation distance between μ and ν. In fact, a striking parallel with (5.1) exists. Indeed, another coupling argument shows that the total variation distance between μ and ν can also be expressed as:

$$d_{\mathrm{TV}}(\mu, \nu) = 2 \inf_{\pi \in \Pi(\mu,\nu)} \int_{E \times E} \mathbf{1}_{\{x \neq y\}} d\pi(x, y).$$

Representation in Terms of Random Variables. Consider now an atomless probability space $(\Omega, \mathcal{F}, \mathbb{P})$. By atomless we mean that for any $A \in \mathcal{F}$ such that $\mathbb{P}(A) > 0$, there exists $B \in \mathcal{F}$, $B \subset A$, such that $0 < \mathbb{P}(B) < \mathbb{P}(A)$. Then, for any distribution $\mu \in \mathcal{P}(E)$, we can construct a random variable $X : \Omega \to E$ with μ as distribution. A proof of this fact can be found for example in Proposition 9.1.11 in Bogachev [64] or Proposition 9.1.2 and Theorem 13.1.1 in Dudley [143]. Applying this result to probability measures on product spaces, say $E \times E$ equipped with any product distance, we see that whenever μ and ν are probability distributions in $\mathcal{P}(E)$ and $\pi \in \Pi(\mu, \nu)$, we can find a pair of two random variables $(X, Y) : \Omega \to E \times E$ such that $\pi = \mathcal{L}(X, Y)$. We then have the two representation formulas:

$$d_{\mathrm{LP}}(\mu, \nu) = \inf \Big\{ \varepsilon > 0 : \inf_{\mathcal{L}(X)=\mu,\, \mathcal{L}(Y)=\nu} \mathbb{P}\big[d(X, Y) > \varepsilon\big] < \varepsilon \Big\},$$

and

$$d_{\mathrm{TV}}(\mu, \nu) = 2 \inf_{\mathcal{L}(X)=\mu,\, \mathcal{L}(Y)=\nu} \mathbb{P}\big[X \neq Y\big].$$

Wasserstein Distances

We now introduce formally a class of metrics which we already used in several instances, and which we will use most frequently in this book. For any $p \geqslant 1$, we denote by $\mathcal{P}_p(E)$ the subspace of $\mathcal{P}(E)$ of the probability measures of order p, namely those probability measures which integrate the p-th power of the distance to a fixed point whose choice is irrelevant in the definition of $\mathcal{P}_p(E)$.

For any $p \geqslant 1$ and $\mu, \nu \in \mathcal{P}_p(E)$, the p-Wasserstein distance $W_p(\mu, \nu)$ is defined by:

$$W_p(\mu, \nu) = \inf_{\pi \in \Pi(\mu,\nu)} \left[\int_{E \times E} d(x, y)^p \, d\pi(x, y) \right]^{1/p}. \tag{5.4}$$

Note that the quantity $W_p(\mu, \nu)$ depends upon the actual distance d in the sense that another distance would lead to different values of $W_p(\mu, \nu)$, even if the topology of E and the space $\mathcal{P}(E)$ were to remain the same. Observe also that, whenever μ and ν belong to $\mathcal{P}_p(E)$, any $\pi \in \Pi(\mu, \nu)$ is also in $\mathcal{P}_p(E \times E)$, $E \times E$ being equipped with any product distance. For convenience, we often use product distances of the form $d_{E \times E}((x_1, x_2), (y_1, y_2)) = (d(x_1, y_1)^q + d(x_2, y_2)^q)^{1/q}$ for some $q \in [1, \infty)$, or $d_{E \times E}((x_1, x_2), (y_1, y_2)) = \max(d(x_1, y_1), d(x_2, y_2))$. The quantity $\int_{E \times E} d(x, y)^p d\pi(x, y)$ is sometimes referred to as a *cost* associated with the coupling π.

The set $\Pi(\mu, \nu)$ is obviously a nonempty compact subset of $\mathcal{P}(E \times E)$ equipped with the Lévy-Prokhorov metric and, by a classical lower semicontinuity argument, the infimum appearing in the definition (5.4) of W_p is always attained by at least one coupling, which we will call an optimal coupling for W_p. For the sake of convenience, we shall use the notation $\Pi_p^{\text{opt}}(\mu, \nu)$ for the set of couplings at which the infimum is attained:

$$\Pi_p^{\text{opt}}(\mu, \nu) = \Big\{\pi \in \Pi_p(\mu, \nu) : \big(W_p(\mu, \nu)\big)^p = \int_{\mathbb{R}^d \times \mathbb{R}^d} d(x, y)^p d\pi(x, y)\Big\}. \qquad (5.5)$$

We shall often use the notation Π^{opt} for Π_p^{opt} when $p = 2$. A measurable map ψ from E into itself is called a transport map from μ to ν if it maps μ into ν, in other words, if the push-forward $\mu \circ \psi^{-1}$ is equal to ν. The associated transport plan is defined as $\mu \circ (I_E, \psi)^{-1}$ where as before I_E denotes the identity map of E, namely the push-forward of the measure μ by the map $E \ni x \mapsto (x, \psi(x)) \in E \times E$. This particular form of transport plan given by a transport map is often called a deterministic transport plan or a deterministic coupling.

The claim that W_p is a distance is not completely obvious. Indeed, the proof of the triangle inequality is not immediate as it requires using disintegration of measures, see Theorem (Vol II)-1.1 for a short remainder. If μ, ν and θ are elements of $\mathcal{P}_p(E)$, and if $\pi^{1,2}$ (resp. $\pi^{2,3}$) is a coupling between μ and ν (resp. ν and θ), we can write:

$$\pi^{1,2}(dx, dy) = \pi^{1,2}(dx, y)\nu(dy), \qquad (\text{resp. } \pi^{2,3}(dy, dz) = \nu(dy)\pi^{2,3}(y, dz)\,)$$

for the disintegration of $\pi^{1,2}$ on its second marginal ν (resp. $\pi^{2,3}$ on its first marginal ν), and then define the measure π on $E \times E \times E$ by:

$$\pi(dx, dy, dz) = \pi^{1,2}(dx, y)\nu(dy)\pi^{2,3}(y, dz).$$

Then the probability measure $\pi^{1,3}$ defined as the projection of π onto its first and third coordinates provides a coupling between μ and θ which can be used to prove the desired inequality:

$$W_p(\mu, \theta) \leqslant W_p(\mu, \nu) + W_p(\nu, \theta).$$

Notice also that Hölder's inequality implies that:

$$W_p(\mu, \nu) \leqslant W_q(\mu, \nu), \qquad \mu, \nu \in \mathcal{P}_p(E), \quad 1 \leqslant p \leqslant q.$$

Remark 5.1 *While the terminology* coupling *is ubiquitous in probability and statistics,* transport plan *is systematically used in the optimal transportation literature (see Subsection 5.1.3 below). We shall use these two terminologies interchangeably. We believe that the simultaneous use of both terms will not be the source of confusion or ambiguity in the sequel.*

Remark 5.2 *Instead of the generic terminology Wasserstein distance, we shall try to use "p-Wasserstein distance" to make clear that we are using the distance* W_p *on* $\mathcal{P}_p(E)$*, for some* $p \geqslant 1$*. Indeed, in general, the term Wasserstein distance is restricted to the distance* W_2 *while the distance* W_1 *is often called the Kantorovich-Rubinstein distance because of the role it plays in optimal transportation. We emphasize this connection in the next few results.*

The next result is known as the *Kantorovich duality* theorem. It is central to the theory of optimal transportation.

Proposition 5.3 *If* (E, d) *is a complete separable metric space,* $p \geqslant 1$ *and* $\mu, \nu \in \mathcal{P}_p(E)$*, then:*

$$W_p(\mu, \nu)^p = \sup_{\phi, \psi,\ \phi(x)+\psi(y) \leqslant d(x,y)^p} \left[\int_E \phi(x)\mu(dx) + \int_E \psi(y)\nu(dy) \right], \tag{5.6}$$

where the supremum is taken over all the real valued bounded continuous functions ϕ *and* ψ *on E. Moreover, if* $\pi \in \Pi_p^{\mathrm{opt}}(\mu, \nu)$ *is an optimal transport plan between* μ *and* ν*, then there exists* $\phi \in L^1(E, \mu)$ *and* $\psi \in L^1(E, \nu)$ *such that, for* π *almost every* $(x, y) \in E \times E$*,*

$$\phi(x) + \psi(y) = d(x, y)^p.$$

Proof.

First Step. Let us denote by $\tilde{W}_p(\mu, \nu)^p$ the right-hand side of (5.6), and prove that $\tilde{W}_p$ satisfies the triangle inequality in the sense that for three probability measures μ, ν, and θ we have:

$$\tilde{W}_p(\mu, \nu) \leqslant \tilde{W}_p(\mu, \theta) + \tilde{W}_p(\theta, \nu).$$

The idea of the proof is borrowed from the classical analysis proof that the norm of L^p spaces satisfies the triangle inequality. For $i = 1, 2$, let $(0, \infty)^2 \ni (s, t) \mapsto c_i(s, t) \in (0, \infty)$ be deterministic functions such that:

$$(a + b)^p = \inf_{s,t>0} [c_1(s, t)a^p + c_2(s, t)b^p], \qquad a, b \geqslant 0. \tag{5.7}$$

Explicit formulas can be found for $c_1(s, t)$ and $c_2(s, t)$. We refrain from giving them because they do not play any role in the proof, but the reader can easily check that, in the case $p = 2$, we can choose $c_1(s, t) = 1 + t$ and $c_2(s, t) = 1 + 1/t$.

Now let ϕ and ψ be two real valued bounded continuous functions on E satisfying $\phi(x) + \psi(y) \leqslant d(x, y)^p$ for all $x, y \in E$. Notice that (5.7) implies that, for any $s, t > 0$ and $x, y, z \in E$, we have:

$$\phi(x) + \psi(y) \leqslant d(x, y)^p \leqslant c_1(s, t)d(x, z)^p + c_2(s, t)d(z, y)^p. \tag{5.8}$$

We define the real valued function ζ by:

$$\zeta(z) = \inf_{x\in E}[c_1(s,t)d(x,z)^p - \phi(x)], \qquad z \in E.$$

By construction,

$$\phi(x) + \zeta(z) \leqslant c_1(s,t)d(x,z)^p, \qquad x, z \in E. \tag{5.9}$$

Moreover,

$$\begin{aligned}\psi(y) - \zeta(z) &\leqslant \phi(x) + \psi(y) - \zeta(z) - \phi(x)\\ &\leqslant c_1(s,t)d(x,z)^p + c_2(s,t)d(y,z)^p - \zeta(z) - \phi(x),\end{aligned}$$

where we used the assumption on the functions ϕ and ψ and inequality (5.8). Since the left-hand side does not depend upon $x \in E$, it is still not greater than the infimum of the right-hand side with respect to x. Consequently, we get:

$$\begin{aligned}\psi(y) - \zeta(z) &\leqslant \inf_{x\in E}[c_1(s,t)d(x,z)^p - \phi(x)] + c_2(s,t)d(y,z)^p - \zeta(z)\\ &= c_2(s,t)d(y,z)^p.\end{aligned} \tag{5.10}$$

Putting together (5.9) and (5.10) we get:

$$\begin{aligned}&\int_E \phi(x)\mu(dx) + \int_E \psi(y)\nu(dy)\\ &\leqslant \left[\int_E \phi(x)\mu(dx) + \int_E \zeta(z)\theta(dz)\right] + \left[\int_E \psi(y)\nu(dx) - \int_E \zeta(z)\theta(dz)\right]\\ &\leqslant c_1(s,t)\tilde{W}_p(\mu,\theta)^p + c_2(s,t)\tilde{W}_p(\theta,\nu)^p,\end{aligned}$$

where we used the definition of $\tilde{W}_p$. Using again the definition of $\tilde{W}_p$, we can take the supremum in the left-hand side over the functions ϕ and ψ satisfying $\phi(x)+\psi(y) \leqslant d(x,y)^p$. We get:

$$\tilde{W}_p(\mu,\nu)^p \leqslant c_1(s,t)\tilde{W}_p(\mu,\theta)^p + c_2(s,t)\tilde{W}_p(\theta,\nu)^p.$$

We can now take the infimum over s and t in the right-hand side and still get an upper bound for $\tilde{W}_p(\mu,\nu)^p$. But by (5.7), this infimum is equal to $[\tilde{W}_p(\mu,\theta) + \tilde{W}_p(\theta,\nu)]^p$ which proves the desired triangle inequality for $\tilde{W}_p$.

Second Step. We now prove that (5.6) holds when the space E is finite, say $E = \{e_1, \cdots, e_n\}$, in which case we use the notation $\mu(i) = \mu(\{e_i\})$ and $\nu(i) = \nu(\{e_i\})$ for $i = 1, \cdots, n$. By definition, we have:

$$W_p(\mu,\nu)^p = \inf_{\substack{\pi(i,j)\geqslant 0,\ \sum_{1\leqslant i\leqslant n}\pi(i,j)=\nu(j),\\ \sum_{1\leqslant j\leqslant n}\pi(i,j)=\mu(i)}} \sum_{1\leqslant i,j\leqslant d} d(e_i,e_j)^p\pi(i,j).$$

If we treat the $n \times n$ matrix $(d(e_i, e_j)^p)_{1 \leqslant i,j \leqslant n}$ given by the distance on the space E as an n^2 vector b, then the p-Wasserstein distance between μ and ν is given by the value of a plain linear program. This program is given by the infimum appearing in the right-hand side of the equality below for the $(2n) \times n^2$ matrix A derived by the equality constraints of the above definition of $W_p(\mu, \nu)^p$, and the $2n$ vector c comprising the values $\mu(1), \cdots, \mu(n), \nu(1), \cdots, \nu(n)$. To sum up, $b = (b_{(i,j)} = d(e_i, e_j)^p)_{1 \leqslant i,j \leqslant n}$, $A = (A_{\ell,(i,j)})_{1 \leqslant \ell \leqslant 2n, 1 \leqslant i,j \leqslant n}$ with $A_{\ell,(i,j)} = \mathbf{1}_{i=\ell}$ if $\ell \leqslant n$ and $A_{\ell,(i,j)} = \mathbf{1}_{j=\ell-n}$ if $\ell > n$, and $c = (c(\ell))_{1 \leqslant \ell \leqslant 2n}$ with $c(\ell) = \mu(\ell)$ if $\ell \leqslant n$ and $c(\ell) = \nu(\ell - n)$ if $\ell > n$. We may think of this problem as the *primal* problem:

$$\inf_{\pi(i,j) \geqslant 0,\, A\pi = c} b \cdot \pi,$$

and then write that its value is given by the value of the corresponding *dual* problem. We recall the classical duality theory for finite dimensional linear programming with obvious notation:

$$\sup_{A^\dagger x \leqslant b} c \cdot x = \inf_{y \geqslant 0,\, Ay = c} b \cdot y.$$

Specializing this duality result to the present situation, we find:

$$W_p(\mu, \nu)^p = \sup_{\phi(i) + \psi(j) \leqslant d(i,j)^p} \sum_{1 \leqslant i \leqslant n} \phi(i)\mu(i) + \sum_{1 \leqslant j \leqslant n} \psi(j)\nu(j),$$

if we denote by $\phi(1), \cdots, \phi(n), \psi(1), \cdots, \psi(n)$ the components of the vector x. This is exactly the Kantorovich's duality (5.6) in the case of the finite set E.

Third Step. Next, we prove, the inequality $W_p(\mu, \nu)^p \geqslant \tilde{W}_p(\mu, \nu)^p$ in full generality. This follows from the fact that if ϕ and ψ are real valued bounded continuous functions on E satisfying $\phi(x) + \psi(y) \leqslant d(x, y)^p$, then for any coupling $\pi \in \Pi(\mu, \nu)$, we have:

$$\begin{aligned}\int_E \phi(x)\mu(dx) + \int_E \psi(y)\nu(dy) &= \int_{E \times E} \phi(x)\pi(dx, dy) + \int_{E \times E} \psi(y)\pi(dx, dy) \\ &\leqslant \int_{E \times E} d(x, y)^p \pi(dx, dy).\end{aligned}$$

Since the left-hand side is independent of the coupling π, one can take the infimum of the right-hand side over all the couplings and get that $W_p(\mu, \nu)^p$ is still an upper bound for the left-hand side. But we can now take the supremum of the left-hand side over all the couples (ϕ, ψ) and obtain the desired inequality.

Fourth Step. Finally, we prove the remaining inequality by an approximation procedure. Let $x_0 \in E$ be fixed. Given $\epsilon > 0$, there exists a compact set $K_\epsilon \subset E$ such that:

$$\int_{K_\epsilon^c} d(x, x_0)^p [\mu(dx) + \nu(dx)] < \epsilon^p.$$

Next we construct a finite partition $(E_\epsilon^j)_{1 \leqslant j \leqslant n}$ of K_ϵ by Borel sets of diameter at most ϵ, and for each $j \in \{1, \cdots, n\}$ we pick an element $x_j \in E_\epsilon^j$. Finally, we define the map ψ from E into E by $\psi(x) = x_j$ whenever $x \in E_\epsilon^j$ and $\psi(x) = x_0$ if $x \in K_\epsilon^c$. Clearly, ψ is a coupling mapping between μ and $\tilde{\mu} = \mu \circ \psi^{-1}$ as well as between ν and $\tilde{\nu} = \nu \circ \psi^{-1}$. Using this coupling to get an upper bound for the distances $W_p(\mu, \tilde{\mu})$ and $W_p(\tilde{\nu}, \nu)$, we find that:

$$W_p(\mu, \tilde{\mu}) \leqslant 2^{1/p}\epsilon \qquad \text{and} \qquad W_p(\tilde{\nu}, \nu) \leqslant 2^{1/p}\epsilon,$$

and, using the triangle inequality for the distance W_p we get:

$$W_p(\mu, \nu) \leqslant W_p(\mu, \tilde{\mu}) + W_p(\tilde{\mu}, \tilde{\nu}) + W_p(\tilde{\nu}, \nu) \leqslant W_p(\tilde{\mu}, \tilde{\nu}) + 2^{1+1/p}\epsilon. \tag{5.11}$$

Using the result proven for probability measures on finite spaces in the *Second Step*, we see that $W_p(\tilde{\mu}, \tilde{\nu}) = \tilde{W}_p(\tilde{\mu}, \tilde{\nu})$ and using the triangle inequality proven in the *First Step*, we get:

$$\begin{aligned}
\tilde{W}_p(\tilde{\mu}, \tilde{\nu}) &\leqslant \tilde{W}_p(\tilde{\mu}, \mu) + \tilde{W}_p(\mu, \nu) + \tilde{W}_p(\nu, \tilde{\nu}) \\
&\leqslant W_p(\tilde{\mu}, \mu) + \tilde{W}_p(\mu, \nu) + W_p(\nu, \tilde{\nu}) \\
&\leqslant \tilde{W}_p(\mu, \nu) + 2^{1+1/p}\epsilon,
\end{aligned} \tag{5.12}$$

where we used once more the fact that $\tilde{W}_p$ is not greater than W_p as proven in *Third Step*. Putting together (5.11) and (5.12) we get:

$$W_p(\mu, \nu) \leqslant \tilde{W}_p(\mu, \nu) + 2^{2+1/p}\epsilon.$$

and letting $\epsilon \searrow 0$ concludes the proof. □

As a direct consequence of the duality identity proven in Proposition 5.3, we have the following characterization of the 1-Wasserstein distance W_1. This result was already mentioned several times, for example when we used the Kantorovich-Rubinstein distance d_{KR} in the proof of Remark 1.5 in Chapter 1, or when we used the Kantorovich-Rubinstein norm $\|\cdot\|_{\mathrm{KR}\star}$ in Subsections 4.3.2 and 4.3.5 in the proofs of Theorems 4.29 and 4.39, and again in the proof of Proposition 4.57.

Corollary 5.4 *If (E, d) is a complete separable metric space, and $\mu, \nu \in \mathcal{P}_1(E)$, then,*

$$W_1(\mu, \nu) = \sup_{\phi:\, |\phi(x)-\phi(y)| \leqslant d(x,y)} \int_E \phi(x)(\mu - \nu)(dx). \tag{5.13}$$

This proves that the Wasserstein distance W_1 coincides with the Kantorovich-Rubinstein distance d_{KR} introduced earlier.

Proof. In the case $p = 1$, the constraint in the supremum of the Kantorovich duality, namely the inequality $\phi(x) + \psi(y) \leqslant d(x, y)$ can be replaced by:

$$\phi(x) = \inf_{y \in E}[d(x, y) - \psi(y)], \qquad x \in E,$$

from which we immediately conclude that ϕ is 1-Lipschitz. So, limiting oneself to functions ϕ that are 1-Lipschitz, the inequality $\phi(x) + \psi(y) \leqslant d(x, y)$ can be replaced by:

$$\psi(y) = \inf_{x \in E}[d(x, y) - \phi(x)] = -\phi(y), \qquad y \in E,$$

so that, in the Kantorovich duality, it is enough to maximize over pairs of Lip-1 functions $(\phi, -\phi)$ which completes the proof. □

Notice that this representation of the 1-Wasserstein distance W_1 implies that $W_1(\mu, \nu)$ depends only upon the difference $\mu - \nu$!

Representation in Terms of Random Variables. Consider now an atomless probability space $(\Omega, \mathcal{F}, \mathbb{P})$. An E-valued random variable X is said to be of order p for a real $p > 0$, if $\mathbb{E}[d(x_0, X)^p] < \infty$ for one, and hence all, $x_0 \in E$. If $p \geqslant 1$ and X and Y are E-valued random variables of order p, then

$$W_p\big(\mathcal{L}(X), \mathcal{L}(Y)\big) \leqslant \mathbb{E}\big[d(X, Y)^p\big]^{1/p}. \tag{5.14}$$

We shall use this simple estimate quite often throughout the book. Moreover, when μ and ν are of order p,

$$W_p(\mu, \nu)^p = \inf \Big\{ \mathbb{E}\big[d(X, Y)^p\big];\ \mathcal{L}(X) = \mu,\ \mathcal{L}(Y) = \nu \Big\}.$$

Connection Between Wasserstein Distances and Weak Convergence

It is natural to wonder how the convergence of measures in $\mathcal{P}_p(E)$ relates to the weak convergence of measures. The answer is quite simple when E is compact. In this case, for any $p \geqslant 1$, if $(\mu_n)_{n\geqslant 0}$ and μ are probability measures on E, $(\mu_n)_{n\geqslant 0}$ converges toward μ for the weak convergence of probability measures if and only if $(W_p(\mu_n, \mu))_{n\geqslant 0}$ converges toward 0. The analog characterization is slightly more involved in the general case.

Theorem 5.5 *For any $p \geqslant 1$, if $(\mu_n)_{n\geqslant 1}$ and μ are in $\mathcal{P}_p(E)$, $\lim_{n\to\infty} W_p(\mu_n, \mu) = 0$ if and only if $(\mu_n)_{n\geqslant 1}$ converges toward μ for the weak convergence of probability measures and:*

$$\lim_{n\to\infty} \int_E d(x_0, x)^p d\mu_n(x) = \int_E d(x_0, x)^p d\mu(x), \tag{5.15}$$

for one (and hence for all) $x_0 \in E$. The latter is also equivalent to the fact that $(\mu_n)_{n\geqslant 1}$ converges toward μ for the weak convergence of probability measures and is p-uniformly integrable, namely

$$\lim_{r\to\infty} \sup_{n\geqslant 1} \int_E d(x_0, x)^p \mathbf{1}_{\{d(x_0,x)\geqslant r\}} d\mu_n(x) = 0. \tag{5.16}$$

A famous theorem of Skorohod states that the weak convergence of $(\mu_n)_{n\geqslant 1}$ toward μ is equivalent to the existence of random variables $(X_n)_{n\geqslant 1}$ and X defined on the same probability space, say $(\Omega, \mathcal{F}, \mathbb{P})$, such that $\mathcal{L}(X) = \mu$ and $\mathcal{L}(X_n) = \mu_n$ for each $n \geqslant 1$, and $\lim_{n\to\infty} X_n = X$ almost surely. We discuss this result later in the chapter and we even provide a proof of a somewhat stronger statement tailor-made to our needs in this book. See Lemma 5.29 and the ensuing discussion. In any case, Skorohod's theorem together with Fatou's lemma imply that:

$$\begin{aligned}\int_E d(x_0,x)^p d\mu(x) &= \mathbb{E}\big[d(x_0,X)^p\big] \\ &\leqslant \liminf_{n\to\infty} \mathbb{E}\big[d(x_0,X_n)^p\big] \\ &\leqslant \liminf_{n\to\infty} \int_E d(x_0,x)^p d\mu_n(x).\end{aligned} \tag{5.17}$$

Proof.

First Step. Let us first assume that $\lim_{n\to\infty} W_p(\mu_n,\mu) = 0$. We then observe that $(\mu_n)_{n\geqslant 1}$ converges in law toward μ.

To do so, we denote by π_n an element of $\Pi_p^{\text{opt}}(\mu_n,\mu)$, for any $n \geqslant 1$. Then, for any bounded and uniformly continuous function f from E to $\mathbb{R}$, we have

$$\int_E f(x)d\mu_n(x) - \int_E f(x)d\mu(x) = \int_{E\times E} \big(f(x) - f(y)\big)d\pi_n(x,y).$$

Splitting the integral in the right-hand side according to the partition of $E \times E$ into the sets $\{(x,y);\ d(x,y) > \epsilon\}$ and $\{(x,y);\ d(x,y) \leqslant \epsilon\}$, for a given $\epsilon > 0$, and using the boundedness and the uniform continuity of f, it is plain to deduce that $(\mu_n)_{n\geqslant 1}$ converges in law toward μ.

We now prove the convergence of the moments. Again, let us fix $\epsilon > 0$ momentarily. There exists a constant $c_\epsilon > 0$ such that for all $a, b \geqslant 0$ we have $(a+b)^p \leqslant (1+\epsilon)a^p + c_\epsilon b^p$. So, for $x, y \in E$, we have:

$$d(x_0,x)^p \leqslant (d(x_0,y) + d(y,x))^p \leqslant (1+\epsilon)d(x_0,y)^p + c_\epsilon d(y,x)^p,$$

and integrating both sides with respect to $\pi_n \in \Pi_p^{\text{opt}}(\mu_n,\mu)$ we get:

$$\int_E d(x_0,x)^p\, d\mu_n(x) \leqslant (1+\epsilon)\int_E d(x_0,y)^p\, d\mu(y) + c_\epsilon \int_{E\times E} d(y,x)^p\, d\pi_n(x,y).$$

By definition, the right most integral is equal to $W_p(\mu_n,\mu)^p$ which goes to 0 as $n \to \infty$. This implies:

$$\limsup_{n\to\infty} \int_E d(x_0,x)^p\, d\mu_n(x) \leqslant (1+\epsilon)\int_E d(x_0,y)^p\, d\mu(y),$$

in which we can take $\epsilon \searrow 0$. The resulting inequality together with (5.17) gives (5.15).

Second Step. Conversely, let us assume that $(\mu_n)_{n\geqslant 1}$ converges weakly toward μ and that (5.15) holds. Invoking Skorohod's representation theorem, we can find a sequence of $\mathbb{R}^d$-valued random variables $(X_n)_{n\geqslant 1}$, constructed on some probability space $(\Omega, \mathcal{F}, \mathbb{P})$ with $\mathcal{L}(X_n) = \mu_n$ for any $n \geqslant 1$ and converging almost surely to some random variable X with $\mathcal{L}(X) = \mu$. By (5.17), $X \in L^p(\Omega, \mathcal{A}, \mathbb{P}; \mathbb{R}^d)$. Of course, $W_p(\mu_n,\mu)^p \leqslant \mathbb{E}[d(X_n,X)^p]$, for any $n \geqslant 1$. Therefore, in order to complete the proof, it suffices to prove that the sequence $(X_n)_{n\geqslant 1}$ is p-uniformly integrable, see (5.16). For any $r > 0$, we have:

$$\mathbb{E}\big[d(x_0,X_n)^p \mathbf{1}_{\{d(x_0,X_n)\geq r\}}\big] = \mathbb{E}\big[d(x_0,X_n)^p\big] - \mathbb{E}\big[\big(d(x_0,X_n)\wedge r\big)^p\big] + r^p\mathbb{P}\big[d(x_0,X_n)\geq r\big].$$

By (5.15) and by the Portemanteau theorem, we have:

$$\begin{aligned}
&\limsup_{n\to\infty} \mathbb{E}\big[d(x_0,X_n)^p \mathbf{1}_{\{d(x_0,X_n)\geq r\}}\big] \\
&\leq \mathbb{E}\big[d(x_0,X)^p\big] - \mathbb{E}\big[\big(d(x_0,X)\wedge r\big)^p\big] + r^p\mathbb{P}\big[d(x_0,X)\geq r\big] \\
&= \mathbb{E}\big[d(x_0,X)^p \mathbf{1}_{\{d(x_0,X)\geq r\}}\big].
\end{aligned}$$

Obviously, the last term can be made as small as we want by taking the limit $r \to \infty$. Uniform integrability easily follows. The last claim in the statement is clear. □

Combining Theorem 5.5 with standard uniform integrability arguments we obtain the following important corollary.

Corollary 5.6 *For any $p \geq 1$, any subset $\mathcal{K} \subset \mathcal{P}_p(E)$, relatively compact for the topology of weak convergence of probability measures, any $x_0 \in E$, and any sequences $(a_n)_{n\geq 1}$ and $(b_n)_{n\geq 1}$ of positive real numbers tending to $+\infty$ with n, the set:*

$$\mathcal{K} \cap \left\{ \mu \in \mathcal{P}_p(E);\ \forall n \geq 1,\ \int_{\{d(x_0,x)\geq a_n\}} d(x_0,x)^p d\mu(x) < \frac{1}{b_n} \right\},$$

is relatively compact for the Wasserstein distance W_p.

Topological Properties of $\mathcal{P}_p(E)$

An important fact is that $\mathcal{P}_p(E)$, when equipped with the Wasserstein distance W_p, is a complete separable metric space whenever, as we have assumed, (E,d) is itself complete and separable. As explained earlier, we are indulging in a slight abuse of terminology, and accordingly, we shall say that $(\mathcal{P}_p(E), W_p)$ is a Polish space. This makes licit the application of weak convergence results for probability measures on $\mathcal{P}_p(E)$, among which Prokhorov and Skorohod theorems. We shall systematically equip $\mathcal{P}_p(E)$ with its Borel σ-field. The following proposition is a useful tool to characterize real valued Borel measurable functions on $\mathcal{P}_p(E)$.

Proposition 5.7 *The Borel σ-field $\mathcal{B}(\mathcal{P}_p(E))$ of $\mathcal{P}_p(E)$ is generated by the family of functions $(\mathcal{P}_p(E) \ni \mu \mapsto \mu(D))_{D\in\mathcal{B}(E)}$, where $\mathcal{B}(E)$ is the Borel σ-field of E. More generally, if $\mathcal{E}$ is a collection of subsets of E which generates $\mathcal{B}(E)$ and is closed under finite intersections, then $\mathcal{B}(\mathcal{P}_p(E))$ is generated by the family of functions $(\mathcal{P}_p(E) \ni \mu \mapsto \mu(D))_{D\in\mathcal{E}}$. In particular, for any Borel measurable function $\psi : E \to \mathbb{R}$ which satisfies $|\psi(x)| \leq C(1 + d(x_0,x)^p)$ for some $C \geq 0$, $x_0 \in E$, and all $x \in E$, the function $\mathcal{P}_p(E) \ni \mu \mapsto \int_E \psi d\mu$ is Borel measurable on $\mathcal{P}_p(E)$.*

Proposition 5.7 remains true with $\mathcal{P}(E)$ instead of $\mathcal{P}_p(E)$, $\mathcal{P}(E)$ being equipped with the Lévy-Prokhorov metric. Indeed, the Borel σ-field $\mathcal{B}(\mathcal{P}(E))$ on $\mathcal{P}(E)$ is generated by the family of mappings $(\mathcal{P}(E) \ni \mu \mapsto \mu(D))_{D\in\mathcal{B}(E)}$. The fact that $\mathcal{B}(\mathcal{P}_p(E)) = \{D\cap\mathcal{P}_p(E);\ D \in \mathcal{B}(\mathcal{P}(E))\}$, which we shall also denote by $\mathcal{B}(\mathcal{P}(E))\cap \mathcal{P}_p(E)$, can be checked directly by inspection. Indeed, for any $\mu_0 \in \mathcal{P}_p(E)$, the mapping $\mathcal{P}_p(E) \ni \mu \mapsto W_p(\mu, \mu_0)$ is lower semicontinuous for the Lévy-Prokhorov distance, which proves that, for any $\varepsilon > 0$, the set $\{\mu \in \mathcal{P}_p(E) : \ W_p(\mu, \mu_0) < \varepsilon\} \in \mathcal{B}(\mathcal{P}(E))$. Therefore, $\mathcal{B}(\mathcal{P}_p(E)) \subset \mathcal{B}(\mathcal{P}(E)) \cap \mathcal{P}_p(E)$. Conversely, for any closed subset $D \subset \mathcal{P}(E)$ for the Lévy-Prokhorov distance, the set $D \cap \mathcal{P}_p(E)$ is a closed subset of $\mathcal{P}_p(E)$ equipped with W_p. We get that $D \cap \mathcal{P}_p(E) \in \mathcal{B}(\mathcal{P}_p(E))$. Since the σ-algebra generated by sets of the form $D \cap \mathcal{P}_p(E)$, with D as above, is $\mathcal{B}(\mathcal{P}(E)) \cap \mathcal{P}_p(E)$, we get the required equality.

As an application of Proposition 5.7, observe that if $[0, T] \ni t \mapsto \mu_t \in \mathcal{P}_p(E)$ is measurable and $\varrho : [0, T] \times E \ni (t, x) \mapsto \varrho(t, x) \in \mathbb{R}$ is jointly measurable and satisfies $|\varrho(t, x)| \leqslant C(1 + d_E(x_0, x)^p)$ for all $(t, x) \in [0, T] \times E$ and for some $C > 0$ and $x_0 \in E$, then the mapping $[0, T] \ni t \mapsto \int_E \psi(t, x)d\mu_t(x)$ is measurable. The proof is a consequence of the monotone class theorem. Moreover, if $\varrho : \mathcal{P}_p(E)\times E \ni (\mu, x) \mapsto \varrho(\mu, x) \in \mathbb{R}$ is jointly measurable and satisfies $\int_E |\varrho(\mu, x)|d\mu(x) < \infty$ for all $\mu \in \mathcal{P}_p(E)$, then the mapping $\mathcal{P}_p(E) \ni \mu \mapsto \int_E \varrho(\mu, x)d\mu(x)$ is measurable.

5.1.2 Glivenko-Cantelli Convergence in the Wasserstein Distance

In this subsection, we analyze the rate of convergence, for the Wasserstein distance, in the Glivenko-Cantelli theorem.

We start with a basic reminder. If $(X_n)_{n\geqslant 1}$ is a sequence of independent identically distributed (i.i.d. for short) random variables in $\mathbb{R}^d$ with common distribution $\mu \in \mathcal{P}(\mathbb{R}^d)$ and if, for each $N \geqslant 1$, we denote by $\bar{\mu}^N$ the empirical measure:

$$\bar{\mu}^N = \frac{1}{N}\sum_{i=1}^{N} \delta_{X_i},$$

the classical Glivenko-Cantelli theorem implies the weak convergence of $\bar{\mu}^N$ toward μ, almost surely and (thus) in probability. Whenever the random variables $(X_n)_{n\geqslant 1}$ are square-integrable, convergence also holds almost surely for the 2-Wasserstein distance:

$$\mathbb{P}\Big[\lim_{N\to+\infty} W_2(\mu, \bar{\mu}^N) = 0\Big] = 1.$$

This follows from Theorem 5.5 and the law of large numbers, which asserts that:

$$\mathbb{P}\Big[\lim_{N\to+\infty} \int_{\mathbb{R}^d} |x|^2 d\bar{\mu}^N(x) = \int_{\mathbb{R}^d} |x|^2 d\mu(x)\Big] = 1. \tag{5.18}$$

Actually, the sequence $(W_2(\mu, \bar{\mu}^N))_{N \geqslant 1}$ is also uniformly square-integrable, since $\mathbb{P}$ almost-surely:

$$W_2\big(\delta_0, \bar{\mu}^N\big)^2 = \frac{1}{N}\sum_{i=1}^{N} |X_i|^2,$$

the above right-hand side converging in L^1. Therefore, provided that X_1 is square-integrable, it holds that:

$$\lim_{N \to +\infty} \mathbb{E}\big[W_2(\mu, \bar{\mu}^N)^2\big] = 0. \tag{5.19}$$

The purpose of this section is to provide a *non-asymptotic* a priori estimate which quantifies the rate of convergence in the above limit when $\mu \in \mathcal{P}_q(\mathbb{R}^d)$ for some $q > 4$. We shall use this estimate repeatedly throughout the book.

Theorem 5.8 *If $\mu \in \mathcal{P}_q(\mathbb{R}^d)$ for some $q > 4$ (that is if $M_q(\mu)^q = \int_{\mathbb{R}^d} |x|^q d\mu(x) < \infty$), then, for each dimension $d \geqslant 1$, there exists a constant $C = C(d, q, M_q(\mu))$ such that, for all $N \geqslant 2$:*

$$\mathbb{E}\big[W_2(\bar{\mu}^N, \mu)^2\big] \leqslant C \begin{cases} N^{-1/2}, & \text{if } d < 4, \\ N^{-1/2}\log N, & \text{if } d = 4, \\ N^{-2/d}, & \text{if } d > 4. \end{cases} \tag{5.20}$$

As far as we know, the above a priori estimate does not exist in book form. It will be used repeatedly throughout the book so we give a detailed proof. It relies on several technical results which we present, for the sake of completeness, in the form of three lemmas. The reader only interested in the applications of the estimate (5.20) *may want to skip these lemmas which will only be needed in the proof of Theorem 5.8.*

Remark 5.9 *Observe that, for an L-Lipschitz-continuous function $\phi : \mathbb{R}^d \to \mathbb{R}^d$ and for two probability distributions μ and ν in $\mathcal{P}_2(\mathbb{R}^d)$, $W_2(\mu \circ \phi^{-1}, \nu \circ \phi^{-1}) \leqslant LW_2(\mu, \nu)$. When $\mu \in \mathcal{P}_q(\mathbb{R}^d)$, with $M_q(\mu) \neq 0$, choose $\phi : \mathbb{R}^d \ni x \mapsto M_q(\mu)x$ and let $\psi : \mathbb{R}^d \ni x \mapsto x/M_q(\mu)$. We deduce that $W_2(\bar{\mu}^N, \mu) \leqslant M_q(\mu)W_2(\bar{\mu}^N \circ \psi^{-1}, \mu \circ \psi^{-1})$. Since*

$$\bar{\mu}^N \circ \psi^{-1} = \frac{1}{N}\sum_{i=1}^{N} \delta_{\psi(X_i)},$$

and $M_q(\mu \circ \psi^{-1}) = 1$, *we obtain:*

$$\begin{aligned}\mathbb{E}\big[W_2(\bar{\mu}^N, \mu)^2\big] &\leqslant M_q(\mu)^2 \mathbb{E}\big[W_2\big(\bar{\mu}^N \circ \psi^{-1}, \mu \circ \psi^{-1}\big)^2\big] \\ &\leqslant C(d, q, 1) M_q(\mu)^2 \begin{cases} N^{-1/2}, & \text{if } d < 4, \\ N^{-1/2} \log N, & \text{if } d = 4, \\ N^{-2/d}, & \text{if } d > 4, \end{cases}\end{aligned}$$

which shows how the constant $C(d, q, M_q(\mu))$ *depends on* $M_q(\mu)$.

First Lemma in the Proof of Theorem 5.8

For each integer $\ell \geqslant 0$, we denote by $\mathcal{P}_\ell$ the partition of the hypercube $(-1, 1]^d$ into $2^{d\ell}$ translations of the hypercube $(-2^{-\ell}, 2^{-\ell}]^d$.

Lemma 5.10 *There exists a universal constant* $c > 0$ *such that, for any pair* (μ, ν) *of probability measures on* $(-1, 1]^d$, *it holds that:*

$$W_2(\mu, \nu)^2 \leqslant c \sum_{\ell \geqslant 0} 2^{-2\ell} \sum_{B \in \mathcal{P}_\ell} \mu(B) \sum_{C \in \mathcal{P}_{\ell+1},\, C \subset B} \left| \frac{\mu(C)}{\mu(B)} - \frac{\nu(C)}{\nu(B)} \right|, \tag{5.21}$$

where $[\mu(C)/\mu(B)]$ *is set to* $1/2^d$ *whenever* $\mu(B) = 0$ *(and similarly for* ν *in lieu of* μ*).*

Proof. We first isolate an argument which will be used repeatedly in the proof.

Preliminary Step. If $(A_k)_{k \geqslant 0}$ is a measurable partition of a Polish space E on which θ and η are probability measures, we define a new probability $\tilde{\theta}$ by its restrictions to each of the $(A_k)_{k \geqslant 0}$ given by $\tilde{\theta}(\cdot \cap A_k) = \eta(A_k)\theta(\cdot | A_k)$ for any $k \geqslant 0$. Of course, we implicitly assume that θ charges all the $(A_k)_{k \geqslant 0}$. Below we use the notation $\tilde{\theta}_{|A_k}(\cdot)$ for $\tilde{\theta}(\cdot \cap A_k)$. We then have:

$$(\theta \wedge \tilde{\theta})_{|A_k}(\cdot) = \big(\theta(A_k) \wedge \eta(A_k)\big)\theta(\cdot | A_k), \quad k \geqslant 0,$$

so that, if we set:

$$\delta = \frac{1}{2} \sum_{k \geqslant 0} |\theta(A_k) - \eta(A_k)| = 1 - \sum_{k \geqslant 0} \big(\theta(A_k) \wedge \eta(A_k)\big),$$

we have:

$$\begin{aligned}(\theta - \tilde{\theta})_+ &= \sum_{k \geqslant 0} \big(\theta(A_k) - \eta(A_k)\big)_+ \theta(\cdot | A_k) \\ &= \sum_{k \geqslant 0} \big(\theta(A_k) - (\theta \wedge \eta)(A_k)\big)\theta(\cdot | A_k) = \theta - \theta \wedge \tilde{\theta}, \\ (\tilde{\theta} - \theta)_+ &= \sum_{k \geqslant 0} \big(\eta(A_k) - \theta(A_k)\big)_+ \theta(\cdot | A_k) \\ &= \sum_{k \geqslant 0} \big(\eta(A_k) - (\theta \wedge \eta)(A_k)\big)_+ \theta(\cdot | A_k) = \tilde{\theta} - \theta \wedge \tilde{\theta},\end{aligned}$$

and thus:

$$(\theta - \tilde{\theta})_+(E) = (\tilde{\theta} - \theta)_+(E) = \delta.$$

In particular, if we use temporarily the notation ψ for the function $E \ni x \mapsto (x, x) \in E \times E$, then the probability measure ξ on $E \times E$ defined by:

$$\begin{aligned}
\xi(B \times C) &= (\theta \wedge \tilde{\theta})(B \cap C) + \delta^{-1}(\theta - \tilde{\theta})_+(B)(\tilde{\theta} - \theta)_+(C) \\
&= \int_{B \times C} \mathbf{1}_{\{x=y\}}\big[(\theta \wedge \tilde{\theta}) \circ \psi^{-1}\big](dx, dy) \\
&\quad + \delta^{-1} \int_{B \times C} \mathbf{1}_{\{x \neq y\}}(\theta - \tilde{\theta})_+(dx)(\tilde{\theta} - \theta)_+(dy),
\end{aligned}$$

for $B, C \in \mathcal{B}(E)$, or equivalently:

$$\begin{aligned}
&\xi(B) \\
&= \int_B \mathbf{1}_{\{x=y\}}\big[(\theta \wedge \tilde{\theta}) \circ \psi^{-1}\big](dx, dy) + \delta^{-1} \int_B \mathbf{1}_{\{x \neq y\}}(\theta - \tilde{\theta})_+(dx)(\tilde{\theta} - \theta)_+(dy) \\
&= \int_B \mathbf{1}_{\{x=y\}}\big[(\theta \wedge \tilde{\theta}) \circ \psi^{-1}\big](dx, dy) + \delta^{-1} \int_B (\theta - \tilde{\theta})_+(dx)(\tilde{\theta} - \theta)_+(dy),
\end{aligned}$$

for $B \in \mathcal{B}(E \times E)$, is a coupling between θ and $\tilde{\theta}$ satisfying $\xi(\{(x, y) \in E^2 : x \neq y\}) = \delta$. Accordingly, if (X, Y) is a couple of random variables with joint distribution ξ, then $\mathcal{L}(X) = \theta$, $\mathcal{L}(Y) = \tilde{\theta}$, and for each $k \geqslant 0$, $\mathbb{P}[X = Y, X \in A_k] = \theta(A_k) \wedge \tilde{\theta}(A_k)$, and $\mathbb{P}[X \neq Y] = \delta$.

Note also that if θ and $\tilde{\theta}$ are probability measures such that $\theta(A_k) = \tilde{\theta}(A_k)$ and ξ_k is a coupling between $\theta(\cdot|A_k)$ and $\tilde{\theta}(\cdot|A_k)$ for each k, then $\xi = \sum_{k \geqslant 0} \theta(A_k)\xi_k$ is a coupling between θ and $\tilde{\theta}$.

Second Step. We now return to the setting of the statement of the lemma, and we let μ and ν be two probability measures on $(-1, 1]^d$. We assume for a while that, for any Borel subset $B \subset (-1, 1]^d$ with positive Lebesgue measure, $\mu(B)$ and $\nu(B)$ are also positive (in other words, the Lebesgue measure on $(-1, 1]^d$ is absolutely continuous with respect to μ and ν). We then construct a sequence $(\mu_n)_{n \geqslant 0}$, with $\mu_0 = \mu$, that converges to ν in the weak sense, and hence in the Wasserstein sense since the supports are bounded.

For each integer $\ell \geqslant 1$, we define the probability measure μ_ℓ by:

$$\mu_\ell(\cdot) = \sum_{B \in \mathcal{P}_\ell} \nu(B)\mu(\cdot|B).$$

For each continuous function $f : \mathbb{R}^d \to \mathbb{R}$, if we denote by w its modulus of continuity on $[-1, 1]^d$ with respect to the Euclidean norm, we have:

$$\begin{aligned}
\int_{\mathbb{R}^d} f(x) d\mu_\ell(x) &= \sum_{B \in \mathcal{P}_\ell} \int_B f(x) d\mu_\ell(x) \\
&= \sum_{B \in \mathcal{P}_\ell} \left(2^{\ell d} \int_B f(x) dx\right) \mu_\ell(B) + O\big(w(\sqrt{d}2^{-\ell})\big)
\end{aligned}$$

$$= \sum_{B \in \mathcal{P}_\ell} \left(2^{\ell d} \int_B f(x) dx \right) \nu(B) + O\big(w(\sqrt{d} 2^{-\ell})\big)$$

$$= \int_{\mathbb{R}^d} f(x) d\nu(x) + O\big(w(\sqrt{d} 2^{-\ell})\big),$$

where we used the Landau notation $O(\cdot)$ for a function $O(\cdot)$ of the form $\mathbb{R} \ni x \mapsto \delta(x) \in \mathbb{R}$ such that $|\delta(x)| \leqslant C|x|$ for some constant $C \geqslant 0$. We deduce that $(\mu_\ell)_{\ell \in \mathbb{N}}$ converges weakly toward ν. By Theorem 5.5, the convergence also holds in the sense of the 2-Wasserstein distance, which implies that:

$$W_2(\mu, \nu) \leqslant \sup_{\ell \geqslant 0} W_2(\mu, \mu_\ell),$$

showing that in order to prove (5.21), it suffices to prove that its right-hand side is an upper bound for $W_2(\mu, \mu_\ell)^2$ for ℓ fixed.

Third Step. Now, for each $\ell \geqslant 0$, we construct a coupling $\xi_\ell \in \Pi(\mu_\ell, \mu_{\ell+1})$. The strategy is to apply the *Preliminary Step* with $\theta = \mu_\ell$, $\eta = \nu$ and then $\tilde{\theta} = \mu_{\ell+1}$. To do so, observe first that, for any $B \in \mathcal{P}_\ell$,

$$\mu_{\ell+1}(B) = \sum_{i=1}^{2^d} \nu(B_i) \mu(B|B_i) = \sum_{i=1}^{2^d} \nu(B_i) = \nu(B),$$

where $B_1, \cdots, B_{2^d}$ form the partition of B into 2^d hypercubes in $\mathcal{P}_{\ell+1}$. Moreover, for each $C \in \mathcal{P}_{\ell+1}$ contained in B, we have:

$$\mu_\ell(C) = \nu(B)\mu(C|B) = \frac{\nu(B)\mu(C)}{\mu(B)},$$

$$\mu_\ell(\cdot|C) = \frac{\nu(B)}{\mu(B)\mu_\ell(C)} \mu_{|C}(\cdot) = \mu(\cdot|C),$$

from which we get:

$$\mu_{\ell+1|C} = \nu(C)\mu(\cdot|C) = \nu(C)\mu_\ell(\cdot|C) = \nu(C)\big(\mu_\ell(\cdot|B)\big)(\cdot|C). \tag{5.22}$$

Dividing both sides by $\nu(B)$, we can reinterpret this equality in the framework of the *Preliminary Step*: It says that, if we start with $\theta = \mu_\ell(\cdot|B)$ and $\eta = \nu(\cdot|B)$, the probability $\mu_{\ell+1}(\cdot|B)$ is nothing but the probability $\tilde{\theta}$ obtained as in the *Preliminary Step* from the partition $\{C \in \mathcal{P}_{\ell+1} : C \subset B\}$ of B. We then denote by $\xi_{\ell,B}$ the resulting coupling, and let:

$$\xi_\ell(\cdot) = \sum_{B \in \mathcal{P}_\ell} \mu_\ell(B) \xi_{\ell,B}(\cdot).$$

Observe that $\mu_\ell(B) = \nu(B)$, so that $\mu_\ell(B) = \mu_{\ell+1}(B)$. Therefore, $\xi_\ell \in \Pi(\mu_\ell, \mu_{\ell+1})$; ξ_ℓ is a coupling constructed from the aggregation prescription. Notice, again from the *Preliminary Step*, that:

$$
\begin{aligned}
\xi_\ell(\{(x,y) : x \neq y\}) &= \sum_{B \in \mathcal{P}_\ell} \mu_\ell(B)\xi_{\ell,B}\big(\{(x,y) : x \neq y\}\big) \\
&= \sum_{B \in \mathcal{P}_\ell} \left[\mu_\ell(B)\left(\frac{1}{2} \sum_{C \in \mathcal{P}_{\ell+1}, C \subset B} \big|\mu_\ell(C|B) - \nu(C|B)\big|\right)\right] \\
&= \frac{1}{2} \sum_{B \in \mathcal{P}_\ell} \sum_{C \in \mathcal{P}_{\ell+1}, C \subset B} \left|\nu(C) - \nu(B)\frac{\mu(C)}{\mu(B)}\right|.
\end{aligned}
$$

Notice also that, for any $B \in \mathcal{P}_\ell$, $\xi_\ell(B \times ((-1,1]^d \setminus B)) = 0$.

Fourth Step. For $\ell \geqslant 0$, we call $K_{\ell+1}$ the conditional law of ξ_ℓ given the first marginal, namely $\xi_\ell(dx_\ell, dx_{\ell+1}) = \mu_\ell(dx_\ell)K_{\ell+1}(x_\ell, dx_{\ell+1})$. In particular,

$$
\mu_{\ell+1}(dx_{\ell+1}) = \int_{(-1,1]^d} \mu_\ell(dx_\ell)K_{\ell+1}(x_\ell, dx_{\ell+1}).
$$

Then, Ionescu-Tulcea's theorem guarantees the existence of a sequence $(Z_\ell)_{\ell \geqslant 1}$ of $(-1,1]^d$-valued random variables constructed on some probability space $(\Omega, \mathcal{F}, \mathbb{P})$ such that:

$$
\begin{aligned}
&\mathbb{P}\big[Z_0 \in A_0, Z_1 \in A_1, \cdots, Z_\ell \in A_\ell\big] \\
&= \int_{A_0}\int_{A_1} \cdots \int_{A_\ell} \mu_0(dx_0)K_1(x_0, dx_1)\cdots K_\ell(x_{\ell-1}, dx_\ell),
\end{aligned}
$$

for every $\ell \geqslant 1$, and any Borel sets $A_0, A_1, \cdots, A_\ell$ in $(-1,1]^d$. For each $\ell \geqslant 1$, the joint distribution of (Z_0, Z_ℓ) is a coupling between $\mu = \mu_0$ and μ_ℓ. Now, if the random variable L is defined by $L = \inf\{\ell \geqslant 0 : Z_\ell \neq Z_{\ell+1}\}$:

$$
\begin{aligned}
W_2(\mu,\nu)^2 \leqslant \sup_{\ell \geqslant 1} W_2(\mu,\mu_\ell)^2 &\leqslant \sup_{\ell \geqslant 1} \mathbb{E}\big[|Z_0 - Z_\ell|^2\big] \\
&\leqslant 2\sup_{\ell \geqslant 1} \mathbb{E}\big[\big(|Z_0 - Z_L|^2 + |Z_L - Z_\ell|^2\big)\mathbf{1}_{\{L \leqslant \ell-1\}}\big] \\
&= 2\sup_{\ell \geqslant 1} \mathbb{E}\big[\big(|Z_{L-1} - Z_L|^2 + |Z_L - Z_\ell|^2\big)\mathbf{1}_{\{L \leqslant \ell-1\}}\big] \\
&\leqslant c\sup_{\ell \geqslant 1} \mathbb{E}\big[2^{-2L}\mathbf{1}_{\{L \leqslant \ell-1\}}\big],
\end{aligned}
$$

for a universal constant $c > 0$. The constant c being allowed to increase from line to line, we get:

$$
\begin{aligned}
W_2(\mu,\nu)^2 &\leqslant c\sum_{\ell \geqslant 0} 2^{-2\ell}\mathbb{P}\big[Z_\ell \neq Z_{\ell+1}\big] \\
&= c\sum_{\ell \geqslant 0} 2^{-2\ell}\xi_{\ell+1}\big(\{(x,y);\, x \neq y\}\big) \\
&= \frac{c}{2}\sum_{\ell \geqslant 0} 2^{-2\ell} \sum_{B \in \mathcal{P}_\ell} \sum_{C \in \mathcal{P}_{\ell+1}, C \subset B} \left|\nu(C) - \nu(B)\frac{\mu(C)}{\mu(B)}\right|,
\end{aligned}
\tag{5.23}
$$

which completes the proof when the Lebesgue measure on $(-1,1]^d$ is absolutely continuous with respect to μ and ν. Observe that we have exchanged the role of μ and ν in (5.21).

Conclusion. In order to complete the proof, it remains to discuss the general case when the Lebesgue measure is no longer absolutely continuous with respect to μ and ν. We then approximate μ and ν by:

$$\mu^\epsilon = (1-\epsilon)\mu + \frac{\epsilon}{2^d}\mathrm{Leb}_{d|(-1,1]^d} \; ; \quad \nu^\epsilon = (1-\epsilon)\nu + \frac{\epsilon}{2^d}\mathrm{Leb}_{d|(-1,1]^d},$$

where Leb_d denotes the d-dimensional Lebesgue measure and $\mathrm{Leb}_{d|(-1,1]^d}(\cdot) = \mathrm{Leb}_d(\cdot \cap (-1,1]^d))$. Then, $(\mu^\epsilon)_{\varepsilon>0}$ and $(\nu^\epsilon)_{\epsilon>0}$ converge in total variation to μ and ν as ϵ tends to 0. Obviously, we can apply the conclusion of the fourth step to each pair $(\mu^\epsilon, \nu^\epsilon)$, $\epsilon > 0$. Letting ϵ tend to 0 in (5.23) completes the proof. □

Second Lemma in the Proof of Theorem 5.8

In order to control $W_2(\mu,\nu)$, we shall introduce two auxiliary quantities based on the right-hand side of (5.21). First, for any $\mu,\nu \in \mathcal{P}((-1,1]^d)$, we let:

$$\delta_2(\mu,\nu) = \sum_{\ell \geqslant 1} 2^{-2\ell} \sum_{B \in \mathcal{P}_\ell} |\mu(B) - \nu(B)|.$$

Next, we introduce a partition $(B_n)_{n\geqslant 0}$ of $\mathbb{R}^d$ with $B_0 = (-1,1]^d$ and

$$B_n = (-2^n, 2^n]^d \backslash (-2^{n-1}, 2^{n-1}]^d, \qquad n \geqslant 1.$$

For $\mu \in \mathcal{P}(\mathbb{R}^d)$ and $n \geqslant 0$ such that $\mu(B_n) > 0$, we define the probability measure μ^{B_n} on $(-1,1]^d$ by:

$$\mu^{B_n}(A) = \mu\big(2^n A | B_n\big),$$

for any Borel set $A \subset (-1,1]^d$. In other words, μ^{B_n} is the push-forward by the map $\mathbb{R}^d \ni x \mapsto x/2^n$ of the probability μ conditioned to be in B_n, namely the probability:

$$B \mapsto \mu(B|B_n) = \frac{\mu(B \cap B_n)}{\mu(B_n)}.$$

Whenever $\mu(B_n) = 0$, we define μ^{B_n} as any arbitrary measure in $\mathcal{P}((-1,1]^d)$. We now let:

$$D_2(\mu,\nu) = \sum_{n \geqslant 0} 2^{2n}\Big(|\mu(B_n) - \nu(B_n)| + (\mu(B_n) \wedge \nu(B_n))\, \delta_2(\mu^{B_n}, \nu^{B_n}) \Big), \tag{5.24}$$

for $\mu,\nu \in \mathcal{P}_2(\mathbb{R}^d)$. Importantly, $D_2(\mu,\nu)$ does not depend upon the (arbitrary) choice of μ^{B_n} or ν^{B_n} when $\mu(B_n) = 0$ or $\nu(B_n) = 0$. The relevance of D_2 to our analysis of convergence in the sense of the Wasserstein distance W_2 is provided by the following estimate.

Lemma 5.11 *There exists a universal constant κ such that, for all pairs of probability measures μ and ν in $\mathcal{P}_2(\mathbb{R}^d)$, we have:*

$$W_2(\mu,\nu)^2 \leqslant \kappa D_2(\mu,\nu). \tag{5.25}$$

Proof. If μ and ν are supported in $(-1,1]^d$, using estimate (5.21) from Lemma 5.10, we have:

$$\begin{aligned}
W_2(\mu,\nu)^2 &\leqslant c\sum_{\ell\geqslant 0} 2^{-2\ell}\sum_{B\in\mathcal{P}_\ell}\mu(B)\sum_{C\in\mathcal{P}_{\ell+1},\,C\subset B}\left|\frac{\mu(C)}{\mu(B)}-\frac{\nu(C)}{\nu(B)}\right| \\
&\leqslant c\sum_{\ell\geqslant 0} 2^{-2\ell}\sum_{B\in\mathcal{P}_\ell}\sum_{C\in\mathcal{P}_{\ell+1},\,C\subset B}\left(\frac{\nu(C)}{\nu(B)}|\mu(B)-\nu(B)|+|\mu(C)-\nu(C)|\right) \\
&\leqslant c\sum_{\ell\geqslant 0} 2^{-2\ell}\left(\sum_{B\in\mathcal{P}_\ell}|\mu(B)-\nu(B)|+\sum_{C\in\mathcal{P}_{\ell+1}}|\mu(C)-\nu(C)|\right) \\
&\leqslant c(1+2^2)\sum_{\ell\geqslant 1} 2^{-2\ell}\sum_{B\in\mathcal{P}_\ell}|\mu(B)-\nu(B)|,
\end{aligned}$$

where we used the fact that $\sum_{B\in\mathcal{P}_0}|\mu(B)-\nu(B)|=0$. The above is nothing but the desired right-hand side of (5.25) which completes the proof when μ and ν are supported in $(-1,1]^d$.

In the general case, for each $n\geqslant 1$, we denote by π_n the optimal coupling of μ^{B_n} and ν^{B_n}, and by ξ_n (that we shall also denote by $\xi_n(dx,dy)$ for pedagogical reasons) the push-forward of π_n by scaling by 2^n, namely by the mapping $(x,y)\mapsto(2^n x, 2^n y)$. Obviously,

$$2^{2n}W_2(\mu^{B_n},\nu^{B_n})^2 = 2^{2n}\int_{\mathbb{R}^d}\int_{\mathbb{R}^d}|x-y|^2 d\pi_n(x,y) = \int_{\mathbb{R}^d}\int_{\mathbb{R}^d}|x-y|^2 d\xi_n(x,y).$$

Next we define the measure ξ on $\mathbb{R}^d\times\mathbb{R}^d$ by:

$$\xi(dx,dy) = \sum_{n\geqslant 0}(\mu(B_n)\wedge\nu(B_n))\xi_n(dx,dy) + a^{-1}\alpha(dx)\beta(dy),$$

where $a=(1/2)\sum_{n\geqslant 0}|\mu(B_n)-\nu(B_n)|$,

$$\alpha(dx) = \sum_{n\geqslant 0}(\mu(B_n)-\nu(B_n))_+\mu(dx|B_n),$$

$$\beta(dy) = \sum_{n\geqslant 0}(\nu(B_n)-\mu(B_n))_+\nu(dy|B_n).$$

Following the proof of the preliminary step in Lemma 5.10, we notice that:

$$a = \sum_{n\geqslant 0}(\mu(B_n)-\nu(B_n))_+ = \sum_{n\geqslant 0}(\nu(B_n)-\mu(B_n))_+ = 1-\sum_{n\geqslant 0}\mu(B_n)\wedge\nu(B_n).$$

We also note that by construction, the marginals of ξ are μ and ν respectively. Indeed, if $A\in\mathcal{B}(\mathbb{R}^d)$ we have:

$$
\begin{aligned}
\xi(A \times \mathbb{R}^d) &= \sum_{n \geqslant 0} (\mu(B_n) \wedge \nu(B_n)) \xi_n(A \times \mathbb{R}^d) + a^{-1}\alpha(A)\beta(\mathbb{R}^d) \\
&= \sum_{n \geqslant 0} (\mu(B_n) \wedge \nu(B_n)) \pi_n(2^{-n}A \times \mathbb{R}^d) + \alpha(A) \\
&= \sum_{n \geqslant 0} (\mu(B_n) \wedge \nu(B_n)) \mu^{B_n}(2^{-n}A) + \sum_{n \geqslant 0} (\mu(B_n) - \nu(B_n))^+ \mu(A|B_n) \\
&= \sum_{n \geqslant 0} \mu(B_n)\mu(A|B_n) = \mu(A),
\end{aligned}
$$

where we used the fact that $a = \beta(\mathbb{R}^d)$. Notice that the proof works correctly even if $\mu(B_n) = 0$ for some $n \geqslant 0$. We argue similarly for the second marginal. Moreover,

$$
\begin{aligned}
\int_{\mathbb{R}^d}\int_{\mathbb{R}^d} |x-y|^2 a^{-1}\alpha(dx)\beta(dy) &\leqslant \frac{2}{a}\int_{\mathbb{R}^d}\int_{\mathbb{R}^d} \Big(|x|^2 + |y|^2\Big)\alpha(dx)\beta(dy) \\
&\leqslant 2\int_{\mathbb{R}^d} |x|^2\alpha(dx) + 2\int_{\mathbb{R}^d} |y|^2\beta(dy) \\
&\leqslant 2\sum_{n \geqslant 0} 2^{2n}\Big((\mu(B_n) - \nu(B_n))^+ + (\nu(B_n) - \mu(B_n))^+\Big) \\
&\leqslant 2\sum_{n \geqslant 0} 2^{2n} |\mu(B_n) - \nu(B_n)|,
\end{aligned}
$$

where we used once more the fact that $a = \alpha(\mathbb{R}^d) = \beta(\mathbb{R}^d)$. Now, using the fact that the marginals of ξ are μ and ν, we have:

$$
\begin{aligned}
W_2(\mu,\nu)^2 &\leqslant \int_{\mathbb{R}^d}\int_{\mathbb{R}^d} |x-y|^2 \xi(dx,dy) \\
&\leqslant \sum_{n \geqslant 0} 2^{2n}\Big(2|\mu(B_n) - \nu(B_n)| + \big(\mu(B_n) \wedge \nu(B_n)\big)\big(W_2(\mu^{B_n}, \nu^{B_n})\big)^2\Big),
\end{aligned}
$$

where we used the fact that $\int_{\mathbb{R}^d}\int_{\mathbb{R}^d} |x-y|^2\xi_n(dx,dy) = 2^{2n}(W_2(\mu^{B_n}, \nu^{B_n}))^2$. Since μ^{B_n} and ν^{B_n} are probability measures on $(-1,1]^d$, we can use the first part of the proof to bound $W_2(\mu^{B_n}, \nu^{B_n})^2$ by $5c(\delta_2(\mu^{B_n}, \nu^{B_n}))^2$, completing the proof in the general case. □

Third Lemma in the Proof of Theorem 5.8

A crucial ingredient in the proof of Theorem 5.8 is the following technical estimate.

Lemma 5.12 *There exists a universal constant C such that, for all probability measures μ and ν in $\mathcal{P}_2(\mathbb{R}^d)$, we have:*

$$
D_2(\mu,\nu) \leqslant C\sum_{n \geqslant 0} 2^{2n}\sum_{\ell \geqslant 0} 2^{-2\ell}\sum_{B \in \mathcal{P}_\ell} \big|\mu\big((2^nB) \cap B_n\big) - \nu\big((2^nB) \cap B_n\big)\big|,
$$

with the same notation as in the definition (5.24) of $D_2(\mu,\nu)$, and where the notation 2^nB stands for the set $\{2^nx \in \mathbb{R}^d;\ x \in B\}$.

Proof. Going back to the definition of $D_2(\mu, \nu)$, we first notice that, for each $n \geqslant 0$:

$$|\mu(B_n) - \nu(B_n)| = \sum_{B \in \mathcal{P}_0} \left|\mu\big((2^n B) \cap B_n\big) - \nu\big((2^n B) \cap B_n\big)\right|,$$

and, whenever $\nu(B_n) > 0$:

$$\begin{aligned}
&\big(\mu(B_n) \wedge \nu(B_n)\big)\delta_2(\mu^{B_n}, \nu^{B_n}) \\
&\leqslant \mu(B_n) \sum_{\ell \geqslant 1} 2^{-2\ell} \sum_{B \in \mathcal{P}_\ell} \left|\mu(2^n B|B_n) - \nu(2^n B|B_n)\right| \\
&\leqslant \sum_{\ell \geqslant 1} 2^{-2\ell} \sum_{B \in \mathcal{P}_\ell} \left|\mu\big((2^n B) \cap B_n\big) - \nu\big((2^n B) \cap B_n\big)\right| \\
&\qquad + \left|1 - \frac{\mu(B_n)}{\nu(B_n)}\right| \sum_{\ell \geqslant 1} 2^{-2\ell} \sum_{B \in \mathcal{P}_\ell} \nu\big((2^n B) \cap B_n\big),
\end{aligned}$$

which completes the proof since the last term is not greater than $|\mu(B_n) - \nu(B_n)|/3$, and the bound remains true whenever $\nu(B_n) = 0$. □

End of the Proof of Theorem 5.8

We are now in a position to prove the main estimate of this subsection.

Proof of Theorem 5.8. Let us assume that $\mu \in \mathcal{P}_q(\mathbb{R}^d)$, for some $q > 4$, and without loss of generality that $\int_{\mathbb{R}^d} |x|^q d\mu(x) = 1$. By Markov inequality, $\mu(B_n) \leqslant 2^{-q(n-1)}$ for all $n \geqslant 0$.

First Step. We shall apply Lemma 5.11 in order to find a bound for $\mathbb{E}[D_2(\bar{\mu}^N, \mu)]$. For a Borel subset $A \subset \mathbb{R}^d$, the random variable $N\bar{\mu}^N(A)$ is Binomial with parameters N and $\mu(A)$ so that:

$$\mathbb{E}\big[|\bar{\mu}^N(A) - \mu(A)|\big] \leqslant \min\big[2\mu(A), \sqrt{\mu(A)/N}\big].$$

Using Cauchy-Schwarz' inequality and the fact that the partition $\mathcal{P}_\ell$ has exactly $2^{d\ell}$ elements, we deduce that, for all $n \geqslant 0$ and $\ell \geqslant 0$,

$$\sum_{B \in \mathcal{P}_\ell} \mathbb{E}\big[\left|\bar{\mu}^N\big((2^n B) \cap B_n\big) - \mu\big((2^n B) \cap B_n\big)\right|\big] \leqslant \min\big[2\mu(B_n), 2^{d\ell/2}\sqrt{\mu(B_n)/N}\big].$$

Using the result of Lemma 5.12 and the fact that $\mu(B_n) \leqslant 2^{-q(n-1)}$, we get, for a universal constant C possibly depending upon q (and whose value is allowed to increase from line to line):

$$\mathbb{E}\big[D_2(\bar{\mu}^N, \mu)\big] \leqslant C \sum_{n \geqslant 0} 2^{2n} \sum_{\ell \geqslant 0} 2^{-2\ell} \min\big[2^{-qn}, 2^{d\ell/2}\sqrt{2^{-qn}/N}\big]. \tag{5.26}$$

Second Step. We first consider the case $d \geqslant 4$. For $N \geqslant 2$ fixed, we estimate the right-hand side of (5.26) by computing the sums in the order they appear. Let $n_0 = \lfloor q^{-1} \log_2 N \rfloor$ where we use the notation $\lfloor x \rfloor$ for the integer part of x, namely the largest integer smaller than or equal to x. For each integer $n \geqslant 0$, we define $\ell(n) = d^{-1}(\log_2 N - qn)$. Notice that $\ell(n) \geqslant 0$ if and only if $n \leqslant n_0$ and

$$\ell \leqslant \ell(n) \Leftrightarrow 2^{-qn} \geqslant 2^{d\ell/2}\sqrt{2^{-qn}/N}$$
$$\Leftrightarrow \min\left[2^{-qn}, 2^{d\ell/2}\sqrt{2^{-qn}/N}\right] = 2^{d\ell/2}\sqrt{2^{-qn}/N}.$$

Similarly,

$$\ell > \ell(n) \Leftrightarrow 2^{-qn} < 2^{d\ell/2}\sqrt{2^{-qn}/N} \Rightarrow \min\left[2^{-qn}, 2^{d\ell/2}\sqrt{2^{-qn}/N}\right] = 2^{-qn}.$$

So if we split the above sum over n into two parts, $\Sigma_1 = \sum_{0\leqslant n\leqslant n_0}\cdots$ and $\Sigma_2 = \sum_{n>n_0}\cdots$,

$$\begin{aligned}
\Sigma_1 &= \sum_{n=0}^{n_0} 2^{2n}\Bigg(\sum_{\ell=0}^{\lfloor \ell(n)\rfloor} 2^{-2\ell}2^{d\ell/2}\sqrt{2^{-qn}/N} + \sum_{\ell>\lfloor \ell(n)\rfloor} 2^{-2\ell}2^{-qn}\Bigg) \\
&= \sum_{n=0}^{n_0} 2^{2n}\Bigg(\sum_{\ell=0}^{\lfloor \ell(n)\rfloor} 2^{-2\ell}2^{d\ell/2}\sqrt{2^{-qn}/N} + 3^{-1}2^{-qn}2^{-2\lfloor \ell(n)\rfloor}\Bigg) \\
&\leqslant N^{-1/2}\sum_{n=0}^{n_0} 2^{(2-q/2)n}\sum_{\ell=0}^{\lfloor \ell(n)\rfloor} 2^{(-2+d/2)\ell} + CN^{-2/d}\sum_{n=0}^{n_0} 2^{(2-q+2q/d)n}.
\end{aligned}$$

Obviously, the second sum above is bounded by a constant times $N^{-2/d}$ since $2-q+2q/d < 0$, recall $d \geq 4$. As for the first sum, when $d > 4$, it is bounded from above by:

$$\begin{aligned}
CN^{-1/2}\sum_{n=0}^{n_0} 2^{(2-q/2)n}2^{(-2+d/2)\ell(n)} &\leqslant CN^{-1/2}N^{(d-4)/(2d)}\sum_{n=0}^{n_0} 2^{(2-q+2q/d)n} \\
&\leqslant CN^{-2/d},
\end{aligned}$$

where we used once more the fact that $2 - q + 2q/d < 0$ because $q > 4$, and where the constant C is allowed to increase from line to line.

Now when $d = 4$ the upper bound on the first term reads:

$$CN^{-1/2}\sum_{n=0}^{n_0} 2^{(2-q/2)n}\ell(n) \leqslant CN^{-1/2}\log_2 N.$$

Similarly, whether $d > 4$ or $d = 4$,

$$\Sigma_2 = \sum_{n>n_0} 2^{2n}\sum_{\ell\geqslant 0} 2^{-2\ell}2^{-qn} = \frac{4}{3}\frac{1}{2^{q-2}-1}2^{-(q-2)n_0} \leqslant CN^{-(1-2/q)}.$$

All together, we proved that, when $d \geqslant 4$ and $q > 4$,

$$\mathbb{E}\big[D_2(\bar{\mu}^N, \mu)\big] \leqslant C\begin{cases} N^{-1/2}\log_2 N & \text{if } d = 4 \\ N^{-2/d} & \text{if } d > 4. \end{cases}$$

Third Step. In order to treat the case $d < 4$, we interchange the order of the summations in the right-hand side of (5.26) and write:

$$\mathbb{E}\big[D_2(\bar{\mu}^N, \mu)\big] \leqslant C \sum_{\ell \geqslant 0} 2^{-2\ell} \sum_{n \geqslant 0} 2^{2n} \min\big[2^{-qn}, 2^{d\ell/2}\sqrt{2^{-qn}/N}\big],$$

and as before assume that N is fixed. Let $\ell_0 = \lfloor d^{-1} \log_2 N \rfloor$ and let us split the above sum into two parts, $\Sigma_1 = \sum_{0 \leqslant \ell \leqslant \ell_0} \cdots$ and $\Sigma_2 = \sum_{\ell > \ell_0} \cdots$. For each integer $\ell \geqslant 0$, we define $n(\ell) = q^{-1}(\log_2 N - d\ell)$. Notice that $n(\ell) \geqslant 0$ if and only if $\ell \leqslant \ell_0$ and

$$\begin{aligned} n \leqslant n(\ell) &\Leftrightarrow 2^{-qn} \geqslant 2^{d\ell/2}\sqrt{2^{-qn}/N} \\ &\Leftrightarrow \min\big[2^{-qn}, 2^{d\ell/2}\sqrt{2^{-qn}/N}\big] = 2^{d\ell/2}\sqrt{2^{-qn}/N}, \end{aligned}$$

and similarly,

$$n > n(\ell) \Leftrightarrow 2^{-qn} < 2^{d\ell/2}\sqrt{2^{-qn}/N} \Rightarrow \min\big[2^{-qn}, 2^{d\ell/2}\sqrt{2^{-qn}/N}\big] = 2^{-qn}.$$

Consequently,

$$\begin{aligned} \Sigma_1 &= \sum_{\ell=0}^{\ell_0} 2^{-2\ell} \Bigg(\sum_{n=0}^{\lfloor n(\ell) \rfloor} 2^{2n} 2^{d\ell/2} \sqrt{2^{-qn}/N} + \sum_{n > \lfloor n(\ell) \rfloor} 2^{2n} 2^{-qn} \Bigg) \\ &\leqslant N^{-1/2} \sum_{\ell=0}^{\ell_0} 2^{-2\ell} 2^{d\ell/2} \sum_{n=0}^{\lfloor n(\ell) \rfloor} 2^{2n} 2^{-qn/2} + C \sum_{\ell=0}^{\ell_0} 2^{-2\ell} 2^{-(q-2)n(\ell)} \\ &\leqslant CN^{-1/2} \sum_{\ell=0}^{\ell_0} 2^{(d/2-2)\ell} + CN^{-(1-2/q)} \sum_{\ell=0}^{\ell_0} 2^{(d-2-2d/q)\ell}. \end{aligned}$$

The second sum on the last line is less than C if $d-2-2d/q < 0$ and is less than $CN^{1-2/q-2/d}$ if $d - 2 - 2d/q > 0$. If $d - 2 - 2d/q = 0$, which happens if and only if $d = 3$ and $q = 6$ since we assumed $d < 4$ and $q > 4$, it is less than $C \log_2(N)$. So, in any case the last term is the right hand side is less than $CN^{-1/2}$ since $d < 4$ and $q > 4$. Similarly,

$$\Sigma_2 = \sum_{\ell > \ell_0} 2^{-2\ell} \sum_{n \geqslant 0} 2^{2n} 2^{-qn} = C \sum_{\ell > \ell_0} 2^{-2\ell} \leqslant CN^{-2/d}.$$

This concludes the proof in all cases. □

5.1.3 Optimal Transportation Tidbits

In preparation for a thorough discussion of notions of differentiability and convexity of functions of measures, we return to the metric spaces $\mathcal{P}_p(E)$ equipped with the Wasserstein distances W_p, and we specialize the analysis to the case $p = 2$ and $E = \mathbb{R}^d$ of crucial importance for the developments of this book.

The geometric nature of our approach to differentiability and convexity suggests that we revisit the elements of optimal transportation introduced earlier. A crucial question in optimal transportation is whether transport plans may be induced by transport maps. This is especially important when the plans are optimal. Recall the definitions of transport plans and transport maps given in the paragraph devoted to Wasserstein distances in Subsection 5.1.1 above.

We usually prefer the notation $\partial\varphi$ to $\nabla\varphi$ to denote the gradient (first derivative) of a differentiable function φ. However, in this subsection, we introduce and use the notion of subdifferential of a function φ which is not necessarily differentiable, and since the most commonly used notation for the subdifferential is $\partial\varphi$, we shall use the notation $\nabla\varphi$ for the gradient of a differentiable function φ. Also, we often denote by I the identity mapping on $\mathbb{R}^d$, which, though it is connected with, should not be confused with the identity matrix which we denote by I_d. We are confident that the context should make it clear.

The next proposition identifies an instance of an optimal transport map that may be easily identified.

Proposition 5.13 *Given a probability measure $\mu \in \mathcal{P}_2(\mathbb{R}^d)$ and a twice continuously differentiable and strictly convex function φ from $\mathbb{R}^d$ to $\mathbb{R}$ satisfying $\int_{\mathbb{R}^d} |\nabla\varphi(x)|^2 d\mu(x) < \infty$, we denote by $\nu = \mu \circ \nabla\varphi^{-1}$ the push-forward image of μ by $\nabla\varphi$. Then, there exists a unique deterministic optimal transport plan in $\Pi_2^{\text{opt}}(\mu,\nu)$, it is induced by the map $\nabla\varphi$, and as a result:*

$$W_2(\mu,\nu)^2 = \int_{\mathbb{R}^d} |x - \nabla\varphi(x)|^2 d\mu(x).$$

Proof. By strict convexity of φ, the gradient $\nabla\varphi$ of φ is increasing in the sense that:

$$\forall x, y \in \mathbb{R}^d, \quad x \neq y \Rightarrow (x-y)\cdot(\nabla\varphi(x) - \nabla\varphi(y)) > 0.$$

In particular, $\nabla\varphi$ is one-to-one from $\mathbb{R}^d$ onto its range (also known as co-domain). Since the Hessian of φ has a strictly positive determinant, the global inversion theorem ensures that $\nabla\varphi$ is a C^1 diffeomorphism from $\mathbb{R}^d$ on its range. We denote the inverse by $(\nabla\varphi)^{-1}$. The remainder of the proof relies on a duality argument. We compute the so-called square-transform of the potential $|x|^2 - 2\varphi(x)$:

$$\phi(y) = \inf_{x\in\mathbb{R}^d} \big\{|x-y|^2 - |x|^2 + 2\varphi(x)\big\} = |y|^2 + \inf_{x\in\mathbb{R}^d} \big\{-2x\cdot y + 2\varphi(x)\big\}, \quad y \in \mathbb{R}^d. \tag{5.27}$$

By strict convexity of φ, we deduce that, when y is in the range of $\nabla\varphi$, the infimum is attained at the unique root $x \in \mathbb{R}^d$ of the equation $\nabla\varphi(x) = y$, so that:

$$\begin{aligned}\phi(y) &= |(\nabla\varphi)^{-1}(y) - y|^2 - |(\nabla\varphi)^{-1}(y)|^2 + 2\varphi\big((\nabla\varphi)^{-1}(y)\big) \\ &= |y|^2 - 2y\cdot(\nabla\varphi)^{-1}(y) + 2\varphi\big((\nabla\varphi)^{-1}(y)\big).\end{aligned} \tag{5.28}$$

The key point is to observe that because of the definition (5.27) of the square transform, we have:

$$\forall x, y \in \mathbb{R}^d, \quad |x-y|^2 \geqslant |x|^2 + \phi(y) - 2\varphi(x).$$

Now, if X and Y are two $\mathbb{R}^d$-valued random variables on some probability space $(\Omega, \mathcal{F}, \mathbb{P})$, such that $\mathcal{L}(X) = \mu$ and $\mathcal{L}(Y) = \nu$,

$$\begin{aligned} \mathbb{E}\big[|X-Y|^2\big] &\geqslant \mathbb{E}\big[|X|^2 + \phi(Y) - 2\varphi(X)\big] \\ &= \mathbb{E}\big[|X|^2 + \phi\big(\nabla\varphi(X)\big) - 2\varphi(X)\big], \end{aligned} \tag{5.29}$$

where we used the fact that Y and $\nabla\varphi(X)$ have the same distribution. A priori, the right-hand side could be $-\infty$, but Inserting (5.28), we get:

$$\begin{aligned} \mathbb{E}\big[|X-Y|^2\big] &\geqslant \mathbb{E}\big[|X|^2 + |\nabla\varphi(X)|^2 - 2\nabla\varphi(X)\cdot X + 2\varphi(X) - 2\varphi(X)\big] \\ &= \mathbb{E}\big[|X - \nabla\varphi(X)|^2\big], \end{aligned}$$

which shows that $\nabla\varphi$ is an optimal transport map. As for the proof of uniqueness, we return to the first line in (5.29). By definition of ϕ, it holds $\mathbb{P}$ -a.s.

$$|X-Y|^2 \geqslant |X|^2 + \phi(Y) - 2\varphi(X).$$

The expectation of the right-hand side only depends on μ and ν and does not depend on the joint law of (X, Y). Therefore, it is equal to $W_2(\mu,\nu)^2$. So, if $\mathbb{E}[|X-Y|^2] = W_2(\mu,\nu)^2$, then the above inequality becomes an equality:

$$|X-Y|^2 = |X|^2 + \phi(Y) - 2\varphi(X),$$

which shows that X reaches the infimum in the definition of $\phi(Y)$. Notice now that Y belongs to the codomain of $\nabla\varphi$ with probability 1 (since ν is the push-forward image of μ by $\nabla\varphi$). Therefore, the minimum in the definition of $\phi(Y)$ is unique and is given by $\nabla\varphi^{-1}(Y)$, in other words $Y = \nabla\varphi(X)$. □

Remark 5.14 *Strict convexity is crucial for the conclusion of Proposition 5.13. Here is a simple counter-example. Let $\mu \in \mathcal{P}_2(\mathbb{R})$ have mean 0 and let ν be the push-forward image of μ by the mapping $\mathbb{R} \ni x \mapsto -x$. We show that the transport map $\mathbb{R} \ni x \mapsto -x$ does not induce an optimal transport plan from μ to ν.*

The cost of the transport map $\mathbb{R} \ni x \mapsto -x$ is:

$$\mathbb{E}\big[|X-(-X)|^2\big] = 4\mathbb{E}\big[|X|^2\big],$$

where X is any random variable with distribution μ. Now, if Y is independent of X and $\mathcal{L}(Y) = \mathcal{L}(-X)$, then,

$$\mathbb{E}\big[|X-Y|^2\big] = 2\mathbb{E}\big[|X|^2\big],$$

which says the transport plan $\mu\otimes\nu$ is of smaller cost than the plan associated with the map $\mathbb{R} \ni x \mapsto -x$.

Remark 5.15 *One can easily push further the analysis in the one-dimensional case. For example if μ and ν are two distributions in $\mathcal{P}_2(\mathbb{R})$, if we denote by F_μ and F_ν their cumulative distribution functions, μ being atomless (i.e., F_μ is continuous), it is well known that ν is the image of μ by $F_\nu^{-1} \circ F_\mu$, where F_ν^{-1} denotes the pseudo-inverse of F_ν. Moreover, if μ and ν have smooth positive densities, then $F_\nu^{-1} \circ F_\mu$ is strictly increasing. So, we can apply Proposition 5.13 by choosing φ as any anti-derivative of $F_\nu^{-1} \circ F_\mu$. We deduce that there exists a unique optimal plan induced by the transport map $F_\nu^{-1} \circ F_\mu$. In fact, this result remains true under the mere assumption that μ has no atom,* i.e., *F_μ is continuous.*

Brenier's Theorem. The remaining of this subsection is devoted to the proof of a celebrated result of Brenier which asserts that in the d-dimensional case, the situation identified in Proposition 5.13 is generic when the measure $\mu \in \mathcal{P}(\mathbb{R}^d)$ is absolutely continuous.

Definition 5.16 *If $\varphi : \mathbb{R}^d \mapsto (-\infty, +\infty]$ is convex and proper in the sense that it is not identically equal to $+\infty$, for each $x \in \mathbb{R}^d$, we define the subdifferential of φ at x as the set:*

$$\partial\varphi(x) = \{u \in \mathbb{R}^d : \ \forall y \in \mathbb{R}^d, \ \varphi(y) \geq \varphi(x) + u \cdot (y - x)\}$$

Obviously, if φ is finite and differentiable at the point x, the subdifferential $\partial\varphi(x)$ is the singleton $\{\nabla\varphi(x)\}$ given by the actual derivative (or gradient) of φ at x.

Definition 5.17 *A set $A \subset \mathbb{R}^d \times \mathbb{R}^d$ is said to be cyclic monotone if for any integer $m \geqslant 1$ and any $(x_1, y_1), \cdots, (x_m, y_m)$ in A we have:*

$$y_1 \cdot (x_2 - x_1) + y_2 \cdot (x_3 - x_2) + \cdots + y_{m-1} \cdot (x_m - x_{m-1}) + y_m \cdot (x_1 - x_m) \leqslant 0.$$

The relevance of cyclic monotonicity to our use of differentiable convex functions is the following classical result of convex analysis due to Rockafellar [320].

Proposition 5.18 *A nonempty set $A \subset \mathbb{R}^d \times \mathbb{R}^d$ is cyclic monotone if and only if it is included in the subdifferential of a lower-semicontinuous proper convex function on $\mathbb{R}^d$ in that sense that $A \subset \partial\varphi = \{(x, y) \in \mathbb{R}^d \times \mathbb{R}^d : \ y \in \partial\varphi(x)\}$.*

Proof. The fact that a subset of the subdifferential of a convex function is cyclic monotone is easy, and can be proven by simple computation from the definition of cyclic monotony by iterating the definition of subdifferential. We do not give the details of the argument because we shall not use this half of the equivalence. We give a detailed proof of the reciprocal which we shall use in the sequel.

Let $(x_0, y_0) \in A$ and let us define the function φ by:

$$\varphi(x) = \sup\big\{y_m \cdot (x - x_m) + y_{m-1} \cdot (x_m - x_{m-1}) + \cdots + y_0 \cdot (x_1 - x_0); \\ m \geqslant 1, \ (x_1, y_1), \cdots, (x_m, y_m) \in A\big\}.$$

The function φ is convex and lower semi-continuous as the supremum of linear functions. Moreover, is it proper because $\varphi(x_0) = 0$. Indeed, $\varphi(x_0) \leqslant 0$ by definition of cyclic monotonicity, and $\varphi(x_0) \geqslant 0$ by choosing $m = 1$ and $x_1 = x_0$, so that $\varphi(x_0) = 0$. Now, if $(x, y) \in A$, for any $z \in \mathbb{R}^d$ we easily get from the very definition of φ that:

$$\varphi(z) \geqslant y \cdot (z - x) + \varphi(x),$$

which proves that $y \in \partial\varphi(x)$, completing the proof. □

Here is the connection with optimal transportation:

Proposition 5.19 *For all measures μ and ν in $\mathcal{P}_2(\mathbb{R}^d)$, the topological support of any optimal transport plan is cyclic monotone.*

Proof. Let us assume that the topological support of an optimal plan π^* is not cyclic monotone. There exist an integer $m \geqslant 1$ and couples $(x_1, y_1), \cdots, (x_m, y_m)$ in $\text{Supp}(\pi^*)$ such that:

$$\sum_{k=1}^{m} \left(|x_{k+1} - y_k|^2 - |x_k - y_k|^2 \right) < 0,$$

where we use the notation x_{m+1} for x_1. By definition of the topological support of a measure, we can find neighborhoods U_i of x_i and V_i of y_i for $i = 1, \cdots, m$ such that $\pi^*(U_i \times V_i) > 0$ for all $i = 1, \cdots, m$ and

$$\sum_{k=1}^{m} \left(|\tilde{x}_{k+1} - \tilde{y}_k|^2 - |\tilde{x}_k - \tilde{y}_k|^2 \right) < 0, \qquad \tilde{x}_k \in U_k,\ \tilde{y}_k \in V_k,\ k = 1, \cdots, m.$$

For $i = 1, \cdots, m$, we define the conditional measures $\pi_i = \pi^*[\cdot | U_i \times V_i] = \pi^*[\cdot \cap U_i \times V_i]/\pi^*(U_i \times V_i)$, and we denote by $\pi_i^{(1)}$ and $\pi_i^{(2)}$ their marginals. Finally we define the measure π by:

$$\pi = \pi^* + \frac{c}{m} \sum_{k=1}^{m} \left(\pi_{k+1}^{(1)} \otimes \pi_k^{(2)} - \pi_k \right),$$

for a positive constant c to be chosen below and with $\pi_{m+1}^{(1)} = \pi_1^{(1)}$. Observe that, for all $A \in \mathcal{B}(\mathbb{R}^d \times \mathbb{R}^d)$,

$$\begin{aligned}
\pi(A) &\geqslant \pi^*(A) - \frac{c}{m} \sum_{k=1}^{m} \frac{\pi^*(A \cap (U_k \times V_k))}{\pi^*(U_k \times V_k)} \\
&\geqslant \frac{1}{m} \sum_{k=1}^{m} \left(\pi^*(A) - \pi^*(A \cap (U_k \times V_k)) \right) \geqslant 0,
\end{aligned}$$

if $c < \min_{1 \leqslant i \leqslant m} \pi^*(U_i \times V_i)$. The measure π is obviously a probability. Its second marginal is the same as the second marginal of π^* which is ν. As for its first marginal, it is μ because of the cyclic summation over k. Hence it is a coupling between μ and ν. We reach the desired contradiction by computing:

$$\begin{aligned}
&\int_{\mathbb{R}^d \times \mathbb{R}^d} |x-y|^2 \pi(dx, dy) - \int_{\mathbb{R}^d \times \mathbb{R}^d} |x-y|^2 \pi^*(dx, dy) \\
&= \frac{c}{m} \int_{(\mathbb{R}^d)^{4m}} \sum_{k=1}^m \left(|x_k - y_k|^2 - |\tilde{x}_k - \tilde{y}_k|^2 \right) \prod_{k=1}^m \pi_{k+1}^{(1)}(dx_{k+1}) \pi_k^{(2)}(dy_k) \pi_k(d\tilde{x}_k, d\tilde{y}_k) \\
&< 0,
\end{aligned}$$

because $x_k, \tilde{x}_k \in U_k$ and $y_k, \tilde{y}_k \in V_k$ for $k = 1, \cdots, m$, and with the convention that $x_{k+1} = x_1$. □

Finally, we state and prove Brenier's theorem.

Theorem 5.20 *If $\mu, \nu \in \mathcal{P}_2(\mathbb{R}^d)$ and if μ is absolutely continuous, there exists an optimal transport map from μ to ν given by the gradient $\nabla\varphi$ of a $(-\infty, \infty]$-valued lower semi-continuous proper convex function φ on $\mathbb{R}^d$ that is μ almost-everywhere differentiable.*

Proof. Let π^* be an optimal transport plan from μ to ν. Proposition 5.19 implies that its support $\text{Supp}(\pi^*)$ is cyclic monotone, and by the part of Rockafellar's characterization which we proved in Proposition 5.18, there exists a $(-\infty, \infty]$-valued lower semi-continuous proper convex function φ on $\mathbb{R}^d$ such that $\text{Supp}(\pi^*) \subset \partial\varphi$, which we can rewrite as:

$$\pi^*\big(\{(x, y) \in \mathbb{R}^d \times \mathbb{R}^d;\ y \in \partial\varphi(x)\}\big) = 1. \tag{5.30}$$

Choose now $(x_0, y_0) \in \text{Supp}(\pi^*)$. Then, for all $x \in \mathbb{R}^d$,

$$\varphi(x) \geqslant \varphi(x_0) + y_0 \cdot (x - x_0).$$

Also, for all $(x, y) \in \text{Supp}(\pi^*)$,

$$\varphi(x_0) \geqslant \varphi(x) + y \cdot (x_0 - x),$$

so that, for all $(x, y) \in \text{Supp}(\pi^*)$,

$$\varphi(x_0) + y_0 \cdot (x - x_0) \leqslant \varphi(x) \leqslant \varphi(x_0) + y \cdot (x - x_0). \tag{5.31}$$

We now observe that the left- and right-hand sides are integrable with respect to π^* and satisfy:

$$\int_{\mathbb{R}^d \times \mathbb{R}^d} \big|\varphi(x_0) + y_0 \cdot (x - x_0)\big| d\pi^*(x, y) \leqslant |\varphi(x_0)| + |y_0| \big(M_1(\mu) + |x_0|\big) < \infty,$$

$$\int_{\mathbb{R}^d \times \mathbb{R}^d} \big|\varphi(x_0) + y \cdot (x - x_0)\big| d\pi^*(x, y) \leqslant |\varphi(x_0)| + M_2(\nu)| \big(M_2(\mu) + |x_0|\big) < \infty.$$

Integrating (5.31) with respect to π^* on $\mathrm{Supp}(\pi^*)$, we get:

$$\int_{\mathbb{R}^d} |\varphi(x)| d\mu(x) < \infty.$$

Therefore, φ is a.s. finite under μ, that is the domain of φ, i.e., $\mathrm{Dom}(\varphi) = \{x \in \mathbb{R}^d : |\varphi(x)| < \infty\}$, is of full measure under μ. Of course, $\mathrm{Dom}(\varphi)$ is a convex subset of $\mathbb{R}^d$, and its boundary has a zero Lebesgue measure, from which get $\mu(\mathrm{Int}(\mathrm{Dom}(\varphi))) = 1$ since μ is absolutely continuous with respect to the Lebesgue measure, where $\mathrm{Int}(\mathrm{Dom}(\varphi))$ is the interior of $\mathrm{Dom}(\varphi)$.

Recall now that a proper convex function is continuous and locally Lipschitz, and thus almost everywhere differentiable, on the interior of its domain. Since $\mu(\mathrm{Int}(\mathrm{Dom}(\varphi))) = 1$, we deduce that $\mu(\{x \in \mathbb{R}^d : \{\nabla\varphi(x)\} = \partial\varphi(x)\}) = 1$, from which we conclude that:

$$\pi^*\big(\{(x, y) \in \mathbb{R}^d \times \mathbb{R}^d : y = \nabla\varphi(x)\}\big) = 1,$$

which completes the proof. □

Remark 5.21 *The above proof shows that any optimal transport plan is of the form* $\mu \circ (I, \nabla\varphi)^{-1}$ *for some proper convex function. A simple duality argument shows that in fact, there is uniqueness, not only of the optimal transport plan, but of the gradient of a convex function transporting* μ *onto* ν*. Moreover, if* ν *is also absolutely continuous, and if we denote by* φ^* *the convex conjugate of* φ*, then* φ^* *is the convex function whose gradient transports* ν *onto* μ *optimally and it is not difficult to see that for* μ*-almost every* $x \in \mathbb{R}^d$ *and* ν*-almost every* $y \in \mathbb{R}^d$*, we have:*

$$\nabla\varphi^* \circ \nabla\varphi(x) = x, \quad \textit{and} \quad \nabla\varphi \circ \nabla\varphi^*(y) = y.$$

5.2 Differentiability of Functions of Probability Measures

There are many notions of differentiability for functions defined on spaces of probability measures, and recent progress in the theory of optimal transportation have put some of their geometric characteristics in the limelight. See Section 5.4 below where we review some of them.

However, the approach which we find convenient for the type of stochastic optimization problems we are interested in is slightly different. It is more of a functional analytic nature rather than of a geometric nature. Our choice is driven by the fact that we need to control infinitesimal perturbations of probability measures induced by infinitesimal variations in a linear space of random variables. For that reason, differentiation is based on the *lifting* of functions $\mathcal{P}_2(\mathbb{R}^d) \ni \mu \mapsto u(\mu)$ to functions $\tilde{u}$ defined on a Hilbert space $L^2(\Omega, \mathcal{F}, \mathbb{P}; \mathbb{R}^d)$ over some probability space $(\Omega, \mathcal{F}, \mathbb{P})$ by setting $\tilde{u}(X) = u(\mathcal{L}(X))$, for $X \in L^2(\Omega, \mathcal{F}, \mathbb{P}; \mathbb{R}^d)$, Ω being a Polish space, $\mathcal{F}$ its Borel σ-field and $\mathbb{P}$ an atomless probability measure (since Ω is Polish, $\mathbb{P}$ is atomless if and only if every singleton has a zero measure).

Throughout the analysis below, we shall use repeatedly the fact that, over an atomless probability space $(\Omega, \mathcal{F}, \mathbb{P})$, for any probability distribution μ on a Polish space E, we can construct an E-valued random variable on Ω with μ as distribution. In this regard, we refer to Remark 5.26 for the properties of the lifting $\tilde{u}$ over general spaces $(\Omega, \mathcal{F}, \mathbb{P})$ that are neither Polish nor atomless.

Definition 5.22 *A function u on $\mathcal{P}_2(\mathbb{R}^d)$ is said to be* L-differentiable *at $\mu_0 \in \mathcal{P}_2(\mathbb{R}^d)$ if there exists a random variable X_0 with law μ_0, in other words satisfying $\mathcal{L}(X_0) = \mu_0$, such that the lifted function $\tilde{u}$ is Fréchet differentiable at X_0.*

The Fréchet derivative of $\tilde{u}$ at X_0 can be viewed as an element of $L^2(\Omega, \mathcal{F}, \mathbb{P}; \mathbb{R}^d)$ by identifying $L^2(\Omega, \mathcal{F}, \mathbb{P}; \mathbb{R}^d)$ and its dual. When studying this form of differentiation, the first item on our agenda is to show that this notion of differentiability is intrinsic.

5.2.1 Structure of the L-Derivative

We first prove that the law of the random variable $D\tilde{u}(\tilde{X}_0)$ does not depend upon the particular choice of the random variable $\tilde{X}_0$ satisfying $\mathcal{L}(\tilde{X}_0) = \mu_0$. See Proposition 5.24 below, whose proof will use the following simple measure theoretical lemma:

Lemma 5.23 *If X and Y are elements of $L^2(\Omega, \mathcal{F}, \mathbb{P}; \mathbb{R}^d)$ with the same law, then for each $\epsilon > 0$ there exist two measurable measure preserving mappings τ and τ^{-1} from Ω into itself such that $\mathbb{P}\{\omega \in \Omega : (\tau \circ \tau^{-1})(\omega) = (\tau^{-1} \circ \tau)(\omega) = \omega\} = 1$ and $\mathbb{P}[|Y - X \circ \tau^{-1}| \leqslant \epsilon] = 1$.*

Proof. Let $(A_n)_{n \geqslant 1}$ be a partition of $\mathbb{R}^d$ by Borel sets of diameter at most ϵ, and, for each $n \geqslant 1$, let us set $B_n = \{X \in A_n\}$ and $C_n = \{Y \in A_n\}$. For each $n \geqslant 1$, $\mathbb{P}(B_n) = \mathbb{P}(C_n)$.

We now use the fact that $\mathbb{P}$ is an atomless probability measure on a Polish space. We denote by $\mathcal{F}^*$ the completion of $\mathcal{F}$ under $\mathbb{P}$ and by $\mathbb{P}^*$ the extension of $\mathbb{P}$ to $\mathcal{F}^*$. Whenever $\mathbb{P}(B_n) = \mathbb{P}(C_n) > 0$, there exist two subsets $M_n \subset B_n$ and $N_n \subset C_n$, both being included in Borel subsets of zero measure under $\mathbb{P}$, and a one-to-one map τ_n from $B_n \setminus M_n$ onto $C_n \setminus N_n$ such that τ_n and τ_n^{-1} are measurable with respect to the restriction of $\mathcal{F}^*$ to $B_n \setminus M_n$ and $C_n \setminus N_n$ and preserve the restrictions of the measure $\mathbb{P}^*$ to $B_n \setminus M_n$ and $C_n \setminus N_n$ (see Corollary 6.6.7 and Theorem 9.2.2 in Bogachev [64]).

Then, we can extend τ_n into a measurable mapping, still denoted by τ_n, from B_n to C_n (measurability being understood with respect to the restrictions of $\mathcal{F}^*$ to B_n and C_n) and then τ_n^{-1} into a measurable mapping, still denoted by τ_n^{-1}, from C_n to B_n (measurability being understood with respect to the restrictions of $\mathcal{F}^*$ to C_n and B_n) in such a way that $\tau_n \circ \tau_n^{-1}$ is the identity on $C_n \setminus N_n$, $\tau_n^{-1} \circ \tau_n$ is the identity on $B_n \setminus M_n$, $\tau_n^{-1}(N_n) \subset M_n$ and $\tau_n(M_n) \subset N_n$. Necessarily, $(\tau_n)^{-1}(N_n) \subset M_n$ and $(\tau_n^{-1})^{-1}(M_n) \subset N_n$. Here, $(\tau_n)^{-1}(\cdot)$ denotes the pre-image by τ_n and, similarly, $(\tau_n^{-1})^{-1}(\cdot)$ denotes the pre-image by τ_n^{-1}. Obviously, for all $A \in \mathcal{F}^*$, with $A \subset C_n$, we have $\mathbb{P}^*(A) = \mathbb{P}^*(A \setminus N_n) = \mathbb{P}^*(\tau_n^{-1}(A \setminus N_n)) = \mathbb{P}^*((\tau_n)^{-1}(A))$. Similarly, for all $A \in \mathcal{F}^*$, with $A \subset B_n$, we have $\mathbb{P}^*(A) = \mathbb{P}^*((\tau_n^{-1})^{-1}(A))$.

Whenever $\mathbb{P}(B_n) = \mathbb{P}(C_n) = 0$, we construct τ_n and τ_n^{-1} according to the same principle, but with $M_n = B_n$ and $N_n = C_n$.

Since $(B_n)_{n\geqslant 1}$ and $(C_n)_{n\geqslant 1}$ are partitions of Ω by measurable sets, the maps $\tau = \sum_{n\geqslant 1}\tau_n \mathbf{1}_{B_n}$ and $\tau^{-1} = \sum_{n\geqslant 1}\tau_n^{-1}\mathbf{1}_{C_n}$ are measurable from $(\Omega,\mathcal{F}^*)$ into itself. Letting $M = \cup_{n\geqslant 1}M_n$ and $N = \cup_{n\geqslant 1}N_n$, we have $\mathbb{P}^*(M) = \mathbb{P}^*(N) = 0$. On $\Omega\setminus N$, $\tau\circ\tau^{-1}$ is the identity and, on $\Omega\setminus M$, $\tau^{-1}\circ\tau$ is also the identity.

Since τ and τ^{-1} are measurable with respect to $\mathcal{F}^*$, we can find two mappings $\tilde\tau$ and $\tilde\tau^{-1}$ from Ω into itself, measurable with respect to $\mathcal{F}$, such that τ and $\tilde\tau$, and similarly τ^{-1} and $\tilde\tau^{-1}$, coincide outside an event A in $\mathcal{F}$ of zero measure under $\mathbb{P}$. In particular, $\tilde\tau$ and $\tilde\tau^{-1}$ preserve the probability measure $\mathbb{P}$. As a by-product, $\mathbb{P}((\tilde\tau)^{-1}(A)) = 0$ and $\mathbb{P}((\tilde\tau^{-1})^{-1}(A)) = 0$. For any $\omega\notin A\cup(\tilde\tau)^{-1}(A)$, we have $\tau(\omega) = \tilde\tau(\omega)\notin A$ and thus $\tau^{-1}(\tau(\omega)) = \tilde\tau^{-1}(\tilde\tau(\omega))$. Therefore, $\mathbb{P}(\{\omega\in\Omega : \tilde\tau^{-1}(\tilde\tau(\omega)) = \omega\}) = \mathbb{P}^*(\{\omega\in\Omega : \tau^{-1}(\tau(\omega)) = \omega\}) = 1$. Similarly, $\mathbb{P}(\{\omega\in\Omega : \tilde\tau(\tilde\tau^{-1}(\omega)) = \omega\}) = 1$.

Finally, observe that, by construction, $\|Y - X\circ\tau^{-1}\|_\infty \leqslant \epsilon$. Since τ and $\tilde\tau$ coincide outside A, we deduce that $\mathbb{P}[|Y - X\circ\tilde\tau^{-1}|\leqslant\epsilon] = 1$. □

Proposition 5.24 *Let u be a real valued function on $\mathcal{P}_2(\mathbb{R}^d)$, $\tilde u$ its lifting to $L^2(\Omega,\mathcal{F},\mathbb{P};\mathbb{R}^d)$. If u is* L*-differentiable at $\mu_0\in\mathcal{P}_2(\mathbb{R}^d)$ in the sense of Definition 5.22, then the lifting $\tilde u$ is differentiable at each $X\in L^2(\Omega,\mathcal{F},\mathbb{P};\mathbb{R}^d)$ such that $\mu_0 = \mathcal{L}(X)$, and the law of the pair $(X, D\tilde u(X))$ does not depend upon the random variable X as long as $\mu_0 = \mathcal{L}(X)$.*

Proof. By definition, there exists a random variable X_0 with law μ_0 such that the lifted function $\tilde u$ is Fréchet differentiable at X_0. Let $X\in L^2(\Omega,\mathcal{F},\mathbb{P};\mathbb{R}^d)$ be such that $\mathcal{L}(X) = \mu_0$. Then Lemma 5.23 implies that, for any $\epsilon>0$, there exist two measurable measure preserving mappings τ_ϵ and τ_ϵ^{-1} from Ω into itself, such that $\mathbb{P}(\{\omega\in\Omega : (\tau_\epsilon\circ\tau_\epsilon^{-1})(\omega) = \omega\}) = \mathbb{P}(\{\omega\in\Omega : (\tau_\epsilon^{-1}\circ\tau_\epsilon)(\omega) = \omega\}) = 1$ and $\mathbb{P}[|X_0 - X\circ\tau_\epsilon|\leqslant\epsilon] = 1$. Using the fact that the lifting $\tilde u$ is differentiable at X_0 and that its values depend only upon the distributions of its arguments, we get, for any $Y\in L^2(\Omega,\mathcal{F},\mathbb{P};\mathbb{R}^d)$:

$$
\begin{aligned}
\tilde u(X+Y) &= \tilde u\big(X\circ\tau_\epsilon + Y\circ\tau_\epsilon\big)\\
&= \tilde u(X_0) + \mathbb{E}\big[D\tilde u(X_0)\cdot\big(X\circ\tau_\epsilon + Y\circ\tau_\epsilon - X_0\big)\big]\\
&\qquad + o\big(\|X\circ\tau_\epsilon + Y\circ\tau_\epsilon - X_0\|_2\big)\\
&= \tilde u(X_0) + \mathbb{E}\big[D\tilde u(X_0)\cdot\big(X\circ\tau_\epsilon + Y\circ\tau_\epsilon - X_0\big)\big]\\
&\qquad + o\big(\|X\circ\tau_\epsilon - X_0\|_2 + \|Y\|_2\big)\\
&= \tilde u(X) + \mathbb{E}\big[D\tilde u(X_0)\circ\tau_\epsilon^{-1}\cdot Y\big] + O\big(\|X\circ\tau_\epsilon - X_0\|_2\big) + o\big(\|Y\|_2\big).
\end{aligned}
\tag{5.32}
$$

It is important to observe that the symbols $O(\cdot)$ and $o(\cdot)$ which we use according to the Landau convention, are here uniform with respect to ϵ. Here $o(\cdot)$ stands for a function $o(\cdot)$ of the form $\mathbb{R}\ni x\mapsto x\delta(x)\in\mathbb{R}$ with $\lim_{x\to 0}\delta(x) = 0$.

Let us assume momentarily that $D\tilde u(X_0)\circ\tau_\epsilon^{-1}$ converges in $L^2(\Omega,\mathcal{F},\mathbb{P};\mathbb{R}^{2d})$ when $\epsilon\searrow 0$, and let us denote by Z its limit. Then, (X,Z) has the same law as $(X_0, D\tilde u(X_0))$ because τ_ϵ is measure preserving. Taking the limit $\epsilon\searrow 0$ in (5.32) we get:

$$
\tilde u(X+Y) = \tilde u(X) + \mathbb{E}\big[Z\cdot Y\big] + o\big(\|Y\|_2\big),
$$

which proves that $D\tilde u(X)$ exists and is equal to Z.

We conclude the proof by proving that $(D\tilde{u}(X_0) \circ \tau_\epsilon^{-1})_{\epsilon>0}$ forms a Cauchy family as $\epsilon \searrow 0$. This follows from the fact that, if we subtract the value of (5.32) for ϵ to its value for $\epsilon' \in (0, \epsilon)$, we find that:

$$\sup_{\epsilon' \in (0,\epsilon)} \left| \mathbb{E}\big[\big(D\tilde{u}(X_0) \circ \tau_\epsilon^{-1} - D\tilde{u}(X_0) \circ \tau_{\epsilon'}^{-1}\big) \cdot Y\big]\right| \leqslant C\big(\epsilon + o(\|Y\|_2)\big),$$

which is enough to conclude by taking $\|Y\|_2 = \sqrt{\epsilon}$, dividing by $\|Y\|_2$, and finally, taking the supremum over these Y's. □

The following result gives the structure of the L-derivative of a function of probability measures, and provides the exact form in which it will be used.

Proposition 5.25 *Let u be a real valued continuously* L*-differentiable function on $\mathcal{P}_2(\mathbb{R}^d)$, and $\tilde{u}$ its lifting to $L^2(\Omega, \mathcal{F}, \mathbb{P}; \mathbb{R}^d)$. Then for any $\mu \in \mathcal{P}_2(\mathbb{R}^d)$, there exists a measurable function $\xi : \mathbb{R}^d \to \mathbb{R}^d$ such that for all $X \in L^2(\Omega, \mathcal{F}, \mathbb{P}; \mathbb{R}^d)$ with $\mathcal{L}(X) = \mu$, it holds that $D\tilde{u}(X) = \xi(X)$ almost surely.*

When we say that u is continuously L-differentiable, we mean that the Fréchet derivative $D\tilde{u}(X)$ of the lifting $\tilde{u}$ is a continuous function of X from the space $L^2(\Omega, \mathcal{F}, \mathbb{P}; \mathbb{R}^d)$ into itself. Also, notice that the function ξ from $\mathbb{R}^d$ into itself given in the above statement is uniquely defined μ-almost everywhere on $\mathbb{R}^d$, and that necessarily $\int_{\mathbb{R}^d} |\xi(x)|^2 d\mu(x) < \infty$. Moreover, notice that in the equality $D\tilde{u}(X) = \xi(X)$, the meaning of the apparently similar evaluations at X are not the same in the left and right sides of the equality. In the left-hand side, $D\tilde{u}$ is seen as a mapping from $L^2(\Omega, \mathcal{F}, \mathbb{P}; \mathbb{R}^d)$ into itself which is evaluated at the random variable X, seen as an element of $L^2(\Omega, \mathcal{F}, \mathbb{P}; \mathbb{R}^d)$, the result of this evaluation being another $\mathbb{R}^d$-valued random variable on $(\Omega, \mathcal{F}, \mathbb{P})$. In the right-hand side, ξ is a mapping from $\mathbb{R}^d$ into itself which is evaluated at each realization of the random variable X. In other words, for almost every $\omega \in \Omega$, it holds that $[D\tilde{u}(X)](\omega) = \xi(X(\omega))$.

Proof. For a given $X \in L^2(\Omega, \mathcal{F}, \mathbb{P}; \mathbb{R}^d)$, the goal is to prove that, as a random variable, $D\tilde{u}(X)$ is measurable with respect to the σ-field generated by X (fact which we denote by $D\tilde{u}(X) \in \sigma\{X\}$), as the existence of ξ such that $\mathbb{P}[D\tilde{u}(X) = \xi(X)] = 1$ then follows from standard measure theory arguments. The fact that ξ is independent of the choice of the random variable X representing the distribution μ then follows from Proposition 5.24.

Without any loss of generality, we may assume that u (and thus $\tilde{u}$) is bounded. Indeed, it suffices to compose u with any smooth bounded function matching the identity on a sufficiently large interval in order to recover the general case. For the time being, we also assume that μ is absolutely continuous with respect to the Lebesgue measure and that:

$$\int_{\mathbb{R}^d} |x|^q d\mu(x) < \infty,$$

for some $q > 4$. For each $\epsilon > 0$, we define the function Ψ on $L^4(\Omega, \mathcal{F}, \mathbb{P}; \mathbb{R}^d)$ by:

$$\Psi(Y) = \tilde{u}(Y) + \frac{1}{2\epsilon}\mathbb{E}\big[|X - Y|^2\big] + \mathbb{E}[|Y|^4].$$

Note that Ψ is Fréchet differentiable on $L^4(\Omega, \mathcal{F}, \mathbb{P}; \mathbb{R}^d)$ and that its Fréchet derivative is given by (or at least can be identified with) $D\Psi(Y) = D\tilde{u}(Y) + \epsilon^{-1}(Y - X) + 4|Y|^2 Y$. Notice also that $\Psi(Y) \to +\infty$ as $\mathbb{E}[|Y|^4] \to +\infty$ since $\tilde{u}$ is bounded. We then call $(Z_n)_{n\geqslant 0}$ a minimizing sequence for Ψ, and for each $n \geqslant 0$, we let $\nu_n = \mathcal{L}(Z_n)$. Since μ is absolutely continuous, we can use Brenier's Theorem 5.20 stating that there exists a real valued convex function ψ_n on $\mathbb{R}^d$, which is differentiable μ-almost everywhere, such that the random variable $Y_n = \nabla\psi_n(X)$ satisfies $\mathcal{L}(Y_n) = \nu_n$ and $\mathbb{E}[|X - Y_n|^2] = W_2(\mu, \nu_n)^2$. These two facts imply that:

$$\begin{aligned}\Psi(Y_n) &= \tilde{u}(Y_n) + \frac{1}{2\epsilon} W_2(\mu, \nu_n)^2 + \mathbb{E}[|Y_n|^4] \\ &= \tilde{u}(Z_n) + \frac{1}{2\epsilon} W_2(\mu, \nu_n)^2 + \mathbb{E}[|Z_n|^4] \leqslant \Psi(Z_n),\end{aligned}$$

proving that $(Y_n)_{n\geqslant 0}$ is also a minimizing sequence of Ψ. Since the lifting $\tilde{u}$ is bounded, we conclude that:

$$\sup_{n\geqslant 0} \int_{\mathbb{R}^d} |x|^4 d\nu_n(x) = \sup_{n\geqslant 0} \mathbb{E}[|Y_n|^4] < \infty,$$

and consequently that the sequence $(\nu_n)_{n\geqslant 0}$ is tight. Extracting a subsequence if necessary, we can assume that this sequence converges (in the sense of weak convergence as well as for the distance W_2 because of the uniform bound on the fourth moments), and we call ν its limit. Notice that:

$$\begin{aligned}u(\nu) &= \lim_{n\to\infty} u(\nu_n) = \lim_{n\to\infty} \tilde{u}(Y_n), \\ W_2(\mu, \nu)^2 &= \lim_{n\to\infty} W_2(\mu, \nu_n)^2 = \lim_{n\to\infty} \mathbb{E}[|X - Y_n|^2],\end{aligned}$$

the first part following from the fact that $\tilde{u}$ is continuous on $L^2(\Omega, \mathcal{F}, \mathbb{P}; \mathbb{R}^d)$. By Fatou's lemma (modulo Skorohod's equivalent form of weak convergence), we also have:

$$\int_{\mathbb{R}^d} |x|^4 d\nu(x) \leqslant \liminf_{n\to\infty} \int_{\mathbb{R}^d} |x|^4 d\nu_n(x) = \liminf_{n\to\infty} \mathbb{E}[|Y_n|^4].$$

Using once more Brenier's Theorem 5.20, we get the existence of a real valued convex function ψ on $\mathbb{R}^d$ such that if we set $Y = \nabla\psi(X)$, then $\mathcal{L}(Y) = \nu$ and $W_2(\mu, \nu)^2 = \mathbb{E}[|X - Y|^2]$. Such a Y is a minimizer of Ψ so that $D\Psi(Y) = 0$ which gives:

$$D\tilde{u}(Y) = -\epsilon^{-1}(Y - X) - 4|Y|^2 Y,$$

which together with the identity $Y = \nabla\psi(X)$, also shows that $D\tilde{u}(Y) \in \sigma\{X\}$. Since the latter is closed in $L^2(\Omega, \mathcal{F}, \mathbb{P}; \mathbb{R}^d)$, we conclude that $D\tilde{u}(X) \in \sigma\{X\}$ by letting $\epsilon \searrow 0$ since Y converges toward X (notice $2\epsilon\Psi(Y) \leqslant 2\epsilon\Psi(X)$), and consequently $D\tilde{u}(Y)$ converges toward $D\tilde{u}(X)$ by continuity of $D\tilde{u}$.

Now if $\mu \in \mathcal{P}_2(\mathbb{R}^d)$ is still absolutely continuous but does not necessarily satisfy the above moment condition, we use X with distribution μ, i.e., such that $\mathcal{L}(X) = \mu$, and we apply the above proof to $X_n = nX/\sqrt{n^2 + |X|^2}$, whose law is absolutely continuous, has moments of all orders, and converges toward X in $L^2(\Omega, \mathcal{F}, \mathbb{P}; \mathbb{R}^d)$. Hence $D\tilde{u}(X_n) \in \sigma\{X_n\} \subset \sigma\{X\}$ and, letting $n \to \infty$, we get $D\tilde{u}(X) \in \sigma\{X\}$, again by continuity of $D\tilde{u}$ and the fact that $\sigma\{X\}$ is closed.

Finally, if $\mu \in \mathcal{P}_2(\mathbb{R}^d)$ is not assumed to be absolutely continuous, we consider a triple of random variables (X, G_1, G_2) with $\mathcal{L}(X) = \mu$, X being independent of (G_1, G_2), G_1 and G_2 being two independent standard d-dimensional Gaussian random variables (recall that we work on an atomless probability space). For each $n \geqslant 1$, we set $X_{i,n} = X + n^{-1}G_i$, $i = 1, 2$. The distribution of $X_{i,n}$ is absolutely continuous so, by what we just saw, $D\tilde{u}(X_{i,n}) \in \sigma\{X, G_i\}$, and taking the limit $n \to \infty$ as before we get $D\tilde{u}(X) \in \sigma\{X, G_i\}$ for $i = 1, 2$. Since G_1 and G_2 are independent, we infer that $D\tilde{u}(X) \in \sigma\{X\}$, which concludes the proof. □

Proposition 5.24 implies that the distribution of the L-derivative of u at μ_0, when viewed as a random variable, depends only upon the law μ_0, and not upon the particular random variable X_0 having distribution μ_0. The Fréchet derivative $[D\tilde{u}](X_0)$ is called the representation of the L-derivative of u at μ_0 along the variable X_0. Since it is viewed as an element of $L^2(\Omega, \mathcal{F}, \mathbb{P}; \mathbb{R}^d)$, by definition,

$$u(\mu) = u(\mu_0) + [D\tilde{u}](X_0) \cdot (X - X_0) + o(\|X - X_0\|_2), \tag{5.33}$$

whenever X and X_0 are random variables with distributions μ and μ_0 respectively, the dot product being the L^2- inner product of $L^2(\Omega, \mathcal{F}, \mathbb{P}; \mathbb{R}^d)$, and $\|\cdot\|_2$ the associated norm. Proposition 5.25 implies that, as a random variable, this Fréchet derivative is of the form $\xi(X_0)$ for some deterministic measurable function $\xi : \mathbb{R}^d \to \mathbb{R}^d$, which is uniquely defined μ_0-almost everywhere on $\mathbb{R}^d$. The equivalence class of ξ in $L^2(\mathbb{R}^d, \mu_0; \mathbb{R}^d)$ being uniquely defined, we can denote it by $\partial_\mu u(\mu_0)$ or $\partial u(\mu_0)$ when no confusion is possible. We shall call $\partial_\mu u(\mu_0)$ the L-derivative of u at μ_0 and most often identify it with a function $\partial_\mu u(\mu_0)(\cdot) : \mathbb{R}^d \ni x \mapsto \partial_\mu u(\mu_0)(x) \in \mathbb{R}^d$ or by $\partial u(\mu_0)(\cdot)$ when no confusion is possible. With this notation, (5.33) takes the form:

$$u(\mu) = u(\mu_0) + \mathbb{E}\big[\partial_\mu u\big(\mathcal{L}(X_0)\big)(X_0) \cdot (X - X_0)\big] + o\big(\|X - X_0\|_2\big). \tag{5.34}$$

Remark 5.26 *The above construction of $\partial_\mu u(\mu_0)$ allows us to express $[D\tilde{u}](X_0)$ as a function of any random variable X_0 with distribution μ_0, wherever this random variable is defined. In particular, the differentiation formulas* (5.33) *and* (5.34) *are invariant under changes of the space $(\Omega, \mathcal{F}, \mathbb{P})$ and the pair of variables (X_0, X) used for the representation of u, in the sense that $[D\tilde{u}](X_0)$ always reads as $\partial_\mu u(\mu_0)(X_0)$, whatever the choices of $(\Omega, \mathcal{F}, \mathbb{P})$ and X_0. So, this construction permits to express $[D\tilde{u}](X_0)$ as a function of any random variable X_0 with distribution μ_0, irrespective of where this random variable is defined, giving a meaning to the* L*-derivative of u at μ_0 independently of the lifting chosen to construct it. In this regard, it is worth noticing that there is no need to assume $(\Omega, \mathcal{F}, \mathbb{P})$ to be Polish and atomless to give a meaning to* (5.34)*, as long as we are able to construct random variables X and X_0 with μ and μ_0 as distributions.*

Remark 5.26 seems pretty obvious. Actually, the proof requires some care since the Landau symbol $o(\cdot)$ in (5.34) a priori depends on the variable X_0 at which the expansion is performed, and thus on the underlying probability space used to construct the lift. In order to proceed, one may let:

$$\vartheta(\pi) = \left(\int_{(\mathbb{R}^d)^2} |x-y|^2 d\pi(x,y)\right)^{-1/2} \times \left(u(\mu) - u(\mu_0) - \int_{(\mathbb{R}^d)^2} \partial_\mu u(\mu_0)(x)\cdot(y-x)\, d\pi(x,y)\right),$$

for any $\pi \in \mathcal{P}_2((\mathbb{R}^d)^2)$ such that $\int_{(\mathbb{R}^d)^2} |x-y|^2 d\pi(x,y) \neq 0$, where μ_0 denotes the first marginal of π on $\mathbb{R}^d$. When $\int_{(\mathbb{R}^d)^2} |x-y|^2 d\pi(x,y) = 0$, we just let $\vartheta(\pi) = 0$.

Expansion (5.34) says that there exists a probability space $(\Omega, \mathcal{F}, \mathbb{P})$ such that, for any $X_0 \in L^2(\Omega, \mathcal{F}, \mathbb{P}; \mathbb{R}^d)$,

$$\lim_{X\to X_0} \vartheta\big(\mathcal{L}(X_0, X)\big) = 0,$$

the limit in the left-hand side holding true in $L^2(\Omega, \mathcal{F}, \mathbb{P}; \mathbb{R}^d)$. We shall prove in Subsection 5.3.1 below, see Lemma 5.30, that this implies:

$$\vartheta(\pi) \to 0 \quad \text{as} \quad \int_{(\mathbb{R}^d)^2} |x-y|^2 d\pi(x,y) \to 0,$$

as long as the first marginal of π on $\mathbb{R}^d$ remains fixed, which shows rigorously that (5.34) may be transferred from one given probability space to any other as explained in Remark 5.26.

The same argument shows that if the Fréchet derivative of the lift $\tilde{u}$, when constructed on some probability space, is continuous, then the same holds true on any other sufficiently rich probability space. Indeed, it suffices to apply the same argument as above but with ϑ given by:

$$\vartheta(\pi) = \int_{(\mathbb{R}^d)^2} \big|\partial_\mu u(\mu)(y) - \partial_\mu u(\mu_0)(x)\big|^2 d\pi(x,y),$$

where μ stands for the second marginal of π on $\mathbb{R}^d$.

Remark 5.27 *Let us assume that $u : \mathcal{P}_2(\mathbb{R}^d) \to \mathbb{R}$ is c-Lipschitz continuous in μ with respect to the 2-Wasserstein distance. If u is continuously* L*-differentiable, then necessarily it holds $\mathbb{E}[|\partial_\mu u(\mu)(X)|^2]^{1/2} \leqslant c$, for any $\mu \in \mathcal{P}_2(\mathbb{R}^d)$ and any random variable X having distribution μ. Indeed, if $\tilde{u}$ is a lifting of u, $\mathcal{L}(X) = \mu$ and $\mathcal{L}(Y) = \nu$ for $\mu, \nu \in \mathcal{P}_2(\mathbb{R}^d)$, then we have:*

$$
\begin{aligned}
|D\tilde{u}(X)\cdot(Y-X)| &\leqslant |\tilde{u}(Y)-\tilde{u}(X)| + o\big(\|Y-X\|_2\big) \\
&= |u(\nu)-u(\mu)| + o\big(\|Y-X\|_2\big) \\
&\leqslant cW_2(\nu,\mu) + o\big(\|Y-X\|_2\big) \\
&\leqslant c\|Y-X\|_2 + o\big(\|Y-X\|_2\big).
\end{aligned}
$$

Dividing by $\|Y-X\|_2$ (assuming that $Y \neq X$) and letting $\|Y-X\|_2$ tend to zero, we get the desired result.

Of course, the argument works the other way around. Indeed, if u is continuously L-differentiable and $D\tilde{u}$ is bounded by C in L^2, then for any two random variables X and X' in $L^2(\Omega, \mathcal{F}, \mathbb{P}; \mathbb{R}^d)$, we can find $t \in [0,1]$ such that

$$
\begin{aligned}
&|u(\mathcal{L}(X)) - u(\mathcal{L}(X'))| \\
&= \big|\mathbb{E}\big[\partial_\mu u\big(\mathcal{L}(tX+(1-t)X')\big)\big(tX+(1-t)X'\big)(X-X')\big]\big| \\
&\leqslant \big\|\partial_\mu u\big(\mathcal{L}(tX+(1-t)X')\big)\big(tX+(1-t)X'\big)\big\|_2 \|X-X'\|_2 \\
&\leqslant C\|X-X'\|_2.
\end{aligned}
$$

Taking the infimum over the random variables X and X' with prescribed marginal distributions, we deduce that u is Lipschitz continuous with respect to the 2-Wasserstein distance.

5.2.2 Examples

We illustrate the structure of this peculiar form of differentiation with a couple of fundamental examples which we shall use throughout the book.

Example 1. It is plain to compute the L-derivative of a linear function, namely when the function u is of the first order form:

$$
u(\mu) = \int_{\mathbb{R}^d} h(x)d\mu(x) = \langle h, \mu\rangle, \tag{5.35}
$$

for some scalar continuously differentiable function h defined on $\mathbb{R}^d$, whose derivative is at most of linear growth. Indeed, in this case, the lifted function $\tilde{u}$ is defined by $\tilde{u}(X) = \mathbb{E}[h(X)]$ and

$$
\begin{aligned}
\tilde{u}(X+Y) &= \tilde{u}(X) + \mathbb{E}\int_0^1 \big[\partial h(X+\lambda Y)\cdot Y\big]d\lambda \\
&= \tilde{u}(X) + \mathbb{E}\big[\partial h(X)\cdot Y\big] + \mathbb{E}\int_0^1 \big[\big(\partial h(X+\lambda Y) - \partial h(X)\big)\cdot Y\big]d\lambda.
\end{aligned}
$$

Observing that the last term in the right-hand side is less than:

$$
\begin{aligned}
&\Big|\mathbb{E}\int_0^1 \big[\big(\partial h(X+\lambda Y) - \partial h(X)\big)\cdot Y\big]d\lambda\Big| \\
&\leqslant \int_0^1 \mathbb{E}\Big[\sup_{|y|\leqslant \|Y\|_2^{1/2}} \big|\partial h(X+\lambda y) - \partial h(X)\big|\,|Y|\Big]d\lambda \\
&\quad + C\mathbb{E}\Big[\mathbf{1}_{\{|Y|\geqslant \|Y\|_2^{1/2}\}}\big(1+|X|+|Y|\big)|Y|\Big] \\
&\leqslant \mathbb{E}\Big[\sup_{|y|\leqslant \|Y\|_2^{1/2}} \big|\partial h(X+y) - \partial h(X)\big|^2\Big]^{1/2}\|Y\|_2 \\
&\quad + C\|Y\|_2\Big(\|Y\|_2 + \sqrt{\|Y\|_2} + \sup_{\mathbb{P}(A)\leqslant \|Y\|_2} \mathbb{E}\big[|X|^2\mathbf{1}_A\big]^{1/2}\Big),
\end{aligned}
$$

where the constant C is connected with the growth of ∂h, it is easy to check that the Fréchet derivative of $\tilde{u}$ at X is given by $\partial h(X)$ (where ∂h is the classical gradient of h) viewed as an element of the dual since $D\tilde{u}(X)\cdot Y = \mathbb{E}[\partial h(X)\cdot Y]$. Consequently, we can think of $\partial_\mu u(\mu)$ as the deterministic function ∂h. Example (5.35) highlights the fact that this notion of L-differentiability is very different from the usual one. Indeed, given the fact that the function u defined by (5.35) is linear in the measure variable μ, when viewed as an element of the dual of a function space, one should expect the derivative to be h and NOT its derivative ∂h! We shall revisit this issue in Section 5.4 where we show that this apparent anomaly is in fact generic. We shall use this particular example to derive, from the general Pontryagin principle for the optimal control of McKean-Vlasov diffusion processes which we prove in Chapter 6, a simple form applicable to scalar interactions which are given by functions of measures of this type.

Example 2. Next, we consider the example of a quadratic function of a measure. This example will be used when we generalize the notion of potential mean field game. To be specific, we assume that the function u is of the form:

$$
u(\mu) = \langle h*\mu, \mu\rangle = \int_{\mathbb{R}^d} [h*\mu](x)d\mu(x) = \int_{\mathbb{R}^d}\int_{\mathbb{R}^d} h(x-y)d\mu(y)d\mu(x), \qquad (5.36)
$$

for some continuously differentiable function h on $\mathbb{R}^d$, whose derivative is at most of linear growth. We find it convenient to lift this function into the function $\tilde{u}$ defined by:

$$
\tilde{u}(X) = \int_{\mathbb{R}^d} \mathbb{E}[h(x-X)]\, d\mathbb{P}_X(x),
$$

where $\mathbb{P}_X$ denotes the distribution of X, which we usually denote by $\mathcal{L}(X)$. Indeed, with this lifting, we can compute the Gâteaux derivative of $\tilde{u}$ as follows. For $U \in L^2(\Omega, \mathcal{F}, \mathbb{P}; \mathbb{R}^d)$ with $\|U\|_2 \leqslant 1$, we have:

$$
\begin{aligned}
&[\tilde{u}(X + \epsilon U) - \tilde{u}(X)] \\
&= \int_{\mathbb{R}^d} \mathbb{E}[h(x - X - \epsilon U)] \, d\mathbb{P}_{X+\epsilon U}(x) - \int_{\mathbb{R}^d} \mathbb{E}[h(x - X)] \, d\mathbb{P}_X(x) \\
&= \int_{\mathbb{R}^d} \mathbb{E}[h(x - X)] \, d\mathbb{P}_{X+\epsilon U}(x) - \int_{\mathbb{R}^d} \mathbb{E}[h(x - X)] \, d\mathbb{P}_X(x) \\
&\quad + \int_{\mathbb{R}^d} \mathbb{E}[h(x - X - \epsilon U) - h(x - X)] \, d\mathbb{P}_{X+\epsilon U}(x) \\
&= (i) + (ii).
\end{aligned}
$$

Using Fubini's theorem, we notice that:

$$
\begin{aligned}
\lim_{\epsilon \searrow 0} \frac{1}{\epsilon}(i) &= \lim_{\epsilon \searrow 0} \frac{1}{\epsilon}\left[\int_{\mathbb{R}^d} \mathbb{E}[h(X + \epsilon U - x)] \, d\mathbb{P}_X(x) - \int_{\mathbb{R}^d} \mathbb{E}[h(x - X)] \, d\mathbb{P}_X(x) \right] \\
&= \int_{\mathbb{R}^d} \mathbb{E}[\partial h(X - x) \cdot U] \, d\mathbb{P}_X(x).
\end{aligned}
$$

Moreover, it is clear that:

$$
\lim_{\epsilon \searrow 0} \frac{1}{\epsilon}(ii) = -\int_{\mathbb{R}^d} \mathbb{E}[\partial h(x - X) \cdot U] \, d\mathbb{P}_X(x),
$$

from which we conclude that the Gâteaux derivative of the function $\tilde{u}$ at X with distribution μ in the direction U is given by $\mathbb{E}\{[(\partial h + \bar{\partial h}) * \mu](X) \cdot U\}$ if we use the notation $\bar{f}$ to denote the function $\bar{f}(x) = -f(-x)$. Since the mapping $L^2(\Omega, \mathcal{F}, \mathbb{P}; \mathbb{R}^d) \ni X \mapsto [(\partial h + \bar{\partial h}) * (\mathcal{L}(X))](X) \in L^2(\Omega, \mathcal{F}, \mathbb{P}; \mathbb{R}^d)$ is continuous, we deduce that $\tilde{u}$ is Fréchet differentiable and that the Fréchet derivative at X with distribution μ is given by $[(\partial h + \bar{\partial h}) * \mu](X)$.

Notice that when h is even (i.e., when $h(x) = h(-x)$), then ∂h is odd (i.e., $\partial h(-x) = -\partial h(x)$) and the derivative is given by $2\partial h * \mu$ or:

$$
\partial_\mu u(\mu)(\cdot) = \big(2[\partial h] * \mu\big)(\cdot).
$$

Example 3. We now consider a slight generalization of the above example:

$$
u(\mu) = \int_{\mathbb{R}^d} v(x, \mu) d\mu(x),
$$

for some continuous function $\mathbb{R}^d \times \mathcal{P}_2(\mathbb{R}^d) \ni (x, \mu) \mapsto v(x, \mu) \in \mathbb{R}$. We assume that v is differentiable in $x \in \mathbb{R}^d$ for μ fixed, the derivative being jointly continuous in (x, μ), and at most of linear growth in x, uniformly in μ in bounded subsets of

$\mathcal{P}_2(\mathbb{R}^d)$. A subset $\mathcal{K}$ of $\mathcal{P}_2(\mathbb{R}^d)$ is said to be bounded if there exists $a > 0$ such that for all $\mu \in \mathcal{K}$, $\int_{\mathbb{R}^d} |x|^2 d\mu(x) \leqslant a$. We also assume that v is L-continuously differentiable in μ for x fixed, and that for each $\mu \in \mathcal{P}_2(\mathbb{R}^d)$, we can choose, for each $x \in \mathbb{R}^d$, a version of $\partial_\mu v(x,\mu)(\cdot)$ in $L^2(\mathbb{R}^d,\mu;\mathbb{R}^d)$ in such a way that the mapping $\mathbb{R}^d \times \mathbb{R}^d \ni (x,x') \mapsto \partial_\mu v(x,\mu)(x')$ is measurable and at most of linear growth, uniformly in μ when restricted to bounded subsets of $\mathcal{P}_2(\mathbb{R}^d)$.

Observe that v is at most of quadratic growth in x, uniformly in μ in bounded subsets, proving that u is well defined. Indeed, for $X \in L^2(\Omega,\mathcal{F},\mathbb{P};\mathbb{R}^d)$ with $\mathcal{L}(X) = \mu$:

$$
\begin{aligned}
&v(x,\mu) \\
&= v(0,\mu) + \int_0^1 \partial_x v(\lambda x,\mu) \cdot x \, d\lambda \\
&= v(0,\delta_0) + \int_0^1 \mathbb{E}\big[\partial_\mu v\big(0,\mathcal{L}(\lambda X)\big)(\lambda X) \cdot X\big] \, d\lambda + \int_0^1 \partial_x v(\lambda x,\mu) \cdot x \, d\lambda .
\end{aligned}
$$

Using the growth condition on $\partial_x v$ and $\partial_\mu v$, the claim easily follows.

In full analogy with the previous example, we now claim that:

$$
\partial_\mu u(\mu)(\cdot) = \partial_x v(\cdot,\mu) + \int_{\mathbb{R}^d} \partial_\mu v(x',\mu)(\cdot) d\mu(x'). \tag{5.37}
$$

For the proof, we introduce an approach which we shall use repeatedly throughout the chapter. We denote by $(\tilde{\Omega},\tilde{\mathcal{F}},\tilde{\mathbb{P}})$ a copy of the space $(\Omega,\mathcal{F},\mathbb{P})$, and we use the following convention. For any random variable $X \in L^2(\Omega,\mathcal{F},\mathbb{P};\mathbb{R}^d)$, we denote by $\tilde{X}$ the copy of X on $(\tilde{\Omega},\tilde{\mathcal{F}},\tilde{\mathbb{P}})$. Then, for $X, Y \in L^2(\Omega,\mathcal{F},\mathbb{P};\mathbb{R}^d)$ with $\mathcal{L}(X) = \mu$ and $\|Y\|_2 \leqslant 1$:

$$
\begin{aligned}
&[\tilde{u}(X+Y) - \tilde{u}(X)] \\
&= \mathbb{E}\big[v\big(X+Y,\mathcal{L}(X+Y)\big) - v\big(X,\mathcal{L}(X+Y)\big)\big] \\
&\quad + \mathbb{E}\big[v\big(X,\mathcal{L}(X+Y)\big) - v\big(X,\mathcal{L}(X)\big)\big] \\
&= \mathbb{E}\big[\partial_x v\big(X,\mathcal{L}(X)\big) \cdot Y\big] + \mathbb{E}\tilde{\mathbb{E}}\Big[\partial_\mu v\big(X,\mathcal{L}(X)\big)(\tilde{X}) \cdot \tilde{Y}\Big] \\
&\quad + \mathbb{E}\big[v\big(X+Y,\mathcal{L}(X+Y)\big) - v\big(X,\mathcal{L}(X+Y)\big) - \partial_x v\big(X,\mathcal{L}(X)\big) \cdot Y\big] \\
&\quad + \mathbb{E}\tilde{\mathbb{E}}\Big[v\big(X,\mathcal{L}(X+Y)\big) - v\big(X,\mathcal{L}(X)\big) - \partial_\mu v\big(X,\mathcal{L}(X)\big)(\tilde{X}) \cdot \tilde{Y}\Big] \\
&= \mathbb{E}\big[\partial_x v\big(X,\mathcal{L}(X)\big) \cdot Y\big] + \mathbb{E}\tilde{\mathbb{E}}\Big[\partial_\mu v\big(X,\mathcal{L}(X)\big)(\tilde{X}) \cdot \tilde{Y}\Big] \\
&\quad + (i) + (ii).
\end{aligned}
$$

By repeating *mutatis mutandis* the proof of the first example, we get:

$$
\begin{aligned}
(i) &= \mathbb{E}\Big[\int_0^1 \Big(\partial_x v\big(X+\lambda Y, \mathcal{L}(X+Y)\big) - \partial_x v\big(X, \mathcal{L}(X)\big)\Big)\cdot Y d\lambda\Big] \\
&\leqslant \mathbb{E}\Big[\sup_{|y| \leqslant \|Y\|_2^{1/2}} \sup_{\nu : W_2(\mu,\nu) \leqslant \|Y\|_2} \big|\partial_x v(X+y,\nu) - \partial_x v(X,\mu)\big|^2\Big]^{1/2} \|Y\|_2 \\
&\quad + C\|Y\|_2\Big(\|Y\|_2 + \sqrt{\|Y\|_2} + \sup_{\mathbb{P}(A)\leqslant \|Y\|_2} \mathbb{E}\big[|X|^2 \mathbf{1}_A\big]^{1/2}\Big),
\end{aligned}
$$

where C may depend on $M_2(\mu)$, which shows that $(i) = o(\|Y\|_2)$. Regarding (ii), we have:

$$
\begin{aligned}
(ii) &= \mathbb{E}\tilde{\mathbb{E}}\Big[\int_0^1 \Big(\partial_\mu v\big(X, \mathcal{L}(X+\lambda Y)\big)(\tilde{X}+\lambda\tilde{Y}) - \partial_\mu v\big(X, \mathcal{L}(X)\big)(\tilde{X})\Big)\cdot \tilde{Y} d\lambda\Big] \\
&\leqslant \|Y\|_2 \mathbb{E}\Big[\sup_{\|Z\|_2 \leqslant \|Y\|_2} \tilde{\mathbb{E}}\Big(\big|\partial_\mu v\big(X, \mathcal{L}(X+Z)\big)(\tilde{X}+\tilde{Z}) - \partial_\mu v\big(X, \mathcal{L}(X)\big)(\tilde{X})\big|^2\Big)\Big]^{1/2}.
\end{aligned}
$$

Without any loss of generality, we may assume that $L^2(\Omega, \mathcal{F}, \mathbb{P}; \mathbb{R}^d)$ is separable, which is the case if $\mathcal{F}$ is a countably generated σ-field or the completion of a countably generated σ-field. Also, by assumption, the function $\tilde{Z} \mapsto \partial_\mu v(x, \mathcal{L}(\tilde{Z}))(\tilde{Z})$ from $L^2(\tilde{\Omega}, \tilde{\mathcal{F}}, \tilde{\mathbb{P}}; \mathbb{R}^d)$ into itself is continuous for each $x \in \mathbb{R}^d$. By the growth condition imposed on $\partial_\mu v$, the above supremum reduces to a supremum over a countable family and is a measurable function of X, and (ii) is in fact $o(\|Y\|_2)$, proving (5.37). The proof is easily completed.

Example 4. As a particular case of the above example, we may choose:

$$
v(x,\mu) = \int_{\mathbb{R}^d} g(x,x') d\mu(x'), \quad x \in \mathbb{R}^d, \ \mu \in \mathcal{P}_2(\mathbb{R}^d),
$$

for a function $g : \mathbb{R}^d \times \mathbb{R}^d \to \mathbb{R}$. If g is continuously differentiable in (x, x'), with $\partial_x g$ and $\partial_{x'} g$ being at most of linear growth in (x, x'), then v satisfies the assumption of Example 3 and the L-derivative of the function $u : \mathcal{P}_2(\mathbb{R}^d) \ni \mu \mapsto u(\mu) = \int_{\mathbb{R}^d} v(x,\mu) d\mu(x)$ writes:

$$
\partial_\mu u(\mu)(x) = \int_{\mathbb{R}^d} \partial_x g(x,x') d\mu(x') + \int_{\mathbb{R}^d} \partial_{x'} g(x',x) d\mu(x'),
$$

for $x \in \mathbb{R}^d$, $\mu \in \mathcal{P}_2(\mathbb{R}^d)$. Indeed, by Lebesgue's differentiation theorem, we have

$$
\partial_x v(x,\mu) = \int_{\mathbb{R}^d} \partial_x g(x,x') d\mu(x'),
$$

which is jointly continuous in (x, μ) and at most of linear growth in x, uniformly in μ in bounded subsets. Moreover,

$$\partial_\mu v(x, \mu)(x') = \partial_{x'} g(x, x'),$$

which also satisfies the prescribed conditions.

Example 5. Finally we consider a slightly different challenge than the four examples considered above. We assume that $\mathbb{R}^d \times \mathcal{P}_2(\mathbb{R}^d) \ni (x, \mu) \mapsto v(x, \mu) \in \mathbb{R}^d$ satisfies the same assumption as in Example 3, and we define the function u by:

$$L^2(\Omega, \mathcal{F}, \mathbb{P}; \mathbb{R}^d) \ni X \mapsto u(X) = v(X, \mathcal{L}(X)) \in L^1(\Omega, \mathcal{F}, \mathbb{P}; \mathbb{R}),$$

and we compute the Fréchet derivative $Du(X)$ of u at $X \in L^2(\Omega, \mathcal{F}, \mathbb{P}; \mathbb{R}^d)$, which should be interpreted as a bounded operator $Du(X)(\cdot)$ from $L^2(\Omega, \mathcal{F}, \mathbb{P}; \mathbb{R}^d)$ into $L^1(\Omega, \mathcal{F}, \mathbb{P}; \mathbb{R})$. For any $Y \in L^2(\Omega, \mathcal{F}, \mathbb{P}; \mathbb{R}^d)$, with $\|Y\|_2 \leqslant 1$, we have:

$$\begin{aligned} u(X + \epsilon Y) - u(X) = v\big(X + \epsilon Y, \mathcal{L}(X + \epsilon Y)\big) - v\big(X, \mathcal{L}(X + \epsilon Y)\big) \\ + v\big(X, \mathcal{L}(X + \epsilon Y)\big) - v\big(X, \mathcal{L}(X)\big). \end{aligned}$$

Repeating the computations of Example 3, it is easy to see that:

$$Du(X)(Y) = \partial_x v(X, \mathcal{L}(X)) \cdot Y + \tilde{\mathbb{E}}\big[\partial_\mu v(X, \mathcal{L}(X))(\tilde{X}) \cdot \tilde{Y}\big].$$

Remark 5.28 *The result of Proposition 5.36 below shows that, under a mild regularity assumption on the Fréchet derivatives, the differentials constructed above can be handled by rather regular versions. Indeed, if the function $\tilde{u}$ is Fréchet differentiable and its Fréchet derivative is uniformly Lipschitz, i.e., there exists a constant $c > 0$ such that $\|D\tilde{u}(X) - D\tilde{u}(X')\|_2 \leqslant c\|X - X'\|_2$ for all X, X' in $L^2(\Omega, \mathcal{F}, \mathbb{P}; \mathbb{R}^d)$, then there exists a function $\partial_\mu u$:*

$$\mathcal{P}_2(\mathbb{R}^d) \times \mathbb{R}^d \ni (\mu, x) \mapsto \partial_\mu u(\mu)(x)$$

such that $|\partial_\mu u(\mu)(x) - \partial_\mu u(\mu)(x')| \leqslant c|x - x'|$ for all $x, x' \in \mathbb{R}^d$ and $\mu \in \mathcal{P}_2(\mathbb{R}^d)$, and for every $\mu \in \mathcal{P}_2(\mathbb{R}^d)$, $\partial_\mu u(\mu)(X) = D\tilde{u}(X)$ almost surely if $\mu = \mathcal{L}(X)$.

5.3 Regularity Properties of the L-Differentials

L-derivatives of functions of probability measures are defined in an L^2-sense and a modicum of care is needed to handle sets of measure zero if one tries to manipulate versions of these derivatives which could be defined everywhere, hopefully keeping some form of regularity. The beginning of this section collects a couple of results in this direction.

Because of the technical nature of some of these results, the reader may want to skip the proofs in a first reading and jump to the more intuitive and enlightening results of this section.

5.3.1 Measure Theoretical Preliminaries

Representation of Random Variables

The first result will be used several times in the book, and its statement is influenced by a couple of applications in Chapter (Vol II)-1.

Lemma 5.29 *There exists a measurable mapping $\psi : [0,1) \times \mathcal{P}_2(\mathbb{R}^d) \ni (\eta,\mu) \mapsto \psi(\eta,\mu) \in \mathbb{R}^d$ such that, for any $\mu \in \mathcal{P}_2(\mathbb{R}^d)$, the image of the Lebesgue measure on $[0,1)$ by $\psi(\cdot,\mu)$ is μ itself. Furthermore, for every atomless probability space $(\Omega,\mathcal{F},\mathbb{P})$, there exists a measurable mapping $\mathcal{P}_2(\mathbb{R}^d) \ni \mu \mapsto X^\mu \in L^2(\Omega,\mathcal{F},\mathbb{P};\mathbb{R}^d)$ such that $X^\mu \sim \mu$ for every $\mu \in \mathcal{P}_2(\mathbb{R}^d)$.*

Recall that $L^2(\Omega,\mathcal{F},\mathbb{P};\mathbb{R}^d)$ is the quotient of the space of square-integrable random variables for $\mathbb{P}$ almost sure equality.

Proof.

First Step. We first construct $\psi(\cdot,\mu)$ for measures μ concentrated on the unit cube in the sense that $\mu([0,1)^d) = 1$. We call $\mathcal{U}$ the set of such measures. Observe from Proposition 5.7 that $\mathcal{U}$ is a Borel subset of $\mathcal{P}_2(\mathbb{R}^d)$.

Given some $n \geqslant 0$, we split the hypercube $[0,1)^d$ into $(2^n)^d$ hypercubes of the form $Q^n(k_1,\cdots,k_d) = \prod_{i=1}^d [k_i/2^n,(k_i+1)/2^n)$, with $(k_1,\cdots,k_d) \in (\mathbb{Z} \cap [0,2^n-1])^d$. For any d-tuple $(k_1,\cdots,k_d)$, we let $M^{n,\mu}(k_1,\cdots,k_d) = \mu(Q^n(k_1,\cdots,k_d))$.

The strategy is to arrange the $Q^n(k_1,\cdots,k_d)$ increasingly according to some order. To this end, we observe that, for any $1 \leqslant i \leqslant d$, $k_i/2^n$ may be uniquely written as:

$$\frac{k_i}{2^n} = \sum_{j=1}^n \frac{\varepsilon_j^n(k_i)}{2^j}, \tag{5.38}$$

with $\varepsilon_j^n(k_i) \in \{0,1\}$. Given $(k_1,\cdots,k_d)$ and $(k'_1,\cdots,k'_d)$ in $(\mathbb{Z} \cap [0,2^n-1])^d$, with $(k_1,\cdots,k_d) \neq (k'_1,\cdots,k'_d)$, we say that $(k_1,\cdots,k_d) \prec_n (k'_1,\cdots,k'_d)$ if, letting:

$$p = \inf\big\{j \in \{1,\cdots,n\} : \big(\varepsilon_j^n(k_1),\cdots,\varepsilon_j^n(k_d)\big) \neq \big(\varepsilon_j^n(k'_1),\cdots,\varepsilon_j^n(k'_d)\big)\big\},$$

$$q = \inf\big\{i \in \{1,\cdots,d\} : \varepsilon_p^n(k_i) \neq \varepsilon_p^n(k'_i)\big)\big\},$$

it holds $0 = \varepsilon_p^n(k_q) < \varepsilon_p^n(k'_q) = 1$. In other words, the order is defined by taking into account first the index j in (5.38) and then the coordinate i. Writing $x \preceq_n y$ if $x \prec_n y$ or $x = y$, $\preceq_n$ is a total order on $\{0,\cdots,2^n-1\}^d$.

We divide the interval $[0,1)$ into a sequence $(I^{n,\mu}(k_1,\cdots,k_d))_{(k_1,\cdots,k_d)\in(\mathbb{Z}\cap[0,2^n-1])^d}$ of $(2^n)^d$ disjoint (possibly empty) intervals, closed on the left and open on the right, of length $M^{n,\mu}(k_1, \cdots, k_d)$, and ordered increasingly according to $\prec_n$. This means that, for any $x \in I^{n,\mu}(k_1, \cdots, k_d)$ and $x' \in I^{n,\mu}(k_1', \cdots, k_d')$, $x < x'$ if $(k_1, \cdots, k_d) \prec_n (k_1', \cdots, k_d')$. Then, we let:

$$\psi_n(\eta, \mu) = \sum_{(k_1,\cdots,k_d) \in (\mathbb{Z} \cap [0,2^n-1])^d} (2^{-n}k_1, \cdots, 2^{-n}k_d)\mathbf{1}_{\{\eta \in I^{n,\mu}(k_1,\cdots,k_d)\}}, \quad \eta \in [0, 1).$$

It is plain to check that the mapping $[0, 1) \times \mathcal{U} \ni (\eta, \mu) \mapsto \psi_n(\eta, \mu) \in [0, 1)^d$ is jointly measurable. Indeed, writing $I^{n,\mu}(k_1, \cdots, k_d)$ as $[a^{n,\mu}(k_1, \cdots, k_d), b^{n,\mu}(k_1, \cdots, k_d))$, the mapping $\mathcal{U} \ni \mu \mapsto (a^{n,\mu}(k_1, \cdots, k_d), b^{n,\mu}(k_1, \cdots, k_d))$ is measurable, since:

$$a^{n,\mu}(k_1, \cdots, k_d) = \mu\Bigl(\bigcup_{(k_1',\cdots,k_d') \prec_n (k_1,\cdots,k_d)} Q^n(k_1', \cdots, k_d') \Bigr),$$

$$b^{n,\mu}(k_1, \cdots, k_d) = \mu\Bigl(\bigcup_{(k_1',\cdots,k_d') \preceq_n (k_1,\cdots,k_d)} Q^n(k_1', \cdots, k_d') \Bigr).$$

Then, we notice that for any bounded and continuous function ℓ:

$$\begin{aligned} \int_0^1 \ell(\psi_n(\eta, \mu))d\eta &= \sum_{(k_1,\cdots,k_d) \in (\mathbb{Z} \cap [0,2^n-1])^d} \ell\Bigl(\frac{k_1}{2^n}, \cdots, \frac{k_d}{2^n}\Bigr)\mu(Q^n(k_1, \cdots, k_d)) \\ &\to \int_{\mathbb{R}^d} \ell(x)d\mu(x), \end{aligned}$$

proving that on $[0, 1)$ equipped with the Lebesgue measure, the sequence of random variables $([0, 1) \ni \eta \mapsto \psi_n(\eta, \mu))_{n \geqslant 0}$ converges in distribution to μ as n tends to $+\infty$. Moreover, because of our choice of ordering, we have:

$$I^{n,\mu}(k_1, \cdots, k_d) = \bigcup_{(\epsilon_1,\cdots,\epsilon_d) \in \{0,1\}^d} I^{n+1,\mu}(2k_1 + \epsilon_1, \cdots, 2k_d + \epsilon_d). \tag{5.39}$$

Indeed, for any $(k_1, \cdots, k_d)$ and $(k_1', \cdots, k_d')$ in $\{0, \cdots, 2^n - 1\}^d$, $(k_1', \cdots, k_d') \prec_n (k_1, \cdots, k_d)$ if and only if $(2k_1' + \epsilon_1', \cdots, 2k_d' + \epsilon_d') \prec_{n+1} (2k_1, \cdots, 2k_d)$ for any $\epsilon_1, \epsilon_1', \cdots, \epsilon_d, \epsilon_d' \in \{0, 1\}$, which implies that:

$$\begin{aligned} a^{n,\mu}(k_1, \cdots, k_d) &= \mu\Bigl(\bigcup_{(k_1',\cdots,k_d') \prec_{n+1} (2k_1,\cdots,2k_d)} Q^{n+1}(k_1', \cdots, k_d') \Bigr) \\ &= a^{n+1,\mu}(2k_1, \cdots, 2k_d) \\ &\leqslant \inf_{(\epsilon_1,\cdots,\epsilon_d) \in \{0,1\}^d} a^{n+1,\mu}(2k_1 + \epsilon_1, \cdots, 2k_d + \varepsilon_d), \end{aligned}$$

where we observed that $(2k_1, \cdots, 2k_d) \preceq_{n+1} (2k_1 + \epsilon_1, \cdots, 2k_d + \epsilon_d)$. Similarly,

$$
\begin{aligned}
b^{n,\mu}(k_1,\cdots,k_d) \\
&= \mu\Big(\Big(\bigcup_{(k_1',\cdots,k_d')\prec_{n+1}(2k_1,\cdots,2k_d)}\Big) \\
&\qquad \cup \Big(\bigcup_{(2k_1,\cdots,2k_d)\preceq_{n+1}(k_1',\cdots,k_d')\preceq_{n+1}(2k_1+1,\cdots,2k_d+1)} Q^n(k_1',\cdots,k_d')\Big)\Big) \\
&\geqslant \sup_{(\epsilon_1,\cdots,\epsilon_d)\in\{0,1\}^d} b^{n+1,\mu}(2k_1+\epsilon_1,\cdots,2k_d+\epsilon_d).
\end{aligned}
$$

This proves that in (5.39), the right-hand side is included in the left-hand side. Observing that both sides are intervals of the same length (closed on the left and open on the right), this proves the equality.

As a by-product, there exists a constant C, independent of n and μ, such that:

$$
\forall \eta \in [0,1), \quad |\psi_n(\eta,\mu) - \psi_{n+1}(\eta,\mu)| \leqslant \frac{C}{2^n}. \tag{5.40}
$$

We deduce that, for each $\mu \in \mathcal{U}$, the random variables $([0,1) \ni \eta \mapsto \psi_n(\eta,\mu))_{n\geqslant 0}$ converge pointwise. So, we can define:

$$
\psi_\infty(\eta,\mu) = \lim_{n\to\infty} \psi_n(\eta,\mu).
$$

The function $[0,1) \times \mathcal{U} \ni (\eta,\mu) \mapsto \psi_\infty(\eta,\mu)$ is jointly measurable. Moreover, for any $\mu \in \mathcal{U}$, the random variable $[0,1) \ni \eta \mapsto \psi_\infty(\eta,\mu)$ has μ as distribution.

Second Step. When the support of μ is general, we define $\mu \circ \phi^{-1}$ as the push-forward (or image) of μ by the mapping:

$$
\phi(x_1,\cdots,x_d) = \Big(\frac{1}{\pi}\arctan(x_1) + \frac{1}{2},\cdots,\frac{1}{\pi}\arctan(x_d) + \frac{1}{2}\Big), \tag{5.41}
$$

for $x_1,\cdots,x_d \in \mathbb{R}^d$. Observing that:

$$
\phi^{-1}(u_1,\cdots,u_d) = \Big(\tan\big(\pi u_1 - \frac{\pi}{2}\big),\cdots,\tan\big(\pi u_d - \frac{\pi}{2}\big)\Big), \quad (u_1,\cdots,u_d) \in (0,1)^d,
$$

we then let:

$$
\psi(\eta,\mu) = \phi^{-1}\Big(\psi_\infty(\eta,\mu\circ\phi^{-1})\Big), \quad \eta \in [0,1),\ \mu \in \mathcal{P}_2(\mathbb{R}^d).
$$

By construction, the random variable $[0,1) \ni \eta \mapsto \psi_\infty(\eta,\mu\circ\phi^{-1})$ has $\mu\circ\phi^{-1}$ as distribution if as before we assume that $[0,1)$ is equipped with the Lebesgue measure to form a probability space. Therefore, $[0,1) \ni \eta \mapsto \psi(\eta,\mu)$ has μ as distribution. Moreover, by using Proposition 5.7 once again, we can easily verify that the mapping $[0,1) \times \mathcal{P}_2(\mathbb{R}^d) \ni (\eta,\mu) \mapsto \psi(\eta,\mu)$ is measurable.

Third Step. In order to complete the proof, we consider an atomless probability space $(\Omega, \mathcal{F}, \mathbb{P})$ equipped with a uniformly distributed random variable $\xi : \Omega \to [0, 1)$. First, from the first two steps of the proof, we know that for any $\mu \in \mathcal{P}_2(\Omega)$, the random variable $\Omega \ni \omega \mapsto \psi(\xi(\omega), \mu)$ has distribution μ. Next, we argue that the mapping $\mathcal{P}_2(\mathbb{R}^d) \ni \mu \mapsto \mathrm{cl}(\Omega \ni \omega \mapsto \psi(\xi(\omega), \mu)) \in L^2(\Omega, \mathcal{F}, \mathbb{P}; \mathbb{R}^d)$ is measurable, where we use the notation $\mathrm{cl}(\zeta)$ for the equivalence class of the random variable ζ for the $\mathbb{P}$-almost sure equality.

For the proof, we shall use the fact that for a bounded continuous function $\ell : \mathbb{R}^d \times \mathbb{R}^d \to \mathbb{R}^d$, the mapping $L^2(\Omega, \mathcal{F}, \mathbb{P}; \mathbb{R}^d) \times \mathbb{R}^d \ni (X, x) \mapsto \mathrm{cl}(\ell(X, x)) \in L^2(\Omega, \mathcal{F}, \mathbb{P}; \mathbb{R}^d)$ is continuous and thus measurable. Measurability is preserved by replacing ℓ by $(\mathbf{1}_{I_1} \circ \ell, \cdots, \mathbf{1}_{I_d} \circ \ell)$ for a collection of d intervals $I_1, \cdots, I_d$. Recalling the definition of ψ_n in the first step of the proof, we deduce that the map $\mathcal{P}_2(\mathbb{R}^d) \ni \mu \mapsto \mathrm{cl}(\psi_n(\xi, \mu \circ \phi^{-1})) \in L^2(\Omega, \mathcal{F}, \mathbb{P}; \mathbb{R}^d)$ is measurable for any $n \geqslant 0$. Since for each $\mu \in \mathcal{P}_2(\mathbb{R}^d)$, $(\mathrm{cl}(\psi_n(\xi, \mu \circ \phi^{-1})))_{n \geqslant 0}$ converges in $L^2(\Omega, \mathcal{F}, \mathbb{P}; \mathbb{R}^d)$ to $\mathrm{cl}(\psi_\infty(\xi, \mu \circ \phi^{-1}))$, we conclude that the map $\mathcal{P}_2(\mathbb{R}^d) \ni \mu \mapsto \mathrm{cl}(\psi_\infty(\xi, \mu \circ \phi^{-1})) \in L^2(\Omega, \mathcal{F}, \mathbb{P}; \mathbb{R}^d)$ is measurable. In order to prove that the map $\mathcal{P}_2(\mathbb{R}^d) \ni \mu \mapsto \mathrm{cl}(\psi(\xi, \mu)) \in L^2(\Omega, \mathcal{F}, \mathbb{P}; \mathbb{R}^d)$ is measurable, we can work component by component. Since the i-th component of $\psi(\xi, \mu)$ is given by the tangent (up to shift) of the i-th component of $\psi_\infty(\xi, \mu \circ \phi^{-1})$, we conclude by approximating the tangent by its power series expansion and using the fact that if $\mu \mapsto \varphi(\mu) \in L^2(\Omega, \mathcal{F}, \mathbb{P}; \mathbb{R})$ is measurable and $\varphi(\mu)$ is a bounded random variable, then any integer power of $\varphi(\mu)$ is also measurable (as a function of μ with values in $L^2(\Omega, \mathcal{F}, \mathbb{P}; \mathbb{R})$). □

Here is an application of Lemma 5.29:

Lemma 5.30 *Let $\vartheta : \mathcal{P}_2((\mathbb{R}^d)^2) \to \mathbb{R}$ satisfy, for some atomless probability space $(\Omega, \mathcal{F}, \mathbb{P})$,*

$$\lim_{X \to X_0} \vartheta\big(\mathcal{L}(X_0, X)\big) = 0,$$

the limit in the left-hand side holding true in $L^2(\Omega, \mathcal{F}, \mathbb{P}; \mathbb{R}^d)$, then, for any $\mu_0 \in \mathcal{P}_2(\mathbb{R}^d)$,

$$\vartheta(\pi) \to 0 \quad \text{as} \quad \int_{(\mathbb{R}^d)^2} |x - y|^2 d\pi(x, y) \to 0,$$

as long as the first marginal of π on $\mathbb{R}^d$ remains equal to μ_0.

Proof. Given a probability measure $\pi \in \mathcal{P}_2((\mathbb{R}^d)^2)$, with μ_0 as first marginal on $\mathbb{R}^d$, we call $(\gamma(x, \cdot))_{x \in \mathbb{R}^d}$ the disintegration of π with respect to μ_0, namely:

$$\pi(dx, dy) = \mu_0(dx)\gamma(x, dy).$$

Recall that, for any $D \in \mathcal{B}(\mathbb{R}^d)$, the mapping $\mathbb{R}^d \ni x \mapsto \gamma(x, D)$ is measurable, see Theorem (Vol II)-1.1 if needed. Notice also that, without any loss of generality, we can assume that, for all $x \in \mathbb{R}^d$, $\gamma(x, \cdot) \in \mathcal{P}_2(\mathbb{R}^d)$. By Proposition 5.7, we deduce that the mapping $\mathbb{R}^d \ni x \mapsto \gamma(x, \cdot) \in \mathcal{P}_2(\mathbb{R}^d)$ is measurable. In particular, the mapping $[0, 1) \times \mathbb{R}^d \ni (\eta, x) \mapsto \psi(\eta, \gamma(x, \cdot))$ is measurable, where ψ is given by Lemma 5.29.

On $(\Omega, \mathcal{F}, \mathbb{P})$, we consider two independent random variables X_0 and Z, X_0 being $\mathbb{R}^d$-valued and having μ_0 as distribution and Z being uniformly distributed on $[0,1)$. We then let $X = \psi(Z, \gamma(X_0, \cdot))$, so that, conditional on X_0, X has $\gamma(X_0, \cdot)$ as distribution. Hence, the pair (X_0, X) has π as joint distribution. In particular,

$$\mathbb{E}\big[|X_0 - X|^2\big] = \int_{(\mathbb{R}^d)^2} |x-y|^2 d\pi(x,y).$$

Observing that X_0 does not depend on π, we deduce that:

$$\left(\int_{(\mathbb{R}^d)^2} |x-y|^2 d\pi(x,y) \to 0\right) \Rightarrow \left(\mathbb{E}\big[|X_0 - X|^2\big] \to 0\right),$$

which implies $\vartheta(\pi) = \vartheta(\mathcal{L}(X_0, X)) \to 0$. □

Connection with Skorohod's Representation Theorem

Lemma 5.29 gives a canonical way to construct a $\mathbb{R}^d$-valued random variable on the probability space $([0,1), \mathcal{B}([0,1)), \text{Leb}_1)$ with a prescribed distribution. It is natural to wonder whether this canonical representation is continuous.

Lemma 5.31 *With ϕ defined in (5.41), for any $\mu \in \mathcal{P}_2(\mathbb{R}^d)$ such that the measure $\mu \circ \phi^{-1}$ satisfies:*

$$(\mu \circ \phi^{-1})\Big(\big\{x \in [0,1)^d : x_i = \frac{k}{2^n}\big\}\Big) = \mu\Big(\big\{x \in \mathbb{R}^d : x_i = \tan\big(\frac{\pi k}{2^n} - \frac{\pi}{2}\big)\big\}\Big) = 0,$$

for all $i \in \{1, \cdots, d\}$, $n \geq 1$ and $k \in \{0, \cdots, 2^n - 1\}$, we can find a Borel subset $C \subset [0,1)$ (obviously depending upon μ), with $\text{Leb}_1(C) = 1$, such that:

$$\forall \eta \in C, \qquad \lim_{\mu' \Rightarrow \mu} \psi(\eta, \mu') = \psi(\eta, \mu),$$

the convergence $\mu' \Rightarrow \mu$ being understood in the sense of the weak convergence of probability measures. In particular, if ξ is a uniformly distributed $[0,1)$-valued random variable constructed on some probability space $(\Omega, \mathcal{F}, \mathbb{P})$, then $\mathbb{P}[\lim_{\mu' \Rightarrow \mu} \psi(\xi, \mu') = \psi(\xi, \mu)] = 1$.

Proof. We use the same notation as in the proof of Lemma 5.29. It suffices to prove that, for every $\mu \in \mathcal{P}_2([0,1)^d)$ satisfying, for all $i \in \{1, \cdots, d\}$, $n \geq 1$ and $k \in \{0, \cdots, 2^n - 1\}$,

$$\mu\Big(\big\{x \in [0,1)^d : x_i = \frac{k}{2^n}\big\}\Big) = 0, \tag{5.42}$$

there exists a Borel subset $C \subset [0,1)$, with $\text{Leb}_1(C) = 1$, such that:

$$\forall \eta \in C, \qquad \lim_{\mu' \in \mathcal{P}_2([0,1)^d):\, \mu' \Rightarrow \mu} \psi_\infty(\eta, \mu') = \psi_\infty(\eta, \mu).$$

For a given $\mu \in \mathcal{P}_2([0,1)^d)$ satisfying (5.42), we thus consider a sequence $(\mu_N)_{N \geqslant 1}$ of probability measures on $[0,1)^d$ weakly converging to μ. We then recall that, for any integer $n \geqslant 1$,

$$a^{n,\mu}(k_1,\cdots,k_d) = \mu\Bigg(\bigcup_{(k'_1,\cdots,k'_d) \prec_n (k_1,\cdots,k_d)} Q^n(k'_1,\cdots,k'_d)\Bigg),$$

$$b^{n,\mu}(k_1,\cdots,k_d) = \mu\Bigg(\bigcup_{(k'_1,\cdots,k'_d) \preceq_n (k_1,\cdots,k_d)} Q^n(k'_1,\cdots,k'_d)\Bigg),$$

where the relationship $x \preceq_n y$ stands for $x \prec_n y$ or $x = y$. For a given value of n, observe that the boundaries of both

$$\bigcup_{(k_1,\cdots,k_d) \prec_n (k'_1,\cdots,k'_d)} Q^n(k_1,\cdots,k_d) \quad \text{and} \quad \bigcup_{(k_1,\cdots,k_d) \preceq_n (k'_1,\cdots,k'_d)} Q^n(k'_1,\cdots,k'_d)$$

are included in $\cup_{i=1}^d \cup_{\ell=0}^{2^n} \{x \in [0,1)^d : x_i = \ell/2^n\}$. Thanks to (5.42), they are of zero measure under μ. We deduce that:

$$\lim_{N\to\infty} a^{n,\mu_N}(k_1,\cdots,k_d) = a^{n,\mu}(k_1,\cdots,k_d),$$

$$\lim_{N\to\infty} b^{n,\mu_N}(k_1,\cdots,k_d) = b^{n,\mu}(k_1,\cdots,k_d).$$

In particular, for any tuple $(k_1,\cdots,k_d) \in \{0,\cdots,2^n-1\}^d$ and any real η in the interval $(a^{n,\mu}(k_1,\cdots,k_d), b^{n,\mu}(k_1,\cdots,k_d))$ (if the interval is not empty), we can find an integer $N^{n,\mu}(k_1,\cdots,k_d)$ such that, for $N \geqslant N^{n,\mu}(k_1,\cdots,k_d)$,

$$a^{n,\mu_N}(k_1,\cdots,k_d) < \eta < b^{n,\mu_N}(k_1,\cdots,k_d),$$

proving that:

$$\psi_n(\eta,\mu_N) = \Big(\frac{k_1}{2^n},\cdots,\frac{k_d}{2^n}\Big) = \psi_n(\eta,\mu).$$

Since

$$\big|\psi_\infty(\eta,\mu_N) - \psi_n(\eta,\mu_N)\big| \leqslant \frac{C}{2^n}, \quad \big|\psi_\infty(\eta,\mu) - \psi_n(\eta,\mu)\big| \leqslant \frac{C}{2^n},$$

we deduce that, for $N \geqslant N^{n,\mu}(k_1,\cdots,k_d)$,

$$\big|\psi_\infty(\eta,\mu_N) - \psi_\infty(\eta,\mu)\big| \leqslant \frac{C}{2^{n-1}},$$

which completes the proof. □

Lemma 5.31 is reminiscent of Skorohod's representation theorem as it provides a way to represent a weakly convergent sequence of probability measures by means of an almost sure convergent sequence of random variables. However, the statement is of a somewhat limited scope since it holds true only for probability measures $\mu\circ\phi^{-1}$ that have no mass on the hyperplanes having a prescribed dyadic coordinate. In order to extend it to any probability measures, we may use an idea found in the proof of Blackwell and Dubins' theorem. For any integer $n \geqslant 1$, we let:

$$\tilde{\psi}(\eta, a, \mu) = \psi\big(\eta, \mu\circ\tau_a^{-1}\big) - a\boldsymbol{e}, \qquad \eta, a \in [0,1),\ \mu \in \mathcal{P}_2(\mathbb{R}^d),$$

where $\tau_a(x) = x + a\boldsymbol{e}$ and $\boldsymbol{e} = (1,\cdots,1) \in \mathbb{R}^d$. Then, for any $a \in [0,1)$ and $\mu \in \mathcal{P}_2(\mathbb{R}^d)$, the measure $\text{Leb}_1 \circ ([0,1) \ni \eta \mapsto \tilde{\psi}(\eta, a, \mu))^{-1}$ is exactly μ.

Now, for a given $\mu \in \mathcal{P}_2(\mathbb{R}^d)$, we can find a countable subset $Q = (c_\ell)_{\ell\in\mathbb{N}}$ of $\mathbb{R}$ such that, for any $i \in \{1,\cdots,d\}$ and any $c \in \mathbb{R}\setminus Q$, $\mu(\{x \in \mathbb{R}^d : x_i = c\}) = 0$. Observe moreover that, for any $a \in \mathbb{R}$, any integer $n \geqslant 1$ and any $k \in \{0,\cdots,2^n-1\}$:

$$\begin{aligned}
&\mu\circ\tau_a^{-1}\Big(\big\{x \in \mathbb{R}^d;\ x_i = \tan\big(\frac{\pi k}{2^n} - \frac{\pi}{2}\big)\big\}\Big) \\
&\quad = \mu\Big(\big\{x \in \mathbb{R}^d;\ x_i = \tan\big(\frac{\pi k}{2^n} - \frac{\pi}{2}\big) - a\big\}\Big),
\end{aligned}$$

so that:

$$\begin{aligned}
&\mu\circ\tau_a^{-1}\Big(\big\{x \in \mathbb{R}^d;\ x_i = \tan\big(\frac{\pi k}{2^n} - \frac{\pi}{2}\big)\big\}\Big) > 0 \\
&\quad \Leftrightarrow \exists \ell \in \mathbb{N},\ a = \tan\big(\frac{\pi k}{2^n} - \frac{\pi}{2}\big) - c_\ell.
\end{aligned}$$

Letting:

$$\bar{Q} = \Big\{\tan\big(\frac{\pi k}{2^n} - \frac{\pi}{2}\big) - c; \qquad c \in Q,\ n \geqslant 0\setminus\{0\},\ k \in \{0,\cdots,2^n-1\}\Big\},$$

we deduce that, for $a \notin \bar{Q}$:

$$\forall n \geqslant 1,\ \forall k \in \{0,\cdots,2^n-1\}, \quad \mu\circ\tau_a^{-1}\Big(\big\{x \in \mathbb{R}^d;\ x_i = \tan\big(\frac{\pi k}{2^n} - \frac{\pi}{2}\big)\big\}\Big) = 0.$$

From Lemma 5.31, we deduce that, for $a \notin \bar{Q}$ and $\mu \in \mathcal{P}_2(\mathbb{R}^d)$, there exists a Borel subset $C \subset [0,1)$ such that, for all $\eta \in C$, the function $\tilde{\psi}(\eta, a, \cdot)$ is continuous at μ. So for any sequence $(\mu_N)_{N\geqslant 1}$ weakly converging to μ,

$$\forall a \notin \bar{Q}, \quad \text{Leb}_1\Big(\big\{\eta \in [0,1) : \lim_{N\to\infty} \tilde{\psi}(\eta, a, \mu_N) = \tilde{\psi}(\eta, a, \mu)\big\}\Big) = 1.$$

And then,

$$\int_{\mathbb{R}} \mathrm{Leb}_1\Big(\{\eta \in [0,1) : \lim_{N\to\infty} \tilde{\psi}(\eta, a, \mu_N) = \tilde{\psi}(\eta, a, \mu)\}\Big)\varphi(a)da = 1,$$

where φ denotes the density of the standard Gaussian probability measure. By Fubini's theorem, for almost every $(\eta, a) \ni [0,1) \times \mathbb{R}$, the function $\tilde{\psi}(\eta, a, \cdot)$ is continuous at μ. This shows that it suffices to add an additional parameter in the definition of ψ in order to make it continuous in $\mu \in \mathcal{P}_2(\mathbb{R}^d)$ for almost every fixed values of the other parameters. Put it differently, if ξ and G denote two independent random variables on some probability space $(\Omega, \mathcal{F}, \mathbb{P})$, with ξ being uniformly distributed on $[0,1)$ and $G \sim N(0,1)$, then $\mathbb{P}[\lim_{N\to\infty} \tilde{\psi}(\xi, G, \mu_N) = \tilde{\psi}(\xi, G, \mu)] = 1$, with $\tilde{\psi}(\xi, G, \mu_N) \sim \mu_N$, for all $N \geqslant 1$, and $\tilde{\psi}(\xi, G, \mu) \sim \mu$. This recovers Blackwell and Dubins' theorem.
Recall that we use freely the notation $X \sim \mu$ in lieu of $\mathcal{L}(X) = \mu$ whenever the expressions for X and μ may render the typesetting too cumbersome.

Remark 5.32 *Blackwell and Dubins' extension of Skorohod's theorem is very important for our purposes. Indeed, when dealing with a weakly convergent sequence of probability measures $(\mu_N)_{N\geqslant 1}$, we often consider an almost surely convergent sequence of random variables $(X_N)_{N\geqslant 1}$ representing the family $(\mu_N)_{N\geqslant 1}$ in the sense that, for all $N \geqslant 1$, $\mathcal{L}(X_N) = \mu_N$. Blackwell and Dubins' theorem asserts that, on a probability space $(\Omega, \mathcal{F}, \mathbb{P})$ equipped with a pair of random variables (ξ, G) as above, it is always possible to construct such a sequence. In particular, whenever the probability space is atomless, it is always possible to construct such a pair (ξ, G) and, subsequently, an almost surely convergent sequence representing the family $(\mu_N)_{N\geqslant 1}$.*

Jointly Measurable Version of the L-Derivative

The next result says that one can always find a reasonable version of the derivative of a continuously L-differentiable function of probability measures.

Proposition 5.33 *Given a continuously* L*-differentiable function $u : \mathcal{P}_2(\mathbb{R}^d) \to \mathbb{R}$, for each $\mu \in \mathcal{P}_2(\mathbb{R}^d)$, one can redefine, $\partial_\mu u(\mu)(\cdot) : \mathbb{R}^d \ni x \mapsto \partial_\mu u(\mu)(x)$ on a μ-negligible set in such a way that the mapping $\mathcal{P}_2(\mathbb{R}^d) \times \mathbb{R}^d \ni (\mu, x) \mapsto \partial_\mu u(\mu)(x)$ is jointly measurable when $\mathcal{P}_2(\mathbb{R}^d)$ is equipped with the Borel σ-field generated by the 2-Wasserstein topology. Whenever $\partial_\mu u(\mu)(\cdot)$ has a continuous version, the version constructed above for measurability reasons coincides with it on the support* $\mathrm{Supp}(\mu)$ *of μ.*

Proof. Let $(\Omega, \mathcal{F}, \mathbb{P})$ be a Polish atomless probability space. We use the fact that for any bounded continuous function $\ell : \mathbb{R}^d \times \mathbb{R}^d \to \mathbb{R}$ and d intervals $I_1, \cdots, I_d$ on the real line, the mapping $L^2(\Omega, \mathcal{F}, \mathbb{P}; \mathbb{R}^d) \times \mathbb{R}^d \ni (X, x) \mapsto \mathrm{cl}(\mathbf{1}_{I_1}(\ell(X,x)), \cdots, \mathbf{1}_{I_d}(\ell(X,x))) \in L^2(\Omega, \mathcal{F}, \mathbb{P}; \mathbb{R}^d)$ is measurable. As before, $L^2(\Omega, \mathcal{F}, \mathbb{P}; \mathbb{R}^d)$ is the quotient of the space of square-integrable random variables for $\mathbb{P}$-almost sure equality, and $\mathrm{cl}(\zeta)$ denotes the equivalence class of ζ in $L^2(\Omega, \mathcal{F}, \mathbb{P}; \mathbb{R}^d)$.

Here is the way we apply this simple fact. Denoting by '$\cdot$' the inner product in $L^2(\Omega, \mathcal{F}, \mathbb{P}; \mathbb{R}^d)$, the mapping $[L^2(\Omega, \mathcal{F}, \mathbb{P}; \mathbb{R}^d)]^2 \ni (X, Y) \mapsto [D\tilde{u}](X) \cdot Y$ is measurable as the pointwise limit of measurable mappings. Therefore, for any vector $e \in \mathbb{R}^d$ and any $\varepsilon > 0$, the mapping $L^2(\Omega, \mathcal{F}, \mathbb{P}; \mathbb{R}^d) \times \mathbb{R}^d \ni (X, x) \mapsto [D\tilde{u}](X) \cdot (e\mathbf{1}_{\{|X-x| \leqslant \varepsilon\}})$ is jointly measurable. Then, the mapping $\psi = (\psi_1, \cdots, \psi_d) : L^2(\Omega, \mathcal{F}, \mathbb{P}; \mathbb{R}^d) \times \mathbb{R}^d \to \mathbb{R}^d$ given by

$$\psi_i(X, x) = \liminf_{\varepsilon \searrow 0} \left[\frac{[D\tilde{u}](X) \cdot (e_i \mathbf{1}_{\{|X-x| \leqslant \varepsilon\}})}{\mathbb{P}(|X - x| \leqslant \varepsilon)} \mathbf{1}_{\{\mathbb{P}(|X-x| \leqslant \varepsilon) > 0\}} \right]$$

is also jointly measurable, where $(e_1, \cdots, e_d)$ is the canonical basis of $\mathbb{R}^d$. By Lebesgue-Besicovitch differentiation theorem, $\psi(X, \cdot)$ is a version of $\partial_\mu u(\mathcal{L}(X))(\cdot)$ in the space $L^2(\mathbb{R}^d, \mathcal{B}(\mathbb{R}^d), \mathcal{L}(X); \mathbb{R}^d)$. If $\partial_\mu u(\mathcal{L}(X))(\cdot)$ admits a continuous version, then it coincides with it on the support of $\mathcal{L}(X)$.

We can now conclude because the chain rule and Lemma 5.29 imply that the mapping $\mathcal{P}_2(\mathbb{R}^d) \times \mathbb{R}^d \ni (\mu, x) \mapsto \psi(X^\mu, x)$ is measurable whenever $(\Omega, \mathcal{F}, \mathbb{P})$ is assumed to be atomless. □

5.3.2 L-Differentiability of Functions of Empirical Measures

The rather special notion of differentiability introduced in this chapter is best understood as differentiation of functions of limits of empirical measures in the directions of the atoms of the measures. We illustrate this fact in the next two propositions.

Definition 5.34 *Given a function $u : \mathcal{P}_2(\mathbb{R}^d) \to \mathbb{R}$ and an integer $N \geqslant 1$, we define the empirical projection of u onto $\mathbb{R}^d$ by:*

$$u^N : (\mathbb{R}^d)^N \ni (x_1, \cdots, x_N) \mapsto u\left(\frac{1}{N}\sum_{i=1}^N \delta_{x_i}\right).$$

The following result connects the L-derivative of a function of probability measures to the standard partial derivatives of its empirical projections. We shall extend these connections to second order derivatives in Proposition 5.91 later in the chapter, after a detailed discussion of the relevant notions of second order differentiability.

Proposition 5.35 *If $u : \mathcal{P}_2(\mathbb{R}^d) \to \mathbb{R}$ is continuously* L*-differentiable, then its empirical projection u^N is differentiable on $(\mathbb{R}^d)^N$ and, for all $i \in \{1, \cdots, N\}$,*

$$\partial_{x_i} u^N(x_1, \cdots, x_N) = \frac{1}{N} \partial_\mu u\left(\frac{1}{N}\sum_{j=1}^N \delta_{x_j}\right)(x_i).$$

In the right-hand side of the above equality, the function $\partial_\mu u(N^{-1}\sum_{j=1}^N \delta_{x_j})(\cdot)$ is uniquely defined in $L^2(\mathbb{R}^d, N^{-1}\sum_{j=1}^N \delta_{x_j}; \mathbb{R}^d)$. In particular, for each $i \in \{1, \cdots, N\}$, $\partial_\mu u(N^{-1}\sum_{j=1}^N \delta_{x_j})(x_i)$ is uniquely defined.

Proof. On an atomless Polish probability space $(\Omega, \mathcal{F}, \mathbb{P})$, we consider a random variable ϑ uniformly distributed over the set $\{1, \cdots, N\}$. Then, for any fixed $\boldsymbol{x} = (x_1, \cdots, x_N) \in (\mathbb{R}^d)^N$, x_ϑ is a random variable having distribution $\bar{\mu}_{\boldsymbol{x}}^N = N^{-1}\sum_{i=1}^N \delta_{x_i}$. In particular, with the same notation as above for $\tilde{u}$,

$$u^N(\boldsymbol{x}) = u^N(x_1, \cdots, x_N) = \tilde{u}(x_\vartheta).$$

Therefore, for $\boldsymbol{h} = (h_1, \cdots, h_N) \in (\mathbb{R}^d)^N$,

$$u^N(\boldsymbol{x} + \boldsymbol{h}) = \tilde{u}(x_\vartheta + h_\vartheta) = \tilde{u}(x_\vartheta) + D\tilde{u}(x_\vartheta) \cdot h_\vartheta + o(|h|),$$

the dot product being here the L^2- inner product over $(\Omega, \mathcal{F}, \mathbb{P})$, from which we deduce:

$$u^N(\boldsymbol{x} + \boldsymbol{h}) = u^N(\boldsymbol{x}) + \frac{1}{N}\sum_{i=1}^N \partial_\mu u(\bar{\mu}_{\boldsymbol{x}}^N)(x_i) \cdot h_i + o(|h|),$$

which is the desired result. □

5.3.3 Lipschitz L-Differentials and Regular Versions

If a real valued function u on $\mathcal{P}_2(\mathbb{R}^d)$ is L-differentiable, the Fréchet derivative $D\tilde{u}$ of its lifting $\tilde{u}$ is a mapping from $L^2(\Omega, \mathcal{F}, \mathbb{P}; \mathbb{R}^d)$ into itself since we identify $L^2(\Omega, \mathcal{F}, \mathbb{P}; \mathbb{R}^d)$ and its dual. It is Lipschitz continuous if there exists a constant $C > 0$ such that, for any identically distributed square integrable random variables X and Y in $L^2(\Omega, \mathcal{F}, \mathbb{P}; \mathbb{R}^d)$, we have:

$$\|D\tilde{u}(X) - D\tilde{u}(Y)\|_2 \leqslant C\|X - Y\|_2.$$

Since the L-derivative of a function has the particular form given by Proposition 5.25, the Lipschitz property can be rewritten as:

$$\mathbb{E}\big[|\partial_\mu u(\mathbb{P}_X)(X) - \partial_\mu u(\mathbb{P}_Y)(Y)|^2\big] \leqslant C^2\mathbb{E}\big[|X - Y|^2\big], \tag{5.43}$$

for any square integrable random variables X and Y in $L^2(\Omega, \mathcal{F}, \mathbb{P}; \mathbb{R}^d)$. Recall that we use the notation $\mathbb{P}_X$ instead of $\mathcal{L}(X)$ when we want to emphasize the probability $\mathbb{P}$ or we do not want to use too many parentheses in the formulas. From our discussion of the construction of $\partial_\mu u$, we know that for each μ, $\partial u(\mu)(\cdot)$ is only uniquely defined μ-almost everywhere. However, the following result says that under the above Lipschitz assumption, there exists a Lipschitz continuous version of $\partial u(\mu)(\cdot)$.

Proposition 5.36 *Assume that $(v(\mu)(\cdot))_{\mu\in\mathcal{P}_2(\mathbb{R}^d)}$ is a family of Borel-measurable mappings from $\mathbb{R}^d$ into itself for which there exists a constant C such that, for any identically distributed square integrable random variables ξ and ξ' in $L^2(\Omega,\mathcal{F},\mathbb{P};\mathbb{R}^d)$ over an atomless probability space $(\Omega,\mathcal{F},\mathbb{P})$, it holds:*

$$\mathbb{E}\big[|v(\mathbb{P}_\xi)(\xi) - v(\mathbb{P}_{\xi'})(\xi')|^2\big] \leqslant C^2\mathbb{E}\big[|\xi - \xi'|^2\big]. \tag{5.44}$$

Then, for each $\mu \in \mathcal{P}_2(\mathbb{R}^d)$, one can redefine $v(\mu)(\cdot)$ on a μ-negligible set in such a way that:

$$\forall x, x' \in \mathbb{R}^d, \quad |v(\mu)(x) - v(\mu)(x')| \leqslant C|x - x'|,$$

for the same C as in (5.44).

Remark 5.37 *Here, the atomless property is just used to guarantee the existence of random variables with a prescribed distribution on any Polish space.*

The proof of Proposition 5.36 is rather long and technical, so the reader mostly interested in the practical applications of the notion of L*-differentiability may want to skip it in a first reading.*

Proof.

First Step. We first consider the case of a bounded function v, and assume that μ has a strictly positive continuous density p on the whole $\mathbb{R}^d$, p and its derivatives being of exponential decay at infinity. We claim that there exists a continuously differentiable one-to-one function U from $(0,1)^d$ onto $\mathbb{R}^d$ such that, whenever $\eta_1,\cdots,\eta_d$ are d independent random variables uniformly distributed on $(0,1)$, $U(\eta_1,\cdots,\eta_d)$ has distribution μ. It satisfies for any $(z_1,\cdots,z_d) \in (0,1)^d$:

$$\frac{\partial U_i}{\partial z_i}(z_1,\cdots,z_d) \neq 0, \quad \frac{\partial U_j}{\partial z_i}(z_1,\cdots,z_d) = 0, \quad 1 \leqslant i < j \leqslant d.$$

The result is well known when $d = 1$. In such a case, U is the inverse of the cumulative distribution function of μ. In higher dimension, U can be constructed by an induction argument on the dimension. Assume indeed that some $\hat{U}$ has been constructed for the first marginal distribution $\hat{\mu}$ of μ on $\mathbb{R}^{d-1}$, that is for the push-forward of μ by the projection mapping $\mathbb{R}^d \ni (x_1,\cdots,x_d) \mapsto (x_1,\cdots,x_{d-1})$. Given $(x_1,\cdots,x_{d-1}) \in \mathbb{R}^{d-1}$, we then denote by $p(\cdot|x_1,\cdots,x_{d-1})$ the conditional density of μ given the $d-1$ first coordinates:

$$p(x_d|x_1,\cdots,x_{d-1}) = \frac{p(x_1,\cdots,x_d)}{\hat{p}(x_1,\cdots,x_{d-1})}, \quad x_1,\cdots,x_{d-1} \in \mathbb{R}^{d-1},$$

where $\hat{p}$ denotes the density of $\hat{\mu}$ (which is continuously differentiable and positive). We then denote by $(0,1) \ni z_d \mapsto U^{(d)}(z_d|x_1,\cdots,x_{d-1})$ the inverse of the cumulative distribution function of the law of density $p(\cdot\,|x_1,\cdots,x_{d-1})$. It satisfies:

$$F_d\big(U^{(d)}(z_d|x_1,\cdots,x_{d-1})|x_1,\cdots,x_{d-1}\big) = z_d,$$

with

$$F_d(x_d|x_1,\cdots,x_{d-1}) = \int_{-\infty}^{x_d} p(y|x_1,\cdots,x_{d-1})dy,$$

which is continuously differentiable in $(x_1,\cdots,x_d)$ (using the exponential decay of the density at infinity). By the implicit function theorem, the mapping:

$$\mathbb{R}^{d-1} \times (0,1) \ni (x_1,\cdots,x_{d-1},z_d) \mapsto U^{(d)}(z_d|x_1,\cdots,x_{d-1})$$

is continuously differentiable. The partial derivative with respect to z_d is given by:

$$\frac{\partial U^{(d)}}{\partial z_d}(z_d|x_1,\cdots,x_{d-1}) = \frac{1}{p(U^{(d)}(z_d|x_1,\cdots,x_{d-1})|x_1,\cdots,x_{d-1})},$$

which is nonzero. We now let:

$$U(z_1,\cdots,z_d) = \big(\hat{U}(z_1,\cdots,z_{d-1}), U^{(d)}(z_d|\hat{U}(z_1,\cdots,z_{d-1}))\big), \quad z_1,\cdots,z_d \in (0,1)^d.$$

By construction, $U(\eta_1,\cdots,\eta_d)$ has distribution μ. Indeed, $\hat{U}(\eta_1,\cdots,\eta_{d-1})$ has distribution $\hat{\mu}$ and the conditional law of $U_d(\eta_1,\cdots,\eta_d)$ given $\eta_1,\cdots,\eta_{d-1}$ is the conditional law of μ given the $d-1$ first coordinates, since $U_d(\eta_1,\cdots,\eta_d) = U^{(d)}(\eta_d|\hat{U}(\eta_1,\cdots,\eta_{d-1}))$. It satisfies $[\partial U_d/\partial z_d](z_1,\cdots,z_d) > 0$ and $[\partial U_i/\partial z_d]\,(z_1,\cdots,z_d) = 0$ for $i < d$. In particular, since the induction assumption implies that $\hat{U}$ is one-to-one and $[\partial U_d/\partial z_d](z_1,\cdots,z_d) > 0$, U must be one-to-one as well. As the Jacobian matrix of U is triangular with nonzero elements on the diagonal, it is invertible. By the global inversion theorem, U is a diffeomorphism. The range of U is the support of μ, that is $\mathbb{R}^d$. This proves that U is one-to-one from $(0,1)^d$ onto $\mathbb{R}^d$.

Second Step. We still assume that v is bounded and μ has a strictly positive continuous density p on the whole $\mathbb{R}^d$, p and its derivatives being of exponential decay at infinity. We will use the mapping U constructed in the first step. For three random variables ξ, ξ' and G in $L^2(\Omega,\mathcal{F},\mathbb{P};\mathbb{R}^d)$, the pair (ξ,ξ') being independent of G, the random variables ξ and ξ' having the same distribution, and G being normally distributed with mean 0 and covariance matrix given by the identity I_d in dimension d, in notation $G \sim N_d(0,I_d)$, then (5.44) implies that, for any integer $n \geqslant 1$:

$$\mathbb{E}\big[|v(\mathbb{P}_{\xi+n^{-1}G})(\xi+n^{-1}G) - v(\mathbb{P}_{\xi+n^{-1}G})(\xi'+n^{-1}G)|^2\big] \leqslant C^2\mathbb{E}\big[|\xi-\xi'|^2\big].$$

In particular, setting:

$$\begin{aligned} v_n(x) &= \mathbb{E}\big[v(\mathbb{P}_{\xi+n^{-1}G})(x+n^{-1}G)\big] \\ &= \frac{n^d}{(2\pi)^{d/2}}\int_{\mathbb{R}^d} v(\mathbb{P}_{\xi+n^{-1}G})(y)\exp\big[-n^2\frac{|x-y|^2}{2}\big]dy, \end{aligned}$$

we have:

$$\mathbb{E}\big[|v_n(\xi)-v_n(\xi')|^2\big] \leqslant C^2\mathbb{E}\big[|\xi-\xi'|^2\big]. \tag{5.45}$$

Notice that v_n is infinitely differentiable with bounded derivatives.

We now choose a specific coupling for ξ and ξ'. Indeed, we know that, for any $\eta = (\eta_1, \cdots, \eta_d)$ and $\eta' = (\eta'_1, \cdots, \eta'_d)$, with uniform distributions on $(0,1)^d$, $U(\eta)$ and $U(\eta')$ have the same distribution as ξ. Without any loss of generality, we may assume that the probability space $(\Omega, \mathcal{F}, \mathbb{P})$ is given by $(0,1)^d \times \mathbb{R}^d$ endowed with its Borel σ-algebra and the product of the Lebesgue measure on $(0,1)^d$ and of the Gaussian measure $N_d(0, I_d)$. The random variables η and G are then chosen as the canonical mappings $\eta : (0,1)^d \times \mathbb{R}^d \ni (z,y) \mapsto z$ and $G : (0,1)^d \times \mathbb{R}^d \ni (z,y) \mapsto y$.

We then define η' as a function of the variable $z \in (0,1)^d$ only. For a given $z^0 = (z^0_1, \cdots, z^0_d) \in (0,1)^d$ and for h small enough so that the open ball $B(z^0, h)$ of center z^0 and radius h is included in $(0,1)^d$, we let:

$$\eta'(z) = \begin{cases} z - 2(z_d - z^0_d)e_d & \text{if } z \in B(z^0, h), \\ z, & \text{outside}, \end{cases}$$

where e_d is the d-th vector of the canonical basis, that is η' matches locally the symmetry with respect to the hyperplane containing z^0 and orthogonal to e_d. Clearly, η' preserves the Lebesgue measure. With this particular choice, we rewrite (5.45) as:

$$\int_{(0,1)^d} \big| v_n\big(U(\eta(z))\big) - v_n\big(U(\eta'(z))\big)\big|^2 dz \leqslant C^2 \int_{(0,1)^d} \big| U(\eta(z)) - U(\eta'(z))\big|^2 dz,$$

or equivalently:

$$\begin{aligned} &\int_{|r|<h} \big| v_n\big[U\big(z^0 + r - 2r_d e_d\big)\big] - v_n\big(U(z^0 + r)\big)\big|^2 dr \\ &\qquad \leqslant C^2 \int_{|r|<h} \big| U\big(z^0 + r - 2r_d e_d\big) - U(z^0 + r)\big|^2 dr. \end{aligned} \tag{5.46}$$

Since U is continuously differentiable, we have:

$$v_n(U(z^0 + r)) = v_n(U(z^0)) + \partial v_n(U(z^0)) \cdot \big[\partial U(z^0) \cdot r\big] + o(r),$$

where $\partial U(z^0)$ is a $d \times d$ matrix. We deduce that:

$$\begin{aligned} v_n\big[U\big(z^0 + r - 2r_d e_d\big)\big] - v_n(U(z^0 + r)) &= -2 \sum_{i=1}^d \frac{\partial v_n}{\partial x_i}(U(z^0)) \frac{\partial U_i}{\partial z_d}(z^0) r_d + o(r) \\ &= -2 \frac{\partial v_n}{\partial x_d}(U(z^0)) \frac{\partial U_d}{\partial z_d}(z^0) r_d + o(r), \end{aligned}$$

since $\partial U_i / \partial z_d = 0$ for $i \neq d$, and

$$\begin{aligned} &\int_{|r|<h} \big| v_n\big[U\big(z^0 + r - 2r_d e_d\big)\big] - v_n(U(z^0 - r))\big|^2 dr \\ &= 4 \Big| \frac{\partial v_n}{\partial x_d}(U(z^0)) \frac{\partial U_d}{\partial z_d}(z^0) \Big|^2 \int_{|r|<h} r_d^2 dr + o(h^{d+2}). \end{aligned} \tag{5.47}$$

Similarly,

$$\int_{|r|<h} \big|U(z^0 + r - 2r_d e_d) - U(z^0 + r)\big|^2 dr = 4\Big|\frac{\partial U_d}{\partial z_d}(z^0)\Big|^2 \int_{|r|<h} r_d^2 dr + o(h^{d+2}), \qquad (5.48)$$

and putting together (5.46), (5.47) and (5.48), we obtain:

$$\Big|\frac{\partial v_n}{\partial x_d}(U(z^0))\frac{\partial U_d}{\partial z_d}(z^0)\Big|^2 \leqslant C^2\Big|\frac{\partial U_d}{\partial z_d}(z^0)\Big|^2.$$

Since $[\partial U_d/\partial z_d](z^0)$ is different from zero, we deduce that:

$$\Big|\frac{\partial v_n}{\partial x_d}(U(z^0))\Big|^2 \leqslant C^2,$$

and since U is a one-to-one mapping from $(0,1)^d$ onto $\mathbb{R}^d$, and $z^0 \in (0,1)^d$ is arbitrary, we conclude that $|[\partial v_n/\partial x_d](x)| \leqslant C$, for any $x \in \mathbb{R}^d$. By changing the basis used for the construction of U (we used the canonical basis but we could use any orthonormal basis as well), we have $|\nabla v_n(x)e| \leqslant C$ for any $x, e \in \mathbb{R}^d$ with $|e| = 1$. This proves that the functions $(v_n)_{n\geqslant 1}$ are uniformly bounded and C-Lipschitz continuous. We then denote by $\hat{v}$ the limit of a subsequence converging for the topology of uniform convergence on compact subsets. For simplicity, we keep the index n to denote the subsequence. Assumption (5.44) implies:

$$\mathbb{E}\big[|v_n(\xi) - v(\mathbb{P}_\xi)(\xi)|^2\big] \leqslant \mathbb{E}\big[|v(\mathbb{P}_{\xi+n^{-1}G})(\xi + n^{-1}G) - v(\mathbb{P}_\xi)(\xi)|^2\big] \leqslant C^2 n^{-2},$$

and taking the limit $n \to +\infty$, we deduce that $\hat{v}$ and $v(\cdot, \mathbb{P}_\xi)$ coincide $\mathbb{P}_\xi$ almost everywhere. This completes the proof when v is bounded and ξ has a continuous positive density p, p and its derivatives being of exponential decay at infinity.

Third Step. When v is bounded and ξ is bounded and has a general distribution, we approximate ξ by $\xi + n^{-1}G$ again. Then, $\xi + n^{-1}G$ has a positive continuous density, the density and its derivatives being of Gaussian decay at infinity, so that, by the second step, the function $\mathbb{R}^d \ni x \mapsto v(\mathbb{P}_{\xi+n^{-1}G})(x)$ can be assumed to be C-Lipschitz continuous for each $n \geqslant 1$. Extracting a convergent subsequence and passing to the limit as above, we deduce that $v(\mathbb{P}_\xi)(\cdot)$ admits a C-Lipschitz continuous version.

When v is bounded and ξ is not assumed to be bounded, we approximate ξ by its orthogonal projection on the ball of center 0 and radius n. We then complete the proof in a similar way.

Finally when v is not bounded, we approximate v by $(\psi_n(v))_{n\geqslant 1}$ where, for each $n \geqslant 1$, ψ_n is a bounded smooth function from $\mathbb{R}$ into itself such that $\psi_n(r) = r$ for $r \in [-n, n]$ and $|[d\psi_n/dr](r)| \leqslant 1$ for all $r \in \mathbb{R}$. Then, for each $n \geqslant 1$, there exists a C-Lipschitz continuous version of $\psi_n(v(\mathbb{P}_\xi))(\cdot)$. Letting n tend to ∞, we complete the proof. □

Under the Lipschitz assumption (5.43) on the Fréchet derivative of the lifting of u, we can use Proposition 5.36 in order to define $\partial_\mu u(\mu)(x)$ for every μ and every x while preserving the Lipschitz property in the variable x. From now on, whenever the derivative $D\tilde{u}$ is Lipschitz, we shall use such a version of $\partial_\mu u(\mu)(\cdot)$. Importantly, observe that this version is uniquely defined on the support of μ only.

So, if $\mu, \nu \in \mathcal{P}_2(\mathbb{R}^d)$ and X and Y are random variables such that $\mathcal{L}(X) = \mu$ and $\mathcal{L}(Y) = \nu$, we have:

$$\begin{aligned}
&\mathbb{E}\big[|\partial_\mu u(\mu)(X) - \partial_\mu u(\nu)(X)|^2\big] \\
&\quad \leqslant 2\Big(\mathbb{E}\big[|\partial_\mu u(\mu)(X) - \partial_\mu u(\nu)(Y)|^2\big] + \mathbb{E}\big[|\partial_\mu u(\nu)(Y) - \partial_\mu u(\nu)(X)|^2\big]\Big) \\
&\quad \leqslant 4C^2 \mathbb{E}\big[|Y - X|^2\big],
\end{aligned}$$

where we used the Lipschitz property (5.43) of the derivative together with the result of Proposition 5.36 applied to the function $\partial_\mu u(\nu)$. Now, taking the infimum over all the couplings (X, Y) with marginals μ and ν, we obtain:

$$\inf_{X, \mathcal{L}(X)=\mu} \mathbb{E}\big[|\partial_\mu u(\mu)(X) - \partial_\mu u(\nu)(X)|^2\big] \leqslant 4C^2 W_2(\mathcal{L}(X), \mathcal{L}(Y))^2,$$

and since the left-hand side depends only upon μ and not on X as long as $\mathcal{L}(X) = \mu$, we get:

$$\mathbb{E}\big[|\partial_\mu u(\mu)(X) - \partial_\mu u(\nu)(X)|^2\big] \leqslant 4C^2 W_2(\mu, \nu)^2. \tag{5.49}$$

As a corollary, we claim a refinement of Proposition 5.33:

Corollary 5.38 *Let $u : \mathcal{P}_2(\mathbb{R}^d) \to \mathbb{R}$ be* L-*differentiable, the Fréchet derivative $D\tilde{u}$ of the lifting of u to $L^2(\Omega, \mathcal{F}, \mathbb{P}; \mathbb{R}^d)$ for an atomless Polish probability space $(\Omega, \mathcal{F}, \mathbb{P})$ being Lipschitz.*

Then for each $\mu \in \mathcal{P}_2(\mathbb{R}^d)$, we can find a Lipschitz version of $\partial_\mu u(\mu)(\cdot) : \mathbb{R}^d \to \mathbb{R}^d$ with a Lipschitz constant independent of μ and such that $\partial_\mu u : \mathcal{P}_2(\mathbb{R}^d) \times \mathbb{R}^d \ni (\mu, x) \mapsto \partial_\mu u(\mu)(x)$ is measurable and continuous at any point (μ, x) such that x belongs to the support of μ.

Proof. We already know that, for any $\mu \in \mathcal{P}_2(\mathbb{R}^d)$, we can find a Lipschitz version of $\partial_\mu u(\mu)(\cdot)$. Whenever the support of μ is the entire space $\mathbb{R}^d$, this Lipschitz version is the unique continuous version of $\partial_\mu u(\mu)(\cdot)$.

First Step. For any $\sigma \in (0, 1]$, we consider the function:

$$\mathcal{U}^\sigma : \mathcal{P}_2(\mathbb{R}^d) \times \mathbb{R}^d \ni (\mu, x) \mapsto \mathcal{U}^\sigma(\mu, x) = \partial_\mu u\big(\mu * N_d(0, \sigma I_d)\big)(x),$$

where as usual, I_d is the d-dimensional identity matrix and $N_d(0, \sigma I_d)$ is the d-dimensional Gaussian law with zero as mean and σI_d as covariance matrix. We observe that $\mathcal{U}^\sigma$ is uniquely defined and that, for any $\mu \in \mathcal{P}_2(\mathbb{R}^d)$, $\mathcal{U}^\sigma(\mu, \cdot)$ is Lipschitz continuous, uniformly in μ and $\sigma > 0$. Moreover, for any $\mu \in \mathcal{P}_2(\mathbb{R}^d)$, any random variable X with μ as distribution, and any random variable G independent of X, with $N_d(0, I_d)$ as distribution, X and G being constructed on $(\Omega, \mathcal{F}, \mathbb{P})$, it holds that:

$$\begin{aligned}
|\mathcal{U}^\sigma(\mu, 0)| &\leqslant C|X + \sigma G| + |\mathcal{U}^\sigma(\mu, X + \sigma G)| \\
&= C|X + \sigma G| + |D\tilde{u}(X + \sigma G)|.
\end{aligned}$$

Taking squares and then expectations, we deduce:

$$|\mathcal{U}^\sigma(\mu,0)|^2 \leqslant C\Big(1+\mathbb{E}\big[|X|^2\big]+\mathbb{E}\big[|D\tilde{u}(X+\sigma G)|^2\big]\Big)$$
$$\leqslant C\Big(1+\mathbb{E}\big[|X|^2\big]+\mathbb{E}\big[|D\tilde{u}(0)|^2\big]\Big),$$

the value of the constant C being allowed to increase from line to line, and where we used the Lipschitz property of $D\tilde{u}$ in the last line. So for a constant C independent of $\sigma \in (0,1]$, we have:

$$|\mathcal{U}^\sigma(\mu,0)|^2 \leqslant C\Big(1+\big(M_2(\mu)\big)^2\Big).$$

We now claim that the mapping $\mathcal{U}^\sigma$ is jointly continuous. It suffices to observe that, for a sequence $(\mu_n)_{n\geqslant 0}$ converging to $\mu \in \mathcal{P}_2(\mathbb{R}^d)$, the family of mappings $(\mathcal{U}^\sigma(\mu_n,\cdot))_{n\geqslant 0}$ is uniformly continuous, the sequence $(\mathcal{U}^\sigma(\mu_n,0))_{n\geqslant 0}$ being bounded. Therefore, the family of functions $(\mathcal{U}^\sigma(\mu_n,\cdot))_{n\geqslant 0}$ is relatively compact for the topology of uniform convergence on compact subsets. Passing to the $L^2(\Omega,\mathcal{F},\mathbb{P};\mathbb{R}^d)$-limit in the identity:

$$D\tilde{u}(X_n+\sigma G) = \mathcal{U}^\sigma(\mu_n, X_n+\sigma G),$$

where, for all $n \geqslant 0$, $X_n \sim \mu_n$ is independent of G, we deduce that any limit of $(\mathcal{U}^\sigma(\mu_n,\cdot))_{n\geqslant 0}$ coincides with $\mathcal{U}^\sigma(\mu,\cdot)$. Joint continuity easily follows.

Second Step. We now let:

$$\mathcal{U}(\mu,x) = \liminf_{n\to\infty} \mathcal{U}^{2^{-n}}(\mu,x), \quad \mu \in \mathcal{P}_2(\mathbb{R}^d), \quad x \in \mathbb{R}^d,$$

the liminf being taken component by component. Clearly, $\mathcal{U}$ is jointly measurable on $\mathcal{P}_2(\mathbb{R}^d)\times\mathbb{R}^d$. Moreover, since, for each $\sigma \in (0,1]$ and $\mu \in \mathcal{P}_2(\mathbb{R}^d)$, the function $\mathbb{R}^d \ni x \mapsto \mathcal{U}^\sigma(\mu,x)$ is C-Lipschitz continuous for a constant C independent of μ and σ, each coordinate of $\mathcal{U}(\mu,\cdot)$ is also C-Lipschitz continuous. Hence, $\mathcal{U}(\mu,\cdot)$ is Lipschitz continuous, uniformly in $\mu \in \mathcal{P}_2(\mathbb{R}^d)$. Similarly, $\mathcal{U}(\mu,0)$ is bounded, uniformly in μ.

We now identify $\mathcal{U}(\mu,\cdot)$ with a version of $\partial_\mu u(\mu)(\cdot)$. Recall that, for any $\sigma > 0$ and any pair of independent random variables (X,G), with $X \sim \mu$ and $G \sim \mathcal{N}_d(0,\sigma I_d)$, it holds that:

$$\mathbb{P}\big[D\tilde{u}(X+\sigma G) = \mathcal{U}^\sigma(\mu, X+\sigma G)\big] = 1.$$

Since $D\tilde{u}$ is C-Lipschitz continuous, we already know that :

$$\big\|D\tilde{u}\big(X+2^{-n}G\big) - D\tilde{u}(X)\big\|_2 \leqslant 2^{-n}C,$$

so that, by a straightforward application of Borel Cantelli lemma, we get:

$$\mathbb{P}\big[\lim_{n\to\infty} D\tilde{u}(X+2^{-n}G) = D\tilde{u}(X)\big] = 1.$$

Moreover, the C-Lipschitz property of $\mathcal{U}^\sigma(\mu,\cdot)$ implies that we also have:

$$\mathbb{P}\big[\lim_{n\to\infty} |\mathcal{U}^{2^{-n}}(\mu, X+2^{-n}G) - \mathcal{U}^{2^{-n}}(\mu,X)|\big] = 1.$$

Therefore:

$$\mathbb{P}\big[\liminf_{n\to\infty} \mathcal{U}^{2^{-n}}(\mu, X + 2^{-n}G) = \mathcal{U}(\mu, X)\big] = 1,$$

and finally:

$$\mathbb{P}\big[D\tilde{u}(X) = \mathcal{U}(\mu, X)\big] = 1.$$

Third Step. It remains to check that $\mathcal{U}$ is continuous at any (μ, x) such that x belongs to the support of μ. We thus consider such a pair (μ, x) together with a sequence $(\mu_n, x_n)_{n\geqslant 0}$ converging to (μ, x). By the same argument as in the first step, we may extract a subsequence of $(\mathcal{U}(\mu_n, \cdot))_{n\geqslant 0}$ converging for the topology of uniform convergence on compact subsets. For simplicity, we still denote this sequence by $(\mathcal{U}(\mu_n, \cdot))_{n\geqslant 0}$ and we call $v(\cdot)$ its limit. Given a sequence $(X_n)_{n\geqslant 0}$ of random variables converging to X in $L^2(\Omega, \mathcal{F}, \mathbb{P}; \mathbb{R}^d)$, with $X_n \sim \mu_n$ and $X \sim \mu$ (see for instance Subsection 5.3.1 for the construction), we can pass to the limit in:

$$\mathbb{P}\big[D\tilde{u}(X_n) = \mathcal{U}(\mu_n, X_n)\big] = 1,$$

and exploiting the fact $\mathcal{U}(\mu_n, \cdot)$ is Lipschitz continuous, uniformly in n, deduce that:

$$\mathbb{P}\big[D\tilde{u}(X) = v(X)\big] = 1,$$

which proves that v coincides with $\mathcal{U}(\mu, \cdot)$ almost everywhere under μ. Since both are continuous, they must coincide on the support of μ. In particular, for the same x as above, $v(x) = \mathcal{U}(\mu, x)$. Since $\lim_{n\to\infty} \mathcal{U}(\mu_n, x_n) = v(x)$, this completes the proof. □

Finite Dimensional Projection

We will use the following consequence of estimate (5.49):

Proposition 5.39 *Let u be an* L*-differentiable function on $\mathcal{P}_2(\mathbb{R}^d)$ with a Lipschitz derivative, and let $\mu \in \mathcal{P}_2(\mathbb{R}^d)$, $\boldsymbol{x} = (x_1, \cdots, x_N) \in (\mathbb{R}^d)^N$ and $\boldsymbol{y} = (y_1, \cdots, y_N) \in (\mathbb{R}^d)^N$. Then, with the usual notation u^N for the empirical projection of u (recall Definition 5.35) and $\bar{\mu}^N_{\boldsymbol{x}}$ for the empirical measure, we have:*

$$\begin{aligned}
\partial u^N(\boldsymbol{x}) \cdot (\boldsymbol{y} - \boldsymbol{x}) &= \frac{1}{N} \sum_{i=1}^N \partial_\mu u(\mu)(x_i) \cdot (y_i - x_i) \\
&\quad + O\bigg[W_2(\bar{\mu}^N_{\boldsymbol{x}}, \mu) \Big(\frac{1}{N} \sum_{i=1}^N |x_i - y_i|^2 \Big)^{1/2}\bigg],
\end{aligned}$$

the dot product in the left-hand side standing for the usual Euclidean inner product.

Proof. Using Proposition 5.35, we get:

$$
\begin{aligned}
&\partial u^N(\boldsymbol{x}) \cdot (\boldsymbol{y}-\boldsymbol{x}) \\
&= \sum_{i=1}^N \partial_{x_i} u^N(\boldsymbol{x}) \cdot (y_i - x_i) \\
&= \frac{1}{N}\sum_{i=1}^N \partial_\mu u(\bar{\mu}^N_{\boldsymbol{x}})(x_i) \cdot (y_i - x_i) \\
&= \frac{1}{N}\sum_{i=1}^N \partial_\mu u(\mu)(x_i) \cdot (y_i - x_i) + \frac{1}{N}\sum_{i=1}^N \big[\partial_\mu u(\bar{\mu}^N_{\boldsymbol{x}})(x_i) - \partial_\mu u(\mu)(x_i)\big] \cdot (y_i - x_i),
\end{aligned}
$$

where '·' in the left-hand side is the inner product in $\mathbb{R}^{dN}$, while '·' in the right-hand side is the inner product in $\mathbb{R}^d$. Now, by Cauchy-Schwarz' inequality,

$$
\begin{aligned}
&\left|\frac{1}{N}\sum_{i=1}^N [\partial_\mu u(\bar{\mu}^N_{\boldsymbol{x}})(x_i) - \partial_\mu u(\mu)(x_i)] \cdot (y_i - x_i)\right| \\
&\qquad \leqslant \left(\frac{1}{N}\sum_{i=1}^N |\partial_\mu u(\bar{\mu}^N_{\boldsymbol{x}})(x_i) - \partial_\mu u(\mu)(x_i)|^2\right)^{1/2} \left(\frac{1}{N}\sum_{i=1}^N |y_i - x_i|^2\right)^{1/2} \\
&\qquad = \left(\mathbb{E}\big[|\partial_\mu u(\bar{\mu}^N_{\boldsymbol{x}})(x_\vartheta) - \partial_\mu u(\mu)(x_\vartheta)|^2\big]\right)^{1/2} \left(\frac{1}{N}\sum_{i=1}^N |y_i - x_i|^2\right)^{1/2} \\
&\qquad \leqslant 2C W_2(\bar{\mu}^N_{\boldsymbol{x}}, \mu) \left(\frac{1}{N}\sum_{i=1}^N |y_i - x_i|^2\right)^{1/2},
\end{aligned}
$$

if we use the same notation for ϑ as in the proof of Proposition 5.35, and apply the estimate (5.49) with $X = x_\vartheta$, $\mu = \bar{\mu}^N_{\boldsymbol{x}}$, and $\nu = \mu$. □

Remark 5.40 *We shall use the estimate of Proposition 5.39 with $x_i = X_i$ when the X_i's are independent $\mathbb{R}^d$-valued random variables constructed on some $(\Omega, \mathcal{F}, \mathbb{P})$ with common distribution μ, in which case the empirical measure $\bar{\mu}^N_{\boldsymbol{x}}$ is the realization of the (random) empirical measure $\bar{\mu}^N$ of the (random) sample $X_1, \cdots, X_N$. Whenever $\mu \in \mathcal{P}_2(\mathbb{R}^d)$, we know from Subsection 5.1.2 that:*

$$
\mathbb{P}\Big[\lim_{n\to+\infty} W_2(\bar{\mu}^N, \mu) = 0\Big] = 1,
$$

and

$$
\lim_{n\to+\infty} \mathbb{E}\Big[W_2(\bar{\mu}^N, \mu)^2\Big] = 0.
$$

Theorem 5.8 provides a sharp estimate of the rate of convergence whenever $\mu \in \mathcal{P}_q(\mathbb{R}^d)$, namely when $\int_{\mathbb{R}^d} |x|^q \mu(dx) < \infty$ for some $q > 4$. This gives an intuitive interpretation of the result of Proposition 5.39 which then says that, when N is large, the gradient of the empirical projection u^N computed at the empirical *sample* $(X_i)_{1 \leqslant i \leqslant N}$ *is close to the sample* $(\partial_\mu u(\mu)(X_i))_{1 \leqslant i \leqslant N}$, *the accuracy of the approximation being specified in the* $L^2(\Omega, \mathcal{F}, \mathbb{P}; \mathbb{R}^d)$ *norm by* (5.20) *when* μ *is sufficiently integrable.*

5.3.4 Joint Differentiability

We often consider functions $h : \mathbb{R}^n \times \mathcal{P}_2(\mathbb{R}^d) \ni (x, \mu) \mapsto h(x, \mu) \in \mathbb{R}$ depending upon a point x in the n-dimensional Euclidean space $\mathbb{R}^n$ and a probability measure μ in $\mathcal{P}_2(\mathbb{R}^d)$.

Joint Differentiability

The notion of joint differentiability is defined according to the same procedure as before: h is said to be jointly differentiable if the *lifting* $\tilde{h} : \mathbb{R}^n \times L^2(\Omega, \mathcal{F}, \mathbb{P}; \mathbb{R}^d) \ni (x, X) \mapsto h(x, \mathbb{P}_X)$ over some atomless Polish probability space $(\Omega, \mathcal{F}, \mathbb{P})$ is jointly differentiable. If $\tilde{h}$ is continuously differentiable in the direction X, we can define the partial derivatives in x and μ. They read $\mathbb{R}^n \times \mathcal{P}_2(\mathbb{R}^d) \ni (x, \mu) \mapsto \partial_x h(x, \mu)$ and $\mathbb{R}^n \times \mathcal{P}_2(\mathbb{R}^d) \ni (x, \mu) \mapsto \partial_\mu h(x, \mu)(\cdot) \in L^2(\mathbb{R}^d, \mu; \mathbb{R}^d)$ respectively. By construction, the partial Fréchet derivative of $\tilde{h}$ in the direction X is given by the mapping $L^2(\Omega, \mathcal{F}, \mathbb{P}; \mathbb{R}^d) \ni (x, X) \mapsto D_X \tilde{h}(x, X) = \partial_\mu h(x, \mathbb{P}_X)(X) \in L^2(\Omega, \mathcal{F}, \mathbb{P}; \mathbb{R}^d)$. The statement and the proof of Proposition 5.33 can be easily adapted to the joint measurability of $\mathbb{R}^n \times \mathcal{P}_2(\mathbb{R}^d) \times \mathbb{R}^d \ni (x, \mu, v) \mapsto \partial_\mu h(x, \mu)(v)$. In order to distinguish the variable x in $h(x, \mu)$ from the variable at which the derivative with respect to μ is computed, we shall often denote the latter by v.

A standard result from classical analysis which we often use says that joint continuous differentiability in the two arguments is equivalent to the partial differentiability in each of the two arguments together with the joint continuity of the partial derivatives. Here, the joint continuity of $\partial_x h$ is understood as the joint continuity with respect to the Euclidean distance on $\mathbb{R}^n$ and the Wasserstein distance on $\mathcal{P}_2(\mathbb{R}^d)$. The joint continuity of $\partial_\mu h$ needs to be understood as the joint continuity of the mapping $(x, X) \mapsto \partial_\mu h(x, \mathbb{P}_X)(X)$ from $\mathbb{R}^n \times L^2(\Omega, \mathcal{F}, \mathbb{P}; \mathbb{R}^d)$ into $L^2(\Omega, \mathcal{F}, \mathbb{P}; \mathbb{R}^d)$. The proof follows from the standard decomposition:

$$
\begin{aligned}
&h(x', \mu') - h(x, \mu) \\
&= h(x', \mu') - h(x, \mu') + h(x, \mu') - h(x, \mu) \\
&= \int_0^1 \partial_x h\big(\lambda x' + (1-\lambda)x, \mu'\big) \cdot (x' - x) \; d\lambda \\
&\quad + \int_0^1 \mathbb{E}\Big[\partial_\mu h\big(x, \mathcal{L}(\lambda X' + (1-\lambda)X)\big)\big(\lambda X' + (1-\lambda)X\big) \cdot (X' - X)\Big] \, d\lambda,
\end{aligned}
$$

for $x, x' \in \mathbb{R}^d$ and $\mu, \mu' \in \mathcal{P}_2(\mathbb{R}^d)$, and X and X' such that $\mathcal{L}(X) = \mu$ and $\mathcal{L}(X') = \mu'$.

Moreover, when the partial derivatives of $\tilde{h}$ in X are Lipschitz continuous in X, we can use the result of Proposition 5.36 which implies that, for any (x, μ), the *representation* $\mathbb{R}^d \ni v \mapsto \partial_\mu h(x, \mu)(v) \in \mathbb{R}^d$ makes sense as a Lipschitz function in v and that an appropriate version of (5.49) holds true. Moreover, we can adapt Corollary 5.38 in the following way: If $D_X\tilde{h}$ is jointly continuous and the Lipschitz constant of $D_X\tilde{h}$ in X is uniform with respect to x in compact subsets of $\mathbb{R}^n$, then, for each $(x, \mu) \in \mathbb{R}^n \times \mathcal{P}_2(\mathbb{R}^d)$, we can find a Lipschitz continuous version of $\mathbb{R}^d \ni v \mapsto \partial_\mu h(x, \mu)(v) \in \mathbb{R}^d$ such that the mapping $\mathbb{R}^n \times \mathcal{P}_2(\mathbb{R}^d) \times \mathbb{R}^d \ni (x, \mu, v) \mapsto \partial_\mu h(x, \mu)(v)$ is measurable and is continuous at any point (x, μ, v) such that v belongs to the support of μ.

Fully Continuous Derivatives

It is sometimes convenient to have, for any $(x, \mu) \in \mathbb{R}^n \times \mathcal{P}_2(\mathbb{R}^d)$, a version of $\mathbb{R}^d \ni v \mapsto \partial_\mu h(x, \mu)(v) \in \mathbb{R}^d$ such that the global map $\mathbb{R}^n \times \mathcal{P}_2(\mathbb{R}^d) \times \mathbb{R}^d \ni (x, \mu, v) \mapsto \partial_\mu h(x, \mu)(v)$ is, not only continuous at any point (x, μ, v) such that v belongs to the support of μ, but is everywhere continuous. Whenever it exists, such a version will be said to be fully regular.

We here provide a simple criterion that ensures the existence of such a fully regular version:

Lemma 5.41 *Let $(u(x, \mu)(\cdot))_{x\in\mathbb{R}^n, \mu\in\mathcal{P}_2(\mathbb{R}^d)}$ be a collection of real-valued functions satisfying, for all $x \in \mathbb{R}^n$ and $\mu \in \mathcal{P}_2(\mathbb{R}^d)$, $u(x, \mu)(\cdot) \in L^\infty(\mathbb{R}^d, \mu; \mathbb{R})$, and for which there exists a constant C such that, for all $x, x' \in \mathbb{R}^n$, and $\xi, \xi', \chi \in L^2(\Omega, \mathcal{F}, \mathbb{P}; \mathbb{R}^d)$,*

$$\begin{aligned} &\mathbb{E}\big[\big(u(x, \mathcal{L}(\xi))(\xi) - u(x', \mathcal{L}(\xi'))(\xi')\big)\chi\big] \\ &\quad \leq C\Big[\|\chi\|_1\big(|x - x'| + \|\xi - \xi'\|_1\big) + \mathbb{E}\big[|\xi - \xi'|\,|\chi|\big]\Big], \end{aligned}$$

where $(\Omega, \mathcal{F}, \mathbb{P})$ is an atomless probability space. Then, for each $(x, \mu) \in \mathbb{R}^n \times \mathcal{P}_2(\mathbb{R}^d)$, we can find a version of $u(x, \mu)(\cdot) \in L^\infty(\mathbb{R}^d, \mu; \mathbb{R})$ such that, for the same constant C as above, for all $x, x' \in \mathbb{R}^n$, $\mu, \mu' \in \mathcal{P}_2(\mathbb{R}^d)$ and $v, v' \in \mathbb{R}^d$,

$$\big|u(x, \mu)(v) - u(x', \mu')(v')\big| \leq C\big(|x - x'| + W_1(\mu, \mu') + |v - v'|\big).$$

Remark 5.42 *Observe that, differently from what we have done so far, we here use the L^1 norm instead of the L^2 norm, and the 1-Wasserstein distance W_1 instead of the 2-Wasserstein distance W_2, in order to characterize continuity with respect to the measure argument. Of course, this is more demanding. This choice is dictated by the argument used below for exhibiting a fully regular version of $\mathcal{U}$, which is based on the duality between L^1 and L^∞.*

Notice also that, although $u(x, \mu)(\cdot) \in L^\infty(\mathbb{R}^d, \mu; \mathbb{R})$, we do not claim that the version provided by the statement is bounded on the whole $\mathbb{R}^d$.

Proof. As a preliminary remark, notice that from the main assumption in the statement, the map $\mathbb{R}^d \times L^2(\Omega, \mathcal{F}, \mathbb{P}; \mathbb{R}^d) \ni (x, \xi) \mapsto u(x, \mathcal{L}(\xi))(\xi) \in L^2(\Omega, \mathcal{F}, \mathbb{P}; \mathbb{R}^d)$ is continuous.

First Step. Consider $x, x' \in \mathbb{R}^n$ and $\xi, \xi' \in L^2(\Omega, \mathcal{F}, \mathbb{P}; \mathbb{R}^d)$. Observe that from the regularity assumption, with probability 1 under $\mathbb{P}$,

$$\big|u\big(x, \mathcal{L}(\xi)\big)(\xi) - u\big(x', \mathcal{L}(\xi')\big)(\xi')\big| \leqslant C\big(|x - x'| + \|\xi - \xi'\|_1 + |\xi - \xi'|\big).$$

In particular, for a Gaussian random variable $Z \sim N_d(0, I_d)$, Z being independent of (ξ, ξ'), it holds, for any integer $p \geqslant 1$,

$$\begin{aligned}&\big|u\big(x, \mathcal{L}(\xi + \tfrac{1}{p}Z)\big)(\xi + \tfrac{1}{p}Z) - u\big(x', \mathcal{L}(\xi' + \tfrac{1}{p}Z)\big)(\xi' + \tfrac{1}{p}Z)\big| \\ &\quad \leqslant C\big(|x - x'| + \|\xi - \xi'\|_1 + |\xi - \xi'|\big).\end{aligned}$$

Integrating with respect to Z only, we get

$$\begin{aligned}&\bigg|\int_{\mathbb{R}^d} u\big(x, \mathcal{L}(\xi + \tfrac{1}{p}Z)\big)(\xi + \tfrac{1}{p}z)\varphi_d(z)dz - \int_{\mathbb{R}^d} u\big(x', \mathcal{L}(\xi' + \tfrac{1}{p}Z)\big)(\xi' + \tfrac{1}{p}z)\varphi_d(z)dz\bigg| \\ &\quad \leqslant C\big(|x - x'| + \|\xi - \xi'\|_1 + |\xi - \xi'|\big).\end{aligned}$$

Observe that the integrals in the left-hand side are well defined: Since $\mathcal{L}(\xi + \frac{1}{p}Z)$ has a positive density, $u(x, \mathcal{L}(\xi + \frac{1}{p}Z))(\cdot) \in L^\infty(\mathbb{R}^d, \mathrm{Leb}_d; \mathbb{R})$, and similarly for the second integral.

Letting, for all $x \in \mathbb{R}^n$, $\mu \in \mathcal{P}_2(\mathbb{R}^d)$ and $v \in \mathbb{R}^d$,

$$u_p(x, \mu)(v) = \int_{\mathbb{R}^d} u\big(x, \mu * (N_d(0, \tfrac{1}{p^2}I_d))\big)(v + \tfrac{1}{p}z)\varphi_d(z)dz,$$

we get that $u_p(x, \mu)(\cdot)$ is continuous and satisfies, for all x, x' in $\mathbb{R}^n$ and ξ, ξ' in $L^2(\Omega, \mathcal{F}, \mathbb{P}; \mathbb{R}^d)$, with probability 1 under $\mathbb{P}$,

$$\big|u_p\big(x, \mathcal{L}(\xi)\big)(\xi) - u_p\big(x, \mathcal{L}(\xi')\big)(\xi')\big| \leqslant C\big(|x - x'| + \|\xi - \xi'\|_1 + |\xi - \xi'|\big). \tag{5.50}$$

Second Step. We now consider $\mu, \mu' \in \mathcal{P}_2(\mathbb{R}^d)$ such that both have a strictly positive smooth density that decays at least exponentially fast at the infinity and whose derivative also decays at least exponentially fact at infinity. We let π be an optimal coupling between μ and μ' for the 1-Wasserstein distance, that is

$$W_1(\mu, \mu') = \int_{\mathbb{R}^d \times \mathbb{R}^d} |v - v'| d\pi(v, v'),$$

and we let (ξ, ξ') be a pair of random variables with π as distribution.

Following the proof of Proposition 5.36, we can find two continuous mappings $\psi : (0,1)^d \to \mathbb{R}^d$ and $\psi' : (0,1)^d \to \mathbb{R}^d$ such that, for any random variable η with uniform distribution on $(0,1)^d$, it holds that $\psi(\eta) \sim \mu$ and $\psi'(\eta) \sim \mu'$. Importantly, both ψ and ψ' are one-to-one from $(0,1)^d$ onto $\mathbb{R}^d$.

Therefore, for given values of v, v' in $\mathbb{R}^d$, we can find $y_0, y_0' \in (0,1)^d$ such that:

$$v = \psi(y_0), \quad v' = \psi(y_0').$$

Then, for a given random variable η with uniform distribution on $(0,1)^d$ and for $\delta > 0$ such that $B(y_0,\delta) \subset (0,1)^d$ and $B(y_0',\delta) \subset (0,1)^d$, where $B(y_0,\delta)$ denotes the d-dimensional open ball of center y_0 and of radius δ, we let:

$$\eta' = \begin{cases} \eta & \text{if } \eta \notin B(y_0,\delta) \cup B(y_0',\delta), \\ \eta + y_0' - y_0 & \text{if } \eta \in B(y_0,\delta), \\ \eta + y_0 - y_0' & \text{if } \eta \in B(y_0',\delta). \end{cases}$$

Then, η' is also uniformly distributed on $(0,1)^d$.

Without any loss of generality, we may assume that (ξ,ξ') is independent of η, and thus of (η,η'). We then let $\bar{\xi} = \psi(\eta)$ and $\bar{\xi}' = \psi'(\eta')$.

Third Step. We now consider a Bernoulli random variable ε with parameter $q \in (0,1)$, independent of (ξ,ξ',η). We let:

$$\xi^\varepsilon = \varepsilon\xi + (1-\varepsilon)\bar{\xi}, \quad \xi^{\varepsilon,\prime} = \varepsilon\xi' + (1-\varepsilon)\bar{\xi}'.$$

Clearly, ξ^ε and $\xi^{\varepsilon,\prime}$ have μ and μ' as respective distributions. Taking advantage of the conclusion of the first step, we deduce that, with probability 1 under $\mathbb{P}$,

$$\left| u_p(x,\mu)(\xi^\varepsilon) - u_p(x',\mu')(\xi^{\varepsilon,\prime}) \right| \leqslant C\big(|x-x'| + \|\xi^\varepsilon - \xi^{\varepsilon,\prime}\|_1 + |\xi^\varepsilon - \xi^{\varepsilon,\prime}|\big).$$

In particular, almost surely on the event $\{\varepsilon = 0\} \cap \{\eta \in B(x_0,\delta)\}$,

$$\begin{aligned} &\left| u_p(x,\mu)\big(\psi(\eta)\big) - u_p(x',\mu')\big(\psi'(\eta + y_0' - y_0)\big) \right| \\ &\quad \leqslant C\big(|x-x'| + \|\xi^\varepsilon - \xi^{\varepsilon,\prime}\|_1 + |\psi(\eta) - \psi'(\eta + y_0' - y_0)|\big). \end{aligned}$$

Therefore, we can find a sequence $(y^m)_{m\geqslant 1}$ converging toward y_0 such that:

$$\begin{aligned} &\left| u_p(x,\mu)\big(\psi(y^m)\big) - u_p(x',\mu')\big(\psi'(y^m + y_0' - y_0)\big) \right| \\ &\quad \leqslant C\big(|x-x'| + \|\xi^\varepsilon - \xi^{\varepsilon,\prime}\|_1 + |\psi(y^m) - \psi'(y^m + y_0' - y_0)|\big). \end{aligned}$$

By continuity of $u_p(x,\mu)(\cdot)$ and $u_p(x,\mu')(\cdot)$ and of ψ and ψ', we get, by letting m tend to ∞,

$$\left| u_p(x,\mu)(v) - u_p(x,\mu')(v') \right| \leqslant C\big(|x-x'| + \|\xi^\varepsilon - \xi^{\varepsilon,\prime}\|_1 + |v-v'|\big),$$

where we used the fact that $\psi(y_0) = v$ and $\psi'(y_0') = v'$.

Letting the parameter q of ε tend to 1 and recalling the choice of (ξ,ξ'), we deduce that:

$$\left| u_p(x,\mu)(v) - u_p(x',\mu')(v') \right| \leqslant C\big(|x-x'| + W_1(\mu,\mu') + |v-v'|\big). \tag{5.51}$$

Inequality (5.51) holds true for probability measures μ,μ' that have a strictly positive smooth density that decays at least exponentially fast at infinity and whose derivative also decays at least exponentially fast at infinity. Since the set of such smooth probability measures is dense in $\mathcal{P}_2(\mathbb{R}^d)$, we deduce that the restriction of u_p to smooth probability measures extends by continuity to the whole $\mathbb{R}^n \times \mathcal{P}_2(\mathbb{R}^d) \times \mathbb{R}^d$. Of course, the continuous extension, denoted by $\bar{u}_p$, satisfies (5.51). By (5.50), for any $(x, \xi) \in \mathbb{R}^n \times L^2(\Omega, \mathcal{F}, \mathbb{P}; \mathbb{R}^d)$, it holds that $\mathbb{P}[u_p(x, \mathcal{L}(\xi))(\xi) = \bar{u}_p(x, \mathcal{L}(\xi))(\xi)] = 1$. Since $u_p(x, \mathcal{L}(\xi))(\cdot)$ and $\bar{u}_p(x, \mathcal{L}(\xi))(\cdot)$ are continuous, we deduce that, for any $\mu \in \mathcal{P}_2(\mathbb{R}^d)$, $\bar{u}_p(x, \mu)(v)$ coincides with $u_p(x, \mu)(v)$ when v belongs to the support of μ. Put differently, $\bar{u}_p(x, \mu)(\cdot)$ provides a version of $u_p(x, \mu)(\cdot)$ in $L^\infty(\mathbb{R}^d, \mu; \mathbb{R})$.

Fourth Step. Actually, inequality (5.51) shows that $\bar{u}_p$ extends to the whole $\mathbb{R}^n \times \mathcal{P}_1(\mathbb{R}^d) \times \mathbb{R}^d$, the extension still satisfying (5.51). The extension is at most of linear growth:

$$|\bar{u}_p(x, \mu)(v)| \leqslant |\bar{u}_p(0, \delta_0)(0)| + C\big(|x| + M_1(\mu) + |v|\big).$$

Since $u_p(0, \delta_0)(0) = \mathbb{E}[u(0, \frac{1}{p}Z)(\frac{1}{p}Z)]$ and since the map $\mathbb{R}^d \times L^2(\Omega, \mathcal{F}, \mathbb{P}; \mathbb{R}^d) \ni (x, \xi) \mapsto u(x, \mathcal{L}(\xi))(\xi) \in L^2(\Omega, \mathcal{F}, \mathbb{P}; \mathbb{R}^d)$ is continuous, we deduce that the sequence $(\bar{u}_p(0, \delta_0)(0) = u_p(0, \delta_0)(0))_{p \geqslant 1}$ is bounded, which shows that the functions $(\bar{u}_p)_{p \geqslant 1}$ are uniformly at most of linear growth.

Recalling that any bounded subset of $\mathcal{P}_2(\mathbb{R}^d)$ is a compact subset of $\mathcal{P}_1(\mathbb{R}^d)$, we deduce from the Arzelà-Ascoli theorem that there exists a subsequence, still denoted by $(\bar{u}_p)_{p \geqslant 1}$, that converges uniformly on any bounded subset of $\mathbb{R}^n \times \mathcal{P}_2(\mathbb{R}^d) \times \mathbb{R}^d$.

It remains to identify the limit of $\bar{u}_p(x, \mu)(\cdot)$ with a version of $u(x, \mu)(\cdot)$ in $L^\infty(\mathbb{R}^d, \mu; \mathbb{R})$. This follows from the fact that, for any bounded and measurable function $g : \mathbb{R}^d \to \mathbb{R}$,

$$\mathbb{E}\big[\bar{u}_p\big(x, \mathcal{L}(\xi)\big)(\xi) g(\xi)\big] = \mathbb{E}\big[u\big(x, \mathcal{L}(\xi + \tfrac{1}{p}Z)\big)(\xi + \tfrac{1}{p}Z) g(\xi)\big],$$

which implies that:

$$\lim_{p \to \infty} \mathbb{E}\big[\bar{u}_p\big(x, \mathcal{L}(\xi)\big)(\xi) g(\xi)\big] = \mathbb{E}\big[u\big(x, \mathcal{L}(\xi)\big)(\xi) g(\xi)\big],$$

where we used the first inequality in the statement of Lemma 5.41. □

Other Forms of Joint Regularity of the Derivative

We shall need other forms of joint regularity of $\partial_\mu h$. Observe for instance that, whenever $\tilde{h} : \mathbb{R}^n \times L^2(\Omega, \mathcal{F}, \mathbb{P}; \mathbb{R}^d) \ni X \mapsto h(x, \mathbb{P}_X)$ is Fréchet differentiable we have:

$$\lim_{\epsilon \searrow 0} \sup_{\|Y\|_2 \leqslant 1} \left| \frac{1}{\epsilon} \Big(\tilde{h}(x, X + \epsilon Y) - \tilde{h}(x, X) \Big) - D_X \tilde{h}(x, X) \cdot Y \right| = 0.$$

So, if h is jointly measurable on $\mathbb{R}^n \times \mathcal{P}_2(\mathbb{R}^d)$, then, for any $Z \in L^2(\Omega, \mathcal{F}, \mathbb{P}; \mathbb{R}^d)$, the mapping $\mathbb{R}^n \times [L^2(\Omega, \mathcal{F}, \mathbb{P}; \mathbb{R}^d)]^2 \ni (x, X, Y) \mapsto \mathbb{E}[(D_X \tilde{h}(x, X) - Z) \cdot Y]$ is measurable and the mapping $\mathbb{R}^n \times L^2(\Omega, \mathcal{F}, \mathbb{P}; \mathbb{R}^d) \ni (x, X) \mapsto \mathbb{E}[(D_X \tilde{h}(x, X) - Z) \cdot \psi(X, Z)]$ is also measurable for any bounded and continuous function ψ from $(\mathbb{R}^d)^2$ into $\mathbb{R}^d$. If $\mathcal{Q}$ is a dense countable subset of the space of continuous functions from $(\mathbb{R}^d)^2 \to \mathbb{R}^d$ converging to 0 at infinity, we see that:

$$\begin{aligned} &\mathbb{R}^n \times L^2(\Omega, \mathcal{F}, \mathbb{P}; \mathbb{R}^d) \ni (x, X) \\ &\quad \mapsto \sup_{\psi \in \mathcal{Q}} \Big\{ \mathbb{E}\big[\big(D_X \tilde{h}(x, X) - Z\big) \cdot \psi(X, Z)\big] \mathbf{1}_{\{\|\psi(X,Z)\|_2 \leqslant 1\}} \Big\} \end{aligned}$$

is also measurable. Recalling that $D_X\tilde{h}(x,X)$ belongs to $L^2(\Omega,\sigma\{X\},\mathbb{P};\mathbb{R}^d)$, we deduce that the right-hand side is exactly $\|D_X\tilde{h}(x,X) - Z\|_2$. This proves that $\mathbb{R}^n \times L^2(\Omega,\mathcal{F},\mathbb{P};\mathbb{R}^d) \ni (x,X) \mapsto D_X\tilde{h}(x,X)$ is measurable when $L^2(\Omega,\mathcal{F},\mathbb{P};\mathbb{R}^d)$ is separable, since the Borel σ-field of $L^2(\Omega,\mathcal{F},\mathbb{P};\mathbb{R}^d)$ is then generated by a countable collection of balls. The space $L^2(\Omega,\mathcal{F},\mathbb{P};\mathbb{R}^d)$ is separable when $\mathcal{F}$ is a countably generated σ-field or the completion of a countably generated σ-field. When $\tilde{h}$ is continuously differentiable in the direction X, $D_X\tilde{h}$ may be represented by means of $\partial_\mu h$ and the mapping $\mathbb{R}^n \times L^2(\Omega,\mathcal{F},\mathbb{P};\mathbb{R}^d) \ni (x,X) \mapsto \partial_\mu h(x,\mathcal{L}(X))(X) \in L^2(\Omega,\mathcal{F},\mathbb{P};\mathbb{R}^d)$ is measurable when $L^2(\Omega,\mathcal{F},\mathbb{P};\mathbb{R}^d)$ is separable.

Continuity may be addressed by using the same procedure. For instance, if $\Psi : \mathbb{R}^n \times \mathcal{P}_2(\mathbb{R}^d) \times \mathbb{R}^d \ni (x,\mu,v) \mapsto \Psi(x,\mu,v) \in \mathbb{R}$ is continuous at any point (x,μ,v) such that $v \in \text{Supp}(\mu)$ and satisfies:

$$\sup_{(x,\mu)\in\mathcal{K}} \int_{\mathbb{R}^d} |\Psi(x,\mu,v)|^2 d\mu(v) < \infty,$$

for any compact subset $\mathcal{K} \subset \mathbb{R}^n \times \mathcal{P}_2(\mathbb{R}^d)$, with $\Psi(x,\mu,\cdot) : \mathbb{R}^d \ni v \mapsto \Psi(x,\mu,v)$ being measurable, then the function $\mathbb{R}^n \times L^2(\Omega,\mathcal{F},\mathbb{P};\mathbb{R}) \ni (x,X) \mapsto \Psi(x,\mathcal{L}(X),X) \in L^p(\Omega,\mathcal{F},\mathbb{P};\mathbb{R})$ is continuous when $p \in [1,2)$ and is measurable when $p = 2$. It is continuous when $p = 2$ if we assume further that, for any compact subset $\mathcal{K} \subset \mathbb{R}^n \times \mathcal{P}_2(\mathbb{R}^d)$,

$$\lim_{a\to\infty} \sup_{(x,\mu)\in\mathcal{K}} \int_{\mathbb{R}^d} |\Psi(x,\mu,v)|^2 \mathbf{1}_{|\Psi(x,\mu,v)| \geqslant a} d\mu(v) = 0.$$

The proof is quite simple. It follows from the fact that, whenever a sequence $(x_\ell, X_\ell)_{\ell\geqslant 1}$ converges to some (x,X) in $\mathbb{R}^n \times L^2(\Omega,\mathcal{F},\mathbb{P};\mathbb{R}^d)$, then the sequence $(\Psi(x_\ell,\mathcal{L}(X_\ell),X_\ell))_{\ell\geqslant 1}$ converges to $\Psi(x,\mathcal{L}(X),X)$ in probability and the proof is completed using uniform integrability. When $p = 2$ and continuity does not hold, the measurability may be proved by approximating Ψ by $(\rho_\ell \circ \Psi)_{\ell\geqslant 1}$ where, for each $\ell \geqslant 1$, $\rho_\ell : \mathbb{R} \to \mathbb{R}$ is a bounded and continuous function satisfying $\rho_\ell(x) = x$ for $|x| \leqslant \ell$ and $|\rho_\ell(x)| \leqslant \ell$ for all $x \in \mathbb{R}^d$.

5.4 Comparisons with Other Notions of Differentiability

We argued repeatedly that the notion of L-differentiability was natural in the context of functions of probability measures appearing as the distributions of random variables which needed to be perturbed. More convincing arguments will come with the discussions of applications in which we need to track the dependence with respect to the marginal distributions of a stochastic dynamical system. See for example Section 5.6 later in this chapter or Chapter 6.

This being said, L-differentiability may not appear as the most intuitive notion, especially from a geometric analysis perspective. The goal of this section is to enlighten the reader as to the relationships between L-differentiability and some of the more traditional approaches which have been advocated in similar contexts, and especially in the theory of optimal transportation.

5.4.1 Smooth Functions of Measures

Natural generalizations of multivariate calculus suggest that differential calculus should be easily set up in a vector space. Fréchet's theory of differentiable functions on an open subset of a Banach space is a case in point. Notice that the notion of L-differentiability is based on an attempt to provide such a vector space structure by lifting functions of probability measures to a flat vector space away from the space $\mathcal{P}_2(\mathbb{R}^d)$. Typically, the lifting lives on a Hilbert space represented as an L^2-space, on which differential calculus can be handled with standard Fréchet's theory. However, since the space of probability measures, say $\mathcal{P}_2(\mathbb{R}^d)$ for the sake of definiteness, is a rather thin manifold in the vector space of measures, it is tempting to work with functions u defined on an open neighborhood of $\mathcal{P}_2(\mathbb{R}^d)$, say in the linear space of finite measures for instance, and use a form of linear differential calculus on this open set. It is instructive to compare this notion of differentiability to L-differentiability.

Linear Functional Derivative

We refrain from dwelling on the exact topological structure put on the space of finite measures. Instead, we start with a rather informal definition of a smooth function on the space $\mathcal{M}(\mathbb{R}^d)$ of finite measures on $\mathbb{R}^d$. We would like to say that a function $u : \mathcal{M}(\mathbb{R}^d) \to \mathbb{R}$ is smooth if there exists a function:

$$\frac{\delta u}{\delta m} : \mathcal{M}(\mathbb{R}^d) \times \mathbb{R}^d \to \mathbb{R}$$

such that, for every measures m and m' in $\mathcal{M}(\mathbb{R}^d)$, it holds:

$$u(m') - u(m) = \int_0^1 \int_{\mathbb{R}^d} \frac{\delta u}{\delta m}\big(tm' + (1-t)m\big)(x) d[m' - m](x)\, dt, \qquad (5.52)$$

with $[\delta u/\delta m](tm' + (1-t)m)(\cdot)$ being close to $[\delta u/\delta m](m)(\cdot)$ as m' gets close to m in a suitable sense. The function $[\delta u/\delta m](m)(\cdot)$ should be understood as the Fréchet or Gâteaux derivative of u at m, but for the sake of definiteness we shall call it the (linear) functional derivative of u with respect to the measure m.

Stated in this way, the definition is not genuine as nothing is said of the well posedness of the right-hand side in (5.52), neither on the topologies needed to quantify the proximity of m' and m, and subsequently $[\delta u/\delta m](m')(\cdot)$ and $[\delta u/\delta m](m)(\cdot)$.

Nevertheless, formula (5.52) is quite appealing. Whenever m and m' are elements of $\mathcal{P}_2(\mathbb{R}^d)$, the derivative $[\delta u/\delta m]$ is evaluated at measures $(tm' + (1-t)m)_{0\leqslant t\leqslant 1}$, which are also probability measures. This suggests how we could define the linear functional derivative of a function u defined only on $\mathcal{P}_2(\mathbb{R}^d)$ as opposed to a neighborhood of the space of probability measures.

Definition 5.43 *A function $u : \mathcal{P}_2(\mathbb{R}^d) \to \mathbb{R}$ is said to have a linear functional derivative if there exists a function:*

$$\frac{\delta u}{\delta m} : \mathcal{P}_2(\mathbb{R}^d) \times \mathbb{R}^d \ni (m,x) \mapsto \frac{\delta u}{\delta m}(m)(x) \in \mathbb{R},$$

continuous for the product topology, $\mathcal{P}_2(\mathbb{R}^d)$ being equipped with the 2-Wasserstein distance, such that, for any bounded subset $\mathcal{K} \subset \mathcal{P}_2(\mathbb{R}^d)$, the function $\mathbb{R}^d \ni x \mapsto [\delta u/\delta m](m)(x)$ is at most of quadratic growth in x uniformly in m for $m \in \mathcal{K}$, and such that for all m and m' in $\mathcal{P}_2(\mathbb{R}^d)$, it holds:

$$u(m') - u(m) = \int_0^1 \int_{\mathbb{R}^d} \frac{\delta u}{\delta m}\big(tm' + (1-t)m\big)(x) d[m' - m](x)\, dt. \tag{5.53}$$

The assumption in Definition 5.43 is tailor-made to guarantee the well posedness of the right-hand side of (5.53). It also ensures that for any $m \in \mathcal{P}_2(\mathbb{R}^d)$,

$$\lim_{m'\to m} \int_{\mathbb{R}^d} \Big[\frac{\delta u}{\delta m}\big(tm' + (1-t)m\big)(x) - \frac{\delta u}{\delta m}(m)(x)\Big] d\big[m' - m\big](x) = 0,$$

which may be proved by writing:

$$\int_{\mathbb{R}^d} \Big[\frac{\delta u}{\delta m}\big(tm' + (1-t)m\big)(x)\Big] dm'(x) = \mathbb{E}\Big[\frac{\delta u}{\delta m}\big(tm' + (1-t)m\big)(X')\Big],$$

for a random variable $X' \sim m'$ and then by letting $X' \to X$ in $L^2(\Omega, \mathcal{F}, \mathbb{P}; \mathbb{R}^d)$, with $X \sim m$, which is possible thanks to the Skorohod representation theorem provided that $(\Omega, \mathcal{F}, \mathbb{P})$ is chosen accordingly. See Subsection 5.3.1. The convergence follows from the continuity of $[\delta u/\delta m]$ together with a domination argument and the fact that $W_2(tm' + (1-t)m, m) \to 0$ as m' tends towards m. In particular, u must be continuous on $\mathcal{P}_2(\mathbb{R}^d)$.

We can provide a first-order expansion of $u(m')$ as m' tends to m under extra assumptions on $[\delta u/\delta m]$.

Proposition 5.44 *Let u have a linear functional derivative in the sense of Definition 5.43. Assume further that, for any $m \in \mathcal{P}_2(\mathbb{R}^d)$, the function $\mathbb{R}^d \ni x \mapsto [\delta u/\delta m](m)(x)$ is differentiable and the derivative*

$$\mathcal{P}_2(\mathbb{R}^d) \times \mathbb{R}^d \ni (m, x) \mapsto \partial_x \Big[\frac{\delta u}{\delta m} \Big](m)(x) \in \mathbb{R}^d$$

is jointly continuous in (m, x) *and is at most of linear growth in* x*, uniformly in* $m \in \mathcal{K}$ *for any bounded subset* $\mathcal{K} \subset \mathcal{P}_2(\mathbb{R}^d)$.

Then, for any $m \in \mathcal{P}_2(\mathbb{R}^d)$*, we can find a function* $\varepsilon : \mathbb{R}_+ \to \mathbb{R}_+$ *such that* $\lim_{r \searrow 0} \varepsilon(r) = 0$ *and for all* $m' \in \mathcal{P}_2(\mathbb{R}^d)$,

$$u(m') - u(m) = \int_{\mathbb{R}^d} \frac{\delta u}{\delta m}(m)(x) d[m' - m](x) + o\big(W_2(m, m')\big), \tag{5.54}$$

where $o(W_2(m, m'))$ *is a term depending on* m *and* m' *and satisfying* $|o(W_2(m, m'))| \leq \varepsilon(W_2(m, m'))W_2(m, m')$.

Clearly, (5.54) is reminiscent of the notion of Fréchet derivative.

Proof. We write:

$$\begin{aligned} u(m') - u(m) &= \int_{\mathbb{R}^d} \frac{\delta u}{\delta m}(m)(x) d[m' - m](x) \\ &+ \int_0^1 \int_{\mathbb{R}^d} \Big[\frac{\delta u}{\delta m}\big(tm' + (1-t)m\big)(x) - \frac{\delta u}{\delta m}(m)(x) \Big] d[m' - m](x)\, dt. \end{aligned}$$

Denoting by π an optimal transport plan from m to m', we have:

$$\begin{aligned} &u(m') - u(m) \\ &= \int_{\mathbb{R}^d} \frac{\delta u}{\delta m}(m)(x) d[m' - m](x) \\ &\quad + \int_0^1 \int_{\mathbb{R}^d \times \mathbb{R}^d} \Big[\Big(\frac{\delta u}{\delta m}\big(tm' + (1-t)m\big)(y) - \frac{\delta u}{\delta m}(m)(y) \Big) \\ &\qquad\qquad - \Big(\frac{\delta u}{\delta m}\big(tm' + (1-t)m\big)(x) - \frac{\delta u}{\delta m}(m)(x) \Big) \Big] d\pi(x, y)\, dt \\ &= \int_{\mathbb{R}^d} \frac{\delta u}{\delta m}(m)(x) d[m' - m](x) \\ &\quad + \int_0^1 \int_0^1 \int_{\mathbb{R}^d \times \mathbb{R}^d} \Big[\Big(\partial_x \frac{\delta u}{\delta m}\big(tm' + (1-t)m\big)\big(\lambda y + (1-\lambda)x\big) \\ &\qquad\qquad - \partial_x \frac{\delta u}{\delta m}(m)\big(\lambda y + (1-\lambda)x\big) \Big) \cdot (y - x) \Big] d\pi(x, y)\, d\lambda\, dt. \end{aligned}$$

In order to complete the proof, it suffices to show that the last term in the above right-hand side is $o(W_2(m', m))$ as $W_2(m', m) \to 0$ while m is fixed. Invoking Lemma 5.30 and letting, with the same notation as in the statement, $\varepsilon(r) = \sup\{\vartheta(\pi)\}$, the supremum being taken over the probability measures $\pi \in \mathcal{P}_2((\mathbb{R}^d)^2)$ having m as first marginal on $\mathbb{R}^d$ and satisfying $\int_{(\mathbb{R}^d)^2} |x - y|^2 d\pi(x, y) \leq r^2$, we just have to prove that:

$$\lim_{X' \to X} \int_0^1 \int_0^1 \mathbb{E}\Big[\Big(\partial_x \frac{\delta u}{\delta m}(tm' + (1-t)m)\big(\lambda X' + (1-\lambda)X\big) \\ - \partial_x \frac{\delta u}{\delta m}(m)\big(\lambda X' + (1-\lambda)X\big)\Big) \cdot \frac{X' - X}{\|X' - X\|_2}\Big] d\lambda dt = 0,$$

the convergence of X' to X being understood in $L^2(\Omega, \mathcal{F}, \mathbb{P}; \mathbb{R}^d)$, for some well-chosen probability space $(\Omega, \mathcal{F}, \mathbb{P})$, with the convention that $X' \sim m'$ and that $X \sim m$. The above limit follows from the growth and continuity properties of $\partial_x[\delta u/\delta m]$ by the same argument as that used to handle Examples 1 and 3 in Subsection 5.2.2. □

Remark 5.45 *In order to distinguish the functional derivative $\delta u/\delta m$ from the L-derivative $\partial_\mu u$, we use the letter δ instead of ∂ for the differential symbol. Moreover, we use the letter m instead of μ for the measure argument.*

Remark 5.46 *Formulas* (5.53) *and* (5.54) *only involve integrals with respect to the measure $m' - m$. As a result, any constant can be added to $\delta u/\delta m$ without affecting either of these formulas as long as the measures m and m' have the same total mass. Consequently, $\delta u/\delta m$ is only defined up to an additive constant.*

Remark 5.47 *Observe also that if there exists a continuous function $[\delta u/\delta m]$ which is at most of quadratic growth in x uniformly in m for m in a bounded subset of $\mathcal{P}_2(\mathbb{R}^d)$ and for which the conclusion* (5.54) *of Proposition 5.44 holds true, then u has $[\delta u/\delta m]$ as linear functional derivative. This follows from the fact that, for any two $m, m' \in \mathcal{P}_2(\mathbb{R}^d)$, under* (5.54)*, the mapping* $[0,1] \ni t \mapsto u(m + t(m' - m))$ *is differentiable with*

$$\frac{d}{dt}\big[u\big(m + t(m' - m)\big)\big] = \int_{\mathbb{R}^d} \frac{\delta u}{\delta m}\big(m + t(m' - m)\big)(x) d\big[m' - m\big](x), \quad t \in [0,1],$$

the right-hand side being continuous in t.

Connection with the L-Derivative

The purpose of the present discussion is to relate the linear functional derivative to the L-derivative whenever they both exist. As one can expect by now, we shall make systematic use of the lifting:

$$\tilde{u} : L^2(\Omega, \mathcal{F}, \mathbb{P}; \mathbb{R}^d) \ni X \mapsto u\big(\mathcal{L}(X)\big),$$

where $(\Omega, \mathcal{F}, \mathbb{P})$ is an atomless Polish probability space.

If u has a linear functional derivative in the sense of Definition 5.43, if X and Y are $\mathbb{R}^d$-valued square integrable random variables, and if we denote by $\tilde{u}$ a lifting of u to the space on which X and Y are defined, we have:

$$
\begin{aligned}
\tilde{u}(X+\epsilon Y)-\tilde{u}(X) &= u\big(\mathcal{L}(X+\epsilon Y)\big)-u\big(\mathcal{L}(X)\big) \\
&= \int_0^1 \left[\int_{\mathbb{R}^d} \frac{\delta u}{\delta m}\big(\mathcal{L}^\epsilon(t)\big)(x)\, d\big[\mathcal{L}(X+\epsilon Y)-\mathcal{L}(X)\big](x)\right] dt \\
&= \int_0^1 \mathbb{E}\left[\frac{\delta u}{\delta m}\big(\mathcal{L}^\epsilon(t)\big)(X+\epsilon Y)-\frac{\delta u}{\delta m}\big(\mathcal{L}^\epsilon(t)\big)(X)\right] dt,
\end{aligned}
$$

where we used the notation $\mathcal{L}^\epsilon(t)$ for $t\mathcal{L}(X+\epsilon Y)+(1-t)\mathcal{L}(X)$. Furthermore, if we assume that the function u satisfies the assumption of Proposition 5.44, then, following the proof of (5.54), we have for $\|Y\|_2 \leqslant 1$:

$$
\begin{aligned}
&\tilde{u}(X+\epsilon Y)-\tilde{u}(X) \\
&= \epsilon \int_0^1 \mathbb{E}\left[\int_0^1 \left[\partial_x \frac{\delta u}{\delta m}\big(\mathcal{L}^\epsilon(t)\big)(X+\epsilon\lambda Y)\cdot Y\right] d\lambda\right] dt \\
&= \epsilon \mathbb{E}\left[\partial_x \frac{\delta u}{\delta m}\big(\mathcal{L}^0(t)\big)(X)\cdot Y\right] \\
&\quad + \epsilon \int_0^1 \mathbb{E}\left[\int_0^1 \left[\left(\partial_x \frac{\delta u}{\delta m}\big(\mathcal{L}^\epsilon(t)\big)(X+\epsilon\lambda Y)-\partial_x \frac{\delta u}{\delta m}\big(\mathcal{L}^0(t)\big)(X)\right)\cdot Y\right] d\lambda\right] dt \\
&= \epsilon \mathbb{E}\left[\partial_x \frac{\delta u}{\delta m}\big(\mathcal{L}(X)\big)(X)\cdot Y\right] + o(\epsilon),
\end{aligned}
\tag{5.55}
$$

where the Landau notation $o(\epsilon)$ in the last line is uniform with respect to $Y \in L^2(\Omega, \mathcal{F}, \mathbb{P}; \mathbb{R}^d)$ with $\|Y\|_2 \leqslant 1$. This shows that the lifting $\tilde{u}$ is Fréchet differentiable at X and that its Fréchet derivative is given by:

$$
D\tilde{u} : L^2(\Omega, \mathcal{F}, \mathbb{P}; \mathbb{R}^d) \ni X \mapsto \partial_x \frac{\delta u}{\delta m}\big(\mathcal{L}(X)\big)(X),
$$

which is continuous because of the continuity and growth properties of $\partial_x[\delta u/\delta m]$. We state the above result in a proposition for later reference.

Proposition 5.48 *Under the assumptions of Proposition 5.44, the function u is* L*-differentiable and*

$$
\boxed{\partial_\mu u(\mu)(\cdot) = \partial_x \frac{\delta u}{\delta m}(\mu)(\cdot), \quad \mu \in \mathcal{P}_2(\mathbb{R}^d).}
$$

Remark 5.49 *The above derivation has the following enlightening interpretation. If a smooth extension of u to the space of measures exists, the* L*-derivative of u at $\mu \in \mathcal{P}_2(\mathbb{R}^d)$, when viewed as a function on $\mathbb{R}^d$, is the derivative (gradient) of the (Fréchet or Gâteaux) derivative of u when considered as a function on the vector* *space of signed measures. This general fact was already encountered with our first example of an* L*-derivative computation. The fact that the* L*-derivative, when viewed as a function from* $\mathbb{R}^d$ *into itself, appears to be the gradient of a scalar function will be argued rigorously in a more general setting in Proposition 5.50 below.*

5.4.2 From the L-Derivative to the Functional Derivative

While we showed in the previous subsection how to reconstruct the L-derivative from the functional derivative, we now attempt to recover the linear functional derivative from the L-derivative.

The L-Derivative as a Gradient

In order to shed more light on the comparison with functional derivatives, we prove that L-derivatives are typically given by gradients of scalar functions on $\mathbb{R}^d$.

Proposition 5.50 *If the scalar function* u *is* L*-differentiable on* $\mathcal{P}_2(\mathbb{R}^d)$ *and the Fréchet derivative of its lifting is uniformly Lipschitz, then for each* $\mu \in \mathcal{P}_2(\mathbb{R}^d)$*, we can find a continuous version of* $\partial_\mu u(\mu)(\cdot)$ *which, when viewed as a function from* $\mathbb{R}^d$ *into itself, is the gradient of a scalar-valued continuously differentiable function* p^μ*. When* p^μ *is chosen to be* 0 *at* 0*, it may be expressed as:*

$$p^\mu(x) = \int_0^1 \partial_\mu u(\mu)(tx) \cdot x dt, \qquad x \in \mathbb{R}^d. \tag{5.56}$$

Proof.

First Step. We use the fact that, if a locally square-integrable vector field $f : \mathbb{R}^d \to \mathbb{R}^d$ is such that $\int_{\mathbb{R}^d} f(x) \cdot b(x)dx = 0$ for every smooth divergence free vector field b with compact support, then there exists a locally square-integrable function $p : \mathbb{R}^d \to \mathbb{R}$ such that:

$$f = \nabla p,$$

in the sense of distributions. See Remark 1.9 in Temam's book [331]. We show that whenever f is continuous, p must be continuously differentiable. Consider indeed a sequence of mollifiers $(\eta^\varepsilon)_{\varepsilon>0}$ with compact support. Then, for all $\varepsilon > 0$,

$$f * \eta^\varepsilon = \nabla\big(p * \eta^\varepsilon\big),$$

where we used the symbol $*$ to denote convolution. Since both sides of the equality are smooth, we have:

$$\forall x \in \mathbb{R}^d, \quad \big(p * \eta^\varepsilon\big)(x) = \big(p * \eta^\varepsilon\big)(0) + \int_0^1 \Big[\big(f * \eta^\varepsilon\big)(tx) \cdot x\Big] dt. \tag{5.57}$$

Since f is continuous, the right-hand side is continuous in x, uniformly in $\varepsilon > 0$ on any compact subset of $\mathbb{R}^d$. This shows that the functions $(\mathbb{R}^d \ni x \mapsto p * \eta^\varepsilon(x) - p * \eta^\varepsilon(0))_{\varepsilon>0}$ are equicontinuous on compact sets. Therefore, the family $(\mathbb{R}^d \ni x \mapsto p * \eta^\varepsilon(x) - p * \eta^\varepsilon(0))_{\varepsilon>0}$ is relatively compact for the topology of uniform convergence on compact subsets of $\mathbb{R}^d$. We call $\tilde{p}$ a limit. Similarly,

$$\forall x, y \in \mathbb{R}^d, \quad (p * \eta^\varepsilon)(y) = (p * \eta^\varepsilon)(x) + \int_0^1 \Big[(f * \eta^\varepsilon)\big(ty + (1-t)x\big) \cdot (y - x) \Big] dt.$$

Letting ε tend to 0, we get:

$$\forall x, y \in \mathbb{R}^d, \quad \tilde{p}(y) = \tilde{p}(x) + \int_0^1 \Big[f\big(ty + (1-t)x\big) \cdot (y - x) \Big] dt,$$

which proves that $\tilde{p}$ is differentiable and that f is its gradient.

Second Step. Let $m \in \mathcal{P}_2(\mathbb{R}^d)$ be fixed and choose an arbitrary smooth divergence free vector field b with a compact support. For every $\epsilon > 0$, we define $m_\epsilon = m * \varphi_\epsilon$ where φ_ϵ denotes the density of the Gaussian distribution with mean 0 and variance $\epsilon^2 I_d$. As defined, m_ϵ is a smooth function which is strictly positive everywhere and, in particular, bounded below by a strictly positive constant on the support of b. Consequently, the ordinary differential equation:

$$\dot{X}_t^\epsilon = \frac{b}{m_\epsilon}(X_t^\epsilon), \quad t \geqslant 0,$$

has a unique solution for every initial condition, and the measure $m_\epsilon(dx) = m_\epsilon(x)dx$ is invariant for this dynamical system, as b is divergence free. The fact that we use the same notation m_ϵ for a measure and its density should not be a source of confusion. In any case, $\mathcal{L}(X_t^\epsilon) = \mathcal{L}(X_0^\epsilon)$ if $\mathcal{L}(X_0^\epsilon) = m_\epsilon$. Note that the randomness in $\boldsymbol{X}^\epsilon$ comes from the initial condition only. Consequently,

$$\begin{aligned}
0 &= \frac{1}{t}\big[u\big(\mathcal{L}(X_t^\epsilon)\big) - u\big(\mathcal{L}(X_0^\epsilon)\big)\big] \\
&= \frac{1}{t}\big[\tilde{u}(X_t^\epsilon) - \tilde{u}(X_0^\epsilon)\big] \\
&= \frac{1}{t} D\tilde{u}(X_0^\epsilon) \cdot (X_t^\epsilon - X_0^\epsilon) + \frac{1}{t} o\big(\|X_t^\epsilon - X_0^\epsilon\|_2\big) \\
&= \mathbb{E}\Big[\partial_\mu u(m_\epsilon)(X_0^\epsilon) \cdot \frac{b}{m_\epsilon}(X_0^\epsilon)\Big] + o(1),
\end{aligned}$$

where $\lim_{t \searrow 0} o(1) = 0$. Taking the limit $t \searrow 0$, we get:

$$\int_{\mathbb{R}^d} \partial_\mu u(m_\epsilon)(x) \cdot b(x)\, dx = 0. \tag{5.58}$$

We can now take the limit $\epsilon \searrow 0$ in the above equality and get the desired result. Recall that Proposition 5.36 and Corollary 5.38 guarantee the existence, for every $\mu \in \mathcal{P}_2(\mathbb{R}^d)$, of a version $x \mapsto \partial_\mu u(\mu)(x)$ of the L-derivative such that $x \mapsto \partial_\mu u(\mu)(x)$ is Lipschitz continuous with a constant independent of μ and such that $|\partial_\mu u(\mu)(0)| \leqslant c(1 + M_2(\mu))$ for a constant c independent of μ. So, the family $(\partial_\mu u(m_\epsilon)(\cdot))_{\epsilon>0}$ is uniformly continuous on compact subsets of $\mathbb{R}^d$, and we can extract a subsequence which converges uniformly on compact sets toward a function providing a version of $\partial_\mu u(m)(\cdot)$. Since this limit satisfies (5.58) for every b, it is a gradient.

Third Step. Since the version constructed in the second step is continuous, we deduce from the first step that its potential p^μ is continuously differentiable. By choosing the version of p^μ which vanishes at 0 and by taking the limit in (5.57), we prove that the representation formula (5.56) holds. □

Recovering the Linear Functional Derivative

We now show how to recover the linear functional derivative from the L-derivative of a function of measures.

Proposition 5.51 *Assume that the scalar function u is* L*-differentiable on $\mathcal{P}_2(\mathbb{R}^d)$ and that the Fréchet derivative of its lifting is uniformly Lipschitz. Assume also that, for each $\mu \in \mathcal{P}_2(\mathbb{R}^d)$, we can find a version of $\mathbb{R}^d \ni x \mapsto \partial_\mu u(\mu)(x)$ such that the mapping $\mathbb{R}^d \times \mathcal{P}_2(\mathbb{R}^d) \ni (x, \mu) \mapsto \partial_\mu u(\mu)(x)$ is continuous. Then, the function u has a linear functional derivative, which satisfies the fundamental relationship in the statement of Proposition 5.48. Moreover, the conclusion of Proposition 5.44 holds.*

Proof.

First Step. The goal is to check that we can write u as in Definition 5.43, with $[\delta u/\delta m]$ being equal to the version of the potential constructed in Proposition 5.50.

We thus consider two square-integrable random variables X and Y together with U a third random variable assumed to be uniform on the segment $[0, 1]$ and independent of (X, Y). As usual, we work on an atomless Polish probability space $(\Omega, \mathcal{F}, \mathbb{P})$. We also denote by Φ (resp. φ) the cumulative distribution (resp. density) function of the univariate standard normal distribution $N(0, 1)$. Given two parameters $\sigma > 0$ and $t \in (0, 1)$, we let:

$$\xi^{t,\sigma} = \Phi\Big(\frac{t-U}{\sigma}\Big)Y + \Big[1 - \Phi\Big(\frac{t-U}{\sigma}\Big)\Big]X.$$

This definition was chosen so that, as σ tends to 0, $\xi^{t,\sigma}$ tends almost surely toward:

$$\xi^{t,0} = \mathbf{1}_{\{U<t\}}Y + \mathbf{1}_{\{U>t\}}X,$$

which has $t\mathcal{L}(Y) + (1-t)\mathcal{L}(X)$ as distribution.

When $\sigma > 0$, the mapping $(0, 1) \ni t \mapsto \xi^{t,\sigma}$ is differentiable with derivative:

$$\frac{d}{dt}\xi^{t,\sigma} = \frac{1}{\sigma}\varphi\Big(\frac{t-U}{\sigma}\Big)(Y - X).$$

Since the derivative is bounded by $C|X-Y|$ for a constant C independent of t, differentiability also holds in the L^2 sense. Therefore, by definition of the L-derivative, we have:

$$u\big(\mathcal{L}(Y)\big) - u\big(\mathcal{L}(X)\big) = \lim_{\sigma \searrow 0} \int_0^1 \mathbb{E}\Big[\partial_\mu u\big(\mathcal{L}(\xi^{t,\sigma})\big)(\xi^{t,\sigma}) \cdot \Big(\frac{1}{\sigma}\varphi\Big(\frac{t-U}{\sigma}\Big)(Y-X)\Big)\Big]dt.$$

Since U is independent of (X, Y), we have:

$$
\begin{aligned}
&u(\mathcal{L}(Y)) - u(\mathcal{L}(X)) \\
&= \lim_{\sigma \searrow 0} \mathbb{E} \int_0^1 \int_0^1 \Big[\partial_\mu u(\mathcal{L}(\xi^{t,\sigma})) \Big(\Phi(\frac{t-\theta}{\sigma}) Y + [1 - \Phi(\frac{t-\theta}{\sigma})] X \Big) \\
&\qquad \cdot \frac{1}{\sigma} \varphi(\frac{t-\theta}{\sigma}) (Y - X) \Big] d\theta \, dt.
\end{aligned}
\tag{5.59}
$$

Second Step. Thanks to Proposition 5.50, we may consider, for any $\mu \in \mathcal{P}_2(\mathbb{R}^d)$, a potential $p^\mu : \mathbb{R}^d \to \mathbb{R}$ of $\partial_\mu u(\mu)(\cdot)$ provided we carefully choose the version of $\partial_\mu u(\mu)(\cdot)$. Actually, since the L-derivative of u is assumed to have a version jointly continuous in the space and measure variables, this jointly continuous version must coincide with the version constructed in the proof of Proposition 5.50. When μ has $\mathbb{R}^d$ as support, $\partial_\mu u(\mu)(\cdot)$ admits a unique continuous version on the entire $\mathbb{R}^d$. When the support of μ is a strict subset of $\mathbb{R}^d$, uniqueness of the continuous version holds true on the support of μ only. However, the density argument used in Proposition 5.50 to construct the continuous version shows that, in that case as well, it coincides with the jointly continuous version of the L-derivative.

Choosing the version of the potential that vanishes at 0, we have the representation formula:

$$
p^\mu(x) = \int_0^1 \partial_\mu u(\mu)(tx) \cdot x dt, \qquad (x, \mu) \in \mathbb{R}^d \times \mathcal{P}_2(\mathbb{R}^d),
$$

which proves that p is jointly continuous in (x, μ). Returning to the conclusion of the first step, we observe that:

$$
\begin{aligned}
&\partial_\mu u(\mathcal{L}(\xi^{t,\sigma})) \Big(\Phi(\frac{t-\theta}{\sigma}) Y + [1 - \Phi(\frac{t-\theta}{\sigma})] X \Big) \cdot \Big(\frac{1}{\sigma} \varphi(\frac{t-\theta}{\sigma}) (Y - X) \Big) \\
&= -\partial_\theta \Big[p^{\mathcal{L}(\xi^{t,\sigma})} \Big(\Phi(\frac{t-\theta}{\sigma}) Y + [1 - \Phi(\frac{t-\theta}{\sigma})] X \Big) \Big].
\end{aligned}
$$

Therefore, using (5.59) and the first step, we get:

$$
\begin{aligned}
u(\mathcal{L}(Y)) - u(\mathcal{L}(X)) = - \lim_{\sigma \searrow 0} \mathbb{E} \int_0^1 \Big[& p^{\mathcal{L}(\xi^{t,\sigma})} \Big(\Phi(\frac{t-1}{\sigma}) Y + [1 - \Phi(\frac{t-1}{\sigma})] X \Big) \\
& - p^{\mathcal{L}(\xi^{t,\sigma})} \Big(\Phi(\frac{t}{\sigma}) Y + [1 - \Phi(\frac{t}{\sigma})] X \Big) \Big] dt.
\end{aligned}
$$

From the proof of Proposition 5.50, we know that $\partial_\mu u(\mu)(x)$ is at most of linear growth in x, uniformly in μ in bounded sets. Therefore, $p^\mu(x)$ is at most of quadratic growth in x, uniformly in μ in bounded sets. By a uniform integrability argument, we can exchange the limit and the integral. We get:

$$u(\mathcal{L}(Y)) - u(\mathcal{L}(X)) = -\mathbb{E}\int_0^1 \Big[p^{t\mathcal{L}(Y)+(1-t)\mathcal{L}(X)}(X) - p^{t\mathcal{L}(Y)+(1-t)\mathcal{L}(X)}(Y)\Big]dt$$

$$= \int_0^1 \int_{\mathbb{R}^d} p^{t\mathcal{L}(Y)+(1-t)\mathcal{L}(X)}(y)\big(\mathcal{L}(Y) - \mathcal{L}(X)\big)(dy)\, dt,$$

which proves that, for all $\mu \in \mathcal{P}_2(\mathbb{R}^d)$, $p^\mu(\cdot) = [\delta u/\delta m](\mu)(\cdot)$ up to an additive constant depending on μ.

Third Step. The last two claims in the statement are easily proven. The fundamental relationship is a straightforward consequence of the fact that $p^\mu(\cdot) = [\delta u/\delta m](\mu)(\cdot)$, while the last claim follows from the fact that the assumptions of Proposition 5.44 are satisfied, the linear growth property of $\partial_\mu u$ being shown as in the proof of Proposition 5.50. □

5.4.3 Geometric Analysis on the Wasserstein Space $\mathcal{P}_2(\mathbb{R}^d)$

In the spirit of the theory of differential manifolds, we introduce a geometric structure on the Wasserstein space $\mathcal{P}_2(\mathbb{R}^d)$ via a natural notion of displacement. Our goal is to connect the notion of Lions' L-derivative introduced earlier, to the notion of Wasserstein gradient issued from this geometric differential structure.

Displacements in $\mathcal{P}_2(\mathbb{R}^d)$

Our earlier discussion of optimal transportation, recall for example Subsection 5.1.3, took place in a static framework. Our goal is now to give a dynamic flavor to this theory. For that purpose, we need a differential geometric structure on the space of probability measures, and its introduction requires the notion of differentiable curves joining elements of this space. So, given two probability measures μ and ν in $\mathcal{P}_2(\mathbb{R}^d)$ for which we can find an optimal transport map ψ from μ to ν, we would like to construct *natural* paths, think for example of the graphs of functions from $[0, 1]$ to $\mathcal{P}_2(\mathbb{R}^d)$, that would go from μ to ν.

A straightforward though rather naive solution would be to use the probability measures μ_t defined by $\mu_t = (1-t)\mu + t\nu$ for $0 \leqslant t \leqslant 1$. This is natural indeed as $\mathcal{P}_2(\mathbb{R}^d)$ can be viewed as embedded in the linear space $\mathcal{M}(\mathbb{R}^d)$ of signed measures on $\mathbb{R}^d$ equipped with its vector space structure. However this *natural* guess ignores the metric structure of $\mathcal{P}_2(\mathbb{R}^d)$ provided by the Wasserstein distance W_2, and offers no insight in the transport of $\mu_0 = \mu$ into $\mu_1 = \nu$.

Inspired by the lifting procedure used in the construction of L-derivatives, one can also search for a path from $[0, 1]$ into $L^2(\Omega, \mathcal{F}, \mathbb{P}; \mathbb{R}^d)$ for an atomless Polish probability space $(\Omega, \mathcal{F}, \mathbb{P})$, which would go from a random variable X with distribution μ to $\psi(X)$ which has ν as distribution if ψ is indeed a transport map from μ to ν. Taking advantage of the flat nature of the linear space L^2, it is tempting to consider:

$$X_t = (1-t)X + t\psi(X) = X + t\big(\psi(X) - X\big), \quad t \in [0, 1],$$

and use the path $\boldsymbol{\mu} = (\mu_t)_{0 \leqslant t \leqslant T}$ in $\mathcal{P}_2(\mathbb{R}^d)$ given by $\mu_t = \mathcal{L}(X_t)$. Notice that in this case:

$$\dot{X}_t = \psi(X) - X, \quad t \in [0, 1].$$

Brenier's Theorem 5.20 suggests, at least when μ is absolutely continuous, that we search for ψ in the form of the gradient of a convex function φ. Notice that, if θ is any smooth real valued function on $\mathbb{R}^d$ with compact support, then for $\epsilon > 0$ small enough, the function $\varphi : \mathbb{R}^d \ni x \mapsto \varphi(x) = (1/2)|x|^2 + \epsilon\theta(x) \in \mathbb{R}$ is a smooth strongly convex function. Using $\psi = \nabla\varphi$ for such a function φ we get:

$$X_t = X + t\epsilon\nabla\theta(X) = (I + t\epsilon\nabla\theta)(X), \qquad \text{and} \qquad \dot{X}_t = \epsilon\nabla\theta(X), \quad t \in [0, 1].$$

By Proposition 5.13, ψ is an optimal transport map from μ to $\nu = \mu \circ \psi^{-1}$. At this stage, the key observation is that, φ being strongly convex, the map $\nabla\varphi = I + t\epsilon\nabla\theta$ is invertible for any $t \in [0, 1]$. Indeed, for $x \in \mathbb{R}^d$, the map $\mathbb{R}^d \ni y \mapsto y \cdot x - \varphi(y)$ has a unique maximizer, say y_x; it satisfies $x = \nabla\varphi(y_x)$. Therefore, the path $[0, 1] \ni t \mapsto X_t$ solves the ordinary differential equation:

$$\dot{X}_t = \epsilon\nabla\theta\Big(\big(I + t\epsilon\nabla\theta\big)^{-1}(X_t)\Big), \quad t \in [0, 1]. \tag{5.60}$$

While it is unclear at this stage that this situation is generic, the above example shows that, in order to go from X to $\psi(X)$ with an optimal transport map, one can simply follow the flow of an Ordinary Differential Equation (ODE) in a linear vector space. As we shall see next, it can be proved that the path $(\mathcal{L}(X_t))_{0 \leqslant t \leqslant 1}$ is in fact a geodesic for the 2-Wasserstein distance.

This is indeed a key idea in the theory of optimal transportation according to which probability measures are *transported* along the flow of an ODE induced by some possibly time-dependent vector field:

$$\dot{\phi}_t^x = b(t, \phi_t^x), \quad t \in [0, 1], \quad \phi_0^x = x, \tag{5.61}$$

for a Borel-measurable mapping $b : [0, 1] \times \mathbb{R}^d \to \mathbb{R}^d$ which plays the role of the vector field. If the ODE (5.61) is well posed for any initial condition $x \in \mathbb{R}^d$, we may indeed transport an initial distribution $\mu_0 \in \mathcal{P}_2(\mathbb{R}^d)$ by pushing it forward with the flow solving the ODE, getting:

$$\mu_t = \mu_0 \circ \big(\mathbb{R}^d \ni x \mapsto \phi_t^x \in \mathbb{R}^d\big)^{-1}.$$

Reformulated in probabilistic terms, μ_t is the distribution of the solution X_t of the ODE at time t when initialized at time $t = 0$ with a random variable X_0 having μ_0 as distribution. In particular,

$$\dot{X}_t = b(t, X_t), \quad t \in [0, 1], \tag{5.62}$$

which should be compared with (5.60). As a result, integration by parts implies that the dynamics of $(\mu_t)_{0\leqslant t\leqslant 1}$ are given by the first-order Fokker-Planck equation:

$$\partial_t \mu_t + \operatorname{div}\bigl(b(t,\cdot)\mu_t\bigr) = 0, \quad t \in [0,1], \tag{5.63}$$

understood in the sense of distributions. Notice that there are plenty more vector fields transporting μ_0 to μ_1 in this way. Indeed, if we add a divergence free (for μ_t) vector field $w(t,\cdot)$ to $b(t,\cdot)$, that is a vector field satisfying:

$$\operatorname{div}\bigl(w(t,\cdot)\mu_t\bigr) = 0, \tag{5.64}$$

in the sense of distributions, then equation (5.63) remains unchanged. Recall that this equation needs to be understood in the sense that:

$$\forall \psi \in \mathcal{C}_c^\infty(\mathbb{R}^d), \quad \int_{\mathbb{R}^d} w(t,x)\cdot \nabla\psi(x) d\mu_t(x) = 0,$$

where $\mathcal{C}_c^\infty(\mathbb{R}^d)$ is the space of real valued smooth functions with compact support in $\mathbb{R}^d$.

Remark 5.52 *The above argument, though informal, will make up half of the proof of an important result of Benamou and Brenier proven below as Theorem 5.53.*

Optimality of the Displacements

When $\mu_0 = \mu$ and $\mu_1 = \nu$ are given, the definition of the Wasserstein distances and the solution of the classical optimal transportation problem involve the construction of couplings (also called transport plans) of the two measures with the goal of minimizing a cost depending only upon the properties of the coupling. This problem is static in nature. As stated earlier, the purpose of this section is to introduce a dynamic component in the transport of μ_0 to μ_1.

As highlighted by the informal discussion above, a major issue is the construction of a vector field $b(t,\cdot)$ transporting $\mu_0 = \mu$ into $\mu_1 = \nu$ and minimizing a cost functional. In the particular case of the quadratic cost $|x-y|^2$, such an optimization problem should be somehow connected with the definition of the 2-Wasserstein distance. This is the object of Theorem 5.53 proven below.

Coming back to the approach introduced above based on the lifting of the optimization problem from the space $\mathcal{P}_2(\mathbb{R}^d)$ to a flat space L^2 of random variables, the search for an optimal displacement can be reformulated as an optimal control problem. Indeed, the desired vector field b can be interpreted as a control in closed-loop feedback form. So we assume that the dynamics of the control system are given by the solution $\boldsymbol{X} = (X_t)_{0\leqslant t\leqslant 1}$ of an equation of the form:

$$\dot{X}_t = \alpha_t,$$

with initial condition $\mathcal{L}(X_0) = \mu$, driven by a control process $\boldsymbol{\alpha} = (\alpha_t)_{0\leqslant t\leqslant 1}$, the goal being to minimize the energy $J(\boldsymbol{\alpha}) = \mathbb{E}\int_0^1 |\alpha_t|^2 dt$ under the constraint $\mathcal{L}(X_1) = \nu$. As explained, we restrict the control process to be in closed loop feedback form, i.e., $\alpha_t = b(t, X_t)$, and we understand that the divergence free component of $b(t,\cdot)$ has no impact on the dynamics of (5.63). Therefore, we may restrict the search for $b(t,\cdot)$ to the subspace orthogonal to the divergence free vector fields for μ_t, namely the closure of the gradients in $L^2(\mathbb{R}^d, \mu_t; \mathbb{R}^d)$, where $\mu_t = \mathcal{L}(X_t)$. Part of the difficulty is the fact that this space is not known since the measure μ_t is unknown. The following result, which is often referred to as Benamou and Brenier's theorem, addresses this quandary.

Theorem 5.53 *For any $\mu, \nu \in \mathcal{P}_2(\mathbb{R}^d)$, the 2-Wasserstein distance between μ and ν satisfies:*

$$W_2(\mu,\nu)^2 = \inf_{(\boldsymbol{\mu},\boldsymbol{b}) \in \mathcal{A}(\mu,\nu)} A(\boldsymbol{\mu},\boldsymbol{b}) \tag{5.65}$$

where the action $A(\boldsymbol{\mu},\boldsymbol{b})$ is defined by:

$$A(\boldsymbol{\mu},\boldsymbol{b}) = \int_0^1 \int_{\mathbb{R}^d} |b(t,x)|^2 d\mu_t(x)\, dt,$$

the infimum being taken over the set $\mathcal{A}(\mu,\nu)$ of pairs $(\boldsymbol{\mu},\boldsymbol{b}) \in C([0,1]; \mathcal{P}_2(\mathbb{R}^d)) \times L^2([0,1] \times \mathbb{R}^d, \mu_t(dx)dt; \mathbb{R}^d)$, where $\mu_0 = \mu$, $\mu_1 = \nu$, and:

$$\partial_t \mu_t + \mathrm{div}\big(b(t,\cdot)\mu_t\big) = 0,$$

in the sense of distributions.

Remark 5.54 *The above version of Benamou and Brenier's theorem is due to Ambrosio, Gigli, and Savaré, but the lines of the proof below are inspired from Villani's monograph; in this latter reference, μ and ν are required to be absolutely continuous and compactly supported. The reader is referred to the Notes & Complements at the end of the chapter for precise citations.*

Proof. We provide a sketch of proof only in the case when μ and ν are absolutely continuous. For a regular enough vector field $\boldsymbol{b}$ (for example locally bounded in (t,x) and Lipschitz in x, uniformly in t), and for each $t \in [0,1]$, let us define $\mu_t = \mu \circ X_t(\cdot)^{-1}$ where $(X_t(x))_{0 \leqslant t \leqslant 1, x \in \mathbb{R}^d}$ is the flow associated with the vector field $\boldsymbol{b}$. Then, $(\boldsymbol{\mu},\boldsymbol{b}) \in \mathcal{A}(\mu,\nu)$. Using successively, the definition of $(\mu_t)_{0 \leqslant t \leqslant 1}$, the definition of the solution $(X_t)_{0 \leqslant t \leqslant 1}$, Fubini's theorem, Hölder inequality, the definition of the set $\mathcal{A}(\mu,\nu)$, and finally the definition of $W_2(\mu,\nu)$, we get:

$$\begin{aligned}
\int_0^1 \int_{\mathbb{R}^d} |b(t,x)|^2\, \mu_t(dx)dt &= \int_0^1 \int_{\mathbb{R}^d} |b(t,X_t(x))|^2\, d\mu(x)dt \\
&= \int_0^1 \int_{\mathbb{R}^d} |\dot{X}_t(x)|^2\, d\mu(x)dt \\
&= \int_{\mathbb{R}^d} \Big(\int_0^1 |\dot{X}_t(x)|^2 dt \Big)\, d\mu(x)
\end{aligned}$$

$$
\begin{aligned}
&\geqslant \int_{\mathbb{R}^d} \left| \int_0^1 \dot{X}_t(x)dt \right|^2 d\mu(x) \\
&= \int_{\mathbb{R}^d} \left|X_1(x) - x\right|^2 d\mu(x) \geqslant W_2(\mu, \nu)^2.
\end{aligned}
$$

Since the right-hand side is independent of $(\boldsymbol{\mu}, \boldsymbol{b}) \in \mathcal{A}(\mu, \nu)$, we can take the infimum of the left-hand side over $\mathcal{A}(\mu, \nu)$ and still preserve the inequality. In order to prove that $W_2(\mu, \nu)^2$ is not greater than the right-hand side of (5.65), we need to consider general $(\boldsymbol{\mu}, \boldsymbol{b}) \in \mathcal{A}(\mu, \nu)$, without assuming that $\boldsymbol{b}$ is Lipschitz in space and make sure that the above inequality still holds. This can be done by a mollifying argument and controlling the limits when removing the mollification. The details are rather involved and we shall not give them here. Details can be found in the references given in the Notes & Complements at the end of the chapter.

We now prove the reverse inequality. For that, we assume that μ is absolutely continuous and we use the Brenier map φ. So $\mu \circ (\nabla \varphi)^{-1} = \nu$ and:

$$
W_2(\mu, \nu)^2 = \int_{\mathbb{R}^d} |\nabla \varphi(x) - x|^2 \mu(dx).
$$

Piggybacking on the informal discussion of the beginning of the section, for $t \in [0, 1]$ and $x \in \mathbb{R}^d$ we set:

$$
\varphi_t(x) = \frac{1-t}{2}|x|^2 + t\varphi(x), \qquad \text{and} \qquad X_t(x) = \nabla \varphi_t(x) = (1-t)x + t\nabla \varphi(x),
$$

the second definition making sense at points x where φ is well defined (which is true almost everywhere under the Lebesgue measure Leb_d on $\mathbb{R}^d$). Since μ is absolutely continuous with respect to Leb_d, we can define the flow $\boldsymbol{\mu}$ by $\mu_t = \mu \circ X_t(\cdot)^{-1}$, for $t \in [0, 1]$. From the definition of $(X_t(x))_{0 \leqslant t \leqslant 1}$ we get:

$$
\dot{X}_t(x) = -x + \nabla \varphi(x), \qquad t \in [0, 1]. \tag{5.66}
$$

Our goal is to find a vector field $\boldsymbol{b}$ for which $b(t, \cdot)$ is defined μ_t-almost surely and in such a way that the above right-hand side can be rewritten as $b(t, X_t(x))$ for μ_0-almost every $x \in \mathbb{R}^d$. Indeed, for such a vector field, we have:

$$
\begin{aligned}
\int_0^1 \int_{\mathbb{R}^d} |b(t,x)|^2 \mu_t(dx)dt &= \int_0^1 \int_{\mathbb{R}^d} |b(t, X_t(x))|^2 \mu_0(dx)dt \\
&= W_2(\mu, \nu)^2,
\end{aligned}
$$

where we used the definition of the flow $\boldsymbol{\mu} = (\mu_t)_{0 \leqslant t \leqslant 1}$ and the definition of $\boldsymbol{b}$. This shows that $(\boldsymbol{\mu}, \boldsymbol{b})$ is optimal. The existence of $\boldsymbol{b}$ follows from Remark 5.21 following Brenier's Theorem 5.20 by setting:

$$
b(t,x) = \nabla \varphi\big(\nabla \varphi_t^*(x)\big) - \nabla \varphi_t^*(x), \qquad t \in [0, 1],\ \mu_t - \text{almost every } x \in \mathbb{R}^d,
$$

where φ_t^* is the convex conjugate of φ_t. When $t = 1$, $\varphi_1 = \varphi^*$ and Remark 5.21 guarantees that $\nabla\varphi(\nabla\varphi^*(y)) = y$ for ν-almost every $y \in \mathbb{R}^d$ and $\nabla\varphi^*(\nabla\varphi(x)) = x$ for μ almost every $x \in \mathbb{R}^d$. When $t \in [0,1)$, φ_t is strongly convex, which shows that $\mu_t = \mu \circ \nabla\varphi_t^{-1}$ is absolutely continuous. Proceeding as in Remark 5.21, we deduce that $\nabla\varphi_t(\nabla\varphi_t^*(y)) = y$ for μ_t-almost every $y \in \mathbb{R}^d$ and $\nabla\varphi_t^*(\nabla\varphi_t(x)) = x$ for μ-almost every $x \in \mathbb{R}^d$. □

Remark 5.55 *The construction of* $\boldsymbol{X}$ *in* (5.66) *is a generalization of the construction of* $\boldsymbol{X}$ *in* (5.60)*, as provided in the introductory discussion of the subsection.*

The optimal flow $\boldsymbol{\mu}$ constructed in the proof of Benamou-Brenier's theorem is given by:

$$\mu_t = \mu \circ \big((1-t)I + t\nabla\varphi\big)^{-1}.$$

It is called the McCann's interpolation between μ and ν. Observe also that $b(t,\cdot)$ may be rewritten:

$$\begin{aligned} b(t,x) &= \nabla\varphi\big(\nabla\varphi_t^*(x)\big) - \nabla\varphi_t^*(x) \\ &= \Big(\frac{1}{t}\nabla\varphi_t - \frac{1-t}{t}I\Big)\big(\nabla\varphi_t^*(x)\big) - \nabla\varphi_t^*(x) \\ &= \frac{1}{t}\big(x - \nabla\varphi_t^*(x)\big), \end{aligned}$$

for μ_t almost every $x \in \mathbb{R}^d$ and for $t \in (0,1]$.

Remark 5.56 *It is possible to* extract *more properties of the McCann's interpolation flow from the above proof. Indeed, in the spirit of the differential geometric discussion we provide next, and even though we will not study geodesics per se, we mention the fact that the infimum in* (5.65) *is achieved along a geodesic path of constant speed* $(\mu_t)_{0\leqslant t\leqslant 1}$*, i.e.* $W_2(\mu_t,\mu_{t+h}) = hW_2(\mu,\nu)$ *for all* $0 \leqslant t < t+h \leqslant 1$*, and the time-dependent vector field* $\boldsymbol{b} : [0,1]\times\mathbb{R}^d \to \mathbb{R}$ *satisfies:*

$$b(t,\cdot) \in \overline{\big\{\nabla\varphi;\ \varphi \in \mathcal{C}_c^\infty(\mathbb{R}^d;\mathbb{R})\big\}}^{L^2(\mathbb{R}^d,\mu_t;\mathbb{R}^d)}, \quad \mathrm{Leb}_1 \text{ a.e. } t \in [0,1],$$

the right-hand side denoting the closure of the set of smooth gradients in the space $L^2(\mathbb{R}^d,\mu_t;\mathbb{R}^d)$.

Prompted by the statement of the above remark, we introduce formally constant speed geodesics in $\mathcal{P}_2(\mathbb{R}^d)$ and we prove basic properties which will help our discussion of the geometric properties of the Wasserstein space $\mathcal{P}_2(\mathbb{R}^d)$.

Definition 5.57 *A curve* $[0,1] \ni t \mapsto \mu_t \in \mathcal{P}_2(\mathbb{R}^d)$ *is called a geodesic path of constant speed if* $W_2(\mu_s,\mu_t) = (t-s)W_2(\mu_0,\mu_1)$ *whenever* $0 \leqslant s \leqslant t \leqslant 1$.

A first important property of these special curves is given in the following proposition.

Proposition 5.58 *If* $[0,1] \ni t \mapsto \mu_t \in \mathcal{P}_2(\mathbb{R}^d)$ *is a geodesic path of constant speed, then for any* $t \in (0,1)$, $\Pi_2^{\text{opt}}(\mu_t, \mu_0)$ *(resp.* $\Pi_2^{\text{opt}}(\mu_t, \mu_1)$*) contains a unique transport plan which is given by a transport map.*

Proof. Let $\gamma \in \Pi_2^{\text{opt}}(\mu_0, \mu_t)$ and $\eta \in \Pi_2^{\text{opt}}(\mu_t, \mu_1)$ be optimal transportation plans, and write their disintegrations with respect to their second and first marginals respectively as:

$$\gamma(dx, dy) = \gamma(dx, y)\mu_t(dy), \qquad \text{and} \qquad \eta(dy, dz) = \mu_t(dy)\eta(y, dz).$$

Finally define the probability measure $\lambda \in \mathcal{P}_2((\mathbb{R}^d)^3)$ by:

$$\lambda(dx, dy, dz) = \gamma(dx, y)\mu_t(dy)\eta(y, dz),$$

and use the notations $\pi^{1,2}$, $\pi^{1,3}$ and $\pi^{2,3}$ for the projections defined by $\pi^{1,2}(x,y,z) = (x,y)$, $\pi^{1,3}(x,y,z) = (x,z)$, and $\pi^{2,3}(x,y,z) = (y,z)$. By construction, we have $\lambda \circ (\pi^{1,2})^{-1} = \gamma$, and $\lambda \circ (\pi^{2,3})^{-1} = \eta$. Moreover, if we define $\mu \in \mathcal{P}_2(\mathbb{R}^d \times \mathbb{R}^d)$ by $\mu = \lambda \circ (\pi^{1,3})^{-1}$, then $\mu \in \Pi(\mu_0, \mu_1)$. We show that in fact, μ is an optimal transport plan in the sense that $\mu \in \Pi_2^{\text{opt}}(\mu_0, \mu_1)$. Indeed:

$$\begin{aligned}
W_2(\mu_0, \mu_1) &\leqslant \left[\int_{(\mathbb{R}^d)^2} |x - z|^2 \mu(dx, dz)\right]^{1/2} \\
&= \left[\int_{(\mathbb{R}^d)^3} |x - z|^2 \lambda(dx, dy, dz)\right]^{1/2} \\
&\leqslant \left[\int_{(\mathbb{R}^d)^3} |x - y|^2 \lambda(dx, dy, dz)\right]^{1/2} + \left[\int_{(\mathbb{R}^d)^3} |y - z|^2 \lambda(dx, dy, dz)\right]^{1/2} \\
&= \left[\int_{(\mathbb{R}^d)^2} |x - y|^2 \gamma(dx, dy)\right]^{1/2} + \left[\int_{(\mathbb{R}^d)^2} |y - z|^2 \eta(dy, dz)\right]^{1/2} \\
&= W_2(\mu_0, \mu_t) + W_2(\mu_t, \mu_1) = tW_2(\mu_0, \mu_1) + (1-t)W_2(\mu_0, \mu_1) \\
&= W_2(\mu_0, \mu_1),
\end{aligned}$$

so that all the above inequalities are in fact equalities. In particular, since the norm of $L^2((\mathbb{R}^d)^3, \lambda; \mathbb{R}^d)$ is strictly convex, this implies that there exists $\alpha > 0$ such that:

$$y - x = \alpha(z - x) \quad \text{for} \quad \lambda - \text{a.e. } (x, y, z). \tag{5.67}$$

Using the fact that $W_2(\mu_0, \mu_t) = tW_2(\mu_0, \mu_1)$, we conclude that $\alpha = t$. Defining the function $\bar{\gamma}$ by $\bar{\gamma}(y) = \int_{\mathbb{R}^d} x\gamma(dx, y)$ for μ_t-a.e. $y \in \mathbb{R}^d$, and integrating both sides of (5.67) with respect to the probability measure $\gamma(dx, y)$ we get:

$$y - \bar{\gamma}(y) = t(z - \bar{\gamma}(y)) \quad \text{for} \quad \eta - \text{a.e. } (y, z),$$

which shows that the map:

$$y \mapsto \varphi_t(y) = \frac{1}{t}y - \frac{1-t}{t}\bar{\gamma}(y)$$

is a transport map giving the transport plan η. Since $\bar{\gamma}$ depends only upon γ, and γ and η were chosen independently of each other, this shows that the transport plan η is unique. This concludes the proof for $\Pi_2^{\text{opt}}(\mu_t, \mu_1)$. We reach the same conclusion for $\Pi_2^{\text{opt}}(\mu_t, \mu_0)$ by exchanging the roles of μ_0 and μ_1 through a simple time reversal $t \mapsto 1 - t$. □

The following simple result provides a large class of constant speed geodesics. In particular, it justifies the claim made in Remark 5.56 about the McCann interpolation.

Proposition 5.59 *If μ_0 and μ_1 belong to $\mathcal{P}_2(\mathbb{R}^d)$, with $\mu_0 \neq \mu_1$, and $\mu \in \Pi_2^{\text{opt}}(\mu_0, \mu_1)$ is an optimal plan, then the curve $[0,1] \ni t \mapsto \mu_t = \mu \circ (\pi_t)^{-1}$ in $\mathcal{P}_2(\mathbb{R}^d)$ where π_t is the projection:*

$$\mathbb{R}^d \times \mathbb{R}^d \ni (x, y) \mapsto \pi_t(x, y) = (1-t)x + ty \in \mathbb{R}^d, \tag{5.68}$$

is a geodesic path of constant speed.

Remark 5.60 *As announced in Remark 5.56, the above proposition says that, when μ_1 is given by $\mu_1 = \mu_0 \circ (\nabla\varphi)^{-1}$, where φ is the Brenier map, the McCann's interpolation between μ_0 and μ_1, as constructed in the proof of Benamou-Brenier's theorem, is a geodesic path of constant speed between μ_0 and μ_1.*

Proof. Let $0 \leqslant s \leqslant t$. Then,

$$\begin{aligned}
W_2(\mu_s, \mu_t) &\leqslant \left[\int_{(\mathbb{R}^d)^2} |\pi_t(x,y) - \pi_s(x,y)|^2 \mu(dx, dy) \right]^{1/2} \\
&\leqslant (t-s) \left[\int_{(\mathbb{R}^d)^2} |x - y|^2 \mu(dx, dy) \right]^{1/2} \\
&\leqslant (t-s) W_2(\mu_0, \mu_1).
\end{aligned}$$

In fact, this inequality is an equality because, if there were a couple (s,t) with $s \leqslant t$ for which this inequality was strict, we would have:

$$\begin{aligned}
W_2(\mu_0, \mu_1) &\leqslant W_2(\mu_0, \mu_s) + W_2(\mu_s, \mu_t) + W_2(\mu_t, \mu_1) \\
&< s W_2(\mu_0, \mu_1) + (t-s) W_2(\mu_0, \mu_1) + (1-t) W_2(\mu_0, \mu_1) \\
&= W_2(\mu_0, \mu_1).
\end{aligned}$$

which is an obvious contradiction. □

Differential Geometry on $\mathcal{P}_2(\mathbb{R}^d)$

As suggested in Remark 5.56, the variational formula for the 2-Wasserstein distance given in Theorem 5.53 suggests that the tangent space to $\mathcal{P}_2(\mathbb{R}^d)$ at a point $\mu \in \mathcal{P}_2(\mathbb{R}^d)$ should consist of gradients, or at least, limits of gradients. We may recast this intuition in the framework of our discussion of the introductory example presented at the beginning of the subsection of displacements in $\mathcal{P}_2(\mathbb{R}^d)$. Therein, we observe that, as the perturbation ϵ in (5.60) goes to zero, namely as μ and ν get closer and closer to each other, the *direction* used to go from μ to ν converges to $\nabla\theta$. Importantly, θ in (5.60) may be any compactly supported smooth function from $\mathbb{R}^d$ to $\mathbb{R}$ so that the *admissible direction* $\nabla\theta$ used for transporting μ locally may be the gradient of any compactly supported smooth function from $\mathbb{R}^d$ to $\mathbb{R}$. More generally, we can check that the function $\mathbb{R}^d \ni x \mapsto \nabla\theta((I+\epsilon\nabla\theta)^{-1}(x))$ which provides the vector field in (5.60) is always a gradient provided that ϵ is small enough, which shows that, at any time $t \in [0,1]$, the law of X_t in (5.60) indeed moves along a gradient vector field. This can be proven by adapting the duality argument used in Proposition 5.13. If we set:

$$\phi_\epsilon(y) = \inf_{x\in\mathbb{R}^d} \{|x-y|^2 + 2\epsilon\theta(x)\},$$

and if we mimic the computations of Proposition 5.13, we see that for any $y \in \mathbb{R}^d$:

$$\phi_\epsilon(y) = \left|y - (I+\epsilon\nabla\theta)^{-1}(y)\right|^2 + 2\epsilon\theta\big((I+\epsilon\nabla\theta)^{-1}(y)\big).$$

Next we notice that the range of $I+\epsilon\nabla\theta$ is the whole space, so that by expanding $\varphi_\epsilon(I+\epsilon\nabla\theta)$, we get for all $y \in \mathbb{R}^d$:

$$\nabla\phi_\varepsilon(y) = 2y - 2(I+\epsilon\nabla\theta)^{-1}(y) = 2\epsilon\nabla\theta\Big(\big(I+\epsilon\nabla\theta\big)^{-1}(y)\Big),$$

which completes the proof of our claim.

The geometric picture sketched above suggests that it may be appropriate to endow the space $\mathcal{P}_2(\mathbb{R}^d)$, equipped with the 2-Wasserstein distance W_2, with a Riemannian structure based on the result of Theorem 5.53. Indeed, the above discussion hints at the following choice for the tangent space $\mathrm{Tan}_\mu(\mathcal{P}_2(\mathbb{R}^d))$ at a point $\mu \in \mathcal{P}_2(\mathbb{R}^d)$. It should be defined as the closure, in $L^2(\mathbb{R}^d,\mu;\mathbb{R}^d)$, of the vector space of gradients of smooth functions with compact supports, namely:

$$\mathrm{Tan}_\mu\big(\mathcal{P}_2(\mathbb{R}^d)\big) = \overline{\{\nabla\varphi;\ \varphi\in\mathcal{C}_c^\infty(\mathbb{R}^d;\mathbb{R})\}}^{L^2(\mathbb{R}^d,\mu;\mathbb{R}^d)}.$$

Since gradients are orthogonal to divergence free fields, $\mathrm{Tan}_\mu(\mathcal{P}_2(\mathbb{R}^d))$ is also equal to:

$$\begin{aligned}\mathrm{Tan}_\mu\big(\mathcal{P}_2(\mathbb{R}^d)\big) = \Big\{v \in L^2(\mathbb{R}^d,\mu;\mathbb{R}^d) :&\\ \forall w \in L^2(\mathbb{R}^d,\mu;\mathbb{R}^d) \text{ with } \mathrm{div}(w\mu)=0,\ \int_{\mathbb{R}^d} v\cdot w\,d\mu = 0\Big\}.\end{aligned}$$

As a first step in our search for connections between L-derivatives and the Wasserstein geometry advocated in this section, we have:

Lemma 5.61 *If the scalar function u is continuously* L*-differentiable on $\mathcal{P}_2(\mathbb{R}^d)$, then it holds that:*

$$\forall \mu \in \mathcal{P}_2(\mathbb{R}^d), \quad \partial_\mu u(\mu)(\cdot) \in \mathrm{Tan}_\mu\big(\mathcal{P}_2(\mathbb{R}^d)\big).$$

Proof.

First Step. We start with the case when the Fréchet derivative of the lifting of u is uniformly Lipschitz. Then, for a given $\mu \in \mathcal{P}_2(\mathbb{R}^d)$, we know from Proposition 5.50 that there exists a continuously differentiable function $p^\mu : \mathbb{R}^d \to \mathbb{R}$, with a Lipschitz continuous gradient ∇p^μ, such that, μ almost everywhere:

$$\partial_\mu u(\mu)(\cdot) = \nabla p^\mu.$$

The Lipschitz property of the L-derivative implies that we can find a constant C such that, for any $x, x' \in \mathbb{R}^d$,

$$\big|\nabla p^\mu(x) - \nabla p^\mu(x')\big| \leqslant C|x - x'|,$$

and

$$\big|\nabla p^\mu(x)\big| \leqslant C\big(1 + |x|\big), \quad \big|p^\mu(x)\big| \leqslant C\big(1 + |x|^2\big).$$

For a sequence $(\rho_n)_{n\geqslant 1}$ of C^∞ mollifiers with supports included in a fixed compact set and converging to δ_0, we set:

$$p_n^\mu(x) = \big(p^\mu * \rho_n\big)(x), \quad x \in \mathbb{R}^d.$$

Since ∇p^μ is continuous, ∇p_n^μ converges to ∇p^μ, uniformly on compact subsets. The convergence also holds in $L^2(\mathbb{R}^d, \mu; \mathbb{R}^d)$ since, up to a possible modification of the constant C:

$$\big|\nabla p_n^\mu(x)\big| \leqslant C\big(1 + |x|\big), \quad x \in \mathbb{R}^d,$$

the right-hand side being obviously square integrable under μ.

Next we show that we can approximate ∇p^μ by gradients of functions in $C_c^\infty(\mathbb{R}^d)$. We start with a smooth cut-off function $\eta : \mathbb{R}^d \to \mathbb{R}$ with $\eta(x) = 1$ for $|x| \leqslant 1$ and $\eta(x) = 0$ for $|x| \geqslant 2$. Then we set $\eta_n(x) = \eta(x/n)$ for $x \in \mathbb{R}^d$ and $n \geqslant 1$, and prove that $\nabla(\eta_n p_n^\mu)$ converges to ∇p^μ in $L^2(\mathbb{R}^d, \mu; \mathbb{R}^d)$. Clearly, it suffices to prove that $\nabla(\eta_n p_n^\mu - p_n^\mu)$ converges to 0 in $L^2(\mathbb{R}^d, \mu; \mathbb{R}^d)$. Expanding the gradient, we obtain:

$$\nabla\big(\eta_n p_n^\mu - p_n^\mu\big) = (\eta_n - 1)\nabla p_n^\mu + p_n^\mu \nabla \eta_n.$$

Since $\eta_n - 1$ converges to 0 pointwise, the same domination argument as above shows that: $(\eta_n - 1)\nabla p_n^\mu$ converges to 0 in $L^2(\mathbb{R}^d; \mu)$. In order to handle the second term in the right-hand side, we observe that $\nabla \eta_n$ converges to 0 pointwise and that:

$$\big|p_n^\mu \nabla \eta_n\big| \leqslant \frac{C}{n}(1 + |x|^2)\mathbf{1}_{\{n \leqslant |x| \leqslant 2n\}} \leqslant C(1 + |x|),$$

which is enough to apply Lebesgue's dominated theorem once again.

Second Step. We now handle the general case. The proof is based on a regularization argument which will be proved in Section 5.6. For that reason, the reader may want to skip this step on a first reading.

Indeed, we shall prove in Lemma 5.94 below that for any smooth function $\rho : \mathbb{R}^d \to \mathbb{R}^d$ with compact support, the function $\mathcal{P}_2(\mathbb{R}^d) \ni \mu \mapsto u(\mu \circ \rho^{-1}) \in \mathbb{R}$ is bounded and continuously L-differentiable; also, its Fréchet derivative is bounded (in L^2). In particular, the lifting of $\mathcal{P}_2(\mathbb{R}^d) \ni \mu \mapsto u(\mu \circ \rho^{-1}) \in \mathbb{R}$ is Lipschitz continuous.

Next, we consider a sequence $(\rho_n)_{n\geqslant 1}$ of compactly supported smooth functions from $\mathbb{R}^d$ into itself such that, for any $n \geqslant 1$ and for all $x \in \mathbb{R}^d$, $\rho_n(x) = x$ for $|x| \leq n$ and $|\rho_n(x)| \leqslant C|x|$ for a constant C independent of n. We then let $u_n(\mu) = u(\mu \circ \rho_n^{-1})$ for all $\mu \in \mathcal{P}_2(\mathbb{R}^d)$. As a byproduct of the proof of Lemma 5.95, it is straightforward to prove that:

$$\forall \mu \in \mathcal{P}_2(\mathbb{R}^d), \quad \lim_{n\to\infty} \int_{\mathbb{R}^d} |\partial_\mu u_n(\mu)(x) - \partial_\mu u(\mu)(x)|^2 d\mu(x) = 0.$$

This shows that $\partial_\mu u \in \mathrm{Tan}_\mu(\mathcal{P}_2(\mathbb{R}^d))$ provided that $\partial_\mu u_n \in \mathrm{Tan}_\mu(\mathcal{P}_2(\mathbb{R}^d))$ for all $n \geqslant 1$. Put differently, we can assume that the lifting $\tilde{u}$ of u is bounded and continuously Fréchet differentiable on $L^2(\Omega, \mathcal{F}, \mathbb{P}; \mathbb{R}^d)$ and that its Fréchet derivative is bounded (in $L^2(\Omega, \mathcal{F}, \mathbb{P}; \mathbb{R}^d)$).

For any $0 < \delta < \epsilon$, we call $\tilde{u}^{\epsilon,\delta}$ the sup-inf convolution of $\tilde{u}$ with parameters (ϵ, δ):

$$\tilde{u}^{\epsilon,\delta}(X) = \sup_{Z\in\mathcal{H}} \inf_{Y\in\mathcal{H}} \Big[\tilde{u}(Y) + \frac{1}{2\epsilon}\mathbb{E}[|Y-Z|^2] - \frac{1}{2\delta}\mathbb{E}[|Z-X|^2]\Big],$$

for $X \in L^2(\Omega, \mathcal{F}, \mathbb{P}; \mathbb{R}^d)$, where $\mathcal{H} = L^2(\Omega, \mathcal{F}, \mathbb{P}; \mathbb{R}^d)$. It is known (see the Notes & Complements for a precise citation) that $\tilde{u}^{\epsilon,\delta}$ is Fréchet differentiable and has a Lipschitz continuous derivative. Also $\tilde{u}^{\epsilon,\delta}$ converges to $\tilde{u}$, uniformly on $\mathcal{H}$. Thanks to the boundedness of u, it is then well checked that there exists a constant $C \geqslant 0$ such that:

$$\sup_{Z\in\mathcal{H}} \inf_{Y\in\mathcal{H}} \Big[\tilde{u}(Y) + \frac{1}{2\epsilon}\mathbb{E}[|Y-Z|^2] - \frac{1}{2\delta}\mathbb{E}[|Z-X|^2]\Big] \leqslant \sup_{Z\in\mathcal{H}} \Big[C - \frac{1}{2\delta}\mathbb{E}[|Z-X|^2]\Big],$$

which shows that the maximization over Z may be restricted to those Z such that $\|Z - X\|_2^2 \leqslant C\delta$, the value of C being allowed to increase from line to line. Therefore, the minimization over Y may be restricted to those Y such that $\|Z - Y\|^2 \leqslant C\varepsilon$, that is $\|Y - X\|_2^2 \leqslant C\varepsilon$ for a new value of C.

Now, for another $W \in \mathcal{H}$, we have:

$$\begin{aligned}
\tilde{u}^{\varepsilon,\delta}(X+W) &= \sup_{Z\in\mathcal{H}} \inf_{Y\in\mathcal{H}} \Big[\tilde{u}(Y) + \frac{1}{2\epsilon}\mathbb{E}[|Y-Z|^2] - \frac{1}{2\delta}\mathbb{E}[|Z-(X+W)|^2]\Big] \\
&= \sup_{Z\in\mathcal{H}} \inf_{Y\in\mathcal{H}} \Big[\tilde{u}(Y) + \frac{1}{2\epsilon}\mathbb{E}[|Y-W-Z|^2] - \frac{1}{2\delta}\mathbb{E}[|Z-X|^2]\Big] \\
&= \sup_{Z\in\mathcal{H}} \inf_{Y\in\mathcal{H}} \Big[\tilde{u}(Y+W) + \frac{1}{2\epsilon}\mathbb{E}[|Y-Z|^2] - \frac{1}{2\delta}\mathbb{E}[|Z-X|^2]\Big].
\end{aligned} \tag{5.69}$$

Since $\tilde{u}$ is continuously Fréchet differentiable, we can find a mapping $\varepsilon_X : \mathbb{R}_+ \to \mathbb{R}_+$ depending upon the variable X such that $\lim_{r\searrow 0} \varepsilon_X(r) = 0$ and

$$\big|\tilde{u}(Y+W) - \tilde{u}(Y) - \langle D\tilde{u}(X), W\rangle_{\mathcal{H}}\big| \leqslant \|W\|_2 \varepsilon_X\big(\|Y-X\|_2 + \|W\|_2\big),$$

where $\langle\cdot,\cdot\rangle_{\mathcal{H}}$ denotes the inner product in $\mathcal{H}$.

Plugging the above expansion in (5.69), we get:

$$\left|\tilde{u}^{\varepsilon,\delta}(X+W) - \tilde{u}^{\epsilon,\delta}(X) - \langle D\tilde{u}(X), W\rangle_{\mathcal{H}}\right| \leqslant \|W\|_2 \varepsilon_X\big(\epsilon + \|W\|_2\big).$$

Recalling that $\tilde{u}^{\varepsilon,\delta}$ has a Lipschitz continuous Fréchet derivative, we deduce that, for all $r > 0$,

$$\left\|D\tilde{u}^{\epsilon,\delta}(X) - D\tilde{u}(X)\right\|_2 \leqslant \varepsilon_X\big(\epsilon + r\big) + C_{\epsilon,\delta} r,$$

where the constant $C_{\epsilon,\delta}$ depends upon ϵ and δ. Letting first r tend to 0 and then ϵ to 0, we deduce that $\|D\tilde{u}^{\epsilon,\delta}(X) - D\tilde{u}(X)\|_2$ tends to 0 as ϵ tends to 0.

Therefore, in order to complete the proof, it suffices to show that $D\tilde{u}^{\epsilon,\delta}(X)$ may be represented in the form $\theta^{\epsilon,\delta}(X)$ for some $\theta^{\epsilon,\delta} \in \mathrm{Tan}_\mu(\mathcal{P}_2(\mathbb{R}^d))$. To do so, it suffices to observe that:

$$\tilde{u}^{\epsilon,\delta}(X) = \sup\inf\left[u(\nu) + \frac{1}{2\delta}\int_{(\mathbb{R}^d)^2} |y-z|^2 d\pi(y,z) - \int_{(\mathbb{R}^d)^2} |x-z|^2 d\varrho(x,z)\right],$$

the infimum being taken over the probability measures $\pi \in \mathcal{P}_2((\mathbb{R}^d)^2)$, the argument ν in u standing for the first marginal of π on $\mathbb{R}^d$, and the supremum being taken over the probability measures $\varrho \in \mathcal{P}_2((\mathbb{R}^d)^2)$ with $\mu = \mathcal{L}(X)$ as first marginal on $\mathbb{R}^d$ and with the same second marginal on $\mathbb{R}^d$ as π. Obviously, the right-hand side only depends on $\mu = \mathcal{L}(X)$, which shows that $\tilde{u}^{\epsilon,\delta}$ may be projected as a function $u^{\epsilon,\delta}$ on $\mathcal{P}_2(\mathbb{R}^d)$. By the first step of the proof, $\partial_\mu u^{\epsilon,\delta}(\mu)(\cdot) \in \mathrm{Tan}_\mu(\mathcal{P}_2(\mathbb{R}^d))$. □

Geometric Differentiability and Lions' L-Derivatives

In the geometric theory of Wasserstein spaces, the notion of differentiability is usually defined in terms of sub- and super-differentials:

Definition 5.62 *Let $u : \mathcal{P}_2(\mathbb{R}^d) \to \mathbb{R}$ and let $\mu \in \mathcal{P}_2(\mathbb{R}^d)$.*

1. *A function $\xi \in \mathrm{Tan}_\mu(\mathcal{P}_2(\mathbb{R}^d))$ is said to belong to the sub-differential of u at μ, and we write $\xi \in \partial^- u(\mu)$, if for all $\mu' \in \mathcal{P}_2(\mathbb{R}^d)$:*

$$u(\mu') \geqslant u(\mu) + \sup_{\pi \in \Pi_2^{\mathrm{opt}}(\mu,\mu')} \int_{\mathbb{R}^d\times\mathbb{R}^d} \xi(x)\cdot\big(y-x\big)d\pi(x,y) + o\big(W_2(\mu,\mu')\big),$$

   where $\Pi_2^{\mathrm{opt}}(\mu,\mu')$ is the set of optimal transport plans from μ to μ' defined in (5.5). If $\partial^- u(\mu)$ is not empty, u is said to be sub-differentiable at μ.
2. *A function $\xi \in \mathrm{Tan}_\mu(\mathcal{P}_2(\mathbb{R}^d))$ is said to belong to the super-differential of u at μ, and we write $\xi \in \partial^+ u(\mu)$, if $-\xi \in \partial^-(-u)(\mu)$. If $\partial^+ u(\mu)$ is not empty, u is said to be super-differentiable at μ.*
3. *The function u is said to be W-differentiable at μ (in other words in the Wasserstein sense) if it is both sub- and super-differentiable, that is if both $\partial^- u(\mu)$ and $\partial^+ u(\mu)$ are not empty.*

The notion of gradient is then given by:

Proposition 5.63 *If $u : \mathcal{P}_2(\mathbb{R}^d) \to \mathbb{R}$ is W-differentiable at μ in the sense of Definition 5.62, then the sets $\partial^- u(\mu)$ and $\partial^+ u(\mu)$ coincide and contain one element only. This element of* $\mathrm{Tan}_\mu(\mathcal{P}_2(\mathbb{R}^d))$ *is called the Wasserstein gradient, or W-gradient, of u at μ and is denoted $\nabla_\mu u(\mu)$.*

Elements in $\mathrm{Tan}_\mu(\mathcal{P}_2(\mathbb{R}^d))$ are identified when equal up to a μ-null Borel set.

Proof. Let us assume that $\xi^- \in \partial^- u(\mu)$ and $\xi^+ \in \partial^+ u(\mu)$. The proof is based on the very same argument we used in our introductory discussion of displacements when we suggested that optimal transport of measures was taking place along gradients. In the definition of the sub-differential, choose $\mu' = \mu \circ (I + \epsilon \nabla \theta)^{-1}$ where θ is a smooth function with compact support from $\mathbb{R}^d$ into $\mathbb{R}$ and $\epsilon \in \mathbb{R}$. For $|\epsilon|$ small enough, $I + \epsilon \nabla \theta$ is the gradient of a strictly convex function, namely $\mathbb{R}^d \ni x \mapsto (1/2)|x|^2 + \epsilon \theta(x)$. By Proposition 5.13, we know that the map $\mathbb{R}^d \ni x \mapsto x + \epsilon \nabla \theta(x) \in \mathbb{R}^d$ is an optimal transport map from μ to μ', and that it defines the unique optimal transport plan from μ to μ'. Therefore, by definition of the sub- and super-differentials, we have that:

$$u(\mu') \geqslant u(\mu) + \epsilon \int_{\mathbb{R}^d} \xi^-(x) \cdot \nabla \theta(x) d\mu(x) + o(\epsilon),$$

and

$$u(\mu') \leqslant u(\mu) + \epsilon \int_{\mathbb{R}^d} \xi^+(x) \cdot \nabla \theta(x) d\mu(x) + o(\epsilon),$$

from which we deduce, by letting ϵ tend to 0 (on both sides 0^+ and 0^-), that:

$$\int_{\mathbb{R}^d} \big(\xi^+(x) - \xi^-(x)\big) \nabla \theta(x) d\mu(x) = 0.$$

The above is true for any smooth function θ from $\mathbb{R}^d$ to $\mathbb{R}$ with a compact support. Since $\xi^+ - \xi^- \in \mathrm{Tan}_\mu(\mathcal{P}_2(\mathbb{R}^d))$, and the latter is equal to the closure in $L^2(\mathbb{R}^d, \mu; \mathbb{R}^d)$ of the gradients of smooth functions with compact supports, we conclude that $\xi^+ = \xi^-$ almost everywhere under μ. □

Finally, we reconcile the notions of Lions L-derivative and Wasserstein W-gradient.

Theorem 5.64 *If $u : \mathcal{P}_2(\mathbb{R}^d) \to \mathbb{R}$ is continuously* L*-differentiable, then u is also W-differentiable in the sense of Definition 5.62 at any $\mu \in \mathcal{P}_2(\mathbb{R}^d)$, and $\partial_\mu u(\mu) = \nabla_\mu u(\mu)$.*

Proof. Given two probability measures $\mu, \mu' \in \mathcal{P}_2(\mathbb{R}^d)$, we assume $\pi \in \Pi_2^{\mathrm{opt}}(\mu, \mu')$ and we call $(\gamma(x, \cdot))_{x \in \mathbb{R}^d}$ the disintegration of π with respect to μ, namely:

$$\pi(dx, dy) = \mu(dx)\gamma(x, dy).$$

On an atomless Polish probability space $(\Omega, \mathcal{F}, \mathbb{P})$, we consider two $\mathbb{R}^d$-valued random variables X and X' such that the pair (X, X') has π as joint distribution. By definition of the L-derivative, we have:

$$
\begin{aligned}
u(\mu') &= u(\mu) + \int_0^1 \mathbb{E}\big[\partial_\mu u(\mathcal{L}((1-t)X + tX'))\big((1-t)X + tX'\big) \cdot (X' - X)\big]dt \\
&= u(\mu) + \mathbb{E}\big[\partial_\mu u(\mathcal{L}(X))(X) \cdot (X' - X)\big] \\
&\quad + \int_0^1 \mathbb{E}\Big[\Big(\partial_\mu u(\mathcal{L}((1-t)X + tX'))\big((1-t)X + tX'\big) - \partial_\mu u(\mathcal{L}(X))(X)\Big) \\
&\qquad \cdot (X' - X)\Big]dt.
\end{aligned}
$$

Following the discussion right after Remark 5.26 based on the application of Lemma 5.30, we can prove that:

$$
\begin{aligned}
&\bigg| \int_0^1 \mathbb{E}\Big[\Big(\partial_\mu u(\mathcal{L}((1-t)X + tX'))\big((1-t)X + tX'\big) - \partial_\mu u(\mathcal{L}(X))(X)\Big) \cdot (X' - X)\Big]dt \bigg| \\
&\leqslant \bigg(\int_{(\mathbb{R}^d)^2} |x-y|^2 d\pi(x,y)\bigg)^{1/2} \varepsilon_\mu\bigg[\bigg(\int_{(\mathbb{R}^d)^2} |x-y|^2 d\pi(x,y)\bigg)^{1/2}\bigg],
\end{aligned}
$$

where $\varepsilon_\mu : \mathbb{R}_+ \to \mathbb{R}_+$ satisfies $\lim_{r \searrow 0} \varepsilon_\mu(r) = 0$. Since the coupling π is optimal, it holds $\int_{(\mathbb{R}^d)^2} |x-y|^2 d\pi(x,y) = W_2(\mu,\mu')^2$, so that:

$$
u(\mu') \geqslant u(\mu) + \int_{\mathbb{R}^d} \partial_\mu u(\mu)(x) \cdot (y - x)\, d\pi(x,y) - W_2(\mu,\mu')\varepsilon_\mu\big(W_2(\mu,\mu')\big).
$$

Therefore, taking the supremum over the optimal plans $\pi \in \Pi_2^{\text{opt}}(\mu,\mu')$, we deduce that:

$$
\begin{aligned}
u(\mu') \geqslant u(\mu) + &\sup_{\pi \in \Pi_2^{\text{opt}}(\mu,\mu')} \int_{\mathbb{R}^d} \partial_\mu u(\mu)(x) \cdot (y - x)\, d\pi(x,y) \\
&- W_2(\mu,\mu')\varepsilon_\mu\big(W_2(\mu,\mu')\big).
\end{aligned}
$$

Lemma 5.61 asserts that $\partial_\mu u(\mu) \in \text{Tan}_\mu(\mathcal{P}_2(\mathbb{R}^d))$. This proves that $\partial_\mu u(\mu) \in \partial^- u(\mu)$. We then prove that $\partial_\mu u(\mu) \in \partial^+ u(\mu)$ in the same way. □

5.4.4 Finite State Spaces

Mean field games with finite state spaces have been studied sporadically, their relevance coming from the importance of some of their applications. Our introductory Chapter 1 contains a couple of such examples. In these models, marginal distributions have a fixed finite support, and functions of these distributions appear as functions on a finite dimensional simplex. As before, when these functions are assumed to be restrictions of smooth functions on the ambient Euclidean space, a natural notion of differentiability can be used. However, this notion differs from the notion of L-differentiability, and we propose to highlight the differences.

In this subsection, we work with a finite state space $E = \{e_1, \cdots, e_d\}$. Without any loss of generality, we shall regard $e_1, \cdots, e_d$ as the canonical basis of the d-dimensional Euclidean space. In other words, we identify E with the subset of $\mathbb{R}^d$ formed by the unit coordinate vectors $e_1 = (1, 0, \cdots, 0)$, $\cdots$, $e_d = (0, \cdots, 0, 1)$.

An Informal Preliminary Discussion

Our starting point is formula (5.52) which holds for finite measures, though not necessarily for functions only defined on the space of probability measures. Here, a finite measure m on E can be identified with the masses it assigns to each of the singletons, in other words, with the element $(m_1, \cdots, m_d)$ of $\mathbb{R}^d$ defined as $m_i = m(\{e_i\})$ for $i = 1, \cdots, d$. Probability measures correspond to the elements $(p_1, \cdots, p_d)$ of the simplex $\mathcal{S}_d$ of $p_i \geqslant 0$ for $i = 1, \cdots, d$ and $p_1 + \cdots + p_d = 1$. Assuming that the function u is defined on the space of measures $\mathcal{M}(E)$, or at least on an open subset of $\mathcal{M}(E)$ containing the space $\mathcal{P}(E)$ of probability measures, and is smooth in the sense of (5.52), we get:

$$
\begin{aligned}
&u\Big(\sum_{i=1}^{d} m_i'\delta_{e_i}\Big) - u\Big(\sum_{i=1}^{d} m_i\delta_{e_i}\Big) \\
&= \int_0^1 \int_{\mathbb{R}^d} \frac{\delta u}{\delta m}\Big(\sum_{i=1}^{d}[tm_i' + (1-t)m_i]\delta_{e_i}\Big)(x)\, d\Big(\sum_{i=1}^{d}[m_i' - m_i]\delta_{e_i}\Big)(x)\, dt \\
&= \sum_{i=1}^{d}\Big(\int_0^1 \frac{\delta u}{\delta m}\Big(\sum_{i=1}^{d}[tm_i' + (1-t)m_i]\delta_{e_i}\Big)(e_i)dt\Big)(m_i' - m_i),
\end{aligned}
\tag{5.70}
$$

from which we conclude that the *derivation* in the space of measures (in the rough sense of (5.52)) should coincide with the usual derivative of the function $\mathbb{R}^d \ni (m_1, \cdots, m_d) \mapsto u\big(\sum_{i=1}^d m_i\delta_{e_i}\big)$, with the identity:

$$
\boxed{\frac{\partial}{\partial m_k}\Big[u\Big(\sum_{i=1}^{d} m_i\delta_{e_i}\Big)\Big] = \frac{\delta u}{\delta m}\Big(\sum_{i=1}^{d} m_i\delta_{e_i}\Big)(e_k),}
$$

that should hold true for all $k \in \{1, \cdots, d\}$ and $(m_1, \cdots, m_d) \in \mathbb{R}^d$. Importantly, this formula is for functions of measures and not only for functions of probability measures. In particular, $(m_1, \cdots, m_d)$ lives in the whole space $\mathbb{R}^d$ and not only in the simplex $\mathcal{S}_d$. We shall prove in Proposition 5.66 below that the formula takes a somewhat different form when the domain of definition of u is restricted to the subset of probability measures.

The above formula says that we can view the d values taken by the function $[\delta u/\delta m](\sum_{i=1}^d m_i\delta_{e_i})$ on E as the d components of the gradient of the function u when viewed as a function on $\mathbb{R}^d$. Clearly, the present discussion is rather informal. Indeed, implicit continuity assumptions were used to derive the above identity from (5.70).

Connection with the Linear Functional Derivative

In order to make things more rigorous, we focus the discussion on functionals u defined on the space $\mathcal{P}_2(\mathbb{R}^d)$. We make use of the sets $\mathcal{S}_{d-1,\leqslant} = \{(p_1, \cdots, p_{d-1}) \in [0,1]^{d-1} : \sum_{i=1}^{d-1} p_i \leqslant 1\}$ and $\mathcal{S}_d = \{(p_1, \cdots, p_{d-1}, p_d) \in [0,1]^d : \sum_{i=1}^{d} p_i = 1\}$. Obviously, $\mathcal{S}_{d-1,\leqslant}$ and $\mathcal{S}_d$ are in one-to-one correspondence and both may be identified with a $(d-1)$-dimensional simplex. Most importantly, $\mathcal{S}_{d-1,\leqslant}$ has a nonempty interior in $\mathbb{R}^{d-1}$ and, of course, $\mathcal{S}_d$ coincides with the set $\mathcal{P}(E)$ of probability measures on E. We first notice that, when viewed as a subset of $\mathbb{R}^d$, the Euclidean and Wasserstein distances are equivalent on $\mathcal{S}_d$.

Lemma 5.65 *For all $\boldsymbol{p}, \boldsymbol{p}' \in \mathcal{S}_d$,*

$$\frac{1}{2}|\boldsymbol{p} - \boldsymbol{p}'| \leqslant W_2(\boldsymbol{p}, \boldsymbol{p}') \leqslant \sqrt{d}|\boldsymbol{p} - \boldsymbol{p}'|^{1/2},$$

where $|\cdot|$ denotes the Euclidean norm.

Proof. We have, for any two $\boldsymbol{p}, \boldsymbol{p}' \in \mathcal{S}_d$,

$$|\boldsymbol{p} - \boldsymbol{p}'|^2 = \sum_{i=1}^{d} |p_i - p_i'|^2,$$

while, for two random variables X and X' (constructed on some probability space $(\Omega, \mathcal{F}, \mathbb{P})$) with values in E and with $\boldsymbol{p}$ and $\boldsymbol{p}'$ as respective distributions, we have (using the fact that $e_1, \cdots, e_d$ are chosen as the vectors of the canonical basis of $\mathbb{R}^d$):

$$\begin{aligned}
\mathbb{E}\big[|X - X'|^2\big] &= 2\mathbb{P}\big[X \neq X'\big] \\
&= 2 - 2\mathbb{P}\big[X = X'\big] \\
&\geqslant 2 - 2\sum_{i=1}^{d} \mathbb{P}\big[X = e_i\big]^{1/2}\mathbb{P}\big[X' = e_i\big]^{1/2} = \sum_{i=1}^{d}\Big(p_i^{1/2} - (p_i')^{1/2}\Big)^2.
\end{aligned}$$

Using the fact that $|p_i - p_i'| = |(p_i^{1/2} - (p_i')^{1/2})(p_i^{1/2} + (p_i')^{1/2})| \leqslant 2|p_i^{1/2} - (p_i')^{1/2}|$, we get:

$$W_2\big(\boldsymbol{p}, \boldsymbol{p}'\big)^2 \geqslant \frac{1}{4}|\boldsymbol{p} - \boldsymbol{p}'|^2.$$

Conversely, for $\boldsymbol{p}, \boldsymbol{p}' \in \mathcal{S}_d$, we can construct two random variables X and X' (on the same probability space as above), with $\boldsymbol{p}$ and $\boldsymbol{p}'$ as respective distributions, such that $\mathbb{P}[X = X' = e_i] = \min(p_i, p_i')$ for all $i \in \{1, \cdots, d\}$. We then have:

$$
\begin{aligned}
W_2(\boldsymbol{p},\boldsymbol{p}')^2 &\leqslant \mathbb{E}\big[|X-X'|^2\big] \\
&= 2\mathbb{P}\big(X \neq X'\big) \\
&\leqslant 2\Big(1-\sum_{i=1}^{d}\min(p_i,p_i')\Big) \\
&= 2\Big[1+\frac{1}{2}\sum_{i=1}^{d}\Big(|p_i-p_i'|-(p_i+p_i')\Big)\Big] = \sum_{i=1}^{d}|p_i-p_i'| \leqslant \sqrt{d}|\boldsymbol{p}-\boldsymbol{p}'|,
\end{aligned}
$$

which completes the proof. □

Repeating (5.70), we get:

Proposition 5.66 *Let $u : \mathcal{P}_2(\mathbb{R}^d) \to \mathbb{R}$ have a linear functional derivative in the sense of Definition 5.43. Then, the function:*

$$
\mathcal{S}_{d-1,\leqslant} \ni (p_1,\cdots,p_{d-1}) \mapsto u\Big(\sum_{i=1}^{d} p_i\delta_{e_i}\Big) \quad \text{with} \quad p_d = 1-\sum_{i=1}^{d-1} p_i,
$$

is continuously differentiable and for all $i \in \{1,\cdots,d-1\}$,

$$
\frac{\partial}{\partial p_i}\Big[u\Big(\sum_{j=1}^{d} p_j\delta_{e_j}\Big)\Big] = \frac{\delta u}{\delta m}\Big(\sum_{j=1}^{d} p_j\delta_{e_j}\Big)(e_i) - \frac{\delta u}{\delta m}\Big(\sum_{j=1}^{d} p_j\delta_{e_j}\Big)(e_d),
$$

for $(p_1,\cdots,p_{d-1}) \in \mathcal{S}_{d-1,\leqslant}$ and $p_d = 1-(p_1+\cdots+p_{d-1})$.

Proof. From (5.70), for $\boldsymbol{p},\boldsymbol{p}' \in \mathcal{S}_d$, we have:

$$
\begin{aligned}
&u\Big(\sum_{i=1}^{d} p_i'\delta_{e_i}\Big) - u\Big(\sum_{i=1}^{d} p_i\delta_{e_i}\Big) \\
&= \sum_{i=1}^{d}\Big[\Big(\int_0^1 \frac{\delta u}{\delta m}\Big(\sum_{j=1}^{d}[tp_j'+(1-t)p_j]\delta_{e_j}\Big)(e_i)dt\Big)(p_i'-p_i)\Big],
\end{aligned}
$$

from which we get:

$$
\begin{aligned}
&u\Big(\sum_{i=1}^d p_i'\delta_{e_i}\Big) - u\Big(\sum_{i=1}^d p_i\delta_{e_i}\Big) \\
&= \sum_{i=1}^{d-1}\Big[\Big(\int_0^1 \Big[\frac{\delta u}{\delta m}\Big(\sum_{j=1}^d [tp_j' + (1-t)p_j]\delta_{e_j}\Big)(e_i) \\
&\qquad\qquad - \frac{\delta u}{\delta m}\Big(\sum_{j=1}^d [tp_j' + (1-t)p_j]\delta_{e_j}\Big)(e_d)\Big]dt\Big)(p_i' - p_i)\Big] \qquad (5.71)\\
&= \sum_{i=1}^{d-1}\Big[\Big(\frac{\delta u}{\delta m}\Big(\sum_{j=1}^d p_j\delta_{e_j}\Big)(e_i) - \frac{\delta u}{\delta m}\Big(\sum_{j=1}^d p_j\delta_{e_j}\Big)(e_d)\Big)(p_i' - p_i)\Big] + o(|\boldsymbol{p}' - \boldsymbol{p}|),
\end{aligned}
$$

the last equality following from the continuity of $[\delta u/\delta m]$ on $\mathcal{P}_2(\mathbb{R}^d)\times\mathbb{R}^d$ assumed as part of the definition of the existence of a linear functional derivative. Notice that we also used the fact that the set $\{\sum_{i=1}^d p_i\delta_{e_i};\ \boldsymbol{p} = (p_1,\cdots,p_d) \in \mathcal{S}_d\}$ is a compact subset of $\mathcal{P}_2(\mathbb{R}^d)$, which guarantees that for any $i \in \{1,\cdots,d\}$, the function:

$$
\mathcal{S}_d \ni \boldsymbol{p} \mapsto \frac{\delta u}{\delta m}\Big(\sum_{j=1}^d p_j\delta_{e_j}\Big)(e_i)
$$

is uniformly continuous with respect to the Wasserstein distance, and hence the Euclidean distance because of Lemma 5.65. □

Corollary 5.67 *Assume that there exists a continuously differentiable function $\bar{u}$ defined on a d-dimensional open neighborhood of the set $\mathcal{S}_d$ such that for all $\boldsymbol{p} \in \mathcal{S}_d$,*

$$
u\Big(\sum_{i=1}^d p_i\delta_{e_i}\Big) = \bar{u}(p_1,\cdots,p_d).
$$

Then, for all $i \in \{1,\cdots,d-1\}$,

$$
\frac{\delta u}{\delta m}\Big(\sum_{j=1}^d p_j\delta_{e_j}\Big)(e_i) - \frac{\delta u}{\delta m}\Big(\sum_{j=1}^d p_j\delta_{e_j}\Big)(e_d) = \frac{\partial \bar{u}}{\partial p_i}(p_1,\cdots,p_d) - \frac{\partial \bar{u}}{\partial p_d}(p_1,\cdots,p_d).
$$

Proof. It suffices to observe that, for two $\boldsymbol{p},\boldsymbol{p}' \in \mathcal{S}_d$:

$$
\begin{aligned}
& u\Big(\sum_{i=1}^{d} p_i'\delta_{e_i}\Big) - u\Big(\sum_{i=1}^{d} p_i\delta_{e_i}\Big) \\
&= \bar{u}(p_1', \cdots, p_d') - \bar{u}(p_1, \cdots, p_d) \\
&= \sum_{i=1}^{d} \frac{\partial \bar{u}}{\partial p_i}(p_1, \cdots, p_d)(p_i' - p_i) + o\big(|\boldsymbol{p}' - \boldsymbol{p}|\big) \\
&= \sum_{i=1}^{d-1} \Big[\frac{\partial \bar{u}}{\partial p_i}(p_1, \cdots, p_d) - \frac{\partial \bar{u}}{\partial p_d}(p_1, \cdots, p_d)\Big](p_i' - p_i) + o\big(|\boldsymbol{p}' - \boldsymbol{p}|\big),
\end{aligned}
$$

where we used the fact that $\sum_{i=1}^{d}(p_i' - p_i) = 0$ in order to derive the last equality. Identifying with the expansion (5.71), we complete the proof. □

Remark 5.68 *Observe that, in contrast with our preliminary discussion in* (5.70) *for functions of signed measures, the statement of Corollary 5.67 identifies the vector* $((\delta u/\delta m)(\sum_{j=1}^{d} p_j\delta_{e_j})(e_i))_{i\in\{1,\cdots,d\}}$ *with* $((\partial\bar{u}/\partial p_i)(p_1, \cdots, p_d))_{i\in\{1,\cdots,d\}}$ *up to an additive constant. This is consistent with our previous observation of the fact that* $\delta u/\delta m$ *is uniquely defined up to an additive constant when* u *is only defined on the subset of probability measures. Recall Remark 5.46.*

Connection with L-Derivatives

Notice that the derivatives computed above catch infinitesimal changes in the values of the function u when its argument changes through infinitesimal variations of the weights $(p_i)_i$. Indeed, the atoms $(e_i)_i$ do not change when we work on a finite state space. This is in sharp contrast with the rationale we gave for the L-differentiation when we considered the empirical projections of a function of probability measures. Indeed we showed that since the weights were fixed to $1/N$ and the locations of the atoms were changing, L-differentials catch infinitesimal changes in the values of the function u when its argument changes through infinitesimal variations of the atoms $(e_i)_i$. As a result, L-differentiation of functions of probability distributions on a finite state space is likely to be less straightforward. This is confirmed by putting together the formulas obtained above and in the Subsection 5.4.1. If the L-derivative has to involve the derivation of the functional derivative with respect to its argument, the formula in Proposition 5.66 seems to indicate that this will be delicate since the $(e_k)_k$ are fixed.

Remark 5.69 *In an attempt to illustrate this difficulty, we introduce a possible lifting for a generic function on a finite set E. As above, we identify E with the subset of* $\mathbb{R}^d$ *formed by the unit coordinate vectors* $e_1 = (1, 0, \cdots, 0)$, $\cdots$, $e_d = (0, \cdots, 0, 1)$ *and for any probability measure* $\boldsymbol{p} = p_1\delta_{e_1} + \cdots + p_d\delta_{e_d}$ *on E, with* $p_i > 0$ *for all* $i \in \{1, \cdots, d\}$, *we consider the* $\mathbb{R}^d$*-valued random variable* $X = (X_1, \cdots, X_d)$ *defined by:*

$$X_1 = \mathbf{1}_{\{U_1 \leqslant p_1\}},$$
$$X_2 = \mathbf{1}_{\{U_1 > p_1, U_2 \leqslant p_2/(1-p_1)\}},$$
$$\cdots\cdots\cdots\cdots$$
$$X_d = \mathbf{1}_{\{U_1 > p_1, U_2 > p_2/(1-p_1), \cdots, U_{d-1} > p_{d-1}/(1-p_1-\cdots-p_{d-2})\}}$$

where $(U_1, \cdots, U_{d-1})$ *are independent random variables uniformly distributed on the unit interval* $[0, 1]$. *It is plain to check that, by construction,* $\mathbf{p} = \mathcal{L}(X)$.

Observe that the vector X encodes the smallest index $i \in \{1, \cdots, d-1\}$ *such that* $U_i \leqslant p_i/(1 - p_1 - \cdots - p_{i-1})$.

Remark 5.69 provides an alternative road for proving Proposition 5.66, at least when $p_i > 0$ for all $i \in \{1, \cdots, d\}$ and when the assumption of Proposition 5.51 is in force so that $[\delta u/\delta m](\mu)(\cdot)$ is a potential of $\partial_\mu u(\mu)(\cdot)$.

Proof of Proposition 5.66 Using L-Derivatives.

First Step. We start with the case $d = 2$, and we use the same strategy as in the proof of Proposition 5.51. For a parameter $\sigma > 0$, we set:

$$\forall p \in (0, 1),\ w \in \mathbb{R}, \quad \zeta^\sigma(p, w) = \Phi\big(\frac{p - w}{\sigma}\big)e_1 + \Big(1 - \Phi\big(\frac{p - w}{\sigma}\big)\Big)e_2,$$

where as usual Φ denotes the cumulative distribution function of the standard normal distribution $N(0, 1)$. Observe that:

$$\forall p \in (0, 1),\ w \in \mathbb{R}, \quad \partial_w \zeta^\sigma(p, w) = -\frac{1}{\sigma}\varphi\big(\frac{p - w}{\sigma}\big)(e_1 - e_2) = -\partial_p \zeta^\sigma(p, w),$$

where as before φ denotes the density of the standard normal distribution. Then, for two p and p' in $(0, 1)$ and any random variable U uniformly distributed in $[0, 1]$, the mapping:

$$[0, 1] \ni t \mapsto \xi_t^\sigma = \zeta^\sigma\big(tp' + (1 - t)p, U\big) \in L^2(\Omega, \mathcal{F}, \mathbb{P}; \mathbb{R}^d)$$

is differentiable, with:

$$\begin{aligned}
\frac{d}{dt}(\xi_t^\sigma) &= \frac{p' - p}{\sigma}\varphi\Big(\frac{tp' + (1 - t)p - U}{\sigma}\Big)(e_1 - e_2) \\
&= -(p' - p)\partial_w \zeta^\sigma\big(tp' + (1 - t)p, U\big).
\end{aligned}$$

We deduce that:

$$\begin{aligned}
&u\big(\mathcal{L}(\xi_1^\sigma)\big) - u\big(\mathcal{L}(\xi_0^\sigma)\big) \\
&= -(p' - p)\mathbb{E}\int_0^1 \partial_\mu u\big(\mathcal{L}(\xi_t^\sigma)\big)(\xi_t^\sigma) \cdot \partial_w \zeta^\sigma\big(tp' + (1 - t)p, U\big)dt \\
&= -(p' - p) \\
&\quad \times \int_0^1 \Big[\int_0^1 \partial_\mu u\big(\mathcal{L}(\xi_t^\sigma)\big)\Big(\zeta^\sigma\big(tp' + (1 - t)p, w\big)\Big) \cdot \partial_w \zeta^\sigma\big(tp' + (1 - t)p, w\big)dw\Big]dt.
\end{aligned}$$

Recalling that $[\delta u/\delta m](\mu)(\cdot)$ is a potential of $\partial_\mu u(\mu)(\cdot)$, we get:

$$u\big(\mathcal{L}(\xi_1^\sigma)\big) - u\big(\mathcal{L}(\xi_0^\sigma)\big) = (p'-p)\int_0^1 \bigg[\frac{\delta u}{\delta m}\big(\mathcal{L}(\xi_t^\sigma)\big)\big(\zeta^\sigma(tp'+(1-t)p,0)\big) - \frac{\delta u}{\delta m}\big(\mathcal{L}(\xi_t^\sigma)\big)\big(\zeta^\sigma(tp'+(1-t)p,1)\big)\bigg]dt.$$

Letting σ tend to 0, we get:

$$\begin{aligned} &u\big(p'\delta_{e_1} + (1-p')\delta_{e_2}\big) - u\big(p\delta_{e_1} + (1-p)\delta_{e_2}\big) \\ &= (p'-p)\int_0^1 \bigg[\frac{\delta u}{\delta m}\Big(\big(tp'+(1-t)p\big)\delta_{e_1} + \big(1-tp'-(1-t)p\big)\delta_{e_2}\Big)(e_1) \\ &\quad - \frac{\delta u}{\delta m}\Big(\big(tp'+(1-t)p\big)\delta_{e_1} + \big(1-tp'-(1-t)p\big)\delta_{e_2}\Big)(e_2)\bigg]dt, \end{aligned}$$

where we used the fact that, for $p \in (0,1)$,

$$\lim_{\sigma \searrow 0} \zeta^\sigma(p,0) = e_1, \quad \lim_{\sigma \searrow 0} \zeta^\sigma(p,1) = e_2.$$

The proof is easily completed in that case.

Second Step. We now consider the case $d \geqslant 3$. We denote by $\mathcal{S}_d^\circ$ the open $(d-1)$-dimensional simplex of the d-tuples $\boldsymbol{p} = (p_1, \cdots, p_d) \in (0,1)^d$ such that $\sum_{i=1}^d p_i = 1$. For any $\sigma > 0$, $\boldsymbol{p} \in \mathcal{S}_d^\circ$ and $w \in \mathbb{R}^{d-1}$, we then let:

$$\zeta_i^\sigma(\boldsymbol{p}, w) = \phi_i^\sigma\big(q(\boldsymbol{p}), w\big), \quad i \in \{1, \cdots, d\},$$

where, for $q = (q_1, \cdots, q_{d-1}) \in (0,1)^{d-1}$,

$$\begin{aligned} \phi_1^\sigma(q,w) &= \Phi\Big(\frac{q_1 - w_1}{\sigma}\Big), \\ \phi_i^\sigma(q,w) &= \bigg[\prod_{j=1}^{i-1}\Big(1 - \Phi\Big(\frac{q_j - w_j}{\sigma}\Big)\Big)\bigg]\Phi\Big(\frac{q_i - w_i}{\sigma}\Big), \quad i \in \{2, \cdots, d-1\}, \\ \phi_d^\sigma(q,w) &= \prod_{j=1}^{d-1}\Big(1 - \Phi\Big(\frac{q_j - w_j}{\sigma}\Big)\Big), \end{aligned}$$

and

$$\begin{aligned} q_1(\boldsymbol{p}) &= p_1, \\ q_i(\boldsymbol{p}) &= \frac{p_i}{1 - (p_1 + \cdots + p_{i-1})} = \frac{p_i}{p_i + \cdots + p_d}, \quad i \in \{2, \cdots, d-1\}. \end{aligned}$$

Below, we use the convention $q_d(\boldsymbol{p}) = 1$. Observe that:

$$\forall q \in (0,1)^{d-1},\ w \in \mathbb{R}^{d-1}, \quad \partial_w \phi_i^\sigma(q,w) = -\partial_q \phi_i^\sigma(q,w).$$

Now, for any two $\boldsymbol{p}$ and $\boldsymbol{p}'$ in $\mathcal{S}_d^\circ$ and any vector $U = (U_1, \cdots, U_{d-1})$ of $d-1$ independent random variables $U_1, \cdots, U_{d-1}$ uniformly distributed on $[0,1]$, the mapping:

$$[0,1] \ni t \mapsto \xi_t^\sigma = \sum_{i=1}^d \zeta_i^\sigma\big(t\boldsymbol{p}' + (1-t)\boldsymbol{p}, U\big) e_i \in L^2(\Omega, \mathcal{F}, \mathbb{P}; \mathbb{R}^d)$$

is differentiable, with:

$$\frac{d}{dt}\xi_t^\sigma = \sum_{i=1}^d \Big(\partial_q \phi_i^\sigma\big(q(t\boldsymbol{p}' + (1-t)\boldsymbol{p}), U\big) \cdot \frac{d}{dt}\big[q\big(t\boldsymbol{p}' + (1-t)\boldsymbol{p}\big)\big]\Big) e_i.$$

We deduce that:

$$\begin{aligned}
&u(\mathcal{L}(\xi_1^\sigma)) - u(\mathcal{L}(\xi_0^\sigma)) \\
&= -\sum_{i=1}^d \mathbb{E} \int_0^1 \partial_\mu u(\mathcal{L}(\xi_t^\sigma))(\xi_t^\sigma) \\
&\qquad \cdot \Big[\Big(\partial_w \phi_i^\sigma\big(q(t\boldsymbol{p}' + (1-t)\boldsymbol{p}), U\big) \cdot \frac{d}{dt}\big[q\big(t\boldsymbol{p}' + (1-t)\boldsymbol{p}\big)\big]\Big) e_i\Big] dt,
\end{aligned}$$

so that:

$$\begin{aligned}
&u(\mathcal{L}(\xi_1^\sigma)) - u(\mathcal{L}(\xi_0^\sigma)) \\
&= -\sum_{i=1}^d \int_0^1 \int_{(0,1)^{d-1}} \Big\{ \partial_\mu u(\mathcal{L}(\xi_t^\sigma))\Big(\sum_{j=1}^d \zeta_j^\sigma\big(t\boldsymbol{p}' + (1-t)\boldsymbol{p}, w\big) e_j \Big) \\
&\qquad \cdot \Big[\Big(\partial_w \zeta_i^\sigma\big(t\boldsymbol{p}' + (1-t)\boldsymbol{p}, w\big) \cdot \frac{d}{dt}\big[q\big(t\boldsymbol{p}' + (1-t)\boldsymbol{p}\big)\big]\Big) e_i\Big]\Big\} dw\, dt.
\end{aligned}$$

Using once again the fact that $[\delta u/\delta m](\mu)(\cdot)$ is a potential of $\partial_\mu u(\mu)(\cdot)$, and in particular the relationship $\partial_{x_i}[\delta u/\delta m](\mu)(x) = \partial_\mu u(\mu)(x) \cdot e_i$, we get:

$$\begin{aligned}
&u(\mathcal{L}(\xi_1^\sigma)) - u(\mathcal{L}(\xi_0^\sigma)) \\
&= -\int_0^1 \int_{(0,1)^{d-1}} \Big[\partial_w \Big(\frac{\delta u}{\delta m}(\mathcal{L}(\xi_t^\sigma))\Big(\sum_{j=1}^d \zeta_j^\sigma\big(t\boldsymbol{p}' + (1-t)\boldsymbol{p}, w\big) e_j \Big)\Big) \\
&\qquad \cdot \frac{d}{dt}\big[q\big(t\boldsymbol{p}' + (1-t)\boldsymbol{p}\big)\big]\Big] dw\, dt.
\end{aligned}$$

By Stokes' theorem,

$$u(\mathcal{L}(\xi_1^\sigma)) - u(\mathcal{L}(\xi_0^\sigma))$$
$$= -\int_0^1 \Big[\int_{\partial(0,1)^{d-1}} \frac{\delta u}{\delta m}(\mathcal{L}(\xi_t^\sigma))\Big(\sum_{j=1}^d \zeta_j^\sigma(t\boldsymbol{p}' + (1-t)\boldsymbol{p}, w)e_j\Big) \times \Big[\frac{d}{dt}\big[q(t\boldsymbol{p}' + (1-t)\boldsymbol{p})\big]\cdot \boldsymbol{n}(w)\Big] ds(w)\Big] dt, \tag{5.72}$$

where $\boldsymbol{n}(w)$ denotes the outward unit normal vector to $\partial(0,1)^{d-1}$ at w and s denotes the surface measure on $\partial(0,1)^{d-1}$.

Third Step. Now, for any $t \in (0,1)$,

$$\lim_{\sigma \searrow 0}\Big[\int_{\partial(0,1)^{d-1}} \frac{\delta u}{\delta m}(\mathcal{L}(\xi_t^\sigma))\Big(\sum_{j=1}^d \zeta_j^\sigma(t\boldsymbol{p}' + (1-t)\boldsymbol{p}, w)e_j\Big) \times \Big[\frac{d}{dt}\big[q(t\boldsymbol{p}' + (1-t)\boldsymbol{p})\big]\cdot \boldsymbol{n}(w)\Big] ds(w)\Big]$$
$$= \sum_{i=1}^{d-1}\sum_{\ell=1}^{d}\Big(\int_{\{w_i=1\}}\Big[\frac{\delta u}{\delta m}\Big(\sum_{j=1}^d (tp_j' + (1-t)p_j)\delta_{e_j}\Big)(e_\ell)$$
$$\times \prod_{k=1}^{\ell-1}\mathbf{1}_{w_k > q_k(t\boldsymbol{p}'+(1-t)\boldsymbol{p})}\mathbf{1}_{w_\ell \le q_\ell(t\boldsymbol{p}'+(1-t)\boldsymbol{p})}\Big[\frac{d}{dt}\big[q(t\boldsymbol{p}' + (1-t)\boldsymbol{p})\big]\cdot e_i\Big] ds(w)\Big)\Big]$$
$$- \sum_{i=1}^{d-1}\sum_{\ell=1}^{d}\Big(\int_{\{w_i=0\}}\Big[\frac{\delta u}{\delta m}\Big(\sum_{j=1}^d (tp_j' + (1-t)p_j)\delta_{e_j}\Big)(e_\ell)$$
$$\times \prod_{k=1}^{\ell-1}\mathbf{1}_{w_k > q_k(t\boldsymbol{p}'+(1-t)\boldsymbol{p})}\mathbf{1}_{w_\ell \le q_\ell(t\boldsymbol{p}'+(1-t)\boldsymbol{p})}\Big[\frac{d}{dt}\big[q(t\boldsymbol{p}' + (1-t)\boldsymbol{p})\big]\cdot e_i\Big] ds(w)\Big)\Big],$$

where we used the convention $w_d = 0$ and the fact that, whenever $w_i = 0$ (resp. 1), $\boldsymbol{n}(w) = -e_i$ (resp. $+e_i$) except at the vertices where the outward normal unit vector is not uniquely defined. Above, we used two main ingredients. First, we used the fact that for any $\boldsymbol{p} \in \mathcal{S}_d^o$ and $\mathbb{P}$-almost surely:

$$\lim_{\sigma \searrow 0}\zeta_i^\sigma(t\boldsymbol{p}' + (1-t)\boldsymbol{p}, U) = \Big[\prod_{j=1}^{i-1}\mathbf{1}_{\{U_j > q_j(t\boldsymbol{p}'+(1-t)\boldsymbol{p})\}}\Big]\mathbf{1}_{\{U_i < q_i(t\boldsymbol{p}'+(1-t)\boldsymbol{p})\}}.$$

By Remark 5.69, we deduce that for any $t \in [0,1]$, the random vector ξ_t^σ converges in law to $\sum_{j=1}^d (tp_j' + (1-t)p_j)\delta_{e_j}$ as σ tends to 0. Also, we used the fact that for $w \neq q(t\boldsymbol{p}' + (1-t)\boldsymbol{p})$, the vector $\sum_{j=1}^d \zeta_j^\sigma(t\boldsymbol{p}' + (1-t)\boldsymbol{p}, w)e_j$ converges to the vector e_ℓ, where ℓ is the smallest index such that $w_\ell \le q_\ell(t\boldsymbol{p}' + (1-t)\boldsymbol{p})$.

Recalling the convention $q_d(t\boldsymbol{p}' + (1-t)\boldsymbol{p}) = 1$ and thus $[d/dt](q_d(t\boldsymbol{p}' + (1-t)\boldsymbol{p})) = 0$, we deduce that for any $t \in (0,1)$:

$$
\begin{aligned}
&\lim_{\sigma \searrow 0} \Big[\int_{\partial(0,1)^{d-1}} \frac{\delta u}{\delta m}\big(\mathcal{L}(\xi_t^\sigma)\big)\Big(\sum_{j=1}^d \zeta_j^\sigma\big(t\boldsymbol{p}' + (1-t)\boldsymbol{p}, w\big) e_j \Big) \\
&\qquad\qquad\qquad \times \Big[\frac{d}{dt}\big[q\big(t\boldsymbol{p}' + (1-t)\boldsymbol{p}\big)\big] \cdot \boldsymbol{n}(w) \Big] ds(w) \Big] \\
&= \sum_{\ell=1}^d \sum_{i \neq \ell} \Big[\frac{\delta u}{\delta m}\Big(\sum_{j=1}^d (t p_j' + (1-t)p_j)\delta_{e_j} \Big)(e_\ell) \\
&\quad \times \Big[\prod_{k \leq \ell-1, k \neq i} \Big(1 - q_k(t\boldsymbol{p}' + (1-t)\boldsymbol{p})\Big) \Big] q_\ell(t\boldsymbol{p}' + (1-t)\boldsymbol{p}) \frac{d}{dt}\big[q_i\big(t\boldsymbol{p}' + (1-t)\boldsymbol{p}\big)\big] \Big] \\
&- \sum_{\ell=1}^d \sum_{i=\ell+1}^d \Big[\frac{\delta u}{\delta m}\Big(\sum_{j=1}^d (t p_j' + (1-t)p_j)\delta_{e_j} \Big)(e_\ell) \\
&\quad \times \Big[\prod_{k \leq \ell-1} \Big(1 - q_k(t\boldsymbol{p}' + (1-t)\boldsymbol{p})\Big) \Big] q_\ell(t\boldsymbol{p}' + (1-t)\boldsymbol{p}) \frac{d}{dt}\big[q_i\big(t\boldsymbol{p}' + (1-t)\boldsymbol{p}\big)\big] \Big] \\
&- \sum_{\ell=1}^d \Big[\frac{\delta u}{\delta m}\Big(\sum_{j=1}^d (t p_j' + (1-t)p_j)\delta_{e_j} \Big)(e_\ell) \\
&\quad \times \Big[\prod_{k \leq \ell-1} \Big(1 - q_k(t\boldsymbol{p}' + (1-t)\boldsymbol{p})\Big) \Big] \frac{d}{dt}\big[q_\ell\big(t\boldsymbol{p}' + (1-t)\boldsymbol{p}\big)\big] \Big],
\end{aligned}
$$

which gives:

$$
\begin{aligned}
&\lim_{\sigma \searrow 0} \Big[\int_{\partial(0,1)^{d-1}} \frac{\delta u}{\delta m}\big(\mathcal{L}(\xi_t^\sigma)\big)\Big(\sum_{j=1}^d \zeta_j^\sigma\big(t\boldsymbol{p}' + (1-t)\boldsymbol{p}, w\big) e_j \Big) \\
&\qquad\qquad\qquad \times \Big[\frac{d}{dt}\big[q\big(t\boldsymbol{p}' + (1-t)\boldsymbol{p}\big)\big] \cdot \boldsymbol{n}(w) \Big] ds(w) \Big] \\
&= \sum_{\ell=1}^d \sum_{i \leq \ell-1} \Big[\frac{\delta u}{\delta m}\Big(\sum_{j=1}^d (t p_j' + (1-t)p_j)\delta_{e_j} \Big)(e_\ell) \\
&\quad \times \Big[\prod_{k \leq \ell-1, k \neq i} \Big(1 - q_k(t\boldsymbol{p}' + (1-t)\boldsymbol{p})\Big) \Big] q_\ell(t\boldsymbol{p}' + (1-t)\boldsymbol{p}) \frac{d}{dt}\big[q_i\big(t\boldsymbol{p}' + (1-t)\boldsymbol{p}\big)\big] \Big] \\
&- \sum_{\ell=1}^d \Big[\frac{\delta u}{\delta m}\Big(\sum_{j=1}^d (t p_j' + (1-t)p_j)\delta_{e_j} \Big)(e_\ell) \\
&\quad \times \Big[\prod_{k \leq \ell-1} \Big(1 - q_k(t\boldsymbol{p}' + (1-t)\boldsymbol{p})\Big) \Big] \frac{d}{dt}\big[q_\ell\big(t\boldsymbol{p}' + (1-t)\boldsymbol{p}\big)\big] \Big].
\end{aligned}
$$

Now, we observe that:

$$
\begin{aligned}
&-\frac{d}{dt}\Big\{\Big[\prod_{k\leqslant \ell-1}\Big(1-q_k(t\boldsymbol{p}'+(1-t)\boldsymbol{p})\Big)\Big]q_\ell\big(t\boldsymbol{p}'+(1-t)\boldsymbol{p}\big)\Big\}\\
&=\sum_{i=1}^{\ell-1}\Big[\prod_{k\leqslant \ell-1,k\neq i}\Big(1-q_k(t\boldsymbol{p}'+(1-t)\boldsymbol{p})\Big)\Big]q_\ell\big(t\boldsymbol{p}'+(1-t)\boldsymbol{p}\big)\frac{d}{dt}\big[q_i\big(t\boldsymbol{p}'+(1-t)\boldsymbol{p}\big)\big]\\
&\quad-\Big[\prod_{k\leqslant \ell-1}\Big(1-q_k(t\boldsymbol{p}'+(1-t)\boldsymbol{p})\Big)\Big]\frac{d}{dt}\big[q_\ell\big(t\boldsymbol{p}'+(1-t)\boldsymbol{p}\big)\big],
\end{aligned}
$$

so that:

$$
\begin{aligned}
&\lim_{\sigma\searrow 0}\Big[\int_{\partial(0,1)^{d-1}}\frac{\delta u}{\delta m}(\mathcal{L}(\xi_t^\sigma))\Big(\sum_{j=1}^d \zeta_j^\sigma\big(t\boldsymbol{p}'+(1-t)\boldsymbol{p},w\big)e_j\Big)\\
&\qquad\qquad\times\Big[\frac{d}{dt}\big[q\big(t\boldsymbol{p}'+(1-t)\boldsymbol{p}\big)\big]\cdot\boldsymbol{n}(w)\Big]ds(w)\Big]\\
&=-\sum_{\ell=1}^d\Big[\frac{\delta u}{\delta m}\Big(\sum_{j=1}^d(tp_j'+(1-t)p_j)\delta_{e_j}\Big)(e_\ell)\\
&\qquad\qquad\times\frac{d}{dt}\Big\{\Big[\prod_{k\leqslant \ell-1}\Big(1-q_k(t\boldsymbol{p}'+(1-t)\boldsymbol{p})\Big)\Big]q_\ell\big(t\boldsymbol{p}'+(1-t)\boldsymbol{p}\big)\Big\}\Big].
\end{aligned}
$$

Meanwhile, we also have, for any $\boldsymbol{p}\in\mathcal{S}_d^\circ$:

$$
\begin{aligned}
&\Big[\prod_{k\leqslant \ell-1}\big(1-q_k(\boldsymbol{p})\big)\Big]q_\ell(\boldsymbol{p})\\
&=\Big[\prod_{k\leqslant \ell-1}\Big(1-\frac{p_k}{1-(p_1+\cdots+p_{k-1})}\Big)\Big]\frac{p_\ell}{1-(p_1+\cdots+p_{\ell-1})}\\
&=\Big[\prod_{k\leqslant \ell-1}\frac{1-(p_1+\cdots+p_k)}{1-(p_1+\cdots+p_{k-1})}\Big]\frac{p_\ell}{1-(p_1+\cdots+p_{\ell-1})}=p_\ell,
\end{aligned}
$$

where we used the convention that $p_0=0$. So that, by differentiating the above identity along the curve $(0,1)\ni t\mapsto t\boldsymbol{p}'+(1-t)\boldsymbol{p}$, we get:

$$
\frac{d}{dt}\Big\{\Big[\prod_{k\leqslant \ell-1}\Big(1-q_k(t\boldsymbol{p}'+(1-t)\boldsymbol{p})\Big)\Big]q_\ell\big(t\boldsymbol{p}'+(1-t)\boldsymbol{p}\big)\Big\}=p_\ell'-p_\ell.
$$

We finally get:

$$\lim_{\sigma \searrow 0} \left[\int_{\partial(0,1)^{d-1}} \frac{\delta u}{\delta m}(\mathcal{L}(\xi_t^\sigma)) \Big(\sum_{j=1}^{d} \zeta_j^\sigma (t\boldsymbol{p}' + (1-t)\boldsymbol{p}, w) e_j \Big) \right.$$

$$\left. \times \left[\frac{d}{dt} [q(t\boldsymbol{p}' + (1-t)\boldsymbol{p})] \cdot \boldsymbol{n}(w) \right] ds(w) \right]$$

$$= - \sum_{\ell=1}^{d} \left(\frac{\delta u}{\delta m} \Big(\sum_{j=1}^{d} (tp_j' + (1-t)p_j) \delta_{e_j} \Big) (e_\ell)(p_\ell' - p_\ell) \right).$$

Plugging into (5.72), we obtain:

$$u\Big(\sum_{i=1}^{d} p_i' \delta_{e_i}\Big) - u\Big(\sum_{i=1}^{d} p_i \delta_{e_i}\Big) = \sum_{\ell=1}^{d} \int_0^1 \left(\frac{\delta u}{\delta m} \Big(\sum_{j=1}^{d} (tp_j' + (1-t)p_j) \delta_{e_j} \Big) (e_\ell)(p_\ell' - p_\ell) \right) dt,$$

from which the conclusion follows. □

5.5 Convex Functions of Probability Measures

Most of the practical applications for which the theoretical results of the book have been developed are concerned with optimizations of functions. So the fact that sufficient conditions will often involve convexity assumptions should not come as a surprise. For functions defined on *flat linear spaces*, the notion of convexity based on the convexity of restrictions of the functions to lines, and the notion of convexity based on the idea of *graph sitting above the tangent hyperplane* are easily seen to be equivalent. This is not clear any longer for functions defined on curved spaces like $\mathcal{P}_2(\mathbb{R}^d)$. In most of the applications considered in this book, we shall use convex functions whose graphs are above their tangents when the latter are defined in terms of L-derivatives. So we first study this class of convex functions. For the sake of completeness, we next define the class of functions which are convex when restricted to geodesic curves, and we compare this form of convexity to the previous one.

5.5.1 L-Convexity for Functions of Probability Measures

We first study the notion of convexity associated with the special notion of L-differentiability introduced in this chapter. Using this notion of differentiability, the notion of convexity coming from the *above the tangent* philosophy can be defined as follows.

Definition 5.70 *A continuously differentiable function h from $\mathcal{P}_2(\mathbb{R}^d)$ into $\mathbb{R}$ is said to be* L-convex *(or just convex if the context is clear), if for every μ and μ' in $\mathcal{P}_2(\mathbb{R}^d)$, we have:*

$$h(\mu') - h(\mu) - \mathbb{E}[\partial_\mu h(\mu)(X) \cdot (X' - X)] \geqslant 0, \tag{5.73}$$

whenever X and X' are square integrable random variables with distributions μ and μ'.

It is easily checked that the lifting of an L-convex function is convex on the L^2 space used for defining the lifting. The converse is true whenever the lifting is continuously Fréchet differentiable.

More generally, a function h on $\mathbb{R}^n \times \mathcal{P}_2(\mathbb{R}^d)$ which is jointly differentiable in the above sense is said to be L-jointly convex, if for every (x, μ) and (x', μ') in $\mathbb{R}^n \times \mathcal{P}_2(\mathbb{R}^d)$, we have:

$$h(x', \mu') - h(x, \mu) - \partial_x h(x, \mu) \cdot (x' - x) - \mathbb{E}[\partial_\mu h(x, \mu)(X) \cdot (X' - X)] \geqslant 0, \tag{5.74}$$

whenever X and X' are square integrable random variables with distributions μ and μ' respectively.

Example 1. Given a nondecreasing convex differentiable function $g : \mathbb{R} \to \mathbb{R}$ and a convex differentiable function $\zeta : \mathbb{R}^d \to \mathbb{R}$, whose derivative is at most of linear growth, the function $\mathcal{P}_2(\mathbb{R}^d) \ni \mu \mapsto h(\mu) = g(\langle \zeta, \mu \rangle)$, is L-convex, where $\langle \zeta, \mu \rangle = \int_{\mathbb{R}^d} \zeta(x) d\mu(x)$.

Indeed, from the first example in Subsection 5.2.2, we know that the function h is L-differentiable, with $\partial_\mu h(\mu)(x) = g'(\langle \zeta, \mu \rangle) \partial \zeta(x)$, for $\mu \in \mathcal{P}_2(\mathbb{R}^d)$ and $x \in \mathbb{R}^d$. Then, for any $\mu, \mu' \in \mathcal{P}_2(\mathbb{R}^d)$,

$$g(\langle \zeta, \mu' \rangle) - g(\langle \zeta, \mu \rangle) = g\big(\mathbb{E}[\zeta(X')]\big) - g\big(\mathbb{E}[\zeta(X)]\big),$$

where $X \sim \mu$ and $X' \sim \mu'$. Since ζ is convex, it holds that $\zeta(X') \geqslant \zeta(X) + \partial \zeta(X) \cdot (X' - X)$. Taking the expectation, we get $\mathbb{E}[\zeta(X')] \geqslant \mathbb{E}[\zeta(X)] + \mathbb{E}[\partial \zeta(X) \cdot (X' - X)]$. Now, using the fact that g is nondecreasing and convex, we get:

$$\begin{aligned} g\big(\mathbb{E}[\zeta(X')]\big) &\geqslant g\big(\mathbb{E}[\zeta(X)] + \mathbb{E}[\partial \zeta(X) \cdot (X' - X)]\big) \\ &\geqslant g\big(\mathbb{E}[\zeta(X)]\big) + g'\big(\mathbb{E}[\zeta(X)]\big)\mathbb{E}[\partial \zeta(X) \cdot (X' - X)]\big), \end{aligned}$$

which completes the proof.

Example 2. Given a convex differentiable function $g : \mathbb{R}^d \times \mathbb{R}^d \to \mathbb{R}$, whose derivative is at most of linear growth, the function:

$$\mathcal{P}_2(\mathbb{R}^d) \ni \mu \mapsto h(\mu) = \int_{\mathbb{R}^d} \int_{\mathbb{R}^d} g(x, x') d\mu(x) d\mu(x')$$

is L-convex. Indeed, we know from Example 4 in Subsection 5.2.2 that the function h is L-differentiable, with:

$$\partial_\mu h(\mu)(x) = \int_{\mathbb{R}^d} \partial_x g(x, x') d\mu(x') + \int_{\mathbb{R}^d} \partial_{x'} g(x', x) d\mu(x').$$

For $\mu, \mu' \in \mathcal{P}_2(\mathbb{R}^d)$, we consider a pair (X, X') of random variables constructed on $L^2(\Omega, \mathcal{F}, \mathbb{P}; \mathbb{R}^d)$ with μ and μ' as respective marginal distributions together with the copy $(\tilde{X}, \tilde{X}')$ constructed on the copy $(\tilde{\Omega}, \tilde{\mathcal{F}}, \tilde{\mathbb{P}})$ along the principles prescribed in Example 3 in Subsection 5.2.2. Then,

$$h(\mu') = \mathbb{E}\tilde{\mathbb{E}}\big[g(X', \tilde{X}')\big], \quad h(\mu) = \mathbb{E}\tilde{\mathbb{E}}\big[g(X, \tilde{X})\big].$$

Since g is convex, we have:

$$g(X', \tilde{X}') \geq g(X, \tilde{X}) + \partial_x g(X, \tilde{X}) \cdot (X' - X) + \partial_{x'} g(X, \tilde{X}) \cdot (\tilde{X}' - \tilde{X}).$$

In order to complete the proof, it suffices to take the expectation $\mathbb{E}\tilde{\mathbb{E}}$ and to notice that:

$$\begin{aligned}
\mathbb{E}\tilde{\mathbb{E}}\big[\partial_x g(X, \tilde{X}) \cdot (X' - X)\big] &= \mathbb{E}\bigg[\bigg(\int_{\mathbb{R}^d} \partial_x g(X, x') d\mu(x')\bigg) \cdot (X' - X)\bigg], \\
\mathbb{E}\tilde{\mathbb{E}}\big[\partial_{x'} g(X, \tilde{X}) \cdot (\tilde{X}' - \tilde{X})\big] &= \tilde{\mathbb{E}}\bigg[\bigg(\int_{\mathbb{R}^d} \partial_{x'} g(x', \tilde{X}) d\mu(x')\bigg) \cdot (\tilde{X}' - \tilde{X})\bigg] \\
&= \mathbb{E}\bigg[\bigg(\int_{\mathbb{R}^d} \partial_{x'} g(x', X) d\mu(x')\bigg) \cdot (X' - X)\bigg].
\end{aligned}$$

Example 3. As a particular case of Example 2, we may take $g : \mathbb{R}^d \to \mathbb{R}$ a convex differentiable function, whose derivative is at most of linear growth, then the function:

$$\mathcal{P}_2(\mathbb{R}^d) \ni \mu \mapsto h(\mu) = \int_{\mathbb{R}^d} \int_{\mathbb{R}^d} g(x - x') d\mu(x) d\mu(x')$$

is L-convex.

A Sobering Counter-Example. We now prove that if $\mu_0 \in \mathcal{P}_2(E)$ is fixed, the square distance function:

$$\mathcal{P}_2(E) \ni \mu \to W_2(\mu_0, \mu)^2 \in \mathbb{R}$$

may not be L-convex or even L-differentiable!

We first notice that if $F : \mathcal{P}_2(\mathbb{R}^d) \to \mathbb{R}$ is L-differentiable (resp. convex), then for any $\mathbb{R}^d$-valued square integrable random variables X_0 and Y_0, the function $[0,1] \ni t \mapsto F(\mathcal{L}((1-t)X_0 + tY_0)) \in \mathbb{R}$ is differentiable (resp. convex).

Next, we remark that, if x^1, x^2, x^3, and x^4 are in $\mathbb{R}^d$, μ and ν are defined by:

$$\mu = \frac{1}{2}\Big(\delta_{x^1} + \delta_{x^2}\Big), \qquad \text{and} \qquad \nu = \frac{1}{2}\Big(\delta_{x^3} + \delta_{x^4}\Big),$$

and X and Y are $\mathbb{R}^d$-valued random variables such that $\mathcal{L}(X) = \mu$ and $\mathcal{L}(Y) = \nu$, then:

$$\begin{aligned} \mathbb{E}[|X-Y|^2] &= \frac{1}{2}\Big(|x^1|^2 + |x^2|^2 + |x^3|^2 + |x^4|^2\Big) \\ &\quad - \big(x^1 + x^2\big) \cdot x^4 - \big(\alpha x^1 + \beta x^2\big) \cdot \big(x^3 - x^4\big), \end{aligned} \tag{5.75}$$

where:

$$\alpha = \mathbb{P}\big[Y = x^3 | X = x^1\big], \qquad \text{and} \qquad \beta = \mathbb{P}\big[Y = x^3 | X = x^2\big],$$

so that:

$$\begin{aligned} W_2(\mu,\nu)^2 = \inf_{\alpha,\beta \in [0,1]} \bigg[&\frac{1}{2}\Big(|x^1|^2 + |x^2|^2 + |x^3|^2 + |x^4|^2\Big) - (x^1 + x^2)\cdot x^4 \\ &- (\alpha x^1 + \beta x^2)\cdot (x^3 - x^4)\bigg]. \end{aligned} \tag{5.76}$$

Now, if $x^1 = (0,0)$, $x^2 = (2,1)$, $x^3 = (-2,1)$ and $x^4 = (0,0)$ and $\mathbb{P}[X_0 = x^1, Y_0 = x^3] = \mathbb{P}[X_0 = x^2, Y_0 = x^4] = 1/2$, then, for any $t \in [0,1]$, we have:

$$\mu^{(t)} = \frac{1}{2}\Big(\delta_{(-2t,t)} + \delta_{(2-2t,1-t)}\Big) \quad \text{with} \quad \mu^{(t)} = \mathcal{L}\big((1-t)X_0 + tY_0\big).$$

Finally, we set $\mu_0 = \frac{1}{2}(\delta_{(0,0)} + \delta_{(0,-2)})$ and if X and Y are two other random variables with distributions μ_0 and $\mu^{(t)}$ respectively, using (5.75) we get:

$$\mathbb{E}[|X-Y|^2] = 5t^2 - 7t + \frac{13}{2} + 2\beta(2t-1),$$

so that minimizing over β we find:

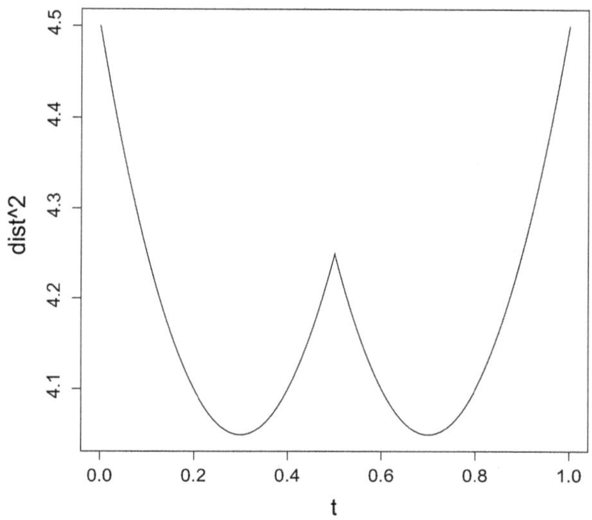

$$W_2(\mu_0, \mu^{(t)})^2 = \begin{cases} 5t^2 - 3t + \dfrac{9}{2} & \text{if} \quad t \leqslant \dfrac{1}{2} \\ 5t^2 - 7t + \dfrac{13}{2} & \text{if } t > \dfrac{1}{2}. \end{cases}$$

Clearly, the plot of the function $t \mapsto W_2(\mu_0, \mu^{(t)})^2$ shows that the latter is not convex, and not even differentiable because of the cusp at $t = 1/2$. This proves our claims.

Of course, it should be noticed that, for any $\nu \in \mathcal{P}_2(\mathbb{R}^d)$, the map $\mathcal{P}_2(\mathbb{R}^d) \ni \mu \mapsto W_2(\mu, \nu)^2$ is always convex for the structure inherited from the linear space of measures. Precisely, for any $t \in [0, 1]$ and any $\mu, \mu' \in \mathcal{P}_2(\mathbb{R}^d)$,

$$W_2\big((1-t)\mu + t\mu', \nu\big)^2 \leqslant (1-t)W_2(\mu, \nu)^2 + tW_2(\mu', \nu)^2.$$

The proof is readily seen. If π and π' belong to $\Pi_2^{\text{opt}}(\mu, \nu)$ and $\Pi_2^{\text{opt}}(\mu', \nu)$ respectively, then the measure $(1-t)\pi + t\pi'$ belongs to $\Pi_2((1-t)\mu + t\mu', \nu)$ and

$$\int_{\mathbb{R}^d \times \mathbb{R}^d} |x - y|^2 d\big((1-t)\pi + t\pi'\big)(x, y) = (1-t)W_2(\mu, \nu)^2 + tW_2(\mu', \nu)^2.$$

In the next paragraph, we discuss an even more striking example in which convexity for the linear structure and L-convexity completely differ.

An Intriguing Example

We return to the second example of L-convex function presented earlier. There $u(\mu) = \langle g * \mu, \mu \rangle$, and we now choose the function g to be $g(x) = |x|^2$, so that $\partial h(x) = 2x$ and $\partial_\mu u(\mu)(x) = 4 \int_{\mathbb{R}^d} (x-y) d\mu(y)$. Notice that:

$$u(\mu) = \mathbb{E}\tilde{\mathbb{E}}[|X - \tilde{X}|^2],$$

where X and $\tilde{X}$ are independent copies, constructed on $(\Omega, \mathcal{F}, \mathbb{P})$ and $(\tilde{\Omega}, \tilde{\mathcal{F}}, \tilde{\mathbb{P}})$ respectively, with the same distribution μ. By developing the squared norm, we get for any $\epsilon > 0$:

$$\begin{aligned}
\tilde{u}(X + \epsilon Y) &= \mathbb{E}\tilde{\mathbb{E}}\big[|X + \epsilon Y - (\tilde{X} + \epsilon \tilde{Y})|^2\big] \\
&= \tilde{u}(X) + 2\epsilon\, \mathbb{E}\tilde{\mathbb{E}}\big[(X - \tilde{X}) \cdot (Y - \tilde{Y})\big] + \epsilon^2 \mathbb{E}\tilde{\mathbb{E}}\big[|Y - \tilde{Y}|^2\big] \\
&= \tilde{u}(X) + 4\epsilon \mathbb{E}\big[(X - \mathbb{E}[X]) \cdot Y\big] + \epsilon^2 \mathbb{E}\tilde{\mathbb{E}}\big[|Y - \tilde{Y}|^2\big] \\
&= \tilde{u}(X) + \epsilon \langle D\tilde{u}(X), Y \rangle_{L^2} + \epsilon^2 \mathbb{E}\tilde{\mathbb{E}}\big[|Y - \tilde{Y}|^2\big],
\end{aligned}$$

where we used Fubini's theorem in order to pass from the second to the third line. The second order term being nonnegative, we recover the fact that the function u is convex in the sense of Definition 5.70.

We now explain why we find this example intriguing. The function u may be rewritten:

$$u(m) = \int_{\mathbb{R}^d} \int_{\mathbb{R}^d} |x - y|^2 dm(x)\, dm(y), \quad m \in \mathcal{P}_2(\mathbb{R}^d).$$

Notice that, for any $m, m' \in \mathcal{P}_2(\mathbb{R}^d)$,

$$\begin{aligned}
u(m') = &\int_{\mathbb{R}^d} \int_{\mathbb{R}^d} |x - y|^2 dm(x) dm(y) \\
&+ 2 \int_{\mathbb{R}^d} \int_{\mathbb{R}^d} |x - y|^2 dm(x) d(m' - m)(y) \\
&+ \int_{\mathbb{R}^d} \int_{\mathbb{R}^d} |x - y|^2 d(m' - m)(x) d(m' - m)(y).
\end{aligned} \tag{5.77}$$

Furthermore, the last integral may be simplified:

$$
\begin{aligned}
&\int_{\mathbb{R}^d}\int_{\mathbb{R}^d} |x-y|^2 d(m'-m)(x)d(m'-m)(y) \\
&= \int_{\mathbb{R}^d}\int_{\mathbb{R}^d} |x|^2\, d(m'-m)(x)d(m'-m)(y) \\
&+ \int_{\mathbb{R}^d}\int_{\mathbb{R}^d} |y|^2\, d(m'-m)(x)d(m'-m)(y) \\
&-2 \int_{\mathbb{R}^d}\int_{\mathbb{R}^d} x\cdot y\, d(m'-m)(x)d(m'-m)(y) \\
&= -2\left|\int_{\mathbb{R}^d} x\, d(m'-m)(x)\right|^2.
\end{aligned}
$$

Obviously the absolute value of this expression is less than $2W_2(m,m')^2$. This says that the absolute value of the last term in the right-hand side in (5.77) is less than $2W_2(m,m')^2$. By Remark 5.47, this implies that u admits the following linear functional derivative:

$$
\frac{\delta u}{\delta m}(m)(x) = 2\int_{\mathbb{R}^d} |x-y|^2 dm(y), \quad (m,x) \in \mathcal{P}_2(\mathbb{R}^d) \times \mathbb{R}^d.
$$

Now, using (5.77) to develop the function u near m along the ray from m to m', we find:

$$
\begin{aligned}
&u\big(m+\epsilon(m'-m)\big) \\
&= \int_{\mathbb{R}^d}\int_{\mathbb{R}^d} |x-y|^2\, dm(x)dm(y) + 2\epsilon \int_{\mathbb{R}^d}\int_{\mathbb{R}^d} |x-y|^2\, dm(x)d(m'-m)(y) \\
&\quad + \epsilon^2 \int_{\mathbb{R}^d}\int_{\mathbb{R}^d} |x-y|^2\, d(m'-m)(x)d(m'-m)(y) \\
&= u(m) + \epsilon \int_{\mathbb{R}^d} \frac{\delta u}{\delta m}(m)(y)\, d(m'-m)(y) - 2\epsilon^2 \left|\int_{\mathbb{R}^d} x\, d(m'-m)(x)\right|^2.
\end{aligned}
$$

The second order correction is now negative, suggesting concavity instead of the convexity argued earlier. The space of finite measures is *flat*, and this concavity can be interpreted using the standard intuition associated with the shapes of surfaces plotted above a flat space. However, the notion of convexity derived from an expansion using L-derivatives cannot be interpreted in terms of Euclidean geometry.

We already encountered this example in Subsection 3.4.3. We shall appeal to it again in the next subsection and in Section 5.7.1, when revisiting the uniqueness of mean field game solutions.

5.5.2 L-Convexity and Monotonicity

Already in Chapter 1, we used the notion of monotonicity in our first *baby steps* toward the existence and uniqueness of Nash equilibria for mean field games. The discussion of this notion culminated in Section 3.4 of Chapter 3 for general uniqueness results, which we shall revisit in Subsection 5.7.1 below. In this subsection, we derive a couple of simple properties relating monotonicity to convexity. First, we define the notion of monotonicity for operators on Hilbert spaces (which will be L^2-spaces in all the examples considered below), which generalizes the Definition 3.31 of L-monotonicity:

Definition 5.71 *A function U from a Hilbert space $\mathcal{H}$ into itself is said to be monotone if, for all $X, X' \in \mathcal{H}$, we have:*

$$\big\langle U(X) - U(X'),\, X - X' \big\rangle_{\mathcal{H}} \geq 0,$$

where $\langle \cdot, \cdot \rangle_{\mathcal{H}}$ denotes the inner product in $\mathcal{H}$.

Lemma 5.72 *If a real valued function F on a Hilbert space $\mathcal{H}$ is Fréchet differentiable with a continuous derivative, then:*

$$F \text{ is convex} \iff DF \text{ is monotone.}$$

Proof. If F is convex, then, for all $X, X' \in \mathcal{H}$, we have:

$$F(X') - F(X) - \langle DF(X), X' - X \rangle_{\mathcal{H}} \geq 0,$$

as well as:

$$F(X) - F(X') - \langle DF(X'), X - X' \rangle_{\mathcal{H}} \geq 0,$$

and summing both inequalities gives the monotonicity condition for DF. Conversely, if DF is assumed to be monotone, since F is continuously differentiable we have:

$$\begin{aligned}
&F(X') - F(X) - \langle DF(X), X' - X \rangle_{\mathcal{H}} \\
&\quad = \int_0^1 \big\langle DF\big((1-t)X + tX'\big), X' - X \big\rangle_{\mathcal{H}} dt - \langle DF(X), X' - X \rangle_{\mathcal{H}} \\
&\quad = \int_0^1 \big\langle DF\big((1-t)X + tX'\big) - DF(X), X' - X \big\rangle_{\mathcal{H}} dt \\
&\quad = \int_0^1 \big\langle DF\big((1-t)X + tX'\big) - DF(X), (1-t)X + tX' - X \big\rangle_{\mathcal{H}} \frac{dt}{t},
\end{aligned}$$

which is nonnegative because of the monotonicity assumption, proving that F is convex. □

The second result of this subsection makes the connection with the Lasry-Lions monotonicity condition introduced in Chapter 3 to ensure uniqueness of mean field game solutions. It is reminiscent of Example 5 of Section 5.2.2.

Lemma 5.73 *If u is a real valued function on $\mathbb{R}^d \times \mathcal{P}_2(\mathbb{R}^d)$ such that:*

1. *u is monotone in the sense of Definition 3.28;*
2. *for each fixed $\mu \in \mathcal{P}_2(\mathbb{R}^d)$, the function $\mathbb{R}^d \ni x \mapsto u(x, \mu)$ is convex;*
3. *for each fixed $\mu \in \mathcal{P}_2(\mathbb{R}^d)$, the function $\mathbb{R}^d \ni x \mapsto u(x, \mu)$ is differentiable and the derivative $\mathbb{R}^d \ni x \mapsto \partial_x u(x, \mu)$ is at most of linear growth, uniformly in μ in bounded subsets of $\mathcal{P}_2(\mu)$;*

then the function U from the Hilbert space $L^2(\Omega, \mathcal{F}, \mathbb{P}; \mathbb{R}^d)$ into itself defined by $U(X) = \partial_x u(X, \mathcal{L}(X))$ is monotone.

Proof. For two random variables X and X' in $L^2(\Omega, \mathcal{F}, \mathbb{P}; \mathbb{R}^d)$, we have:

$$\langle U(X) - U(X'), X - X' \rangle_{L^2} = \langle \partial_x u(X, \mathcal{L}(X)) - \partial_x u(X', \mathcal{L}(X')), X - X' \rangle_{L^2}.$$

We use the convexity property:

$$u(X, \mathcal{L}(X')) \geq u(X', \mathcal{L}(X')) + \langle \partial_x u(X', \mathcal{L}(X')), X - X' \rangle_{L^2}.$$

Similarly,

$$u(X', \mathcal{L}(X)) \geq u(X, \mathcal{L}(X)) + \langle \partial_x u(X, \mathcal{L}(X)), X' - X \rangle_{L^2}.$$

Summing term by term, we get:

$$\begin{aligned} u(X, \mathcal{L}(X')) + u(X', \mathcal{L}(X)) \geq\ & u(X', \mathcal{L}(X')) + u(X, \mathcal{L}(X)) \\ & + \langle \partial_x u(X', \mathcal{L}(X')) - \partial_x u(X, \mathcal{L}(X)), X - X' \rangle_{L^2}. \end{aligned}$$

Rearranging the terms and taking expectations, we get:

$$\begin{aligned} &\langle \partial_x u(X, \mathcal{L}(X)) - \partial_x u(X', \mathcal{L}(X')), X - X' \rangle_{L^2} \\ &\quad \geq \mathbb{E}\big[u(X, \mathcal{L}(X)) - u(X, \mathcal{L}(X'))\big] - \mathbb{E}\big[u(X', \mathcal{L}(X)) - u(X', \mathcal{L}(X'))\big] \\ &\quad = \int_{\mathbb{R}^d} \big[u(x, \mathcal{L}(X)) - u(x, \mathcal{L}(X'))\big] d\big[\mathcal{L}(X) - \mathcal{L}(X')\big](x), \end{aligned}$$

which is nonnegative by the monotonicity assumption. This completes the proof of the monotonicity of U. □

Remark 5.74 *The converse of the implication proven in the above lemma is not true, as shown by the following example. Indeed, consider the function*

$$u(x, \mu) = \frac{1}{2}\left(x - \bar{\mu}\right)^2, \quad x \in \mathbb{R}, \ \mu \in \mathcal{P}_2(\mathbb{R}),$$

where $\bar{\mu}$ denotes the mean of μ. Then, $\partial_x u(x, \mu) = x - \bar{\mu}$, so that, with $U(X) = \partial_x u(X, \mathcal{L}(X)) = X - \mathbb{E}[X]$, we get:

$$\begin{aligned}\langle U(X) - U(X'), X - X' \rangle_{L^2} &= \mathbb{E}\Big[\big(U(X) - U(X')\big)(X - X')\Big] \\ &= \mathbb{E}\Big[\big(X - X' - \mathbb{E}(X - X')\big)(X - X')\Big] \\ &= \mathrm{Var}(X - X') \geqslant 0,\end{aligned}$$

which proves that U is monotone as an operator on a Hilbert space. Moreover, u is obviously convex in x for μ fixed. However,

$$\begin{aligned}\int_{\mathbb{R}} \Big[u(x, \mu) - u(x, \mu')\Big] d\big(\mu - \mu'\big)(x) &= \frac{1}{2} \int_{\mathbb{R}} \Big[\big(x - \bar{\mu}\big)^2 - \big(x - \bar{\mu}'\big)^2\Big] d\big(\mu - \mu'\big)(x) \\ &= -\big(\bar{\mu} - \bar{\mu}'\big) \int_{\mathbb{R}} \big(x - \frac{\bar{\mu} + \bar{\mu}'}{2}\big) d\big(\mu - \mu'\big)(x) \\ &= -\big(\bar{\mu} - \bar{\mu}'\big)^2 \leqslant 0,\end{aligned}$$

which shows that u is not monotone in the sense of Definition 3.28!

Remark 5.75 *If the notion of convexity for smooth functions of measures is understood with respect to the linear functional derivative, then convexity of a function of measures should be defined by the property:*

$$u(m') - u(m) - \int_{\mathbb{R}^d} \frac{\delta u}{\delta m}(m)(x) d(m' - m)(x) \geqslant 0,$$

for all $m, m' \in \mathcal{P}_2(\mathbb{R}^d)$. Then the result of Lemma 5.72 still holds in the sense that u is convex in this sense if and only if $[\delta u / \delta m]$ is monotone in the sense of Definition 3.28. The proof is exactly the same.

5.5.3 Displacement Convexity

We now introduce the notion of convexity most popular in the theory of optimal transportation of measures, and we connect it to the notion of L-convexity studied above.

Definition 5.76 *A subset $\mathcal{P}$ of $\mathcal{P}_2(\mathbb{R}^d)$ is said to be displacement convex if for any μ_0 and μ_1 in $\mathcal{P}$, any optimal transport plan $\mu \in \Pi_2^{\mathrm{opt}}(\mu_0, \mu_1)$, and any $t \in [0, 1]$, $\mu \circ (\pi_t)^{-1} \in \mathcal{P}$.*

Recall the definition (5.68) of the projection π_t. With this definition of displacement convex sets, the definition of displacement convex functions follows naturally.

Definition 5.77 *If $\mathcal{P} \subset \mathcal{P}_2(\mathbb{R}^d)$ is displacement convex, a function $h : \mathcal{P} \to \mathbb{R}$ is said to be displacement convex (resp. strictly displacement convex) if for every $\mu_0, \mu_1 \in \mathcal{P}$ and $\mu \in \Pi_2^{\text{opt}}(\mu_0, \mu_1)$, the function $[0,1] \ni t \mapsto h(\mu \circ (\pi_t)^{-1})$ is convex (resp. strictly convex). If $\lambda > 0$, h is said to be λ-uniformly displacement convex if for every $\mu_0, \mu_1 \in \mathcal{P}$, we have:*

$$h(\mu \circ (\pi_t)^{-1}) \leqslant (1-t)h(\mu_0) + th(\mu_1) - \frac{\lambda}{2} t(1-t) W_2(\mu_0, \mu_1)^2. \tag{5.78}$$

Notice that (5.78) can be proven by showing:

$$\frac{d^2}{dt^2} h(\mu \circ (\pi_t)^{-1}) \geqslant \lambda W_2(\mu_0, \mu_1)^2.$$

We now revisit Example 3 introduced earlier in this section in the new framework of displacement convexity.

Proposition 5.78 *If $g : \mathbb{R}^d \to \mathbb{R}$ is continuously differentiable and convex, then the function $\mathcal{P}_2(\mathbb{R}^d) \ni \mu \mapsto h(\mu) = \int_{\mathbb{R}^d} \int_{\mathbb{R}^d} g(x-y) d\mu(x) d\mu(y)$ is displacement convex. If g is strictly convex, then for every $m \in \mathbb{R}^d$, the restriction of h to the set $\mathcal{P}^{(m)} = \{\theta \in \mathcal{P}_2(\mathbb{R}^d) : \int_{\mathbb{R}^d} x \theta(dx) = m\}$ is strictly convex. Finally, if g is λ-uniformly convex (in the sense that the Hessian of g is bounded below by λI_d, with $\lambda > 0$, when g is twice differentiable), then the restriction of h to the set $\mathcal{P}^{(m)}$ is λ-uniformly displacement convex.*

Proof. The first claim is easy to check. Indeed, if $\mu_0, \mu_1 \in \mathcal{P}_2(\mathbb{R}^d)$, $\mu \in \Pi_2^{\text{opt}}(\mu_0, \mu_1)$, and we use the notation μ_t for $\mu \circ (\pi_t)^{-1}$, then:

$$\begin{aligned}
h(\mu_t) &= \int_{\mathbb{R}^d} \int_{\mathbb{R}^d} g(z-z') d\mu_t(z) d\mu_t(z') \\
&= \int_{(\mathbb{R}^d)^2} \int_{(\mathbb{R}^d)^2} g\big(\pi_t(x,y) - \pi_t(x',y')\big) d\mu(x,y) d\mu(x',y') \\
&= \int_{(\mathbb{R}^d)^2} \int_{(\mathbb{R}^d)^2} g\big((1-t)(x-x') + t(y-y')\big) d\mu(x,y) d\mu(x',y') \\
&\leqslant \int_{(\mathbb{R}^d)^2} \int_{(\mathbb{R}^d)^2} \big[(1-t)g(x-x') + tg(y-y')\big] d\mu(x,y) d\mu(x',y') \\
&= (1-t) \int_{\mathbb{R}^d} \int_{\mathbb{R}^d} g(x-x') d\mu_0(x) d\mu_0(x') + t \int_{\mathbb{R}^d} \int_{\mathbb{R}^d} g(y-y') d\mu_1(y) d\mu_1(y') \\
&= (1-t)h(\mu_0) + th(\mu_1).
\end{aligned}$$

If we now assume that g is strictly convex, the above inequality can only be an equality if and only if $\mu^{\otimes 2}(\{(x,y),(x',y') : x-y = x'-y'\}) = \mu^{\otimes 2}(\{(x,y),(x',y') : x-x' = y-y'\}) = 1$, which is equivalent to the fact that there exists an element $a \in \mathbb{R}^d$ such that $\mu(\{(x,y);\ x-y = a\}) = 1$, which implies that μ_1 is merely a shift of μ_0. In that case, for each fixed $m \in \mathbb{R}^d$, the restriction of the function h to $\mathcal{P}^{(m)}$ is strictly convex.

Finally, if $t \in [0,1]$ is fixed, the above computation shows that:

$$
\begin{aligned}
&(1-t)h(\mu_0) + th(\mu_1) - h(\mu_t) \\
&= \int_{(\mathbb{R}^d)^2}\int_{(\mathbb{R}^d)^2} \big[(1-t)g(x-x') + tg(y-y') \\
&\qquad\qquad - g\big((1-t)(x-x') + t(y-y')\big)\big] d\mu(x,y)d\mu(x',y').
\end{aligned}
$$

For the sake of convenience, set $\Delta x = x - x'$ and $\Delta y = y - y'$. We have:

$$
\begin{aligned}
&(1-t)g(\Delta x) + tg(\Delta y) - g\big((1-t)\Delta x + t\Delta y\big) \\
&= (1-t)\big[g(\Delta x) - g\big((1-t)\Delta x + t\Delta y\big)\big] + t\big[g(\Delta y) - g\big((1-t)\Delta x + t\Delta y\big)\big] \\
&= (1-t)\int_0^1 \nabla g\big(s\Delta x + (1-s)((1-t)\Delta x + t\Delta y)\big) \cdot \big(\Delta x - ((1-t)\Delta x + t\Delta y)\big) ds \\
&\quad + t\int_0^1 \nabla g\big(s\Delta y + (1-s)((1-t)\Delta x + t\Delta y)\big) \cdot \big(\Delta y - ((1-t)\Delta x + t\Delta y)\big) ds \\
&= t(1-t)\int_0^1 [\nabla g\big([s + (1-s)(1-t)]\Delta x + (1-s)t\Delta y\big) \\
&\qquad\qquad - \nabla g\big((1-s)(1-t)\Delta x + [s + (1-s)t\Delta y]\big)] \cdot (\Delta x - \Delta y) ds \\
&\geqslant t(1-t)\frac{\lambda}{2}|\Delta x - \Delta y|^2,
\end{aligned}
$$

because of the λ-uniform convexity of g. Consequently,

$$
\begin{aligned}
&(1-t)h(\mu_0) + th(\mu_1) - h(\mu_t) \\
&\geqslant \frac{\lambda}{2}t(1-t)\int_{(\mathbb{R}^d)^2}\int_{(\mathbb{R}^d)^2} |x - x' - (y-y')|^2 d\mu(x,y)d\mu(x',y').
\end{aligned}
$$

Now, if we assume that $\int_{\mathbb{R}^d} x d\mu_0(x) = \int_{\mathbb{R}^d} y d\mu_1(y)$, we have:

$$
\begin{aligned}
\int_{(\mathbb{R}^d)^2}\int_{(\mathbb{R}^d)^2} |x - x' - (y-y')|^2 d\mu(x,y)d\mu(x',y') &= 2\int_{(\mathbb{R}^d)^2} |x-y|^2 d\mu(x,y) \\
&= 2W_2(\mu_0,\mu_1)^2.
\end{aligned}
$$

This is exactly what we needed to complete the proof. □

We close this subsection with a comparison of the notion of displacement convexity and the concept of L-convexity introduced earlier.

Proposition 5.79 *Let us assume that the real valued function h is continuously* L-*differentiable on $\mathcal{P}_2(\mathbb{R}^d)$. Then h is* L-*convex if and only if it is displacement convex.*

Proof. As usual we denote by $\tilde{h}$ the lifting of h to an L^2-space. Recall that saying that h is L-convex means that the graph of $\tilde{h}$ is above its tangent as given by its Fréchet derivative. So if μ_0 and μ_1 are given in $\mathcal{P}_2(\mathbb{R}^d)$, if $\mu \in \Pi_2^{\text{opt}}(\mu_0, \mu_1)$, we denote by (X, Y) a couple of random variables in the L^2-space over $\mathcal{P}_2(\mathbb{R}^d)$ with joint distribution μ, that is $\mathcal{L}(X, Y) = \mu$. If we use the same notation $\mu_t = \mu \circ (\pi_t)^{-1}$ for the optimal displacement from μ_0 to μ_1, then $\mu_t = \mathcal{L}(X_t)$ with $X_t = (1-t)X + tY$. Now:

$$\begin{aligned} h(\mu_1) - h(\mu_0) - \frac{d}{dt}h(\mu_t)\Big|_{t=0} &= h(\mu_1) - h(\mu_0) - \frac{d}{dt}\tilde{h}(X_t)\Big|_{t=0} \\ &= h(\mu_1) - h(\mu_0) - D\tilde{h}(X)\cdot(Y-X) \\ &= h(\mu_1) - h(\mu_0) - \mathbb{E}\big[\partial_\mu h(\mu_0)(X)\cdot(Y-X)\big]. \end{aligned}$$

From the definition of the L-convexity and the *above the tangent* formulation of displacement convexity, we see that h is L-convex if and only if it is displacement convex. □

Furthermore, if we assume that the Fréchet derivative of the lifting $\tilde{h}$ is Lipschitz continuous, we can use the result of Theorem 5.64 identifying the L-derivative and the W-derivative (in the sense of Definition 5.62), namely that $\partial_\mu u(\mu) = \nabla_\mu u(\mu)$ for any $\mu \in \mathcal{P}_2(\mathbb{R}^d)$, and in this way, we can also identify the notion of displacement convexity to the notion of convexity of functions whose graphs are above the tangents when the latter are determined by the differential calculus on Wasserstein space.

5.6 Itô's Formula Along a Flow of Measures

The goal of this section is to provide a chain rule for the differentiation of functions of t of the form $(u(\mu_t))_{t\geqslant 0}$ when u is an $\mathbb{R}$-valued smooth function defined on the space $\mathcal{P}_2(\mathbb{R}^d)$ of probability measures of order 2 on $\mathbb{R}^d$, and $\boldsymbol{\mu} = (\mu_t)_{t\geqslant 0}$ is the flow of marginal distributions of an $\mathbb{R}^d$-valued Itô process $\boldsymbol{X} = (X_t)_{t\geqslant 0}$. We shall sometimes use the terminology Itô's formula instead of chain rule because the dynamics are driven by an Itô process.

5.6.1 Possible Road Maps for the Proof

There are two obvious strategies to expand $u(\mu_t)$ and derive an infinitesimal chain rule for the differential in time.

For each given $t > 0$, a first strategy consists in dividing the interval $[0, t]$ into small intervals of length $\Delta t = t/N$ for some integer $N \geqslant 1$, and writing the difference $u(\mu_t) - u(\mu_0)$ as a telescoping sum:

$$u(\mu_t) - u(\mu_0) = \sum_{i=0}^{N-1} \big[u(\mu_{i\Delta t}) - u(\mu_{(i-1)\Delta t})\big].$$

One could then use an appropriate form of Taylor's formula at the order 2 for functions of probability measures and expand each difference in the above summation. Since the remainder terms are expected to be smaller than the step size Δt, one should be able to derive the chain rule by collecting the terms and letting Δt tend to 0. This strategy fits the original proof of Itô's formula in classical stochastic differential calculus.

A different strategy consists in another approximation of the Itô dynamics. Instead of discretizing in time as in the previous approach, it is tempting to reduce the space dimension by approximating the flow $\boldsymbol{\mu} = (\mu_t)_{t\geqslant 0}$ by a flow of empirical measures:

$$\left(\bar{\mu}_t^N = \frac{1}{N}\sum_{\ell=1}^{N} \delta_{X_t^\ell}\right)_{t\geq 0},$$

for $N \geqslant 1$, where $\boldsymbol{X}^1 = (X_t^1)_{t\geqslant 0}, \cdots, \boldsymbol{X}^N = (X_t^N)_{t\geqslant 0}$ are N independent copies of $X = (X_t)_{t\geqslant 0}$ (constructed on an appropriate probability space $(\Omega, \mathcal{F}, \mathbb{P})$). Using the empirical projection of u defined as the real valued function u^N on $\mathbb{R}^{Nd}$ by:

$$u^N(x^1, \cdots, x^N) = u\left(\frac{1}{N}\sum_{\ell=1}^{N} \delta_{x^\ell}\right), \tag{5.79}$$

the strategy is then to expand $u^N(X_t^1, \cdots, X_t^N)$ using the standard version of Itô's formula in finite dimension, and try to control the limit when N tends to infinity. Obviously, we should recover the same chain rule as the one obtained by the first approach.

Whatever the strategy, it is necessary to pay special attention to the regularity conditions needed to expand $(u(\mu_t))_{t\geqslant 0}$ infinitesimally. As evidenced by (5.79), we may expect that, not only, u has to be once differentiable in the measure argument, but also to have second order derivatives in order to allow for the application of Itô's formula to the empirical projection u^N. For instance, it would be quite tempting to require the lifting $\tilde{u}$ to be twice (continuously) Fréchet differentiable. However, as we show in Remark 5.80 below, this is a very restrictive condition for our purpose. We shall spend quite a bit of time in the next subsections identifying the right notion of second order differentiability needed to perform the desired chain rule expansion. See Remark 5.81 for a first account.

Remark 5.80 *If u is a function on $\mathcal{P}_2(\mathbb{R}^d)$ and if we assume that the lifting $\tilde{u}$ on some $(\Omega, \mathcal{F}, \mathbb{P})$ is twice continuously Fréchet differentiable at X, then the second Fréchet derivative $D^2\tilde{u}(X)$ is a symmetric operator on $L^2(\Omega, \mathcal{F}, \mathbb{P}; \mathbb{R}^d)$, and we can* *use the* quadratic form *notation to denote the second order directional derivatives* $D^2\tilde{u}(X)[Y,Z]$ *in the directions* Y *and* Z *of* $L^2(\Omega,\mathcal{F},\mathbb{P};\mathbb{R}^d)$.

However, this notion of differentiability can have serious shortcomings for some of our purpose. Indeed as we are about to show, there exist smooth functions $h : \mathbb{R}^d \to \mathbb{R}$ *with compact supports for which the function* $\tilde{u} : L^2(\Omega,\mathcal{F},\mathbb{P};\mathbb{R}^d) \ni X \mapsto \mathbb{E}[h(X)]$ *is not twice continuously Fréchet differentiable. For such a* $\tilde{u}$, *we already know that* $D\tilde{u}(X) = \partial h(X)$. *Therefore, for any* $Y \in L^2(\Omega,\mathcal{F},\mathbb{P};\mathbb{R}^d)$,

$$\lim_{\epsilon\to 0} \frac{D\tilde{u}(X+\epsilon Y) - D\tilde{u}(X)}{\epsilon} = \partial^2 h(X)Y,$$

where $\partial^2 h(X)Y$ *is understood as* $(\sum_{j=1}^d \partial^2_{x_i x_j} h(X)Y_j)_{1\leqslant i\leqslant d}$, *the limit being understood pointwise or in* $L^2(\Omega,\mathcal{F},\mathbb{P};\mathbb{R}^d)$ *sense. Therefore, if it exists, the second order Fréchet derivative must be given by the mapping* $L^2(\Omega,\mathcal{F},\mathbb{P};\mathbb{R}^d) \ni X \mapsto (L^2(\Omega,\mathcal{F},\mathbb{P};\mathbb{R}^d) \ni Y \mapsto \partial^2 h(X)Y)$. *In particular, this second order Fréchet derivative will be continuous if*

$$\lim_{X'\to_{L^2} X} \sup_{\|Y\|_2\leqslant 1} \|(\partial^2 h(X') - \partial^2 h(X))Y\|_2 = 0.$$

Now, choose $d = 1$, $X \equiv 0$, $\mathbb{P}[X' = 1] = 1 - \mathbb{P}[X' = 0] = \epsilon$ *for some* $\epsilon \in (0,1)$, *and assume that* $\partial^2 h$ *equals to the identity on* $[0,1]$. *Then,*

$$\sup_{\|Y\|_2\leqslant 1} \|(\partial^2 h(X') - \partial^2 h(X))Y\|_2 = \sup_{\|Y\|_2\leqslant 1} \mathbb{E}\big[\mathbf{1}_{\{X'=1\}}Y^2\big]^{1/2}.$$

Furthermore, if we choose $Y = \varepsilon^{-1/2}X'$, *so that* $\|Y\|_2 = 1$ *and* $\mathbb{E}[\mathbf{1}_{\{X'=1\}}Y^2] = 1$, *we see that the above right-hand side cannot tend to* 0 *as* ε *tend* 0, *while obviously,* $X' \to X$ *in* L^2 *as* ε *tends to* 0*!*

Remark 5.81 *The first strategy mentioned above may appear to be most natural as it mimics the proof of the standard Itô formula. However, the second approach seems to be right in line with our desire to apply our tools to models of large populations of individuals interacting through empirical measures. Indeed, the new perspective provided by the second strategy of proof enlightens the choice we made for a form of differential calculus on the space of probability measures. In any case, both strategies require some smoothness conditions on* u. *As we just accounted for,* u *must be twice differentiable (in some sense) in both cases. However, the strategy based on approximations by empirical projections ends up being less demanding in terms of assumptions. Indeed, by taking full advantage of the finite dimensional stochastic calculus chain rule, it allows us to apply standard finite dimensional mollification arguments, and in so doing, weaken the smoothness conditions required on the coefficients. In particular, this approach works under a weak notion of second order differentiability. See Theorem 5.99 below.*

5.6.2 Full $\mathcal{C}^2$-Regularity

We first establish the chain rule for $(u(\mu_t))_{t\geq 0}$ when u satisfies a strong notion of $\mathcal{C}^2$ - regularity. We enumerate the properties we require for this notion to hold.

Assumption (Full $\mathcal{C}^2$ Regularity).

(A1) The function u is $\mathcal{C}^1$ in the sense of L-differentiation, and its first derivative has a jointly continuous version $\mathcal{P}_2(\mathbb{R}^d) \times \mathbb{R}^d \ni (\mu, v) \mapsto \partial_\mu u(\mu)(v) \in \mathbb{R}^d$.

(A2) For each fixed $\mu \in \mathcal{P}_2(\mathbb{R}^d)$, the version of $\mathbb{R}^d \ni v \mapsto \partial_\mu u(\mu)(v) \in \mathbb{R}^d$ used in **(A1)** is differentiable on $\mathbb{R}^d$ in the classical sense and its derivative is given by a jointly continuous function $\mathcal{P}_2(\mathbb{R}^d) \times \mathbb{R}^d \ni (\mu, v) \mapsto \partial_v \partial_\mu u(\mu)(v) \in \mathbb{R}^{d\times d}$.

(A3) For each fixed $v \in \mathbb{R}^d$, the version of $\mathcal{P}_2(\mathbb{R}^d) \ni \mu \mapsto \partial_\mu u(\mu)(v) \in \mathbb{R}^d$ in **(A1)** is continuously L-differentiable component by component, with a derivative given by a function $(\mu, v, v') \mapsto \partial_\mu^2 u(\mu)(v)(v') \in \mathbb{R}^{d\times d}$ such that for any $\mu \in \mathcal{P}_2(\mathbb{R}^d)$ and any $X \in L^2(\Omega, \mathcal{F}, \mathbb{P}; \mathbb{R}^d)$ with $\mathcal{L}(X) = \mu$ over a probability space $(\Omega, \mathcal{F}, \mathbb{P})$, $\partial_\mu^2 u(\mu)(v)(X)$ gives the Fréchet derivative at X of $L^2(\Omega, \mathcal{F}, \mathbb{P}; \mathbb{R}^d) \ni X' \mapsto \partial_\mu u(\mathcal{L}(X'))(v)$, for every $v \in \mathbb{R}^d$. Denoting $\partial_\mu^2 u(\mu)(v)(v')$ by $\partial_\mu^2 u(\mu)(v, v')$, the map $\mathcal{P}_2(\mathbb{R}^d) \times \mathbb{R}^d \times \mathbb{R}^d \ni (\mu, v, v') \mapsto \partial_\mu^2 u(\mu)(v, v')$ is also assumed to be continuous for the product topology.

Remark 5.82 *The following observations may be useful.*

1. *Under (**A1**), there exists one and only one version of $\partial_\mu u(\mu)(\cdot) \in L^2(\mathbb{R}^d, \mu; \mathbb{R}^d)$ for each $\mu \in \mathcal{P}_2(\mathbb{R}^d)$ such that the mapping $\mathcal{P}_2(\mathbb{R}^d) \times \mathbb{R}^d \ni (\mu, v) \mapsto \partial_\mu u(\mu)(v) \in \mathbb{R}^d$ is jointly continuous.*
2. *Under (**A2**), there exists one and only one version of $\partial_\mu u(\mu)(\cdot)$ for each $\mu \in \mathcal{P}_2(\mathbb{R}^d)$ such that $\mathbb{R}^d \ni v \mapsto \partial_\mu u(\mu)(v)$ is differentiable for each $\mu \in \mathcal{P}_2(\mathbb{R}^d)$ and the mapping $\mathcal{P}_2(\mathbb{R}^d) \times \mathbb{R}^d \ni (\mu, v) \mapsto \partial_v \partial_\mu u(\mu)(v)$ is jointly continuous. In particular, the values of the derivatives $\partial_\mu u(\mu)(v)$ and $\partial_v \partial_\mu u(\mu)(v)$, for $\mu \in \mathcal{P}_2(\mathbb{R}^d)$ and $v \in \mathbb{R}^d$, are uniquely determined.*
3. *Under (**A3**), there exists one and only one continuous version of $\partial_\mu u(\mu)(\cdot)$ for each $\mu \in \mathcal{P}_2(\mathbb{R}^d)$ such that for each fixed $v \in \mathbb{R}^d$, the mapping $\mathcal{P}_2(\mathbb{R}^d) \ni \mu \mapsto \partial_\mu u(\mu)(v)$ is L-continuously differentiable and the derivative $\mathcal{P}_2(\mathbb{R}^d) \times \mathbb{R}^d \times \mathbb{R}^d \ni (\mu, v, v') \mapsto \partial_\mu^2 u(\mu)(v, v')$ is jointly continuous. Also, the values of $\partial_\mu u(\mu)(v)$ and $\partial_\mu^2 u(\mu)(v, v')$ are uniquely determined.*

Proof.

First Step. The proof of the first claim in Remark 5.82 is straightforward. When the support of μ is the entire $\mathbb{R}^d$, there exists at most one continuous version of the mapping $\mathbb{R}^d \ni v \mapsto \partial_\mu u(\mu)(v) \in \mathbb{R}^d$. By approximating any $\mu \in \mathcal{P}_2(\mathbb{R}^d)$ by a sequence of probability measures with full supports, we deduce that $\partial_\mu u(\mu)(\cdot)$ is also uniquely determined when the mapping $(\mu, v) \mapsto \partial_\mu u(\mu)(v)$ is jointly continuous, as required in (**A1**).

Second Step. The proof of the second claim is pretty similar. By the same argument as above, we observe that the mapping $(\mu, v) \mapsto \partial_v \partial_\mu u(\mu)(v)$ is uniquely determined under (**A2**). Then, we use the fact that for any $\mu \in \mathcal{P}_2(\mathbb{R}^d)$, any $v_0 \in \text{Supp}(\mu)$ and any $v \in \mathbb{R}^d$,

$$\partial_v u(\mu)(v) = \partial_v u(\mu)(v_0) + \int_0^1 \partial_v \partial_\mu u(\mu)\big(tv + (1-t)v_0\big) \cdot (v - v_0)dt.$$

The second term in the right-hand side is uniquely determined. Since $\partial_v u(\mu)(\cdot)$ is differentiable, it is also continuous. Hence, the value of $\partial_v u(\mu)(v_0)$ is also uniquely determined since v_0 belongs to the support of μ. As a result, the left-hand side is uniquely determined under (**A2**).

Third Step. Proceeding as in the first two steps, we deduce that $(\mu, v, v') \mapsto \partial^2_\mu u(\mu)(v, v')$ is uniquely determined under (**A3**). Then, we use the fact that for any $v \in \mathbb{R}^d$ and any $\mu \in \mathcal{P}_2(\mathbb{R}^d)$,

$$\partial_\mu u\big(\mu * N_d(0, I_d)\big)(v) - \partial_\mu u(\mu)(v) = \int_0^1 \mathbb{E}\big[\partial^2_\mu u\big(\mathcal{L}(X + tZ)\big)(v, X + tZ) \cdot Z\big]dt,$$

where $N_d(0, I_d)$ denotes the d-dimensional Gaussian distribution with I_d as covariance matrix and X and Z are two independent random variables with values in $\mathbb{R}^d$, with $X \sim \mu$ and $Z \sim N_d(0, I_d)$. In the above equality, the right-hand side is uniquely determined while the first term in the left-hand side is also uniquely determined since $\mu * N_d(0, I_d)$ has the entire $\mathbb{R}^d$ as support. We easily deduce that $\partial_\mu u(\mu)(v)$ is uniquely determined. □

As usual, the space $\mathcal{P}_2(\mathbb{R}^d)$ is endowed with the 2-Wasserstein distance.

Definition 5.83 *We say that a real valued function u on $\mathcal{P}_2(\mathbb{R}^d)$ is fully $\mathcal{C}^2$ if it satisfies the three conditions (**A1**), (**A2**) and (**A3**) of assumption* **Full $\mathcal{C}^2$ Regularity** *above.*

Notice that when the first derivative $D\tilde{u}$ exists and is Lipschitz, Proposition 5.36 (see also Corollary 5.38) guarantees the existence of a version $\mathcal{P}_2(\mathbb{R}^d) \times \mathbb{R}^d \ni (\mu, v) \mapsto \partial_\mu u(\mu)(v)$ which is Lipschitz in v uniformly in μ. And if this function is differentiable in μ with a Lipschitz derivative, the same Proposition 5.36 guarantees the existence of a regular version of the second derivative.

However, neither Proposition 5.36 nor Corollary 5.38 ensure the existence of a jointly continuous version of $\partial_\mu u$ (and *a fortiori* of $\partial_v \partial_\mu u$ or of $\partial^2_\mu u$). Indeed, Corollary 5.38 just provides sufficient conditions ensuring that $\partial_\mu u$ is jointly continuous at any point $(\mu, v) \in \mathcal{P}_2(\mathbb{R}^d) \times \mathbb{R}^d$ such that v is in the support of μ. In this regard, only Lemma 5.41 may be useful. So, the regularity conditions required in the three bullet points of assumption **Full $\mathcal{C}^2$ Regularity** above appear as very strong. Part of the objectives of this section will be precisely to relax them.

Remark 5.84 *As announced in Subsection 5.3.4, we used the letter v (and not the more common letter x) in order to denote the Euclidean variable in the derivative $\partial_\mu u$. This is especially useful when u is defined on the larger space $\mathbb{R}^d \times \mathcal{P}_2(\mathbb{R}^d)$ (as it will be often the case below) and thus reads $u : \mathbb{R}^d \times \mathcal{P}_2(\mathbb{R}^d) \ni (x, \mu) \mapsto u(x, \mu)$: In that case, our convention permits to distinguish $\partial_x \partial_\mu u(x, \mu)(v)$ (which is the partial derivative of $\partial_\mu u$ with respect to the original Euclidean variable x) from $\partial_v \partial_\mu u(x, \mu)(v)$ (which is the partial derivative of $\partial_\mu u$ with respect to the auxiliary Euclidean variable appearing in the* L*-derivative).*

Notation. Throughout the section, we use the following notations. For $\mu \in \mathcal{P}_2(\mathbb{R}^d)$, $v, v' \in \mathbb{R}^d$,

$$\partial_v \partial_\mu u(\mu)(v) = \Big(\partial_{v_j} [\partial_\mu u(\mu)]_i(v) \Big)_{1 \leqslant i,j \leqslant d},$$

$$\partial^2_\mu u(\mu)(v, v') = \Big(\big(\partial_\mu [\partial_\mu u(\mu)]_i(v)\big)_j (v') \Big)_{1 \leqslant i,j \leqslant d}.$$

Moreover, for any $y, z \in \mathbb{R}^d$,

$$\begin{aligned}
\partial_v \partial_\mu u(\mu)(v) y &= \Big(\sum_{j=1}^d \partial_{v_j} [\partial_\mu u(\mu)]_i(v) y_j \Big)_{1 \leqslant i \leqslant d} \in \mathbb{R}^d, \\
\partial^2_\mu u(\mu)(v, v') y &= \Big(\sum_{j=1}^d \big(\partial_\mu [\partial_\mu u(\mu)]_i(v)\big)_j (v') y_j \Big)_{1 \leqslant i \leqslant d} \in \mathbb{R}^d, \\
\partial_v \partial_\mu u(\mu)(v) \cdot (y \otimes z) &= \sum_{i,j=1}^d \partial_{v_j} [\partial_\mu u(\mu)]_i(v) z_j y_i \in \mathbb{R}, \\
\partial^2_\mu u(\mu)(v, v') \cdot (y \otimes z) &= \sum_{i,j=1}^d \big(\partial_\mu [\partial_\mu u(\mu)]_i(v)\big)_j (v') z_j y_i \in \mathbb{R}.
\end{aligned} \tag{5.80}$$

For a $d \times d$ matrix a, we shall also write:

$$\partial_v \partial_\mu u(\mu)(v) \cdot a = \sum_{i,j=1}^d \partial_{v_j} [\partial_\mu u(\mu)]_i(v) a_{i,j} = \mathrm{trace}\big\{ \partial_v \partial_\mu u(\mu)(v) a^\dagger \big\},$$

$$\partial^2_\mu u(\mu)(v, v') \cdot a = \sum_{i,j=1}^d \big(\partial_\mu [\partial_\mu u(\mu)]_i(v)\big)_j a_{i,j} = \mathrm{trace}\big\{ \partial^2_\mu u(\mu)(v, v') a^\dagger \big\}.$$

Connection with the Second-Order Differentiability of the Lifting

As we already alluded to in Remark 5.80, an assumption of the type "the lifting $\tilde{u}$ is twice continuously Fréchet differentiable" may not be well suited to our needs for the purpose of the proof of the chain rule. However, whenever u is fully $\mathcal{C}^2$ and its second-order derivatives satisfy suitable boundedness conditions, the first-order Fréchet derivative $D\tilde{u} : L^2(\Omega, \mathcal{F}, \mathbb{P}; \mathbb{R}^d) \ni X \mapsto \partial_\mu u(\mathcal{L}(X))(X) \in L^2(\Omega, \mathcal{F}, \mathbb{P}; \mathbb{R}^d)$ is continuously Gâteaux differentiable:

Proposition 5.85 *Assume that u is fully $\mathcal{C}^2$ and satisfies, for any compact subset $\mathcal{K} \subset \mathcal{P}_2(\mathbb{R}^d)$:*

$$\sup_{\mu \in \mathcal{K}} \left[\sup_{v \in \mathbb{R}^d} \left(\left|\partial_v \partial_\mu u(\mu)(v)\right|^2 + \int_{\mathbb{R}^d} \left|\partial_\mu^2 u(\mu)(v, v')\right|^2 d\mu(v') \right) \right] < +\infty.$$

Then, for any $X, Y \in L^2(\Omega, \mathcal{F}, \mathbb{P}; \mathbb{R}^d)$, the mapping:

$$\mathbb{R} \ni t \mapsto D\tilde{u}(X + tY) \in L^2(\Omega, \mathcal{F}, \mathbb{P}; \mathbb{R}^d)$$

is differentiable with:

$$\begin{aligned} &\frac{d}{dt}_{|t=0} \big[D\tilde{u}(X + tY) \big] \\ &\quad = \partial_v \partial_\mu u\big(\mathcal{L}(X)\big)(X) Y + \int_{\mathbb{R}^d \times \mathbb{R}^d} \partial_\mu^2 u\big(\mathcal{L}(X)\big)(X, x) y \, d\mathbb{P}_{(X,Y)}(x, y), \end{aligned}$$

the right-hand side being linear in the variable Y and defining a continuous function from $L^2(\Omega, \mathcal{F}, \mathbb{P}; \mathbb{R}^d) \times L^2(\Omega, \mathcal{F}, \mathbb{P}; \mathbb{R}^d)$ into $L^2(\Omega, \mathcal{F}, \mathbb{P}; \mathbb{R}^d)$.

Proposition 5.85 allows us to say that $D\tilde{u}$ is continuously Gâteaux differentiable.

Remark 5.86 *Observe that the linearity of the Gâteaux derivative in the direction Y is best seen if we use a copy $(\tilde{\Omega}, \tilde{\mathcal{F}}, \tilde{\mathbb{P}})$ of the space $(\Omega, \mathcal{F}, \mathbb{P})$ and we write:*

$$\frac{d}{dt}_{|t=0} \big[D\tilde{u}(X + tY) \big] = \partial_v \partial_\mu u\big(\mathcal{L}(X)\big)(X) Y + \tilde{\mathbb{E}}\big[\partial_\mu^2 u\big(\mathcal{L}(X)\big)(X, \tilde{X}) \tilde{Y} \big],$$

where, by convention, $\tilde{X}$ and $\tilde{Y}$ are copies of X and Y on $(\tilde{\Omega}, \tilde{\mathcal{F}}, \tilde{\mathbb{P}})$.

Proof of Proposition 5.85. Consider $X, Y \in L^2(\Omega, \mathcal{F}, \mathbb{P}; \mathbb{R}^d)$. Recalling that, for any $t \in \mathbb{R}$, $D\tilde{u}(X + tY) = \partial_\mu u(\mathcal{L}(X + tY))(X + tY)$, it is clear, that for any realization $\omega \in \Omega$, even if we choose not to write ω like probabilists do:

$$\begin{aligned} &\frac{d}{dt} \big[D\tilde{u}(X + tY) \big] \\ &\quad = \partial_v \partial_\mu u\big(\mathcal{L}(X + tY)\big)(X + tY) Y + \tilde{\mathbb{E}}\big[\partial_\mu^2 u\big(\mathcal{L}(X + tY)\big)(X + tY, \tilde{X} + t\tilde{Y}) \tilde{Y} \big]. \end{aligned}$$

Under the growth conditions assumed in the statement of the proposition, the right-hand side is in $L^2(\Omega, \mathcal{F}, \mathbb{P}; \mathbb{R}^d)$.

In order to prove that differentiability holds in $L^2(\Omega, \mathcal{F}, \mathbb{P}; \mathbb{R}^d)$, it suffices to prove that the mapping:

$$\begin{aligned}&[L^2(\Omega, \mathcal{F}, \mathbb{P}; \mathbb{R}^d)]^2 \ni (X, Y) \mapsto \\ &\qquad \partial_v \partial_\mu u(\mathcal{L}(X))(X)Y + \tilde{\mathbb{E}}\big[\partial^2_\mu u(\mathcal{L}(X))(X, \tilde{X})\tilde{Y}\big] \in L^2(\Omega, \mathcal{F}, \mathbb{P}; \mathbb{R}^d)\end{aligned}$$

is continuous. For two pairs (X, Y) and (X', Y') in $[L^2(\Omega, \mathcal{F}, \mathbb{P}; \mathbb{R}^d)]^2$, we have:

$$\begin{aligned}&\partial_v \partial_\mu u(\mathcal{L}(X'))(X')Y' + \tilde{\mathbb{E}}\big[\partial^2_\mu u(\mathcal{L}(X'))(X', \tilde{X}')\tilde{Y}'\big] \\ &\quad - \partial_v \partial_\mu u(\mathcal{L}(X))(X)Y + \tilde{\mathbb{E}}\big[\partial^2_\mu u(\mathcal{L}(X))(X, \tilde{X})\tilde{Y}\big] \\ &= \Big\{\partial_v \partial_\mu u(\mathcal{L}(X'))(X')(Y' - Y) + \tilde{\mathbb{E}}\big[\partial^2_\mu u(\mathcal{L}(X'))(X', \tilde{X}')(\tilde{Y}' - \tilde{Y})\big]\Big\} \\ &\quad + \Big\{\Big(\partial_v \partial_\mu u(\mathcal{L}(X'))(X') - \partial_v \partial_\mu u(\mathcal{L}(X))(X)\Big)Y \\ &\qquad + \tilde{\mathbb{E}}\Big[\Big(\partial^2_\mu u(\mathcal{L}(X'))(X', \tilde{X}') - \partial^2_\mu u(\mathcal{L}(X))(X, \tilde{X})\Big)\tilde{Y}\Big]\Big\} \\ &= (i) + (ii).\end{aligned}$$

From the boundedness conditions imposed on $\partial_v \partial_\mu u$ and on $\partial^2_\mu u$, it is clear that $L^2 - \lim_{(X',Y') \to_{L^2} (X,Y)} (i) = 0$. As for (ii), observe that, as $X' \to_{L^2} X$, the first term in (ii) converges to 0 in probability. By a standard uniform integrability argument (using the fact that $\partial_v \partial_\mu u(\mathcal{L}(X'))(\cdot)$ remains uniformly bounded as $X' \to_{L^2} X$), the convergence also holds in $L^2(\Omega, \mathcal{F}, \mathbb{P}; \mathbb{R}^d)$. Consider now the second term in (ii). Observe that, without any loss of generality, we may assume that $\mathcal{L}(X')$ and $\mathcal{L}(X)$ belong to a compact subset $\mathcal{K}$ of $\mathcal{P}_2(\mathbb{R}^d)$. Then, for any $R > 0$, the function $\partial^2_\mu u$ is bounded and uniformly continuous on $\mathcal{K} \times \{(v, v') \in \mathbb{R}^d;\ |v|, |v'| \leqslant R\}$. Consequently, by a new uniform integrability argument, as X' tends to X, we have:

$$\mathbb{E}\tilde{\mathbb{E}}\Big[\Big|\Big(\partial^2_\mu u(\mathcal{L}(X'))(X', \tilde{X}') - \partial^2_\mu u(\mathcal{L}(X))(X, \tilde{X})\Big)\tilde{Y}\Big|^2 \mathbf{1}_{\{|X'|+|\tilde{X}'|+|X|+|\tilde{X}| \leqslant R\}}\Big] \to 0.$$

In order to complete the proof, we notice that from Cauchy-Schwarz' inequality and from the a priori bound on $\partial^2_\mu u$, we have:

$$\begin{aligned}&\mathbb{E}\Big[\Big|\tilde{\mathbb{E}}\Big[\Big(\partial^2_\mu u(\mathcal{L}(X'))(X', \tilde{X}') - \partial^2_\mu u(\mathcal{L}(X))(X, \tilde{X})\Big)\tilde{Y}\mathbf{1}_{\{|X'|+|\tilde{X}'|+|X|+|\tilde{X}|>R\}}\Big]\Big|^2\Big] \\ &\quad \leqslant \mathbb{E}\tilde{\mathbb{E}}\Big[|\tilde{Y}|^2 \mathbf{1}_{\{|X'|+|\tilde{X}'|+|X|+|\tilde{X}|>R\}}\Big],\end{aligned}$$

which tends to 0 as R tends to $+\infty$, uniformly in $\mathcal{L}(X'), \mathcal{L}(X) \in \mathcal{K}$. □

The boundedness conditions imposed on the second-order derivatives of u in the statement of Proposition 5.85 are rather strong. However, weaker forms of differentiability of the lifting $\tilde{u}$ may be derived under weaker conditions. For instance, the following result can be established using the arguments used in the proof of Proposition 5.85.

Proposition 5.87 *Assume that u is fully $\mathcal{C}^2$ and satisfies, for any compact subset $\mathcal{K} \subset \mathcal{P}_2(\mathbb{R}^d)$,*

$$\sup_{\mu \in \mathcal{K}} \left[\int_{\mathbb{R}^d} \left| \partial_v \partial_\mu u(\mu)(v) \right|^2 d\mu(v) + \int_{\mathbb{R}^d} \int_{\mathbb{R}^d} |\partial_\mu^2 u(\mu)(v, v')|^2 d\mu(v) d\mu(v') \right] < \infty.$$

Then, for any $X \in L^2(\Omega, \mathcal{F}, \mathbb{P}; \mathbb{R}^d)$ and $Y, Z \in L^\infty(\Omega, \mathcal{F}, \mathbb{P}; \mathbb{R}^d)$, the mapping:

$$\mathbb{R}^2 \ni (s, t) \mapsto \tilde{u}(X + sY + tZ),$$

is twice continuously differentiable. Moreover,

$$\begin{aligned}
&\frac{\partial^2}{\partial s \partial t}_{|(s,t)=(0,0)} \left[\tilde{u}(X + sY + tZ) \right] \\
&= \frac{\partial}{\partial s}_{|s=0} \left\{ \mathbb{E}\left[\partial_\mu u\big(\mathcal{L}(X + sY)\big)(X + sY) \cdot Z \right] \right\} \\
&= \mathbb{E}\left[\partial_v \partial_\mu u\big(\mathcal{L}(X)\big)(X) \cdot Z \otimes Y \right] + \mathbb{E}\tilde{\mathbb{E}}\left[\partial_\mu^2 u\big(\mathcal{L}(X)\big)(X, \tilde{X}) \cdot Z \otimes \tilde{Y} \right],
\end{aligned}$$

where, as usual, we use the tilde notation to denote copies of the various objects at hand.

Remark 5.88 *Notice that, in the statement of Proposition 5.87, the variables Y and Z are required to be in $L^\infty(\Omega, \mathcal{F}, \mathbb{P}; \mathbb{R}^d)$. Obviously, this requirement is stronger than the condition used so far for defining the Gâteaux and Fréchet derivatives of the lifting of u, which have been computed along directions in $L^2(\Omega, \mathcal{F}, \mathbb{P}; \mathbb{R}^d)$.*

Observe also that we here address the differentiability of $\mathbb{R} \ni s \mapsto \mathbb{E}[D\tilde{u}(X + sY)(X + sY) \cdot Z] \in \mathbb{R}$. This is in contrast with the statement of Proposition 5.85, in which we addressed the differentiability of the mapping $\mathbb{R} \ni s \mapsto D\tilde{u}(X + sY)(X + sY) \in L^2(\Omega, \mathcal{F}, \mathbb{P}; \mathbb{R}^d)$.

Symmetry of the Second-Order Derivatives. By Schwarz' theorem, we can exchange the roles of Y and Z in the above identity. With the same notation as in the statement of Proposition 5.87, this shows that:

$$\begin{aligned}
&\mathbb{E}\left[\partial_v \partial_\mu u\big(\mathcal{L}(X)\big)(X) \cdot Z \otimes Y \right] + \mathbb{E}\tilde{\mathbb{E}}\left[\partial_\mu^2 u\big(\mathcal{L}(X)\big)(X, \tilde{X}) \cdot Z \otimes \tilde{Y} \right] \\
&= \mathbb{E}\left[\partial_v \partial_\mu u\big(\mathcal{L}(X)\big)(X) \cdot Y \otimes Z \right] + \mathbb{E}\tilde{\mathbb{E}}\left[\partial_\mu^2 u\big(\mathcal{L}(X)\big)(X, \tilde{X}) \cdot Y \otimes \tilde{Z} \right].
\end{aligned}$$

Choosing Y of the form $\varepsilon\psi(X)$ and Z of the form $\varepsilon\zeta(X)$, for two bounded Borel measurable functions from $\mathbb{R}^d$ to $\mathbb{R}^d$, with $\mathbb{P}[\varepsilon = 1] = \mathbb{P}[\varepsilon = -1] = 1/2$ and

ε independent of X, we get rid of the second-order derivatives $\partial_\mu^2 u$ in the above identity, so that:

$$\mathbb{E}\Big[\partial_v \partial_\mu u\big(\mathcal{L}(X)\big)(X) \cdot \zeta(X) \otimes \psi(X)\Big] = \mathbb{E}\Big[\partial_v \partial_\mu u\big(\mathcal{L}(X)\big)(X) \cdot \psi(X) \otimes \zeta(X)\Big],$$

from which we deduce that $\partial_v \partial_\mu u(\mathcal{L}(X))(X)$ takes values in the set of symmetric matrices of size d. By continuity of $\partial_\mu u(\mu)(\cdot)$ in the variable v, where $\mu = \mathcal{L}(X)$, this shows that $\partial_v \partial_\mu u(\mu)(v)$ is a symmetric matrix for any $v \in \mathbb{R}^d$ when the support of μ is the entire $\mathbb{R}^d$. By continuity in μ, we deduce that $\partial_v \partial_\mu u(\mu)(v)$ is a symmetric matrix for any $\mu \in \mathcal{P}_2(\mathbb{R}^d)$ and any $v \in \mathbb{R}^d$.

Choosing Y and Z of the form $\psi(X)$ and $\zeta(X)$ respectively, for two bounded Borel measurable functions from $\mathbb{R}^d$ to $\mathbb{R}$, we get, making use of the symmetry of $\partial_v \partial_\mu u$:

$$\begin{aligned}
&\mathbb{E}\tilde{\mathbb{E}}\Big[\partial_\mu^2 u\big(\mathcal{L}(X)\big)(X,\tilde{X}) \cdot \zeta(X) \otimes \psi(\tilde{X})\Big] \\
&= \mathbb{E}\tilde{\mathbb{E}}\Big[\partial_\mu^2 u\big(\mathcal{L}(X)\big)(X,\tilde{X}) \cdot \psi(X) \otimes \zeta(\tilde{X})\Big] \\
&= \mathbb{E}\tilde{\mathbb{E}}\Big[\partial_\mu^2 u\big(\mathcal{L}(X)\big)(\tilde{X},X) \cdot \psi(\tilde{X}) \otimes \zeta(X)\Big] \\
&= \mathbb{E}\tilde{\mathbb{E}}\Big[\Big(\partial_\mu^2 u\big(\mathcal{L}(X)\big)(\tilde{X},X)\Big)^\dagger \cdot \zeta(X) \otimes \psi(\tilde{X})\Big],
\end{aligned}$$

from which we deduce that $\partial_\mu^2 u(\mathcal{L}(X))(X,\tilde{X}) = (\partial_\mu^2 u(\mathcal{L}(X))(\tilde{X},X))^\dagger$. By the same argument as above, we conclude that $\partial_\mu^2 u(\mu)(v,v') = (\partial_\mu^2 u(\mu)(v',v))^\dagger$, for any $v, v' \in \mathbb{R}^d$ and $\mu \in \mathcal{P}_2(\mathbb{R}^d)$.

The following corollary summarizes our discussion.

Corollary 5.89 *Assume that the function u is fully $\mathcal{C}^2$. Then, for any $\mu \in \mathcal{P}_2(\mathbb{R}^d)$ and any $v, v' \in \mathbb{R}^d$, $\partial_v \partial_\mu u(\mu)(v)$ is a symmetric matrix and $\partial_\mu^2 u(\mu)(v,v') = (\partial_\mu^2 u(\mu)(v',v))^\dagger$.*

Proof. The proof is essentially identical to the above argument except for the integrability assumptions on $\partial_v \partial_\mu u$ and $\partial_\mu^2 u$. However, we can repeat the argument when μ has a bounded support. Indeed, by continuity, $\partial_v \partial_\mu u(\mu)$ and $\partial_\mu^2 u(\mu)$ are bounded on the support of μ whenever it is bounded, and the bounds remain uniform as long as the support of μ remains included in a prescribed bounded subset of $\mathbb{R}^d$.

This shows that the symmetry relationships hold on the support of μ when the latter is bounded. By approximating any probability measure with $\mathbb{R}^d$ as support by a sequence of probability measures with bounded supports, using the continuity of the derivatives, we show that the symmetry relationships hold on the whole $\mathbb{R}^d$ when the support of μ is the whole $\mathbb{R}^d$. By continuity again, we complete the proof as above in the case when the support of μ is merely a subset of $\mathbb{R}^d$. □

Remark 5.90 *Symmetry of $\partial_v \partial_\mu u$ should not come as a surprise since Lemma 5.61 asserts that $\partial_\mu u$ is somehow a gradient.*

Connection with the Differentiability of the Empirical Projection

Finally, we enlighten the notion of full $\mathcal{C}^2$-regularity by extending the connection between the derivatives of u^N and those of u given in Proposition 5.35 to the second order. Recall that given a function $u : \mathcal{P}_2(\mathbb{R}^d) \to \mathbb{R}$ and an integer $N \geqslant 1$, the empirical projection of u onto $\mathbb{R}^d$ was defined as the function:

$$u^N : (\mathbb{R}^d)^N \ni (x_1, \cdots, x_N) \mapsto u\left(\frac{1}{N}\sum_{i=1}^N \delta_{x_i}\right).$$

Proposition 5.91 *Assume that u is fully $\mathcal{C}^2$. Then, for any $N \geqslant 1$, the empirical projection u^N is $\mathcal{C}^2$ on $(\mathbb{R}^d)^N$ and, for all $x^1, \cdots, x^N \in \mathbb{R}^d$,*

$$\begin{aligned} &\partial^2_{x^i x^j} u^N(x^1, \cdots, x^N) \\ &= \frac{1}{N}\partial_v \partial_\mu u\left(\frac{1}{N}\sum_{\ell=1}^N \delta_{x^\ell}\right)(x^i)\mathbf{1}_{i=j} + \frac{1}{N^2}\partial^2_\mu u\left(\frac{1}{N}\sum_{\ell=1}^N \delta_{x^\ell}\right)(x^i, x^j). \end{aligned} \tag{5.81}$$

Proof. The proof piggybacks on the computation of the first order derivatives given in Proposition 5.35. When $i \neq j$, (5.81) can be obtained by applying the result of Proposition 5.35 twice. A modicum of care is required when $i = j$ as differentiability is performed simultaneously in the directions of μ and v in the first order derivative $\partial_\mu u(\mu)(v)$. However, the assumption of joint continuity of the second-order derivatives $\partial^2_\mu u(\mu)(v)$ and $\partial_v \partial_\mu u(\mu)(v, v')$ can be used to handle this minor difficulty. □

5.6.3 Chain Rule Under Full $\mathcal{C}^2$-Regularity

Statement of the Chain Rule

On a complete probability space $(\Omega, \mathcal{F}, \mathbb{P})$ equipped with a right-continuous and complete filtration $\mathbb{F} = (\mathcal{F}_t)_{t \geqslant 0}$, we now consider an $\mathbb{R}^d$-valued Itô process:

$$dX_t = b_t dt + \sigma_t dW_t, \quad X_0 \in L^2(\Omega, \mathcal{F}, \mathbb{P}), \tag{5.82}$$

where $\boldsymbol{W} = (W_t)_{t \geqslant 0}$ is an $\mathbb{F}$-Brownian motion with values in $\mathbb{R}^d$, and $(b_t)_{t \geqslant 0}$ and $(\sigma_t)_{t \geqslant 0}$ are $\mathbb{F}$-progressively measurable processes with values in $\mathbb{R}^d$ and $\mathbb{R}^{d \times d}$ respectively. We assume that they satisfy:

$$\forall T > 0, \quad \mathbb{E}\left[\int_0^T \left(|b_t|^2 + |\sigma_t|^4\right) dt\right] < +\infty. \tag{5.83}$$

The main result of this section is the form of Itô's formula given by the following chain rule.

Theorem 5.92 *Under the above conditions, in particular assuming that* (5.82) *and* (5.83) *hold, let us further assume that u is fully $\mathcal{C}^2$ and that, for any compact subset $\mathcal{K} \subset \mathcal{P}_2(\mathbb{R}^d)$,*

$$\sup_{\mu\in\mathcal{K}}\left[\int_{\mathbb{R}^d} \left|\partial_v \partial_\mu u(\mu)(v)\right|^2 d\mu(v)\right] < +\infty. \tag{5.84}$$

Then, if for any $t \geqslant 0$ we denote by μ_t the marginal distribution $\mu_t = \mathcal{L}(X_t)$ and we let $a_t = \sigma_t\sigma_t^\dagger$, it holds that:

$$\begin{aligned} u(\mu_t) = u(\mu_0) &+ \int_0^t \mathbb{E}\big[\partial_\mu u(\mu_s)(X_s)\cdot b_s\big]ds \\ &+ \frac{1}{2}\int_0^t \mathbb{E}\big[\partial_v\big(\partial_\mu u(\mu_s)\big)(X_s)\cdot a_s\big]ds. \end{aligned} \tag{5.85}$$

Remark 5.93 *By symmetry, observe that:*

$$\partial_v\partial_\mu u(\mu_s)(X_s)\cdot a_s = \text{trace}\big\{\partial_v\partial_\mu u(\mu_s)(X_s)a_s^\dagger\big\} = \text{trace}\big\{\partial_v\partial_\mu u(\mu_s)(X_s)a_s\big\}.$$

Moreover, the lifting $\tilde{u}$ of u being continuously Fréchet differentiable, the bound (5.84) *is obviously satisfied by $\partial_\mu u$, namely:*

$$\sup_{\mu\in\mathcal{K}}\left[\int_{\mathbb{R}^d} \left|\partial_\mu u(\mu)(v)\right|^2 d\mu(v)\right] < +\infty. \tag{5.86}$$

Proof of the Chain Rule

The proof of Theorem 5.92 relies on a mollification argument of independent interest.

Lemma 5.94 *Let $u : \mathcal{P}_2(\mathbb{R}^d) \to \mathbb{R}$ be a fully $\mathcal{C}^2$ function and $\rho : \mathbb{R}^d \to \mathbb{R}^d$ be a smooth function with compact support (i.e., equal to* 0 *outside of a bounded subset of $\mathbb{R}^d$). Define the function $u \star \rho$ on $\mathcal{P}_2(\mathbb{R}^d)$ by:*

$$\big(u\star\rho\big)(\mu) = u\big(\mu\circ\rho^{-1}\big), \qquad \mu\in\mathcal{P}_2(\mathbb{R}^d).$$

Then, $u \star \rho$ is fully $\mathcal{C}^2$. Moreover, $u \star \rho$ and its first and second order derivatives are bounded and uniformly continuous on the whole space.

Proof. The lifted version of $u \star \rho$ is nothing but $\tilde{u} \circ \tilde{\rho}$, where $\tilde{\rho}$ is defined as $\tilde{\rho} : L^2(\Omega, \mathcal{F}, \mathbb{P}; \mathbb{R}^d) \ni X \mapsto \tilde{\rho}(X) = \rho(X)$. It is then plain to check that:

$$
\begin{aligned}
&\partial_\mu\big[u \star \rho\big](\mu)(v) \\
&\quad = \Big(\sum_{k=1}^d \big[\partial_\mu u(\mu \circ \rho^{-1})(\rho(v))\big]_k \frac{\partial \rho_k}{\partial v_i}(v) \Big)_{i=1,\cdots,d}, \\
&\partial^2_\mu\big[u \star \rho\big](\mu)(v, v') \\
&\quad = \Big(\sum_{k,\ell=1}^d \big[\partial^2_\mu u(\mu \circ \rho^{-1})(\rho(v), \rho(v'))\big]_{k,\ell} \frac{\partial \rho_k}{\partial v_i}(v) \frac{\partial \rho_\ell}{\partial v_j}(v') \Big)_{i,j=1,\cdots,d}, \\
&\partial_v \partial_\mu\big[u \star \rho\big](\mu)(v) \\
&\quad = \Big(\sum_{k=1}^d \big[\partial_\mu u(\mu \circ \rho^{-1})(\rho(v))\big]_k \frac{\partial^2 \rho_k}{\partial v_i \partial v_j}(v) \\
&\qquad\quad + \sum_{k,\ell=1}^d \big[\partial_v \partial_\mu u(\mu \circ \rho^{-1})(\rho(v))\big]_{k,\ell} \frac{\partial \rho_k}{\partial v_i}(v) \frac{\partial \rho_\ell}{\partial v_j}(v) \Big)_{i,j=1,\cdots,d}.
\end{aligned}
\tag{5.87}
$$

Since ρ is compactly supported, the mapping $\mathcal{P}_2(\mathbb{R}^d) \ni \mu \mapsto \mu \circ \rho^{-1}$ has a relatively compact range in $\mathcal{P}_2(\mathbb{R}^d)$ by Theorem 5.5. By continuity of u and its derivatives, we deduce that $u \star \rho$ and its first and second order derivatives are bounded and uniformly continuous on the whole space. □

As a consequence of Lemma 5.94, we get:

Lemma 5.95 *Assume that the chain rule holds for any fully $\mathcal{C}^2$ function u with bounded and uniformly continuous (with respect to the space and measure arguments) first and second order derivatives. Then, it holds for any function u satisfying the assumptions of Theorem 5.92.*

Proof. Assume that the chain rule has been proved for any bounded and uniformly continuous function u with bounded and uniformly continuous derivatives of orders 1 and 2. Then, for u satisfying the assumption of Theorem 5.92, we can apply the chain rule to $u \star \rho$, for any ρ as in the statement of Lemma 5.94. In particular, we can apply the chain rule to $u \star \rho_n$ for any $n \geqslant 1$, where $(\rho_n)_{n \geqslant 1}$ is a sequence of compactly supported smooth functions such that $(\rho_n, \partial \rho_n, \partial^2 \rho_n)(v) \to (v, I_d, 0)$ uniformly on compact sets as $n \to \infty$, where I_d denotes the identity matrix of size d and 0 is here the zero of $\mathbb{R}^{d \times d}$. In order to pass to the limit in the chain rule (5.85), the only thing we need to do is to verify some almost sure (or pointwise) convergence in the underlying expectations, and to check that the relevant uniform integrability arguments can be used.

Without any loss of generality, we can assume that there exists a constant C such that $|\rho_n(v)| \leqslant C|v|$, $|\partial \rho_n(v)| \leqslant C$ and $|\partial^2 \rho_n(v)| \leqslant C$ for any $n \geqslant 1$ and $v \in \mathbb{R}^d$, and that $\rho_n(v) = v$ for any $n \geqslant 1$ and v with $|v| \leqslant n$. Then, for any $\mu \in \mathcal{P}_2(\mathbb{R}^d)$ and any random variable X with μ as distribution, it holds:

$$
W_2\big(\mu \circ \rho_n^{-1}, \mu\big)^2 \leqslant \mathbb{E}\big[|\rho_n(X) - X|^2 \mathbf{1}_{\{|X| \geqslant n\}}\big] \leqslant C\mathbb{E}\big[|X|^2 \mathbf{1}_{\{|X| \geqslant n\}}\big],
$$

which tends to 0 as $n \to \infty$. By continuity of u and its partial derivatives, and by (5.87), it is plain to deduce that almost surely,

$$\begin{aligned} &u \star \rho_n(\mu) \to u(\mu), \quad \partial_\mu[u \star \rho_n](\mu)(X) \to \partial_\mu u(\mu)(X), \\ &\partial_v \partial_\mu[u \star \rho_n](\mu)(X) \to \partial_v \partial_\mu u(\mu)(X). \end{aligned} \tag{5.88}$$

Moreover, we notice that:

$$\sup_{n \geqslant 1} \mathbb{E}\Big[\big|\partial_\mu[u \star \rho_n](\mu)(X)\big|^2 + \big|\partial_v \partial_\mu[u \star \rho_n](\mu)(X)\big|^2 \Big] < \infty. \tag{5.89}$$

Indeed, by (5.87), it is enough to check that:

$$\begin{aligned} \sup_{n \geqslant 1} \bigg[&\int_{\mathbb{R}^d} \big|\partial_\mu u(\mu \circ \rho_n^{-1})(v)\big|^2 d(\mu \circ \rho_n^{-1})(v) \\ &+ \int_{\mathbb{R}^d} \big|\partial_v \partial_\mu u(\mu \circ \rho_n^{-1})(v)\big|^2 d(\mu \circ \rho_n^{-1})(v) \bigg] < \infty, \end{aligned}$$

which follows directly from (5.84) and (5.86), noticing that the sequence $(\mu \circ \rho_n^{-1})_{n \geqslant 1}$ lives in a compact subset of $\mathcal{P}_2(\mathbb{R}^d)$ as it is convergent.

By (5.88) and (5.89) and by a standard uniform integrability argument, we deduce that, for any $t \geqslant 0$ and any $s \in [0, t]$ such that $\mathbb{E}[|b_s|^2 + |\sigma_s|^4] < \infty$,

$$\begin{aligned} &\lim_{n \to +\infty} \mathbb{E}\big[\partial_\mu[u \star \rho_n](\mathcal{L}(X))(X) \cdot b_s\big] = \mathbb{E}\big[\partial_\mu u(\mathcal{L}(X))(X) \cdot b_s\big], \\ &\lim_{n \to +\infty} \mathbb{E}\big[\partial_v \partial_\mu[u \star \rho_n](\mathcal{L}(X))(X) \cdot a_s\big] = \mathbb{E}\big[\partial_v \partial_\mu u(\mathcal{L}(X))(X) \cdot a_s\big]. \end{aligned}$$

Recall that the above is true for any $\mu \in \mathcal{P}_2(\mathbb{R}^d)$ and any $X \in L^2(\Omega, \mathcal{F}, \mathbb{P}; \mathbb{R}^d)$ with μ as distribution. We then choose $X = X_s$ in the above limits. As the bound $\mathbb{E}[|b_s|^2 + |\sigma_s|^4] < \infty$ is satisfied for almost every $s \in [0, t]$, we can pass to the limit inside the integrals appearing in the chain rule applied to each of the $(u \star \rho_n)_{n \geqslant 1}$. In order to pass to the limit in the chain rule itself, we must exchange the pathwise limit which holds for almost every $s \in [0, t]$ and the integral with respect to s. The argument is the same as in (5.89). Indeed, since the flow of measures $(\mathcal{L}(X_s))_{0 \leqslant s \leqslant t}$ is continuous for the 2-Wasserstein distance, the family of measures $((\mathcal{L}(\rho_n(X_s)))_{0 \leqslant s \leqslant t})_{n \geqslant 1}$ is relatively compact and thus:

$$\sup_{n \geqslant 1} \sup_{s \in [0,t]} \mathbb{E}\Big[\big|\partial_\mu[u \star \rho_n](\mathcal{L}(X_s))(X_s)\big|^2 + \big|\partial_v \partial_\mu[u \star \rho_n](\mathcal{L}(X_s))(X_s)\big|^2 \Big] < \infty,$$

which is enough to prove that the functions:

$$\Big([0, t] \ni s \mapsto \mathbb{E}\Big[\partial_\mu[u \star \rho_n](\mathcal{L}(X_s))(X_s) \cdot b_s\Big] + \mathbb{E}\Big[\partial_v \partial_\mu[u \star \rho_n]\big(\mathcal{L}(X_s)\big)(X_s) \cdot a_s\Big] \Big)_{n \geqslant 1}$$

are uniformly integrable on $[0, t]$. □

We now turn to the proof of Theorem 5.92. We just give a sketch as a complete proof will be given for a refined version in Theorem 5.99 later in this section.

Proof of Theorem 5.92. By Lemma 5.95, we can replace u by $u \star \rho$ for some compactly supported smooth function ρ or equivalently, we can replace $(X_t)_{t \geqslant 0}$ by $(\rho(X_t))_{t \geqslant 0}$. In other words, we can assume without any loss of generality that u and its first and second order derivatives are bounded and uniformly continuous, and that u and its derivatives are uniformly continuous and that $(X_t)_{t\geqslant 0}$ is a bounded Itô process.

Repeating the proof of Lemma 5.95, we can even assume that $(b_t)_{t\geqslant 0}$ and $(\sigma_t)_{t\geqslant 0}$ are also bounded. Indeed, it suffices to prove the chain rule when $(X_t)_{t\geqslant 0}$ is driven by truncated processes and pass to the limit along a sequence of truncations converging to $(X_t)_{t\geqslant 0}$.

Let us denote by $((X_t^\ell)_{t\geqslant 0})_{\ell\geqslant 1}$ a sequence of i.i.d. copies of $(X_t)_{t\geqslant 0}$. That is, for any $\ell \geqslant 1$,

$$dX_t^\ell = b_t^\ell dt + \sigma_t^\ell dW_t^\ell, \quad t \geqslant 0,$$

where $((b_t^\ell, \sigma_t^\ell, W_t^\ell)_{t\geqslant 0}, X_0^\ell)_{\ell\geqslant 1}$ are i.i.d copies of $((b_t, \sigma_t, W_t)_{t\geqslant 0}, X_0)$ constructed on an extension of $(\Omega, \mathcal{F}, \mathbb{P})$. Recalling the definition of the flow of marginal empirical measures:

$$\bar{\mu}_t^N = \frac{1}{N}\sum_{\ell=1}^N \delta_{X_t^\ell},$$

the classical Itô formula yields together with Proposition 5.91, $\mathbb{P}$-a.s., for any $t \geqslant 0$:

$$\begin{aligned}
u^N(X_t^1, \cdots, X_t^N) &= u^N(X_0^1, \cdots, X_0^N) \\
&\quad + \frac{1}{N}\sum_{\ell=1}^N \int_0^t \partial_\mu u(\bar{\mu}_s^N)(X_s^\ell) \cdot b_s^\ell ds \\
&\quad + \frac{1}{N}\sum_{\ell=1}^N \int_0^t \partial_\mu u(\bar{\mu}_s^N)(X_s^\ell) \cdot (\sigma_s^\ell dW_s^\ell) \\
&\quad + \frac{1}{2N}\sum_{\ell=1}^N \int_0^t \mathrm{trace}\{\partial_v \partial_\mu u(\bar{\mu}_s^N)(X_s^\ell) a_s^\ell\} ds \\
&\quad + \frac{1}{2N^2}\sum_{\ell=1}^N \int_0^t \mathrm{trace}\{\partial_\mu^2 u(\bar{\mu}_s^N)(X_s^\ell, X_s^\ell) a_s^\ell\} ds,
\end{aligned}$$

with $a_s^\ell = \sigma_s^\ell(\sigma_s^\ell)^\dagger$. We take expectations on both sides of this equality and obtain, using the fact that the stochastic integral has zero expectation thanks to the boundedness of the coefficients:

$$\begin{aligned}
\mathbb{E}[u(\bar{\mu}_t^N)] &= \mathbb{E}[u(\bar{\mu}_0^N)] + \frac{1}{N}\sum_{\ell=1}^N \mathbb{E}\Big[\int_0^t \partial_\mu u(\bar{\mu}_s^N)(X_s^\ell) \cdot b_s^\ell ds\Big] \\
&\quad + \frac{1}{2N}\sum_{\ell=1}^N \mathbb{E}\Big[\int_0^t \mathrm{trace}\Big[\partial_v \partial_\mu u(\bar{\mu}_s^N)(X_s^\ell) a_s^\ell\Big] ds\Big] \\
&\quad + \frac{1}{2N^2}\sum_{\ell=1}^N \mathbb{E}\Big[\int_0^t \mathrm{trace}\Big[\partial_\mu^2 u(\bar{\mu}_s^N)(X_s^\ell, X_s^\ell) a_s^\ell\Big] ds\Big].
\end{aligned}$$

All the expectations are finite, thanks to the boundedness of the coefficients. Using the fact that the processes $((a_s^\ell, b_s^\ell, X_s^\ell)_{0\leqslant s\leqslant t})_{\ell\in\{1,\cdots,N\}}$ are i.i.d., we deduce that:

$$
\begin{aligned}
\mathbb{E}\big[u(\bar{\mu}_t^N)\big] &= \mathbb{E}\big[u(\bar{\mu}_0^N)\big] + \int_0^t \mathbb{E}\Big\{\partial_\mu u(\bar{\mu}_s^N)(X_s^1)\cdot b_s^1\Big\} ds \\
&\quad + \frac{1}{2}\int_0^t \mathbb{E}\Big\{\text{trace}\Big[\partial_v \partial_\mu u(\bar{\mu}_s^N)(X_s^1) a_s^1\Big]\Big\} ds \\
&\quad + \frac{1}{2N}\int_0^t \mathbb{E}\Big\{\text{trace}\Big[\partial_\mu^2 u(\bar{\mu}_s^N)(X_s^1, X_s^1) a_s^1\Big]\Big\} ds \\
&= \mathbb{E}\big[u(\bar{\mu}_0^N)\big] + (i) + (ii) + (iii).
\end{aligned}
$$

In particular, because of the additional $1/N$, (iii) converges to 0. Moreover, we know from (5.19) that, for any $s \in [0, t]$,

$$
\mathbb{P}\Big[\lim_{N\to\infty} W_2\big(\bar{\mu}_s^N, \mu_s\big)^2 = 0\Big] = 1. \tag{5.90}
$$

This implies, together with the continuity of u with respect to the distance W_2, that $\mathbb{E}[u(\bar{\mu}_t^N)]$ (respectively $\mathbb{E}[u(\bar{\mu}_0^N)]$) converges to $u(\mu_t)$ (respectively $u(\mu_0)$). Combining the boundedness and the uniform continuity of $\partial_\mu u$ on $\mathcal{P}_2(\mathbb{R}^d) \times \mathbb{R}^d$ with the boundedness of $(b_s)_{0\leqslant s\leqslant t}$, we prove in the same way that (i) converges to the first integral appearing in the right-hand side of (5.85). Similar arguments lead to the convergence of (ii). □

5.6.4 Partial $\mathcal{C}^2$-Regularity

One of the most remarkable feature of equation (5.85) is the fact that the second order derivative $\partial_\mu^2 u$ does not appear in the final form of the chain rule provided by Theorem 5.92. Thus, it is quite natural to wonder if the chain rule could still hold without assuming the existence of $\partial_\mu^2 u$. Motivated by this quandary, we prove that the chain rule does indeed hold under a weaker set of assumptions not requiring the existence of $\partial_\mu^2 u$. We shall refer to this set of assumptions as *partial $\mathcal{C}^2$ regularity*.

Assumption (Partial $\mathcal{C}^2$ Regularity). The lifting $\tilde{u}$ is continuously Fréchet differentiable, and, for any $\mu \in \mathcal{P}_2(\mathbb{R}^d)$, we can find a continuous version of the mapping $\mathbb{R}^d \ni v \mapsto \partial_\mu u(\mu)(v)$ such that:

(A1) The mapping $\mathcal{P}_2(\mathbb{R}^d) \times \mathbb{R}^d \ni (\mu, v) \mapsto \partial_\mu u(\mu)(v)$ is locally bounded (namely is bounded on any compact subset) and is continuous at any (μ, v) such that $v \in \text{Supp}(\mu)$.

(A2) For any $\mu \in \mathcal{P}_2(\mathbb{R}^d)$, the mapping $\mathbb{R}^d \ni v \mapsto \partial_\mu u(\mu)(v) \in \mathbb{R}^d$ is continuously differentiable and its derivative is locally bounded and is jointly continuous with respect to (μ, v) at any point (μ, v) such that $v \in \text{Supp}(\mu)$, the derivative being denoted by $\mathbb{R}^d \ni v \mapsto \partial_v \partial_\mu u(\mu)(v) \in \mathbb{R}^{d\times d}$.

Above, we use the notation $\mathrm{Supp}(\mu)$ for the support of μ.

Definition 5.96 *We say that u is partially $\mathcal{C}^2$ if it satisfies assumption* **Partial $\mathcal{C}^2$ Regularity**.

Remark 5.97 *Observe that, contrary to what we did in the Definition 5.83 of the full $\mathcal{C}^2$ regularity, joint continuity of the first and second order derivatives is only required at pairs (μ, v) such that $v \in \mathrm{Supp}(\mu)$. According to our discussion in Corollary 5.38, this is much more satisfactory. Notice also that, on the support of μ, $\partial_\mu u(\mu)(\cdot)$ and $\partial_v \partial_\mu u(\mu)(\cdot)$ are uniquely defined provided that they are continuous.*

Remark 5.98 *Following our discussion of the fully $\mathcal{C}^2$ case, we argue the symmetry of $\partial_v \partial_\mu u(\mu)(\cdot)$ whenever u is partially $\mathcal{C}^2$. The only difficulty is that we cannot repeat the proof of Corollary 5.89 (which applies to the fully $\mathcal{C}^2$ case) since it requires the existence of $\partial^2_\mu u(\mu)$ explicitly.*

We shall prove next (see the proof of Theorem 5.99 below) that whenever u, $\partial_\mu u$ and $\partial_v \partial_\mu u$ are bounded and uniformly continuous, there exists a family of twice continuously differentiable functions $(u_n^N : (\mathbb{R}^d)^N \to \mathbb{R})_{n,N \geq 1}$ together with a sequence of reals $(\varepsilon_p)_{p \geq 1}$ converging to 0 as p tends to ∞, such that, for any $\mu \in \mathcal{P}_2(\mathbb{R}^d)$ and any N-tuple $(X^1, \cdots, X^N)$ of independent random variables constructed on some auxiliary probability space $(\Omega, \mathcal{F}, \mathbb{P})$, with common distribution μ, it holds for any $i \in \{1, \cdots, N\}$:

$$\mathbb{E}\Big[\big|N\partial^2_{x^i x^i} u_n^N(X^1, \cdots, X^N) - \partial_v \partial_\mu u(\mu)(X^i)\big|\Big] \leqslant \varepsilon_n + n\varepsilon_N.$$

Since u_n^N is twice continuously differentiable, we have:

$$\big[\partial^2_{x^i x^i} u_n^N(X^1, \cdots, X^N)\big]^\dagger = \partial^2_{x^i x^i} u_n^N(X^1, \cdots, X^N).$$

Therefore,

$$\mathbb{E}\Big[\Big|\Big(\partial_v \partial_\mu u(\mu)(X^1)\Big)^\dagger - \partial_v \partial_\mu u(\mu)(X^1)\Big|\Big] \leqslant 2\big(\varepsilon_n + n\varepsilon_N\big).$$

Letting N and then n tend to ∞, we deduce that $\partial_v \partial_\mu u(\mu)(\cdot)^\dagger = \partial_v \partial_\mu u(\mu)(\cdot)$ μ-almost everywhere and thus everywhere on the support of μ since $\partial_v \partial_\mu u(\mu)(\cdot)$ is continuous.

This shows that $\partial_v \partial_\mu u$ is symmetric whenever u and its derivatives are bounded and uniformly continuous. In order to complete the proof, we need a new approximation argument. It relies on the fact that, as shown in the proof of Theorem 5.99, for any $\mu \in \mathcal{P}_2(\mathbb{R}^d)$ and $v \in \mathrm{Supp}(\mu)$, $\partial_v \partial_\mu u(\mu)(v)$ is the limit of a sequence $(\partial_v \partial_\mu u_n(\mu)(v))_{n \geq 1}$, where for any $n \geqslant 1$, $u_n : \mathcal{P}_2(\mathbb{R}^d) \to \mathbb{R}$ is partially $\mathcal{C}^2$ and u_n and its derivatives are bounded and uniformly continuous. This completes the proof that, for $v \in \mathrm{Supp}(\mu)$, $\partial_v \partial_\mu u(\mu)(v)$ is a symmetric matrix.

Chain Rule Under Partial $\mathcal{C}^2$-Regularity

We now show that the chain rule still holds under the weaker assumption of partial $\mathcal{C}^2$ regularity.

Theorem 5.99 *Assume that u is partially $\mathcal{C}^2$ and that, for any compact subset $\mathcal{K} \subset \mathcal{P}_2(\mathbb{R}^d)$, the bound (5.84) holds true. Then, the chain rule (5.85) holds for any Itô process of the form (5.82) satisfying (5.83).*

Remark 5.100 *If we had to combine Remarks 5.97 and 5.98 in an informal statement, we would say that "everything works exactly as in the fully $\mathcal{C}^2$ case for $\partial_v \partial_\mu u(\mu)(v)$ as long as v is in the support of μ." Clearly, this suffices for our purpose since, in Itô's formula (5.85), the second-order derivative is only evaluated at points (μ, v) such that $v \in \text{Supp}(\mu)$.*

Remark 5.101 *Notice that, in (5.85), the mappings $[0,T] \ni s \mapsto \mathbb{E}[\partial_\mu u(\mu_s)(X_s) \cdot b_s]$ and $[0,T] \ni s \mapsto \mathbb{E}[\partial_v \partial_\mu u(\mu_s)(X_s) \cdot a_s]$ are measurable if we assume without any loss of generality that a_s and b_s are square integrable for any $s \in [0,T]$. This follows from the fact that for any $t \in [0,T]$, the mappings:*

$$L^2(\Omega, \mathcal{F}, \mathbb{P}; \mathbb{R}^d) \ni X \mapsto \mathbb{E}\big[\partial_\mu u\big(\mathcal{L}(X)\big)(X) \cdot b_t\big],$$
$$L^2(\Omega, \mathcal{F}, \mathbb{P}; \mathbb{R}^d) \ni X \mapsto \mathbb{E}\big[\partial_v \partial_\mu u\big(\mathcal{L}(X)\big)(X) \cdot a_t\big],$$

are continuous, which implies that:

$$\begin{aligned}
&\mathbb{E}\Big[\partial_\mu u\big(\mathcal{L}(X_t)\big)(X_t) \cdot b_t\Big] \\
&= \lim_{N \to \infty} \sum_{k=0}^{N-1} \mathbb{E}\Big[\partial_\mu u\big(\mathcal{L}(X_{Tk/N})\big)(X_{Tk/N}) \cdot b_t\Big] \mathbf{1}_{Tk/N \leqslant t < T(k+1)/N} \qquad (5.91) \\
&\quad + \mathbb{E}\Big[\partial_\mu u\big(\mathcal{L}(X_T)\big)(X_T) \cdot b_T\Big] \mathbf{1}_{t=T},
\end{aligned}$$

and similarly for $(\partial_v \partial_\mu u(\mathcal{L}(X_t))(X_t))_{0 \leqslant t \leqslant T}$ and $(a_t)_{0 \leqslant t \leqslant T}$.

The proof can be completed by noticing that for any $Z \in L^2(\Omega, \mathcal{F}, \mathbb{P}; \mathbb{R}^d)$, the mapping $[0,T] \ni t \mapsto \mathbb{E}[Z \cdot b_t]$ is measurable by Fubini's theorem. We deduce that the right-hand side in (5.91) is measurable in t. Letting N tend to the infinity, we complete the proof for $[0,T] \ni t \mapsto \mathbb{E}[\partial_\mu u(\mathcal{L}(X_t))(X_t) \cdot b_t]$. The same argument works for $[0,T] \ni t \mapsto \mathbb{E}[\partial_v \partial_\mu u(\mathcal{L}(X_t))(X_t) \cdot a_t]$.

Proof of Theorem 5.99.

First Step. We start with the same mollification procedure as in the proof of Theorem 5.92, see (5.87).

We consider again $u \star \rho$. By local boundedness of $\partial_\mu u$ and $\partial_v \partial_\mu u$, the functions $\partial_\mu(u \star \rho)$ and $\partial_v \partial_\mu (u \star \rho)$ are bounded on the whole space. However, contrary to the argument in the proof of Theorem 5.92, we cannot claim here that $\partial_\mu(u \star \rho)$ and $\partial_v \partial_\mu (u \star \rho)$ are continuous on the whole space since $\partial_\mu u$ and $\partial_v \partial_\mu u$ are only continuous at points (μ, v) such that v is in the support of μ. Still, from formulas (5.87), we notice that $\partial_\mu (u \star \rho)$ and $\partial_v \partial_\mu (u \star \rho)$ are also continuous at points (μ, v) such that v is in the support of μ, the reason being that $v \in \mathrm{Supp}(\mu)$ implies $\rho(v) \in \mathrm{Supp}(\mu \circ \rho^{-1})$.

Next, we replace $\mathcal{P}_2(\mathbb{R}^d) \ni \mu \mapsto (u \star \rho)(\mu)$ by $\mathcal{P}_2(\mathbb{R}^d) \ni \mu \mapsto (u \star \rho)(\mu * \varphi)$ where φ is the density of the standard normal (Gaussian) distribution $N_d(0, I_d)$ on $\mathbb{R}^d$, and $\mu * \varphi$ is the usual convolution product given by:

$$\mathbb{R}^d \ni x \mapsto \int_{\mathbb{R}^d} \varphi(x - y) d\mu(y).$$

Using the fact that a lifting of the map $\mu \mapsto u(\mu * \varphi)$ is given by $X \mapsto \tilde{u}(X + \xi)$ where $\tilde{u}$ is the lifting of u and ξ is an $N_d(0, I_d)$ Gaussian vector independent of X, we see that:

$$\partial_\mu \big[u(\mu * \varphi) \big](v) = \int_{\mathbb{R}^d} \partial_\mu u(\mu * \varphi)(v - v')\varphi(v')dv'.$$

Applying this formula to $u \star \rho$ instead of u, we get:

$$\partial_\mu \big[(u \star \rho)(\mu * \varphi) \big](v) = \int_{\mathbb{R}^d} \partial_\mu (u \star \rho)(\mu * \varphi)(v - v')\varphi(v')dv'.$$

Similarly, we get:

$$\partial_v \partial_\mu \big[(u \star \rho)(\mu * \varphi) \big](v) = \int_{\mathbb{R}^d} \partial_v \partial_\mu (u \star \rho)(\mu * \varphi)(v - v')\varphi(v')dv'.$$

Since the support of $\mu * \varphi$ is the whole $\mathbb{R}^d$, for any $v \in \mathbb{R}^d$, $(\mu * \varphi, v)$ is a continuity point of both $\partial_\mu (u \star \rho)$ and $\partial_v \partial_\mu (u \star \rho)$. Therefore, the mappings $\mathcal{P}_2(\mathbb{R}^d) \times \mathbb{R}^d \ni (\mu, v) \mapsto \partial_\mu (u \star \rho)(\mu * \varphi)(v)$ and $\mathcal{P}_2(\mathbb{R}^d) \times \mathbb{R}^d \ni (\mu, v) \mapsto \partial_v \partial_\mu (u \star \rho)(\mu * \varphi)(v)$ are continuous. Since they are bounded, we deduce from Lebesgue's theorem that the maps $(\mu, v) \mapsto \partial_\mu[(u \star \rho)(\mu * \rho)](v)$ and $(\mu, v) \mapsto \partial_v \partial_\mu[(u \star \rho)(\mu * \rho)](v)$ are continuous on the whole $\mathcal{P}_2(\mathbb{R}^d) \times \mathbb{R}^d$.

Moreover, whenever φ is replaced by the density φ_ϵ of $N_d(0, \epsilon I_d)$ which converges to the Dirac mass at 0 for the W_2 distance when $\epsilon \searrow 0$, it is easy to check that, for any $\mu \in \mathcal{P}_2(\mathbb{R}^d)$ and any $v \in \mathrm{Supp}(\mu)$, $\partial_\mu[(u \star \rho)(\mu * \varphi_\epsilon)](v)$ and $\partial_v \partial_\mu[(u \star \rho)(\mu * \varphi_\epsilon)](v)$ converge to $\partial_\mu (u \star \rho)(\mu)(v)$ and $\partial_v \partial_\mu (u \star \rho)(\mu)(v)$. In particular, if Itô's formula holds true for functionals of the type $\mathcal{P}_2(\mathbb{R}^d) \ni \mu \mapsto (u \star \rho)(\mu * \varphi_\epsilon)$, it also holds true for functionals of the type $\mathcal{P}_2(\mathbb{R}^d) \ni \mu \mapsto (u \star \rho)(\mu)$ and then for functionals of the type $\mathcal{P}_2(\mathbb{R}^d) \ni \mu \mapsto u(\mu)$ by the same approximation argument as in the proof of Theorem 5.92, noticing in particular that (5.88) remains true.

Therefore, without any loss of generality, we can assume that u and its first and partial second order derivatives are bounded and continuous on the whole space. Then, repeating once again the argument from Theorem 5.92, we can also assume that u and its derivatives are uniformly continuous and that $(X_t)_{t \geq 0}$ is a bounded Itô process.

Second Step. As before, we use a mollification argument. For a smooth compactly supported density ρ on $\mathbb{R}^d$, and using the same notations as above, for each integer $n \geq 1$, we define the mollified version u_n^N of u^N by:

$$
\begin{aligned}
u_n^N(x^1,\cdots,x^N) &= n^{Nd}\int_{(\mathbb{R}^d)^N} u^N(x^1-y^1,\cdots,x^N-y^N)\prod_{\ell=1}^N \rho(ny^\ell)\prod_{\ell=1}^N dy^\ell \\
&= \mathbb{E}\Big[u\Big(\frac{1}{N}\sum_{i=1}^N \delta_{x^i-Y^i/n}\Big)\Big],
\end{aligned}
\tag{5.92}
$$

where $Y^1,\cdots,Y^N$ are N i.i.d. random variables with density ρ. From the estimate:

$$
W_2\Big(\frac{1}{N}\sum_{i=1}^N \delta_{x^i-Y^i/n},\frac{1}{N}\sum_{i=1}^N \delta_{x^i}\Big)^2 \leqslant \frac{1}{N}\sum_{i=1}^N \Big|\frac{Y^i}{n}\Big|^2,
$$

we deduce that:

$$
W_2\Big(\frac{1}{N}\sum_{i=1}^N \delta_{x^i-Y^i/n},\frac{1}{N}\sum_{i=1}^N \delta_{x^i}\Big)^2 \leqslant \frac{C}{n^2},
\tag{5.93}
$$

the constant C depending upon the size of the support of ρ. Above and in the rest of the proof, the constant C is a general constant which is allowed to increase from line to line. Importantly, it does not depend on n or N.

Recalling that the function $\mathcal{P}_2(\mathbb{R}^d)\times\mathbb{R}^d \ni (\mu,x)\mapsto \partial_\mu u(\mu)(x)$ is assumed to be bounded, we deduce from Remark 5.27, 5.92, and 5.93 that:

$$
\begin{aligned}
&\big|u_n^N(x^1,\cdots,x^N)-u^N(x^1,\cdots,x^N)\big| \\
&= \Big|\mathbb{E}\Big[u\Big(\frac{1}{N}\sum_{i=1}^N \delta_{x^i-Y^i/n}\Big)-u\Big(\frac{1}{N}\sum_{i=1}^N \delta_{x^i}\Big)\Big]\Big| \leqslant Cn^{-1}.
\end{aligned}
\tag{5.94}
$$

Given a bounded random variable X with distribution μ, we know from Theorem 5.8 that $\mathbb{E}[W_2(\mu,\bar{\mu}^N)^2]$ tends to 0 as N tends to infinity, $\bar{\mu}^N$ denoting the empirical measure of a sample of N independent random variables with the same law as X. Moreover, the rate of convergence of $(\mathbb{E}[W_2(\mu,\bar{\mu}^N)^2])_{N\geqslant 1}$ towards 0 only depends upon the moments of X. Together with (5.94), this implies that we can find a sequence $(\varepsilon_\ell)_{\ell\geqslant 1}$ independent of t, converging to 0 as ℓ tends to ∞, and such that, for any $n,N\geqslant 1$ and $t\geqslant 0$,

$$
\begin{aligned}
&\mathbb{E}\Big[\big|u_n^N(X_t^1,\cdots,X_t^N)-u(\mu_t)\big|\Big] \\
&\leqslant \mathbb{E}\Big[\big|u_n^N(X_t^1,\cdots,X_t^N)-u^N(X_t^1,\cdots,X_t^N)\big|\Big]+\mathbb{E}\Big[\big|u(\bar{\mu}_t^N)-u(\mu_t)\big|\Big] \\
&\leqslant \varepsilon_n+\varepsilon_N.
\end{aligned}
\tag{5.95}
$$

By boundedness of u, we deduce that, for any $p\geqslant 1$ and any $t\geqslant 0$,

$$
\mathbb{E}\Big[\big|u_n^N(X_t^1,\cdots,X_t^N)-u(\mu_t)\big|^p\Big]^{1/p} \leqslant \varepsilon_n^{(p)}+\varepsilon_N^{(p)},
\tag{5.96}
$$

for a sequence $(\varepsilon_\ell^{(p)})_{\ell\geqslant 1}$ which tends to 0 as ℓ tends to ∞.

Using Proposition 5.91, we get:

$$\begin{aligned}
\partial_{x^i} u_n^N(x^1,\cdots,x^N) &= n^{Nd}\int_{(\mathbb{R}^d)^N} \partial_{x^i} u^N(x^1-y^1,\cdots,x^N-y^N)\prod_{\ell=1}^N \rho(ny^\ell)\prod_{\ell=1}^N dy^\ell \\
&= \frac{n^{Nd}}{N}\int_{(\mathbb{R}^d)^N} \partial_\mu u\Big(\frac{1}{N}\sum_{\ell=1}^N \delta_{x^\ell-y^\ell}\Big)(x^i-y^i)\prod_{\ell=1}^N \rho(ny^\ell)\prod_{\ell=1}^N dy^\ell \\
&= \frac{1}{N}\mathbb{E}\Big[\partial_\mu u\Big(\frac{1}{N}\sum_{\ell=1}^N \delta_{x^\ell-Y^\ell/n}\Big)\Big(x^i-\frac{Y^i}{n}\Big)\Big].
\end{aligned}$$

Using the boundedness and the uniform continuity of $\partial_\mu u$ on the whole space and following the proof of (5.95), we deduce that, for any $t \geqslant 0$,

$$\mathbb{E}\Big[\big|N\partial_{x^i} u_n^N(X_t^1,\cdots,X_t^N)-\partial_\mu u(\mu_t)(X_t^i)\big|\Big] \leqslant \varepsilon_n+\varepsilon_N. \tag{5.97}$$

Again, by boundedness of $\partial_\mu u$, we deduce that, for any $p \geqslant 1$ and any $t \geqslant 0$,

$$\mathbb{E}\Big[\big|N\partial_{x^i} u_n^N(X_t^1,\cdots,X_t^N)-\partial_\mu u(\mu_t)(X_t^i)\big|^p\Big]^{1/p} \leqslant \varepsilon_n^{(p)}+\varepsilon_N^{(p)}. \tag{5.98}$$

Now, we differentiate once more with respect to x^i:

$$\begin{aligned}
&\partial^2_{x^i x^i} u_n^N(x^1,\cdots,x^N) \\
&= \frac{n^{Nd+1}}{N}\int_{(\mathbb{R}^d)^N}\Big\{\partial_\mu u\Big(\frac{1}{N}\sum_{\ell=1}^N \delta_{x^\ell-y^\ell}\Big)(x^i-y^i)\Big\}\otimes\nabla\rho(ny^i)\prod_{\ell\neq i}\rho(ny^\ell)\prod_{\ell=1}^N dy^\ell,
\end{aligned}$$

the tensor product operating on elements of $\mathbb{R}^d$. We then rewrite the derivative as:

$$N\partial^2_{x^i x^i} u_n^N(x^1,\cdots,x^N) = T_{n,i}^{1,N}(x^1,\cdots,x^N)+T_{n,i}^{2,N}(x^1,\cdots,x^N),$$

with:

$$\begin{aligned}
&T_{n,i}^{1,N}(x^1,\cdots,x^N) \\
&= n^{Nd+1}\int_{(\mathbb{R}^d)^N}\Big\{\partial_\mu u\Big(\frac{1}{N}\sum_{\ell\neq i}\delta_{x^\ell-y^\ell}+\frac{1}{N}\delta_{x^i}\Big)(x^i-y^i)\Big\}\otimes\nabla\rho(ny^i)\prod_{\ell\neq i}\rho(ny^\ell)\prod_{\ell=1}^N dy^\ell,
\end{aligned}$$

and

$$
\begin{aligned}
&T^{2,N}_{n,i}(x^1,\cdots,x^N)\\
&= n^{Nd+1}\int_{(\mathbb{R}^d)^N}\biggl\{\biggl[\biggl(\partial_\mu u\Bigl(\frac{1}{N}\sum_{\ell=1}^N \delta_{x^\ell-y^\ell}\Bigr)-\partial_\mu u\Bigl(\frac{1}{N}\sum_{\ell\neq i}\delta_{x^\ell-y^\ell}+\frac{1}{N}\delta_{x^i}\Bigr)\biggr](x^i-y^i)\biggr\}\\
&\qquad\qquad\otimes \nabla\rho(ny^i)\prod_{\ell\neq i}\rho(ny^\ell)\prod_{\ell=1}^N dy^\ell .
\end{aligned}
$$

By integration by parts (recall that $\mathbb{R}^d \ni x \mapsto \partial_\mu u(\mu)(x)$ is differentiable), we can split $T^{1,N}_{n,i}$ into:

$$
T^{1,N}_{n,i}(x^1,\cdots,x^N) = T^{11,N}_{n,i}(x^1,\cdots,x^N) + T^{12,N}_{n,i}(x^1,\cdots,x^N),
$$

with:

$$
T^{11,N}_{n,i}(x^1,\cdots,x^N) = n^{Nd}\int_{(\mathbb{R}^d)^N}\biggl\{\partial_v\partial_\mu u\Bigl(\frac{1}{N}\sum_{\ell=1}^N\delta_{x^\ell-y^\ell}\Bigr)(x^i-y^i)\biggr\}\prod_{\ell=1}^N\rho(ny^\ell)\prod_{\ell=1}^N dy^\ell
$$

and

$$
\begin{aligned}
&T^{12,N}_{n,i}(x^1,\cdots,x^N)\\
&= n^{Nd}\int_{(\mathbb{R}^d)^N}\biggl\{\partial_v\partial_\mu u\Bigl(\frac{1}{N}\sum_{\ell\neq i}\delta_{x^\ell-y^\ell}+\frac{1}{N}\delta_{x^i}\Bigr)\\
&\qquad\qquad -\partial_v\partial_\mu u\Bigl(\frac{1}{N}\sum_{\ell=1}^N\delta_{x^\ell-y^\ell}\Bigr)(x^i-y^i)\biggr\}\prod_{\ell=1}^N\rho(ny^\ell)\prod_{\ell=1}^N dy^\ell .
\end{aligned}
$$

The first term is treated as (5.95) and (5.97). Namely, we argue that, because of the uniform continuity of $\partial_v\partial_\mu u$, we have for any $t \geqslant 0$:

$$
\mathbb{E}\bigl[\bigl|T^{11,N}_{n,i}(X^1_t,\cdots,X^N_t) - \partial_v\partial_\mu u(\mu_t)(X^i_t)\bigr|\bigr] \leqslant \varepsilon_n + \varepsilon_N, \tag{5.99}
$$

from which we get that for any $p \geqslant 1$ and any $t \geqslant 0$,

$$
\mathbb{E}\bigl[\bigl|T^{11,N}_{n,i}(X^1_t,\cdots,X^N_t) - \partial_v\partial_\mu u(\mu_t)(X^i_t)\bigr|^p\bigr]^{1/p} \leqslant \varepsilon^{(p)}_n + \varepsilon^{(p)}_N. \tag{5.100}
$$

To handle the second term, we use once more the uniform continuity of $\partial_v\partial_\mu u$. Indeed, we have:

$$
|T^{12,N}_{n,i}(x^1,\cdots,x^N)| \leqslant \varepsilon_N,
$$

as implied by:

$$W_2\left(\frac{1}{N}\sum_{\ell\neq i}\delta_{x^\ell-y^\ell}+\frac{1}{N}\delta_{x^i},\frac{1}{N}\sum_{\ell=1}^N\delta_{x^\ell-y^\ell}\right)^2\leqslant\frac{1}{N}|y^i|^2\leqslant\frac{C}{N},$$

if, as in the definition of $T^{12,N}_{n,i}(x^1,\cdots,x^N)$, the quantity ny_i is restricted to the compact support of ρ. This implies that, for any $t\geqslant 0$,

$$\mathbb{E}\Big[\big|T^{12,N}_{n,i}(X^1_t,\cdots,X^N_t)\big|\Big]\leqslant\varepsilon_N, \tag{5.101}$$

and consequently that, for any $p\geqslant 1$ and $t\geqslant 0$,

$$\mathbb{E}\Big[\big|T^{12,N}_{n,i}(X^1_t,\cdots,X^N_t)\big|^p\Big]^{1/p}\leqslant\varepsilon^{(p)}_N. \tag{5.102}$$

We finally handle $T^{2,N}_{n,i}$. Following the proof of (5.102), we have, for any $p\geqslant 1$ and any $t\geqslant 0$,

$$\mathbb{E}\big[\big|T^{2,N}_{n,i}(X^1_t,\cdots,X^N_t)\big|^p\big]^{1/p}\leqslant n\varepsilon^{(p)}_N, \tag{5.103}$$

the additional n coming from the differentiation of the regularization kernel.

Third Step. In order to complete the proof, we apply Itô's formula to $(u^N_n(X^1_t,\cdots,X^N_t))_{t\geqslant 0}$ for given values of n and N. We obtain:

$$\begin{aligned}0&=u^N_n(X^1_t,\cdots,X^N_t)-u^N_n(X^1_0,\cdots,X^N_0)\\&\quad-\frac{1}{N}\sum_{\ell=1}^N\int_0^t\partial_{x^\ell}u^N_n(X^1_s,\cdots,X^N_s)\cdot b^\ell_s ds\\&\quad-\frac{1}{N}\sum_{\ell=1}^N\int_0^t\partial_{x^\ell}u^N_n(X^1_s,\cdots,X^N_s)\cdot(\sigma^\ell_s dW^\ell_s)\\&\quad-\frac{1}{2N}\sum_{\ell=1}^N\int_0^t\text{trace}\big[\partial^2_{x^\ell x^\ell}u^N_n(X^1_s,\cdots,X^N_s)a^\ell_s\big]ds,\end{aligned}$$

with $a^\ell_s=\sigma^\ell_s(\sigma^\ell_s)^\dagger$. We compare with the expected result, by computing the difference:

$$\begin{aligned}\Delta^N_t&=u(\mu_t)-u(\mu_0)-\frac{1}{N}\sum_{\ell=1}^N\int_0^t\partial_\mu u(\mu_s)(X^\ell_s)\cdot b^\ell_s ds\\&\quad-\frac{1}{N}\sum_{\ell=1}^N\int_0^t\partial_\mu u(\mu_s)(X^\ell_s)\cdot(\sigma^\ell_s dW^\ell_s)\\&\quad-\frac{1}{2N}\sum_{\ell=1}^N\int_0^t\text{trace}\big[\partial_v\partial_\mu u(\mu_s)(X^\ell_s)a^\ell_s\big]ds.\end{aligned} \tag{5.104}$$

From (5.96), (5.98), (5.100), (5.102), and (5.103), we obtain, for any $T > 0$,

$$\sup_{0 \leqslant t \leqslant T} \left|\mathbb{E}\big[\Delta_t^N\big]\right| \leqslant \varepsilon_n + (1+n)\varepsilon_N,$$

the sequence $(\varepsilon_\ell)_{\ell \geqslant 1}$ now depending on T. Using a straightforward exchangeability argument and letting N tend to ∞, we deduce that:

$$\sup_{0 \leqslant t \leqslant T} |\Delta_t| \leqslant \varepsilon_n, \tag{5.105}$$

where:

$$\begin{aligned}\Delta_t = u(\mu_t) - u(\mu_0) &- \int_0^t \mathbb{E}\big[\partial_\mu u(\mu_s)(X_s) \cdot b_s\big] ds \\ &- \frac{1}{2}\int_0^t \mathbb{E}\Big[\text{trace}\big[\partial_v \partial_\mu u(\mu_s)(X_s) a_s\big]\Big] ds.\end{aligned}$$

Letting n tend ∞ in (5.105), we conclude that $\Delta \equiv 0$, which completes the proof. □

Extension of the Chain Rule

The chain rule formula (5.85) stated and proven in Theorem 5.92 and extended in Theorem 5.99 was given for time independent functions u for the sake of simplicity. Clearly a similar chain rule holds if u also depends upon time and the state of a diffusion process. An even more general form will be proven in Theorem 4.17 in Chapter 4 (second volume) to include conditional diffusions and flows of conditional probability distributions. In order to avoid having to quote ahead results from Chapter 4 (second volume), we state under the assumption below a simpler form of this generalization which will be sufficient for the purpose of the applications discussed in this chapter:

Assumption (Joint Chain Rule). For a given $T > 0$, the function u is a continuous function from $[0,T] \times \mathbb{R}^d \times \mathcal{P}_2(\mathbb{R}^d)$ to $\mathbb{R}$ such that:

(A1) For any $\mu \in \mathcal{P}_2(\mathbb{R}^d)$, the function $[0,T] \times \mathbb{R}^d \ni (t,x) \mapsto u(t,x,\mu)$ is of class $\mathcal{C}^{1,2}$, the functions $\partial_t u$, $\partial_x u$ and $\partial^2_{xx} u$ being (jointly) continuous in (t,x,μ).

(A2) For any $(t,x) \in [0,T] \times \mathbb{R}^d$, the function $\mathcal{P}_2(\mathbb{R}^d) \ni \mu \mapsto u(t,x,\mu)$ is continuously L-differentiable and, for any $\mu \in \mathcal{P}_2(\mathbb{R}^d)$, we can find a version of the mapping $\mathbb{R}^d \ni v \mapsto \partial_\mu u(t,x,\mu)(v)$ such that the mapping $[0,T] \times \mathbb{R}^d \times \mathcal{P}_2(\mathbb{R}^d) \times \mathbb{R}^d \ni (t,x,\mu,v) \mapsto \partial_\mu u(t,x,\mu)(v)$ is locally bounded and is continuous at any (t,x,μ,v) such that $v \in \text{Supp}(\mu)$.

(continued)

(A3) For the version of $\partial_\mu u$ mentioned above and for any $(t,x,\mu) \in [0,T] \times \mathbb{R}^d \times \mathcal{P}_2(\mathbb{R}^d)$, the mapping $\mathbb{R}^d \ni v \mapsto \partial_\mu u(t,x,\mu)(v) \in \mathbb{R}^d$ is continuously differentiable and its derivative, denoted by $\mathbb{R}^d \ni v \mapsto \partial_v \partial_\mu u(t,x,\mu)(v) \in \mathbb{R}^{d\times d}$, is locally bounded and is jointly continuous in (t,x,μ,v) at any point (t,x,μ,v) such that $v \in \mathrm{Supp}(\mu)$.

Proposition 5.102 *If u satisfies assumption* **Joint Chain Rule** *and if for every compact subset $\mathcal{K} \subset \mathbb{R}^d \times \mathcal{P}_2(\mathbb{R}^d)$, it holds that:*

$$\sup_{(t,x,\mu)\in[0,T]\times\mathcal{K}} \left[\int_{\mathbb{R}^d} \left|\partial_\mu u(t,x,\mu)(v)\right|^2 d\mu(v) + \int_{\mathbb{R}^d} \left|\partial_v \partial_\mu u(t,x,\mu)(v)\right|^2 d\mu(v) \right] < \infty, \tag{5.106}$$

if we set $\mu_t = \mathcal{L}(X_t)$ for $t \in [0,T]$ for an Itô process $(X_t)_{0\leqslant t\leqslant T}$ of the form (5.82) *satisfying* (5.83) *at time T, and if $(\xi_t)_{t\in[0,T]}$ is another d-dimensional Itô process on the same filtered probability space $(\Omega,\mathcal{F},\mathbb{F},\mathbb{P})$ with similar dynamics $d\xi_t = \eta_t dt + \gamma_t dW_t$, for two $\mathbb{F}$-progressively measurable processes $(\eta_t)_{0\leqslant t\leqslant T}$ and $(\gamma_t)_{0\leqslant t\leqslant T}$ with values in $\mathbb{R}^d$ and $\mathbb{R}^{d\times d}$ respectively such that:*

$$\mathbb{P}\left[\int_0^T \left(|\eta_t| + |\gamma_t|^2\right)dt < \infty\right] = 1,$$

then, $\mathbb{P}$ almost surely, for all $t \in [0,T]$, it holds:

$$\begin{aligned} u(t,\xi_t,\mu_t) &= u(0,\xi_0,\mu_0) + \int_0^t \partial_x u(s,\xi_s,\mu_s)\cdot\left(\gamma_s dW_s\right) \\ &+ \int_0^t \left(\partial_t u(s,\xi_s,\mu_s) + \partial_x u(s,\xi_s,\mu_s)\cdot\eta_s\right)ds \\ &+ \frac{1}{2}\int_0^t \mathrm{trace}\left[\partial_{xx}^2 u(s,\xi_s,\mu_s)\gamma_s\gamma_s^\dagger\right]\Big)ds \\ &+ \int_0^t \tilde{\mathbb{E}}\left[\partial_\mu u(s,\xi_s,\mu_s)(\tilde{X}_s)\cdot\tilde{b}_s\right]ds \\ &+ \frac{1}{2}\int_0^t \tilde{\mathbb{E}}\left[\mathrm{trace}\left(\partial_v\partial_\mu u(s,\xi_s,\mu_s)(\tilde{X}_s)\tilde{\sigma}_s\tilde{\sigma}_s^\dagger\right)\right]ds, \end{aligned} \tag{5.107}$$

where the process $(\tilde{X}_t,\tilde{b}_t,\tilde{\sigma}_t)_{0\leqslant t\leqslant T}$ is a copy of the process $(X_t,b_t,\sigma_t)_{0\leqslant t\leqslant T}$, on a copy $(\tilde{\Omega},\tilde{\mathcal{F}},\tilde{\mathbb{P}})$ of the probability space $(\Omega,\mathcal{F},\mathbb{P})$.

Remark 5.103 *Importantly, in full analogy with Remark 5.101, the processes*

$$[0,T] \times \Omega \ni (s,\omega) \mapsto \tilde{\mathbb{E}}\big[\partial_\mu u\big(s,\xi_s(\omega),\mu_s\big)(\tilde{X}_s)\cdot \tilde{b}_s\big],$$
$$[0,T] \times \Omega \ni (s,\omega) \mapsto \tilde{\mathbb{E}}\big[\mathrm{trace}\big(\partial_v \partial_\mu u\big(s,\xi_s(\omega),\mu_s\big)(\tilde{X}_s)\tilde{\sigma}_s\tilde{\sigma}_s^\dagger\big)\big],$$

are progressively measurable if we assume that a_s and $\sigma_s\sigma_s^\dagger$ are square integrable for any $s \in [0,T]$. This is due to the fact that the functions:

$$[0,T] \times \mathbb{R}^d \ni (s,x) \mapsto \tilde{\mathbb{E}}\big[\partial_\mu u\big(s,x,\mu_s\big)(\tilde{X}_s)\cdot \tilde{b}_s\big],$$
$$[0,T] \times \mathbb{R}^d \ni (s,x) \mapsto \tilde{\mathbb{E}}\big[\mathrm{trace}\big(\partial_v \partial_\mu u\big(s,x,\mu_s\big)(\tilde{X}_s)\tilde{\sigma}_s\tilde{\sigma}_s^\dagger\big)\big],$$

are jointly measurable. For the first of them, the measurability follows from the fact that we can find a jointly measurable version of $\partial_\mu u : [0,T] \times \mathbb{R}^d \times \mathcal{P}_2(\mathbb{R}^d) \times \mathbb{R}^d \ni (t,x,\mu,v) \mapsto \partial_\mu u(t,x,\mu)(v)$, as explained in Subsection 5.3.4. For the second, arguing the measurability is less straightforward. Still, we can use the fact (see again Subsection 5.3.4) that the mapping:

$$\begin{aligned}&[0,T] \times \mathbb{R}^d \times L^2(\tilde{\Omega},\tilde{\mathcal{F}},\tilde{\mathbb{P}};\mathbb{R}^d) \ni (t,x,\tilde{X}) \\ &\qquad \mapsto \varrho_{d\times d}\Big(\partial_v\partial_\mu u\big(t,x,\mathcal{L}(\tilde{X})\big)(\tilde{X})\Big) \in L^2(\tilde{\Omega},\tilde{\mathcal{F}},\tilde{\mathbb{P}};\mathbb{R}^{d\times d})\end{aligned}$$

is continuous for any compactly supported smooth function $\varrho_{d\times d}$ from $\mathbb{R}^{d\times d}$ into itself, see again Subsection 5.3.4. Then, the proof can be completed as in Remark 5.101.

Proof of Proposition 5.102. As a preliminary remark, we observe that for any $(t,x) \in [0,T]\times\mathbb{R}^d$, the mapping $\mathcal{P}_2(\mathbb{R}^d) \ni \mu \mapsto u(t,x,\mu)$ is partially $\mathcal{C}^2$.

First Step. We first assume that the processes $(b_t)_{0\leqslant t\leqslant T}$ and $(\sigma_t)_{0\leqslant t\leqslant T}$ have continuous paths and satisfy:

$$\mathbb{E}\Big[\sup_{0\leqslant t\leqslant T}\big(|b_t|^2 + |\sigma_t|^4\big)\Big] < \infty.$$

We then define the function:

$$U(t,x) = u(t,x,\mu_t), \quad t \in [0,T],\ x \in \mathbb{R}^d.$$

Since the path $[0,T] \ni t \mapsto \mu_t \in \mathcal{P}_2(\mathbb{R}^d)$ is continuous for the Wasserstein distance, U is continuous. By a similar argument, U is twice differentiable in space and $\partial_x U$ and $\partial^2_{xx} U$ are jointly continuous in time and space.

We now prove that U is differentiable with respect to the time variable and that $\partial_t U$ is continuous. For any $t \in [0,T)$ and $h > 0$ such that $t+h \in [0,T]$ and for any $x \in \mathbb{R}^d$, we have:

$$\begin{aligned}
&U(t+h,x) - U(t,x)\\
&= \Big(u\big(t+h,x,\mu_{t+h}\big) - u\big(t,x,\mu_{t+h}\big)\Big) + \Big(u(t,x,\mu_{t+h}) - u(t,x,\mu_t)\Big)\\
&= \int_t^{t+h} \partial_t u(s,x,\mu_{t+h}) ds + \int_t^{t+h} \mathbb{E}\big[\partial_\mu u(t,x,\mu_s)(X_s)\cdot b_s\big] ds\\
&\quad + \frac{1}{2}\int_t^{t+h} \mathbb{E}\big[\partial_v \partial_\mu u(t,x,\mu_s)(X_s)\cdot a_s\big] ds\\
&= (i) + (ii) + (iii),
\end{aligned}$$

where we used the chain rule for partially $\mathcal{C}^2$ functions. Using the joint continuity of $\partial_t u$, we clearly have $\lim_{h\searrow 0}(i)/h = \partial_t u(t,x,\mu_t)$. Using the joint continuity of $\partial_\mu u$ (at points (t,x,μ,v) such that $v \in \mathrm{Supp}(\mu)$) and the pathwise continuity of $(b_s)_{0\leqslant s\leqslant T}$, we have, in probability,

$$\lim_{s\searrow t} \partial_\mu u(t,x,\mu_s)(X_s)\cdot b_s = \partial_\mu u(t,x,\mu_t)(X_t)\cdot b_t.$$

Obviously, by Cauchy-Schwarz' inequality, we have, for any event $A \in \mathcal{F}$,

$$\mathbb{E}\big[\mathbf{1}_A|\partial_\mu u(t,x,\mu_s)(X_s)\cdot b_s|\big] \leqslant \mathbb{E}\big[|\partial_\mu u(t,x,\mu_s)(X_s)|^2\big]^{1/2}\mathbb{E}\big[\mathbf{1}_A \sup_{0\leqslant s\leqslant T}|b_s|^2\big]^{1/2},$$

so that, by (5.106) the family $(\partial_\mu u(t,x,\mu_s)(X_s))_{0\leqslant s\leqslant T}$ is uniformly integrable. We deduce that:

$$\lim_{s\searrow t} \mathbb{E}\big[\partial_\mu u(t,x,\mu_s)(X_s)\cdot b_s\big] = \mathbb{E}\big[\partial_\mu u(t,x,\mu_t)(X_t)\cdot b_t\big],$$

and, subsequently:

$$\lim_{h\searrow 0}\frac{1}{h}(ii) = \mathbb{E}\big[\partial_\mu u(t,x,\mu_t)(X_t)\cdot b_t\big].$$

Similarly, we also have:

$$\lim_{h\searrow 0}\frac{1}{h}(iii) = \frac{1}{2}\mathbb{E}\big[\partial_v\partial_\mu u(t,x,\mu_t)(X_t)\cdot a_t\big],$$

showing that U is right-differentiable in time with:

$$\partial_t U(t,x) = \partial_t u(t,x,\mu_t) + \mathbb{E}\big[\partial_\mu u(t,x,\mu_t)(X_t)\cdot b_t\big] + \frac{1}{2}\mathbb{E}\big[\partial_v\partial_\mu u(t,x,\mu_t)(X_t)\cdot a_t\big].$$

Using the same argument as the one used above to investigate the last two limits, we can prove that $\partial_t U$ is jointly continuous in time and space. This shows that U is of class $\mathcal{C}^{1,2}$ on $[0,T]\times\mathbb{R}^d$. Applying standard Itô's formula to $(U(t,\xi_t) = u(t,\xi_t,\mu_t))_{0\leqslant t\leqslant T}$, we obtain (5.107).

Second Step. We now get rid of the continuity assumption on the processes $(b_t)_{0\leqslant t\leqslant T}$ and $(\sigma_t)_{0\leqslant t\leqslant T}$. Let $((b_t^n)_{0\leqslant t\leqslant T})_{n\geqslant 0}$ and $((\sigma_t^n)_{0\leqslant t\leqslant T})_{n\geqslant 0}$ be sequences of $\mathbb{F}$-progressively measurable processes such that, for each $n \geqslant 0$, $(b_t^n)_{0 \leqslant t \leqslant T}$ and $(\sigma_t^n)_{0 \leqslant t \leqslant T}$ satisfy the assumptions used in the first step, together with:

$$\lim_{n \to \infty} \mathbb{E} \int_0^T \big(|b_t - b_t^n|^2 + |\sigma_t - \sigma_t^n|^4 \big) dt = 0.$$

In particular, if for $n \geqslant 0$ and for $t \in [0, T]$, we set:

$$X_t^n = X_0 + \int_0^t b_s^n ds + \int_0^t \sigma_s^n dW_s,$$

and $\mu_t^n = \mathcal{L}(X_t^n)$, then we have:

$$\lim_{n \to \infty} \sup_{0 \leqslant t \leqslant T} W_2(\mu_t, \mu_t^n)^2 = 0.$$

The goal is then to pass to the limit in (5.107). To do so, we use repeatedly (5.106) together with the fact that the sequence $((\mu_t^n)_{0 \leq t \leq T})_{n \geq 0}$ is relatively compact in $\mathcal{P}_2(\mathbb{R}^d)$. By using a localization sequence for the process $(\xi_s)_{0 \leqslant s \leqslant T}$, we may assume that it lives in a bounded subset of $\mathbb{R}^d$. Considering the penultimate line in (5.107), observe by Cauchy Schwarz' inequality that:

$$\begin{aligned} &\sup_{0 \leqslant t \leqslant T} \left| \int_0^t \tilde{\mathbb{E}}\big[\partial_\mu u(s, \xi_s, \mu_s^n)(\tilde{X}_s^n) \cdot \tilde{b}_s^n\big] ds - \int_0^t \tilde{\mathbb{E}}\big[\partial_\mu u(s, \xi_s, \mu_s^n)(\tilde{X}_s^n) \cdot \tilde{b}_s\big] ds \right| \\ &\quad \leqslant c \left[\mathbb{E} \int_0^T |b_t^n - b_t|^2 dt \right]^{1/2}. \end{aligned}$$

Therefore, in order to pass to the limit (in the pathwise sense, uniformly in time) in the first of the last two terms of (5.107), it suffices to focus on the limit of:

$$\int_0^t \tilde{\mathbb{E}}\big[\partial_\mu u(s, \xi_s, \mu_s^n)(\tilde{X}_s^n) \cdot \tilde{b}_s\big] ds.$$

Repeating the uniform integrability arguments used in the first step of the proof, we claim that $\mathbb{P}$ almost surely, for almost every $s \in [0, T]$ (namely those for which $\mathbb{E}[|b_s|^2] < \infty$),

$$\lim_{n \to \infty} \tilde{\mathbb{E}}\big[\partial_\mu u(s, \xi_s, \mu_s^n)(\tilde{X}_s^n) \cdot \tilde{b}_s\big] = \tilde{\mathbb{E}}\big[\partial_\mu u(s, \xi_s, \mu_s)(\tilde{X}_s) \cdot \tilde{b}_s\big],$$

which, after we take advantage of the fact that $(\xi_t)_{0 \leqslant t \leqslant T}$ takes values in a bounded subset of $\mathbb{R}^d$, of (5.106), and after we apply Lebesgue's dominated convergence theorem, proves that $\mathbb{P}$ almost surely:

$$\begin{aligned} &\limsup_{n \to \infty} \sup_{0 \leqslant t \leqslant T} \left| \int_0^t \tilde{\mathbb{E}}\big[\partial_\mu u(s, \xi_s, \mu_s^n)(\tilde{X}_s^n) \cdot \tilde{b}_s\big] ds - \int_0^t \tilde{\mathbb{E}}\big[\partial_\mu u(s, \xi_s, \mu_s)(\tilde{X}_s) \cdot \tilde{b}_s\big] ds \right| \\ &\quad \leqslant \limsup_{n \to \infty} \int_0^T \left| \tilde{\mathbb{E}}\big[\partial_\mu u(s, \xi_s, \mu_s^n)(\tilde{X}_s^n) \cdot \tilde{b}_s\big] - \tilde{\mathbb{E}}\big[\partial_\mu u(s, \xi_s, \mu_s)(\tilde{X}_s) \cdot \tilde{b}_s\big] \right| ds = 0. \end{aligned}$$

The last term of (5.107) is handled in the same way. The terms appearing in the second and third lines of (5.107) are easily handled. Regarding the stochastic integral, it suffices to notice that, for a universal constant $C \geqslant 0$,

$$\mathbb{E}\Big[\sup_{0\leqslant t\leqslant T}\Big|\int_0^t \partial_x u(s,\xi_s,\mu_s^n)\cdot(\gamma_s dW_s) - \int_0^t \partial_x u(s,\xi_s,\mu_s)\cdot(\gamma_s dW_s)\Big|^2\Big]$$

$$\leqslant \mathbb{E}\int_0^T \big|\big(\partial_x u(t,\xi_t,\mu_t^n) - \partial_x u(t,\xi_t,\mu_t)\big)\big|^2 |\gamma_t|^2 dt.$$

Since $\partial_x u$ is assumed to be (jointly) continuous and $(\xi_t)_{0\leqslant t\leqslant T}$ is assumed to take values in a bounded subset of $\mathbb{R}^d$, the right-hand side tends to 0 as n tends to ∞, which completes the proof. □

5.6.5 Sufficient Condition for Partial $\mathcal{C}^2$-Regularity

The following set of assumptions provides a sufficient condition for $\mathcal{C}^2$ partial regularity, which will be very useful in the sequel:

Assumption (Sufficiency for Partial $\mathcal{C}^2$). The function $u : \mathcal{P}_2(\mathbb{R}^d) \to \mathbb{R}$ is L-continuously differentiable and, on an atomless probability space $(\Omega,\mathcal{F},\mathbb{P})$, its lifted version $\tilde{u} : L^2(\Omega,\mathcal{F},\mathbb{P};\mathbb{R}^d) \ni X \mapsto u(\mathcal{L}(X)) \in \mathbb{R}$ satisfies:

(A1) For any $\chi \in L^2(\Omega,\mathcal{F},\mathbb{P};\mathbb{R}^d)$ and any continuously differentiable map $\mathbb{R} \ni \lambda \mapsto X^\lambda \in L^2(\Omega,\mathcal{F},\mathbb{P};\mathbb{R}^d)$ with the property that all the $(X^\lambda)_{\lambda\in\mathbb{R}}$ have the same distribution and that $\mathbb{P}[|[d/d\lambda]X^\lambda| \leqslant 1] = 1$, the mapping:

$$\mathbb{R} \ni \lambda \mapsto D\tilde{u}(X^\lambda)\cdot\chi = \mathbb{E}\big[\partial_\mu u(\mathcal{L}(X^\lambda))(X^\lambda)\cdot\chi\big] \in \mathbb{R}$$

is continuously differentiable, the derivative at $\lambda = 0$ only depending upon the family $(X^\lambda)_{\lambda\in\mathbb{R}}$ through the values of X^0 and $[d/d\lambda]_{|\lambda=0}X^\lambda$, and being denoted by:

$$\partial^2_{\zeta,\chi}\tilde{u}(X) = \frac{d}{d\lambda}_{|\lambda=0}\big[D\tilde{u}(X^\lambda)\cdot\chi\big],$$

whenever $X = X^0$ and $\zeta = \dfrac{d}{d\lambda}_{|\lambda=0} X^\lambda$.

(A2) There exists a constant C such that, for any X, X', χ and ζ in $L^2(\Omega,\mathcal{F},\mathbb{P};\mathbb{R}^d)$, with $X \sim X'$ and $|\zeta| \leqslant 1$ (with probability 1), it holds:

(continued)

$$(i) \quad |D\tilde{u}(X) \cdot \chi| + |\partial^2_{\xi,\chi} u(X)| \leqslant C\|\chi\|_2,$$
$$(ii) \quad |D\tilde{u}(X) \cdot \chi - D\tilde{u}(X') \cdot \chi| + |\partial^2_{\xi,\chi}\tilde{u}(X) - \partial^2_{\xi,\chi}\tilde{u}(X')|$$
$$\leqslant C\|X - X'\|_2\|\chi\|_2.$$

Actually, the fact that the derivative of $D\tilde{u}(X^\lambda) \cdot \chi$ at $\lambda = 0$ only depends on X^0 and χ may be seen as a consequence of (**A2**). Indeed, by the Lipschitz property (*ii*), it holds that $|D\tilde{u}(X^\lambda) \cdot \chi - D\tilde{u}(X^0 + \lambda\xi) \cdot \chi| \leq \|X^\lambda - X^0 - \lambda\xi\|_2\|\chi\|_2 = o(\lambda)\|\chi\|_2$, which proves that $[d/d\lambda]_{|\lambda=0} D\tilde{u}(X^\lambda) \cdot \chi = [d/d\lambda]_{|\lambda=0} D\tilde{u}(X^0 + \lambda\xi) \cdot \chi$.

Theorem 5.104 *Under assumption* **Sufficiency for Partial** $\mathcal{C}^2$, *u is partially* $\mathcal{C}^2$ *and, for any compact subset* $\mathcal{K} \subset \mathcal{P}_2(\mathbb{R}^d)$, *we have:*

$$\sup_{\mu \in \mathcal{K}} \left[\int_{\mathbb{R}^d} \left|\partial_v \partial_\mu u(\mu)(v)\right|^2 d\mu(v) \right] < \infty,$$

so that the chain rule applies to any Itô process satisfying (5.83).

Remark 5.105 *The thrust of Theorem 5.104 is to focus on the smoothness of the mapping* $\mathbb{R}^d \ni v \mapsto \partial_\mu u(\mu)(v)$ *independently of the smoothness in the direction* μ *by restricting the test random variables* $(X^\lambda)_{\lambda \in \mathbb{R}}$ *to an identically distributed family. One of the issue in the proof is precisely to construct such a family of test random variables.*

As a warm-up to the proof of Theorem 5.104, we discuss what Proposition 5.36 says in the framework of Theorem 5.104. Rewriting $D\tilde{u}(X)\cdot\chi$ as $\mathbb{E}[\partial_\mu u(\mathcal{L}(X))(X)\cdot\chi]$, we can write, for any $X, X' \in L^2(\Omega, \mathcal{F}, \mathbb{P}; \mathbb{R}^d)$ with $\mathcal{L}(X) = \mathcal{L}(X') = \mu$ for some $\mu \in \mathcal{P}_2(\mathbb{R}^d)$,

$$\left|\mathbb{E}\big[\big(\partial_\mu u(\mu)(X') - \partial_\mu u(\mu)(X)\big) \cdot \chi\big]\right| \leqslant C\mathbb{E}\big[|X - X'|^2\big]^{1/2}\mathbb{E}\big[|\chi|^2\big]^{1/2},$$

which implies that, for any $\mu \in \mathcal{P}_2(\mathbb{R}^d)$, we can find a Lipschitz continuous version of the map $\mathbb{R}^d \ni x \mapsto \partial_\mu u(\mu)(x)$, with C as Lipschitz constant (in particular, the Lipschitz property holds true uniformly with respect to μ). Therefore, for any $X \in L^2(\Omega, \mathcal{F}, \mathbb{P}; \mathbb{R}^d)$ with $X \sim \mu$, $|\partial_\mu u(\mu)(0)| \leqslant C\mathbb{E}[|X|] + \mathbb{E}[|\partial_\mu u(\mu)(X)|]$, the last term being bounded thanks to (*i*). Obviously the right-hand side is uniformly bounded for μ in bounded subsets of $\mathcal{P}_2(\mathbb{R}^d)$, from which we deduce that ∂_μ is locally bounded.

Actually, on the model of Corollary 5.38, we can say a little bit more. Indeed, by the same arguments as in the proof of Corollary 5.38, we can prove that the function $\mathcal{P}_2(\mathbb{R}^d) \times \mathbb{R}^d \ni (\mu, v) \mapsto \partial_\mu u(\mu)(v)$ is jointly continuous at any point (μ, v) such that $v \in \mathrm{Supp}(\mu)$.

We now proceed with the proof of Theorem 5.104. It relies on two main ingredients. The first one is a new mollification argument. The second is a coupling argument which permits to choose relevant versions of the random variables along which the differentiation is performed.

Proof of Theorem 5.104.

First Step. Given a distribution μ and a random variable X with distribution μ, we introduce, for each integer $n \geqslant 1$, the mollified version μ^n of μ defined as:

$$\mu^n = \mu * \varphi_{d,n},$$

where the function $\varphi_{d,n}$ denotes the density of the mean-zero d-dimensional Gaussian distribution $N_d(0,(1/n)I_d)$ with covariance matrix $(1/n)I_d$, where as usual I_d is the identity matrix of dimension d. We then define the mapping:

$$\mathcal{V}^n(\mu, v) = \int_{\mathbb{R}^d} \partial_\mu u(\mu^n)(v - x) n^{d/2} \varphi_d\big(n^{1/2}x\big) dx, \tag{5.108}$$

where the function $\varphi_d = \varphi_{d,1}$ denotes the density of the standard d-dimensional Gaussian distribution. The mapping $\mathcal{V}^n$ is given by the convolution of $\partial_\mu u(\mu^n)(\cdot)$ with the measure $N_d(0,(1/n)I_d)$. According to the warm-up preceding the proof, the sequence $(\partial_\mu u(\mu^n)(0))_{n\geqslant 1}$ is bounded and the functions $(\partial_\mu u(\mu^n) : \mathbb{R}^d \ni v \mapsto \partial_\mu u(\mu^n)(v) \in \mathbb{R}^d)_{n\geqslant 1}$ are uniformly Lipschitz continuous. Thus, the sequence of functions $(\mathcal{V}^n(\mu,\cdot))_{n\geqslant 1}$ is relatively compact for the topology of uniform convergence on compact subsets. Any limit must coincide with $\partial_\mu u(\mu)(\cdot)$ at points v in the support of μ or, put it differently, any limit provides a version of $\partial_\mu u(\mu)(\cdot)$ which is Lipschitz continuous, the Lipschitz constant being uniform in μ. When μ has full support, the sequence $(\mathcal{V}^n(\mu,\cdot))_{n\geqslant 1}$ converges to the unique continuous version of $\partial_\mu u(\mu)$, the convergence being uniform on compact subsets.

Let $X^n = X + n^{-1/2}G$, where G is an $N_d(0, I_d)$ Gaussian variable independent of X, so that $\mathcal{L}(X^n) = \mu^n$. We then observe that, for any $\mathbb{R}^d$-valued square integrable random variable χ such that the pair (X,χ) is independent of G,

$$\begin{aligned} D\tilde{u}(X^n)\cdot\chi &= \mathbb{E}\big[\partial_\mu u(\mu^n)(X^n)\cdot\chi\big] \\ &= \mathbb{E}\bigg[\bigg(\int_{\mathbb{R}^d} \partial_\mu u(\mu^n)(X - x) n^{d/2}\varphi_d(n^{1/2}x)dx\bigg)\cdot\chi\bigg] \\ &= \mathbb{E}\big[\mathcal{V}^n(\mu, X)\cdot\chi\big]. \end{aligned} \tag{5.109}$$

The main advantage of this formula is the fact that the mapping $\mathbb{R}^d \ni v \mapsto \mathcal{V}^n(\mu, v)$ is differentiable with respect to v, which is not known for $\mathbb{R}^d \ni x \mapsto \partial_\mu u(\mu)(v)$ at this stage of the proof.

Second Step. We construct now, independently of the measure μ considered above, a family $(Y^\lambda)_{\lambda\in\mathbb{R}}$ that is differentiable with respect to λ in $L^2(\Omega,\mathcal{F},\mathbb{P};\mathbb{R})$ but which is, at the same time, invariant in law, all the Y^λ, for $\lambda \in \mathbb{R}$, being uniformly distributed on $[-\pi/2, \pi/2]$.

Given two independent $N(0,1)$ random variables Z and Z', for any $\lambda \in \mathbb{R}$, we set:

$$Z^\lambda = \cos(\lambda)Z + \sin(\lambda)Z', \quad Z'^{,\lambda} = -\sin(\lambda)Z + \cos(\lambda)Z'.$$

For any $\lambda \in \mathbb{R}$, the pair $(Z^\lambda, Z'^{,\lambda})$ has the same law as (Z, Z') (because of the invariance of the Gaussian distribution by rotation). Next, we define the random variables Y^λ by:

$$Y^\lambda = \arcsin\Big(\frac{Z^\lambda}{\sqrt{(Z^\lambda)^2 + (Z'^{,\lambda})^2}}\Big) = \arcsin\Big(\frac{Z^\lambda}{\sqrt{Z^2 + (Z')^2}}\Big).$$

For any $\lambda \in \mathbb{R}$, Y^λ is uniformly distributed over $[-\pi/2, \pi/2]$. Pointwise (that is to say for $\omega \in \Omega$ fixed), the mapping $\mathbb{R} \ni \lambda \mapsto Y^\lambda$ is differentiable at any λ such that $Z'^{,\lambda} \neq 0$. Noticing that $[d/d\lambda]Z^\lambda = Z'^{,\lambda}$ pointwise, we get in that case:

$$\frac{d}{d\lambda}Y^\lambda = \frac{Z'^{,\lambda}}{\sqrt{Z^2 + (Z')^2}}\Big(1 - \frac{(Z^\lambda)^2}{(Z^\lambda)^2 + (Z'^{,\lambda})^2}\Big)^{-1/2} = \mathrm{sign}(Z'^{,\lambda}).$$

On the event $\{Z'^{,0} \neq 0\} = \{Z' \neq 0\}$, which is of probability 1, the set of λ's such that $Z'^{,\lambda} = 0$ is locally finite. The above derivative being bounded by 1, this says that pointwise, the mapping $\mathbb{R} \ni \lambda \mapsto Y^\lambda$ is 1-Lipschitz continuous. Therefore, the random variables $(Y^\lambda - Y^0)/\lambda$, $\lambda \neq 0$, are bounded by 1. Moreover, still on the event $\{Z' \neq 0\}$, the above computation shows that:

$$\lim_{\lambda \to 0} \frac{Y^\lambda - Y^0}{\lambda} = \mathrm{sign}(Z'). \tag{5.110}$$

Therefore, by Lebesgue's dominated convergence theorem, the mapping $\mathbb{R} \ni \lambda \mapsto Y^\lambda \in L^2(\Omega, \mathcal{F}, \mathbb{P}; \mathbb{R})$ is differentiable at $\lambda = 0$ with $\mathrm{sign}(Z')$ as derivative. In the sequel, we will denote Y^0 by Y.

Actually, by a rotation argument, differentiability holds at any $\lambda \in \mathbb{R}$, with $[d/d\lambda]Y^\lambda = \mathrm{sign}(Z'^{,\lambda})$. It is then clear that $\mathbb{R} \ni \lambda \mapsto \mathrm{sign}(Z'^{,\lambda}) \in L^2(\Omega, \mathcal{A}, \mathbb{P}; \mathbb{R}^d)$ is continuous. Indeed, the path $\mathbb{R} \ni \lambda \mapsto Z'^{,\lambda}$ is continuous. Composition by the function *sign* preserves continuity since, for any $\lambda \in \mathbb{R}$, the set of zeroes of $Z'^{,\lambda}$ is of zero probability.

Third Step. Assume now that μ is a given distribution and that X is a random variable with distribution μ, X being independent of the pair (Z, Z'). Given the same $(Y^\lambda)_{\lambda \in \mathbb{R}}$ as above, for $\delta > 0$, we let:

$$X^\lambda = (\delta \times Y^\lambda)e + X,$$

for each $\lambda \in \mathbb{R}$, where e is an arbitrary deterministic unit vector in $\mathbb{R}^d$. We omit the dependence upon δ in the notation X^λ. The mapping $\mathbb{R} \ni \lambda \mapsto X^\lambda$ is continuously differentiable in $L^2(\Omega, \mathcal{F}, \mathbb{P}; \mathbb{R}^d)$, with:

$$\frac{d}{d\lambda}_{|\lambda=0} X^\lambda = (\delta \times \mathrm{sign}(Z'))e.$$

Going back to (5.109), we get, for another random variable $\chi \in L^2(\Omega, \mathcal{F}, \mathbb{P}; \mathbb{R}^d)$, with (X, χ, Z, Z') independent of G,

$$D\tilde{u}\Big(X^\lambda + \frac{1}{\sqrt{n}}G\Big) \cdot \chi = \mathbb{E}\big[\mathcal{V}^n\big(\mathcal{L}(X^\lambda), X^\lambda\big) \cdot \chi\big].$$

As the mapping $\mathbb{R} \ni \lambda \mapsto X^\lambda$ is continuously differentiable in $L^2(\Omega, \mathcal{F}, \mathbb{P}; \mathbb{R}^d)$ and since all the random variables $(X^\lambda)_{\lambda\in\mathbb{R}}$ have the same distribution, we deduce from **(A1)** in the standing assumption that (for (X, χ, Z, Z') independent of G):

$$\begin{aligned}
&\partial^2_{\mathrm{sign}(Z')e,\chi}\tilde{u}\Big(X + \delta Y e + \frac{1}{\sqrt{n}}G\Big) \\
&= \frac{d}{d\lambda}_{|\lambda=0}\Big[D\tilde{u}\Big(X^{\lambda/\delta} + \frac{1}{\sqrt{n}}G\Big)\cdot\chi\Big] \\
&= \frac{1}{\delta}\frac{d}{d\lambda}_{|\lambda=0}\Big[D\tilde{u}\Big(X^{\lambda} + \frac{1}{\sqrt{n}}G\Big)\cdot\chi\Big] \\
&= \mathbb{E}\Big[\mathrm{trace}\Big\{\partial_v\mathcal{V}^n\Big(\mathcal{L}(X+\delta Y e), X+\delta Y e\Big)\big((\mathrm{sign}(Z')\chi)\otimes e\big)\Big\}\Big].
\end{aligned}$$

Observe in the above formula that $\partial_v\mathcal{V}^n$ takes values in the set of symmetric $d \times d$ matrices since $\partial_\mu u(\mu)(\cdot)$ derives from a potential for all $\mu \in \mathcal{P}_2(\mathbb{R}^d)$, see Proposition 5.50.

Noticing that the random variable $|\mathrm{sign}(Z')|$ is equal to 1 almost surely, we can replace χ by $\mathrm{sign}(Z')\chi$ with (X, χ) independent of (Z, Z'), so that:

$$\begin{aligned}
&\partial^2_{\mathrm{sign}(Z')e,\mathrm{sign}(Z')\chi}\tilde{u}\Big(X + \delta Y e + \frac{1}{\sqrt{n}}G\Big) \\
&\quad= \mathbb{E}\Big[\mathrm{trace}\Big\{\partial_v\mathcal{V}^n\Big(\mathcal{L}(X+\delta Y e), X+\delta Y e\Big)\big(\chi\otimes e\big)\Big\}\Big].
\end{aligned}$$

Finally, we let:

$$\mathcal{W}^{n,\delta}(\mu, v) = \int_{\mathbb{R}} \partial_v\mathcal{V}^n(\mu * p^\delta, v + \delta r e)p(r)dr, \tag{5.111}$$

where p is the uniform density on $[-\pi/2, \pi/2]$ and $p^\delta(\cdot) = p(\cdot/\delta)/\delta$ is the uniform density on $[-\delta\pi/2, \delta\pi/2]$. As usual, $\mu * p^\delta$ is an abbreviated notation for denoting the convolution of μ with the uniform distribution on the segment $[-(\delta\pi/2)e, (\delta\pi/2)e]$. Since the pair (X, χ) is independent of (Z, Z'), we end up with the duality formula:

$$\partial^2_{\mathrm{sign}(Z')e,\mathrm{sign}(Z')\chi}\tilde{u}\Big(X + \delta Y e + \frac{1}{\sqrt{n}}G\Big) = \mathbb{E}\Big[\mathrm{trace}\Big\{\mathcal{W}^{n,\delta}(\mu, X)\big(\chi\otimes e\big)\Big\}\Big]. \tag{5.112}$$

By the smoothness assumption on $\partial^2_{\xi,\chi}\tilde{u}$ (see *(ii)* in **(A2)** in assumption **Sufficiency for Partial** $\mathcal{C}^2$), we deduce that, for another X', with distribution μ as well, such that the triple (X, X', χ) is independent of (Z, Z') and the 5-tuple (X, X', χ, Z, Z') is independent of G,

$$\Big|\mathbb{E}\Big[\mathrm{trace}\Big\{\big(\mathcal{W}^{n,\delta}(\mu, X) - \mathcal{W}^{n,\delta}(\mu, X')\big)\big(\chi\otimes e\big)\Big\}\Big]\Big| \leqslant C\mathbb{E}\big[|X - X'|^2\big]^{1/2}\mathbb{E}\big[|\chi|^2\big]^{1/2}, \tag{5.113}$$

the constant C being independent of μ, δ and n. The above is true for any $\sigma\{X, X'\}$-measurable $\chi \in L^2(\Omega, \mathcal{F}, \mathbb{P}; \mathbb{R}^d)$. We deduce that, for any other $e' \in \mathbb{R}^d$ with $|e'| = 1$,

$$\mathbb{E}\Big[\Big|\mathrm{trace}\Big\{\big(\mathcal{W}^{n,\delta}(\mu,X)-\mathcal{W}^{n,\delta}(\mu,X')\big)\big(e'\otimes e\big)\Big\}\Big|^2\Big]\leqslant C\mathbb{E}\big[|X-X'|^2\big]^{1/2}.$$

By Proposition 5.36, this says that $\mathbb{R}^d \ni v \mapsto \mathrm{trace}\{(\mathcal{W}^{n,\delta}(\mu,v))(e'\otimes e)\}$ has a C-Lipschitz continuous version.

Fourth Step. From (5.108) and (5.111), we know that:

$$\begin{aligned}\mathcal{W}^{n,\delta}(\mu,v) &= \int_{\mathbb{R}} \partial_v \mathcal{V}^n\big(\mu * p^\delta, v+\delta re\big)p(r)dr \\ &= n^{(d+1)/2}\int_{\mathbb{R}\times\mathbb{R}^d} \partial_\mu u\big(\mu * p^\delta * N_d(0,\tfrac{1}{n}I_d), w+\delta re\big)p(r)\partial\varphi_d\big(n^{1/2}(v-w)\big)drdw.\end{aligned}$$

Since $\mu * N_d(0,(1/n)I_d)$ has full support, we know that $\partial_\mu u(\mu * p^\delta * N_d(0,(1/n)I_d),\cdot)$ converges towards $\partial_\mu u(\mu * N_d(0,(1/n)I_d),\cdot)$ as δ tends to 0, uniformly on compact subsets (see the warm-up). We deduce that, as δ tends to 0, $\mathcal{W}^{n,\delta}(\mu,v)$ converges to:

$$\begin{aligned}\mathcal{W}^n(\mu,v) &= n^{(d+1)/2}\int_{\mathbb{R}^d} \partial_\mu u\big(\mu * N_d(0,\tfrac{1}{n}I_d), w\big)\partial\varphi_d\big(n^{1/2}(v-w)\big)dw \\ &= \partial_v\bigg(n^{d/2}\int_{\mathbb{R}^d} \partial_\mu u\big(\mu * N_d(0,\tfrac{1}{n}I_d), w\big)\varphi_d\big(n^{1/2}(v-w)\big)dw\bigg) = \partial_v \mathcal{V}^n(\mu,v).\end{aligned}$$

Therefore, we deduce that the mappings $(\mathbb{R}^d \ni v \mapsto \mathrm{trace}\{(\partial_v \mathcal{V}^n(\mu,v))(e'\otimes e)\})_{n\geqslant 1}$ are Lipschitz continuous, uniformly in μ. Since $\partial_v \mathcal{V}^n(\mu,v)$ is independent of e and e', this implies that the mappings $(\mathbb{R}^d \ni v \mapsto \partial_v \mathcal{V}^n(\mu,v))_{n\geqslant 1}$ are Lipschitz continuous, uniformly with respect to μ.

By (5.112) and *(ii)* in **(A2)** in assumption **Sufficiency for Partial** C^2,

$$\sup_{n\geqslant 1,\delta\in[0,1]} \mathbb{E}\big[\big|\mathrm{trace}\big\{\big(\mathcal{W}^{n,\delta}(\mu,X)\big)(e'\otimes e)\big\}\big|^2\big]\leqslant C,$$

for a possibly new value of C. Above, $X \sim \mu$. Letting δ tend to 0, we deduce from Fatou's lemma that:

$$\sup_{n\geqslant 1}\mathbb{E}\Big[\big|\mathrm{trace}\big\{\big(\partial_v \mathcal{V}^n(\mu,X)\big)(e'\otimes e)\big\}\big|^2\Big]\leqslant C,$$

and thus that $\sup_{n\geqslant 1}\mathbb{E}[|\partial_v \mathcal{V}^n(\mu,X)|^2] \leqslant C$, which implies by Lipschitz property of $\partial_v \mathcal{V}^n(\mu,\cdot)$, that:

$$\forall n\geqslant 1,\quad |\partial_x \mathcal{V}^n(\mu,0)|\leqslant C\big(1+\mathbb{E}[|X|^2]\big). \tag{5.114}$$

This says that the sequence of mappings $(\mathbb{R}^d \ni v \mapsto \partial_v \mathcal{V}^n(\mu,v))_{n\geqslant 1}$ is relatively compact for the topology of uniform convergence. Therefore, we can extract a convergent subsequence. As the limit of $\mathcal{V}^n(\mu,\cdot)$ is $\partial_\mu u(\mu)(\cdot)$, we deduce that $\mathbb{R}^d \ni v \mapsto \partial_\mu u(\mu)(v)$ is differentiable with respect to v. Passing to the limit in (5.112) (first on δ and then on n), we deduce that

$$\partial^2_{\text{sign}(Z')e,\text{sign}(Z')\chi}\tilde{u}(X) = \mathbb{E}\Big[\text{trace}\Big\{\big(\partial_v\partial_\mu u(\mu)(X)\big)\big(\chi \otimes e\big)\Big\}\Big]. \tag{5.115}$$

Again, by *(ii)* in (**A2**) in assumption **Sufficiency for Partial** $\mathcal{C}^2$, there exists a constant C such that, for any $\mu \in \mathcal{P}_2(\mathbb{R}^d)$ and any $X \in L^2(\Omega, \mathcal{F}, \mathbb{P}; \mathbb{R}^d)$ with μ as distribution, $\mathbb{E}[|\partial_v\partial_\mu u(\mu)(X)|^2] \leqslant C$, which is a required condition for applying the chain rule. In order to complete the proof, it remains to prove that the mapping $\mathcal{P}_2(\mathbb{R}^d) \times \mathbb{R}^d \ni (\mu, v) \mapsto \partial_v\partial_\mu u(\mu)(v)$ is jointly continuous at any point (μ, v) such that $v \in \text{Supp}(\mu)$. We already know that it is Lipschitz continuous with respect to v, uniformly in μ. For a sequence $(\mu^n)_{n\geqslant 1}$ in $\mathcal{P}_2(\mathbb{R}^d)$ converging for the 2-Wasserstein distance to some $\mu \in \mathcal{P}_2(\mathbb{R}^d)$, we deduce from the Lipschitz property and by the same argument as in (5.114) that the sequence of functions $(\mathbb{R}^d \ni v \mapsto \partial_v\partial_\mu u(\mu^n)(v))_{n\geqslant 1}$ is relatively compact for the topology of uniform convergence on compact subsets. By means of the bound $\sup_{n\geqslant 1}\mathbb{E}[|\partial_v\partial_\mu u(\mu^n)(X^n)|^2] \leqslant C$, with $X^n \sim \mu^n$, it is quite easy to pass to the limit in the right-hand side of (5.115). By *(ii)* in (**A2**) in assumption **Sufficiency for Partial** $\mathcal{C}^2$, we can also pass to the limit in the left-hand side. Equation (5.115) then permits to identify any limit with $\partial_v\partial_\mu u(\mu)(\cdot)$ on the support of μ. Since the mappings $(\partial_v\partial_\mu u(\mu^n)(\cdot))_{n\geqslant 1}$ are uniformly continuous on compact subsets, we deduce that, for an additional sequence $(v^n)_{n\geqslant 1}$, with values in $\mathbb{R}^d$, that converges to some $v \in \text{Supp}(\mu)$, the sequence $(\partial_v\partial_\mu u(\mu^n)(v^n))_{n\geqslant 1}$ converges, up to a subsequence, to $\partial_v\partial_\mu u(\mu)(v)$. Now, by relative compactness of the sequence $(\mathbb{R}^d \ni v \mapsto \partial_v\partial_\mu u(\mu^n)(v))_{n\geqslant 1}$, the sequence $(\partial_v\partial_\mu u(\mu^n)(v^n))_{n\geqslant 1}$ is bounded. By a standard compactness argument, the sequence $(\partial_v\partial_\mu u(\mu^n)(v^n))_{n\geqslant 1}$ must be convergent with $\partial_v\partial_\mu u(\mu)(v)$ as limit. Arguing as we did for $\partial_\mu u$ in the warm-up preceding the proof, we easily prove that $\partial_v\partial_\mu u$ is locally bounded. □

5.7 Applications

5.7.1 Revisiting Uniqueness in MFG

We now comment more on Remark 5.75 within the framework of mean field games.

Revisiting the Lasry-Lions Monotonicity Condition

On a given complete probability space $(\Omega, \mathcal{F}, \mathbb{P})$, equipped with a right-continuous and complete filtration $\mathbb{F}$, with a $\mathbb{F}$-Brownian motion $(W_t)_{0\leqslant t\leqslant T}$ with values in $\mathbb{R}^d$, consider the mean field game associated with the parameterized stochastic control problem:

$$\inf_{\alpha\in\mathbb{A}} \mathbb{E}\bigg[\int_0^T f(t, X_t, \mu_t, \alpha_t)dt + g(X_T, \mu_T)\bigg],$$

$$\text{subject to} \quad dX_t = b(t, X_t, \alpha_t)dt + \sigma(t, X_t)dW_t; \quad X_0 \in L^2(\Omega, \mathcal{F}_0, \mathbb{P}; \mathbb{R}^d),$$

where $b : [0,T] \times \mathbb{R}^d \times A \to \mathbb{R}^d$, $\sigma : [0,T] \times \mathbb{R}^d \to \mathbb{R}^{d\times d}$, $f : [0,T] \times \mathbb{R}^d \times \mathcal{P}_2(\mathbb{R}^d) \times A \to \mathbb{R}$ and $g : \mathbb{R}^d \times \mathcal{P}_2(\mathbb{R}^d) \to \mathbb{R}$, with A being a closed convex subset of $\mathbb{R}^k$ and f having the same separated structure as in assumption **Lasry-Lions Monotonicity** from Section 3.4, namely:

$$f(t, x, \mu, \alpha) = f_0(t, x, \mu) + f_1(t, x, \alpha),$$
$$(t, x, \mu, \alpha) \in [0, T] \times \mathbb{R}^d \times \mathcal{P}_2(\mathbb{R}^d) \times A.$$

Assume further that there exist two functions $F_0 : [0, T] \times \mathcal{P}_2(\mathbb{R}^d) \to \mathbb{R}$ and $G : \mathcal{P}_2(\mathbb{R}^d) \to \mathbb{R}$ such that, for any $t \in [0, T]$, $F_0(t, \cdot)$ and G are differentiable for the linear functional differentiation defined in Subsection 5.4.1, such that, for all $(t, x, \mu) \in [0, T] \times \mathbb{R}^d \times \mathcal{P}_2(\mathbb{R}^d)$,

$$f_0(t, x, \mu) = \frac{\delta F_0}{\delta \mu}(t, \mu)(x),$$
$$g(x, \mu) = \frac{\delta G}{\delta \mu}(t, \mu)(x).$$

Then, f_0 and g satisfy the Lasry-Lions monotonicity property in Definition 3.28 if F_0 and G are convex in the direction μ in the sense of Remark 5.75. In other words, the Lasry-Lions monotonicity condition (used in Theorem 3.29 for guaranteeing uniqueness) can be interpreted as a convexity property (of F_0 and G) in the direction of the measure argument for the linear functional differentiation introduced in Subsection 5.4.1.

Using the L-Differential Calculus

The connection between convexity and monotonicity is quite appealing as it provides another interpretation of the Lasry-Lions condition. Actually, it becomes even more intriguing if we recall from the discussion in Subsection 5.4.1 that the notion of convexity for the L-differential calculus may differ from the notion of convexity for the linear functional differentiation. It is thus a natural question to wonder whether convexity, when regarded in the L-sense, could help for uniqueness.

In order to address this problem, we must recall the notion of L-monotonicity introduced in Definition 3.31. From Lemma 5.72, we know that, if h is the L-derivative of an L-differentiable and L-convex function $H : \mathcal{P}_2(\mathbb{R}^d) \to \mathbb{R}$, namely:

$$\forall (x, \mu) \in \mathbb{R}^d \times \mathcal{P}_2(\mathbb{R}^d), \quad h(x, \mu) = \partial_\mu H(\mu)(x),$$

then h is L-monotone. Whenever Proposition 5.51 applies, this says that $\partial_x[\delta H/\delta \mu]$ is L-monotone when H is L-convex.

Returning to the mean field game described above, we deduce that $\partial_x f_0(t, \cdot)$ and $\partial_x g$ are L-monotone if $F_0(t, \cdot)$ and G are L-convex in the direction μ. Therefore, in full analogy with the above interpretation of the Lasry-Lions monotonicity condition, the L-monotonicity condition in the statement of Theorem 3.32 (which provides another sufficient condition for guaranteeing uniqueness) can be also interpreted as a convexity condition (of F_0 and G) in the direction of the measure argument but for the L-differentiation!

5.7.2 A Primer on the Master Equation

As an application of the tools introduced in this chapter, we now provide a short initiation to the seminal notion of master equation for mean field games. We give a primer only since we shall dedicate the entire Part I of the second volume of the book to it.

In short, the master equation is a partial differential equation set on the enlarged state space $[0,T]\times\mathbb{R}^d\times\mathcal{P}_2(\mathbb{R}^d)$. Generally speaking, it is associated with a McKean-Vlasov forward-backward stochastic differential equation of the same type as those introduced in Chapters 3 and 4 for characterizing equilibria to mean field games, see for instance (4.49). The connection between the partial differential equation and the forward-backward stochastic system proceeds in the same way as that between a partial differential equation on the Euclidean space $[0,T]\times\mathbb{R}^d$ and a classical Markovian forward-backward stochastic differential equation and thus works along the lines exposed in Subsection 4.1.2. Basically, the solution of the master equation is the decoupling field (the so-called *master field* according to the terminology introduced in Subsection 4.2.4) of the corresponding McKean-Vlasov forward-backward stochastic differential equation.

On a complete filtered probability space $(\Omega,\mathcal{F},\mathbb{F}=(\mathcal{F}_t)_{t\geqslant 0},\mathbb{P})$, we thus consider a system of the same type as (4.49) with the slight difference that $m=1$:

$$\begin{cases} dX_t = B\big(t,X_t,\mathcal{L}(X_t),Y_t,Z_t\big)dt + \Sigma\big(t,X_t,\mathcal{L}(X_t)\big)dW_t \\ dY_t = -F\big(t,X_t,\mathcal{L}(X_t),Y_t,Z_t\big)dt + Z_t\cdot dW_t, \qquad t\in[0,T], \end{cases} \tag{5.116}$$

with $Y_T=G(X_T,\mathcal{L}(X_T))$ as terminal condition. Here, we assume that $B:[0,T]\times\mathbb{R}^d\times\mathcal{P}_2(\mathbb{R}^d)\times\mathbb{R}\times\mathbb{R}^d\rightarrow\mathbb{R}^d$, $F:[0,T]\times\mathbb{R}^d\times\mathcal{P}_2(\mathbb{R}^d)\times\mathbb{R}\times\mathbb{R}^d\rightarrow\mathbb{R}$, $\Sigma:[0,T]\times\mathbb{R}^d\times\mathcal{P}_2(\mathbb{R}^d)\rightarrow\mathbb{R}^{d\times d}$ and $G:\mathbb{R}^d\times\mathcal{P}_2(\mathbb{R}^d)\rightarrow\mathbb{R}$ are Borel-measurable, locally bounded and Lipschitz-continuous in the variables $(x,\mu,y,z)\in\mathbb{R}^d\times\mathcal{P}_2(\mathbb{R}^d)\times\mathbb{R}\times\mathbb{R}^d$, uniformly in the time parameter $t\in[0,T]$.

With (5.116), we associate the following partial differential equation on $[0,T]\times\mathbb{R}^d\times\mathcal{P}_2(\mathbb{R}^d)$, with $u:[0,T]\times\mathbb{R}^d\times\mathcal{P}_2(\mathbb{R}^d)\rightarrow\mathbb{R}$ as unknown:

$$\begin{aligned} &\partial_t u(t,x,\mu) + B\big(t,x,\mu,u(t,x,\mu),(\Sigma^\dagger\partial_x u)(t,x,\mu)\big)\cdot\partial_x u(t,x,\mu) \\ &\quad + \frac{1}{2}\mathrm{trace}\Big[(\Sigma\Sigma^\dagger)(t,x,\mu)\partial^2_{xx}u(t,x,\mu)\Big] \\ &\quad + \int_{\mathbb{R}^d} B\big(t,v,\mu,u(t,v,\mu),(\Sigma^\dagger\partial_x u)(t,v,\mu)\big)\cdot\partial_\mu u(t,x,\mu)(v)d\mu(v) \\ &\quad + \frac{1}{2}\int_{\mathbb{R}^d}\mathrm{trace}\Big[(\Sigma\Sigma^\dagger)(t,v,\mu)\partial_v\partial_\mu u(t,x,\mu)(v)\Big]d\mu(v) \\ &\quad + F\big(t,x,\mu,u(t,x,\mu),(\Sigma^\dagger\partial_x u)(t,x,\mu)\big) = 0, \end{aligned} \tag{5.117}$$

for $(t,x,\mu)\in[0,T]\times\mathbb{R}^d\times\mathcal{P}_2(\mathbb{R}^d)$, with $u(T,x,\mu)=G(x,\mu)$ as terminal condition.

The next statement consists of a verification argument that makes the connection between (5.116) and (5.117):

Proposition 5.106 *Assume that there exists a function* $u : [0, T] \times \mathbb{R}^d \times \mathcal{P}_2(\mathbb{R}^d) \to \mathbb{R}$ *satisfying assumption* **Joint Chain Rule** *together with* (5.106) *and such that* u *and* $\partial_x u$ *are Lipschitz continuous in* (x, μ) *uniformly in time. Assume further that* Σ *is bounded.*

Then, for any initial condition $X_0 \in L^2(\Omega, \mathcal{F}_0, \mathbb{P}; \mathbb{R}^d)$*, the system* (5.116) *admits a unique solution* $(X_t, Y_t, Z_t)_{0 \leqslant t \leqslant T}$ *satisfying:*

$$\mathbb{E}\left[\sup_{0 \leqslant t \leqslant T} \big(|X_t|^2 + |Y_t|^2\big) + \int_0^T |Z_t|^2 dt\right] < \infty.$$

It satisfies, $\mathbb{P}$ *almost surely,*

$$\forall t \in [0, T], \; Y_t = u\big(t, X_t, \mathcal{L}(X_t)\big),$$

and, $\mathrm{Leb}_1 \otimes \mathbb{P}$ *almost everywhere,*

$$Z_t = \big(\Sigma^\dagger \partial_x u\big)\big(t, X_t, \mathcal{L}(X_t)\big).$$

Remark 5.107 *The master equation for mean field games is obtained by choosing* B*,* Σ*,* F *and* G *as in the statement of Theorem 4.44. We let the reader write the corresponding form of* (5.117)*. In that case, Proposition 5.106 shows that the function* $[0, T] \times \mathbb{R}^d \ni (t, x) \mapsto u(t, x, \mathcal{L}(X_t))$ *should coincide with the solution of the HJB equation in the mean field game system* (3.12)*, while* $(\mu_t)_{0 \leqslant t \leqslant T}$ *therein should coincide with* $(\mathcal{L}(X_t))_{0 \leqslant t \leqslant T}$.

Proof.

First Step. We first prove the existence of a solution. To do so, we notice that, under our assumption, the stochastic differential equation:

$$dX_t = B\Big(t, X_t, \mathcal{L}(X_t), u\big(t, X_t, \mathcal{L}(X_t)\big), \big(\Sigma^\dagger \partial_x u\big)\big(t, X_t, \mathcal{L}(X_t)\big)\Big) dt + \Sigma\big(t, X_t, \mathcal{L}(X_t)\big) dW_t,$$

for $t \in [0, T]$, with $X_0 \in L^2(\Omega, \mathcal{F}_0, \mathbb{P}; \mathbb{R}^d)$ as initial condition, is uniquely solvable, see Theorem 4.21. The solution satisfies $\mathbb{E}[\sup_{0 \leqslant t \leqslant T} |X_t|^2] < \infty$.

Let now:

$$Y_t = u\big(t, X_t, \mathcal{L}(X_t)\big), \quad Z_t = \big(\Sigma^\dagger \partial_x u\big)\big(t, X_t, \mathcal{L}(X_t)\big), \quad t \in [0, T].$$

Combining the PDE (5.117) with Proposition 5.102, we deduce that $(X_t, Y_t, Z_t)_{0 \leqslant t \leqslant T}$ is a solution of (5.116).

Second Step. We now consider another solution $(X'_t, Y'_t, Z'_t)_{0 \leqslant t \leqslant T}$ to (5.116), with the same initial condition $X'_0 = X_0$ as in the first step. We let:

$$\mathcal{Y}_t = u\big(t, X'_t, \mathcal{L}(X'_t)\big), \quad \mathcal{Z}_t = \big(\Sigma^\dagger \partial_x u\big)\big(t, X'_t, \mathcal{L}(X'_t)\big), \quad t \in [0, T].$$

Again, we can combine the PDE (5.117) with Proposition 5.102. We deduce that:

$$\begin{aligned} d\mathcal{Y}_t &= \Big(B\big(t, \mathcal{L}(X'_t), X'_t, Y'_t, Z'_t\big) - B\big(t, \mathcal{L}(X'_t), X'_t, \mathcal{Y}_t, \mathcal{Z}_t\big)\Big) \cdot \partial_x u\big(t, X'_t, \mathcal{L}(X'_t)\big)dt \\ &\quad + \tilde{\mathbb{E}}\Big[\Big(B\big(t, \mathcal{L}(X'_t), \tilde{X}'_t, \tilde{Y}'_t, \tilde{Z}'_t\big) - B\big(t, \mathcal{L}(X'_t), \tilde{X}'_t, \tilde{\mathcal{Y}}_t, \tilde{\mathcal{Z}}_t\big)\Big) \cdot \partial_\mu u\big(t, X'_t, \mathcal{L}(X'_t)\big)(\tilde{X}'_t)\Big]dt \\ &\quad - F\big(t, \mathcal{L}(X'_t), X'_t, \mathcal{Y}_t, \mathcal{Z}_t\big)dt + \mathcal{Z}_t \cdot dW_t, \qquad t \in [0, T], \end{aligned}$$

with the terminal boundary condition $\mathcal{Y}'_T = G(X'_T, \mathcal{L}(X'_T))$, and where we used the same convention as above for the variables labeled with a tilde: They denote copies of the original variables that are constructed on a copy of the original probability space.

In order to complete the proof, it suffices to regard the difference $(Y'_t - \mathcal{Y}_t, Z'_t - \mathcal{Z}_t)_{0 \leqslant t \leqslant T}$ as the solution of a backward SDE with random coefficients with 0 as terminal condition. Notice then from the fact that u is Lipschitz continuous in μ, that $D_X \tilde{u}$ takes values in a bounded subset of $L^2(\Omega, \mathcal{F}, \mathbb{P}; \mathbb{R}^d)$, see Remark 5.27. In particular,

$$\begin{aligned} &\Big|\tilde{\mathbb{E}}\Big[\Big(B\big(t, \mathcal{L}(X'_t), \tilde{X}'_t, \tilde{Y}'_t, \tilde{Z}'_t\big) - B\big(t, \mathcal{L}(X'_t), \tilde{X}'_t, \tilde{\mathcal{Y}}_t, \tilde{\mathcal{Z}}_t\big)\Big) \cdot \partial_\mu u\big(t, X'_t, \mathcal{L}(X'_t)\big)(\tilde{X}'_t)\Big]\Big| \\ &\leqslant C\tilde{\mathbb{E}}\Big[|\tilde{Y}'_t - \tilde{\mathcal{Y}}_t|^2 + |\tilde{Z}'_t - \tilde{\mathcal{Z}}_t|^2\Big]^{1/2} \tilde{\mathbb{E}}\Big[|\partial_\mu u(t, X'_t, \mathcal{L}(X'_t))(\tilde{X}'_t)|^2\Big]^{1/2} \\ &\leqslant C\tilde{\mathbb{E}}\Big[|\tilde{Y}'_t - \tilde{\mathcal{Y}}_t|^2 + |\tilde{Z}'_t - \tilde{\mathcal{Z}}_t|^2\Big]^{1/2}, \end{aligned}$$

for a value of C, independent of t, that is allowed to increase from line to line. By adapting the stability arguments used in the proof of Theorem 4.23, we get that:

$$\sup_{0 \leqslant t \leqslant T} \mathbb{E}\big[|Y'_t - \mathcal{Y}_t|^2\big] + \mathbb{E}\int_0^T |Z'_t - \mathcal{Z}_t|^2 dt = 0.$$

Therefore, $(X'_t)_{0 \leqslant t \leqslant T}$ solves the same SDE as $(X_t)_{0 \leqslant t \leqslant T}$, which proves uniqueness. □

5.7.3 Application to a Nonstandard Control Problem

The purpose of this short section is to present an application of the chain rule to a nonstandard control problem. In full analogy with the classical case and with the previous subsection, we use a verification argument whereby the classical solution of a partial differential equation, if it exists, provides a solution to the optimal control problem. However, the control problem has to be of a very special nature to justify the need for such a sophisticated form of the chain rule. Case in point, the application we propose to investigate is an example of the optimal control of McKean-Vlasov dynamics studied in full generality in Chapter 6.

For the purpose of the current application, we seek to minimize the cost $J(\boldsymbol{\alpha})$ defined as

$$
\begin{aligned}
J(\boldsymbol{\alpha}) &= \mathbb{E}\biggl[\int_0^T \bigl[f\bigl(\mathcal{L}(X_t^{\boldsymbol{\alpha}})\bigr) + \frac{1}{2}|\alpha_t|^2\bigr]dt\biggr] \\
&= \int_0^T f\bigl(\mathcal{L}(X_t^{\boldsymbol{\alpha}})\bigr)dt + \mathbb{E}\biggl[\int_0^T \frac{1}{2}|\alpha_t|^2 dt\biggr],
\end{aligned}
\tag{5.118}
$$

over the set $\mathbb{A} = \mathbb{H}^{2,d}$ of admissible controls for the controlled dynamics:

$$
dX_t^{\boldsymbol{\alpha}} = \alpha_t dt + dW_t, \quad t \in [0,T], \tag{5.119}
$$

where the initial condition X_0 is in $L^2(\Omega, \mathcal{F}_0, \mathbb{P}; \mathbb{R}^d)$, and $\boldsymbol{W} = (W_t)_{0 \leqslant t \leqslant T}$ is a standard Wiener process in $\mathbb{R}^d$ defined on a filtered complete probability space $(\Omega, \mathcal{F}, \mathbb{F} = (\mathcal{F}_t)_{0 \leqslant t \leqslant T}, \mathbb{P})$. The McKean-Vlasov nature of the problem comes from the fact that the running cost is a function of the marginal distribution of the controlled state. The function f can be quite general, but, for the sake of definiteness, we shall assume that it is continuous on $\mathcal{P}_2(\mathbb{R}^d)$ for the 2-Wasserstein distance W_2.

Proposition 5.108 *Let us assume that there exists a function $u : [0,T] \times \mathcal{P}_2(\mathbb{R}^d) \to \mathbb{R}$, differentiable in t, with $\partial_t u$ being continuous in (t,μ), partially $\mathcal{C}^2$ in the measure variable, with $D_X\tilde{u} : [0,T] \times L^2(\Omega, \mathcal{F}, \mathbb{P}; \mathbb{R}^d) \ni (t,X) \mapsto D_X\tilde{u}(t,X) \in L^2(\Omega, \mathcal{F}, \mathbb{P}; \mathbb{R}^d)$ being Lipschitz continuous with respect to X, uniformly in time, $\partial_\mu u : [0,T] \times \mathcal{P}_2(\mathbb{R}^d) \times \mathbb{R}^d \ni (t,\mu,v) \mapsto \partial_\mu u(t,\mu)(v)$ and $\partial_v \partial_\mu u : [0,T] \times \mathcal{P}_2(\mathbb{R}^d) \times \mathbb{R}^d \ni (t,\mu,v) \mapsto \partial_v \partial_\mu u(t,\mu)(v)$ being continuous at any point (t,μ,v) such that $v \in \mathrm{Supp}(\mu)$, satisfying $u(T,\cdot) \equiv 0$ and*

$$
\sup_{t \in [0,T]} \sup_{\mu \in \mathcal{K}} \biggl[\int_{\mathbb{R}^d} \bigl|\partial_v \partial_\mu u(t,\mu)(v)\bigr|^2 d\mu(v)\biggr] < \infty, \tag{5.120}
$$

for all compact $\mathcal{K} \subset \mathcal{P}_2(\mathbb{R}^d)$. Furthermore, if we assume that u satisfies the infinite-dimensional PDE*:*

$$
\begin{aligned}
&\partial_t u(t,\mu) - \frac{1}{2}\int_{\mathbb{R}^d} |\partial_\mu u(t,\mu)(v)|^2 d\mu(v) \\
&\quad + \frac{1}{2}\mathrm{trace}\biggl[\int_{\mathbb{R}^d} \partial_v \partial_\mu u(t,\mu)(v) d\mu(v)\biggr] + f(\mu) = 0,
\end{aligned}
\tag{5.121}
$$

then, the McKean-Vlasov SDE

$$
d\hat{X}_t = -\partial_\mu u\bigl(t, \mathcal{L}(\hat{X}_t)\bigr)(\hat{X}_t)dt + dW_t, \quad 0 \leqslant t \leqslant T, \tag{5.122}
$$

with $X_0 \in L^2(\Omega, \mathcal{F}_0, \mathbb{P}; \mathbb{R}^d)$ *as initial condition has a unique solution* $(\hat{X}_t)_{0 \leqslant t \leqslant T}$ *satisfying* $\mathbb{E}[\sup_{0 \leqslant t \leqslant T} |\hat{X}_t|^2] < \infty$ *and this solution is the unique optimal path in the sense that the control* $\hat{\boldsymbol{\alpha}} = (\hat{\alpha}_t)_{0 \leqslant t \leqslant T}$ *defined by* $\hat{\alpha}_t = -\partial_\mu u(t, \mathcal{L}(\hat{X}_t))(\hat{X}_t)$ *minimizes the cost:*

$$J(\hat{\boldsymbol{\alpha}}) = \inf_{\boldsymbol{\alpha} \in \mathbb{A}} J(\boldsymbol{\alpha}).$$

Of course, u must be interpreted as a value function. In particular, the terminal condition $u(T, \cdot)$ is null because we did not include a terminal cost in the expression for the cost functional J. The terminal condition would be set equal to g if, in the definition $J(\boldsymbol{\alpha})$, we added $g(\mathcal{L}(X_T^{\boldsymbol{\alpha}}))$.

Proof.

First Step. We first prove that (5.122) is uniquely solvable. We first recall from Subsection 5.3.4 that we can a find a version of each $\partial_\mu u(t, \mu)(\cdot) \in L^2(\mathbb{R}^d, \mu)$ such that the mapping $[0, T] \times \mathcal{P}_2(\mathbb{R}^d) \times \mathbb{R}^d \ni (t, \mu, v) \mapsto \partial_\mu u(t, \mu)(v)$ is measurable. Of course, for any $t \in [0, T]$ and any random variable $X \in L^2(\Omega, \mathcal{F}, \mathbb{P}; \mathbb{R}^d)$, $\partial_\mu u(t, \mathcal{L}(X))(X)$ is almost surely equal to $D_X\tilde{u}(t, X)$. In particular, the Lipschitz property of $D_X\tilde{u}$ in the variable X shows that, for any $X, Y \in L^2(\Omega, \mathcal{F}, \mathbb{P}; \mathbb{R}^d)$,

$$\sup_{0 \leqslant t \leqslant T} \mathbb{E}\Big[\big|\partial_\mu u\big(t, \mathcal{L}(X)\big)(X) - \partial_\mu u\big(t, \mathcal{L}(Y)\big)(Y)\big|^2 \Big] \leqslant C\|X - Y\|_2^2.$$

Moreover, by choosing $\mu = \delta_0$ (Dirac point mass at 0), we have

$$\sup_{0 \leqslant t \leqslant T} |\partial_\mu u(t, \delta_0)(0)|^2 < \infty.$$

This suffices to implement Picard's fixed point theorem along the lines of the proof of Theorem 4.21. As a byproduct of the above estimates, we also get that:

$$\sup_{t \in [0,T]} \sup_{\mu \in \mathcal{K}} \bigg[\int_{\mathbb{R}^d} \big|\partial_\mu u(t, \mu)(v)\big|^2 d\mu(v) \bigg] < \infty,$$

for all compact $\mathcal{K} \subset \mathcal{K}_2(\mathbb{R}^d)$, which is a necessary condition to apply the chain rule in the second step below.

Second Step. Consider now a generic admissible control $\boldsymbol{\alpha} = (\alpha_t)_{0 \leqslant t \leqslant T}$, denote by $\boldsymbol{X}^{\boldsymbol{\alpha}} = (X_t^{\boldsymbol{\alpha}})_{0 \leqslant t \leqslant T}$ the corresponding controlled state given by (5.119), and let us apply the time dependent form of the chain rule of Theorem 5.99 discussed in Proposition 5.102 to $(u(t, \mathcal{L}(X_t^{\boldsymbol{\alpha}})))_{0 \leqslant t \leqslant T}$. We get:

$$
\begin{aligned}
&du\big(t, \mathcal{L}(X_t^{\alpha})\big) \\
&= \Big[\partial_t u\big(t, \mathcal{L}(X_t^{\alpha})\big) + \mathbb{E}\Big[\partial_\mu u\big(t, \mathcal{L}(X_t^{\alpha})\big)(X_t^{\alpha}) \cdot \alpha_t\Big] \\
&\qquad\qquad + \frac{1}{2}\mathbb{E}\Big[\text{trace}\big[\partial_v \partial_\mu u\big(t, \mathcal{L}(X_t^{\alpha})\big)(X_t^{\alpha})\big]\Big]\Big] dt \\
&= \Big[-f\big(\mathcal{L}(X_t^{\alpha})\big) + \frac{1}{2}\mathbb{E}\Big[\big|\partial_\mu u\big(t, \mathcal{L}(X_t^{\alpha})\big)(X_t^{\alpha})\big|^2\Big] + \mathbb{E}\Big[\partial_\mu u\big(t, \mathcal{L}(X_t^{\alpha})\big)(X_t^{\alpha}) \cdot \alpha_t\Big]\Big] dt \\
&= \Big[-f\big(\mathcal{L}(X_t^{\alpha})\big) - \frac{1}{2}\mathbb{E}\big[|\alpha_t|^2\big] + \frac{1}{2}\mathbb{E}\Big[\big|\alpha_t + \partial_\mu u\big(t, \mathcal{L}(X_t^{\alpha})\big)(X_t^{\alpha})\big|^2\Big]\Big] dt
\end{aligned}
$$

where we used the PDE (5.121) satisfied by u before identifying a *perfect square*. If we integrate both sides and use the definition of the cost $J(\boldsymbol{\alpha})$, we get:

$$
J(\boldsymbol{\alpha}) = u\big(0, \mathcal{L}(X_0)\big) + \frac{1}{2}\mathbb{E}\Big[\int_0^T \Big[\big|\alpha_t + \partial_\mu u\big(t, \mathcal{L}(X_t^{\alpha})\big)(X_t^{\alpha})\big|^2\Big] dt\Big],
$$

which shows that $(\hat{X}_t)_{0 \le t \le T}$ is the unique optimal path. □

Remark 5.109 *Equation* (5.121) *is a simple form of the master equation for the optimal control of McKean-Vlasov dynamics which we shall derive in Chapter 6.*

Remark 5.110 *Benamou and Brenier's Theorem 5.53 provides a first variational formula for the* 2*-Wasserstein distance* W_2*. We shall show in Subsection 6.7.3 that the* 2*-Wasserstein distance* W_2 *can be viewed (up to a slight modification) as the solution of an optimization problem of the type considered in this section.*

5.7.4 Application to McKean-Vlasov SDEs

As another application of the chain rule for the flow of marginals measures of an Itô process, we revisit the propagation of chaos for McKean-Vlasov processes.

Semi-group Generated by a McKean-Vlasov SDE

Given a complete probability space $(\Omega, \mathcal{F}, \mathbb{P})$, equipped with a filtration $\mathbb{F} = (\mathcal{F}_t)_{t \geqslant 0}$ satisfying the usual assumptions, and with an $\mathbb{R}^d$-valued $\mathbb{F}$-Brownian motion $\boldsymbol{W} = (W_t)_{t \geqslant 0}$, we consider the McKean-Vlasov SDE:

$$
dX_t = b\big(t, X_t, \mathcal{L}(X_t)\big)dt + \sigma\big(t, X_t, \mathcal{L}(X_t)\big)dW_t, \quad t \geqslant 0, \tag{5.123}
$$

where the coefficients:

$$
(b, \sigma) : [0, T] \times \mathbb{R}^d \times \mathcal{P}_2(\mathbb{R}^d) \to \mathbb{R}^d \times \mathbb{R}^{d \times d}
$$

are (jointly) continuous and satisfy the Cauchy-Lipschitz assumptions:

$$|b(t,x,\mu)| \leqslant c\big(1+|x|+M_2(\mu)\big), \quad |\sigma(t,x,\mu)| \leqslant c\big(1+M_2(\mu)\big),$$
$$|b(t,x,\mu)-b(t,x',\mu')|+|\sigma(t,x,\mu)-\sigma(t,x',\mu')| \leqslant c\big(|x-x'|+W_2(\mu,\mu')\big),$$

for all $t \in [0,T]$, $x,x' \in \mathbb{R}^d$ and $\mu,\mu' \in \mathcal{P}_2(\mathbb{R}^d)$, for some constant $c \geqslant 0$. We assume that $(\Omega,\mathcal{F}_0,\mathbb{P})$ is atomless so that, for any $\mu \in \mathcal{P}_2(\mathbb{R}^d)$, there exists a random variable $X_0 \in L^2(\Omega,\mathcal{F}_0,\mathbb{P};\mathbb{R}^d)$ such that $X_0 \sim \mu$.

Under these Cauchy-Lipschitz assumptions, Theorem 4.21 implies that, for any initial random variable $X_0 \in L^2(\Omega,\mathcal{F}_0,\mathbb{P};\mathbb{R}^d)$, the McKean-Vlasov SDE (5.123) is uniquely solvable. We emphasize that uniqueness must also hold in law. Indeed, for any two solutions constructed on possibly different spaces, the standard Yamada-Watanabe theorem permits to construct on the same probability space, two new solutions, driven by the same initial condition and by the same random noise, each one being distributed according to one of the two original laws. These two new solutions also satisfy the McKean-Vlasov SDE, but on the same space. Consequently, they must be equal. In particular, for any $t \geqslant 0$, the law of X_t only depends upon the initial distribution of X_0. This makes it possible to define, for any $t \geqslant 0$ and any function $\phi : \mathcal{P}_2(\mathbb{R}^d) \to \mathbb{R}$, the function $\mathscr{P}_t\phi$ by:

$$\big[\mathscr{P}_t\phi\big](\mu) = \phi\big(\mathcal{L}(X_t^{X_0})\big), \qquad \mu \in \mathcal{P}_2(\mathbb{R}^d).$$

Here we denote by $X_t^{X_0}$ the solution at time t of (5.123) with initial condition X_0, and we assume that X_0 has distribution μ.

As a side effect of the proof of existence and uniqueness of a solution of the McKean-Vlasov SDE (5.123), we get that, for any time $T > 0$, there exists a constant $C \geqslant 0$, such that, for any $X_0, X_0' \in L^2(\Omega,\mathcal{F}_0,\mathbb{P};\mathbb{R}^d)$,

$$\mathbb{E}\big[\sup_{0\leqslant t\leqslant T} |X_t^{X_0} - X_t^{X_0'}|^2\big] \leqslant C^2\mathbb{E}\big[|X_0 - X_0'|^2\big].$$

This implies that, for any $t \geqslant 0$, $\mathscr{P}_t\phi : \mathcal{P}_2(\mathbb{R}^d) \to \mathbb{R}$ is bounded and continuous whenever ϕ is bounded and continuous for the 2-Wasserstein distance W_2. It also shows that $\mathscr{P}_t\phi$ is Lipschitz continuous for the same Wasserstein distance W_2 whenever ϕ is Lipschitz continuous, and more generally, that $\mathscr{P}_t\phi$ is bounded and uniformly continuous whenever ϕ is bounded and uniformly continuous. In particular, whenever b and σ are time-independent, uniqueness implies that $(\mathscr{P}_t)_{t\geqslant 0}$ is a one-parameter semigroup of operators on the Banach space $C_b(\mathcal{P}_2(\mathbb{R}^d);\mathbb{R})$ of bounded continuous functions on $\mathcal{P}_2(\mathbb{R}^d)$ (equipped with the 2-Wasserstein distance). It is also a one-parameter semigroup of operators on the Banach space $\mathcal{UC}_b(\mathcal{P}_2(\mathbb{R}^d);\mathbb{R})$ of bounded uniformly continuous functions on $\mathcal{P}_2(\mathbb{R}^d)$ and on the Banach space $\mathcal{UC}_0(\mathcal{P}_2(\mathbb{R}^d);\mathbb{R})$ of bounded uniformly continuous functions ϕ on $\mathcal{P}_2(\mathbb{R}^d)$ such that $\phi(\mu) \to 0$ as $M_2(\mu) \to \infty$. The last claim follows from the fact that, for any $t > 0$, there exists a constant $C_t > 0$ such that, for any $X_0 \in L^2(\Omega, \mathcal{F}_0, \mathbb{P}; \mathbb{R}^d)$, $\mathbb{E}[|X_t^{X_0}|^2] \geqslant \exp(-C_t)(\mathbb{E}[|X_0|^2] - C_t)$.

Notice that, for any $\phi \in \mathcal{UC}_0(\mathcal{P}_2(\mathbb{R}^d);\mathbb{R})$, $\sup_{\mu \in \mathcal{P}_2(\mathbb{R}^d)} |\mathscr{P}_t\phi(\mu) - \phi(\mu)|$ tends to 0 as t tends to 0, proving that, whenever b and σ are time-homogeneous, the semi-group $(\mathscr{P}_t)_{t\geqslant 0}$ is strongly continuous on $\mathcal{UC}_0(\mathcal{P}_2(\mathbb{R}^d);\mathbb{R})$. Indeed, there exists a constant $C > 0$ such that, for any $X_0 \in L^2(\Omega, \mathcal{F}, \mathbb{P}; \mathbb{R}^d)$ and any $t \in [0, 1]$:

$$\mathbb{E}\big[|X_t^{X_0} - X_0|^2\big] \leqslant Ct\big(1 + \mathbb{E}\big[|X_0|^2\big]\big),$$
$$\mathbb{E}[|X_t^{X_0}|^2] \geqslant \exp(-C)\big(\mathbb{E}[|X_0|^2] - C\big).$$

Therefore, for a given $\phi \in \mathcal{UC}_0(\mathcal{P}_2(\mathbb{R}^d);\mathbb{R})$, we have, for any initial condition $X_0 \in L^2(\Omega, \mathcal{F}_0, \mathbb{P}; \mathbb{R}^d)$ and for any $R > 0$,

$$\begin{aligned}
\Big|\phi\big(\mathcal{L}(X_t^{X_0})\big) - \phi\big(\mathcal{L}(X_0)\big)\Big| &= \Big|\phi\big(\mathcal{L}(X_t^{X_0})\big) - \phi\big(\mathcal{L}(X_0)\big)\Big| \mathbf{1}_{\{\mathbb{E}[|X_0|^2] \leqslant R^2\}} \\
&\quad + \Big|\phi\big(\mathcal{L}(X_t^{X_0})\big) - \phi\big(\mathcal{L}(X_0)\big)\Big| \mathbf{1}_{\{\mathbb{E}[|X_0|^2] > R^2\}} \\
&\leqslant \sup_{W_2(\mu,\nu) \leqslant Ct(1+R)^2} \big|\phi(\mu) - \phi(\nu)\big| + \sup_{M_2(\mu) \geqslant R} \phi(\mu) \\
&\quad + \sup_{M_2(\mu)^2 \geqslant \exp(-C)R^2 - C} \phi(\mu).
\end{aligned}$$

Choosing R large enough first, and then t small enough, we complete the proof of the strong continuity of the semi-group.

Generator of the Semi-group

Now let us assume that $\phi : \mathcal{P}_2(\mathbb{R}^d) \to \mathbb{R}$ is partially $\mathcal{C}^2$ and satisfy, for any compact subset $\mathcal{K} \subset \mathcal{P}_2(\mathbb{R}^d)$,

$$\sup_{\mu \in \mathcal{K}} \bigg[\int_{\mathbb{R}^d} \big|\partial_v \partial_\mu \phi(\mu)(v)\big|^2 d\mu(v) \bigg] < \infty.$$

Then, for any initial condition X_0 for the McKean-Vlasov SDE (5.123), we may expand the function $\mathbb{R}_+ \ni t \mapsto \big(\mathscr{P}_t\phi\big)(\mu)$ with $\mathcal{L}(X_0) = \mu$ using the chain rule. From the bound $\mathbb{E}[\sup_{0\leqslant t\leqslant T} |X_t^{X_0}|^2] < \infty$, which holds true for any $T > 0$, and from the growth conditions on σ and b we see that:

$$\sup_{0\leqslant t\leqslant T} \mathbb{E}\Big[\big|b\big(t, X_t^{X_0}, \mathcal{L}(X_t^{X_0})\big)\big|^2 + \big|\sigma\big(t, X_t^{X_0}, \mathcal{L}(X_t^{X_0})\big)\big|^4 \Big] < \infty,$$

where we used the fact that σ is bounded in x. Therefore, Theorem 5.99 implies that for all $t \geqslant 0$:

$$\begin{aligned}\frac{d}{dt}\big[(\mathscr{P}_t\phi)(\mu)\big] &= \frac{d}{dt}\big[\phi(\mu_t)\big]\\ &= \int_{\mathbb{R}^d} b(t,v,\mu_t)\cdot\partial_\mu\phi(\mu_t)(v)d\mu_t(v)\\ &\quad+\frac{1}{2}\int_{\mathbb{R}^d}\mathrm{trace}\big[\big(\sigma\sigma^\dagger\big)(t,v,\mu_t)\partial_v\partial_\mu\phi(\mu_t)(v)\big]d\mu_t(v),\end{aligned}$$

with $\mu_t = \mathcal{L}(X_t^{X_0})$. Actually, Theorem 5.99 just provides the right-differentiability of the function $\mathbb{R}_+ \ni t \mapsto \mathscr{P}_t\phi)(\mu)$. Then, differentiability follows from the fact that $\mathbb{R}_+ \ni t \mapsto \mu_t \in \mathcal{P}_2(\mathbb{R}^d)$ is continuous and that, by the same argument as in the first step of the proof of Proposition 5.102, the above right-hand side is also continuous in time. In particular, if we set:

$$\begin{aligned}\big[\mathscr{L}_t\phi\big](\mu) &= \int_{\mathbb{R}^d} b(t,v,\mu)\cdot\partial_\mu\phi(\mu)(v)d\mu(v)\\ &\quad+\frac{1}{2}\int_{\mathbb{R}^d}\mathrm{trace}\big[\big(\sigma\sigma^\dagger\big)(t,v,\mu)\partial_v\partial_\mu\phi(\mu)(v)\big]d\mu(v),\end{aligned} \tag{5.124}$$

for $\mu \in \mathcal{P}_2(\mathbb{R}^d)$, we get:

$$\frac{d}{dt}\big[(\mathscr{P}_t\phi)(\mu)\big] = \big[\mathscr{P}_t\big(\mathscr{L}_t\phi\big)\big](\mu), \qquad \mu\in\mathcal{P}_2(\mathbb{R}^d), \tag{5.125}$$

and thus:

$$\frac{d}{dt}\big[(\mathscr{P}_t\phi)(\mu)\big]_{|t=0} = \big(\mathscr{L}_0\phi\big)(\mu), \qquad \mu\in\mathcal{P}_2(\mathbb{R}^d). \tag{5.126}$$

Whenever b and σ do not depend upon time, $\mathscr{L}_t$ is also independent of t, and will be denoted by $\mathscr{L}$. The above application of the chain rule then says that the domain of the infinitesimal generator of the strongly continuous semi-group $(\mathscr{P}_t)_{t\geqslant 0}$ on $\mathcal{UC}_0(\mathcal{P}_2(\mathbb{R}^d))$ contains the intersection of $\mathcal{UC}_0(\mathcal{P}_2(\mathbb{R}^d))$ with the space of partially $\mathcal{C}^2$ functions, intersection on which this generator coincides with $\mathscr{L}$. As a result, identity (5.125) can be interpreted as a forward Kolmogorov equation on $\mathcal{P}_2(\mathbb{R}^d)$. Observe also that the intersection of $\mathcal{UC}_0(\mathcal{P}_2(\mathbb{R}^d))$ and the space of partially $\mathcal{C}^2$ functions is not empty. Indeed, for any compactly supported smooth function $\rho : \mathbb{R}_+ \to \mathbb{R}$, the function $\phi : \mathcal{P}_2(\mathbb{R}^d) \ni \mu \mapsto \rho(M_2(\mu)^2)$ is in this intersection, with $\partial_\mu\phi(\mu)(v) = 2v\rho'(M_2(\mu)^2)$ and $\partial_v\partial_\mu\phi(\mu)(v) = 2\rho'(M_2(\mu)^2)$. Any multiplication of ϕ with a function of the same type as those described in the first step of the proof of Theorem 5.99 also belongs to the intersection of $\mathcal{UC}_0(\mathbb{R}^d)$ with the space of partially $\mathcal{C}^2$ functions.

Backward Equation

Assume now that there exists a subspace $\mathscr{C}$ of bounded smooth functions $\phi : \mathcal{P}_2(\mathbb{R}^d) \to \mathbb{R}$ for which the mapping:

$$\Phi : \mathbb{R}_+ \times \mathcal{P}_2(\mathbb{R}^d) \ni (t, \mu) \mapsto \phi(\mathcal{L}(X_T^{T-t,\xi\sim\mu})) \tag{5.127}$$

where $X_T^{T-t,\xi\sim\mu}$ is the value of the solution at time T starting from a random variable $\xi \sim \mu$ at time $T - t$, satisfies assumption **Joint Chain Rule** together with (5.106). Notice that we shall only need the simplest form of the chain rule since Φ is independent of the space variable x.

Then, for any initial condition $X_0 \in L^2(\Omega, \mathcal{F}_0, \mathbb{P}; \mathbb{R}^d)$ of the McKean-Vlasov SDE (5.123), we get from the extended chain rule Proposition 5.102 that, for any $T > 0$ and $t \in [0, T]$,

$$\begin{aligned} &\frac{d}{dt}\big[\Phi(T-t,\mu_t)\big] \\ &= -\partial_t \Phi(T-t,\mu_t) + \int_{\mathbb{R}^d} b(t,v,\mu_t)\cdot \partial_\mu \Phi(T-t,\mu_t)(v)d\mu_t(v) \\ &\quad + \frac{1}{2}\int_{\mathbb{R}^d} \text{trace}\big[(\sigma\sigma^\dagger)(t,v,\mu_t)\partial_v \partial_\mu \Phi(T-t,\mu_t)(v)\big]d\mu_t(v), \end{aligned} \tag{5.128}$$

where we used $\mu_t = \mathcal{L}(X_t^{X_0})$. Observe that the left-hand side must be zero since, for any $t \in [0, T]$,

$$\Phi(T-t,\mu_t) = \phi\big(\mathcal{L}(X_T^{t,\xi\sim\mu_t})\big) = \phi\big(\mathcal{L}(X_T^{X_0})\big). \tag{5.129}$$

As a result, the right-hand side in (5.128) must be zero. Noticing that the dynamics of the McKean-Vlasov SDE may be initialized from any $\mu \in \mathcal{P}_2(\mathbb{R}^d)$ at any time $t \in [0, T]$, we get the backward equation:

$$\begin{aligned} &\partial_t\big[\Phi(T-t,\mu)\big] + \int_{\mathbb{R}^d} b(t,v,\mu)\cdot \partial_\mu \Phi(T-t,\mu)(v)d\mu(v) \\ &\quad + \frac{1}{2}\int_{\mathbb{R}^d} \text{trace}\big[(\sigma\sigma^\dagger)(t,v,\mu)\partial_v \partial_\mu \Phi(T-t,\mu)(v)\big]d\mu(v) = 0, \end{aligned} \tag{5.130}$$

with the terminal condition $\Phi(T-t,\mu)|_{t=T} = \phi(\mu)$, for $\mu \in \mathcal{P}_2(\mathbb{R}^d)$. Observe that the above PDE reads as a simplified (linear) version of the (nonlinear) master equation (5.117).

Propagation of Chaos Revisited

We now consider the standard particle system approximating the McKean-Vlasov SDE (5.123), namely:

$$dX_t^i = b\big(t, X_t^i, \bar{\mu}_t^N\big)dt + \sigma\big(t, X_t^i, \bar{\mu}_t^N\big)dW_t^i, \qquad i = 1, \cdots, N$$

for some integer $N \geqslant 1$, where $(X_0^i)_{i=1,\cdots,N}$ are independent $\mathcal{F}_0$-measurable random variables with the same distribution as some $X_0 \in L^2(\Omega, \mathcal{F}_0, \mathbb{P}; \mathbb{R}^d)$ and $(\boldsymbol{W}^i)_{i=1,\cdots,N}$ are independent Wiener processes of dimension d. Without any loss of generality, we can work on the same probability space as above. As before, we use the empirical measure:

$$\bar{\mu}_t^N = \frac{1}{N}\sum_{i=1}^N \delta_{X_t^i}, \quad t \in [0, T].$$

Assume, in addition to the previous assumptions, that σ is bounded and that, for any function ϕ in the class $\mathscr{C}$, the function Φ also satisfies:

1. the functions $[0,T] \times \mathcal{P}_2(\mathbb{R}^d) \times \mathbb{R}^d \ni (t,\mu,v) \mapsto \partial_\mu \Phi(t,\mu)(v)$ and $[0,T] \times \mathcal{P}_2(\mathbb{R}^d) \times \mathbb{R}^d \ni (t,\mu,v) \mapsto \partial_v[\partial_\mu \Phi(t,\mu)](v)$ are continuous;
2. the function $\mathcal{P}_2(\mathbb{R}^d) \ni \mu \mapsto \partial_\mu \Phi(t,\mu)(v)$ is L-differentiable for any $(t,v) \in [0,T] \times \mathbb{R}^d$, the mapping:

$$[0,T] \times \mathcal{P}_2(\mathbb{R}^d) \times \mathbb{R}^d \times \mathbb{R}^d \ni (t,\mu,v,v') \mapsto \partial_\mu^2 \Phi(t,\mu)(v,v')$$

   being continuous;
3. it holds that:

$$\sup_{t\in[0,T]} \sup_{\mu \in \mathcal{P}_2(\mathbb{R}^d)} \left[\int_{\mathbb{R}^d} \left(\left|\partial_\mu \Phi(t,\mu)(v)\right|^2 + \left|\partial_\mu^2 \Phi(t,\mu)(v,v)\right|^2 \right) d\mu(v) \right] < \infty.$$

Notice in particular that, for any $t \in [0,T]$, the function $\Phi(t,\cdot)$ is fully $\mathcal{C}^2$.

Then, Propositions 5.35 and 5.91 imply that the empirical projection function:

$$[0,T] \times (\mathbb{R}^d)^N \ni \big(t,(x^1,\cdots,x^N)\big) \mapsto \Phi\Big(T-t, \frac{1}{N}\sum_{i=1}^N \delta_{x^i}\Big)$$

is of class $\mathcal{C}^{1,2}$ with specific partial derivatives, so that we can apply standard Itô's formula. We get:

$$\begin{aligned}
&d\, \Phi\big(T-t, \bar{\mu}_t^N\big) \\
&= -\partial_t \Phi\big(T-t,\cdot\big)\big(\bar{\mu}_t^N\big)dt \\
&\quad + \frac{1}{N}\sum_{i=1}^N b\big(t, X_t^i, \bar{\mu}_t^N\big) \cdot \partial_\mu \Phi(T-t, \bar{\mu}_t^N)(X_t^i)dt \\
&\quad + \frac{1}{N}\sum_{i=1}^N \partial_\mu \Phi(T-t, \bar{\mu}_t^N)(X_t^i) \cdot \Big(\sigma\big(t, X_t^i, \bar{\mu}_t^N\big)dW_t^i\Big)
\end{aligned}$$

$$+\frac{1}{2N}\sum_{i=1}^{N}\mathrm{trace}\Big[\big(\sigma\sigma^\dagger\big)\big(t,X_t^i,\bar\mu_t^N\big)\partial_v\partial_\mu\Phi(T-t,\bar\mu_t^N)(X_t^i)\Big]dt$$

$$+\frac{1}{2N^2}\sum_{i=1}^{N}\mathrm{trace}\Big[\big(\sigma\sigma^\dagger\big)\big(t,X_t^i,\bar\mu_t^N\big)\partial_\mu^2\Phi(T-t,\bar\mu_t^N)(X_t^i,X_t^i)\Big]dt.$$

Notice that the second and fourth terms in the right-hand side can be rewritten as:

$$\frac{1}{N}\sum_{i=1}^{N}b\big(t,X_t^i,\bar\mu_t^N\big)\cdot\partial_\mu\Phi(T-t,\bar\mu_t^N)(X_t^i)$$
$$=\int_{\mathbb{R}^d}b\big(t,v,\bar\mu_t^N\big)\cdot\partial_\mu\Phi(T-t,\bar\mu_t^N)(v)d\bar\mu_t^N(v),$$

$$\frac{1}{2N}\sum_{i=1}^{N}\mathrm{trace}\Big[\big(\sigma\sigma^\dagger\big)\big(t,X_t^i,\bar\mu_t^N\big)\partial_v\partial_\mu\Phi(T-t,\bar\mu_t^N)(X_t^i)\Big]$$
$$=\frac{1}{2}\int_{\mathbb{R}^d}\mathrm{trace}\Big[\big(\sigma\sigma^\dagger\big)\big(t,v,\bar\mu_t^N\big)\partial_v\partial_\mu\Phi(T-t,\bar\mu_t^N)(v)\Big]d\bar\mu_t^N(v),$$

so that, using the PDE (5.130) satisfied by Φ at $(t,\bar\mu_t^N)$, and as before, the notation $\mu_t=\mathcal{L}(X_t^{X_0})$, we get:

$$\begin{aligned}
&d\big[\Phi\big(T-t,\bar\mu_t^N\big)-\Phi\big(T-t,\mu_t\big)\big]\\
&=d\,\Phi\big(T-t,\bar\mu_t^N\big)\\
&=\frac{1}{N}\sum_{i=1}^{N}\sigma\big(t,X_t^i,\bar\mu_t^N\big)\cdot\Big(\partial_\mu\Phi(T-t,\bar\mu_t^N)(X_t^i)dW_t^i\Big)\\
&\quad+\frac{1}{2N^2}\sum_{i=1}^{N}\mathrm{trace}\Big[\big(\sigma\sigma^\dagger\big)\big(t,X_t^i,\bar\mu_t^N\big)\partial_\mu^2\Phi(T-t,\bar\mu_t^N)(X_t^i,X_t^i)\Big]dt,
\end{aligned}\tag{5.131}$$

where we used the fact that $\Phi(T-t,\mu_t)$ is constant in t, see (5.129). By assumption 3 above, there exists a constant C such that:

$$\mathbb{E}\Big[\sup_{0\leqslant t\leqslant T}\big|\Phi\big(T-t,\bar\mu_t^N\big)-\Phi\big(T-t,\mu_t\big)\big|^2\Big]\leqslant\frac{C}{N}+C\mathbb{E}\Big[\big|\Phi\big(T,\bar\mu_0^N\big)-\Phi\big(T,\mu_0\big)\big|^2\Big].$$

Since $\Phi(T,\cdot)$ is continuous and bounded, the right-hand side tends to 0. Choosing $t=T$ in the left-hand side, we get:

$$\forall\phi\in\mathcal{C},\quad\lim_{N\to\infty}\mathbb{E}\Big[\big|\phi\big(\bar\mu_T^N\big)-\phi\big(\mu_T\big)\big|^2\Big]=0,\tag{5.132}$$

which is another form of the propagation of chaos, at least when the class $\mathcal{C}$ is rich enough.

Remark 5.111 *We can specify the rate of convergence in* (5.132). *Basically, we get, under the standing assumption on ϕ, and on $\Phi(T,\cdot)$, the latter being Lipschitz continuous in μ thanks to assumption 3, that:*

$$\mathbb{E}\Big[\big|\phi(\bar{\mu}_T^N) - \phi(\mu_T)\big|^2\Big]^{1/2} \leqslant \frac{C}{\sqrt{N}} + \mathbb{E}\big[W_2(\bar{\mu}_0^N, \mu_0)^2\big]^{1/2},$$

the normalization by the root of N in the right-hand side being reminiscent of that appearing in the statement of the central limit theorem whilst the last term may be estimated by means of Theorem 5.8.

Another way of quantifying the convergence is to take the expectation in (5.131). *Under the same assumption as above, we get:*

$$\big|\mathbb{E}[\phi(\bar{\mu}_T^N)] - \phi(\mu_T)\big| \leqslant \frac{C}{N} + \big|\mathbb{E}[\Phi(T,\bar{\mu}_0^N)] - \Phi(T,\mu_0)\big|, \tag{5.133}$$

which provides a bound on the rate of convergence of the semi-group generated by the system of particles towards the limiting McKean-Vlasov flow. Obviously, the order of the last term should depend on the smoothness of $\Phi(T,\cdot)$.

Back to the Chain Rule

We conclude this section with a remark regarding the statement of the chain rule. Indeed, the family of differential operators $(\mathscr{L}_t)_{t\geqslant 0}$ introduced in (5.124) allows for a condensed form of the chain rule stated in Proposition 5.102. We state it here for the sake of later reference.

Using the same kind of notation as in the statement of Proposition 5.102, assume that $(\xi_t)_{0\leqslant t\leqslant T}$ is the solution of another d-dimensional stochastic differential equation:

$$d\xi_t = \eta(t,\xi_t)dt + \gamma(t,\xi_t)dW_t, \quad t\in[0,T]\,; \quad \xi_0 \in L^2(\Omega,\mathcal{F}_0,\mathbb{P};\mathbb{R}^d), \tag{5.134}$$

where $\eta : \mathbb{R}_+ \times \mathbb{R}^d \to \mathbb{R}^d$ and $\gamma : \mathbb{R}_+ \times \mathbb{R}^d \to \mathbb{R}^{d\times d}$ are Borel functions satisfying the usual growth and Lipschitz continuity conditions which guarantee existence and uniqueness of the solutions for the above equation. As we already did several times in the book, to the diffusion equation (5.134), we associate the family of differential operators $(L_t)_{t\geqslant 0}$ serving as infinitesimal generators. They act on twice continuously differentiable functions φ on $\mathbb{R}^d$ in the following way:

$$L_t\varphi : \mathbb{R}^d \ni x \mapsto [L_t\varphi](x) = \eta(t,x)\cdot\partial\varphi(x) + \frac{1}{2}\mathrm{trace}\Big[\big(\gamma_t\gamma_t^\dagger\big)(t,x)\partial^2\varphi(x)\Big], \tag{5.135}$$

for $t \in [0, T]$. With these notations and definition (5.124) for $(\mathscr{L}_t)_{t\geq 0}$, the chain rule (5.107) now reads:

$$
\begin{aligned}
d\, u(t, \xi_t, \mu_t) &= \Big[\partial_t u(t, \xi_t, \mu_t) + \big[L_{t,x} u\big](t, \xi_t, \mu_t) + \big[\mathscr{L}_{t,\mu} u\big](t, \xi_t, \mu_t) \Big]\, dt \\
&\quad + \partial_x u(t, \xi_t, \mu_t) \cdot \big(\gamma_t dW_t\big), \qquad t \in [0, T],
\end{aligned}
$$

where we used the convenient conventions:

$$
[L_{t,x} u](t, x, \mu) = [L_t u(t, \cdot, \mu)](x), \quad \text{and} \quad [\mathscr{L}_{t,\mu} u](t, x, \mu) = [\mathscr{L}_t u(t, x, \cdot)](\mu).
$$

5.8 Notes & Complements

The Lévy-Prokhorov distance d_{LP} between probability measures on a separable metric space was first introduced by Prokhorov in 1956 as a generalization to metric spaces of a distance introduced by Lévy for probability measures on the real line. This distance is not easy to compute in general. It is theoretically important because it provides a metric for the weak convergence of probability measures. The fact that $d_{\mathrm{LP}}(\mu, \nu)$ is precisely the minimum distance in probability between random variables with distributions μ and ν respectively was proven by Strassen for complete separable metric spaces and extended by Dudley to arbitrary separable metric spaces, see Theorems 11.6.2 and 11.6.3 in Dudley's book [143]. The equality between the Wasserstein distance W_1 and the Kantorovich-Rubinstein distance d_{KR} is known as the Kantorovich-Rubinstein theorem.

For more properties of these distances, as well as detailed references and a better historical perspective, we refer the interested reader to Dudley's book [143] and Villani's monograph [338]. The proof of Proposition 5.7 may be found in Bertsekas and Shreve [56, Chapter 7].

Throughout the book, we need to control the distance between the empirical distribution of a sample of independent identically distributed (i.i.d. for short) random variables, and the common distribution of the variables in the sample. This is painfully apparent, not only in this chapter, but in many other instances throughout the text. This time honored problem is central both in probability, statistics and theoretical computer science. For us, relevant applications include Monte Carlo methods and estimates for particle systems, and approximations of partial differential equations. While many distances can be used, the Wasserstein distance is natural for particle approximations of PDEs. As for the moment estimates of Theorem 5.8, preliminary results by Horowitz and Kandarikar can be found in [198], and in book form in the text of Rachev and Rüschendorf [317]. While far from optimal, the form of these results could have been sufficient for the purpose of the book. However, we chose to give a sharper version in a self-contained presentation. We followed the idea of Dereich, Scheutzow, and Schottstedt [135] as presented in the recent work of Fournier and Guillin [161]. The upper bounds provided in the text for the rate of convergence of the empirical measures toward the true distribution are essentially optimal. Indeed, the lower bound proved in [36] by Barthe and Bordenave shows that the rate $N^{-1/d}$ cannot be improved for the 1-Wasserstein distance W_1 in dimension $d \geqslant 3$.

There are many notions of differentiability for functions defined on spaces of probability measures, and recent progress in the theory of optimal transportation have put some of them in the limelight. We refer the interested reader to the books of Ambrosio, Gigli and Savaré [21] and Villani [337, 338] for detailed exposés of these geometric approaches in textbook form. The notion of differentiability used in the text was introduced by P.L. Lions in his lectures at the *Collège de France* [265], hence our terminology of L-differentiability. Our presentation benefited from Cardaliaguet's lucid and readable account [83]. In particular, the statements and the strategies of the proofs of Propositions 5.24 and 5.25 are borrowed from [83]. The idea behind the connection between the L-derivative and the linear functional derivative is from the paper by Cardaliaguet, Delarue, Lasry, and Lions [86]. Proposition 5.51 was already proven in that paper, though using a different strategy. As for the various notions of differentiability in normed spaces used in the text, we refer to any textbook on analysis and differential calculus, see for instance [308].

Subsections 5.3.1, 5.3.2, and 5.3.3 are essentially borrowed from the paper by Carmona and Delarue [98]. The discussion of the Blackwell and Dubins' theorem is inspired from the original note by Blackwell and Dubins [63]. The connection between monotonicity and convexity, as exposed in Subsection 5.5.2, was discussed in Lions' lectures at the *Collège de France* [265].

The reader interested in the theory of optimal transportation can complement the results presented in this chapter, including Brenier's theorem whose original proof can be found in [69], and the transport along vector fields like in the Benamou and Brenier's theorem, with Villani's book [337], and Ambrosio, Gigli and Savaré textbook [21].

The sobering counter-example showing the lack of differentiability and convexity of the square of the Wasserstein distance to a fixed measure is borrowed from the book [21] by Ambrosio, Gigli, and Savaré. The remaining discussion about the Wasserstein gradients is modeled after this book. Theorem 5.53 is a version of a famous result by Benamou and Brenier [41] and Otto [295]. Once again, the statement is taken from [21], but the proof given in the text, which requires the probability measures to be absolutely continuous, follows the arguments used in Villani's monograph [337]. The properties of the inf-sup convolution, as used in the proof of Lemma 5.61, may be found in the paper by Lasry and Lions [259].

The notions of full and partial $\mathcal{C}^2$ regularity presented in Section 5.6 are taken from the papers by Buckdahn, Li, Peng, and Rainer [79] and by Chassagneux, Crisan, and Delarue [114]. The various versions of Itô's formula are taken from the same work. The sufficient condition for ensuring the partial $\mathcal{C}^2$ regularity can also be found in [114]. We shall use it in Chapter (Vol II)-5 in order to prove the existence of a classical solution to the master equation.

The verification argument for the master equation established in Subsection 5.7.2 is inspired by a similar result for classical FBSDEs which can be found in the paper [271] by Ma, Protter, and Young, and from the introductory paper of Carmona and Delarue [97] on the master equation. We shall provide an in-depth analysis of the master equation in Chapters (Vol II)-4 and (Vol II)-5. The optimization problem introduced in Subsection 5.7.3 will be revisited in Chapter 6. We refer to the paper by Gangbo, Nguyen, and Tudorascu [167] for an analytic treatment of this example. The analysis of the semi-group generated by a McKean-Vlasov diffusion process on $\mathcal{C}_b(\mathcal{P}_2(\mathbb{R}^d); \mathbb{R}^d)$ as well as the corresponding propagation chaos were inspired by the analysis of the convergence of finite games toward mean field games, as provided in the paper by Cardaliaguet, Delarue, Lasry, and Lions. We address these questions of convergence in Chapter (Vol II)-6. Similar ideas to that exposed in Subsection 5.7.4 have been developed by Kolokoltsov for investigating propagation of chaos, see for instance the earlier article [233] together with the monograph [234]; in particular, the inequality (5.133) in Remark 5.111 plays a key role in [233, 234] and in the subsequent works by the same author on mean field games, see the Notes & Complements of Chapter (Vol II)-6 for precise references. Propagation of chaos for standard McKean-Vlasov SDEs will be revisited in Chapter (Vol II)-2.

Optimal Control of SDEs of McKean-Vlasov Type

6

Abstract
The purpose of this chapter is to provide a detailed probabilistic analysis of the optimal control of nonlinear stochastic dynamical systems of McKean-Vlasov type. We tackle the characterization and construction of solutions of this special type of optimal control problem by means of forward-backward stochastic differential equations. Because of the presence of the distribution of the controlled state in the coefficients, the approach based on the Pontryagin stochastic maximum principle requires special attention. We provide a version of this maximum principle based on the differential calculus for functions of probability measures introduced and developed in Chapter 5. We test the results of the analysis on linear quadratic models and a few other models already considered in the framework of mean field games. Finally, we highlight the similarities and the differences between this problem and MFG problems with which it is often confused.

6.1 Introduction

Stochastic Differential Equations (SDEs) of McKean-Vlasov type are usually referred to as *nonlinear* SDEs, the term nonlinear emphasizing the possible dependence of the coefficients upon the marginal distributions of the solutions. This terminology has no bearing on a possible nonlinear dependence of the coefficients of the equations upon the state variable. This special feature of the coefficients, even when the latter are nonrandom, creates nonlocal feedback effects which rule out the standard Markov property. Including the marginal distribution in the state of the system could restore the Markov property at the cost of a leap in complexity of the state of the process. The latter would have to include a probability measure, and subsequently become infinite dimensional. While the analysis of the infinitesimal

R. Carmona, F. Delarue, *Probabilistic Theory of Mean Field Games with Applications I*, Probability Theory and Stochastic Modelling 83,
DOI 10.1007/978-3-319-58920-6_6

generator could be done with tools developed for infinite dimensional differential operators, the standard differential calculus, even in infinite dimension, would have a hard time capturing the fact that the second component of the state process has to match the marginal distribution of the first component. The chain rules developed in Chapter 5 were introduced to handle these difficulties.

Because of the crucial role they play in the probabilistic approach to mean field games, existence and uniqueness results for forward, backward, and forward-backward stochastic differential equations of the McKean-Vlasov (MKV for short) type were given in Section 4.2 and Section 4.3 of Chapter 4. While most of the proofs benefited from an understanding of the metric structure of the spaces of probability measures, they did not require any form of differential calculus on these spaces. Here we concentrate on the optimal control of stochastic systems whose dynamics are given by equations of the McKean-Vlasov type. This is where the special differential calculus introduced in Chapter 5 comes handy. Strangely enough, the optimal control of dynamics driven by McKean-Vlasov SDEs seems to be a brand new problem, to a great degree ignored in the standard stochastic control literature. See nevertheless the Notes & Complements at the end of the chapter for exceptions.

As we saw in Chapter 4, solving a McKean-Vlasov SDE is done by a fixed point argument. First one fixes a set of candidates for the distribution of the solution, then one solves the resulting standard SDE, the fixed point argument being to demand that the distribution of the solution be equal to the distribution we started from. A stochastic control problem adds an extra optimization layer to the fixed point. This formulation bears a lot of resemblance to the approach to mean field game problems as formulated in Chapters 1 and 3, and it is of the utmost importance to understand the extent of the similarities and differences between the two problems.

SDEs of McKean-Vlasov type were introduced to describe the asymptotic behavior of a generic element of a large population of particles with *mean field* interactions. The adjective *large* underscores the fact that the analysis is intended to describe the asymptotic regime when the number of particles tends to infinity. In this asymptotic regime, particles become independent of each others, and the state of each single particle satisfies an SDE of McKean-Vlasov type. Such a phenomenon is usually referred to as *propagation of chaos*. We already alluded to it in the applications of Chapter 5 and we shall revisit it in a more detailed fashion in Chapter (Vol II)-1. To see the relevance of this theory to the models of stochastic differential games where players interact in a mean field way, we assume for the sake of definiteness that the private states $\boldsymbol{X}^{N,i} = (X_t^{N,i})_{0 \leqslant t \leqslant T}$ of N players satisfy the system of SDEs:

$$dX_t^{N,i} = b\big(t, X_t^{N,i}, \bar{\mu}^N_{X_t^N}, \alpha_t^i\big)dt + \sigma\big(t, X_t^{N,i}, \bar{\mu}^N_{X_t^N}, \alpha_t^i\big)dW_t^i, \quad t \in [0,T], \tag{6.1}$$

for some time horizon $T > 0$ and with a common (deterministic) initial condition. Such a model is similar to those introduced in Chapter 2. However, differently from Chapter 2, we assume that all the players use distributed controls

$(\alpha_t^i = \phi(t, X_t^{N,i}))_{0 \leqslant t \leqslant T}$ given by the same feedback function ϕ in order to minimize an expected cost from running and terminal costs:

$$J^i(\phi) = \mathbb{E}\Big[\int_0^T f\big(t, X_t^{N,i}, \bar{\mu}^N_{X_t^N}, \alpha_t^i\big) dt + g\big(X_T^{N,i}, \bar{\mu}^N_{X_T^N}\big) \Big]. \tag{6.2}$$

Actually, $J^i(\phi)$ is in fact independent of i and is thus common to all the players since all of them use the same feedback function.

In contrast with our approach to MFG problems, instead of optimizing over the control right away, we assume that the common feedback function ϕ is momentarily kept fixed, and we first consider the large population limit. The theory of propagation of chaos states that, in the limit $N \to \infty$, for any fixed integer k, the joint distribution of the k-dimensional process $(X_t^{N,1}, \cdots, X_t^{N,k})_{0 \leqslant t \leqslant T}$ converges to a product distribution (in other words the k processes $\boldsymbol{X}^{N,i} = (X_t^{N,i})_{0 \leqslant t \leqslant T}$ for $i = 1, \cdots, k$ become independent in the limit) and the distribution of each single marginal process converges toward the distribution of the unique solution $\boldsymbol{X} = (X_t)_{0 \leqslant t \leqslant T}$ of the McKean–Vlasov evolution equation:

$$\begin{aligned} dX_t &= b\big(t, X_t, \mathcal{L}(X_t), \phi(t, X_t)\big)dt \\ &\quad + \sigma\big(t, X_t, \mathcal{L}(X_t), \phi(t, X_t)\big)dW_t, \quad t \in [0, T], \end{aligned} \tag{6.3}$$

where $\boldsymbol{W} = (W_t)_{0 \leqslant t \leqslant T}$ is a standard Wiener process. So if the common feedback control function ϕ is fixed, in the limit $N \to \infty$, the private states of the players become independent of each other, and for each given i, the distribution of the private state process $\boldsymbol{X}^{N,i} = (X_t^{N,i})_{0 \leqslant t \leqslant T}$ evolving according to (6.1) converges toward the distribution of the solution of (6.3). In this limit, the objective functions that the players try to minimize become:

$$J(\alpha) = \mathbb{E}\Big[\int_0^T f\big(t, X_t, \mathcal{L}(X_t), \phi(t, X_t)\big) dt + g\big(X_T, \mathcal{L}(X_T)\big) \Big], \tag{6.4}$$

so if we choose to perform the optimization after taking the limit $N \to \infty$, i.e., assuming that the limit has already been taken, the objective of each player becomes the minimization of the functional (6.4) over a class of admissible feedback control functions ϕ under the dynamical constraint (6.3). This is a form of optimal stochastic control of a state evolving according to the stochastic differential equation (6.3) when the admissible controls are in closed loop feedback form. More generally, such a problem can be stated for controls $\boldsymbol{\alpha} = (\alpha_t)_{0 \leqslant t \leqslant T}$ adapted to any specific information structure. Such a formulation amounts to finding:

$$\begin{aligned} &\boldsymbol{\alpha}^* = \arg\min_{\boldsymbol{\alpha}} \mathbb{E}\Big[\int_0^T f\big(t, X_t, \mathcal{L}(X_t), \alpha_t\big) dt + g\big(X_T, \mathcal{L}(X_T)\big) \Big] \\ &\quad \text{subject to} \\ &dX_t = b\big(t, X_t, \mathcal{L}(X_t), \alpha_t\big)dt + \sigma\big(t, X_t, \mathcal{L}(X_t), \alpha_t\big)dW_t, \quad t \in [0, T]. \end{aligned} \tag{6.5}$$

We call this kind of problem the *optimal control of stochastic McKean-Vlasov dynamics* or the *mean field stochastic optimal control.*

Summary. While similar in purpose to the MFG strategy used to identify equilibria for large symmetric games, the above plan of action may lead to a different notion of equilibrium. To emphasize this point, we rely on the following diagram which illustrates the fact that there are two different paths to go from the North West corner corresponding to the statement of a N player stochastic differential game, to the South East corner where one should expect to find equilibria. For better or worse, this diagram is not *commutative* and this chapter provides insight on the matter. Accordingly, we demonstrate by examples that the choice of a particular path has drastic consequences on the properties of the resulting equilibria.

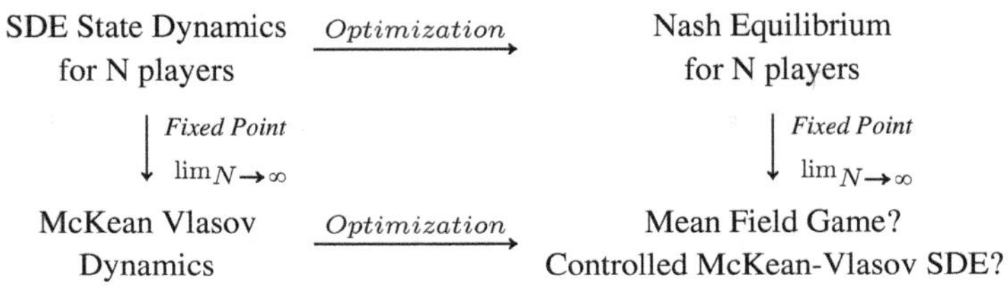

It is important to emphasize what we mean by the limit $N \to \infty$. Our goal is to identify properties of the solutions of the limiting problem which, when re-injected into the formulation of the game with finitely many players, give approximate solutions to the problem with N players. This interpretation suggests that taking this limit is essentially the same as solving for a fixed point, hence the labels used in the above diagram.

As we explain later on in Chapter (Vol II)-6, the solutions to both problems provide approximate equilibrium states for large populations of individuals whose interactions and objective functions are of mean field type. The differences between these notions of equilibrium are subtle and depend upon the formulation of the optimization component of the equilibrium model. We provide simple examples of linear quadratic models illustrating these differences in Section 6.7.1 below.

Despite the fact that the above diagram is not commutative, mean field games and mean field stochastic control problems are in fact connected. This is part of the objectives of our introductory (and mostly heuristic) Section 6.2 below to show that, in some cases, the optimal trajectories of a control problem of the McKean-Vlasov type are given by the solution of a mean field game, possibly driven by different coefficients. The typical example when this is indeed the case is the so-called class of *potential games*, which we already alluded to in Chapters 1 and 2 and which we revisit in Section 6.7 below.

Another objective of Section 6.2 is to provide a general review of the conceivable strategies that may be implemented to solve mean field control problems. As for mean field games, both analytic and probabilistic approaches may be used. However, since the dynamics described by McKean-Vlasov SDEs are genuinely non-Markovian, it is natural to approach the optimization problem using suitable probabilistic reformulations of the problem, as opposed to a contrived adaptation of the Hamilton-Jacobi-Bellman paradigm. This is exactly what we shall do in most of the remaining sections of the chapter. In full analogy with our discussion of the MFG problem, our probabilistic approach is based upon a characterization of the optimal trajectories as the solutions of a forward-backward SDE of the McKean-Vlasov type. Precisely, we shall develop an appropriate version of the Pontryagin maximum principle. As we explained in the previous chapters, the stochastic maximum principle approach is based on the introduction (and the manipulation) of adjoint processes defined as solutions of BSDEs. The formulation of the adjoint equations requires the computation of partial derivatives of the Hamiltonian function with respect to the variables defining the state of the system. In the case of McKean-Vlasov dynamics, the marginal distributions of the solutions are full-fledged state variables. As a consequence, the Hamiltonian function needs to be differentiated with respect to these marginal distributions as well, and the Pontryagin system takes the form of forward-backward SDE of the McKean-Vlasov type driven by the usual derivative of the Hamiltonian with respect to the state variable but also by its L-derivative, notion introduced in Chapter 5. We believe that the need for a differential calculus over the Wasserstein space is the main reason why it took so long for the problem to be studied in its full generality. Indeed, the early attempts at tackling this problem were restricted to scalar interactions in which the dynamics depend solely upon moments of the marginal distributions. For in those cases, differentiability with respect to the measure can be done by standard calculus chain rules. In this chapter, we consider the problem in its full generality taking full advantage of the notion of differentiability presented in Chapter 5. It is worth mentioning that, in comparison, the linear functional derivative is preferred to the L-derivative when the Pontryagin principle is applied under the analytic formulation; as we shall see later on, the resulting forward-backward system then takes a form which is reminiscent of the MFG system (3.12).

The Pontryagin maximum principle is a very powerful tool. However, as we already saw in Chapters 2 and 3, the insights it provides come at the price of restrictive assumptions on the models, especially its sufficient condition for optimality. So quite expectedly, the results of this chapter based on the Pontryagin principle rely on a set of technical assumptions which limit the class of models to dynamics given by coefficients which are essentially linear in the state, control and measure variables, and costs which are convex in the state and the control variables. While seemingly restrictive, these assumptions are typical in the applications of the Pontryagin maximum principle to standard optimization problems. This prompts us to provide in Section 6.6 a road map to a less demanding procedure for solving the optimization problem by means of compactness arguments based upon the notion of relaxed controls.

The necessary part of the Pontryagin stochastic maximum principle suggests that one searches the control set for a candidate minimizing the Hamiltonian function, while the sufficient part points to the insertion of the found minimizer into both the forward equation governing the dynamics, and the backward equation providing the adjoint processes. The presence of the minimizer in both of these equations creates a strong coupling between the forward and backward equations, and the solution of the control problem reduces de facto to the solution of a Forward-Backward Stochastic Differential Equation (FBSDE). In the present situation, the marginal distributions of the solutions appear in the coefficients. These FBSDEs of McKean-Vlasov type were studied in Section 4.3, but one of the assumptions used there, typically the boundedness of the coefficients with respect to the state variable, precludes the application of this result to the Linear Quadratic (LQ) models often used as benchmarks in stochastic control. Like in Chapters 3 and 4, we extend the basic results of Section 4.3 by taking advantage of the convexity of the Hamiltonian. A strong form of this convexity assumption can be used to apply the *continuation method*, providing existence and uniqueness of the solution of the FBSDE at hand. Restoring the Markov property by extending the state space as alluded to earlier, we identify the backward component of the solution of this FBSDE to a function of the forward component and its marginal distribution. This function is known as the *decoupling field* of the FBSDE. In the classical cases, it can be found by solving a PDE. In the present set-up, such a PDE is infinite dimensional as it involves differentiation with respect to the state of the forward dynamics as well as its distribution. Precisely, it reads as the derivative of an infinite dimensional Hamilton-Jacobi-Bellman equation, similar to that presented in the applications of Chapter 5. This PDE is formulated in Section 6.5. Somehow, it is related to the *master equation* for mean field games, which we shall address with care in the second volume of the book.

6.2 Probabilistic and Analytic Formulations

In analogy with our presentation of mean field games in Chapter 3, we address the optimal control of McKean-Vlasov diffusion processes in two different ways, based on a direct analytic formulation and a probabilistic methodology respectively. In the case of the analytic approach, we develop the Pontryagin and the HJB strategies for the optimization problem, while we concentrate mostly on the Pontryagin approach in the probabilistic case. Interestingly, we shall see that in both cases, the Pontryagin systems can be linked to forward/backward systems of the types of those identified in Chapters 3 and 4 for MFG problems. We shall take advantage of this unexpected connection when we revisit the class of potential mean field game problems later in the chapter.

6.2.1 Probabilistic Formulation of the Optimization Problem

As in Chapter 4, we assume that $\boldsymbol{W} = (W_t)_{0\leqslant t\leqslant T}$ is a d-dimensional standard Wiener process defined on a complete filtered probability space $(\Omega, \mathcal{F}, \mathbb{F}, \mathbb{P})$, where the filtration $\mathbb{F} = (\mathcal{F}_t)_{0\leqslant t\leqslant T}$ is complete and right-continuous. As before, for each random variable/vector or stochastic process X, we denote by $\mathbb{P} \circ X^{-1}$, $\mathbb{P}_X$ or $\mathcal{L}(X)$ the law (alternatively called the distribution) of X.

Controlled Dynamics

For a given initial condition $\xi \in L^2(\Omega, \mathcal{F}_0, \mathbb{P}; \mathbb{R}^d)$, the stochastic dynamics of the controlled state are given by a stochastic process $\boldsymbol{X} = (X_t)_{0 \leqslant t \leqslant T}$ satisfying a nonlinear SDE of the form:

$$\begin{aligned} dX_t &= b(t, X_t, \mathcal{L}(X_t), \alpha_t)dt \\ &\quad + \sigma(t, X_t, \mathcal{L}(X_t), \alpha_t)dW_t, \quad t \in [0, T]\,; \quad X_0 = \xi, \end{aligned} \tag{6.6}$$

where the drift and diffusion coefficients of the state X_t of the system at time t are given by a pair of deterministic (measurable) functions $(b, \sigma) : [0, T] \times \mathbb{R}^d \times \mathcal{P}_2(\mathbb{R}^d) \times A \to \mathbb{R}^d \times \mathbb{R}^{d \times d}$, see Proposition 5.7 for the description of the σ-field on $\mathcal{P}_2(\mathbb{R}^d)$, and $\boldsymbol{\alpha} = (\alpha_t)_{0 \leqslant t \leqslant T}$ is a progressively measurable process with values in a measurable space $(A, \mathcal{A})$. Typically, A will be a closed convex subset of the Euclidean space $\mathbb{R}^k$, for $k \geq 1$, and $\mathcal{A}$ the σ-field induced by the Borel σ-field of this Euclidean space. The fact that $\boldsymbol{W}$ and $\boldsymbol{X}$ are required to have the same dimension d is for convenience only. As already explained in the introduction, the term *nonlinear* used to qualify (6.6), does not refer to the fact that the coefficients b and σ could be nonlinear functions of x, but instead to the fact that they depend not only on the value of the unknown process X_t at time t, but also on its marginal distribution $\mathcal{L}(X_t)$. Using the terminology introduced in Section 4.2, we call (6.6) a *controlled stochastic differential equation of the McKean-Vlasov type*. Sometimes the McKean-Vlasov dynamics posited in (6.6) are also called of *mean field type*, in which case we use the terminology *mean field stochastic control problem*. This is justified by the fact that the uncontrolled stochastic differential equations of McKean-Vlasov type first appeared as the infinite particle limits of large systems of particles with mean field interactions (see for instance Chapter 1 of Volume II).

Throughout the chapter, we assume that the drift coefficient b and the volatility σ satisfy the following assumptions:

Assumption (MKV Lipschitz Regularity).

(A1) The function $[0, T] \ni t \mapsto (b, \sigma)(t, 0, \delta_0, 0) \in \mathbb{R}^d \times \mathbb{R}^{d \times m}$ is bounded.

(A2) There exists $c > 0$ such that for all $t \in [0, T]$, $\alpha, \alpha' \in A$, $x, x' \in \mathbb{R}^d$ and $\mu, \mu' \in \mathcal{P}_2(\mathbb{R}^d)$,

$$\begin{aligned} &|b(t, x, \mu, \alpha) - b(t, x', \mu', \alpha')| + |\sigma(t, x, \mu, \alpha) - \sigma(t, x', \mu', \alpha')| \\ &\leqslant c\big[|x - x'| + |\alpha - \alpha'| + W_2(\mu, \mu')\big], \end{aligned}$$

where $W_2(\mu, \mu')$ denotes the 2-Wasserstein distance on the space $\mathcal{P}_2(\mathbb{R}^d)$ (recall Section 5.1 of Chapter 5 for a definition).

The set $\mathbb{A}$ of *admissible* control processes $\boldsymbol{\alpha}$ is defined as the set of A-valued progressively measurable processes $\boldsymbol{\alpha} = (\alpha_t)_{0\leqslant t\leqslant T}$ satisfying:

$$\mathbb{E}\int_0^T |\alpha_t|^2 dt < \infty. \tag{6.7}$$

Remark 6.1 *The above choice of admissibility is motivated by the search for optimal controls in open loop forms. This class of controls is especially well suited to the probabilistic approach based on the Pontryagin stochastic maximum principle. However, we shall consider other classes of admissible controls from time to time. Indeed, it will be convenient to work with Markovian controls in closed loop feedback form* $\alpha_t = \phi(t, X_t)$ *for a deterministic function* $\phi : [0, T] \times \mathbb{R}^d \to A$ *when we introduce the value function of the optimization problem, and we search for equations (typically PDEs) satisfied by this value function. We refer to the next section for a first account in that direction.*

Cost Functional

Under the above assumptions (**A1**) and (**A2**), Theorem 4.21 of Section 4.2.1 implies that for every admissible control $\boldsymbol{\alpha} \in \mathbb{A}$, there exists a unique solution $\boldsymbol{X} = \boldsymbol{X}^{\boldsymbol{\alpha}}$ of (6.6). Moreover for every $p \in [1, 2]$, this solution satisfies:

$$\mathbb{E}\Big[\sup_{0\leqslant t\leqslant T} |X_t|^p\Big] < \infty. \tag{6.8}$$

In this chapter, we are interested in the minimization of the objective function:

$$J(\boldsymbol{\alpha}) = \mathbb{E}\Big[\int_0^T f\big(t, X_t, \mathcal{L}(X_t), \alpha_t\big)dt + g\big(X_T, \mathcal{L}(X_T)\big)\Big], \tag{6.9}$$

over the set $\mathbb{A}$ of admissible control processes. The *running cost* function f is given by a real valued deterministic (measurable) function on $[0, T]\times\mathbb{R}^d\times\mathcal{P}_2(\mathbb{R}^d)\times A$, and the *terminal cost* function g by a real valued deterministic (measurable) function on $\mathbb{R}^d \times \mathcal{P}_2(\mathbb{R}^d)$. Assumptions on the cost functions f and g will be spelled out later on, but typically we shall assume:

Assumption (MKV Quadratic Growth). There exists a constant C such that:

$$|f(t, x, \mu, \alpha)| \leqslant C\big(1 + |x| + M_2(\mu) + |\alpha|\big)^2,$$
$$|g(x, \mu)| \leqslant C\big(1 + |x| + M_2(\mu)\big)^2,$$

for all $(t, x, \mu, \alpha) \in [0, T]\times\mathbb{R}^d\times\mathcal{P}_2(\mathbb{R}^d)\times A$, where $M_2(\mu)$ denotes the square root of the second moment of μ, see (3.26).

For the sake of simplicity we assume that all the coefficients are deterministic. Some of the results can be extended to random coefficients. See the Notes & Complements at the end of the chapter for a discussion of some of these possible generalizations.

First Comparison with MFG

We emphasize one more time that the optimization problem (6.9) differs from the optimization problem encountered in the theory of mean field games. Indeed, when solving a mean field game problem, the optimization of the cost functional (6.9) is performed for a fixed flow of probability measures. In other words, the argument $(\mathcal{L}(X_t))_{0\leqslant t\leqslant T}$ in (6.6) and (6.9) is kept fixed as $\boldsymbol{\alpha}$ varies, and the resulting controlled processes are driven by the same flow of measures, which is not necessarily the flow of marginal distributions of the process $(X_t)_{0\leqslant t\leqslant T}$, but merely an input. Solving the mean field game then consists in identifying an input flow of probability measures such that the optimal states have precisely the input as flow of marginal distributions. This is different from the problem considered in this chapter since, throughout the optimization process, the flow of probability measures used in (6.6) and (6.9) need to be equal to the flow of marginal distributions $(\mathcal{L}(X_t))_{0\leqslant t\leqslant T}$ of the solution of (6.6).

6.2.2 Reformulation as a Deterministic Optimal Control Problem

As announced, we now reformulate the optimization problem analytically.

Reformulation of the Problem Over Controls in Closed Loop Feedback Form

As suggested by Remark 6.1, we may restrict ourselves to Markovian controls $\boldsymbol{\alpha} = (\phi(t, X_t))_{0\leqslant t\leqslant T}$ given by feedback control functions, in which case (6.6) becomes (at least formally):

$$\begin{aligned} dX_t &= b\big(t, X_t, \mathcal{L}(X_t), \phi(t, X_t)\big)dt \\ &\quad + \sigma\big(t, X_t, \mathcal{L}(X_t), \phi(t, X_t)\big)dW_t, \quad t \in [0, T]\,; \quad X_0 = \xi. \end{aligned} \tag{6.10}$$

As before, we use a bold face character to indicate that we are working with a function of t, and we use the notation $(\cdot)$ to emphasize that we are dealing with a function of $x \in \mathbb{R}^d$. In this way, we can distinguish the notation $\boldsymbol{\alpha}(\cdot) = (\phi(t, \cdot))_{0\leqslant t\leqslant T}$ for the feedback function from the values $\alpha \in A$ in the range of this function and, also, from the control process $\boldsymbol{\alpha} = (\alpha_t = (\phi(t, X_t)))_{0\leqslant t\leqslant T}$ obtained by implementing the feedback function $\boldsymbol{\alpha}(\cdot)$ in the SDE (6.6). Since we keep the discussion in this section at an informal level only, we do not discuss the well posedness of (6.10).

We now rewrite the optimal control problem (6.3)–(6.4) in the form of a deterministic control problem. Since the coefficients f and g are deterministic, we can view the objective $J(\boldsymbol{\alpha})$ as a deterministic function of the marginal distributions $\boldsymbol{\mu} = (\mu_t = \mathcal{L}(X_t))_{0 \leqslant t \leqslant T}$ by writing:

$$J(\phi) = \int_0^T \langle f(t, \cdot, \mu_t, \phi(t, \cdot)), \mu_t \rangle dt + \langle g(\cdot, \mu_T), \mu_T \rangle, \tag{6.11}$$

if we use the notation $\langle \varphi, \nu \rangle$ for the integral of the function φ with respect to the measure ν.

Motivated by this (deterministic) form of the objective function, we rewrite the dynamics of the controlled state in terms of its distribution only, replacing the dynamical equation (6.3) given by a stochastic differential equation, by a deterministic dynamical equation for the marginal distributions themselves. A simple application of the classical form of Itô's formula to the McKean-Vlasov SDE (if well posed) shows that, similar to (3.12), the dynamics of the marginal distributions are given by the nonlinear Kolmogorov-Fokker-Planck's equation:

$$\partial_t \mu_t = L_t^{\phi(t,\cdot),\dagger} \mu_t, \quad t \in [0, T] \; ; \quad \mu_0 \sim \xi, \tag{6.12}$$

where the action of the operator $L_t^{\phi(t,\cdot),\dagger}$ on measures (in the sense of distributions) is given by:

$$\begin{aligned} L_t^{\phi(t,\cdot),\dagger} \mu &= -\text{div}_x \big[b(t, \cdot, \mu, \phi(t, \cdot)) \mu \big] \\ &\quad + \frac{1}{2} \text{trace} \big[\partial_{xx}^2 \big(\sigma(t, \cdot, \mu, \phi(t, \cdot)) \big(\sigma(t, \cdot, \mu, \phi(t, \cdot)) \big)^\dagger \mu \big) \big]. \end{aligned} \tag{6.13}$$

So, instead of considering (6.3)–(6.4), we may want to focus on the deterministic optimal control problem:

$$\inf \bigg[\int_0^T \langle f(t, \cdot, \mu_t, \phi(t, \cdot)), \mu_t \rangle \, dt + \langle g(\cdot, \mu_T), \mu_T \rangle \bigg], \tag{6.14}$$

the infimum being taken over the set of pairs $(\mu_t, \phi(t, \cdot))_{0 \leqslant t \leqslant T}$, for which the mappings $[0, T] \ni t \mapsto \mu_t \in \mathcal{P}_2(\mathbb{R}^d)$ and $[0, T] \times \mathbb{R}^d \ni (t, x) \mapsto \phi(t, x)$ are measurable, and satisfy:

$$\int_0^T \| \phi(t, \cdot) \|^2_{L^2(\mathbb{R}^d, \mu_t; \mathbb{R}^k)} dt < \infty,$$

and (6.12) in the sense of distributions. We refer to the discussion following Proposition 5.7 for the time measurability properties of the integrand in (6.11).

Example: Optimal Transportation as an MKV Control Problem
It is worth mentioning that Benamou-Brenier's Theorem 5.53 of optimal transportation proven in Chapter 5, may be recast as a McKean-Vlasov control problem of the form (6.14). Indeed, for a given target measure $\nu_1 \in \mathcal{P}_2(\mathbb{R}^d)$, choose $T = 1$ and the coefficients:

$$
\begin{aligned}
&b(t,x,\mu,\alpha) = \alpha, \quad \sigma(t,x,\mu,\alpha) = 0, \\
&f(t,x,\mu,\alpha) = |\alpha|^2, \quad g(x,\mu) = \begin{cases} 0 & \text{if } \mu = \nu_1 \\ +\infty & \text{otherwise} \end{cases},
\end{aligned}
\tag{6.15}
$$

for $(t,x,\mu,\alpha) \in [0,T] \times \mathbb{R}^d \times \mathcal{P}_2(\mathbb{R}^d) \times A$. Notice that this control problem is of the McKean-Vlasov type because the terminal cost function depends upon the distribution of the controlled process. Now, for a given initial distribution ν_0, the special form of the terminal cost function g should enforce the terminal value $\mu_1 = \nu_1$ for the optimal flow $\boldsymbol{\mu} = (\mu_t)_{0 \leqslant t \leqslant 1}$. If we optimize over closed loop controls, the feedback functions ϕ can be identified to the vector fields of the theorem of Benamou and Brenier, and consequently, the minimal cost in (6.14) (with $\mu_0 = \nu_0$) should coincide with $W_2(\nu_0,\nu_1)^2$. We shall come back to this example in Subsection 6.7.3 below.

6.2.3 The State Variable, the Hamiltonian, and the Adjoint Variables

The present discussion being merely intended to compare the probabilistic and the analytic approaches to the problem, we shall keep part of the presentation at a heuristic level.

Following the approach to standard control problems reviewed in Chapter 3 we introduce a Hamiltonian with finite-dimensional variables. To wit, we call Hamiltonian the function H defined by:

$$
H(t,x,\mu,y,z,\alpha) = b(t,x,\mu,\alpha) \cdot y + \sigma(t,x,\mu,\alpha) \cdot z + f(t,x,\mu,\alpha), \tag{6.16}
$$

for $(t,x,\mu,y,z,\alpha) \in [0,T] \times \mathbb{R}^d \times \mathcal{P}_2(\mathbb{R}^d) \times \mathbb{R}^d \times \mathbb{R}^{d\times d} \times A$, where the dot notation stands for the inner product in the Euclidean space unambiguously determined by the context. Although this definition will turn out to be useful throughout the analysis, the function H so defined cannot stand for the Hamiltonian associated with the original control problem stated in Subsection 6.2.1, nor can it stand for the Hamiltonian of the reformulation as a deterministic control problem given in Subsection 6.2.2. Indeed, the state variable is infinite dimensional in both cases, and it is natural to expect the adjoint variable to be of infinite dimension as well.

Therefore, in order to understand the connection between H and the McKean-Vlasov optimal control problem, we must discuss the choice of the state variable and then discuss the form of the Hamiltonian.

Using the Probabilistic Approach

We first start with the probabilistic approach proposed in Subsection 6.2.1 and provide the relevant form of the corresponding Hamiltonian. Due to the very nature of McKean-Vlasov stochastic differential equations, it is quite tempting to choose $(X_t, \mathcal{L}(X_t))$ as the variable describing the state of the system at time t, where the first coordinate is understood as the realization $X_t(\omega)$ of the random variable X_t and the second as its distribution. In so doing, the state variable lives in $\mathbb{R}^d \times \mathcal{P}_2(\mathbb{R}^d)$. However, given the development of the notion of L-differentiability on the Wasserstein space in Chapter 5, a more convenient strategy could be to lift the measure component of the state variable into a random variable belonging to $L^2(\Omega, \mathcal{F}, \mathbb{P}; \mathbb{R}^d)$. Of course, such a lifting procedure is absolutely trivial in the probabilistic approach as we may choose the random variable X_t itself as the lift of $\mathcal{L}(X_t)$. This prompts us to define the Hamiltonian $\tilde{H}$ by:

$$\tilde{H}(t, x, \tilde{X}, y, z, \alpha) = H(t, x, \mu, y, z, \alpha) \tag{6.17}$$

for any random variable $\tilde{X}$ with distribution μ. Below, we shall use the following convention: Whenever X is a random variable constructed on $(\Omega, \mathcal{F}, \mathbb{P})$, $\tilde{X}$ denotes an independent copy on a clone $(\tilde{\Omega}, \tilde{\mathcal{F}}, \tilde{\mathbb{P}})$ of the space $(\Omega, \mathcal{F}, \mathbb{P})$, according to the same principle as in Chapter 5. Now, for any $t \in [0, T]$, we may regard the random variable X_t, seen as an element of $L^2(\Omega, \mathcal{F}, \mathbb{P}; \mathbb{R}^d)$, as the state variable itself as it encodes both the realization $X_t(\omega)$ of X_t and the copy $\tilde{X}_t$.

So we may regard the full-fledged adjoint variables of the mean field control problem as random variables $Y \in L^2(\Omega, \mathcal{F}, \mathbb{P}; \mathbb{R}^d)$ and $Z \in L^2(\Omega, \mathcal{F}, \mathbb{P}; \mathbb{R}^{d\times d})$, in which case the Hamiltonian should be given by:

$$\tilde{\mathcal{H}}(t, X, Y, Z, \beta) = \mathbb{E}\big[\tilde{H}(t, X, \tilde{X}, Y, Z, \beta)\big],$$

for $t \in [0, T]$, $X, Y \in L^2(\Omega, \mathcal{F}, \mathbb{P}; \mathbb{R}^d)$, $Z \in L^2(\Omega, \mathcal{F}, \mathbb{P}; \mathbb{R}^{d\times d})$ and $\beta \in L^2(\Omega, \mathcal{F}, \mathbb{P}; A)$.

The following two observations are in order:

1. First, if $\hat{\alpha}(t, x, \mu, y, z)$ is a minimizer of $A \ni \alpha \mapsto H(t, x, \mu, y, z, \alpha)$ in the sense that:

$$\hat{\alpha}(t, x, \mu, y, z) = \operatorname{argmin}_{\alpha \in A} H(t, x, \mu, y, z, \alpha), \tag{6.18}$$

then, for any $\beta \in L^2(\Omega, \mathcal{F}, \mathbb{P}; A)$,

$$\tilde{\mathcal{H}}\Big(t, X, Y, Z, \hat{\alpha}\big(t, X, \mathcal{L}(X), Y, Z\big)\Big) \leq \tilde{\mathcal{H}}(t, X, Y, Z, \beta), \tag{6.19}$$

proving that $\hat{\alpha}(t, X, \mathcal{L}(X), Y, Z)$, when it is square integrable, is a minimizer of $L^2(\Omega, \mathcal{F}, \mathbb{P}; A) \ni \beta \mapsto \tilde{\mathcal{H}}(t, X, Y, Z, \beta)$.

2. Second, whenever the coefficients are smooth, the Fréchet derivative of $\tilde{\mathcal{H}}$ in X is given by:

$$
\begin{aligned}
D_X\tilde{\mathcal{H}}(t,X,Y,Z,\beta) &= \partial_x H(t,X,\mathcal{L}(X),Y,Z) \\
&\quad + \tilde{\mathbb{E}}\big[\partial_\mu H(t,\tilde{X},\mathcal{L}(X),\tilde{Y},\tilde{Z},\tilde{\beta})(X)\big],
\end{aligned}
\tag{6.20}
$$

the second term in the right-hand side resulting from the definition of the L-differential and from Fubini's theorem, see Example 3 in Subsection 5.2.2.

Using the Analytic Approach

We now provide a similar discussion for the analytic approach. This will allow us to compare both approaches.

Returning to the analytic formulation (6.12)–(6.14), we thus consider the deterministic optimal control problem:

$$
\inf_{\phi,\ \partial_t\mu_t = L_t^{\phi(t,\cdot),\dagger}\mu_t} J(\phi),
$$

the partial differential operator $L_t^{\phi(t,\cdot),\dagger}$ being defined in (6.13). Obviously, this partial differential operator is unbounded but, for pedagogical reasons, we choose to ignore the technical issues related to the definition of its domain and keep the discussion rather informal.

First, we observe that the state variable is a probability measure in $\mathcal{P}_2(\mathbb{R}^d)$ and we shall regard it as an element of the space of signed measures ν on $\mathbb{R}^d$ such that $\int_{\mathbb{R}^d}(1+|x|)^2 d|\nu|(x) < \infty$. So by duality, the adjoint variable should be a continuous function $u \in C(\mathbb{R}^d;\mathbb{R})$ with sub-quadratic growth. It is then natural to introduce the (formal) Hamiltonian:

$$
\begin{aligned}
\mathscr{H}\big(t,\mu,u,\beta\big) &= \Big\langle u(\cdot), -\mathrm{div}_x\big[b(t,\cdot,\mu,\beta(\cdot))\mu\big] + \frac{1}{2}\mathrm{trace}\Big[\partial^2_{xx}\Big(a\big(t,\cdot,\mu,\beta(\cdot)\big)\mu\Big)\Big]\Big\rangle \\
&\quad + \big\langle f\big(t,\cdot,\mu,\beta(\cdot)\big),\mu\big\rangle,
\end{aligned}
$$

for $u \in C(\mathbb{R}^d;\mathbb{R})$ with sub-quadratic growth as dual variable of the state variable $\mu \in \mathcal{P}_2(\mathbb{R}^d)$ and for $\beta \in L^2(\mathbb{R}^d,\mu;A)$. As always, we use the short notation $a = \sigma\sigma^\dagger$. Whenever u is smooth enough, say C^2, we can use integration by parts. We get:

$$
\begin{aligned}
\mathscr{H}\big(t,\mu,u,\beta\big) &= \Big\langle b\big(t,\cdot,\mu,\beta(\cdot)\big)\cdot\partial_x u(\cdot) + \frac{1}{2}\mathrm{trace}\Big[a\big(t,\cdot,\mu,\beta(\cdot)\big)\partial^2_{xx}u(\cdot)\Big],\mu\Big\rangle \\
&\quad + \big\langle f\big(t,\cdot,\mu,\beta(\cdot)\big),\mu\big\rangle.
\end{aligned}
$$

Writing out the integration with respect to μ, we obtain:

$$\begin{aligned}
&\mathscr{H}\big(t, \mu, u, \beta\big) \\
&= \int_{\mathbb{R}^d} \Big[b\big(t, x, \mu, \beta(x)\big) \cdot \partial_x u(x) \\
&\qquad + \frac{1}{2}\mathrm{trace}\big[a\big(t, x, \mu, \beta(x)\big)\partial^2_{xx}u(x)\big] + f\big(t, x, \mu, \beta(x)\big)\Big] d\mu(x) \\
&= \int_{\mathbb{R}^d} K\big(t, x, \mu, \partial_x u(x), \partial^2_{xx}u(x), \beta(x)\big) d\mu(x),
\end{aligned} \tag{6.21}$$

if we use the notation K for the operator symbol:

$$\begin{aligned}
K(t, x, \mu, y, z, \alpha) &= b(t, x, \mu, \alpha) \cdot y + \frac{1}{2}a(t, x, \mu, \alpha) \cdot z + f(t, x, \mu, \alpha) \\
&= H\Big(t, x, \mu, y, \tfrac{1}{2}z\sigma(t, x, \mu, \alpha), \alpha\Big),
\end{aligned} \tag{6.22}$$

for $(t, x, \mu, y, z, \alpha) \in [0, T] \times \mathbb{R}^d \times \mathcal{P}_2(\mathbb{R}^d) \times \mathbb{R}^d \times \mathbb{R}^{d\times d} \times A$, with H as in (6.16). In order to minimize the Hamiltonian $\mathscr{H}$ with respect to the control variable β, namely in order to compute:

$$\mathscr{H}^*(t, \mu, u) = \inf_{\beta \in L^2(\mathbb{R}^d, \mu; A)} \mathscr{H}(t, \mu, u, \beta),$$

we assume that the operator symbol K has a minimum in $\alpha \in A$ for each t, x, μ, y, and z fixed. More precisely, we assume the existence of a function:

$$[0, T] \times \mathbb{R}^d \times \mathcal{P}_2(\mathbb{R}^d) \times \mathbb{R}^d \times \mathbb{R}^{d\times d} \ni (t, x, \mu, y, z) \mapsto \alpha^*(t, x, \mu, y, z) \in A$$

such that:

$$K\big(t, x, \mu, y, z, \alpha^*(t, x, \mu, y, z)\big) = \inf_{\alpha \in A} K(t, x, \mu, y, z, \alpha), \tag{6.23}$$

for all t, x, μ, y and z fixed. Then,

$$\begin{aligned}
&\mathscr{H}^*(t, \mu, u) \\
&= \int_{\mathbb{R}^d} K\big(t, x, \mu, \partial_x u(x), \partial^2_{xx}u(x), \alpha^*(t, x, \mu, \partial_x u(x), \partial^2_{xx}u(x))\big) d\mu(x),
\end{aligned}$$

implying that the minimum in $\mathscr{H}^*(t, \mu, u)$ is achieved at the function $\beta(\cdot) = \alpha^*(t, \cdot, \mu, \partial_x u(\cdot), \partial^2_{xx}u(\cdot))$, when the latter is square-integrable. It is worth noticing that the last two equations are the analogues of (6.18) and (6.19).

We now mimic (6.20) and compute the derivative of the Hamiltonian $\mathscr{H}$ with respect to the measure argument μ standing for the state variable of the optimization problem. Since we are using the standard duality between function spaces and spaces of measures, we should use the functional derivative introduced in Subsection 5.4.1 instead of the L-derivative in order to compute this derivative. We get:

$$\begin{aligned}
&\frac{\delta \mathscr{H}}{\delta m}(t, \mu, u, \beta)(\bullet) \\
&= b(t, \bullet, \mu, \beta(\bullet)) \cdot \partial_x u(\bullet) \\
&\quad + \frac{1}{2}\text{trace}\big[a(t, \bullet, \mu, \beta(\bullet)) \cdot \partial^2_{xx} u(\bullet)\big] + f(t, \bullet, \mu, \beta(\bullet)) \\
&\quad + \Big\langle \frac{\delta b}{\delta m}(t, \cdot, \mu, \beta(\cdot))(\bullet) \cdot \partial_x u(\cdot) \\
&\quad + \frac{1}{2}\text{trace}\Big[\frac{\delta a}{\delta m}(t, \cdot, \mu, \beta(\cdot))(\bullet) \cdot \partial^2_{xx} u(\cdot)\Big] + \frac{\delta f}{\delta m}(t, \cdot, \mu, \beta(\cdot))(\bullet), \mu\Big\rangle,
\end{aligned} \tag{6.24}$$

for t, μ, u and β as above. In the duality products, the integration variable is $\cdot$ and $\bullet$ is fixed.

In order to be consistent with the notations introduced in Chapter 5, we denote by $\delta/\delta m$ the linear functional derivative with respect to the measure argument, although the latter one is denoted by μ. In this way, we avoid any confusion with the notation ∂_μ for the L-derivative.

Remark 6.2 *It is worth observing that $\hat{\alpha}$ in* (6.18) *and α^* in* (6.23) *coincide whenever σ is independent of the control variable.*

Application: Writing an HJB Equation for the McKean-Vlasov Control Problem

The value function of the present deterministic control problem is the function:

$$v(t, \mu) = \inf_{\phi}\bigg[\int_t^T \big\langle f\big(s, \cdot, \mu_s, \phi(s, \cdot)\big), \mu_s\big\rangle ds + \big\langle g(\cdot, \mu_T), \mu_T\big\rangle\bigg], \tag{6.25}$$

where $(\mu_s)_{t \leqslant s \leqslant T}$ has the dynamics (6.12) with the initial condition $\mu_t = \mu \in \mathcal{P}_2(\mathbb{R}^d)$ at time $t \in [0, T]$. The dynamic programming principle says that it should satisfy (at least in some generalized sense) the HJB equation:

$$\partial_t v(t, \mu) + \mathscr{H}^*\Big(t, \mu, \frac{\delta v}{\delta m}(t, \mu)\Big) = 0, \quad (t, \mu) \in [0, T] \times \mathcal{P}_2(\mathbb{R}^d). \tag{6.26}$$

The third argument in the Hamiltonian $\mathscr{H}^*$ is the function $u(\cdot) = [\delta v/\delta m](t, \mu)(\cdot)$ and the terminal condition is $v(T, \mu) = \langle g(\cdot, \mu), \mu\rangle$. For exactly the same reasons as above, we use the standard Fréchet or Gateaux differentiation introduced in Subsection 5.4.1 instead of the L-derivative in order to compute the derivative of the value function with respect to the state variable μ.

Example 6.3 Let us assume that the matrix σ is equal to the identity, that the drift is equal to the control, i.e., $b(t,x,\mu,\alpha) = \alpha \in A$, with $A = \mathbb{R}^d$, and that the running cost is separable in the sense that:

$$f(t,x,\mu,\alpha) = \frac{1}{2}|\alpha|^2 + f_0(t,x,\mu),$$

for some smooth function f_0 defined on $[0,T] \times \mathbb{R}^d \times \mathcal{P}_2(\mathbb{R}^d)$. The minimizer of $K(t,x,\mu,y,z,\cdot)$ is simply given by $\alpha^*(t,x,\mu,y,z) = -y$ so that:

$$\mathscr{H}^*(t,\mu,u) = \int_{\mathbb{R}^d} \Big[-\frac{1}{2}|\partial_x u(x)|^2 + \frac{1}{2}\Delta_x u(x) + f_0(t,x,\mu)\Big] d\mu(x),$$

and the HJB equation (6.26) reads:

$$\begin{aligned}\partial_t v(t,\mu) + \int_{\mathbb{R}^d} \Big[-\frac{1}{2}\Big|\partial_x \frac{\delta v}{\delta m}(t,\mu)(x)\Big|^2 \\ + \frac{1}{2}\Delta_x \frac{\delta v}{\delta m}(t,\mu)(x) + f_0(t,x,\mu)\Big] d\mu(x) = 0,\end{aligned}$$

or equivalently:

$$\begin{aligned}\partial_t v(t,\mu) + \int_{\mathbb{R}^d} \Big[-\frac{1}{2}\Big|\partial_\mu v(t,\mu)(x)\Big|^2 \\ + \frac{1}{2}\mathrm{trace}\big[\partial_v \partial_\mu v(t,\mu)(x)\big] + f_0(t,x,\mu)\Big] d\mu(x) = 0,\end{aligned} \tag{6.27}$$

if we use the relationship between the L-derivative and the functional derivative proven in Subsection 5.4.1 of Chapter 5.

Example 6.4 We now assume that the matrix σ is equal to the identity, that the drift does not depend upon the measure μ, i.e., $b(t,x,\mu,\alpha) = b(t,x,\alpha)$, and that the running cost is separable in the sense that $f(t,x,\mu,\alpha) = f_0(t,x,\mu) + f_1(t,x,\alpha)$ for some smooth functions f_0 and f_1 defined on $[0,T] \times \mathbb{R}^d \times \mathcal{P}_2(\mathbb{R}^d)$ and $[0,T] \times \mathbb{R}^d \times A$ respectively. In this case, the minimizer α^* of $K(t,x,\mu,y,z,\cdot)$ is the minimizer of $b(t,x,\alpha) \cdot y + f_1(t,x,\alpha)$ (and also of $H(t,x,\mu,y,z)$), and for that reason, it depends neither on the measure μ nor on z, i.e., $\alpha^*(t,x,\mu,y,z) = \hat{\alpha}(t,x,\mu,y,z) = \alpha^*(t,x,y)$ (which we shall also denote by $\hat{\alpha}(t,x,y)$). Consequently:

$$\begin{aligned}\mathscr{H}^*(t,\mu,u) = \int_{\mathbb{R}^d} \Big[b\Big(t,x,\hat{\alpha}\big(t,x,\partial_x u(t,x)\big)\Big) \cdot \partial_x u(t,x) + \frac{1}{2}\Delta_x u(t,x) \\ + f_0(t,x,\mu) + f_1\Big(t,x,\hat{\alpha}\big(t,x,\partial_x u(t,x)\big)\Big)\Big] d\mu(x),\end{aligned}$$

and if we replace the dual variable u by $\delta v/\delta m$, as before, we can use the fact that $\partial_x u$ becomes $\partial_\mu v$ and Δu becomes trace$[\partial_x \partial_\mu v]$. Therefore, the HJB equation reads:

$$\begin{aligned}\partial_t v(t,\mu) + \int_{\mathbb{R}^d} \Big[b\Big(t, x, \hat{\alpha}\big(t, x, \partial_\mu v(t,\mu)(x)\big)\Big) \cdot \partial_\mu v(t,\mu)(x) \\ + \frac{1}{2}\mathrm{trace}\big[\partial_v \partial_\mu v(t,\mu)(x)\big] + f_0(t,x,\mu) \\ + f_1\Big(t, x, \hat{\alpha}\big(t, x, \partial_\mu v(t,\mu)(x)\big)\Big)\Big] d\mu(x) = 0,\end{aligned} \tag{6.28}$$

for $(t,\mu) \in [0,T] \times \mathcal{P}_2(\mathbb{R}^d)$. Making use of the shorten notation $H(t,x,\mu,y,\alpha)$ for the *reduced* Hamiltonian associated with the full Hamiltonian (6.16), (6.28) can also be written:

$$\begin{aligned}\partial_t v(t,\mu) + \int_{\mathbb{R}^d} \Big[\frac{1}{2}\mathrm{trace}\big[\partial_x \partial_\mu v(t,\mu)(x)\big] \\ + H\Big(t, x, \mu, \partial_\mu v(t,\mu)(x), \hat{\alpha}\big(t, x, \partial_\mu v(t,\mu)(x)\big)\Big)\Big] d\mu(x) = 0,\end{aligned}$$

for $(t,\mu) \in [0,T] \times \mathcal{P}_2(\mathbb{R}^d)$.

6.2.4 Pontryagin Maximum Principles for Both Formulations

As in the previous subsection, the present discussion remains mostly at a heuristic level. It is intended to compare the probabilistic and analytic approaches to the problem of optimal control of McKean-Vlasov dynamics.

The thrust of this chapter is to develop and implement appropriate forms of the Pontryagin maximum principle for the optimal control of McKean-Vlasov equations. Since the discussion of the previous subsection led to two different formulations of the problem, we shall need to work with two versions of the maximum principle depending on whether the original control problem is recast as a deterministic or stochastic control problem. In the present subsection, we provide a brief account of the two approaches, discussing them side by side.

The Case of the Probabilistic Formulation

The following definition of the adjoint processes is imposed on us by the expression (6.20) obtained earlier for the derivative of the finite dimensional Hamiltonian.

Definition 6.5 *In addition to assumption* **MKV Lipschitz Regularity**, *assume that the coefficients b, σ, f and g are jointly differentiable with respect to x and μ.*

Then, for any admissible control $\boldsymbol{\alpha} = (\alpha_t)_{0 \leqslant t \leqslant T} \in \mathbb{A}$, *we denote by* $\boldsymbol{X} = \boldsymbol{X}^{\alpha}$ *the corresponding controlled state process, and whenever*

$$\mathbb{E} \int_0^T \Big[\big| \partial_x f(t, X_t, \mathcal{L}(X_t), \alpha_t) \big|^2 + \tilde{\mathbb{E}} \Big[\big| \partial_\mu f(t, X_t, \mathcal{L}(X_t), \alpha_t)(\tilde{X}_t) \big|^2 \Big] \Big] dt < \infty, \tag{6.29}$$

and

$$\mathbb{E} \Big[\big| \partial_x g(X_T, \mathcal{L}(X_t)) \big|^2 + \tilde{\mathbb{E}} \big[|\partial_\mu g(X_T, \mathcal{L}(X_t))(\tilde{X}_T)|^2 \big] \Big] < \infty, \tag{6.30}$$

we call adjoint processes of $\boldsymbol{X}$ *(or of the admissible control* $\boldsymbol{\alpha}$*), any couple* $(\boldsymbol{Y}, \boldsymbol{Z})$ *of progressively measurable stochastic processes* $\boldsymbol{Y} = (Y_t)_{0 \leqslant t \leqslant T}$ *and* $\boldsymbol{Z} = (Z_t)_{0 \leqslant t \leqslant T}$ *in* $\mathbb{S}^{2,d} \times \mathbb{H}^{2,d \times d}$ *satisfying the equation:*

$$\begin{cases} dY_t = -\Big[\partial_x H\big(t, X_t, \mathcal{L}(X_t), Y_t, Z_t, \alpha_t\big) \\ \qquad\qquad + \tilde{\mathbb{E}}\big[\partial_\mu H\big(t, \tilde{X}_t, \mathcal{L}(X_t), \tilde{Y}_t, \tilde{Z}_t, \tilde{\alpha}_t\big)(X_t) \big] \Big] dt \\ \qquad + Z_t dW_t, \quad t \in [0, T], \\ Y_T = \partial_x g\big(X_T, \mathcal{L}(X_T)\big) + \tilde{\mathbb{E}}\big[\partial_\mu g\big(\tilde{X}_T, \mathcal{L}(X_T)\big)(X_T) \big], \end{cases} \tag{6.31}$$

where $(\tilde{\boldsymbol{X}}, \tilde{\boldsymbol{Y}}, \tilde{\boldsymbol{Z}}, \tilde{\boldsymbol{\alpha}})$ *is an independent copy of* $(\boldsymbol{X}, \boldsymbol{Y}, \boldsymbol{Z}, \boldsymbol{\alpha})$ *defined on the space* $(\tilde{\Omega}, \tilde{\mathcal{F}}, \tilde{\mathbb{P}})$ *and* $\tilde{\mathbb{E}}$ *denotes the expectation on* $(\tilde{\Omega}, \tilde{\mathcal{F}}, \tilde{\mathbb{P}})$.

Equation (6.31) is referred to as the adjoint equation.

Remark 6.6 *Beyond the mathematical rationale* (6.20) *for the form of the above adjoint equation, its intuitive justification can be argued as follows. The search for optimality goes through the computation of the variations of the cost* $J(\boldsymbol{\alpha})$ *for infinitesimal variations in the control process* $\boldsymbol{\alpha}$. *The main difference with the classical case is that when* $\boldsymbol{\alpha}$ *varies, the variations of both the state* $(X_t^{\alpha})_{0 \leqslant t \leqslant T}$ *and its distribution* $(\mathcal{L}(X_t^{\alpha}))_{0 \leqslant t \leqslant T}$ *have to be controlled. So it should not come as a surprise that derivatives with respect to the state variable* x *and the measure* μ *appear in the right-hand sides of both the equation for* $(Y_t)_{0 \leqslant t \leqslant T}$ *and its terminal condition* Y_T. *The specific ways in which these derivatives enter formula* (6.31) *directly follow from* (6.20). *However, they may not appear transparent in a first reading. They are best understood from the proofs of the necessary and sufficient conditions of the stochastic maximum principle given below. As in the proof of* (6.20), *these derivatives are manipulated inside an expectation, and an interchange of this expectation and the expectation over the space introduced for the lifting of the functions of measures is the source of the special form of formula* (6.31). *Fubini's theorem is the reason for the special role played by the independent copies appearing in* (6.31).

Notice that $\tilde{\mathbb{E}}[\partial_\mu H(t, \tilde{X}_t, \mathcal{L}(X_t), \tilde{Y}_t, \tilde{Z}_t, \tilde{\alpha}_t)(X_t)]$ is a (measurable) function of the random variable X_t as it stands for $\tilde{\mathbb{E}}[\partial_\mu H(t, \tilde{X}_t, \mathcal{L}(X_t), \tilde{Y}_t, \tilde{Z}_t, \tilde{\alpha}_t)(x)]_{|x=X_t}$. Measurability can be proved by following the discussion in Subsection 5.3.4, which shows that we can find a measurable version of the mapping $(t, x, \mu, y, z, v) \mapsto \partial_\mu H(t, x, \mu, y, z)(v)$; measurability of $\tilde{\mathbb{E}}[\partial_\mu H(t, \tilde{X}_t, \mathcal{L}(X_t), \tilde{Y}_t, \tilde{Z}_t, \tilde{\alpha}_t)(x)]_{|x}$ with respect to (t, x) is then a consequence of Fubini's theorem. The same remark applies to $\tilde{\mathbb{E}}[\partial_\mu g(\tilde{X}_T, \mathcal{L}(X_T))(X_T)]$. Notice also that, when b, σ, f and g do not depend upon the marginal distributions of the controlled state process, the extra terms appearing in the adjoint equation and its terminal condition disappear, and this equation coincides with the classical adjoint equation of stochastic control.

We now expand the derivative of the Hamiltonian in (6.31). To do so, we make the following observation: If $h : \mathcal{P}_2(\mathbb{R}^d) \to \mathbb{R}^\ell$ is differentiable, then, for each $p \in \mathbb{R}^\ell$, the real valued function $\mu \mapsto h(\mu) \cdot p$ is differentiable; its differential is a function $\partial_\mu[h(\mu) \cdot p](\cdot)$ defined μ-almost everywhere on $\mathbb{R}^d$, and $\partial_\mu[h(\mu) \cdot p](x)$ is equal to $(\partial_\mu h(\mu)(x))^\dagger p$, where $\partial_\mu h(\mu)(x)$ is understood as a matrix of size $\ell \times d$. For the sake of convenience, we shall use the notation $\partial_\mu h(\mu)(x) \odot p$ for $\partial_\mu[h(\mu) \cdot p](x) = (\partial_\mu h(\mu)(x))^\dagger p$, with a similar notation for $\partial_x h(x) \odot p = \partial_x[h(x) \cdot p]$ when $h : \mathbb{R}^d \to \mathbb{R}^\ell$. Now, the adjoint equation rewrites:

$$
\begin{aligned}
dY_t = -\Big[&\partial_x b\big(t, X_t, \mathcal{L}(X_t), \alpha_t\big) \odot Y_t + \partial_x \sigma\big(t, X_t, \mathcal{L}(X_t), \alpha_t\big) \odot Z_t \\
&+ \partial_x f\big(t, X_t, \mathcal{L}(X_t), \alpha_t\big) \\
&+ \tilde{\mathbb{E}}\Big[\partial_\mu b(t, \tilde{X}_t, \mathcal{L}(X_t), \tilde{\alpha}_t)(X_t) \odot \tilde{Y}_t \\
&+ \partial_\mu \sigma(t, \tilde{X}_t, \mathcal{L}(X_t), \tilde{\alpha}_t)(X_t) \odot \tilde{Z}_t \\
&+ \partial_\mu f(t, \tilde{X}_t, \mathcal{L}(X_t), \tilde{\alpha}_t)(X_t) \Big]\Big] dt \\
&+ Z_t dW_t, \quad t \in [0, T],
\end{aligned}
\tag{6.32}
$$

with the terminal condition $Y_T = \partial_x g(X_T, \mathcal{L}(X_T)) + \tilde{\mathbb{E}}[\partial_\mu g(\tilde{X}_T, \mathcal{L}(X_T))(X_T)]$. Notice that $\partial_x b$ and $\partial_x \sigma$ are bounded since b and σ are assumed to be Lipschitz continuous in the variable x by (**A2**). Also, $\mathbb{E}[|\partial_\mu b(t, \tilde{X}_t, \mathcal{L}(X_t), \tilde{\alpha}_t)(X_t)|^2]^{1/2}$ and $\mathbb{E}[|\partial_\mu \sigma(t, \tilde{X}_t, \mathcal{L}(X_t), \tilde{\alpha}_t)(X_t)|^2]^{1/2}$ are bounded by c since b and σ are assumed to be c-Lipschitz continuous in the variable μ with respect to the 2-Wasserstein distance. Indeed, Remark 5.27 ensures that, for a differentiable function $h : \mathcal{P}_2(\mathbb{R}^d) \to \mathbb{R}$, which is c-Lipschitz continuous in μ with respect to the 2-Wasserstein distance, it holds $\mathbb{E}[|\partial_\mu h(X)|^2]^{1/2} \leqslant c$, for any $\mu \in \mathcal{P}_2(\mathbb{R}^d)$ and any random variable X having distribution μ.

Remark 6.7 *The notation $\odot$ is especially convenient for terms of the form $\partial_x \sigma \odot z$ or $\partial_\mu \sigma \odot z$. In contrast, we did not introduce it in Chapter 4 since, in the various applications of the stochastic maximum principle addressed therein, σ is assumed to be constant.*

Existence of the Adjoint Processes. Given an admissible control $\boldsymbol{\alpha} \in \mathbb{A}$ and the corresponding controlled state process $\boldsymbol{X} = \boldsymbol{X}^{\boldsymbol{\alpha}}$, equation (6.32) is a BSDE of the McKean-Vlasov type, as those introduced in Subsection 4.2.2. Whenever the filtration $\mathbb{F}$ is generated by $\mathcal{F}_0$ and by $\boldsymbol{W}$, the existence and uniqueness result for mean field BSDEs given by Theorem 4.23 in Chapter 4 can be easily adapted to the present needs up to a slight generalization for allowing the McKean-Vlasov interaction to include the variable z and the dimension of ξ to differ from d. In particular, the bound:

$$\max\Big\{\mathbb{E}\tilde{\mathbb{E}}\big[|\partial_\mu b(t,\tilde{X}_t,\mathcal{L}(X_t),\tilde{\alpha}_t)(X_t)|^2\big],\\ \mathbb{E}\tilde{\mathbb{E}}\big[|\partial_\mu \sigma(t,\tilde{X}_t,\mathcal{L}(X_t),\tilde{\alpha}_t)(X_t)|^2\big]\Big\} < \infty,$$

guarantees existence and uniqueness of a solution of equation (6.32).

Optimality Condition. From the standard version of the stochastic Pontryagin maximum principle, we infer that, if $\boldsymbol{\alpha} = (\alpha_t)_{0\leqslant t\leqslant T}$ is an optimal control process with $\boldsymbol{X} = (X_t)_{0\leqslant t\leqslant T}$ as optimal path, then it should be of the form:

$$\alpha_t = \hat{\alpha}\big(t, X_t, \mathcal{L}(X_t), Y_t, Z_t\big), \quad t \in [0,T], \tag{6.33}$$

where $\hat{\alpha}$ is the minimizer of H, as defined in (6.18) and $(\boldsymbol{Y},\boldsymbol{Z}) = (Y_t,Z_t)_{0\leqslant t\leqslant T}$ solves (6.31). The goal of Sections 6.3 and 6.4 below is to prove this claim rigorously.

The Case of the Deterministic Formulation

We now return to the deterministic formulation of the McKean-Vlasov optimal control problem introduced in Subsection 6.2.2 and we aim at providing a heuristic for the Pontryagin principle in that case. A natural infinite dimensional generalization of the deterministic Pontryagin maximum principle suggests that, for each admissible control in closed loop feedback form $\boldsymbol{\alpha} = (\phi(t,\cdot))_{0\leqslant t\leqslant T}$, the associated adjoint variable $\boldsymbol{u} = (u(t,\cdot))_{0\leqslant t\leqslant T}$ should be defined as the solution of the adjoint equation:

$$\begin{aligned} du(t,\bullet) &= -\frac{\delta \mathscr{H}}{\delta m}\big(t,\mu_t,u(t,\cdot),\phi(t,\cdot)\big)(\bullet)\,dt,\\ u_T(\bullet) &= \frac{\delta}{\delta m}\bigg[\int_{\mathbb{R}^d} g(x,\mu)d\mu(x)\bigg]_{|\mu=\mu_T}(\bullet), \end{aligned}$$

where $(\mu_t)_{0\leqslant t\leqslant T}$ satisfies (6.12). Again, the fact that we use the standard duality between function spaces and spaces of measures prompts us to compute the derivative of the Hamiltonian and the terminal condition using functional derivatives. So the adjoint equation should read:

$$\begin{aligned}\frac{du_t}{dt}(\bullet) = -\Big[& b\big(t, \bullet, \mu_t, \phi(t, \bullet)\big) \cdot \partial_x u_t(\bullet) \\ & + \frac{1}{2}\mathrm{trace}\big[a\big(t, \bullet, \mu_t, \phi(t, \bullet)\big) \cdot \partial^2_{xx} u_t(\bullet)\big] + f\big(t, \bullet, \mu_t, \phi(t, \bullet)\big) \\ + \Big\langle & \frac{\delta b}{\delta m}\big(t, \cdot, \mu_t, \phi(t, \cdot)\big)(\bullet) \cdot \partial_x u_t(\cdot) \\ & + \frac{1}{2}\mathrm{trace}\Big[\frac{\delta a}{\delta m}\big(t, \cdot, \mu_t, \phi(t, \cdot)\big)(\bullet) \cdot \partial^2_{xx} u_t(\cdot)\Big] \\ & + \frac{\delta f}{\delta m}\big(t, \cdot, \mu_t, \phi(t, \cdot)\big)(\bullet), \mu_t\Big\rangle\Big],\end{aligned} \tag{6.34}$$

where we denoted $\boldsymbol{u}$ by $(u_t(\bullet))_{0 \leqslant t \leqslant T}$ instead of $(u(t, \bullet))_{0 \leqslant t \leqslant T}$ and where we used $\cdot$ as integration variable in the duality products. The terminal condition should be:

$$u_T(\bullet) = g(\bullet, \mu_T) + \int_{\mathbb{R}^d} \frac{\delta g}{\delta m}(x, \mu_T)(\bullet) d\mu_T(x).$$

Optimality Condition. Now, the maximum principle suggests that, if $\boldsymbol{\alpha}$ is an optimal control with $\boldsymbol{\mu} = (\mu_t)_{0 \leqslant t \leqslant T}$ as optimal path, then it should be of the form:

$$\phi(t, x) = \alpha^*\big(t, x, \mu_t, \partial_x u(t, x), \partial^2_{xx} u(t, x)\big), \quad (t, x) \in [0, T] \times \mathbb{R}^d,$$

α^* being the minimizer of K defined in (6.23) and $\boldsymbol{u} = (u(t, \cdot))_{0 \leqslant t \leqslant T}$ solving (6.34). Therefore, combining the Kolmogorov-Fokker-Planck equation (6.12) for $\boldsymbol{\mu} = (\mu_t)_{0 \leqslant t \leqslant T}$ with the equation (6.34) for $\boldsymbol{u} = (u(t, \cdot))_{0 \leqslant t \leqslant T}$, we deduce that the pair $(\boldsymbol{\mu}, \boldsymbol{u})$ should necessarily solve the forward-backward system:

$$\left\{\begin{aligned} & \partial_t \mu_t = L_t^{\alpha^*(t, \cdot, \mu_t, \partial_x u(t, \cdot), \partial^2_{xx} u(t, \cdot)), \dagger} \mu_t, \quad t \in [0, T]\,; \quad \mu_0 \sim \xi, \\ & \partial_t u(t, x) = -\frac{\delta \mathscr{H}}{\delta m}\Big(t, \mu_t, u(t, \cdot), \alpha^*\big(t, \cdot, \mu_t, \partial_x u(t, \cdot), \partial^2_{xx} u(t, \cdot)\big)\Big)(x), \\ & \qquad\qquad\qquad\qquad\qquad\qquad\qquad\qquad (t, x) \in [0, T] \times \mathbb{R}^d, \\ & u(T, x) = g(x, \mu_T) + \int_{\mathbb{R}^d} \frac{\delta g}{\delta m}(y, \mu_T)(x) d\mu_T(y), \quad x \in \mathbb{R}^d, \end{aligned}\right. \tag{6.35}$$

where $\delta \mathscr{H}/\delta m$ can be computed as in (6.24), namely, with the same notation as in (6.21):

$$\begin{aligned}\frac{\delta \mathscr{H}}{\delta m}(t, \mu, u, \beta)(x) &= K\big(t, x, \mu, \partial_x u(x), \partial^2_{xx} u(x), \beta(x)\big) \\ &\quad + \int_{\mathbb{R}^d} \frac{\delta K}{\delta m}\big(t, y, \mu, \partial_x u(y), \partial^2_{xx} u(y), \beta(y)\big)(x) d\mu(y).\end{aligned}$$

In full analogy with Remark 3.26 for stochastic control in finite dimension, the adjoint variable $\boldsymbol{u}$ should coincide with the derivative of the value function $\boldsymbol{v} = (v(t,\cdot))_{0\leqslant t\leqslant T}$ in (6.25) computed along the optimal path $(\mu_t)_{0\leqslant t\leqslant T}$, namely we should have:

$$u(t,x) = \frac{\delta v}{\delta m}(t,\mu_t)(x), \quad (t,x) \in [0,T] \times \mathbb{R}^d. \tag{6.36}$$

Using the relationship between the L-derivative and the functional derivative, we derive the following (formal) identification:

$$\partial_x u(t,x) = \partial_\mu v(t,\mu_t)(x), \quad \partial^2_{xx} u(t,x) = \partial_v \partial_\mu v(t,\mu_t)(x),$$

for $(t,x) \in [0,T] \times \mathbb{R}^d$. Relationship (6.36) is nothing but the analogue of the equation (4.8) for the decoupling field of a forward-backward system. Here, the decoupling field is the mapping:

$$U : [0,T] \times \mathbb{R}^d \times \mathcal{P}_2(\mathbb{R}^d) \ni (t,x,\mu) \mapsto \frac{\delta v}{\delta m}(t,\mu)(x).$$

Differentiating the HJB equation (6.26) with respect to μ, we deduce that U should satisfy (at least formally):

$$\partial_t U(t,x,\mu) + \frac{\delta}{\delta m}\Big[\mathscr{H}^*\Big(t,\mu,U(t,\cdot,\mu)\Big)\Big](x) = 0, \tag{6.37}$$

for $(t,x,\mu) \in [0,T] \times \mathbb{R}^d \times \mathcal{P}_2(\mathbb{R}^d)$, which is sometimes called the master equation of the McKean-Vlasov control problem (6.12)–(6.14). Unfortunately, it is only fair to say that the jury is still out on the use of the terminology *master equation*. See nevertheless the discussion in Section 6.5. Here, the terminal boundary condition for $U(T,\cdot,\cdot)$ is:

$$U(T,x,\mu) = g(x,\mu) + \int_{\mathbb{R}^d} \frac{\delta g}{\delta m}(y,\mu)(x)d\mu(y), \quad x \in \mathbb{R}^d, \ \mu \in \mathcal{P}_2(\mathbb{R}^d).$$

We conclude this subsection by revisiting the two examples for which we derived the HJB equations, and we write explicitly the corresponding master equations.

Example 6.8 Under the assumptions of Example 6.3 above, the master equation takes the form:

$$\begin{aligned}
&\partial_t U(t,x,\mu) + \frac{1}{2}\Delta_x U(t,x,\mu) - \frac{1}{2}\big|\partial_x U(t,x,\mu)\big|^2 + f_0(t,x,\mu) \\
&\quad + \int_{\mathbb{R}^d} \Big[\frac{1}{2}\Delta_y \frac{\delta U}{\delta m}(t,y,\mu)(x) - \partial_y \frac{\delta U}{\delta m}(t,y,\mu)(x) \cdot \partial_y U(t,y,\mu) \\
&\qquad + \frac{\delta f_0}{\delta m}(t,y,\mu)(x)\Big] d\mu(y) = 0,
\end{aligned}$$

for $(t, x, \mu) \in [0, T] \times \mathbb{R}^d \times \mathcal{P}_2(\mathbb{R}^d)$, where $\partial_y U(t, y, \mu)$ and $\partial_y[\delta U/\delta m](t, y, \mu)(x)$ are seen as column vectors of dimension d. Notice that, on the first line, we recognize the structure of a finite dimensional HJB equation.

Example 6.9 Under the assumptions of Example 6.4 above with $A = \mathbb{R}^k$, we have:

$$
\begin{aligned}
&\partial_t U(t, x, \mu) + \frac{1}{2}\Delta_x U(t, x, \mu) + H\Big(t, x, \mu, \partial_x U(t, x, \mu), \hat{\alpha}\big(t, x, \partial_x U(t, x, \mu)\big)\Big) \\
&\quad + \int_{\mathbb{R}^d} \Big[\frac{1}{2}\Delta_y \frac{\delta U}{\delta m}(t, y, \mu)(x) \\
&\qquad - b\Big(t, y, \hat{\alpha}\big(t, y, \partial_y U(t, y, \mu)\big)\Big) \cdot \partial_y \frac{\delta U}{\delta m}(t, y, \mu)(x) \\
&\qquad + \frac{\delta f_0}{\delta m}(t, y, \mu)(x)\Big] d\mu(y) = 0,
\end{aligned}
$$

for $(t, x, \mu) \in [0, T] \times \mathbb{R}^d \times \mathcal{P}_2(\mathbb{R}^d)$, where we used the fact that $\hat{\alpha}(t, x, y)$ is a zero of $\partial_\alpha H(t, x, \mu, y, \cdot)$ since $A = \mathbb{R}^k$. As in the case of Example 6.8, the first part of this master equation has the structure of a finite dimensional HJB equation.

6.2.5 Connecting the Two Approaches with Mean Field Games

As in the two previous subsections, we try to keep most of the discussions at a heuristic level.

In the previous subsections, we gave two distinct formulations for the optimal control of stochastic differential equations of the McKean-Vlasov type, and for each of them, we developed a dedicated form of the Pontryagin maximum principle. However, it is not clear how these two forms relate to each other. Here, still in an informal way, we explain the connections between these two versions of the Pontryagin maximum principle. Then, we highlight how they can be linked to mean field game problems.

For simplicity, and since we only accounted for mean field games in this specific case, **we shall assume that σ does not depend upon the control** α. In particular, due to the relationship (6.22) between H and K, $\hat{\alpha}$ in (6.18) and α^* in (6.23) coincide, see Remark 6.2. Moreover, $\hat{\alpha}$ and thus α^* as well, only depend upon the variables (t, x, μ, y).

A New Optimization Problem

Starting from the deterministic formulation of the control problem, the strategy is to regard the backward equation in (6.35) as an HJB equation. To do so, we assume throughout this subsection that the solution $\boldsymbol{u} = (u(t, \cdot))_{0 \leq t \leq T}$ of the equation:

$$
\begin{aligned}
&\partial_t u(t,x) \\
&= -K\Big(t,x,\mu_t,\partial_x u(t,x),\partial^2_{xx}u(t,x),\hat{\alpha}\big(t,x,\mu_t,\partial_x u(t,x)\big)\Big) \qquad (6.38)\\
&\quad - \int_{\mathbb{R}^d}\frac{\delta K}{\delta m}\Big(t,y,\mu_t,\partial_y u(t,y),\partial^2_{yy}u(t,y),\alpha^*\big(t,y,\mu_t,\partial_y u(t,y)\big)\Big)(x)d\mu_t(y),
\end{aligned}
$$

for $(t,x) \in [0,T] \times \mathbb{R}^d$, is smooth, where (see (6.23)):

$$
\begin{aligned}
&K\Big(t,x,\mu_t,\partial_x u(t,x),\partial^2_{xx}u(t,x),\alpha^*\big(t,x,\mu_t,\partial_x u(t,x)\big)\Big) \\
&= \inf_{\alpha\in A}\Big[b(t,x,\mu_t,\alpha)\cdot\partial_x u(t,x) \qquad (6.39)\\
&\quad + \frac{1}{2}\mathrm{trace}\big[a(t,x,\mu_t)\partial^2_{xx}u(t,x)\big] + f(t,x,\mu_t,\alpha)\Big].
\end{aligned}
$$

Therefore, this solution $\boldsymbol{u} = (u(t,\cdot))_{0\leqslant t\leqslant T}$ may be understood as the value function of a new stochastic control problem, set over controlled *standard* diffusion processes of the form:

$$
dX_t^{\boldsymbol{\alpha}} = b\big(t,X_t^{\boldsymbol{\alpha}},\mu_t,\alpha_t\big)dt + \sigma\big(t,X_t^{\boldsymbol{\alpha}},\mu_t\big)dW_t, \quad t\in[0,T]\,; \quad X_0^{\boldsymbol{\alpha}} = \xi, \qquad (6.40)
$$

$\boldsymbol{\alpha}$ standing for an A-valued control process as in (6.6), and associated with the cost functional:

$$
\begin{aligned}
J(\boldsymbol{\alpha}) = \mathbb{E}\bigg[& g(X_T^{\boldsymbol{\alpha}},\mu_T) + \int_{\mathbb{R}^d}\frac{\delta g}{\delta m}(y,\mu_T)(X_T^{\boldsymbol{\alpha}})d\mu_T(y) \\
&+ \int_0^T f\big(t,X_t^{\boldsymbol{\alpha}},\mu_t,\alpha_t\big)dt \\
&+ \int_0^T\int_{\mathbb{R}^d}\frac{\delta K}{\delta m}\Big(t,y,\mu_t,\partial_y u(t,y),\partial^2_{yy}u(t,y), \qquad (6.41)\\
&\qquad \alpha^*\big(t,y,\mu_t,\partial_y u(t,y)\big)\Big)(X_t^{\boldsymbol{\alpha}})d\mu_t(y)dt\bigg].
\end{aligned}
$$

It is very important to observe that the flow of measures $\boldsymbol{\mu} = (\mu_t)_{0\leqslant t\leqslant T}$ appearing in the dynamics (6.40) of $\boldsymbol{X}$ and in the definition (6.41) of $J(\boldsymbol{\alpha})$ is not required to coincide with the flow of marginal distributions of $\boldsymbol{X}$. Instead, $\boldsymbol{\mu}$ is merely the forward component of the solution to (6.35). This is in stark contrast with (6.6). In particular, we stress the fact that the control problem (6.40)–(6.41) is not a McKean-Vlasov control problem but a standard control problem.

Optimal Path of the New Optimization Problem

Return to (6.35) and consider an optimal flow $\boldsymbol{\mu} = (\mu_t)_{0\leqslant t\leqslant T}$. Assume also that we can construct a process $\boldsymbol{X} = (X_t)_{0\leqslant t\leqslant T}$ having $\boldsymbol{\mu} = (\mu_t)_{0\leqslant t\leqslant T}$ as flow of marginal laws and solving the McKean-Vlasov SDE associated with the Kolmogorov-Fokker-Planck equation (see the references at the end of the subsection) in (6.35), namely:

$$\begin{aligned} dX_t &= b\Big(t, X_t, \mathcal{L}(X_t), \alpha^*\big(t, X_t, \mathcal{L}(X_t), \partial_x u(t, X_t)\big)\Big)dt \\ &\quad + \sigma\big(t, X_t, \mathcal{L}(X_t)\big)dW_t, \end{aligned} \tag{6.42}$$

for all $t \in [0, T]$, with $X_0 = \xi$ and with:

$$\forall t \in [0, T], \quad \mathcal{L}(X_t) = \mu_t. \tag{6.43}$$

Since $\mathcal{L}(X_t) = \mu_t$ for all $t \in [0, T]$, we may regard $\boldsymbol{X}$ as a controlled diffusion process of the same form as in (6.40), with $\boldsymbol{\alpha}$ equal to:

$$\boldsymbol{\alpha} = \Big(\alpha_t = \alpha^*\big(t, X_t, \mu_t, \partial_x u(t, X_t)\big)\Big)_{0\leqslant t\leqslant T}, \quad t \in [0, T].$$

Since u is the value function of the HJB equation (6.38) and α^* is a minimizer of the Hamiltonian K in α, see (6.39), the process $\boldsymbol{X} = (X_t)_{0\leqslant t\leqslant T}$ is an optimal path for the optimal control problem (6.40)–(6.41), see Lemma 4.47.

Stochastic Pontryagin Maximum Principle for the New Optimization Problem

We now apply the standard stochastic Pontryagin maximum principle to the problem (6.40)–(6.41). The Hamiltonian associated with (6.40)–(6.41) is given by:

$$\begin{aligned} &H'(t, x, y, z, \alpha) \\ &= H(t, x, \mu_t, y, z, \alpha) \\ &\quad + \int_{\mathbb{R}^d} \frac{\delta K}{\delta m}\Big(t, y, \mu_t, \partial_y u(t, y), \partial^2_{yy} u(t, y), \hat{\alpha}\big(t, y, \mu_t, \partial_y u(t, y)\big)\Big)(x) d\mu_t(y), \end{aligned}$$

for $(t, x, y, z, \alpha) \in [0, T] \times \mathbb{R}^d \times \mathbb{R}^d \times \mathbb{R}^{d\times d} \times A$, where we used the identity $\hat{\alpha} = \alpha^*$. Clearly, the minimizer of $H'(t, x, y, z, \cdot)$ with respect to α is the minimizer $\hat{\alpha}(t, x, \mu_t, y)$ of $H(t, x, \mu_t, y, z, \cdot)$. See (6.18) and recall that σ is independent of the control. Therefore, making use of the connection between the L-derivative and the functional derivative, the stochastic Pontryagin forward-backward system associated with (6.40)–(6.41) is:

$$
\left\{
\begin{aligned}
dX'_t &= b\big(t, X'_t, \mu_t, \hat{\alpha}(t, X'_t, \mu_t, Y'_t)\big)dt \\
&\qquad + \sigma\big(t, X'_t, \mu_t\big)dW_t, \quad t \in [0,T]\,; \qquad X'_0 = \xi, \\
dY'_t &= -\Big[\partial_x H\big(t, X'_t, \mu_t, Y'_t, Z'_t, \hat{\alpha}(t, X'_t, \mu_t, Y'_t)\big) \\
&\qquad + \int_{\mathbb{R}^d} \partial_\mu K\big(t, y, \mu_t, \partial_y u(t,y), \partial^2_{yy} u(t,y), \\
&\qquad\qquad \hat{\alpha}\big(t, y, \mu_t, \partial_y u(t,y)\big)\big)(X'_t) d\mu_t(y)\Big] dt \\
&\qquad + Z'_t dW_t, \qquad t \in [0,T],
\end{aligned}
\right.
\tag{6.44}
$$

with terminal condition:

$$
Y'_T = \partial_x g\big(X'_T, \mu_T\big) + \int_{\mathbb{R}^d} \partial_\mu g(y, \mu_T)(X'_T) d\mu_T(y).
$$

Since $\boldsymbol{X}$ in (6.42) is already known to be an optimal path of the optimal control problem (6.40)–(6.41), the standard connection between the HJB equation (6.38) and the Pontryagin system (6.44) prompts us to consider:

$$
\big(X'_t, Y'_t, Z'_t\big)_{0 \leqslant t \leqslant T} = \Big(X_t, \partial_x u(t, X_t), \partial^2_{xx} u(t, X_t) \sigma\big(t, X_t, \mu_t\big)\Big)_{0 \leqslant t \leqslant T}, \tag{6.45}
$$

as a possible candidate for the solution of (6.44). See Subsection 3.3.2 for the case σ constant, the generalization to the current setting being straightforward. The fact that this triple is indeed a solution of (6.44) may be easily checked by writing the PDE satisfied by $\partial_x u$ by differentiation of (6.38), and then by expanding $(\partial_x u(t, X_t))_{0 \leqslant t \leqslant T}$ using Itô's formula.

Recovering the Stochastic Pontryagin System

Plugging the form of $(\boldsymbol{X}', \boldsymbol{Y}', \boldsymbol{Z}')$ given by (6.45) into the backward equation in (6.44), we get:

$$
\left\{
\begin{aligned}
dX'_t &= b\big(t, X'_t, \mu_t, \hat{\alpha}(t, X'_t, \mu_t, Y'_t)\big)dt \\
&\qquad + \sigma\big(t, X'_t, \mu_t\big)dW_t, \qquad t \in [0,T]\,; \quad X'_0 = \xi, \\
dY'_t &= -\Big[\partial_x H\big(t, X'_t, \mu_t, Y'_t, Z'_t, \hat{\alpha}(t, X'_t, \mu_t, Y'_t)\big) \\
&\qquad + \tilde{\mathbb{E}}\Big[\partial_\mu K\big(t, \tilde{X}'_t, \mu_t, \tilde{Y}'_t, \partial^2_{yy} u(t, \tilde{X}'_t), \hat{\alpha}(t, \tilde{X}'_t, \mu_t, \tilde{Y}'_t)\big)(X'_t)\Big]\Big] dt \\
&\qquad + Z'_t dW_t, \qquad t \in [0,T],
\end{aligned}
\right.
\tag{6.46}
$$

with terminal condition:

$$
\begin{aligned}
Y'_T &= \partial_x g\big(X'_T, \mu_T\big) + \int_{\mathbb{R}^d} \partial_\mu g(y, \mu_T)(X'_T) d\mu_T(y) \\
&= \partial_x g\big(X'_T, \mu_T\big) + \tilde{\mathbb{E}}\big[\partial_\mu g(\tilde{X}'_T, \mu_T)(X'_T)\big].
\end{aligned}
$$

Using (6.22), we notice that:

$$
\begin{aligned}
&\partial_\mu K(t,x,\mu,y,z,\alpha)(x') \\
&\quad = \partial_\mu H\big(t,x,\mu,y,\alpha\big)(x') + \frac{1}{2}\partial_\mu\big[\mathrm{trace}\big(\sigma(t,x,\mu)\sigma(t,x,\mu)^\dagger z^\dagger\big)\big](x'),
\end{aligned}
\tag{6.47}
$$

for $(t,x,x',\mu,y,z,\alpha) \in [0,T] \times \mathbb{R}^d \times \mathbb{R}^d \times \mathcal{P}_2(\mathbb{R}^d) \times \mathbb{R}^d \times \mathbb{R}^{d\times d} \times A$, and with $H(t,x,\mu,y,\alpha) = b(t,x,\mu,\alpha)\cdot y + f(t,x,\mu,\alpha)$ denoting the reduced Hamiltonian. For a symmetric matrix z of size $d \times d$ and an integer $i \in \{1,\dots,d\}$, it holds:

$$
\begin{aligned}
&\partial^i_\mu\big[\mathrm{trace}\big(\sigma(t,x,\mu)\sigma(t,x,\mu)^\dagger z^\dagger\big)\big](x') \\
&= \mathrm{trace}\big[\big(\partial^i_\mu\sigma(t,x,\mu)(x')\big)\sigma(t,x,\mu)^\dagger z^\dagger\big] \\
&\qquad + \mathrm{trace}\big[\sigma(t,x,\mu)\big(\partial^i_\mu\sigma(t,x,\mu)^\dagger(x')\big)z^\dagger\big] \\
&= \mathrm{trace}\big[\big(\partial^i_\mu\sigma(t,x,\mu)(x')\big)\sigma(t,x,\mu)^\dagger z^\dagger\big] \\
&\qquad + \mathrm{trace}\big[\sigma(t,x,\mu)\big(\partial^i_\mu\sigma(t,x,\mu)^\dagger(x')\big)z\big] \\
&= \mathrm{trace}\big[\big(\partial^i_\mu\sigma(t,x,\mu)(x')\big)\sigma(t,x,\mu)^\dagger z^\dagger\big] \\
&\qquad + \mathrm{trace}\big[z\sigma(t,x,\mu)\big(\partial^i_\mu\sigma(t,x,\mu)^\dagger(x')\big)\big] \\
&= 2\mathrm{trace}\big[\big(\partial^i_\mu\sigma(t,x,\mu)(x')\big)\big(z\sigma(t,x,\mu)\big)^\dagger\big] \\
&= 2\big(\partial^i_\mu\sigma(t,x,\mu)(x')\big)\cdot\big(z\sigma(t,x,\mu)\big),
\end{aligned}
\tag{6.48}
$$

where ∂^i_μ denotes the i coordinate of the derivative ∂_μ, that is $\partial^i_\mu = (\partial_\mu\cdot)_i$ with the notations used in Chapter 5. Plugging this expression into (6.47), we deduce:

$$
\partial_\mu K(t,x,\mu,y,z,\alpha)(x') = (\partial_\mu H)\big(t,x,\mu,y,z\sigma(t,x,\mu),\alpha\big)(x'),
$$

for $(t,x,x',\mu,y,z,\alpha) \in [0,T] \times \mathbb{R}^d \times \mathbb{R}^d \times \mathcal{P}_2(\mathbb{R}^d) \times \mathbb{R}^d \times \mathbb{R}^{d\times d} \times A$, with z being symmetric. Returning to (6.46), we finally obtain:

$$
\left\{
\begin{aligned}
dX'_t &= b\big(t,X'_t,\mu_t,\hat\alpha(t,X'_t,\mu_t,Y'_t)\big)dt \\
&\qquad +\sigma\big(t,X'_t,\mu_t\big)dW_t, \quad t\in[0,T]\,; \quad X'_0 = \xi, \\
dY'_t &= -\Big[\partial_x H\big(t,X'_t,\mu_t,Y'_t,Z'_t,\hat\alpha(t,X'_t,\mu_t,Y'_t)\big) \\
&\qquad +\tilde{\mathbb{E}}\Big[\partial_\mu H\big(t,\tilde X'_t,\mu_t,\tilde Y'_t,\tilde Z'_t,\hat\alpha\big(t,\tilde X'_t,\mu_t,\tilde Y'_t\big)\big)(X'_t)\Big]\Big]dt \\
&\qquad +Z'_t dW_t, \quad t\in[0,T].
\end{aligned}
\right.
$$

This is remarkable as it shows that, starting from the deterministic formulation of the optimal control of McKean-Vlasov dynamics, we were able to completely recover the forward/backward system (6.32) given by the stochastic Pontryagin principle applied to the probabilistic formulation as described in Subsection 6.2.4. We summarize the above result.

The FBSDE system of McKean-Vlasov type derived from the version of the stochastic Pontryagin maximum principle for the optimal control of McKean-Vlasov dynamics can be identified with the FBSDE system for a standard optimal control problem derived from the Pontryagin maximum principle for the deterministic formulation of the original McKean-Vlasov optimal control problem!

Connection with Mean Field Games

The connection with mean field games should now be clear. The process $\boldsymbol{X}$ defined in (6.42) is an optimal path of the stochastic control problem (6.40)–(6.41) set in the environment $\boldsymbol{\mu} = (\mu_t = \mathcal{L}(X_t))_{0 \leq t \leq T}$. Because of the identity (6.43), it is a solution of the mean field game defined over controlled processes of the form (6.40) with respect to the cost functional I defined in (6.41). The flow $\boldsymbol{\mu} = (\mu_t)_{0 \leq t \leq T}$ therein is now understood as an input. As above, we can reformulate this statement as follows:

The forward/backward PDE system issued from the application of the Pontryagin maximum principle to the deterministic formulation of the McKean-Vlasov problem may be identified with an auxiliary MFG problem. In fact, the FBSDE system given by the Pontryagin principle for the McKean-Vlasov problem may be identified with the FBSDE system given by the application of the Pontryagin principle (in the sense described in Chapters 3 and 4) to this auxiliary MFG problem.

Although the above formulation is quite appealing, it is actually somewhat deceiving. Indeed, the cost functional of the auxiliary MFG problem (6.40)–(6.41) involves the value function u of the primary McKean-Vlasov control problem. Its formulation is thus rather implicit.

However, it is possible to identify interesting cases for which the auxiliary MFG problem has an explicit structure. For example, if we revisit Example 6.4 and assume that b and σ do not depend upon the measure parameter μ, namely:

$$b(t, x, \mu, \alpha) = b(t, x, \alpha), \quad \sigma(t, x, \mu) = \sigma(t, x),$$

and that the running cost f is separable in the sense that:

$$f(t, x, \mu, \alpha) = f_0(t, x, \alpha) + f_1(t, x, \mu),$$

for $(t, x, \mu, \alpha) \in [0, T] \times \mathbb{R}^d \times \mathcal{P}_2(\mathbb{R}^d) \times A$ then, by (6.22), it holds that:

$$\frac{\delta K}{\delta m}(t, x, \mu, y, z, \alpha) = \frac{\delta f}{\delta m}(t, x, \mu, \alpha) = \frac{\delta f_1}{\delta m}(t, x, \mu).$$

Therefore, the auxiliary MFG problem consists in optimizing the cost functional:

$$
\begin{aligned}
I(\boldsymbol{\alpha}) = \mathbb{E}\bigg[& g(X_T^{\boldsymbol{\alpha}}, \mu_T) + \int_{\mathbb{R}^d} \frac{\delta g}{\delta m}(y, \mu_T)(X_T^{\boldsymbol{\alpha}}) d\mu_T(y) \\
& + \int_0^T \Big(f\big(t, X_t^{\boldsymbol{\alpha}}, \mu_t, \alpha_t\big) + \frac{\delta f_1}{\delta m}(t, y, \mu_t)(X_t^{\boldsymbol{\alpha}}) \Big) d\mu_t(y) dt \bigg],
\end{aligned}
\tag{6.49}
$$

where $\boldsymbol{\mu} = (\mu_t)_{0 \leqslant t \leqslant T}$ is a continuous trajectory with values in $\mathcal{P}_2(\mathbb{R}^d)$, under the dynamic constraint:

$$
dX_t^{\boldsymbol{\alpha}} = b\big(t, X_t^{\boldsymbol{\alpha}}, \alpha_t\big)dt + \sigma\big(t, X_t^{\boldsymbol{\alpha}}\big)dW_t, \quad t \in [0, T] \,; \quad X_0^{\boldsymbol{\alpha}} = \xi.
$$

We shall revisit this example in Subsection 6.7.2 through the lenses of potential games.

Direct Comparison of the Two Approaches

Instead of dealing with two different forms of the Pontryagin maximum principle applied to the stochastic and deterministic formulations of the optimal control of McKean-Vlasov dynamics, we may also invoke a direct argument in order to connect the two approaches. This argument could be based on the following two steps:

1. First, we could associate, with any controlled process $\boldsymbol{X} = \boldsymbol{X}^{\boldsymbol{\alpha}}$ as in (6.6), a controlled process driven by a control in closed loop feedback form as in (6.10), with the same marginal distributions as $\boldsymbol{X}^{\boldsymbol{\alpha}}$, and a cost smaller than $J(\boldsymbol{\alpha})$ (recall formula (6.9) for the cost). Since any controlled process, with a square-integrable control in closed loop feedback form, induces a solution to the Kolmogorov-Fokker-Planck equation (6.12), this would prove that the optimal cost under the probabilistic approach is always greater than the optimal cost under the deterministic formulation.

   Such a construction can be achieved in the typical cases considered in Chapters 3 and 4 where the drift b is linear in α and the cost functional f is convex in α. It suffices to consider, as a control in closed loop feedback form, the conditional expectation, given the σ-field $\sigma\{X_t\}$, of the original control α_t, at any time $t \in [0, T]$.
2. In order to prove the converse bound, we could associate, with any controlled flow of distributions $(\mu_t)_{0 \leqslant t \leqslant T}$ solving the Kolmogorov-Fokker-Planck equation (6.12) for some Markovian control $(\phi(t, \cdot))_{0 \leqslant t \leqslant T}$ with the appropriate integrability properties, a diffusion process solving (6.10). This may be achieved in the strong or weak sense, depending on whether the solution to (6.10) can be constructed on the original probability space, or merely on some probability space, say for instance the canonical setup. In the first case, we would deduce that the optimal cost under the probabilistic approach cannot exceed the optimal cost under the deterministic formulation, proving the desired converse bound. In

the second case, we could reach the same conclusion provided that we allow, in the probabilistic formulation, for controlled processes constructed on possibly different spaces. This requires to redefine the cost functional as a function of the joint law of the Brownian motion $\boldsymbol{W}$, of the controlled process $\boldsymbol{X} = \boldsymbol{X}^{\alpha}$ and of the control $\boldsymbol{\alpha}$ (instead of a function of the realizations of $\boldsymbol{\alpha}$). We provide a short account in that direction in Section 6.6.

Motivated by the example of the connection between optimal transportation and stochastic control of McKean-Vlasov diffusion processes given in Subsection 6.2.1, we shall implement the two steps described above in Subsection 6.7.3 in the particular case when the coefficients are given by (6.15).

6.3 Stochastic Pontryagin Principle for Optimality

In this section, we derive a form of the stochastic Pontryagin maximum principle for the optimal control of McKean-Vlasov dynamics in the framework introduced in Subsection 6.2.1. We establish necessary conditions for optimality as well as sufficient conditions when the Hamiltonian satisfies appropriate assumptions of convexity. The role of convexity is twofold. On the one hand, convexity allows for an elegant proof of the necessary part of the Pontryagin principle. On the other hand, convexity plays a major role in the proof of the sufficient condition for optimality. It is indeed mandatory to require the Hamiltonian H to be convex in x, μ and α in order to establish the sufficient part of the Pontryagin principle. As explained in Section 6.4 below, typical examples of these assumptions include cases of drift and volatility functions b and σ being linear functions of x, μ and α, these seemingly restrictive assumptions imposing the same kind of limitation as in the classical (non-McKean-Vlasov) case.

For the time being, we state two forms of the regularity assumptions which will be used in this chapter. Given a filtered complete probability space $(\Omega, \mathcal{F}, \mathbb{F}, \mathbb{P})$ as in Subsection 6.2.1, they are:

Assumption (Pontryagin Optimality).

(A1) The functions b, σ and f are differentiable with respect to (x, α), the mappings $(x, \mu, \alpha) \mapsto \partial_x(b, \sigma, f)(t, x, \mu, \alpha)$ and $(x, \mu, \alpha) \mapsto \partial_\alpha(b, \sigma, f)(t, x, \mu, \alpha)$ being continuous for each $t \in [0, T]$. The functions b, σ and f are also differentiable with respect to the variable μ, the mapping $\mathbb{R}^d \times L^2(\Omega, \mathcal{F}, \mathbb{P}; \mathbb{R}^d) \times A \ni (x, X, \alpha) \mapsto \partial_\mu(b, \sigma, f)(t, x, \mathcal{L}(X), \alpha)(X) \in L^2(\Omega, \mathcal{F}, \mathbb{P}; \mathbb{R}^{d\times d} \times \mathbb{R}^{(d\times d)\times d} \times \mathbb{R}^d)$ being continuous for each $t \in [0, T]$. Similarly, the function g is differentiable with respect to x, the mapping $(x, \mu) \mapsto \partial_x g(x, \mu)$ being continuous.

(continued)

The function g is also differentiable with respect to the variable μ, the mapping $\mathbb{R}^d \times L^2(\Omega, \mathcal{F}, \mathbb{P}; \mathbb{R}^d) \ni (x, X) \mapsto \partial_\mu g(x, \mathcal{L}(X))(X) \in L^2(\Omega, \mathcal{F}, \mathbb{P}; \mathbb{R}^d)$ being continuous.

(A2) The function $[0, T] \ni t \mapsto (b, \sigma, f)(t, 0, \delta_0, 0)$ is uniformly bounded. The derivatives $\partial_x(b, \sigma)$ and $\partial_\alpha(b, \sigma)$ are uniformly bounded and the mapping $x' \mapsto \partial_\mu(b, \sigma)(t, x, \mu, \alpha)(x')$ has an $L^2(\mathbb{R}^d, \mu; \mathbb{R}^d)$-norm which is also uniformly bounded (i.e., uniformly in (t, x, μ, α)). There exists a constant L such that, for any $R \geqslant 0$ and any (t, x, μ, α) such that $|x|, M_2(\mu), |\alpha| \leqslant R$, $|\partial_x f(t, x, \mu, \alpha)|$, $|\partial_x g(x, \mu)|$, and $|\partial_\alpha f(t, x, \mu, \alpha)|$ are bounded by $L(1 + R)$ and the $L^2(\mathbb{R}^d, \mu; \mathbb{R}^d)$-norms of $x' \mapsto \partial_\mu f(t, x, \mu, \alpha)(x')$ and $x' \mapsto \partial_\mu g(x, \mu)(x')$ are bounded by $L(1 + R)$.

Observe that our formulation of the joint differentiability is very much in the spirit of Subsection 5.3.4. Also notice that we used the notation $M_2(\mu)^2$ introduced in (3.7) for the second moment of a measure:

$$M_2(\mu)^2 = \int_{\mathbb{R}^d} |x|^2 d\mu(x), \quad \mu \in \mathcal{P}_2(\mathbb{R}^d).$$

Note that assumption **Pontryagin Optimality** implies assumption **MKV Lipschitz Regularity**.

Throughout the section, $(\tilde{\Omega}, \tilde{\mathcal{F}}, \tilde{\mathbb{P}})$ denotes a copy of $(\Omega, \mathcal{F}, \mathbb{P})$. It will be used for expanding the L-derivatives of the coefficients driving the optimal control problem. The expectation under $\tilde{\mathbb{P}}$ is denoted by $\tilde{\mathbb{E}}$.

6.3.1 A Necessary Condition

As before, we assume that the set A of control values is a closed convex subset of $\mathbb{R}^k$ and we denote by $\mathbb{A}$ the set of admissible control processes and by $\boldsymbol{X} = \boldsymbol{X}^{\boldsymbol{\alpha}}$ the controlled state process, namely the solution of (6.6) with a given initial condition $X_0 = \xi \in L^2(\Omega, \mathcal{F}_0, \mathbb{P}; \mathbb{R}^d)$. The filtration $\mathbb{F}$ is assumed to be generated by $\mathcal{F}_0$ and by $\boldsymbol{W}$. Our first task is to compute the Gâteaux derivative of the cost functional J at $\boldsymbol{\alpha}$ in all directions. In order to do so, we choose $\boldsymbol{\beta} \in \mathbb{H}^{2,k}$ such that $\boldsymbol{\alpha} + \epsilon\boldsymbol{\beta} \in \mathbb{A}$ for $\epsilon > 0$ small enough. We then compute the variation of J at $\boldsymbol{\alpha}$ in the direction of $\boldsymbol{\beta}$ (think of $\boldsymbol{\beta}$ as the difference between another element of $\mathbb{A}$ and $\boldsymbol{\alpha}$).

Derivative of the Controlled Process

Using the notation $\boldsymbol{\theta} = (\theta_t = (X_t, \mathcal{L}(X_t), \alpha_t))_{0 \leqslant t \leqslant T}$, we define the variation process $V = (V_t)_{0 \leqslant t \leqslant T}$ as the solution of the stochastic differential equation:

$$dV_t = \big[\gamma_t V_t + \rho_t(\mathcal{L}(X_t, V_t)) + \eta_t\big]dt + \big[\hat{\gamma}_t V_t + \hat{\rho}_t(\mathcal{L}(X_t, V_t)) + \hat{\eta}_t\big]dW_t, \qquad (6.50)$$

with $V_0 = 0$, where the coefficients γ_t, δ_t, ρ_t, $\hat{\gamma}_t$, $\hat{\rho}_t$, and $\hat{\eta}_t$ are defined by:

$$\gamma_t = \partial_x b(t, \theta_t), \quad \hat{\gamma}_t = \partial_x \sigma(t, \theta_t), \quad \eta_t = \partial_\alpha b(t, \theta_t)\beta_t, \quad \hat{\eta}_t = \partial_\alpha \sigma(t, \theta_t)\beta_t.$$

They are progressively measurable bounded processes with values in the spaces $\mathbb{R}^{d\times d}$, $\mathbb{R}^{(d\times d)\times d}$, $\mathbb{R}^d$, and $\mathbb{R}^{d\times d}$ respectively (the parentheses around $d \times d$ indicating that $\hat{\gamma}_t \cdot u$ is seen as an element of $\mathbb{R}^{d\times d}$ whenever $u \in \mathbb{R}^d$), and:

$$\begin{aligned}
\rho_t &= \tilde{\mathbb{E}}\big[\partial_\mu b(t, \theta_t)(\tilde{X}_t)\tilde{V}_t\big] = \tilde{\mathbb{E}}\big[\partial_\mu b(t, x, \mathcal{L}(X_t), \alpha)(\tilde{X}_t)\tilde{V}_t\big]\Big|_{\substack{x=X_t \\ \alpha=\alpha_t}}, \\
\hat{\rho}_t &= \tilde{\mathbb{E}}\big[\partial_\mu \sigma(t, \theta_t)(\tilde{X}_t)\tilde{V}_t\big] = \tilde{\mathbb{E}}\big[\partial_\mu \sigma(t, x, \mathcal{L}(X_t), \alpha)(\tilde{X}_t)\tilde{V}_t\big]\Big|_{\substack{x=X_t \\ \alpha=\alpha_t}},
\end{aligned} \tag{6.51}$$

which are progressively measurable processes with values in $\mathbb{R}^d$ and $\mathbb{R}^{d\times d}$ respectively, and where $(\tilde{X}_t, \tilde{V}_t)$ is a copy of (X_t, V_t) defined on $(\tilde{\Omega}, \tilde{\mathcal{F}}, \tilde{\mathbb{P}})$. We refer to Subsection 5.3.4 for a complete account of the measurability properties. As expectations of functions of $(\tilde{X}_t, \tilde{V}_t)$, ρ_t, and $\hat{\rho}_t$ depend upon the joint distribution of X_t and V_t. In (6.50) we wrote $\rho_t(\mathcal{L}(X_t, V_t))$ and $\hat{\rho}_t(\mathcal{L}(X_t, V_t))$ in order to stress this dependence upon the joint distribution of X_t and V_t. Even though we are dealing with possibly random coefficients, the existence and uniqueness of the variation process are guaranteed by a suitable version of Theorem 4.21 applied to the couple (X, V) and the system formed by (6.6) and (6.50). Our assumption on the boundedness of the partial derivatives of the coefficients implies that $\boldsymbol{V}$ satisfies $\mathbb{E}[\sup_{0\leqslant t\leqslant T} |V_t|^p] < \infty$ for every finite $p \geqslant 1$. In particular $(\rho_t)_{0\leq t\leq T}$ and $(\hat{\rho}_t)_{0\leq t\leq T}$ are bounded.

Lemma 6.10 *For $\epsilon > 0$ small enough, we denote by $\boldsymbol{\alpha}^\epsilon$ the admissible control defined by $\alpha_t^\epsilon = \alpha_t + \epsilon\beta_t$, and by $\boldsymbol{X}^\epsilon = \boldsymbol{X}^{\boldsymbol{\alpha}^\epsilon}$ the corresponding controlled state. We have:*

$$\lim_{\epsilon \searrow 0} \mathbb{E}\left[\sup_{0\leqslant t\leqslant T} \left|\frac{X_t^\epsilon - X_t}{\epsilon} - V_t\right|^2\right] = 0. \tag{6.52}$$

Proof. For the purpose of this proof we set $\theta_t^\epsilon = (X_t^\epsilon, \mathcal{L}(X_t^\epsilon), \alpha_t^\epsilon)$ and $V_t^\epsilon = \epsilon^{-1}(X_t^\epsilon - X_t) - V_t$. Notice that $V_0^\epsilon = 0$ and that:

$$\begin{aligned}
dV_t^\epsilon &= \Big[\frac{1}{\epsilon}\big[b(t, \theta_t^\epsilon) - b(t, \theta_t)\big] - \partial_x b(t, \theta_t) \cdot V_t - \partial_\alpha b(t, \theta_t) \cdot \beta_t \\
&\quad - \tilde{\mathbb{E}}\big[\partial_\mu b(t, \theta_t)(\tilde{X}_t) \cdot \tilde{V}_t\big]\Big] dt \\
&\quad + \Big[\frac{1}{\epsilon}\big[\sigma(t, \theta_t^\epsilon) - \sigma(t, \theta_t)\big] - \partial_x \sigma(t, \theta_t) \cdot V_t - \partial_\alpha \sigma(t, \theta_t) \cdot \beta_t \\
&\quad - \tilde{\mathbb{E}}\big[\partial_\mu \sigma(t, \theta_t)(\tilde{X}_t) \cdot \tilde{V}_t\big]\Big] dW_t \\
&= V_t^{\epsilon,1} dt + V_t^{\epsilon,2} dW_t.
\end{aligned} \tag{6.53}$$

Now for each $t \in [0, T]$ and each $\epsilon > 0$, we have:

$$\frac{1}{\epsilon}\big[b(t,\theta_t^\epsilon) - b(t,\theta_t)\big] = \int_0^1 \partial_x b\big(t,\theta_t^{\lambda,\epsilon}\big)\cdot(V_t^\epsilon + V_t)\,d\lambda + \int_0^1 \partial_\alpha b\big(t,\theta_t^{\lambda,\epsilon}\big)\cdot\beta_t\,d\lambda$$
$$+ \int_0^1 \tilde{\mathbb{E}}\big[\partial_\mu b\big(t,\theta_t^{\lambda,\epsilon}\big)(\tilde{X}_t^{\lambda,\epsilon})\cdot(\tilde{V}_t^\epsilon + \tilde{V}_t)\big]\,d\lambda,$$

where, in order to simplify the notation, we set $X_t^{\lambda,\epsilon} = X_t + \lambda\epsilon(V_t^\epsilon + V_t)$, $\alpha_t^{\lambda,\epsilon} = \alpha_t + \lambda\epsilon\beta_t$ and $\theta_t^{\lambda,\epsilon} = (X_t^{\lambda,\epsilon}, \mathcal{L}(X_t^{\lambda,\epsilon}), \alpha_t^{\lambda,\epsilon})$. Computing the '$dt$'-term, we get:

$$\begin{aligned}
V_t^{\epsilon,1} &= \int_0^1 \partial_x b\big(t,\theta_t^{\lambda,\epsilon}\big)\cdot V_t^\epsilon\,d\lambda + \int_0^1 \tilde{\mathbb{E}}\big[\partial_\mu b\big(t,\theta_t^{\lambda,\epsilon}\big)(\tilde{X}_t^{\lambda,\epsilon})\cdot\tilde{V}_t^\epsilon\big]\,d\lambda \\
&\quad + \int_0^1 \big[\partial_x b\big(t,\theta_t^{\lambda,\epsilon}\big) - \partial_x b(t,\theta_t)\big]\cdot V_t\,d\lambda + \int_0^1 \big[\partial_\alpha b\big(t,\theta_t^{\lambda,\epsilon}\big) - \partial_\alpha b(t,\theta_t)\big]\cdot\beta_t\,d\lambda \\
&\quad + \int_0^1 \tilde{\mathbb{E}}\big[\big(\partial_\mu b\big(t,\theta_t^{\lambda,\epsilon}\big)(\tilde{X}_t^{\lambda,\epsilon}) - \partial_\mu b(t,\theta_t)(\tilde{X}_t)\big)\cdot\tilde{V}_t\big]\,d\lambda \\
&= \int_0^1 \partial_x b\big(t,\theta_t^{\lambda,\epsilon}\big)\cdot V_t^\epsilon\,d\lambda + \int_0^1 \tilde{\mathbb{E}}\big[\partial_\mu b\big(t,\theta_t^{\lambda,\epsilon}\big)(\tilde{X}_t^{\lambda,\epsilon})\cdot\tilde{V}_t^\epsilon\big]\,d\lambda \\
&\quad + I_t^{\epsilon,1} + I_t^{\epsilon,2} + I_t^{\epsilon,3}.
\end{aligned}$$

By (**A2**) in assumption **Pontryagin Optimality**, the last three terms of the above right-hand side are bounded in $L^2([0,T]\times\Omega)$, uniformly in ϵ. Next, we treat the diffusion part $V^{\epsilon,2}$ in the same way using Jensen's inequality and Burkholder-Davis-Gundy's inequality to control the quadratic variation of the stochastic integrals. Consequently, going back to (6.53), we see that, for any $S \in [0,T]$,

$$\mathbb{E}\big[\sup_{0\leqslant t\leqslant S}|V_t^\epsilon|^2\big] \leqslant c' + c'\int_0^S \mathbb{E}\big[\sup_{0\leqslant s\leqslant t}|V_s^\epsilon|^2\big]dt,$$

where as usual $c' > 0$ is a generic constant whose value can change from line to line, as long as it remains independent of ϵ. Applying Gronwall's inequality, we deduce that $\mathbb{E}[\sup_{0\leqslant t\leqslant T}|V_t^\epsilon|^2] \leqslant c'$. Therefore, we have:

$$\lim_{\epsilon\searrow 0}\mathbb{E}\big[\sup_{0\leqslant\lambda\leqslant 1}\sup_{0\leqslant t\leqslant T}\big|X_t^{\lambda,\epsilon} - X_t\big|^2\big] = 0.$$

We then prove that $I^{\epsilon,1}$, $I^{\epsilon,2}$ and $I^{\epsilon,3}$ converge to 0 in $L^2([0,T]\times\Omega)$ as $\epsilon \searrow 0$. Indeed,

$$\begin{aligned}
\mathbb{E}\int_0^T |I_t^{\epsilon,1}|^2 dt &= \mathbb{E}\int_0^T \bigg|\int_0^1 \big([\partial_x b\big(t,\theta_t^{\lambda,\epsilon}\big) - \partial_x b(t,\theta_t)]\cdot V_t\big)d\lambda\bigg|^2 dt \\
&\leqslant \mathbb{E}\int_0^T\int_0^1 |\partial_x b\big(t,\theta_t^{\lambda,\epsilon}\big) - \partial_x b(t,\theta_t)|^2|V_t|^2 d\lambda dt.
\end{aligned}$$

Since the function $\partial_x b$ is bounded and continuous in x, μ, and α, the above right-hand side converges to 0 as $\epsilon \searrow 0$. A similar argument applies to $I_t^{\epsilon,2}$ and $I_t^{\epsilon,3}$. Again, we treat the diffusion part $V^{\epsilon,2}$ in the same way using Jensen's inequality and Burkholder-Davis-Gundy's inequality. Consequently, going back to (6.53), we finally see that, for any $S \in [0, T]$,

$$\mathbb{E}[\sup_{0\leqslant t\leqslant S} |V_t^\epsilon|^2] \leqslant c' \int_0^S \mathbb{E}[\sup_{0\leqslant s\leqslant t} |V_s^\epsilon|^2]dt + c_\epsilon,$$

where $\lim_{\epsilon\searrow 0} c_\epsilon = 0$. Finally, we get the desired result applying Gronwall's inequality. □

Gâteaux Derivative of the Objective Function

We now compute the Gâteaux derivative of the objective function.

Lemma 6.11 *The function* $\mathbb{A} \ni \boldsymbol{\alpha} \mapsto J(\boldsymbol{\alpha})$ *is Gâteaux differentiable in the direction* $\boldsymbol{\beta}$ *and its derivative is given by:*

$$\begin{aligned}
&\frac{d}{d\epsilon}J(\boldsymbol{\alpha}+\epsilon\boldsymbol{\beta})\big|_{\epsilon=0} \\
&= \mathbb{E}\int_0^T \big[\partial_x f(t,\theta_t)\cdot V_t + \tilde{\mathbb{E}}[\partial_\mu f(t,\theta_t)(\tilde{X}_t)\cdot \tilde{V}_t] + \partial_\alpha f(t,\theta_t)\cdot \beta_t\big]dt \\
&\qquad + \mathbb{E}\big[\partial_x g(X_T,\mathcal{L}(X_T))\cdot V_T + \tilde{\mathbb{E}}[\partial_\mu g(X_T,\mathcal{L}(X_T))(\tilde{X}_T)\cdot \tilde{V}_T]\big].
\end{aligned} \tag{6.54}$$

Proof. We use freely the notation introduced in the proof of the previous lemma.

$$\begin{aligned}
&\frac{d}{d\epsilon}J(\boldsymbol{\alpha}+\epsilon\boldsymbol{\beta})\big|_{\epsilon=0} \\
&= \lim_{\epsilon\searrow 0}\frac{1}{\epsilon}\mathbb{E}\bigg[\int_0^T \big[f(t,\theta_t^\epsilon) - f(t,\theta_t)\big]dt + \big[g(X_T^\epsilon,\mathcal{L}(X_T^\epsilon)) - g(X_T,\mathcal{L}(X_T))\big]\bigg].
\end{aligned} \tag{6.55}$$

Computing the two limits separately we get:

$$\begin{aligned}
\lim_{\epsilon\searrow 0}\frac{1}{\epsilon}\mathbb{E}\int_0^T \big[f(t,\theta_t^\epsilon) - f(t,\theta_t)\big]dt &= \lim_{\epsilon\searrow 0}\frac{1}{\epsilon}\mathbb{E}\int_0^T\int_0^1 \frac{d}{d\lambda}\{f(t,\theta_t^{\lambda,\epsilon})\}\,d\lambda dt \\
&= \lim_{\epsilon\searrow 0}\mathbb{E}\int_0^T\int_0^1 \big[\partial_x f(t,\theta_t^{\lambda,\epsilon})\cdot(V_t^\epsilon + V_t) \\
&\qquad + \tilde{\mathbb{E}}[\partial_\mu f(t,\theta_t^{\lambda,\epsilon})(\tilde{X}_t^{\lambda,\epsilon})\cdot(\tilde{V}_t^\epsilon + \tilde{V}_t)] + \partial_\alpha f(t,\theta_t^{\lambda,\epsilon})\cdot\beta_t\big]\,d\lambda dt \\
&= \mathbb{E}\int_0^T \big[\partial_x f(t,\theta_t)\cdot V_t + \tilde{\mathbb{E}}\big[\partial_\mu f(t,\theta_t)(\tilde{X}_t)\cdot\tilde{V}_t\big] + \partial_\alpha f(t,\theta_t)\cdot\beta_t\big]dt,
\end{aligned}$$

where we used the hypothesis on the continuity and growth of the derivatives of f, the uniform convergence proven in the previous lemma, and standard uniform integrability arguments. The second term in (6.55) is handled in a similar way. □

Since conditions (6.29)–(6.30) are satisfied under assumption **Pontryagin Optimality**, the duality relationship is given by:

Lemma 6.12 *Given $(Y_t, Z_t)_{0 \leqslant t \leqslant T}$ as in Definition 6.5, it holds:*

$$\mathbb{E}[Y_T \cdot V_T] = \mathbb{E}\int_0^T \big[Y_t \cdot \big(\partial_\alpha b(t,\theta_t)\beta_t\big) + Z_t \cdot \big(\partial_\alpha \sigma(t,\theta_t)\beta_t\big) \\ - \partial_x f(t,\theta_t) \cdot V_t - \tilde{\mathbb{E}}\big[\partial_\mu f(t,\theta_t)(\tilde{X}_t) \cdot \tilde{V}_t\big]\big]\,dt. \tag{6.56}$$

Proof. Letting $\Theta_t = (X_t, \mathcal{L}(X_t), Y_t, Z_t, \alpha_t)$ and using the definitions (6.50) of the variation process $\boldsymbol{V}$, and (6.31) or (6.32) of the adjoint process $\boldsymbol{Y}$, integration by parts gives:

$$\begin{aligned}
&Y_T \cdot V_T \\
&= Y_0 \cdot V_0 + \int_0^T Y_t \cdot dV_t + \int_0^T dY_t \cdot V_t + \int_0^T d[Y,V]_t \\
&= M_T + \int_0^T \Big[Y_t \cdot \big(\partial_x b(t,\theta_t)V_t\big) + Y_t \cdot \tilde{\mathbb{E}}\big[\partial_\mu b(t,\theta_t)(\tilde{X}_t)\tilde{V}_t\big] + Y_t \cdot \big(\partial_\alpha b(t,\theta_t)\beta_t\big) \\
&\qquad - \partial_x H(t,\Theta_t) \cdot V_t - \tilde{\mathbb{E}}\big[\partial_\mu H(t,\tilde{\Theta}_t)(X_t) \cdot V_t\big] \\
&\qquad + Z_t \cdot \big(\partial_x \sigma(t,\theta_t)V_t\big) + Z_t \cdot \tilde{\mathbb{E}}\big[\partial_\mu \sigma(t,\theta_t)(\tilde{X}_t)\tilde{V}_t\big] \\
&\qquad + Z_t \cdot \big(\partial_\alpha \sigma(t,\theta_t)\beta_t\big)\Big]dt,
\end{aligned}$$

where $(M_t)_{0 \leqslant t \leqslant T}$ is a mean zero integrable martingale. By taking expectations on both sides and applying Fubini's theorem:

$$\begin{aligned}
&\mathbb{E}\tilde{\mathbb{E}}\big[\partial_\mu H(t,\tilde{\Theta}_t)(X_t) \cdot V_t\big] \\
&\quad = \mathbb{E}\tilde{\mathbb{E}}\big[\partial_\mu H(t,\Theta_t)(\tilde{X}_t) \cdot \tilde{V}_t\big] \\
&\quad = \mathbb{E}\tilde{\mathbb{E}}\big[\big(\partial_\mu b(t,\theta_t)(\tilde{X}_t)\tilde{V}_t\big) \cdot Y_t + \big(\partial_\mu \sigma(t,\theta_t)(\tilde{X}_t)\tilde{V}_t\big) \cdot Z_t + \partial_\mu f(t,\theta_t)(\tilde{X}_t) \cdot \tilde{V}_t\big],
\end{aligned}$$

we get the desired equality (6.56) by handling in a similar way the derivatives in x. □

Putting together the duality relation (6.56) and (6.54) we get:

Corollary 6.13 *The Gâteaux derivative of J at $\boldsymbol{\alpha}$ in the direction $\boldsymbol{\beta}$ can be written as:*

$$\frac{d}{d\epsilon}J(\boldsymbol{\alpha} + \epsilon\boldsymbol{\beta})\big|_{\epsilon=0} = \mathbb{E}\int_0^T \partial_\alpha H(t, X_t, \mathcal{L}(X_t), Y_t, Z_t, \alpha_t) \cdot \beta_t\,dt. \tag{6.57}$$

Proof. Using Fubini's theorem, the second expectation appearing in the expression (6.54) of the Gâteaux derivative of J given in Lemma 6.11 can be rewritten as:

$$\begin{aligned}
&\mathbb{E}\big[\partial_x g(X_T, \mathcal{L}(X_T)) \cdot V_T + \tilde{\mathbb{E}}\big(\partial_\mu g(X_T, \mathcal{L}(X_T))(\tilde{X}_T) \cdot \tilde{V}_T\big)\big] \\
&\qquad = \mathbb{E}\big[\partial_x g(X_T, \mathcal{L}(X_T)) \cdot V_T\big] + \mathbb{E}\tilde{\mathbb{E}}\big[\partial_\mu g(\tilde{X}_T, \mathcal{L}(X_T))(X_T) \cdot V_T\big] \\
&\qquad = \mathbb{E}[Y_T \cdot V_T],
\end{aligned}$$

and using the expression derived in Lemma 6.12 for $\mathbb{E}[Y_T \cdot V_T]$ in (6.54) we get the desired result. □

Main Statement

The main result of this subsection is the following:

Theorem 6.14 *Under assumption* **Pontryagin Optimality**, *if we assume further that $\mathbb{F}$ is generated by $\mathcal{F}_0$ and $\boldsymbol{W}$, that the Hamiltonian H is convex in $\alpha \in A$, that the admissible control $\boldsymbol{\alpha} = (\alpha_t)_{0 \leqslant t \leqslant T} \in \mathbb{A}$ is optimal, that $\boldsymbol{X} = (X_t)_{0 \leqslant t \leqslant T}$ is the associated (optimally) controlled state, and that $(\boldsymbol{Y}, \boldsymbol{Z}) = (Y_t, Z_t)_{0 \leqslant t \leqslant T}$ are the associated adjoint processes solving* (6.31), *then we have:*

$$\forall \alpha \in A, \quad H\big(t, X_t, \mathcal{L}(X_t), Y_t, Z_t, \alpha_t\big) \leqslant H\big(t, X_t, \mathcal{L}(X_t), Y_t, Z_t, \alpha\big), \tag{6.58}$$

$Leb_1 \otimes \mathbb{P}$ almost everywhere.

Proof. Since A is convex, given $\boldsymbol{\beta} \in \mathbb{A}$ we can choose the perturbation $\alpha_t^\epsilon = \alpha_t + \epsilon(\beta_t - \alpha_t)$ which is still in $\mathbb{A}$ for $0 \leqslant \epsilon \leqslant 1$. Since $\boldsymbol{\alpha}$ is optimal, we have the inequality

$$\frac{d}{d\epsilon} J(\boldsymbol{\alpha} + \epsilon(\boldsymbol{\beta} - \boldsymbol{\alpha}))\big|_{\epsilon=0} = \mathbb{E}\int_0^T \big[\partial_\alpha H\big(t, X_t, \mathcal{L}(X_t), Y_t, Z_t, \alpha_t\big) \cdot (\beta_t - \alpha_t)\big]\, dt \geqslant 0.$$

By convexity of the Hamiltonian with respect to the control variable $\alpha \in A$, we conclude that

$$\mathbb{E}\int_0^T \big[H\big(t, X_t, \mathcal{L}(X_t), Y_t, Z_t, \beta_t\big) - H\big(t, X_t, \mathcal{L}(X_t), Y_t, Z_t, \alpha_t\big)\big] dt \geqslant 0,$$

for all $\boldsymbol{\beta}$. Now, if for a given (deterministic) $\alpha \in A$ we choose $\boldsymbol{\beta}$ in the following way:

$$\beta_t(\omega) = \begin{cases} \alpha & \text{if } (t, \omega) \in C, \\ \alpha_t(\omega) & \text{otherwise,} \end{cases}$$

for an arbitrary progressively measurable set $C \subset [0, T] \times \Omega$ (that is $C \cap [0, t] \in \mathcal{B}([0, t]) \otimes \mathcal{F}_t$ for any $t \in [0, T]$), we see that:

$$\mathbb{E}\int_0^T \mathbf{1}_C \big[H\big(t, X_t, \mathcal{L}(X_t), Y_t, Z_t, \alpha\big) - H\big(t, X_t, \mathcal{L}(X_t), Y_t, Z_t, \alpha_t\big)\big] dt \geqslant 0,$$

from which we conclude that:

$$H\big(t, X_t, \mathcal{L}(X_t), Y_t, Z_t, \alpha\big) - H\big(t, X_t, \mathcal{L}(X_t), Y_t, Z_t, \alpha_t\big) \geqslant 0 \qquad \mathrm{Leb}_1 \otimes \mathbb{P}\ a.e.,$$

which is the desired conclusion. □

When convexity of the set A does not hold, the following weaker version of the necessary part of the stochastic Pontryagin principle holds:

Proposition 6.15 *Under the same assumption* **Pontryagin Optimality** *as before, without assuming that H is a convex function of α, but requiring now that A is an open subset (not necessarily convex), if we still assume that the admissible control $\boldsymbol{\alpha} = (\alpha_t)_{0 \leqslant t \leqslant T} \in \mathbb{A}$ is optimal, that $\boldsymbol{X} = (X_t)_{0 \leqslant t \leqslant T}$ is the associated (optimally) controlled state, and that $(\boldsymbol{Y}, \boldsymbol{Z}) = (Y_t, Z_t)_{0 \leqslant t \leqslant T}$ are the associated adjoint processes, then, we have:*

$$\partial_\alpha H\big(t, X_t, \mathcal{L}(X_t), Y_t, Z_t, \alpha_t\big) = 0 \qquad \mathrm{Leb}_1 \otimes \mathbb{P}\ a.e..$$

Proof. Given $\epsilon_0 > 0$, $\beta \in \mathbb{R}^k$ with $|\beta| = 1$, and a progressively measurable set $C \subset [0, T] \times \Omega$, we let

$$\beta_t = \beta \mathbf{1}_{C \cap \{\mathrm{dist}(\alpha_t, A^\complement) > \epsilon_0\}},$$

for $t \in [0, T]$. By construction, $\alpha_t + \epsilon \beta_t \in A$ for all $t \in [0, T]$ and $\epsilon \in (0, \epsilon_0)$. Following the proof of Theorem 6.14, we claim:

$$\mathbb{E} \int_0^T \big[\partial_\alpha H\big(t, X_t, \mathcal{L}(X_t), Y_t, Z_t, \alpha_t\big) \cdot \beta_t\big]\, dt \geqslant 0,$$

from which we deduce that:

$$\mathbf{1}_{\{\mathrm{dist}(\alpha_t, A^\complement) > \epsilon_0\}} \partial_\alpha H\big(t, X_t, \mathcal{L}(X_t), Y_t, Z_t, \alpha_t\big) \cdot \beta \geqslant 0 \qquad \mathrm{Leb}_1 \otimes \mathbb{P}\ a.e..$$

As β and ϵ_0 are arbitrary, we finally get:

$$\mathbf{1}_{\{\mathrm{dist}(\alpha_t, A^\complement) > 0\}} \partial_\alpha H\big(t, X_t, \mathcal{L}(X_t), Y_t, Z_t, \alpha_t\big) = 0 \qquad \mathrm{Leb}_1 \otimes \mathbb{P}\ a.e..$$

Recalling that A is open, the result follows. □

6.3.2 A Sufficient Condition

The necessary condition for optimality identified in the previous subsection can be turned into a sufficient condition for optimality under some technical assumptions.

Theorem 6.16 *Under assumption* **Pontryagin Optimality** *as before, let* $\boldsymbol{\alpha} \in \mathbb{A}$ *be an admissible control,* $\mathbf{X} = (X_t)_{0 \leqslant t \leqslant T}$ *the corresponding controlled state process, and* $(\mathbf{Y}, \mathbf{Z}) = (Y_t, Z_t)_{0 \leqslant t \leqslant T}$ *the corresponding adjoint processes. Let us also assume that:*

1. $\mathbb{R}^d \times \mathcal{P}_2(\mathbb{R}^d) \ni (x, \mu) \mapsto g(x, \mu)$ *is convex;*
2. $\mathbb{R}^d \times \mathcal{P}_2(\mathbb{R}^d) \times A \ni (x, \mu, \alpha) \mapsto H(t, x, \mu, Y_t, Z_t, \alpha)$ *is convex* $\mathrm{Leb}_1 \otimes \mathbb{P}$ *almost everywhere.*

If

$$H(t, X_t, \mathcal{L}(X_t), Y_t, Z_t, \alpha_t) = \inf_{\alpha \in A} H(t, X_t, \mathcal{L}(X_t), Y_t, Z_t, \alpha), \tag{6.59}$$

$\mathrm{Leb}_1 \otimes \mathbb{P}$ *a.e., then* $\boldsymbol{\alpha}$ *is an optimal control, i.e.,* $J(\boldsymbol{\alpha}) = \inf_{\boldsymbol{\alpha}' \in \mathbb{A}} J(\boldsymbol{\alpha}')$.

We refer to Section 5.5 for definitions and properties of L-convex functionals. Here, in analogy with (5.74), the convexity property of H takes the form:

$$\begin{aligned} H(t, x', \mu', Y_t, Z_t, \alpha') \geqslant\ & H(t, x, \mu, Y_t, Z_t, \alpha) \\ & + \partial_x H(t, x, \mu, Y_t, Z_t, \alpha) \cdot (x' - x) \\ & + \partial_\alpha H(t, x, \mu, Y_t, Z_t, \alpha) \cdot (\alpha' - \alpha) \\ & + \tilde{\mathbb{E}}\big[\partial_\mu H(t, x, \mu, Y_t, Z_t) \cdot (\tilde{X}' - \tilde{X})\big], \end{aligned} \tag{6.60}$$

$\mathrm{Leb}_1 \otimes \mathbb{P}$ almost surely, for all $t \in [0, T]$, $x, x' \in \mathbb{R}^d$, $\mu, \mu' \in \mathcal{P}_2(\mathbb{R}^d)$ and $\alpha, \alpha' \in A$, when $\tilde{X} \sim \mu$ and $\tilde{X}' \sim \mu'$. A similar definition holds for g.

Remark 6.17 *As made clear by the proof below, the optimal control* $\boldsymbol{\alpha}$ *is unique if* H *is* λ*-strongly convex in* α *for some* $\lambda > 0$*, namely if there is an extra* $+\lambda|\alpha - \alpha'|^2$ *in the right-hand side of* (6.60).

Also, the proof shows that $J(\boldsymbol{\alpha}) \leqslant J(\boldsymbol{\alpha}')$ *for control processes* $\boldsymbol{\alpha}'$ *that are progressively measurable for a larger filtration* $\mathbb{F}'$ *containing* $\mathbb{F}$ *such that* $\mathbf{W}$ *is an* $\mathbb{F}'$*-Brownian motion. The fact that* $\mathbb{F}$ *is generated by* $\mathcal{F}_0$ *and* $\mathbf{W}$ *is just needed to guarantee that the adjoint BSDE* (6.31) *is solvable when* $\boldsymbol{\alpha}$ *is given.*

Proof. Let $\boldsymbol{\alpha}' \in \mathbb{A}$ be a generic admissible control, and $\mathbf{X}' = \mathbf{X}^{\boldsymbol{\alpha}'}$ the corresponding controlled state. By definition of the objective function of the control problem we have:

$$
\begin{aligned}
&J(\boldsymbol{\alpha}) - J(\boldsymbol{\alpha}') \\
&= \mathbb{E}\big[g(X_T, \mathcal{L}(X_T)) - g(X'_T, \mathcal{L}(X'_T))\big] + \mathbb{E}\int_0^T \big[f(t,\theta_t) - f(t,\theta'_t)\big]dt \\
&= \mathbb{E}\big[g(X_T, \mathcal{L}(X_T)) - g(X'_T, \mathcal{L}(X'_T))\big] + \mathbb{E}\int_0^T \big[H(t,\Theta_t) - H(t,\Theta'_t)\big]dt \\
&\quad - \mathbb{E}\int_0^T \big\{\big[b(t,\theta_t) - b(t,\theta'_t)\big]\cdot Y_t + \big[\sigma(t,\theta_t) - \sigma(t,\theta'_t)\big]\cdot Z_t\big\}dt,
\end{aligned}
\tag{6.61}
$$

by definition of the Hamiltonian, where $\theta_t = (X_t, \mathcal{L}(X_t), \alpha_t)$ and $\Theta_t = (X_t, \mathcal{L}(X_t), Y_t, Z_t, \alpha_t)$ (and similarly for θ'_t and Θ'_t, Θ'_t relying on the same Y_t and Z_t as Θ_t). The function g being convex, we have:

$$
g(x,\mu) - g(x',\mu') \leqslant (x - x')\cdot \partial_x g(x,\mu) + \tilde{\mathbb{E}}\big[\partial_\mu g(x,\mu)(\tilde{X})\cdot(\tilde{X} - \tilde{X}')\big],
$$

so that:

$$
\begin{aligned}
&\mathbb{E}\big[g\big(X_T, \mathcal{L}(X_T)\big) - g\big(X'_T, \mathcal{L}(X'_T)\big)\big] \\
&\leqslant \mathbb{E}\big[\partial_x g(X_T, \mathcal{L}(X_T))\cdot (X_T - X'_T) \\
&\qquad\qquad + \tilde{\mathbb{E}}\big[\partial_\mu g(X_T, \mathcal{L}(X_T))(\tilde{X}_T)\cdot(\tilde{X}_T - \tilde{X}'_T)\big]\big] \\
&= \mathbb{E}\big[\big(\partial_x g(X_T, \mathcal{L}(X_T)) + \tilde{\mathbb{E}}[\partial_\mu g(\tilde{X}_T, \mathcal{L}(X_T))(X_T)]\big)\cdot (X_T - X'_T)\big] \\
&= \mathbb{E}\big[Y_T\cdot (X_T - X'_T)\big] = \mathbb{E}\big[(X_T - X'_T)\cdot Y_T\big],
\end{aligned}
\tag{6.62}
$$

where we used Fubini's theorem and the fact that the "tilde random variables" are independent copies of the "non-tilde variables". Using the adjoint equation and taking expectation, we get:

$$
\begin{aligned}
&\mathbb{E}\big[(X_T - X'_T)\cdot Y_T\big] \\
&= \mathbb{E}\bigg[\int_0^T (X_t - X'_t)\cdot dY_t + \int_0^T Y_t\cdot d[X_t - X'_t] + \int_0^T [\sigma(t,\theta_t) - \sigma(t,\theta'_t)]\cdot Z_t dt\bigg] \\
&= -\mathbb{E}\int_0^T \big[\partial_x H(t,\Theta_t)\cdot (X_t - X'_t) + \tilde{\mathbb{E}}\big[\partial_\mu H(t,\tilde{\Theta}_t)(X_t)\big]\cdot (X_t - X'_t)\big]dt \\
&\quad + \mathbb{E}\int_0^T \big[[b(t,\theta_t) - b(t,\theta'_t)]\cdot Y_t + [\sigma(t,\theta_t) - \sigma(t,\theta'_t)]\cdot Z_t\big]dt,
\end{aligned}
$$

where we used integration by parts and the fact that $\boldsymbol{Y} = (Y_t)_{0\leqslant t\leqslant T}$ solves the adjoint equation. Using Fubini's theorem and the fact that $\tilde{\Theta}_t$ is an independent copy of Θ_t, the expectation of the second term in the second line can be rewritten as:

$$
\begin{aligned}
&\mathbb{E}\int_0^T \tilde{\mathbb{E}}\big[\partial_\mu H(t,\tilde{\Theta}_t)(X_t)\big]\cdot(X_t - X_t')dt \\
&= \mathbb{E}\tilde{\mathbb{E}}\int_0^T \big[[\partial_\mu H(t,\Theta_t)(\tilde{X}_t)]\cdot(\tilde{X}_t - \tilde{X}_t')\big]dt \\
&= \mathbb{E}\int_0^T \tilde{\mathbb{E}}\big[\partial_\mu H(t,\Theta_t)(\tilde{X}_t)\cdot(\tilde{X}_t - \tilde{X}_t')\big]dt.
\end{aligned}
\tag{6.63}
$$

Consequently, using (6.61), (6.62), and (6.63), we obtain:

$$
\begin{aligned}
&J(\boldsymbol{\alpha}) - J(\boldsymbol{\alpha}') \\
&\leqslant \mathbb{E}\int_0^T [H(t,\Theta_t) - H(t,\Theta_t')]dt \\
&- \mathbb{E}\int_0^T \Big[\partial_x H(t,\Theta_t)\cdot(X_t - X_t') + \tilde{\mathbb{E}}\big[\partial_\mu H(t,\tilde{\Theta}_t)(\tilde{X}_t)\cdot(\tilde{X}_t - \tilde{X}_t')\big]\Big]dt \leqslant 0,
\end{aligned}
\tag{6.64}
$$

because of the convexity assumption on H, see in particular (6.60), and because of the criticality of the admissible control $\boldsymbol{\alpha} = (\alpha_t)_{0\leqslant t\leqslant T}$, see (6.59), which says $(\alpha_t - \beta)\cdot \partial_\alpha H(t, X_t, \mu_t, Y_t, Z_t, \alpha_t) \leqslant 0$ for all $\beta \in A$, see (3.11) if needed. □

6.3.3 Special Cases

We apply the general formalism developed in this chapter to a set of particular cases which already appeared in the literature, and we provide the special forms of the stochastic Pontryagin principle which apply in these cases. We discuss only sufficient conditions for optimality for the sake of definiteness. The corresponding necessary conditions can easily be derived from the results of Subsection 6.3.1.

Scalar Interactions

We first consider scalar interactions for which the dependence upon the probability measure comes through functions of scalar moments of the measure. More specifically, we assume that:

$$
\begin{aligned}
b(t,x,\mu,\alpha) &= \hat{b}(t,x,\langle\psi,\mu\rangle,\alpha), &\quad \sigma(t,x,\mu,\alpha) &= \hat{\sigma}(t,x,\langle\phi,\mu\rangle,\alpha),\\
f(t,x,\mu,\alpha) &= \hat{f}(t,x,\langle\gamma,\mu\rangle,\alpha), &\quad g(x,\mu) &= \hat{g}(x,\langle\zeta,\mu\rangle),
\end{aligned}
$$

for some scalar continuously differentiable functions ψ, ϕ, γ and ζ with derivatives at most of linear growth, functions $\hat{b}$, $\hat{\sigma}$ and $\hat{f}$ defined on $[0,T]\times\mathbb{R}^d\times\mathbb{R}\times A$ with values in $\mathbb{R}^d$, $\mathbb{R}^{d\times d}$ and $\mathbb{R}$ respectively, and a real valued function $\hat{g}$ defined on $\mathbb{R}^d\times\mathbb{R}$. As before, we use the bracket notation $\langle h,\mu\rangle$ to denote the integral of the function h with respect to the measure μ. The functions $\hat{b}$, $\hat{\sigma}$, $\hat{f}$ and $\hat{g}$ are similar in spirit to the coefficients b, σ, f and g except for the fact that the measure variable μ is now replaced by a numeric variable for which we shall use the notation r for the sake of definiteness. Under these conditions, the Hamiltonian function H reads:

$$H(t,x,\mu,y,z,\alpha) = \hat{b}(t,x,\langle\psi,\mu\rangle,\alpha)\cdot y + \hat{\sigma}(t,x,\langle\phi,\mu\rangle,\alpha)\cdot z + \hat{f}(t,x,\langle\gamma,\mu\rangle,\alpha).$$

We derive the particular form taken by the adjoint equation in the present situation. We start with the terminal condition as it is easier to identify. According to (6.31), it reads:

$$Y_T = \partial_x g(X_T, \mathcal{L}(X_T)) + \tilde{\mathbb{E}}[\partial_\mu g(\tilde{X}_T, \mathcal{L}(\tilde{X}_T))(X_T)].$$

Since the terminal cost is of the form $g(x,\mu) = \hat{g}(x,\langle\zeta,\mu\rangle)$, given our definition of differentiability with respect to the variable μ, we know, as a generalization of (5.35), that $\partial_\mu g(x,\mu)(\cdot)$ reads:

$$\partial_\mu g(x,\mu)(x') = \partial_r \hat{g}\big(x,\langle\zeta,\mu\rangle\big)\partial\zeta(x'), \quad x' \in \mathbb{R}^d.$$

Therefore, the terminal condition Y_T can be rewritten as:

$$Y_T = \partial_x \hat{g}\big(X_T, \mathbb{E}[\zeta(X_T)]\big) + \tilde{\mathbb{E}}\big[\partial_r \hat{g}\big(\tilde{X}_T, \mathbb{E}[\zeta(X_T)]\big)\big]\partial\zeta(X_T).$$

Notice that the 'tildes' can be removed at this stage since $\tilde{X}_T$ has the same distribution as X_T. Note also that if g and $\hat{g}$ do not depend upon x, the function $\mathcal{P}_2(\mathbb{R}^d) \ni \mu \mapsto g(\mu) = \hat{g}(\langle\zeta,\mu\rangle)$ is convex if ζ is convex and $\hat{g}$ is nondecreasing and convex, see Example 1 in Subsection 5.5.1.

Similarly, $\partial_\mu H(t,x,\mu,y,z,\alpha)$ can be identified to the $\mathbb{R}^d$-valued function defined by:

$$\begin{aligned}\partial_\mu H(t,x,\mu,y,z,\alpha)(x') &= \big[\partial_r \hat{b}(t,x,\langle\psi,\mu\rangle,\alpha)\odot y\big]\partial\psi(x')\\ &\quad + \big[\partial_r \hat{\sigma}(t,x,\langle\phi,\mu\rangle,\alpha)\odot z\big]\partial\phi(x')\\ &\quad + \partial_r \hat{f}(t,x,\langle\gamma,\mu\rangle,\alpha)\ \partial\gamma(x'),\end{aligned}$$

and the dynamic part of the adjoint equation (6.31) rewrites:

$$\begin{aligned}dY_t = &-\Big[\partial_x \hat{b}(t,X_t,\mathbb{E}[\psi(X_t)],\alpha_t)\odot Y_t + \partial_x \hat{\sigma}(t,X_t,\mathbb{E}[\phi(X_t)],\alpha_t)\odot Z_t\\ &\quad + \partial_x \hat{f}(t,X_t,\mathbb{E}[\gamma(X_t)],\alpha_t)\Big]dt + Z_t dW_t\\ &-\Big[\tilde{\mathbb{E}}\big[\partial_r \hat{b}(t,\tilde{X}_t,\mathbb{E}[\psi(X_t)],\tilde{\alpha}_t)\odot \tilde{Y}_t\big]\partial\psi(X_t)\\ &\quad + \tilde{\mathbb{E}}\big[\partial_r \hat{\sigma}(t,\tilde{X}_t,\mathbb{E}[\phi(X_t)],\tilde{\alpha}_t)\odot \tilde{Z}_t\big]\partial\phi(X_t)\\ &\quad + \tilde{\mathbb{E}}\big[\partial_r \hat{f}(t,\tilde{X}_t,\mathbb{E}[\gamma(X_t)],\tilde{\alpha}_t)\big]\partial\gamma(X_t)\Big]dt,\end{aligned}$$

which again, can be slightly simplified by *removing the 'tildes'*.

First Order Interactions

In the case of first order interactions, the dependence upon the probability measure is linear in the sense that the coefficients b, σ, f, and g are given in the form:

$$
\begin{aligned}
b(t,x,\mu,\alpha) &= \langle \hat{b}(t,x,\cdot,\alpha), \mu\rangle, & \sigma(t,x,\mu,\alpha) &= \langle \hat{\sigma}(t,x,\cdot,\alpha), \mu\rangle, \\
f(t,x,\mu,\alpha) &= \langle \hat{f}(t,x,\cdot,\alpha), \mu\rangle, & g(x,\mu) &= \langle \hat{g}(x,\cdot), \mu\rangle.
\end{aligned}
$$

for some functions $\hat{b}$, $\hat{\sigma}$, and $\hat{f}$ defined on $[0,T]\times\mathbb{R}^d\times\mathbb{R}^d\times A$ with values in $\mathbb{R}^d$, $\mathbb{R}^{d\times d}$ and $\mathbb{R}$ respectively and continuously differentiable with respect to (x,x',α) with derivatives at most of linear growth, and a real valued function $\hat{g}$ defined on $\mathbb{R}^d\times\mathbb{R}^d$ and continuously differentiable with derivatives at most of linear growth. This form of dependence comes from the original derivation of the McKean-Vlasov equation as limit of large systems of particles whose dynamics are given by stochastic differential equations with *mean field* interactions as in:

$$
\begin{aligned}
dX_t^i &= \frac{1}{N}\sum_{j=1}^N \hat{b}(t,X_t^i,X_t^j)dt \\
&\quad + \frac{1}{N}\sum_{j=1}^N \hat{\sigma}(t,X_t^i,X_t^j)dW_t^j, \quad i=1,\cdots,N, \quad 0\leqslant t\leqslant T,
\end{aligned}
\tag{6.65}
$$

where the $(W^j)_{1\leqslant j\leqslant N}$'s are N independent standard Wiener processes in $\mathbb{R}^d$. In the present situation the linearity in μ implies that $\partial_\mu g(x,\mu)(x') = \partial_{x'}\hat{g}(x,x')$ and similarly:

$$
\begin{aligned}
&\partial_\mu H(t,x,\mu,y,z,\alpha)(x') \\
&\quad = \partial_{x'}\hat{b}(t,x,x',\alpha)\odot y + \partial_{x'}\hat{\sigma}(t,x,x',\alpha)\odot z + \partial_{x'}\hat{f}(t,x,x',\alpha),
\end{aligned}
$$

and the dynamic part of the adjoint equation (6.31) rewrites:

$$
dY_t = -\tilde{\mathbb{E}}\big[\partial_x \hat{H}(t,X_t,\tilde{X}_t,Y_t,Z_t,\alpha_t) + \partial_{x'}\hat{H}(t,\tilde{X}_t,X_t,\tilde{Y}_t,\tilde{Z}_t,\tilde{\alpha}_t)\big]dt + Z_t dW_t,
$$

if we use the obvious notation:

$$
\hat{H}(t,x,x',y,z,\alpha) = \hat{b}(t,x,x',\alpha)\cdot y + \hat{\sigma}(t,x,x',\alpha)\cdot z + \hat{f}(t,x,x',\alpha),
$$

and the terminal condition is given by:

$$
Y_T = \tilde{\mathbb{E}}\big[\partial_x \hat{g}(X_T,\tilde{X}_T) + \partial_{x'}\hat{g}(\tilde{X}_T,X_T)\big].
$$

Notice that g is convex if $\hat{g}$ is convex in the usual sense, see Example 2 in Subsection 5.5.1. Similarly, for all $t\in[0,T]$, $y\in\mathbb{R}^d$ and $z\in\mathbb{R}^{d\times d}$, $\mathbb{R}^d\times\mathcal{P}_2(\mathbb{R}^d)\times A\ni(x,\mu,\alpha)\mapsto H(t,x,\mu,y,z,\alpha)$ is convex if $\mathbb{R}^d\times\mathbb{R}^d\times A\ni(x,x',\alpha)\mapsto \hat{H}(t,x,x',y,z,\alpha)$ is convex in the usual sense.

6.4 Solvability of the Pontryagin FBSDE

We now turn to the application of the Pontryagin stochastic maximum principle to the solution of the optimal control of McKean-Vlasov dynamics. The strategy is to identify a minimizer of the Hamiltonian, and to use it in the forward dynamics and the adjoint equation. This creates a coupling between these equations, leading to the study of an FBSDE of mean field type. As explained in the introduction, the existence results of Chapters 3 and 4 do not cover some of the solvable models such as the linear quadratic (LQ) models. Here we establish existence and uniqueness by taking advantage of the specific structure of the equations inherited from the underlying optimization problem. Assuming that the terminal cost and the Hamiltonian satisfy the same convexity assumptions as in the statement of Theorem 6.16, we prove that unique solvability holds by applying the *continuation method* for FBSDEs.

Throughout the section, we use the same notation as in the previous one: $(\tilde{\Omega}, \tilde{\mathcal{F}}, \tilde{\mathbb{P}})$ denotes a copy of $(\Omega, \mathcal{F}, \mathbb{P})$. The expectation under $\tilde{\mathbb{P}}$ is denoted by $\tilde{\mathbb{E}}$.

6.4.1 Technical Assumptions

We state the conditions we shall use from now on. These assumptions subsume assumption **Pontryagin Optimality** introduced in Section 6.3. As it is most often the case in applications of the stochastic maximum principle, we choose the set A of control values to be a closed convex subset of $\mathbb{R}^k$ and we consider a *linear* model for the forward dynamics of the state.

Assumption (Control of MKV Dynamics). There exist two constants $L \geqslant 0$ and $\lambda > 0$ such that:

(A1) The drift and volatility functions b and σ are linear in x, μ and α. To wit, for all $(t, x, \mu, \alpha) \in [0, T] \times \mathbb{R}^d \times \mathcal{P}_2(\mathbb{R}^d) \times A$, we assume that:

$$b(t, x, \mu, \alpha) = b_0(t) + b_1(t)x + \bar{b}_1(t)\bar{\mu} + b_2(t)\alpha,$$

$$\sigma(t, x, \mu, \alpha) = \sigma_0(t) + \sigma_1(t)x + \bar{\sigma}_1(t)\bar{\mu} + \sigma_2(t)\alpha,$$

for some bounded measurable deterministic functions b_0, b_1, $\bar{b}_1$ and b_2 with values in $\mathbb{R}^d$, $\mathbb{R}^{d\times d}$, $\mathbb{R}^{d\times d}$ and $\mathbb{R}^{d\times k}$, and σ_0, σ_1, $\bar{\sigma}_1$ and σ_2 with values in $\mathbb{R}^{d\times d}$, $\mathbb{R}^{(d\times d)\times d}$, $\mathbb{R}^{(d\times d)\times d}$ and $\mathbb{R}^{(d\times d)\times k}$ (as usual, the parentheses around $d \times d$ indicate that $\sigma(t)u$ is seen as an element of $\mathbb{R}^{d\times d}$ whenever $u \in \mathbb{R}^{\ell}$, with $\ell = k, d$), and where we use the notation $\bar{\mu} = \int x \, d\mu(x)$ for the mean of a measure μ.

Regarding the regularity, we shall assume:

(A2) The functions f and g satisfy the same assumptions as in assumption **Pontryagin Optimality** in Section 6.3. In particular, for all $(t, x, \mu, \alpha) \in [0, T] \times \mathbb{R}^d \times \mathcal{P}_2(\mathbb{R}^d) \times A$,

$$
\begin{aligned}
&\big|f(t,x',\mu',\alpha') - f(t,x,\mu,\alpha)\big| + \big|g(x',\mu') - g(x,\mu)\big| \\
&\quad \leqslant L\big[1 + |x'| + |x| + |\alpha'| + |\alpha| + M_2(\mu) + M_2(\mu')\big] \\
&\qquad \times \big[|(x',\alpha') - (x,\alpha)| + W_2(\mu',\mu)\big].
\end{aligned}
$$

(A3) The derivatives of f and g with respect to (x, α) and x respectively are L-Lipschitz continuous with respect to (x, α, μ) and (x, μ) respectively, the Lipschitz property in the variable μ being understood in the sense of the 2-Wasserstein distance. Moreover, for any $t \in [0, T]$, any $x, x' \in \mathbb{R}^d$, any $\alpha, \alpha' \in \mathbb{R}^k$, any $\mu, \mu' \in \mathcal{P}_2(\mathbb{R}^d)$, and any $\mathbb{R}^d$-valued random variables X and X' having μ and μ' as distributions,

$$
\begin{aligned}
&\mathbb{E}\big[|\partial_\mu f(t,x',\mu',\alpha')(X') - \partial_\mu f(t,x,\mu,\alpha)(X)|^2\big] \\
&\qquad \leqslant L\big(|(x',\alpha') - (x,\alpha)|^2 + \mathbb{E}\big[|X' - X|^2\big]\big), \\
&\mathbb{E}\big[|\partial_\mu g(x',\mu')(X') - \partial_\mu g(x,\mu)(X)|^2\big] \\
&\qquad \leqslant L\big(|x' - x|^2 + \mathbb{E}\big[|X' - X|^2\big]\big).
\end{aligned}
$$

Finally,

(A4) The function f satisfies the L-convexity property:

$$
\begin{aligned}
&f(t,x',\mu',\alpha') - f(t,x,\mu,\alpha) - \partial_{(x,\alpha)} f(t,x,\mu,\alpha) \cdot (x' - x, \alpha' - \alpha) \\
&\qquad - \mathbb{E}\big[\partial_\mu f(t,x,\mu,\alpha)(X) \cdot (X' - X)\big] \geqslant \lambda |\alpha' - \alpha|^2,
\end{aligned}
$$

for $t \in [0, T]$, $(x, \mu, \alpha) \in \mathbb{R}^d \times \mathcal{P}_2(\mathbb{R}^d) \times A$ and $(x', \mu', \alpha') \in \mathbb{R}^d \times \mathcal{P}_2(\mathbb{R}^d) \times A$, whenever $X, X' \in L^2(\Omega, \mathcal{F}, \mathbb{P}; \mathbb{R}^d)$ with distributions μ and μ' respectively. The function g is also assumed to be L-convex in (x, μ).

Comparing (5.43) with (**A3**), we notice that the liftings $L^2(\Omega, \mathcal{F}, \mathbb{P}; \mathbb{R}^d) \ni X \mapsto \tilde{f}(t, x, \mathcal{L}(X), \alpha)$ and $L^2(\Omega, \mathcal{F}, \mathbb{P}; \mathbb{R}^d) \ni X \mapsto g(x, \mathcal{L}(X))$ have Lipschitz continuous derivatives. As a consequence of Proposition 5.36, for any $t \in [0, T]$, $x \in \mathbb{R}^d$, $\mu \in \mathcal{P}_2(\mathbb{R}^d)$ and $\alpha \in A$, there exist versions of $\mathbb{R}^d \ni x' \mapsto \partial_\mu f(t, x, \mu, \alpha)(x')$ and $\mathbb{R}^d \ni x' \mapsto \partial_\mu g(x, \mu)(x')$ which are L-Lipschitz continuous.

6.4.2 The Hamiltonian and the Adjoint Equations

The drift and the volatility being linear, the Hamiltonian takes the particular form:

$$\begin{aligned} H(t, x, \mu, y, z, \alpha) = &\big[b_0(t) + b_1(t)x + \bar{b}_1(t)\bar{\mu} + b_2(t)\alpha\big] \cdot y \\ &+ \big[\sigma_0(t) + \sigma_1(t)x + \bar{\sigma}_1(t)\bar{\mu} + \sigma_2(t)\alpha\big] \cdot z + f(t, x, \mu, \alpha), \end{aligned}$$

for $t \in [0, T]$, $x, y \in \mathbb{R}^d$, $z \in \mathbb{R}^{d\times d}$, $\mu \in \mathcal{P}_2(\mathbb{R}^d)$ and $\alpha \in A$. Given $(t, x, \mu, y, z) \in [0, T] \times \mathbb{R}^d \times \mathcal{P}_2(\mathbb{R}^d) \times \mathbb{R}^d \times \mathbb{R}^{d\times d}$, the function $A \ni \alpha \mapsto H(t, x, \mu, y, z, \alpha)$ is strictly convex so that there exists a unique minimizer $\hat{\alpha}(t, x, \mu, y, z)$:

$$\hat{\alpha}(t, x, \mu, y, z) = \mathrm{argmin}_{\alpha\in A} H(t, x, \mu, y, z, \alpha). \tag{6.66}$$

Assumption **Control of MKV Dynamics** being slightly stronger than the assumptions used in Chapter 3, we can strengthen the conclusions of Lemma 3.3 while still using the same arguments.

Lemma 6.18 *The function* $[0, T] \times \mathbb{R}^d \times \mathcal{P}_2(\mathbb{R}^d) \times \mathbb{R}^d \times \mathbb{R}^{d\times d} \ni (t, x, \mu, y, z) \mapsto \hat{\alpha}(t, x, \mu, y, z) \in A$ *is measurable, locally bounded and Lipschitz continuous with respect to* (x, μ, y, z), *uniformly in* $t \in [0, T]$, *the Lipschitz constant depending only upon* λ, *the supremum norms of* b_2 *and* σ_2 *and the Lipschitz constant of* $\partial_\alpha f$ *in* (x, μ).

Proof. Except maybe for the Lipschitz property with respect to the measure argument, these facts were explicitly proved in Lemma 3.3. Lemma 3.3 applies when $\sigma = 0$, but the proof may be easily adapted to the current setting. The regularity of $\hat{\alpha}$ with respect to μ follows from the following remark. If $(t, x, y, z) \in [0, T]\times\mathbb{R}^d\times\mathbb{R}^d\times\mathbb{R}^{d\times d}$ is fixed and μ, μ' are generic elements in $\mathcal{P}_2(\mathbb{R}^d)$, $\hat{\alpha} = \hat{\alpha}(t, x, \mu, y, z)$ and $\hat{\alpha}' = \hat{\alpha}(t, x, \mu', y, z)$ denoting the associated minimizers, we deduce from the convexity condition (**A4**) in assumption **Control of MKV Dynamics**:

$$\begin{aligned} 2\lambda|\hat{\alpha}' - \hat{\alpha}|^2 &\leqslant (\hat{\alpha}' - \hat{\alpha}) \cdot \big[\partial_\alpha f\big(t, x, \mu, \hat{\alpha}'\big) - \partial_\alpha f\big(t, x, \mu, \hat{\alpha}\big)\big] \\ &= (\hat{\alpha}' - \hat{\alpha}) \cdot \big[\partial_\alpha H\big(t, x, \mu, y, z, \hat{\alpha}'\big) - \partial_\alpha H\big(t, x, \mu, y, z, \hat{\alpha}\big)\big] \\ &\leqslant (\hat{\alpha}' - \hat{\alpha}) \cdot \big[\partial_\alpha H\big(t, x, \mu, y, z, \hat{\alpha}'\big) - \partial_\alpha H\big(t, x, \mu', y, z, \hat{\alpha}'\big)\big] \\ &= (\hat{\alpha}' - \hat{\alpha}) \cdot \big[\partial_\alpha f\big(t, x, \mu, \hat{\alpha}'\big) - \partial_\alpha f\big(t, x, \mu', \hat{\alpha}'\big)\big] \\ &\leqslant C|\hat{\alpha}' - \hat{\alpha}|\ W_2(\mu', \mu), \end{aligned} \tag{6.67}$$

the passage from the second to the third line following from the two inequalities:

$$\forall \beta \in A, \quad (\beta - \hat{\alpha}) \cdot \partial_\alpha H\big(t, x, \mu, y, z, \hat{\alpha}\big) \geqslant 0,$$
$$(\beta - \hat{\alpha}') \cdot \partial_\alpha H\big(t, x, \mu', y, z, \hat{\alpha}'\big) \geqslant 0,$$

from which we get:

$$-(\hat{\alpha}' - \hat{\alpha}) \cdot \partial_\alpha H\big(t, x, \mu, y, z, \hat{\alpha}\big) \leqslant 0 \leqslant -\big(\hat{\alpha}' - \hat{\alpha}\big) \cdot \partial_\alpha H\big(t, x, \mu', y, z, \hat{\alpha}'\big).$$

This concludes the proof. □

For each admissible control $\boldsymbol{\alpha} = (\alpha_t)_{0 \leqslant t \leqslant T}$, if we denote the corresponding solution of the state equation by $\boldsymbol{X} = (X_t)_{0 \leqslant t \leqslant T}$, then the adjoint BSDE (6.31) introduced in Definition 6.5 reads:

$$\begin{aligned} dY_t &= -\partial_x f\big(t, X_t, \mathcal{L}(X_t), \alpha_t\big)dt - b_1(t)^\dagger Y_t dt - \sigma_1(t)^\dagger Z_t dt \\ &\quad - \tilde{\mathbb{E}}\big[\partial_\mu f\big(t, \tilde{X}_t, \mathcal{L}(X_t), \tilde{\alpha}_t\big)(X_t)\big]dt - \bar{b}_1(t)^\dagger \mathbb{E}[Y_t]dt - \bar{\sigma}_1(t)^\dagger \mathbb{E}[Z_t]dt \\ &\quad + Z_t dW_t. \end{aligned} \tag{6.68}$$

Given the necessary and sufficient conditions proven in the previous section, our goal is to use the control $\hat{\boldsymbol{\alpha}} = (\hat{\alpha}_t)_{0 \leqslant t \leqslant T}$ defined by $\hat{\alpha}_t = \hat{\alpha}(t, X_t, \mathcal{L}(X_t), Y_t, Z_t)$ where $\hat{\alpha}$ is the minimizer function constructed above and $(X_t, Y_t, Z_t)_{0 \leqslant t \leqslant T}$ is a solution of the FBSDE:

$$\left\{ \begin{aligned} dX_t &= \big[b_0(t) + b_1(t)X_t + \bar{b}_1(t)\mathbb{E}[X_t] \\ &\qquad\qquad + b_2(t)\hat{\alpha}(t, X_t, \mathcal{L}(X_t), Y_t, Z_t)\big]dt \\ &\quad + \big[\sigma_0(t) + \sigma_1(t)X_t + \bar{\sigma}_1(t)\mathbb{E}[X_t] \\ &\qquad\qquad + \sigma_2(t)\hat{\alpha}(t, X_t, \mathcal{L}(X_t), Y_t, Z_t)\big]dW_t, \\ dY_t &= -\big[\partial_x f\big(t, X_t, \mathcal{L}(X_t), \hat{\alpha}(t, X_t, \mathcal{L}(X_t), Y_t, Z_t)\big) \\ &\qquad\qquad + b_1(t)^\dagger Y_t + \sigma_1(t)^\dagger Z_t\big]dt \\ &\quad - \Big[\tilde{\mathbb{E}}\big[\partial_\mu f\big(t, \tilde{X}_t, \mathcal{L}(X_t), \hat{\alpha}(t, \tilde{X}_t, \mathcal{L}(X_t), \tilde{Y}_t, \tilde{Z}_t)\big)(X_t)\big] \\ &\qquad\qquad + \bar{b}_1(t)^\dagger \mathbb{E}[Y_t] + \bar{\sigma}_1(t)^\dagger \mathbb{E}[Z_t]\Big]dt \\ &\quad + Z_t dW_t, \end{aligned} \right. \tag{6.69}$$

with initial condition $X_0 = \xi \in L^2(\Omega, \mathcal{F}_0, \mathbb{P}; \mathbb{R}^d)$ and terminal condition $Y_T = \partial_x g(X_T, \mathcal{L}(X_T)) + \tilde{\mathbb{E}}[\partial_\mu g(\tilde{X}_T, \mathcal{L}(X_T))(X_T)]$.

6.4.3 Main Existence and Uniqueness Result

The system (6.69) is a McKean-Vlasov FBSDE of the same type as (4.30) in Chapter 4. We shall prove the following specific solvability result:

Theorem 6.19 *Under assumption* **Control of MKV Dynamics** *and for any initial condition* $\xi \in L^2(\Omega, \mathcal{F}_0, \mathbb{P}; \mathbb{R}^d)$*, the forward-backward system* (6.69) *is uniquely solvable.*

Observe from Theorem 6.16 that, under the above assumption, the solution of (6.69) is the unique optimal path of the mean field stochastic control problem defined in (6.6)–(6.9).

The proof is an adaptation of the *continuation method* for FBSDEs. The idea is to prove that existence and uniqueness are preserved under small perturbations of the coefficients. Starting from a case for which existence and uniqueness are known to hold, we then establish Theorem 6.19 by modifying iteratively the coefficients so that (6.69) is eventually shown to belong to the class of uniquely solvable systems. A simple strategy is to modify the coefficients in a linear way. The notations becoming quickly unruly, we use the following conventions.

Parameterized Solutions

Like in Subsection 6.3.1, the notation $(\Theta_t)_{0\leqslant t\leqslant T}$ stands for stochastic processes of the form $(X_t, \mathcal{L}(X_t), Y_t, Z_t, \alpha_t)_{0\leqslant t\leqslant T}$ with values in $\mathbb{R}^d \times \mathcal{P}_2(\mathbb{R}^d) \times \mathbb{R}^d \times \mathbb{R}^{d\times d} \times A$. We will denote by $\mathscr{S}$ the space of processes $\boldsymbol{\Theta} = (\Theta_t)_{0\leqslant t\leqslant T}$ such that $(X_t, Y_t, Z_t, \alpha_t)_{0\leqslant t\leqslant T}$ is $\mathbb{F}$–progressively measurable, $\boldsymbol{X} = (X_t)_{0\leqslant t\leqslant T}$ and $\boldsymbol{Y} = (Y_t)_{0\leqslant t\leqslant T}$ have continuous sample paths, and

$$\|\boldsymbol{\Theta}\|_{\mathscr{S}} = \mathbb{E}\biggl[\sup_{0\leqslant t\leqslant T}\bigl[|X_t|^2 + |Y_t|^2\bigr] + \int_0^T \bigl[|Z_t|^2 + |\alpha_t|^2\bigr]dt\biggr]^{1/2} < \infty. \tag{6.70}$$

Similarly, the notation $(\theta_t)_{0\leqslant t\leqslant T}$ is generic for processes $(X_t, \mathcal{L}(X_t), \alpha_t)_{0\leqslant t\leqslant T}$ with values in $\mathbb{R}^d \times \mathcal{P}_2(\mathbb{R}^d) \times A$. All the processes $(\theta_t)_{0\leqslant t\leqslant T}$ considered below will be restrictions of *extended* processes $(\Theta_t)_{0\leqslant t\leqslant T} \in \mathscr{S}$.

An input for (6.69) will be a four-tuple $\mathcal{I} = ((\mathcal{I}_t^b, \mathcal{I}_t^\sigma, \mathcal{I}_t^f)_{0\leqslant t\leqslant T}, \mathcal{I}_T^g)$, $(\mathcal{I}_t^b)_{0\leqslant t\leqslant T}$, $(\mathcal{I}_t^\sigma)_{0\leqslant t\leqslant T}$ and $(\mathcal{I}_t^f)_{0\leqslant t\leqslant T}$ being three square-integrable progressively measurable processes with values in $\mathbb{R}^d$, $\mathbb{R}^{d\times d}$ and $\mathbb{R}^d$ respectively, and $\mathcal{I}_T^g$ denoting a square-integrable $\mathcal{F}_T$-measurable random variable with values in $\mathbb{R}^d$. Such an input is specifically designed to be injected into the dynamics of (6.69), $\mathcal{I}^b$ being plugged into the drift of the forward equation, $\mathcal{I}^\sigma$ into the volatility of the forward equation, $\mathcal{I}^f$ into the bounded variation term of the backward equation and $\mathcal{I}^g$ into the terminal condition of the backward equation. The space of inputs is denoted by $\mathbb{I}$. This justifies their respective dimensions. It is endowed with the norm:

$$\|\mathcal{I}\|_{\mathbb{I}} = \mathbb{E}\left[|\mathcal{I}_T^g|^2 + \int_0^T \left[|\mathcal{I}_t^b|^2 + |\mathcal{I}_t^\sigma|^2 + |\mathcal{I}_t^f|^2 \right] dt \right]^{1/2}. \tag{6.71}$$

Definition 6.20 *For any $\gamma \in [0,1]$, $\xi \in L^2(\Omega, \mathcal{F}_0, \mathbb{P}; \mathbb{R}^d)$ and input $\mathcal{I} \in \mathbb{I}$, the FBSDE:*

$$\begin{cases} dX_t = \big(\gamma b(t,\theta_t) + \mathcal{I}_t^b\big)dt + \big(\gamma\sigma(t,\theta_t) + \mathcal{I}_t^\sigma\big)dW_t, \\ dY_t = -\big(\gamma\{\partial_x H(t,\Theta_t) + \tilde{\mathbb{E}}[\partial_\mu H(t,\tilde{\Theta}_t)(X_t)]\} + \mathcal{I}_t^f\big)dt \\ \qquad +Z_t dW_t, \quad t \in [0,T], \end{cases} \tag{6.72}$$

with:

$$\alpha_t = \hat{\alpha}(t, X_t, \mathcal{L}(X_t), Y_t, Z_t), \quad t \in [0,T], \tag{6.73}$$

as optimality condition, $X_0 = \xi$ as initial condition, and

$$Y_T = \gamma\{\partial_x g(X_T, \mathcal{L}(X_T)) + \tilde{\mathbb{E}}[\partial_\mu g(\tilde{X}_T, \mathcal{L}(X_T))(X_T)]\} + \mathcal{I}_T^g$$

as terminal condition, is referred to as $\mathcal{E}(\gamma,\xi,\mathcal{I})$.

Whenever $(X_t,Y_t,Z_t)_{0\leqslant t\leqslant T}$ is a solution, the full process $\boldsymbol{\Theta} = (\Theta_t)_{0\leqslant t\leqslant T} = (X_t, \mathcal{L}(X_t), Y_t, Z_t, \alpha_t)_{0\leqslant t\leqslant T}$ is referred to as the associated extended solution.

Remark 6.21 *The way the coupling between the forward and backward equations enters (6.72) is a bit different from the way Equation (6.69) is written. In the formulation used in the statement of Definition 6.20, the coupling between the forward and the backward equations follows from the optimality condition (6.73). Because of that optimality condition, the two formulations are equivalent in the sense that, when $\gamma = 1$ and $\mathcal{I} \equiv 0$, the pair (6.72)–(6.73) coincides with (6.69). The formulation used above matches the one used in the statements of Theorems 6.14 and 6.16.*

Induction Argument

The following definition is stated for the sake of convenience only. It will help articulate concisely the induction step of the proof of Theorem 6.19.

Definition 6.22 *For any $\gamma \in [0,1]$, we say that property $(\mathcal{S}_\gamma)$ holds if, for any $\xi \in L^2(\Omega, \mathcal{F}_0, \mathbb{P}; \mathbb{R}^d)$ and any $\mathcal{I} \in \mathbb{I}$, the FBSDE $\mathcal{E}(\gamma,\xi,\mathcal{I})$ has a unique extended solution in $\mathscr{S}$.*

Lemma 6.23 *Let $\gamma \in [0,1]$ such that $(\mathcal{S}_\gamma)$ holds. Then, there exists a constant C, independent of γ, such that, for any $\xi, \xi' \in L^2(\Omega, \mathcal{F}_0, \mathbb{P}; \mathbb{R}^d)$ and $\mathcal{I}, \mathcal{I}' \in \mathbb{I}$, the respective extended solutions $\boldsymbol{\Theta}$ and $\boldsymbol{\Theta}'$ of $\mathcal{E}(\gamma,\xi,\mathcal{I})$ and $\mathcal{E}(\gamma,\xi',\mathcal{I}')$ satisfy:*

$$\|\boldsymbol{\Theta} - \boldsymbol{\Theta}'\|_{\mathscr{S}} \leqslant C\big(\mathbb{E}\big[|\xi - \xi'|^2\big]^{1/2} + \|\mathcal{I} - \mathcal{I}'\|_{\mathbb{I}}\big).$$

Proof. We use a mere variation on the proof of the classical stochastic maximum principle. With the same notation as in the statement, and using $\boldsymbol{\Theta}$ for $(X_t, \mathcal{L}(X_t), Y_t, Z_t, \alpha_t)_{0 \leqslant t \leqslant T}$ and $(\theta_t = (X_t, \mathcal{L}(X_t), \alpha_t))_{0 \leqslant t \leqslant T}$, we compute:

$$\begin{aligned}
&\mathbb{E}\big[(X_T' - X_T) \cdot Y_T\big] \\
&= \mathbb{E}\big[(\xi' - \xi) \cdot Y_0\big] \\
&\quad - \gamma\Big\{\mathbb{E}\int_0^T \big[\partial_x H(t, \Theta_t) \cdot (X_t' - X_t) + \tilde{\mathbb{E}}\big[\partial_\mu H(t, \tilde{\Theta}_t)(X_t)\big] \cdot (X_t' - X_t)\big] dt \\
&\qquad - \mathbb{E}\int_0^T \big[[b(t, \theta_t') - b(t, \theta_t)] \cdot Y_t + [\sigma(t, \theta_t') - \sigma(t, \theta_t)] \cdot Z_t\big] dt\Big\} \\
&\quad - \Big\{\mathbb{E}\int_0^T \big[(X_t' - X_t) \cdot \mathcal{I}_t^f + (\mathcal{I}_t^b - \mathcal{I}_t^{b,\prime}) \cdot Y_t + (\mathcal{I}_t^\sigma - \mathcal{I}_t^{\sigma,\prime}) \cdot Z_t\big] dt\Big\} \\
&= T_0 - \gamma T_1 - T_2.
\end{aligned}$$

Following (6.62),

$$\begin{aligned}
&\mathbb{E}\big[(X_T' - X_T) \cdot Y_T\big] \\
&\quad = \gamma \mathbb{E}\big[\big(\partial_x g(X_T, \mathcal{L}(X_T)) + \tilde{\mathbb{E}}[\partial_\mu g(\tilde{X}_T, \mathcal{L}(X_T))(X_T)]\big) \cdot (X_T' - X_T)\big] \\
&\qquad + \mathbb{E}\big[(\mathcal{I}_T^{g,\prime} - \mathcal{I}_T^g) \cdot Y_T\big] \\
&\quad \leqslant \gamma \mathbb{E}\big[g(X_T', \mathcal{L}(\tilde{X}_T')) - g(X_T, \mathcal{L}(X_T))\big] + \mathbb{E}\big[(\mathcal{I}_T^{g,\prime} - \mathcal{I}_T^g) \cdot Y_T\big].
\end{aligned}$$

Identifying the two expressions above and repeating the proof of Theorem 6.16, we obtain:

$$\gamma J(\boldsymbol{\alpha}') - \gamma J(\boldsymbol{\alpha}) \geqslant \gamma \lambda \mathbb{E}\int_0^T |\alpha_t - \alpha_t'|^2 dt + T_0 - T_2 + \mathbb{E}\big[(\mathcal{I}_T^g - \mathcal{I}_T^{g,\prime}) \cdot Y_T\big]. \tag{6.74}$$

Now, we can reverse the roles of $\boldsymbol{\alpha}$ and $\boldsymbol{\alpha}'$ in (6.74). Denoting by T_0' and T_2' the corresponding terms in the inequality and summing both inequalities, we deduce that:

$$2\gamma\lambda \mathbb{E}\int_0^T |\alpha_t - \alpha_t'|^2 dt + T_0 + T_0' - (T_2 + T_2') + \mathbb{E}\big[(\mathcal{I}_T^g - \mathcal{I}_T^{g,\prime}) \cdot (Y_T - Y_T')\big] \leqslant 0.$$

The sum $T_2 + T_2'$ reads:

$$T_2 + T_2' = \mathbb{E}\int_0^T \big[-(\mathcal{I}_t^f - \mathcal{I}_t^{f,\prime}) \cdot (X_t - X_t') + (\mathcal{I}_t^b - \mathcal{I}_t^{b,\prime}) \cdot (Y_t - Y_t') + (\mathcal{I}_t^\sigma - \mathcal{I}_t^{\sigma,\prime}) \cdot (Z_t - Z_t')\big] dt.$$

Similarly,

$$T_0 + T_0' = -\mathbb{E}\big[(\xi - \xi') \cdot (Y_0 - Y_0')\big].$$

Therefore, using Young's inequality, there exists a constant C (the value of which may change from line to line), C being independent of γ, such that, for any $\varepsilon > 0$,

$$\gamma \mathbb{E} \int_0^T |\alpha_t - \alpha'_t|^2 dt \leqslant \varepsilon \|\boldsymbol{\Theta} - \boldsymbol{\Theta}'\|^2_{\mathscr{S}} + \frac{C}{\varepsilon}\big(\mathbb{E}[|\xi - \xi'|^2] + \|\mathcal{I} - \mathcal{I}'\|^2_{\mathbb{I}}\big). \tag{6.75}$$

From standard estimates for BSDEs, there exists a constant C, independent of γ, such that:

$$\begin{aligned} &\mathbb{E}\bigg[\sup_{0\leqslant t\leqslant T} |Y_t - Y'_t|^2 + \int_0^T |Z_t - Z'_t|^2 dt\bigg] \\ &\qquad \leqslant C\gamma \mathbb{E}\bigg[\sup_{0\leqslant t\leqslant T} |X_t - X'_t|^2 + \int_0^T |\alpha_t - \alpha'_t|^2 dt\bigg] + C\|\mathcal{I} - \mathcal{I}'\|^2_{\mathbb{I}}. \end{aligned} \tag{6.76}$$

Similarly,

$$\mathbb{E}[\sup_{0\leqslant t\leqslant T} |X_t - X'_t|^2] \leqslant \mathbb{E}[|\xi - \xi'|^2] + C\gamma \mathbb{E}\int_0^T |\alpha_t - \alpha'_t|^2 dt + C\|\mathcal{I} - \mathcal{I}'\|^2_{\mathbb{I}}. \tag{6.77}$$

From (6.76), (6.77), and (6.75), we deduce that:

$$\begin{aligned} &\mathbb{E}\bigg[\sup_{0\leqslant t\leqslant T} |X_t - X'_t|^2 + \sup_{0\leqslant t\leqslant T} |Y_t - Y'_t|^2 + \int_0^T |Z_t - Z'_t|^2 dt\bigg] \\ &\qquad \leqslant C\gamma \mathbb{E}\int_0^T |\alpha_t - \alpha'_t|^2 dt + C\big(\mathbb{E}[|\xi - \xi'|^2] + \|\mathcal{I} - \mathcal{I}'\|^2_{\mathbb{I}}\big) \\ &\qquad \leqslant C\varepsilon \|\boldsymbol{\Theta} - \boldsymbol{\Theta}'\|^2_{\mathscr{S}} + \frac{C}{\varepsilon}\big(\mathbb{E}[|\xi - \xi'|^2] + \|\mathcal{I} - \mathcal{I}'\|^2_{\mathbb{I}}\big). \end{aligned} \tag{6.78}$$

Using (6.75) again and choosing ε small enough, we complete the proof. □

Lemma 6.24 *There exists $\delta_0 > 0$ such that, if $(\mathcal{S}_\gamma)$ holds for some $\gamma \in [0, 1)$, then $(\mathcal{S}_{\gamma+\eta})$ also holds for any $\eta \in (0, \delta_0]$ satisfying $\gamma + \eta \leqslant 1$.*

Proof. The proof follows from a standard Picard's contraction argument. Indeed, if γ is such that $(\mathcal{S}_\gamma)$ holds, for $\eta > 0$, $\xi \in L^2(\Omega, \mathcal{F}_0, \mathbb{P}; \mathbb{R}^d)$ and $\mathcal{I} \in \mathbb{I}$, we then define a mapping Φ from $\mathscr{S}$ into itself whose fixed points coincide with the solutions of $\mathcal{E}(\gamma + \eta, \xi, \mathcal{I})$. The definition of Φ is as follows. Given a process $\boldsymbol{\Theta} \in \mathscr{S}$, we denote by $\boldsymbol{\Theta}'$ the extended solution of the FBSDE $\mathcal{E}(\gamma, \xi, \mathcal{I}')$ with:

$$\begin{aligned} \mathcal{I}^{b,\prime}_t &= \eta b(t, \theta_t) + \mathcal{I}^b_t, \\ \mathcal{I}^{\sigma,\prime}_t &= \eta \sigma(t, \theta_t) + \mathcal{I}^b_t, \\ \mathcal{I}^{f,\prime}_t &= \eta \partial_x H(t, \Theta_t) + \eta \tilde{\mathbb{E}}\big[\partial_\mu H(t, \tilde{\Theta}_t)(X_t)\big] + \mathcal{I}^f_t, \\ \mathcal{I}^{g,\prime}_T &= \eta \partial_x g(X_T, \mathcal{L}(X_T)) + \eta \tilde{\mathbb{E}}\big[\partial_\mu g(\tilde{X}_T, \mathcal{L}(X_T))(X_T)\big] + \mathcal{I}^g_T. \end{aligned}$$

By assumption, it is uniquely defined and it belongs to $\mathscr{S}$, so that the mapping $\Phi : \boldsymbol{\Theta} \mapsto \boldsymbol{\Theta}'$ maps $\mathscr{S}$ into itself. It is then clear that a process $\boldsymbol{\Theta} \in \mathscr{S}$ is a fixed point of Φ if and only if Θ is an extended solution of $\mathcal{E}(\gamma + \eta, \xi, \mathcal{I})$. So we only need to prove that Φ is a contraction when η is small enough. This is a consequence of Lemma 6.23 . Indeed, given $\boldsymbol{\Theta}^1$ and $\boldsymbol{\Theta}^2$ two processes in $\mathscr{S}$, if we denote by $\boldsymbol{\Theta}'^{,1}$ and $\boldsymbol{\Theta}'^{,2}$ their respective images by Φ, Lemma 6.23 implies that:

$$\|\boldsymbol{\Theta}'^{,1} - \boldsymbol{\Theta}'^{,2}\|_{\mathscr{S}} \leqslant C\eta \|\boldsymbol{\Theta}^1 - \boldsymbol{\Theta}^2\|_{\mathscr{S}},$$

which is enough to conclude he proof. □

The proof of Theorem 6.19 follows a straightforward induction argument based on Lemma 6.24 as $(\mathcal{S}_0)$ obviously holds.

The Master Field

Now that existence and uniqueness have been proven, the master field is constructed following the argument used for Lemma 4.25 in Subsection 4.2.4.

Lemma 6.25 *For any $t \in [0,T]$ and $\xi \in L^2(\Omega, \mathcal{F}_t, \mathbb{P}; \mathbb{R}^d)$, there exists a unique solution $(X_s^{t,\xi}, Y_s^{t,\xi}, Z_s^{t,\xi})_{t \leqslant s \leqslant T}$, of the Pontryagin forward/backward system (6.69) on $[t,T]$ with $X_t^{t,\xi} = \xi$ as initial condition. Moreover, for any $\mu \in \mathcal{P}_2(\mathbb{R}^d)$, there exists a measurable mapping $\mathcal{U}(t,\cdot,\mu) : \mathbb{R}^d \ni x \mapsto \mathcal{U}(t,x,\mu)$ such that:*

$$\mathbb{P}\Big[Y_t^{t,\xi} = \mathcal{U}\big(t,\xi,\mathcal{L}(\xi)\big)\Big] = 1. \tag{6.79}$$

Finally, there exists a constant C, depending only on the parameters in assumption **Control of MKV Dynamics**, *such that, for any $t \in [0,T]$ and any $\xi^1, \xi^2 \in L^2(\Omega, \mathcal{F}_t, \mathbb{P}; \mathbb{R}^d)$,*

$$\mathbb{E}\Big[\big|\mathcal{U}\big(t,\xi^1,\mathcal{L}(\xi^1)\big) - \mathcal{U}\big(t,\xi^2,\mathcal{L}(\xi^2)\big)\big|^2\Big]^{1/2} \leqslant C\mathbb{E}\big[|\xi^1 - \xi^2|^2\big]^{1/2}. \tag{6.80}$$

We here use the letter $\mathcal{U}$ instead of U as in the statement of Lemma 4.25 in order to distinguish from U in (6.37).

Proof. Given $t \in [0,T)$ and $\xi \in L^2(\Omega, \mathcal{F}_t, \mathbb{P}; \mathbb{R}^d)$, existence and uniqueness of a solution of (6.69) on $[t,T]$ with ξ as initial condition is a direct consequence of Theorem 6.19. The construction of the decoupling field is done exactly, mutatis mutandis, as in the proof of Lemma 4.25. Finally, the Lipschitz property (6.80) of $\mathcal{U}(0,\cdot,\cdot)$ is a direct consequence of Lemma 6.23 with $\gamma = 1$. Shifting time if necessary, the same argument applies to $\mathcal{U}(t,\cdot,\cdot)$. □

Remark 6.26 *We shall revisit the notion of master field in the next section, see Subsection 6.5.2. The notion of master field for mean field games will be addressed in detail in Chapters (Vol II)-4 and (Vol II)-5.*

Also, observe that, from Proposition 5.36, for any $\mu \in \mathcal{P}_2(\mathbb{R}^d)$, there exists a version of $\mathbb{R}^d \ni x \mapsto \mathcal{U}(t,x,\mu)$ in $L^2(\mathbb{R}^d;\mu)$ that is Lipschitz-continuous with respect to x, for the same Lipschitz constant C as in (6.80).

6.5 Several Forms of the Master Equation

Our goal is now to provide a probabilistic derivation of the master equation (6.37) obtained in Subsection 6.2.4 and to connect its solution U with the decoupling field $\mathcal{U}$ identified in the statement of Lemma 6.25. In order to do so, we assume throughout the section that assumption **Control of MKV Dynamics** introduced in the previous section is in force, but **with the restriction that σ is uncontrolled**. In particular, Theorem 6.19 applies and guarantees that for any given initial condition $\xi \in L^2(\Omega, \mathcal{F}_0, \mathbb{P}; \mathbb{R}^d)$, the mean field stochastic control problem (6.6)–(6.9) has a unique optimal path, which is characterized by the solution of the forward-backward system (6.69).

Most of the discussion below could be generalized to more general forms of drift and volatility coefficients, and running and terminal cost functions, as long as existence and uniqueness of an optimal path remain true.

6.5.1 Dynamic Programming Principle

Recall that the objective is to minimize the quantity:

$$\mathbb{E}\bigg[\int_0^T f\big(t, X_t^{\boldsymbol{\alpha}}, \mathcal{L}(X_t^{\boldsymbol{\alpha}}), \alpha_t\big)dt + g\big(X_T^{\boldsymbol{\alpha}}, \mathcal{L}(X_T^{\boldsymbol{\alpha}})\big)\bigg], \tag{6.81}$$

over the space of A-valued square integrable $\mathbb{F}$-adapted controls $\boldsymbol{\alpha} = (\alpha_t)_{0 \leqslant t \leqslant T}$ under the dynamic constraint:

$$dX_t^{\boldsymbol{\alpha}} = b\big(t, X_t^{\boldsymbol{\alpha}}, \mathcal{L}(X_t^{\boldsymbol{\alpha}}), \alpha_t\big)dt + \sigma\big(s, X_t^{\boldsymbol{\alpha}}, \mathcal{L}(X_t^{\boldsymbol{\alpha}})\big)dW_t, \quad t \in [0,T], \tag{6.82}$$

with the initial condition $X_0^{\boldsymbol{\alpha}} = \xi \in L^2(\Omega, \mathcal{F}, \mathbb{P}; \mathbb{R}^d)$.

The Value Function of the Optimal Control Problem

Our analysis relies on manipulations of several forms of the value function associated with the optimal control problem.

Inspired by the analytic point of view introduced in (6.25) for the control of the Fokker-Planck equation, we let:

Definition 6.27 *Under the assumption prescribed above, for any $t \in [0,T]$ and $\xi \in L^2(\Omega, \mathcal{F}_t, \mathbb{P}; \mathbb{R}^d)$, the quantity:*

$$\inf_{\boldsymbol{\alpha}_{|[t,T]}} \mathbb{E}\left[\int_t^T f(s, X_s^{\boldsymbol{\alpha}}, \mathcal{L}(X_s^{\boldsymbol{\alpha}}), \alpha_s)ds + g(X_T^{\boldsymbol{\alpha}}, \mathcal{L}(X_T^{\boldsymbol{\alpha}}))\right], \tag{6.83}$$

under the prescription $X_t^{\boldsymbol{\alpha}} = \xi$, where the infimum is taken over A-valued square-integrable $(\mathcal{F}_s)_{t\leqslant s\leqslant T}$-progressively measurable processes $\boldsymbol{\alpha}_{|[t,T]} = (\alpha_s)_{t\leqslant s\leqslant T}$, only depends upon t and the distribution of ξ. For that reason, we can denote it by $v(t,\mu)$ if μ is the law of ξ.

The function v is called the value function of the mean field stochastic control problem.

Proof. Without any loss of generality, we can assume that $t = 0$. By Theorem 6.19, the mean field stochastic control problem (6.6)–(6.9) with ξ as initial condition has a unique optimal path, which is characterized by the solution of the forward-backward system (6.69).

We shall prove in Theorem (Vol II)-1.33 a relevant version of the Yamada and Watanabe theorem for McKean-Vlasov forward-backward SDEs. It says that, the forward-backward system (6.69) being uniquely solvable in the strong sense, the law of its solution only depends on the law of the initial condition. □

Remark 6.28 *The definition of v in (6.83) is quite similar to that given in (6.25), except that (6.83) is based upon the probabilistic formulation of the control problem while (6.25) is based upon the analytic approach. In order to identify the two definitions rigorously, it is necessary to connect first the two formulations of the optimal control problem. We refer to the final discussion in Subsection 6.2.5 for a short account.*

We claim that v satisfies the following dynamic programming principle:

Proposition 6.29 *The value function v, as defined above, satisfies, for all $t \in [0,T]$, $h \in [0, T-t]$ and $\mu \in \mathcal{P}_2(\mathbb{R}^d)$:*

$$\begin{aligned} &v(t,\mu) \\ &= \inf_{\boldsymbol{\alpha}_{|[t,t+h]}} \left\{\mathbb{E}\left[\int_t^{t+h} f(s, X_s^{\boldsymbol{\alpha}}, \mathcal{L}(X_s^{\boldsymbol{\alpha}}), \alpha_s)ds\right] + v\big(t+h, \mathcal{L}(X_{t+h}^{\boldsymbol{\alpha}})\big)\right\}, \end{aligned} \tag{6.84}$$

under the prescription $X_t^{\boldsymbol{\alpha}} = \xi \in L^2(\Omega, \mathcal{F}, \mathbb{P}; \mathbb{R}^d)$ with $\xi \sim \mu$, where the infimum is taken over A-valued square-integrable $(\mathcal{F}_s)_{t\leqslant s\leqslant t+h}$-progressively measurable processes $\boldsymbol{\alpha}_{|[t,t+h]} = (\alpha_s)_{t\leqslant s\leqslant t+h}$.

Notice that, despite the presence of the expectation in (6.84), the dynamic programming principle is deterministic. Indeed, the underlying state variable is the marginal law of the controlled process $\boldsymbol{X}^{\boldsymbol{\alpha}}$.

Proof. The proof of (6.84) is pretty standard. For t, h and μ as in the statement, we first prove that $v(t,\mu)$ is greater than the right-hand side in (6.84). To do so, it suffices to start from the definition of $v(t,\mu)$ in Definition 6.27 and observe that for any $\boldsymbol{\alpha}_{|[t,T]}$:

$$\begin{aligned}
&\mathbb{E}\left[\int_t^T f\big(s, X_s^{\alpha}, \mathcal{L}(X_s^{\alpha}), \alpha_s\big)ds + g\big(X_T^{\alpha}, \mathcal{L}(X_T^{\alpha})\big)\right] \\
&= \mathbb{E}\left[\int_t^{t+h} f\big(s, X_s^{\alpha}, \mathcal{L}(X_s^{\alpha}), \alpha_s\big)ds\right] \\
&\quad + \mathbb{E}\left[\int_{t+h}^T f\big(s, X_s^{\alpha}, \mathcal{L}(X_s^{\alpha}), \alpha_s\big)ds + g\big(X_T^{\alpha}, \mathcal{L}(X_T^{\alpha})\big)\right],
\end{aligned}$$

but the last term in the right-hand side is greater than $v(t+h, \mathcal{L}(X_{t+h}^{\alpha}))$, which proves that $v(t, \mu)$ is greater than the right-hand side in (6.84).

To prove the converse inequality, we consider a control $\boldsymbol{\alpha} = (\alpha_s)_{t \leqslant s \leqslant t+h}$ defined over $[t, t+h]$ and a control $\boldsymbol{\alpha}' = (\alpha'_s)_{t+h \leqslant s \leqslant T}$ defined over $[t+h, T]$. We patch them together by letting $\beta_s = \alpha_s$ for $s \in [t, t+h]$ and $\beta_s = \alpha'_s$ for $s \in (t+h, T]$. Clearly, $\boldsymbol{\beta}$ is an admissible control over $[t, T]$. Therefore,

$$v(t, \mu) \leqslant \mathbb{E}\left[\int_t^T f\big(s, X_s^{\beta}, \mathcal{L}(X_s^{\beta}), \beta_s\big)ds + g\big(X_T^{\beta}, \mathcal{L}(X_T^{\beta})\big)\right],$$

with $X_t^{\beta} = \xi \in L^2(\Omega, \mathcal{F}_t, \mathbb{P}; \mathbb{R}^d)$, with $\xi \sim \mu$. It is pretty clear that:

$$X_s^{\beta} = X_s^{\alpha', X_{t+h}^{\alpha}}, \quad s \in [t+h, T],$$

where $(X_s^{\alpha', X_{t+h}^{\alpha}})_{t+h \leqslant s \leqslant T}$ is the solution of (6.6) with X_{t+h}^{α} as initial condition at time $t+h$. By freezing $\boldsymbol{\alpha}$ and by minimizing over $\boldsymbol{\alpha}'$, we get:

$$v(t, \mu) \leqslant \mathbb{E}\left[\int_t^{t+h} f\big(s, X_s^{\alpha}, \mathcal{L}(X_s^{\alpha}), \alpha_s\big)ds\right] + v\big(t+h, \mathcal{L}(X_{t+h}^{\alpha})\big),$$

which completes the proof. □

Whenever v is smooth enough, this weak form of the DPP is sufficient to recover the HJB equation (6.26), with the difference that computations are then based upon the L-differential calculus instead of the linear functional derivative. Indeed, the use of the L-derivative is very natural as $(v(t+h, \mathcal{L}(X_{t+h}^{\alpha})))_{t \leqslant t+h \leqslant T}$ may be expanded in h by means of the chain rule proven in Chapter 5. We refer to Subsection 5.7.3 for a preliminary discussion of the same kind. We shall address this question again below.

Whenever v is not smooth, the DPP may be used to derive the HJB equation in the viscosity sense. We shall do so in Chapter (Vol II)-4, but for the master equation associated with mean field games. As for mean field stochastic control, we refer to citations in the Notes & Complements below.

The Value Function Over the Enlarged State Space

We now define another form of value function. Our goal is indeed to duplicate at any time $t \in [0, T)$ the form taken by the value function $v(T, \cdot)$ at time T. The latter one writes:

$$v(T,\mu) = \int_{\mathbb{R}^d} g(x,\mu)d\mu(x), \quad \mu \in \mathcal{P}_2(\mathbb{R}^d),$$

and, in a similar fashion, we would like to write:

$$v(t,\mu) = \int_{\mathbb{R}^d} V(t,x,\mu)d\mu(x), \quad t \in [0,T], \quad \mu \in \mathcal{P}_2(\mathbb{R}^d),$$

for some function V, in which case V would read as a *value function* defined over the *enlarged* state space $\mathbb{R}^d \times \mathcal{P}_2(\mathbb{R}^d)$.

A natural way to do so is to proceed by conditioning. Roughly speaking, for each $(t,x,\mu) \in [0,T] \times \mathbb{R}^d \times \mathcal{P}_2(\mathbb{R}^d)$, we should define $V(t,x,\mu)$ as the conditional expected future costs:

$$\begin{aligned} &V(t,x,\mu) \\ &= \mathbb{E}\Big[\int_t^T f\big(s, X_s^{\hat{\alpha}}, \mathcal{L}(X_s^{\hat{\alpha}}), \hat{\alpha}_s\big)ds + g\big(X_T^{\hat{\alpha}}, \mathcal{L}(X_T^{\hat{\alpha}})\big) \,|\, X_t^{\hat{\alpha}} = x\Big], \end{aligned} \tag{6.85}$$

where $\hat{\boldsymbol{\alpha}}$ minimizes the quantity (6.83) under the constraint $X_t^{\boldsymbol{\alpha}} = \xi \sim \mu$. Notice that, with this definition of the value function, for each $t \in [0,T]$ and $\mu \in \mathcal{P}_2(\mathbb{R}^d)$, $V(t,x,\mu)$ is only defined for μ-almost every $x \in \mathbb{R}^d$. Below, the 'hat' symbol always refers to optimal quantities, and $(X_s^{\hat{\alpha}})_{t\leqslant s\leqslant T}$ is sometimes denoted by $\hat{\boldsymbol{X}} = (\hat{X}_s)_{t\leqslant s\leqslant T}$. Put differently, $\boldsymbol{X}^{\hat{\boldsymbol{\alpha}}}$ in (6.85) is understood as the optimal path minimizing the cost functional (6.83) over McKean-Vlasov diffusion processes satisfying (6.82) with the initial condition $X_t^{\boldsymbol{\alpha}} = \xi \in L^2(\Omega, \mathcal{F}_t, \mathbb{P}; \mathbb{R}^d)$, where $\xi \sim \mu$. We shall prove below that the definition (6.85) is consistent in the sense that the right-hand side in (6.85) is independent of the choice of the random variable ξ representing the distribution μ.

In order to reformulate (6.85) in a more proper fashion, we use the fact that the minimizer $(\hat{\alpha}_s)_{t\leqslant s\leqslant T}$ has a feedback form, given by Lemma 6.25. Namely $\hat{\alpha}_s$ reads as $\bar{\alpha}(s, X_s^{\hat{\alpha}}, \mathcal{L}(X_s^{\hat{\alpha}}))$, where:

$$\bar{\alpha}(t,x,\mu) = \hat{\alpha}\big(t,x,\mu,\mathcal{U}(t,x,\mu)\big). \tag{6.86}$$

Above, we wrote $\hat{\alpha}(t,x,\mu,y)$ for $\hat{\alpha}(t,x,\mu,y,z)$. Indeed, $\hat{\alpha}$ is independent of z since σ is uncontrolled, which is a crucial fact at this stage of the proof. Therefore, for $t \in [0,T]$ and $\xi \in L^2(\Omega, \mathcal{F}_t, \mathbb{P}; \mathbb{R}^d)$, the optimal path $\hat{\boldsymbol{X}}^{t,\xi}$, with ξ as initial condition at time t, is the solution of the McKean-Vlasov SDE:

$$d\hat{X}_s = b\big(s, X_s, \mathcal{L}(\hat{X}_s), \bar{\alpha}(s, X_s, \mathcal{L}(\hat{X}_s))\big)ds + \sigma\big(s, X_s, \mathcal{L}(X_s)\big)dW_s,$$

for $s \in [t,T]$, with the initial condition $\hat{X}_t = \xi \in L^2(\Omega, \mathcal{F}_t, \mathbb{P}; \mathbb{R}^d)$. By Lemma 6.25, $L^2(\Omega, \mathcal{F}_t, \mathbb{P}; \mathbb{R}^d) \ni X \mapsto \mathcal{U}(t,X,\mathcal{L}(X))$ is Lipschitz continuous in X. Following the last argument in the proof of Lemma 4.56, it is bounded on bounded subsets. Hence, the above SDE is uniquely solvable. As already explained, strong uniqueness for McKean-Vlasov SDEs implies weak uniqueness, see the proof of Definition 6.27. Therefore, the law of the path $\hat{X}$ only depends on (t, μ), where $\mu = \mathcal{L}(\xi)$. In particular, we can write $\mathcal{L}(\hat{X}_s^{t,\mu})$ for $\mathcal{L}(\hat{X}_s^{t,\xi})$, since the latter only depends on ξ through μ.

Now, in order to well define the conditioning in (6.85), we may define, for any $(t, x, \mu) \in [0, T] \times \mathbb{R}^d \times \mathcal{P}_2(\mathbb{R}^d)$, $\hat{\boldsymbol{X}}^{t,x,\mu} = (\hat{X}_s^{t,x,\mu})_{t \leqslant s \leqslant T}$ as the solution of the SDE:

$$\begin{aligned} d\hat{X}_s^{t,x,\mu} &= b\Big(s, \hat{X}_s^{t,x,\mu}, \mathcal{L}(\hat{X}_s^{t,\mu}), \bar{\alpha}\big(s, \hat{X}_s^{t,x,\mu}, \mathcal{L}(\hat{X}_s^{t,\mu})\big)\Big)dt \\ &\quad + \sigma\big(s, \hat{X}_s^{t,x,\mu}, \mathcal{L}(\hat{X}_s^{t,\mu})\big)dW_s, \quad s \in [t, T], \end{aligned} \tag{6.87}$$

with the initial condition $\hat{X}_t^{t,x,\mu} = x$ at time t. Observe that (6.87) is not a McKean-Vlasov equation! Indeed, the measure argument input is not $\mathcal{L}(\hat{X}_s^{t,x,\mu})$ but $\mathcal{L}(\hat{X}_s^{t,\mu})$. Notice also that the use of the notation $\hat{\boldsymbol{X}}^{t,\mu}$ in this definition is perfectly legitimate since only the law of $\hat{\boldsymbol{X}}^{t,\mu}$ is needed in (6.87), regardless of the choice of the representative ξ of μ. Also by combining Lemma 6.25 and Proposition 5.36, $\mathcal{U}$ is Lipschitz in x, uniformly in t and μ, which guarantees that (6.87) is uniquely solvable.

As a conclusion, we have a more satisfactory definition of V.

Definition 6.30 *If assumption* **Control of MKV Dynamics** *holds and σ is uncontrolled, we call extended value function the function V defined by:*

$$\begin{aligned} V(t, x, \mu) = \mathbb{E}\bigg[\int_t^T & f\Big(s, \hat{X}_s^{t,x,\mu}, \mathcal{L}(\hat{X}_s^{t,\mu}), \bar{\alpha}\big(s, \hat{X}_s^{t,x,\mu}, \mathcal{L}(\hat{X}_s^{t,\mu})\big)\Big)ds \\ & + g\big(\hat{X}_T^{t,x,\mu}, \mathcal{L}(\hat{X}_T^{t,\mu})\big)\bigg], \end{aligned} \tag{6.88}$$

for $(t, x, \mu) \in [0, T] \times \mathbb{R}^d \times \mathcal{P}_2(\mathbb{R}^d)$. It satisfies:

$$v(t, \mu) = \int_{\mathbb{R}^d} V(t, x, \mu) d\mu(x), \quad (t, \mu) \in [0, T] \times \mathbb{R}^d.$$

Proof. The identification of v is a mere consequence of the Markov property for the SDE (6.87), which holds since the equation is well posed. □

Dynamic Programming Principle

Our goal is to characterize the function V as the solution of a partial differential equation (PDE) on the space $[0, T] \times \mathbb{R}^d \times \mathcal{P}_2(\mathbb{R}^d)$. In the classical theory of optimal control problems, when the Dynamic Programming Principle (DPP for short) holds, and the value function is computed along optimal paths, up to the accumulated running costs, this value function is a martingale. So using the chain rule stated in

Proposition 5.102, one may try to guess the kind of relationship it should satisfy in order for the bounded variation part to vanish. The implementation of this approach depends only upon the availability of a form of the dynamic programming principle which is the basis for the martingale property of the value function along optimal paths.

Following the approach used in finite dimension, a natural strategy is to use (6.88) as a basis for the derivation of a dynamic programming principle for V. We shall use the fact that the pair $(\hat{X}_s^{t,x,\mu}, \mathcal{L}(\hat{X}_s^{t,\mu}))_{t \leqslant s \leqslant T}$ is Markovian with values in $\mathbb{R}^d \times \mathcal{P}_2(\mathbb{R}^d)$. The Markov property states that for any $t \leqslant t+h \leqslant T$, the future states of $(\mathcal{L}(\hat{X}_s^{t,\mu}))_{t+h \leqslant s \leqslant T}$ are uniquely determined by $\mathcal{L}(\hat{X}_{t+h}^{t,\mu})$, namely:

$$\mathcal{L}(\hat{X}_s^{t,\mu}) = \mathcal{L}\Big(\hat{X}_s^{t+h, \mathcal{L}(X_{t+h}^{t,\mu})}\Big), \quad s \in [t+h, T],$$

and the conditional law of the future states of $(\hat{X}_s^{t,x,\mu})_{t+h \leqslant s \leqslant T}$ given the past before $t+h$ is uniquely determined by $\hat{X}_{t+h}^{t,x,\mu}$ and $\mathcal{L}(\hat{X}_{t+h}^{t,\mu})$, namely:

$$\hat{X}_s^{t,x,\mu} = \hat{X}_s^{t+h, \hat{X}_{t+h}^{t,\mu}, \mathcal{L}(\hat{X}_{t+h}^{t,\mu})}, \quad s \in [t+h, T],$$

It follows that:

Under the same assumption as before, V satisfies the dynamic programming principle:

$$\begin{aligned} &V\big(t+h, \hat{X}_{t+h}^{t,x,\mu}, \mathcal{L}(\hat{X}_{t+h}^{t,\mu})\big) \\ &\quad = \mathbb{E}\bigg[\int_{t+h}^{T} f\Big(s, \hat{X}_s^{t,x,\mu}, \mathcal{L}(\hat{X}_s^{t,\mu}), \bar{\alpha}\big(s, \hat{X}_s^{t,x,\mu}, \mathcal{L}(\hat{X}_s^{t,\mu})\big)\Big) ds \\ &\qquad\qquad + g\big(\hat{X}_T^{t,x,\mu}, \mathcal{L}(\hat{X}_T^{t,\mu})\big) \,\big|\, \mathcal{F}_{t+h}\bigg], \end{aligned}$$

for $t \leqslant t+h \leqslant T$.

Taking expectations on both sides and using the definition (6.88), this shows that:

Proposition 6.31 *Under the above assumption, V satisfies the following version of the dynamic programming principle:*

$$\begin{aligned} V(t, x, \mu) = \mathbb{E}\bigg[&\int_t^{t+h} f\Big(s, \hat{X}_s^{t,x,\mu}, \mathcal{L}(\hat{X}_s^{t,\mu}), \bar{\alpha}\big(s, \hat{X}_s^{t,x,\mu}, \mathcal{L}(\hat{X}_s^{t,\mu})\big)\Big) ds \\ &+ V\big(t+h, \hat{X}_{t+h}^{t,x,\mu}, \mathcal{L}(\hat{X}_{t+h}^{t,\mu})\big)\bigg], \end{aligned} \tag{6.89}$$

for $t \leqslant t+h \leqslant T$ and $\mu \in \mathcal{P}_2(\mathbb{R}^d)$.

6.5.2 Derivation of the Master Equation for the Value Function

According to the strategy we hinted at earlier, our derivation of the master equation is based on the application of the chain rule stated in Proposition 5.102 to the dynamics of V along optimal paths. Consequently, for the purpose of this derivation, we assume that the value function V defined in (6.88) is smooth enough so that we can apply the chain rule (5.107). In order to satisfy the integrability constraints in the chain rule, we also require σ to be bounded.

It would be possible to carry out the complete analysis of the regularity of V, but the proof would be lengthy and would require a lot of technicalities. The interested reader may have a look at the references at the end of the chapter. Also, she/he may find in Chapter (Vol II)-5 a similar analysis, but for mean field games instead of mean field stochastic control problems.

Form of the Equation

From (6.87), we get, for any $(t,x,\mu) \in [0,T] \times \mathbb{R}^d \times \mathcal{P}_2(\mathbb{R}^d)$:

$$
\begin{aligned}
& dV\big(s, \hat{X}_s^{t,x,\mu}, \mathcal{L}(\hat{X}_s^{t,\mu})\big) \\
&= \Big(\partial_t V\big(s, \hat{X}_s^{t,x,\mu}, \mathcal{L}(\hat{X}_s^{t,\mu})\big) + \partial_x V\big(s, \hat{X}_s^{t,x,\mu}, \mathcal{L}(\hat{X}_s^{t,\mu})\big) \cdot b_s \\
&\qquad + \frac{1}{2} \mathrm{trace}\Big[\partial_{xx}^2 V\big(s, \hat{X}_s^{t,x,\mu}, \mathcal{L}(\hat{X}_s^{t,\mu})\big) \sigma_s \sigma_s^\dagger \Big] \\
&\qquad + \tilde{\mathbb{E}}\Big[\partial_\mu V\big(s, \hat{X}_s^{t,x,\mu}, \mathcal{L}(\hat{X}_s^{t,\mu})\big)\big(\widetilde{\{\hat{X}_s^{t,\mu}\}} \big) \cdot \tilde{b}_s \Big] \\
&\qquad + \frac{1}{2} \tilde{\mathbb{E}}\Big[\mathrm{trace}\Big(\partial_v \partial_\mu V\big(s, \hat{X}_s^{t,x,\mu}, \mathcal{L}(\hat{X}_s^{t,\mu})\big)\big(\widetilde{\{\hat{X}_s^{t,\mu}\}} \big) \tilde{\sigma}_s \tilde{\sigma}_s^\dagger \Big) \Big] \Big) ds \\
&\quad + \partial_x V\big(s, \hat{X}_s^{t,x,\mu}, \mathcal{L}(\hat{X}_s^{t,\mu})\big) \cdot \big(\sigma_s dW_s\big),
\end{aligned}
$$

where we wrote b_s for $b(s, \hat{X}_s^{t,x,\mu}, \mathcal{L}(\hat{X}_s^{t,\mu}), \hat{\alpha}_s)$, σ_s for $\sigma(s, \hat{X}_s^{t,x,\mu}, \mathcal{L}(\hat{X}_s^{t,\mu}))$, and $\hat{\alpha}_s$ for $\tilde{\alpha}(s, \hat{X}_s^{t,x,\mu}, \mathcal{L}(\hat{X}_s^{t,\mu}))$, where $\tilde{\alpha}$ is the optimal feedback function, as defined in (6.86). As usual, $(\tilde{b}_s, \tilde{\sigma}_s)$ is a copy of (b_s, σ_s) on the space $(\tilde{\Omega}, \tilde{\mathcal{F}}, \tilde{\mathbb{P}})$, the expectation over which is denoted by $\tilde{\mathbb{E}}$. Similarly,

$$
\widetilde{\{\hat{X}_s^{t,\mu}\}}
$$

is a copy of $\hat{X}_s^{t,\mu}$.

Inserting the above expansion in (6.89) in order to get the limit form of the left-hand side as h tends to 0, we deduce that V satisfies the equation:

$$
\begin{aligned}
&\partial_t V(t,x,\mu) + b\big(t,x,\mu,\bar{\alpha}(t,x,\mu)\big)\cdot\partial_x V(t,x,\mu) \\
&\quad + \frac{1}{2}\mathrm{trace}\Big[a(t,x,\mu)\partial^2_{xx}V(t,x,\mu)\Big] + f\big(t,x,\mu,\bar{\alpha}(t,x,\mu)\big) \\
&\quad + \int_{\mathbb{R}^d}\Big[b\big(t,x',\mu,\bar{\alpha}(t,x',\mu)\big)\cdot\partial_\mu V(t,x,\mu)(x') \\
&\qquad\qquad + \frac{1}{2}\mathrm{trace}\Big(a\big(t,x',\mu\big)\partial_v\partial_\mu V(t,x,\mu)(x')\Big)\Big]d\mu(x') = 0,
\end{aligned}
\tag{6.90}
$$

for $(t,x,\mu)\in[0,T]\times\mathbb{R}^d\times\mathcal{P}_2(\mathbb{R}^d)$, with the terminal condition $V(T,x,\mu)=g(x,\mu)$. We call this equation the master equation for the value function of the problem.

Formal Identification of the Feedback Function

We conclude this subsection with the identification of the special form of the minimizer $\bar{\alpha}$. We go back to (6.84) and rewrite it as:

$$
\begin{aligned}
v(t,\mu) = \inf_{\boldsymbol{\alpha}|[t,t+h]}\bigg\{\mathbb{E}\bigg[\int_t^{t+h} f\big(s,X_s^{\boldsymbol{\alpha}},\mathcal{L}(X_s^{\boldsymbol{\alpha}}),\alpha_s\big)ds \\
+ V\big(t+h,X_{t+h}^{\boldsymbol{\alpha}},\mathcal{L}(X_{t+h}^{\boldsymbol{\alpha}})\big)\bigg]\bigg\},
\end{aligned}
$$

with the prescription $X_t^{\boldsymbol{\alpha}} = \xi \in L^2(\Omega,\mathcal{F}_t,\mathbb{P};\mathbb{R}^d)$, where $\xi\sim\mu$. If V satisfies the assumption of Proposition 5.102, then, for any admissible control $\boldsymbol{\alpha}$, we can expand the quantity $V(t+h,X_{t+h}^{\boldsymbol{\alpha}},\mathcal{L}(X_{t+h}^{\boldsymbol{\alpha}}))$ appearing in the right-hand side by using Itô's formula (5.107) and the PDE (6.90). Dividing by h, letting h tend to 0 and assuming that $\boldsymbol{\alpha}$ is continuous at time t, we obtain:

$$
\begin{aligned}
&\mathbb{E}\Big[b\big(t,X_t^{\boldsymbol{\alpha}},\mathcal{L}(X_t^{\boldsymbol{\alpha}}),\alpha_t\big)\cdot\partial_x V\big(t,X_t^{\boldsymbol{\alpha}},\mathcal{L}(X_t^{\boldsymbol{\alpha}})\big) \\
&\quad + \tilde{\mathbb{E}}\big[b\big(t,\tilde{X}_t^{\boldsymbol{\alpha}},\mathcal{L}(X_t^{\boldsymbol{\alpha}}),\tilde{\alpha}_t\big)\cdot\partial_\mu V\big(t,X_t^{\boldsymbol{\alpha}},\mathcal{L}(X_t^{\boldsymbol{\alpha}})\big)(\tilde{X}_t^{\boldsymbol{\alpha}})\big] \\
&\quad + f\big(t,X_t^{\boldsymbol{\alpha}},\mathcal{L}(X_t^{\boldsymbol{\alpha}}),\alpha_t\big)\Big] \\
&\geqslant \mathbb{E}\Big[b\big(t,X_t^{\boldsymbol{\alpha}},\mathcal{L}(X_t^{\boldsymbol{\alpha}}),\bar{\alpha}(t,X_t^{\boldsymbol{\alpha}},\mathcal{L}(X_t^{\boldsymbol{\alpha}}))\big)\cdot\partial_x V\big(t,X_t^{\boldsymbol{\alpha}},\mathcal{L}(X_t^{\boldsymbol{\alpha}})\big) \\
&\quad + \tilde{\mathbb{E}}\big[b\big(t,\tilde{X}_t^{\boldsymbol{\alpha}},\mathcal{L}(X_t^{\boldsymbol{\alpha}}),\bar{\alpha}(t,\tilde{X}_t^{\boldsymbol{\alpha}},\mathcal{L}(X_t^{\boldsymbol{\alpha}}))\big)\cdot\partial_\mu V\big(t,X_t^{\boldsymbol{\alpha}},\mathcal{L}(X_t^{\boldsymbol{\alpha}})\big)(\tilde{X}_t^{\boldsymbol{\alpha}})\big] \\
&\quad + f\big(t,X_t^{\boldsymbol{\alpha}},\mathcal{L}(X_t^{\boldsymbol{\alpha}}),\bar{\alpha}(t,X_t^{\boldsymbol{\alpha}},\mathcal{L}(X_t^{\boldsymbol{\alpha}}))\big)\Big].
\end{aligned}
$$

We deduce that, for any random variable $\chi \in L^2(\Omega, \mathcal{F}_t, \mathbb{P}; \mathbb{R}^d)$, $\bar{\alpha}(t, \chi, \mathcal{L}(\chi))$ should satisfy:

$$
\begin{aligned}
&\bar{\alpha}\big(t, \chi, \mathcal{L}(\chi)\big) \\
&= \underset{\alpha \in L^2(\Omega, \mathcal{F}_t, \mathbb{P}; A)}{\operatorname{argmin}} \mathbb{E}\Big[b\big(t, \chi, \mathcal{L}(\chi), \alpha\big) \cdot \partial_x V\big(t, \chi, \mathcal{L}(\chi)\big) \\
&\qquad + \tilde{\mathbb{E}}\big[b\big(t, \tilde{\chi}, \mathcal{L}(\chi), \tilde{\alpha}\big) \cdot \partial_\mu V\big(t, \chi, \mathcal{L}(\chi)\big)(\tilde{\chi})\big] + f\big(t, \chi, \mathcal{L}(\chi), \alpha\big)\Big].
\end{aligned}
$$

Now, by Fubini's theorem, the minimization can be reformulated as:

$$
\begin{aligned}
\bar{\alpha}\big(t, \chi, \mathcal{L}(\chi)\big) = \underset{\alpha \in L^2(\Omega, \mathcal{F}_t, \mathbb{P}; A)}{\operatorname{argmin}} &\mathbb{E}\Big[b\big(t, \chi, \mathcal{L}(\chi), \alpha\big) \cdot \Big(\partial_x V\big(t, \chi, \mathcal{L}(\chi)\big) \\
&\qquad + \tilde{\mathbb{E}}\big[\partial_\mu V\big(t, \tilde{\chi}, \mathcal{L}(\chi)\big)(\chi)\big]\Big) \\
&+ f\big(t, \chi, \mathcal{L}(\chi), \alpha\big)\Big].
\end{aligned}
\tag{6.91}
$$

Recalling that the mapping $\hat{\alpha} : [0, T] \times \mathbb{R}^d \times \mathcal{P}_2(\mathbb{R}^d) \times \mathbb{R}^d \ni (t, x, \mu, y) \mapsto \hat{\alpha}(t, x, \mu, y)$ minimizes the reduced Hamiltonian H, the minimizer in (6.91) must be:

$$
\begin{aligned}
\bar{\alpha}\big(t, \chi, \mu\big) &= \hat{\alpha}\Big(t, \chi, \mu, \partial_x V(t, \chi, \mu) + \tilde{\mathbb{E}}\big[\partial_\mu V\big(t, \tilde{\chi}, \mu\big)(\chi)\big]\Big) \\
&= \hat{\alpha}\bigg(t, \chi, \mu, \partial_x V\big(t, \chi, \mu\big) + \int_{\mathbb{R}^d} \partial_\mu V\big(t, x', \mu\big)(\chi) d\mu(x')\bigg),
\end{aligned}
$$

with $\mu = \mathcal{L}(\chi)$, showing that:

$$
\bar{\alpha}(t, x, \mu) = \hat{\alpha}\Big(t, x, \mu, \partial_x V(t, x, \mu) + \int_{\mathbb{R}^d} \partial_\mu V(t, x', \mu)(x) d\mu(x')\Big) \tag{6.92}
$$

is an optimal feedback. Plugging this relationship into (6.90), we obtain the full-fledged form to the master equation:

$$
\begin{aligned}
&\partial_t V(t,x,\mu) + \frac{1}{2}\mathrm{trace}\Big[a(t,x,\mu)\partial^2_{xx}V(t,x,\mu)\Big] \\
&\quad + b\bigg(t,x,\mu,\hat{\alpha}\Big(t,x,\mu,\partial_x V(t,x,\mu) + \int_{\mathbb{R}^d}\partial_\mu V(t,x',\mu)(x)d\mu(x')\Big)\bigg) \\
&\qquad\qquad\qquad\qquad\qquad \cdot \partial_x V(t,x,\mu) \\
&\quad + f\bigg(t,x,\mu,\hat{\alpha}\Big(t,x,\mu,\partial_x V(t,x,\mu) + \int_{\mathbb{R}^d}\partial_\mu V(t,x',\mu)(x)d\mu(x')\Big)\bigg) \\
&\quad + \int_{\mathbb{R}^d}\bigg[b\bigg(t,\tilde{x},\mu,\hat{\alpha}\Big(t,\tilde{x},\mu,\partial_x V(t,\tilde{x},\mu) + \int_{\mathbb{R}^d}\partial_\mu V(t,x',\mu)(\tilde{x})d\mu(x')\Big)\bigg) \\
&\qquad\qquad\qquad\qquad\qquad \cdot \partial_\mu V(t,x,\mu)(\tilde{x}) \\
&\qquad\qquad + \frac{1}{2}\mathrm{trace}\Big(a(t,\tilde{x},\mu)\partial_v\partial_\mu V(t,x,\mu)(\tilde{x})\Big)\bigg]d\mu(\tilde{x}) = 0,
\end{aligned}
\tag{6.93}
$$

for $(t,x,\mu) \in [0,T]\times\mathbb{R}^d\times\mathcal{P}_2(\mathbb{R}^d)$, with the terminal condition $V(T,x,\mu) = g(x,\mu)$. Also, the optimal path solving the optimal control of the McKean-Vlasov dynamics is given by:

$$
\begin{aligned}
d\hat{X}_s = b\bigg(&s,\hat{X}_s,\hat{\mu}_s,\hat{\alpha}\Big(s,\hat{X}_s,\hat{\mu}_s,\partial_x V(s,\hat{X}_s,\hat{\mu}_s) \\
&\qquad + \int_{\mathbb{R}^d}\partial_\mu V(s,x',\hat{\mu}_s)(\hat{X}_s)d\hat{\mu}_s(x')\Big)\bigg)ds \\
&+ \sigma\big(s,\hat{X}_s,\hat{\mu}_s\big)dW_s,
\end{aligned}
\tag{6.94}
$$

subject to the constraint $\hat{\mu}_s = \mathcal{L}(\hat{X}_s)$ for $s \in [t,T]$, with $\mathcal{L}(\hat{X}_t) = \mu$, for some initial distribution $\mu \in \mathcal{P}_2(\mathbb{R}^d)$. Moreover, by comparing (6.86) and (6.92), we may conjecture that:

$$
\begin{aligned}
\mathcal{U}(t,x',\mu) = \partial_x V(t,x,\mu) + \int_{\mathbb{R}^d}\partial_\mu V(t,x',\mu)(x)d\mu(x'),& \\
(t,x,\mu) \in [0,T]\times\mathbb{R}^d\times\mathcal{P}_2(\mathbb{R}^d).&
\end{aligned}
\tag{6.95}
$$

The fact that the right-hand side contains two different terms is a perfect reflection of the backward propagation of the terminal condition of the FBSDE (6.69). Indeed, this terminal condition has two terms corresponding to the partial derivatives of the terminal cost function g with respect to the state variable x and the distribution μ. Recalling Definition 6.30, this leads to us the formal identification:

$$\mathcal{U}(t,x,\mu) = \partial_\mu v(t,\mu)(x), \quad (t,x,\mu) \in [0,T] \times \mathbb{R}^d \times \mathcal{P}_2(\mathbb{R}^d).$$

The proof of the above relationship can be made rigorous. Observing from (6.88) that:

$$\begin{aligned} v(t,\mu) = \mathbb{E}\bigg[\int_t^T f\Big(s, \hat{X}_s^{t,\xi}, \mathcal{L}(\hat{X}_s^{t,\xi}), \bar{\alpha}\big(s, \hat{X}_s^{t,\xi}, \mathcal{L}(\hat{X}_s^{t,\xi})\big)\Big) ds \\ + g\big(\hat{X}_T^{t,\xi}, \mathcal{L}(\hat{X}_T^{t,\xi})\big)\bigg], \end{aligned} \tag{6.96}$$

for $\xi \sim \mu$, the identity between $\mathcal{U}$ and $\partial_\mu v$ can be established under appropriate regularity properties by differentiating the above right-hand side with respect to ξ, seen as an element of $L^2(\Omega, \mathcal{F}_t, \mathbb{P}; \mathbb{R}^d)$. This requires to investigate first the derivative of $\hat{X}^{t,\xi}$ with respect to ξ and more generally to address the differentiability of the solution of (6.69) with respect to its initial condition, when seen as a random variable. Such a program will be carried out in Chapter (Vol II)-5 in order to analyze the master equation for mean field games.

Verification Argument

The relevance of the master equation (6.90) is contained in the following verification result, which is an extension of Proposition 5.108 and which does not require **Control of MKV Dynamics** to hold.

Proposition 6.32 *On top of the assumptions* **MKV Lipschitz Regularity** *and* **MKV Quadratic Growth**, *assume that σ is uncontrolled and satisfy $|\sigma(t,x,\mu)| \leqslant C(1 + M_2(\mu))$, for some $C \geqslant 0$ and for all $(t,x,\mu) \in [0,T] \times \mathbb{R}^d \times \mathcal{P}_2(\mathbb{R}^d)$. Assume also that the reduced Hamiltonian has a unique minimizer $\hat{\alpha} : [0,T] \times \mathbb{R}^d \times \mathcal{P}_2(\mathbb{R}^d) \times \mathbb{R}^d \ni (t,x,\mu,y) \mapsto \hat{\alpha}(t,x,\mu,y)$ at most of linear growth in (x,μ,y) uniformly in $t \in [0,T]$ and that there exists a solution V to* (6.90), *satisfying the assumption of Proposition 5.102 together with:*

$$\begin{aligned} |\partial_x V(t,x,\mu)| + \bigg(\int_{\mathbb{R}^d} |\partial_\mu V(t,x',\mu)(x)|^2 d\mu(x') \bigg)^{1/2} \\ \leqslant C\big(1 + |x| + M_2(\mu)\big). \end{aligned} \tag{6.97}$$

Assume finally that, for any initial condition $\xi \in L^2(\Omega, \mathcal{F}_0, \mathbb{P}; \mathbb{R}^d)$, equation (6.94) *with $\hat{X}_0 = \xi$ as initial condition has a unique solution $\hat{X}^{0,\xi}$. Then $(\hat{\alpha}_s^{0,\xi} = \bar{\alpha}(s, \hat{X}_s^{0,\xi}, \mathcal{L}(\hat{X}_s^{0,\xi})))_{0 \leqslant s \leqslant T}$, with $\bar{\alpha}$ as in* (6.92), *solves the minimization problem* (6.6)–(6.9).

Proof. We first notice that because of the linear growth assumption and (6.97), the supremum over $[0,T]$ of the solution of (6.94) is square integrable and that, for any square integrable control $\boldsymbol{\alpha}$, the supremum of X^{α} (with $X_0^{\alpha} = \xi$) is also square integrable. Next, we replace g by $V(T,\cdot,\cdot)$ in (6.9), and apply the chain rule (5.107), the integrability condition (6.97) ensuring that the expectation of the martingale part is zero. Using the same Fubini argument as in (6.91), we deduce that the right-hand side is indeed greater than $\int_{\mathbb{R}^d} V(0,x,\mu)d\mu(x)$, with $\mu = \mathcal{L}(\xi)$. Choosing $\boldsymbol{\alpha} = \hat{\boldsymbol{\alpha}}^{0,\xi}$, we see that equality must hold. □

6.5.3 A Master Equation for the Derivative of the Value Function

As explained earlier, the decoupling field $\mathcal{U}$ of the FBSDE (6.69). can be identified with the L-derivative of the value function v, as defined in Definition 6.27. In full analogy with the previous subsection, but also with Subsections 4.1.2 and 5.7.2, the goal of this subsection is to derive informally an equation (most likely a PDE, though possibly in infinite dimension) satisfied by $\mathcal{U}$.

To do so, we assume that $\mathcal{U}$ satisfies the assumptions of the Itô chain rule stated in Proposition 5.102. We start from the equality:

$$Y_t = \mathcal{U}\big(t, X_t, \mathcal{L}(X_t)\big), \tag{6.98}$$

where $(X_t, Y_t, Z_t)_{0 \le t \le T}$ solves (6.32). Since $\mathcal{U}$ is assumed to be jointly continuous in all the variables, the above relationship should hold, with probability 1, for all $t \in [0,T]$.

Then, we identify the time-differentials of both sides, using the backward equation in (6.32) for dY_t, the optimality condition (6.33), and Itô's chain rule (5.107) for $d\mathcal{U}(t, X_t, \mathcal{L}(X_t))$. Using (6.31), we get:

$$\begin{aligned} dY_t = &- \Big[\partial_x H\big(t, X_t, \mathcal{L}(X_t), Y_t, Z_t, \hat{\alpha}_t\big) \\ &- \tilde{\mathbb{E}}\big[\partial_\mu H\big(t, \tilde{X}_t, \mathcal{L}(X_t), \widetilde{\{\hat{\alpha}_t\}}, \tilde{Y}_t, \tilde{Z}_t\big)(X_t)\big]\Big]dt \\ &+ Z_t dW_t, \quad t \in [0,T], \end{aligned} \tag{6.99}$$

with the terminal condition $Y_T = \partial_x g(X_T, \mathcal{L}(X_T)) + \tilde{\mathbb{E}}[\partial_\mu g(\tilde{X}_T, \mathcal{L}(X_T))(X_T)]$, and with the optimality constraint:

$$\hat{\alpha}_t = \hat{\alpha}\big(t, X_t, \mathcal{L}(X_t), Y_t\big) = \hat{\alpha}\big(t, X_t, \mathcal{L}(X_t), \mathcal{U}(t, X_t, \mathcal{L}(X_t))\big), \quad t \in [0,T].$$

In (6.99), $\widetilde{\{\hat{\alpha}_t\}}$ denotes the copy of $\hat{\alpha}_t$ on $(\tilde{\Omega}, \tilde{\mathcal{F}}, \tilde{\mathbb{P}})$. By the chain rule (5.107), we get:

$$
\begin{aligned}
d\mathcal{U}(t, X_t, \mu_t) = \Big[&\partial_t \mathcal{U}\big(t, X_t, \mathcal{L}(X_t)\big) + \partial_x \mathcal{U}\big(t, X_t, \mathcal{L}(X_t)\big) b_t \\
&+ \frac{1}{2}\big[\partial^2_{xx} \mathcal{U}\big(t, X_t, \mathcal{L}(X_t)\big)\sigma_t \sigma_t^\dagger\big] \\
&+ \tilde{\mathbb{E}}\big[\partial_\mu \mathcal{U}(t, X_t, \mathcal{L}(X_t))(\tilde{X}_t)\tilde{b}_t\big] \\
&+ \frac{1}{2}\tilde{\mathbb{E}}\Big[\partial_v \partial_\mu \mathcal{U}(t, X_t, \mathcal{L}(X_t))(\tilde{X}_t)\tilde{\sigma}_t \tilde{\sigma}_t^\dagger\Big]\Big] dt \\
&+ \partial_x \mathcal{U}\big(t, X_t, \mathcal{L}(X_t)\big)\big(\sigma_t dW_t\big),
\end{aligned}
\tag{6.100}
$$

where we used the notation b_t for $b(t, X_t, \mathcal{L}(X_t), \hat{\alpha}_t)$ and σ_t for $\sigma(t, X_t, \mathcal{L}(X_t))$. As usual, $(\tilde{X}_t, \tilde{b}_t, \tilde{\sigma}_t)$ is a copy of (X_t, b_t, σ_t), the expectation over which is denoted by $\tilde{\mathbb{E}}$. Notice also that $\partial_x \mathcal{U}(t, X_t, \mathcal{L}(X_t))$ and $\partial_\mu \mathcal{U}(t, X_t, \mathcal{L}(X_t))(\tilde{X}_t)$ are $d \times d$ matrices acting on the d-dimensional drifts b_t and $\tilde{b}_t$. Similarly, $\partial^2_{xx} \mathcal{U}(t, X_t, \mathcal{L}(X_t))$ and $\partial_v \partial_\mu \mathcal{U}(t, X_t, \mathcal{L}(X_t))(\tilde{X}_t)$ are of dimension $d \times (d \times d)$. The $d \times d$ components act on $\sigma_t \sigma_t^\dagger$ and $\tilde{\sigma}_t \tilde{\sigma}_t^\dagger$.

Identifying the quadratic variation terms in (6.99) and (6.100), we get:

$$
Z_t = \partial_x \mathcal{U}\big(t, X_t, \mathcal{L}(X_t)\big)\sigma\big(t, X_t, \mathcal{L}(X_t)\big), \quad t \in [0, T]. \tag{6.101}
$$

We can now identify the bounded variation parts of the differentials of both sides of (6.98) after replacing $\hat{\alpha}_t$ by the argument of the minimization of the Hamiltonian, namely $\hat{\alpha}(t, X_t, \mathcal{L}(X_t), Y_t)$, and Y_t and Z_t by (6.98) and (6.101) respectively. We get:

$$
\begin{aligned}
0 = &\partial_t \mathcal{U}\big(t, X_t, \mathcal{L}(X_t)\big) + \partial_x \mathcal{U}\big(t, X_t, \mathcal{L}(X_t)\big) b_t \\
&+ \frac{1}{2}\partial^2_{xx} \mathcal{U}\big(t, X_t, \mathcal{L}(X_t)\big)\sigma_t \sigma_t^\dagger \\
&+ \tilde{\mathbb{E}}\Big[\partial_\mu \mathcal{U}\big(t, X_t, \mathcal{L}(X_t)\big)(\tilde{X}_t)\tilde{b}_t\Big] \\
&+ \frac{1}{2}\tilde{\mathbb{E}}\Big[\partial_v \partial_\mu \mathcal{U}\big(t, X_t, \mathcal{L}(X_t)\big)(\tilde{X}_t)\tilde{\sigma}_t \tilde{\sigma}_t^\dagger\Big] \\
&+ \partial_x H\big(t, X_t, \mathcal{L}(X_t), Y_t, Z_t, \hat{\alpha}_t\big) \\
&+ \tilde{\mathbb{E}}\Big[\partial_\mu H\big(t, \tilde{X}_t, \mathcal{L}(X_t), \tilde{Y}_t, \tilde{Z}_t, \widetilde{\{\hat{\alpha}_t\}}\ \big)(X_t)\Big].
\end{aligned}
\tag{6.102}
$$

Since $(\tilde{X}_t, \tilde{Y}_t, \tilde{Z}_t, \widetilde{\{\hat{\alpha}_t\}})$ is an independent copy of $(X_t, Y_t, Z_t, \hat{\alpha}_t)$, if we use formulas (6.98) and (6.101), since b_t, σ_t and α_t are also functions of X_t only, the above expectations $\tilde{\mathbb{E}}$ are nothing but mere integrals with respect to the measure $\mathcal{L}(X_t)$. So equality (6.102) will be satisfied along all the optimal paths $[0, T] \ni t \mapsto (X_t, \mathcal{L}(X_t))$ if the function $\mathcal{U}$ satisfies the following system of PDEs:

$$
\begin{aligned}
0 &= \partial_t \mathcal{U}(t,x,\mu) + \partial_x \mathcal{U}(t,x,\mu) b\Big(t,x,\mu,\hat{\alpha}\big(t,x,\mu,\mathcal{U}(t,x,\mu)\big)\Big) \\
&\quad + \frac{1}{2}\partial^2_{xx}\mathcal{U}(t,x,\mu) a(t,x,\mu) \\
&\quad + \int_{\mathbb{R}^d} \Big[\partial_\mu \mathcal{U}(t,x,\mu)(x') b\Big(t,x',\mu,\hat{\alpha}\big(t,x',\mu,\mathcal{U}(t,x',\mu)\big)\Big) \\
&\qquad\qquad + \frac{1}{2}\partial_v \partial_\mu \mathcal{U}(t,x,\mu)(x') a(t,x',\mu)\Big] d\mu(x') \\
&\quad + \partial_x H\Big(t,x,\mu,\mathcal{U}(t,x,\mu), \partial_x \mathcal{U}(t,x,\mu)\sigma(t,x,\mu), \hat{\alpha}\big(t,x,\mu,\mathcal{U}(t,x,\mu)\big)\Big) \\
&\quad + \int_{\mathbb{R}^d} \partial_\mu H\Big(t,x',\mu,\mathcal{U}(t,x',\mu), \partial_x \mathcal{U}(t,x',\mu)\sigma(t,x',\mu), \\
&\qquad\qquad \hat{\alpha}\big(t,x',\mu,\mathcal{U}(t,x',\mu)\big)\Big)(x) d\mu(x'),
\end{aligned}
\tag{6.103}
$$

for $(t,x,\mu) \in [0,T] \times \mathbb{R}^d \times \mathcal{P}_2(\mathbb{R}^d)$, with the terminal condition:

$$
\mathcal{U}(T,x,\mu) = \partial_x g(x,\mu) + \int_{\mathbb{R}^d} \partial_\mu g(x',\mu)(x) d\mu(x'), \quad x \in \mathbb{R}^d, \; \mu \in \mathcal{P}_2(\mathbb{R}^d),
$$

where as usual, we used:

$$
a(t,x,\mu) = \big(\sigma\sigma^\dagger\big)(t,x,\mu), \quad (t,x,\mu) \in [0,T] \times \mathbb{R}^d \times \mathcal{P}_2(\mathbb{R}^d).
$$

Example 6.33 If we revisit Example 6.3 for which $b(t,x,\mu,\alpha) = \alpha$, $\sigma = I_d$, and the running cost function f is of the form $\frac{1}{2}|\alpha|^2 + f_0(t,x,\mu)$, since $\hat{\alpha}(t,x,\mu,y) = -y$ in that case, the above master equation becomes:

$$
\begin{aligned}
0 &= \partial_t \mathcal{U}(t,x,\mu) - \partial_x \mathcal{U}(t,x,\mu)\mathcal{U}(t,x,\mu) + \frac{1}{2}\Delta_x \mathcal{U}(t,x,\mu) \\
&\quad + \int_{\mathbb{R}^d} \Big[- \partial_\mu \mathcal{U}(t,x,\mu)(x')\mathcal{U}(t,x',\mu) \\
&\qquad\qquad + \frac{1}{2}\mathrm{trace}\Big(\partial_v \partial_\mu \mathcal{U}(t,x,\mu)(x')\Big)\Big] d\mu(x') \\
&\quad + \partial_x f_0(t,x,\mu) + \int_{\mathbb{R}^d} \partial_\mu f_0(t,x',\mu)(x) d\mu(x').
\end{aligned}
$$

6.6 A Weak Formulation Bypassing Pontryagin Principle

The full analysis based on the stochastic maximum principle we provided earlier in the chapter is very robust, in part because it describes the optimizers of the control problem in quite a simple way. However, since it relies on a strong joint convexity assumption in x, μ and α, the conditions under which it can be applied may not be satisfied in some practical examples.

The purpose of this section is to provide, at least in a particular case, an alternative formulation of the control of McKean-Vlasov stochastic differential equations for which existence of an optimizer can be proven without requiring the Hamiltonian to be convex in the state variable.

6.6.1 Introduction of the Weak Formulation

In order to circumvent the lack of joint convexity in the cost functions, we introduce a new formulation of the minimization problem $\inf_{\boldsymbol{\alpha}} J(\boldsymbol{\alpha})$ defined in (6.6) and (6.9).

Statement of the Problem

Throughout this section, we shall focus on the case where the dynamics of $\boldsymbol{X}^{\boldsymbol{\alpha}}$ are linearly controlled through the drift. More precisely, we assume that:

$$dX_t^{\boldsymbol{\alpha}} = (\Lambda X_t^{\boldsymbol{\alpha}} + B\alpha_t)dt + \Sigma dW_t, \quad t \in [0, T], \tag{6.104}$$

with a prescribed initial condition $X_0^{\boldsymbol{\alpha}} = X_0$. Here, $\boldsymbol{X}^{\boldsymbol{\alpha}} = (X_t^{\boldsymbol{\alpha}})_{0 \leqslant t \leqslant T}$ takes values in $\mathbb{R}^d$, Λ, B and Σ are constant matrices of dimensions $d \times d$, $d \times k$ and $d \times m$ respectively, and $\boldsymbol{W} = (W_t)_{0 \leqslant t \leqslant T}$ is a m-dimensional Wiener process. Motivated by practical applications addressed in the next section, we assume that $m \neq d$, which differs from what we have done so far. Without any loss of generality we can assume $d \geqslant m$, which is the only interesting case, and that Σ is of rank m. The complete probability space carrying $\boldsymbol{W}$ is denoted by $(\Omega, \mathcal{F}, \mathbb{P})$. It is equipped with a complete and right-continuous filtration $\mathbb{F}$ and $\boldsymbol{W}$ is assumed to be a Brownian motion with respect to $\mathbb{F}$. Above, the control $\boldsymbol{\alpha} = (\alpha_t)_{0 \leqslant t \leqslant T}$ is a square-integrable and $\mathbb{F}$-progressively measurable process with values in a closed convex subset $A \subset \mathbb{R}^k$. Of course, whenever X_0 is deterministic, $\mathbb{F}$ may be the complete (and thus right-continuous) augmentation of the filtration generated by $\boldsymbol{W}$, but, as we shall see next, it may also be a larger filtration. In any case, for the purpose of the present discussion, we remark that the dynamics (6.104) imply that:

$$dX_t^{\boldsymbol{\alpha}} = (\Lambda X_t^{\boldsymbol{\alpha}} + B\alpha_t)dt + dM_t, \quad t \in [0, T], \tag{6.105}$$

where $\boldsymbol{M} = (M_t)_{t \geqslant 0}$ is a continuous martingale with quadratic variation $[M, M]_t = \Sigma\Sigma^\dagger t$. Notice also that, the state $X_t^{\boldsymbol{\alpha}}$ in (6.104) satisfies:

$$X_t^{\boldsymbol{\alpha}} = e^{t\Lambda} X_0^{\boldsymbol{\alpha}} + \int_0^t e^{(t-s)\Lambda} B\alpha_s \, ds + Y_t, \quad t \in [0, T], \tag{6.106}$$

where the process $\boldsymbol{Y} = (Y_t)_{t\geqslant 0}$ is given by $Y_t = \int_0^t e^{(t-s)\Lambda} \Sigma \, dW_s$ and satisfies:

$$dY_t = \Lambda Y_t dt + \Sigma dW_t = \Lambda Y_t dt + dM_t, \qquad Y_0 = 0. \tag{6.107}$$

For the analysis, it is important to observe that, starting from a state process given by (6.106) with a continuous process $\boldsymbol{Y}$ with $Y_0 = 0$ such that $(M_t = Y_t - \int_0^t \Lambda Y_s ds)_{0\leqslant t\leqslant T}$ is a martingale with quadratic variation $([M,M]_t = \Sigma\Sigma^\dagger t)_{0\leqslant t\leqslant T}$, we can recover the dynamics (6.104) in the following way. If $\Sigma = UDV$ is the singular value decomposition of the matrix Σ, the matrix U (resp. V) is an orthogonal $d \times d$ (resp. $m \times m$) matrix, and the matrix D is a $d \times m$ diagonal matrix (i.e., the entries $D_{i,i}$ for $i = 1, \cdots, m$ are the only nonzero entries). We then define the process $\boldsymbol{W} = (W_t)_{t\geqslant 0}$ by:

$$W_t = V^\dagger D^{-1} U^\dagger M_t, \quad t \in [0,T],$$

where D^{-1} denotes the $m \times d$ diagonal matrix with $[D^{-1}]_{i,i} = [D_{i,i}]^{-1}$ for $i = 1, \cdots, m$. The process $\boldsymbol{W}$ is an m-dimensional Brownian motion for the same filtration, because it is a continuous martingale with quadratic variation $[W,W]_t = tI_m$. Moreover, by construction $M_t = \Sigma W_t$. While the computation of the quadratic variation of $\boldsymbol{W}$ is straightforward given its definition, the fact that $M_t = \Sigma W_t$ is less obvious. It can be proven by checking that $U^\dagger M_t = U^\dagger \Sigma W_t$ which is equivalent to:

$$U^\dagger M_t = U^\dagger UDVV^\dagger D^{-1} U^\dagger M_t = [DD^{-1}][U^\dagger M_t], \quad t \in [0,T].$$

Since $[DD^{-1}]$ is a $d \times d$ diagonal matrix with ones on the first m entries of the diagonal, the desired inequality can be proven by showing that the last $d-m$ entries of the random vector $U^\dagger M_t$ are identically zero. This is indeed the case because:

$$\mathrm{cov}(U^\dagger M_t) = U^\dagger \mathbb{E}[M_t M_t^\dagger] U = U^\dagger \Sigma \Sigma^\dagger U \, t = DD^\dagger t, \quad t \in [0,T].$$

Remark 6.34 *For pedagogical reasons, we could have chosen a simpler model for the dynamics of the state, typically something of the form $dX_t^{\boldsymbol{\alpha}} = \alpha_t dt + dW_t$. Despite the fact that it increases the technicalities of the proofs, the choice of* (6.104) *(and in particular the assumption $d \geqslant m$) was made to accommodate applications such as the flocking model discussed at the end of Subsection* 6.7.2 *below on potential mean field games.*

We begin our discussion of the weak formulation with a cost functional of the type:

$$J(\boldsymbol{\alpha}) = \mathbb{E}\bigg[\int_0^T f\big(t, X_t^{\boldsymbol{\alpha}}, \mathcal{L}(X_t^{\boldsymbol{\alpha}}), \alpha_t\big)dt + g\big(X_T^{\boldsymbol{\alpha}}, \mathcal{L}(X_T^{\boldsymbol{\alpha}})\big)\bigg], \tag{6.108}$$

where f and g satisfy:

Assumption (MKV Weak Formulation). The coefficients f and g are real valued measurable functions on $[0, T] \times \mathbb{R}^d \times \mathcal{P}_2(\mathbb{R}^d) \times A$ and $\mathbb{R}^d \times \mathcal{P}_2(\mathbb{R}^d)$ respectively satisfying:

(A1) g is continuous and, for any $t \in [0, T]$, the function $f(t, \cdot, \cdot, \cdot)$ is also continuous;

(A2) there exist two constants $C \geqslant 0$ and $\lambda > 0$ such that, for all $(t, x, \mu, \alpha) \in [0, T] \times \mathbb{R}^d \times \mathcal{P}_2(\mathbb{R}^d) \times A$,

$$
\begin{aligned}
&|f(t,x,\mu,\alpha)| + |g(x,\alpha)| \leqslant C\big(1 + |x|^2 + M_2(\mu)^2 + |\alpha|^2\big),\\
&f(t,x,\mu,\alpha) \geqslant \lambda|\alpha|^2 - C\big(1 + |x| + M_2(\mu)\big),\\
&g(x,\alpha) \geqslant -C\big(1 + |x| + M_2(\mu)\big).
\end{aligned}
$$

(A3) For any $t \in [0, T]$, $x \in \mathbb{R}^d$ and $\mu \in \mathcal{P}_2(\mathbb{R}^d)$, the function $A \ni \alpha \mapsto f(t, x, \mu, \alpha)$ is convex.

In order to establish the existence of a control process minimizing the expected cost, we could try to prove that J is lower-semicontinuous in $\boldsymbol{\alpha}$ with respect to some topology and that, for the same topology, the sublevel sets of J are compact. Of course, the main difficulty with such an approach is to find a convenient topology for the investigation of the properties of J. This is where a new formulation may help. This motivates us to reformulate the minimization problem in order to ease the analysis of the cost functional.

Reformulation

A first step is to regard the cost functional as a mere function of the law of the control process $\boldsymbol{\alpha}$ as opposed to a function of the actual realizations of $\boldsymbol{\alpha}$. This requires a new definition of the optimization problem to dissociate it from the specific choice of the probabilistic space $(\Omega, \mathcal{F}, \mathbb{P})$. Observe indeed that, instead of constructing control processes on a prescribed probability space (as we did above), we may consider a triple $(X_0, \boldsymbol{Y}, \boldsymbol{\alpha})$, defined on some complete probability space, still denoted by $(\Omega, \mathcal{F}, \mathbb{P})$ for simplicity, such that:

1. X_0 is distributed according to μ_0, originally given for the initial distribution;
2. $\boldsymbol{\alpha} = (\alpha_t)_{0 \leqslant t \leqslant T}$ is an A-valued process jointly measurable in (t, ω) such that:
   a) for any $\omega \in \Omega$, $\int_0^T |\alpha_t(\omega)|^2 dt$ is finite,
   b) $\mathbb{E} \int_0^T |\alpha_t|^2 dt < \infty$;

3. $\boldsymbol{Y}$ is a d-dimensional continuous process such that $Y_0 = 0$ and the process $(M_t = Y_t - A\int_0^t Y_s ds)_{0\leqslant t\leqslant T}$ is a martingale with quadratic variation $([M,M]_t = \Sigma\Sigma^\dagger t)_{0\leqslant t\leqslant T}$ for the complete and right-continuous augmentation of the filtration generated by the process $(X_0, \boldsymbol{Y}, (\int_0^t \alpha_s ds)_{0\leqslant t\leqslant T})$.

The triple $(X_0, \boldsymbol{Y}, \boldsymbol{\alpha})$ is then said to be *admissible*. Quite remarkably, it is not attached to an *a priori* prescribed probability space. This fact is in full analogy with the notion of weak solution in the theory of stochastic differential equations. Notice that 2.*b*) implies 2.*a*) on an event of probability 1. In particular, whenever 2.*b*) holds, we can modify $\boldsymbol{\alpha}$ on a negligible event in such a way that 2.*a*) holds as well. The rationale for defining the filtration in terms of the indefinite integral of $\boldsymbol{\alpha}$ will be made clear below.

Under prescription 2, $\boldsymbol{\alpha}$ can be viewed as a random variable with values in the Polish space $L^2([0,T];A)$. On the probability space $(\Omega, \mathcal{F}, \mathbb{P})$ equipped with the filtration defined in item 3 above, we can define $\boldsymbol{X}$ as in (6.106). By construction this process satisfies the dynamics (6.105), and because of the reconstruction of $\boldsymbol{W}$ from $\boldsymbol{M}$, the dynamics (6.104) as well. We can then define the cost $J(\boldsymbol{\alpha})$ as in (6.108). Clearly, the value of the cost only depends upon the law of the triple $(X_0, \boldsymbol{Y}, \boldsymbol{\alpha})$ on the canonical space $\Omega_{\text{canon}} = \mathbb{R}^d \times \mathcal{C}([0,T];\mathbb{R}^d) \times L^2([0,T];A)$. Indeed, by construction, two admissible triples having the same law share the same cost. This suggests to define directly the cost functional as a function on the space $\mathcal{P}(\Omega_{\text{canon}})$ and subsequently, to transfer any admissible triple onto the canonical space Ω_{canon}.

For this reason, we introduce the following definition.

Definition 6.35 *A probability measure $\mathbb{P}$ on the space $\Omega_{canon} = \mathbb{R}^d \times \mathcal{C}([0,T];\mathbb{R}^d) \times L^2([0,T];A)$ equipped with the $\mathbb{P}$-completion of the Borel σ-field is said to be admissible for the optimal control problem* (6.104)–(6.108) *in the weak sense if under $\mathbb{P}$, the canonical process $(\xi, \boldsymbol{y}, \boldsymbol{a})$ on Ω_{canon} satisfies:*

1. *ξ is distributed according to μ_0,*
2. *the process $\left(y_t - A\int_0^t y_s ds\right)_{0\leqslant t\leqslant T}$ is a martingale with quadratic variation $(\Sigma\Sigma^\dagger t)_{0\leqslant t\leqslant T}$ for the complete and right-continuous augmentation $\mathbb{F} = (\mathcal{F}_t)_{0\leqslant t\leqslant T}$ of the filtration generated by the process $(\xi, y_t, \int_0^t a_s ds)_{0\leqslant t\leqslant T}$,*
3. *$\mathbb{E}\int_0^T |a_t|^2 dt$ is finite.*

On Ω_{canon}, we define:

$$x_t = e^{tA}\xi + \int_0^t e^{(t-s)A} B a_s ds + y_t, \quad t \in [0,T],$$

and then:

$$\mathcal{J}(\mathbb{P}) = \mathbb{E}^{\mathbb{P}}\left[\int_0^T f\left(t, x_t, \mathbb{P}\circ x_t^{-1}, a_t\right)dt + g\left(x_T, \mathbb{P}\circ x_T^{-1}\right)\right],$$

where $\mathbb{E}^{\mathbb{P}}$ denotes the expectation under $\mathbb{P}$. The collection of admissible probabilities is denoted by $\mathscr{A}$ (which should not be confused with $\mathcal{A}$ which is the notation we use for the Borel σ-field of A).

Notice that neither the definition of $\boldsymbol{x} = (x_t)_{0\leqslant t\leqslant T}$ nor that of $\mathscr{J}(\mathbb{P})$ depend upon the choice of the function representing the equivalence class of the realization of $\boldsymbol{a} = (a_t)_{0\leqslant t\leqslant T} \in L^2([0,T];A)$. However, if needed, we may choose, as a canonical representative of the equivalence class of $\boldsymbol{a}$, the process:

$$\tilde{a}_t(\omega) = \begin{cases} \lim_{n\to+\infty} n \displaystyle\int_{(t-1/n)_+}^t a_s(\omega)ds & \text{if the limit exists,} \\ 0 & \text{otherwise,} \end{cases}$$

for all $(t,\omega) \in [0,T] \times \Omega$. By Lebesgue's differentiation theorem, we know that, in the above definition, the limit in the right-hand side exists for all $\omega \in \Omega$ and for almost every $t \in [0,T]$. Moreover, for any $\omega \in \Omega$, the function $[0,T] \ni t \mapsto \tilde{a}_t(\omega)$ is a representative of the equivalence class $\boldsymbol{a}(\omega) \in L^2([0,T];A)$. It is progressively measurable with respect to the filtration generated by the process $(\int_0^t a_s ds)_{0\leqslant t\leqslant T}$. It is plain to see that this filtration contains the filtration generated by the process $(\int_0^t e^{(t-s)A} B a_s ds)_{0\leqslant t\leqslant T}$.

Observe finally that, for any admissible triple $(X_0, \boldsymbol{Y}, \boldsymbol{\alpha})$ irrespective of the probability space $(\Omega, \mathcal{F}, \mathbb{P})$ on which it is defined, the law of $(X_0, \boldsymbol{Y}, \boldsymbol{\alpha})$ is admissible in the sense of Definition 6.35.

Remark 6.36 *Any controlled trajectory in the original sense as described in equations* (6.104)–(6.108) *induces an admissible probability measure in the sense of Definition 6.35.*

Main Statement

Here is the main result of this section.

Theorem 6.37 *Under assumption* **MKV Weak Formulation**, *there exists a probability $\mathbb{P}^\star$ on Ω_{canon} such that $\mathscr{J}(\mathbb{P}^\star)$ is equal to the infimum of $\mathscr{J}$ over the set of admissible probability measures on Ω_{canon}.*

6.6.2 Relaxed Controls

The advantage of the weak formulation is quite clear: it is much easier to establish the relative compactness of a family of laws of random variables than the relative compactness of the family formed by the random variables themselves. However, although this fact makes the approach more appealing, it still does not answer the need for a topology on the space of control processes for which one can prove tightness for sequences of controls whose costs are uniformly bounded. Indeed, this is more or less what we should prove in order to establish the relative compactness of the sublevel sets of $\mathcal{J}$.

A way to bypass this difficulty is to relax the formulation of the control problem using the notion of *relaxed control*, which has been widely used in the theory of stochastic control. Roughly speaking, a *relaxed control* is a random measure on $[0, T] \times A$ satisfying some conditions that are described below. Obviously, standard controls ought to be *relaxed controls*. A natural way to make sure that this is indeed the case is to associate, with each control process $\boldsymbol{\alpha} = (\alpha_t)_{0 \leqslant t \leqslant T}$ defined on a complete filtered probability space $(\Omega, \mathcal{F}, \mathbb{F}, \mathbb{P})$, the random measure $Q(\omega, dt, d\alpha) = dt\delta_{\alpha_t(\omega)}(d\alpha)$ rigorously defined as:

$$Q(\omega, B) = \int_0^T \mathbf{1}_B\big(t, \alpha_t(\omega)\big)dt, \quad B \in \mathcal{B}([0, T] \times A), \quad \omega \in \Omega.$$

Notice that the marginal distribution of Q on $[0, T]$ is the Lebesgue measure. This prompts us to introduce the following definition.

Definition 6.38 *If we denote by $\mathcal{Q}$ the set of finite positive measures on $[0, T] \times A$ whose first projection is the Lebesgue measure on $[0, T]$, then a relaxed control on $(\Omega, \mathcal{F})$ is a random variable $Q : \Omega \ni \omega \mapsto Q(\omega, \cdot) \in \mathcal{Q}$.*

Recall also from Proposition 5.7 that, for any $B \in \mathcal{B}([0, T] \times A)$, the mapping $\Omega \ni \omega \mapsto Q(\omega, B)$, which we sometimes denote by $Q(B)$, is a random variable. A key observation is that every element $q \in \mathcal{Q}$ may be normalized into a probability measure $q/T \in \mathcal{P}([0, T] \times A)$. In this respect, the following simple result will come handy in what follows.

Proposition 6.39 *The set $\{q/T,\ q \in \mathcal{Q}\}$ is a closed subset of $\mathcal{P}([0, T] \times A)$ equipped with the topology of weak convergence.*

Proof. Without any loss of generality, we assume that $T = 1$. We then consider a sequence $(q^n)_{n \geqslant 0}$, with values in $\mathcal{Q}$, which converges in the weak sense to some $q \in \mathcal{P}([0, T] \times A)$, and prove that for any $B \in \mathcal{B}([0, T])$, $q(B \times A) = \text{Leb}_1(B)$, which will show that q belongs to $\mathcal{Q}$. Consider a continuous function ℓ from $[0, T]$ to $\mathbb{R}$. We have:

$$\lim_{n \to \infty} \int_0^T \int_A \ell(t) q^n(dt, da) = \int_0^T \int_A \ell(t) q(dt, da).$$

Obviously, the fact that the left-hand side is equal to $\int_0^T \ell(t)dt$ concludes the proof. □

The fact that the first marginal of a relaxed control is fixed plays a key role. For instance, we shall appeal several times to the following property, sometimes referred to as *stable convergence in law*, see replace by Lemma (Vol II)-7.34 for a proof:

Proposition 6.40 *Let ℓ be a bounded jointly measurable function from $[0, T] \times A$ such that, for any $t \in [0, T]$, the mapping $A \ni a \mapsto \ell(t, a)$ is continuous, and $(q^n)_{n\geqslant 0}$ be a sequence with values in $\mathcal{Q}$ converging weakly to $q \in \mathcal{Q}$. Then,*

$$\lim_{n\to+\infty} \int_0^T \int_A \ell(t, a)q^n(dt, da) = \int_0^T \int_A \ell(t, a)q(dt, da).$$

Disintegration of Relaxed Controls

For each $t \in [0, T]$, we say that a relaxed control Q is $\mathbb{F}$-adapted if, for any $t \in [0, T]$ and for any $B \in \mathcal{B}([0, T] \times A)$, the random variable $Q(B \cap ([0, t] \times A))$ is $\mathcal{F}_t$-measurable.

Equivalently, denoting by $\mathbb{F}^{Q,\text{nat}}$ the filtration generated by the $\mathcal{P}([0, T] \times A)$-valued process $((1/t)Q(\cdot \cap ([0, t] \times A)))_{0<t\leqslant T}$ (see Proposition 5.7 for the meaning if needed), Q is $\mathbb{F}$-adapted if $\mathcal{F}_t^{Q,\text{nat}} \subset \mathcal{F}_t$ for all $t \in [0, T]$.

We shall learn from Theorem (Vol II)-1.1 that if Q is a relaxed control, for each $\omega \in \Omega$, we can associate with $Q(\omega, \cdot)$ a kernel $(Q_t(\omega, \cdot))_{t\in[0,T]}$, each $Q_t(\omega, \cdot)$ being a probability measure on A, such that $Q(\omega, \cdot) = dtQ_t(\omega, \cdot)$. Equivalently, for any $B \in \mathcal{B}([0, T] \times A)$,

$$Q(\omega, B) = \int_{[0,T]\times A}\int \mathbf{1}_B(t, a)Q_t(\omega, da)dt.$$

With the above notation, we can prove the following properties which will be helpful in manipulating relaxed controls.

Proposition 6.41 *If Q is an $\mathbb{F}$-adapted relaxed control, we can redefine the process $(Q_t(\omega, \cdot))_{0\leqslant t\leqslant T}$ in such a way that:*

1. *for any $t \in [0, T]$, the mapping $[0, t] \times \Omega \ni (s, \omega) \mapsto Q_s(\omega, \cdot) \in \mathcal{P}(A)$ is measurable when its domain $[0, t] \times \Omega$ is equipped with the σ-field $\mathcal{B}([0, t]) \otimes \mathcal{F}_t$ and its range $\mathcal{P}(A)$ with the Borel σ-field of the Lévy-Prokhorov metric;*
2. *for $\mathbb{P}$-almost every $\omega \in \Omega$, $Q(\omega, \cdot) = dtQ_t(\omega, \cdot)$.*

Proof. For any $\omega \in \Omega$, $t \in [0, T]$ and $h > 0$, let:

$$Q_t^h(\omega, B) = h^{-1}Q(\omega, [(t-h)_+, t] \times B), \qquad B \in \mathcal{B}(A).$$

For any $t \in [0, T]$, $Q_t^h : \Omega \ni \omega \mapsto Q_t^h(\omega, B)$ is a random variable. Moreover, for $0 \leqslant s < t \leqslant T$,

$$\begin{aligned}|Q_t^h(\omega, B) - Q_s^h(\omega, B)| &\leqslant h^{-1}\Big(Q(\omega, [s, t] \times A) + Q(\omega, [(s-h)_+, (t-h)_+] \times A)\Big) \\ &\leqslant 2\frac{t-s}{h},\end{aligned}$$

so that $Q_\cdot^h(\cdot, B)$ is continuous in t and is thus jointly measurable in (t, ω). We deduce that $[h, T] \times \Omega \ni (t, \omega) \mapsto Q_t^h(\omega, \cdot) \in \mathcal{P}(A)$ is measurable.

Observe now that, for any $\omega \in \Omega$ and for almost every $t \in [0, T]$, for any ℓ in a countable dense subset of the space of real valued continuous functions on $\mathbb{R}$ converging to 0 at infinity, we have:

$$\int_A \ell(a) Q_t(\omega, da)$$
$$= \lim_{h \searrow 0} h^{-1} \int_{(t-h)+}^{t} \int_A \ell(a) Q\big(\omega, (ds, da)\big) = \lim_{h \searrow 0} \int_A \ell(a) Q_t^h(\omega, da).$$

This follows from the fact that the path:

$$[0, T] \ni t \mapsto \int_0^t \int_A \ell(a) Q_s(\omega, da) ds = \int_0^t \int_A \ell(a) Q\big(\omega, (ds, da)\big)$$

is Lipschitz continuous. In particular, for any $\omega \in \Omega$, $\varepsilon > 0$ and almost every $t \in (\varepsilon, T]$, the family $(Q_t^h(\omega, \cdot))_{0<h<\varepsilon}$ converges weakly to $Q_t(\omega, \cdot)$ as $h \searrow 0$.

For each fixed $\varepsilon > 0$, let $D_\varepsilon \subset [\varepsilon, T] \times \Omega$ be the set of points (t, ω) at which the sequence $(Q_t^{1/n}(\omega, \cdot))_{n>\varepsilon^{-1}}$ has a limit in $\mathcal{P}(A)$ as n tends to ∞. By joint measurability of the map $[\varepsilon, T] \times \Omega \ni (t, \omega) \mapsto Q_t^{1/n}(\omega, \cdot)$, we have $D_\varepsilon \in \mathcal{B}([\varepsilon, T]) \otimes \mathcal{F}$. Moreover, for all $\omega \in \Omega$, the set $\{t \in [\varepsilon, T] : (t, \omega) \not\in D_\varepsilon\}$ has zero Lebesgue measure. Therefore, D_ε has full measure in $[\varepsilon, T] \times \Omega$. In particular, we can redefine $Q_t(\omega, \cdot)$ as the limit of $(Q_t^{1/n}(\omega, \cdot))_{n \geq 1}$ when $(t, \omega) \in \cup_{\varepsilon>0} D_\varepsilon$ (each D_ε being regarded as an element of $\mathcal{B}([0, T]) \otimes \mathcal{F}$), and a fixed arbitrary value otherwise.

By the same argument as above, for any $\varepsilon \leqslant h \leqslant t \leqslant T$, the mapping $[\varepsilon, t] \times \Omega \ni (s, \omega) \mapsto Q_s^h(\omega, \cdot)$ is $\mathcal{B}([\varepsilon, t]) \otimes \mathcal{F}_t$-measurable. In particular, the set $\{(s, \omega) \in [\varepsilon, t] \times \Omega : (s, \omega) \in D_\varepsilon\}$ belongs to $\mathcal{B}([\varepsilon, t]) \otimes \mathcal{F}_t$. Therefore, $[0, t] \times \Omega \ni (s, \omega) \mapsto Q_s(\omega, \cdot)$ is $\mathcal{B}([0, t]) \otimes \mathcal{F}_t$-measurable. □

6.6.3 Minimization Under the Weak Relaxed Formulation

We now provide a new formulation of the optimal control problem which accommodates relaxed controls.

Strong Relaxed Formulation of the Optimization Problem

We first implement the notion of relaxed control under the original strong formulation (6.104)–(6.108) of the optimization problem. Given an underlying complete filtered probability space $(\Omega, \mathcal{F}, \mathbb{F}, \mathbb{P})$ supporting a $\mathbb{F}$-Wiener process $\boldsymbol{W}$, we define $Y_t = \int_0^t e^{(t-s)\Lambda} \Sigma dW_s$ and $M_t = Y_t - \int_0^t \Lambda Y_s ds$ as before, for $t \in [0, T]$. Next, to any $\mathbb{F}$-adapted relaxed control Q such that:

$$\mathbb{E} \int_0^T \int_A |a|^2 Q(dt, da) < \infty,$$

we associate the process:

$$X_t^Q = e^{tA}X_0 + \int_0^t \int_A e^{(t-s)A} Ba\, Q(ds, da) + Y_t, \quad t \in [0, T], \tag{6.109}$$

for a given initial condition X_0, as well as the cost functional:

$$J(Q) = \mathbb{E}\left[\int_0^T \int_A f\big(t, X_t^Q, \mathcal{L}(X_t^Q), a\big)Q(dt, da) + g\big(X_T^Q, \mathcal{L}(X_T^Q)\big)\right]. \tag{6.110}$$

Observe that the two integrals with respect to Q may be rewritten in terms of $(Q_t)_{0\le t\le T}$. Of course, the resulting writing does not depend upon the choice of the kernel $(Q_t)_{0\le t\le T}$ in Proposition 6.41.

Weak Relaxed Formulation

As in the introduction of the weak formulation in Subsection 6.6.1, observe that instead of constructing the relaxed control on a prescribed probability space, we may consider a triple $(X_0, \boldsymbol{Y}, Q)$, defined on some complete probability space $(\Omega, \mathcal{F}, \mathbb{P})$, such that:

1. the law of X_0 is the probability measure μ_0 originally chosen for the initial distribution;
2. Q is a random variable with values in $\mathcal{Q}$ such that $\mathbb{E}\int_0^T \int_A |a|^2 Q(dt, da)$ is finite;
3. $\boldsymbol{Y}$ is a d-dimensional continuous process such that $Y_0 = 0$ and the process $\boldsymbol{M}$ defined by $M_t = Y_t - \int_0^t AY_s ds$ for $0 \leqslant t \leqslant T$ is a martingale for the complete and right-continuous augmentation of the filtration generated by $(X_0, \boldsymbol{Y}, Q)$ (see the paragraph preceding the statement of Proposition 6.41 for the definition of the filtration generated by Q) with quadratic variation $[M, M]_t = \Sigma\Sigma^\dagger t$.

According to the discussion of Subsection 6.6.1, the triple $(X_0, \boldsymbol{Y}, Q)$ is called *admissible*. On the resulting filtered space, we can define $\boldsymbol{X}^Q$ as in (6.109) and the cost $J(Q)$ as in (6.110). Clearly, the value of the cost functional $J(Q)$ only depends upon the law of $(X_0, \boldsymbol{Y}, Q)$ on $\mathbb{R}^d \times \mathcal{C}([0, T]; \mathbb{R}^d) \times \mathcal{Q}$. We now state the analog of Definition 6.35 for relaxed controls.

Definition 6.42 *A probability measure $\mathbb{P}$ on $\Omega_{canon}^{relax} = \mathbb{R}^d \times \mathcal{C}([0, T]; \mathbb{R}^d) \times \mathcal{Q}$, equipped with the $\mathbb{P}$-completion of the Borel σ-field is said to be admissible for the optimal control problem* (6.104)–(6.108) *in the weak relaxed sense if under $\mathbb{P}$, the canonical process $(\xi, \boldsymbol{y}, q)$ on Ω_{canon}^{relax} satisfies:*

1. *ξ is distributed according to μ_0;*
2. *$\boldsymbol{y} = (y_t)_{0\leqslant t\leqslant T}$ is such that $y_0 = 0$ and the process $\boldsymbol{m} = (m_t)_{0\leqslant t\leqslant T}$ defined by $m_t = y_t - \int_0^t Ay_s ds$ is a martingale for the complete and right-continuous augmentation $\mathbb{F}$ of the canonical filtration generated by the process $(\xi, y_t, q(\cdot \cap ([0, t] \times A)))_{0\leqslant t\leqslant T}$ with quadratic variation $[m, m]_t = \Sigma\Sigma^\dagger t$;*
3. *$\mathbb{E}\int_0^T |a|^2 q(dt, da) < \infty$.*

We denote by $\mathscr{A}^{\mathrm{relax}}$ the set of probability measures on $\Omega_{\mathrm{canon}}^{\mathrm{relax}}$ which are admissible in the weak relaxed sense. To any such $\mathbb{P}$, we associate the cost $\mathscr{J}^{\mathrm{relax}}(\mathbb{P})$ defined as the value of $J(q)$ in (6.110) computed on $\Omega_{\mathrm{canon}}^{\mathrm{relax}}$.

Proof Theorem 6.37

We prove Theorem 6.37 assuming that the same result holds for the relaxed formulation. So, we momentarily assume the following result which will be proved later on.

Theorem 6.43 *Under assumption* **MKV Weak Formulation**, *the functional $\mathscr{J}^{relax}$ has a minimizer over the set $\mathscr{A}^{relax}$. In other words, there exists a probability $\mathbb{P}^{\star,relax}$ on Ω_{canon}^{relax} such that $\mathscr{J}^{relax}(\mathbb{P}^{\star,relax})$ is equal to the infimum of $\mathscr{J}^{relax}$ over the set $\mathscr{A}^{relax}$ of admissible probability measures on Ω_{canon}^{relax}.*

Taking for granted the conclusion of Theorem 6.43, the proof of Theorem 6.37 may be completed as follows:

Proof of Theorem 6.37: Consider a minimizing probability measure $\mathbb{P}^{\star,\mathrm{relax}}$ identified in the statement of Theorem 6.43. It is a probability measure on the canonical space $\Omega_{\mathrm{canon}}^{\mathrm{relax}}$. With each $(\xi, \boldsymbol{y}, q) \in \Omega_{\mathrm{canon}}^{\mathrm{relax}}$ such that:

$$\int_0^T \int_A |a| q(dt, da) < \infty,$$

we associate the process $\boldsymbol{x} = (x_t)_{0 \leqslant t \leqslant T}$ defined by:

$$x_t = e^{tA}\xi + \int_0^t \int_A e^{(t-s)A} Ba\, q(ds, da) + y_t, \quad t \in [0, T]. \tag{6.111}$$

Under $\mathbb{P}^{\star,\mathrm{relax}}$, the integral $\int_0^T \int_A |a|^2 q(ds, da)$ is almost surely finite. In particular, $\boldsymbol{x}$ has continuous sample paths. We call $(q_t(\omega, \cdot))_{0 \leqslant t \leqslant T}$ the kernel given by Proposition 6.41 satisfying $q(\omega, \cdot) = dt q_t(\omega, \cdot)$ for $\mathbb{P}^{\star,\mathrm{relax}}$ almost every $\omega \in \Omega_{\mathrm{canon}}^{\mathrm{relax}}$. Then, by convexity of the running cost in the variable α, we have:

$$\begin{aligned}
&\mathscr{J}^{\mathrm{relax}}\big(\mathbb{P}^{\star,\mathrm{relax}}\big) \\
&= \mathbb{E}^{\mathbb{P}^{\star,\mathrm{relax}}}\bigg[g\big(x_T, \mathbb{P}^{\star,\mathrm{relax}} \circ x_T^{-1}\big) + \int_0^T \int_A f\big(t, x_t, \mathbb{P}^{\star,\mathrm{relax}} \circ x_t^{-1}, a\big) q(dt, da)\bigg] \\
&= \mathbb{E}^{\mathbb{P}^{\star,\mathrm{relax}}}\bigg[g\big(x_T, \mathbb{P}^{\star,\mathrm{relax}} \circ x_T^{-1}\big) + \int_0^T \int_A f\big(t, x_t, \mathbb{P}^{\star,\mathrm{relax}} \circ x_t^{-1}, a\big) q_t(da) dt\bigg] \\
&\geqslant \mathbb{E}^{\mathbb{P}^{\star,\mathrm{relax}}}\bigg[g\big(x_T, \mathbb{P}^{\star,\mathrm{relax}} \circ x_T^{-1}\big) + \int_0^T f\Big(t, x_t, \mathbb{P}^{\star,\mathrm{relax}} \circ x_t^{-1}, \int_A a q_t(da)\Big) dt\bigg].
\end{aligned} \tag{6.112}$$

Notice that, $\mathbb{P}^{\star,\text{relax}}$-almost surely, for almost every $t \in [0,T]$, $\int_A a q_t(da)$ makes sense and belongs to A. Letting:

$$\alpha_t(\omega) = \begin{cases} \displaystyle\int_A a q_t(\omega, da) & \text{if } \displaystyle\int_A |a| q_t(\omega, da) < \infty, \\ 0 & \text{otherwise,} \end{cases}$$

for $(t,\omega) \in [0,T] \times \Omega$, we have:

$$\mathbb{E}\int_0^T |\alpha_t|^2 dt \leqslant \mathbb{E}\int_0^T \int_A |a|^2 q(dt, da) < \infty.$$

Notice also that (6.111) may be rewritten:

$$x_t = e^{tA}\xi + \int_0^t e^{(t-s)A} B\alpha_s ds + y_t, \quad t \in [0,T]. \tag{6.113}$$

Now we define $\mathbb{P}^\star$ as the law of $(\xi, \boldsymbol{y}, \boldsymbol{\alpha})$ on $\mathbb{R}^d \times \mathcal{C}([0,T];\mathbb{R}^d) \times L^2([0,T];A)$ under $\mathbb{P}^{\star,\text{relax}}$. Also, with the same notations as in Definition 6.35 and by the very definition of $\boldsymbol{x} = (x_t)_{0 \leqslant t \leqslant T}$ in (6.111), the law of $(\xi, \boldsymbol{y}, \boldsymbol{a}, \boldsymbol{x})$ under $\mathbb{P}^\star$ is the same as the law of $(\xi, \boldsymbol{y}, \boldsymbol{\alpha}, \boldsymbol{x})$ under $\mathbb{P}^{\star,\text{relax}}$. By (6.112), we then have:

$$\mathscr{J}(\mathbb{P}^\star) \leqslant \mathscr{J}^{\text{relax}}(\mathbb{P}^{\star,\text{relax}}).$$

Since any admissible probability measure for the weak formulation (see Definition 6.35) is also an admissible probability measure for the weak relaxed formulation (see Definition 6.42), or equivalently since $\mathscr{A} \subset \mathscr{A}^{\text{relax}}$, we deduce that, for any $\mathbb{P} \in \mathscr{A}^{\text{relax}}$,

$$\mathscr{J}(\mathbb{P}^\star) \leqslant \mathscr{J}^{\text{relax}}(\mathbb{P}^{\star,\text{relax}}) \leqslant \mathscr{J}(\mathbb{P}),$$

which completes the proof. □

6.6.4 Proof of the Solvability Under the Weak Relaxed Formulation

The rest of the section is devoted to the proof of Theorem 6.43. Without any loss of generality we assume that $T = 1$ which allows us to identify $\mathcal{Q}$ with a subset of $\mathcal{P}([0,1] \times A)$.

In order to prove that the weak relaxed formulation admits a minimizer, we proceed in two steps. First we prove that any nonempty sublevel set of the form $\{\mathbb{P} \in \mathscr{A}^{\text{relax}} : \mathscr{J}^{\text{relax}}(\mathbb{P}) \leqslant K\}$ for $K \in \mathbb{R}$, is relatively compact for the weak-topology. Next, we show that $\mathscr{J}^{\text{relax}}$ is lower semi-continuous for the weak topology.

Compactness of the Sublevel Sets

In order to investigate the compactness of the sublevel sets, we start with the following observation: any admissible probability measure $\mathbb{P}$ in the sense of Definition 6.42 may be regarded as a probability measure on $\mathbb{R}^d \times \mathcal{C}([0,1];\mathbb{R}^d) \times (\mathcal{Q} \cap \mathcal{P}_\eta([0,1] \times A))$, for $\eta \in [1,2]$.

Lemma 6.44 *Let $K \in \mathbb{R}$ such that the sublevel set $\{\mathbb{P} \in \mathscr{A}^{relax} : \mathscr{J}^{relax}(\mathbb{P}) \leqslant K\}$ is not empty. Then, for any $\eta \in [1,2)$, $\{\mathbb{P} \in \mathscr{A}^{relax} : \mathscr{J}^{relax}(\mathbb{P}) \leqslant K\}$ is relatively compact for the weak topology on $\mathcal{P}(\mathbb{R}^d \times \mathcal{C}([0,1];\mathbb{R}^d) \times \mathcal{Q}^\eta)$, where $\mathcal{Q}^\eta = \mathcal{Q} \cap \mathcal{P}_\eta([0,1] \times A)$ is equipped with the η-Wasserstein distance on $\mathcal{P}_\eta([0,1] \times A)$. Moreover, any weak limit of sequences with values in $\{\mathbb{P} \in \mathscr{A}^{relax} : \mathscr{J}^{relax}(\mathbb{P}) \leqslant K\}$ belongs to $\mathscr{A}^{relax}$.*

Notice that Proposition 6.39 implies that $\mathcal{Q}^\eta$ is a closed subset of $\mathcal{P}_\eta([0,1] \times A)$ for the η-Wasserstein distance.

Proof. Recall that we assume that $T = 1$ for simplicity. Throughout the proof, we work on the canonical space $\Omega^{\text{relax}}_{\text{canon}}$ and the canonical random variable is denoted by $(\xi, \boldsymbol{y}, q)$. With each $(\xi, \boldsymbol{y}, q) \in \Omega^{\text{relax}}_{\text{canon}}$ such that:

$$\int_0^1 \int_A |a| q(dt, da) < \infty,$$

we associate the process $\boldsymbol{x} = (x_t)_{0 \leqslant t \leqslant 1}$ defined by (6.111) for $t \in [0,1]$. When $\mathbb{P} \in \mathscr{A}^{\text{relax}}$, the integral $\int_0^1 \int_A |a|^2 q(ds, da)$ is finite $\mathbb{P}$-almost surely. In particular, $\boldsymbol{x}$ has continuous sample paths.

First Step. Let us consider a sequence $(\mathbb{P}^n)_{n \geqslant 1}$ with values in $\mathscr{A}^{\text{relax}}$ such that $\mathscr{J}^{\text{relax}}(\mathbb{P}^n) \leqslant K$ for all $n \geqslant 1$. By assumption (**A2**), we have, for a possibly new value of the constant C whose value is allowed to increase from line to line:

$$\lambda \mathbb{E}^n \int_0^1 \int_A |a|^2 q(dt, da) \leqslant \mathscr{J}^{\text{relax}}(\mathbb{P}^n) + C\Big(1 + \mathbb{E}^n\big[\sup_{0 \leqslant t \leqslant 1} |x_t|\big]\Big).$$

We use the notation $\mathbb{E}^n$ to denote the expectation with respect to $\mathbb{P}^n$. Observe now that by definition (6.111) of $\boldsymbol{x}$, we have:

$$\mathbb{E}^n\big[\sup_{0 \leqslant t \leqslant 1} |x_t|\big] \leqslant C\Big(1 + \mathbb{E}^n \int_0^1 \int_A |a|^2 q(dt, da)\Big)^{1/2}.$$

Consequently, there exists a constant c_K, depending on K, such that:

$$\mathbb{E}^n \int_0^1 \int_A |a|^2 q(dt, da) \leqslant c_K, \tag{6.114}$$

from which we deduce that the sequence $(\mathbb{P}^n \circ q^{-1})_{n\geqslant 1}$ is tight on $\mathcal{P}([0,1]\times A)$. In particular, for any $\varepsilon, c > 0$:

$$\sup_{n\geqslant 1} \mathbb{P}^n\Big[q\big(|a| \geqslant \frac{c}{\varepsilon}\big) \geqslant \frac{1}{c}\Big] \leqslant c \sup_{n\geqslant 1} \mathbb{E}^n\Big[q\big(|a| \geqslant \frac{c}{\varepsilon}\big)\Big] \leqslant \frac{c_K \varepsilon^2}{c}.$$

Therefore, for any $\varepsilon > 0$ and any $p \in \mathbb{N}$:

$$\sup_{n\geqslant 1} \mathbb{P}^n\Big[q\big(|a| \geqslant \frac{2^p}{\varepsilon}\big) \geqslant \frac{1}{2^p}\Big] \leqslant \frac{c_K \varepsilon^2}{2^p}.$$

Finally, by summing over $p \in \mathbb{N}$, we obtain:

$$\sup_{n\geqslant 1} \mathbb{P}^n\Big[\bigcup_{p\in\mathbb{N}} \Big\{q\big(|a| \geqslant \frac{2^p}{\varepsilon}\big) \geqslant \frac{1}{2^p}\Big\}\Big] \leqslant \sum_{p\in\mathbb{N}} \frac{c_K \varepsilon^2}{2^p} = 2c_K\varepsilon^2. \tag{6.115}$$

Since, for any $\varepsilon > 0$, the set:

$$\Big\{q \in \mathcal{P}([0,1]\times A);\ \forall p \in \mathbb{N},\ q\big(|a| \geqslant \frac{2^p}{\varepsilon}\big) \leqslant \frac{1}{2^p}\Big\} \tag{6.116}$$

is a relatively compact subset of $\mathcal{P}([0,1]\times A)$ by Prokhorov's theorem, we conclude that the sequence $(\mathbb{P}^n \circ q^{-1})_{n\geqslant 1}$ is tight on $\mathcal{P}([0,1]\times A)$. Since $\mathcal{Q}$ is a closed subset of $\mathcal{P}([0,1]\times A)$, it is also tight on $\mathcal{Q}$.

Second Step. We now prove that, for any $\eta \in [1,2)$, the family $(\mathbb{P}^n \circ q^{-1})_{n\geqslant 1}$ is tight on $\mathcal{P}_\eta([0,1]\times A)$ equipped with the η-Wasserstein distance. In order to do so, we make use of Corollary 5.6.

We first recall the bound (6.114), from which we get, for every $R > 0$:

$$\mathbb{E}^n \int_0^1 \int_A \mathbf{1}_{\{|a| \geqslant R\}} |a|^\eta q(dt,da) \leqslant \frac{1}{R^{2-\eta}} \mathbb{E}^n \int_0^1 \int_A |a|^2 q(dt,da) \leqslant \frac{c_K}{R^{2-\eta}}.$$

Therefore, for any $\varepsilon > 0$ and $p \in \mathbb{N}$, Markov's inequality yields:

$$\begin{aligned}\sup_{n\geqslant 1} \mathbb{P}^n\Big(\int_0^1 \int_A \mathbf{1}_{\{|a|\geqslant 2^p\}} |a|^\eta q(dt,da) \geqslant \frac{1}{2^{(2-\eta)p/2}\varepsilon}\Big) &\leqslant c_K\varepsilon \frac{2^{(2-\eta)p/2}}{2^{p(2-\eta)}} \\ &= c_K\varepsilon 2^{-(2-\eta)p/2}.\end{aligned}$$

Summing over $p \in \mathbb{N}$, we finally have:

$$\sup_{n\geqslant 1} \mathbb{P}^n\Big(\bigcup_{p\in\mathbb{N}} \Big\{\int_0^1 \int_A \mathbf{1}_{\{|a|\geqslant 2^p\}} |a|^\eta q(dt,da) \geqslant \frac{1}{2^{(2-\eta)p/2}\varepsilon}\Big\}\Big) \leqslant \frac{c_K\varepsilon}{1-2^{-(2-\eta)/2}}. \tag{6.117}$$

Now, using Corollary 5.6, the fact that the time component t runs through the compact interval $[0,1]$, and the relative compactness of the set (6.116) for the topology of weak convergence, we conclude that, for any $\varepsilon > 0$, the set:

$$\Big\{q \in \mathcal{P}([0,1]\times A);\ \forall p \in \mathbb{N},\ q\big(|a| \geq \frac{2^p}{\varepsilon}\big) \leq \frac{1}{2^p},$$
$$\int_0^1 \int_A \mathbf{1}_{\{|a|\geq 2^p\}} |a|^\eta q(dt,da) < \frac{1}{2^{(2-\eta)p/2}\varepsilon}\Big\}$$

is relatively compact in $\mathcal{P}_\eta([0,1]\times A)$. Since the bounds (6.115) and (6.117) control the mass that $\mathbb{P}^n \circ q^{-1}$ puts outside of this relatively compact set, the proof of the tightness of $(\mathbb{P}^n \circ q^{-1})_{n\geq 1}$ on $\mathcal{Q}^\eta$ is complete.

Third Step. Obviously, $(\mathbb{P}^n)_{n\geq 1}$ is tight on $\Omega^{\text{relax}}_{\text{canon}}$. Therefore, we can extract a subsequence converging in the weak sense. Let $\mathbb{P}$ be a limit point. In order to complete the proof, we must show that $\mathbb{P}$ satisfies the properties of Definition 6.42.

It is clear that under $\mathbb{P}$, ξ has distribution μ_0. Moreover, it is standard to prove that under $\mathbb{P}$, the process $\boldsymbol{m}$ defined by $(m_t = y_t - \int_0^t \Lambda y_s ds)_{0\leq t\leq 1}$ is a martingale for the complete and right-continuous augmentation of the filtration of the canonical process $(\xi, \boldsymbol{y}, q)$ with quadratic variation $([m,m]_t = \Sigma\Sigma^\dagger t)_{0\leq t\leq 1}$.

Finally, observe that if we denote by $\mathbb{E}$ the expectation with respect to $\mathbb{P}$ we have:

$$\mathbb{E}\int_0^1 \int_A |a|^2 q(dt,da) = \lim_{p\to\infty} \mathbb{E}\int_0^1 \int_A |a|^2 \varphi_p(|a|) q(dt,da),$$

where $(\varphi_p)_{p\in\mathbb{N}}$ is a nondecreasing sequence of continuous functions with values in $[0,1]$, equal to 1 on $[-p,p]$ and vanishing outside $[-2p,2p]$. Since the mapping $\mathcal{Q} \ni Q \mapsto \int_0^1 \int_A |a|^2 \varphi_p(|a|) Q(dt,da)$ is continuous, we deduce that, for all $p \in \mathbb{N}$,

$$\begin{aligned}\mathbb{E}\int_0^1 \int_A |a|^2 \varphi_p(|a|) q(dt,da) &= \lim_{n\to\infty} \mathbb{E}^n \int_0^1 \int_A |a|^2 \varphi_p(|a|) q(dt,da)\\ &\leq \liminf_{n\to\infty} \mathbb{E}^n \int_0^1 \int_A |a|^2 q(dt,da) \leq C_K.\end{aligned}$$

We conclude that:

$$\mathbb{E}\int_0^1 \int_A |a|^2 q(dt,da) \leq C_K,$$

which completes the proof □

Lower Semicontinuity of the Cost Functional

We now prove that the cost functional $\mathscr{J}^{\text{relax}}$ is lower semicontinuous. The proof relies on the following fact about Polish spaces, which is a variant of Proposition 6.40; for the sake of completeness, we show how to derive it as a corollary of Proposition 6.40.

Lemma 6.45 *Let $(\mathcal{S}, d)$ be a Polish space and $\varphi : [0,1]\times\mathcal{S}\times A \to \mathbb{R}$ be a bounded measurable function, such that for any $t \in [0,1]$, the function $\mathcal{S}\times A \ni (\varsigma, a) \mapsto \varphi(t,\varsigma,a)$ is continuous.*

Then, for any sequence $(\varsigma^n, q^n)_{n\geqslant 1}$ converging to some (ς, q) for the product topology on $\mathcal{C}([0,1];\mathcal{S}) \times \mathcal{P}([0,1]\times A)$, we have:

$$\liminf_{n\to\infty} \int_0^1 \int_A \varphi(t, \varsigma_t^n, a) q^n(dt, da) = \int_0^1 \int_A \varphi(t, \varsigma_t, a) q(dt, da).$$

Proof.

First Step. We first assume that there exists $c > 0$ such that $\varphi(t, \varsigma, a) = 0$ if $|a| \geqslant c$. Then, we write:

$$\begin{aligned}
&\int_0^1 \int_A \varphi(t, \varsigma_t^n, a) q^n(dt, da) - \int_0^1 \int_A \varphi(t, \varsigma_t, a) q(dt, da) \\
&= \int_0^1 \int_A \Big(\varphi(t, \varsigma_t^n, a) - \varphi(t, \varsigma_t, a)\Big) q^n(dt, da) + \int_0^1 \int_A \varphi(t, \varsigma_t, a)\big(q^n - q\big)(dt, da).
\end{aligned} \tag{6.118}$$

Since for any $t \in [0,1]$, the function $\mathcal{S} \times A \ni (\varsigma, a) \mapsto \varphi(t, \varsigma, a)$ is continuous in (ς, a) and null when $|a| \geqslant c$, we conclude that for any $t \in [0,1]$:

$$\lim_{n\to\infty} \sup_{a\in A} \big|\varphi(t, \varsigma_t^n, a) - \varphi(t, \varsigma_t, a)\big| = 0.$$

Now,

$$\left| \int_0^1 \int_A \Big(\varphi(t, \varsigma_t^n, a) - \varphi(t, \varsigma_t, a)\Big) q^n(dt, da) \right| \leqslant \int_0^1 \sup_{a\in A} \Big|\varphi(t, \varsigma_t^n, a) - \varphi(t, \varsigma_t, a)\Big| dt,$$

and by Lebesgue's dominated convergence theorem, we get:

$$\lim_{n\to\infty} \int_0^1 \int_A \Big(\varphi(t, \varsigma_t^n, a) - \varphi(t, \varsigma_t, a)\Big) q^n(dt, da) = 0,$$

which takes care of the first term in the right-hand side of (6.118). As for the second term, it tends to 0 as n tends to ∞ when φ is jointly continuous in its three arguments. In order to overcome the lack of continuity in the variable t, we use Proposition 6.40.

Second Step. We now turn to the general case when φ does not vanish anymore when $|a|$ is large. We consider a sequence of continuous functions $(\psi_p)_{p\in\mathbb{N}}$, with values in $[0,1]$, such that each $\psi_p(x) = 1$ when $|x| \leqslant p$ and $\psi_p(x) = 0$ whenever $|x| > 2p$. Then, the first step applies to each function $[0,1] \times \mathcal{S} \times A \ni (t, \varsigma, a) \mapsto \varphi(t, \varsigma, a)\psi_p(a)$. Therefore, in order to complete the proof, it suffices to notice that:

$$\begin{aligned}
\sup_{n\geqslant 1} \left| \int_0^1 \int_A \varphi(t, \varsigma_t, a)\big(\psi_p(a) - 1\big) q^n(dt, da) \right| &\leqslant C \sup_{n\geqslant 1} \int_0^1 \int_A \big|\psi_p(a) - 1\big| q^n(dt, da) \\
&\leqslant C \sup_{n\geqslant 1} \int_0^1 \int_A \mathbf{1}_{\{|a|\geqslant 2p\}} q^n(dt, da),
\end{aligned}$$

where C depends upon the sup norm of φ. Since the sequence of measures $(q^n)_{n\geqslant 1}$ is convergent, it is tight, so that the last term tends to 0 as p tends to ∞. □

We now prove the desired semi-continuity result.

Lemma 6.46 *For any $\eta \in (1,2)$, the cost functional $\mathscr{J}^{relax} : \mathscr{A}^{relax} \to \mathbb{R}$ given in Definition 6.42 is lower semicontinuous on any sublevel set with respect to the weak topology on $\mathcal{P}(\mathbb{R}^d \times \mathcal{C}([0,T];\mathbb{R}^d) \times \mathcal{Q}^\eta)$, where $\mathcal{Q}^\eta = \mathcal{Q} \cap \mathcal{P}_\eta([0,1] \times A)$ is equipped with the η-Wasserstein distance.*

Proof. Let us consider a sequence $(\mathbb{P}^n)_{n\geqslant 1}$ in $\mathscr{A}^{\text{relax}}$ such that $\sup_{n\geqslant 1} \mathscr{J}^{\text{relax}}(\mathbb{P}^n) \leqslant K$, for some $K \in \mathbb{R}$, and converging to $\mathbb{P}$ with respect to the weak topology on $\mathcal{P}(\mathbb{R}^d \times \mathcal{C}([0,T];\mathbb{R}^d) \times \mathcal{Q}^\eta)$. By Lemma 6.44, $\mathbb{P}$ belongs to $\mathscr{A}^{\text{relax}}$.

First Step. When $\boldsymbol{x}$ is defined on the canonical space $\Omega_{\text{canon}}^{\text{relax},\eta} = \mathbb{R}^d \times \mathcal{C}([0,1];\mathbb{R}^d) \times \mathcal{Q}^\eta$, with $\mathcal{Q}^\eta = \mathcal{Q} \cap \mathcal{P}_\eta([0,1] \times A)$ via formula (6.111), we view it as the image of $(\xi, \boldsymbol{y}, q)$ under the mapping $\mathcal{X}$ defined by:

$$\begin{aligned}
\Omega_{\text{canon}}^{\text{relax},\eta} \ni (\xi, \boldsymbol{y}, q) &\mapsto \mathcal{X}(\xi, \boldsymbol{y}, q) \\
&= \Big(e^{tA}\xi + \int_0^t \int_A e^{(t-s)A} Ba\, q(ds, da) + y_t \Big)_{0\leqslant t\leqslant 1} \in \mathcal{C}([0,1];\mathbb{R}^d).
\end{aligned}$$

We claim that $\mathcal{X}$ is continuous as the sum of three continuous functions. Obviously, it is enough to check that $(\int_0^t \int_A e^{(t-s)A} Ba\ q(ds, da))_{0\leqslant t\leqslant 1} \in \mathcal{C}([0,1];\mathbb{R}^d)$ is a continuous function of $q \in \mathcal{Q}^\eta$. Proposition 6.40 implies continuity for each fixed $t \in [0,1]$, the uniformity following from the fact that, if $0 \leqslant s \leqslant t$, we have:

$$\begin{aligned}
&\Big| \int_0^t \int_A e^{(t-u)A} Ba\, q(du, da) - \int_0^s \int_A e^{(s-u)A} Ba\, q(du, da) \Big| \\
&\leqslant \Big| \int_0^s \int_A [e^{tA} - e^{sA}] e^{-uA} Ba\, q(du, da) \Big| + \Big| \int_s^t \int_A e^{(t-u)A} Ba\, q(du, da) \Big| \\
&\leqslant C\bigg(|t-s| \int_0^1 \int_A |a|\, q(du, da) + |t-s|^{(\eta-1)/\eta} \Big[\int_0^1 \int_A |a|^\eta\, q(dt, da) \Big]^{1/\eta} \bigg).
\end{aligned}$$

for some constant C depending only upon the norms of the matrices A and B.

Second Step. From the first step, we have that $(\mathbb{P}^n \circ (\xi, \boldsymbol{y}, q, \boldsymbol{x})^{-1})_{n\geqslant 1}$ converges to $\mathbb{P} \circ (\xi, \boldsymbol{y}, q, \boldsymbol{x})^{-1}$ on $\Omega_{\text{canon}}^{\text{relax},\eta} \times \mathcal{C}([0,1];\mathbb{R}^d)$. Therefore, letting, for each $t \in [0,1]$ and $n \geqslant 1$, $\mu_t^n = \mathbb{P}^n \circ x_t^{-1}$, we deduce that the sequence $(\mu_t^n)_{n\geqslant 1}$ converges weakly to $\mu_t = \mathbb{P} \circ x_t^{-1}$. Also,

$$\sup_{n\geqslant 1} \sup_{0\leqslant t\leqslant 1} |x_t| \leqslant C\Big(|\xi| + \int_0^1 \int_A |a| q(dt, da) \Big) + \sup_{0\leqslant t\leqslant 1} |y_t|.$$

Now, (6.114) implies the existence of a constant C_K such that:

$$\sup_{n\geqslant 1} \mathbb{E}^n \bigg[\Big| \int_0^1 \int_A |a| q(dt, da) \Big|^2 \bigg] \leqslant \sup_{n\geqslant 1} \bigg[\mathbb{E}^n \int_0^1 \int_A |a|^2 q(dt, da) \bigg] \leqslant C_K, \tag{6.119}$$

and allowing the value of C_K to increase from line to line if needed,

$$\sup_{n\geqslant 1} \mathbb{E}^n\big[\sup_{0\leqslant t\leqslant 1} |x_t|^2\big] \leqslant C_K.$$

Therefore, for any $\eta \in [1,2)$, the sequence of measures $(\mathbb{P}^n \circ (\sup_{0\leqslant t\leqslant 1} |x_t|^\eta)^{-1})_{n\geqslant 1}$ is η-uniformly integrable. We deduce that, for any $t \in [0,1]$, the sequence $(\mu_t^n)_{n\geqslant 1}$ converges to μ_t in $\mathcal{P}_\eta(\mathbb{R}^d)$. Moreover:

$$\mathbb{E}\int_0^1\int_A |a|^2 q(dt,da) \leqslant C_K, \quad \text{and} \quad \sup_{0\leqslant t\leqslant 1} \big(M_2(\mu_t)\big)^2 \leqslant \mathbb{E}\big[\sup_{0\leqslant t\leqslant 1} |x_t|^2\big] \leqslant C_K. \tag{6.120}$$

Third Step. By Skorohod's representation theorem on the Polish space $\Omega_{\text{canon}}^{\text{relax},\eta} \times \mathcal{C}([0,1];\mathbb{R}^d)$, we get a sequence of random variables $(\xi^n, \boldsymbol{Y}^n, \mathcal{Q}^n, \boldsymbol{X}^n)_{n\geqslant 1, n=\infty}$, defined on the same probability space $(\Xi, \mathcal{G}, \mathbf{P})$, such that, for all $n \geqslant 1$, $(\xi^n, \boldsymbol{Y}^n, \mathcal{Q}^n, \boldsymbol{X}^n)$ is distributed according to $\mathbb{P}^n \circ (\xi, \boldsymbol{y}, q, \boldsymbol{x})^{-1}$, $(\xi^\infty, \boldsymbol{Y}^\infty, \mathcal{Q}^\infty, \boldsymbol{X}^\infty)$ is distributed according to $\mathbb{P} \circ (\xi, \boldsymbol{w}, q, \boldsymbol{x})^{-1}$ and, $\mathbf{P}$-almost surely,

$$\lim_{n\to\infty} (\xi^n, \boldsymbol{Y}^n, \mathcal{Q}^n, \boldsymbol{X}^n) = (\xi^\infty, \boldsymbol{Y}^\infty, \mathcal{Q}^\infty, \boldsymbol{X}^\infty).$$

By Lemma 6.45, we deduce that for any $p \in \mathbb{N}$ and $\mathbf{P}$-almost surely:

$$\begin{aligned}
&\int_0^1\int_A f(t, X_t^\infty, \mu_t^\infty, a)\psi_p(a)\mathcal{Q}^\infty(dt,da) \\
&\quad = \lim_{n\to\infty} \int_0^1\int_A f(t, X_t^n, \mu_t^n, a)\psi_p(a)\mathcal{Q}^n(dt,da),
\end{aligned}$$

where ψ_p is a continuous cut-off function from $\mathbb{R}^k$ to $[0,1]$, which is equal to 1 on the ball of center 0 and radius p, and 0 outside the ball of center 0 and radius $2p$.

We now apply Fatou's lemma to the sequence $(\int_0^1\int_A f(t, X_t^n, \mu_t^n, a)\psi_p(a)\mathcal{Q}^n(dt,da) + C_p\int_0^1 (1 + |X_t^n| + M_2(\mu_t^n))dt)_{n\geq 1}$, which is non-negative for a well-chosen constant C_p. By uniform integrability of the random variables $(\int_0^1 (1 + |X_t^n| + M_2(\mu_t^n))dt)_{n\geq 1}$, we get for any $p \in \mathbb{N}$:

$$\begin{aligned}
&\mathbf{E}\int_0^1\int_A f(t, X_t^\infty, \mu_t^\infty, a)\psi_p(a)\mathcal{Q}^\infty(dt,da) \\
&\quad \leqslant \liminf_{n\to\infty} \mathbf{E}\int_0^1\int_A f(t, X_t^n, \mu_t^n, a)\psi_p(a)\mathcal{Q}^n(dt,da),
\end{aligned}$$

where we used the notation $\mathbf{E}$ for the expectation with respect to $\mathbf{P}$ over Ξ. Returning to the original sequence $(\mathbb{P}^n \circ (\xi, \boldsymbol{y}, q, \boldsymbol{x})^{-1})_{n\geqslant 1}$, we deduce that:

$$\begin{aligned}
&\mathbb{E}\int_0^1\int_A f(t, x_t, \mu_t, a)\psi_p(a)q(dt,da) \\
&\quad \leqslant \liminf_{n\to\infty} \mathbb{E}^n\int_0^1\int_A f(t, x_t, \mu_t^n, a)\psi_p(a)q(dt,da).
\end{aligned}$$

Fourth Step. Since for any $t \in [0,1]$ we have:

$$|f(t, x_t, \mu_t, a)| \leqslant \Big(1 + \sup_{0\leqslant t\leqslant 1} |x_t|^2 + \sup_{0\leqslant t\leqslant 1} \big(M_2(\mu_t)\big)^2 + |a|^2\Big),$$

and since (6.120) implies:

$$\mathbb{E}\int_0^1 \int_A \Big(1 + \sup_{0\leqslant t\leqslant 1} |x_t|^2 + \sup_{0\leqslant t\leqslant 1} \big(M_2(\mu_t)\big)^2 + |a|^2\Big) q(dt, da) < \infty,$$

we have:

$$\lim_{p\to\infty} \mathbb{E}\int_0^1 \int_A f(t, x_t, \mu_t, a)\psi_p(a) q(dt, da) = \mathbb{E}\int_0^1 \int_A f(t, x_t, \mu_t, a) q(dt, da).$$

The assumption on the running cost function implies that, for all $n \geqslant 1$, $t \in [0,1]$ and $a \in A$:

$$\begin{aligned} f(t, x_t, \mu_t^n, a) &\geqslant \lambda |a|^2 - C\big(1 + |x_t| + M_2(\mu_t^n)\big) \\ &\geqslant \lambda |a|^2 - C\big(1 + |x_t|\big), \end{aligned}$$

where we used the fact the sequence $(\sup_{0\leqslant t\leqslant 1} M_2(\mu_t^n))_{n\geqslant 1}$ is bounded, and allowed the constant C to vary from line to line. Therefore,

$$\begin{aligned} &\mathbb{E}^n \int_0^1 \int_A f(t, x_t, \mu_t^n, a)\psi_p(a) q(dt, da) \\ &= \mathbb{E}^n \int_0^1 \int_A f(t, x_t, \mu_t^n, a) q(dt, da) + \mathbb{E}^n \int_0^1 \int_A f(t, x_t, \mu_t^n, a)\big(\psi_p(a) - 1\big) q(dt, da) \\ &\leqslant \mathbb{E}^n \int_0^1 \int_A f(t, x_t, \mu_t^n, a) q(dt, da) + C\mathbb{E}^n \int_0^1 \int_A \big(1 + \sup_{0\leqslant t\leqslant 1} |x_t|\big) \mathbf{1}_{\{|a|\geqslant p\}} q(dt, da). \end{aligned}$$

By Cauchy Schwarz' inequality, we get:

$$\begin{aligned} &\mathbb{E}^n \int_0^1 \int_A \big(1 + \sup_{0\leqslant t\leqslant 1} |x_t|\big) \mathbf{1}_{\{|a|\geqslant p\}} q(dt, da) \\ &\leqslant \bigg(\mathbb{E}^n \int_0^1 \int_A \big(1 + \sup_{0\leqslant t\leqslant 1} |x_t|\big)^2 q(dt, da)\bigg)^{1/2} \bigg(\mathbb{E}^n \int_0^1 \int_A \mathbf{1}_{\{|a|\geqslant p\}} q(dt, da)\bigg)^{1/2} \\ &\leqslant \frac{C}{p}, \end{aligned}$$

where we also used (6.120). Finally:

$$\limsup_{p\to\infty}\liminf_{n\to\infty}\mathbb{E}^n\int_0^1\int_A f(t,x_t,\mu_t^n,a)\psi_p(a)q(dt,da)$$

$$\leqslant \liminf_{n\to\infty}\mathbb{E}^n\int_0^1\int_A f(t,x_t,\mu_t^n,a)q(dt,da).$$

Taking the limit as p tends to ∞ in the conclusion of the third step, we obtain:

$$\mathbb{E}\int_0^1\int_A f(t,x_t,\mu_t,a)q(dt,da) \leqslant \liminf_{n\to\infty}\mathbb{E}^n\int_0^1\int_A f(t,x_t,\mu_t^n,a)q(dt,da).$$

Handling the terminal cost as the running cost in the third step, we complete the proof. □

Conclusion

The proof of Theorem 6.43 is easily completed, using the relative compactness of the sublevel sets and the lower semi-continuity of the cost functional $\mathcal{J}^{\text{relax}}$.

6.7 Examples

This section may be viewed as a hodgepodge of extensions of the theory presented in this chapter and discussions of models already introduced in earlier chapters.

6.7.1 Linear Quadratic (LQ) McKean Vlasov (Mean Field) Control

The treatment of this subsection parallels the discussion of Section 3.5 of Chapter 3 where we solved linear quadratic mean field games. As before, we start with the multidimensional models before focusing on the one-dimensional case. We use the same notation as in Section 3.5 for the purpose of comparison.

The Linear Quadratic (LQ) Model

As before, the drift is given by:

$$b(t,x,\mu,\alpha) = b_1(t)x + \bar{b}_1(t)\bar{\mu} + b_2(t)\alpha,$$

the running cost function by:

$$f(t,x,\mu,\alpha) = \frac{1}{2}\Big(x^\dagger q(t)x + \big(x-s(t)\bar{\mu}\big)^\dagger\bar{q}(t)\big(x-s(t)\bar{\mu}\big) + \alpha^\dagger r(t)\alpha\Big),$$

and the terminal cost by:

$$g(x,\mu) = \frac{1}{2}\Big(x^\dagger qx + \big(x-s\bar{\mu}\big)^\dagger\bar{q}\big(x-s\bar{\mu}\big)\Big).$$

The volatility σ is assumed to be constant.

We assume that b_1, $\bar{b}_1$ and b_2 are deterministic continuous function on $[0, T]$ with values in $\mathbb{R}^{d\times d}$, $\mathbb{R}^{d\times d}$, and $\mathbb{R}^{d\times k}$ respectively. Similarly, we assume that q, r, s and $\bar{q}$ are deterministic continuous function on $[0, T]$ with values in $\mathbb{R}^{d\times d}$, $\mathbb{R}^{k\times k}$, $\mathbb{R}^{d\times d}$ and $\mathbb{R}^{d\times d}$ respectively, $q(t)$ and $\bar{q}(t)$ being symmetric and nonnegative semi-definite, and $r(t)$ symmetric and (strictly) positive definite (hence invertible). We also assume that the $d \times d$ matrices q and $\bar{q}$ are symmetric and nonnegative semi-definite. The set A is taken as the entire $\mathbb{R}^k$.

In the present set-up, the stochastic control problem is to solve the optimization problem:

$$
\begin{aligned}
\inf_{\alpha\in\mathbb{A}} \mathbb{E}\bigg[\frac{1}{2}\Big(X_T^\dagger q X_T + (X_T - s\mathbb{E}[X_T])^\dagger \bar{q}(X_T - s\mathbb{E}[X_T])\Big) \\
+ \frac{1}{2}\int_0^T \Big(X_t^\dagger q(t) X_t + \big(X_t - s(t)\mathbb{E}[X_t]\big)^\dagger \bar{q}(t)\big(X_t - s(t)\mathbb{E}[X_t]\big) \\
+ \alpha_t^\dagger r(t)\alpha_t\Big) dt\bigg],
\end{aligned}
\tag{6.121}
$$

subject to

$$
dX_t = \Big[b_1(t)X_t + b_2(t)\alpha_t + \bar{b}_1(t)\mathbb{E}[X_t]\Big]dt + \sigma dW_t, \quad X_0 = x_0.
$$

It is easy to check that assumption **Control of MKV Dynamics** under which Theorem 6.19 was proven is satisfied. However, like in the case of linear quadratic mean field games studied in Section 3.5 of Chapter 3, we use the very special structure of the model to show how a simple *pedestrian* approach can lead directly to the solution of the problem. The reduced Hamiltonian is given by:

$$
\begin{aligned}
H(t, x, \mu, y, \alpha) &= \big[b_1(t)x + \bar{b}_1(t)\bar{\mu} + b_2(t)\alpha\big] \cdot y \\
&\quad + \frac{1}{2}\Big(x^\dagger q(t) x + \big(x - s(t)\bar{\mu}\big)^\dagger \bar{q}(t)\big(x - s(t)\bar{\mu}\big) + \alpha^\dagger r(t)\alpha\Big),
\end{aligned}
$$

for $(t, x, \mu, y, \alpha) \in [0, T] \times \mathbb{R}^d \times \mathcal{P}_2(\mathbb{R}^d) \times \mathbb{R}^d \times \mathbb{R}^k$. This Hamiltonian is minimized for:

$$
\hat{\alpha} = \hat{\alpha}(t, x, \mu, y) = -r(t)^{-1} b_2(t)^\dagger y,
\tag{6.122}
$$

which is independent of the measure argument μ. Its derivative with respect to μ is given by:

$$
\partial_\mu H(t, x, \mu, y)(x') = \bar{b}_1(t)^\dagger y - s(t)^\dagger \bar{q}(t)(x - s(t)\bar{\mu}),
$$

which is a constant function of the variable x'.

The Mean Field FBSDE

The McKean-Vlasov FBSDE derived from the Pontryagin stochastic maximum principle reads:

$$\begin{cases} dX_t = \Big(b_1(t)X_t - b_2(t)r(t)^{-1}b_2(t)^\dagger Y_t + \bar{b}_1(t)\mathbb{E}[X_t]\Big)dt + \sigma dW_t, \\ dY_t = -\Big(b_1(t)^\dagger Y_t + [q(t) + \bar{q}(t)]X_t - \bar{q}(t)s(t)\mathbb{E}[X_t]\Big)dt \\ \qquad - \Big(\bar{b}_1(t)^\dagger \mathbb{E}[Y_t] - s(t)^\dagger \bar{q}(t)[I_d - s(t)]\mathbb{E}[X_t]\Big)dt + Z_t dW_t, \end{cases} \tag{6.123}$$

for $t \in [0, T]$, with initial condition X_0 and terminal condition $Y_T = [q + \bar{q}]X_T + [s^\dagger \bar{q}s - (\bar{q}s + s^\dagger \bar{q})]\mathbb{E}[X_T]$. Above, I_d denotes the identity matrix of dimension d. By Theorems 6.14 and 6.16, the above system characterizes the optimal trajectories of (6.121).

In order to proceed with the analysis of (6.123), we take the expectations of both sides. Using the notation $\bar{x}_t$ and $\bar{y}_t$ for the expectations $\mathbb{E}[X_t]$ and $\mathbb{E}[Y_t]$ respectively, we find that:

$$\begin{cases} d\bar{x}_t = \Big(\big[b_1(t) + \bar{b}_1(t)\big]\bar{x}_t - b_2(t)r(t)^{-1}b_2(t)^\dagger \bar{y}_t\Big)dt, \\ d\bar{y}_t = \Big(\big[-q(t) - \bar{q}(t) + \bar{q}(t)s(t) + s(t)^\dagger \bar{q}(t) - s(t)^\dagger \bar{q}(t)s(t)\big]\bar{x}_t \\ \qquad - \big[b_1(t)^\dagger + \bar{b}_1(t)^\dagger\big]\bar{y}_t\Big)dt, \qquad t \in [0, T], \\ x_0 = \mathbb{E}[X_0], \qquad \bar{y}_T = \big[q + \bar{q} + s^\dagger \bar{q}s - (\bar{q}s + s^\dagger \bar{q})\big]\bar{x}_T. \end{cases} \tag{6.124}$$

Of course, any solution of (6.123) provides a solution to (6.124). Conversely, if (6.124) is uniquely solvable and (6.123), when $(\mathbb{E}[X_t])_{0 \leqslant t \leqslant T}$ and $(\mathbb{E}[Y_t])_{0 \leqslant t \leqslant T}$ are replaced by frozen inputs $(\bar{x}_t)_{0 \leqslant t \leqslant T}$ and $(\bar{y}_t)_{0 \leqslant t \leqslant T}$ and the system (6.123) is thus regarded as a non-McKean-Vlasov system, the original McKean-Vlasov system (6.123) is also uniquely solvable. We already used similar arguments in Section 3.5. In particular, we learnt from the analysis provided therein that, under the standing assumption, the system (6.123), with $(\mathbb{E}[X_t])_{0 \leqslant t \leqslant T}$ and $(\mathbb{E}[Y_t])_{0 \leqslant t \leqslant T}$ regarded as frozen inputs, is uniquely solvable. Therefore, the analysis of (6.121) boils down to that of (6.124).

As in the MFG case, we rewrite (6.124) in the form:

$$\begin{cases} \dot{\bar{x}}_t = a_t \bar{x}_t + b_t \bar{y}_t, \\ \dot{\bar{y}}_t = c_t \bar{x}_t + d_t \bar{y}_t, \qquad t \in [0, T], \\ \bar{x}_0 = \mathbb{E}[X_0] \qquad \bar{y}_T = e\bar{x}_T, \end{cases} \tag{6.125}$$

with

$$a_t = b_1(t) + \bar{b}_1(t), \qquad b_t = -b_2(t)r(t)^{-1}b_2(t)^\dagger,$$
$$c_t = -q(t) - \bar{q}(t) + \bar{q}(t)s(t) + s(t)^\dagger\bar{q}(t) - s(t)^\dagger\bar{q}(t)s(t),$$
$$d_t = -b_1(t) - \bar{b}_1(t), \qquad e = q + \bar{q} + s^\dagger\bar{q}s - (\bar{q}s + s^\dagger\bar{q}),$$

for $t \in [0,T]$. Pay attention to the fact that the notation b_t does not stand for a drift term. The drift coefficients are denoted by $b_1(t)$, $\bar{b}_1(t)$ and $b_2(t)$. We can solve the system (6.125) if we are able to solve the matrix Riccati equation:

$$\dot{\bar{\eta}}_t + \bar{\eta}_t a_t - d_t\bar{\eta}_t + \bar{\eta}_t b_t \bar{\eta}_t - c_t = 0, \qquad \bar{\eta}_T = e, \tag{6.126}$$

in which case $\bar{y}_t = \bar{\eta}_t\bar{x}_t$, for $t \in [0,T]$. Clearly, (6.126) is similar to the MFG case (see Section 3.5), except for the fact that the coefficients c_t and d_t are different, and the terminal condition requires $e = q + \bar{q} + s^\dagger\bar{q}s - (\bar{q}s + s^\dagger\bar{q})$. Assuming momentarily that the matrix Riccati equation (6.126) has a unique solution $\bar{\eta}_t$, the solution of (6.125) is obtained by solving:

$$\dot{\bar{x}}_t = [a_t + b_t\bar{\eta}_t]\bar{x}_t, \qquad \bar{x}_0 = \mathbb{E}[X_0], \tag{6.127}$$

and setting $\bar{y}_t = \bar{\eta}_t\bar{x}_t$.

The solvability of the Riccati equation (6.126) may be addressed by the same arguments as in Section 3.5. Indeed, as in the case of the linear quadratic mean field game models, existence and uniqueness for a solution of this matrix Riccati equation are equivalent to the unique solvability of a deterministic control problem. This deterministic control problem has a convex Hamiltonian (with strong convexity in α) whenever the matrix coefficients are continuous, q, $\bar{q}$, $q(t)$ and $\bar{q}(t)$ are nonnegative definite, and $r(t)$ is strictly positive definite. This suffices to prove the solvability of (6.126).

Returning to (6.123) and plugging $\mathbb{E}[X_t] = \bar{x}_t$ and $\mathbb{E}[Y_t] = \bar{y}_t$ into the McKean-Vlasov FBSDE (6.123), we reduce the problem to the solution of the affine FBSDE:

$$\begin{cases} dX_t &= [\mathfrak{a}_t X_t + \mathfrak{b}_t Y_t + \mathfrak{c}_t]dt + \sigma dW_t, \qquad X_0 = x_0, \\ dY_t &= [\mathfrak{m}_t X_t - \mathfrak{a}_t^\dagger Y_t + \mathfrak{d}_t]dt + Z_t dW_t, \qquad Y_T = \mathfrak{q}X_T + \mathfrak{r}, \end{cases} \tag{6.128}$$

with:

$$\begin{cases} \mathfrak{a}_t = b_1(t), \qquad \mathfrak{b}_t = -b_2(t)r(t)^{-1}b_2(t)^\dagger, \qquad \mathfrak{c}_t = \bar{b}_1(t)\bar{x}_t, \\ \mathfrak{m}_t = -[q(t) + \bar{q}(t)], \\ \mathfrak{d}_t = [\bar{q}(t)s(t) + s(t)^\dagger\bar{q}(t) - s(t)^\dagger\bar{q}(t)s(t)]\bar{x}_t - \bar{b}_1(t)^\dagger\bar{y}_t, \\ \mathfrak{q} = q + \bar{q}, \qquad \mathfrak{r} = [s^\dagger\bar{q}s - (\bar{q}s + s^\dagger\bar{q})]\bar{x}_T. \end{cases}$$

As usual, the affine structure of the FBSDE suggests that the decoupling field will be an affine function, so we search for deterministic differentiable functions $t \mapsto \eta_t$ and $t \mapsto \chi_t$ such that:

$$Y_t = \eta_t X_t + \chi_t, \qquad t \in [0, T].$$

Computing dY_t from this ansatz, using the expression of dX_t given by the first equation of (6.128), and identifying term by term with the expression of dY_t given in (6.128) we get:

$$\begin{cases} \dot{\eta}_t + \eta_t \mathfrak{b}_t \eta_t + \mathfrak{a}_t^\dagger \eta_t + \eta_t \mathfrak{a}_t - \mathfrak{m}_t = 0, & \eta_T = \mathfrak{q}, \\ \dot{\chi}_t + (\mathfrak{a}_t^\dagger + \eta_t \mathfrak{b}_t)\chi_t - \mathfrak{d}_t + \eta_t \mathfrak{c}_t = 0, & \chi_T = \mathfrak{r}, \\ Z_t = \eta_t \sigma. \end{cases} \tag{6.129}$$

As before, the first equation is a matrix Riccati equation. If and when it can be solved, the third equation becomes solved automatically, and the second equation becomes a first order linear ODE, though not homogenous this time, which can be solved by standard methods. Notice that the quadratic terms of the two Riccati equations (6.126) and (6.129) are the same since $b_t = \mathfrak{b}_t = -b_2(t)r(t)^{-1}b_2(t)^\dagger$. However, the terminal conditions are different since the terminal condition in (6.129) is given by $\mathfrak{q} = q + \bar{q}$, while it is given by $e = q + \bar{q} + s^\dagger \bar{q} s - (\bar{q} s + s^\dagger \bar{q})$ in (6.125). Notice also that the first order terms are different as well. Anyway, although it differs from (6.126), (6.129) may be proved to be uniquely solvable by the same argument, since existence and uniqueness of a solution to (6.129) are equivalent to the unique solvability of a deterministic control problem with a convex Hamiltonian. This proves once more that we have existence and uniqueness of a solution to the LQ McKean-Vlasov control problem under the standing assumption.

When X_0 is deterministic, the optimally controlled state is Gaussian despite the nonlinearity due to the McKean-Vlasov nature of the dynamics, and because of the linearity of the ansatz, the adjoint process $\boldsymbol{Y} = (Y_t)_{0 \leqslant t \leqslant T}$ is also Gaussian. Also, using again the form of the ansatz, we see that the optimal control $\boldsymbol{\alpha} = (\alpha_t)_{0 \leqslant t \leqslant T}$ which was originally expected to be an open loop control, is in fact in closed loop feedback form $\hat{\alpha}_t = \varphi(t, X_t)$ since it can be rewritten as:

$$\hat{\alpha}_t = -r(t)^{-1}b_2(t)^\dagger \eta_t X_t - r(t)^{-1}b_2(t)^\dagger \chi_t, \quad t \in [0, T],$$

which incidentally shows that the optimal control is also a Gaussian process.

Remark 6.47 *The reader might notice that, within the linear-quadratic framework, the conditions for the unique solvability of the adjoint equations are not the same in the MFG approach and in the Control of MKV Dynamics. On the one hand, optimization over controlled MKV dynamics reads as an optimization problem of purely (strictly) convex nature, for which existence and uniqueness of an optimal state are expected. On the other hand, the optimal states in the MFG approach appear as the* *fixed points of a matching problem. Without any additional assumptions, there is no reason why this matching problem should derive from a convex optimization, even if the original coefficients of the game are linear-quadratic.*

Back to the Simple Example of Mean Field Game

We revisit the one-dimensional model introduced at the end of Section 3.5 in Chapter 3 for the purpose of a mean field game analysis. The drift was chosen to be $b(t,x,\mu,\alpha) = \alpha$, so that $b_1(t) = \bar{b}_1(t) = 0$, $b_2(t) = 1$, and the state equation read:

$$dX_t = \alpha_t dt + \sigma dW_t, \qquad t \in [0,T]; \qquad X_0 = x_0.$$

The running cost function was of the simple form $f(t,x,\mu,\alpha) = \alpha^2/2$, so that $r(t) = 1$ and $q(t) = \bar{q}(t) = 0$. As before, we assume that the terminal cost function is given by $g(x,\mu) = \bar{q}(x - s\bar{\mu})^2/2$ for some $\bar{q} > 0$ and $s \in \mathbb{R}$. Using the notation and the results above, we see that the McKean-Vlasov FBSDE derived from the Pontryagin stochastic maximum principle is the same as (3.67), except for its terminal conditions. Indeed, while $\mathfrak{q}$ is the same since $\mathfrak{q} = \bar{q}$, $\mathfrak{r}$ is now given by $\mathfrak{r} = \bar{q}s(s-2)\mathbb{E}[x_T]$. Postulating again an affine relationship $Y_t = \eta_t X_t + \chi_t$ and solving for the two deterministic functions η and χ, we find the same expression as in (3.68):

$$X_t = x_0 \frac{1 + \mathfrak{q}(T-t)}{1 + \mathfrak{q}T} - \frac{\mathfrak{r}t}{1 + \mathfrak{q}T} + \sigma[1 + \mathfrak{q}(T-t)] \int_0^t \frac{dW_s}{1 + \mathfrak{q}(T-s)}.$$

This makes the computation of $\mathbb{E}[X_T]$ very simple. We find:

$$\mathbb{E}[X_T] = \frac{x_0}{1 + \bar{q}(1-s)^2 T},$$

which always makes sense, and which is different from (3.69).

The Master Field

Quite remarkably, the master field $\mathcal{U}$ introduced in Lemma 6.25 has a very nice structure in the linear-quadratic case. From the identity $Y_t = \eta_t X_t + \chi_t$, $t \in [0,T]$, it is tempting to identify $\mathcal{U}(t,x,\mu)$ with $\eta_t x + \chi_t$. However, this guess is overly naive (and false) since χ_t depends upon the distribution of X_t. Indeed, going back to (6.129), one sees that $\mathfrak{d}_t$ depends on $\bar{y}_t$. Obviously, the mapping $[0,T] \times \mathbb{R}^d \ni x \mapsto \eta_t x + \chi_t$ identifies with u in the statement of Theorem 4.53.

However, it is worth mentioning that $(\eta_t)_{0 \leqslant t \leqslant T}$ is defined autonomously. In particular, we can rewrite the relationship between Y_t and X_t under the form:

$$\forall t \in [0,T], \quad Y_t - \bar{y}_t = \eta_t\big(X_t - \bar{x}_t\big) + \chi_t - \bar{y}_t + \eta_t \bar{x}_t.$$

Taking expectations, we deduce that, necessarily,

$$\forall t \in [0,T], \quad \chi_t - \bar{y}_t + \eta_t \bar{x}_t = 0,$$

so that:

$$\forall t \in [0,T], \quad Y_t = \eta_t X_t + \bar{y}_t - \eta_t \bar{x}_t.$$

Recalling from (6.126) and (6.127) that:

$$\bar{y}_t = \bar{\eta}_t \bar{x}_t, \quad t \in [0,T],$$

where $(\bar{\eta}_t)_{0 \leqslant t \leqslant T}$ is also defined autonomously, we finally end up with the relationship:

$$\forall t \in [0,T], \quad Y_t = \eta_t X_t + \big(\bar{\eta}_t - \eta_t\big)\mathbb{E}[X_t],$$

proving that the decoupling $\mathcal{U}$ is in fact given by:

$$\forall (t,x,\mu) \in [0,T] \times \mathbb{R}^d \times \mathcal{P}_2(\mathbb{R}^d), \quad \mathcal{U}(t,x,\mu) = \eta_t x + \big(\bar{\eta}_t - \eta_t\big)\bar{\mu}.$$

When $t = T$, the above decomposition is consistent with the writing of Y_T under the form $Y_T = [q + \bar{q}]X_T + [e - (q + \bar{q})]\mathbb{E}[X_T]$.

As an exercise, the reader may check that it solves the master equation (6.103).

6.7.2 Potential MFGs and MKV Optimization

We first revisit the discussion initiated in Subsection 2.3.3. Accordingly, our goal is to identify the notion of potential game appropriate in the setting of mean field games.

Informal Discussion

MFG Problem. We consider a mean field game of the form:

$$\inf \mathbb{E}\bigg[\int_0^T \Big[\frac{1}{2}|\alpha_t|^2 + f\big(t, X_t^{\boldsymbol{\alpha}}, \mu_t\big)\Big]dt + g\big(X_T^{\boldsymbol{\alpha}}, \mu_T\big)\bigg], \tag{6.130}$$

the infimum being taken over control processes $\boldsymbol{\alpha} = (\alpha_t)_{0 \leqslant t \leqslant T}$, the process $\boldsymbol{X}^{\boldsymbol{\alpha}} = (X_t^{\boldsymbol{\alpha}})_{0 \leqslant t \leqslant T}$ denoting the controlled diffusion process:

$$X_t = \xi + \int_0^t \alpha_s ds + \sigma W_t, \quad t \in [0,T]. \tag{6.131}$$

The problem is defined on a complete probability space $(\Omega, \mathcal{F}, \mathbb{P})$ equipped with an $\mathbb{R}^d$-valued $\mathbb{F}$-Brownian motion $\boldsymbol{W} = (W_t)_{0 \leqslant t \leqslant T}$ for a complete and right-continuous filtration $\mathbb{F}$, and with an initial condition $\xi \in L^2(\Omega, \mathcal{F}_0, \mathbb{P}; \mathbb{R}^d)$. The control process $\boldsymbol{\alpha} = (\alpha_t)_{0 \leqslant t \leqslant T}$ is assumed to take values in the whole Euclidean space $A = \mathbb{R}^d$. As usual, it is required to be $\mathbb{F}$-progressively measurable and to be square-integrable with respect to the product measure $\mathrm{Leb}_1 \otimes \mathbb{P}$. The coefficients $f : [0,T] \times \mathbb{R}^d \times \mathcal{P}_2(\mathbb{R}^d) \to \mathbb{R}$ and $g : \mathbb{R}^d \times \mathcal{P}_2(\mathbb{R}^d) \to \mathbb{R}$ are assumed to satisfy suitable conditions, which will be specified next. The volatility is given by a positive real number σ.

As usual, $\boldsymbol{\mu} = (\mu_t)_{0 \leqslant t \leqslant T}$ denotes a continuous path with values in $\mathcal{P}_2(\mathbb{R}^d)$. According to the discussions of Chapters 3 and 4, the goal is to find a flow $\boldsymbol{\mu} = (\mu_t)_{0 \leqslant t \leqslant T}$ such that the optimal path of the above optimal control problem has exactly $\boldsymbol{\mu}$ as flow of marginal distributions.

MKV Optimization Problem. We further assume the existence of functions $F : [0,T] \times \mathcal{P}_2(\mathbb{R}^d) \to \mathbb{R}$ and $G : \mathcal{P}_2(\mathbb{R}^d) \to \mathbb{R}$, L-differentiable in the measure argument, such that:

$$\partial_x f(t,x,\mu) = \partial_\mu F(t,\mu)(x) \quad \text{and} \quad \partial_x g(x,\mu) = \partial_\mu G(\mu)(x) \tag{6.132}$$

for μ-almost every $x \in \mathbb{R}^d$. We then consider the central planner optimization problem:

$$\inf \left\{ \int_0^T \left(\frac{1}{2}\mathbb{E}\big[|\alpha_t|^2\big] + F\big(t, \mathcal{L}(X_t)\big) \right) dt + G\big(\mathcal{L}(X_T)\big) \right\}, \tag{6.133}$$

the infimum being taken over the same class of control processes $\boldsymbol{\alpha} = (\alpha_t)_{0 \leqslant t \leqslant T}$ as above, and $\boldsymbol{X}^{\boldsymbol{\alpha}}$ denoting the same controlled diffusion process as in (6.131). Obviously, the optimization problem (6.133) is a special case of the class of McKean-Vlasov optimal control problems considered in this chapter.

Comparison of the Two Problems. Pursuing the discussion initiated in Subsection 6.2.5, we compare the two problems (6.130) and (6.133).

Assume now that F and G admit linear functional derivatives with respect to the measure argument, see for instance Proposition 5.51 for conditions under which the existence of the L-derivative implies the existence of the linear functional derivative. Using the same notation as in (6.133), we notice that in the present situation, the cost functional I which we introduced in (6.49) to associate an auxiliary MFG problem to an MKV optimal control problem, takes the form:

$$\begin{aligned} I(\boldsymbol{\alpha}) = \mathbb{E}\bigg[\int_0^T \left(\frac{1}{2}|\alpha_t|^2 + F(t,\mu_t) + \frac{\delta F}{\delta m}(t,\mu_t)(X_t^{\boldsymbol{\alpha}}) \right) dt \\ + \left(G(\mu_T) + \frac{\delta G}{\delta m}(\mu_T)(X_T^{\boldsymbol{\alpha}}) \right) \bigg], \end{aligned}$$

for a given continuous path $\boldsymbol{\mu} = (\mu_t)_{0 \leqslant t \leqslant T}$ with values in $\mathcal{P}_2(\mathbb{R}^d)$, where, as above,

$$X_t^{\boldsymbol{\alpha}} = \xi + \int_0^t \alpha_s ds + \sigma W_t, \quad t \in [0,T].$$

Consequently, we can find a nonrandom quantity $C(\boldsymbol{\mu})$, depending upon the flow $\boldsymbol{\mu}$ and independent of the control $\boldsymbol{\alpha}$, such that:

$$I(\boldsymbol{\alpha}) = \mathbb{E}\bigg[\int_0^T \Big(\frac{1}{2}|\alpha_t|^2 + \frac{\delta F}{\delta m}(t,\mu_t)(X_t^{\boldsymbol{\alpha}})\Big)dt + \frac{\delta G}{\delta m}(\mu_T)(X_T^{\boldsymbol{\alpha}})\bigg] + C(\boldsymbol{\mu}).$$

By Proposition 5.48 and (6.132), we know that, up to an additive constant depending on t and $\mu_t, f(t,\cdot,\mu_t)$ is equal to $[\delta F/\delta m](t,\mu_t)(\cdot)$ and, up to an additive constant depending on μ_T, $g(\cdot,\mu_T)$ is equal to $[\delta G/\delta m](\mu_T)(\cdot)$. Therefore, up to a modification of $C(\boldsymbol{\mu})$, we have:

$$I(\boldsymbol{\alpha}) = \mathbb{E}\bigg[\int_0^T \Big(\frac{1}{2}|\alpha_t|^2 + f\big(t,X_t^{\boldsymbol{\alpha}},\mu_t\big)\Big)dt + g\big(X_T^{\boldsymbol{\alpha}},\mu_T\big)\bigg] + C(\boldsymbol{\mu}).$$

Up to the constant $C(\boldsymbol{\mu})$, we recognize the same cost functional as in the definition of the MFG problem (6.130). Since the constant $C(\boldsymbol{\mu})$ plays no role when optimizing with respect to $\boldsymbol{\alpha}$ while keeping $\boldsymbol{\mu}$ frozen, we deduce that the auxiliary MFG problem associated with (6.133) through the procedure defined in Subsection 6.2.5 coincides with the MFG problem (6.130). Moreover, the analysis provided in Subsection 6.2.5 shows that the Pontryagin forward-backward systems associated with the mean field game (6.130) and with the mean field stochastic control problem (6.133) are the same.

Generic Model

We now specialize the choice of F and G and make the above discussion rigorous in that case. To do so, let us assume that ℓ is an even and continuously differentiable function on $\mathbb{R}^d$ and that its derivative is at most of linear growth, and similarly that for any $t \in [0,T]$, $h(t,\cdot)$ is also even and continuously differentiable on $\mathbb{R}^d$, with a derivative at most of linear growth. Then, we define the functions F and G by:

$$\begin{aligned} F(t,\mu) &= \frac{1}{2}\int_{\mathbb{R}^d}\int_{\mathbb{R}^d} h(t,x-x')d\mu(x)d\mu(x'), \\ G(\mu) &= \frac{1}{2}\int_{\mathbb{R}^d}\int_{\mathbb{R}^d} \ell(x-x')d\mu(x)d\mu(x'), \end{aligned} \tag{6.134}$$

for $(t,\mu) \in [0,T] \times \mathcal{P}_2(\mathbb{R}^d)$. By Example 2 in Subsection 5.2.2 and by Proposition 5.51, these functions have functional derivatives with respect to the measure argument which are given by:

$$\begin{aligned} \frac{\delta F}{\delta m}(t,\mu)(x) &= \int_{\mathbb{R}^d} h(t,x-x')d\mu(x'), \\ \frac{\delta G}{\delta m}(\mu)(x) &= \int_{\mathbb{R}^d} \ell(x-x')d\mu(x'), \end{aligned} \tag{6.135}$$

for $(t,x,\mu) \in [0,T] \times \mathbb{R}^d \times \mathcal{P}_2(\mathbb{R}^d)$, so they can be used in the present setup.

Rigorous Existence and Uniqueness Result

Proposition 6.48 *Let $h : [0,T] \times \mathbb{R}^d \to \mathbb{R}$ be a jointly measurable function, even, twice continuously differentiable and convex in the space variable, with bounded second-order derivatives uniformly in time, and $\ell : \mathbb{R}^d \to \mathbb{R}$ be an even, twice continuously differentiable and convex function with bounded second-order derivatives. For F and G defined via* (6.134) *above, the McKean-Vlasov optimal control problem* (6.131)–(6.133) *has a unique optimal path $\hat{\boldsymbol{X}} = \boldsymbol{X}^{\hat{\boldsymbol{\alpha}}} = (X_t^{\hat{\boldsymbol{\alpha}}})_{0 \leqslant t \leqslant T}$ given by a specific control process $\hat{\boldsymbol{\alpha}} = (\hat{\alpha}_t)_{0 \leqslant t \leqslant T}$.*

Moreover, its flow of marginal distributions $\boldsymbol{\mu} = (\mu_t = \mathcal{L}(\hat{X}_t))_{0 \leqslant t \leqslant T}$ is the unique solution to the MFG problem (6.130)–(6.131), *with:*

$$f(t,x,\mu) = \int_{\mathbb{R}^d} h(t, x - x') d\mu(x'), \quad g(x,\mu) = \int_{\mathbb{R}^d} \ell(x - x') d\mu(x'),$$

for $(t,x) \in [0,T] \times \mathbb{R}^d$.

Proof. From (6.135), we have:

$$\partial_\mu F(t,\mu)(x) = \int_{\mathbb{R}^d} \partial_x h(t, x - x') d\mu(x'), \quad \partial_\mu G(\mu)(x) = \int_{\mathbb{R}^d} \partial_x \ell(x - x') d\mu(x').$$

It is straightforward to check that assumption **Control of MKV Dynamics** holds, the convexity property of F and G in (**A4**) following from the convexity property of h and ℓ.

Therefore, by Theorem 6.19, the Pontryagin system for the McKean-Vlasov control problem has a unique solution. Also, the unique solution of the Pontryagin system is the unique optimal path of the McKean-Vlasov control problem. In the present situation, the Pontryagin system has the form:

$$\begin{cases} dX_t = -Y_t dt + \sigma dW_t, \\ dY_t = -\tilde{\mathbb{E}}\big[\partial_x h\big(t, X_t - \tilde{X}_t\big)\big] dt + Z_t dW_t, \quad t \in [0,T], \\ X_0 = \xi, \quad Y_T = \tilde{\mathbb{E}}\big[\partial_x \ell\big(X_T - \tilde{X}_T\big)\big]. \end{cases} \tag{6.136}$$

Letting $\boldsymbol{\mu} = (\mu_t = \mathcal{L}(X_t))_{0 \leqslant t \leqslant T}$, we may rewrite the above system as:

$$\begin{cases} dX_t = -Y_t dt + \sigma dW_t, \\ dY_t = -\partial_x f(t, X_t, \mu_t) dt + Z_t dW_t, \quad t \in [0,T], \\ X_0 = \xi, \quad Y_T = \partial_x g(X_T, \mu_T), \end{cases}$$

which is exactly the Pontryagin system for the standard optimal control problem (6.130). Observing that f and g are convex in the Euclidean space variable, we deduce from Theorem 3.17 that $\boldsymbol{X} = (X_t)_{0 \leqslant t \leqslant T}$ is an optimal path of the optimal control problem (6.130). Therefore, $\boldsymbol{\mu} = (\mu_t)_{0 \leqslant t \leqslant T}$ is an MFG equilibrium, and Theorem 6.19 again shows that it is the unique one since the MFG equilibria are characterized by the McKean-Vlasov FBSDE (6.136). □

The property of potential mean field games which we identified above says that the solution (via the Pontryagin stochastic maximum principle) of the mean field game problem (6.130)–(6.131) reduces to the solution of a central planner optimization problem. This much was expected on the basis of previous discussions of potential games. However, what is remarkable is the fact that this central planner optimization problem is in fact an optimal control of McKean-Vlasov dynamics.

Flocking as a Potential Mean Field Game

The MFG flocking model introduced in Chapter 1 was solved in Chapter 4 under the extra assumption that the running cost function was bounded. We revisit the model based on the cost function (4.180) in light of the present discussion of potential games. Recall that the model is based on the assumption that the dynamics of a typical bird in the population are given by controlled kinetic equations:

$$dx_t = v_t dt, \qquad dv_t = \alpha_t dt + \sigma dW_t, \qquad t \geqslant 0,$$

where x_t denotes the position of a bird at time t, and v_t its velocity. Here, $\boldsymbol{W} = (W_t)_{t\geqslant 0}$ is a three-dimensional Wiener process and $\boldsymbol{\alpha} = (\alpha_t)_{t\geqslant 0}$ is a three-dimensional progressively measurable process with respect to the filtration generated by the initial position and by $\boldsymbol{W}$. The process $\boldsymbol{\alpha}$ plays the role of the control of the bird on its velocity.

With the rationale of mean field games as a framework for the search for a large population consensus, the finite player game formulation of Chapter 1 suggests that we consider the following cost functional:

$$J^{\mu}(\boldsymbol{\alpha}) = \mathbb{E}\biggl[\int_0^T \bigl[\frac{1}{2}|\alpha_t|^2 + (h * \mu_t)(x_t, v_t)\bigr]dt\biggr], \tag{6.137}$$

for a given time horizon $T > 0$, a continuous flow $\boldsymbol{\mu} = (\mu_t)_{t\geqslant 0}$ of measures on $\mathbb{R}^6$ and an even function h of the variables x and v. Notice that the special convolution form of the running cost function is covered by the second example of N-player potential game in Subsection 2.3.3. In the context of the limit $N \to \infty$ of large populations, the convolution form appearing in the cost functional (6.137) is reminiscent of that used to define f and g in the statement of Proposition 6.48.

Unfortunately, Proposition 6.48, as stated above, cannot apply directly to (6.137) because of the degeneracy of the equation for the position $(x_t)_{0\leqslant t\leqslant T}$.

However, we can easily recast (6.137) in a more tractable setting, as done in Section 6.6. Indeed, with $(X_t^{\boldsymbol{\alpha}} = (x_t^{\boldsymbol{\alpha}}, v_t^{\boldsymbol{\alpha}}))_{0\leqslant t\leqslant T}$, the flocking controlled dynamics can be written as:

$$dX_t^{\boldsymbol{\alpha}} = (AX_t^{\boldsymbol{\alpha}} + B\alpha_t)dt + \Sigma dW_t, \quad t \in [0, T], \tag{6.138}$$

with:

$$\Lambda = \begin{pmatrix} 0_d & I_d \\ 0_d & 0_d \end{pmatrix} \quad B = \begin{pmatrix} 0_d \\ I_d \end{pmatrix} \quad \text{and} \quad \Sigma = \begin{pmatrix} 0_d & 0_d \\ 0_d & \sigma I_d \end{pmatrix}.$$

Then, it is pretty easy to reformulate the statement of Proposition 6.48 by using (6.138) in lieu of (6.131) as controlled dynamics. For instance, (6.136) in the proof of the proposition becomes:

$$\begin{cases} dX_t = \Big(\Lambda X_t - B^\dagger Y_t\Big)dt + \Sigma dW_t, \\ dY_t = -\Big(\Lambda^\dagger Y_t + \tilde{\mathbb{E}}\big[\partial_x h(X_t - \tilde{X}_t)\big]\Big)dt + Z_t dW_t, \quad t \in [0,T], \\ X_0 = \xi, \quad Y_T = \tilde{\mathbb{E}}\big[\partial_x \ell(X_T - \tilde{X}_T)\big]. \end{cases}$$

So, when h in (6.137) is convex, Proposition 6.48 applies up to a slight modification of the argument in the proof.

Therefore, the flocking problem reduces to the minimization of the functional:

$$J(\boldsymbol{\alpha}) = \int_0^T \Big[\frac{1}{2}\mathbb{E}\big[|\alpha_t|^2\big] + F\big(\mathcal{L}(x_t, v_t)\big)\Big]dt, \tag{6.139}$$

where:

$$F(\mu) = \frac{1}{2}\int_{\mathbb{R}^6} (h * \mu)(x,v)d\mu(x,v), \tag{6.140}$$

and, when h is convex, the mean field game can be solved by solving the central planner optimization problem, the latter being an optimal control problem of the McKean-Vlasov type.

Still, the typical form for h in the flocking model is:

$$h(x,v) = \frac{|v|^2}{(1+|x|^2)^\beta}, \tag{6.141}$$

for some $\beta \geqslant 0$, see (4.180) with $\kappa = \sqrt{2}$ therein. Except for the case $\beta = 0$, the function h is not convex! Therefore Proposition 6.48 does not apply. Still we can solve the *central planner* optimization and regard its solutions as possible candidates for solving the MFG problem. Indeed, although the running cost function F is not convex – which prevents us from applying the results of Section 6.4 – assumption **MKV Weak Formulation** is satisfied and we can appeal to Theorem 6.37 to prove existence of a solution of the McKean-Vlasov central planner optimization in the weak formulation.

6.7.3 Optimal Transportation as an MKV Control Problem

The purpose of this section is to revisit the example introduced at the very end of Subsection 6.2.2. There, starting from the statement of Benamou-Brenier's Theorem 5.53, we gave an informal argument to show that the 2-Wasserstein distance could appear as the value function of an optimal control problem of the McKean-Vlasov type formulated in an analytic way. By analytic formulation, we mean that the controlled trajectories were not regarded as controlled stochastic processes, as we did in most of the chapter, but as deterministic flows of probability measures satisfying a Kolmogorov-Fokker-Planck equation of the form (6.12).

The fact that two approaches, a probabilistic one and an analytic one, are conceivable for handling mean field stochastic control problems was already explained in the introductory Section 6.2. However, there, we just gave a few indications on the strategies that could be used to pass from one formulation to another. Motivated by the statement of Benamou-Brenier's theorem, our goal is here to address these questions more properly in the framework of optimal transportation.

As mentioned at the end of Section 6.2, one difficulty for passing from the analytic to the probabilistic approach is to reconstruct, for a given flow $\boldsymbol{\mu}$ of probability measures satisfying a continuity equation of the Kolmogorov-Fokker-Planck type, a stochastic process $\boldsymbol{X}$ admitting $\boldsymbol{\mu}$ as flow of marginal laws. This is exactly the purpose of the following statement, which is taken from Ambrosio-Gigli-Savaré's monograph:

Theorem 6.49 *Let $\boldsymbol{\mu} = (\mu_t)_{0\leqslant t\leqslant T}$ be a flow of probability measures on $\mathbb{R}^d$, continuous with respect to the topology of weak convergence, such that:*

$$\int_0^T \int_{\mathbb{R}^d} \big(\partial_t \rho(t,x) + b(t,x)\cdot \partial_x \rho(t,x)\big) d\mu_t(x) dt = 0, \tag{6.142}$$

for all real valued smooth function ρ from $[0,T]\times\mathbb{R}^d$ with compact support included in $(0,T)\times\mathbb{R}^d$, and for some measurable vector field $b : [0,T]\times\mathbb{R}^d \to \mathbb{R}^d$ satisfying

$$\int_0^T \int_{\mathbb{R}^d} |b(t,x)|^2 d\mu_t(x) dt < \infty.$$

Then, there exists a probability measure $\mathbb{P}$ on the canonical space $\Omega = \mathbb{R}^d \times \mathcal{C}([0,T];\mathbb{R}^d)$ such that under $\mathbb{P}$, the canonical process $(\xi, \boldsymbol{x} = (x_t)_{0\leqslant t\leqslant T})$ satisfies:

$$x_t = \xi + \int_0^t b(s,x_s)ds, \quad t \in [0,T], \tag{6.143}$$

and, for all $t \in [0,T]$, $\mu_t = \mathcal{L}(x_t)$. In particular, the trajectories of $(x_t)_{0\leqslant t\leqslant T}$ belong to the so-called Cameron-Martin space of absolutely continuous trajectories whose derivative is square integrable on $[0,T]$; they satisfy:

$$\mathbb{E}^{\mathbb{P}} \int_0^T |\dot{x}_t|^2 dt < \infty,$$

where $\mathbb{E}^{\mathbb{P}}$ *denotes the expectation under* $\mathbb{P}$.

Conversely, any probabilistic measure $\mathbb{P}$ *under which the canonical process has trajectories in the Cameron-Martin space such that* (6.143) *holds for some measurable velocity field* $b : [0, T] \times \mathbb{R}^d \to \mathbb{R}^d$ *satisfying:*

$$\mathbb{E}^{\mathbb{P}} \int_0^T |b(t, x_t)| dt < \infty,$$

induces, via its flow of marginal measures $\boldsymbol{\mu} = (\mathbb{P} \circ x_t^{-1})_{0 \leqslant t \leqslant T}$, *a solution to the Kolmogorov-Fokker-Planck equation* (6.142).

In our terminology, the process $(x_t)_{0 \leqslant t \leqslant T}$, as written in (6.143), is a controlled process with a Markovian control. Obviously, Theorem 6.49 permits to reformulate Benamou-Brenier's Theorem 5.53 in a more probabilistic fashion. We deduce that the square Wasserstein distance $W_2(\mu_0, \mu_1)^2$ between two measures μ_0 and μ_1 in $\mathcal{P}_2(\mathbb{R}^d)$ reads as the infimum of the kinetic energy:

$$\inf \mathbb{E}^{\mathbb{P}} \int_0^1 |\dot{x}_t|^2 dt,$$

the infimum being taken over probability measures $\mathbb{P}$ on the canonical space Ω under which $\xi \sim \mu_0$, $x_1 \sim \mu_1$, and $(x_t)_{0 \leqslant t \leqslant 1}$ satisfies (6.143) for $T = 1$ and for some Borel vector field b satisfying $\mathbb{E} \int_0^1 |b(t, x_t)| dt < \infty$.

In this regard, observe that the fact that the infimum is taken over probability measures and not over control processes is reminiscent of the weak formulation used in Section 6.6.

Interestingly, we may wonder about a similar formulation of Benamou-Brenier's theorem using open loop instead of Markovian controls in (6.143). This prompts us to quote another result, which we already alluded to at the end of Section 6.2:

Theorem 6.50 *For some time horizon* $T > 0$, *let* $\boldsymbol{X} = (X_t)_{0 \leqslant t \leqslant T}$ *be an* $\mathbb{R}^d$*-valued absolutely continuous process defined on some probability space* $(\Xi, \mathcal{G}, \mathbf{P})$ *with dynamics of the form:*

$$X_t = X_0 + \int_0^t \alpha_s ds, \tag{6.144}$$

where $\mathbf{E}[|X_0|^2] < \infty$ *and* $\boldsymbol{\alpha} = (\alpha_t)_{0 \leqslant t \leqslant T}$ *is a jointly measurable process satisfying* $\mathbf{E} \int_0^T |\alpha_t|^2 dt < \infty$, *where* $\mathbf{E}$ *is the expectation under* $\mathbf{P}$.

Then, there exist a vector field $b : [0,T] \times \mathbb{R}^d \to \mathbb{R}^d$ and a probability measure $\mathbf{P}$ on the canonical space $\Omega = \mathbb{R}^d \times \mathcal{C}([0,T];\mathbb{R}^d)$ satisfying:

$$\mathbb{E} \int_0^T |b(s,x_s)|^2 ds < \infty,$$

and $\mathbb{P}$ almost-surely:

$$x_t = \xi + \int_0^t b(s,x_s)ds, \quad t \in [0,T],$$

where $(\xi, \boldsymbol{x} = (x_t)_{0\leqslant t \leqslant T})$ is the canonical process on Ω, such that for any $t \in [0,T]$, the law of x_t under $\mathbb{P}$ is the same as the law of X_t under $\mathbf{P}$.

We refer to the Notes & Complements for precise references.

Combining Theorems 6.49 and 6.50, we can replace the closed loop formulation in Benamou-Brennier's theorem by an open loop version. We obtain a new formulation of Benamou-Brennier's theorem which reads as follows. Given two probability measures μ_0 and μ_1 in $\mathcal{P}_2(\mathbb{R}^d)$, the square Wasserstein distance between μ_0 and μ_1 may be expressed as:

$$W_2(\mu_0,\mu_1)^2 = \inf \mathbb{E}^{\mathbb{P}} \int_0^1 |\dot{x}_t|^2 dt,$$

where the infimum is taken over all the probability measures $\mathbb{P}$ on the canonical space $\Omega = \mathbb{R}^d \times \mathcal{C}([0,1];\mathbb{R}^d)$ under which the canonical process $(\xi, \boldsymbol{x} = (x_t)_{0\leqslant t\leqslant 1})$ satisfies:

1. $x_0 = \xi$,
2. $\boldsymbol{x}$ is absolutely continuous and $\mathbb{E}^{\mathbb{P}} \int_0^1 |\dot{x}_t|^2 dt < \infty$,
3. the law of x_0 is μ_0,
4. the law of x_1 is μ_1.

Of course, this may be rewritten as a mean field stochastic control problem, but formulated in the weak sense as in Section 6.6. To any probability $\mathbb{P}$ on the canonical space such that the three (and not four) first items above are satisfied, we may indeed associate the cost:

$$\mathcal{J}(\mathbb{P}) = \mathbb{E}^{\mathbb{P}} \int_0^1 |\dot{x}_t|^2 dt + g\big(\mathcal{L}(x_1)\big), \tag{6.145}$$

where:

$$g(\mu) = \begin{cases} 0 & \text{if } \mu = \mu_1, \\ +\infty & \text{if } \mu \neq \mu_1. \end{cases}$$

We now face a control problem of the McKean-Vlasov type as the cost functional depends upon the marginal distribution of the process $\boldsymbol{x}$ at the terminal time.

Mollification of the Control Problem

In contrast with (6.144), we often considered in the text nondegenerate dynamics of the form:

$$dX_t = \alpha_t dt + \sigma dW_t, \quad t \in [0,T], \tag{6.146}$$

where $(W_t)_{0 \leqslant t \leqslant T}$ is a d-dimensional Brownian motion on a filtered probability space $(\Omega, \mathcal{F}, \mathbb{F}, \mathbb{P})$ satisfying the usual assumption, $\boldsymbol{\alpha}$ is an $\mathbb{R}^d$-valued and square-integrable progressively measurable process, and σ is a $d \times d$-matrix. In what follows, we assume that σ is a diagonal matrix of the form σI_d for $\sigma > 0$,

In such a framework, it is natural to wonder about the analogue of (6.145), in which case the minimization problem takes the form:

$$\inf J(\boldsymbol{\alpha}),$$

with:

$$J(\boldsymbol{\alpha}) = \mathbb{E}\left[\int_0^1 |\alpha_t|^2 dt\right] + g\big(\mathcal{L}(X_1)\big),$$

and as above:

$$g(\mu) = \begin{cases} 0 & \text{if } \mu = \mu_1, \\ +\infty & \text{if } \mu \neq \mu_1. \end{cases}$$

In (6.146), the volatility σ provides some form of mollification. For that reason, it may sound convenient to regularize μ_1 as well in the terminal condition and to replace g by:

$$g^\sigma(\mu) = \begin{cases} 0 & \text{if } \mu = \mu_1 * N_d(0, \sigma^2 I_d), \\ +\infty & \text{if } \mu \neq \mu_1 * N_d(0, \sigma^2 I_d). \end{cases}$$

We denote the new cost functional by:

$$J^\sigma(\boldsymbol{\alpha}) = \mathbb{E}\left[\int_0^1 |\alpha_t|^2 dt\right] + g^\sigma\big(\mathcal{L}(X_1)\big),$$

Under these conditions one can prove:

Theorem 6.51 *Let* $\mu_0, \mu_1 \in \mathcal{P}_2(\mathbb{R}^d)$, μ_0 *being absolutely continuous with respect to the Lebesgue measure, then,*

$$\lim_{\sigma \searrow 0} \inf_{\boldsymbol{\alpha}} J^\sigma(\boldsymbol{\alpha}) = W_2(\mu_0, \mu_1)^2.$$

See the Notes & Complements below for precise references.

6.7.4 Back to the Diffusion Form of Aiyagari's Growth Model

We revisit the diffusion version of Aiyagari's growth model solved in Subsection 3.6.3 of Chapter 3 as a mean field game problem. We remark that the argument we used in Subsection 3.6.3 of Chapter 3 to make the connection between the MFG solution and a standard control problem strongly suggests that the equilibrium we found may also solve the optimal control problem for the corresponding McKean-Vlasov dynamics. The purpose of this subsection is to revisit this fact.

The MKV analogue of the diffusion form of the Aiyagari's growth model solved in Subsection 3.6.3 of Chapter 3 requires the analysis of the forward dynamical system:

$$\begin{cases} dZ_t = -(Z_t - 1)\,dt + dW_t, \\ dA_t = \big[(1-\alpha)\mathbb{E}[A_t]^\alpha Z_t + \big(\alpha \mathbb{E}[A_t]^{\alpha-1} - \delta\big)A_t - c_t\big]dt, \quad t \in [0,T], \end{cases}$$

driven by a square-integrable and $\mathbb{F}$-adapted control process $\boldsymbol{c} = (c_t)_{0 \leqslant t \leqslant T}$ with nonnegative values. Notice that this is exactly the forward system used in (3.74) for the formulation of the MFG model provided $\bar{\mu}_t$ is replaced by $\mathbb{E}[A_t]$. As explained several times, the fact that the flow of measures $\boldsymbol{\mu} = (\mu_t)_{0 \leqslant t \leqslant T}$ is not frozen before the optimization is the main difference between the MFG and the MKV problems. Whenever the solution is well defined (which is equivalent to the fact that $\mathbb{E}[A_t] > 0$ for any $t \in [0,T]$), we can consider the same cost functional:

$$J(\boldsymbol{c}) = \mathbb{E}\bigg[\int_0^T (-U)(c_t)dt - \tilde{U}(A_T)\bigg],$$

where $U(c) = (c^{1-\gamma}-1)/(1-\gamma)$ for $\gamma > 0$, with $U(c) = \ln(c)$ if $\gamma = 1$, and $\tilde{U}$ is the identity function. As in the MFG version of the problem, the (time-homogeneous) reduced Hamiltonian has the form:

$$\begin{aligned} &H(z, a, \mu, y_z, y_a, c) \\ &\quad = [1 - z]y_z + \Big[(1-\alpha)\bar{\mu}^\alpha z + \big(\alpha\bar{\mu}^{\alpha-1} - \delta\big)a - c\Big]y_a - U(c), \end{aligned}$$

where $\bar{\mu} = \int_{\mathbb{R}^2} a\mu(dz, da)$ is the mean of the second marginal of μ. Also, we denoted the control by a; here, α is a constant exponent. Obviously, $H(z, a, \mu, y_z, y_a, c)$ makes sense only if $\bar{\mu} \geqslant 0$. A first difference with the analysis performed for the MFG version of the problem is the fact that we keep the variable z in the expression of the Hamiltonian, but, as in the MFG case, its adjoint process does enter the equation for the optimal trajectories. An obvious reason for keeping z is that the measure argument is now part of the state of the forward dynamics, and we need to compute exactly the value of $\partial_\mu H$. In the present situation,

$$\partial_\mu H(a, z, \mu, y_a, y_z, c)(v) = \left[(1-\alpha)\alpha\bar{\mu}^{\alpha-1}z + \alpha(\alpha-1)\bar{\mu}^{\alpha-2}a\right]y_a,$$

if $\bar{\mu} > 0$. Since the first adjoint process $\boldsymbol{Y}_z = (Y_{z,t})_{0\leqslant t\leqslant T}$ has no influence on the value of the optimal trajectory, we can focus on the dynamics of the adjoint of the wealth process $\mathbf{A} = (A_t)_{0\leqslant t\leqslant T}$ for which we use the notation $\boldsymbol{Y} = (Y_t)_{0\leqslant t\leqslant T}$ instead of $\boldsymbol{Y}_a = (Y_{a,t})_{0\leqslant t\leqslant T}$ for the sake of simplicity. The McKean-Vlasov forward-backward system derived from the Pontryagin maximum principle proved in this chapter for MKV diffusion processes (see Definition 6.5) writes:

$$\begin{cases} dA_t = \left[(1-\alpha)\mathbb{E}[A_t]^\alpha Z_t + [\alpha\mathbb{E}[A_t]^{\alpha-1} - \delta]A_t - (-Y_t)^{-1/\gamma}\right]dt \\ dY_t = -Y_t\left[\alpha\mathbb{E}[A_t]^{\alpha-1} - \delta\right]dt - (1-\alpha)\alpha\mathbb{E}[A_t]^{\alpha-1}\mathbb{E}[Z_tY_t]dt \\ \qquad\qquad -\alpha(\alpha-1)\mathbb{E}[A_t]^{\alpha-2}\mathbb{E}[A_tY_t]dt + \tilde{Z}_t dW_t, \quad t \in [0,T], \\ Y_T = -1, \end{cases} \tag{6.147}$$

and it requires $(\mathbb{E}[A_t])_{0\leqslant t\leqslant T}$ to be positive-valued.

Like in the analysis of the MFG equilibrium, the solution of the backward equation is deterministic. Therefore, recalling that $\mathbb{E}[Z_t] = 1$ for any $t \in [0,T]$,

$$\begin{aligned} dY_t &\\ &= -Y_t\left[\alpha\mathbb{E}[A_t]^{\alpha-1} - \delta\right]dt - (1-\alpha)\alpha\mathbb{E}[A_t]^{\alpha-1}Y_t dt - \alpha(\alpha-1)\mathbb{E}[A_t]^{\alpha-1}Y_t dt \\ &= -Y_t\left[\alpha\mathbb{E}[A_t]^{\alpha-1} - \delta\right]dt, \end{aligned}$$

which is the same backward equation as in (3.76). Taking the mean in the forward equation of (6.147), we then deduce that the pair $(\mathbb{E}[A_t], Y_t)_{0\leqslant t\leqslant T}$ solves (3.76), which makes it possible to repeat the analysis of the case of the MFG version of the problem, provided that $\bar{\mu}_0$ is large enough. In such a case, the system (6.147) has a unique solution such that $\mathbb{E}[A_t] > 0$ for all $t \in [0,T]$.

6.8 Notes & Complements

We claimed in the introduction that the problem of the optimal control of SDEs of McKean-Vlasov type has *notoriously* been ignored in the mathematical literature. However, it is fair to mention that some special cases such as the mean variance portfolio selection problem have been considered by Anderson and Djehiche in [24] and Fischer and Livieri in [155] in the spirit of this chapter. The linear quadratic case was discussed (though quite recently) by two groups of authors in [53] and [99]. Bensoussan, Sung, Yam, and Yung on one hand and Carmona, Delarue, and Lachapelle on the other hand, simultaneously and independently of each other, discussed the linear quadratic case. However, the technical analysis presented in this chapter follows the approach of Carmona and Delarue as originally developed in [98] which is similar to, though different from, the one presented in the monograph [50] by Bensoussan, Frehse, and Yam. Also, inspired by the surge of interest for the theory of optimal control, several works have been published on the analysis of Hamilton-Jacobi-Bellman equations on the Wasserstein space, see for instance Feng and Katsoulakis [152], Gangbo, Nguyen, and Tudorascu [167] and Pham and Wei [311]. As explained in the chapter, see also the additional comments right below, HJB equations on the Wasserstein space play a central role in the deterministic analysis of mean field stochastic control problems.

Several versions of the stochastic maximum principle for optimization problems over systems with mean field interactions exist in the literature. For example, Hosking derives in [201] a maximum principle for a finite player game with mean field interactions. Also, Brandis-Meyer, Oksendal, and Zhou [281] use Malliavin calculus to derive a stochastic maximum principle for a mean field control problem including jumps.

The discussion of models with scalar interactions in Subsection 5.2.2 shows how the model treated by Anderson and Djehiche in [24] appears as an example of our more general formulation of the Pontryagin stochastic maximum principle. In fact, the mean variance portfolio optimization example discussed in [24] as well as the solution proposed in [53] and [99] of the optimal control of linear-quadratic (LQ) McKean-Vlasov dynamics are based on the general form of the Pontryagin principle proven in this chapter as applied to models with scalar interactions.

The continuation method alluded to in the proof of Theorem 6.19 was originally introduced for the analysis of forward-backward stochastic differential equations, by Peng and Wu in [307].

Throughout the chapter, we assumed that the space A of controls was convex. This assumption was only made for the sake of simplicity. More general spaces can be handled at the cost of using spike variation techniques, and adding one extra adjoint equation. See for example [343, Chapter 3] for a discussion of the classical (i.e., non-McKean-Vlasov) case. Without the motivation from specific applications, we chose to refrain from providing this level of generality and avoid an excessive overhead in notation and technicalities.

The interpretation of mean field games as a criticality for some McKean-Vlasov control problem was noticed in the earlier articles by Lasry and Lions [260–262]. It was used in theoretical works on mean field games, see for instance the papers [85, 87, 88, 92] cited in the Notes & Complements of Chapter 3, but also for numerical purposes, see Benamou and Carlier [42]. The similarities and the differences between mean field games and McKean-Vlasov problems when driven by the same dynamics and the same cost functionals were first identified and discussed by Carmona, Delarue, and Lachapelle in [99], where it is clearly emphasized that optimizing first and searching for a fixed point afterwards leads to the solution of a mean field game problem, while finding the fixed point first and then optimizing afterwards leads to the solution of the optimal control of McKean-Vlasov SDEs.

The discussion of Section 6.2 comparing the probabilistic approach to the control of McKean-Vlasov dynamics advocated and developed in this chapter, to a possible deterministic infinite dimensional control problem, is part of the folklore on the subject, see for instance Bensoussan, Frehse, and Yam [50], Gangbo, Nguyen, and Tudorascu [167], Laurière and Pironneau [263], and Pham and Wei [311]. We here provide a systematic presentation based on the differential calculus developed in Chapter 5 and with a strong bent on the use of *master equations* as they will play a crucial role in the second volume of the book. The argument proposed at the end of Section 6.2 to condition open loop controls to end up with controls in closed loop feedback form is similar to the argument used in the construction of diffusion processes with marginal laws given by an Itô process, as in Gyongy's original paper [191]. Motivated by problems in financial mathematics, the same idea was used more recently, for instance in the paper [72] by Brunick and Shreve. We refer to Carlen [93], Mikami [282] and Quastel and Varadhan [316] for the construction of weak solutions with marginal laws satisfying the Kolmogorov-Fokker-Planck equation (6.12). We may invoke the paper by Zvonkin [346] and Veretennikov [336] in order to guarantee (under suitable assumptions) that the solutions are strong.

An interesting challenge is to investigate rigorously the infinite-dimensional Hamilton-Jacobi-Bellman equation (6.28). Existence and uniqueness of a viscosity solution were addressed by Pham and Wei [311]. As highlighted by Section 6.5, the problem can be formulated in the enlarged space $\mathbb{R}^d \times \mathcal{P}_2(\mathbb{R}^d)$, the resulting equation (6.90) being of the same nature as the *master equation* for mean field games introduced in Chapter (Vol II)-4, except that $\tilde{\alpha}$ therein has a different interpretation. The differences between the two classes of master equations are borne by the following fundamental distinction. The *master equation* for controlled McKean-Vlasov processes should be a form of Hamilton-Jacobi-Bellman equation as it derives from an optimization problem. However, the *master equation* for mean field games does not have to be an equation of the Hamilton-Jacobi-Bellman type since the fixed point condition describing equilibria in a mean field game does not derive from an optimization criterion. This important distinction was first emphasized by Carmona and Delarue in [97]. Existence and uniqueness of a classical solution to (6.90) were investigated by Chassagneux, Crisan, and Delarue [114]. The arguments developed in [114] will be revisited in Chapter (Vol II)-5 when addressing the existence and uniqueness of a classical solution to the master equation for mean field games.

As suggested in the text, a natural question concerns the possible extension of the results of this chapter to equations with random coefficients. It is indeed well known that the classical Pontryagin stochastic maximum principle also applies to systems with random coefficients. The same should hold in the McKean-Vlasov case, though proofs should require a *modicum of care*.

To be more specific, Theorems 6.14 (necessary condition in the Pontryagin principle), 6.16 (sufficient condition in the Pontryagin principle) and 6.19 are still valid for random coefficients. Allowing random coefficients means that b, σ and f may depend upon the realization $\omega \in \Omega$ in a progressively measurable way with respect to the filtration $\mathbb{F} = (\mathcal{F}_t)_{0 \leqslant t \leqslant T}$, and that g may also depend upon the randomness in a measurable way with respect to the σ-field $\mathcal{F}_T$. The various assumptions, which may differ from one theorem to another, are then supposed to hold *path by path*. The form of the adjoint BSDE is the same as (6.31), provided that the *independent copy* made from the space $(\Omega, \mathcal{F}, \mathbb{P})$ to the space $(\tilde{\Omega}, \tilde{\mathcal{F}}, \tilde{\mathbb{P}})$ also takes into account the dependence upon the randomness. This means that, in (6.31), the value of $\partial_\mu H$ in $\partial_\mu H(t, \tilde{X}_t, \mathcal{L}(X_t), \tilde{Y}_t, \tilde{Z}_t, \tilde{\alpha}_t)$ must be computed along the corresponding realization $\tilde{\omega} \in \tilde{\Omega}$ (as $\tilde{X}_t$, $\tilde{Y}_t$, $\tilde{Z}_t$ and $\tilde{\alpha}_t$ actually stand for $\tilde{X}_t(\tilde{\omega})$, $\tilde{Y}_t(\tilde{\omega})$, $\tilde{Z}_t(\tilde{\omega})$ and $\tilde{\alpha}_t(\tilde{\omega})$). This principle holds for all the *copies* considered in the computations reported in the text.

Still, it is not clear how the results on the decoupling field can be extended to the case of random coefficients. Indeed $Y_t^{t,\xi}$ is already random, so that the decoupling field u_ξ introduced in the proof of Lemma 6.25 is also random. Following the proof of Lemma 6.25, we can write it as $u_\xi(\omega, t, x)$, but u_ξ cannot be entirely determined by the law of ξ, as it must also depend upon the joint law of ξ and (b, f, σ, g). Unfortunately, the proofs of Lemma 6.25 and Proposition 6.32 rely on a coupling argument which fails when the decoupling field depends on the joint distribution of ξ and (b, f, σ, g).

We already alluded to the theory of relaxed controls, as used in Section 6.6, in the Notes & Complements of Chapter 3. As already explained, the theory goes back to the earlier works by Young [344] and Fleming [158] and we refer to the survey by Borkar [65] and to the monograph by Yong and Zhou [343] for an overview of controlled diffusion processes with relaxed controls. Relaxed controls for mean field games were introduced by Lacker [254], but to the best of our knowledge, the theory has never been used for solving mean field stochastic control problems. Section 6.6 is a first step in this direction. The notion of stable convergence used in Section 6.6, see Lemma 6.40, is due to Jacod and Mémin [216], see the original paper together with the monograph [196] by Häusler and Luschgy.

The goal of the discussion of Subsection 6.7.3 is to highlight the possible interpretation of the Monge-Kantorovich problem of optimal transportation as an optimal control problem of the McKean-Vlasov type. Theorem 6.49 is proved in Section 8.2 of the monograph by Ambrosio, Gigli, and Savaré [21]. As mentioned earlier, switching from the closed loop to the open loop formulation can be justified by Gyongy's conditioning trick which can be found in the papers [191] and [72] already cited above. In particular, Theorem 6.50 is explicitly taken from [72]. Theorem 6.51 in this subsection is taken from the paper [283] of Mikami, in which the connection between optimal transportation and optimal control is already present. The terminal constraint which we view as a McKean-Vlasov feature in this book was circumvented in [283] by a form of duality provided by classical HJB equations. Motivated by applications to finance, a lot of attention has recently been paid to similar stochastic optimization problems with a prescribed constraint for the law of the terminal condition but in the case when not only the drift but also the volatility is controlled. These problems are usually called *semimartingale transportation problems*, see for instance Tan and Touzi [328]. Last, observe that in the reference [2] by Achdou, Camilli, and Capuzzo-Dolcetta, the authors address mean field games in which both the initial and terminal states of the population are prescribed; these are referred to as *mean field planning problems*.

Our discussion of the differentiability of functions of measures on a finite state space in Subsection 5.4.4 of Chapter 5 was motivated in part by the second half of Guéant's analysis [185] of MFGs on a finite graph. There, the author shows that, in the case of potential games on a finite graph, the master equation is solved by the derivative of the value function of the deterministic control problem naturally associated with the dynamic evolution of the distribution given by the forward Kolmogorov equation. By considering Markov controls in feedback form and rewriting the optimization problem of the central planner as a deterministic control problem for the distribution of the state instead of a stochastic control problem for the state, the author does not have to acknowledge the fact that the optimization problem of the central planner is in fact a stochastic control problem of the McKean-Vlasov type.

In Chapter (Vol II)-6, we provide a particle interpretation of mean field stochastic control problems. In contrast with mean field games for which players choose individually their response to other players' actions, mean field stochastic control corresponds to controlled particle systems obeying a central planner, who decides of a common strategy in order to minimize some common cost. Somehow, this justifies the alternative terminology *central planner optimization problem* used throughout the chapter for naming mean field stochastic control problems.

Similar to mean field games, other types of mean field stochastic control problems may be considered. For instance, the analogue of mean field stochastic control, but for games, is presented in Chapter 7; it consists of games between a finite number of infinite homogenous populations, each population having its own central planner. We also refer to Djehiche, Tembine, and Tempone [138] for control problems with risk-sensitive cost functionals.

Epilogue to Volume I

Extensions for Volume I

7

Abstract
The goal of this chapter is to follow-up on some of the examples introduced in Chapter 1, especially those which are not directly covered by the probabilistic theory of stochastic differential mean field games developed so far. Indeed, Chapter 1 included a considerable amount of applications hinting at mathematical models with distinctive features which were not accommodated in the previous chapters. We devote this chapter to presentations, even if only informal, of extensions of the Mean Field Game paradigm to these models. They include extensions to several homogenous populations, infinite horizon optimization, and finite state space models. These mean field game models have a great potential for the quantitative analysis of very important practical applications, and we show how the technology developed in this book can be brought to bear on their solutions.

7.1 First Extensions

To start with, we present in this section two natural extensions of the class of mean field games we have studied so far.

The first extension concerns mean field games with several populations or, equivalently, with multiclass agents. The second one is about mean field games with infinite time horizon.

7.1.1 Mean Field Games with Several Populations

From a modeling perspective, one of the major shortcomings of the standard mean field game theory is the strong symmetry requirement that all the players in the game are *statistically identical*. A first way to break this symmetry is to assume that the

R. Carmona, F. Delarue, *Probabilistic Theory of Mean Field Games with Applications I*, Probability Theory and Stochastic Modelling 83,
DOI 10.1007/978-3-319-58920-6_7

players belong to a finite number of homogeneous groups in which the mean field asymptotic theory can be carried out, group by group. Another way is to assume that one of the players dominates the others in the sense that it directly influences the states of the other players while it only feels the others through their collective state. This subsection is devoted to the former approach only. The latter, which is more demanding from the technical point of view, will be addressed in the last chapter of the next volume.

For the sake of simplicity, we restrict ourselves to the case of two homogeneous subgroups in the population. Clearly, the discussion below can be easily adapted to cover cases with a higher number of subgroups.

Finite-Player Game

The generic N-player stochastic differential games leading to mean field games were introduced in Chapter 2. See in particular Subsection 2.3. A key feature was the fact that the dynamics of the states of the players were driven by the same drift and volatility coefficients b and σ from $[0,T] \times \mathbb{R}^d \times \mathcal{P}(\mathbb{R}^d) \times A$ into $\mathbb{R}^d$ and $\mathbb{R}^{d\times m}$, where A is the set of admissible actions, d the dimension of the state space of the players, and m the dimension of the noise. Recall that m was chosen equal to d in Chapters 3 and 4.

Accordingly, when the population is divided into two homogeneous subgroups, we shall consider two sets of drift and volatility coefficients (b_1,σ_1) and (b_2,σ_2). Obviously, (b_1,σ_1) denotes the drift and volatility coefficients of players from the first group, and similarly for (b_2,σ_2). Observe that the set A and the dimension parameters d and m can be chosen to be proper to each of the two subgroups, in which case (A,d,m) becomes (A_1,d_1,m_1) and (A_2,d_2,m_2). Also, due to the mean field hypothesis, we now require that each (b_l,σ_l), for $l \in \{1,2\}$, is a function defined on $[0,T] \times \mathbb{R}^{d_l} \times \mathcal{P}(\mathbb{R}^{d_1}) \times \mathcal{P}(\mathbb{R}^{d_2}) \times A_l$, which accounts for the fact that the dynamics of the state of any player in the game now feel the collective states of the two subgroups. In the end, the dynamics of players from the group 1 take the form:

$$dX_t^{1,i} = b_1\big(t, X_t^{1,i}, \bar{\mu}^{N_1-1}_{X_t^{1,-i}}, \bar{\mu}^{N_2}_{X_t^2}, \alpha_t^{1,i}\big)dt + \sigma_1\big(t, X_t^{1,i}, \bar{\mu}^{N_1-1}_{X_t^{1,-i}}, \bar{\mu}^{N_2}_{X_t^2}, \alpha_t^{1,i}\big)dW_t^{1,i},$$

for $i \in \{1,\cdots,N_1\}$, while, for players from the second group, they take the form:

$$dX_t^{2,i} = b_2\big(t, X_t^{2,i}, \bar{\mu}^{N_1}_{X_t^1}, \bar{\mu}^{N_2-1}_{X_t^{2,-i}}, \alpha_t^{2,i}\big)dt + \sigma_2\big(t, X_t^{2,i}, \bar{\mu}^{N_1}_{X_t^1}, \bar{\mu}^{N_2-1}_{X_t^{2,-i}}, \alpha_t^{2,i}\big)dW_t^{2,i},$$

for $i \in \{1,\cdots,N_2\}$, where N_1 and N_2 denote the respective numbers of players in populations 1 and 2 and $(\boldsymbol{W}^{1,i})_{i=1,\cdots,N_1}$ and $(\boldsymbol{W}^{2,i})_{i=1,\cdots,N_2}$ are independent families of independent $\mathbb{R}^{m_1}$ and $\mathbb{R}^{m_2}$-valued Wiener processes constructed on some filtered probability space $(\Omega,\mathcal{F},\mathbb{F},\mathbb{P})$. Each $\boldsymbol{\alpha}^{1,i}$ (resp. $\boldsymbol{\alpha}^{2,i}$) is required to be $\mathbb{F}$-progressively measurable and to take values in A_1 (resp. A_2). Depending upon the nature of the equilibria we are looking for, we may demand that these controls satisfy more restrictive measurability properties, or even be of a special feedback form. However, for the purpose of the present discussion, we shall not address these technicalities here. Also, we use the same notations as in Chapter 2 for the empirical distributions, namely:

$$\bar{\mu}^{N_1-1}_{X^{1,-i}_t} = \frac{1}{N_1-1}\sum_{j=1,j\neq i}^{N_1} \delta_{X^{1,j}_t}, \quad \bar{\mu}^{N_2}_{X^2_t} = \frac{1}{N_2}\sum_{j=1}^{N_2} \delta_{X^{2,j}_t},$$

$$\bar{\mu}^{N_1}_{X^1_t} = \frac{1}{N_1}\sum_{j=1}^{N_1} \delta_{X^{1,j}_t}, \quad \bar{\mu}^{N_2-1}_{X^{2,-i}_t} = \frac{1}{N_2-1}\sum_{j=1,j\neq i}^{N_2} \delta_{X^{2,j}_t}.$$

The costs to players of each of the subgroups are defined in a similar manner. For tuples of strategies $(\boldsymbol{\alpha}^{1,i})_{i=1,\cdots,N_1}$ and $(\boldsymbol{\alpha}^{2,i})_{i=1,\cdots,N_2}$, the cost to any player $i \in \{1,\dots,N_1\}$ from the first group is:

$$\begin{aligned}
&J^{1,i}\big((\boldsymbol{\alpha}^{1,j})_{j=1,\cdots,N_1}, (\boldsymbol{\alpha}^{2,j})_{j=1,\cdots,N_2}\big) \\
&= \mathbb{E}\bigg[\int_0^T f_1\big(t, X^{1,i}_t, \bar{\mu}^{N_1-1}_{X^{1,-i}_t}, \bar{\mu}^{N_2}_{X^2_t}, \alpha^{1,i}_t\big)dt + g_1\big(X^{1,i}_T, \bar{\mu}^{N_1-1}_{X^{1,-i}_T}, \bar{\mu}^{N_2}_{X^2_T}\big)\bigg],
\end{aligned}$$

and, similarly, for any player $i \in \{1,\cdots,N_2\}$ from the second group:

$$\begin{aligned}
&J^{2,i}\big((\boldsymbol{\alpha}^{1,j})_{j=1,\cdots,N_1}, (\boldsymbol{\alpha}^{2,j})_{j=1,\cdots,N_2}\big) \\
&= \mathbb{E}\bigg[\int_0^T f_2\big(t, X^{2,i}_t, \bar{\mu}^{N_1}_{X^1_t}, \bar{\mu}^{N_2-1}_{X^{2,-i}_t}, \alpha^{2,i}_t\big)dt + g_2\big(X^{2,i}_T, \bar{\mu}^{N_1}_{X^1_T}, \bar{\mu}^{N_2-1}_{X^2_{T,-i}}\big)\bigg].
\end{aligned}$$

Obviously, f_1 is a function from $[0,T] \times \mathbb{R}^{d_1} \times \mathcal{P}(\mathbb{R}^{d_1}) \times \mathcal{P}(\mathbb{R}^{d_2}) \times A_1$ into $\mathbb{R}$ and g_1 is a function $\mathbb{R}^{d_1} \times \mathcal{P}(\mathbb{R}^{d_1}) \times \mathcal{P}(\mathbb{R}^{d_2})$ into $\mathbb{R}$, and similarly for f_2 and g_2.

Remark 7.1 *Obviously, the model is not as general as what we could think of. For instance, we could incorporate common noises, in analogy with mean field games with a common noise investigated in Part I in the second volume. In that case, we could use either the same common noise for the two subgroups or two different common noises, one for each subgroup.*

Also, in this presentation, the way the coefficients are required to depend upon the empirical distributions of the two subgroups is somewhat restrictive. Indeed, it would be more realistic to allow the coefficients to depend upon the proportions of players from each of the two subgroups in the population. However, although this would make sense from a modeling standpoint, the mathematical significance would be rather limited. Indeed, the proportions $N_1/(N_1+N_2)$ and $N_2/(N_1+N_2)$ of each of the two subgroups would be regarded as fixed, since there is no way, in this model, for a player to switch from one subgroup to another. This observation will become especially clear in the next paragraph: We shall take the mean field limit $N_1, N_2 \to \infty$ with the prescription that:

$$\frac{N_1}{N_1 + N_2} \to \pi_1, \qquad \frac{N_2}{N_1 + N_2} \to \pi_2, \tag{7.1}$$

π_1 and π_2 representing the limiting proportion of players from each subgroup in the limiting population. In this approach, π_1 and π_2 are a priori *prescribed.*

Asymptotic Formulation

We now present the limiting formulation of the game when N_1 and N_2 tend to infinity in such a way that (7.1) is satisfied, π_1 and π_2 representing the limiting proportion of players from each subgroup in the limiting population.

The intuition leading to the definition of an asymptotic Nash equilibrium is exactly the same as in standard mean field games. Asymptotically, any unilateral change of strategy decided by one of the players cannot affect the global states of any of the two populations. So, in the limiting framework, everything works as if the best response of any player was computed as the solution of a standard optimal control problem within the environment determined by the equilibrium distributions of the two populations. So, the search for an asymptotic equilibrium should comprise the following two-steps:

1. For any two deterministic flows of probability measures $\boldsymbol{\mu}^1 = (\mu_t^1)_{0 \leqslant t \leqslant T}$ and $\boldsymbol{\mu}^2 = (\mu_t^2)_{0 \leqslant t \leqslant T}$ given on $\mathbb{R}^{d_1}$ and $\mathbb{R}^{d_2}$ respectively, solve the two optimal control problems:

$$\inf_{\boldsymbol{\alpha}^1} J^{1,\boldsymbol{\mu}^1,\boldsymbol{\mu}^2}(\boldsymbol{\alpha}^1) \qquad \text{and} \qquad \inf_{\boldsymbol{\alpha}^2} J^{2,\boldsymbol{\mu}^1,\boldsymbol{\mu}^2}(\boldsymbol{\alpha}^2),$$

over $\mathbb{F}^1$-progressively measurable processes $\boldsymbol{\alpha}^1$ and $\mathbb{F}^2$-progressively measurable processes $\boldsymbol{\alpha}^2$, where $\mathbb{F}^1$ and $\mathbb{F}^2$ are the complete filtrations generated by two independent $\mathbb{R}^{d_1}$ and $\mathbb{R}^{d_2}$-valued Brownian motions $\boldsymbol{W}^1$ and $\boldsymbol{W}^2$ constructed on some probability space $(\Omega, \mathcal{F}, \mathbb{P})$, and where:

$$J^{1,\boldsymbol{\mu}^1,\boldsymbol{\mu}^2}(\boldsymbol{\alpha}^1) = \mathbb{E}\biggl[\int_0^T f_1\bigl(t, X_t^1, \mu_t^1, \mu_t^2, \alpha_t^1\bigr)dt + g_1\bigl(X_T^1, \mu_T^1, \mu_T^2\bigr)\biggr],$$

$$J^{2,\boldsymbol{\mu}^1,\boldsymbol{\mu}^2}(\boldsymbol{\alpha}^2) = \mathbb{E}\biggl[\int_0^T f_2\bigl(t, X_t^2, \mu_t^1, \mu_t^2, \alpha_t^2\bigr)dt + g_2\bigl(X_T^2, \mu_T^1, \mu_T^2\bigr)\biggr],$$

with

$$dX_t^1 = b_1\bigl(t, X_t^1, \mu_t^1, \mu_t^2, \alpha_t^1\bigr)dt + \sigma_1\bigl(t, X_t^1, \mu_t^1, \mu_t^2, \alpha_t^1\bigr)dW_t^1,$$

$$dX_t^2 = b_2\bigl(t, X_t^2, \mu_t^1, \mu_t^2, \alpha_t^2\bigr)dt + \sigma_2\bigl(t, X_t^2, \mu_t^1, \mu_t^2, \alpha_t^2\bigr)dW_t^2,$$

for $t \in [0, T]$, and $X_0^1 = x_0^1 \in \mathbb{R}^{d_1}$ and $X_0^2 = x_0^2 \in \mathbb{R}^{d_2}$ as initial conditions.

2. Find flows $\boldsymbol{\mu}^1 = (\mu_t^1)_{0 \leqslant t \leqslant T}$ and $\boldsymbol{\mu}^2 = (\mu_t^2)_{0 \leqslant t \leqslant T}$ such that:

$$\forall t \in [0, T], \quad \mu_t^1 = \mathcal{L}(\hat{X}_t^{1,\boldsymbol{\mu}^1,\boldsymbol{\mu}^2}), \quad \mu_2^1 = \mathcal{L}(\hat{X}_t^{2,\boldsymbol{\mu}^1,\boldsymbol{\mu}^2}),$$

if $\hat{\boldsymbol{X}}^{1,\boldsymbol{\mu}^1,\boldsymbol{\mu}^2}$ and $\hat{\boldsymbol{X}}^{2,\boldsymbol{\mu}^1,\boldsymbol{\mu}^2}$ are solutions of the above optimal control problems.

We assume that the initial conditions X_0^1 and X_0^2 are deterministic for convenience only. The above procedure can be easily extended to cases when X_0^1 and X_0^2 are random.

Each optimization problem articulated in step 1 above can be handled using the tools introduced for the analysis of standard mean field games. For instance, both problems can be reformulated by means of either: *(i)* a Hamilton-Jacobi-Bellman equation, *(ii)* or an FBSDE for the value function, *(iii)* or the stochastic maximum principle, which relies on another FBSDE.

Below, we review these three approaches when σ_1 and σ_2 are independent of the control variables.

PDE Formulation. In this approach, the optimization problems of step 1 are formulated in terms of HJB equations and the fixed point conditions of step 2 in terms of Fokker-Planck-Kolmogorov equations. Consequently, the PDE approach to mean field games with two populations consists in solving a system of four coupled PDEs: two backward Hamilton-Jacobi-Bellman equations describing the value functions of the two optimization problems, and two forward Fokker-Planck-Kolmogorov equations describing the state laws of each of the two subgroups. In analogy with (3.12), the resulting system takes the form:

$$\begin{cases} \partial_t V_l(t,x) + \dfrac{1}{2}\mathrm{trace}\Big[\big(\sigma_l\sigma_l^\dagger\big)(t,x,\mu_t^1,\mu_t^2)\partial_{xx}^2 V_l(t,x)\Big] \\ \qquad\qquad + H_l^{(r)}\Big(t,x,\mu_t^1,\mu_t^2,\partial_x V_l(t,x),\hat{\alpha}_l(t,x,\mu_t^1,\mu_t^2,\partial_x V_l(t,x))\Big) = 0, \\ \partial_t \mu_t^l - \dfrac{1}{2}\mathrm{trace}\Big[\partial_{xx}^2\Big(\big(\sigma_l\sigma_l^\dagger\big)(t,x,\mu_t^1,\mu_t^2)\mu_t^l\Big)\Big] \\ \qquad\qquad + \mathrm{div}_x\Big(b_l\big(t,x,\mu_t^1,\mu_t^2,\hat{\alpha}_l(t,x,\mu_t^1,\mu_t^2,\partial_x V_l(t,x))\big)\mu_t^l\Big) = 0, \end{cases}$$

in $[0, T] \times \mathbb{R}^{d_l}$, with $V_l(T, \cdot) = g_l(\cdot, \mu_T^1, \mu_T^2)$ as terminal condition for the first equation, and $\mu_0^l = \delta_{x_0^l}$ as initial condition for the second, and for $l = 1, 2$. Above $H_l^{(r)}$ is the reduced Hamiltonian associated with (b_l, σ_l, f_l):

$$H_l^{(r)}(t, x_l, \mu_1, \mu_2, y_l, \alpha_l) = b_l(t, x_l, \mu_1, \mu_2, \alpha_l) \cdot y_l + f_l(t, x_l, \mu_1, \mu_2, \alpha_l), \tag{7.2}$$

for $t \in [0, T]$, $x_l \in \mathbb{R}^{d_l}$, $\mu_l \in \mathcal{P}(\mathbb{R}^{d_l})$, $y_l \in \mathbb{R}^{d_l}$ and $\alpha_l \in A_l$, while $\hat{\alpha}_l(t, x_l, \mu_1, \mu_2, y_l)$ is a minimizer:

$$\hat{\alpha}_l(t, x_l, \mu_1, \mu_2, y_l) \in \mathrm{argmin}_{\alpha_l \in A_l} H_l^{(r)}(t, x_l, \mu_1, \mu_2, y_l, \alpha_l).$$

First FBSDE Formulation. When using an FBSDE, instead of a HJB equation, for describing the value function, we end up with a McKean-Vlasov FBSDE of the same type as (4.54):

$$\left\{\begin{array}{l} dX_t^l = b_l\big(t, X_t^l, \mathcal{L}(X_t^1), \mathcal{L}(X_t^2), \hat{\alpha}_t^l\big) + \sigma_l\big(t, X_t^l, \mathcal{L}(X_t^1), \mathcal{L}(X_t^2)\big) dW_t^l, \\ dY_t^l = -f_l\big(t, X_t^l, \mathcal{L}(X_t^1), \mathcal{L}(X_t^2), \hat{\alpha}_t^l\big) dt + Z_t^l dW_t^l, \\ \text{where } \hat{\alpha}_t^l = \hat{\alpha}_l\big(t, X_t^l, \mathcal{L}(X_t^1), \mathcal{L}(X_t^2), \sigma_l(t, X_t^l, \mathcal{L}(X_t^1), \mathcal{L}(X_t^2))^{-1\dagger} Z_t^l\big), \end{array}\right. \tag{7.3}$$

for $t \in [0, T]$, with $Y_T^l = g_l(X_T^l, \mathcal{L}(X_T^1), \mathcal{L}(X_T^2))$ as terminal condition, for $l = 1, 2$.

Second FBSDE Formulation. When using the stochastic Pontryagin principle, we obtain an FBSDE of the same type as (4.70):

$$\left\{\begin{array}{l} dX_t^l = b_l\big(t, X_t^l, \mathcal{L}(X_t^1), \mathcal{L}(X_t^2), \hat{\alpha}_t^l\big) dt \\ \qquad\qquad\qquad + \sigma_l\big(t, X_t^l, \mathcal{L}(X_t^1), \mathcal{L}(X_t^2)\big) dW_t^l, \\ dY_t^l = -\partial_x H_l\big(t, X_t^l, \mathcal{L}(X_t^1), \mathcal{L}(X_t^2), Y_t^l, Z_t^l, \hat{\alpha}_t^l\big) dt + Z_t^l dW_t^l, \\ \text{where } \hat{\alpha}_t^l = \hat{\alpha}_l\big(t, X_t^l, \mathcal{L}(X_t^1), \mathcal{L}(X_t^2), Y_t^l\big), \end{array}\right. \tag{7.4}$$

for $t \in [0, T]$, with the terminal condition $Y_T^l = \partial_x g_l(X_T^l, \mathcal{L}(X_T^1), \mathcal{L}(X_T^2))$, for $l = 1, 2$. Above, H_l stands for the full-fledged Hamiltonian:

$$\begin{aligned} H_l(t, x_l, \mu_1, \mu_2, y_l, z_l, \alpha_l) &= b_l(t, x_l, \mu_1, \mu_2, \alpha_l) \cdot y_l + f_l(t, x_l, \mu_1, \mu_2, \alpha_l) \\ &\quad + \mathrm{trace}\big[\sigma_l(t, x_l, \mu_1, \mu_2) z_l^\dagger\big]. \end{aligned}$$

In contrast with (4.70), observe that σ_l may not be constant, which explains why the backward equation involves the full Hamiltonian instead of the reduced one.

The analysis of FBSDEs of the McKean-Vlasov type (7.3) and (7.4) can be carried out as the analysis of (4.54) and (4.70). In the latter case, σ_1 and σ_2 need to be assumed to be constant. We shall revisit the Pontryagin principle for processes with nonconstant volatility coefficients in Chapter (Vol II)-1, see Subsection (Vol II)-1.4.4.

A Practical Example

As a practical application, we revisit the crowd congestion model discussed in Subsection 1.5.3.

Instead of one representative individual, we consider two sub-populations with $\mathbb{R}$ as state space for the individuals, and with the same dynamics as in (1.50):

$$\begin{aligned} dX_t^1 &= \alpha_t^1 dt + \sigma dW_t^1, \\ dX_t^2 &= \alpha_t^2 dt + \sigma dW_t^2, \end{aligned} \tag{7.5}$$

for $t \in [0, T]$, where $\sigma > 0$ and $\boldsymbol{W}^1$ and $\boldsymbol{W}^2$ are two independent Wiener processes. Here, $\boldsymbol{\alpha}^1$ and $\boldsymbol{\alpha}^2$ are $\mathbb{F}^1$ and $\mathbb{F}^2$-progressively measurable square integrable processes with values in $A_1 = A_2 = \mathbb{R}$. As in (1.51), we then choose:

$$
\begin{aligned}
& f_1(t, x, \mu_1, \mu_2, \alpha) \qquad (7.6) \\
&= \frac{1}{2}|\alpha|^2 \Big(\int_{\mathbb{R}} \rho(x - x')d\mu_1(x') \Big)^{a_1} \Big(\int_{\mathbb{R}} \rho(x - x')d\mu_2(x') \Big)^{a_2} + e^{-rt}k(t, x),
\end{aligned}
$$

as running cost for the representative player of the first group, and similarly for the representative player of the second group. Here ρ is a smooth density with a support concentrated around 0.

As explained in Chapter 1, the function k models the effect of panic depending upon where the player is. Also, powers $a_1 \geqslant 0$ and $a_2 \geqslant 0$ are intended to penalize congestion. Whenever $a_2 > a_1$, individuals from group 1 primarily avoid congestion with people from group 2, which may be typical of a xenophobic behavior.

We refer to the Notes & Complements below for references on this model.

Potential Games

As another example, we consider the analogue of potential games, but for models with two homogeneous subpopulations. We consider two $\mathbb{R}^{d_1}$ and $\mathbb{R}^{d_2}$-valued representative players with state dynamics of the form:

$$
\begin{aligned}
dX_t^1 &= \alpha_t^1 dt + \sigma dW_t^1, \\
dX_t^2 &= \alpha_t^2 dt + \sigma dW_t^2, \quad t \in [0, T],
\end{aligned}
$$

where, as above, $\boldsymbol{W}^1$ and $\boldsymbol{W}^2$ are two independent Wiener processes with values in $\mathbb{R}^{d_1}$ and $\mathbb{R}^{d_2}$ respectively, and $\boldsymbol{\alpha}^1$ and $\boldsymbol{\alpha}^2$ are two $\mathbb{F}^1$ and $\mathbb{F}^2$ square-integrable progressively measurable processes with values in $A_1 \subset \mathbb{R}^{d_1}$ and $A_2 \subset \mathbb{R}^{d_2}$. Their cost functionals are of the form:

$$
J^l(\boldsymbol{\alpha}^1, \boldsymbol{\alpha}^2) = \int_0^T \Big(F_l\big(t, \mathcal{L}(X_t^1), \mathcal{L}(X_t^2)\big) + \frac{1}{2}\mathbb{E}[|\alpha_t^l|^2] \Big) dt + G_l\big(\mathcal{L}(X_T^1), \mathcal{L}(X_T^2)\big),
$$

where $F_l : [0, T] \times \mathcal{P}(\mathbb{R}^{d_1}) \times \mathcal{P}(\mathbb{R}^{d_2}) \to \mathbb{R}$ and $G_l : \mathcal{P}(\mathbb{R}^{d_1}) \times \mathcal{P}(\mathbb{R}^{d_2}) \to \mathbb{R}$. A pair of strategies $(\hat{\boldsymbol{\alpha}}^1, \hat{\boldsymbol{\alpha}}^2)$ is said to be a Nash equilibrium if for any $l = 1, 2$ and any other admissible strategy $\boldsymbol{\alpha}^l$, it holds:

$$
J^l(\boldsymbol{\alpha}^l, \hat{\boldsymbol{\alpha}}^{-l}) \geqslant J^l(\hat{\boldsymbol{\alpha}}^1, \hat{\boldsymbol{\alpha}}^2),
$$

where, as usual, $(\boldsymbol{\alpha}^l, \hat{\boldsymbol{\alpha}}^{-l}) = (\boldsymbol{\alpha}^1, \hat{\boldsymbol{\alpha}}^2)$ if $l = 1$ and $(\boldsymbol{\alpha}^l, \hat{\boldsymbol{\alpha}}^{-l}) = (\hat{\boldsymbol{\alpha}}^1, \boldsymbol{\alpha}^2)$ if $l = 2$.

Remark 7.2 *As explained above, whenever X_0^1 and X_0^2 are deterministic, $\mathbb{F}^1$ and $\mathbb{F}^2$ may be chosen as the complete filtration generated by $\boldsymbol{W}^1$ and $\boldsymbol{W}^2$. Whenever X_0^1 and X_0^2 are random, both $\mathbb{F}^1$ and $\mathbb{F}^2$ have to be augmented in an obvious manner.*

Notice that in contrast with the notion of equilibrium defined for mean field games with two subgroups, the Nash equilibrium is here regarded as an equilibrium between the two populations. The formulation is in fact reminiscent of the mean field stochastic control problems investigated in Chapter 6, since the marginal laws appearing in the cost functionals are directly influenced by the strategies.

In order to find a Nash equilibrium, we can follow the arguments developed in Chapter 6 and implement the stochastic Pontryagin principle, except that we have to use the version of the stochastic maximum principle for games instead of the version for control problems. Based upon our experience from Chapter 6, we expect that the resulting adjoint equations depend upon the differential calculus used on the space of probability measures of order 2.

If we choose the L-differential calculus introduced in Chapter 5 and if we assume that the coefficients satisfy suitable differentiability assumptions, then in full analogy with Definition 6.5, we associate with each couple $(\boldsymbol{\alpha}^1, \boldsymbol{\alpha}^2)$ two pairs of backward SDEs:

$$\begin{cases} dY_t^{i,j} = -\tilde{\mathbb{E}}\big[\partial_{\mu_j} H^i\big(t, \mathcal{L}(X_t^1), \mathcal{L}(X_t^2), \tilde{Y}_t^{i,1}, \tilde{Y}_t^{i,2}, \tilde{\alpha}_t^1, \tilde{\alpha}_t^2\big)(X_t^j)\big]dt \\ \qquad\qquad + Z_t^{i,j} dW_t^j, \quad t \in [0,T], \\ Y_T^{i,j} = \partial_{\mu_j} G_i\big(\mathcal{L}(X_T^1), \mathcal{L}(X_T^2)\big)(X_T^j), \end{cases} \tag{7.7}$$

for $i,j \in \{1,2\}$, where, as usual, the tuple $(\tilde{\boldsymbol{Y}}^{i,1}, \tilde{\boldsymbol{Y}}^{i,2}, \tilde{\boldsymbol{\alpha}}^1, \tilde{\boldsymbol{\alpha}}^2)$ is a copy of $(\boldsymbol{Y}^{i,1}, \boldsymbol{Y}^{i,2}, \boldsymbol{\alpha}^1, \boldsymbol{\alpha}^2)$ defined on $(\tilde{\Omega}, \tilde{\mathcal{F}}, \tilde{\mathbb{P}})$ and $\tilde{\mathbb{E}}$ denotes the expectation on $(\tilde{\Omega}, \tilde{\mathcal{F}}, \tilde{\mathbb{P}})$. Here, the Hamiltonians H^1 and H^2 are defined as reduced Hamiltonians for games (observe that the index is in exponent in order to distinguish these equations from (7.3)):

$$H^i(t, \mu^1, \mu^2, y_1, y_2, \alpha_1, \alpha_2) = \sum_{j=1}^{2} \alpha_j \cdot y_j + \frac{1}{2}|\alpha_i|^2 + F_i(t, \mu_1, \mu_2).$$

We here use the reduced Hamiltonians because the volatility is uncontrolled. Also, in (7.7), we represent the martingale part with respect to the sole $\boldsymbol{W}^j$ because the randomness in the equation only comes from $\boldsymbol{X}^j$, which is $\mathbb{F}^j$-progressively measurable.

Computing $\partial_{\mu_j} H^i$ given the current assumptions, we get:

$$\begin{cases} dY_t^{i,j} = -\partial_{\mu_j} F_i\big(t, \mathcal{L}(X_t^1), \mathcal{L}(X_t^2)\big)(X_t^j)dt + Z_t^{i,j} dW_t^j, \quad t \in [0,T], \\ Y_T^{i,j} = \partial_{\mu_j} G_i\big(\mathcal{L}(X_T^1), \mathcal{L}(X_T^2)\big)(X_T^j). \end{cases}$$

Since the first order optimality condition given by the stochastic maximum principle takes the form $\alpha_t^i = -Y_t^{i,i}$, this yields the following McKean-Vlasov FBSDE:

$$\begin{cases} dX_t^i = -Y_t^{i,i}dt + \sigma dW_t^i, \\ dY_t^{i,j} = -\partial_{\mu_j}F_i\big(t, \mathcal{L}(X_t^1), \mathcal{L}(X_t^2)\big)(X_t^j)dt + Z_t^{i,j}dW_t^j, \quad t \in [0,T], \\ Y_T^{i,j} = \partial_{\mu_j}G_i\big(\mathcal{L}(X_T^1), \mathcal{L}(X_T^2)\big)(X_T^j). \end{cases}$$

In particular, if we can find functions $f_l : [0,T] \times \mathbb{R}^{d_l} \times \mathcal{P}_2(\mathbb{R}^{d_1}) \times \mathcal{P}_2(\mathbb{R}^{d_2}) \to \mathbb{R}$ and $g_l : \mathbb{R}^{d_l} \times \mathcal{P}_2(\mathbb{R}^{d_1}) \times \mathcal{P}_2(\mathbb{R}^{d_2}) \to \mathbb{R}$ for $l = 1, 2$, such that:

$$\begin{aligned} \partial_x f_l(t, x, \mu_1, \mu_2) &= \partial_{\mu_l} F_l(t, \mu_1, \mu_2)(x), \\ \partial_x g_l(x, \mu_1, \mu_2) &= \partial_{\mu_l} G_l(\mu_1, \mu_2)(x), \qquad l = 1, 2, \end{aligned}$$

then we can identify $(\boldsymbol{Y}^{1,1}, \boldsymbol{Y}^{2,2})$ with $(\boldsymbol{Y}^1, \boldsymbol{Y}^2)$ in (7.4) with $(t, x_l, \alpha_l) \mapsto f_l(t, x_l, \mu_1, \mu_2) + \frac{1}{2}|\alpha_l|^2$ as running cost for player l. Formally, this identifies the Nash equilibrium between the two subgroups with the solution of a mean field game set over the two subpopulations.

Note that we used $\mathcal{P}_2$ instead of $\mathcal{P}$ when we specified the domains of f_l and g_l as L-differentiation requires to work on $\mathcal{P}_2$,

Example. In the spirit of the congestion cost functions (7.6), we may choose, with the same dynamics as in (7.5),

$$F_l = F_0(\mu_l) + \lambda \int_{\mathbb{R}} \Big(\int_{\mathbb{R}} \rho(x - x') d\mu_1(x') \Big) \Big(\int_{\mathbb{R}} \rho(x - x') d\mu_2(x') \Big) dx,$$

for $\lambda \geqslant 0$ and a smooth density ρ with a small support containing 0. As above, λ can be interpreted as a *xenophobia* parameter. In particular, we may expect some forms of *segregation* for λ large, suggesting that the supports of the state distributions of the two populations might separate from one another in equilibrium. We refer to the bibliography in the Notes & Complements for references to further discussions on this issue.

7.1.2 Infinite Horizon MFGs

Some of the examples discussed in Chapter 1 were presented with an infinite time horizon: see for instance the economic growth model in Subsection 1.4.2, the model of production of exhaustible resources in Subsection 1.4.4 and the Cucker-Smale model of flocking in Subsection 1.5.1. However, the mathematical theory covered in the book has been limited to mean field games with a finite time horizon. The goal of this subsection is to provide information on the methodology which could be implemented to solve infinite horizon models.

In order to do so, we distinguish two cases: *(i)* Mean field games with an infinite time horizon and a discounted running cost, which cover the aforementioned economic growth model and model of production of exhaustible resources; *(ii)* Ergodic mean field games, which appeared in the presentation of the Cucker-Smale model.

Mean Field Games with Infinite Time Horizon and Discounted Running Cost

Following the presentation of mean field games given in Chapter 3, we consider a player in interaction with a homogeneous population and with controlled dynamics of the form:

$$dX_t = b(t, X_t, \mu_t, \alpha_t)dt + \sigma(t, X_t, \mu_t, \alpha_t)dW_t, \quad t \geqslant 0,$$

with $X_0 = x_0$ as initial condition for some $x_0 \in \mathbb{R}^d$. Here, $\boldsymbol{W} = (W_t)_{t \geqslant 0}$ is an m-dimensional Brownian motion constructed on a probability space $(\Omega, \mathcal{F}, \mathbb{P})$, $\boldsymbol{\mu} = (\mu_t)_{t \geqslant 0}$ is a flow of probability measures on $\mathbb{R}^d$ accounting for the state of the population and $\boldsymbol{\alpha} = (\alpha_t)_{t \geqslant 0}$ is an A-valued $\mathbb{F}$-progressively measurable control process. The set A is a Borel subset of $\mathbb{R}^k$ and $\mathbb{F}$ is the complete filtration generated by $\boldsymbol{W}$. The functions b and σ are defined on $[0, \infty) \times \mathbb{R}^d \times \mathcal{P}_2(\mathbb{R}^d) \times A$ and take values in $\mathbb{R}^d$ and $\mathbb{R}^{d \times m}$ respectively. Roughly speaking, the set-up is the same as in Chapter 3 except for the fact that the dynamics are now defined on the entire time interval $[0, \infty)$.

Accordingly, we associate with each $\boldsymbol{\alpha}$ an expected cost given by the integral from 0 to ∞ of some running cost. With an instantaneous cost function $f : [0, \infty) \times \mathbb{R}^d \times \mathcal{P}(\mathbb{R}^d) \times A \to \mathbb{R}$ of the type used in Chapter 3, we let:

$$J^{\boldsymbol{\mu}}(\boldsymbol{\alpha}) = \mathbb{E}\left[\int_0^\infty e^{-\beta t} f(t, X_t, \mu_t, \alpha_t) dt \right],$$

where $\beta > 0$ is an actualization factor, most often a discount factor as in the financial applications. In some sense, one may think of a zero terminal cost function g.

Formally, the search for an equilibrium within the population should follow the same procedure as that defined for mean field games with a finite time horizon:

1. For any deterministic flow of probability measures $\boldsymbol{\mu} = (\mu_t)_{t \geqslant 0}$ on $\mathbb{R}^d$, solve the infinite horizon optimal control problem:

$$\inf_{\boldsymbol{\alpha}} J^{\boldsymbol{\mu}}(\boldsymbol{\alpha}),$$

over $\mathbb{F}$-progressively measurable A-valued processes $\boldsymbol{\alpha}$.
2. Find a flow $\boldsymbol{\mu}$ and a solution $\hat{\boldsymbol{X}}^{\boldsymbol{\mu}}$ to the above optimization problem such that:

$$\forall t \geqslant 0, \quad \mu_t = \mathcal{L}(\hat{X}_t^{\boldsymbol{\mu}}). \tag{7.8}$$

Remark 7.3 *Obviously, we could consider a more general version of the model, including for example a common noise, or a random initial condition. We leave it to the reader to adapt the above definition of an equilibrium accordingly.*

When comparing with the analysis of mean field games with a finite time horizon presented in Chapters 3 and 4, the main difference lies in the optimal control problem defined in step 1. Indeed, since the cost functional is defined via an integral over an unbounded interval, this integral may not make sense under the regularity and integrability assumptions used in Chapters 3 and 4, and additional conditions may be needed to make the whole machinery work.

In this subsection, we do not address this question in detail, though we provide references in the Notes & Complements to works where results addressing this issue can be found. Still, we observe that whenever f is bounded, the cost is obviously well defined. However, since it is often convenient to assume that f is strictly convex in α, in order to accommodate these two seemingly contradictory constraints, it then makes sense to require that A is bounded. Also, notice that more generally, when f is bounded from below, the cost is well defined, although it may be infinite. Furthermore, whenever f is neither bounded from above nor from below, special properties of the drift function b and the volatility σ can still make it possible for the cost to still be well defined. Indeed, some of these properties can be used to control the growth of the solution $(X_t)_{t\geqslant 0}$ and its moments which may grow exponentially, polynomially, or could be bounded.

As oftentimes in this book, we assume that σ is uncontrolled.

Value Function and HJB Equation. When the cost functional is well defined for a sufficiently large class $\mathbb{A}$ of admissible control processes $\boldsymbol{\alpha}$, one can try to characterize the solutions to the optimal control problem $\inf_{\boldsymbol{\alpha}\in\mathbb{A}} J^{\mu}(\boldsymbol{\alpha})$ by means of similar equations to those used when the time horizon is finite. A common way to do so is to introduce the analogue of the value function:

$$V(t,x) = e^{\beta t} \inf_{\boldsymbol{\alpha}\in\mathbb{A}^t} \mathbb{E}\left[\int_t^{\infty} e^{-\beta s} f(s, X_s, \mu_s, \alpha_s) ds \,|\, X_t = x \right], \tag{7.9}$$

where $\mathbb{A}^t$ is the class of admissible controls starting from time t. Here, the exponential pre-factor is a normalization accounting for the fact that the system is initialized at time t. By a formal application of the dynamic programming principle, we expect that:

$$V(t,x) = \inf_{\boldsymbol{\alpha}\in\mathbb{A}^t} \mathbb{E}\left[e^{\beta t} \int_t^{t+h} e^{-\beta s} f(s, X_s, \mu_s, \alpha_s) ds + e^{-\beta h} V(t+h, X_{t+h}) \,|\, X_t = x \right]. \tag{7.10}$$

Whenever V is smooth, Itô's formula yields:

$$\begin{aligned}\partial_t V(t,x) + \frac{1}{2}\text{trace}\Big[\big(\sigma\sigma^\dagger\big)(t,x,\mu_t)\partial^2_{xx}V(t,x)\Big] \\ - \beta V(t,x) + \inf_{\alpha\in A} H^{(r)}\big(t,x,\mu_t,\partial_x V(t,x),\alpha\big) = 0,\end{aligned} \tag{7.11}$$

for $(t,x) \in [0,\infty) \times \mathbb{R}^d$, where $H^{(r)}$ is the reduced Hamiltonian:

$$H^{(r)}(t,x,\mu,y,\alpha) = b(t,x,\mu,\alpha)\cdot y + f(t,x,\mu,\alpha).$$

Obviously, (7.11) has the same form as the HJB equation appearing in the statement of Lemma 4.47, except for the presence of an additional zero-order term and the apparent lack of a terminal condition. The terminal condition should be replaced by a condition on the asymptotic behavior of $V(t,\cdot)$ as t tends to ∞. The need for such an additional condition on the growth of $V(t,\cdot)$ for $t \to \infty$, becomes especially clear when implementing the analog of the verification argument used in Lemma 4.47 in the case of finite horizon models. Following the statement of this lemma, assume indeed that V is a classical solution to (7.11), and for an admissible control process $\boldsymbol{\alpha}$, expand $(e^{-\beta t}V(t,X_t))_{t\geqslant 0}$ using Itô's formula. Taking expectation in the resulting expansion (provided this is permissible), we get for any $t \geqslant 0$,

$$\begin{aligned}&\mathbb{E}\big[e^{-\beta t}V(t,X_t)\big] + \mathbb{E}\Big[\int_0^t e^{-\beta s} f(s,X_s,\mu_s,\alpha_s)ds\Big] \\ &= V(0,x_0) + \mathbb{E}\Big[\int_0^t e^{-\beta s}\Big[H^{(r)}\big(s,X_s,\mu_s,\partial_x V(s,X_s),\alpha_s\big) \\ &\qquad - \inf_{\alpha\in A} H^{(r)}\big(s,X_s,\mu_s,\partial_x V(s,X_s),\alpha\big)\Big]ds\Big].\end{aligned}$$

If $\mathbb{E}[e^{-\beta t}V(t,X_t)]$ tends to 0 as t tends to ∞, we obtain:

$$\mathbb{E}\Big[\int_0^\infty e^{-\beta s} f(s,X_s,\mu_s,\alpha_s)ds\Big] \geqslant V(0,x_0),$$

with equality if

$$\forall t \geqslant 0, \quad \alpha_t = \hat{\alpha}\big(t,X_t,\partial_x V(t,X_t)\big),$$

where:

$$\hat{\alpha}(t,x,\mu,y) \in \text{argmin}_{\alpha\in A} H^{(r)}(t,x,\mu,y,\alpha),$$

and with strict inequality if the above identity for $\boldsymbol{\alpha}$ is not satisfied on some measurable subset of $[0, \infty) \times \Omega$ with a nonzero measure for $\text{Leb}_1 \otimes \mathbb{P}$. Therefore, if the minimizer $\hat{\alpha}$ is well defined, if the SDE:

$$dX_t = b\big(t, X_t, \mu_t, \hat{\alpha}(t, X_t, \mu_t, \partial_x V(t, X_t))\big)dt + \sigma(t, X_t, \mu_t)dW_t, \quad t \geqslant 0,$$

with $X_0 = x_0$, is solvable, and if $\hat{\boldsymbol{\alpha}} = (\hat{\alpha}(t, X_t, \mu_t, \partial_x V(t, X_t)))_{0 \leqslant t \leqslant T}$ is admissible, then V is the value function of the optimal control problem defined by $\inf_{\boldsymbol{\alpha} \in \mathbb{A}} J^{\mu}(\boldsymbol{\alpha})$, and $\boldsymbol{X}$ is an optimal path. If the minimizer $\hat{\alpha}$ is strict and the above SDE is uniquely solvable, then $\boldsymbol{X}$ is the unique optimal path.

When the running cost f is bounded, it makes sense to require V to be uniformly bounded in lieu of a terminal condition for (7.11). In this case $e^{-\beta t}\mathbb{E}[V(t, X_t)]$ tends to 0 as $t \to \infty$ for all $\mathbb{F}$-progressively measurable processes $\boldsymbol{\alpha}$ with values in A.

A standard procedure to construct a solution to (7.11) is to implement the following: 1) for each integer $n \geqslant 1$, find a solution V_n to the equation restricted to $[0, n] \times \mathbb{R}^d$ with $V_n(n, \cdot) = 0$ as explicit terminal condition; 2) prove that the sequence $(V_n)_{n \geqslant 1}$ converges in some sense towards a solution V to (7.11) with an admissible behavior at infinity.

Uniqueness of classical solutions may be proved by implementing the above verification argument that demonstrates the optimality of the feedback function $[0, \infty) \times \mathbb{R}^d \ni (t, x) \mapsto \hat{\alpha}(t, x, \mu_t, \partial_x V(t, x))$.

FBSDE Formulation. We just argued that the optimal control problem $\inf_{\boldsymbol{\alpha} \in \mathbb{A}} J^{\mu}(\boldsymbol{\alpha})$ could be handled by means of the same HJB equation as in the finite horizon case, but with an additional zero-order term, and with an asymptotic terminal condition. In fact, it can also be handled with the same FBSDEs as those used in the finite horizon case.

Based on the above discussion on the HJB equation (7.11), we can reasonably expect that the FBSDE representing the value function (7.9) should be:

$$\begin{cases} dX_t = b\big(t, X_t, \mu_t, \hat{\alpha}(t, X_t, \mu_t, \sigma(t, X_t, \mu_t)^{-1\dagger} Z_t)\big)dt \\ \qquad\qquad\qquad\qquad\qquad + \sigma(t, X_t, \mu_t)dW_t, \\ dY_t = -\big[f\big(t, X_t, \mu_t, \hat{\alpha}(t, X_t, \mu_t, \sigma(t, X_t, \mu_t)^{-1\dagger} Z_t)\big) - \beta Y_t\big]dt \\ \qquad\qquad\qquad\qquad\qquad + Z_t dW_t, \end{cases} \tag{7.12}$$

for all $t \geqslant 0$, with $X_0 = x_0$ as initial condition. Like (7.11), (7.12) has no explicit terminal condition. Instead, it is necessary to impose conditions on the behavior of Y_t as t tends to ∞. As above, a standard strategy for constructing a solution is to solve the approximating problem on the interval $[0, n]$ instead of $[0, \infty)$, with 0 as explicit terminal condition at time n, and then let n tend to ∞. Also, uniqueness of the solution may be proved by combining, as in the above verification argument, the condition imposed on the asymptotic behavior of the solution together with the strategy used in Proposition 4.51 to prove uniqueness of the optimal paths on finite intervals.

The strategy is the same when dealing with the stochastic Pontryagin principle. The corresponding FBSDE should be of the form:

$$\begin{cases} dX_t = b\big(t, X_t, \mu_t, \hat{\alpha}(t, X_t, \mu_t, Y_t)\big)dt + \sigma(t, X_t, \mu_t)dW_t, \\ dY_t = -\big[\partial_x H\big(t, X_t, \mu_t, Y_t, Z_t, \hat{\alpha}(t, X_t, \mu_t, Y_t)\big) - \beta Y_t\big]dt + Z_t dW_t, \end{cases} \tag{7.13}$$

for $t \geqslant 0$, with appropriate conditions on the behavior of the solution as t tends to ∞. Above, H stands for the full Hamiltonian of the problem.

We refer to the Notes & Complements at the end of the chapter for further references on these kinds of equations, including results on the choice of the asymptotic boundary condition, and on practical applications.

The Fixed Point Condition. Generally speaking, step 2 in the definition of an equilibrium can be solved by means of the same fixed point argument as in the finite horizon case. For instance, one may look for a fixed point on each interval $[0, n]$, and extract a converging subsequence as n tends to ∞, convergence being then understood as convergence on each $\mathcal{C}([0, T], \mathcal{P}_2(\mathbb{R}^d))$, $T > 0$, equipped with the uniform topology.

Ergodic Mean Field Games

Revisiting Infinite Horizon MFGs with a Discounted Running Cost. Within the same infinite horizon framework as above, another fixed point condition is conceivable when the coefficients are time independent. In such a case, the dynamics take the following form:

$$dX_t = b(X_t, \mu_t, \alpha_t)dt + \sigma(X_t, \mu_t)dW_t, \quad t \geqslant 0\,; \quad X_0 = x_0,$$

with

$$J^{\mu}(\boldsymbol{\alpha}) = \mathbb{E}\bigg[\int_0^{\infty} e^{-\beta t} f(X_t, \mu_t, \alpha_t)dt\bigg]$$

as cost functional. If we now require that the flow $\boldsymbol{\mu} = (\mu_t)_{t \geqslant 0}$ remains constant over time, that is $\mu_t = \mu$ for all $t \geqslant 0$ for some $\mu \in \mathcal{P}(\mathbb{R}^d)$, then the optimization problem becomes the minimization of the cost functional:

$$J^{\mu}(\boldsymbol{\alpha}) = \mathbb{E}\bigg[\int_0^{\infty} e^{-\beta t} f(X_t, \mu, \alpha_t)dt\bigg], \tag{7.14}$$

under the dynamical constraint:

$$dX_t = b(X_t, \mu, \alpha_t)dt + \sigma(X_t, \mu)dW_t, \quad t \geqslant 0.$$

Notice that we denoted $J^{\boldsymbol{\mu}}$ by J^{μ} with the superscript μ in a regular font (as opposed to the boldface $\boldsymbol{\mu}$) to emphasize the fact that the flow is now constant. Accordingly the value function in (7.9) is expected to become time independent. It reads:

$$V(x) = \inf_{\boldsymbol{\alpha}\in\mathbb{A}} \mathbb{E}\left[\int_0^\infty e^{-\beta s} f(X_s, \mu, \alpha_s) ds \,|\, X_0 = x\right].$$

In this case, the HJB equation (7.11) becomes stationary:

$$\frac{1}{2}\text{trace}\Big[\big(\sigma\sigma^\dagger\big)(x,\mu)\partial^2_{xx}V(x)\Big] - \beta V(x) + \inf_{\alpha\in A} H^{(r)}\big(x,\mu,\partial_x V(x),\alpha\big) = 0, \qquad (7.15)$$

$H^{(r)}$ being now independent of t. Then, minimizers $\hat{\alpha}$ of $H^{(r)}$ merely write $\hat{\alpha}(x,\mu,y)$ and optimal paths are given by time-homogeneous diffusion processes:

$$d\hat{X}^\mu_t = b\big(\hat{X}^\mu_t, \mu, \hat{\alpha}(\hat{X}^\mu_t, \mu, \partial_x V(\hat{X}^\mu_t))\big)dt + \sigma(\hat{X}^\mu_t, \mu)dW_t, \quad t \geqslant 0.$$

As a new fixed point condition in the definition of an equilibrium, we now require in step 2 that μ is an invariant measure of $(\hat{X}^\mu_t)_{t\geqslant 0}$ instead of (7.8).

Of course, there is no real reason why $(\hat{X}^\mu_t)_{t\geqslant 0}$ should be in stationary regime. In other words, we should not expect that $\mathcal{L}(\hat{X}^\mu_t) = \mu$ for all $t \geqslant 0$. The rationale behind the choice of μ as the (or an) invariant distribution of $(\hat{X}^\mu_t)_{t\geqslant 0}$ is the fact that in this framework, and under suitable assumptions, $\mathcal{L}(\hat{X}^\mu_t)$ should be close to μ when t is large. So everything works as if, in the original cost functional $J^{\boldsymbol{\mu}}$ with a time-dependent flow $\boldsymbol{\mu}$, the large-time limit $\mu = \lim_{t\to\infty}\mu_t$, the limit being for instance taken in the sense of weak convergence, was substituted for the entire flow $\boldsymbol{\mu} = (\mu_t)_{t\geqslant 0}$.

Ergodic Cost. In order to fully legitimize the substitution of $\boldsymbol{\mu}$ by its long-run limit μ, a convenient strategy is to provide a formulation of the cost functional $J^{\boldsymbol{\mu}}$ which is independent of the initial condition of $\boldsymbol{X} = (X_t)_{t\geqslant 0}$. A natural candidate is:

$$J^{\boldsymbol{\mu},\text{erg}}(\boldsymbol{\alpha}) = \lim_{T\to\infty}\frac{1}{T}\mathbb{E}\left[\int_0^T f(X_t, \mu_t, \alpha_t)dt\right], \qquad (7.16)$$

for a flow $\boldsymbol{\mu} = (\mu_t)_{t\geqslant 0}$ with values in $\mathcal{P}(\mathbb{R}^d)$, where as above, we assume that b and σ are independent of t and also of μ, that is:

$$dX_t = b(X_t, \alpha_t)dt + \sigma(X_t)dW_t, \quad t \geqslant 0; \quad X_0 = x_0. \qquad (7.17)$$

When μ_t converges to some $\mu \in \mathcal{P}(\mathbb{R}^d)$ as t tends to ∞ and f satisfies suitable regularity assumptions, $J^{\boldsymbol{\mu}}(\boldsymbol{\alpha})$ should be the same as:

$$J^{\mu,\text{erg}}(\boldsymbol{\alpha}) = \lim_{T\to\infty}\frac{1}{T}\mathbb{E}\left[\int_0^T f(X_t, \mu, \alpha_t)dt\right]. \qquad (7.18)$$

Actually, if we restrict (7.17) to control processes $\boldsymbol{\alpha} = (\alpha_t)_{t\geqslant 0}$ in stationary Markov feedback form $\boldsymbol{\alpha} = (\alpha_t = \phi(X_t))_{t\geqslant 0}$ for functions ϕ such that the diffusion process $(X_t)_{t\geqslant 0}$ is an irreducible strong Feller Markov process with an invariant measure (in which case the latter is unique), then $J^{\mu,\mathrm{erg}}(\boldsymbol{\alpha})$ becomes:

$$J^{\mu,\mathrm{erg}}(\boldsymbol{\alpha}) = \int_{\mathbb{R}^d} f\big(x, \mu, \phi(x)\big) d\nu(x), \tag{7.19}$$

where ν is the invariant measure of $\boldsymbol{X} = (X_t)_{t\geqslant 0}$.

Now, for a given $\mu \in \mathcal{P}(\mathbb{R}^d)$, the optimization problem $\inf_{\boldsymbol{\alpha}\in\mathbb{A}} J^{\mu,\mathrm{erg}}(\boldsymbol{\alpha})$ is expected to be connected with $\inf_{\boldsymbol{\alpha}\in\mathbb{A}} J^{\mu,\beta}(\boldsymbol{\alpha})$, for $\beta > 0$, where $J^{\mu,\beta}$ stands for the same cost functional as before, but with the additional superscript β in the notation in order to distinguish it from the new $J^{\mu,\mathrm{erg}}$:

$$J^{\mu,\beta}(\boldsymbol{\alpha}) = \mathbb{E}\bigg[\int_0^\infty e^{-\beta t} f(X_t, \mu, \alpha_t) dt\bigg].$$

This is the same cost functional as in (7.14). We now emphasize the dependence upon the actualization rate β. It is natural to expect:

$$\inf_{\boldsymbol{\alpha}\in\mathbb{A}} J^{\mu,\mathrm{erg}}(\boldsymbol{\alpha}) = \lim_{\beta\to 0} \beta \inf_{\boldsymbol{\alpha}\in\mathbb{A}} J^{\mu,\beta}(\boldsymbol{\alpha}),$$

or, reformulated in terms of value functions,

$$\inf_{\boldsymbol{\alpha}\in\mathbb{A}} J^{\mu,\mathrm{erg}}(\boldsymbol{\alpha}) = \lim_{\beta\to 0} \beta V^\beta(x_0),$$

where:

$$V^\beta(x) = \inf_{\boldsymbol{\alpha}\in\mathbb{A}} \mathbb{E}\bigg[\int_0^\infty e^{-\beta s} f(X_s, \mu, \alpha_s) ds \,|\, X_0 = x\bigg].$$

Since $\inf_{\boldsymbol{\alpha}\in\mathbb{A}} J^{\mu,\mathrm{erg}}(\boldsymbol{\alpha})$ is expected to be independent of the starting point x_0, we should have:

$$\inf_{\boldsymbol{\alpha}\in\mathbb{A}} J^{\mu,\mathrm{erg}}(\boldsymbol{\alpha}) = \lim_{\beta\to 0} \beta V^\beta(x),$$

for any $x \in \mathbb{R}^d$.

Letting $\lambda = \inf_{\boldsymbol{\alpha}\in\mathbb{A}} J^{\mu,\mathrm{erg}}(\boldsymbol{\alpha})$ and passing to the limit in a formal way in (7.15), we derive the HJB equation:

$$\frac{1}{2}\mathrm{trace}\Big[\big(\sigma\sigma^\dagger\big)(x)\partial^2_{xx} V(x)\Big] - \lambda + \inf_{\alpha\in A} H^{(r)}\big(x, \mu, \partial_x V(x), \alpha\big) = 0. \tag{7.20}$$

From this equation, V can only be determined up to an additive constant, and the constant λ needs to be part of the solution. For this reason, V is usually constructed as the limit of $V^\beta - V^\beta(0)$. In that case, optimal paths are given by:

$$d\hat{X}_t^\mu = b\big(\hat{X}_t^\mu, \hat{\alpha}(\hat{X}_t^\mu, \mu, \partial_x V(\hat{X}_t^\mu))\big)dt + \sigma(\hat{X}_t^\mu)dW_t, \quad t \geq 0,$$

and, at least formally, λ is given by the integral:

$$\lambda = \int_{\mathbb{R}^d} f\big(x, \mu, \hat{\alpha}(x, \mu, \partial_x V(x))\big)d\hat{\mu}(x),$$

where $\hat{\mu}$ is the unique invariant measure of $\hat{\boldsymbol{X}}^\mu$. This identification of λ follows from (7.19) when $\hat{\boldsymbol{X}}^\mu$ is irreducible, strong Feller and has an invariant measure.

In this framework, the fixed point condition for mean field games merely writes $\hat{\mu} = \mu$, that is μ is the invariant measure of $\hat{\boldsymbol{X}}^\mu$. Therefore, for an ergodic mean field game, the search for an equilibrium consists in the following two-step procedure:

1. For any deterministic probability measure μ on $\mathbb{R}^d$, solve the ergodic optimal control problem:

$$\inf_{\boldsymbol{\alpha}} J^{\mu,\text{erg}}(\boldsymbol{\alpha}),$$

   over $\mathbb{F}$-progressively measurable A-valued processes $\boldsymbol{\alpha}$.
2. Find a probability measure μ such that μ is an invariant measure of the diffusion process $\hat{\boldsymbol{X}}^\mu$ where $\hat{\boldsymbol{X}}^\mu$ is a solution of the optimal control problem in step 1.

Observe that in step 1, we can choose whether or not to reduce the analysis to controls $\boldsymbol{\alpha}$ in stationary Markov feedback form. In any case, a reasonable guess is that the optimal control should be in stationary Markov feedback form. The requirement that $\hat{\boldsymbol{X}}^\mu$ has an invariant measure for any $\mu \in \mathcal{P}(\mathbb{R}^d)$ is a very restrictive condition. In order to satisfy it, one usually imposes specific conditions on the coefficients b and σ. For example, denoting by ϕ the optimal stationary feedback function, one may want to require that the diffusion process $\hat{\boldsymbol{X}}^\mu$ solving the SDE:

$$d\hat{X}_t^\mu = b\big(\hat{X}_t^\mu, \phi(\hat{X}_t^\mu)\big)dt + \sigma(\hat{X}_t^\mu)dW_t, \quad t \geq 0,$$

has suitable non-degeneracy and positive recurrence properties, this being the case if the set A is bounded, the coefficient σ is bounded from above and uniformly elliptic, and the drift b is dissipative in the x-direction. We refer to the Notes & Complements below for standard references on the subject.

Obviously, if the analysis is reduced to controlled processes which are irreducible and strong Feller, the measure μ in step 2 is the unique invariant measure of $\hat{\boldsymbol{X}}^\mu$. Also, once a fixed point μ in step 2 is found, the two formulations (7.16) and (7.18) coincide if the marginals $(\mu_t = \mathcal{L}(\hat{X}_t^\mu))_{t\geq 0}$ converge to μ, which is an invariant measure of $\hat{X}^\mu$ because of the fixed point condition. For instance, if $\hat{X}^\mu$ is irreducible and strong Feller, in which case μ is the unique invariant measure of $\hat{X}^\mu$, then μ_t converges in law to μ as $t \to \infty$. Depending on the smoothness of f in the measure argument, convergence may be investigated with respect to other topologies or distances, such as the 2-Wasserstein distance.

PDE Formulation. Accounting for the HJB equation (7.20) together with the standard Poisson equation for the invariant measure of a diffusion process, we end up describing ergodic mean field games with a system of two stationary PDEs:

$$\begin{cases} \frac{1}{2}\text{trace}\Big[(\sigma\sigma^\dagger)(x)\partial^2_{xx}V(x)\Big] - \lambda + \inf_{\alpha\in A} H\big(x,\mu,\partial_x V(x),\alpha\big) = 0, \\ -\frac{1}{2}\text{trace}\Big[\partial^2_{xx}\Big((\sigma\sigma^\dagger)(x)\mu\Big)\Big] + \text{div}_x\Big(b\big(x,\hat{\alpha}(x,\mu,\partial_x V(t,x))\big)\mu\Big) = 0, \end{cases}$$

for $x \in \mathbb{R}^d$, with the constraint that μ is a probability distribution. As noticed for (7.20), we stress the fact that λ is part of the solution.

We shall not address the solvability of the above system. We refer to the Notes & Complements below for references on that question.

Ergodic BSDE. In analogy with (7.12), the stationary HJB equation may be represented by means of an ergodic BSDE, which can be obtained by replacing βY_t in (7.12) by λ. Once again, we refer to the end of the chapter for further discussion of that point.

The *N*-Player Game. At the risk of indulging in anticipation of Chapter (Vol II)-6, we cannot resist the temptation to describe the connection with finite-players games. The reason is that, as explained in the Notes & Complements of Chapter (Vol II)-6, the first published results on the convergence of finite-player games equilibria to mean field games were obtained for ergodic mean field games. In this last paragraph, we follow the presentation used in these early papers to explain the differences with the strategy that we shall adopt in Chapter (Vol II)-6.

With coefficients of the same form as above, consider N states with dynamics:

$$dX_t^{N,i} = b(X_t^{N,i}, \alpha_t^{N,i})dt + \sigma(X_t^{N,i})dW_t^i, \quad t \geqslant 0, \quad X_0^i = x_0^i \in \mathbb{R}^d,$$

for $i = 1, \cdots, N$, where as usual, $\boldsymbol{W}^1, \cdots, \boldsymbol{W}^N$ are N independent Wiener processes with values in $\mathbb{R}^m$ and $\boldsymbol{\alpha}^{N,1}, \cdots, \boldsymbol{\alpha}^{N,N}$ are progressively measurable processes with values in A. With each player $i \in \{1, \cdots, N\}$, we associate the ergodic cost:

$$J^{N,i}\big(\boldsymbol{\alpha}^{N,1}, \cdots, \boldsymbol{\alpha}^{N,N}\big) = \lim_{T\to\infty} \frac{1}{T}\mathbb{E}\Big[\int_0^T f(X_t^i, \bar{\mu}_t^{N,i}, \alpha_t^i)dt\Big],$$

where:

$$\bar{\mu}_t^{N,i} = \frac{1}{N-1} \sum_{j=1,j\neq i}^{N} \delta_{X_t^{N,j}}, \quad t \geqslant 0.$$

If we force player $i \in \{1, \cdots, N\}$ to use a strategy $\boldsymbol{\alpha}^i$ adapted to the sole Wiener process $\boldsymbol{W}^i$, then the states of the players are independent. In particular, if we restrict ourselves to stationary Markovian equilibria of the form $\boldsymbol{\alpha}^{N,i} = (\alpha_t^{N,i} = \phi^{N,i}(X_t^{N,i}))_{t\geq 0}$, for some function $\phi^{N,i} : \mathbb{R}^d \to A$, then the best response of player $i \in \{1, \cdots, N\}$ to the other players using feedback functions $\phi^{N,1}, \cdots, \phi^{N,i-1}, \phi^{N,i+1}, \cdots, \phi^{N,N}$, consists, at least under reasonable assumptions, in minimizing the ergodic cost:

$$\begin{aligned} J^{N,i}(\boldsymbol{\alpha}^i) &\\ = \lim_{T\to\infty} \frac{1}{T} \mathbb{E}\biggl[\int_0^T \biggl(\int_{\mathbb{R}^{(N-1)d}} f\Bigl(X_t^i, \frac{1}{N} \sum_{j=1,j\neq i}^{N} \delta_{z^j}, \alpha_t^i\Bigr) \prod_{j=1,j\neq i}^{N} d\mu^j(z^j) \biggr) dt \biggr], \end{aligned}$$

where $\mu^1, \cdots, \mu^{i-1}, \mu^{i+1}, \cdots, \mu^N$ are the invariant measures of the diffusion processes:

$$dX_t^j = b\bigl(X_t^j, \phi^{N,j}(X_t^j)\bigr)dt + \sigma(X_t^j)dW_t^j, \quad t \geqslant 0\,; \quad X_0^j = x_0^j,$$

for $j = 1, \cdots, i-1, i+1, \cdots, N$ respectively. Letting:

$$\mu^{-i} = (\mu^1, \cdots, \mu^{i-1}, \mu^{i+1}, \cdots, \mu^N),$$

the reduced Hamiltonian reads:

$$\begin{aligned} &H^i(x, \mu^{-i}, y, \alpha) \\ &\quad = b(x, \alpha) \cdot y + \int_{\mathbb{R}^{(N-1)d}} f\Bigl(x, \frac{1}{N} \sum_{j=1,j\neq i}^{N} \delta_{z^j}, \alpha\Bigr) \prod_{j=1,j\neq i}^{N} d\mu^j(z^j), \end{aligned}$$

for $x, y \in \mathbb{R}^d$, and $\alpha \in A$. This prompts us to let, as the analogue of the Nash system (2.17) for the N-player game with a finite time horizon, the following system of equations:

$$\begin{cases} \frac{1}{2}\mathrm{trace}\Bigl[\bigl(\sigma\sigma^\dagger\bigr)(x)\partial_{xx}^2 v^{N,i}(x)\Bigr] - \lambda^i + \inf_{\alpha\in A} H^i\bigl(x, \mu^{-i}, \partial_x v^{N,i}(x), \alpha\bigr) = 0, \\ -\frac{1}{2}\mathrm{trace}\Bigl[\partial_{xx}^2\Bigl(\bigl(\sigma\sigma^\dagger\bigr)(x)\mu^i\Bigr)\Bigr] \\ \qquad\qquad +\mathrm{div}_x\Bigl(b\bigl(x, \hat{\alpha}^i(x, \mu^{-i}, \partial_x v^{N,i}(t,x))\bigr)\mu^i\Bigr) = 0, \end{cases} \tag{7.21}$$

for $x \in \mathbb{R}^d$ and $i = 1, \cdots, N$, each μ^i being a probability measure on $\mathbb{R}^d$, and $\hat{\alpha}^i(x, \mu^{-i}, y)$ denoting $\text{argmin}_{\alpha \in A} H^i(x, \mu^{-i}, y, \alpha)$. The second equation guarantees that μ^i is the invariant measure of $\boldsymbol{X}^{N,i}$ when computed over the feedback function $\phi^{N,i} : \mathbb{R}^d \ni x \mapsto \hat{\alpha}^i(x, \mu^{-i}, \partial_x v^{N,i}(x)) \in A$. The finite time horizon analogue of (7.21) will be studied in Chapter (Vol II)-6, see (Vol II)-(6.94).

The system (7.21) is a system of N forward-backward equations over $\mathbb{R}^d$, these equations being coupled by the measures $\mu^1, \cdots, \mu^N$. In contrast with the aforementioned Nash system (Vol II)-(6.94) that we shall study in Chapter (Vol II)-6, the functions $(v^{N,i})_{i=1,\cdots,N}$ are defined over $\mathbb{R}^d$ and not over $\mathbb{R}^{Nd}$. Obviously, this makes a big difference when investigating the convergence of the equilibria as N tends to ∞. It is indeed much easier to prove compactness properties of the functions $(v^{N,i})_{i=1,\cdots,N}$ when defined on $\mathbb{R}^d$. Of course, the crucial point here is the fact that each control $\boldsymbol{\alpha}^{N,i}$ is required to be $\boldsymbol{W}^i$-adapted. As already emphasized, this forces the states of the different players to be independent! In Chapter (Vol II)-6, the system (Vol II)-(6.94) corresponds to the more general case when the controls are allowed to be adapted with respect to all the noises. This is one of the main objective of Chapter (Vol II)-6 to address the convergence of the solution of the system in this much more challenging regime.

We refer to the Notes & Complements below for references.

7.2 Mean Field Games with Finitely Many States

This section is devoted to the analysis of models for which the state space of the system is finite. The methods can be adjusted to handle countable discrete state spaces like the set of integers $\mathbb{N}$ used in the example *Searching for Knowledge* of Section 1.6 of Chapter 1. We refrain from working at this level of generality to avoid having to deal with heavier notations and most importantly, to add technical conditions to guarantee that all the quantities (which would typically involve infinite sums) are actually finite and well defined. Moreover, and even though we still work with continuous time, the stochastic analysis tools developed throughout the book for solutions of stochastic differential mean field games can no longer be used in their original forms. The solutions and implementations need to be ported from the framework of stochastic differential games to a discrete game set-up.

In this section, we assume that the possible states of the system comprise a finite set $E = \{e_1, \cdots, e_d\}$. Even though this will not play a major role in this section, we can always assume, as we did in Subsection 5.4.4 of Chapter 5, that the set E is embedded in $\mathbb{R}^d$ by regarding its elements as the vectors of the canonical basis of $\mathbb{R}^d$ formed by the unit coordinate vectors $e_1 = (1, 0, \cdots, 0), \cdots, e_d = (0, \cdots, 0, 1)$.

Before we introduce the specifics of the finite player games from which we derive mean field game models, we review some of the features of the basic stochastic differential mean field games studied in the first part of the book. At any given time, players choose their actions from a Borel set $A \subset \mathbb{R}^k$. The dynamics of the state of the system are completely determined by the drift and volatility functions b and σ which depend upon time, the value of the state of a given player, a probability measure serving as a proxy for the distribution of the states of the other players, and a possible action. For each $\mu \in \mathcal{P}_2(\mathbb{R}^d)$ and $\alpha \in A$, the functions $[0,T] \times \mathbb{R}^d \ni (t,x) \mapsto b(t,x,\mu,\alpha)$ and $[0,T] \times \mathbb{R}^d \ni (t,x) \mapsto \sigma(t,x,\mu,\alpha)$ determine the local means and standard deviations of the changes of the state in infinitesimally small time intervals. Put more mathematically, they determine the infinitesimal generator $(L_t^{\mu,\alpha})_{0\leqslant t\leqslant T}$ of a Markov diffusion process for the dynamics of the states from time t on. Replacing μ by a deterministic flow $\boldsymbol{\mu} = (\mu_t)_{0\leqslant t\leqslant T}$ of measures, or α by a feedback function $\phi(t,x)$ would change the dynamics without affecting the Markovian character of these dynamics. However, replacing μ by a stochastic flow of measures, or α by an adapted process with values in A, would prevent us from constructing the controlled dynamics of the states as a Markov process. Nevertheless, such dynamics could still be constructed from the local mean and standard deviation characteristics (now random processes) by solving stochastic differential equations with random coefficients. We now explain why and how the situation is different, and possibly more delicate, when the state space is a finite set E instead of $\mathbb{R}^d$.

We first recall the characterization of the infinitesimal generators of continuous time Markovian dynamics on a finite state space E:

Definition 7.4 *A real valued function q on $E \times E$ satisfying the two properties:*

(i) $q(x,x') \geqslant 0, \qquad x,x' \in E,\ x \neq x';$
(ii) $q(x,x) = -\sum_{x'\neq x} q(x,x'), \qquad x \in E.$

is called a Q-matrix.

Such Q-matrices determine the infinitesimal generators of E-valued continuous time Markov chains by providing the rates of change of the state in infinitesimally small time intervals. For this reason, they will play the roles played by the drift and volatility coefficients b and σ. So in analogy with the framework of stochastic differential mean field games recalled above, we fix the set-up of our game models in the following way.

Assumption (Discrete MFG Rates).

(A1) For each $\alpha \in A$, $\mu \in \mathcal{P}(E)$, and $t \in [0,T]$, $\big(\lambda_t(x,x',\mu,\alpha)\big)_{x,x'\in E}$ is a Q-matrix and there exists a positive constant c_λ such that:

$$|\lambda_t(x,x',\mu,\alpha)| \leq c_\lambda(1+|\alpha|), \qquad t \in [0,T],\ x,x' \in E,\ \mu \in \mathcal{P}(E),\ \alpha \in A.$$

(continued)

Also, for each $x, x' \in E$, the mapping $[0, T] \times \mathcal{P}(E) \times A \ni (t, \mu, \alpha) \mapsto \lambda_t(x, x', \mu, \alpha)$ is continuous, $\mathcal{P}(E)$ being equipped with the weak convergence topology, which is here equivalent to the Wasserstein topology since E is finite.

(A2) The rates $(\lambda_t)_{0 \leqslant t \leqslant T}$ are linear in α, namely:

$$\lambda_t(x, x', \mu, \alpha) = \Lambda_t(x, x', \mu) \cdot \alpha, \quad x, x' \in E, \ \mu \in \mathcal{P}(E),$$

where, for each $x, x' \in E'$, the mapping $[0, T] \times \mathcal{P}(E) \ni (t, \mu) \mapsto \Lambda_t(x, x', \mu) \in \mathbb{R}^k$ is bounded and continuous and satisfies:

$$\begin{cases} \Lambda_t(x, x', \mu) \cdot \alpha \geqslant 0, & \text{for } x \neq x' \text{ and } \alpha \in A, \\ \Lambda_t(x, x, \mu) = - \sum_{x' \in E, x' \neq x} \Lambda_t(x, x', \mu). \end{cases}$$

The linearity assumption **(A2)** will only be needed later on for some specific technical results. We included it here for the sake of completeness as it is an assumption on the form of the jump rates after all. Observe however that it is rather restrictive as the sign constraint $\lambda_t(x, x', \mu, \alpha) \geqslant 0$ for $x \neq x'$ becomes $\Lambda_t(x, x', \mu) \cdot \alpha \geqslant 0$, which precludes A to contains opposite vectors. We let the reader check that a more general affine, instead of linear, condition would in fact suffice to implement the arguments used below.

Our first task is to show that the rates $\lambda_t(x, x', \mu, \alpha)$ can be used to construct dynamics controlled by the players when μ is replaced by μ_t for a flow $\boldsymbol{\mu} = (\mu_t)_{0 \leqslant t \leqslant T}$ of measures which is either deterministic or Markovian in the sense that μ_t is a function of time and the state X_t at time t, and similarly when α is replaced by a feedback function $\phi(t, X_t)$ of the state X_t at time t.

Constructing state dynamics when the flow of measures and the controls are only assumed to be adapted to a given filtration is much more involved mathematically. It requires the construction of point processes from their local characteristics as given by their dual predictable projections. Indeed, the jump intensities are the only characteristics which can be controlled. More details on what is actually needed, at least in the mean field game formulation, will be given at the beginning of Subsection 7.2.2 below. So for the sake of simplicity, we restrict ourselves to models *without common noise*, and we only consider Markovian control strategies. References to specific texts containing elements of the control theory of point processes are given in the Notes & Complements at the end of the chapter.

7.2.1 N Player Symmetric Games in a Finite State Space

For pedagogical reasons, we first describe, though in a rather informal way, the finite player games leading to the mean field game models considered in this section. Complete hypotheses and precise and rigorous statements will be given in the following subsections.

The state of player $i \in \{1, \cdots, N\}$ is given at time t by an element X_t^i of the finite set E. Intuitively, the way player i acts on the system is by choosing, or at least influencing, the rate at which its own state will switch from its current value $X_t^i = x$ to another possible value $x' \in E$. So even if this choice were to depend upon other quantities such as the current or past values X_s^j for $j \neq i$ and $0 \leqslant s \leqslant t$ of the states of the other players, the action of player i at time t should only affect directly the value of the rates at which its own state will jump to other states $x' \in E$. Again, these rates can and will depend upon the current value of the state of player i, as well as other quantities like the current values of the states of the other players.

Uncontrolled Transition Rates with Mean Field Interaction

In order to model the symmetry and the *mean field* nature of the interactions, we ask the rates of change of the state of a given player i to depend upon the state of the system only through the current value of the state of player i and the empirical distribution of the states of the other players. To be more specific, if we assume that the dynamics of the states of the players are not controlled, and that the state of the system at time t is $X_t = (X_t^1, \cdots, X_t^N)$, then the individual states of the N players jump independently of each other with transition probabilities of the form:

$$\mathbb{P}\Big[X_{t+\Delta t}^i = x' \mid X_t^i = x,\ \bar{\mu}_{X_t^{-i}}^{N-1} = \mu\Big] = \lambda_t^i(x, x', \mu)\Delta t + o(\Delta t), \tag{7.22}$$

for $i = 1, \cdots, N$, whenever $x' \neq x$, where as usual, $\bar{\mu}_{X_t^{-i}}^{N-1} = \frac{1}{N-1}\sum_{j=1, j\neq i}^{N} \delta_{X_t^j}$ is the empirical distribution of the states X_t^{-i} of the other players.

The rate $\lambda_t^i(x, x', \mu)$ is here to give the small Δt asymptotic behavior of the probability at time t that the state of player i changes from x to x' when the empirical distribution of the states of all the other players is μ. The formulation of this requirement is inspired by the properties of continuous time Markov chains in finite state spaces.

Controlled Transition Rates

Let us now see how we could include the control strategies of the players in our list of desiderata.

We fix a filtered probability space $(\Omega, \mathcal{F}, \mathbb{F}, \mathbb{P})$ together with an adapted process $\boldsymbol{\alpha} = (\boldsymbol{\alpha}^1, \cdots, \boldsymbol{\alpha}^N)$ with values in $A^N = A \times \cdots \times A$. Here, $\boldsymbol{\alpha}^i = (\alpha_t^i)_{0 \leqslant t \leqslant T}$ for $i = 1, \cdots, N$, and we should think of α_t^i as the control exerted at time t by player i. Notice that we require the control strategies to be adapted despite the presence of jumps in the sample paths of the state process.

If we denote by $\boldsymbol{X}^i = (X^i_t)_{0\leqslant t\leqslant T}$ the time evolution of the state of player i for $i = 1,\cdots,N$, as a result of the strategy profile $\boldsymbol{\alpha}$, we ask the state of the system to be a $E^N = E \times \cdots \times E$-valued process $\boldsymbol{X} = (\boldsymbol{X}^1,\cdots,\boldsymbol{X}^N)$ with a law characterized by the transition probabilities $\mathbb{P}[X_{t+\Delta t} = \boldsymbol{x}'|\mathcal{F}_t, X_t = \boldsymbol{x}]$, which we would like to specify from the transition rates $\lambda_t(x^i, x'^i, \bar{\mu}^{N-1}_{\boldsymbol{x}^{-i}}, \alpha^i_t)$ whenever $\boldsymbol{x} = (x^1,\cdots,x^N)$ and $\boldsymbol{x}' = (x'^1,\cdots,x'^N)$ are elements of E^N. This would allow us to work with control strategies which depend upon the history of the system in a non-anticipative way, opening the door to searches for open loop Nash equilibria based on appropriate forms of the stochastic maximum principle. However, as we explained in the informal introduction to this section, the construction of such state processes is rather involved, and in order to avoid having to develop and present too many technical tools which would distract us from our main objective, we shall restrict ourselves to Markovian control strategies.

To be more specific, we shall only consider control strategies of the form $\alpha^i_t = \phi^i(t, X_t)$ for some deterministic measurable A-valued function ϕ^i on $[0,T]\times E^N$. We denote by $\mathbb{A}^i = \mathbb{A}$ this set of strategies. For the time being, it will serve as the set of admissible control strategies for each player i. We may add extra conditions for a control strategy to be admissible later on.

In any case, the controlled state evolves as a Markov process in E^N, with *càd-làg* trajectories, and its distribution is entirely determined by the transitions:

$$\mathbb{P}[X_{t+\Delta t} = \boldsymbol{x}'|X_t = \boldsymbol{x},\ \mathcal{F}_t]$$
$$= \begin{cases} 1 - \sum_{j=1}^{N} \lambda_t\big(x^j, x'^j, \mu^{N-1}_{\boldsymbol{x}^{-j}}, \phi^j(t,\boldsymbol{x})\big)\Delta t + o(\Delta t) & \text{if } \ \boldsymbol{x}' = \boldsymbol{x}, \\ \lambda_t\big(x^i, x'^i, \mu^{N-1}_{\boldsymbol{x}^{-i}}, \phi^i(t,\boldsymbol{x})\big)\Delta t + o(\Delta t) & \text{if } \ x'^i \neq x^i \text{ and } \boldsymbol{x}'^{-i} = \boldsymbol{x}^{-i}, \\ & \quad \text{for some } i \in \{1,\dots,N\}, \end{cases} \tag{7.23}$$

for $\boldsymbol{x}, \boldsymbol{x}' \in E^N$. Clearly, the above definition implies that in an infinitesimal time interval of length Δt, at most one player's state will change.

Recall from Subsection 5.4.4 of Chapter 5 that a probability measure $\mu \in \mathcal{P}(E)$ can be identified with the set of weights $\boldsymbol{p} = (p_\ell)_{1\leqslant \ell\leqslant d} \in \mathcal{S}_d$ it gives to each of the elements of the set E. In other words, we can identify the measure μ with the element $\boldsymbol{p}$ of the simplex given by its components $p_\ell = \mu(\{e_\ell\})$ for $\ell = 1,\cdots,d$ so that $\mu = \sum_{1\leqslant \ell\leqslant d} p_\ell \delta_{e_\ell}$. Notice that when $\mu = \bar{\mu}^N_{\boldsymbol{x}}$ is the empirical measure of a sample $\boldsymbol{x} = (x^1,\cdots,x^N)$ of elements of E, we have:

$$\bar{\mu}^N_{\boldsymbol{x}} = \frac{1}{N}\sum_{i=1}^{N} \delta_{x^i} = \sum_{\ell=1}^{d} p_\ell \delta_{e_\ell}$$

where $p_\ell = \#\{i;\ 1 \leqslant i \leqslant N,\ x^i = e_\ell\}/N$ is the proportion of elements x^i of the sample which are equal to e_ℓ.

Cost Functionals

The goal of player i is to minimize its expected cost as given by:

$$J^i(\boldsymbol{\alpha}^1, \cdots, \boldsymbol{\alpha}^N) = \mathbb{E}\left[\int_0^T f(t, X_t^i, \bar{\mu}_{X_t^{-i}}^{N-1}, \alpha_t^i)\, dt + g(X_T^i, \bar{\mu}_{X_T^{-i}}^{N-1})\right],$$

where the strategies $\boldsymbol{\alpha}^j$ belong to $\mathbb{A}$ for $j = 1, \cdots, N$, and where the running cost function f and the terminal cost function g are real valued functions defined on the sets $[0, T] \times E \times \mathcal{P}(E) \times A$ and $E \times \mathcal{P}(E)$ respectively.

7.2.2 Mean Field Game Formulation

We emphasize the similarities and the differences between the present analysis and the stochastic differential models studied throughout the book by using similar notations and formulating hypotheses and results in as close a manner as possible.

General Formulation

The reader may skip this short subsection in a first reading. Indeed, it will not be used in the remainder part of this section where we consider only Markovian games. Still, we thought it would be useful to hint at the technicalities we would have to face should we want to consider models with common noise or games with major and minor players as we do in Subsection (Vol II)-7.1.9.

In its most abstract form, a general formulation of the problem would involve a filtered probability space $(\Omega, \mathcal{F}, \mathbb{F}, \mathbb{P})$ together with two predictable processes $\boldsymbol{\alpha} = (\alpha_t)_{0 \leq t \leq T}$ and $\boldsymbol{\mu} = (\mu_t)_{0 \leq t \leq T}$ with values in A and $\mathcal{P}(E)$ respectively. We should think of μ_t as a random input serving as proxy for the conditional distribution of the state of a generic player, the conditioning being with respect to a random environment external to the game and common to all the players. We should think of α_t as the control exerted at time t by a generic player. Given the environment $\boldsymbol{\mu}$ and the control strategy $\boldsymbol{\alpha}$, the state of the process should be a continuous time process $\boldsymbol{X} = (X_t)_{0 \leq t \leq T}$ with values in E. We shall assume that it is *càd-làg* by which we mean right continuous and with left limits. Because E is finite, this process determines a random point measure on $[0, T] \times E$:

$$\nu(dt, dx) = \sum_{\substack{0 \leq t \leq T \\ X_{t-} \neq X_t}} \delta_{(t, X_t)}(dt, dx) = \sum_{n \geq 0} \delta_{(T_n, X_n)}(dt, dx),$$

if we think of $\boldsymbol{X}$ as a marked point process $(T_n, X_n)_{n \geq 0}$ where T_n denotes the time of the n-th jump of the process and X_n the landing point of the jump. The mathematical formulation of the fact that a generic player controls the state in the random environment $\boldsymbol{\mu}$ with the strategy $\boldsymbol{\alpha}$ would state that the dual predictable projection $\bar{\nu}$ of the measure ν is given by:

$$\bar{\nu}(dt, \{x'\}) = \lambda_t(X_{t-}, x', \mu_t, \alpha_t)dt,$$

which appears as the data from which the state evolution needs to be constructed. Such a level of generality raises questions which could distract us from the main thrust of this section. Indeed, reconstructing the point measure ν from its dual predictable projection $\bar{\nu}$ is rather involved, and even when this can be done, the measure ν does not always determine uniquely the state process $\boldsymbol{X}$ if the filtration $\mathbb{F}$ is larger than the filtration $\mathbb{F}^X$ generated by the process $\boldsymbol{X}$. This difficulty can be resolved by working on the canonical space on which we can construct the marked point process, but this can prevent us from capturing natural models for important applications. See the Notes & Complements at the end of the chapter for references to textbooks and articles providing the necessary material needed to tackle these subtle issues.

Markovian Mean Field Games

Given the technical issues raised in the above discussion, we restrict ourselves to the case of deterministic flows $\boldsymbol{\mu} = (\mu_t)_{0t \leqslant T}$ in $\mathcal{P}(E)$ and Markovian control strategies $\boldsymbol{\alpha} = (\alpha_t)_{0 \leqslant t \leqslant T}$ given by feedback functions in the form $\alpha_t = \phi(t, X_t)$. As above, we denote by $\mathbb{A}$ the set of these strategies. Notice that the predictable version of $\boldsymbol{\alpha}$ we alluded to in the previous paragraph should be thought of as $(\phi(t-, X_{t-}))_{0 \le t \le T}$, which coincides $\mathrm{Leb}_1 \otimes \mathbb{P}$ almost-everywhere with $(\phi(t, X_t))_{0 \le t \le T}$, where as usual Leb_1 is the Lebesgue measure on $[0, T]$.

Naturally, the mean field game problem associated with the finite player game formulated above can be stated as the following set of two successive steps:

1. For each fixed deterministic flow $\boldsymbol{\mu} = (\mu_t)_{0 \leqslant t \leqslant T}$ of probability measures on E, solve the optimal control problem:

$$\inf_{\boldsymbol{\alpha} \in \mathbb{A}} \mathbb{E}\left[\int_0^T f(t, X_t, \mu_t, \alpha_t)dt + g(X_T, \mu_T)\right]$$

where for each $\boldsymbol{\alpha} \in \mathbb{A}$, the process $\boldsymbol{X} = (X_t)_{0 \leqslant t \leqslant T}$ is a nonhomogeneous E-valued Markov chain with transition probabilities determined by the Q-matrix of rates q_t given by the formula:

$$q_t(x, x') = \lambda_t(x, x', \mu_t, \phi(t, x)), \qquad t \in [0, T], \ x, x' \in E.$$

2. Find $\boldsymbol{\mu} = (\mu_t)_{0 \leqslant t \leqslant T}$ such that the flow of marginal distributions of a solution $\boldsymbol{X} = (X_t)_{0 \leqslant t \leqslant T}$ to the above optimal control problem coincides with the flow we started from in the sense that $\mathcal{L}(X_t) = \mu_t$ for all $t \in [0, T]$.

Before we even start formulating assumptions and specifying regularity properties, we reiterate the fact that, since the space E is finite, the space $\mathcal{P}(E)$ can be identified to the simplex $\mathcal{S}_d$, and because of the equivalence of the norms proved in Lemma 5.65, the regularity properties such as continuity or Lipschitz regularity can be equivalently formulated in terms of the ambient Euclidean norm of $\mathcal{S}_d$ instead of the Wasserstein distance used in the previous chapters. For this reason, we shall not specify which metric structure we use on $\mathcal{P}(E)$. Differentiability issues will not come up until we discuss the master equation later in this section.

Assumptions on the Cost Functions

We first identify the set of cost functions we shall use in this section.

Assumption (Discrete MFG Cost Functions).

(A1) The function f is jointly continuous in (t, μ, α). For each fixed $(t, x, \mu) \in [0, T] \times E \times \mathcal{P}(E)$, it is differentiable with respect to the control parameter α, with a derivative $\partial_\alpha f$ which is jointly continuous in (t, μ, α).

(A2) The set A is closed and convex and the function f is uniformly γ-convex in $\alpha \in A$ for some strictly positive constant $\gamma > 0$, uniformly in $(t, x, \mu) \in [0, T] \times E \times \mathcal{P}(E)$, in the sense that:

$$f(t, x, \mu, \alpha') - f(t, x, \mu, \alpha) - \partial_\alpha f(t, x, \mu, \alpha) \cdot (\alpha' - \alpha) \geq \gamma |\alpha - \alpha'|^2, \tag{7.24}$$

for all $(t, x, \mu) \in [0, T] \times E \times \mathcal{P}(E)$ and α and α' in A.

(A3) The terminal cost function $g : E \times \mathcal{P}(E) \ni (x, \mu) \mapsto g(x, \mu) \in \mathbb{R}$ is continuous in μ for each $x \in E$.

Notice that the assumptions made above are automatically uniform with respect to the state variable x since the state space E is finite.

The Hamiltonian

For each $x \in E$ we define the difference operator Δ_x acting on functions on E by the formula:

$$[\Delta_x h](x') = h(x') - h(x), \qquad x, x' \in E, \tag{7.25}$$

whenever h is a real valued function on E. If $E = \{e_1, \cdots, e_d\}$, such a function h can be identified to the element $y \in \mathbb{R}^d$ defined by $y_i = h(e_i)$ for $i = 1, \cdots, d$, and in this way, if $x = e_i$, the difference operator Δ_x can be viewed as well as the transformation Δ_i of $\mathbb{R}^d$ given by:

$$\Delta_i y = (y_1 - y_i, \cdots, y_d - y_i), \qquad y = (y_1, \cdots, y_d) \in \mathbb{R}^d, \ i = 1, \cdots, d. \tag{7.26}$$

The introduction of this definition is justified by the fact that the action of the infinitesimal generators $(L_t)_{t\geq 0}$ of a nonhomogeneous continuous time Markov chain in E with Q-matrix (or transition rates) $(q_t)_{t\geq 0}$ is given by, for each function h on E:

$$\begin{aligned} [L_t h](x) = q_t(x,\cdot)\cdot h = \sum_{x'\in E} q_t(x,x')h(x') &= \sum_{x'\in E} q_t(x,x')[h(x') - h(x)] \\ &= q_t(x,\cdot)\cdot \Delta_x h. \end{aligned} \tag{7.27}$$

We shall use the notation Δ_x for the difference operator when convenient.

In any case, the Kolmogorov or Fokker-Planck equation for a flow of probability measures $\boldsymbol{\nu} = (\nu_t)_{t\geq 0}$ governed by $(q_t)_{t\geq 0}$ is given by $\partial_t \nu_t = L_t^\dagger \nu_t$ where we use the notation $L_t^\dagger$ to denote the adjoint/transpose of the operator/matrix L_t. In developed form, the Kolmogorov equation reads:

$$\partial_t \nu_t(\{x\}) = \sum_{x'\in E} q_t(x',x)\nu_t(\{x'\}), \quad x\in E,\ t\geqslant 0. \tag{7.28}$$

Notice in particular that, for any $x\in E$, $(\nu_t(\{x\}))_{t\geq 0}$ is time continuous.

Coming back to our control problem, and using $h\in\mathbb{R}^E$ for the adjoint variable, the Hamiltonian is defined as the real valued function H on $[0,T]\times E\times\mathcal{P}(E)\times\mathbb{R}^E\times A$ given by:

$$\begin{aligned} H(t,x,\mu,h,\alpha) &= [L_t^{\mu,\alpha} h](x) + f(t,x,\mu,\alpha) \\ &= \sum_{x'\in E} \lambda_t(x,x',\mu,\alpha)h(x') + f(t,x,\mu,\alpha). \end{aligned} \tag{7.29}$$

We use the obvious notation $L_t^{\mu,\alpha}$ for the infinitesimal generator L_t defined above for the Q-matrix $q_t(x,x') = \lambda_t(x,x',\mu,\alpha)$.

Using **(A2)** in **Discrete MFG Rates** and **(A2)** in **Discrete MFG Cost Functions** and following the steps of the proof of Lemma 3.3 in Chapter 3, we can prove the existence of a unique minimizer $\hat{\alpha}$ defined by:

$$\hat{\alpha}(t,x,\mu,h) = \text{argmin}_{\alpha\in A} H(t,x,\mu,h,\alpha). \tag{7.30}$$

It is jointly continuous in (t,μ,h) and Lipschitz continuous and at most of linear growth in h.

In this framework, the minimized Hamiltonian is the real valued function H^* defined on $[0,T]\times E\times\mathcal{P}(E)\times\mathbb{R}^E$ by:

$$H^*(t,x,\mu,h) = \inf_{\alpha\in A} H(t,x,\mu,h,\alpha) = H\big(t,x,\mu,h,\hat{\alpha}(t,x,\mu,h)\big). \tag{7.31}$$

It is jointly continuous and locally Lipschitz continuous in h.

For the sake of convenience, we implicitly assume below that **(A1)** and **(A2)** in assumption **Discrete MFG Cost Functions** and assumption **Discrete MFG Rates** are satisfied. However, in some specific cases, we shall still use the notations $\hat{\alpha}$ and H^* even though λ_t is not linear in α.

The HJB Equation and a Verification Theorem

The flow $\boldsymbol{\mu} = (\mu_t)_{0 \leqslant t \leqslant T}$ being fixed, we define the value function of the optimization problem of the generic player by:

$$u^{\boldsymbol{\mu}}(t,x) = \inf_{\alpha \in \mathbb{A}_t} \mathbb{E}\left[\int_t^T f(s, X_s, \mu_s, \alpha_s) ds + g(X_T, \mu_T) \big| \, X_t = x \right], \tag{7.32}$$

where the set of admissible controls $\mathbb{A}_t$ comprises only feedback functions on $[t, T] \times E$. Since we are in a fully Markovian setup, the dynamic programming principle holds and the HJB equation reads:

$$\partial_t u^{\boldsymbol{\mu}}(t,x) + H^*\big(t, x, \mu_t, u^{\boldsymbol{\mu}}(t,\cdot)\big) = 0, \qquad t \in [0,T], \ x \in E, \tag{7.33}$$

with terminal condition $u^{\boldsymbol{\mu}}(T,x) = g(x, \mu_T)$. Thanks to (7.27), (7.33) can be rewritten in a form very similar to that used before in the book, compare for instance with (3.12):

$$\partial_t u^{\boldsymbol{\mu}}(t,x) + H^*\big(t, x, \mu_t, \Delta_x u^{\boldsymbol{\mu}}(t,\cdot)\big) = 0, \qquad t \in [0,T], \ x \in E. \tag{7.34}$$

Since E is finite, observe that (7.33) appears as a vector ordinary differential equation. Under the above assumptions on the running cost function, the coefficients of this ordinary differential equation are locally Lipschitz continuous, which guarantees existence and uniqueness of a local solution.

As usual, this local solution can be extended in a unique manner to the entire $[0, T]$ whenever we can prove an *a priori* bound. In order to establish such an *a priori* bound, we proceed in two steps. The first one reads as an analog of the verification argument proven in Lemma 4.47:

Proposition 7.5 *If the function $u : [0,T] \times E \ni (t,x) \mapsto u(t,x) \in \mathbb{R}$ is a continuously differentiable solution of the HJB equation (7.33) with terminal condition $u(T,x) = g(x, \mu_T)$ for $x \in E$, then u is the value function of the optimal control problem when the flow $\boldsymbol{\mu}$ is given, and the feedback function:*

$$\hat{\phi}(t,x) = \hat{\alpha}\big(t, x, \mu_t, u(t,\cdot)\big) \tag{7.35}$$

gives the unique optimal Markovian control.

Proof. As usual, the verification argument can be proved by writing:

$$\mathbb{E}\left[u(t,X_t) + \int_0^t f(s,X_s,\mu_s,\alpha_s)ds\right],$$

as the integral of its derivative between t and T, when $\boldsymbol{X} = (X_t)_{0\leqslant t\leqslant T}$ is driven by a control $\boldsymbol{\alpha}$. This can be achieved by means of (7.27). We then get that the cost to $\boldsymbol{\alpha}$ is greater than $\mathbb{E}[u(0,X_0)]$, with equality if $\phi = \hat{\phi}$. □

Obviously, the verification argument still applies when u is just defined on $[S,T]\times E$, for some $S \in [0,T]$, instead of the whole $[0,T]\times E$. In particular, it applies to the local solution u to (7.33); in such a case, $[S,T]$ is the largest subinterval of $[0,T]$ on which the local solution is known to be defined. As a byproduct, the local solution coincides with the value function u^{μ} on $[S,T]\times E$ and it suffices to prove that $\sup_{t\in[0,T],x\in E}|u^{\mu}(t,x)|$ is finite to have an *a priori* bound for the local solution to (7.33). The latter follows from:

Lemma 7.6 *Under the standing assumptions, there exists a constant C such that, for any $S \in [0,T]$ for which the HJB equation* (7.33) *has a continuously differentiable solution u on $[S,T]\times E$ with terminal condition $u(T,x) = g(x,\mu_T)$, it holds:*

$$\sup_{t\in[S,T],x\in E}|u^{\mu}(t,x)| = \sup_{t\in[S,T],x\in E}|u(t,x)| \leqslant C.$$

Proof. We consider S as in the statement. Thanks to the preliminary discussion before the statement of Lemma 7.6, we know that u^{μ} and u coincides on $[S,T]$.

For $t \in [0,T]$ and $\alpha_0 \in A$, we can use $(\alpha_s = \alpha_0)_{t\leqslant s\leqslant T}$ in the right-hand side of (7.32), we easily get an upper bound for u^{μ} and thus for u on $[S,T]\times E$, the bound being independent of S.

In order to establish the lower bound, we make use of the convexity of f:

$$\begin{aligned}
&u^{\mu}(t,x)\\
&\geqslant \mathbb{E}\biggl[\int_t^T \Bigl(f(s,X_s,\mu_s,\alpha_0) + \partial_\alpha f(s,X_s,\mu_s,\alpha_0)\cdot\bigl(\hat{\alpha}(s,X_s,\mu_s,u^{\mu}(s,\cdot)) - \alpha_0\bigr)\Bigr)ds\\
&\qquad + g(X_T,\mu_T)\,\big|\,X_t = x\biggr]\\
&\geqslant -C\biggl(1 + \mathbb{E}\biggl[\int_t^T \bigl|\hat{\alpha}\bigl(s,X_s,\mu_s,u^{\mu}(s,\cdot)\bigr)\bigr|ds\,\big|\,X_t = x\biggr]\biggr),
\end{aligned}$$

where $C \geqslant 0$ depends upon α_0, and the three maxima $\max_{t\in[0,T],x'\in E}|f(t,x',\mu_t,\alpha_0)|$, $\max_{t\in[0,T],x'\in E}|\partial_\alpha f(t,x',\mu_t,\alpha_0)|$, and $\max_{x'\in E}|g(x',\mu_T)|$.

Allowing C to increase from line to line and using the fact $|\hat{\alpha}(s, X_s, \mu_s, u^{\mu}(s,\cdot))| \leqslant C(1 + |u^{\mu}(s,\cdot)|)$, see Lemma 3.3, we finally get:

$$u^{\mu}(t,x) \geqslant -C\left(1 + \int_t^T |u^{\mu}(s,\cdot)|ds\right).$$

Recalling that we already have an upper bound for u^{μ}, we deduce that:

$$|u^{\mu}(t,\cdot)| \leqslant C\left(1 + \int_t^T |u^{\mu}(s,\cdot)|ds\right).$$

We conclude by Gronwall's lemma. □

The Fixed-Point Step and Kolmogorov's Equation

The second step of the mean field game problem as articulated above in the introduction of the subsection corresponds to the search for a fixed point of the best response function in the classical case of Nash equilibria for finite player games. A flow $\boldsymbol{\mu} = (\mu_t)_{0\leqslant t\leqslant T}$ provides such a fixed point if $\mu_0 = \mathcal{L}(X_0)$, which is usually prescribed as the state initial condition, and if $(\mu_t)_{0\leqslant t\leqslant T}$ satisfies the forward Kolmogorov equation for the marginal law $(\mathcal{L}(X_t))_{0\leqslant t\leqslant T}$ of the optimally controlled path $(X_t)_{0\leqslant t\leqslant T}$ in environment $\boldsymbol{\mu}$. Indeed, because of the Markov property, the forward Kolmogorov equation characterizes the marginal law $(\mathcal{L}(X_t))_{0\leqslant t\leqslant T}$. In the present situation, if we recall the form (7.28) of the Kolmogorov equation, and if we use the optimal control identified in Proposition 7.5 above, this equation reads:

$$\begin{aligned}\partial_t \mu_t(\{x\}) &= \sum_{x'\in E} \mu_t(\{x'\})\hat{q}_t(x',x) \\ &= \sum_{x'\in E, x'\neq x} \mu_t(\{x'\})\hat{q}_t(x',x) - \sum_{x'\in E, x'\neq x} \mu_t(\{x\})\hat{q}_t(x,x'),\end{aligned} \tag{7.36}$$

where we use the notation $(\hat{q}_t)_{t\geqslant 0}$ for the Q-matrix of transition rates of the optimal continuous time Markov chain identified in the above verification theorem. In other words:

$$\hat{q}_t(x,x') = \hat{q}_t^{\boldsymbol{\mu}}(x,x') = \lambda_t\big(x, x', \mu_t, \hat{\alpha}(t, x, \mu_t, u^{\mu}(t,\cdot))\big). \tag{7.37}$$

As for the stochastic differential games studied in this book, mean field game Nash equilibria are then identified to the solutions of the system of Kolmogorov and Hamilton-Jacobi-Bellman equations (7.33) and (7.36).

Existence of a Mean Field Game Equilibrium

Let us denote by $\mathcal{C}_d = C([0,T];\mathcal{P}(E))$ the space of continuous flows $\boldsymbol{\mu} = (\mu_t)_{0\leqslant t\leqslant T}$ of probability measures on E. Clearly, $C([0,T];\mathcal{P}(E))$ can be identified with $C([0,T];\mathcal{S}_d)$, and can be naturally embedded into $C([0,T];\mathbb{R}^d)$. For each fixed $\boldsymbol{\mu} \in \mathcal{C}_d$, we denote by $u^{\boldsymbol{\mu}}$ the unique solution of the HJB equation (7.33). Using the optimal control identified in the verification result of Proposition 7.5, we construct the flow (7.37) of Q-matrices. Using this flow of transition rates, we solve the forward Kolmogorov equation (7.36) with initial condition μ_0. We denote the resulting flow of measure by $\Psi(\boldsymbol{\mu})$. It is plain to see that $\Psi(\boldsymbol{\mu}) \in \mathcal{C}_d$ and proving existence of a solution of the mean field game problem as stated above can now be done by proving that this map Ψ has a fixed point in $\mathcal{C}_d$. The continuity of Ψ is due to the fact that the minimizer $\hat{\alpha}$ is continuous in the variables μ and h. Notice that the transition rates $(\hat{q}_t^{\mu})_{0\leq t\leq T}$ given by the minimizer $\hat{\alpha}$ are uniformly bounded so that the solution of the corresponding Kolmogorov equation is time Lipschitz with a Lipschitz constant independent of $\boldsymbol{\mu}$. Using Arzela-Ascoli's theorem to identify relatively compact subsets of $\mathcal{C}_d$ we conclude using Schauder's fixed-point theorem.

Monotonicity and Uniqueness of the Solution

In order to prove uniqueness of the solution, we may follow the same strategy as for stochastic differential mean field games. Indeed, as in the case of assumption **Lasry-Lions Monotonicity** of Section 3.4, we may assume that the terminal cost function g is monotone in the sense that:

$$\int_E [g(x,\mu) - g(x,\mu')]d(\mu - \mu')(x)$$
$$= \sum_{x\in E}[g(x,\mu) - g(x,\mu')]\big(\mu(\{x\}) - \mu'(\{x\})\big) \geq 0,$$

for all $\mu, \mu' \in \mathcal{P}(E)$, and that the running cost function f has a decomposition of the form:

$$f(t,x,\mu,\alpha) = f_0(t,x,\mu) + f_1(t,x,\alpha)$$

with $f_0(t,\cdot,\cdot)$ monotone for each fixed $t \in [0,T]$. A crucial ingredient in the proof of Theorem 3.29 is the fact that the drift and the volatility are independent of the flow $\boldsymbol{\mu}$, so that, given an admissible control strategy $\boldsymbol{\alpha}$, the law of the control process is entirely determined, irrespective of the flow $\boldsymbol{\mu}$. This is still the case in the present situation if, on the top of the above conditions, we assume that:

$$\text{the } Q\text{-matrices } (\lambda_t)_{0\leq t\leq T} \text{ are independent of } \mu.$$

Under these conditions, we can repeat the proof of Theorem 3.29 mutatis mutandis in the present situation and prove uniqueness.

The Master Equation

The master equation is the subject of the second part of the second volume of the book. However, instances of this equation were already mentioned in Chapter 5 and Chapter 6. The self-contained nature of this section on mean field games with finite states makes it possible to discuss the master equation without the *heavy lifting* of the second part of the second volume.

We consider a real valued function $\mathcal{U}$ defined on $[0, T] \times E \times \mathcal{P}(E)$ which is assumed to be functionally differentiable in the variable $\mu \in \mathcal{P}(E)$ for every fixed $(t, x) \in [0, T] \times E$. Recall that $\mathcal{P}(E)$ can be identified with the simplex $\mathcal{S}_d$ via the correspondence $\mu \leftrightarrow \boldsymbol{p} = (p_1, \cdots, p_d)$ with $p_i = \mu(\{e_i\})$ for $i = 1, \cdots, d$, or in other words, $\mu = \sum_{i=1}^d p_i \delta_{e_i}$. Given this identification, Proposition 5.66 explains how one goes from the standard Euclidean partial derivatives $\partial \mathcal{U}/\partial p_i$ to the linear functional derivative $\delta \mathcal{U}/\delta \mu$ when $\mathcal{U}$ is defined on the whole $\mathcal{P}_2(\mathbb{R}^d)$. If that were to be the case, we would be able to use this identification to express the master equation in terms of this functional derivative.

In the present context of finite state mean field games, and in analogy with the master equation (5.117) for stochastic differential mean field games without common noise introduced in Chapter 5, the master equation here takes the form:

$$\begin{aligned} &\partial_t \mathcal{U}(t, x, \mu) + H^*\big(t, x, \mu, \mathcal{U}(t, \cdot, \mu)\big) \\ &\qquad + \sum_{x' \in E} h^*\big(t, \mu, \mathcal{U}(t, \cdot, \mu)\big)(x') \frac{\partial \mathcal{U}(t, x, \mu)}{\partial \mu(\{x'\})} = 0, \end{aligned} \tag{7.38}$$

for $(t, x, \mu) \in [0, T] \times E \times \mathcal{P}(E)$, with $\mathcal{U}(T, x, \mu) = g(x, \mu)$ as terminal condition. The $\mathbb{R}^E$-valued function h^* is defined on $[0, T] \times \mathcal{P}(E) \times \mathbb{R}^E \times E$ by:

$$\begin{aligned} h^*(t, \mu, u)(x') &= \int_E \lambda_t\big(x, x', \mu, \hat{\alpha}(t, x, \mu, u)\big) d\mu(x) \\ &= \sum_{x \in E} \lambda_t\big(x, x', \mu, \hat{\alpha}(t, x, \mu, u)\big) \mu(\{x\}). \end{aligned} \tag{7.39}$$

While it is convenient to identify $\mathcal{P}(E)$ with the simplex $\mathcal{S}_d$, we need to specify the kind of derivative we use when we take derivatives with respect to measure arguments. In (7.38), $\partial \mathcal{U}(t, x, \mu)/\partial \mu(\{x'\})$ is understood as the derivative of $\mathcal{U}$ with respect to the weight $\mu(\{x'\})$ whenever $\mathcal{U}(t, x, \cdot)$ itself is regarded as a differentiable function of the d-tuple $(\mu(\{x'\}))_{x' \in E}$ on an open neighborhood of $\mathcal{S}_d$. However, it is important to notice that the choice of the derivative is not an issue in the statement of the master equation (7.38). Indeed, we can think of $h^*(t, \mu, u)$ as a function of $x' \in E$. Since it is an integral of a Q-matrix with respect to its first variable, the values of $h^*(t, \mu, u)(x')$ sum up to 0. Consequently, we can add a constant to the partial derivatives of $\mathcal{U}$ in the right-hand side of (7.38) without changing its value. Namely, (7.38) may be rewritten as:

$$\begin{aligned} &\partial_t \mathcal{U}(t, x, \mu) + H^*\big(t, x, \mu, \mathcal{U}(t, \cdot, \mu)\big) \\ &\qquad + \sum_{x' \in E} h^*\big(t, \mu, \mathcal{U}(t, \cdot, \mu)\big)(x') \Big(\frac{\partial \mathcal{U}(t, x, \mu)}{\partial \mu(\{x'\})} - \frac{\partial \mathcal{U}(t, x, \mu)}{\partial \mu(\{x\})} \Big) = 0. \end{aligned}$$

It is worth noticing that the above summation is also equal to:

$$\sum_{x'\neq x} h^*\big(t,\mu,\mathcal{U}(t,\cdot,\mu)\big)(x')\Big(\frac{\partial\mathcal{U}(t,x,\mu)}{\partial\mu(\{x'\})}-\frac{\partial\mathcal{U}(t,x,\mu)}{\partial\mu(\{x\})}\Big). \tag{7.40}$$

Now, using Proposition 5.66 and Corollary 5.67, we can identify:

$$\frac{\partial\mathcal{U}(t,x,\mu)}{\partial\mu(\{x'\})}-\frac{\partial\mathcal{U}(t,x,\mu)}{\partial\mu(\{x\})},$$

for $x'\neq x$, with the partial derivative of $\mathcal{U}(t,x,\cdot)$ with respect to $\mu(\{x'\})$ whenever $\mathcal{U}(t,x,\cdot)$ is regarded as a smooth function of the $(d-1)$ tuple $(\mu(\{x'\}))_{x'\in E\setminus\{x\}}$, which we can see as an element of the $(d-1)$-dimensional domain:

$$\mathcal{S}_{d-1,\leqslant}=\Big\{(p_1,\cdots,p_{d-1})\in[0,1]^{d-1}:\sum_{i=1}^{d-1}p_i\leqslant 1\Big\}.$$

This implies that whether we use partial derivatives on the $(d-1)$-dimensional domain $\mathcal{S}_{d-1,\leqslant}$ or linear functional derivatives, the master equation remains the same. From a numerical point of view, this implies that, in order to compute the sum in (7.40), it suffices to compute the derivatives with respect to $(\mu(\{x'\}))_{x'\in E\setminus\{x\}}$, regarded as an element of the $(d-1)$-dimensional domain $\mathcal{S}_{d-1,\leqslant}$, and then compute the inner product with $(h^*(t,\mu,\mathcal{U}(t,\cdot,\mu))(x'))_{x'\in E}$ with the convention that the entry $h^*(t,\mu,\mathcal{U}(t,\cdot,\mu))(x)$ is multiplied by zero.

As in the general theory of stochastic differential mean field games, the interest in the master equation (7.38), when it can be solved, is to encapsulate both the Kolmogorov and HJB equations, (7.36) and (7.33) respectively, in one single equation, see for instance Remark 5.107. In the present situation, the general procedure used to check this claim goes as follows:

Proposition 7.7 *Let us assume that $\mathcal{U}$ is a real valued function defined on $[0,T]\times E\times\mathcal{P}(E)$ which solves equation (7.38) with terminal condition $\mathcal{U}(T,x,\mu)=g(x,\mu)$ for $(x,\mu)\in E\times\mathcal{P}(E)$. If $\boldsymbol{\mu}:[0,T]\ni t\mapsto\mu_t\in\mathcal{P}(E)$ is the solution of the ordinary differential equation:*

$$\partial_t\mu_t(\{x\})=h^*\big(t,\mu_t,\mathcal{U}(t,\cdot,\mu_t)\big)(x),\quad x\in E, \tag{7.41}$$

with a given initial condition μ_0, then the function $u:[0,T]\times E\ni(t,x)\mapsto u(t,x)=\mathcal{U}(t,x,\mu_t)\in\mathbb{R}$ solves the HJB equation (7.33) for the flow $\boldsymbol{\mu}$, and appears as the value function of the optimization problem in the environment $\boldsymbol{\mu}$. Also, (7.41) can be identified with the Kolmogorov equation (7.36). Consequently, $\boldsymbol{\mu}$ is an equilibrium.

The proof is immediate in the present situation. We can use the chain rule to compute $\partial_t u(t,x)$, and the fact that $\mathcal{U}$ satisfies the master equation (7.38), and $\boldsymbol{\mu}$

satisfies (7.41), imply the desired result. The claim that (7.41) identifies with the Kolmogorov equation (7.36) is justified by (7.37).

7.2.3 A First Form of the Cyber-Security Model

For the sake of illustration, we propose a first analysis of the Botnet defense model alluded to in Subsection 1.6.2 of Chapter 1. There, we introduced the model as an instance of a game with major and minor players. Here, we exogenize the behavior of the major player and we concentrate on the population of potential victims assuming that the hacker has already chosen its strategy, and that all the players know it. We shall consider the full model with an attacker and a field of targets in Subsection (Vol II)-7.1.9 of Chapter (Vol II)-7.

We first specify the state space and the transition rates for the dynamics of the states of the potential victims, as well as their cost functions. We refer to the bibliography cited in the Notes & Complements below for a complete account on this model. We assume that each vulnerable computer can be in one of the following $d = 4$ states:

- DI: defended infected;
- DS: defended and susceptible to infection;
- UI: unprotected infected;
- US: unprotected and susceptible to infection.

So $E = \{\mathrm{DI}, \mathrm{DS}, \mathrm{UI}, \mathrm{US}\}$. In this simplistic model, the rate λ_t is independent of t and each network computer owner can choose one of two actions, that is $A = \{0, 1\}$. Action 0 means that the computer owner is happy with its level of defense (Defended or Unprotected) and does not try to change its own state, while action 1 means that the computer owner is willing to update the level of protection of its computer and switch to the other state (Unprotected or Defended). In the latter case, updating occurs after an exponential time with parameter $\lambda > 0$, which accounts for the speed of the response of the defense system.

When infected, a computer may recover at a rate depending on its protection level: the recovery rate is denoted by $q^{\mathrm{D}}_{\mathrm{rec}}$ for a protected computer and by $q^{\mathrm{U}}_{\mathrm{rec}}$ for an unprotected one.

Conversely, a computer may become infected in two ways, either directly from the attacks of the hacker or indirectly from infected computers that spread out the infection. The rate of direct infection depends upon the intensity of the attacks, as fixed by the botnet herder. This intensity is denoted by v_{H} and the rate of direct infection of a protected computer is $v_{\mathrm{H}} q^{\mathrm{D}}_{\mathrm{inf}}$ while the rate of direct infection of an unprotected computer is $v_{\mathrm{H}} q^{\mathrm{U}}_{\mathrm{inf}}$. Also, the rates of infection spreading from infected to susceptible computers depend upon the distribution of states μ within the population of computers. The rate of infection of an unprotected susceptible computer by other unprotected infected computers is $\beta_{\mathrm{UU}} \mu\{\mathrm{UI}\}$, the rate of infection of a protected susceptible computer by other unprotected infected computers is

$\beta_{\text{UD}}\mu\{\text{UI}\}$, the rate of infection of an unprotected susceptible computer by other protected infected computers is $\beta_{\text{DU}}\mu\{\text{DI}\}$, and the rate of infection of a protected susceptible computer by other protected infected computers is $\beta_{\text{DD}}\mu\{\text{DI}\}$.

Finally, the rate of transition $\lambda_t(x, x', \mu, v_{\text{H}}, \alpha)$ for the state of a computer in the network is given by:

$$\lambda_t(\cdot,\cdot,\mu,v_{\text{H}},0) =$$

$$\begin{array}{c} \text{DI} \\ \text{DS} \\ \text{UI} \\ \text{US} \end{array}\begin{bmatrix} \cdots & q^{\text{D}}_{\text{rec}} & 0 & 0 \\ v_{\text{H}}q^{\text{D}}_{\text{inf}} + \beta_{\text{DD}}\mu(\{\text{DI}\}) + \beta_{\text{UD}}\mu(\{\text{UI}\}) & \cdots & 0 & 0 \\ 0 & 0 & \cdots & q^{\text{U}}_{\text{rec}} \\ 0 & 0 & v_{\text{H}}q^{\text{U}}_{\text{inf}} + \beta_{\text{UU}}\mu(\{\text{UI}\}) + \beta_{\text{DU}}\mu(\{\text{DI}\}) & \cdots \end{bmatrix}$$

with columns labelled DI, DS, UI, US, and:

$$\lambda_t(\cdot,\cdot,\mu,v_{\text{H}},1) =$$

$$\begin{array}{c} \text{DI} \\ \text{DS} \\ \text{UI} \\ \text{US} \end{array}\begin{bmatrix} \cdots & q^{\text{D}}_{\text{rec}} & \lambda & 0 \\ v_{\text{H}}q^{\text{D}}_{\text{inf}} + \beta_{\text{DD}}\mu(\{\text{DI}\}) + \beta_{\text{UD}}\mu(\{\text{UI}\}) & \cdots & 0 & \lambda \\ \lambda & 0 & \cdots & q^{\text{U}}_{\text{rec}} \\ 0 & \lambda & v_{\text{H}}q^{\text{U}}_{\text{inf}} + \beta_{\text{UU}}\mu(\{\text{UI}\}) + \beta_{\text{DU}}\mu(\{\text{DI}\}) & \cdots \end{bmatrix}$$

where all the instances of $\cdots$ should be replaced by the negative of the sum of the entries of the row in which $\cdots$ appears on the diagonal.

As explained earlier, we do not specify the dynamics nor the state of the attacker in this first form of the model. For the present purposes, it suffices to known the value of the intensity of the attacks, here given by v_{H}. Notice also that, in the current form of the model, the rate λ_t not only depends on the action of the typical computer owner, but also on the intensity of the attacks. Each computer owner pays a fee k_{D} per unit of time for the defense of its system, and k_{I} per unit of time for losses resulting from infection. So, if we denote by X_t the state of its computer at time t, and by $\boldsymbol{\alpha} = (\alpha_t)_{0 \leqslant t \leqslant T}$ its control, the expected cost to a typical computer owner is given by:

$$J(\boldsymbol{\alpha}) = \mathbb{E}\left[\int_0^T \big(k_{\text{D}}\mathbf{1}_{\text{D}} + k_{\text{I}}\mathbf{1}_{\text{I}}\big)(X_t)dt\right],$$

where we use the notation $\text{D} = \{\text{DI}, \text{DS}\}$ and $\text{I} = \{\text{DI}, \text{UI}\}$.

Notice that, like for the counter-example investigated in Subsection 7.2.5 below, the control of the minor player is of *bang-bang* type.

Numerical Implementation

For the purpose of illustration, we provide numerical results from a straightforward implementation of the solution of the mean field game of cyber security described above. We chose a time interval $[0, T]$ with $T = 10$ (see comments below for a discussion of this particular choice) which we covered by a regular mesh

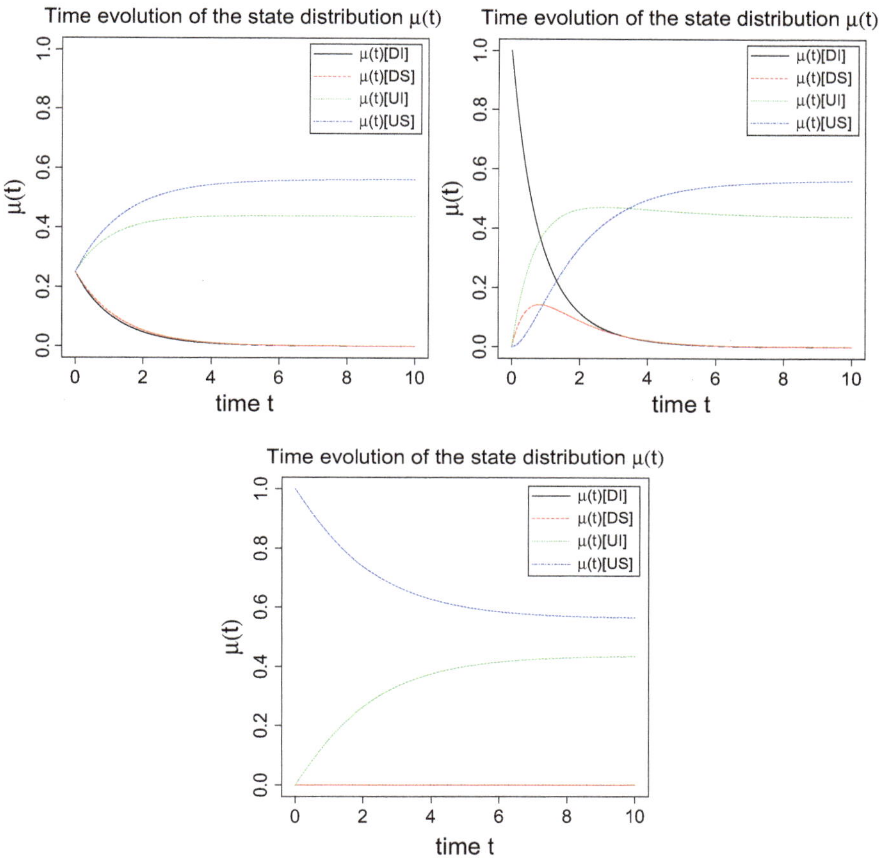

Fig. 7.1 Time evolution in equilibrium, of the distribution μ_t of the states of the computers in the network for different initial conditions μ_0: $\mu_0 = (0.25, 0.25, 0.25, 0.25)$ in the left plot on the top row; $\mu_0 = (1, 0, 0, 0)$ in the right plot on the top row; and $\mu_0 = (0, 0, 0, 1)$ in the bottom plot.

$\{t_i\}_{i=0,\cdots,N_s}$ with $N_s = 10^4$ and $t_i = i\Delta t$ for $\Delta t = 10^{-3}$. We implemented the solutions of the HJB equation (7.33) and the Kolmogorov equation (7.36)–(7.37) with straightforward explicit Euler schemes, and we iterated the solutions of these equations to find the fixed point. In the numerical experiments we conducted, the process converged in a very small number of iterations.

We used the following parameters to produce the plots of Figures 7.1 and 7.2: $\beta\{UU\} = 0.3$, $\beta\{UD\} = 0.4$, $\beta\{DU\} = 0.3$, and $\beta\{DD\} = 0.4$ for the rates of infection; $v_H = 0.6$ for the attack intensity of the hacker, and $\lambda = 0.8$ for the speed of response, $q^D_{\text{rec}} = 0.5$ and $q^U_{\text{rec}} = 0.4$, for the rates of recovery, and $q^D_{\text{inf}} = 0.4$ and $q^U_{\text{inf}} = 0.3$ for the rates of infection.

Finally, the constants appearing in the definition of the expected cost were chosen as $k_D = 0.3$ for the cost of being defended, and $k_I = 0.5$ for the cost of being infected.

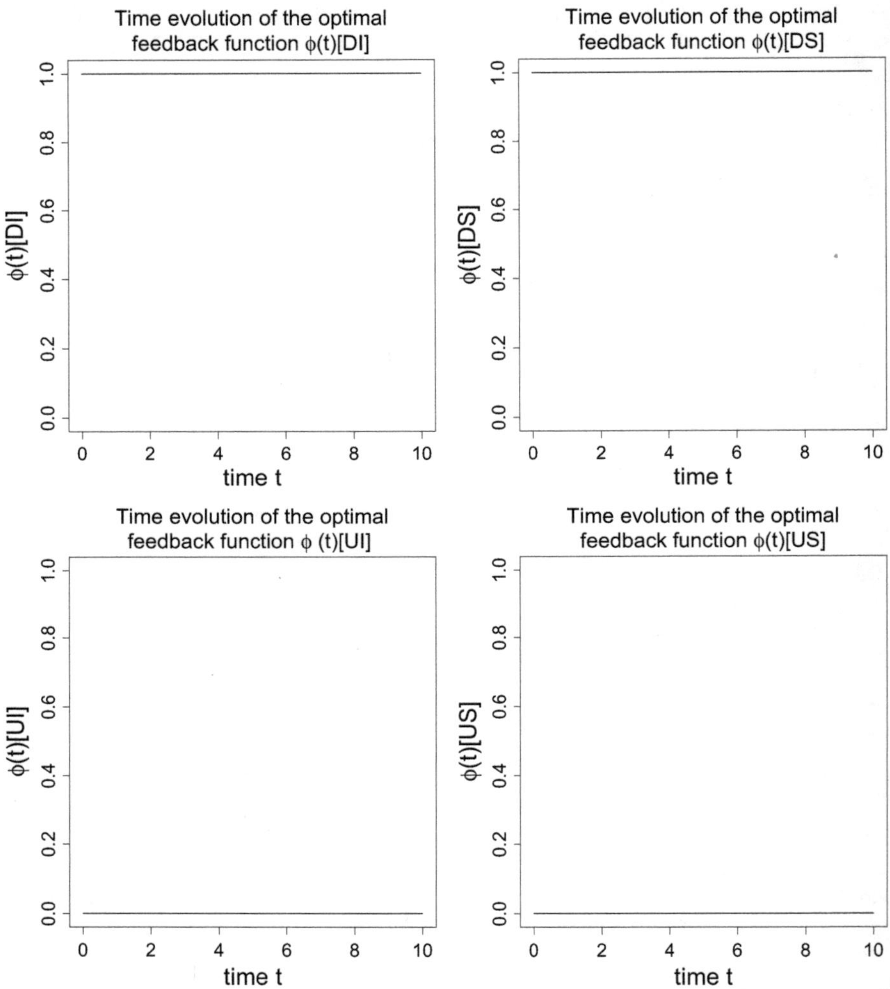

Fig. 7.2 Time evolution of the optimal feedback function $\phi(t, \cdot)$ in equilibrium. From left to right and from top to bottom, $\phi(t, \mathrm{DI})$, $\phi(t, \mathrm{DS})$, $\phi(t, \mathrm{UI})$, and $\phi(t, \mathrm{US})$.

These numerical experiments seem to indicate that, no matter which initial distribution μ_0 we start from, in equilibrium, the distribution of the states converges as time evolves toward a specific distribution in which the proportion of computers being defended is zero, the proportion of computers being unprotected and infected is approximately 0.44 and the proportion of computers unprotected and susceptible to be infected is 0.56. In order to check that $\hat{\mu} = (0, 0, 0.44, 0.56)$ is indeed an invariant measure one can use it as initial condition by choosing $\mu_0 = \hat{\mu}$ and check that the graphs of the four functions $t \mapsto \mu_t(i)$ for $i \in E = \{\mathrm{DI}, \mathrm{DS}, \mathrm{UI}, \mathrm{US}\}$ are horizontal. We do not reproduce the plots, but this is indeed the case. We chose the value of T in order to see that the proportions become constant. Larger values of T confirm this fact. We kept T to a reasonably small value to still see the patterns in the left-hand side of the plots. Varying the parameters of the model gives different values for the limiting levels of $\mu_t(\mathrm{UI})$ and $\mu_t(\mathrm{US})$ while $\mu_t(\mathrm{DI})$ and $\mu_t(\mathrm{DS})$ remain constant and equal to 0.

While the interpretation of this invariant distribution is not clear, its existence, and the strong evidence for the convergence suggest that a strong ergodicity exists in the model and that a search for stationary solutions in an analysis of an infinite horizon model in the spirit of Subsection 7.1.2 is reasonable. See the Notes & Complements at the end of the chapter for references to such an analysis.

For the sake of completeness, we computed and plotted the time evolution of the optimal feedback function $\phi(t,\cdot)$. Interestingly, irrespective of our choice of initial condition μ_0, or even of the parameters of the model, we found that the function $\phi(t,\cdot)$ is constant over time, and given by $\phi(t,\mathrm{DI}) = \phi(t,\mathrm{DS}) = 1$ and $\phi(t,\mathrm{UI}) = \phi(t,\mathrm{US}) = 0$ for all $t \in [0,T]$.

The strong ergodicity which made us believe in the convergence over time of the distribution μ_t and the optimal feedback control function $\phi(t,\cdot)$ toward unique well specified limits, does not always hold. There exist combinations of parameters for which several stationary limits are possible, or no stationary limit exists.

For the purpose of illustration, we used the following parameters: $\beta\{\mathrm{UU}\} = \beta\{\mathrm{UD}\} = 5$, $\beta\{\mathrm{DU}\} = \beta\{\mathrm{DD}\} = 2$, for the rates of infection; $q^{\mathrm{D}}_{\mathrm{rec}} = q^{\mathrm{U}}_{\mathrm{rec}} = 0.3$, for the rates of recovery, and $q^{\mathrm{D}}_{\mathrm{inf}} = 0.3$ and $q^{\mathrm{U}}_{\mathrm{inf}} = 0.4$ for the rates of infection. Finally, we kept the same attack intensity $v_{\mathrm{H}} = 0.6$, the same time horizon $T = 10$, $k_{\mathrm{I}} = 1$ for the cost of being infected, but we chose $k_{\mathrm{D}} = 0.5385$ for the cost of being defended and $\lambda = 1000$ for the speed of response. We explain these choices in the Notes & Complements at the end of the chapter.

With these parameters, the iterations of the successive solutions of the HJB equation and the Kolmogorov Fokker Planck equation do not behave the same way. Instead of a very fast convergence to what we took as the equilibrium measure flow, we see oscillations questioning the convergence toward a unique equilibrium accepted before as a fact. We do not plot the time evolution of the optimal feedback function $\phi(t,\cdot)$ because it appears to be the same as before. However, we illustrate the behavior of the measure flow through these iterations in Figure 7.3.

For the sake of completeness, we also solved numerically the master equation (7.38). For the purpose of comparison, we still work on the interval $[0,T]$ with $T = 10$, but we now use a coarser grid with $N_s = 10$ time steps only. We discretize the simplex S_3 with the grid:

$$\mathcal{G} = \{(\frac{k_1}{M}, \frac{k_2}{M}, \frac{k_3}{M});\ k_i \text{ integer }, 0 \leqslant k_i \leqslant M,\ \text{for } i = 1,2,3,\ \frac{k_1}{M} + \frac{k_2}{M} + \frac{k_3}{M} \leqslant 1\},$$

with $M = 25$ and we use backward derivatives. While we do not know how to provide an instructive plot of the values of the solution of the master equation (7.38), we checked the consistency of our results by using this solution to derive the equilibrium measure flow $\boldsymbol{\mu} = (\mu_t)$ by solving the ordinary differential equation (7.41) from Proposition 7.7. The results are reproduced in Figure 7.4. They were obtained with the parameters used to produce Figure 7.1.

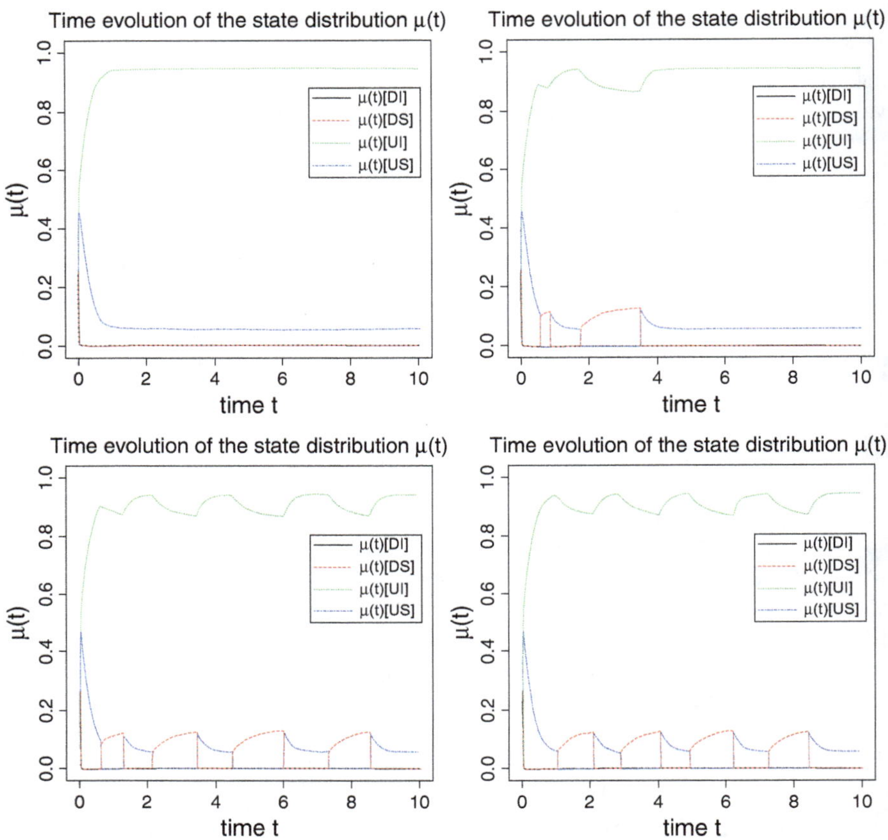

Fig. 7.3 From left to right and from top to bottom, time evolution of the distribution μ_t for the parameters given in the text, after 1, 5, 20, and 100 iterations of the successive solutions of the HJB equation and the Kolmogorov Fokker Planck equation.

7.2.4 Special Cases of MFGs with Finite State Spaces

We now concentrate on the finite state space analog of a model frequently considered in the book. To do so, recall the formulation of the L-monotonicity property above, and assume accordingly that the contributions to the running cost function of the control and the marginal distribution are split apart in the sense that:

$$f(t, x, \mu, \alpha) = f_0(t, x, \mu) + f_1(t, x, \alpha),$$

where for each $t \in [0, T]$ and $x \in E$, the function $\alpha \mapsto f_1(t, x, \alpha)$ is a strictly convex function for a convexity constant independent of t (and x since E is finite).

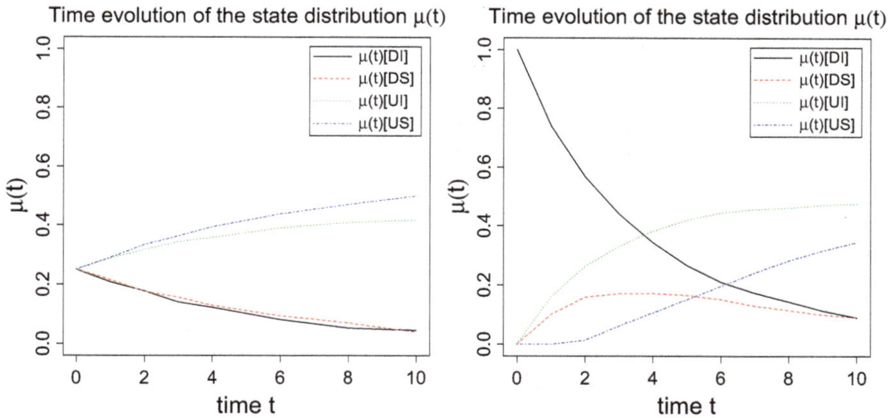

Fig. 7.4 Time evolution of the distribution μ_t as computed as the solution of the ordinary differential equation (7.41) from Proposition 7.7 using the numerical solution of the master equation. For the purpose of comparison, we used the initial conditions μ_0: $\mu_0 = (0.25, 0.25, 0.25, 0.25)$ on the left and $\mu_0 = (1, 0, 0, 0)$ on the right.

As above, we assume that A is a closed convex subset of the Euclidean space $\mathbb{R}^k$. The typical example we have in mind is the analog of the model used in Subsection 6.7.2 provided by:

$$f_1(t, x, \alpha) = \frac{1}{2}|\alpha|^2, \qquad t \in [0, T], \; x \in E, \; \alpha \in A. \tag{7.42}$$

For the sake of convenience, we recall the important definitions and equations. The Hamiltonian H is still given by the same formula:

$$H(t, x, \mu, h, \alpha) = [L_t^{\mu,\alpha} h](x) + f_0(t, x, \mu) + f_1(t, x, \alpha),$$

and the minimizer $\hat{\alpha}$ can be defined as:

$$\hat{\alpha}(t, x, \mu, h) = \text{argmin}_{\alpha \in A} H_1(t, x, \mu, h, \alpha), \tag{7.43}$$

which is uniquely defined when the rate $\lambda_t(x, x', \mu, \alpha)$ is linear in α, where the reduced Hamiltonian H_1 is defined by:

$$H_1(t, x, \mu, h, \alpha) = [L_t^{\mu,\alpha} h](x) + f_1(t, x, \alpha).$$

So, for a given flow $\boldsymbol{\mu} = (\mu_t)_{0 \leqslant t \leqslant T}$ of measures on E, the HJB equation (7.33) reads:

$$0 = \partial_t u(t, x) + H_1^*(t, x, \mu_t, u(t, \cdot)) + f_0(t, x, \mu_t),$$

where the minimized Hamiltonian H_1^* is defined as:

$$H_1^*(t, x, \mu, h) = \lambda_t\big(x, \cdot, \mu, \hat{\alpha}(t, x, \mu, h)\big) \cdot h + f_1\big(t, x, \hat{\alpha}(t, x, \mu, h)\big).$$

Now, the master equation (7.38) can be rewritten as:

$$\partial_t \mathcal{U}(t,x,\mu) + H_1^*\big(t,x,\mu_t,\mathcal{U}(t,\cdot,\mu)\big) + f_0(t,x,\mu) + \sum_{x',x''\in E} \lambda_t\big(x'',x',\mu,\hat{\alpha}\big(t,x'',\mu,\mathcal{U}(t,\cdot,\mu)\big)\big)\mu(\{x''\})\frac{\partial \mathcal{U}(t,x,\mu)}{\partial \mu(\{x'\})} = 0,$$

with the same terminal condition as before. Now, since the sum of the jump rates $(\lambda_t(x'',x',\mu,\hat{\alpha}(t,x'',\mu,\mathcal{U}(t,\cdot,\mu))))_{x'\in E}$ is null, this may be also written:

$$\begin{aligned}&\partial_t \mathcal{U}(t,x,\mu) + H_1^*\big(t,x,\mu_t,\mathcal{U}(t,\cdot,\mu)\big) + f_0(t,x,\mu)\\&\quad + \sum_{x',x''\in E} \lambda_t\big(x'',x',\mu,\hat{\alpha}\big(t,x'',\mu,\mathcal{U}(t,\cdot,\mu)\big)\big)\mu(\{x''\})\Big(\Big[\Delta_{x''}\frac{\partial \mathcal{U}(t,x,\mu)}{\partial \mu(\{\cdot\})}\Big](x')\Big)\\&= 0.\end{aligned}$$

A Form of Linear Quadratic Model

In order to get more explicit formulas, we consider the special case (7.42) to which we add an assumption on the dependence of the jump rates upon the control α. Although all the arguments extend to other cases, we shall assume, in order to simplify the notation, that $k = d-1$ and $A = [0,\infty)^{d-1}$. Assume also that for any $x \in E$, we are given a one-to-one mapping τ_x from $E \setminus \{x\}$ onto $\{1,\dots,d-1\}$ and, for any $\alpha \in A$, let:

$$\alpha(x,x') = \begin{cases} \alpha\big(\tau_x(x')\big) & \text{if } x' \neq x, \\ -\displaystyle\sum_{x''\in E, x''\neq x} \alpha\big(\tau_x(x'')\big) & \text{if } x' = x, \end{cases} \tag{7.44}$$

where $\alpha(i)$ is the ith coordinate of $\alpha = (\alpha(1),\cdots,\alpha(d-1)) \in [0,\infty)^{d-1}$, for $i \in \{1,\dots,d-1\}$.

In agreement with assumption **Discrete MFG Rates**, we then assume that the rate function has the following linear form:

$$\lambda_t(x,x',\mu,\alpha) = \alpha(x,x'), \qquad t\in[0,T],\ x,x'\in E,\ \mu\in\mathcal{P}(E),\ \alpha\in A.$$

while f_1 in (7.42) is equal to:

$$f_1(t,x,\alpha) = \frac{1}{2}\sum_{x'\in E, x'\neq x} |\alpha(x,x')|^2, \qquad t\in[0,T],\ x\in E,\ \alpha\in A.$$

Hence,

$$H_1(t,x,\mu,h,\alpha) = \sum_{x'\in E, x'\neq x} \alpha(x,x')[\Delta_x h](x') + \frac{1}{2}\sum_{x'\in E, x'\neq x} |\alpha(x,x')|^2,$$

and the minimizer $\hat{\alpha}$ is given by:

$$\hat{\alpha}(t,x,\mu,h) = \big([[\Delta_x h](x')]_-\big)_{x'\in\tau_x^{-1}(\{1,\cdots,d-1\})},$$

where as usual we use the notation $y_- = (-y)_+ = \max(-y,0)$ for $y \in \mathbb{R}$. The minimized Hamiltonian is now equal to:

$$\begin{aligned}
H_1^*(t,x,\mu,h) &= -\frac{1}{2}\sum_{x'\in E, x'\neq x} \big([[\Delta_x h](x')]_-\big)^2 \\
&= -\frac{1}{2}\sum_{x'\in E, x'\neq x} \big([h(x)-h(x')]_+\big)^2 \\
&= -\frac{1}{2}\sum_{x'\in E} \big([h(x)-h(x')]_+\big)^2.
\end{aligned}$$

As a result, for any given flow $\boldsymbol{\mu} = (\mu_t)_{0\leqslant t\leqslant T}$ of measures on E, the HJB equation (7.33) reads:

$$\partial_t u^{\mu}(t,x) - \frac{1}{2}\sum_{x'\in E}\big([u^{\mu}(t,x)-u^{\mu}(t,x')]_+\big)^2 + f_0(t,x,\mu_t) = 0,$$

and the master equation becomes:

$$\begin{aligned}
&\partial_t \mathcal{U}(t,x,\mu) - \frac{1}{2}\sum_{x'\in E}\big([\mathcal{U}(t,x,\mu)-\mathcal{U}(t,x',\mu)]_+\big)^2 + f_0(t,x,\mu_t) \\
&\quad + \sum_{x',x''\in E}\mu(\{x''\})\Big(\Big[\Delta_{x''}\frac{\partial \mathcal{U}(t,x,\mu)}{\partial \mu(\{\cdot\})}\Big](x')\Big)\Big[[\Delta_{x''}\mathcal{U}(t,\cdot,\mu)](x')\Big]_- = 0,
\end{aligned}$$

or, equivalently,

$$\begin{aligned}
&\partial_t \mathcal{U}(t,x,\mu) - \frac{1}{2}\sum_{x'\in E}\big([\mathcal{U}(t,x,\mu)-\mathcal{U}(t,x',\mu)]_+\big)^2 + f_0(t,x,\mu_t) \\
&- \sum_{x',x''\in E}\mu(\{x''\})\Big(\frac{\partial \mathcal{U}(t,x,\mu)}{\partial \mu(\{x''\})} - \frac{\partial \mathcal{U}(t,x,\mu)}{\partial \mu(\{x'\})}\Big)\big[\mathcal{U}(t,x'',\mu)-\mathcal{U}(t,x',\mu)\big]_+ = 0.
\end{aligned} \tag{7.45}$$

Potential Mean Field Games

As in the case of stochastic differential mean field games, we identify potential games through special characteristics of the running and terminal cost functions. Motivated by the analysis in Subsection 6.7.2, we assume the existence of real valued functions F and G defined on $[0,T]\times\mathcal{P}(E)$ and $\mathcal{P}(E)$ respectively, which admit linear functional derivatives giving the running and terminal cost functions of the game in the sense that:

$$f_0(t, x, \mu) = \frac{\partial F(t, \mu)}{\partial \mu(\{x\})}, \quad \text{and} \quad g(x, \mu) = \frac{\partial G(\mu)}{\partial \mu(\{x\})},$$

for $t \in [0, T]$, $x \in E$ and $\mu \in \mathcal{P}(E)$. In the present context, the central planner problem is the following optimal control problem of the McKean-Vlasov type:

$$\inf_{\alpha \in \mathbb{A}} \left[\int_0^T \left[\mathbb{E}\big(f_1(t, X_t, \alpha_t)\big) + F\big(t, \mathcal{L}(X_t)\big) \right] dt + G\big(\mathcal{L}(X_T)\big) \right],$$

with f_1 as in (7.42) and where the infimum is taken over the set of admissible Markov strategies $\alpha_t = \phi(t, X_t) = (\phi_1(t, X_t), \cdots, \phi_{d-1}(t, X_t))$ with values in $[0, \infty)^{d-1}$ and $X = X^\alpha = (X_t)_{0 \leqslant t \leqslant T}$ is the inhomogeneous continuous time E-valued Markov process with transition rates given by (7.44), namely by the Q-matrices:

$$q_t(x, x') = \begin{cases} \phi_{\tau_x(x')}(t, x) & \text{if } x' \neq x, \\ -\sum_{x'' \in E} \phi_{\tau_x(x'')}(t, x) & \text{if } x' = x, \end{cases}$$

which we denote by $\phi(t, x, x')$. Here, we do not invoke the machinery developed in Chapter 6 for the optimal control of stochastic McKean-Vlasov equations. Indeed, since we use only Markovian strategies given in feedback forms, we rewrite the problem as a deterministic control problem over $\mathcal{P}(E)$ involving the minimization of the functional:

$$\begin{aligned} J(\boldsymbol{\phi}) &= \int_0^T \left[\langle f_1(t, \cdot, \phi(t, \cdot)), \mu_t \rangle + F(t, \mu_t) \right] dt + G(\mu_T) \\ &= \int_0^T \left[\frac{1}{2} \sum_{x \in E} \sum_{x' \in E, x' \neq x} \mu_t(\{x\}) |\phi(t, x, x')|^2 + F(t, \mu_t) \right] dt + G(\mu_T), \end{aligned}$$

under the dynamic constraint:

$$d\mu_t(\{x\}) = \left[(L_t^{\phi})^\dagger \mu_t \right](x) = \sum_{x' \in E} \mu_t(\{x'\}) \phi(t, x', x), \qquad x \in E.$$

Motivated by the result of Subsection 6.7.2, we try to prove that the solution of this optimal control problem provides a solution to the mean field game problem considered in this section.

We solve the McKean-Vlasov control problem by writing its HJB equation and identifying the optimal control. Once this is done, we check that the optimal control provides the solution of the original mean field game problem defined by the running and terminal cost functions f and g. The Hamiltonian $\mathscr{H}$ of the McKean-Vlasov control problem is given by:

$$\begin{aligned}
&\mathscr{H}(t,\mu,h,\phi) \\
&= \big[(L_t^{\phi})^{\dagger}\mu\big]\cdot h + \langle f_1(t,\cdot,\phi(\cdot)),\mu\rangle + F(t,\mu) \\
&= \sum_{x\in E}\mu(\{x\})\Big[\sum_{x'\in E,x'\neq x}\Big(\phi(t,x,x')[\Delta_x h](x') + \frac{1}{2}|\phi(t,x,x')|^2\Big)\Big] + F(t,\mu),
\end{aligned}$$

and, since we can minimize term by term, the minimized Hamiltonian is given by:

$$\mathscr{H}^*(t,\mu,h) = \sum_{x\in E}\mu(\{x\})\Big[-\frac{1}{2}\sum_{x'\in E,x'\neq x}[h(x)-h(x')]_+^2\Big] + F(t,\mu). \tag{7.46}$$

Using the notation $v(t,\mu)$ for the value function of the deterministic control problem, the HJB equation is given by:

$$\partial_t v(t,\mu) - \frac{1}{2}\sum_{x\in E}\mu(\{x\})\sum_{x'\in E}\Big[\frac{\partial v(t,\mu)}{\partial\mu(\{x\})} - \frac{\partial v(t,\mu)}{\partial\mu(\{x'\})}\Big]_+^2 + F(t,\mu) = 0. \tag{7.47}$$

Again, using Proposition 5.66 and Corollary 5.67, we see that, instead of the standard derivative of functions defined on $\mathbb{R}^d$, we could as well use the linear functional derivatives of functions defined on $\mathcal{P}_2(\mathbb{R}^d)$ if we assume that the function $v(t,\cdot)$ has a smooth extension to $\mathcal{P}_2(\mathbb{R}^d)$.

We now prove that the solution of this deterministic control problem can provide a solution to the original mean field game problem.

Proposition 7.8 *Let us assume that the function* $[0,T]\times\mathcal{P}(E)\ni(t,\mu)\mapsto v(t,\mu)\in\mathbb{R}$ *is twice differentiable with respect to the weights* $(\mu(\{x\}))_{x\in E}$ *and solves equation* (7.47) *with terminal condition* $v(T,\mu)=G(\mu)$. *Then the function* $[0,T]\times E\times\mathcal{P}(E)\ni(t,x,\mu)\mapsto\mathcal{U}(t,x,\mu) = [\partial v/\partial\mu(\{x\})](t,\mu)\in\mathbb{R}$ *solves the master equation* (7.38) *with terminal condition* $\mathcal{U}(T,x,\mu)=g(x,\mu)$.

Here and below, we use the same rules of differentiation as in the writing of the master equation (7.38).

Proof. We give the proof in broad strokes only. The interested reader can easily fill in the technical details. Notice first that a standard verification argument can be used to show that v, as a solution of equation (7.47), is indeed the value function of the deterministic McKean-Vlasov control problem of interest here. Next, exchanging freely the order of partial derivatives, we differentiate both sides of (7.47) with respect to $\mu(\{x_0\})$ for some $x_0\in E$ and obtain:

$$\partial_t \frac{\partial v(t,\mu)}{\partial \mu(\{x_0\})} - \frac{1}{2}\sum_{x'\in E}\left[\frac{\partial v(t,\mu)}{\partial \mu(\{x_0\})} - \frac{\partial v(t,\mu)}{\partial \mu(\{x'\})}\right]_+^2 + \frac{\partial F(t,\mu)}{\partial \mu(\{x_0\})}$$

$$- \sum_{x\in E}\mu(\{x\})\sum_{x'\in E}\left[\left(\frac{\partial^2 v(t,\mu)}{\partial \mu(\{x\})\partial \mu(\{x_0\})} - \frac{\partial^2 v(t,\mu)}{\partial \mu(\{x'\})\partial \mu(\{x_0\})}\right)\right.$$

$$\left. \times \left[\frac{\partial v(t,\mu)}{\partial \mu(\{x\})} - \frac{\partial v(t,\mu)}{\partial \mu(\{x'\})}\right]_+\right] = 0.$$

Setting:

$$\mathcal{U}(t,x,\mu) = \frac{\partial v(t,\mu)}{\partial \mu(\{x\})}, \qquad f_0(t,x,\mu) = \frac{\partial F(t,\mu)}{\partial \mu(\{x\})}, \qquad \text{and} \qquad x = x_0,$$

we recover (7.45). □

Mean Field Games on a Directed Graph

In many practical applications, specific forms for the state transitions are suggested by the very nature of the problem at hand. In particular, the transition from a given state can often be restricted to a predetermined subset of states identified as the set of states which can be reached from the current state. As we explain below, this restriction is typical of Markovian evolutions on directed graphs. Even though the notations and the statements are different, we argue that the spirit of the less specific model studied above can easily be recast in the framework of Markov processes on a directed graph.

To be more specific, let us assume that the state space of the system is the set of nodes of a directed graph. To be consistent with the previous discussion, we denote by $E = \{e_1, \cdots, e_d\}$ the set of nodes, and for each node $x \in E$, we define $V_+(x)$ as the subset of $E \setminus \{x\}$ of nodes x' for which there exists a directed edge from x to x'. The number of these nodes is denoted by d_x and is called the out-degree of x. The main difference with the assumptions used so far is that we now force the dynamics of the state to respect the directed graph structure by demanding that the transition rate Q-matrices satisfy:

$$q_t(x,x') = 0 \qquad \text{whenever} \quad x' \notin V_+(x). \tag{7.48}$$

Similarly, we denote by $V_-(x)$ the subset of $E \setminus \{x\}$ of nodes x' for which there exists a directed edge from x' to x. Notice that the case studied earlier corresponds to $V_+(x) = V_-(x) = E\setminus\{x\}$ for all $x \in E$. It is plain to port all the results proven above to the set-up of mean field games on directed graphs (including potential games and central planer optimization problems) using (7.48). For example, the Kolmogorov equation (7.36) rewrites:

$$\partial_t \mu_t(\{x\}) = \sum_{x'\in V_-(x)} \mu_t(\{x'\})q_t(x',x) - \sum_{x'\in V_+(x)} \mu_t(\{x\})q_t(x,x'). \tag{7.49}$$

We leave the details to the interested reader.

7.2.5 Counter-Example to the Convergence of Markovian Equilibria

The goal of this subsection is to provide an example of a mean field game with finitely many states for which the associated N-player game has a Nash equilibrium that does not converge to the solution of the mean field game.

Throughout the subsection, we consider the following example with $T = 1$, $E = A = \{0, 1\}$ and, for $t \in [0, 1]$, $x, x', \alpha \in E$ and $\mu \in \mathcal{P}(E)$,

$$\begin{cases} \lambda_t(x, x', \mu, \alpha) = \beta \mathbf{1}_{\{x'=\alpha\}}, & x' \neq x, \\ \lambda_t(x, x, \mu, \alpha) = -\beta \mathbf{1}_{\{x=1-\alpha\}}, \end{cases}$$

for some $\beta > 0$, which has a simple interpretation: the next state x' coincides with the action α. Also, we let:

$$f(t, x, \mu, \alpha) = \mathbf{1}_{\{x=0\}} \text{ and } g = 2\mu(\{1\}).$$

In that case, the dynamics of the representative player may be easily described by means of a Poisson process with intensity β. At any occurrence t of the Poisson process, the player jumps to the state α_{t-} if different from its current position and, otherwise, stays in its current state. Here, $(\alpha_t)_{0 \leqslant t \leqslant 1}$ denotes the control process chosen by the representative player.

Equilibria to the Mean Field Game

For a flow of distributions $\boldsymbol{\mu} = (\mu_t)_{0 \leqslant t \leqslant 1}$ in $\mathcal{P}(E)$ and for the state of a representative player $(X_t)_{0 \leqslant t \leqslant 1}$, the cost functional reads:

$$J(\boldsymbol{\alpha}) = \mathbb{E}\bigg[\int_0^1 \mathbf{1}_{\{X_t=0\}} dt + 2\mu_1(\{1\})\bigg] = \mathbb{E}\bigg[\int_0^1 \mathbf{1}_{\{X_t=0\}} dt\bigg] + 2\mu_1(\{1\}).$$

Obviously, for a given flow of probability distributions, the minimum of J is attained for strategies $\boldsymbol{\alpha}$ spending the least amount of time in state 0. So, the best strategy is to choose $\hat{\alpha}_t = 1$, for all $t \in [0, 1]$. In that case, the transition rates have the form:

$$\begin{pmatrix} \lambda_t(0, 0, \mu_t, \hat{\alpha}_t) & \lambda_t(0, 1, \mu_t, \hat{\alpha}_t) \\ \lambda_t(1, 0, \mu_t, \hat{\alpha}_t) & \lambda_t(1, 1, \mu_t, \hat{\alpha}_t) \end{pmatrix} = \begin{pmatrix} -\beta & \beta \\ 0 & 0 \end{pmatrix}.$$

Denoting by $(\hat{X}_t)_{0 \leqslant t \leqslant T}$ the corresponding path, we have:

$$\mathbb{P}[\hat{X}_t = 1] = \mathbb{P}[\hat{X}_0 = 0]\big(1 - \exp(-\beta t)\big) + \mathbb{P}[\hat{X}_0 = 1],$$

which shows that, for a given initial condition $\mu_0 \in \mathcal{P}(E)$, the flow $\boldsymbol{\mu} = (\mu_t)_{0 \leqslant t \leqslant 1}$ is solution to the mean field problem if and only if

$$\begin{aligned} \mu_t(\{1\}) &= \mu_0(\{0\})\big(1 - \exp(-\beta t)\big) + \mu_0(\{1\}) \\ &= 1 - \mu_0(\{0\}) \exp(-\beta t), \quad t \in [0, 1]. \end{aligned} \tag{7.50}$$

Then, the value function, under the equilibrium $\boldsymbol{\mu}$, reads:

$$u^{\mu}(t,0) = \frac{1}{\beta}\big(1 - \exp(-\beta(1-t))\big) + 2\mu_1(\{1\}),$$
$$u^{\mu}(t,1) = 2\mu_1(\{1\}).$$

HJB Equation. The value function may be retrieved by writing down the HJB equation. Here the Hamiltonian is independent of μ and reads:

$$H(t,0,h,\alpha) = \beta\big(h(1) - h(0)\big)\mathbf{1}_{\{\alpha=1\}} + 1,$$
$$H(t,1,h,\alpha) = \beta\big(h(0) - h(1)\big)\mathbf{1}_{\{\alpha=0\}}.$$

Therefore,

$$\hat{\alpha}(t,0,h) = \begin{cases} 1 & \text{if} \quad h(1) - h(0) < 0, \\ 0 & \text{if} \quad h(1) - h(0) > 0, \end{cases} \tag{7.51}$$

and

$$\hat{\alpha}(t,1,h) = \begin{cases} 0 & \text{if} \quad h(0) - h(1) < 0, \\ 1 & \text{if} \quad h(0) - h(1) > 0, \end{cases} \tag{7.52}$$

so that, using the same notation as in (7.31),

$$\begin{aligned} H^*(t,0,h) &= -\beta\big(h(0) - h(1)\big)_+ + 1, \\ H^*(t,1,h) &= -\beta\big(h(1) - h(0)\big)_+. \end{aligned} \tag{7.53}$$

Since $u^{\mu}(t,0) \geqslant u^{\mu}(t,1)$, it is easily checked that:

$$\begin{aligned} &\frac{d}{dt}u^{\mu}(t,0) + H^*\big(t,0,u^{\mu}(t,\cdot)\big) \\ &\qquad = \frac{d}{dt}u^{\mu}(t,0) - \beta\big(u^{\mu}(t,0) - u^{\mu}(t,1)\big) + 1 \\ &\qquad = -\exp(-\beta(1-t)) - \big(1 - \exp(-\beta(1-t))\big) + 1 \\ &\qquad = 0, \end{aligned}$$

and

$$\frac{d}{dt}u^{\mu}(t,1) + H^*\big(t,1,u^{\mu}(t,\cdot)\big) = \frac{d}{dt}u^{\mu}(t,1) = 0.$$

Observing in particular that $u^{\mu}(t,0) > u^{\mu}(t,1)$ for all $t \in [0,T)$, we recover from (7.51) and (7.52) the fact that, whatever the initial point and the initial distribution, the optimal strategy is given by the constant control strategy 1.

Master Equation. By adapting (7.50) to games initialized at arbitrary times $t \in [0, 1]$, we compute the master field:

$$\begin{aligned}\mathcal{U}(t,0,\mu) &= \frac{1}{\beta}\big(1-\exp(-\beta(1-t))\big)+2-2\mu(\{0\})\exp(-\beta(1-t)),\\ \mathcal{U}(t,1,\mu) &= 2-2\mu(\{0\})\exp(-\beta(1-t)),\end{aligned} \tag{7.54}$$

for $(t,\mu) \in [0,1]\times\mathcal{P}(E)$, so that:

$$\begin{aligned}\frac{\partial\mathcal{U}(t,0,\mu)}{\partial\mu(\{0\})}-\frac{\partial\mathcal{U}(t,0,\mu)}{\partial\mu(\{1\})} &= -2\exp(-\beta(1-t)),\\ \frac{\partial\mathcal{U}(t,1,\mu)}{\partial\mu(\{0\})}-\frac{\partial\mathcal{U}(t,1,\mu)}{\partial\mu(\{1\})} &= -2\exp(-\beta(1-t)),\end{aligned}$$

while, with the same notation as in (7.38),

$$\begin{aligned}h^*(\mu)(0) &= -\beta\mu(\{0\}),\\ h^*(\mu)(1) &= \beta\mu(\{0\}).\end{aligned}$$

Therefore,

$$\begin{aligned}&h^*(\mu)(0)\frac{\partial\mathcal{U}(t,0,\mu)}{\partial\mu(\{0\})}+h^*(\mu)(1)\frac{\partial\mathcal{U}(t,0,\mu)}{\partial\mu(\{1\})}\\ &= h^*(\mu)(0)\Big(\frac{\partial\mathcal{U}(t,0,\mu)}{\partial\mu(\{0\})}-\frac{\partial\mathcal{U}(t,0,\mu)}{\partial\mu(\{1\})}\Big)=2\beta\mu(\{0\})\exp(-\beta(1-t)),\end{aligned}$$

and, similarly,

$$\begin{aligned}&h^*(\mu)(0)\frac{\partial\mathcal{U}(t,1,\mu)}{\partial\mu(\{0\})}+h^*(\mu)(1)\frac{\partial\mathcal{U}(t,1,\mu)}{\partial\mu(\{1\})}\\ &= h^*(\mu)(0)\Big(\frac{\partial\mathcal{U}(t,1,\mu)}{\partial\mu(\{0\})}-\frac{\partial\mathcal{U}(t,1,\mu)}{\partial\mu(\{1\})}\Big)=2\beta\mu(\{0\})\exp(-\beta(1-t)),\end{aligned}$$

and then, with the same computation as above, we have:

$$\begin{aligned}&\frac{d}{dt}\mathcal{U}(t,0,\mu)+H^*\big(t,0,\mathcal{U}(t,\cdot,\mu)\big)\\ &\quad+h^*(\mu)(0)\frac{\partial\mathcal{U}(t,0,\mu)}{\partial\mu(\{0\})}+h^*(\mu)(1)\frac{\partial\mathcal{U}(t,0,\mu)}{\partial\mu(\{1\})}\\ &= -\exp(-\beta(1-t))-2\beta\mu(\{0\})\exp(-\beta(1-t))\\ &\quad-\big(1-\exp(-\beta(1-t))\big)+1+2\beta\mu(\{0\})\exp(-\beta(1-t))\\ &= 0,\end{aligned}$$

and, similarly,

$$\begin{aligned}
&\frac{d}{dt}\mathcal{U}(t,1,\mu) + H^*\big(t,1,\mathcal{U}(t,\cdot,\mu)\big) \\
&\qquad + h^*(\mu)(0)\frac{\partial \mathcal{U}(t,1,\mu)}{\partial \mu(\{0\})} + h^*(\mu)(1)\frac{\partial \mathcal{U}(t,1,\mu)}{\partial \mu(\{1\})} \\
&= -2\beta\mu(\{0\})\exp(-\beta(1-t)) + 2\beta\mu(\{0\})\exp(-\beta(1-t)) \\
&= 0,
\end{aligned}$$

which shows that the master equation is indeed satisfied.

Equilibria for the N-Player Game

We now turn to the analysis of the N-player game and search for Nash equilibria over Markovian strategies.

The dynamics of the N players are given by (7.23). They can also be described in a simple way: assuming that players choose Markov feedback functions $(\phi^{N,i})_{i=1,\cdots,N}$, each of them may jump at the occurrences of a Poisson process with intensity β, the Poisson processes being denoted by $((T^i_n)_{n\geqslant 1})_{i=1,\cdots,N}$ and being assumed to be independent. At each $t = T^i_n \in [0,1]$, for some $n \geqslant 1$, the i-th player switches to $\phi^{N,i}(t-, X^{(N)}_{t-})$ if different from $X^{N,i}_{t-}$ and stays in $X^{N,i}_{t-}$ if not, where $(X^{(N)}_t = (X^{N,1}_t, \cdots, X^{N,N}_t))_{0\leqslant t\leqslant 1}$ is the state of the system.

The cost functional to player i reads:

$$J^{N,i}(\phi^{N,1}, \cdots, \phi^{N,N}) = \mathbb{E}\bigg[\int_0^1 \mathbf{1}_{\{X^{N,i}_t = 0\}}dt + 2\bar{X}^{N,-i}_1\bigg],$$

where $\bar{X}^{N,-i}_t = \frac{1}{N-1}\sum_{j=1,j\neq i}^N X^{N,j}_t$.

Notice first that the constant strategies $\phi^{N,i} \equiv 1$ for $i = 1,\cdots,N$ form a Markovian Nash equilibrium. Indeed, if all the players $j \neq i$ use such a constant strategy, then $(\mathbb{E}[\bar{X}^{N,-i}_t])_{0\leqslant t\leqslant 1}$ is independent of $\phi^{N,i}$, so that the expected cost to player i is, up to an additive constant which is independent of $\phi^{N,i}$, its expected time spent in state 0 which is minimized by using the strategy $\phi^{N,i} \equiv 1$, proving that we indeed have identified a Nash equilibrium.

However, this example is especially interesting because one can identify in some cases, another explicit Nash equilibrium. Consider the strategy profiles:

$$\phi^{*N,i}(x^1, \cdots, x^N) = \begin{cases} 1 & \text{if } \bar{\boldsymbol{x}}^{N,-i} > 0, \\ 0 & \text{if } \bar{\boldsymbol{x}}^{N,-i} = 0, \end{cases}$$

for $\boldsymbol{x} = (x^1, \ldots, x^N) \in \{0,1\}^N$, where, as usual,

$$\bar{x}^{N,-i} = \frac{1}{N-1} \sum_{j=1, j \neq i}^{N} x^j,$$

which gives the proportion of players different from i in state 1 in the system described by the state $(x^1, \ldots, x^N) \in \{0, 1\}^N$.

We claim:

Lemma 7.9 *There exists $\beta_0 > 0$ such that, for $\beta \geqslant \beta_0$, there exists $N_0(\beta) \geqslant 2$ such that, for $N \geqslant N_0(\beta)$, the functions $(\phi^{*N,i})_{1 \leqslant i \leqslant N}$ form a Markovian Nash equilibrium.*

Proof: Throughout the proof, β is assumed to be greater than 1.

Our first goal is to compute the cost to player 1 when all the players including player 1 itself use the feedback functions $(\phi^{*N,i})_{i=1,\cdots,N}$. In order to do so, we shall compute the cost when the game is initialized at any time $t \in [0, 1)$. The controlled state of the system should then be denoted by $((X_s^{*N,i})_{i=1,\cdots,N})_{t \leqslant s \leqslant 1}$, but, for simplicity, we shall use the notation $((X_s^{N,i})_{i=1,\cdots,N})_{t \leqslant s \leqslant 1}$. Letting $\bar{X}_t^{N,-1} = \frac{1}{N-1} \sum_{i=2}^{N} X_t^{N,i}$, for $t \in [0, 1]$, the cost to player 1 reads:

$$J^{N,1}(\phi^{*N}) = \mathbb{E}\bigg[\int_t^1 \mathbf{1}_{\{X_s^{N,1}=0\}} ds + 2\bar{X}_1^{N,-1} \bigg].$$

We write $v^{N,1}(t, \boldsymbol{x})$ for $J^{N,1}(\phi^{*N})$ when the initial condition at time t of the system is a deterministic tuple $\boldsymbol{x} = (x^1, \cdots, x^N) \in \{0, 1\}^N$.

Throughout the proof, we denote by $(\varrho_n)_{n \geqslant 0}$ a Poisson process with intensity $N\beta$ and modeling the possible jumping times in the game, at least up until time 1. Precisely, players cannot jump at any time $s \notin \{\varrho_n, n \geqslant 1\}$. At any time ϱ_n, an index $I_n \in \{1, \cdots, N\}$ is selected and player I_n jumps if allowed by its own control strategy.

Obviously, the sequences $(I_n)_{n \geqslant 1}$ and $(\varrho_n)_{n \geqslant 0}$ are independent and the variables $(I_n)_{n \geqslant 1}$ are also independent. For any $n \geqslant 1$, I_n is uniformly distributed on $\{1, \cdots, N\}$.

Also, throughout the proof, we denote by $((Y_s^{N,i})_{1 \leqslant i \leqslant N})_{t \leqslant s \leqslant 1}$ a family of N independent Markov processes with values in $\{0, 1\}$, with the prescription that each $(Y_s^{N,i})_{t \leqslant s \leqslant 1}$ starts from $X_t^{N,i}$, cannot exit from state 1, and jumps from state 0 to state 1 at the first time ρ_n with $I_n = i$. In words, the jumping times of $\boldsymbol{Y}^{N,i}$, for $i = 1, \cdots, N$, are dictated by the same Poisson process as the jumping times of $\boldsymbol{X}^{N,i}$. For any $s \in [t, 1]$, we let $\bar{Y}_s^{N,-1} = \frac{1}{N-1} \sum_{j=2}^{N} Y_s^{N,j}$.

The strategy is to compare $v^{N,1}(t, (1, \boldsymbol{x}^{-1}))$ and $v^{N,1}(t, (0, \boldsymbol{x}^{-1}))$ for all the possible values of $\boldsymbol{x}^{-1}$, where $\boldsymbol{x}^{-1}$ is a shorter notation for $\boldsymbol{x}^{N,-1}$.

First Case. Assume that there are at least two players in $\{2, \cdots, N\}$ starting from state 1 at time t, that is $\bar{\boldsymbol{x}}^{-1} \geqslant \frac{2}{N}$. Then all the players implement strategy 1 up until the end of the game.

a. If player 1 starts from 1 at time t, that is $x^1 = 1$, then:

$$v^{N,1}(t, (1, \boldsymbol{x}^{-1})) = 2\mathbb{E}[\bar{X}_1^{N,-1}] = 2\mathbb{E}[\bar{Y}_1^{N,-1}].$$

b. If player 1 starts from 0 at time t, it will stay in 0 up until τ^1, where τ^1 is the first jumping time of player 1. Since all the players implement strategy 1, we get:

$$v^{N,1}\big(t,(0,\boldsymbol{x}^{-1})\big) = \mathbb{E}\bigg[\int_t^1 \mathbf{1}_{\{X_s^{N,1}=0\}} ds + 2\bar{X}_1^{N,-1}\bigg] \geqslant \mathbb{E}\Big[(1\wedge\tau^1 - t) + 2\bar{Y}_1^{N,-1}\Big],$$

which is strictly greater than $v^{N,1}(t,(1,\boldsymbol{x}^{-1}))$. As a result, we have:

$$\phi^{*N,1}(1,\boldsymbol{x}^{-1}) = 1 = \text{sign}\big(v^{N,1}\big(t,(0,\boldsymbol{x}^{-1})\big) - v^{N,1}\big(t,(1,\boldsymbol{x}^{-1})\big)\big),$$
$$\phi^{*N,1}(0,\boldsymbol{x}^{-1}) = 1 = \text{sign}\big(v^{N,1}\big(t,(0,\boldsymbol{x}^{-1})\big) - v^{N,1}\big(t,(1,\boldsymbol{x}^{-1})\big)\big),$$

where $\text{sign}(r)$ is 1 if $r > 0$, -1 if $r < 0$, and 0 if $r = 0$, and then:

$$\phi^{*N,1}(x^1,\boldsymbol{x}^{-1}) = 1 - x^1 \Leftrightarrow v^{N,1}\big(t,(1-x^1,\boldsymbol{x}^{-1})\big) - v^{N,1}\big(t,(x^1,\boldsymbol{x}^{-1})\big) < 0, \tag{7.55}$$

at least when $\boldsymbol{x}^{-1} \geqslant \frac{2}{N}$.

Second Case. Assume now that there is exactly one player, say I_0, different from 1 that starts from state 1 at time t, that is $\bar{\boldsymbol{x}}^{-1} = 1$.

a. If player 1 starts from 1, that is $x^1 = 1$, then all the players implement strategy 1 up until the end of the game. As a result, we have as before:

$$v^{N,1}\big(t,(1,\boldsymbol{x}^{-1})\big) = 2\mathbb{E}\big[\bar{X}_1^{N,-1}\big] = 2\mathbb{E}\big[\bar{Y}_1^{N,-1}\big],$$

which, in that case, is equal to:

$$v^{N,1}\big(t,(1,\boldsymbol{x}^{-1})\big) = \frac{2}{N-1} + 2\frac{N-2}{N-1}\big(1-\exp(-\beta(1-t))\big).$$

b. If player 1 starts from 0 at time t, then all the players except player I_0 implement strategy 1 up until the first jumping time ϱ_1, while player I_0 implements strategy 0. At time ϱ_1, the system restarts with two particles in state 1 provided that $\varrho_1 < 1$ and $I_1 \neq I_0$. So, on the event $\{\varrho_1 < 1,\ I_1 \neq I_0\}$, the process $(\bar{X}_s^{N,-1})_{t\leqslant s\leqslant T}$ coincides with $(\bar{Y}_s^{N,-1})_{t\leqslant s\leqslant T}$. Therefore,

$$v^{N,1}\big(t,(0,\boldsymbol{x}^{-1})\big) \geqslant 2\mathbb{E}\big[\frac{1}{N-1}\mathbf{1}_{\{\varrho_1>1\}} + \mathbf{1}_{\{\varrho_1<1,\ I_1\neq I_0\}}\bar{Y}_1^{N,-1}\big] + \mathbb{E}\big[1\wedge\tau_1 - t\big],$$

where we used that $\bar{X}_1^{N,-1} = \frac{1}{N-1}$ on the event $\{\varrho_1 > 1\}$. Importantly, since $\mathbb{P}[\varrho_1 = 1] = 0$ and since $Y_1^j = 0$ for all $j \neq I_0$ when $\varrho_1 > 1$, we have with probability 1:

$$\begin{aligned}&\frac{1}{N-1}\mathbf{1}_{\{\varrho_1>1\}} + \mathbf{1}_{\{\varrho_1<1,\ I_1\neq I_0\}}\bar{Y}_1^{N,-1}\\ &= \frac{1}{N-1}\big(1-\mathbf{1}_{\{\varrho_1<1, I_1=I_0\}}\big) + \big(1-\mathbf{1}_{\{\varrho_1<1,\ I_1=I_0\}}\big)\frac{1}{N-1}\sum_{j=2,j\neq I_0}^{N-1} Y_1^j.\end{aligned} \tag{7.56}$$

Then, we have the following bounds:

$$\frac{1}{N-1}\mathbb{P}\big[\varrho_1 < 1, I_1 = I_0\big] = \frac{1}{N(N-1)}\big(1-\exp(-N\beta(1-t))\big) = (1-t)O\Big(\frac{\beta}{N}\Big),$$

and

$$\begin{aligned}\mathbb{E}\bigg[\mathbf{1}_{\{\varrho_1<1,\, I_1=I_0\}}\frac{1}{N-1}\sum_{j=2,j\neq I_0}^{N-1} Y_1^j\bigg] &\leqslant \frac{1}{N}\mathbb{E}\bigg[\frac{1}{N-1}\sum_{j=2,j\neq I_0}^{N-1} Y_1^j \,|\, \varrho_1 < 1,\ I_1 = I_0\bigg]\\ &\leqslant \frac{1}{N-1}\big(1-\exp(-\beta(1-t))\big) = (1-t)O\Big(\frac{\beta}{N}\Big),\end{aligned}$$

where we used the Markov property for the Poisson process in the second line.

Inserting the two previous bounds in (7.56) and comparing with $v^{N,1}(t,(1,\boldsymbol{x}^{-1}))$ obtained in part a, we deduce that:

$$2\mathbb{E}\Big[\frac{1}{N-1}\mathbf{1}_{\{\varrho_1>1\}} + \mathbf{1}_{\{\varrho_1<1,\, I_1\neq I_0\}}\bar{Y}_1^{N,-1}\Big] \geqslant v^{N,1}\big(t,(1,\boldsymbol{x}^{-1})\big) - (1-t)O\Big(\frac{\beta}{N}\Big).$$

Thus, returning to the lower bound for $v^{N,1}(t,(0,\boldsymbol{x}^{-1}))$, we get:

$$\begin{aligned}v^{N,1}\big(t,(0,\boldsymbol{x}^{-1})\big) &\geqslant v^{N,1}\big(t,(1,\boldsymbol{x}^{-1})\big) - (1-t)O\Big(\frac{\beta}{N}\Big) + \int_0^\infty \beta\big((1-t)\wedge r\big)\exp\big(-\beta r\big)dr\\ &\geqslant v^{N,1}\big(t,(1,\boldsymbol{x}^{-1})\big) - (1-t)O\Big(\frac{\beta}{N}\Big) + (1-t)\exp\big(-\beta\big),\end{aligned}$$

which shows that, for each fixed $\beta \geqslant 1$, we can choose N large enough so that $v^{N,1}(t,(0,\boldsymbol{x}^{-1})) > v^{N,1}(t,(1,\boldsymbol{x}^{-1}))$. Hence the conclusion is the same as in the first case, see (7.55).

Third Case. We now assume that $\bar{\boldsymbol{x}}^{-1} = 0$.

a. If player 1 starts from 0, then all the players implement strategy 0 and thus remain in 0. In particular, $v^{N,1}(t,(0,\boldsymbol{x}^{-1})) = 1-t$.

b. Now, if player 1 starts from 1, then all the players except player 1 implement strategy 1, at least up until ϱ_1. If $\varrho_1 < 1$ and $I_1 \geqslant 2$, then there are two players in state 1 at time ϱ_1; after ϱ_1, all of them implement strategy 1. If $\varrho_1 < 1$ and $I_1 = 1$, then player 1 switches to 0 at time ϱ_1 and then all the players remain in 0. So, the cost to player 1 is greater than:

$$\begin{aligned}2\mathbb{E}\Big[\mathbf{1}_{\{\varrho_1<1,\, I_1\geqslant 2\}}\bar{Y}_1^{N,-1}\Big] &= 2\mathbb{E}\Big[\mathbf{1}_{\{I_1\geqslant 2\}}\bar{Y}_1^{N,-1}\Big]\\ &= 2\mathbb{E}\Big[\bar{Y}_1^{N,-1}\Big] - 2\mathbb{E}\Big[\mathbf{1}_{\{\varrho_1<1,\, I_1=1\}}\bar{Y}_1^{N,-1}\Big],\end{aligned}$$

where, as in the second case, we used the fact that $\mathbf{1}_{\{\varrho_1<1\}}\bar{Y}_1^{N,-1} = \bar{Y}_1^{N,-1}$ with probability 1. Now, following the second case again,

$$2\mathbb{E}\Big[\mathbf{1}_{\{\varrho_1<1,\, I_1=1\}}\bar{Y}_1^{N,-1}\Big] \leqslant \frac{2}{N}\mathbb{E}\Big[\bar{Y}_1^{N,-1} \,|\, \varrho_1 < 1,\ I_1 = 1\Big] \leqslant \frac{2}{N}\big(1-\exp(-\beta(1-t))\big).$$

So, we end up with:

$$\begin{aligned} v^{N,1}(t,(1,\boldsymbol{x}^{-1})) &\geqslant 2\big(1-\frac{1}{N}\big)\big(1-\exp(-\beta(1-t))\big) \\ &= 2\big(1-\frac{1}{N}\big)\int_0^{\beta(1-t)} \exp(-r)dr. \end{aligned}$$

If $\beta(1-t) \geqslant \ln 3$, then:

$$v^{N,1}(t,(1,\boldsymbol{x}^{-1})) \geqslant 2\big(1-\frac{1}{N}\big)\int_0^{\ln 3} \exp(-r)dr = 2\frac{2}{3}\big(1-\frac{1}{N}\big),$$

which is greater than 1 and thus than $v^{N,1}(t,(0,\boldsymbol{x}^{-1}))$ for N large enough. If $\beta(1-t) < \ln 3$, then:

$$v^{N,1}(t,(1,\boldsymbol{x}^{-1})) \geqslant 2\big(1-\frac{1}{N}\big)\int_0^{(1-t)} \exp(-r)dr \geqslant 2\big(1-\frac{1}{N}\big)\exp\big(-\frac{\ln 3}{\beta}\big)(1-t).$$

So, choosing β and N large enough, it is also greater than $v^{N,1}(t,(0,\boldsymbol{x}^{-1}))$. In any case, the conclusion (7.55) of the first step remains true for well-chosen values of β and N.

Conclusion. For β and N as in the statement, we get that (7.55) is always satisfied, whatever the value of $\boldsymbol{x}$. Also, recall from its definition that $v^{N,1}$ satisfies the backward ODE:

$$\begin{aligned} &\frac{d}{dt}v^{N,1}(t,\boldsymbol{x}) + \sum_{i=1}^{N}\beta\big(v^{N,1}(t,(1-x^i,\boldsymbol{x}^{-i})) - v^{N,1}(t,(x^i,\boldsymbol{x}^{-i}))\big)\mathbf{1}_{\{\phi^{*N,i}(x^i,\boldsymbol{x}^{-i})=1-x^i\}} \\ &\quad + \mathbf{1}_{\{x^1=0\}} = 0, \end{aligned}$$

with $v^{N,1}(1,\boldsymbol{x}) = 2\bar{x}^{-1}$ as terminal condition. Using (7.55), this may be rewritten as:

$$\begin{aligned} &\frac{d}{dt}v^{N,1}(t,\boldsymbol{x}) + \mathbf{1}_{\{x^1=0\}} \\ &\quad + \beta\big(v^{N,1}(t,(1-x^1,\boldsymbol{x}^{-1})) - v^{N,1}(t,(x^1,\boldsymbol{x}^{-1}))\big)\mathbf{1}_{\{v^{N,1}(t,(1-x^1,\boldsymbol{x}^{-1}))-v(t,(x^1,\boldsymbol{x}^{-1}))<0\}} \\ &\quad + \sum_{i=2}^{N}\beta\big(v^{N,1}(t,(1-x^i,\boldsymbol{x}^{-i})) - v^{N,1}(t,(x^i,\boldsymbol{x}^{-i}))\big)\mathbf{1}_{\{\phi^{*N,i}(x^i,\boldsymbol{x}^{-i})=1-x^i\}} \\ &\quad = 0. \end{aligned}$$

We then recognize on the first two lines the Hamiltonian structure generated by the Hamiltonian H^* in (7.53). This shows that $v^{N,1}$ is the value function of the optimal control problem characterizing the best response of player 1 when all the others play the strategies $(\phi^{*N,i})_{2\leqslant i\leqslant N}$. □

Conclusion

Remarkably, Lemma 7.9 shows that, whenever the N-player game is initialized with $(0,\cdots,0)$, the strategy profiles $(\phi^{*N,i})_{1\leqslant i\leqslant N}$ force all the players to stay in state 0, while, in the limiting mean field game, the Dirac mass δ_0 at point 0 is not an equilibrium.

Obviously, what happens is that the mean field limit captures the trivial Nash equilibrium obtained by letting all the players in the N-player game play strategy 1. In contrast, the mean field formulation cannot keep track of the strategies $(\phi^{*N,i})_{1\leq i\leq N}$ because, in the asymptotic regime, it is no longer possible for the whole population to optimize its response when a single player is deviating. In this regard, it is worth observing that the limiting optimal cost in the mean field problem, when initialized at 0, is $2 - O(1/\beta)$, while, when all the players start from 0 and play strategy 0 in the N-player game, the cost to any player is 1, which shows that the equilibrium captured by the mean field limit is not the one with the minimal cost.

It is also worth mentioning that, in appearance, the limiting mean field game possesses all the properties that we shall use in Chapter (Vol II)-6 to prove the convergence of games with finitely many players to mean field games. Notice in particular that the mean field game is uniquely solvable for any initial distribution and that the master equation has a smooth solution. This may seem contradictory with the existence of the extra Nash equilibrium $(\phi^{*N,i})_{1\leqslant i\leqslant N}$ since the latter facts are basically the main ingredients used in the analysis performed in Chapter (Vol II)-6.

We now explain why the master equation cannot capture the additional Nash equilibrium exhibited in the statement of Lemma 7.9.

The Nash System. Following the analysis performed in Chapter (Vol II)-6, we first write down the Nash system for the N-player game.

Similar to (2.17), see also (Vol II)-(6.94) together with the proof of Lemma 7.9, the Nash system reads as a system of differential equations with a tuple of functions $(v^{N,i} : [0,T] \times E^N \to \mathbb{R})_{i=1,\cdots,N}$ as unknown:

$$\frac{d}{dt} v^{N,i}(t,\boldsymbol{x}) + H^*\big(t, x^i, v^{N,i}(t,\cdot,\boldsymbol{x}^{-i})\big) \\ + \sum_{j=1, j\neq i}^{N} \beta\big(v^{N,i}\big(t,(1-x^j,\boldsymbol{x}^{-j})\big) - v^{N,i}(t,\boldsymbol{x})\big)\mathbf{1}_{\{\phi^{N,j}(t,\boldsymbol{x})=1-x^j\}} = 0, \tag{7.57}$$

where $\boldsymbol{x} = (x^1, \cdots, x^N)$ and $(1-x^j,\boldsymbol{x}^{-j}) = (x^1,\cdots,x^{j-1},1-x^j,x^{j+1},\cdots)$, with the property, inherited from (7.51) and (7.52), that:

$$\phi^{N,j}(t,\boldsymbol{x}) = \begin{cases} 1-x^j & \text{if} \quad v^{N,j}\big(t,(1-x^j,\boldsymbol{x}^{-j})\big) - v^{N,j}(t,\boldsymbol{x}) < 0, \\ x^j & \text{if} \quad v^{N,j}\big(t,(1-x^j,\boldsymbol{x}^{-j})\big) - v^{N,j}(t,\boldsymbol{x}) > 0. \end{cases}$$

Now, the analysis performed in Chapter (Vol II)-6 prompts us to let:

$$u^{N,i}(t,\boldsymbol{x}) = \mathcal{U}(t, x^i, \bar{\mu}^{N-1}_{\boldsymbol{x}^{-i}}).$$

The goal is to check:

Lemma 7.10 *The functions $(u^{N,i})_{1 \leqslant i \leqslant N}$ are solutions of the Nash system* (7.57).

Of course, Lemma 7.10 should not come as a surprise: It is a just way to rephrase the fact that the constant strategy profile 1 is a Nash equilibrium of the N-player game. Actually, its interest is mostly pedagogical as it provides a clear parallel with the statement of Proposition (Vol II)-(6.31) in Chapter (Vol II)-6.

Proof. By (7.54),

$$
\begin{aligned}
u^{N,i}\big(t,(0,\boldsymbol{x}^{-i})\big) &= \frac{1}{\beta}\big(1-\exp(-\beta(1-t))\big) \\
&\quad + 2 - \frac{2}{N-1}\sum_{j=1,j\neq i}^{N}(1-x^j)\exp(-\beta(1-t)), \\
u^{N,i}\big(t,(1,\boldsymbol{x}^{-i})\big) &= 2 - \frac{2}{N-1}\sum_{j=1,j\neq i}^{N}(1-x^j)\exp(-\beta(1-t)).
\end{aligned}
$$

Therefore, for $j \neq i$,

$$
u^{N,i}\big(t,(1-x^j,\boldsymbol{x}^{-j})\big) - u^{N,i}\big(t,(x^j,\boldsymbol{x}^{-j})\big) = \frac{2}{(N-1)}(1-2x^j)\exp(-\beta(1-t)),
$$

while, for $j = i$,

$$
u^{N,i}\big(t,(0,\boldsymbol{x}^{-i})\big) - u^{N,i}\big(t,(1,\boldsymbol{x}^{-i})\big) = \frac{1}{\beta}\big(1-\exp(-\beta(1-t))\big),
$$

so that, letting:

$$
\psi^{N,i}(t,\boldsymbol{x}) = \begin{cases} 1-x^i & \text{if} \quad u^{N,i}\big(t,(1-x^i,\boldsymbol{x}^{-i})\big) - u^{N,i}(t,\boldsymbol{x}) < 0, \\ x^i & \text{if} \quad u^{N,i}\big(t,(1-x^i,\boldsymbol{x}^{-i})\big) - u^{N,i}(t,\boldsymbol{x}) > 0, \end{cases}
$$

we get that $\psi^{N,i}(t,\boldsymbol{x}) = 1$.

Therefore, for a given $i \in \{1,\cdots,N\}$, the last term in (7.57) reads:

$$
\begin{aligned}
&\sum_{j=1,j\neq i}^{N} \beta\big(u^{N,i}\big(t,(1-x^j,\boldsymbol{x}^{-j})\big) - u^{N,i}(t,\boldsymbol{x})\big)\mathbf{1}_{\{\psi^{N,j}(t,\boldsymbol{x})=1-x^j\}} \\
&= \frac{2\beta}{N-1}\exp(-\beta(1-t))\sum_{j=1,j\neq i}^{N}(1-2x^j)\mathbf{1}_{\{x^j=0\}} \\
&= \frac{2\beta}{N-1}\exp(-\beta(1-t))\sum_{j=1,j\neq i}^{N}(1-x^j).
\end{aligned}
$$

Now, thanks to (7.53), the Hamiltonian H^* in (7.57) reads:

$$H^*\big(t,0,u^{N,i}(t,(\cdot,\boldsymbol{x}^{-i}))\big) = -(1-\exp(-\beta(1-t)))+1 = \exp\big(\beta(1-t)\big),$$
$$H^*\big(t,1,u^{N,i}(t,(\cdot,\boldsymbol{x}^{-i}))\big) = 0,$$

Computing the time derivatives of $(u^{N,i})_{1\leqslant i\leqslant N}$, we easily complete the proof. $\square$

Now, if we perform the same computation with the strategy $(\phi^{*N,i})_{1\leqslant i\leqslant N}$ given by Lemma 7.9, then we find that:

$$\phi^{*N,i}(t,\boldsymbol{x}) = 0 \quad \text{if } \boldsymbol{x} = 0,$$

which amounts to say that:

$$v^{*N,i}\big(t,(1,0^{-i})\big) - v^{*N,i}(t,0) > 0,$$

$(v^{*N,i})_{1\leqslant i\leqslant N}$ being defined as the corresponding value functions. Actually, the above inequality is precisely what we checked in the proof of Lemma 7.9. Of course, it is false if $v^{*N,i}$ is replaced by $u^{N,i}$.

So, the explanation is now clear: The control strategy profile is of *bang-bang* type as it oscillates from state 0 or 1 to state 1 or 0 according to the sign of the discrete derivative of the value function of the game. Rephrased with the notation used in the book, the minimizer $\hat{\alpha}(t,x,h)$ in (7.51) and (7.52) is not continuous with respect to h and this explains why the master equation fails to capture the equilibrium identified in Lemma 7.10. In comparison, the minimizer of the Hamiltonian appearing in the analysis of Chapter (Vol II)-6 is regular, which makes a big difference when investigating the convergence property.

7.3 Notes & Complements

Mean field games models with several groups of players were already part of the original contribution of Huang, Caines, and Malhamé, see [211]. Another earlier paper on the subject is due to Lachapelle and Wolfram [253], who introduced models for groups of pedestrians with crowd aversion and xenophobia, that is aversion of an ingroup towards outgroups. One of this model was revisited by Cirant and Verzini in the more recent work [119], with a more detailed analysis of the segregation phenomenon. General well posedness of stationary mean field games with two populations was addressed by Feleqi in [151] under periodic boundary conditions and by Cirant in [117] under Neumann boundary conditions. Convergence of the N-player game was investigated by Feleqi in [151]. A synthetic presentation is also given in Chapter 8 of the textbook by Bensoussan, Frehse, and Yam [50].

The derivation of the HJB equation for stochastic optimal control problems with an infinite horizon and a discounted running cost may be found in Chapter III of the monograph by Fleming and Soner [157]. However, the exposition therein is limited to time homogeneous coefficients, in which case the resulting HJB equation becomes stationary. Obviously, the stationary case is not well fitted to infinite horizon mean field games unless we modify the fixed point condition as explained in the introduction of ergodic mean field games: Under the standard fixed point condition, the distribution of the population may depend on time and, subsequently, the coefficients of the underlying control problem are not time homogeneous. It is only when addressing mean field games with an ergodic cost that the problem becomes time independent.

Regarding the probabilistic approach to this kind of optimal control problems, earlier results on decoupled FBSDEs in infinite horizon were obtained by Peng [303] and Buckdahn and Peng [80] under appropriate monotonicity conditions, which were relaxed by Briand and Hu [70] and Royer [321]. Peng and Shi, in [306], implemented a continuation argument to prove an existence and uniqueness result for fully coupled FBSDEs in infinite horizon. Connection between FBSDEs and optimal control problems in infinite horizon was addressed by Fuhrman and Tessitore [166] and Hu and Tessitore [205]. The corresponding version of the stochastic maximum principle was investigated by Haadem, Øksendal, and Proske [192] and by Maslowski and Veverka [276].

Examples of infinite horizon mean field games may be found in Huang, Caines, and Malhamé [212], Huang [208] and Huang [209]. We refer to Chapter 7 in the monograph by Bensoussan, Frehse, and Yam [50] and to the article by Priuli [314] for other considerations on mean field games with an infinite horizon and a discounted running cost.

Ergodic mean field games, including the convergence of the N-player game, were addressed by Lasry and Lions in their first works on the subject. We refer to the two seminal papers [260, 262] for a complete account of the available results. Refinements were obtained by Cirant in [118], where special attention is paid to cost functionals favoring congestion phenomena, in [153] by Ferreira and Gomes whose analysis allows for degenerate cases, and in [313] by Pimentel and Voskanyan who address the existence of classical solutions. The extended mean field games models introduced and studied in Subsection 4.6 of Chapter 4 were studied in the ergodic case by Gomes, Patrizi, and Voskanyan in [127] where they are called extended ergodic mean fields games. The linear quadratic case was considered by Bardi and Priuli in [35]. The note by Borkar [66] provides a pedagogical introduction to the theory of ergodic optimal control (without mean field interactions). The HJB equations used to solve ergodic control problems and ergodic games were investigated by Bensoussan and Frehse [46, 47]. We refer to the monographs [228] by Khasminskii and [128] by Da Prato and Zabczyk for a general presentation of the ergodic properties of Markov and diffusion processes. Ergodic BSDEs, which we alluded to in Subsection 7.1.2, were investigated by Debussche, Hu, and Tessitore in [131] and Fuhrman, Hu, and Tessitore in [270].

The connection between ergodic mean field games and the large time limit of mean field games with finite time horizon was addressed by Cardaliaguet, Lasry, Lions, and Porretta in [90, 91] for mean field games with nondegenerate diffusion coefficients, and by Cardaliaguet in [84] for degenerate first order mean field games. For numerical methods on ergodic mean field games, we refer the interested reader to the paper [19] of Almulla, Ferreira, and Gomes and to [4] by Achdou and Capuzzo-Dolcetta. In [82], Camilli and Marchi study a class of stationary mean field games on networks.

Our introductory discussion of mean field games with finite state spaces emphasizes the fact that we refrain from considering the equivalent of a common noise, and that we restrict our presentation to Markovian dynamics as given by control strategies in feedback closed loop form. For the construction of state dynamics as marked point processes from their dual predictable projections and the discussion of a few examples of control of queuing systems, the interested reader is referred to the book [231] of Kitaev and Rykov. For the theory of point processes we refer the reader to Brémaud's monograph [67], and to Cinlar's textbook [116] for a clear pedagogical introduction. All these treatises rely heavily on the fundamental work of Jacod [214] on the theory of the predictable projection of a random measure. We also refer to the monograph [173] of Gihman and Skorohod for a general overview of the theory of controlled processes and to the textbook [190] of Guo and Hernández-Lerma for a more specific focus on Markov decision processes.

The bulk of the technical results presented in our discussion of the mean field games with finite state spaces was inspired by the works of Guéant [185, 188] and Gomes, Mohr, and Souza [175]. The earlier publication [174] by the same authors investigates the discrete time case which we did not consider in the text. These works are very similar. The dynamics of the states are given for an evolution on a directed graph in [185, 188], emphasizing the possibility to restrict transitions to and from specific sets of nodes. However, except for a clear emphasis in the notation which could help the intuition regarding the time evolution of the state, this directed graph structure is not really used when it comes to theoretical results and proofs, so the set-up of [175] could be used as well to describe the dynamics. The main difference is that from the start of [185, 188], the contributions of the control α and the marginal distribution μ to the running cost function are split apart and appear in two different additive components of the cost. Even though the form of the contribution of the control is more general than the quadratic function which we use in the text, the strict convexity assumption of [185, 188] ends up playing the same role as our quadratic assumption, and for the sake of exposition, we decided to use the quadratic function instead of carrying around Legendre transforms. Finally, while [175] does not assume that the running cost function splits into parts containing the contributions of the control and the marginal distribution, technical assumptions and dedicated a priori estimates lead to similar arguments and proofs.

The four state model for the behavior of computer owners facing cyber attacks by hackers which we chose for the purpose of illustration is borrowed from the paper [235] by Kolokoltsov and Bensoussan. There, the authors consider the infinite time horizon version of the model, and search for stationary equilibria. They give a complete characterization of the state of affair, i.e., *nonexistence*, *existence and uniqueness*, and even *existence of two* equilibria in the asymptotic regime $\lambda = \infty$. Obviously, the numerical illustrations given in the text are for T and λ finite. We chose the parameters of the model for our numerical results to be consistent with the asymptotic properties expected from their results. This cyber-security model will be revisited in Chapter 7 of Volume II where we extend the framework of finite state mean field games to include major and minor players, generalization which makes the model more realistic for the analysis of cyber attack applications.

The counter-example presented in Subsection 7.2.5 is inspired by the note [139] of Doncel, Gast, and Gaujal. See also the expanded version [140] by the same authors. It may be regarded as a dynamic variant of the classical prisoner's dilemma game. In the finite player version, a strategy for constructing an equilibrium is given by the *tit-for-tat* principle: any player which defects from the cooperation is punished by the others. Intuitively, such a strategy cannot be implemented anymore in the limiting setting since the mass of one defecting player is zero.

We refer to Gast and Gaujal [169, 170], Gast, Gaujal, and LeBoudec [169, 170], and Kolokolstov [239] for a study of mean field control problems, very much in the spirit of Chapter 6, on a finite state space. Some of these articles treat only the discrete time case. Finally, we refer to Basna, Hilbert, and Kolokoltsov [37] for a discussion of mean field games when the dynamics of the states of the players are given by pure jump Markov processes in a continuous state space.

References

1. Y. Achdou, F. Buera, J.M. Lasry, P.L. Lions, and B. Moll. Partial differential equation models in macroeconomics. *Philosophical Transactions of the Royal Society, A*, 372, Oct. 2014.
2. Y. Achdou, F. Camilli, and I. Capuzzo-Dolcetta. Mean field games: numerical methods for the planning problem. *SIAM Journal on Control and Optimization*, 50:77–109, 2010.
3. Y. Achdou, F. Camilli, and I. Capuzzo-Dolcetta. Mean field games: convergence of a finite difference method. *SIAM Journal on Numerical Analysis*, 51:2585–2612, 2013.
4. Y. Achdou and I. Capuzzo-Dolcetta. Mean field games: numerical methods. *SIAM Journal on Numerical Analysis*, 48:1136–1162, 2010.
5. Y. Achdou and M. Laurière. On the system of partial differential equations arising in mean field type control. *Discrete and Continuous Dynamical Systems, A*, 35:3879–3900, 2015.
6. Y. Achdou and V. Perez. Iterative strategies for solving linearized discrete mean field games systems. *Networks and Heterogeneous Media*, 7:197–217, 2012.
7. Y. Achdou and A. Porretta. Convergence of a finite difference scheme to weak solutions of the system of partial differential equations arising in mean field games. *SIAM Journal on Numerical Analysis*, 54:161–186.
8. S. Adlakha and R. Johari. Mean field equilibrium in dynamic games with strategic complementarities. *Operations Research*, 61:971–989, 2013.
9. N. Aghbal and R. Carmona. A solvable mean field game with interactions through the control. Technical report, Princeton University, 2014.
10. P. Aghion and P. Howitt. A model of growth through creative destruction. *Econometrica*, 60:323–352, 1992.
11. S. Ahuja. Wellposedness of mean field games with common noise under a weak monotonicity condition. *SIAM Journal on Control and Optimization*, 54:30–48, 2016.
12. S.R. Aiyagari. Uninsured idiosyncratic risk and aggregate saving. *The Quarterly Journal of Economics*, 109:659–684, 1994.
13. M. Aizenman and B. Simon. Brownian motion and Harnack inequality for Schrödinger operators. *Communications in Pure and Applied Mathematics*, 35:209–273, 1982.
14. S. Alanko. *Regression-based Monte Carlo methods for solving nonlinear PDEs*. PhD thesis, New York University, 2015.
15. D. Aldous. Weak convergence and the general theory of processes. Unpublished notes. http://www.stat.berkeley.edu/~aldous/Papers/weak-gtp.pdf, 1983.
16. D. Aldous. Exchangeability and related topics. In *Ecole d'Eté de Probabilités de Saint Flour 1983*. Volume 1117 of Lecture Notes in Mathematics, pages 1–198. Springer-Verlag Berlin Heidelberg, 1985.
17. C. D. Aliprantis and K. Border. *Infinite Dimensional Analysis. Third Edition*. Springer-Verlag Berlin Heidelberg, 2006.
18. R. Almgren and N. Chriss. Optimal execution of portfolio transactions. *Journal of Risk*, 3:5–39, 2001.

R. Carmona, F. Delarue, *Probabilistic Theory of Mean Field Games with Applications I*, Probability Theory and Stochastic Modelling 83,
DOI 10.1007/978-3-319-58920-6

19. N. Almulla, R. Ferreira, and D. Gomes. Two numerical approaches to stationary mean-field games. *Dynamic Games and Applications*, 7:657–682, 2016.
20. L. Ambrosio and J. Feng. On a class of first order Hamilton-Jacobi equations in metric spaces. *Journal of Differential Equations*, 256:2194–2245, 2014.
21. L. Ambrosio, N. Gigli, and G. Savaré. *Gradient flows in metric spaces and in the Wasserstein space of probability measures*. Birkhäuser Basel, 2004.
22. T.T.K. An and B. Oksendal. Maximum principle for stochastic differential games with partial information. *Journal of Optimization Theory and Applications*, 139:463–483, 2008.
23. T.T.K. An and B. Oksendal. A maximum principle for stochastic differential games with g-expectations and partial information. *Stochastics*, 84:137–155, 2012.
24. D. Andersson and B. Djehiche. A maximum principle for SDEs of mean-field type. *Applied Mathematics & Optimization*, 63:341–356, 2010.
25. F. Antonelli. Backward-forward stochastic differential equations. *Annals of Applied Probability*, 3:777–793, 1993.
26. F. Antonelli and J. Ma. Weak solutions of forward-backward SDE's. *Stochastic Analysis and Applications*, 21(3):493–514, 2003.
27. R. Aumann. Markets with a continuum of traders. *Econometrica*, 32:39–50, 1964.
28. R. J. Aumann. Existence of competitive equilibrium in markets with continuum of traders. *Econometrica*, 34:1–17, 1966.
29. X. Tan C. Zhou B. Bouchard, D. Possamaï. A unified approach to a priori estimates for supersolutions of BSDEs in general filtrations. *Annales de l'institut Henri Poincaré, Probabilités et Statistiques*, to appear.
30. R. Bafico and P. Baldi. Small random perturbations of Peano phenomena. *Stochastics*, 6:279–292, 1982.
31. F. Baghery and B. Oksendal. A maximum principle for stochastic control with partial information. *Stochastic Analysis and Applications*, 25:705–717, 2007.
32. K. Bahlali, B. Mezerdi, M. N'zi, and Y. Ouknine. Weak solutions and a Yamada-Watanabe theorem for FBSDEs. *Random Operators and Stochastic Equations*, 15:271–285, 2007.
33. M. Bardi. Explicit solutions of some linear quadratic mean field games. *Networks and Heterogeneous Media*, 7:243–261, 2012.
34. M. Bardi and E. Feleqi. Nonlinear elliptic systems and mean field games. *Nonlinear Differential Equations and Applications NoDEA*, 23:44, 2016.
35. M. Bardi and F. Priuli. Linear-quadratic N-person and mean-field games with ergodic cost. *SIAM Journal on Control and Optimization*, 52:3022–3052, 2014.
36. F. Barthe and C. Bordenave. Combinatorial optimization over two random point sets. In C. Donati-Martin et al., editors, *Séminaire de Probabilités XLV*. Volume 2046 of Lecture Notes in Mathematics, pages 483–536. Springer International Publilshing, 2013.
37. R. Basna, A. Hilbert, and V.N. Kolokolstov. An approximate Nash equilibrium for pure jump Markov games of mean-field-type on continuous state space. *Stochastics*, 89:967–993, 2017.
38. R.F. Bass. *Diffusions and Elliptic Operators*. Springer-Verlag New York, 1998.
39. R.F. Bass and P. Hsu. Some potential theory for reflecting Brownian motion in Hölder and Lipschitz domains. *Annals of Probability*, 19:486–508, 1991.
40. J.R. Baxter and R.V. Chacon. Compactness of stopping times. *Zeitschrift für Wahrscheinlichkeitstheorie und verwandte Gebiete*, 40:169–181, 1977.
41. J.-D. Benamou and Y. Brenier. A computational fluid mechanics solution to the Monge-Kantorovich mass transfer problem. *Numerische Mathematik*, 84:375–393, 2000.
42. J.D. Benamou and G. Carlier. Augmented Lagrangian methods for transport optimization, mean field games and degenerate elliptic equations. *Journal of Optimization Theory and Applications*, 167:1–26, 2015.
43. J.D. Benamou, G. Carlier, and N. Bonne. An augmented Lagrangian numerical approach to solving mean field games. Technical report, INRIA, 2013. https://hal.inria.fr/hal-00922349/
44. A. Bensoussan, M.H.M. Chau, and S.C.P. Yam. Mean field games with a dominating player. *Applied Mathematics & Optimization*, 74:91–128, 2016.

45. A. Bensoussan and J. Frehse. Nonlinear elliptic systems in stochastic game theory. *Journal für die reine und angewandte Mathematik*, 350:23–67, 1984.
46. A. Bensoussan and J. Frehse. On Bellman equations of ergodic control in $\mathbf{R}^n$. *Journal für die reine und angewandte Mathematik*, 492:125–160, 1992.
47. A. Bensoussan and J. Frehse. Ergodic Bellman systems for stochastic games in arbitrary dimension. *Proceedings of the Royal Society of Edinburgh, A*, 449:65–77, 1995.
48. A. Bensoussan and J. Frehse. Stochastic games for *N* players. *Journal of Optimization Theory and Applications*, 105:543–565, 2000.
49. A. Bensoussan and J. Frehse. Smooth solutions of systems of quasilinear parabolic equations. *ESAIM: Control, Optimisation and Calculus of Variations*, 8:169–193, 2010.
50. A. Bensoussan, J. Frehse, and P. Yam. *Mean Field Games and Mean Field Type Control Theory*. SpringerBriefs in Mathematics. Springer-Verlag New York, 2013.
51. A. Bensoussan, J. Frehse, and S. C. P. Yam. The master equation in mean field theory. *Journal de Mathématiques Pures et Appliquées*, 2014.
52. A. Bensoussan, J. Frehse, and S. C. P. Yam. On the interpretation of the master equation. *Stochastic Processes and their Applications*, 127:2093–2137, 2016.
53. A. Bensoussan, K.C.J. Sung, S.C.P. Yam, and S.P. Yung. Linear quadratic mean field games. *Journal of Optimization Theory and Applications*, 169:469–529, 2016.
54. A. Bensoussan and S. C. P. Yam. Control problem on space of random variables and master equation. Technical report, http://arxiv.org/abs/1508.00713, 2015.
55. D.P. Bertsekas. *Nonlinear Programming*. Athena Scientific, 1995.
56. D.P. Bertsekas and S.E. Shreve. *Stochastic Optimal Control: The Discrete Time Case*. Academic Press, 1978.
57. P. Billingsley. *Convergence of Probability Measures. Third edition.* John Wiley & Sons, Inc., 1995.
58. P. Billingsley. *Probability and Measure. Second edition.* John Wiley & Sons, Inc., 1999.
59. A. Bisin, U. Horst, and O. Özgür. Rational expectations equilibria of economies with local interactions. *Journal of Economic Theory*, 127:74–116, 2006.
60. J.-M. Bismut. Théorie probabiliste du contrôle des diffusions, *Memoirs of the American Mathematical Society*, 167(4), 1976.
61. J.-M. Bismut. Conjugate convex functions in optimal stochastic control. *Journal of Mathematical Analysis and Applications*, 44:384–404, 1973.
62. J.M. Bismut. An introductory approach to duality in optimal stochastic control. *SIAM Review*, 20:62–78, 1978.
63. D. Blackwell and L.E. Dubins. An extension of Skorohod's almost sure convergence theorem. *Proceedings of the American Mathematical Society*, 89:691–692, 1983.
64. V.I. Bogachev. *Measure Theory, Volume 2*. Springer-Verlag Berlin Heidelberg, 2007.
65. V.S. Borkar. Controlled diffusion processes. *Probability Surveys*, 2:213–244, 2005.
66. V.S. Borkar. Ergodic control of diffusion processes. In Marta Sanz-Solé et al., editors, *Proceedings of the International Congress of Mathematics, Madrid, Spain*, pages 1299–1309. European Mathematical Society, 2006.
67. P. Brémaud. *Point Processes and Queues: Martingale Dynamics*. Springer Series in Statistics. Springer-Verlag New York, 1981.
68. P. Brémaud and M. Yor. Changes of filtrations and of probability measures. *Zeitschrift für Wahrscheinlichkeitstheorie und verwandte Gebiete*, 45:269–295, 1978.
69. Y. Brenier. Polar factorization and monotone rearrangement of vector-valued functions. *Communications on Pure and Applied Mathematics*, 44:375–417, 1991.
70. P. Briand and Y. Hu. Stability of BSDEs with random terminal time and homogenization of semilinear elliptic PDEs. *Journal of Functional Analysis*, 155:455–494, 1998.
71. P. Briand and Y. Hu. BSDE with quadratic growth and unbounded terminal value. *Probability Theory and Related Fields*, 136:604–618, 2006.
72. G. Brunick and S. Shreve. Mimicking an Itô process by a solution of a stochastic differential equation. *Annals of Applied Probability*, 23:1584–1628, 2013.

73. J. Bryant. A model of reserves, bank runs and deposit insurance. *Journal of Banking and Finance*, 4:335–344, 1980.
74. R. Buckdahn, B. Djehiche, and J. Li. Mean field backward stochastic differential equations and related partial differential equations. *Stochastic Processes and their Applications*, 119:3133–3154, 2007.
75. R. Buckdahn, B. Djehiche, J. Li, and S. Peng. Mean field backward stochastic differential equations: A limit approach. *Annals of Probability*, 37:1524–1565, 2009.
76. R. Buckdahn and H.-J. Engelbert. A backward stochastic differential equation without strong solution. *Theory of Probability and its Applications*, 50:284–289, 2006.
77. R. Buckdahn and H.-J. Engelbert. On the continuity of weak solutions of backward stochastic differential equations. *Theory of Probability and its Applications*, 52:152–160, 2008.
78. R. Buckdahn, H. J. Engelbert, and A. Răşcanu. On weak solutions of backward stochastic differential equations. *Theory of Probability and its Applications*, 49:16–50, 2005.
79. R. Buckdahn, J. Li, S. Peng, and C. Rainer. Mean-field stochastic differential equations and associated PDEs. *Annals of Probability*, 45:824–878, 2017.
80. R. Buckdahn and S. Peng. Stationary backward stochastic differential equations and associated partial differential equations. *Probability Theory and Related Fields*, 115:383–399, 1999.
81. K. Burdzy, W. Kang, and K. Ramanan. The Skorokhod problem in a time-dependent interval. *Stochastic Processes and their Applications*, 119:428–452, 2009.
82. F. Camilli and C. Marchi. Stationary mean field games systems defined on networks. *SIAM Journal on Control and Optimization*, 54:1085–1103, 2016.
83. P. Cardaliaguet. Notes from P.L. Lions' lectures at the Collège de France. Technical report, https://www.ceremade.dauphine.fr/$\sim$cardalia/MFG100629.pdf, 2012.
84. P. Cardaliaguet. Long time average of first order mean field games and weak KAM theory. *Dynamic Games and Applications*, 3:473–488, 2013.
85. P. Cardaliaguet. Weak solutions for first order mean field games with local coupling. In P. Bettiol et al., editors, *Analysis and Geometry in Control Theory and its Applications*. Springer INdAM Series, pages 111–158. Springer International Publishing, 2015.
86. P. Cardaliaguet, F. Delarue, J.-M. Lasry, and P.-L. Lions. The master equation and the convergence problem in mean field games. Technical report, http://arxiv.org/abs/1509.02505, 2015.
87. P. Cardaliaguet and J. Graber. Mean field games systems of first order. *ESAIM: Control, Optimisation and Calculus of Variations*, 21:690–722, 2015.
88. P. Cardaliaguet, J. Graber, A. Porretta, and D. Tonon. Second order mean field games with degenerate diffusion and local coupling. *Nonlinear Differential Equations and Applications NoDEA*, 22:1287–1317, 2015.
89. P. Cardaliaguet and S. Hadikhanloo. Learning in mean field games: the fictitious play. *ESAIM: Control, Optimisation and Calculus of Variations*, 23:569–591, 2017.
90. P. Cardaliaguet, J.M. Lasry, P.L. Lions, and A. Porretta. Long time average of mean field games. *Networks and Heterogeneous Media*, 7:279–301, 2012.
91. P. Cardaliaguet, J.M. Lasry, P.L. Lions, and A. Porretta. Long time average of mean field games with a nonlocal coupling. *SIAM Journal on Control and Optimization*, 51:3558–3591, 2013.
92. P. Cardaliaguet, A. R. Mészáros, and F. Santambrogio. First order mean field games with density constraints: Pressure equals price. *SIAM Journal on Control and Optimization*, 54:2672–2709, 2016.
93. E. A. Carlen. Conservative diffusions. *Communication in Mathematical Physics*, 94:293–315, 1984.
94. R. Carmona. *Lectures on BSDEs, Stochastic Control and Stochastic Differential Games*. SIAM, 2015.
95. R. Carmona and F. Delarue. Mean field forward-backward stochastic differential equations. *Electronic Communications in Probability*, 2013.

96. R. Carmona and F. Delarue. Probabilistic analysis of mean field games. *SIAM Journal on Control and Optimization*, 51:2705–2734, 2013.
97. R. Carmona and F. Delarue. The master equation for large population equilibriums. In D. Crisan, B. Hambly, T. Zariphopoulou, editors, *Stochastic Analysis and Applications 2014: In Honour of Terry Lyons*, pages 77–128. Springer Cham, 2014.
98. R. Carmona and F. Delarue. Forward-backward stochastic differential equations and controlled Mckean Vlasov dynamics. *Annals of Probability*, 43:2647–2700, 2015.
99. R. Carmona, F. Delarue, and A. Lachapelle. Control of McKean-Vlasov versus mean field games. *Mathematics and Financial Economics*, 7:131–166, 2013.
100. R. Carmona, F. Delarue, and D. Lacker. Mean field games with common noise. *Annals of Probability*, 44:3740–3803, 2016.
101. R. Carmona, J.P. Fouque, M. Moussavi, and L.H. Sun. Systemic risk and stochastic games with delay. Technical report, 2016. https://arxiv.org/abs/1607.06373
102. R. Carmona, J.P. Fouque, and L.H. Sun. Mean field games and systemic risk: a toy model. *Communications in Mathematical Sciences*, 13:911–933, 2015.
103. R. Carmona and D. Lacker. A probabilistic weak formulation of mean field games and applications. *Annals of Applied Probability*, 25:1189–1231, 2015.
104. R. Carmona, F. Delarue, and D. Lacker. Mean field games of timing and models for bank runs. *Applied Mathematics & Optimization*, 76:217–260, 2017.
105. R. Carmona and D. Nualart. *Nonlinear Stochastic Integrators, Equations and Flows*. Gordon & Breach, 1990.
106. R. Carmona and P. Wang. An alternative approach to mean field game with major and minor players, and applications to herders impacts. *Applied Mathematics & Optimization*, 76:5–27, 2017.
107. R. Carmona and K. Webster. The self financing condition in high frequency markets. *Finance Stochastics*, to appear.
108. R. Carmona and K. Webster. A Stackelberg equilibrium for the Limit Order Book. *Mathematical Finance*, to appear.
109. R. Carmona and W.I. Zheng. Reflecting Brownian motions and comparison theorems for Neumann heat kernels. *Journal of Functional Analysis*, 123:109–128, 1994.
110. R. Carmona and G. Zhu. A probabilistic approach to mean field games with major and minor players. *Annals of Applied Probability*, 26:1535–1580, 2016.
111. C. Ceci, A. Cretarola, and F. Russo. BSDEs under partial information and financial applications. *Stochastic Processes and their Applications*, 124:2628–2653, 2014.
112. U. Cetin, H.M. Soner, and N. Touzi. Options hedging under liquidity costs. *Finance Stochastics*, 14:317–341, 2010.
113. P. Chan and R. Sircar. Bertrand and Cournot mean field games. *Applied Mathematics & Optimization*, 71:533–569, 2015.
114. J.F. Chassagneux, D. Crisan, and F. Delarue. McKean-vlasov FBSDEs and related master equation. Technical report, http://arxiv.org/abs/1411.3009, 2015.
115. P. G. Ciarlet. *Introduction to Numerical Linear Algebra and Optimisation*. Cambridge Texts in Applied Mathematics. Cambridge University Press, 1989.
116. E. Çinlar. *Probability and Stochastics*. Graduate Texts in Mathematics. Springer-Verlag New York, 2011.
117. M. Cirant. Multi-population mean field games systems with Neumann boundary conditions. *Journal de Mathématiques Pures et Appliquées*, 103:1294–1315, 2015.
118. M. Cirant. Stationary focusing mean-field games. *Communications in Partial Differential Equations*, 41:1324–1346, 2016.
119. M. Cirant and G. Verzini. Bifurcation and segregation in quadratic two-populations mean field games systems. *ESAIM: Control, Optimisation and Calculus of Variations*, 23:1145–1177, 2017.
120. M. Coghi and F. Flandoli. Propagation of chaos for interacting particles subject to environmental noise. *Annals of Applied Probability*, 26:1407–1442, 2016.

121. M.G Crandall and P.L. Lions. Viscosity solutions of Hamilton-Jacobi equations in infinite dimensions. i. Uniqueness of viscosity solutions. *Journal of Functional Analysis*, 62:379–396, 1985.
122. M.G Crandall and P.L. Lions. Viscosity solutions of Hamilton-Jacobi equations in infinite dimensions. ii. Existence of viscosity solutions. *Journal of Functional Analysis*, 65:368–405, 1986.
123. M.G Crandall and P.L. Lions. Viscosity solutions of Hamilton-Jacobi equations in infinite dimensions. iii. *Journal of Functional Analysis*, 68:214–247, 1986.
124. M.G Crandall and P.L. Lions. Viscosity solutions of Hamilton-Jacobi equations in infinite dimensions. iv. Hamiltonians with unbounded linear terms. *Journal of Functional Analysis*, 90:3237–283, 1990.
125. F. Cucker and E. Mordecki. Flocking in noisy environments. *Journal de Mathématiques Pures et Appliquées*, 89:278–296, 2008.
126. F. Cucker and S. Smale. Emergent behavior in flocks. *IEEE Transactions on Automatic Control*, 52:852–862, 2007.
127. V. Voskanyana D.A. Gomes, S. Patrizi. On the existence of classical solutions for stationary extended mean field games. *Nonlinear Analysis: Theory, Methods & Applications*, 99:49–79, 2014.
128. G. Da Prato and J. Zabczyk. *Ergodicity for Infinite Dimensional Systems*. Cambridge University Press, 1996.
129. E.B. Davies. Spectral properties of compact manifolds and changes of metric. *American Journal of Mathematics*, 112:15–39, 1990.
130. D. Dawson and J. Vaillancourt. Stochastic McKean-Vlasov equations. *NoDEA. Nonlinear Differential Equations and Applications*, 2(2):199–229, 1995.
131. A. Debussche, Y. Hu, and G. Tessitore. Ergodic BSDEs under weak dissipative assumptions. *Stochastic Processes and their Applications*, 121:407–426, 2011.
132. F. Delarue. On the existence and uniqueness of solutions to FBSDEs in a non-degenerate case. *Stochastic Processes and their Applications*, 99:209–286, 2002.
133. F. Delarue. Estimates of the solutions of a system of quasi-linear PDEs. a probabilistic scheme. In J. Azéma et al., editors, *Séminaire de Probabilités XXXVII*, pages 290–332. Springer-Verlag Berlin Heidelberg, 2003.
134. F. Delarue and G. Guatteri. Weak existence and uniqueness for FBSDEs. *Stochastic Processes and their Applications*, 116:1712–1742, 2006.
135. S. Dereich, M. Scheutzow, and R. Schottstedt. Constructive quantization: approximation by empirical measures. *Annales Institut Henri Poincaré, Probabilités Statistiques*, 49:1183–1203, 2013.
136. D.W. Diamond and P.H. Dybvig. Bank runs, deposit insurance, and liquidity. *The Journal of Political Economy*, (91):401–419, 1983.
137. B. Djehiche, A. Tcheukam Siwe, and H. Tembine. Mean field-type games in engineering. Technical report, https://arxiv.org/abs/1605.03281, 2016.
138. B. Djehiche, H. Tembine, and R. Tempone. A stochastic maximum principle for risk-sensitive mean-field type control. *IEEE Transactions on Automatic Control*, 60:2640–2649, 2015.
139. J. Doncel, N. Gast, and B. Gaujal. Are mean-field games the limits of finite stochastic games? *SIGMETRICS Performance Evaluation Review*, 44:18–20, 2016.
140. J. Doncel, N. Gast, and B. Gaujal. Mean-field games with explicit interactions. Technical report, https://hal.inria.fr/hal-01277098/file/main.pdf, 2016.
141. G. Dos Reis. *Some advances on quadratic BSDE: Theory - Numerics - Applications*. LAP LAMBERT Academic Publishing, 2011.
142. R. Duboscq and A. Réveillac. Stochastic regularization effects of semi-martingales on random functions. *Journal de Mathématiques Pures et Appliquées*, 106:1141–1173, 2016.
143. R.M. Dudley. *Real Analysis and Probability*. Wadsworth & Brooks/Cole, 1989.
144. D. Duffie, G. Giroux, and G. Manso. Information percolation. *American Economics Journal: Microeconomic Theory*, 2:1, 2010.

145. D. Duffie, S. Malamud, and G. Manso. Information percolation with equilibrium search dynamics. *Econometrica*, 77:1513–1574, 2009.
146. D. Duffie, S. Malamud, and G. Manso. Information percolation in segmented markets. *Journal of Economic Theory*, 153:1–32, 2014.
147. D. Duffie and G. Manso. Information percolation in large markets. *American Economic Review, Papers and Proceedings*, 97:203–209, 2007.
148. D. Duffie and Y. Sun. Existence of independent random matching. *Annals of Applied Probability*, 17:385–419, 2007.
149. S. Ethier and T. Kurtz. *Markov Processes: Characterization and Convergence*. John Wiley & Sons, Inc., 2005.
150. G. Fabbri, F. Gozzi, and A. Swiech. *Stochastic Optimal Control in Infinite Dimensions: Dynamic Programming and HJB Equations*. Probability Theory and Stochastic Modelling. Springer International Publishing, 2017.
151. E. Feleqi. The derivation of ergodic mean field game equations for several population of players. *Dynamic Games and Applications*, 3:523–536, 2013.
152. J. Feng and M. Katsoulakis. A comparison principle for Hamilton-Jacobi equations related to controlled gradient flows in infinite dimensions. *Archive for Rational Mechanics and Analysis*, 192:275–310, 2009.
153. R. Ferreira and D. Gomes. Existence of weak solutions to stationary mean field games through variational inequalities. Technical report, http://arxiv.org/abs/1512.05828, 2015.
154. M. Fischer. On the connection between symmetric N-player games and mean field games. *Annals of Applied Probability*, 27:757–810, 2017.
155. M. Fischer and G. Livieri. Continuous time mean-variance portfolio optimization through the mean field approach. *ESAIM: Probability and Statistics*, 20:30–44, 2016.
156. F. Flandoli. *Random Perturbation of PDEs and Fluid Dynamics: Ecole d'été de probabilités de Saint-Flour XL*. Volume 2015 of Lecture Notes in Mathematics. Springer-Verlag Berlin Heidelberg, 2011.
157. W.H. Fleming and M. Soner. *Controlled Markov Processes and Viscosity Solutions*. Stochastic Modelling and Applied Probability. Springer-Verlag, New York, 2010.
158. W.H. Fleming. Generalized solutions in optimal stochastic control. In *Proceedings of the Second Kingston Conference on Differential Games*, pages 147–165. Marcel Dekker, 1977.
159. R. Foguen Tchuendom. Restoration of uniqueness of Nash equilibria for a class of linear-quadratic mean field games with common noise. *Dynamic Games and Applications*, to appear.
160. M. Fornasier and F. Solombrino. Mean-field optimal control. *ESAIM: Control, Optimisation and Calculus of Variations*, 20:1123–1152, 2014.
161. N. Fournier and A. Guillin. On the rate of convergence in the Wasserstein distance of the empirical measure. *Probability Theory and Related Fields*, 162:707–738, 2015.
162. A. Friedman. *Partial differential equations of parabolic type*. Prentice-Hall, Englewood Cliffs, N.J., first edition, 1964.
163. A. Friedman. Stochastic differential games. *Journal of Differential Equations*, 11:79–108, 1972.
164. D. Fudenberg and D. Levine. Open-loop and closed-loop equilibria in dynamic games with many players. *Journal of Economic Theory*, 44:1–18, 1988.
165. D. Fudenberg and J. Tirole. *Game Theory*. MIT Press, 1991.
166. M. Fuhrman and G. Tessitore. Infinite horizon backward stochastic differential equations and elliptic equations in Hilbert spaces. *Annals of Probability*, 30:607–660, 2004.
167. W. Gangbo, T. Nguyen, and A. Tudorascu. Hamilton-Jacobi equations in the Wasserstein space. *Methods and Applications of Analysis*, 15:155–184, 2008.
168. W. Gangbo and A. Swiech. Existence of a solution to an equation arising from the theory of mean field games. *Journal of Differential Equations*, 259:6573–6643, 2015.
169. N. Gast and B. Gaujal. A mean field approach for optimization in discrete time. *Journal of Discrete Event Dynamic Systems*, 21:63–101, 2011.

170. N. Gast, B. Gaujal, and J.-Y. Le Boudec. Mean field for Markov decision processes: from discrete to continuous optimization. *IEEE Transactions on Automatic Control*, 57:2266–2280, 2012.
171. J. Gatheral, A. Schied, and A. Slynko. Transient linear price impact and Fredholm integral equations. *Mathematical Finance*, 22:445–474, 2012.
172. R. Gayduk and S. Nadtochiy. Liquidity effects of trading frequency. *Mathematical Finance*, to appear.
173. I.I. Gihman and A.V. Skorohod. *Controlled Stochastic Processes*. Springer-Verlag Berlin Heidelberg New York, 1979.
174. D.A. Gomes, J. Mohr, and R.R. Souza. Discrete time, finite state space mean field games. *Journal de Mathématiques Pures et Appliquées*, 93:308–328, 2010.
175. D.A. Gomes, J. Mohr, and R.R. Souza. Continuous time finite state mean field games. *Applied Mathematics & Optimization*, 68:99–143, 2013.
176. D.A. Gomes, L. Nurbekyan, and E. Pimentel. *Economic Models and Mean-field Games Theory*. Publicaões Matemáticas, IMPA, Rio, Brazil, 2015.
177. D.A. Gomes and E. Pimentel. Time-dependent mean-field games with logarithmic nonlinearities. *SIAM Journal of Mathematical Analysis*, 47:3798–3812, 2015.
178. D.A. Gomes and E. Pimentel. Local regularity for mean-field games in the whole space. *Minimax Theory and its Applications*, 1:65–82, 2016.
179. D.A. Gomes, E. Pimentel, and H. Sánchez-Morgado. Time-dependent mean-field games in the sub- quadratic case. *Communications in Partial Differential Equations*, 40:40–76, 2015.
180. D.A. Gomes, E. Pimentel, and H. Sánchez-Morgado. Time-dependent mean-field games in the superquadratic case. *ESAIM: Control, Optimisation and Calculus of Variations*, 22:562–580, 2016.
181. D.A. Gomes, E. Pimentel, and V. Voskanyan. *Regularity Theory for Mean-Field Game Systems*. SpringerBriefs in Mathematics Springer International Publishing, 2016.
182. D.A. Gomes and J. Saude. Mean field games models - a brief survey. *Dynamic Games and Applications*, 4:110–154, 2014.
183. D.A. Gomes and V. Voskanyan. Extended mean field games. *SIAM Journal on Control and Optimization*, 54:1030–1055, 2016.
184. A. Granas and J. Dugundji. *Fixed point theory*. Springer Monographs in Mathematics. Springer-Verlag New York, 2003.
185. O. Guéant. From infinity to one: The reduction of some mean field games to a global control problem. *Cahier de la Chaire Finance et Développement Durable*, 42, 2011.
186. O. Guéant. Mean field games equations with quadratic Hamiltonian: A specific approach. *Mathematical Models and Methods in Applied Sciences*, 22:291–303, 2012.
187. O. Guéant. New numerical methods for mean field games with quadratic costs. *Networks and Heterogeneous Media*, 2:315–336, 2012.
188. O. Guéant. Existence and uniqueness result for mean field games with congestion effect on graphs. *Applied Mathematics & Optimization*, 72:291–303, 2015.
189. O. Guéant, J.M. Lasry, and P.L. Lions. Mean field games and applications. In R. Carmona et al., editors, *Paris Princeton Lectures on Mathematical Finance 2010*. Volume 2003 of Lecture Notes in Mathematics. Springer-Verlag Berlin Heidelberg, 2010.
190. X. Guo and O. Hernández-Lerma. *Continuous-Time Markov Decision Processes*. Stochastic Modelling and Applied Probability. Springer-Verlag Berlin Heidelberg, 2009.
191. I. Gyöngy. Mimicking the one-dimensional marginal distributions of processes having an Itô differential. *Probability Theory and Related Fields*, 71:501–516, 1986.
192. S. Haadem, B. Øksendal, and F. Proske. Maximum principles for jump diffusion processes with infinite horizon. *Automatica*, 49:2267–2275, 2013.
193. S. Hamadène. Backward-forward SDE's and stochastic differential games. *Stochastic Processes and their Applications*, 77:1–15, 1998.
194. S. Hamadène. Nonzero-sum linear quadratic stochastic differential games and backward forward equations. *Stochastic Analysis and Applications*, 17:117–130, 1999.

195. S. Hamadène and J.P. Lepeltier. Backward equations, stochastic control and zero-sum stochastic differential games. *Stochastics and Stochastic Reports*, 54:221–231, 1995.
196. E. Häusler and H. Luschgy. *Stable Convergence and Stable Limit Theorems*. Probability Theory and Stochastic Modelling. Springer International Publishing, 1995.
197. Z. He and W. Xiong. Dynamic debt runs. *Review of Financial Studies*, 25:1799–1843, 2012.
198. J. Horowitz and R.L. Karandikar. Mean rates of convergence of empirical measures in the Wasserstein metric. *Journal of Computational and Applied Mathematics*, 55:261–273, 1994.
199. U. Horst. Ergodic fluctuations in a stock market model with interacting agents: the mean field case. *Discussion paper No. 106, Sonderforschungbereich 373, Humboldt Universität, Berlin*, 1999.
200. U. Horst. Stationary equilibria in discounted stochastic games with weakly interacting players. *Games and Economic Behavior*, 51:83–108, 2005.
201. J.A. Hosking. A stochastic maximum principle for a stochastic differential game of a mean-field type. *Applied Mathematics & Optimization*, 66:415–454, 2012.
202. Y. Hu. Stochastic maximum principle. In John Baillieul, Tariq Samad, editors, *Encyclopedia of Systems and Control*, pages 1347–1350. Springer-Verlag London, 2015.
203. Y. Hu and S. Peng. Maximum principle for semilinear stochastic evolution control systems. *Stochastics and Stochastic Reports*, 33:159–180, 1990.
204. Y. Hu and S. Tang. Multi-dimensional backward stochastic differential equations of diagonally quadratic generators. *Stochastic Processes and their Applications*, 126:1066–1086, 2016.
205. Y. Hu and G. Tessitore. BSDE on an infinite horizon and elliptic PDEs in infinite dimension. *Nonlinear Differential Equations and Applications NoDEA*, 14:825–846, 2007.
206. C-F. Huang and L. Li. Continuous time stopping games with monotone reward structures. *Mathematics of Operations Research*, 15:496–507.
207. M. Huang. Large-population LQG games involving a major player: the Nash equivalence principle. *SIAM Journal on Control and Optimization*, 48:3318–3353, 2010.
208. M. Huang. A mean field accumulation game with HARA utility. *Dynamics Games and Applications*, 3:446–472, 2013.
209. M. Huang. Mean field capital accumulation games: the long time behavior. In *Proceedings of the 52nd IEEE Conference on Decision and Control*, pages 2499–2504. 2013.
210. M. Huang, P.E. Caines, and R.P. Malhamé. Individual and mass behavior in large population stochastic wireless power control problems: centralized and Nash equilibrium solutions. In *Proceedings of the 42nd IEEE International Conference on Decision and Control*, pages 98–103. 2003.
211. M. Huang, P.E. Caines, and R.P. Malhamé. Large population stochastic dynamic games: closed-loop McKean-Vlasov systems and the Nash certainty equivalence principle. *Communications in Information and Systems*, 6:221–252, 2006.
212. M. Huang, P.E. Caines, and R.P. Malhamé. Large population cost coupled LQG problems with nonuniform agents: individual mass behavior and decentralized ϵ-Nash equilibria. *IEEE Transactions on Automatic Control*, 52:1560–1571, 2007.
213. M. Huang, R.P. Malhamé, and P.E. Caines. Nash equilibria for large population linear stochastic systems with weakly coupled agents. In R.P. Malhamé, E.K. Boukas, editors, *Analysis, Control and Optimization of Complex Dynamic Systems*, pages 215–252. Springer-US, 2005.
214. J. Jacod. Multivariate point processes: predictable projections, Radon-Nykodym derivatives, representation of martingales. *Zeitschrift für Wahrscheinlichkeitstheorie und verwandte. Gebiete*, 31:235–253, 1975.
215. J. Jacod. Weak and strong solutions of stochastic differential equations. *Stochastics*, 3:171–191, 1980.
216. J. Jacod and J. Mémin. Weak and strong solutions of stochastic differential equations: Existence and stability. In D. Williams, editor, *Stochastic Integrals*. Volume 851 of Lecture Notes in Mathematics, pages 169–212. Springer-Verlag Berlin Heidelberg, 1981.

217. M. Huang, P.E. Caines, and R.P. Malhamé. Social optima in mean field LQG control: centralized and decentralized strategies. *IEEE Transactions on Automatic Control*, 57(7):1736–1751, 2012.
218. X. Huang, S. Jaimungal and M. Nourian. Mean-field game strategies for optimal execution. Technical report, University of Toronto, https://papers.ssrn.com/sol3/papers.cfm?abstract_id=2578733, 2017.
219. M. Jeanblanc and Y. Le Cam. Immersion property and credit risk modelling. In F. Delbaen, M. Rásonyi, C. Stricker, editors, *Optimality and Risk-Modern Trends in Mathematical Finance*, pages 99–132. Springer-Verlag Berlin Heidelberg, 2010.
220. R. Jordan, D. Kinderlehrer, and F. Otto. The variational formulation of the Fokker-Planck equation. *SIAM Journal on Mathematical Analysis*, 29:1–17, 1998.
221. B. Jourdain, S. Meleard, and W. Woyczynski. Nonlinear SDEs driven by Lévy processes and related PDEs. *ALEA, Latin American Journal of Probability*, 4:1–29, 2008.
222. J. Kallsen and C. Kühn. Pricing derivatives of american and game type in incomplete markets. *Finance and Stochastics*, 8:261–284, 2004.
223. J. Kallsen and C. Kühn. Convertible bonds: Financial derivatives of game type. In A. Kyprianou, W. Schoutens, P. Wilmott, editors, *Exotic Option Pricing and Advanced Lévy Models*, pages 277–288. John Wiley & Sons, Inc., 2005.
224. N. El Karoui and S.J. Huang. A general result of existence and uniqueness of backward stochastic differential equations. In N. El Karoui, L. Mazliak, editors, *Backward stochastic differential equations*, Research Notes in Mathematics, pages 27–36. Pitman, Longman, Harlow, 1997.
225. N. El Karoui, D.H. Nguyen, and M. Jeanblanc-Picqué. Compactification methods in the control of degenerate diffusions: existence of an optimal control. *Stochastics*, 20:169–219, 1987.
226. N. El Karoui, S. Peng, and M.C. Quenez. Backward stochastic differential equations in finance. *Mathematical Finance*, 7:1–71, 1997.
227. N. Kazamaki. *Continuous Exponential Martingales and BMO*. Volume 1579 of Lecture Notes in Mathematics. Springer-Verlag Berlin Heidelberg, 1994.
228. R.Z. Khasminskii. *Stochastic Stability of Differential Equations*. Sijthoff & Noordhoff, 1980.
229. Y. Kifer. Game options. *Finance and Stochastics*, 4:443–463, 2000.
230. J.F.C. Kingman. Uses of exchangeability. *Annals of Probability*, 6:183–197, 1978.
231. M.Y. Kitaev and V. Rykov. *Controlled Queuing Systems*. CRC Press, 1995.
232. M. Kobylanski. Backward stochastic differential equations and partial differential equations with quadratic growth. *Annals of Probability*, 28:558–602, 2000.
233. V.N. Kolokolstov. Nonlinear Markov semigroups and interacting Lévy processes. *Journal of Statistical Physics*, 126:585–642, 2007.
234. V.N. Kolokolstov. *Nonlinear Markov processes and kinetic equations*. Cambridge University Press, Cambridge, 2010.
235. V.N. Kolokolstov and A. Bensoussan. Mean-field-game model for botnet defense in cybersecurity. *Applied Mathematics & Optimization*, 74:669–692, 2016.
236. V.N. Kolokolstov, J. Li, and W. Yang. Mean field games and nonlinear Markov processes. Technical report, http://arxiv.org/abs/1112.3744, 2011.
237. V.N. Kolokolstov and M. Troeva. On the mean field games with common noise and the McKean-Vlasov SPDEs. Technical report, http://arxiv.org/abs/1506.04594, 2015.
238. V.N. Kolokolstov, M. Troeva, and W. Yang. On the rate of convergence for the mean field approximation of controlled diffusions with large number of players. *Dynamic Games and Applications*, 4:208–230, 2013.
239. V.N. Kolokoltsov. Nonlinear Markov games on a finite state space (mean-field and binary interactions). *International Journal of Statistics and Probability*, 1:77–91, 2012.
240. T. Kruse and A. Popier. BSDEs with monotone generator driven by Brownian and Poisson noises in a general filtration. *Stochastics*, 88:491–539, 2016.
241. P. Krusell and Jr. A. Smith. Income and wealth heterogeneity in the macroeconomy. *Journal of Political Economy*, 106:867–896, 1998.

242. N. Krylov. *Controlled Diffusion Processes*. Stochastic Modelling and Applied Probability. Springer-Verlag Berlin Heidelberg, 1980.
243. H. Kunita. *Stochastic Flows and Stochastic Differential Equations*. Cambridge Studies in Advanced Mathematics. Cambridge University Press, 1990.
244. H. Kunita and S. Watanabe. On square integrable martingales. *Nagoya Mathematical Journal*, 30:209–245, 1967.
245. T.G. Kurtz. Random time changes and convergence in distribution under the Meyer-Zheng conditions. *Annals of Applied Probability*, 19:1010–1034, 1991.
246. T.G. Kurtz. The Yamada-Watanabe-Engelbert theorem for general stochastic equations and inequalities. *Electronic Journal of Probability*, 12:951–965, 2007.
247. T.G. Kurtz. Weak and strong solutions of general stochastic models. *Electronic Journal of Probability*, 19:1–16, 2014.
248. T.G. Kurtz and P. Protter. Weak limit theorems for stochastic integrals and stochastic differential equations. *Annals of Probability*, 19:1035–1070, 1991.
249. T.G. Kurtz and J. Xiong. Particle representations for a class of nonlinear SPDEs. *Stochastic Processes and their Applications*, 83(1):103–126, 1999.
250. T.G. Kurtz and J. Xiong. A stochastic evolution equation arising from the fluctuations of a class of interacting particle systems. *Communications in Mathematical Sciences*, 2(3):325–358, 2004.
251. A. Lachapelle, J.M. Lasry, C.A. Lehalle, and P.L. Lions. Efficiency of the price formation process in the presence of high frequency participants: a mean field games analysis. *Mathematics and Financial Economics*, 10:223–262, 2016.
252. A. Lachapelle, J. Salomon, and G. Turinici. Computation of mean field equilibria in economics. *Mathematical Models and Methods in Applied Sciences*, 20:567–588, 2010.
253. A. Lachapelle and M.T. Wolfram. On a mean field game approach modeling congestion and aversion in pedestrian crowds. *Transportation Research Part B: Methodological*, 45:1572–1589, 2011.
254. D. Lacker. Mean field games via controlled martingale problems: Existence of markovian equilibria. *Stochastic Processes and their Applications*, 125:2856–2894, 2015.
255. D. Lacker. A general characterization of the mean field limit for stochastic differential games. *Probability Theory and Related Fields*, 165:581–648, 2016.
256. D. Lacker. Limit theory for controlled McKean-Vlasov dynamics. http://arxiv.org/1609.08064, 2016.
257. D. Lacker and K. Webster. Translation invariant mean field games with common noise. *Electronic Communications in Probability*, 20, 2015.
258. O.A. Ladyzenskaja, V.A. Solonnikov, and N. N. Ural'ceva. *Linear and Quasi-linear Equations of Parabolic Type*. Translations of Mathematical Monographs. American Mathematical Society, 1968.
259. J.M. Lasry and P.L. Lions. A remark on regularization in Hilbert spaces. *Israël Journal of Mathematics*, 55, 1986.
260. J.M. Lasry and P.L. Lions. Jeux à champ moyen I. Le cas stationnaire. *Comptes Rendus de l'Académie des Sciences de Paris, ser. I*, 343:619–625, 2006.
261. J.M. Lasry and P.L. Lions. Jeux à champ moyen II. Horizon fini et contrôle optimal. *Comptes Rendus de l'Académie des Sciences de Paris, ser. I*, 343:679–684, 2006.
262. J.M. Lasry and P.L. Lions. Mean field games. *Japanese Journal of Mathematics*, 2:229–260, 2007.
263. M. Laurière and O. Pironneau. Dynamic programming for mean field type control. *Comptes Rendus Mathematique, ser. I*, 352:707–713, 2014.
264. G M Lieberman. *Second Order Parabolic Differential Equations*. World Scientific, 1996.
265. P.L. Lions. Théorie des jeux à champs moyen et applications. Lectures at the Collège de France. http://www.college-de-france.fr/default/EN/all/equ_der/cours_et_seminaires.htm, 2007–2008.

266. P.L. Lions. Estimées nouvelles pour les équations quasilinéaires. Seminar in Applied Mathematics at the Collège de France. http://www.college-de-france.fr/site/pierre-louis-lions/seminar-2014-11-14-11h15.htm, 2014.
267. P.L. Lions and A.S. Sznitman. Stochastic differential equations with reflecting boundary conditions. *Communications on Pure and Applied Mathematics*, 37:511–537, 1984.
268. K. Lye and J. Wing. Game strategies in network security. *International Journal on Information Security*, 4:71–86, 2005.
269. J. Komlós M. Ajtai and G. Tusnàdy. On optimal matchings. *Combinatorica*, 4:259–264, 1983.
270. Y. Hu M. Fuhrman and G. Tessitore. Ergodic BSDEs and optimal ergodic control in Banach spaces. *SIAM Journal on Control and Optimization*, 48:1542–1566, 2009.
271. J. Ma, P. Protter, and J. Yong. Solving forward-backward stochastic differential equations explicitly – a four step scheme. *Probability Theory and Related Fields*, 98:339–359, 1994.
272. J. Ma, Z. Wu, D. Zhang, and J. Zhang. On well-posedness of forward-backward SDEs - a unified approach. *Annals of Applied Probability*, 25:2168–2214, 2015.
273. J. Ma, H. Yin, and J. Zhang. On non-Markovian forward backward SDEs and backward stochastic PDEs. *Stochastic Processes and their Applications*, 122:3980–4004, 2012.
274. J. Ma and J. Yong. *Forward-Backward Stochastic Differential Equations and their Applications*. Volume 1702 of Lecture Notes in Mathematics. Springer-Verlag Berlin Heidelberg, 2007.
275. J. Ma and J. Zhang. Path regularity for solutions of backward stochastic differential equations. *Probability Theory and Related Fields*, 122:163–190, 2002.
276. B. Maslowski and P. Veverka. Sufficient stochastic maximum principle for discounted control problem. *Applied Mathematics & Optimization*, 70:225–252, 2014.
277. H.P. McKean. A class of Markov processes associated with nonlinear parabolic equations. *Proceedings of the National Academy of Science*, 56:1907–1911, 1966.
278. H.P. McKean. Propagation of chaos for a class of nonlinear parabolic equations. *Lecture Series in Differential Equations*, 7:41–57, 1967.
279. S. Méléard. Asymptotic behaviour of some interacting particle systems; McKean-Vlasov and Boltzmann models. In D. Talay, L. Denis, L. Tubaro, editors, *Probabilistic Models for Nonlinear Partial Differential Equations*. Volume 1627 of Lecture Notes in Mathematics, pages 42–95. Springer-Verlag Berlin Heidelberg, 1996.
280. P.-A. Meyer and Zheng. Tightness criteria for laws of semimartingales. *Annales de l'Institut Henri Poincaré, Probabilités et Statistiques*, 20:353–372, 1984.
281. T. Meyer-Brandis, B. Oksendal, and X.Y. Zhou. A mean field stochastic maximum principle via Malliavin calculus. *Stochastics*, 84:643–666, 2012.
282. T. Mikami. Markov marginal problems and their applications to Markov optimal control. In W.M. McEneaney, G.G. Yin, Q., Zhang, editors, *Stochastic Analysis, Control, Optimization and Applications, A Volume in Honor of W.H. Fleming*, pages 457–476. Boston, Birkhäuser, 1999.
283. T. Mikami. Monge's problem with a quadratic cost by the zero-noise limit of h-path processes. *Probability Theory and Related Fields*, 29:245–260, 2004.
284. P. Milgrom and J. Roberts. Rationalizability, learning, and equilibrium in games with strategic complementarities. *Econometrica*, 58:1255–1277, 1990.
285. M. Nourian, P.E. Caines, and R.P. Malhamé. Mean field analysis of controlled Cucker-Smale type flocking: Linear analysis and perturbation equations. In S. Bittanti, editor, *Proceedings of the 18th IFAC World Congress, Milan, August 2011*, pages 4471–4476. Curran Associates, Inc., 2011.
286. S. Morris and H.S. Shin. Unique equilibrium in a model of self-fulfilling currency attacks. *American Economic Review*, 88:587–597, 1998.
287. M. Safonov N. Krylov. An estimate for the probability of a diffusion process hitting a set of positive measure. *Doklady Akademii nauk SSSR*, 245:18–20, 1979.
288. J. Nash. Equilibrium points in n-person games. *Proceedings of the National Academy of Sciences of the USA*, 36:48–49, 1950.
289. J. Nash. Non-cooperative games. *Annals of Mathematics*, 54:286–295, 1951.

290. K. Nguyen, T. Alpcan, and T. Basar. Stochastic games for security in networks with interdependent nodes. In *Proceedings of the 2009 International Conference on Game Theory for Networks, 13–15 May, 2009, Istanbul* pages 697–703, 2009.
291. S. Nguyen and M. Huang. Linear-quadratic-Gaussian mixed games with continuum-parametrized minor players. *SIAM Journal on Control and Optimization*, 50:2907–2937, 2012.
292. S. Nguyen and M. Huang. Mean field LQG games with mass behavior responsive to a major player. In *Proceedings of the 51st IEEE Conference on Decision and Control*, pages 5792–5797, 2012.
293. M. Nourian and P. Caines. ϵ-Nash mean field game theory for nonlinear stochastic dynamical systems with major and minor agents. *SIAM Journal on Control and Optimization*, 51:3302–3331, 2013.
294. M. Nutz. A mean field game of optimal stopping. Technical report, Columbia University, https://arxiv.org/abs/1605.09112, 2016.
295. F. Otto. The geometry of dissipative evolution equations: the porous medium equation. *Communications in Partial Differential Equations*, 26:101–174, 2001.
296. E. Pardoux. Homogenization of linear and semilinear second order parabolic PDEs with periodic coefficients: A probabilistic approach. *Journal of Functional Analysis*, 167:469–520, 1999.
297. E. Pardoux and S. Peng. Adapted solution of a backward stochastic differential equation. *Systems & Control Letters*, 14:55–61, 1990.
298. E. Pardoux and S. Peng. Backward SDEs and quasilinear PDEs. In B. L. Rozovskii and R. B. Sowers, editors, *Stochastic Partial Differential Equations and Their Applications*. Volume 176 of Lecture Notes in Control and Information Sciences. Springer-Verlag Berlin Heidelberg, 1992.
299. E. Pardoux and A. Răşcanu. *Stochastic Differential Equations, Backward SDEs, Partial Differential Equations*. Stochastic Modelling and Applied Probability. Springer International Publishing, 2014.
300. E. Pardoux and S. Tang. Forward-backward stochastic differential equations and quasilinear parabolic PDEs. *Probability Theory and Related Fields*, 114:123–150, 1999.
301. K.R. Parthasarathy. *Probability on Metric Spaces*. Chelsea AMS Publishing, 1967.
302. S. Peng. A general stochastic maximum principle for optimal control problems. *SIAM Journal on Control and Optimization*, 2:966–979, 1990.
303. S. Peng. Probabilistic interpretation for systems of quasilinear parabolic partial differential equations. *Stochastics and Stochastics Reports*, 37:61–74, 1991.
304. S. Peng. A generalized dynamic programming principle and Hamilton-Jacobi-Bellman equation. *Stochastics and Stochastics Reports*, 38:119–134, 1992.
305. S. Peng. Stochastic Hamilton Jacobi Bellman equations. *SIAM Journal on Control and Optimization*, 30:284–304, 1992.
306. S. Peng and Y. Shi. Infinite horizon forward-backward stochastic differential equations. *Stochastic Processes and their Applications*, 85:75–92, 2000.
307. S. Peng and Z. Wu. Fully coupled forward-backward stochastic differential equations and applications to optimal control. *SIAM Journal on Control and Optimization*, 37:825–843, 1999.
308. J.P. Penot. *Calculus Without Derivatives*. Graduate Texts in Mathematics. Springer-Verlag New York, 2012.
309. H. Pham. On some recent aspects of stochastic control and their applications. *Probability Surveys*, 2:506–549, 2005.
310. H. Pham. *Continuous-time Stochastic Control and Optimization with Financial Applications*. Stochastic Modelling and Applied Probability. Springer-Verlag Berlin Heidelberg, 2009.
311. H. Pham and X. Wei. Bellman equation and viscosity solutions for mean field stochastic control problem. *ESAIM: Control, Optimisation and Calculus of Variations*, to appear.
312. H. Pham and X. Wei. Dynamic programming for optimal control of stochastic McKean-Vlasov dynamics. *SIAM Journal on Control and Optimization*, 55:1069–1101, 2017.

313. E. Pimentel and V. Voskanyan. Regularity theory for second order stationary mean-field games. *Indiana University Mathematics Journal*, 66:1–22, 2017.
314. F. Priuli. Linear-quadratic N-person and mean-field games: Infinite horizon games with discounted cost and singular limits. *Dynamic Games and Applications*, 5:397–419, 2015.
315. P. Protter. *Stochastic Integration and Differential Equations. A New Approach.* Stochastic Modelling and Applied Probability. Springer-Verlag Berlin Heidelberg, 1990.
316. J. Quastel and S.R.S. Varadhan. Diffusion semigroups and diffusion processes corresponding to degenerate divergence form operators. *Communications on Pure and Applied Mathematics*, 50:667–706, 1997.
317. S.T. Rachev and L. Ruschendorf. *Mass Transportation Problems I: Theory.* Probability and Its Applications. Springer-Verlag New York, 1998.
318. M. Reed and B. Simon. *Methods of Modern Mathematical Physics. I. Functional analysis.* Academic Press, Inc. San Diego, 1980.
319. J.C. Rochet and X. Vives. Coordination failures and the lender of last resort. *Journal of the European Economic Associateion*, 2:1116–1148, 2004.
320. R. T. Rockafellar. *Convex Analysis.* Princeton University Press, 1970.
321. M. Royer. BSDEs with a random terminal time driven by a monotone generator and their links with PDEs. *Stochastics and Stochastic Reports*, 76:281–307, 2004.
322. W. Rudin. *Real and Complex Analysis.* McGraw-Hill, New York, 1966.
323. J. Schauder. Der fixpunktsatz in funktionalräumen. *Studia Mathematica*, 2:171–180, 1930.
324. Y. Sun. The exact law of large numbers via Fubini extension and characterization of insurable risks. *Journal of Economic Theory*, 126:31–69, 2006.
325. A.S. Sznitman. Topics in propagation of chaos. In P-L Hennequin, editor, *Ecole de Probabilités de Saint Flour, XIX-1989.* Volume 1464 of Lecture Notes in Mathematics, pages 165–251. Springer-Verlag Berlin Heidelberg, 1989.
326. M. Jeanblanc T. Bielecki and M. Rutkowski. Hedging of defaultable claims. In R. Carmona et al., editors, *Paris Princeton Lectures on Mathematical Finance 2003.* Volume 1847 of Lecture Notes in Mathematics, pages 1–132. Springer-Verlag Berlin Heidelberg, 2004.
327. M. Jeanblanc T. Bielecki, S. Crepey and M. Rutkowski. Arbitrage pricing of defaultable game options with applications to convertible bonds. *Quantitative Finance*, 8:795–810, 2008.
328. X. Tan and N. Touzi. Optimal transportation under controlled stochastic dynamics. *Annals of Probability*, 41:3201–3240, 2013.
329. H. Tanaka. Stochastic differential equations with reflecting boundary condition in convex regions. *Hiroshima Mathematical Journal*, 9:163–177, 1979.
330. A. Tarski. A lattice theoretic fixed point theorem. *Pacific Journal of Mathematics*, 5:285–309, 1955.
331. R. Temam. *Navier-Stokes Equations.* AMS Chelsea, 1984.
332. H. Tembine, Q. Zhu, and T. Basar. Risk-sensitive mean-field stochastic differential games. *IEEE Transactions on Automatic Control*, 59:835–850, 2014.
333. R. Tevzadze. Solvability of backward stochastic differential equations with quadratic growth. *Stochastic Processes and their Applications*, 118:503–515, 2008.
334. N. Touzi. *Optimal Stochastic Control, Stochastic Target Problems, and Backward SDE.* Fields Institute Monographs. Springer-Verlag New York, 2012.
335. J. Vaillancourt. On the existence of random McKean-Vlasov limits for triangular arrays of exchangeable diffusions. *Stochastic Analysis and Applications*, 6(4):431–446, 1988.
336. A. Y. Veretennikov. Strong solutions and explicit formulas for solutions of stochastic integral equations. *Matematicheskii Sbornik*, 111:434–452, 1980.
337. C. Villani. *Topics in Optimal Transportation.* Graduate Studies in Mathematics. American Mathematical Society, 2003.
338. C. Villani. *Optimal Transport, Old and New.* Grundlehren der mathematischen Wissenschaften. Springer-Verlag Berlin Heidelberg, 2009.
339. X. Vives. Nash equilibrium with strategic complementarities. *Journal of Mathematical Economics*, 19:305–321, 1990.

340. H. Xing and G. Žitković. A class of globally solvable Markovian quadratic BSDE systems and applications. *The Annals of Probability*, to appear.
341. J. Yong. Linear forward backward stochastic differential equations. *Applied Mathematics & Optimization*, 39:93–119, 1999.
342. J. Yong. Linear forward backward stochastic differential equations with random coefficients. *Probability Theory and Related Fields*, 135:53–83, 2006.
343. J. Yong and X. Zhou. *Stochastic Controls: Hamiltonian Systems and HJB Equations*. Stochastic Modelling and Applied Probability. Springer-Verlag New York, 1999.
344. L.C. Young. *Calculus of variations and control theory*. W.B. Saunders, Philadelphia, 1969.
345. E. Zeidler. *Nonlinear Functional Analysis and its Applications I: Fixed-Point Theorems*. Springer-Verlag New York, 1986.
346. A. K. Zvonkin. A transformation of the phase space of a diffusion process that will remove the drift. *Matematicheskii Sbornik*, 93:129–149, 1974.

Assumption Index

Note: Page numbers in Roman refer to Volume I and those in *italics* refer to *Volume II*

R. Carmona, F. Delarue, *Probabilistic Theory of Mean Field Games with Applications I*, Probability Theory and Stochastic Modelling 83,
DOI 10.1007/978-3-319-58920-6

Notation Index

Note: Page numbers in Roman refer to Volume I and those in *italics* refer to *Volume II*

Symbols

R. Carmona, F. Delarue, *Probabilistic Theory of Mean Field Games with Applications I*, Probability Theory and Stochastic Modelling 83,
DOI 10.1007/978-3-319-58920-6

Author Index

Note: Page numbers in Roman refer to Volume I and those in *italics* refer to *Volume II*

R. Carmona, F. Delarue, *Probabilistic Theory of Mean Field Games with Applications I*, Probability Theory and Stochastic Modelling 83,
DOI 10.1007/978-3-319-58920-6

Subject Index

Note: Page numbers in Roman refer to Volume I and those in *italics* refer to *Volume II*

R. Carmona, F. Delarue, *Probabilistic Theory of Mean Field Games with Applications I*, Probability Theory and Stochastic Modelling 83,
DOI 10.1007/978-3-319-58920-6

D

GPSR Compliance
The European Union's (EU) General Product Safety Regulation (GPSR) is a set of rules that requires consumer products to be safe and our obligations to ensure this.

If you have any concerns about our products, you can contact us on

ProductSafety@springernature.com

In case Publisher is established outside the EU, the EU authorized representative is:

Springer Nature Customer Service Center GmbH
Europaplatz 3
69115 Heidelberg, Germany

www.ingramcontent.com/pod-product-compliance
Lightning Source LLC
Chambersburg PA
CBHW071919110625
28065CB00003B/127
* 9 7 8 3 0 3 0 1 3 2 6 0 6 *